AQUATIC AND WETLAND PLANTS OF
SOUTHEASTERN UNITED STATES

AQUATIC AND WETLAND PLANTS OF SOUTHEASTERN UNITED STATES

DICOTYLEDONS

Robert K. Godfrey
Jean W. Wooten

Athens
THE UNIVERSITY OF GEORGIA PRESS

Library of Congress Cataloging in Publication Data

Godfrey, Robert K
    Aquatic and wetland plants of southeastern United States.

    Bibliography.
    Includes indexes.

    1. Aquatic plants—Southern States—Identification.   2. Wet-
land flora—Southern States—Identification.   3. Dicotyledons—
Identification.   I. Wooten, Jean W., joint author.   II. Title.

QK125.G59   583'.0976   80-16452
                    ISBN 0-8203-0532-4

Endpaper drawing courtesy of Wilderness Graphics, Tallahassee

TO
ELEANOR NIERNESEE GODFREY
WIFE AND FRIEND

Contents

Preface

This treatment of dicotyledonous aquatic and wetland flowering plants of southeastern United States is a companion work to our treatment of monocotyledons.* The geographic area of coverage includes North Carolina, South Carolina, Georgia, Florida, Tennessee, Alabama, Mississippi, Arkansas, and Louisiana. This area abuts the eastern boundary of the range encompassed by Donovan S. Correll and Helen B. Correll, *Aquatic and Wetland Plants of Southwestern United States* (1972).

The senior author began research on this work more than twenty years ago at Florida State University with a grant from the National Science Foundation (G-4321) for study of aquatic plants of Florida. Later, funding was obtained from the Institute of General Sciences, National Institutes of Health (RG-6305). While these grants were in force much field exploration was accomplished, voucher specimens were collected, illustrations were prepared, and preliminary drafts of keys and descriptions for some taxa were written.

From 1963 to 1973, owing to changes in the senior author's departmental duties at Florida State University, further research was restricted to field exploration and collection of voucher specimens. Duplicates of the specimens were exchanged with numerous institutions, chiefly those in the southeastern states, so that the specimen holdings of the university herbarium were very significantly augmented and diversified.

Upon his retirement from Florida State University in 1973, the senior author became a Beadel Research Fellow of Tall Timbers Research, Tallahassee, and research on the project was reactivated under its auspices. The junior author, then in residence in Tallahassee, vigorously encouraged renewal of the effort, with expansion of the geographic area of coverage and inclusion of wetland plants, and her collaboration in the project began.

Whatever degree of success may be attributed to our completed contribution, a very large measure of credit goes to Tall Timbers Research, for providing the senior author with a research fellowship, aid of staff (particularly in the onerous task of final typing of the manuscript), use of physical facilities, and, not least, the stimulative interest of the former and present directors, E. V. Komarek, Sr., and D. Bruce Means. The Department of Biological Science, Florida State University, very generously provided work space in its herbarium for the senior author after his retirement and encouraged him to continue using all herbarium facilities. Much appreciation is expressed to the Department of Biology, University of Southern Mississippi, for its support and encouragement of the junior author's participation during the last six years.

To Grady W. Reinert, Barbara Culbertson, and Melanie Darst, artists who drew the illustrations under our supervision, we express admiration for their talents and appreciation for pleasurable associations.

We gratefully acknowledge the privilege of using plates from the following publications: Clyde Reed, *Selected Weeds of the United States* (1970), Elbert E. Little, Jr., and Frank H. Wadsworth, *Common Trees of Puerto Rico* (1964), Elbert E. Little, Jr., et al., *Trees of Puerto Rico* (1974), Fred J. Hermann, *Vetches of the United States*, U.S. Dept. Agr., Handb. No. 168 (1960), Mildred Mathias and Lincoln C. Constance in Lundell's *Flora of Texas* (1951), and T. G. Yuncker in Lundell's *Flora of Texas* (1943).

We extend our thanks to the following persons (and to the journals noted) for permission to use illustrations from their published works: Ernest O. Beal (*Brittonia*), C. R. Gunn (*Brittonia*), Robert Kral (*Brittonia, Sida*), Richard W. Lowden (*Aquatic Botany*), Richard S. Mitchell (*Rhodora*), Robert R. Smith and Daniel B. Ward (*Sida*), and John W. Thieret (*Castanea*).

Carroll E. Wood, Jr., graciously permitted use of certain illustrations from several of the

Aquatic and Wetland Plants of Southeastern United States: Monocotyledons. Athens: University of Georgia Press (1979).

papers of the series, under his general direction, contributing to a generic flora of the southeastern United States, all published in the *Journal of the Arnold Arboretum*; we are grateful to him and to several other authors of papers in the series. Inestimable and sincere gratitude goes to Donovan S. Correll and Helen B. Correll for use of numerous illustrations by Vivian Frazer from their book, *Aquatic and Wetland Plants of Southwestern United States* (1972). Finally, we acknowledge our appreciation to the University of Florida Presses for permitting the use of illustrations from Kurz and Godfrey, *Trees of Northern Florida* (1962).

The formulation of concepts of taxa and the construction of keys in a floristic work of the scope of the present one (and the preceding companion volume) is to no small extent dependent upon use of information in many published works: manuals, floras, floristic studies, and taxonomic revisions. Our indebtedness to the authors of these is of incalculable dimensions and cannot be repaid.

Three colleagues, some of whose ongoing research has not yet been published, have generously communicated with us respecting their work, either by letter or by furnishing to us parts of unfinalized drafts of manuscript, or both. These are Daniel F. Austin (Convolvulaceae), Philip Cantino (*Physostegia*), and Walter S. Judd (*Lyonia*). While we very much appreciate their sharing some of their specialized knowledge with us, they are by no means to be held responsible for whatever deficiencies or inaccuracies occur in our treatments of those groups; these will have been our own.

The publication of this volume was made possible in part by a grant from the Department of Biological Science, Florida State University. The authors and the publisher gratefully acknowledge this support.

The introduction which appeared in the preceding volume treating monocotyledonous plants was written as almost wholly pertinent to this one for dicotyledonous plants. It is, with very slight modification, included in this volume, inasmuch as some persons who secure this one may not possess the other.

ROBERT K. GODFREY
Tall Timbers Research, Inc.
and
Florida State University

JEAN W. WOOTEN
Department of Biology
University of Southern Mississippi

Introduction

The aim of this work is to aid in identifying native and naturalized dicotyledonous plants that inhabit aquatic and wetland places, places permanently or seasonally wet. It is not intended to be a study of ecology of hydrophytes. It is hoped that, as a taxonomic-floristic work, it will have value to ecologists, to governmental agencies concerned with environmental problems, to contractual firms engaged in environmental impact analyses, to individuals with general ecological interests, and to students in colleges and universities. Recognizing that these prospective users will have had diverse or limited experience in plant identification, we have sought to use, whenever possible, nontechnical language. We trust that we may have achieved this goal to a limited but useful extent. A glossary is included to help with the technical terminology necessary to describe features for reasonable understanding.

Those states of the southeastern United States for which this manual affords botanical coverage (see preface) encompass parts of the following physiographic provinces: Atlantic and Gulf coastal plains, southern Appalachian highlands, flanked on the east by the Piedmont plateau, and on the west by the Appalachian plateau, the interior lowlands, and the interior highlands. The region has a diverse topography, a wide range of habitat types, varied and variable organismal communities, thus diverse ecosystems. In many instances our ecosystems, complex and not readily delimited to begin with, are becoming increasingly blurred and degraded both by man's overt direct manipulation and by the vicissitudinous alterations attendant upon one or another of his activities.

FORMAT

The text for each species includes a statement indicating the habitat or habitats in which the plant usually grows and a statement of its geographic distribution. Most of the statements must be taken as generalities, subject in many instances to exception and incompleteness.

For the stated geographic ranges, especially for widely distributed species, we relied upon relatively recent taxonomic revisions whenever possible. Otherwise, we followed or adapted from ranges given in floras and manuals, particularly Small (1933), Fernald (1950), Gleason (1952), and Correll and Correll (1972). The range for each of the wide-ranging species is expressed as nearly as feasible according to the geographic areas which mark the limits of range. In most cases the northern boundary is given first, beginning at the northeast, in the style of the widely used floras by Fernald and Gleason.

The reference list following the glossary includes, for the most part, general references and appropriate papers from the series prepared for a biologically oriented generic flora of the southeastern United States, the latter a joint project of the Arnold Arboretum and the Gray Herbarium of Harvard University, chiefly under the direction of Carroll E. Wood, Jr. and Reed C. Rollins. Numerous literature citations are placed in the text with the treatment to which they most specifically apply.

DISTRIBUTION OF AQUATIC AND WETLAND PLANTS

The subject of this work, aquatic and wetland plants, would seem to carry with it the assumption that some definitive physical boundaries exist so that one can attribute a given plant's occurrence to an aquatic habitat or to a wetland habitat. This is true, however, only to a very limited extent, for the physical boundaries are generally not precise, neither are the habitat requirements or tolerances of a great many species. Some sites, which have waterlogged soils during times of high precipitation, may dry out completely during intervening drought periods.

Aquatic and wetland plants commonly are present sporadically in their habitats. Plants of

1

a given species may be present in great abundance at one time, then disappear or become greatly diminished in number, and reappear later, oftentimes several or many years later. On the shores of karst lakes and ponds, for example, where water levels fluctuate greatly, the wet sands, peats, or sandy peats during times of low water are sometimes inhabited by dense populations of certain species, by populations of other species during other times of low or high water. In karst areas such lakes or ponds probably fluctuate not only in relation to surface water draining into them but also in relation to underground water courses. Some sometimes fluctuate considerably from year to year, sometimes remain at high or low water levels continuously for relatively long intervals. It appears plausible that populations of some species appearing on their exposed shores may have underground parts remaining viable but dormant when submersed; others may have seeds which have long dormant periods when submersed. In the latter case, germination of the seeds may not only require an exposed substrate, but also may require a particular set of environmental conditions prevailing simultaneously in order that germination be induced.

Populations of species characteristically inhabiting boggy pine savannas and flatwoods also vary greatly in number present at a given time. Their periodicity is, in our judgment, attuned to periodicity of fire. Many kinds have underground parts which may remain wholly or partially dormant during fire-free intervals, or may have seeds with extended dormancy. In either case, they appear to be fire adapted.

TAXONOMIC COVERAGE

With respect to taxonomic coverage, we admit to having exercised many arbitrary choices as to whether certain species should be considered as inhabitants of wetlands. Anyone who must decide what plants of a given area are associated with excessive water during at least a part of their life spans will sympathize with the mental anguish accompanying such selection. Some species which are clearly inhabitants of aquatic or wetland places were excluded on other grounds—for example, certain ones of rare and local occurrence not likely to be found anywhere in any abundance. In addition, a few species are in the account only because their inclusion was considered helpful to the problems of identification. Since, after all, the treatment is for an arbitrarily selected portion of a region's total flora, we indulged personal biases which will show us to have been somewhat inconsistent. Criticism for having included certain plants, and for not having included others, may confidently be expected.

HABITATS

Natural lakes and ponds in our region are much more prevalent in the coastal plain than elsewhere and they are especially numerous in Florida. Man-made farm ponds occur more or less throughout, as do larger impoundments associated with flood control, hydroelectric power production, and navigation. Other habitats with open water include streams of assorted sizes, swamps, springs, ocean waters and estuaries, bays, lagoons, drainage and irrigation ditches and canals. It is in the foregoing principally that aquatic plants occur, plants with their parts wholly submersed, or partially submersed and partially floating, or with only their flowering and fruiting parts emersed.

There is, of course, no sharp line of demarcation between open-water aquatic and marsh vegetation. We conceive of marshes as places with fairly permanent and stable surface water (or at least with water regularly flowing over the surface in tidal situations) inhabited by herbaceous plants rooted in the substrate but with photosynthetic and reproductive organs principally emersed. Marshes or marshy situations prevail throughout our region, in coastal and estuarine tidal areas, along shores or in shallow water of ponds, lakes, streams, lagoons, sloughs, backwaters, wet meadows, drainage and irrigation ditches and canals.

Swamps are wetlands dominated by woody plants, the substrate flooded for one or more longish periods during each year, sometimes more or less permanently flooded, but usually without surface water part of the time. They occur on the floodplains of streams and in small to large depressions. During maximum flooding both aquatic- and marshy-type shade-tolerant herbs may abound beneath the trees; during periods of maximal drying out essentially terrestrial but shade-tolerant herbs may flourish there. Wet woodlands are tree-dominated as well but their substrates generally have water standing over them for short intervals during times of high precipitation.

Bogs are generally open or semiopen areas having a grass-sedge-dominated "ground flora" with a varying admixture of plants with showy flowers. Bogs occur throughout the region of our coverage but are most conspicuous and most extensive in the coastal plain, where pine savannas or flatwoods and evergreen shrub or shrub-tree bogs (bays) are distinctive.

Pine savannas or flatwoods, which occur on substrates which are waterlogged or even have some surface water during seasons of heavy precipitation, may become dry during times of low precipitation or "baked dry" during extended droughts. If periodically burned, the pines are usually scattered, the ground flora herbaceous, the latter principally grasses and sedges but with numerous kinds of showy flowering plants as well. If periodic burning is excluded, however, hardwood trees and shrubs, whose top growth is continually arrested by burning, take over and relatively quickly form a closed canopy, which causes the herbaceous flora to become repressed and eventually to die out.

Shrub or shrub-tree bogs, commonly called bays, are flat, poorly drained, boggy areas inhabited by evergreen shrubs and trees, sometimes including conifers tolerant of long hydroperiods. Herbaceous plants are few, or scarcely any, in the bays except during short intervals after severe burning. Bay shrubs and trees quickly regenerate from underground parts after fire and again repress the herbs. Bays occur in the outer or lower coastal plain on poorly drained, broad, flat heads of streams (branch bays), on now relatively elevated flat bottoms of areas that were estuarine when the sea level was much above its present position (estuary bays), and Carolina bays, elliptical shallow basins of variable size which dot the coastal plain of the two Carolinas.

The nature and extent of aquatic and wetland plants occurring in and along rivers and streams are markedly affected by flooding and subsidence of flooding during and between periods of high precipitation. Fluctuations in river and stream flow greatly influence aquatic and marsh plant populations in their main courses. In addition, the kinds of plant populations on the floodplains relate to both long-range and short-term hydrologic phenomena. Pools and backwaters fill and subside, floodwaters inundate floodplains to varying depths at times, and much of the floodplain is relatively dry for varying periods between floods. The kinds and quantities of herbaceous vegetation, especially, differ very greatly on the floodplain through the year or from one year to the next.

Attention is drawn especially to two extraordinary wetland areas lying within the geographic boundaries of our coverage, namely, the Okefenokee Swamp and the Everglades.

The Okefenokee Swamp covers parts of six counties of southeastern Georgia and extends just into Florida. Its area is somewhat over 600 square miles, its greatest width 26 miles, its greatest length 39 miles. Its principal effluent is the Suwannee River, although the St. Mary's River drains some of its eastern part. Persons interested in a detailed account of habitats and vegetation of the Okefenokee are referred to Wright and Wright (1932). For a very interestingly written account of early exploration, of exploitation of its timber resources, of schemes put forth to have the Swamp yield up some of its other riches, and not least for the rich perspective of a man who spent most of his working life in it and about it, we recommend "Forty-five Years with the Okefenokee Swamp—1900–1945" by John M. Hopkins (Georgia Society of Naturalists Bulletin No. 4 (undated)).

An outstanding natural feature of southern Florida (and one of the unique regions of the earth) is (or was!) the Everglades. The Everglades is a great flat expanse, a rock-filled basin with uneven topography essentially filled with peat and muck to a nearly level plain. It extends in a broad belt roughly from Lake Okeechobee southward and southwestward nearly to Cape Sable, its boundaries encompassing approximately 750 square miles of area. It is prevailingly marshy but does not by any means exhibit a uniform vegetational cover. Davis

3

(1943) describes six main vegetational types, and the reader is referred to his work for detail (and for detail about southern Florida's extensive cypress swamps as well). Highly recommended to the reader is the beautifully written popular book by Marjory Stoneman Douglas, *The Everglades: River of Grass*, Hurricane House, Coral Gables, Florida (1947).

The Everglades is now greatly changed from its pristine condition, owing to an extraordinary system of drainage canals, dikes, etc., chiefly related to flood control on the one hand and "reclamation" of lands for agriculture on the other. Some parts, notably the so-called water management areas, reflect a reasonable image of what much of it may have been like before the hand of man changed its face so drastically.

ENVIRONMENTAL PERSPECTIVES

Presently we are acutely aware of diverse problems deriving from utilization of water and management of water resources. Wetlands generally, and coastal wetlands particularly, are recognized as significantly related to the welfare of wildlife and fisheries. Rivers and impoundments are increasingly important in both navigation and recreation. In some places, Florida, for example, wetlands relate importantly to subterranean water supplies upon which both agricultural and urban interests rely. Drainage, flood control, extension of waterways to form "dead-end" watercourses in residential areas create an admixture of aquatic plant (and other) problems. Commercial interests vending aquatic plants to the aquarium trade, apparently a big business, allegedly "plant" exotic aquatics in streams, ponds, lakes, drainage canals, and ditches, where some of them, at least, become a menace to navigation, to water quality, and to drainage—that is, they become pollutants.

A direct economic value of wetland plants may be perceived as virtually nil, other than that they contribute aesthetically to the general landscape. However, wetlands, conceived not just as physical features but as ecosystems, relate to the general economics in a many-faceted, intricately webbed fashion. Water resources, water utilization, water quality, and management of wildlife and fisheries resources are notable among the considerations that give wetlands importance.

Clogging of impoundments, drainage and irrigation ditches and canals, channelized streams, or other water areas contrived by man is commonly attributable to aquatic and marsh plants introduced from floras other than our own, that is, to exotics. Some of the most notable and pestiferous (including monocotyledons) are water hyacinth (*Eichhornia crassipes*), alligator weed (*Alternanthera philoxeroides*), hydrilla (*Hydrilla verticillata*), Brazilian waterweed (*Egeria densa*), and Asian water milfoil (*Myriophyllum spicatum*).

In such waters, and in addition in natural waterways or bodies of water where accidental or purposeful manipulations—pollution, for example—in one way or another alter natural environmental regimes, various native plants contribute to weed problems. Some among the many with such potential are water willow (*Justicia americana*), cattails (*Typha* spp.), frog's bit (*Limnobium spongia*), torpedo grass (*Panicum repens*), smartweeds (*Polygonum* spp.), *Ludwigia* spp., water-lilies (*Nymphaea* spp.), spatter-docks (*Nuphar* spp.), lotus (*Nelumbo lutea*), fanwort (*Cabomba caroliniana*), water shield (*Brasenia schreberi*), water-celery or tape-grass (*Vallisneria americana*), canna (*Canna flaccida*), rushes (*Juncus* spp.), spike-rushes (*Eleocharis* spp.), bulrushes (*Scirpus* spp.), and arrow roots (*Sagittaria* spp.).

The aggressive weedy action of aquatic and marsh plants in waters naturally suitable for or presumably devised to be suitable for human benefit poses economic problems of immense magnitude. We do not consider this a proper forum in which to elaborate upon the intricacies inherent in reaching measures of prevention or control both economically and environmentally sound.

On the coasts of Florida, as conditions become increasingly tropical, coastal salt marshes give way in considerable part to mangrove swamps. On the Atlantic coast this commences at the north at about St. Augustine, on the Gulf coast at about Cedar Key, the greatest development of mangrove communities being along the southwest coast of the peninsula. Mangroves play an important geologic role in building the coasts farther out into the water, in forming islands, in the accumulation of peat and marl deposits. They afford significant protection to the coasts from storms and are vital to fisheries and wildlife (Davis, 1943). Husbanding of the mangrove resource is one of those wetland concerns relating to environmental quality over which the state and the nation must be ever watchful.

In recent years vast acreages of pine savanna and flatwoods (and some other coastal plain community types) have been greatly altered, in most cases veritably destroyed. Large areas have been drained, clear-cut, then converted to permanent pasture or to agricultural crops such as soybeans. Probably the greatest acreages have been drained, clear-cut, the sites then mechanically "cleaned" by gargantuan machines and converted to pine farming, the slash pine (*Pinus elliottii*) being the one mostly used. In the short term following mechanical alteration of the site and the planting of pine seedlings, there is usually an explosion of populations of some, at least, of the native herbaceous plants, but most of these are soon greatly diminished in number or eliminated by competition and shade, the general practice resulting in what is in essence a monoculture.

Individual persons (both lay and professional), various local and national organizations, governmental agencies, and others presently show much concern about environmental deterioration. There is plainly evident a vigorous determination to curtail excessive destructive forces, to place at least sample ecosystems or community types in the public domain, and thus to some extent to preserve them. There is, in addition, much activity, principally on the part of taxonomically oriented persons or organizations, directed toward identifying what individual species are rare, threatened, or endangered so that steps can be taken to ensure a modicum of protection to them. It would seem self-evident, since species are component parts of ecosystems, that the two sets of concerns would in large measure be one, yet one is wont to think sometimes that the twain shall never meet. Apropos of this, it is urged that both the species lovers and the ecosystem or community lovers focus additional attention on finding what manipulative management practices, imitative of natural conditions, may be necessary and useful in maintenance of that which is set aside.

Artificial Keys to the Higher Taxa of Aquatic and Wetland Dicotyledonous Plants of Southeastern United States

Key to Specialized Taxa and to Subsidiary Keys

1. Plants woody.
    2. Leaves or leaf-scars opposite or whorled.   Key 1, p. 7
    2. Leaves or leaf-scars alternate.
        3. The leaves compound.   Key 2, p. 9
        3. The leaves simple.   Key 3, p. 10
1. Plants herbaceous.
    4. The plants with special characteristics as stated in the four entries numbered 5 below.
        5. Plant bearing small, urnlike, or bladderlike traps on at least some of its vegetative parts, the latter partly or wholly free in water in aquatic species (unless stranded on the substrate at times of receding water levels), or largely or wholly within a wet substrate; flowers bilaterally symmetrical, calyx strongly 2-lobed, corolla 2-lipped, lower lip usually 3-lobed and with a basal spur, upper lip usually entire (*Utricularia*).   **Lentibulariaceae,** p. 681
        5. Plant with flowers aggregated on a common receptacle into a head subtended by an involucre, the latter somewhat calyxlike, the whole head commonly resembling a single flower.
                                                                                      Compositae, p. 751
        5. Plant with milky sap; inflorescence small, closely resembling an individual flower; each inflorescence comprised of a cuplike or urnlike structure (cyathium) bearing around its rim 1–5 nectariferous glands each sometimes having petallike appendages; each cyathium bearing within it staminate flowers each consisting of a single stamen, these disposed in fascicles, and a solitary pistillate flower inserted centrally within at the base, the pistillate flower stipitate and becoming exserted from the cyathium; ovary 3-locular, 1 ovule in each locule, styles 3, each usually bifid (*Chamaesyce*, *Euphorbia*).   **Euphorbiaceae,** p. 276
        5. Plant parasitic, leafless, without green pigment, appearing yellow, cream-colored, brownish, or whitish; stem slender, twining or looping from one host plant to another (*Cuscuta*).
                                                                                      Convolvulaceae, p. 555
    4. The plants not having any of the special sets of characteristics in the four entries numbered 5 above.
        6. Flowers, some or all of them on a given plant, unisexual.   Key 4, p. 14
        6. Flowers bisexual.
            7. The flowers without a perianth.   Key 5, p. 15
            7. The flowers having a perianth of at least one series of parts.
                8. Perianth of one series of parts, or if more than one series then not wholly or distinctly differentiated as calyx and corolla.   Key 6, p. 15
                8. Perianth of distinctly differentiated calyx and corolla.
                    9. Pistils more than one in each flower (at least their ovaries wholly separate).   Key 7, p. 16
                    9. Pistil one in each flower (ovary united at least basally, the styles may be two or more).
                        10. Stamens more numerous than the petals or lobes of the corolla.   Key 8, p. 17
                        10. Stamens just as many as the petals or lobes of the corolla or fewer.
                            11. Corolla of separate petals.   Key 9, p. 18
                            11. Corolla partially or wholly united (in some united only at the extreme base).
                                                                                                       Key 10, p. 19

Key 1

Plants Woody, Leaves Simple or Compound and with Leaves or the Leaf-scars Opposite or Whorled

1. Plants trees or shrubs, stems erect.
    2. Leaves compound.

3. Lateral buds enclosed by the bases of the petioles; leaf-scars forming a ring around the twig (*Acer negundo*).   **Aceraceae**, p. 311

3. Lateral buds not enclosed by the bases of the petioles but occurring in the leaf axils; leaf-scars not forming a ring around the stem.

    4. Trees; flowering before new growth commences in spring; flowers unisexual (rarely bisexual), staminate and pistillate on separate plants; fruit a single-winged samara; leaflets entire or if toothed the teeth usually relatively obscure (*Fraxinus*).   **Oleaceae**, p. 511

    4. Shrubs; flowering as growth of the season occurs or later; flowers bisexual; fruit a berrylike drupe or an inflated capsule; margins of leaflets finely serrate, teeth many.

        5. Flowers very small and numerous, disposed in flattish-topped cymes 2–4 dm across; fruit a juicy, purplish black, berrylike drupe 4–6 mm long, containing 3–5 stone-covered seeds (*Sambucus*).   **Caprifoliaceae**, p. 727

        5. Flowers few, borne in a drooping panicle; fruit 3–5 cm long, with a thin, papery, inflated, and bladderlike wall.   **Staphyleaceae**, p. 309

2. Leaves simple.

  6. Leaves evergreen, some present on twigs of a previous season.

    7. Bark thick, spongy, whitish at first, later exfoliating in buffish to pale cinnamon-colored, many papery layers; flowers whitish, densely crowded in terminal spikes or panicles of spikes, the stamens especially numerous and conspicuous, making the inflorescence bottle-brush-like.

            Myrtaceae, p. 387

    7. Bark not as described above (or if so then plant with bright yellow flowers in cymes); inflorescence not at all bottle-brush-like.

        8. Plants of maritime situations.

          9. Stipules present in the form of two green scales around a pair of developing leaves, quickly deciduous leaving a ring-scar on the twig.   **Rhizophoraceae**, p. 387

          9. Stipules none and no ring-scars around the twigs.

            10. Leaves without glands on their petioles; flowers or fruits sessile and compacted at the termini of a few-branched cyme; leaf blades glabrous above, pubescent beneath.

                Avicenniaceae, p. 589

            10. Leaves with a pair of glands on their petioles; flowers or fruits in small, ball-like masses spicately disposed on inflorescence branches, or individually loosely spicate on inflorescence branches; leaf blades glabrous on both surfaces (*Laguncularia*).   **Combretaceae**, p. 384

        8. Plants not of maritime situations.

          11. Leaves punctate; petals separate, bright yellow; stamens numerous (*Hypericum*).

                Guttiferae, p. 334

          11. Leaves not punctate; petals partially to almost wholly united, not bright yellow; stamens 2 or 10.

            12. Corolla cream-yellow; stamens 2 (*Osmanthus*).   **Oleaceae**, p. 513

            12. Corolla pink to rose-purple; stamens 10 (*Kalmia carolina*).   **Ericaceae**, p. 472

  6. Leaves deciduous, none present on twigs of the previous season.

    13. Major veins of leaf palmate.   **Aceraceae**, p. 310

    13. Major veins of leaf pinnate.

      14. Opposite petioles connected by stipules or, when these are shed, by stipular line-scars (*Cephalanthus* and *Pinckneya*).   **Rubiaceae**, p. 714

      14. Opposite petioles not connected by stipules or stipular line-scars.

        15. Leaf blades broadly ovate, 1–3 dm long, longer ones about half as broad basally as long, shorter ones about as broad basally as long (*Catalpa*).   **Bignoniaceae**, p. 708

        15. Leaf blades, none of them broadly ovate.

          16. Buds, petioles, young twigs, and often lower leaf surfaces, more or less scurfy-scaly-pubescent (*Viburnum*).   **Caprifoliaceae**, p. 729

          16. Buds, petioles, etc., glabrous or pubescent but not scurfy-scaly-pubescent.

             17. Flowers and fruits borne in flat-topped cymes terminating branchlets; torn leaves exposing hairlike vascular strands.   **Cornaceae**, p. 430

             17. Flowers and fruits not in terminal, flat-topped cymes; torn leaves not exposing hairlike vascular strands.

                18. Margins of leaves shallowly and usually indistinctly toothed, mostly on their upper halves, only occasional leaves on a given plant entire (*Forestiera*).   **Oleaceae**, p. 513

                18. Margins of all leaves entire.

                  19. Leaves with a spicy aroma when bruised or crushed; flowers borne singly and terminally on short branches of new growth; perianth parts numerous, separate, strap-shaped parts purplish green or purplish brown. **Calycanthaceae**, p. 233

    19. Leaves not spicy-aromatic when bruised or crushed; flowers borne in panicles from buds on wood of the previous season; perianth of a minute 4-lobed calyx and a white or cream-white corolla tubular at least at the very base and 4-lobed (*Chionanthus, Ligustrum*). **Oleaceae**, p. 509

1. Plants woody vines, or scramblers with more or less prostrate or creeping stems, or with at least some stems arching and rooting at the tips.

  20. Leaves compound.

    21. The leaves with 2 leaflets and a tendril (*Bignonia*). **Bignoniaceae**, p. 709

    21. The leaves with 3 or more leaflets, tendrils none.

      22. Climbing by aerial, adventitious roots from the stem (*Campsis*). **Bignoniaceae**, p. 709

      22. Climbing by bending or twining of petioles or stalks of the leaflets (*Clematis*).

      Ranunculaceae, p. 127

  20. Leaves simple.

    23. The leaves half- to subterete and succulent. **Bataceae**, p. 107

    23. The leaves flat, not succulent.

      24. Stems, some of them at least, arching so that their tips reach the ground, take root, and produce new shoots (*Decodon*). **Lythraceae**, p. 375

      24. Stems scrambling or climbing.

        25. Climbing by prolongation of the branch system, the branches clambering upon and through each other or upon and through other vegetation, and in part by developing branches looping or twisting around other branches or around stems of other vegetation. **Hippocrateaceae**, p. 309

        25. Climbing by twining of stems or by means of aerial, adventitious roots on the stems.

          26. Stems climbing by twining.

            27. Sap milky. **Apocynaceae**, p. 538

            27. Sap not milky.

              28. Ovary inferior, the floral tube surmounted by minute calyx segments; fruit a several-seeded berry (*Lonicera*). **Caprifoliaceae**, p. 728

              28. Ovary superior, calyx united below and with distinct and evident lobes; fruit a much-flattened, 2-valved capsule (*Gelsemium*). **Loganiaceae**, p. 514

          26. Stems climbing by means of aerial, adventitious roots (*Decumaria*).

          Saxifragaceae, p. 210

Key 2

Plants Woody, with Compound Leaves and the Leaves or Leaf-scars Alternate

1. Leaves evenly pinnately once or twice compound (terminal leaflets or terminal pinnae paired). **Leguminosae**, p. 235

1. Leaves odd-pinnately compound (terminal leaflets solitary)—occasionally a leaf even-pinnate because of abortion of the terminal leaflet during development.

  2. Stem bearing prickles or spines or both, at least on younger portions.

    3. Leaves 2- or 3-pinnately compound. **Araliaceae**, p. 433

    3. Leaves once pinnately compound (*Rosa, Rubus*). **Rosaceae**, p. 217

  2. Stems not spinescent or prickly.

    4. Leaves evergreen, as evidenced by presence of some on twigs of a previous season (*Metopium, Schinus*). **Anacardiaceae**, p. 293

    4. Leaves deciduous as evidenced by lack of any on twigs of a previous season.

      5. Plant a low, colonial shrub, the stems to about 8 dm tall, the wood of both stems and roots yellow; flowers and fruits in flexuous, drooping racemes or narrow panicles; leaflets very strongly serrate, or lobulate and with the lobes strongly serrate (*Xanthorhiza*). **Ranunculaceae**, p. 125

      5. Plant not having the above combination of characteristics.

        6. The plant a low trailing shrub or more commonly a vine climbing by means of aerial, adventitious roots, *or* a twining vine, *or* a vine climbing by means of tendrils.

          7. Plant a low trailing shrub or more commonly a climbing vine; leaves trifoliolate—take warning: poison ivy! (*Toxicodendron radicans*). **Anacardiaceae**, p. 295

          7. Plant a twining vine with once-pinnately compound leaves, *or* a vine climbing by means of tendrils, leaves mostly bipinnately compound.

8. Leaves once-pinnately compound, leaflets 9–15 per leaf (*Wisteria*).

Leguminosae, p. 244

    8. Leaves mostly bipinnately compound (*Ampelopsis arborea*).   **Vitaceae,** p. 317

  6. The plant erect.

    9. Margins of leaflets entire; plants shrubs or small, slender trees not over 8 dm tall.

      10. Flowers radially symmetrical, they or the fruits borne in broad, open panicles—take warning: poison sumac! (*Toxicodendron vernix*).   **Anacardiaceae,** p. 297

      10. Flowers bilaterally symmetrical, they or the fruits borne in racemes (*Amorpha*).

Leguminosae, p. 240

    9. Margins of leaflets toothed; plants (potentially) relatively large trees (crushed foliage with a distinctive pungent aroma) (*Carya*).   **Juglandaceae,** p. 21

Key 3

Plants Woody, their Leaves or Leaf-scars Alternate, Leaf Blades Simple

1. Leaf blades, at least some of them on a given plant, as broad as long, very nearly so, or even broader than long.

  2. Plant a vine, the main stems, at least, woody.

    2a. Climbing by tendrils opposite at least some of the leaves.   **Vitaceae,** p. 316

    2a. Climbing by twining.   **Aristolochiaceae,** p. 64

  2. Plant erect.

    3. Stipular line-scars present as rings around the nodes of twigs of younger branches.

      4. Leaf blades coarsely and irregularly toothed, or, more commonly, coarsely toothed and lobed; bark of older stems (trunks) sloughing in thin, irregular plates leaving the stem mottled or dappled in several colors; flowers unisexual and minute, those of each sex borne in separate globular heads on the same axis, the staminate on short, stiffish stalks, the pistillate much larger and dangling on longish stalks.   **Platanaceae,** p. 217

      4. Leaf blades 4–6-lobed, the margins otherwise entire, the leaf apices truncate or broadly **V**-notched, bark of older stems (trunks) ridged and furrowed; flowers bisexual, large and conspicuous, borne singly terminating branchlets after the leaves have unfolded (*Liriodendron*).

Magnoliaceae, p. 153

    3. Stipule scars, if present, not forming rings around the twigs at their nodes.

      5. Principal veins of leaf blades palmate.

        6. Leaf blades on a given plant all conspicuously palmately 5-lobed giving much the effect of a star (*Liquidambar*).   **Hamamelidaceae,** p. 215

        6. Leaf blades unlobed or if lobed then not at all giving the effect of a star.

          6a. Leaf and stem pubescence conspicuously stellate (*Pavonia, Urena*).   **Malvaceae,** p. 321

          6a. Leaf pubescence and that of the stem, if any, of unbranched hairs (*Morus*).

Moraceae, p. 59

      5. Principal veins of leaf blades very definitely pinnate.

        7. Margins of leaf blades irregularly doubly serrate; bark of older stems (trunks) exfoliating in curly, thin patches (*Betula*).   **Betulaceae,** p. 38

        7. Margins of leaves not doubly serrate; bark of older stems (trunks) ridged and furrowed or smoothish.

          8. Leaves evergreen, some present on twigs of a previous season; distribution (in our range) only in s. Fla.   **Chrysobalanaceae,** p. 232

          8. Leaves deciduous, none present on twigs of a previous season; distribution other than in s. Fla.

            9. Leaves broadly ovate or rhombic-ovate, more than 3-ranked on the twigs, that is alternating in such a way that they are in more than 3 vertical rows; principal lateral veins arching-ascending or angling-ascending but branching before reaching leaf edges.

              9a. Plant with milky sap; larger leaves not exceeding 7 (rarely as much as 9) cm long (*Sapium*).   **Euphorbiaceae,** p. 289

              9a. Plant not having milky sap; larger leaves 10 cm long or more, mostly more (*Populus*).

Salicaceae, p. 33

           9. Leaves variable in shape, 2-ranked, that is, alternating on either side of the twigs and in 2 vertical rows; principal lateral veins angling-ascending parallel to each other and ending at the leaf margins (*Fothergilla, Hamamelis*).   **Hamamelidaceae,** p. 215

1. Leaf blades markedly longer than broad.

  10. Stems trailing along the ground or climbing by means of tendrils, by twining, or by scrambling.

    11. Stems trailing along the ground; leaf blades not over 1.5 cm long.

      12. Leaves sessile, needlelike, leafy branchlets mosslike; stems rooting at the nodes.

        Diapensiaceae, p. 487

      12. Leaves short-petioled, not needlelike, leafy branchets not mosslike; stems not rooting at the nodes (certain species of *Vaccinium*).   **Ericaceae,** p. 485

    11. Stems climbing by means of tendrils, by twining, or by scrambling; leaf blades much exceeding 1.5 cm in length.

      13. Climbing by means of tendrils at the tips of branchlets (*Brunnichia*).   **Polygonaceae,** p. 65

      13. Climbing by means of twining or scrambling.

        14. Leaf blades lanceolate; inflorescence an involucrate, radiate head (*Aster carolinianus*).

        Compositae, p. 827

        14. Leaf blades elliptic to elliptic-oblong, oblong, or ovate; inflorescence axillary or terminating branchlets, not an involucrate, radiate head.

          15. Principal lateral veins 8–10 to either side of the midrib.

            16. Upper leaf surfaces glossy green and glabrous; twigs glabrous (*Berchemia*).

            Rhamnaceae, p. 315

            16. Upper leaf surfaces dull green, glabrous or pubescent; young twigs pubescent (*Dalbergia*).   **Leguminosae,** p. 248

          15. Principal lateral veins 2 or 3, usually 3, to either side of the midrib (*Colubrina*).

            Rhamnaceae, p. 316

  10. Stems erect.

    17. Leaf blades palmately veined (*Pavonia*).   **Malvaceae,** p. 321

    17. Leaf blades pinnately veined.

      18. Flowers individually small, aggregated on a common receptacle in heads subtended by an involucre.   **Compositae,** p. 751

      18. Flowers not in involucrate heads.

        19. Leaf blades narrowly oblanceolate.

          20. Stem and leaf surfaces glabrous; tips of at least some of the branchlets rigid and sharply thorn-tipped (*Lycium*).   **Solanaceae,** p. 578

          20. Stems and at least the lower leaf surfaces pubescent; branchlets, none of them thorn-tipped.   **Surianiaceae,** p. 260

        19. Leaf blades not narrowly oblanceolate.

          21. Buds conspicuously clustered at or near the ends of the twigs; fruit a nut (acorn) borne in a cup of imbricated scales (*Quercus*).   **Fagaceae,** p. 40

          21. Buds not clustered at or near the ends of twigs; fruit other than a nut borne in a scaly cup.

            22. Stipular scars present as a line around the stem of younger branchlets.

              23. Sap milky (*Ficus*).   **Moraceae,** p. 60

              23. Sap not milky (*Magnolia*).   **Magnoliaceae,** p. 154

            22. Stipular scars, if present, not forming a line all the way around the twigs.

              24. Leaves with a pair of glands on their petioles or at about the bases of their blades (*Conocarpus*).   **Combretaceae,** p. 384

              24. Leaves without glands on their petioles or at about the bases of their blades.

                25. Leaf blades punctate on one or both surfaces (very minutely and inconspicuously so beneath in *Litsia*).

                  26. Lateral veins of the leaf blades evident.

                    27. Young twigs zigzagging; leaves deciduous, none present on twigs of the previous season (*Litsea*).   **Lauraceae,** p. 359

                    27. Young twigs not zigzagging; leaves evergreen, some present on twigs of a previous season.

                      28. Each leaf scar with one vascular bundle scar; stipules or stipule scars present (minute) (*Ilex glabra, I. coriacea*).   **Aquifoliaceae,** p. 301

                      28. Each leaf scar with three vascular bundle scars; stipules none.

                      Myricaceae, p. 27

                  26. Lateral veins of the leaf blades not at all evident or scarcely perceptible.

                  Illiciaceae, p. 151

                25. Leaf blades not punctate.

29. Fresh foliage and twigs notably aromatic when crushed or bruised.
30. Leaves deciduous, none present on twigs of the previous season.
  31. Aroma of crushed leaves and twigs notably spicy (*Lindera*) **Lauraceae,** p. 359
  31. Aroma of crushed twigs and leaves various but not pleasantly spicy.

  Annonaceae, p. 147

30. Leaves evergreen, some present on twigs of the previous season (*Persea*).

  Lauraceae, p. 357

29. Fresh foliage and twigs not aromatic when crushed.
  32. Plants armed with thorns on at least some of the branches.
  33. Margins of leaf blades entire; sap milky (this particularly noticeable in lush shoots and petioles). **Sapotaceae,** p. 505
  33. Margins of leaf blades toothed or lobed and toothed; sap watery (*Crataegus*).

  Rosaceae, p. 229

  32. Plants without thorns.
  34. Leaves 2-ranked, i.e., alternating on either side of the twig and thus in two vertical rows, or leaves 3-ranked, i.e., alternating in such a way that they appear in three vertical rows on the twig.
    35. The leaves 2-ranked.
    36. Principal lateral veins of the leaf blades equally prominent and extending straight and parallel to each other from the midrib.
      37. The principal lateral veins ending at the leaf edge.
        38. Margins of the leaf blades coarsely undulate to coarsely crenate-dentate, *or* crenate-dentate to serrate-dentate only toward the apices of the blades (*Fothergilla, Hamamelis*). **Hamamelidaceae,** p. 215
        38. Margins of the leaf blades closely serrate or doubly serrate distally from very near the bases of the blades.
          39. Bark of older stems (trunks) exfoliating in papery-curly scales; leaf blades mostly deltoid-ovate or subrhomic (*Betula*). **Betulaceae,** p. 38
          39. Bark of older stems (trunks) ridged and grooved or exfoliating in irregular shreds or scales; shape of leaf blades other than as above.
            40. Leaf blades symmetrical; fruiting inflorescences usually present on a given tree throughout the growing season, the individual fruits subtended by foliaceous bracts or enclosed in saclike bracts (*Carpinus, Ostrya*).

  Betulaceae, p. 40

            40. Leaf blades, some of them at least on a given branch, asymmetrical basally; fruiting inflorescences or fruits shed by the time the leaves are fully developed or shortly thereafter, *or* flowering and fruiting occurring in autumn (*Ulmus*). **Ulmaceae,** p. 61
      37. The principal lateral veins of the leaf blades branching well before they reach the leaf edge, the branches ending at the leaf edge (*Planera*).

  Ulmaceae, p. 61

    36. Principal lateral veins of the leaf blades neither equally prominent nor extending essentially straight and parallel to each other from the midrib (*Celtis*).

  Ulmaceae, p. 61

    35. The leaves 3-ranked (somewhat irregularly so in *Itea*).
      41. Principal lateral veins of the leaf blades straight-ascending parallel to each other from the midrib and ending in the serrulations; stipules or stipule scars present (*Alnus*). **Betulaceae,** p. 37
      41. Principal lateral veins of the leaf blades somewhat arcuate-ascending from the midrib, eventually running almost parallel with the leaf margin and anastomosing; stipules or stipule scars none (*Itea*). **Saxifragaceae,** p. 213
  34. Leaves more than 3-ranked on the twigs.
    42. The larger of the lateral veins of the leaf blades prominent and extending approximately parallel to each other from the midvein.
      43. Wood relatively soft, not dense, extremely light in weight; leaf margins entire; flowering occurring before or as the leaves emerge in early spring.

  Leitneriaceae, p. 27

      43. Wood dense, relatively hard, not notably light in weight; margins of most leaves on a given plant serrated distally from about the middle; flowering racemes produced terminally on branchlets of the season well after the leaves are fully matured. **Clethraceae,** p. 469

42. The larger lateral veins neither equally prominent nor extending parallel to each other from the midrib.

44. Vestiture of various parts wholly, or at least in part, of stellate hairs, branched hairs, or scales.

45. Leaves evergreen, some present on twigs of a previous season.

46. Twigs and leaves glabrous. **Myrsinaceae,** p. 502

46. Twigs pubescent or scurfy-scaly; at least the lower leaf surfaces pubescent or pubescent and scaly.

47. Twigs and lower leaf surfaces sparsely pubescent, the hairs mostly with 2–4 short, straight branches from the base; flowers large and showy, corolla of 5 separate white petals, the fully open corolla about 8 cm across. **Theaceae,** p. 350

47. Twigs and at least the lower leaf surfaces abundantly scurfy-scaly; flowers 3–4 mm long, corolla urceolate (*Lyonia ferruginea, L. fruticosa*). **Ericaceae,** p. 469

45. Leaves deciduous, none present on twigs of the previous season.

48. Leaf blades pinnately veined. **Styracaceae,** p. 507

48. Leaf blades palmately veined (*Physocarpus*). **Rosaceae,** p. 219

44. Vestiture of various parts, if any, not of stellate hairs, branched hairs, or scales.

49. Leaf blades with narrow, dark reddish glands along the midribs of their upper surfaces (*Aronia*). **Rosaceae,** p. 226

49. Leaf blades not having glands along the midribs of their upper surfaces.

50. Plant a "woody herb" with woody lower stems and herbaceous upper stems, growing more or less continuously throughout the season and as new growth proceeds producing axillary, showy flowers, the petals 4, bright yellow, lasting but a day (shattering very readily) (Certain species of *Ludwigia*). **Onagraceae,** p. 389

50. Plant not as described above.

51. Plants shrubs to about 2 m tall and with oblanceolate or elliptic leaves 2–7 cm long; flowers small, numerous, borne in terminal panicles or panicled corymbs, each flower with a campanulate floral tube bearing at its summit 5 small sepals and 5 small petals, numerous stamens, pistils usually 5 and free within the floral tube (*Spiraea*). **Rosaceae,** p. 220

51. Plants not having the above combination of characteristics.

52. Stipules or stipule scars present.

53. Plants shrubs up to about 2 m tall; inflorescences narrowly spikelike and terminating branchlets of the season, their axes thickish and somewhat fleshy; flowers unisexual, pistillate borne on the lower portion of a given axis, staminate above; inflorescences or infructescences present much of the season after early summer (*Sebastiana, Stillingia*). **Euphorbiaceae,** p. 283

53. Plants not having the foregoing combination of characteristics.

54. Leaf scars with one vascular bundle scar. **Aquifoliaceae,** p. 301

54. Leaf scars with three vascular bundle scars.

55. Leaf blades rounded and short-mucronate apically; buds with several scales (*Amelanchier*). **Rosaceae,** p. 228

55. Leaf blades acute or acuminate apically; buds with a single, caplike or mittenlike scale (*Salix*). **Salicaceae,** p. 34

52. Stipules or stipule scars none.

56. Pith of the twigs (from the beginning of their development) transversely diaphragmed, solid and homogeneous between the diaphragms. **Nyssaceae,** p. 427

56. Pith of the twigs continuous and homogeneous or transversely diaphragmed and chambered between the diaphragms.

57. The pitch transversely diaphragmed and chambered between the diaphragms. **Symplocaceae,** p. 505

57. The pith in both young and older twigs continuous and homogeneous *or* continuous in the first-year twigs and becoming transversely diaphragmed and chambered between the diaphragms by the second year.

58. Pith of both young and older twigs continuous and homogeneous.

    59. Fruit a septicidally or loculicidally dehiscing capsule or a fleshy berry or drupe; anthers attached basally, inverted at anthesis or during development, dehiscing at the base (the apparent apex) by terminal slits, clefts, or pores and often having awns or spurs; petals at least partially united (except in *Befaria*).

                                                    Ericaceae, p. 469

    59. Fruit a dry, indehiscent drupe; anthers attached near their middles, erect, dehiscing longitudinally.   **Cyrillaceae,** p. 297

58. Pith of first-year twigs continuous and homogeneous, becoming diaphragmed and chambered between the diaphragms as they age.   **Ebenaceae,** p. 502

Key 4

Plants Herbaceous, with all or some of the Flowers on a given Plant Unisexual

1. Plant a vine.

    2. Climbing achieved by means of tendrils, or stems trailing and bearing tendrils.

        3. Ovary inferior.   **Cucurbitaceae,** p. 730

        3. Ovary superior (*Cissus*).   **Vitaceae,** p. 317

    2. Climbing achieved by bending or twining of the stalks of the leaves or of the leaflets (Certain species of *Clematis*).   **Ranunculaceae,** p. 127

1. Plant not a vine.

    4. Leaves compound, leaflets distinctly differentiated as such (*Thalictrum*).   **Ranunculaceae,** p. 130

    4. Leaves simple, entire to variously lobed or divided; if divided, then not having distinctly differentiated leaflets.

        5. The leaves dissected into narrow or filiform segments, or reduced to scales.

            6. Leaves, at least those below the inflorescence, dissected into narrow or filiform segments.

                7. The leaves pinnately dissected; ovary inferior (*Myriophyllum*).   **Haloragaceae,** p. 421

                7. The leaves dichotomously forked; ovary superior.   **Ceratophyllaceae,** p. 121

            6. Leaves represented by opposite scales (*Salicornia*).   **Chenopodiaceae,** p. 93

        5. The leaves neither dissected nor reduced to scales.

            8. Surfaces of leaf blades notably scurfy-scaly (*Atriplex*).   **Chenopodiaceae,** p. 99

            8. Surfaces of leaf blades not scurfy-scaly.

                9. Leaves all or chiefly opposite.

                    10. Plants diminutive; leaves small, not exceeding 6 mm broad, mostly narrower, and mostly 12 mm long or less; flowers minute and inconspicuous, scarcely visible to the unaided eye, borne in the leaf axils.   **Callitrichaceae,** p. 289

                    10. Plants not diminutive; leaves much larger than as described above; flowers small but borne in clearly evident, axillary spikes, cymes, or panicles.

                        11. Leaves sessile, half-terete to subterete, thickened-succulent.   **Bataceae,** p. 107

                        11. Leaves petiolate, their blades flat, not thickened-succulent.   **Urticaceae,** p. 50

                9. Leaves all alternate.

                    12. The leaves terete or subterete, succulent or subsucculent (*Suaeda*).

                                      Chenopodiaceae, p. 95

                    12. The leaves neither approximately terete nor succulent.

                        13. Leaves stellate-pubescent (*Croton*).   **Euphorbiaceae,** p. 280

                        13. Leaves glabrous or, if pubescent, pubescence not stellate.

                            14. Plants dioecious.

                                15. Each flower subtended by (usually 3) chaffy-scarious bracts; perianth chaffy-scarious.   **Amaranthaceae,** p. 101

                                15. Each flower not subtended by chaffy-scarious bracts; perianth not chaffy-scarious (certain species of *Rumex*).   **Polygonaceae,** p. 65

                            14. Plants monoecious or polygamous.

                              16. Plants monoecious.

                                17. Stem (and petioles) clothed with whitish, terete-subulate, stinging spicules dilated-jointed at the base (*Laportea*).   **Urticaceae,** p. 53

17. Stem glabrous or pubescent with nonstinging hairs (*Acalypha*, *Caperonia*, *Phyllanthus*). **Euphorbiaceae**, p. 276
16. Plants polygamous (bisexual, staminate, and pistillate flowers on the same plant) (*Parietaria*). **Urticaceae**, p. 53

Key 5
Plants Herbaceous, Flowers Bisexual, lacking a Perianth

- Plant thalluslike, aquatic, attached to rocks by fleshy discs. **Podostemaceae**, p. 203
- Plant not thalluslike, with relatively large, petiolate leaves; plant rooted in soil in wet places.
  Saururaceae, p. 21

Key 6
Plants Herbaceous, with Bisexual Flowers, Perianths of one Series of Parts *or*, if of more than one Series of Parts, then not distinctly differentiated as Sepals and Petals

1. Pistils more than one in each flower and wholly separate (ovaries separately embedded in pits near the top of a fleshy, spongy, flat-topped receptacle, styles protruding, in Nelumbonaceae).
  2. Leaves peltate (only the leaves subtending the flowers peltate in *Cabomba*, other leaves finely dissected).
    3. Leaves, most of them, finely dichotomously dissected. **Cabombaceae**, p. 157
    3. Leaves all peltate.
      4. Leaf blades orbicular, petioles and lower leaf surfaces not having a gelatinous sheath; flowers large and showy, 15–25 cm across; perianth parts numerous, yellow. **Nelumbonaceae**, p. 158
      4. Leaf blades broadly elliptical, petioles and lower leaf surfaces with a gelatinous sheath; flowers about 2–2.5 cm across; perianth parts 6 (–8), dull purple (*Brasenia*). **Cabombaceae**, p. 157
  2. Leaves, none of them peltate.
    5. The leaves compound, leaflets distinctly differentiated as such (*Clematis, Thalictrum clavatum*).
      Ranunculaceae, p. 123
    5. The leaves simple, more or less divided in some but not with definitely differentiated leaflets.
      6. Plants diminutive, with numerous, narrow leaves closely set basally; flowers solitary terminating scapes, the pistils many and the receptacle bearing them elongating greatly as the fruits mature; sepals long-spurred basally, petals clawed, a nectariferous pit at the summit of the claw (*Myosurus*). **Ranunculaceae**, p. 125
      6. Plants not having the combination of characteristics described above.
        7. Leaves stipulate, stipules subulate, about 5 mm long (*Dalibarda*). **Rosaceae**, p. 225
        7. Leaves not stipulate.
          8. Blades of principal leaves orbicular to broadly ovate, cordate basally; sepals bright yellow (*Caltha*). **Ranunculaceae**, p. 133
          8. Blades of principal leaves variable but not with shape as described above; sepals either white or at least partially purplish.
            9. Sepals obovate, membranous, 3–5 mm long, white; styles short-subulate, glabrous (*Trautvetteria*). **Ranunculaceae**, p. 133
            9. Sepals narrowly oblong on their proximal halves, acuminate apically, 25–40 mm long or a little more, thickish-spongy; styles plumose, greatly elongating as the fruits mature (*Clematis*). **Ranunculaceae**, p. 127
1. Pistil one in each flower (if evidently of more than one carpel, then carpels at least partly united and each with a separate style).
  10. Flower with 5 persistent sepals, the inner two (termed *wings*) much larger than the other three and usually colored like petals; corolla tubular basally, with 3 principal lobes above, the central lobe clawed, often somewhat boatlike, often sublobed, in some fringed apically, lateral lobes similar to each other (*Polygala*). **Polygalaceae**, p. 264
  10. Flower with perianth other than as described above.
    11. Leaves compound (*Sanguisorba*). **Rosaceae**, p. 225
    11. Leaves simple (if divided then not having distinctly differentiated leaflets).
      12. The leaves, at least those on the lower stem, opposite or in "apparent whorls."

15

13. Leaves opposite.
  14. Style one in each flower.
    15. Floral tube fully coalesced with the wall of the ovary and indistinguishable from it (*Ludwigia*).   **Onagraceae**, p. 389
    15. Floral tube separate from the ovary even if closely surrounding all or the basal part of it (some specimens of *Ammania latifolia*, *Didiplis*).   **Lythraceae**, p. 373
  14. Styles 2–5 in each flower.
    16. Stems and leaves not succulent.
      17. Styles 2 (*Chrysosplenium*).   **Saxifragaceae**, p. 206
      17. Styles 4 or 5 (*Sagina*).   **Caryophyllaceae**, p. 120
    16. Stems and leaves succulent.   **Aizoaceae**, p. 111
13. Leaves in "apparent whorls."
  18. Sepals 5; petals none.   **Molluginaceae**, p. 110
  18. Sepals none; petals 3 or 4, usually 4 (*Galium*).   **Rubiaceae**, p. 716
12. The leaves alternate.
  19. Plants with rhizomes on or within a soil substrate beneath water, the long-petiolate leaves borne from the rhizomes, blades usually floating; flowers similarly borne from the rhizomes, long-stalked, floating or emersed; perianth parts many (*Nymphaea*).   **Nymphaeaceae**, p. 160
  19. Plant not having attributes as described above.
    20. Leaves with tubular stipules sheathing the stem.   **Polygonaceae**, p. 65
    20. Leaves with or without stipules, if stipulate the stipules not tubular around the stem [note that in many Umbelliferae dilated bases of petioles (not stipules) sheathe the stem].
      21. Flowers each with 2 or more styles.
        22. Styles (or stigmas) 3 or 2.
          23. Flowers solitary and sessile in the axils of leaves or bracteal leaves; sepals 3; petals none; styles (or stigmas) 3 (*Proserpinaca*).   **Haloragaceae**, p. 419
          23. Flowers usually borne in umbels, simple or compound, sometimes in compact, head-like inflorescences; sepals none (in some genera or species); petals 5; styles (or stigmas) 2.   **Umbelliferae**, p. 433
        22. Styles 10 or 5.
          24. Styles 10; flowers or fruits in racemes and evenly distributed around their glabrous axes; racemes borne on the stems oppositely or suboppositely to the leaves.   **Phytolaccaceae**, p. 107
          24. Styles 5; flowers or fruits loosely spiked along the upper side of 2 to several subequal branches of cymes; cymes terminating branches, axes of cymes stipitate-glandular (*Penthorum*).   **Saxifragaceae**, p. 206
      21. Flowers each with one style (this may be branched above the base).
        25. Leaves terete or subterete, succulent or subsucculent (*Suaeda*).   **Chenopodiaceae**, p. 95
        25. Leaves flat, rarely if ever subsucculent.
          26. Stamens 6, 2 shorter than the other 4 (*Rorippa sessiliflora*).   **Cruciferae**, p. 171
          26. Stamens 5 or fewer, of equal length.
            27. Calyx and bracts subtending the flowers chaffy-scarious.   **Amaranthaceae**, p. 101
            27. Calyx and bracts, if any subtending the flowers, green, not scarious.
              Urticaceae, p. 50

Key 7

Plants Herbaceous, with Bisexual Flowers, Perianths with definitely Distinguishable Sepals and Petals, each Flower with more than one Pistil, Ovaries wholly Separate

1. Leaves alternate.
  2. Flowers hypogynous, the perianth and stamens inserted below the ovaries.
    3. Plant with milky sap (*Amsonia*).   **Apocynaceae**, p. 541
    3. Plant not having milky sap.
      4. Stems and leaves succulent; leaves sessile (*Diamorpha*, *Sedum*).   **Crassulaceae**, p. 203
      4. Stems and leaves not succulent; leaves, at least the lower ones, petiolate (*Ranunculus*).
        Ranunculaceae, p. 133

2. Flowers perigynous or epigynous, the perianth and stamens inserted on or near the rim of a floral tube. **Rosaceae, p. 217**
1. Leaves opposite.
    5. Ovaries 4 in each flower, each with a very short style; leaves 4–6 mm long, sessile bases of opposite pairs united around the nodes; sap not milky (*Tillaea*). **Crassulaceae, p. 203**
    5. Ovaries 2 in each flower, their style or styles conspicuous; leaves much longer than 6 mm, their bases not united around the nodes; sap milky.
        6. Style one in each flower. **Apocynaceae, p. 538**
        6. Styles two in each flower, their stigmas united. **Asclepiadaceae, p. 543**

Key 8

Plants Herbaceous, with Bisexual Flowers, Perianths with definitely Distinguishable Sepals and Petals, each Flower with one Pistil (Ovary United at least basally), Stamens more numerous than the Petals or Lobes of the Corolla

1. Flowers hypogynous or perigynous.
  2. The flowers radially symmetrical.
    3. Leaves uniquely modified and unlike "usual" leaves.
      4. The leaves pitcherlike or trumpetlike. **Sarraceniaceae, p. 191**
      4. The leaves with blades hinged lengthwise in the middle, each of the two halves nearly kidney-shaped, their margins each equipped with conspicuous bristles which interlock when the leaf "closes" (*Dionaea*). **Droseraceae, p. 185**
    3. Leaves not uniquely modified as described in couplet numbered 4 above.
      5. Plants with robust rhizomes on or within a soil substrate and beneath water, the long-petioled leaves and long-stalked flowers both arising from the rhizomes and the leaf blades and flowers floating or somewhat emersed (*Nuphar*). **Nymphaeaceae, p. 166**
      5. Plants not as described above.
        6. Leaves opposite, rarely whorled.
          7. Flowers hypogynous.
            8. The flowers in the leaf axils sessile or nearly so. **Elatinaceae, p. 328**
            8. The flowers in cymes.
              9. Petals white (*Arenaria, Stellaria*). **Caryophyllaceae, p. 119**
              9. Petals yellow, or flesh-colored to mauve-purple or pink. **Guttiferae, p. 330**
          7. Flowers perigynous (*Cuphea*). **Lythraceae, p. 375**
        6. Leaves alternate, at least those on the lower stem.
          10. Stamens 6, 2 shorter than the other 4; flowers in bractless racemes. **Cruciferae, p. 170**
          10. Stamens not as above; flowers in leaf axils or subtended by bracts.
            11. Inflorescence essentially scapose, a leafless or nearly leafless cyme or panicled cyme (*Saxifraga*). **Saxifragaceae, p. 206**
            11. Inflorescence not at all scapose.
              12. Flowers perigynous (*Lythrum*). **Lythraceae, p. 380**
              12. Flowers hypogynous. **Malvaceae, p. 320**
  2. The flowers bilaterally symmetrical.
    13. Calyx wholly petaloid.
      14. Sepals (in ours) 3, the 2 upper ones (as the flower dangles on its stalk) small, the lower one relatively large, saccate, open in front and the saclike portion spurred. **Balsaminaceae, p. 260**
      14. Sepals 5, the inner 2 (termed wings) much larger than the other 3 and colored like petals. **Polygalaceae, p. 264**
    13. Calyx not at all petaloid.
      15. Leaves palmately compound, bearing a stipular prickle at either side of the leaf base; petioles bearing both long and short gland-tipped hairs and usually scattered, opaque, sharp prickles as well; corolla of 4 separate, unequal, long-clawed petals. **Capparaceae, p. 184**
      15. Leaves various but without stipular prickles, petioles not prickly; corolla papilionaceous or petal 1 and with 4 petaloid staminodia alternating with the stamens. **Leguminosae, p. 235**
1. Flowers epigynous.
  16. Plant succulent; styles 3 or more. **Portulacaceae, p. 116**

16. Plant not succulent; style 1.
    17. Floral tube at full maturity flask-shaped; anthers dehiscing by terminal pores.
                                            Melastomataceae, p. 360
    17. Floral tube not at all flask-shaped; anthers dehiscing longitudinally.   **Onagraceae**, p. 389

Key 9

Plants Herbaceous, with Bisexual Flowers, Perianths definitely Distinguishable as Calyx and Corolla, Pistil one in each Flower (Ovary United at least basally), Stamens just as many as the wholly Separate Petals or fewer

1. Diminutive plant, in its entirety rosettelike and only 1–3 cm across (easily overlooked); leaves spatulate, 3–10 mm long (*Lepuropetalon*).   **Saxifragaceae**, p. 205
1. Plant not as described above.
  2. Flowers hypogynous, perigynous, or with a floral tube fused to the ovary only basally.
    3. The flowers perigynous (floral tube may closely surround all or part of the ovary but is free from it).   **Lythraceae**, p. 373
    3. The flowers hypogynous or with only the base of the ovary fused with the floral tube.
      4. Flowers bilaterally symmetrical.   **Violaceae**, p. 350
      4. Flowers radially symmetrical.
        5. Petals 5.
          6. Flower or inflorescence on a scape.
            7. Scape bearing a single flower terminally; plant wholly glabrous (*Parnassia*).
                              Saxifragaceae, p. 209
            7. Scape bearing an essentially 1-sided raceme (flowers short-stalked) terminally; leaves bearing gland-tipped hairs each exuding a droplet of clear, glutinous fluid (which glistens in bright light) (*Drosera*).   **Droseraceae**, p. 185
          6. Flowers or inflorescences not scapose.
            8. The flowers 1–3, very short-stalked, borne in the leaf axils (*Bergia*). **Elatinaceae**, p. 330
            8. The flowers in cymes or cymose clusters or in compact, headlike glomerules terminating branches.
              9. Flowers in cymes or cymose clusters.
                10. Leaves punctate (*Hypericum*).   **Guttiferae**, p. 330
                10. Leaves not punctate.
                    11. Petals white.
                      12. Plants slender, their leaves sessile or subsessile and with entire margins (*Drymaria, Spergularia*).   **Caryophyllaceae**, p. 118
                      12. Coarse plant, lower leaves, at least, long-petiolate, blades large and broad, deeply palmately lobed (*Boykinia*).   **Saxifragaceae**, p. 210
                    11. Petals yellow.   **Linaceae**, p. 257
              9. Flowers in compact, headlike clusters terminating branches (*Melochia*).
                              Sterculiaceae, p. 328
        5. Petals 3 or 4.
          13. Petals and stamens 3 each (*Elatine*).   **Elatinaceae**, p. 330
          13. Petals and stamens 4 each (*Cardamine hirsuta*).   **Cruciferae**, p. 183
  2. Flowers epigynous, ovary wholly inferior.
    14. Style 1.   **Onagraceae**, p. 389
    14. Styles 2.   **Umbelliferae**, p. 433

18

Key 10

Plants Herbaceous, with Bisexual Flowers, Perianths definitely Distinguishable as Calyx and Corolla, Pistil one in each Flower, Stamens just as many as the Lobes of the wholly or partially United Corolla, or fewer

1. Leaves closely set on the stem basally, essentially in a rosette, flower or inflorescence scapose.
  2. Scape bearing a single flower terminally (*Pinguicula*).   **Lentibulariaceae,** p. 671
  2. Scape bearing an inflorescence.
    3. Inflorescence a narrow, cylindrical spike.   **Plantaginaceae,** p. 710
    3. Inflorescence an open panicle.   **Plumbaginaceae,** p. 489
1. Leaves distributed on the stem above the base even if there is a basal rosette.
  4. Leaves pinnately dissected into numerous, narrowly linear segments and appearing distinctly feathery; plant aquatic (*Hottonia*).   **Primulaceae,** p. 490
  4. Leaves not pinnately dissected into narrowly linear segments and appearing feathery or, if so, then plant definitely terrestrial.
    5. Plant with a rhizome in a soil substrate beneath water; long petiolelike stems arising from the rhizome and reaching to or near the water surface, each stem bearing 1 (rarely more than 1) leaf, the single node evident or sometimes obscure; flowers in umbels from the nodes, anthesis of the individual flowers occurring over a considerable period of time.   **Menyanthaceae,** p. 537
    5. Plant not as described above.
      6. Ovary wholly or almost wholly inferior.
        7. Corolla 5-lobed.
          8. Leaves alternate.   **Campanulaceae,** p. 734
          8. Leaves opposite (*Pentodon*).   **Rubiaceae,** p. 723
        7. Corolla 4-lobed (very rarely 3–5-lobed in *Mitchella*).   **Rubiaceae,** p. 712
      6. Ovary superior or only slightly inferior.
        9. Leaves all scalelike, 5 mm long or less (*Bartonia*).   **Gentianaceae,** p. 523
        9. Leaves not scalelike.
          10. Corolla radially symmetrical or nearly so *and* the stamens as many as the lobes of the corolla.
            11. Stamens inserted opposite to the lobes of the corolla.   **Primulaceae,** p. 489
            11. Stamens inserted alternate to the lobes of the corolla.
              12. Ovary deeply 2- or 4-lobed, lobes rounded laterally.
                13. Styles 2; ovary 2-lobed (*Dichondra*).   **Convolvulaceae,** p. 560
                13. Style 1; ovary 4-lobed.
                  14. Leaves alternate.   **Boraginaceae,** p. 580
                  14. Leaves opposite.
                    15. Style cleft at the apex.   **Labiatae,** p. 589
                    15. Style capitate at the summit or with 2 obscure, rounded lobes.   **Verbenaceae,** p. 585
            12. Ovary not conspicuously lobed, or, if lobed, lobes flattened.
              16. The ovary 1-locular or 3-locular.
                17. Ovary 1-locular; leaves not needlelike or awllike.   **Gentianaceae,** p. 523
                17. Ovary 3-locular; leaves needlelike or awllike.   **Diapensiaceae,** p. 487
               16. The ovary 2-locular or 4-locular.
                18. Leaves opposite.
                  19. The leaves with stipules or with stipular line-scars connecting the bases of the pairs.   **Loganiaceae,** p. 514
                  19. The leaves without stipules.   **Verbenaceae,** p. 585
               16. Leaves all alternate or at least those distally on the stem alternate.
                20. Corolla lobed to its middle.   **Hydrophyllaceae,** p. 573
                20. Corolla virtually unlobed or very shallowly lobed.   **Convolvulaceae,** p. 554
          10. Corolla *either* distinctly bilaterally symmetrical *or* the stamens fewer than the lobes of the corolla.
            21. Plant a viscid-pubescent, strong-scented annual, the stem prostrate-spreading, to 1 m

19

long, branches sometimes ascending; ovary 2-locular; fruit a 2-valved capsule with a conspicuous beak, the beak splitting and elastically spreading as the fruit matures, the split halves strongly incurving giving a markedly 2-horned effect (*Proboscidea*). **Martyniaceae,** p. 709
21. Plant without the combination of characteristics described above.
    22. Ovary deeply 4-lobed (from the summit).
        23. Style cleft apically.   **Labiatae,** p. 589
        23. Style capitate apically, sometimes with 2 obscure, rounded lobes.
        Verbenaceae, p. 585
    22. Ovary not deeply 4-lobed.
        24. Anthers united into a tube around the style (*Lobelia*).   **Campanulaceae,** p. 734
        24. Anthers separate.
            25. Leaves with stony concretions in the epidermis, these evident as short lines or dots; fruit a compressed, 2-valved capsule splitting and elastically spreading above the base at dehiscence, the seeds borne on curved or hooked projections from the placentae.
            Acanthaceae, p. 696
            25. Leaves without stony concretions in the epidermis; fruit splitting at maturity into 2 or 4 nutlets or a 2-valved capsule, seeds in which are not borne on curved or hooked projections from the placentae.
                26. Fruit splitting at maturity into 2 or 4 nutlets.   **Verbenaceae,** p. 585
                26. Fruit a 2-locular, 2- or 4-valved capsule containing few to many seeds (capsule in *Micranthemum* usually 1-locular by full maturity and dehiscing by irregular rupturing of its thin wall).   **Scrophulariaceae,** p. 625

Saururaceae (LIZARD'S-TAIL FAMILY)

Saururus cernuus L. LIZARD'S-TAIL. Fig. 1

Perennial herb, rhizomatous and forming colonies. Aerial stems erect, to about 12 dm tall, unbranched below, simple or with few ascending branches above, pubescent. Leaves alternate, petioles shorter than the blades, dilated basally and slightly sheathing the stem, blades ovate-cordate, subpalmately veined, margins entire; lower leaf surface sparsely short-pubescent, more heavily pubescent on the principal veins, usually becoming glabrous with age. Inflorescence a stalked, slender white raceme, curved-nodding distally, to about 3 dm long. Flowers without a perianth but each with a small subtending bract; stamens 6–8 (fewer by abortion), the filaments long and slender, much exceeding the carpels. Carpels 3–5, usually united only at the base, at maturity carpels appressed forming a depressed-globose "fruit," the outer surface of each part wrinkled, parts separating as shed, each indehiscent.

In shallow water of streams, marshy margins of bodies of water, swamps, wet woodlands, ditches. s.w. Que. and s. Ont., Minn., Ill., Mo., Kan., e. Okla., s.e. Tex., generally eastward.

Juglandaceae (WALNUT AND HICKORY FAMILY)

Carya (HICKORIES)

Trees with hard, heavy wood, continuous pith in the twigs. Leaves deciduous, alternate, odd-pinnately compound, without stipules, leaflets pinnately veined, glandular dotted beneath; herbage pungently aromatic when crushed or bruised. Flowers unisexual, plants monoecious, the staminate in pendent catkins, usually in a fascicle of 3 borne from the summit of the twig of the previous year or from the base of the developing shoot of the season; pistillate flowers solitary or in 2–6-flowered spikes terminal on developing shoots of the season. Staminate flower with a 2- or 3-lobed calyx subtended and usually exceeded by a bract, stamens 3–8, anthers sessile, pubescent. Pistillate flower with 1 pistil, the ovary surrounded by and adnate to an involucre 4-lobed at the summit; involucre together with the exocarp of the ovary developing into the husk of the fruit, which at maturity separates from the enclosed hard shell of a 1-seeded nut, the husk wholly or partially splitting into 4 valves.

1. Scales of overwintering buds 4 or 6, arranged margin to margin (valvate); scars left by bud scales wider medially than on the ends, staggered and separated from each other, not forming a band around the twig; valves of the husk of the fruit keeled or somewhat winged on the sutures, at least distally.
  2. Leaflets 11–15, sometimes 9 or 7 on individual leaves of a given tree.
    3. Lower surfaces of leaflets essentially glabrous; nut round in cross-section; kernel of nut sweet, each half barely notched apically; fascicle of staminate catkins sessile.   1. *C. illinoensis*
    3. Lower surfaces of leaflets pubescent at least along the midrib and in the axils of the lateral veins; nut angled in cross-section; kernel of nut bitter, each half deeply 2-lobed apically; fascicle of staminate catkins stalked.   2. *C. aquatica*
  2. Leaflets 5–9.
    4. Overwintering bud rusty orange to yellow, husk of fruit not keeled or winged along the sutures proximally, at maturity splitting only to just below the middle; kernel of the nut bitter; lower surfaces of leaflets pale green to copper-colored but dull and not with a metallic lustre.   3. *C. cordiformis*
    4. Overwintering buds brownish; husk of the fruit keeled or ridged along the full length of the sutures, at maturity splitting to the base; kernel of the nut sweet; lower surfaces of leaflets with a silvery metallic lustre early in the season, with a conspicuously coppery metallic lustre late in the season.
                             4. *C. myristicaeformis*
1. Scales of overwintering buds imbricate; scars left by the bud scales very narrow, collectively forming a band of scars around the twig; valves of the husk of the fruit not keeled or winged.

21

Fig. 1. **Saururus cernuus:** a, base of plant; b, top of plant; c, flower and subtending bract; d, mature fruit; e, seed.

5. Margins of young leaflets densely ciliate and with tufts of hairs on the serrations, the tufts of hairs persistent on older leaflets.   5. *C. ovata*

5. Margins of young leaflets if pubescent without hairs in tufts.
     6. Bark shaggy.   6. *C. laciniosa*
     6. Bark tight.   7. *C. glabra*

1. Carya illinoensis (Wang.) K. Koch. PECAN.   Fig. 2

Potentially a large tree, in natural habitats often with an enlarged buttressed base, the bark grayish, narrowly fissured and separating freely into platelike scales. Overwintering buds valvate, flattened, covered with jointed hairs. Leaflets usually 11–17, sometimes 9, lanceolate to oblanceolate, the lateral inequilateral basally and falcate, apices acuminate, median leaflets usually the largest, margins finely serrate or crenate-serrate, at maturity both surfaces essentially glabrous; terminal leaflet with a stalk 1–2 cm long, laterals subsessile. Fascicles of staminate catkins sessile. Fruit more or less ellipsoid or ovoid, sutures of the husk keeled or narrowly winged, splitting nearly to the base. Nut round in cross-section; kernel sweet, each half barely notched apically. (*Hicoria pecan* (Marsh.) Britt.)

Floodplain forests. s.w. Ohio to Iowa, southward to Ala. and Tex., adjacent Mex.

Many selected varieties widely cultivated in the southern states for the nuts, a valuable article of commerce. Naturalized in hedgerows, old fields, open woodlands and woodland borders, in areas where cultivated.

2. Carya aquatica (Michx. f.) Nutt. WATER HICKORY, BITTER PECAN.   Fig. 3

A medium to large tree, the bark similar to that of *C. illinoensis*. Overwintering buds valvate, reddish brown, clothed with yellow glands which are shed early. Leaflets 11–15, rarely fewer, mostly lanceolate to narrowly oblong, lateral ones equilateral to inequilateral, usually falcate, apices acuminate, upper lateral usually largest, margins serrate, at maturity glabrous above, pubescent at least along the midrib and in the axils of the lateral veins below; terminal leaflet with a stalk about 1 cm long, laterals sessile or subsessile. Fascicle of staminate catkins stalked. Fruit angled, the sutures somewhat keeled or winged, flattened between the sutures, the husks splitting along the full length of the sutures. Kernel bitter, each half deeply 2-lobed apically. (*Hicoria aquatica* (Michx. f.) Britt.)

In floodplains where inundated only temporarily, high river banks, on natural levees along rivers and streams. Chiefly coastal plain, Va. to s.cen. pen. Fla., westward to Tex. and e. Okla., northward in the interior to s.e. Mo. and s. Ill.

Fig. 2. **Carya illinoensis.** (From Kurz and Godfrey, *Trees of Northern Florida.* 1962)

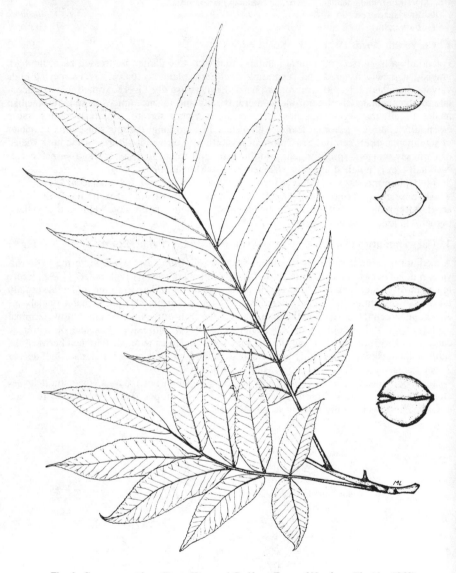

Fig. 3. **Carya aquatica.** (From Kurz and Godfrey, *Trees of Northern Florida*. 1962)

24

3. Carya cordiformis (Wang.) K. Koch. BITTERNUT HICKORY.

Medium-sized tree, the brownish, scaly bark sloughing in small flaky ridges. Overwintering buds valvate, rusty orange to yellow, persistently scurfy. Leaflets 7–9, lanceolate, oblanceolate, elliptic, or obovate, some of them, especially the lanceolate or oblanceolate ones, falcate, cuneate to rounded basally, sometimes inequilateral, apices acute to acuminate, margins finely serrate; upper surface glabrous and dark green, lower paler, dull, evenly pubescent, pubescent only along the veins, varying to glabrous, often clothed with resinous, scurfy, brown to dull coppery scales; terminal leaflet sessile or very short stalked, laterals sessile or subsessile. Fascicles of staminate catkins stalked. Fruits subglobose to ovoid, beaked apically, husk thin, its sutures rounded-keeled above the middle, splitting from the apex to about the middle. Kernel of the nut reddish brown, very bitter. (*Hicoria cordiformis* (Wang.) Britt.)

River and stream banks, floodplain forest where only temporarily inundated, and on well-drained upland forest sites. s. Maine and Que. to Minn., southward to the Fla. Panhandle, e. Okla. and e. Tex.

4. Carya myristicaeformis (Michx. f.) Nutt. NUTMEG HICKORY.

Medium-sized tree, the bark brown tinged with red, irregularly broken into small, thin scales. Overwintering buds valvate, covered with dense, brown, scurfy pubescence. Leaflets 7–9, sometimes 5, the laterals lanceolate, lance-ovate, elliptic, or oblong-elliptic, usually somewhat inequilateral and rounded to cuneate basally, acute apically, terminal one usually larger than the others, acuminate basally and apically; leaflets sessile or nearly so, margins serrate, upper surface dark green, lower at first with a silvery metallic lustre, later becoming coppery metallic, glabrous or nearly so, with resinous, scurfy scales. Fruit ellipsoid to somewhat obovoid, the sutures of the husk ridged or keeled their full length, the thin husk at maturity splitting nearly to the base, surface densely clothed with brown-coppery scales. Nutshell thick and hard, reddish brown, apiculate at both ends. Kernel dark brown, sweet. (*Hicoria myristicaeformis* (Michx. f.) Britt.)

Swamps, river banks, rich woodlands, usually underlain by marl. Rare and local in much of the range, e. S.C., cen. Ala. to Ark., e. Okla., La. and e. Tex.

5. Carya ovata (Mill.) K. Koch. SHELLBARK HICKORY, SHAGBARK HICKORY.

Medium-sized tree, bark light gray, separating into elongate, shaggy plates. Overwintering buds with 10–12 imbricated scales. Leaflets usually 5, sometimes 7, broadly lanceolate, ovate, or obovate, relatively large, the terminal one commonly largest, becoming 1–2 dm long and 5–11 cm wide, lowermost leaflets much smaller than those above; margins at first densely ciliate and with tufts of hairs on the serrations, the tufts usually persistent on older leaflets, surfaces of mature leaves glabrous or nearly so, dark yellowish green above, paler below. Fascicles of staminate catkins stalked. Fruits subglobose to obovoid, the summit flattish, husk thickish, reddish brown to nearly black, not keeled on the sutures, at maturity splitting to the base. Nut 4-angled, shell whitish to pinkish brown, thin. Kernel sweet. (*Hicoria ovata* (Mill.) Britt.)

Rich woodlands, stream banks, sometimes bottomland woodlands, only marginally if at all a wetland plant. s. Maine, s.w. Que., s. Ont. to s. Minn., generally southward to e. N.C., cen. S.C., cen. Ga., s. Ala., Miss., La. and e. Tex.

6. Carya laciniosa (Michx. f.) Loud. BIG SHELLBARK HICKORY, KING-NUT.

Similar to *C. ovata*. Leaflets usually 7, sometimes 9, mature ones glabrous above, permanently pubescent beneath, the serrations without tufts of hairs. (*Hicoria laciniosa* (Michx. f.) Sarg.)

Floodplain forest where inundated during periods of high water. N.Y. to w. Pa., Ohio, s. Mich., Ind., s.e. Iowa, e. Neb., e. Kan., southward to cen. N.C., s.cen. Ala., Miss., La., Ark., e. Okla.

7. Carya glabra (Mill.) Sweet. PIGNUT HICKORY. Fig. 4

A small to large tree, bark gray, tight, shallowly furrowed and with low interlacing ridges forming more or less of a diamond pattern. Overwintering buds with scales imbricated, closely silky-pubescent. Leaflets 5 or 7, sometimes 3 on individual leaves of a given tree, showing much variation in size and shape; lanceolate, elliptic, oblong, oblong-lanceolate, ovate, or lance-ovate, terminal and upper pair equal to subequal, lowest pair smallest; terminal leaflet usually acuminate at both extremities, others variably rounded to short-cuneate basally, acuminate apically, margins finely serrate; mature leaflets glabrous above, varyingly pubescent to glabrous below. Fruit globose, ellipsoid, obpyriform, smooth on the sutures, splitting tardily only part way from the summit, husks olive green to dark brown, varying in thickness. Shell of the nut hard and thick, kernel sweet to acrid. (*Hicoria glabra* (Mill.) Britt.; incl. var. *megacarpa* Sarg., *H. austrina* Small)

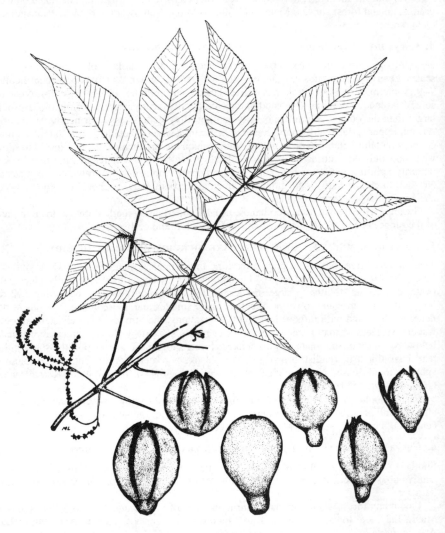

Fig. 4. **Carya glabra.** (From Kurz and Godfrey, *Trees of Northern Florida.* 1962)

26

Inhabiting a wide variety of sites, rich upland woodlands, mixed woodlands on deep, well-drained sands, mixed wet hammocks underlain by limestone, old stabilized dunes, sand pine–oak scrub, bottomlands or floodplain woodlands. e. Mass. and Vt. to s. Mich., s. Ill., and e. Mo., generally southward to cen. pen. Fla., s. Ala., e. and n. Miss., n.e. Ark.

Leitneriaceae (CORKWOOD FAMILY)

Leitneria floridana Chapm. CORKWOOD.   Figs. 5 and 6

Wandlike or treelike shrub to 6 m tall, spreading by shoots from the extensive roots. Bark reddish brown, smooth, with buff-colored lenticels. Twigs at first densely pubescent, becoming smooth with age. Leaves with petioles 2–4 cm long, blades elliptic or lanceolate-elliptic, 5–17 cm long, 2–5 cm wide, bases and apices acute; emerging leaves sparsely silky-pubescent above between the veins, more densely so on the veins, densely silky-pubescent on the lower surface, at maturity the upper surfaces glabrous, the lower softly pubescent, chiefly along the larger veins, margins entire; venation pinnate, the larger lateral veins prominent and extending approximately parallel to each other from the midrib, veinlets forming a fine network visible from either surface but more distinct beneath. Flowers borne in catkins produced from the previous year's wood before the leaves emerge, staminate and pistillate usually on separate plants, plants thus dioecious (rarely the plants polygamodioecious). Staminate catkins erect to spreading, flexuous, often nodding distally, brownish, 2–5 cm long, 1–1.5 cm broad; pistillate catkins stiffly erect, much less conspicuous than the staminate, reddish, mostly 1–2 cm long, about 1 cm broad. Stamens free, mostly in clusters of 10–12, sometimes fewer or more in a cluster, in the axils of spirally arranged, close-set, deltoid-ovate scales, perianth lacking. Pistillate flowers sessile, usually solitary in the axils of spirally arranged primary bracts, each with 2 bractlets at the base and surrounded by a perianthlike structure of (3)4 (–8) segments, two often larger than the others. Fruit an erect, smooth, long-elliptic or oblong-elliptic drupe, usually brownish at maturity.

Sporadic in occurrence. Sawgrass or sawgrass-cabbage palmetto marshes, brackish tidal shores, swampy woodlands, swampy prairies. s. Ga., n. Fla., westward to e. Tex., e. Ark., s.e. Mo.

Reference: Godfrey, R. K., and André F. Clewell. "Polygamodioecious *Leitneria floridana* (Leitneriaceae)." *Sida* 2: 172–173. 1965.

Myricaceae (WAX-MYRTLE FAMILY)

Myrica

Shrubs or small trees. Leaves alternate, simple, punctate on one or both surfaces; stipules none (in ours). Inflorescence a short, scaly, erect catkin. Flowers unisexual, rarely bisexual, plants usually dioecious; perianth none. Staminate flowers solitary in the axils of primary, usually ovate, bracts, stamens 2–20, usually 4–8, the short filaments free or united near the base. Pistillate flowers subtended by an ovate primary bract and usually by 2–4 secondary bractlets in addition, these sometimes absent; pistil 1, 2- (rarely 3-) carpellate, ovary superior, 1-locular, 1-ovulate, style short, stigmas 2. Fruit a nutlike drupe, in most covered by protuberances, infrequently glandular-dotted, usually with a waxy coating.

1. Bracts subtending the staminate flowers longer than the stamens, closely imbricate and hiding the stamens; pistillate flower with 2 bractlets enlarging as the fruit matures, persistent and clasping the fruit; mature fruiting catkins burlike.   1. *M. gale*
1. Bracts subtending the staminate flowers loosely spreading, shorter than the stamens and not at all obscuring them; bractlets of pistillate flowers deciduous after anthesis, the mature fruits naked, not in burlike catkins.

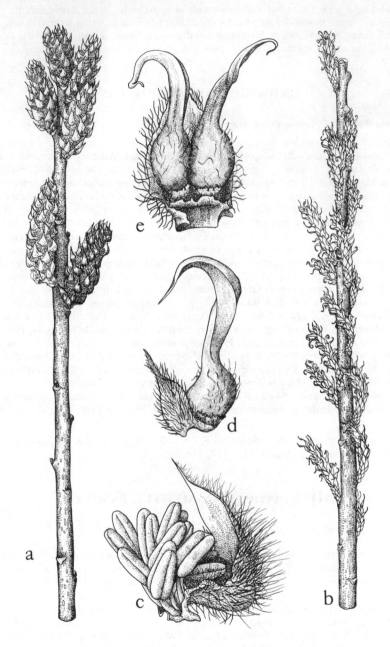

Fig. 5. **Leitneria floridana:** a, twig with staminate catkins; b, twig with pistillate catkins; c, staminate flower with subtending bract; d, usual type of pistillate flower with subtending bract; e, unusual pistillate flower with subtending bract.

Fig. 6. **Leitneria floridana:** a, branch with mature leaves and mature fruiting clusters; b, small section of twig; c, leaf, with section of lower surface enlarged.

2. Leaves not aromatic when crushed, glabrous on both surfaces, margins entire; mature fruits somewhat longer than broad, about 6–7 mm across.   2. *M. inodora*

2. Leaves aromatic when crushed, usually at least sparsely pubescent on one or both surfaces, usually the margins of some leaves on a given plant toothed; fruits globose or short-ovoid, 4.5 mm across or less.

    3. The leaves glandular-punctate on both surfaces.   3. *M. cerifera*

    3. The leaves glandular-punctate on their lower surfaces.

        4. Leaves evergreen; fruits, in early stages of development, not pubescent; in our range widely distributed, chiefly in the coastal plain and not rare.   4. *M. heterophylla*

        4. Leaves deciduous; fruits, in early stages of development, pubescent; in our range occurring only in n.e. N.C., where rare.   5. *M. pensylvanica*

1. Myrica gale L. SWEET GALE.

Shrub to about 2 m tall. Leaves deciduous, sessile, oblanceolate, mostly 2–4 cm long, 0.6–1.8 cm wide, cuneate-attenuate basally, rounded to obtuse apically, margins serrate above the middle, upper surfaces glabrous or pubescent along the midrib, the lower pubescent or glabrous, both surfaces with amber, glandular dots. Catkins at anthesis before shoots of the season emerge; staminate with closely imbricate primary bracts essentially concealing the flowers; fruiting catkins burlike, 2 bractlets below each flower enlarging as the fruit matures, persistent, eventually clasping the flattish fruit. (*Gale palustris* (Lam.) Chev.)

Nfld. to Alaska, southward to N.Y., s. Ont., Mich., Wisc., Minn., southward in the west to Ore., in the east southward in the Appalachian mts. to w. N.C. and e. Tenn.; Euras.

2. Myrica inodora Bartr. ODORLESS WAX-MYRTLE.   Fig. 7

Evergreen, glabrous shrub or small tree with a smooth, light gray bark on older stems. Leaves leathery, cuneately narrowed to short petiolar or subpetiolar bases, expanded portion of leaf elliptic, oblanceolate, or narrowly obovate, mostly 4–8 cm long and to 3.5 cm wide, rounded or obtuse apically, margins entire, narrowly revolute, surfaces very minutely punctate, sometimes showing tiny resinous dots of exudate. Catkins borne in early spring both on leafless wood and in axils of the persistent leaves, not necessarily both on the same twig. Fruits rarely globose, usually oblong-oval, somewhat longer than broad, mostly 6–7 mm across, covered by vertical, more or less platelike tubercles, overall surface of the latter dark brown or nearly black, sometimes lacking white-waxy exudate in the interstices between the tubercles, or with variable amounts of wax between them and on their outer surfaces, the fruits rarely completely white-waxy. (*Cerothamnus inodorus* (Bartr.) Small)

Shrub-tree bogs and bays, nonalluvial acid swamps. Fla. Panhandle, from Leon Co. westward, thence to s. Miss.

3. Myrica cerifera L. WAX-MYRTLE, SOUTHERN BAYBERRY.   Fig. 8

Evergreen, aromatic shrub or small tree, main stems to about 7 m tall and 2 dm in diameter near their bases; sometimes, where burned over every year or every few years, spreading by suckering from subterranean runners and forming colonies of low, bushy-branched stems with smaller leaves than on larger plants. Leaves cuneately narrowed to petiolar or subpetiolar bases, expanded portions of leaves oblanceolate (or linear-oblanceolate on dwarfed plants), mostly acute apically, margins usually irregularly few-toothed from about their middles upwardly, edges not revolute; leaves variable in size on a given plant, toward the tips of the branchlets commonly smaller than below; larger leaves 5–10 cm long, 1.5–2 cm wide; both surfaces punctate, commonly very abundantly so, usually with amber to brown dots of resinous exudate, lower surfaces sometimes short-pubescent, mostly along the midribs, sometimes on the principal lateral veins as well. Catkins borne in early spring, both on the leafless wood and in the axils of the persistent leaves, not necessarily both on the same twig. Mature fruits globose or short-ovoid, 2–3 mm across, the surficial tubercles usually compacted, the entire outer surface white-waxy. (*Cerothamnus cerifera* (L.) Small; incl. *M. pusilla* Raf., *M. cerifera* var. *pumila* Michx., *C. pumilus* (Michx.) Small)

Inhabiting a wide variety of sites: fresh to slightly brackish banks, shores, flats, interdune

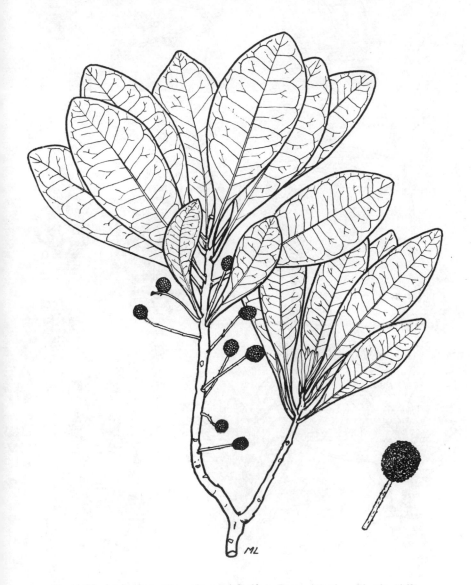

Fig. 7. **Myrica inodora.** (From Kurz and Godfrey, *Trees of Northern Florida*. 1962)

31

Fig. 8. a–g, **Myrica cerifera:** a, branch with staminate catkins; b, branch with pistillate catkins; c, fruiting branch; d, portion of upper leaf surface; e, portion of lower leaf surface; f, staminate flowers; g, fruit; h–q, **Myrica heterophylla:** h, branch with staminate catkins; i, branch with pistillate catkins; j, fruiting branch; k, portion of upper leaf surface; l, portion of lower leaf surface; m, staminate catkin; n, pistillate catkin; o, bract with pistillate flower; p, pistil; q, fruit. (From Correll and Correll. 1972)

swales, pine savannas and flatwoods, cypress-gum ponds and swamps, wet hammocks, upland mixed woodlands, fence and hedge rows. Coastal plain, s. N.J. to s. Fla., westward to e. Tex., s.e. Okla., Ark.; W.I.

4. Myrica heterophylla Raf.   Fig. 8

Evergreen, aromatic shrub to about 3 m tall. Leaves cuneately narrowed to petiolar or subpetiolar bases, expanded portion of leaf elliptic, oblanceolate, or narrowly obovate, leathery, apices obtuse or rounded, infrequently acute, often minutely apiculate at the tips, margins usually irregularly, shallowly few-toothed distally, the edges not revolute; upper surfaces glabrous, usually not glandular-punctate, lower surfaces punctate and with amber to brownish dots of exudate in the punctations or above them, often pubescent on the principal veins; leaf size variable, to 12 cm long and 3 cm wide, not tending to be smaller toward the tips of branchlets. Catkins borne in early spring on leafless twigs or in axils of the persistent leaves, not necessarily both on the same twig. Young fruits not pubescent; mature fruits globose or short-ovoid, 2–4.5 mm across, the tubercles compacted or not, sometimes completely covered with white wax, sometimes with little wax, the tubercles then dark brown; wax tending to be finely stringy and with magnification giving the effect of pubescence. (*Cerothamnus carolinensis* sensu Small (1933) in part)

Evergreen shrub-tree bogs and bays, hillside bogs in pinelands, wet pine savannas and flatwoods. Coastal plain and piedmont, Va. to cen. pen. Fla., westward to e. Tex.

5. Myrica pensylvanica Loisel. BAYBERRY.

Similar to *M. heterophylla*. Leaves more membranous, deciduous, elliptic or oblanceolate, less frequently narrowly obovate, often pubescent above and on the veins below; young fruits pubescent on the tubercles.

In our range, coastal sand flats and dunes, rare, n.e. N.C.; more common northward to Nfld.; locally shores of Lake Erie, isolated stations inland.

Salicaceae (WILLOW FAMILY)

Trees or shrubs. Leaves simple, alternate, stipulate, pinnately veined. Flowers in catkins, without perianth, each subtended by a small, scalelike bract, unisexual, the staminate and pistillate on different plants. Staminate flowers with 1 to many stamens, filaments slender, distinct or partly united. Pistillate flowers of a single pistil, the ovary superior, 1-locular, ovules several to numerous, borne on 2–4 basal or parietal placentae, style short or none, stigmas 2–4 (as many as the lobes of the placentae). Fruit a longitudinally 2–4-valved capsule bearing numerous seeds each bearing a terminal tuft of soft, silky hairs.

● Buds several, imbricated; leaves (in ours) ovate, less than 3 times as long as broad; inflorescences pendulous.   1. *Populus*
● Buds with a single scale; leaves (of ours) not ovate, more than 3 times as long as broad; inflorescences erect or spreading.   2. *Salix*

1. Populus (POPLARS, ASPENS, COTTONWOODS)

Fast-growing trees with bitter-astringent bark and light wood. Buds with several imbricated scales and gummy. Leaves simple, long-petiolate, alternate, deciduous, leaf scars prominent, with 3 vascular bundle scars, blades (in ours) often as broad as long, or broader than long, usually not exceeding 2 times as long as broad; petioles with nectariferous glands at the summit. Stipules present but falling early. Flowers individually small, borne in pendulous catkins before the leaves emerge, unisexual, the plants dioecious, each flower inserted on a shallow or cuplike disc and subtended by a bract. Perianth none. Staminate flowers with 4–12 (–60) stamens; pistillate with 1 pistil, the ovary 1-locular, with 2–4 parietal placentae, style short, stigmas same in number as the placentae. Fruit an ovate 2–4-valved capsule which matures before the leaves reach maturation. Seeds abundant, minute, with a basal tuft of long, silky, ascending hairs.

● Leaf blades acuminate apically, the tips distinctly pointed; petiole flattened and somewhat dilated distally; 2–5 glands at the very base of the blade or on the summit of the petiole, these often erect and somewhat foliar; blades and petioles glabrous or occasionally with sparse, short pubescence near the base, along the lower margins, or on the distal part of the petiole; twigs glabrous and with low, narrow, longitudinal ridges.   1. *P. deltoides*
● Leaf blades bluntly obtuse apically; petioles essentially terete throughout and not dilated distally; glands, if any, on the edges of the blades at either side of the petiole; young leaf blades, petioles, and twigs cottony-pubescent, most of the pubescence often sloughing during maturation leaving only a patch persisting around the veins at the base; twigs not longitudinally ridged.   2. *P. heterophylla*

1. Populus deltoides Bartr. ex Marsh. var. deltoides. EASTERN COTTONWOOD.

Tree to 30 m tall or more, bark grayish, roughly and irregularly fissured and ridged. Twigs yellowish green, glabrous, lustrous, with low, narrow ridges running longitudinally, older branches roughened by the raised leaf scars. Leaf blades at maturity leathery, stiffish, prominently pinnately veined, mostly 5–15 cm long (longer on sprouts and saplings), deltoid, deltoid-ovate, to suborbicular-ovate, mostly truncate basally, sometimes broadly cordate but infrequently with small, cordate-lobing near the summit of the petiole, acuminate apically, the tips distinctly pointed; margins crenate-dentate or crenate-serrate, surfaces glabrous or less frequently with sparse, short pubescence near the base, on the lower margins, or distally on the petioles; petioles flattened and usually dilated distally, the flattening at right angles to the surface of the blades; commonly a pair, sometimes 3, 4, or 5, often but not always somewhat foliose, glands at the base of the blade or some or all of the glands from the petiole just below the blade.

Floodplain and lowland wet woodlands and swamps. Que. and N. Eng. to s. Man., generally southward to n. pen. Fla. and Tex.

2. Populus heterophylla L. SWAMP COTTONWOOD, BLACK COTTONWOOD.

Small to large tree, the bark brownish tinged with red, irregularly fissured and furrowed, on old trunks sloughing in long, narrow plates. Twigs stoutish, somewhat reddish, cottony-pubescent at first, eventually nearly glabrous, without longitudinal ridges, eventually roughened by the raised leaf scars. Mature leaf blades thick and stiffish, prominently pinnately veined, 10–20 cm long, ovate, from about as broad as long to twice as long as broad, their bases truncate to broadly rounded, often with small cordate lobing to either side of the summit of the petiole, sometimes covering the summit of the petiole, apices bluntly obtuse; margins shallowly and usually evenly crenate-serrate, surfaces cottony-pubescent upon expanding, upper surface becoming dark green and glabrous save for a persistent patch of cottony pubescence around the veins at the base, lower surface becoming grayish or whitish, commonly hairy in the vein angles and nearly always with a patch of pubescence around the veins at the base; petioles at first cottony-pubescent, later mostly glabrous, essentially terete and not dilated distally; edges of the blade just next to the petiole usually glandular.

Swamps and lowland wet woodlands, sometimes in sluggish woodland streams. Conn. southward on or near the coastal plain to n. Fla. thence westward to La., northward in the interior to Ohio, s. Mich., Ill., Mo.

2. Salix (WILLOWS)

Trees or shrubs, with bitter, astringent, aromatic bark, soft, brittle wood. Buds covered by a single, usually nonresinous scale. Twigs terete, flexible, often jointed at the base and snapping off easily. Leaves simple, alternate, deciduous, short-petiolate, more than 3 times as long as broad, leaf scars small, with 3 vascular bundle scars; stipules falling quickly, or sometimes shortly persistent, sometimes persistent for a considerable part of the growing season. Flowers individually small, borne in erect to spreading catkins, the catkins mostly developing before or as the leaves emerge; flowers unisexual, the plants dioecious, the flowers set on nectariferous discs, without a perianth, each subtended by a densely pubescent bract. Staminate flower of (1)2 or 3–12 distinct or partially united stamens, 1 or 2 (–4) glands at the base of the flower stalk. Pistillate flowers with 1 pistil, 1–4 glands at the base of the flower stalks, style short or none, stigmas 2, sometimes each 2-cleft. Fruit a 2-valved capsule, the valves recurving at dehiscence. Seeds numerous, minute, with cottony or silky hairs from the base.

1. Leaf blades green and glabrous beneath at maturity.
  2. Mature leaf blades lanceolate or linear-lanceolate, short-tapered basally; catkins mostly produced before or as the leaves emerge.   1. *S. nigra*
  2. Mature leaf blades linear-lanceolate to linear, mostly gradually long-tapered to both extremities; catkins mostly produced when leaves are fully or nearly fully developed.   3. *S. interior*
1. Leaf blades grayish to whitish beneath at maturity, glaucous and relatively sparsely pubescent, or copiously densely pubescent and felty, or whitish with silky hairs.
  3. Larger, mature leaf blades broadly lanceolate to broadly elliptic or elliptic-oblong, 10–15 cm long and 3–5 cm broad, broadly rounded at the base, obtuse apically.   6. *S. floridana*
  3. Larger, mature leaf blades lanceolate to linear-lanceolate, tapered at both extremities.
    4. Leaf margins entire, revolute; lower surfaces with densely matted, felty pubescence.
                                                                                              5. *S. humilis*
    4. Leaf margins toothed, not revolute; lower surfaces glabrous, sparsely pubescent, or silky-pubescent.
      5. Fully grown leaf blades glabrous, glaucous beneath, if pubescent, the pubescence not silky.
                                                                                           2. *S. caroliniana*
      5. Fully grown leaf blades with whitish silky pubescence beneath.   4. *S. sericea*

1. Salix nigra Marsh. BLACK WILLOW.

Tree to 20 m tall, often shrubby and with several trunks, bark ridged and furrowed, dark brown; twigs and branchlets brown or reddish brown, glabrous at maturity, usually pubescent at first. Leaf blades lanceolate, green on both surfaces, sparsely pubescent at first, usually glabrous at maturity, variable in size, the larger to 12 cm long and 1–2 cm wide, cuneate basally, acute to acuminate apically, often falcate, margins serrulate, the teeth with reddish glandular tips, petioles 4–10 mm long. Catkins produced before or as the leaves emerge. Capsule glabrous, ovoid-conical, 3–5 mm long, maturing before or as the leaves mature.

   Swamps, stream banks, sand and gravel bars, shores of bodies of water, meadows, commonly abundantly colonizing wet disturbed areas. s. N.B. to cen. Minn., generally southward to Ga., n. Fla. from Jefferson Co. westward, and Tex.

2. Salix caroliniana Michx.

Similar in general features to *S. nigra*. Mature leaves whitish-glaucous beneath, usually sparsely pubescent, especially along the midribs; margins finely serrulate, the glandular tips of the teeth usually yellowish; large, broadly reniform stipules often persistent much of the growing season on vigorous shoots or sprouts. (*S. longipes* Anders.; incl. *S. marginata* Wimmer (?), *S. amphibia* Small)

   River banks, sand and gravel bars, swamps, open, low, wet places. Md., D.C., westward to s. Ind., s. Ill., Mo. and Kan., southward to s. Fla. and s.cen. Tex.; Cuba.

3. Salix interior Rowlee. SANDBAR WILLOW.

A densely clumped, many-stemmed shrub, mostly 2–5 m tall; twigs reddish to brown, usually glabrous. Stipules small and subulate, or lacking. Branchlets usually very leafy. Petioles very short. Leaf blades linear-lanceolate to linear, the former long-tapering to both extremities, 5–10 (–14) cm long, 3–10 mm wide, at maturity green on both surfaces, glabrous or sparsely very short-pubescent beneath, often thinly silky-pubescent when young; margins with well-separated, small, subdentate teeth, teeth sometimes narrow and appearing much longer than wide. Catkins appearing after the leaves are mature. Ovaries thinly silvery-pubescent, mature capsules 7–10 mm long, glabrous to white-pubescent.

   Alluvial sands, mud bars, sand hills, river and stream banks, floodplain woodlands, borders of wet woodlands, ditches, shores of bodies of water. N.B. and s. Que. to Alaska; southward in the East to Potomac and Ohio valleys, thence to La., from the Miss. River across the Great Plains.

4. Salix sericea Marsh. SILKY WILLOW.

Shrub or small tree, mostly 1–3 m tall, occasionally to 7 m; twigs brown to dark reddish brown, glabrous finally but the young branchlets short-pubescent. Leaf blades narrowly lanceolate, bases obtuse or nearly so, apices acute to acuminate, mostly 6–10 cm long, 1–2 cm broad, margins irregularly serrulate, upper surface dark green and short-pubescent to glabrous, lower surface densely to thinly silvery-silky-pubescent. Stipules lanceolate, 3–7 mm long, mostly falling early. Catkins mostly emerging before the leaves. Capsules ovate, 3–5 mm long, silvery-pubescent.

Marshes, low, wet or swampy woodlands, stream banks, commonly in or near running water of rocky places. N.S. and N.B. to Wisc., e. Iowa and s.e. Mo., southward to N.C., n. Ga., and n. Ala.

5. Salix humilis Marsh. PRAIRIE WILLOW.

Shrub, to about 3 m tall, commonly with spreading underground parts and colonial. Branchlets and twigs with short, ashy gray, soft pubescence. Leaf blades very variable from clone to clone, narrowly to broadly oblanceolate to broadly elliptic, upper surface at first short-pubescent, later becoming dark green and glabrous but usually remaining pubescent on the veins, lower surface usually glaucous beneath a felty, grayish pubescence, margins entire and revolute (sometimes appearing irregularly wavy but then still revolute), bases mostly obtuse, apices acute, obtuse, or abruptly short-acuminate. Catkins develop mostly before the leaves emerge. Capsules pubescent, lanceolate to lance-ovate, 6–9 mm long. (Incl. *S. tristis* Ait.)

Boggy swales, prairies, barrens, thickets, mt. balds. Nfld. and s. Que. to e. N.D., generally southward to n. Fla. and n.e. Tex.

6. Salix floridana Chapm. FLORIDA WILLOW.

A shrub or tree (trees known to us about 30 m tall and 35 cm d.b.h.). Bark of large specimens dark brown, very coarsely ridged and furrowed. Branchlets and young twigs densely short- and soft-pubescent, becoming dark reddish brown and glabrous, or with patches of persistent pubescence. Emerging leaf blades thinly silky-pubescent above, densely so beneath, soon becoming glabrous and green above, glaucous and sparsely pubescent beneath, broadly lanceolate to oblong-elliptic, bases broadly rounded, apices obtuse; larger mature blades to 18 cm long, 4–5 cm broad, finely and irregularly dentate to finely serrate. Catkins developing as the leaves emerge. Capsules ovate-oblong, somewhat leathery, 5–8 mm long, glabrous.

Spring runs in woodland, banks, marshy borders, and floodplains of large spring-fed streams, commonly in water much of the time. cen. Ga to n. Fla.

Betulaceae (BIRCH FAMILY)

Deciduous trees or shrubs. Leaves petiolate, alternate, with deciduous stipules; blades pinnately veined, principal lateral veins straight-ascending parallel to each other from the midrib. Flowers unisexual (plants monoecious), borne in catkins, at anthesis in early spring before or as the new shoots emerge. Staminate catkins (emerging in autumn, dormant over winter) pendulous, long-cylindric, scales subtending a pair of small bractlets and 1–3 naked flowers each having 1–10 stamens; pistillate catkins erect or pendulous, short relative to the staminate, the imbricate scales more or less concealing the flowers, each flower subtended by a small bract and 2 bractlets. Ovary inferior, the floral tube bearing minute calyx segments, or none, on its rim; ovary 2-locular, usually 2 ovules in each locule; styles 2. Fruit a 1-seeded nutlet, with or without an involucre comprised of bracts that enlarged during maturation.

1. Buds stalked; leaves in 3 ranks; pistillate catkins becoming hard and woody, persisting through the winter and into the next season. 1. *Alnus*

1. Buds sessile; leaves in 2 ranks; pistillate catkins not becoming woody and not persisting over winter.
  2. Outer bark buff to coppery, sloughing in thin, usually curly patches; lower surfaces of young leaf blades densely woolly-pubescent, in age remaining woolly-pubescent or with sparse pubescence along the veins and with wooly tufts in the vein angles; bases of leaf blades broadly tapering to truncate.
2. *Betula*

  2. Outer bark gray, smoothish and not sloughing; lower surfaces of young and older leaf blades sparsely pubescent only along the veins and with tufts of hairs in the vein angles; bases of leaf blades rounded. 3. *Carpinus*

1. Alnus (ALDERS)

Alnus serrulata (Ait.) Willd. Fig. 9

Usually a shrub, sometimes attaining the dimensions and form of a small tree. Buds stalked. Young branchlets brown, pilose, sometimes densely so, later the twigs sparsely pubescent, and finally glabrous, reddish brown with a waxy bloom sloughing irregularly; light brown lenticels moderately conspicuous; leaf scars usually raised, triangular, with 3 vascular bundle scars. Pith triangular in cross-section. Leaves deciduous, alternate, 3-ranked, stipulate, the

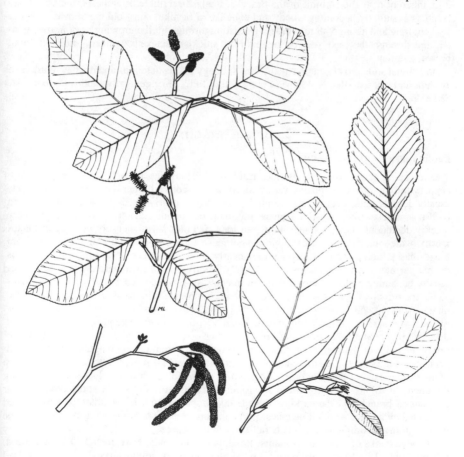

Fig. 9. **Alnus serrulata.** (From Kurz and Godfrey, *Trees of Northern Florida.* 1962)

stipules to about 1 cm long, mostly elliptic to oblanceolate, pubescent, quickly deciduous; petioles short, pubescent; blades 5–10 cm long, 2.5–5 cm broad, obovate, elliptic, or oblong, apices rounded, obtuse, or occasionally short-acuminate, bases obtuse to rounded (usually varying in these characteristics on a single branch); venation pinnate, the ascending lateral veins prominent, especially on the lower surface; margins often slightly wavy, finely irregularly toothed, sometimes with larger primary teeth and finely toothed on and between them; upper surface of mature blades glabrous or sparsely pubescent, usually sparsely hairy on the principal veins, lower surface usually sparsely pubescent, at least on the veins and often with matted hairs in the principal vein axils; expanding leaves sticky and aromatic. Flowers unisexual, borne in catkins, staminate and pistillate catkins on the same plant; catkins borne singly or usually in clusters of 2 to several near the ends of branchlets of the season, evident late in the season, well before leaf-fall, reaching anthesis in very early spring considerably before leaf emergence, the staminate even shedding before leaf emergence; fully developed staminate catkins, at anthesis, dangling, long-cylindrical, conspicuous; staminate flowers in clusters of 3 (–6) subtended by a short-stalked peltate bract, 4 or 5 bractlets, each flower with a 3–5-parted calyx and 3–5 stamens. Pistillate catkins stiffly erect, sticky, much smaller and shorter than the staminate, linear-oblong, 5–10 mm long, about 2 mm wide; pistillate bracts fleshy, obpyramidal, each subtending 2 pistils; bractlets 2–5 adherent near the summit of the bract, scarcely discernible at anthesis, the bracts and bractlets eventually becoming woody and conelike or burlike, short-oblong to ovate-oblong, 1–2 cm long and about 5–8 mm broad, persisting throughout the growing season, over winter, and through the next year or longer. Fruit thickish, laterally winged, shiny, chestnut brown, a nutlet.

In alluvial soils mostly, open stream banks, boggy swales, sloughs and ditches, borders of swamps and wet woodlands. Maine to N.Y., Ohio, Ind., Mo., generally southward to n. Fla. and e. Tex.

2. Betula (BIRCHES)

Betula nigra L. RIVER BIRCH.   Fig. 10

Tree, reaching about 30 m tall. Bark of trunk freely sloughing in curly thin patches of dark coppery brown to buff, exposing yellowish to salmon-colored inner bark. Young branchlets usually densely pubescent, twigs eventually glabrous or nearly so, dark reddish brown, with buffish lenticels. Buds sessile. Leaves alternate, deciduous, 2-ranked, with narrow, very quickly deciduous, hairy stipules 4–5 mm long; petioles short, densely pubescent; blades mostly ovate-triangular or subrhombic, sometimes nearly elliptic, 3–10 cm long, 1.5–3 cm broad, bases broadly cuneate, sometimes nearly truncate, apices obtuse to acute, margins doubly serrate; emerging leaves very densely woolly beneath, moderately woolly above, usually becoming nearly glabrous beneath in age save for pubescence on the principal veins and in the principal vein axils, upper surfaces becoming glabrous save for sparse pubescence on the principal veins; venation pinnate, the lateral veins ascending, especially conspicuous on the lower surface. Flowers unisexual, in catkins, both staminate and pistillate catkins on the same plant. Staminate catkins form in summer, singly or in short clusters at or near the tips of the twigs of the season, remaining dormant over winter, developing fully, elongate and pendent, as the leaves are emerging or just before. Staminate flowers 3 to each peltate bract of the catkin, each of 2 lateral ones subtended by a bractlet, each with a 2–4-lobed calyx and 4 stamens. Pistillate catkins woolly, developing in spring, usually singly at the ends of short new lateral branchlets, the catkins stalked, thick, short-cylindric; pistillate flowers 2 or 3 subtended by a bract with 3 long lobes, perianth none. Fruit a laterally winged nutlet. Scales of the catkins shedding as the nutlets fall early in the season.

Wooded banks of rivers and streams, floodplain woodlands. N.H. to N.J., Pa., westward to Ohio, Ind., Ill., Iowa, s.w. Wisc., s.e. Minn., generally southward except in the s. Appalachians, to n. Fla. and e. Tex.

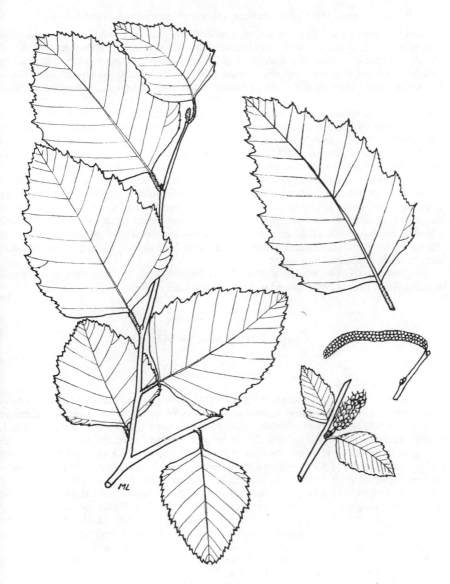

Fig. 10. **Betula nigra.** (From Kurz and Godfrey, *Trees of Northern Florida*. 1962)

3. Carpinus

Carpinus caroliniana Walt. BLUE-BEECH, AMERICAN HORNBEAM, IRONWOOD.

A small understory tree. Bark of the trunk smoothish, granular, grayish, the trunks often somewhat fluted, appearing in places somewhat like flexed muscles. Young branchlets usually at least sparsely pilose, the slender twigs becoming glabrous and reddish with a patchy sloughing of grayish bloom. Leaf scars horizontally oval or half-moon-shaped, with 3 vascular bundle scars. Buds sessile. Leaves simple, deciduous, 2-ranked, with short, slender, pubescent petioles; blades oblong-ovate to oblong, 2–6 cm long, 1–3 (–4) cm broad, rounded basally, apices acute to acuminate; margins doubly serrate; venation pinnate, the ascending laterals not conspicuous seen from above, moderately conspicuous seen from below; upper surfaces mostly glabrous, lower sparsely pubescent on the principal veins and with tufts of hairs in the vein axils. Flowers in catkins, unisexual, both staminate and pistillate catkins on the same plant. Catkins develop before and as the leaves emerge, staminate solitary at various nodes along the woody twigs, elongate, pendulous, with broadly ovate scales each subtending a solitary flower composed of several divided filaments each bearing 2 apically pilose half-anthers; pistillate catkins solitary, terminating short, leafy branchlets of the season; pistillate flowers in pairs, each with a minute bract adnate at base to 2 minute bractlets. Bracts and bractlets greatly enlarge, becoming foliaceous as fruits develop, the catkins then pendulous. Fruit a small ribbed nutlet. Foliaceous fruiting catkins evident through much of the growing season.

Low woodlands, wet woodlands, floodplain forests. N.S., N.B., s. Maine to Minn., Iowa, e. Neb., generally southward to cen. pen. Fla. and e. Tex.

Fagaceae (OAK FAMILY)

Quercus (OAKS)

Shrubs or trees, the trees of some species potentially of large size. Leaves alternate, simple, at least short-petioled, deciduous in autumn or in some overwintering and falling just before or as new growth commences, pinnately veined, margins entire, toothed, or pinnately lobed, stipules present but usually quickly deciduous as the leaves develop; overwintering buds tending to be clustered at or near the tips of the twigs. Flowers unisexual, the plants monoecious. Staminate flowers on pendent catkins borne singly or in clusters from leaf axils near the bases of developing shoots of the season; each flower consisting of a more or less cup- or bowl-shaped calyx united below, 5-lobed above, with 3–12 stamens seated within. Pistillate flowers borne singly or 2 to several in short spikes in axils of developing leaves, usually near the tips of branchlets of the season; each flower consisting of a single pistil with 3 styles, the ovary seated within and adherent to a 6-lobed calyx which in turn is enclosed in an involucre of overlapping scales, these by subsequent development eventually forming a cup around the base of, or wholly around, the matured ovary, the nut or acorn. Ovary 3-locular, with 6 ovules only one of which develops into a seed.

Oaks of a given species exhibit a considerable degree of variability of leaves, those of sprouts or sucker shoots often differing considerably from those on branches of the crown; leaves of lower branches may differ notably from those of exposed crown branches; leaves of seedling or sapling trees commonly differ from those of mature trees. In the descriptions of the various species below, distinctive characteristics of the leaves are derived (unless otherwise noted) from those of mature branches in the crowns.

Identification of oaks is further complicated because hybridization occurs between plants of various species growing in proximity, yielding hybrid progeny with characteristics intermediate between those of the parents. Hybrids are perhaps not so frequently produced as is sometimes alleged. Their apparent frequency is to a considerable extent consequent upon the relatively long life of individual plants. In general, hybrid individuals tend to be more prevalent on sites where mechanical disturbance of the soil had occurred prior to establishment of the seedlings. The following treatment does not take hybrids into account.

In some cases, to distinguish between plants representing pairs or groups of species (especially until one attains sufficient familiarity with subtle differences in leaf and correlated characteristics) it is essential to know the distinctive features of involucral cups and acorns. At other times of the year than that at which mature fruits are present, or in those seasons when a given tree does not produce fruit, one can often find cups and acorns of a previous season beneath the tree and must of necessity rely upon them.

1. Leaves, some of them at least, conspicuously lobed, sinuses between the major lobes reaching ½ the distance to the midrib or a little more.
  2. Lobes blunt or rounded, none of them bristle-tipped.
    3. Upper, inner scales of involucral cups with long, slender, attenuate tips making a conspicuous fringe around the rim.   1. *Q. macrocarpa*
    3. Upper, inner scales of involucral cups not elongate, not forming a fringe around the rim.
      4. Involucral cups wholly or almost enclosing the acorns.   2. *Q. lyrata*
      4. Involucral cups shallowly bowllike, enclosing about ⅓ of the acorn.   3. *Q. similis*
  2. Lobes pointed and they and their marginal teeth bristle-tipped.
    5. Mature leaves with a dense, close, grayish or tawny pubescence on lower surfaces.
                  8. *Q. falcata* var. *pagodaefolia*
    5. Mature leaves with pubescence restricted to tufts in the axils of lateral veins (lower surfaces).
      6. Involucral cups shallowly saucerlike, enclosing little more than the bases of the acorns; acorn subglobose, 1.0–1.2 cm long and broad, evenly buff-colored.   9. *Q. palustris*
      6. Involucral cups bowllike, enclosing about ⅓ of the acorn; acorn oval-elliptic, about 2.5 cm long, 1.5 cm broad, with alternating, obscurely evident, brown and buff longitudinal stripes.
                  10. *Q. nuttallii*
1. Leaf blades entire, toothed, or shallowly lobed, the lobes not reaching ½ the distance to the midrib.
  7. Leaf blades undulate-toothed or undulate-lobed essentially evenly along both margins.
    8. Lateral veins of the leaf all ending at the tips of the marginal lobes or teeth; stalks bearing fruits equaling or shorter than the petioles of adjacent leaves.   4. *Q. michauxii*
    8. Lateral veins of the leaf, some of them, ending at the tips of the marginal teeth or lobes, some ending in the sinuses, some sometimes ending somewhat back of the tips of the teeth or back of the sinuses; stalks bearing the fruits considerably longer than the petioles of adjacent leaves.
                  5. *Q. bicolor*
  7. Leaf blades entire (toothed or lobed only on vigorous sprouts, root sprouts, or some leaves on second phase of shoot growth in summer and then not evenly crenate- or undulate-toothed or -lobed all along each margin), *or* obovate and commonly 3-lobed with the edges entire.
    9. Lower surfaces of mature leaves persistently pubescent throughout.
      10. Pubescence on lower leaf surface grayish, dense or tightly compacted; leaves overwintering, deciduous in spring just before or as new growth commences.   6. *Q. virginiana*
      10. Pubescence of lower leaf surfaces tawny, loose, the surface of the blade showing through more or less as an olive green color; leaves deciduous in autumn.   7. *Q. oglethorpensis*
    9. Lower surfaces of mature leaves glabrous or the pubescence restricted to tufts in the vein axils and along the veins.
      11. Leaf blades rounded to obtuse-angled basally, exhibiting little variation, predominantly lanceolate, narrowly elliptic-lanceolate, or lance-oblong, predominantly broadest below the middle or at the middle.   11. *Q. phellos*
      11. Leaf blades acute-angled to attenuate-cuneate basally, exhibiting considerable variation, predominantly broadest above the middle, spatulate to rhombic or obovate.
        12. Leaves cuneate basally, mostly oblanceolate, rarely spatulate-obovate, usually some leaves on any given tree rhombic.   12. *Q. laurifolia*
        12. Leaves attenuate-cuneate basally, commonly abruptly dilated distally and obovate overall, the dilated apex commonly 3-lobed or shallowly lobulate, some leaves sometimes spatulate, exhibiting not a little variation from tree to tree.   13. *Q. nigra*

1. Quercus macrocarpa Michx. BUR OAK, MOSSY-CUP OAK.

Large tree in our range, shrubby in some parts of the range. Bark of first-year twigs pubescent at first, becoming glabrous and yellowish gray, sometimes corky, that of older twigs and branches gray, on older trunks ashy gray, rough, ridged and furrowed. Leaves deciduous in autumn, mostly obovate in overall outline, 1–3 dm long, irregularly 5–9-lobed, lobes rounded apically, without bristle tips, base cuneate, the center sinus often reaching nearly to the midrib, the 2 lobes above longest, those below progressively shorter, terminal portion broad, oblongish to ovate-lobulate; upper surface glabrous and lustrous green at maturity, lower grayish with close, dense stellate pubescence, soft to the touch. Cup large, to 5 cm across, deeply bowllike, enclosing ⅓ to nearly all the acorn, scales thick and woody, pubescent, medial ones broad-based and bases more or less united, their blunt tips somewhat elevated, humped on the back, upper scales with long, slender, attenuate tips making a conspicuous fringe around the rim. Acorns maturing in one season, 3–5 cm long, 2–4 cm broad, the rounded apices often depressed centrally and with a coarse apicule, powdery-pubescent without, glabrous within.

Floodplain forests and bottomlands, adjacent lower slopes (in our range). s. Maine to s.e. Sask., easterly boundary running southwestwardly to w. Va., Ky., w. Tenn., Ala., Ark., n.w. La., e. and n.cen. Tex., westerly boundary southward to s.e. Mont., n.e. Wyo., Neb., Okla.

2. Quercus lyrata Walt. OVERCUP OAK. Fig. 11

Large tree. Bark of first-year twigs very sparsely pubescent to glabrous at first, brown, that of older twigs and branches grayish, glabrous, on older trunks light brownish gray, forming thick, irregular plates or broad ridges. Leaves clustered near the tips of the branchlets, deciduous in autumn, obovate to oblong in overall outline, 7–20 cm long and 3–12 cm wide, mostly 5–9-lobed, the 2 lobes below the terminal one usually largest and spreading nearly laterally, downwardly the lobes smaller, the lowest triangular; lobes not bristle-tipped; apices of lobes varying from rounded to acute; upper surfaces dark green and glabrous, lower surfaces tightly grayish-tomentose when young, varying from grayish-tomentose to green and essentially glabrous at maturity. Cups oblate to subglobose, about wholly enclosing the acorns, scales thick, pubescent, the medial ones, especially, with subterminal, flattish, pronounced lateral protuberances. Acorns maturing in one season, short-ovoid to subglobose, apices broadly rounded, mostly about 2.5 cm broad, shell appressed-pubescent without, glabrous within.

River and stream banks, floodplain forests. Coastal plain and piedmont, s. N.J. to Fla. Panhandle, westward to e. Tex. and e. Okla., northward in the interior to s. Ind., s. Ill., s.e. Iowa, and e.cen. Mo.

3. Quercus similis Ashe. BOTTOMLAND POST OAK.

Medium-sized to large tree. First-year twigs reddish brown, persistently grayish-pubescent, older trunks grayish brown, fissured and flaky. Leaves deciduous in autumn, obovate in overall outline, 5–16 cm long, mostly about 12 cm, cuneate to rounded basally, with 2 or 3 prominent, lateral, round-tipped lobes on a side, the medial ones largest, oblongish, ascending, the sinuses between them and the next pair above reaching ½–⅔ of the distance to the midrib; upper surfaces glabrous and lustrous green at maturity, the lower grayish, stellate-pubescent. Cup shallowly bowllike, nearly 1.5 cm wide, enclosing about ⅓ of the acorn, scales small, ovate-triangular, blunt apically, closely appressed. Acorns maturing in one season, broadly oval, about 1.5 cm long, rounded apically and apiculate, surface powdery-pubescent without, glabrous within. (*Q. stellata* Wang. var. *paludosa* Sarg.)

Floodplain forests subject to inundation during flooding. La., s. Ark., e. Tex.

4. Quercus michauxii Nutt. SWAMP CHESTNUT OAK. Fig. 12

Large tree. First-year twigs at first copiously pubescent, soon becoming glabrous and reddish brown, bark of older trunks light or silvery gray, scaly. Leaves deciduous in autumn, obovate to oval, regularly or uniformly crenate, crenate-dentate, or shallowly crenate- or dentate-lobed, mostly with 10–15 nearly uniform teeth or lobes on a side; upper surfaces of mature leaves glabrous and dark green, not lustrous, lower grayish or rusty, with pubescence of spreading hairs throughout or only on the veins, commonly soft to the touch; size of leaves

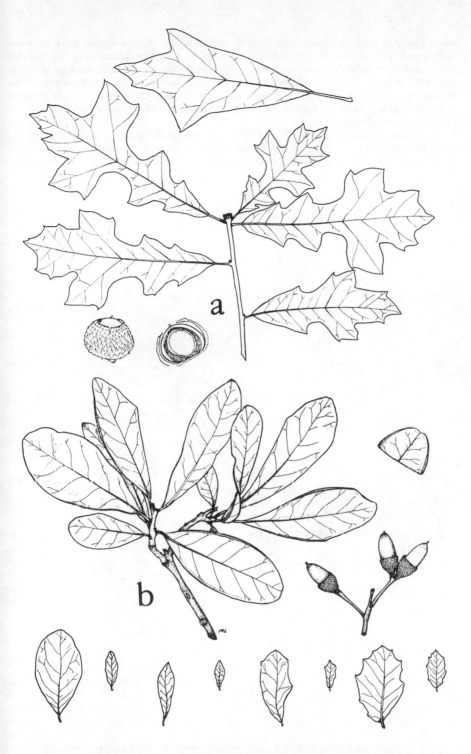

Fig. 11. a, **Quercus lyrata;** b, **Quercus virginiana.** (From Kurz and Godfrey, *Trees of Northern Florida*. 1962)

varying greatly, even on a single tree, from about 6 to nearly 30 cm long and 3–18 cm wide at their widest places. Cup bowllike, relatively large, 3–4 cm broad, enclosing ⅓–½ of the acorn, scales rhombic, pubescent, mostly very thick and hard, blunt apically, strongly keeled-tuberculate on their backs, the innermost scales relatively thin and with subulate tips. Acorns maturing in one season, broadly oval to subglobose, outer surface of the nutshell glabrous or nearly so oh the lower half, powdery-pubescent distally, inner surface glabrous. (*Q. prinus* sensu Small, 1933)

Floodplain forests and wet woodlands subject to periodic flooding and on adjacent mesic, richly wooded slopes, wooded ravine bottoms and slopes. N.J., southward chiefly on the coastal plain but with scattered stations on the piedmont, to about cen. pen. Fla., westward to e. Tex., northward in the interior, s.e. Okla., Ark., s.e. Mo., s. Ill., Tenn., and Ky.

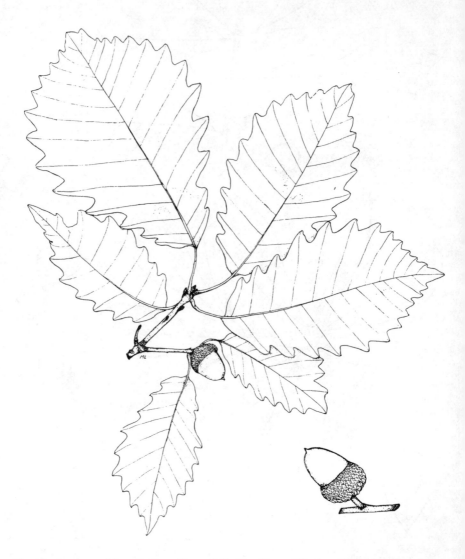

Fig. 12. **Quercus michauxii.** (From Kurz and Godfrey, *Trees of Northern Florida* (1962) as *Q. prinus*)

44

5. **Quercus bicolor** Willd. SWAMP WHITE OAK.

Medium-sized tree. Bark of first-year twigs brown, pubescent, that of upper limbs peeling off in ragged scales, on the trunks thick, grayish brown, markedly ridged and furrowed, or checked. Leaves deciduous in autumn, oblong-obovate in outline, 10–15 cm long, 5–10 cm wide, cuneate basally, blunt apically, sinuate-crenate with 6–10 lobes all along each side, lobes mostly rounded, sometimes pointed, without bristle tips; upper surface glabrous and dark lustrous green at maturity, lower pubescent with both short and longish stellate hairs, gray to pale green depending upon the density of pubescence. Cups bowllike, enclosing ⅓–½ the acorn, scales pubescent, lower and median ones thick, appressed, prominently tuberculate on the back, upper scales acuminate to long-pointed, sometimes making a short, slight fringe around the rim. Acorns maturing in one season, 2–3 cm long, ovoid, rounded at the summit, strongly apiculate, nutshell glabrous without and within.

River and stream banks, swamps, bottomland woods, sometimes locally abundant but in scattered localities in much of the range. Maine and s. Que. to Mont., southward to piedmont, N.C., Tenn., Ala., Mo.

6. **Quercus virginiana** Mill. LIVE OAK.   Fig. 11

Tree, commonly attaining very large girth of a relatively short trunk and with large, very wide-spreading limbs; if growing in relatively dense forests, especially where seasonally wet, often with a high trunk and relatively narrow crown. Wood very hard and durable. Bark of first-year twigs grayish brown and scurfy, that of older twigs and branches grayish, on large trunks dark brown, thick, roughly ridged and furrowed, becoming blocky. Leaves persistent over winter, stiff-leathery, oblong, oblanceolate, or narrowly obovate, not bristle-tipped, prevailingly 4–10 cm long and 1–4 cm broad, considerably larger on occasional trees; unlobed, margins entire, sometimes somewhat revolute; upper surfaces glabrous, dark green, lower grayish, clothed with a tight tomentum, usually so tight as not to be evident to the unaided eye. Not infrequently a second phase of shoot growth occurring in summer, leaves of which are commonly sharply dentate or dentate-lobed as is the case with leaves of vigorous sprouts or root-sprouts. Cups turbinate, enclosing about ⅓–½ of the acorns, scales ovate-triangular, blunt apically, pubescent. Acorns maturing in one season, mostly oblong-ellipsoid, narrowly rounded apically and prominently apiculate, nutshell glabrous without and within.

Predominantly a tree of well-drained, relatively heavy soils, sometimes in almost pure stands, in mixed woodlands, often around ponds and lakes above the level of highest water, also in wet hammocks and wet woodlands where surface water stands seasonally. Outer coastal plain, s.e. Va. through Ga., throughout Fla., westward to e. and s. Tex.

Quercus geminata Small (*Q. virginiana* var. *maritima* (Chapm.) Sarg.) is a closely similar tree, generally more scrubby, which occurs on deep, well-drained sands. In its "pure" form its leaves are much more rugose-veiny, their sides conspicuously rolled downward, the edges markedly revolute, the lower surfaces densely and loosely tomentose. In parts of the overlapping ranges of *Q. virginiana* and *Q. geminata*, many trees are of intermediate character and are presumed to be of hybrid origin. They have the potential for colonizing, are long-lived, and where the landscape has for a long time been subject to much disturbance the intergradation is so rampant that a significant proportion of individual trees cannot be identified as one or the other. This is notably true throughout Florida. We are not aware, however, of the occurrence of intermediates in wetland habitats; there the trees are identifiable as *Q. virginiana*.

7. **Quercus oglethorpensis** Duncan.

Medium-sized tree. Bark of first-year twigs brown, reddish- or purplish-tinged, that of older twigs and branches grayish, on older trunks light gray and broken into thin, loosely appressed scales. Leaves deciduous in autumn, 5–13 cm long, 1.5–4 cm broad, narrowly elliptic or elliptic-oblong to somewhat obovate, some leaves tending to be falcate, entire marginally (or on vigorous shoots undulate to shallowly sinuate distally), rounded to short-cuneate basally, obtuse or broadly acute apically, without bristle-tips; upper surfaces green and glabrous, lower tawny with evenly distributed, yellowish, stellate hairs, somewhat velvety to the touch. Cups saucerlike or shallowly bowllike, enclosing about ⅓ of the acorn, scales small, thinnish, appressed, tips blunt, appressed-pubescent and short-ciliate. Acorns maturing in one season, oval-ovoid, about 1 cm long, depressed at the summit and apiculate, nutshell exteriorly pubescent apically, glabrous within.

Level wooded bottoms astride small streams, on heavy soils relatively impervious to water, the terrain relatively seldom flooded by the streams but with small to large temporary pools. Piedmont, S.C. and Ga., in Saluda, Savannah, and Altamaha River drainage systems.

8. Quercus falcata Michx. var. **pagodaefolia** Ell. CHERRYBARK OAK.   Fig. 13

Large tree. Bark of first-year twigs closely and tightly tomentose, sometimes becoming nearly glabrous by autumn, that of older twigs and limbs reddish to gray and glabrous, on older trunks shallowly fissured and forming narrow, flaky, or scaly ridges, reddish-tinged. Leaves 10–15 cm long, with 5–11 pronounced, subequal lobes mostly spreading nearly at right angles to the midsection, the sinuses broadly U-shaped, base below the lowest lobes truncate, subtruncate, or broadly short-tapered, principal lobes usually (not always) toothed, apices of lobes or teeth bristle-tipped; upper surfaces dark, glossy green and glabrous, the lower grayish to tawny, densely and closely soft tomentose. Cups saucerlike, enclosing about ¼–⅓ of the acorn, scales ovate-triangular, blunt to truncate apically, closely appressed, densely appressed-pubescent except marginally. Acorns maturing the second season, ovoid to subglobose, about 12 mm wide, nutshell evenly buff-colored or brown, pubescent without and within. (*Q. pagoda* Raf.)

Floodplain and bottomland woodlands, adjacent slopes and bluffs. Coastal plain, N.J. to Fla. Panhandle, westward to e. Tex., northward in the interior, s. and e. Ark., s.e. Mo., w. Tenn. and Ky., southernmost Ill. and Ind.

Quercus falcata var. *falcata* is a tree of well-drained uplands. Its leaves are mostly 3–5-lobed, the lobing pattern very variable, those of fertile branches almost always U-shaped in outline below the lowest lobes.

9. Quercus palustris Muenchh. PIN OAK.

Medium-sized tree, the limbs usually somewhat declining. Bark of first-year twigs reddish brown, that of older branches dark brown, on older trunks grayish brown and relatively smooth but producing small scales. Leaves mostly 8–14 cm long and usually about as wide across the largest pair of lobes; deeply lobed, with 2 or 3 pairs of lateral lobes longer than the width of the central body of the blade, the sinuses broadly U-shaped; pair of lobes below the terminal one largest, principal lobes usually with oblongish, nearly parallel sides, distally irregularly 2- or 3-sublobed and toothed, the teeth prominently bristle-tipped; base of leaf sometimes short-tapered below the triangular lowermost lobes, more commonly truncate or subtruncate across the base of spreading, more oblongish lowermost lobes; principal terminal lobe with an oblongish more or less parallel-sided base below its 3 sublobes; upper leaf surfaces lustrous green and glabrous, the lower paler and dull, with tufts of hairs in the vein axils. Cup shallowly saucerlike, 1–1.5 cm broad, 2–3 mm deep, enclosing little more than the base of the acorn, scales triangular-ovate, grayish-dusty-pubescent, closely appressed, blunt apically. Acorns subglobose, about 10–12 mm broad, short-apiculate at the broadly rounded apex, evenly buff-colored, powdery-pubescent without, pubescent within.

Floodplain and bottomland wet woodlands. Mass. to s. Mich. and s.e. Iowa, southward to N.C., Tenn., n. Ark., and n.e. Okla.

10. Quercus nuttallii E. J. Palmer.

Very similar to *Q. palustris*. Leaves similarly lobed but the main lobes narrower, less frequently oblongish and parallel-sided at the base, more frequently triangular. Cups and acorns

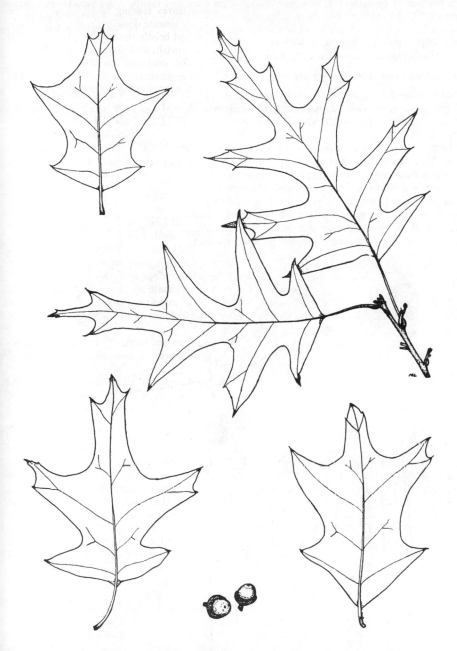

Fig. 13. **Quercus falcata** var. **pagodaefolia.** (From Kurz and Godfrey, *Trees of Northern Florida*. 1962)

larger than in *Q. palustris*, bowllike, about 1 cm deep and nearly 2 cm broad, enclosing about ⅓ of the acorn, with a short, thick stipe, scales similar but coarser. Acorn oval-elliptic, about 2.5 cm long, 1.5 cm broad or a little more, with alternating brown and buff longitudinal stripes, powdery-pubescent without, pubescent within.

Alluvial lowland woodlands. w. Miss., La., Ark., s.e. Mo., e. Tex.

11. Quercus phellos L. WILLOW OAK.   Fig. 14

Medium-sized to large tree. Bark of first-year twigs reddish brown, that of older twigs and limbs grayish, on older trunks dark gray, shallowly fissured and split into irregular small plates or checks. Leaves deciduous in autumn, mostly 6–12 cm long and 0.7–2 (–4) cm broad, varying relatively little in shape, lanceolate, narrowly elliptic-lanceolate, lance-oblong, an occasional leaf narrowly oblanceolate, margins entire, mostly short-tapered basally, acute apically and commonly bristle-tipped; upper surfaces glabrous, dark green, lower surfaces somewhat paler, silky-pubescent when young, when mature with tufts of tri-chomes in the axils of the lateral veins, sometimes lightly floccose or with tufts of hairs irreg-ularly scattered along the larger veins. Cup shallowly saucerlike, 9–12 mm broad, enclosing the base of the acorn, scales grayish-pubescent, narrowly rhombic, blunt apically, closely appressed. Acorns maturing the second season, subglobose, about 1 cm broad, nutshells closely and finely pubescent without, pubescent within.

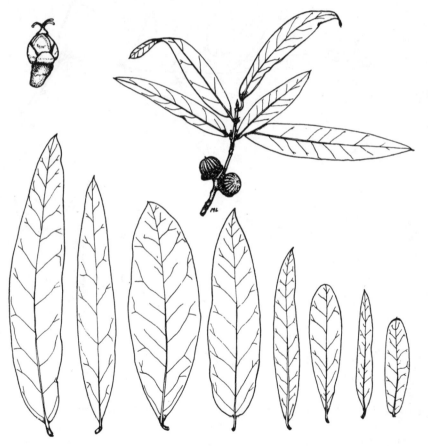

Fig. 14. **Quercus phellos.** (From Kurz and Godfrey, *Trees of Northern Florida.* 1962)

Bottomland and lowland woods, sometimes spreading to hedgerows and into upland woodlands developing on old fields. s. N.J. and s.e. Pa., southward mainly on the coastal plain and piedmont to n. Fla., westward to e. Tex., northward in the interior, s.e. Okla., s. and e. Ark., s.e. Mo., southernmost Ill., Tenn. and Ky.

12. Quercus laurifolia Michx. DIAMOND-LEAF OAK, SWAMP LAUREL OAK.   Fig. 15

Large tree. Bark of first-year twigs brown or reddish brown, gray on older twigs and branches, becoming dark, shallowly ridged and furrowed and somewhat scaly on large trunks. Leaves tardily deciduous in autumn to overwintering, exhibiting considerable variation on a given tree; prevailingly oblanceolate with cuneate bases and rounded to blunt apices, almost always some leaves rhombic as well, these blunt to acute apically, leaves of occasional individual trees obovate, occasional leaves on a given tree 3-lobed toward the apex; some leaves with short bristle tips, mostly without, at least at maturity; edges entire,

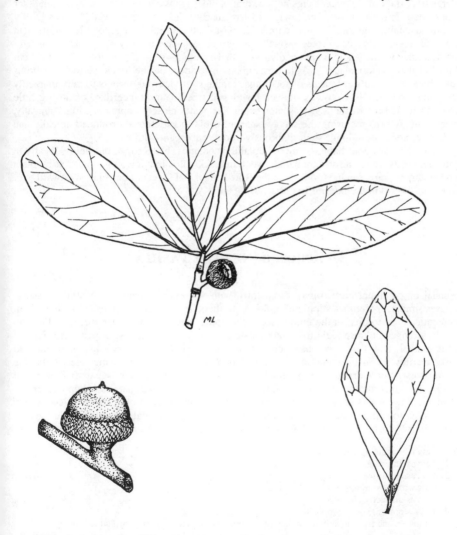

Fig. 15. **Quercus laurifolia.** (From Kurz and Godfrey, *Trees of Northern Florida.* 1962)

49

yellowish-opaque, upper surfaces glabrous, lower glabrous or sometimes with tufts of pubescence in the axils of the principal lateral veins, conspicuous reticulate-venation evident with suitable magnification or seen against transmitted light. Leaves of seedlings, saplings, and suckers usually several-lobed and with little resemblance to those of mature branches of a well-developed tree. Acorn cups shallowly bowllike, enclosing about ⅓ of the acorn, triangular-ovate, apices blunt, more or less appressed-pubescent, margins ciliate. Acorns maturing the second season, ovoid to hemispheric, apically broadly rounded and apiculate, shell minutely pubescent without, pubescent within. (*Q. obtusa* (Willd.) Ashe)

Floodplain forests, river and stream banks, cypress swamps, bayheads, borders of mangrove swamps, wet hammocks. s, N.J. to s. pen Fla., westward to s.e. Tex.

13. Quercus nigra L. WATER OAK.   Figs. 16 and 17

Large tree. Bark of first-year twigs brown, that of older twigs and branches grayish, on older trunks dark grayish brown, shallowly and irregularly ridged and furrowed. Leaves tardily deciduous, falling gradually over winter, thickish-leathery, exhibiting notable variation in size and shape from tree to tree; basically, in outline, broadest distally, spatulate to cuneate-obovate, the broadened apex unlobed varying to 3-lobed, 5–10 (–15) cm long, 2–5 (–7) cm broad at their widest places; bristle tips sometimes evident on the lobes; surfaces glabrous and green, edges entire. Leaves of seedlings, saplings, or vigorous sprouts usually markedly differing from those of crowns of well-developed trees. Cups saucerlike, enclosing little more than the base of the acorn, scales triangular-ovate, closely appressed, blunt apically, pubescent. Acorns maturing the second season, ovoid to subglobose, rounded apically and apiculate, about 1 cm long, shell glabrous without, pubescent within.

In a wide variety of habitats, mixed upland woodlands, hedgerows, old fields, locally in pine flatwoods, river and stream banks, wet hammocks, floodplain woodlands. Coastal plain and piedmont, rarely in the foothills of the Blue Ridge, Del. to cen. pen. Fla., westward to e. Tex., northward in the interior to Okla., Mo., and Ky.

Urticaceae (NETTLE FAMILY)

Annual or perennial herb (ours), frequently with watery sap, some with stinging hairs. Leaves simple, sometimes stipulate, opposite or alternate, usually petiolate, with punctiform concretions (cystoliths) in the epidermis of the upper, lower, or both leaf surfaces. Flowers small, arranged in loose to tight axillary clusters, or in spikes, racemes, or panicles; bisexual or unisexual, plants monoecious, dioecious, or polygamous; if monoecious, staminate and pistillate flowers in different inflorescences or intermingled in the same one. Staminate flowers sessile or short-stalked, greenish, falling after shedding pollen; sepals free or united below, 4–6-merous; petals none; stamens of equal number to sepals and opposite them. Pistillate flowers sessile or nearly so, greenish or reddish; sepals 3 or 4, more or less united; staminodia sometimes present; petals none. Pistil 1, ovary 1-locular, 1-ovuled, style 1 with a capitate or filiform stigma. Fruit an achene, often enclosed by the persistent calyx.

1. Plants with stinging hairs.
    2. Leaves alternate.   1. *Laportea*
    2. Leaves opposite.   2. *Urtica*
1. Plants without stinging hairs.
    3. Leaves distinctly alternate.   3. *Parietaria*
    3. Leaves opposite or subopposite.
        4. Flowers and fruits in axillary spikes; stems not watery-succulent but firm, hard, opaque.
                                                                                 4. *Boehmeria*
        4. Flowers and fruits in axillary panicles; stems watery-succulent, nearly translucent.   5. *Pilea*

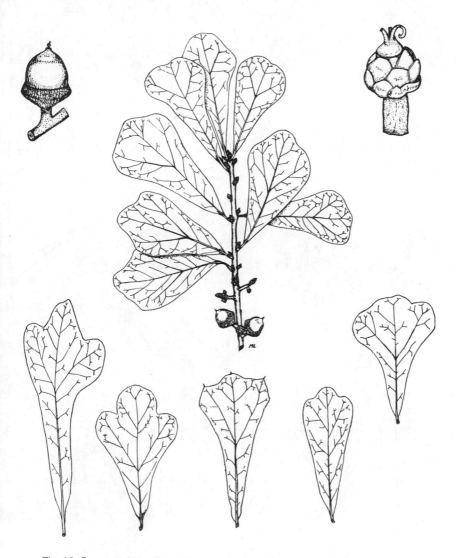

Fig. 16. **Quercus nigra.** (From Kurz and Godfrey, *Trees of Northern Florida*. 1962)

51

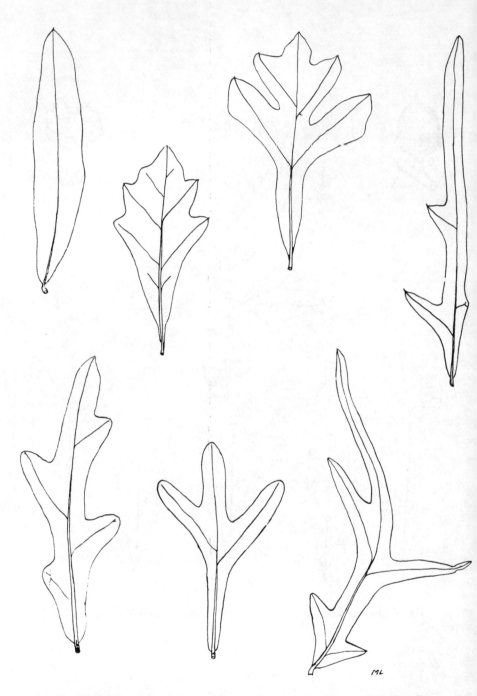

Fig. 17. **Quercus nigra:** leaves of saplings and sprouts. (From Kurz and Godfrey, *Trees of Northern Florida*. 1962)

1. Laportea

Laportea canadensis (L.) Wedd. WOOD NETTLE.

Erect perennial herb with somewhat tuberous roots. Stems usually solitary, variable in height to about 1 m, copiously clothed with whitish, terete-subulate, stinging spicules dilated-jointed at the base. Leaves with stipules that are soon shed, with slender petioles about as long as the blades, the petioles with stinging spicules as on the stem, blades ovate, very variable in size from 3–4 to about 15 cm long and 2 to about 12 cm wide, pinnately veined, mostly truncate or rounded basally, sometimes broadly cuneate, apices acuminate, margins strongly serrate, both surfaces with appressed, whitish hairs. Flowers in axillary and terminal panicles, unisexual, plants mostly monoecious, staminate and pistillate panicles separate, the staminate in axils of leaves lower on the stem than the pistillate, sometimes only a terminal pistillate panicle present. Staminate flowers with 5 sepals, 5 stamens. Pistillate flowers with 4 sepals, 2 enlarging as the fruit matures. Achene flattish, asymmetrical, one side nearly straight, the other broadly rounded, the persistent, often recurved style positioned about at the juncture of the straight and curved sides. (*Urticastrum divaricatum* (L.) Kuntze)

Mesic forests, floodplain forests, stream banks, often locally abundant. N.S. to Man., generally southward to the Fla. Panhandle and n.e. Okla.

2. Urtica

Urtica dioica L. STINGING NETTLE.   Fig. 18

A rhizomatous, colonial perennial, polymorphic in its overall range, segregate species, varieties or subspecies not here distinguished. Stems erect, 1–2 m tall or more, with stinging hairs of variable size, the larger with longish pedestallike bases. Leaves with petioles much shorter than the blades, with prominent stipules 10–15 mm long; blades mostly ovate to lance-ovate, pinnately to subpalmately veined, basally rounded to subcordate, sometimes cuneate, apically acuminate, margins strongly serrate, both surfaces with appressed stinging hairs, sometimes the upper surface glabrate, the hairs on the lower surface chiefly along the veins. Flowers in axillary panicles, the pistillate usually uppermost. Staminate flowers with 4 sepals and 4 stamens inserted around a cuplike rudiment of a pistil. Pistillate flowers with 4 sepals in 2 pairs, the outer smaller, the inner loosely enclosing the achene at maturity. Achene flattened, ovate.

Floodplain forests, wet woodlands, stream banks, wet thickets, clearings, waste places. In parts of most of our range (except Fla.) and very widely ranging beyond.

3. Parietaria (PELLITORIES)

Taprooted annuals (ours), sparsely to densely pubescent with soft, straight, or hooked hairs. Stems erect, ascending, often reclining, commonly densely multistemmed, translucent, watery-succulent. Leaves petiolate, blades pinnately or palmately veined, alternate, without stipules. Bisexual, staminate, and pistillate flowers on the same plant (polygamous). Flowers in short, sessile, axillary cymes, one cymule on each side of the petiole base; flowers subtended by 1–3 bracts; first-opening flower of cyme usually bisexual, others pistillate (staminate apparently rarely, if ever, present in our species). Pistillate flowers sessile, with 4 equal pubescent perianth parts fused toward the base, free above; bisexual flowers with 5 stamens. Perianth enlarging during maturation of the fruit. Fruit a symmetrical or somewhat asymmetrical, hard-walled achene.

1. Leaves palmately veined.
  2. Achene usually 1.2 (–1.4) mm long, without a flanged stipe, apically asymmetrical, with an apicule positioned well to one side of the apex.   1. *P. praetermissa*

53

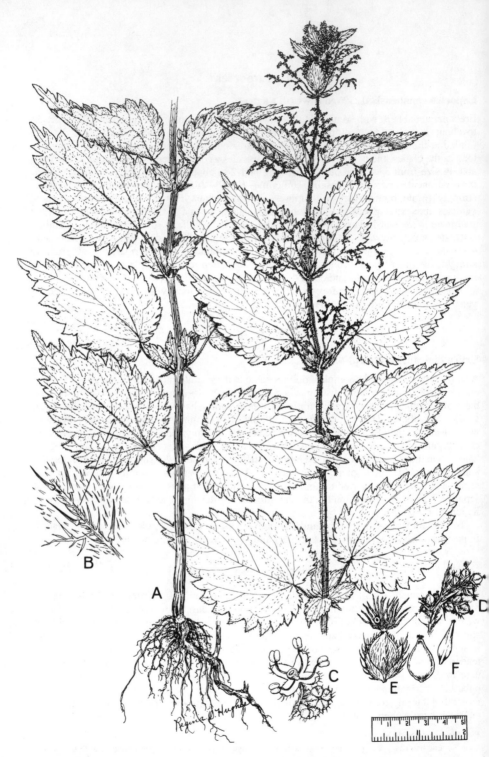

Fig. 18. **Urtica dioica:** A, habit; B, stinging hairs, enlarged; C, flower, closed and open; D, part of fruiting spike; E, fruit; F, achene, face and edge views. (From Reed, *Selected Weeds of the United States* (1970) Fig. 53)

2. Achene less than 1 mm long, with a short, flanged stipe, symmetrical apically and with an apicule central or nearly so at the apex.   2. *P. floridana*
1. Leaves pinnately veined.   3. *P. pensylvanica*

1. **Parietaria praetermissa** Hinton.   Fig. 19

Stems terete, simple or more commonly many from near the base, usually reclining and straggly, finely to densely pubescent with short, hooked hairs. Petioles translucent, watery-succulent, about as long as the blades, pubescent; blades ovate, deltoid, broadly rhombic, or ovate, with 3 veins from the base, thus palmate, principal veins markedly raised beneath (when fresh), surfaces smooth or pubescent, margins entire but short-ciliate, variable in size and in length-width ratio, on some plants the larger about 2 cm long with broadly tapering bases and apices, on some plants to about 6 cm long, the bases often truncate, the apices attenuate; cystoliths abundant in intervein tissues, not evident in fresh leaves, clearly evident as raised dots in dried leaves. Bracts lanceolate to linear, shorter than to a little longer than the flowers, clothed with long, straight and short, hooked hairs. Calyx lobes on fruits brownish, ovate to linear-oblong, acute, loosely connivent below the tips, tips somewhat free. Achene oval, 1.2–1.4, mostly 1.2, mm long, apically asymmetrical, with a short, blunt apiculus positioned well to one side of the rounded apex, the basal short stipe not flanged. (*P. floridana* of authors, not Nutt.)

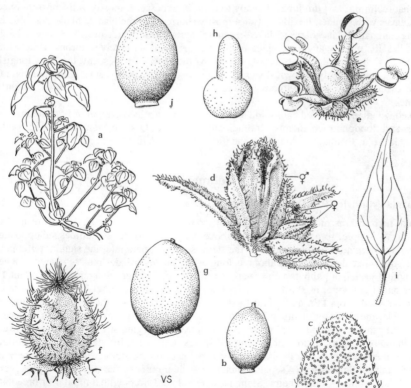

Fig. 19. a,b, **Parietaria floridana:** a, habit of branch; b, achene (note flanged stipe); c–h, **Parietaria praetermissa:** c, adaxial surface of leaf apex with cystoliths; d, inflorescence with bisexual flower, the perianth connivent, and two pistillate flowers; e, bisexual flower just after pollen discharge; f, pistillate flower; g, achene; h, embryo; i,j, **Parietaria pensylvanica:** i, leaf; j, achene. (From Miller in *Jour. Arn. Arb.* 52: 65. 1971)

In mesic to dryish mixed woodlands, sometimes locally abundant on mounds of debris washed up on seashores, often abundant in floodplain swamps and cypress depressions between periods of flooding. Coastal plain, N.C. to s. Fla., westward to La., northward to Mo.

2. Parietaria floridana Nutt. Fig. 19

Vegetatively like *P. praetermissa*, leaves averaging smaller, but size not reliable in distinguishing the two. Achene less than 1 mm long, with a short flanged stipe at base, apically symmetrical, with an apiculus centrally positioned or nearly so. (*P. nummularia* Small)

In habitats similar to those of *P. praetermissa*. Coastal plain from Del. to Fla., westward to Tex.; Calif.; Cuba.

Reference: Hinton, B. D. "*Parietaria praetermissa* (Urticaceae). A New Species from the Southeastern United States." *Sida* 3: 191–194. 1968.

We follow Hinton in distinguishing between *P. praetermissa* and *P. floridana* but without much confidence that they deserve separate recognition. Stature of plants and size of leaves vary greatly in both species (if there are two) depending upon moisture and nutrient content of the substrate and whether growing in sun or shade. If not considered distinct, the name to be applied is *P. floridana*.

3. Parietaria pensylvanica Muhl. ex Willd. Fig. 19

Stems slender, commonly simple, sometimes with a few branches from the base, erect or somewhat reclining, to 4 dm long, sparsely to densely pubescent, mostly with short, hooked hairs. Leaves with slender, flexible petioles mostly shorter than the blades; blades variable in size, predominantly lanceolate to lance-rhombic, sometimes ovate, mostly 3–6 cm long, cuneate basally, usually attenuate apically but the tips blunt, minutely scabrous. Bracts linear, obtuse, 4–5 mm long, considerably exceeding the flowers, pubescent, the hairs longish and straight. Fruiting calyx brown, lobes linear-oblong, acute, pubescent, tightly connivent. Achene ovoid, 1 mm long, without a flanged stipe, symmetrical, shiny brown, slightly apiculate.

Mixed woodlands, shaded banks and ledges, moist waste places, apparently sometimes in marshes and floodplain woodlands. Sporadically, s. Maine, Que. to Man. and B.C., southward to Ga., Fla. Panhandle, Ala., La. and Tex.; B.C. to Calif.; Mex.

4. Boehmeria

Boehermia cylindrica (L.) Sw. BOG-HEMP, FALSE-NETTLE. Fig. 20

Perennial herb. Stems more or less pubescent, sometimes scabrid, to 12 dm tall or a little more, mostly simple, sometimes branched above. Leaves opposite, sometimes subopposite, more rarely alternate near the summit of the stem, stipulate, petiolate, the slender pubescent petioles somewhat shorter than to about as long as the blades. Blades mostly ovate to lance-ovate, with 3 principal palmate veins, variable in size, e.g., larger, longer ones to about 15 cm long and 4 cm wide, or shorter, broader ones 10 cm long and about 7 cm wide, various sizes in between; from thin and pliant to stiff and thickish-rugose, the former on plants in heavy shade, the latter on plants in full sun; usually somewhat pubescent on both surfaces, broadly cuneate to rounded basally, acuminate apically, margins strongly serrate. Flowers clustered on axillary continuous or interrupted spikes, unisexual, the staminate and pistillate variously intermingled or wholly staminate spikes lower on the stem, wholly pistillate ones above. Staminate flowers with 4 sepals united below, their outer surfaces copiously pubescent with both hooked and straight hairs. Stamens 4. Pistillate flowers with a tubular or urceolate calyx, usually 4-lobed, the lobes ciliate or toothed, pubescent with hooked or straight hairs on the outer surface. Style 1, stigmatic down one side. Achene flattish, elliptic, closely invested by the persistent calyx. (Incl. *B. drummondii* Wedd. and *B. decurrens* Small)

Marshes, marshy shores, wet woodlands, swamps, floodplain forests, wet thickets and

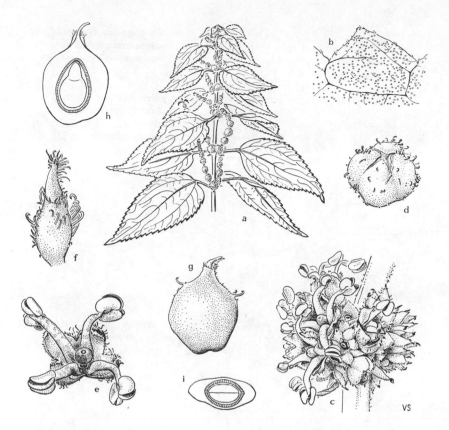

Fig. 20. **Boehmeria cylindrica:** a, plant apex with axillary inflorescences; b, cystoliths at edge of adaxial leaf surface; c, cluster of carpellate and staminate flowers from inflorescence; d, staminate flower bud showing valvate perianth segments; e, postanthesis in staminate flower; f, carpellate flower; g, mature fruit covered by persistent perianth; h, fruit in longitudinal section, endosperm white, endocarp hatched (semidiagrammatic); i, fruit in cross-section, endosperm and endocarp as in h (semidiagrammatic). (From Miller in *Jour. Arn. Arb.* 52: 60. 1971)

clearings, stream banks, cypress-gum ponds and depressions, drainage or irrigation ditches or canals. Que. and Ont. to Minn., generally southward to s. Fla. and Tex.

5. Pilea

Pilea pumila (L.) Gray. CLEARWEED, RICHWEED.   Fig. 21

Low, glabrous annual. Stem solitary or more rarely several from the base, often decumbent below, watery-succulent and nearly translucent. Leaves with slender, flexuous petioles as long as the blades or nearly so; blades thin, flexuous, ovate to broadly elliptic or oval, palmately veined, variable in size, to about 10 cm long and 6 cm wide, basally broadly cuneate to rounded, apically acuminate, marginally mostly crenate-serrate, the tips of the teeth blunt or rounded. Flowers in axillary panicles, unisexual, the staminate and pistillate commonly intermingled (rarely on separate plants). Staminate flowers with calyx united in the bud but separating into 4 (or 3) sepals at anthesis. Stamens 4 (or 3). Pistillate flowers with 3 some-

57

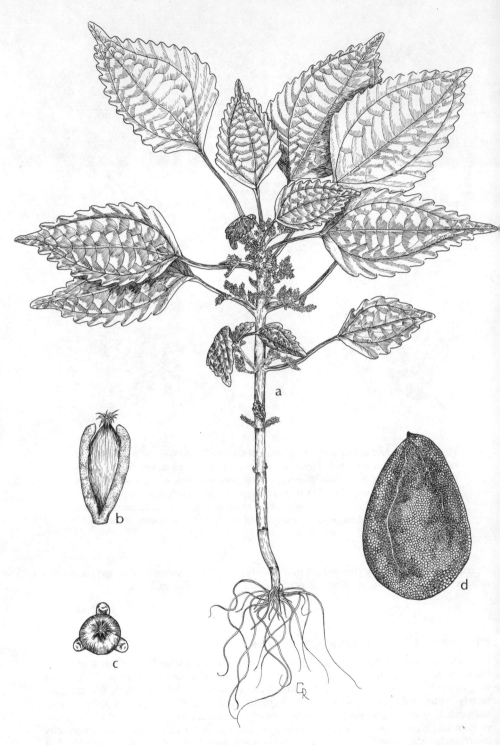

Fig. 21. **Pilea pumila:** a, habit; b, pistillate flower, side view; c, pistillate flower, from above; d, fruit.

what unequal sepals, each subtending a scalelike staminode. Stigma sessile. Achene ovate, longer than broad, flattened, light brown, with slightly raised darker lines or mottling, subtended by the persistent calyx. (*Adicea pumila* (L.) Raf.)

Swamps, floodplain forests, wet woodlands, springy places in woodlands, wet clearings. N.B. to N.D., generally southward to n. Fla. and e. Tex.

A segregate species, *P. fontana* (Lunnell) Rydb., is essentially similar save for characteristics of the fruit. Its fruit is little longer than broad, dark brown to black, with paler margins, slightly roughened and without lines or mottling. It occurs from Pr.Ed.I. to N.C., southward in the Midwest to Neb. and Ind., the Ridge and Valley Province in Va., and along the coastal plain from N.C. to n.e. Fla.

Moraceae (MULBERRY FAMILY)

Trees or shrubs (rarely herbs) with milky or watery sap. Leaves mostly alternate, stipulate, stipules quickly falling as new shoots develop, leaving scars. Flowers unisexual, plants monoecious or dioecious, very small, variously disposed in heads, catkins, or within hollow, fruitlike receptacles. Calyx of 2–6, usually 4, separate or partly united sepals. Corolla none. Staminate flowers with as many stamens as sepals or calyx lobes (rarely fewer) and opposite them. Pistillate flowers 2-carpellate (or 1 carpel aborted), usually 1-locular, styles 2. Fruits diverse.

● Leaves membranous, deciduous; buds with 3–6 scales; major veins of leaf blades palmate, margins of blades toothed; stipules subulate, quickly deciduous and leaving a short line-scar at either side of the petiole base.   1. *Morus*
● Leaves somewhat leathery, evergreen; buds naked but enclosed by a stipule; venation of leaf blades definitely pinnate, margins of blades entire; stipules leaving a stipular ring-scar around the twig at the nodes.   2. *Ficus*

1. Morus (MULBERRIES)

Morus rubra L. RED MULBERRY.

Tree to at least 20 m tall and to 6 or more dm d.b.h. Bark of twigs brown, dotted with round to elliptic, somewhat raised, paler lenticels; bark of older trunks irregularly, narrowly ridged and furrowed. Buds with 3–6 scales. Stipules subulate, soft, 6–8 mm long, very quickly deciduous as the new shoots of the season develop leaving a stipular line-scar at either side of the petiole base. Leaves petiolate, petioles shorter than the blades; blades usually varying considerably in size on a given tree, broadly ovate or oval to nearly orbicular, smaller ones commonly as broad as long, larger ones usually up to twice as long as broad, some sometimes oblique basally, mostly equilateral with truncate, rounded, or slightly cordate bases, apices mostly abruptly short-acuminate, the 3 major veins palmate, larger lateral veins not reaching the leaf margins but pairs interconnecting by a vein arch, margins crenate-serrate, upper surfaces rather dark, dull green, usually slightly scabrous, lower surfaces paler, sparsely to densely pubescent, especially when young, usually retaining at least a little pubescence along the veins in age; some leaf blades of saplings and vigorous sprouts usually lobed, often some of them mittenlike, others 2–5 (–7)-lobed, blades of crowns commonly all unlobed. Flowers in stalked catkins, the staminate and pistillate on different branches of the same plant or on different plants; staminate catkins slenderly cylindric, loosely flowered, the pistillate short-oblong, compactly flowered. Ovaries of pistillate flowers developing into small, juicy, dark red to black drupelets aggregated together to form the mulberry.

Floodplain woodlands of rivers and streams, adjacent wooded slopes. Vt. to S.D., generally southward to s. Fla. and e. and cen. Tex.

Note: *Morus alba* L., a tree attaining much less stature, is similar in its general features. Its leaves are glabrous and somewhat lustrous on both surfaces or pubescent only in the vein axils beneath. Native of Asia, formerly cultivated, now naturalized and growing mostly in places of disturbance near human habitations, apparently rarely in wet places.

2. Ficus (FIGS)

(Ours) evergreen trees with milky sap, commonly epiphytal as seedlings. Leaves simple, alternate, short-petioled, pinnately veined, the larger lateral veins angling-ascending more or less parallel to each other from the midrib, not reaching the leaf margins but interconnecting somewhat away from the margin; leaves leathery, glabrous, margins entire; stipules enclosing naked buds and leaving stipular ring scars around the nodes after shedding. Flowers tiny, unisexual, both staminate and pistillate borne on the inner walls of a fruitlike receptacle (synconium), the pistillate forming very small fruits (achenes) immersed in the thickened receptacle, the entire receptacle or synconium usually referred to as the fruit (fig).

- Inflorescences or infructescences sessile on twigs.   1. *F. aurea*
- Inflorescences or infructescences stalked.   2. *F. citrifolia*

1. Ficus aurea Nutt. STRANGLER FIG.

Potentially a tree to at least 20 m tall, forming descending aerial, adventitious roots which eventually become stout, those developing close to its own main stem or that of the plant upon which it is epiphytic, eventually, to some degree, coalescing and "strangling"; also producing from the widely spreading branches descending roots which become stemlike or proplike. Leaves mostly oblong or oval, their bases broadly short-tapered, to rounded, apices mostly abruptly and bluntly short-acuminate; larger mature leaves 10–15 cm long and 5–8 cm broad. "Fruits" spheroidal or depressed-globose, mostly about 8–10 mm in diameter.

    Bayheads, cypress swamps, banks of sloughs and drainage canals, borders of mangrove swamps, tropical hammocks. s.cen. pen. Fla., Fla. Keys; Bah.Is., Cuba, Jam., Hisp., Cayman Is.

2. Ficus citrifolia Mill. SHORTLEAF FIG.

Shrub or tree to 12 to 15 m tall, sometimes epiphytic, usually eventually producing descending aerial roots some of which coalesce to form a trunk. Similar to *F. aurea* in general appearance (and not easily distinguished without "fruits"); larger mature leaves usually not exceeding 10 cm long, many of them short-oblong, some usually ovate, bases truncate or slightly cordate, sometimes broadly tapered, apices as for *F. aurea*. Inflorescences or infructescences on stalks 5–20 mm long, spheroidal, 1–1.5 cm in diameter. (*F. brevifolia* Nutt., *F. laevigata* Vahl)

    Commonly growing with *F. aurea* (in our range), probably less frequently in wet places. s. pen. Fla., Fla. Keys; W.I.; Mex. to Parag.

Ulmaceae (ELM FAMILY)

Trees or shrubs. Leaves simple, alternate, pinnately veined, 2-ranked, deciduous, with quickly deciduous stipules. Flowers small, bisexual or unisexual. Calyx united at least basally, 3–9- (usually 4- or 5-) lobed. Corolla none. Stamens as many as the calyx lobes and opposite them. Pistil 1, ovary superior, 2-carpellate, 1-locular and 1-ovulate, style 2-parted. Fruit a samara, nut, or drupe.

1. Leaf blades with 4–8 principal lateral veins, the lowest pair arising from the base of the blade and forming there a 3-nerved, or a **V**-nerved part with the midrib; pith of the branchlets chambered; fruit a drupe.   1. *Celtis*
1. Leaf blades with 6–20 principal lateral veins, the lowest arising above the base of the blade; pith of the branchlets solid; fruit a burlike nut or samara.

2. Bark scaly and flaky, exposing reddish brown inner bark as it sloughs; flowers appearing in fascicles on emerging shoots of the season and as the leaves are partially developed; fruit a soft, burlike nut.

2. Planera

2. Bark ridged and furrowed or scaly, not exposing reddish brown inner bark as it sloughs; flowers appearing in fascicles or racemes either from buds on twigs of a previous season before new shoot development in spring *or* from axillary buds in autumn; fruit a samara.

3. Ulmus

1. Celtis

Celtis laevigata Willd. HACKBERRY, SUGARBERRY.

Medium to large tree. Bark gray, smooth or more frequently bearing corky-warty outgrowths, these sometimes in masses. Leaves short-petiolate, blades mostly ovate but varying to lanceolate, with a pair of principal lateral veins arising from the base, variable in size from 4 to about 12 cm long and 1.5–6 cm wide, often somewhat inequilateral, broadly short-tapered to rounded or subtruncate basally, long-acuminate apically; upper surface essentially glabrous at maturity, the lower usually pubescent on the veins, entire or with few teeth on their upper halves. Flowers unisexual, greenish, the staminate in small fascicles on lower portions of emerging branchlets of the season, the pistillate solitary, less frequently in pairs, distally on emerging branchlets of the season. Drupes orange, pinkish orange, or brownish red at maturity, glaucous, oval to subglobose, 6–9 mm across, reaching nearly full size early in the season, ripening in late summer. (*C. mississippiensis* Bosc.; incl. *C. smallii* Beadle)

Floodplain forests, wet woodlands, river banks, rich, moist woodlands, fence rows. Va. to Fla. (except extreme southern part), westward to Tex., northward in the interior, Okla., Mo., s. Ill., s. Ind., w. Ky., Tenn.

2. Planera

Planera aquatica Walt. ex. J. F. Gmel. PLANER-TREE, WATER-ELM.   Fig. 22

Shrub or small tree, rarely to 18 m tall, often with a short trunk and a low, spreading crown; often with considerable sprouting from the base of the trunk, occasional specimens with several stocky short trunks, these having developed from a few such sprouts after the main trunk died. Bark scaly and flaky, sloughing in long, grayish brown plates and exposing reddish brown inner bark. Leaves short-petiolate, blades mostly ovate, sometimes inequilateral, acute to short-acuminate apically, dark green and glabrous above, paler beneath, glabrous or sparsely pubescent mostly on the veins, the veins becoming brownish, margins somewhat irregularly serrate to somewhat doubly serrate. Flowers bisexual or unisexual, usually both on the same plant; staminate flowers with 4 or 5 stamens, borne from the axils of opening bud scales; bisexual flowers, usually with 1 stamen, or pistillate flowers in the axils of the first leaves of emerging shoots. Fruit a soft, stipitate nut about 1 cm long, its surface with irregular, fleshy projections and somewhat burlike, maturing in spring.

Floodplain forests and river shores subject to periodic flooding, sand and gravel bars. Coastal plain, s.e. N.C. to n. Fla., westward to e. Tex. and s.e. Okla., northward in the interior to s.e. Mo., s. Ill., Ky.

3. Ulmus (ELMS)

Trees. Bark ridged and furrowed or scaly. Leaves short-petiolate, lateral veins of the blades essentially equal, straight, and parallel from the midrib outwardly. Flowers bisexual, appearing in fascicles or racemes from buds on twigs of a previous season before new shoot emergence in spring, or in autumn from buds in the leaf axils. Fruit a samara.

1. Leaf blades rounded-obtuse apically, mostly less than 5 cm long; flowering in autumn from buds in the leaf axils.

1. U. crassifolia

Fig. 22. **Planera aquatica:** a, branch; b, enlargement of leaves; c, flowering/fruiting branch, before leaves fully grown; d, fruit showing fleshy processes, persisting stamens below; e, staminate flower. (From Correll and Correll. 1972)

1. Leaf blades acute or acuminate apically, the larger ones, at least, 5 cm long or longer; flowering in spring from buds on twigs of a previous season and before new shoot emergence.
2. The leaf blades lanceolate to narrowly elliptic, or narrowly oblong, the larger ones 3 cm wide or less, the bases often only slightly if at all asymmetric; lower surfaces more or less hairy on the veins but not with tufts of hairs in the axils of the principal lateral veins and the midrib; surfaces of the samaras pubescent, not rugose-veiny, the margins ciliate.   2. *U. alata*
2. The leaf blades, the larger ones at least, oblong-obovate, broadly elliptic, oval, or ovate, 4 cm wide or more, the bases mostly noticeably asymmetric, lower surfaces sparsely hairy to glabrous on the veins but with tufts or webs of hairs in the axils of the principal lateral veins and the midrib; surface of samaras glabrous and conspicuously rugose-veiny, margins ciliate.   3. *U. americana*

1. Ulmus crassifolia Nutt. CEDAR ELM.

Tree to about 25 m tall, the bark scaly-ridged, light brown. Branchlets often with corky-ridged wings. Leaf blades elliptic, oval, or ovate, symmetric or somewhat asymmetric basally, obtuse to rounded apically, mostly 2.5 to about 5 cm long and to about 3 cm wide, glabrous above, pubescent mostly on the veins beneath, margins irregularly shallowly serrate or doubly serrate. Flowering in autumn from buds in the leaf axils. Samaras strongly asymmetrical, about 1 cm long, their surfaces sparsely pubescent and somewhat rugose-veiny, margins ciliate.

Floodplain and lowland woodlands. s.w. Tenn., Ark., e. Okla., La. to cen. and s. Tex.; Suwannee River, Fla.

2. Ulmus alata Michx. WINGED ELM, CORK ELM, WAHOO.   Fig. 23

Medium-sized tree with light brown, irregularly fissured bark. Branchlets, especially of young specimens or sprouts, commonly with corky-ridged wings. Leaf blades lanceolate, narrowly elliptic, or narrowly oblong, 3–10 cm long and to 3, rarely 4, cm wide, equilateral or slightly inequilateral, bases mostly rounded, apices usually acute; margins doubly serrate, upper surfaces glabrous or somewhat scabrous, the lower more or less pubescent on the veins but without tufts of hairs in the axils of the lateral veins and the midrib. Flowering from buds on twigs of a previous season before the new shoots emerge in spring. Samaras elliptic in outline, symmetrical, their surfaces pubescent, not rugose-veiny, margins conspicuously ciliate, 5–8 mm long.

In woodlands of various mixtures and on various sites, both upland and lowland but not where flooded except temporarily; often in floodplain woodlands and on stream banks. Va. to Ky., s. Ind., s. Ill., Mo., generally southward to cen. pen. Fla. and e. Tex.

3. Ulmus americana L. AMERICAN ELM, WHITE ELM.

Medium-sized to large tree with grayish brown, fissured and ridged or flaky bark. Leaf blades vary considerably in size, even on an individual specimen, in degree of asymmetry of their bases, and in marginal serrations. Blades oblong-obovate, broadly elliptic, oval, or ovate, 3–15 cm long (the larger on a given specimen usually not less than 6 cm long) and 5–8 cm wide (the larger on a given specimen usually not less than 4 cm wide); bases varying from equilateral to markedly inequilateral, apices usually acuminate, sometimes acute, or obtuse on occasional smaller leaves; margins doubly serrate for the most part, very variable as to size of teeth; upper surfaces mostly essentially glabrous, sometimes scabrous, the lower more or less pubescent on the veins and with tufts or webs of pubescence in the axils of the lateral veins and the midrib. Flowering from buds on twigs of a previous season before new shoot emergence in spring. Samaras about 1 cm long, symmetrical or somewhat asymmetrical, elliptic in outline, their surfaces glabrous, rugose-veiny, the margins conspicuously ciliate. (Incl. *U. floridana* Chapm.)

Floodplain forests, wet to moist rich woodlands. Nfld. to Sask., generally southward to cen. pen. Fla. and Tex.

Plants of *U. americana* are subject to Dutch elm disease and phloem necrosis, and the populations in the northern part of the range have largely died out.

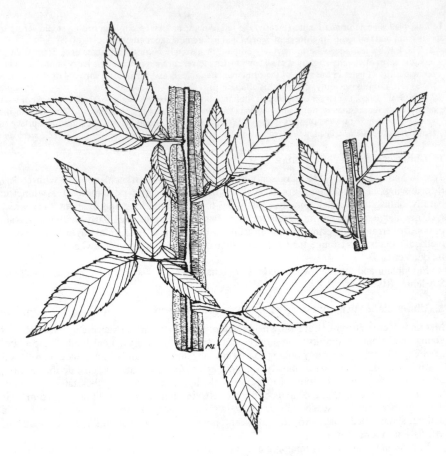

Fig. 23. **Ulmus alata.** (From Kurz and Godfrey, *Trees of Northern Florida.* 1962)

Aristolochiaceae (BIRTHWORT FAMILY)

Aristolochia tomentosa Sims. PIPE-VINE, WOOLLY DUTCHMAN'S PIPE.

A twining vine, the main stems woody, the branchlets of the season herbaceous, sometimes to 25 m into the crowns of trees. Leaves with densely soft-pubescent petioles usually some- what shorter than the blades; blades broadly cordate-ovate to orbicular-reniform, 5–15 cm long or a little more, mostly about as broad as long, margins entire, apices broadly obtuse to rounded, sparsely pubescent above, especially along the veins, softly pubescent, velvety to the touch beneath, notably so when young. Flowers solitary or paired on bractless stalks from leaf axils, the stalks and exterior of floral tube densely soft-pubescent; flower strongly bent a little below the middle, with an oblique, nearly closed purple orifice and a spreading to re- flexed, greenish yellow to purple, 3-lobed limb; flower (straightened out) about 5 cm long fully developed. Fruit a septicidal, longitudinally ribbed capsule, oblong, truncate to rounded at both extremities (rarely beaked apically owing to failure to produce seeds in the distal portion), 4–8 cm long. Seeds numerous, flat, triangular, about 1 cm long.

Floodplain forests, river and creek banks, often on natural levees. s. Ind. to s.e. Kan., generally southward to w. half of Fla. Panhandle and e. and n.cen. Tex.

Note: *A. macrophylla* Lam. (*A. durior* Hill), occurring in rich mountain woodlands in parts of our range, is of closely similar general appearance. Its petioles, lower leaf surfaces, flower stalks, and exterior of the flowers are glabrous or only sparsely pubescent, not velvety to the touch, even when young; its flower stalks bear a conspicuous, foliaceous bract somewhat below their middles.

Polygonaceae (SMARTWEED FAMILY)

(Ours) herbs, or (one of ours) a semiwoody vine. Leaves simple, in most with tubelike, sheathing stipules (ocreae) above the swollen joints of the stem. Flowers small, bisexual or unisexual, variously arranged in inflorescences of fascicled clusters of flowers which may be sheathed at their bases by leafless stipules (ocreolae). Calyx united below, 3–6-lobed, or of 3–6 separate sepals, sometimes with keels or wings, in some petallike, the calyx usually more or less persistent. Corolla none. Stamens 4–9 (rarely fewer or more), the filaments in some dilated basally. Ovary superior, styles 2 or 3, ovule 1. Fruit an achene, usually falling with the remains of the calyx.

1. Plant a vine, climbing by means of tendrils at the tips of branchlets; stipules (ocreae) none; mature sepals decurrent-winged down the fruiting stalks. 1. *Brunnichia*
1. Plant not a vine, without tendrils; stipular sheaths (ocreae) present; mature sepals rarely winged down the fruiting stalks.
  2. Inner perianth segments larger than the outer and (in all except *R. acetosella*) enlarging as the fruit matures. 2. *Rumex*
  2. Inner perianth segments somewhat smaller than the outer, both essentially unchanged in size as the fruit matures *or* the outer enlarging and becoming winged as the fruit matures. 3. *Polygonum*

1. Brunnichia

Brunnichia ovata (Walt.) Shinners. LADIES'-EARDROPS, EARDROP-VINE. Fig. 24

Vine, woody below, herbaceous on new growth, climbing by tendrils terminating lateral branchlets. Bark of woody stems reddish, with grayish stripes and abundant pustulate lenticels; herbaceous young stems with a few scattered hairs. Leaves alternate, short-petiolate, petioles more or less pubescent, a band of short pubescence at the juncture with the stem, ocreae none; blades ovate, variable in size, to 10 cm long and 3–5 cm wide at the base, the bases mostly truncate, occasionally slightly cordate or broadly cuneate, margins entire, pubescent beneath, glabrous or nearly so above; lateral veins arcuate. Flowers stalked, in sessile bracteate fascicles, these in panicled spikes; individual flowers from ocreolae. Flower stalks jointed near the base, gradually dilated above, becoming winged below the fruit during maturation. Flowers bisexual. Calyx segments somewhat petallike, oblong, spreading at anthesis, greenish white above, becoming erect and connivent in fruit and apically somewhat boat-shaped, increasing greatly in size as the fruit matures. Stamens 8. Styles 3, elongate, recurved above at anthesis. Achene obtusely triangular, enclosed in the hardened calyx. (*B. cirrhosa* Banks ex Gaertn.)

River banks, borders of and clearings of floodplain forests, bayous, sometimes in thickets bordering ponds. Coastal plain, S.C. to n. Fla., westward to e. Tex., northward in the interior to s.e. Mo., w. Ky., s. Ill.

2. Rumex (DOCKS AND SORRELS)

Herbaceous perennials, biennials, or winter annuals. Leaves alternate, with sheathing stipules (ocreae). Flowers fascicled, the fascicles in panicled racemes. Flowers small, on jointed stalks, bisexual or unisexual, if unisexual then the plants of some monoecious, others dioecious. Sepals 6, the outer 3 smaller than the inner 3, the latter (in most) enlarging as the

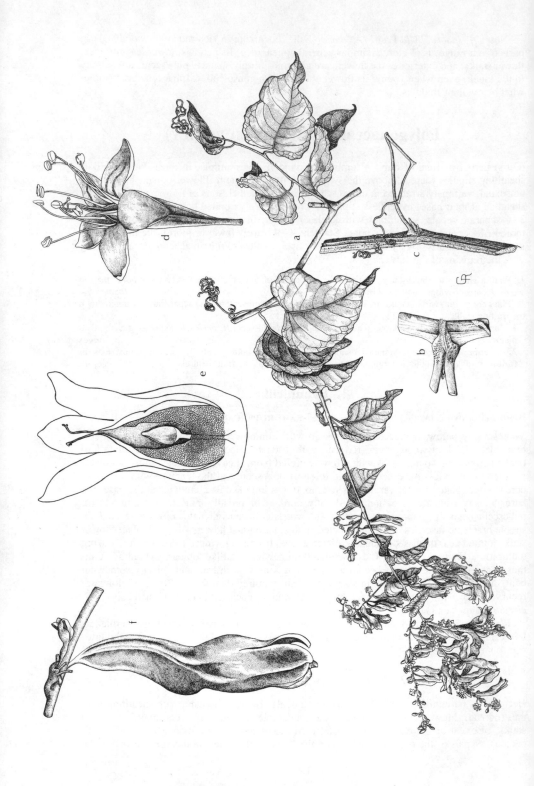

fruit matures (in fruit called valves) and convergent over the 3-angled achene. Petals none. Stamens 6. Stigmas 3, feathery.

Reference: Rechinger, K. H., Jr. "The North American Species of *Rumex*." *Fieldiana*: Botanical Series: 17: 1–151. 1937.

1. Leaves (some of the lower ones at least) sagittate or hastate; flowers unisexual, the plants usually dioecious; inner fruiting sepals (valves) without a tubercle on the midrib.
  2. Plants with slender rhizomes, perennial; valves ovate, basally about 1 mm broad, a little over 1 mm long, not much enlarged in fruit.   1. *R. acetocella*
  2. Plants not rhizomatous, winter annuals; valves reniform, broader than long, 4–5 mm broad, much enlarged in fruit, very broadly winged, the wings usually conspicuously purplish pink.
        2. *R. hastatulus*
1. Leaves not sagittate or hastate; flowers bisexual or if some of them unisexual, then both kinds on the same plant; at least one valve (in ours) with a tubercle on the midrib of at least one of them, usually a tubercle on each of the 3 valves.
  3. Margins of the valves with subulate teeth (best observed on mature fruiting calyx).
    4. Principal leaves broadest below the middle, dark to pale green; surface of tubercle on the valve smooth to rough or wrinkled, not alveolate-reticulate.
      5. Stalks of flowers or fruits (including portions above and below the joints) slender and flexuous, considerably longer than the fruiting calyx.   3. *R. obtusifolius*
      5. Stalks of the flowers or fruits stoutish and stiff, little if any longer than the fruiting calyx.
        4. *R. pulcher*
    4. Principal leaves broadest above the base, usually at or above the middle, grayish green; surface of tubercle on mature valves finely alveolate-reticulate.
      6. Mature valves 4–5 mm long, about 3–4 mm wide, mostly with 4 or 5 marginal teeth.
        5. *R. obovatus*
      6. Mature valves about 3 mm long and 2 mm wide, mostly with 2 or 3 marginal teeth.
        6. *R. paraguayensis*
  3. Margins of the valves entire or wavy, not subulate-toothed.
    7. Stalks of the fruits (including portions both below and above the joints) very slender and much longer than the fruiting calyx.
      8. Leaf margins irregular and markedly crisped or undulate; portion of the fruiting stalk below the joint slender and flexuous, 3–4 mm long; mature fruiting stalks spreading laterally or recurved (bent elbowlike near the joint); tubercle on the mature valve not projecting below the base of the valve.   7. *R. crispus*
      8. Leaf margins flat, entire; portion of the fruiting stalks below the joint very short and stiff, 1 mm long or less, the upper portion of the stalk deflexed at or above the joint; tubercle on the mature valve projecting somewhat below the base of the valve.   8. *R. verticillatus*
    7. Stalk of the fruit (including portions both below and above the joint) shorter than to about equaling the length of the fruiting calyx.
      9. A well-developed tubercle on only 1 of the valves, or if 1 on each valve then 2 of them smaller.
        9. *R. altissimus*
      9. A well-developed tubercle on each of the 3 valves.
        10. Tubercle markedly "puffy," large enough to obscure much of the surface of the valve.
        10. *R. conglomeratus*
        10. Tubercle not "puffy" and not obscuring much of the surface of the valve.
        11. *R. chrysocarpus*

1. Rumex acetosella L. SHEEP SORREL.   Fig. 25

Slenderly rhizomatous perennial with 1 to several stems from the rhizome at a given place, these usually unbranched below the panicle, 2–4 dm tall, glabrous, commonly suffused with reddish purple pigment throughout. Ocreae thin-hyaline, loose and friable. Lower and median leaves petiolate, petioles mostly about as long as the blades; blades mostly hastate basally (some on a given plant sometimes not lobed), the portions of the blades above the lobes linear-oblong to elliptic-oblong or oblong-ovate, often variable in size even on a single plant, 5–20 mm wide, 2–5 cm long; panicles with few, if any, usually unlobed leaves. Flowers mostly unisexual, the plants dioecious (occasionally polygamous). Flower stalks about as long as the calyx. Valves of fruiting calyx ovate, blunt apically, not membranous-

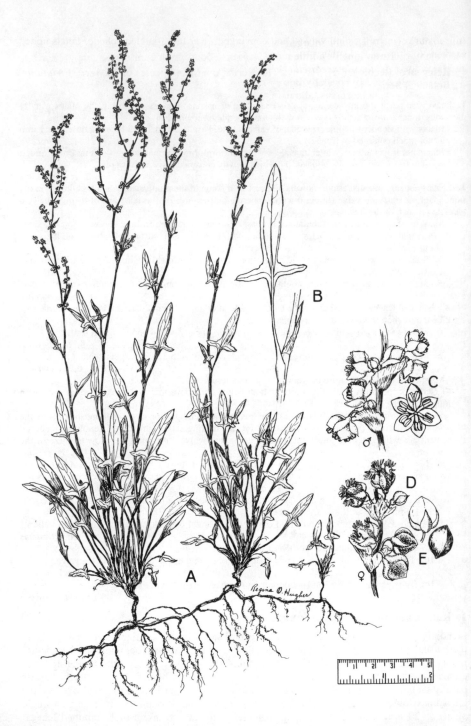

Fig. 25. **Rumex acetosella:** A, habit; B, leaf; C, staminate flowers; D, pistillate flowers; E, fruiting calyx and achene. (From Reed, *Selected Weeds of the United States* (1970) Fig. 62)

winged, not or only slightly enlarged during maturation of the fruit, reticulate-nerved, without tubercles, 1 mm long or a little more. Achene about 1 mm long and as broad, brown to red, lustrous. (*Acetosella acetosella* (L.) Small)

Commonly weedy, especially in sour soils of fallow fields, lawns and waste places, railroad gravels; also in meadows, alluvial outwash, clearings of swampy places, muddy shores. Native of Euras., widely naturalized in temp. N.Am.

2. **Rumex hastatulus** Baldw. ex Ell.

Winter annual with a taproot. Stems 1 to several from the base and with numerous leaves about the base, those on the stem above much reduced and relatively distant from each other. Stem glabrous, to about 8 dm tall. Ocreae thin-hyaline, loose, friable. Lower leaves petiolate, the petioles mostly about as long as the blades; blades mostly hastate basally, the portion of the blades above the lobes variable in size and shape even on a single plant, linear-oblong, oblong, ovate, or elliptic to oblanceolate or spatulate, blunt apically, mostly not over 1.5 cm wide and to about 8 cm long. Panicles with few, if any, usually unlobed bracteal leaves. Flowers unisexual, the plants dioecious. Flower stalks slender, considerably longer than the mature calyx. Valves of the calyx becoming much enlarged during maturation of the fruit, usually as broad as long or broader, commonly cordate basally, subreniform to reniform in overall outline, broadly membranous-winged, without tubercles, about 3 mm across, usually pinkish purple, masses of fruiting plants very colorful. Achene elliptical in overall outline, about 1.5 mm long, brown, lustrous.

Coastal dunes and interdune swales, very abundantly colonizing sandy, fallow fields and clearings, often on exposed shores or bottoms of ponds or lakes at times of low water. Chiefly coastal plain, Mass. to cen. pen. Fla., westward to e., s.e. and n.cen. Tex. and N.Mex., northward in the interior to Mo., Kan. and s. Ill.

3. **Rumex obtusifolius** L. BITTER DOCK.

Coarse perennial with a stout taproot. Stems stout, usually unbranched below the panicle, to 12 dm tall, glabrous, leafy-bracted in the lower part of the panicle. Ocreae conspicuous, membranous but longitudinally striate, friable. Lower and median leaves petiolate, the petioles usually shorter than the blades; blades dark green, usually broadest at the base, ovate-oblong, cordate to truncate basally, obtuse apically, margins usually somewhat undulate, slightly crisped, veins usually reddish, surfaces papillate, short-pubescent on the veins beneath, 10–35 cm long and up to 15 cm broad. Flowers bisexual. Fascicles not crowded, the lower usually separated from one another. Stalks of flowers or fruits, including portion below and above the joints, slender and flexuous, considerably longer than the fruiting calyx, usually arched-recurving. Valves of fruiting calyx triangular-ovate, reticulate-nerved, margins each with 3 or 4 subulate teeth, only 1 valve with a tubercle (or sometimes the other 2 valves with small tubercles). Achene 2 mm long, ovate in overall outline, acute apically, dark reddish brown, shiny.

Usually in moist to wet soils, stream banks, ditches, clearings, marshy shores, floodplains. Native of Eur., widely naturalized, Que. and N.S. to B.C., generally southward to n. Fla. and Calif.

4. **Rumex pulcher** L. FIDDLE DOCK.   Fig. 26

Perennial with a stout taproot, somewhat less coarse than *R. obtusifolius*. Stem usually solitary, to about 8 dm tall, the panicle branches usually widely divergent. Ocreae conspicuous, thin-membranous, friable. Lower leaves petioled, the petioles mostly shorter than the blades, pubescent; blades broadest near the base, often somewhat contracted just above the base, thus fiddle-shaped, pale green, basally cordate to truncate, obtuse apically, more or less ovate-oblong, to 20 cm long and 6 cm wide, pubescent on the veins beneath, margins usually somewhat undulate, slightly crisped. Flowers bisexual. Fascicles (except at tips of branches) well separated from each other. Stalks of flowers or fruits stoutish, stiff, little if any longer than the fruiting calyx. Valves ovate, 4–6 mm long, strongly reticulate-nerved, the tubercle

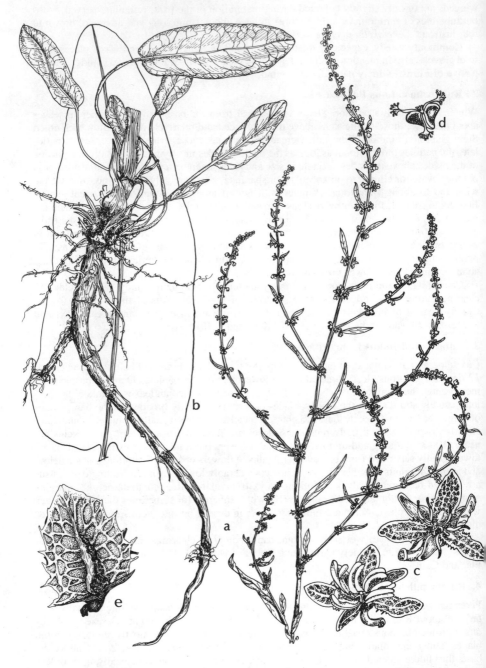

Fig. 26. **Rumex pulcher:** a, basal and upper part of plant; b, outline of leaf blade, from median leaf on stem; c, two flowers; d, pistil; e, fruit. (From Correll and Correll. 1972)

on one valve usually more developed than on the others, tubercle conspicuously "puffy," roughish to rugose or wrinkled, not finely alveolate-reticulate, margin of each valve with 2–6 subulate teeth (rarely entire). Achene 2–2.5 mm long, ovate in overall outline, acute apically, dark reddish brown, lustrous.

Seasonally moist to wet, usually open places, stream banks, drainage and irrigation ditches, bogs, muddy shores, clearings, waste places. Native of Eur., widely naturalized, L.I. to n. Fla., westward to e. half of Tex., n. to Mo.; Calif. and Ore.; Mex.

5. Rumex obovatus Danser.

Annual with a stoutish taproot. Stems 1 to several from the base, commonly 1, mostly 4–7 dm tall, glabrous, panicle branches ascending to widely divergent. Ocreae hyaline, friable, mostly disintegrated by the time fruits are mature. Lower leaves petiolate, the petioles mostly as long as the blades, sometimes shorter, glabrous; blades grayish green, usually broadest above the middle, oblong-obovate, obovate, or elliptic, (the range of shapes even on a single plant), 4–15 cm long, glabrous, bases truncate, subcordate, or cuneate, apices obtuse. Flowers bisexual. Fascicles mostly well separated from each other except near the tips of the panicle branches. Stalks of flowers or fruits, including portion below and above the joints, longer than the mature fruiting calyx, moderately slender and stiffish. Mature valves triangular-ovate, about 4–5 mm long, strongly reticulate-nerved, a prominent "puffy" tubercle on each valve, 1 larger than the other 2, surfaces of tubercles finely alveolate-reticulate, margins of the valves each with 4 or 5 subulate teeth. Achene ovate in overall outline, acute apically, a little over 2 mm long, nearly as broad, surface brown, shiny.

River shores, low wet places in fallow fields, banks of drainage and irrigation ditches, exposed shores or bottoms of ponds at times of low water. Native of s. S.Am., sporadically naturalized, n. Fla. and La.

6. Rumex paraguayensis Parodi.

In general aspect very similar to *R. obovatus*. Fascicles of flowers or fruits mostly crowded. Mature valves smaller, 3 mm long, triangular, apices long-acute, margins with 2 or 3 triangular or subulate teeth; tubercles "puffy," 1 larger than the other 2, their surfaces alveolate-reticulate. Achene broadly elliptic in overall outline, about 2 mm long, surface brown, shiny.

Native of s. S.Am., sporadically adventive in n. Fla. and La.: chiefly wet sands of upper strand and drier sands of shores, St. George Sound, Fla. Panhandle; weedy area, Fontainbleu State Park, La.

7. Rumex crispus L. CURLY DOCK, SOUR DOCK, YELLOW DOCK.   Fig. 27

Coarse, stout perennial, the stem unbranched below the panicle, to about 15 dm tall, glabrous. Ocreae conspicuous, membranous but longitudinally striate, friable. Lower leaves petiolate, the petioles shorter than the blades, glabrous; blades dark green, long-elliptical to oblong-lanceolate, 1–3 dm long, markedly undulate and crisped near the margins, cuneate basally, acute apically, glabrous. Panicle branches ascending-erect, the fascicles of flowers or fruits crowded. Flowers bisexual. Flower stalks, including portions below and above the joints, slender, much longer than the fruiting calyx, portion below the joint 3–4 mm long, stalk mostly bent elbowlike near the joint. Valves ovate, blunt apically, truncate to subcordate basally, reticulate-nerved, 4–5 mm long, margins essentially entire; a tubercle on each of the valves, 1 usually larger than the other 2, sometimes only 1 well developed, tubercle surfaces smooth to slightly wrinkled; tubercle not projecting below the base of the valve. Achene ovate in overall outline, apically acute, about 3 mm long, brown.

Weedy, cultivated fields, drainage and irrigation ditches, marshy shores, bottomland clearings, waste places. Native of Eur., widely naturalized in temp. areas.

8. Rumex verticillatus L. SWAMP DOCK.   Fig. 28

Perennial, usually in water and commonly with many fibrous roots from the lower nodes. Stem often with short axillary branches below the panicle, glabrous, purplish, to about 10

Fig. 27. **Rumex crispus:** A, plant; B, fruit: a, surrounded by persistent calyx, b, showing 3 valves; C, achene. (From Reed, *Selected Weeds of the United States* (1970) Fig. 63)

Fig. 28. **Rumex verticillatus:** a, basal part of plant; b, node; c, top of plant; d, flower; e, fruit.

dm tall. Lower leaves petiolate, the petioles somewhat shorter than the blades, glabrous; blades usually pale green, glabrous, thinnish, elliptic to lance-elliptic, flat, cuneate at both extremities, margins entire, 5–20 cm long and 2–5 cm broad. Ocreae conspicuous, thin-membranous but with longitudinal striae, friable. Flowers bisexual. Fascicles of flowers or fruits mostly crowded, sometimes well separated from each other on the lower parts of the panicle branches. Portion of the flowering or fruiting stalks below the joints short, 1 mm long or less, stiff; portion of the stalks above the joint slender, considerably longer than the fruiting calyx, deflexed at or somewhat above the joint. Valves ovate, blunt apically, 4 mm long or a little more, sometimes broader than long, nerved-reticulate either side of the tubercle but not to the entire or wavy margin; a well-developed, ridgelike tubercle on each valve, this projecting somewhat below the base of the valve. Achene ovate in overall outline, acute apically, about 3 mm long, brown, shiny. (Incl. *R. floridanus* Meisn.)

Usually in water, or in places flooded at times, streams and river floodplains, swamps, marshy shores, drainage and irrigation ditches or canals. Que. and Ont. to Minn., generally southward to s. Fla. and e. and s.e. Tex.

9. Rumex altissimus Wood. PALE DOCK.

Perennial. In general aspect similar to *R. verticillatus*. Stalks of the fruit, including short portion below the joint and longer portion above, shorter than to about equaling the fruiting calyx. Valves broadly ovate, 4–6 mm long, as wide as long or wider than long, blunt apically, truncate to subcordate basally, reticulate-nerved either side of the midrib or tubercle quite to or very nearly to the entire margin; 1 valve with a conspicuous "puffy" tubercle, the other 2 with smaller ones or more commonly with none; surface of the tubercle smooth or slightly wrinkled. Achene ovate in overall outline, short-acute apically, about 3 mm long, brown, shiny.

Floodplain woodlands, marshy shores, wet meadows, drainage and irrigation ditches. N.Y. to Minn., Neb., Colo., generally southward to Ga., Ala., Miss., Tex., and Ariz.

10. Rumex conglomeratus Murr.

Similar in general aspect to *R. crispus*. Lower leaves rounded to cordate basally. Panicle with numerous bracteal leaves, the fascicles of flowers or fruits well separated from each other. Flower or fruit stalks, including portions below and above the joints, shorter than to equaling the fruiting calyx. Valves 2.5–3 mm long, each with a large, plump tubercle which nearly obscures the wings of the valve.

Floodplain woodlands, drainage and irrigation ditches, stream banks, marshy shores. Native of Eur., naturalized on the coastal plain, e. Va. to Ga., sporadically, Ohio, Ind., Mich., e. Tex., Ariz., Pacific states.

11. Rumex chrysocarpus Moris.

Perennial. Stem often with short lateral branches below the inflorescence. Leaf blades pale green to bluish green, lanceolate, 5–12 cm long, 1–3 cm wide. Flowers bisexual. Fascicles of flowers or fruits usually well separated from each other, sometimes approximate, panicle branches not leafy-bracted. Stalks of flowers or fruits, including portions below and above the joints, shorter than to equaling the fruiting calyx. Valves ovate, 4–5 mm long, acute to obtuse apically, wings nerved-reticulate from the midrib or tubercle to the margin; an oblongish tubercle on each valve, its surface usually finely alveolate-reticulate when fully developed. Achene ovate in overall outline, abruptly pointed apically, 2.5 mm long, dark brown, lustrous. (*R. berlandieri* Meisn.)

Swamps and seasonally wet places. La., Tex.; Fla. Panhandle; Mex.

3. Polygonum (SMARTWEEDS AND KNOTWEEDS)

(Ours) annual or perennial herbs, some viny. Leaves simple, alternate, with tubelike, sheathing stipules (ocreae). Stems with swollen nodes. Flowers with jointed stalks, arranged in axillary clusters or in axillary or terminal spikelike racemes, the latter sometimes in panicles. Calyx united below, 4–6-lobed above, pink, green, or white, the lobes spreading at anthesis and then petallike in some, usually with glandular disks lining the lower parts, after anthesis closely embracing the fruit as it matures. Petals none. Stamens 3–9, often unequally inserted, sometimes some in the sinuses of the lobes, others below on the tube, occasionally some on the margins of the glands. Styles 2 or 3 and 2- or 3-divided above, stigmas capitate. Achene lenticular or 3-angled.

1. Stems of mature flowering specimens retrorsely barbed.
  2. Leaves petiolate.
    3. Leaf blades sagittate, strongly retrorsely barbed on the midvein below; achene 3-angled.
                                                           1. *P. sagittatum*
    3. Leaf blades hastate, not barbed on the midvein below but pubescent; achene lenticular.
                                                           2. *P. arifolium*
  2. Leaves sessile.   3. *P. meisnerianum*
1. Stems not barbed.
  4. Stems twining (viny).
    5. Outer calyx lobes enlarging as the fruit matures and becoming strongly winged. 4. *P. scandens*
    5. Outer calyx lobes essentially unchanged as the fruit matures, narrowly keeled on the back.
                                                           5. *P. convolvolus*
  4. Stems not twining or viny.
    6. Flowers in elongate spikelike racemes mostly 15–30 cm long or more, mostly widely separated from each other; styles 2, becoming hard and persistent on the fruit, hooked terminally.
                                                           6. *P. virginianum*
    6. Flowers in relatively short, spikelike racemes 6 cm long or less, commonly crowded at least toward the summit of the raceme, if remote then the racemes not nearly 15 cm long; styles not becoming hard, usually withering as the fruit matures, eventually deciduous.
      7. Stipular sheaths of the leaves (ocreae) entire at the summit (note, however, that the ocreae often fracture and tear and that in those whose ocreae have striate nerves the membranous portions between the nerves may be fractured and shed leaving the striate nerves free and appearing like bristles).
        8. Raceme 1 and terminal, or a terminal unequal pair.   7. *P. amphibium*
        8. Racemes usually several to numerous, both terminal and axillary.
          9. Calyx glandular-dotted; achene lenticular, both faces convex; plant perennial.
                                                   8. *P. densiflorum*
          9. Calyx without glandular dots; achene lenticular, one or both faces flat or concave; plant annual.
            10. Stalks of the raceme glabrous or (usually) with sessile or subsessile glands; racemes mostly nodding.   9. *P. lapathifolium*
            10. Stalks of the racemes pubescent, sometimes stipitate-glandular, sometimes with stipitate glands intermixed with short, stiff pubescence, sometimes with only short, stiff pubescence.
                                               10. *P. pensylvanicum*
      7. Stipular sheaths of the leaves (ocreae) with bristles fringing their summits.
        11. Internodes and stipular leaf sheaths both copiously and conspicuously pubescent with laterally spreading hairs.   11. *P. hirsutum*
        11. Internodes and stipular leaf sheaths not both with spreading hairs.
          12. Achene black, its surface very finely granular and dull; racemes nodding; calyx usually 4-parted.   12. *P. hydropiper*
          12. Achene black, its surface smooth and lustrous; calyx 5-parted.
            13. Plant perennial with horizontal rhizomes or stolons.
              14. Fruiting calyx spherical or nearly so in outline, the lobes incurved toward the slightly exserted tip of the achene.   16. *P. opelousanum*
              14. Fruiting calyx oval in outline and completely enclosing the achene.
                15. Calyces with numerous glandular dots randomly spaced (drying yellowish to brownish and then more evident).   13. *P. punctatum*
                15. Calyces without glands, or with a few scattered, pale, scalelike structures.
                  16. Leaves 1.5 cm broad or more; hairs on the tube of the stipular sheath loose and spreading, not stiff.   14. *P. setaceum*
                  16. Leaves mostly 1 cm wide or less; hairs on the tube of the stipular sheath strongly ascending-appressed, stiff, their bases attached to the tube (adnate).
                                              15. *P. hydropiperoides*

13. Plant annual with a taproot.
  17. Bristles fringing the summit of the sheath around the fascicles of flowers ⅔ as long as to somewhat longer than the tube bearing them.   **18.** *P. caespitosum*
  17. Bristles fringing the summit of the sheath around the fascicles of flowers less than half as long as the tube bearing them, or absent.   **17.** *P. persicaria*

1. Polygonum sagittatum L. TEAR-THUMB, ARROW-VINE.   Fig. 29

Stems slender, erect at first, later loosely and divergently branching, to 2 m long or more, the branches weak, mostly leaning upon or intertwined through other vegetation; stems 4-angled, the angles sharply retrorsely barbed. Leaves short-petiolate, the petioles sharply retrorsely barbed; blades sagittate at base, broadest at base thence gradually tapering to an acute tip, long-triangular in outline, 3–10 cm long and 1–3 cm wide at the base, strongly retrorsely barbed on the midvein beneath. Flowers pink or white, in capitate clusters on smooth, slender stalks, these solitary or in terminal and axillary few-branched, open panicles much exceeding the leaves. Stamens 8. Style 3-divided above. Achene 3-angled, dark brown to black, smooth, ovate in outline, 3 mm long or a little more, shortly beaked. (*Tracaulon sagittatum* (L.) Small)

Marshes, wet woodlands, wet ditches, wet thickets, brushy ditch and stream banks. Nfld. to Sask., generally southward to n. Fla. and e. Tex.

2. Polygonum arifolium L. HALBERD-LEAVED TEAR-THUMB.

Stem slender, weak, to 2 m long or more, erect at first, later reclining on or intertwined with other vegetation, ribbed-angled, retrorsely barbed. Leaves slenderly petiolate, the petioles ½–⅔ as long as the blades; principal blades with ovate-acuminate, horizontally divergent (hastate) basal lobes, broadly oblong or ovate-oblong above the lobes, to 15 cm long and 7 cm broad, pubescent on both surfaces. Flowers few, pink to purplish to white, in short racemes on slender stalks, these retrorsely barbed below and with intermixed glandular and nonglandular hairs distally, the racemes in loose, open, few-branched panicles. Stamens 6. Styles 2-divided above. Achene plumply lenticular, about 4 mm long, dark brown to black, ovate, not beaked. (*Tracaulon arifolium* (L.) Raf.)

Wet meadows, marshes, marshy shores, wet woodlands. N.B. to Minn., generally southward to Ga. and s. Mo.

3. Polygonum meisnerianum Cham. & Schlecht.
var. **beyrichianum** (Cham. & Schlecht.) Meisn. in Mart.   Fig. 30

Stems slender, weak, reclining, usually intertwined amongst each other or upon and within other vegetation, ribbed-angled, unevenly retrorsely barbed. Leaves sessile, linear-oblong, rounded, shortly tapered, or sometimes subcordate basally, long-acute apically, larger ones to 10 cm long or a little more, glabrous on both surfaces, margins with numerous short, sharp spicules. Seedling plants or sprouts arising from cut-off plants may have leaves slightly sagittate at the base. Flowers pinkish, in subcapitate, slender-stalked racemes, on few and widely divergent, terminal and axillary, glandular-pubescent panicle branches. Stamens 5. Style short and thickish below, with 3 short slender divisions above. Achene ovate in outline, obscurely 3-angled, 3 mm long, nearly as broad, brown.

Swamps, wet ditches and thickets. Coastal plain, S.C.; n. Fla.; La.; Mex.; e. S.Am.

Reference: Mitchell, R. S. "A Re-evaluation of *Polygonum meisnerianum* in North America." *Rhodora* 72: 182–188. 1970.

4. Polygonum scandens L. CLIMBING FALSE-BUCKWHEAT.

Herbaceous, perennial twining vine, sometimes climbing well into small trees and shrubs, to 5 m long or more. Stems obscurely angled to terete, usually glabrous, sometimes short-pubescent. Leaves slenderly petiolate, the petioles variable in length, much shorter than the blades to ½ or ⅔ as long as the blades; blades ovate, basally short-cordate, short-hastate, or

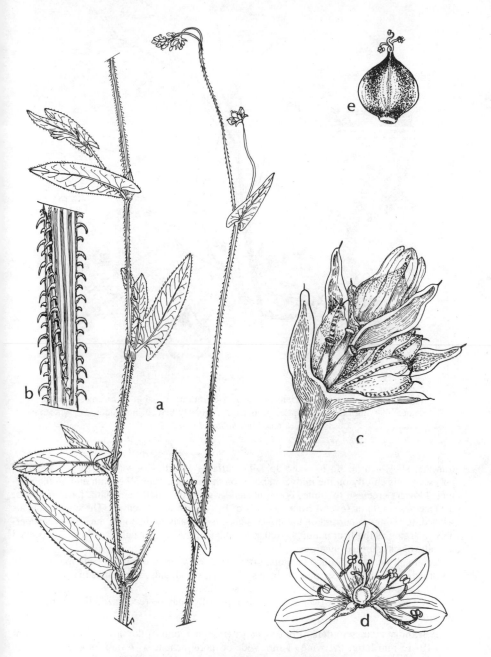

Fig. 29. **Polygonum sagittatum:** a, portions of plant; b, portion of stem; c, flower cluster; d, flower, opened out; e, achene. (From Correll and Correll. 1972)

Fig. 30. **Polygonum meisnerianum** var. **beyrichianum:** A, habit; B, raceme; C, nodal region showing ocrea, leaf base, portions of retrorsely barbed internodes; D, achene. (From Mitchell in *Rhodora* 72: 185. 1970)

truncate, short-acuminate to acute apically, surfaces glabrous or very short-pubescent, if pubescent this chiefly on the midvein above, on the principal veins beneath and on the margins. Flowers greenish to white, borne in slender-stalked, usually elongate, loose racemes, the racemes single or several from the axils of leaves or bracteal leaves. Three of the sepals winged, the wings decurrent on the flower stalks to or nearly to the joint, the sepals considerably enlarging as the fruit matures. Achenes 3-angled, the faces usually concave, not beaked apically, black, surfaces lustrous.

Commonly in well-drained habitats, often in places where the soil has been mechanically disturbed, also fence rows, roadside banks, thickets, woodlands, sometimes in swamps, bottomland woodlands, sand and gravel bars of streams.

Two varieties, partially overlapping geographically, may be recognized for our range, as follows:

P. scandens var. **scandens.** Fruiting calyx (from the joint of the flower stalk to the summit) 10–15 mm long, the wings 1 mm wide or more; achene 4–6 mm long. (*Bilderdykia scandens* (L.) Greene)

Que. to N.D., generally southward to S.C., Ga., and Okla.

P. scandens var. **cristatum** (Engelm. & Gray) Gl. Fruiting calyx (from the joint of the flower stalk to the summit) 7–10 mm long, the wings less than 1 mm wide; achene 2.5–3 mm long. (*P. cristatum* Engelm. & Gray, *Bilderdykia cristata* (Engelm. & Gray) Greene; incl. *P. dumetorum* L., *B. dumetorum* (L.) Dum.)

N.Eng. to Iowa, generally southward to n. Fla., Ark., Okla., e. and n.cen. Tex.

5. Polygonum convolvulus L. BLACK BINDWEED. Fig. 31

Annual, twining or procumbent vine. Closely similar in general habital features to *P. scandens*. Sepals enlarging little as the fruit matures, three of them with pale, opaque, narrow keels that are somewhat decurrent on the flower stalks; fruiting calyx about 5 mm long, strongly 3-angled, the faces roughish or farinose. Achene black, obovate in overall outline, 3-angled, the faces somewhat convex, not beaked, surface finely granular and dull, about 4–5 mm long. (*Bilderdykia convolvulus* (L.) Dum.)

Weedy in cultivated and disturbed grounds, sometimes in clearings of swamps or wet woodlands. Native of Eur., widely naturalized in temp. N.Am.

6. Polygonum virginianum L. JUMPSEED. Fig. 32

Perennial with a hard, knotty rhizome. Stems commonly solitary, sometimes several from the rhizome, usually unbranched. Ocreae appressed-bristly-pubescent, their summits fringed with bristles. Leaves short-petiolate; blades broadly elliptic, ovate-elliptic, varying to lanceolate, bases rounded to broadly short-tapered, apices usually acuminate, both surfaces and margins strigose; very variable in size, from about 5 cm long and 2 cm wide to about 15 cm long and 8–10 cm wide. Flowers in elongate racemes, mostly 15–60 cm long, usually widely spaced on the axis, greenish to white, sometimes pinkish; racemes commonly solitary, sometimes several from a common ocreola, sometimes few-branched. Fruiting calyx about 3 mm long, the stalk reflexed. Stamens 5. Styles 2, becoming hard and rigid, persistent on the fruit, about as long as the achene, each hooked terminally. Achene lenticular, ovate, 3.5–4 mm long, brown, surface lustrous. (*Tovara virginiana* (L.) Raf.; *Antenoron virginianum* (L.) Roberty & Vautier)

Floodplain forests, rich, mesic forests, seepage areas in woodlands, moist thickets, stream banks, clearings of lowland woodlands. Que. and Ont., generally southward to n. Fla. and Tex.; cen. Mex.

7. Polygonum amphibium L. WATER SMARTWEED.

An aquatic and amphibious perennial exhibiting much variability relative both to genetic heterogeneity and phenotypic modification where water levels are relatively stable and where they fluctuate both cyclically and sporadically. Mitchell (1968) presents the results of a very thorough study of the variation based upon manipulative procedures for clones in cultivation together with observations of populations under field conditions.

Plants with floating tips and spreading or floating leaves, sometimes with erect, branched, emersed stems eventually produced. Plants on moist soil with decumbent lower stems rooting at the nodes, upper stems spreading to erect. Leaves short-petiolate, the blades basally cordate to tapered, apices rounded to acuminate, floating leaves more obtuse or rounded apically than aerial leaves, the latter usually acuminate; blade shape variable (even on a single shoot), oblong-elliptic, ovate, or lanceolate, margins entire, surfaces strigose; larger blades to 15 cm long and 4–6 cm wide. Ocreae of aerial shoots of aquatic plants, early in the season, flared, green and herbaceous; those of terrestrial plants striate, membranous between the striae, their surfaces strigose, eventually tending to shatter. Inflorescence of a single, dense, terminal raceme, or often with a smaller second one laterally from just below it, the racemes of aquatic plants 1–4 cm long, those of terrestrial forms 4–11 (–15) cm long, larger racemes 1–1.5 cm thick, the stalk of the raceme glandular-pubescent if arising from an aerial apex, glabrous if arising from a submersed apex. Calyx bright rose-pink. Stamens 5. Style pale, 2-divided at about the middle, the capitate stigmas bright red. Achene lenticular, broadly elliptical to obovate, short-beaked at the summit, the surface minutely granular, about 3–4 mm long. (Incl. *Polygonum natans* Eat., *Polygonum coccineum* (Muhl. ex Willd.) Greene, *Persicaria muhlenbergii* (Meisn.) Small)

79

Fig. 31. **Polygonum convolvulus:** A, habit; B, portion of raceme; C, flower; D, achenes.
(From Reed, *Selected Weeds of the United States* (1970) Fig. 58)

Fig. 32. **Polygonum virginianum:** a, plant; b, ocrea; c, perianth after anthesis and during development of achene within it; d, achene; e, cross-sectional outline of achene.

Ours is the var. *emersum* Michx. Richard S. Mitchell (personal communication) courteously suggests we use the following notes appertaining to the variety as they appear in Richard S. Mitchell and J. Kenneth Dean, *Polygonaceae (Buckwheat Family) of New York State*, Bull. No. 431, New York State Museum (1978).

Typical populations flower from strongly erect or emergent plants which are often quite robust; floating shoots and leaves are not produced; inflorescences are from 4–15 cm long, and leaves are acuminate tipped, frequently with undulating margins; peduncles are glandular-pubescent . . . ; stipules are not flanged with a collar, but turn brown at the margin and shatter; the most poorly adapted hydrophyte, but well adapted to temporarily moist situations: ditches, sloughs, rice fields as a weed.

8. Polygonum densiflorum Meisn.   Fig. 33

Relatively coarse perennial. Stems decumbent below, copiously rooting at the nodes then erect-ascending, glabrous, reddish brown to purple, in the more robust specimens 8–15 mm in diameter. Ocreae usually glabrous. Leaves short-petiolate, blades glabrous, lanceolate to oblong-lanceolate, cuneate basally, long-acute apically, variable in size, the larger ones mostly 10–25 cm long, 2–4 cm wide. Racemes erect, several in loose open panicles, 4–8 cm long, usually relatively loosely flowered, about 1 cm wide, the flower stalks well exserted from the ocreolae, the joints of the stalks slightly below the calyx. Calyx green to greenish white or pinkish, 3 mm long, glandular-dotted. Styles 2. Achene lenticular, broadly oval to orbicular in outline, shortly beaked at the summit, dark brown or nearly black, lustrous, 3 mm long including the beak. (*Persicaria portoricensis* (Bertero) Small)

Swamps, wet thickets, marshy shores, commonly in water. Chiefly coastal plain, s. N.Y. to s. Fla., westward to Tex., northward in the interior to s.e. Mo. and Tenn.; C.Am., W.I., S.Am.

9. Polygonum lapathifolium L.

Annual with erect, commonly branching, glabrous stems; of variable stature but robust specimens to 15 dm tall and with stout stems. Ocreae glabrous, striate-nerved, membranous between the nerves, usually the membranous portions shattering distally leaving the striae which then may give the appearance of bristles. Leaves short-petiolate, blades lanceolate, variable in size, up to 30 cm long and 5 cm wide toward their bases, short-cuneate basally, long-acute apically, sometimes glabrous on both surfaces, or with short spiculelike teeth on the principal veins, especially beneath and on the margins; frequently glandular-punctate beneath. Racemes usually relatively numerous in loose, open panicles, usually rather densely flowered, 3–8 cm long and mostly nodding, 6–8 mm wide, stalks of the racemes with sessile or subsessile glands, notably distally, these sometimes deciduous. Calyx white or pink, 2–3 mm long, not glandular-dotted. Stamens 6. Style 2-divided very nearly to the base. Achene lenticular, broadly oval to suborbicular, usually 1 face flat, the other concave, or both concave, nipplelike at the summit, dark brown, 2–2.5 mm long. (*Persicaria lapathifolia* (L.) Small)

Marshy shores, wet meadows, sand and gravel bars of streams, moist to wet clearings, often weedy in moist to wet cultivated lands or in drainage or irrigation ditches. More or less throughout temp. N.Am., probably introduced from Eur.

10. Polygonum pensylvanicum L. PINKWEED.   Fig. 34

Annual with erect, simple to freely branched stems of variable stature to 2 m tall. Ocreae glabrous, membranous, often soon fracturing. Leaves short-petiolate, blades lanceolate, bases sometimes rounded, usually short-cuneate, long-acute apically, strigose or glabrous beneath, glabrous above, frequently sparsely punctate on both surfaces. Racemes mostly in few-branched, loose, open panicles, the stalks of the racemes sometimes stipitate-glandular, sometimes with short, stiff pubescence intermixed with stipitate glands, *or* with only short, stiff pubescence (rarely glabrous, apparently so occasionally on specimens in Ohio, Mo., Ark.). Calyx bright rose-pink, infrequently white, 3–4 mm long, not glandular-dotted. Style 2-divided to somewhat below the middle, sometimes heterostylous and either the stamens *or*

Fig. 33. **Polygonum densiflorum:** a, habit; b, ocrea; c, portion of inflorescence; d, perianth enclosing mature achene; e, achene; f, cross-sectional outline of achene.

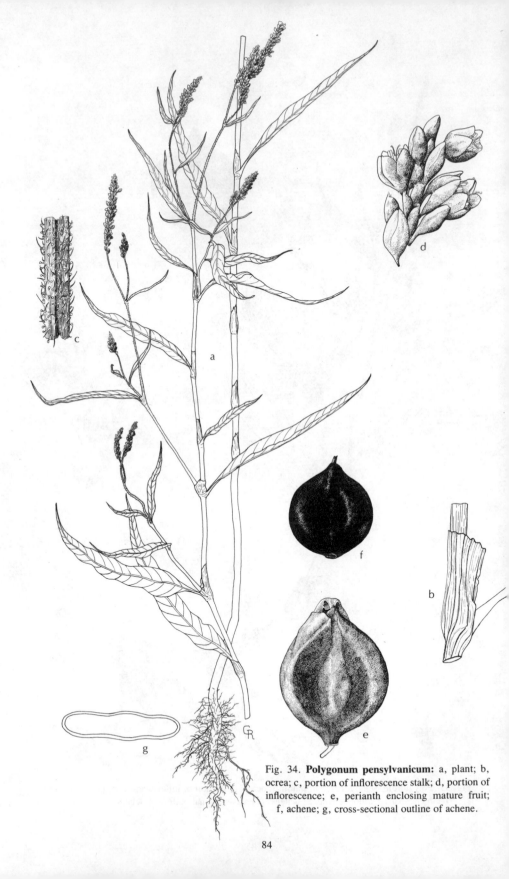

Fig. 34. **Polygonum pensylvanicum:** a, plant; b,
ocrea; c, portion of inflorescence stalk; d, portion of
inflorescence; e, perianth enclosing mature fruit;
f, achene; g, cross-sectional outline of achene.

the styles conspicuously exserted from the calyx. Achene lenticular, broadly oval to sub-reniform, very dark brown or black, surface lustrous. (*Persicaria pensylvanica* (L.) Small; incl. *Polygonum bicorne* Raf., *Polygonum longistylum* Small = *Persicaria longistyla* (Small) Small)

Floodplain forests, marshy shores, wet clearings, swales, shores of streams, commonly weedy in moist to wet cultivated lands and in drainage and irrigation ditches. More or less throughout temp. N.Am.

11. Polygonum hirsutum Walt. Fig. 35

A relatively slender perennial. Lower stems decumbent and copiously rooting at the nodes, then weakly erect. Both internodes and ocreae conspicuously and rather densely clothed with long, spreading pubescence. Leaves sessile, their bases 1–2 cm wide, truncate to very slightly cordate, gradually tapering to acute apices, the longer ones usually not exceeding 10 cm long, both surfaces clothed with long, stiffish, appressed hairs. Racemes erect, 3–6 cm long, 8 mm wide or less, long-stalked, few in open, loose panicles, the stalks of the racemes hirsute, at least below, sometimes glabrous distally. Calyx pink, infrequently whitish, 2 mm long or a little more, not glandular-dotted. Stamens 8. Style 3-divided near the summit. Achene 3-angled, broadly oval or ovate, a little over 2 mm long including the short beak, dark brown, lustrous. (*Persicaria hirsuta* (Walt.) Small)

Marshy shores of ponds and lakes or on shores or bottoms exposed at times of low water, "grassy," relatively open, cypress-gum ponds and depressions, occasionally in drainage ditches. Coastal plain, N.C. to pen. Fla., Fla. Panhandle and s.w. Ga.

12. Polygonum hydropiper L. Fig. 36

Relatively slender, simple to branched annual of variable stature, usually not exceeding 6 dm tall but sometimes to 1 m, distal branches or branchlets flexuous and curving. Stem glabrous. Ocreae membranous, the tube sparsely appressed-pubescent or glabrous, fringing bristles at the summit few and usually 6 mm long or less. Leaves sessile or the blades decurrent to short petioles, blades lanceolate, cuneate basally, acute apically, the larger mostly not over 10 cm long and 2 cm wide, surfaces glabrous or sparsely appressed-pubescent, finely punctate. Flowers sometimes 1 to few in the axils of bracteal leaves, or in more elongate panicled racemes, the racemes narrow, commonly arched. Calyx usually 4-parted, white or greenish white (often tipped with pink in the bud), glandular-dotted, 3–4 mm long. Stamens 6. Style usually 3-divided from about the middle, the undivided portion short and thickish. Achene usually 3-angled, infrequently lenticular, ovate in outline, its surface finely granular, dull black, 2–3 mm long. (*Persicaria hydropiper* (L.) Opiz)

Weedy, wet meadows and pastures, marshy shores, alluvial bottomlands, ditches, moist to wet cultivated grounds. Que. to B.C., generally southward to N.C., w. S.C., thence to Ala., Tex. and Calif.; Euras.

13. Polygonum punctatum Ell. Fig. 37

Vigorous perennial, stems becoming long-decumbent below and rooting at the nodes, the flowering stems erect, to 1 m tall, commonly forming extensive colonies. (Rarely, in our experience, have we seen this plant with the appearance of an annual, that is without decumbent perennial bases; we surmise occasional plants may be flowering the first year from seed.) Stem glabrous or short-pubescent, simple to much branched. Ocreae membranous, the tube glabrous to sparsely short-pubescent, their summits fringed with bristles 5–15 mm long. Leaf blades lanceolate, shortly decurrent basally to short petioles, long-acute distally, variable in size, to 15 cm long and 2 cm wide, surfaces glabrous or short-pubescent on the midvein and sometimes on the principal laterals, usually with short, stiff pubescence on, or on and near, the margins. Racemes several in loose, open panicles. Fascicles of flowers relatively loosely arranged on narrow racemes, sometimes contiguous, more often remote, especially toward the bases of the racemes. Calyx 5-parted, white to greenish white, glandular-dotted (glandular dots 30 or more borne on both the lobes and the tube and drying brown),

85

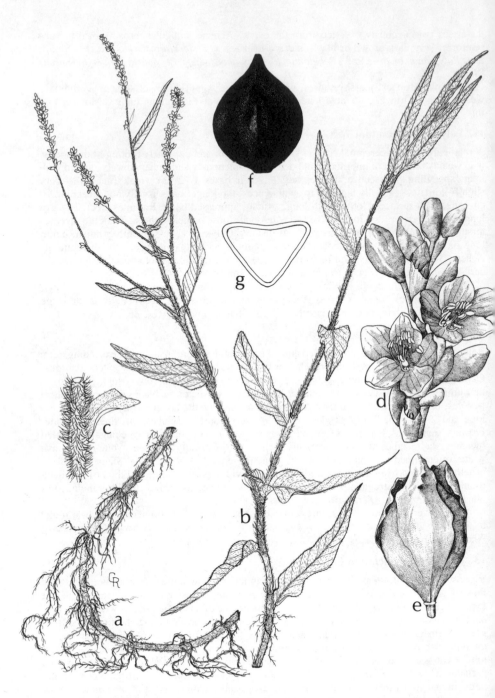

Fig. 35. **Polygonum hirsutum:** a, section of rooting base of stem; b, top of plant; c, short section of stem, showing pubescence; d, portion of inflorescence; e, perianth after anthesis and during development of achene within; f, achene; g, cross-sectional outline of achene.

Fig. 36. **Polygonum hydropiper:** A, habit; B, node showing leaf and ocrea; C, raceme;
D, flower; E, achenes. (From Reed, *Selected Weeds of the United States* (1970) Fig. 59)

Fig. 37. **Polygonum punctatum:** a, lower part of stem with portion of decumbent base; b, upper part of plant; c, ocrea; d, portion of inflorescence; e, perianth surrounding mature achene; f, achene; g, cross-sectional outline of achene.

about 3 mm long in fruit. Stamens 6–8. Style 3- (rarely 2-) parted from just above the middle. Achene usually 3-angled, infrequently lenticular, oval in outline, 2.5–3 mm long, surface black, lustrous. (*Persicaria punctata* (Ell.) Small)

Commonly in water, marshes, marshy shores of ponds, lakes, streams, exposed shores and bottoms of ponds and lakes during times of low water, swamps, floodplain forests, drainage and irrigation ditches and canals, moist to wet disturbed places and cultivated fields. Throughout most of temp. and subtrop. N.Am.; trop. Am.

14. Polygonum setaceum Baldw. ex Ell. Fig. 38

Relatively coarse perennial, the lower stems decumbent and rooting at the nodes, flowering stems ascending-erect, 7–15 dm tall. Ocreae with loose and spreading, not stiff, pubescence, the hairs enlarged at the base but not adnate to the tube, bristles fringing the summit 6–10 mm long; internodes glabrous to sparsely pubescent with spreading hairs. Leaves very short-petiolate or the upper sessile, lanceolate, the larger ones mostly 6–10, rarely to 15 cm long and 1.5–2.5 cm, occasionally to 4 cm, wide, both surfaces and the margin pubescent with spreading (rarely subappressed) hairs. Racemes usually in loose, open panicles, narrow, 5 mm wide or a little more, 2–6 cm long, fascicles of flowers not congested. Calyx usually white or greenish white, occasionally pinkish, 5-lobed, about 3 mm long in fruit, rarely glandular-dotted, if so, then only on the lobes and not on the tube, the dots faint. Achene 3-angled, brown to black, lustrous, 2–2.5 mm long including the short beak. (*Persicaria setacea* (Baldw. ex Ell.) Small; *Polygonum hydropiperoides* var. *setaceum* (Baldw. ex Ell.) Gl.)

Borders of swamps, marshy shores, moist to wet hammocks, wet woodlands and clearings, floodplain forest, drainage and irrigation ditches and canals, sloughs. Coastal plain and piedmont, s. N.J. to s. Fla., westward to Tex., Ark. and Okla.

15. Polygonum hydropiperoides Michx. Fig. 39

With the habit of *P. setaceum* but more slender and of lower stature, mostly about 5 dm tall, sometimes a little taller. Ocreae with short, stiff, strongly appressed-ascending pubescence on the tube, the bases of the hairs flat, enlarged and adnate to the tube, stiff bristles fringing the summit to 7 mm long; internodes mostly with pubescence as on the ocreae. Leaves very short-petiolate to sessile, blades lanceolate to nearly linear-lanceolate, the larger commonly 1 cm wide or less, less frequently to 2 cm wide, about 7–10 cm long; surfaces more or less pubescent with short-appressed, stiff hairs and with short, spiculelike hairs on and near the margins. Racemes in loose, open panicles, narrow, 5 mm wide or a little more, 2–6 cm long, fascicles of flowers not congested. Bristles fringing the summit of the sheaths surrounding the fascicles of flowers less than half as long as the tubes bearing them, or infrequently to ⅔ as long or a little more. Calyx oval in outline, roseate below, pink or cream-white at the lobe apices, rarely greenish, purplish, or white, without glandular dots. Achene 2–3 mm long, 3-angled, dark brownish black, lustrous, not at all exserted from the fruiting calyx. (*Persicaria hydropiperoides* (Michx.) Small)

Often in extensive colonies, marshes, marshy shores, exposed shores and bottoms of ponds or lakes during times of low water, marl prairies, borders of swamps, wet clearings, drainage and irrigation ditches and canals, swales. N.S. to B.C., southward more or less throughout the U.S.

16. Polygonum opelousanum Ridd. ex Small. Fig. 40

Similar in general appearance to *P. hydropiperoides*. Leaf blades glandular beneath with blue-green to yellow, scalelike glands. Calyx oval to spherical in outline, the outer lobes incurved at maturity, greenish at the base, creamy distally or rose- to purple-tinged (not pink), sometimes with a few, pale, scalelike glands unevenly spaced. Achene brownish black, lustrous, usually about as broad as long, 1.5–2.5 mm long, 1.3–2.3 mm broad, its tip slightly exserted from the fruiting calyx. (*Persicaria opelousana* (Ridd. ex Small) Small, *Polygonum hydropiperoides* var. *opelousanum* (Ridd. ex Small) Stone)

In habitats similar to those for *Polygonum hydropiperoides*. Chiefly coastal plain, e. Mass. to s. Fla., w. Tex. and Okla.

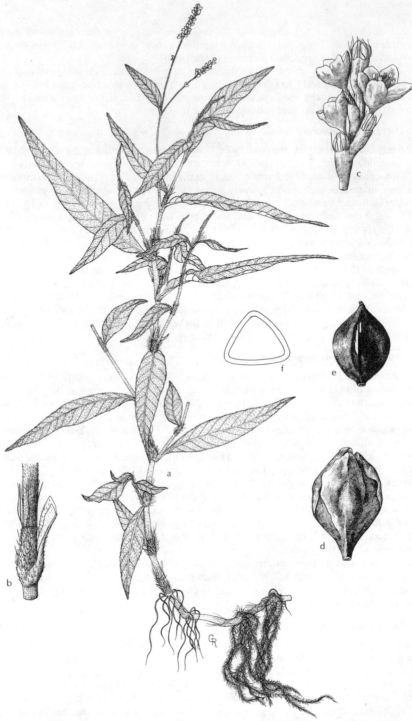

Fig. 38. **Polygonum setaceum:** a, habit; b, ocrea; c, portion of inflorescence; d, perianth enclosing mature fruit; e, achene; f, cross-sectional outline of achene.

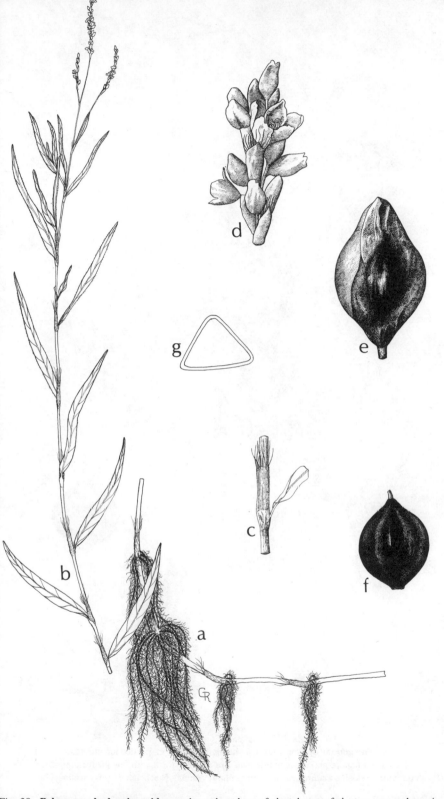

Fig. 39. **Polygonum hydropiperoides:** a, decumbent base of plant; b, top of plant; c, ocrea; d, portion of inflorescence; e, perianth surrounding mature achene; f, achene; g, cross-sectional outline of achene.

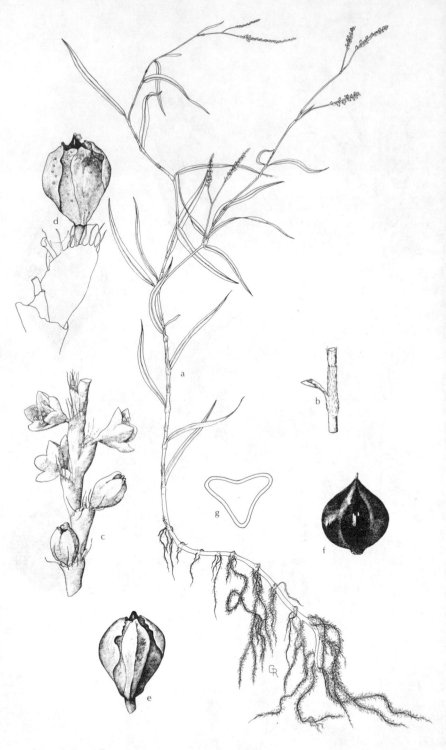

Fig. 40. **Polygonum opelousanum:** a, habit; b, ocrea; c, portion of inflorescence; d, ocreola and mature perianth enclosing mature achene (form with few glands present); e, perianth enclosing mature achene (form without glands); f, achene; g, cross-sectional outline of achene.

17. Polygonum persicaria L. Fig. 41

Plant annual, the stems erect-ascending, simple to much branched, to about 8 dm tall, smooth or sparsely pubescent. Ocreae membranous, sometimes striate-nerved, usually loose, sparsely short-pubescent to smooth, the bristles fringing the summit of the tube about 2 mm long, much shorter than the tube. Leaves short-petiolate to sessile, blades lanceolate, variable in size, 3–15 cm long, 5–18 mm wide, tapering to both extremities, glabrous or sparsely short-pubescent, with appressed spiculelike hairs on the margins. Racemes short, essentially erect, in loose panicles, the fascicles of flowers congested. Fringe of bristles at the summit of the sheaths around the fascicles of flowers less than half as long as the tube bearing them, or absent. Calyx pink or rose-pink. Achene lenticular (infrequently 3-angled), oval, short-beaked, dark brown to black, lustrous, about 2.5 mm long. (*Persicaria persicaria* (L.) Small)

Open banks of streams and ponds, swales, gravel or sand bars, alluvial outwash, moist to wet areas in cultivated fields, disturbed grounds, commonly weedy. Native of Eur., widely naturalized in N.Am.

18. Polygonum caespitosum Bl. var. **longisetum** (DeBruyn) Stewart.

Plant annual, similar in general appearance to *P. persicaria*, stems usually more divergently branched, often sprawling. Leaf blades elliptic to lanceolate, glabrous above, short-pubescent on the veins below and with appressed spiculelike hairs on the margins. Bristles fringing the summit of the ocrea and the sheaths surrounding the flower fascicles ⅔ as long to longer than the tube bearing them. Achene 3-angled, ovate in outline, black, lustrous, about 2 mm long.

Habitats similar to those of *P. persicaria*. Native of Asia, naturalized from Mass. to Ill., generally southward to n. Fla. and La.

Chenopodiaceae (GOOSEFOOT FAMILY)

Annual or perennial herbs with alternate or opposite leaves without stipules, those of our wetlands succulent or subsucculent. Flowers small, individually inconspicuous, bisexual, or unisexual and plants monoecious. Sepals (1) 3 5 (or none), somewhat united. Petals none. Stamens of the same number as the sepals and opposite them. Pistil 1, ovary superior, 1-locular; styles 1–5. Fruit a 1-seeded utricle, in some enclosed by the persistent calyx.

1. Leaves represented by opposite scales.   1. *Salicornia*
1. Leaves not scalelike.
  2. The leaves terete or nearly so.   2. *Suaeda*
  2. The leaves flat.
    3. Leaf surfaces scurfy with branlike scales or farinaceous; fruit completely enclosed by toothed and tuberculate bracts.   3. *Atriplex*
    3. Leaf surfaces cottony-pubescent beneath when young, usually glabrous in age; fruit somewhat wheellike and with a thin peripheral wing.   4. *Cycloloma*

1. Salicornia (GLASSWORTS)

Succulent herbs with opposite, jointed branches. Leaves consisting of fleshy perfoliate scales fused to each other below and more or less fused to the stem which they surround, only the tips, if any, free. Inflorescence a fleshy spike the scales of which are not unlike those of the stem below except that they are shorter and closer together. Flowers bisexual (or some pistillate and the plants polygamous), in groups of 3 and sunken in pits in the axis of a fleshy spike above the axils of scalelike bracts. Calyx a fleshy sac with a 3- or 4-toothed margin around a slitlike opening, spongy in fruit. Corolla none. Stamens 1 or 2, exserted. Styles 2–4-branched. Ovary membranous, even in fruit. Seed thick, hairpinlike, pubescent.

Fig. 41. **Polygonum persicaria:** A, habit; B, raceme; C, node showing leaf base and ocrea; D, achene (two views). (From Reed, *Selected Weeds of the United States* (1970) Fig. 61)

1. Plant perennial with main stems horizontal and mat-forming.   1. *S. virginica*
1. Plant annual, stems erect.
  2. Scale-leaves sharp pointed apically.   2. *S. bigelovii*
  2. Scale-leaves blunt apically.   3. *S. europaea*

1. Salicornia virginica L. PERENNIAL GLASSWORT.   Fig. 42

Plant perennial with horizontal main stems and branches rooting at the nodes and with erect or ascending lateral branches. Single plants may form mats of various sizes, and by interlacing of stems of several to many plants extensive mats may develop. Horizontal branches usually becoming embedded in the substrate and more or less woody. Secondary or lateral shoots about 1–3 dm tall. Leaves oblique-truncate or with minute, broad, blunt tips; on older dried-up portions of the stem the summit of the leaf pair is cuplike. Joints of the flowering spike as broad as long. (*S. perennis* of authors not Mill.)

Salt and brackish marshes and flats. Atlantic coast, N.H. southward to s. Fla., westward along the Gulf coast to Tex.; L.Calif. to B.C.; W.I.; w. Eur.; n. Afr.

2. Salicornia bigelovii Torr. ANNUAL GLASSWORT.   Fig. 43

Plant annual, unbranched or more commonly with several to numerous ascending branches, 1–4 dm tall (often giving the appearance of a miniature succulent tree), green or variously suffused with red pigment. Leaves wholly fused to each other and fused around the stem making it appear comprised of thick naked joints. Scale-leaves very short and narrowed abruptly to sharp points. Joints of the flowering spike about as thick as long, 3–4 mm across.

Salt and brackish marshes and flats. Atlantic coast from N.S. to Fla., Gulf coast from Fla. to Tex. and Yucatan; s. Calif.; W.I. and Bah.Is.

3. Salicornia europaea L.

In general features closely similar to *S. bigelovii*. Scale-leaves blunt apically. Fruiting spikes about 2 mm across.

Salt marshes and flats, Atlantic coast, N.S. to Ga.; in salt-licks and salt marshes inland to Mich., Wisc. and Ill.; Pacific coast; Euras.; Afr.

2. Suaeda (SEA BLITES)

Annual or perennial herbs, the latter sometimes somewhat shrubby. Leaves sessile, succulent, nearly terete, alternate, entire, in size gradually reduced upward, in the upper inflorescence bractlike. Flowers in sessile, mostly compact clusters in the axils of the leaves or bracts, commonly 3–5 in a cluster but sometimes fewer. Clusters more or less surrounding the leaf bases and about half encircling the stem, each flower subtended by a minute, thin, scalelike bract. Calyx of 5 succulent, short sepals fused at the base, closely enveloping the pistil, only the stigmas exserted. Flowers mostly bisexual, occasional ones pistillate. Stamens 5, filaments short, the whole stamen tightly invested within the calyx segment and not exserted. Ovary depressed-globose, styles 2–5. Fruit with a thin-membranous, nearly transparent wall loosely enveloping a single horizontal seed. Seed black, shiny, nearly lenticular, oblique-orbicular in outline, 1–2 mm wide.

- Leaves glaucous or glaucesent; calyx segments essentially equal, not keeled, not hoodlike.
      1. *S. maritima*
- Leaves green, not glaucous; calyx segments unequal, the larger 3 hoodlike on the back. 2. *S. linearis*

1. Suaeda maritima (L.) Dum.   Fig. 44

Annual, sparingly to much branched, the branches spreading-ascending, ascending or depressed, sometimes forming depressed mats; main stems mostly up to 4 dm tall, rarely exceeding 6 dm. Leaves glaucous, to 5 cm long, semiterete. Sepals rounded on the back, essentially equal.

Salt marshes and sea strands. Native to Euras. Naturalized along the Atlantic coast, Que. to Va.; sporadic in Fla.

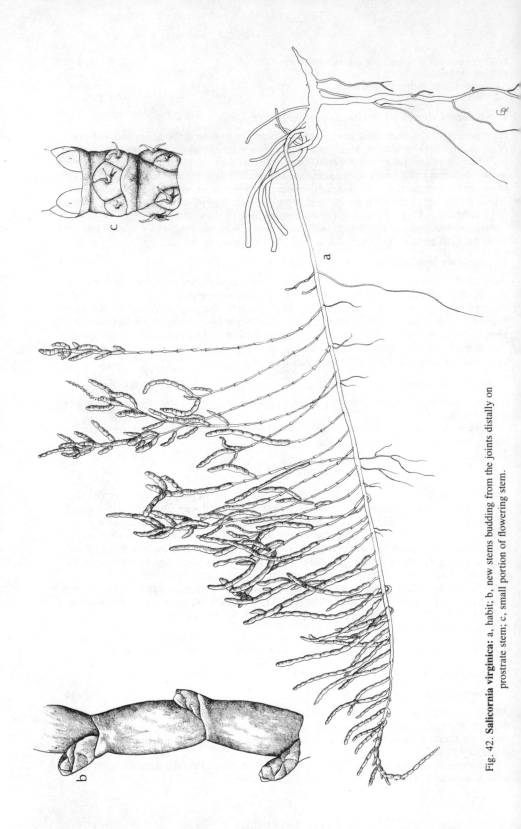

Fig. 42. **Salicornia virginica**: a, habit; b, new stems budding from the joints distally on prostrate stem; c, small portion of flowering stem.

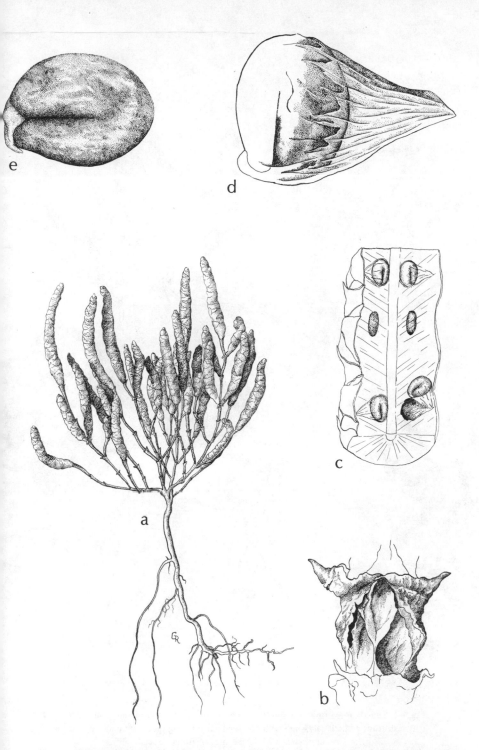

Fig. 43. **Salicornia bigelovii:** a, habit; b, joint after fruits have dehisced; c, longitudinal section, showing, in upper and lower portions of drawing, fruits containing seeds embedded in the fleshy axis; d, fruit; e, seed.

97

Fig. 44. **Suaeda maritima:** a, habit; b, flower; c, fruit, almost completely surrounded by calyx; d, fruit, without calyx and with one seed germinating on parent plant; e, short section of stem with aggregates of fruits; f, seed.

2. Suaeda linearis (Ell.) Moq.   Fig. 45

Annual, or sometimes perennial in the southern parts of the range. Stems commonly much branched, the branches elongate, sometimes reclining but not mat-forming; principal stems very variable in length, becoming as much as 1 m long; green, not glaucous. Leaves semiterete, the larger to about 5 cm long. Lower 2 sepals rounded on the back, the upper 3 larger than the lower and with hoodlike backs. (*Dondia linearis* (Ell.) Millsp.)

Salt marshes, sea strands, often weedy in marly soils near the coast. Atlantic coast, Maine to Fla., Gulf coast, Fla. to Tex.; W.I., Bah.Is.

3. Atriplex (ORACHS)

Annual herbs with angulate or channeled stems. Leaves opposite, alternate, or both opposite and alternate, flat, more or less scurfy with branlike scales, or farinose. Flowers unisexual, plants monoecious. Staminate flowers in terminal spicate-clusters, with 5 sepals, 5 stamens. Pistillate flowers in axillary clusters or spicate clusters, completely enclosed by a pair of persistent bracts, (ours) without a calyx.

● Leaves silvery-gray; fruiting bracts with triangular teeth across the broad summit, crested-tuberculate on both sides of the faces.   1. *A. arenaria*
● Leaves green or purplish green; fruiting bracts triangular-ovate, sometimes hastate, not triangular-toothed on the tapering summit, the faces smooth or with a few tubercles on the faces.   2. *A. patula*

1. Atriplex arenaria Nutt. SEABEACH ORACH.

Highly variable. Plant usually much branched, usually not much over 5 dm tall, often as broad. Leaves alternate, densely silvery-gray, scaly beneath, grayish green above, ovate and rounded at the base to narrowly lance-ovate and tapering basally, to about 3 cm long and to about 1 cm wide, sessile or nearly so. Staminate spikes 1–2 cm long. Fruiting bracts mostly wider than long, united to the middle, with 3–5 triangular teeth across the broad summit, the faces of each with a tuberculate crest either side of the midrib. (Incl. *A. pentandra* (Jacq.) Standl.)

Upper beach strands and dunes, borders of salt marshes. Atlantic coast, s. N.H. to s. Fla., Gulf coast, Fla. to Tex.; W.I.

2. Atriplex patula L. var. **hastata** (L.) Gray.

Stem erect or spreading, to 1 m long. Leaves usually opposite below, alternate above, green or purplish green, sparsely to densely scaly beneath when young; lower leaves, at least, usually petiolate, the blades truncate to hastate at base, more or less triangular to rhombic, obtuse apically; upper leaves lanceolate to oblanceolate or lance-ovate, mostly tapering basally. Fruiting bracts ovate to rhombic-ovate, united at the base, entire or finely toothed on the margins, somewhat longer than broad, each face smooth or with 1 to few tubercles on the sides. (*A. hastata* L.)

Saline or brackish places near the coast, in rich soils inland. Nfld. to B.C., southward to S.C., Ohio, Ind., Ill., Mo., and Tex.; Ore. and Calif.; Euras.

4. Cycloloma

Cycloloma atriplicifolium (Spreng.) Coult. TUMBLE RINGWEED, WINGED PIGWEED.

Annual herb, 1–8 dm tall, usually bushy-branched and about as broad as tall, the stems commonly leafless by the latter stages of maturation and with tumbleweed fitness. Stems mostly striate-angled, loosely cottony-pubescent when young, glabrous or nearly so in age. Leaves alternate, sessile or narrowed to short, subpetiolar bases, medial leaves largest, blades lanceolate or lance-ovate, 2–8 cm long, 6–15 mm broad, acute apically, margins sinuately dentate-toothed or toothed-lobed, surfaces cottony-pubescent when young, usually glabrous in age. Flowers very small, 1–1.5 mm across, sessile and remote from each other on all branch-

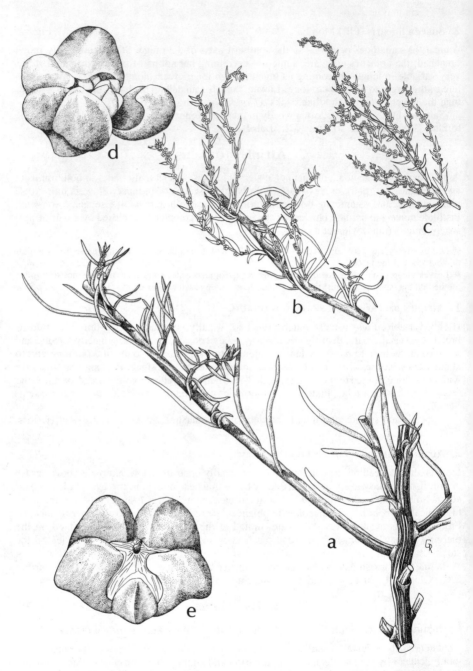

Fig. 45. **Suaeda linearis:** a, small part of principal stem with base of branch; b, median portion of branch below; c, distal portion of branch; d, flower, just opening; e, fruit encompassed by calyx.

lets, subtending bracts minute, bisexual (or sometimes a few pistillate). Calyx 5-lobed, segments keeled and incurved over the ovary, by maturity developing below its lobes a continuous, thin, entire, erose, or toothed, circular wing. Stamens 5. Ovary circular-depressed, densely pubescent, styles 2 or 3. Fruit a depressed-circular, purplish utricle partially to almost wholly enclosed by the winged calyx. Seed circular-lenticular, 1.5–2 mm broad, lustrous black, finely papillose.

Bars and exposed shores of rivers and streams, beds of ephemeral streams, interior dunes; also weedy in sandy, fallow fields and other places. Apparently native from Ind. to Man., southward to Tex. and Ariz., now naturalized in many places eastward.

Amaranthaceae (AMARANTH FAMILY)

(Ours) annual or perennial herbs. Leaves alternate or opposite, petioled or sessile, without stipules, usually with entire margins. Flowers bisexual or unisexual (plants sometimes monoecious, sometimes dioecious, sometimes polygamous), usually with 3 subtending scarious bracts. Calyx of (2–) 5 free or basally united parts, these in most cases scarious. Petals none. Stamens (2–) 5, filaments free or basally united, sometimes with staminodes as well. Pistil 1, ovary superior. Fruit a 1-seeded utricle, circumscissily dehiscent, irregularly dehiscent, or indehiscent, or a 1- to several-seeded capsule.

1. Leaves alternate.   1. *Amaranthus*
1. Leaves opposite.
  2. Flowers in short, headlike spikes, spikes sessile or stalked, borne from the leaf axils or terminally on the branches.
    3. Leaves succulent; stamens 5, staminodes none; style absent; stigmas 2; plants maritime.
                                        2. *Philoxerus*
    3. Leaves not succulent; stamens 5 and alternating staminodes present as well; style 1, stigma capitate; plants of freshwater habitats.   3. *Alternanthera*
  2. Flowers in diffuse panicles.   4. *Iresine*

1. Amaranthus (AMARANTHS, PIGWEEDS)

Plants annual (in some species of extraordinarily variable stature). Leaves alternate, petiolate, edges sometimes irregularly wavy but margin itself entire in most. Flowers usually in axillary clusters or in spikelike panicles or thyrses; subtended by 3 bracts, bisexual or unisexual, in some both kinds on the same plant, or one or the other on separate plants. Perianth parts absent, or 2–5, membranous or scarious, not greatly dissimilar to the subtending bracts. Stamens 5 or 2 or 3. Style none, stigmas 2 or 3, ovary 1-locular, 1-ovuled. Fruit a 1-seeded utricle, indehiscent, irregularly dehiscent, or circumscissily dehiscent.

A group difficult of interpretation for us who have no expertise in it. We exclude from this work those species which are almost exclusively weeds of cultivated grounds and waste places. With the exception of *A. crassipes* and *A. pumilus*, the species here treated belong to what some authors place in the segregate genus *Acnida* (plants of which are dioecious) and we have in considerable part adapted from the work of Jonathan Sauer, "Revision of the Dioecious Amaranths," *Madrono* 13: 5–46 (1955).

1. Flowers in small axillary clusters.
  2. Leaf blades as broad as long to not over twice as long as broad, from suborbicular to short-oblong.
                                        1. *A. pumilis*
  2. Leaf blades significantly longer than broad, chiefly oblanceolate or spatulate.   2. *A. crassipes*
1. Flowers mostly on elongate spikelike branches, or diffusely paniculate or thyrsiform.
  3. In the pistillate plants:
    4. Fruit circumscissily dehiscent.   3. *A. tamariscina*
    4. Fruit indehiscent or irregularly dehiscent.
      5. The fruit angled.

101

    6. Fruit 2.5–4 mm long; seed 2–3 mm long; bracts less than 1 mm long.   **4.** *A. cannabinus*
    6. Fruit 2 mm long or less; seed about 1 mm long; bracts 1.5 mm long or more. **5.** *A. australis*
    5. The fruit not angled.   **6.** *A. tuberculatus*
  3. In the staminate plants:
    7. Bracts a little over 2 mm long.   **3.** *A. tamariscina*
    7. Bracts 1–1.8 mm long.
      8. The bracts about 1 mm long.   **4.** *A. cannabinus*
      8. The bracts 1.5–1.8 mm long.
        9. Bracts with moderately heavy midrib; outer perianth parts with excurrent midveins.
                5. *A. australis*
        9. Bracts with slender midrib; midveins of outer perianth parts not excurrent. **6.** *A. tuberculatus*

1. Amaranthus pumilus Raf. DWARF AMARANTH.

Plant low, with sprawling branches. Leaf blades succulent, little longer than broad, suborbicular to short-oblong, 1–1.5 cm long, retuse at the summit. Flowers in short axillary fascicles, mostly bisexual. Fruit smooth, indehiscent, 4–5 mm long.

    Local, beach sands of strands and spits. e. Mass. to S.C.

2. Amaranthus crassipes Schlecht.

Plant usually with numerous branches from the base, stems prostrate or decumbent below then erect-ascending, somewhat succulent. Leaves oblanceolate to spatulate, the blades narrowed to petiolar bases, blades sometimes ovate, 1–4 cm long, mostly significantly longer than broad, rounded or retuse at the summit. Inflorescences short axillary clusters. Flowers mostly unisexual, both sexes on the same plant, occasionally some bisexual. Fruit with a papillate surface, indehiscent.

    Naturalized in waste places and adventive at seaports, Fla. to Tex.; in Tex. on mud and gravel bars and in playa lakes. Native in trop. Am.

    Specimens of the following dioecious species of amaranths are for us very difficult to identify, especially if from staminate plants, the pistillate ones having diagnostic characters somewhat more readily discerned. We suggest that one interested in these plants seek, in the field, to find mature pistillate plants (these are usually fewer in number in a given area than the staminate); if plants are extremely large, as may be the case with specimens of *A. cannabinus* and *A. australis* particularly, the shape and dimensions of the midstem leaves may be helpful. The geographic area and habitat of a given plant or population may aid in identification by a process of elimination.

3. Amaranthus tamariscina Nutt.   Fig. 46

Plants erect, with ascending branches to 1–2 m tall, or sometimes all branches procumbent and 1 to several dm long. Leaf blades rhombic-oblong, lanceolate, or ovate-lanceolate, to 1 dm long, attenuate at the base, rounded to obtuse apically, sometimes notched at the apex; upper leaves reduced and narrowly oblong. Inflorescence branches stiff, 1–2 dm long, each subtended by a bracteal leaf (unless this has already fallen). Bracts 1.5–2 mm long, with moderately heavy excurrent midrib in the staminate, with a heavy excurrent midrib in the pistillate. Staminate flowers with 5 perianth parts and 5 stamens, the inner perianth parts about 2.5 mm long, the outer about 3 mm, acuminate with excurrent midribs. Pistillate flowers with 1 or 2 perianth parts, the shorter but a rudiment, the longer about 2 mm long. Fruit about 1.5 mm long, circumscissily dehiscent at about the middle, the wall thin, rugose, sometimes with faint ridges, not angled. Seed lenticular, nearly circular in outline, about 1 mm in diameter, dark reddish brown. (*Acnida tamariscina* (Nutt.) Wood)

    Sandy fields, waste places, chiefly in moist to wet soil, shallow water of ponds and lakes, along streams, sloughs, exposed shores of impoundments, swamps. Ind. to Wisc. and S.D., southward to w. Tenn., La., Tex., N.M., occasional as an adventive in the eastern states.

4. Amaranthus cannabinus (L.) Sauer. WATER-HEMP.

Plants of variable stature; plants having commenced growth early in the season usually stout, erect, 1–3 m tall, the branches ascending; plants having commenced growth late in the sea-

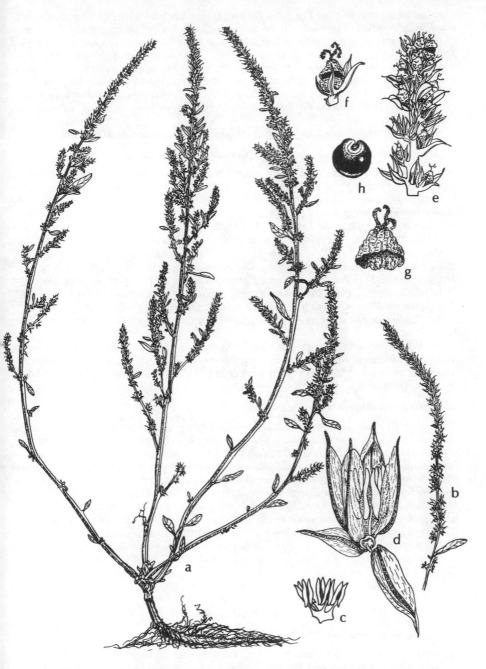

Fig. 46. **Amaranthus tamariscina:** a, staminate plant; b, staminate branchlet; c, stamens; d, staminate flower; e, pistillate branchlet; f, fruit; g, distal portion of circumscissile capsule; h, seed. (From Correll and Correll (1972), as *Acnida tamariscina*)

son often blooming when but 1 or a few dm tall. Larger leaf blades mostly lanceolate, sometimes ovate; smaller leaf blades from narrowly lanceolate to nearly linear. Flowers in elongate, simple to paniculate, slender often leafy-bracted subcontinuous or interrupted spikes. Bracts about 1 mm long. Staminate flowers with 5 nearly equal perianth parts 2.5–3 mm long, the inner emarginate, the outer acute, the midveins not excurrent. Stamens 5. Pistillate flowers without a perianth or with mere rudiments. Fruit 2.5–4 mm long, obovoid, angled, usually with subsidiary wing-angles between the 3 principal ones, sometimes appearing rugose. Seeds 2–3 mm in diameter, flattened, dark reddish brown to black. (*Acnida cannabina* L.)

Salt to fresh tidal shores and marshes. s. Maine to s. Ga., possibly in n.e. Fla.

5. Amaranthus australis (Gray) Sauer. SOUTHERN WATER-HEMP.

Similar to *A. cannabinus* in general features but potentially much more robust; plants having commenced growth early in the season sometimes reach a height of as much as 9 m with a basal stem diameter of 15 cm, perhaps more; plants having commenced growth late in the season may be flowering when relatively diminutive, 1 to a few dm tall. Larger leaf blades ovate, usually long-attenuate distally, petioles to 20 cm long, blades to 30 cm long; smaller leaf blades lanceolate. Inflorescences much as in *A. cannabinus* but with potential for larger dimensions. Bracts 1.5–1.8 (–2) mm long, midrib heavy in the staminate, moderately heavy in the pistillate, not greatly excurrent in either. Staminate flowers with 5 approximately equal perianth parts 2.5–3 mm long, the inner emarginate, the outer acuminate with excurrent midveins. Stamens 5. Pistillate flowers without perianths. Fruit obovoid to turbinate, 1.5–2 mm long, not rugose, stramineous. Seed flattened, 1 mm wide or a little more, lustrous, dark reddish brown to black. (*Acnida cuspidata* Bert. ex Spreng; incl. *Acnida alabamensis* Standl.)

In salt to fresh tidal marshes, borders of mangrove swamps, drainage and irrigation ditches and canals, cultivated mucklands, lagoons, bayous, sloughs, lakeshores. Mostly relatively near the coasts, Fla. to Tex., and from coastal Tex. northwestward to the Plains country; W.I.; Yucatan; Trin. and Venez.

6. Amaranthus tuberculatus (Moq.) Sauer.

Plants of very variable stature, prostrate, ascending, or erect, potentially 3 m tall. Larger leaf blades ovate to lanceolate, the smaller oblong or spatulate. In general features similar to *A. cannabinis*. Bracts 1–1.5 mm long, midrib very slender in the staminate, somewhat less so in the pistillate, excurrent far beyond the body. Staminate flowers with 5 approximately equal perianth parts, 2.5–3 mm long, the inner obtuse or emarginate, the outer acuminate and the midveins not excurrent. Pistillate flowers usually without perianth parts, occasionally with 1 or 2 rudimentary ones. Fruit not angled, ovoid to globose, smooth or tuberculate, 1.5–2 mm long. Seed lenticular, roundish to obovate in outline, about 1 mm wide, dark reddish brown. (*Acnida altissima* Ridd.; *Acnida tuberculata* Moq.; incl. *Acnida concatenata* (Moq.) Small)

Marshy or exposed shores of rivers, creeks, lakes, ponds and in marshes and bogs; weedy in low grounds of fields, gardens, and roadsides close to its natural habitats. w. N.Eng. and Ont. to N.D., southwestward in the eastern part of the range to s. Ohio, n.cen. Tenn., Ark. and La., in the western part of the range to Kan. and Iowa.

2. Philoxerus

Philoxerus vermicularis (L.) R. Br. SILVERHEAD.

Herbaceous perennial, somewhat succulent, stems up to 2 m long, repent, with ascending tips or branches; principal stems much branched, forming mats on wet substrates. Leaves opposite, thick and fleshy, 1.5–6 cm long, 2–12 mm wide, linear-oblong to oblanceolate, apices acute or blunt, each tapered to a sessile base which joins that of the other leaf to form a narrow sheath which is villous within the axils of the leaves. Inflorescence of stalked head-like axillary and terminal spikes. Flowers bisexual, borne in the axils of ovate, acute or obtuse bracts; bractlets ovate-oblong; perianth of 5 oblanceolate sepals 3–5 mm long. Stamens 5, filaments united below. Fruit a compressed, indehiscent, ovoid utricle. Seeds about 1 mm long, orbicular, dark brown.

Sandy beaches, wet sands and marls, coastal dunes, essentially maritime in our area. Fla. to Tex.; W.I., C.Am., S.Am.; Afr.

3. Alternanthera (CHAFF-FLOWERS)

(Ours) perennial herbs with stem usually decumbent at least below and rooting at the nodes. Leaves opposite, sessile. Inflorescence sessile or stalked headlike spikes borne from the leaf axils or terminal on the branches. Flowers bisexual, a chaffy bract and 2 chaffy bractlets closely subtending the flower, perianth parts 4 or 5, unequal, chaffy. Stamens 5, alternating with staminodes. Style 1, stigma capitate.

- Flower heads markedly stalked.   1. *A. philoxeroides*
- Flower heads sessile.   2. *A. sessilis*

1. Alternanthera philoxeroides (Mart.) Griseb. ALLIGATOR WEED.   Fig. 47

Stems forming a dense, tangled mass in the water or occasionally on shore, decumbent, to 1 m long and 0.8 mm thick, glabrous except for a narrow band or strip of multicellular trichomes within the leaf bases, stem becoming hollow and slightly flattened with age, rooting at the nodes, somewhat pink when fresh. Leaves linear-elliptic, to 9 cm long, 1.5 cm wide, apices acute, tipped with a tiny spine, tapering at base to form a very short petiolelike base which dilates and clasps the stem, joining with that of the other leaf to form a narrow sheath; with scattered hairs on both sides when young, becoming glabrous with age, margins hyaline, with minute dentations which bear long trichomes, midvein prominent on both sides. Inflorescence a several-flowered whitish head, axillary or terminal on the branches, their stalks to at least 5.5 cm long and 1.5 mm thick. Stamens 5, filaments partially connate to form a membranous tube, the versatile anthers alternating with fringed petaloid or staminodial enations. Style 1, stigma capitate, papillate. Mature fruit not seen. (*Achyranthes philoxeroides* (Mart.) Standl.)

Ponds, lakes, streams, canals, ditches. Coastal plain, Va. to Fla., westward to Tex. Native in S.Am., naturalized, C.Am. and elsewhere. A noxious aquatic weed.

2. Alternanthera sessilis (L.) R. Br. ex DC.

Plant weakly erect or more commonly with lower stems prostrate and rooting at the nodes then with weakly erect branches about 3–5 dm tall, stems sometimes somewhat pubescent in lines, pubescent at the nodes. Leaves sessile, narrowed to a short petiolelike base, narrowly oblong or lanceolate, the larger about 5 cm long, apically acute. Inflorescence narrowly spikelike, becoming short-cylindrical as it develops, sessile in the leaf axils.

Boggy or marshy clearings, shores of streams or bodies of water. Fla. Panhandle, probably recently introduced. Pantropical.

4. Iresine

Iresine rhizomatosa Standl. BLOODLEAF.

Rhizomatous perennial. Stem erect, procumbent, or more or less clambering, usually with several, subequal, ascending branches, 5–15 dm tall, glabrous or sparsely pubescent, pilose at the somewhat swollen nodes. Leaves opposite, 4–15 cm long overall, with definite petioles 1–4 cm long, or the blades narrowed proximally to winged subpetioles, often, especially on upper leaves, winged-flanged at either side of the petiole base; blades thin, lanceolate to lance-ovate, apically acuminate, surfaces finely very short-pubescent, margins entire, but often short-pubescent. Flowers unisexual (plants dioecious), very small and numerous, in panicled, narrow spikes, the panicles terminating the main stem and its branches, commonly short, subsidiary panicles present along with the principal branches in the axils on

Fig. 47. **Alternanthera philoxeroides:** a, habit; b, flower, opened out; c, flower, showing bracts and perianth parts; d, flower, perianth removed, staminodia and stamens spread out, pistil in center; e–h, variation respecting stamens, staminodia, and pistils sometimes observed.

106

the main stem. Each flower subtended by 3 short, ovate, acute, very thin bracts; sepals 5, lance-acute, a little longer than the subtending bracts, thin and somewhat silvery when fresh, drying stramineous; in the pistillate flowers, the perianth subtended by a ring of long, woolly hairs, the sepals 1-nerved; corolla none; in the staminate flower, 5 stamens, each a little over half the length of the subtending sepal; in the pistillate flower, ovary subglobose, stigmas 2, filiform, sessile. Fruit a globular, thin-walled, indehiscent utricle bearing a single small, lustrous, brownish red seed. (*I. paniculata*, misapplied)

Floodplain forests, adjacent slopes, ·wet woodlands, thickets near streams, interdune swales. Md. to s. Ill. and Kan., southward to Fla. Panhandle and e. half of Tex.

Phytolaccaceae (POKEWEED FAMILY)

Phytolacca americana L. POKEWEED, POKEBERRY.   Fig. 48

Coarse, glabrous, perennial herb with a large, thick and fleshy, poisonous root, purplish stems of diverse stature, commonly widely branched and to 3 m tall. Leaves simple, alternate; petioles 1–5 cm long; blades varying greatly in size, 9–30 cm long, 3–11 cm broad, elliptic, ovate-elliptic, or ovate-lanceolate, bases cuneate to rounded, apices acute to acuminate, margins entire, sometimes crisped; stipules none. Flowers in stalked racemes borne apparently opposite or subopposite to a leaf, the racemes 5–30 cm long, nodding or erect in flower or fruit; flower stalks 2–15 mm long, each subtended by a bract 1–4 mm long and bearing alternate bractlets more or less medially; occasionally a few of the lowest stalks of the raceme have short, lateral flowering branches in the axils of the bractlets, thus the inflorescence paniculate at base. Flowers radially symmetrical, bisexual. Tepals (sepals?) whitish or pinkish, broadly obovate or suborbicular, 2–3 mm long and broad, persistent. Stamens 10, about as long as the sepals. Pistil 1, ovary superior, usually comprised of 10 united carpels, each with a separate style. Fruit a 10-seeded, juicy, more or less oblate-lobed berry, cratcriform at the summit, dark purple-black at maturity, 7–12 mm in diameter. Seed thinly lenticular, subreniform to rotund in outline, lustrous black, 3 mm long. (Incl. *P. rigida* Small)

In diverse, well-drained to wet, usually, but not always, disturbed habitats including floodplain forests and clearings subject to temporary flooding, exposed shores and bars of rivers, moist woodlands and fields. Maine to Minn., Iowa, Kan., generally southward to s. Fla., Tex., and Ariz.; Mex.; introduced into Eur.

Bataceae (SALTWORT FAMILY)

Batis maritima L. SALTWORT.   Fig. 49

Somewhat woody herb with thick fusiform roots, young stems succulent-herbaceous, old stems woody and with a pale buff, very soft, corky bark readily shredding in irregular flaky pieces. Plant pale green or yellow-green, strongly scented, stems elongate-spreading, prostrate, arching, or creeping, the tips of arching branches commonly rooting then forming a cluster of new branches, eventually forming large dense clones. Leaves opposite, sessile, succulent, commonly curved, half- to semiterete, margins entire, without stipules. Flowers unisexual (plants dioecious), small, crowded in fleshy, conelike, axillary, sessile or stalked spikes solitary in the leaf axils, with imbricate, fleshy scales subtending the flowers. Staminate spikes oblong- or ovate-cylindric, sessile, reniform to suborbicular, bracts persistent; calyx cuplike, membranous, 2-lobed. Corolla none. Stamens 4 or 5, inserted on the base of the calyx, alternating with staminodia. Pistillate spikes on short stalks, 4–12-flowered, bracts deciduous; perianth none, the flower thus of a single pistil; ovary 4-locular, 1 ovule in

107

Fig. 48. **Phytolacca americana:** A, portions of plant; B, branchlet with fruiting racemes; C, flower; D, berry; E, seeds. (From Reed, *Selected Weeds of the United States* (1970) Fig. 72)

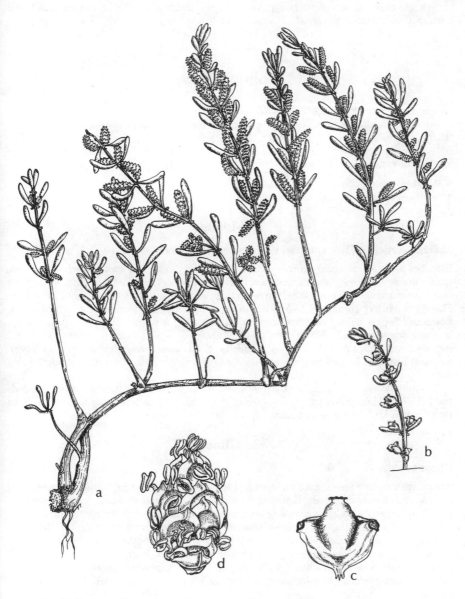

Fig. 49. **Batis maritima:** a, part of a staminate plant; b, small branch of pistillate plant; c, fruit; d, staminate inflorescence. (From Correll and Correll. 1972)

each locule; stigma sessile, capitate. Fruit a fleshy spike consisting of cohering berrylike pistils.

Salt marshes and salt flats, mangrove swamps, muddy tidal shores and flats. Atlantic coast, S.C. southward through the Fla. Keys, Gulf coast to s. Tex.; trop. Am.; Haw.Is.

Molluginaceae

Annual herbs, usually branched from the base, branches prostrate (not rooting at the nodes), or prostrate proximally, erect-ascending distally. Leaves (in ours) opposite or in apparent whorls, petiolate. Flowers in axillary clusters. Sepals 5 (or 4), free (in ours), persistent. Corolla none. Stamens 3–10, rarely more. Pistil 1, ovary superior, 3–5-locular, styles or stigmas the same number as the locules. Fruit capsular, usually with numerous seeds.

- Stems and leaves glabrous; flowers stalked.   1. *Mollugo*
- Stems and leaves pubescent; flowers sessile.   2. *Glinus*

1. Mollugo

Mollugo verticillata L. CARPETWEED, INDIAN-CHICKWEED.   Fig. 50

Branches prostrate or weakly erect-ascending, glabrous. Leaves in apparent whorls of 3–6, mostly unequal in size, glabrous, oblanceolate, spatulate, or linear-oblanceolate, the broader ones with rounded or obtuse apices, narrower ones acute, margins entire, to about 3 cm long. Flowers in axillary clusters of 2–5, with filiform stalks to about 15 mm long, reflexed after flowering. Sepals oblong or elliptic, nearly translucent, 3-veined, margins hyaline-membranous. Stamens usually 3, sometimes 4 or 5. Ovary 3- (–5-) locular, styles 3 (–5), short, stigmas linear. Capsule dehiscing loculicidally. Seeds numerous, strongly curvate-asymmetrical, lustrous chestnut brown, with darker brown ridges along the crest, about 0.6 mm long.

Commonly weedy in cultivated grounds and waste places, on exposed shores of ponds and lakes, sand and gravel bars and shores of rivers, alluvial outwash. Throughout much of temperate N.Am.; native in trop. Am.

2. Glinus

Glinus lotoides L.

Annual with stellate pubescence on stem and leaves, this on some parts cottony-tomentose, branches commonly numerous, radiating from the base, prostrate or weakly erect-ascending. Leaves apparently whorled, unequal in size, petioles about as long as the blades; blades obovate to broadly spatulate, sometimes orbicular, apices broadly rounded to obtuse, the larger to about 4 cm long overall. Flowers in dense axillary clusters. Sepals 5, stellate-tomentose, about 7 mm long, keeled, lanceolate, narrowly long-acute apically. Stamens 3–10, rarely more. Capsule 3-locular, loculicidally dehiscent. Seeds numerous, minute, curvate-asymmetrical, with a whitish bladderlike appendage at the point of attachment, surface reddish brown.

Alluvial muds and sands of streams and lakes, low woodland clearings. Mo. and Okla. to La., westward to Calif.; native in Eur.

Another species of *Glinus*, *G. radiatus* (Ruiz & Pav.) Rohrb. in Mart., native in tropical America, is said to be naturalized in La. and Tex. (perhaps elsewhere in the s.w. U.S.). The two appear to be distinguished primarily on the basis of the texture of the seed coat, that of *G. lotoides* being tuberculate, that of *G. radiatus* smooth. There appears to be no consensus as to whether plants naturalized in the U.S. are representatives of 2 species.

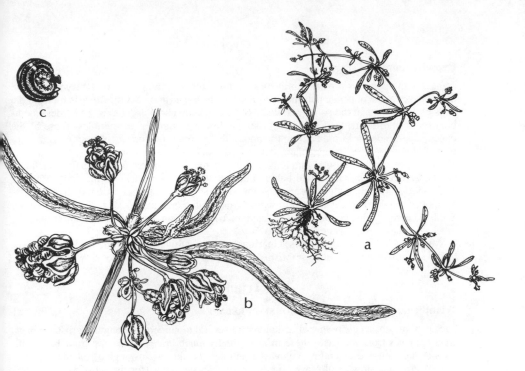

Fig. 50. **Mollugo verticillata:** a, habit; b, node with flowers; c, seed. (From Correll and Correll. 1972)

Aizoaceae (CARPET-WEED FAMILY)

A very large family, many of the species in southern Africa, and exhibiting considerable diversity of structure. Description here abbreviated ("tailored") using the more limited number of characters attributable to the representatives of the few genera in our flora.

Annual or perennial herbs, succulent or subsucculent, commonly with numerous stems from near the base, these often prostrate, in some perennial ones mat-forming. Leaves opposite, with or without stipules. Flowers bisexual, appearing solitary, paired or clustered in the leaf axils; in some with a short floral tube, this adnate to the lower portion of the ovary and thus the ovary partly inferior; calyx segments 5. Petals absent, petallike staminodes present in some. Fertile stamens 5 to numerous, they and the staminodes inserted on the floral tube. Pistil 1, ovary 1–5-locular, styles or stigmas as many as the locules. Fruit a capsule, dehiscing loculicidally, septicidally, or circumscissily. Seeds usually numerous, rarely 1 per locule.

1. Leaves stipulate.
  2. Larger leaf blades mostly not over 6 mm long and 4 mm wide; stipules fimbriate.   1. *Cypselea*
  2. Larger leaf blades 20–30 mm long, mostly as long as wide or wider than long; stipules not fimbriate.   2. *Trianthema*
1. Leaves not stipulate although the petioles are dilated proximally and bear wide membranous wings.
           3. *Sesuvium*

1. Cypselea

Cypselea humifusa Turpin.

Annual with numerous prostrate branches, a single plant forming a small "leafy mat." One of the leaves of the opposite pair much larger than the other (branches mostly arising from the axils of the smaller of the pair), petiolate, petioles mostly about as long as the blades, mem-branous or scarious-fringed stipules border the lower portions of the petioles, those of the opposite pair more or less sheathing the node, blades small, the longer to about 6 mm, elliptic to oblong, oval, or oblanceolate, cuneate to rounded basally, rounded apically, margins entire. Flowers small, about 2 mm long, solitary in the axils, subtended by a pair of laciniate bracts. Sepals 5, unequal, united below, margins scarious, persistent on the fruit. Stamens 1–5, shorter than the calyx, inserted at the sinuses. Ovary superior, stigmas 2(3). Fruit a subglobose, 1-locular capsule, thin-walled, dehiscing circumscissily below the middle. Seeds numerous, tiny, brown, smooth, asymmetrical-curvate, one end much larger than the other, the shape reminiscent of that of a *bota*.

Muddy or drying-muddy places, clearings, shores, cultivated lands, depressions in pinelands. Locally introduced and established in Fla., La., and Calif. Native in W.I.

2. Trianthema

Trianthema portulacastrum L. HORSE-PURSLANE.   Fig. 51

Succulent or subsucculent annual, usually with several to numerous elongate, prostrate principal branches from the base, these in turn usually much branched. Young branchlets with lines of hairs on and extending below the petiolar sheaths. Stems angled-ribbed, pustular. One of the opposite pair of leaves much larger than the other (for the most part branches arising from the axil of the smaller of the pair); petiolate, the petioles to nearly as long as the blades, dilated proximally and with winging stipules, those of a pair sheathing the nodes; blades suborbicular to obovate, sometimes somewhat rhombic and broader than long, the larger 20–30 mm long, 15–30 mm wide, mostly broadly cuneate basally, rounded, truncate, or retuse apically, often shortly apiculate. Flowers solitary in the axils, sessile. Calyx united below, 5-lobed, 2 bracts fused with the tube, lobes green externally, pinkish to whitish within, somewhat hoodlike, with a hornlike protuberance on the back just below the summit. Stamens 5–10, inserted on the floral tube in 2 series alternate and opposite to the calyx lobes. Ovary superior, stigmas 2. Fruit capsular, short-cylindric or turbinate, closely invested by the floral tube, the walls membranous below, thick above, truncate apically, circumscissily dehiscent at or a little below the middle. Seeds 1 to several, roundish-reniform, somewhat asymmetrical, papillate, reddish brown to black.

Sandy and marly shores, marly spoil banks and flats, along streams and irrigation ditches and canals. s. pen. Fla., Fla. Keys, locally northward to N.J., locally westward to Calif., northward in the interior to Okla. and Mo.; pantropical.

3. Sesuvium (SEA-PURSLANES)

Annual or perennial, succulent, glabrous herbs (ours). Leaves of a pair equal to subequal, entire, narrowed to subpetiolar bases, these dilated proximally and membranous-winged, not clasping the node; blades (in ours) mostly linear to oblanceolate or spatulate (rarely obovate). Flowers solitary, appearing axillary. Calyx united below, 5-lobed, margins of the lobes scarious, hooded above, bearing a subapical appendage on the back, pink to lavender within, persistent. Stamens 5 to numerous. Ovary partially inferior, styles 2–5. Capsule circumscissile.

• Flowers and fruits sessile; annual, stems erect-ascending, or if branches prostrate then not rooting at the nodes; leaf tips rounded.   1. *S. maritimum*

Fig. 51. **Trianthema portulacastrum:** a, habit; b, tip of stem; c, node, d, capsule, e, seed. (From Correll and Correll. 1972)

● Flowers and fruits short-stalked; perennial, stems prostrate, rooting at the nodes, commonly forming extensive dense mats; leaf tips acute.   2. *S. portulacastrum*

1. Sesuvium maritimum (Walt.) BSP.   Fig. 52

Annual, varying greatly in stature, sometimes flowering when diminutive, sometimes with mostly erect-ascending branches 1 to few dm tall, sometimes with widely spreading prostrate branches (usually not rooting at the nodes). Leaves of a pair sometimes unequal but generally not conspicuously so, mostly oblanceolate or spatulate, to about 2.5 cm long, cuneate basally, tips rounded. Flowers sessile.

Damp to wet sands, peats, or muck, sea beaches, marshes, semiswampy relatively open places, usually near the coast. Local, N.Y. to s. Fla., westward to Tex.; W.I.

2. Sesuvium portulacastrum L.   Fig. 53

Perennial, producing elongate prostrate branches in great profusion, rooting at the nodes, mat-forming (or mound-forming where sands build up around the growing plants). Leaves of a pair usually approximately equal, linear or mostly oblanceolate, cuneate below, apices acute, 1–2.5 cm long. Flowers and fruits distinctly stalked.

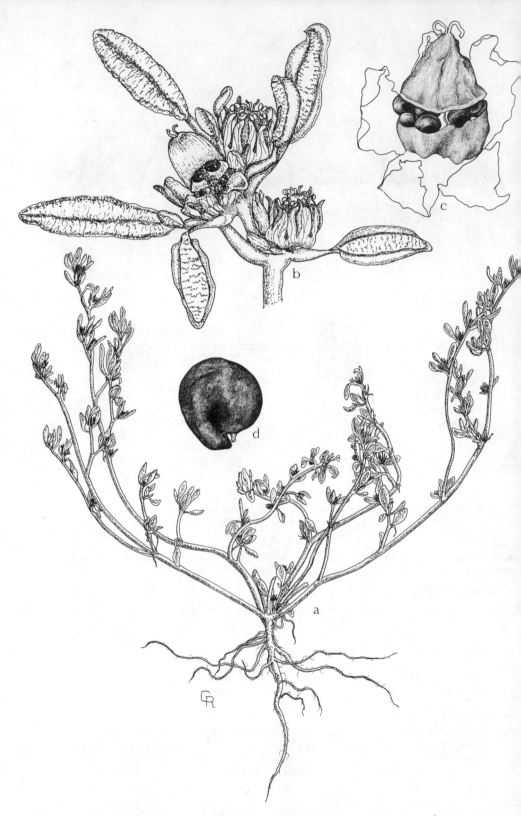

Fig. 52. **Sesuvium maritimum:** a, habit; b, small portion of branch; c, capsule at time of dehiscence; d, seed. (b, from Correll and Correll. 1972)

Fig. 53. a–c, **Sesuvium portulacastrum:** a, habit; b, flower; c, pistil; d–f, **Sesuvium verrucosum:** d, habit; e, flower; f, pistil. (From Correll and Correll. 1972)

Upper beach strands, dunes, interdune swales, banks of marshes, sometimes in salt flats of marshes, spoil banks. N.C. to s. Fla., westward to Tex.; trop. Am.

Sesuvium verrucosum Raf. (Fig. 53), a plant of the interior, similar to *S. portulacastrum*, apparently occurs in our range only in Arkansas. Plants freely branched, branches sometimes prostrate but not rooting at the nodes. Flowers and fruits subsessile or with stoutish, short, inconspicuous stalks.

In saline and alkaline places, about lakes, in creek bottoms, on mud flats. Mo. and Ark. to w. and n.w. Tex., N.Mex., Ariz., Calif.; n. Mex.

Portulacaceae (PURSLANE FAMILY)

Portulaca (PURSLANES)

Annual or perennial succulent herbs, commonly diffusely branching, the branches erect, ascending, or procumbent. Leaves alternate, subopposite, or essentially opposite, flat, subterete, or terete, margins entire, the uppermost often congested below the flowers, with scarious stipules or stipules in the form of hairs. Flowers sessile or subsessile, solitary and axillary, or few in terminal clusters or compact cymes, ovary partly or wholly inferior. Calyx segments 2, the abaxial larger than the adaxial. Petals 4 or 5, usually 5, separate or united basally, inserted on the floral tube, becoming mushy-gelatinous after anthesis. Stamens 7 to many, inserted on the floral tube with the corolla, filaments usually pubescent below. Ovary 1-locular, placentation basal. Fruit a capsule, dehiscing circumscissily, many-seeded.

- Petals yellow; stipular hairs at the leaf axils few and inconspicuous, few and inconspicuous as well around the flower clusters.   1. *P. oleracea*
- Petals purplish pink; stipular hairs at the leaf axils long, abundant, and conspicuous, very abundant and conspicuous around the flower clusters.   2. *P. pilosa*

1. Portulaca oleracea L. COMMON PURSLANE.   Fig. 54

Conspicuously fleshy annual, usually purplish red throughout, with stout, procumbent, radially spreading branches, the principal branches short or to nearly 10 dm long, glabrous with the exception of the few and inconspicuous stipular hairs. Leaves alternate to essentially opposite, flat, cuneate-obovate, apices rounded to nearly truncate, 6–30 mm long, occasionally longer. Flowers solitary or in terminal clusters, the hairs around them few and inconspicuous. Sepals broadly ovate to orbicular, keeled on the back, 3–4.5 mm long. Petals yellow, somewhat longer than the sepals, notched at the summit, narrowly oblong-obovate. Stamens 6–10. Capsule circumscissile at about the middle. Seeds dark reddish brown to black, surface pebbled, lustrous.

Weedy in a wide variety of open habitats, rock exposures, fields, sandy coastal beaches and mangrove flats, salt flats and marshes. Native of the Old World, widely naturalized in temp. and trop. regions of the Americas.

2. Portulaca pilosa L.

Annual or sometimes persisting for more than one year, with few to numerous ascending-erect branches to about 2 dm tall. Leaves alternate, mostly subterete, linear to narrowly oblanceolate or spatulate in outline, with numerous, long, persistent stipular hairs at the leaf axils, leaving the stems after leaf-fall appearing floccose. Flowers mostly in terminal clusters, copiously woolly because of the copious, long stipular hairs of the bracteal leaves. Sepals ovate or triangular-ovate, 2–3 mm long. Petals purplish pink, obovate, apices truncate to rounded and mucronate, about twice as long as the sepals. Capsule circumscissile at about the middle. Seeds black, surface rough, lustrous.

Sandy exposed shores of ponds and lakes, estuarine shores, mangrove flats, weedy in gardens, sandy roadsides, waste places. Chiefly coastal plain, N.C. to s. Fla., westward to La.; trop. Am.

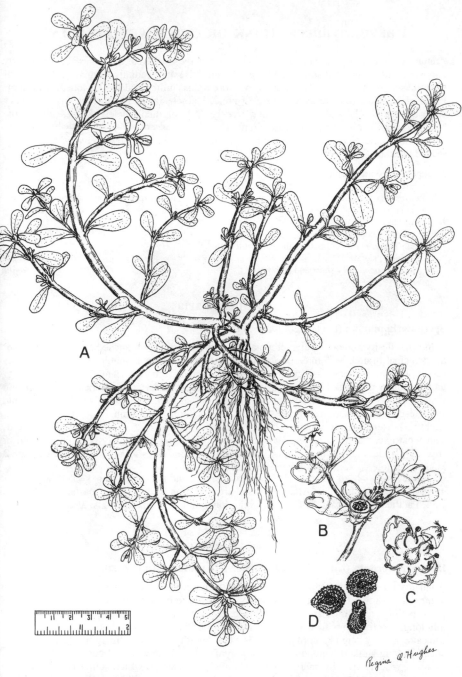

A

B

C

D

Fig. 54. **Portulaca oleracea:** A, habit; B, flowers and capsules; C, open flower; D, seeds. (From Reed, *Selected Weeds of the United States* (1970) Fig. 74)

Caryophyllaceae (PINK OR CHICKWEED FAMILY)

Annual, biennial, or perennial herbs, the stems usually swollen at the nodes. Leaves mostly opposite, with or without stipules. Flowers usually bisexual, radially symmetrical. Calyx of separate sepals, usually 5, sometimes 4, or more or less united and 4- or 5-lobed, persistent. Petals of the same number as the sepals or none. Stamens up to 10, distinct. Pistil 1, ovary superior, usually 1-locular, placentation free central, or incompletely 2–5-locular at the base; styles 2–5, rarely united into 1. Fruit a capsule, dehiscent at least apically by valves or teeth of the same number or twice the number of the styles, seeds few to many, or fruit a 1-seeded utricle.

1. Stipules present, scarious.
    2. Leaves narrowly linear, stipules broadly triangular.   1. *Spergularia*
    2. Leaves suborbicular, short-ovate, or reniform, mostly about as broad as long or broader than long, stipules fimbriate.   2. *Drymaria*
1. Stipules none.
    3. Petals cleft nearly to the base.   3. *Stellaria*
    3. Petals shallowly emarginate at the summit to entire.
        4. Styles 3, slender.   4. *Arenaria*
        4. Styles 5 or 4, very short and stubby.   5. *Sagina*

1. Spergularia

Spergularia marina (L.) Griseb. SAND SPURRY.

Somewhat fleshy annual, rarely perennial, commonly diffusely branched from near the base, the principal branches from 5–35 cm long. Stem glabrous or glandular-pubescent. Leaves sessile, narrowly linear, 1–3 cm long, stipules scarious, broadly triangular, 2–4 mm long, united around the node. Flowers in open cymes, their stalks stipitate-glandular, 2–5 mm long. Sepals 5, separate, lance-ovate to elliptic-oblong, green centrally on the outside and stipitate-glandular, margins broadly scarious, apices blunt, 2.5–4 mm long. Petals usually 5, slightly shorter than the sepals, white or pinkish. Stamens 2–5. Styles 3. Capsule ovoid, somewhat exceeding the calyx, 3-valved to the base. Seeds obliquely obovoid, smooth or stipitate-glandular, without wings or with hyaline, erose wings all around the margin, the 2 kinds often intermixed. (*Tissa marina* (L.) Britt.)

Saline, brackish, or alkaline soils, coastal marshes and flats, interdune swales, sometimes in wet, fresh-water habitats. Atlantic coastal areas, e. Can. to Fla., Gulf coastal areas, Fla. to Tex., Pacific coastal areas, B.C. to L.Calif., local in inland alkaline flats; also Euras.

2. Drymaria

Drymaria cordata (L.) Willd. ex R. & S. WEST INDIAN CHICKWEED.

Annual, with several principal branches from the base, the lower stems prostrate and rooting at the nodes, flowering branches ascending, branches to 6 dm long. Leaves with short, margined petioles and fimbriate stipules, blades suborbicular, short-ovate, or reniform, 5–20 mm long, mostly as broad as long or a little broader than long, faintly nerved, bases mostly truncate or very slightly cordate, apices very broadly rounded, margins entire. Flowers small, in long-stalked, compound, axillary cymes. Sepals separate, about 4 mm long, elliptic-oblong, acute apically, margins narrowly hyaline. Petals 5, white, somewhat shorter than the sepals, clawed basally, the blades deeply cleft, the lobes oblong or linear-oblong. Stamens 5. Styles 3. Capsule oval or ovate-oval in outline, about 3 mm long.

Lawns, gardens, ditches, pastures where the soil is moist to wet, and on exposed sand or gravel bars in rivers. Fla. to La. Native of trop. Am.

3. Stellaria (CHICKWEEDS)

Annuals or perennials, commonly diffusely branched, without stipules. Flowers in terminal or axillary cymes, or both. Sepals 5. Petals 5, cleft or bifid almost to the base. Stamens 10 or fewer. Styles 3–5. Capsule dehiscing by twice as many valves as there are styles.

- Leaves ovate or lance-ovate; styles 5.   1. *S. aquatica*
- Leaves linear to narrowly lanceolate; styles 3.   2. *S. graminea*

1. Stellaria aquatica (L.) Scop.

Perennial. Stems weak, decumbent below and rooting at the nodes, usually glabrous below, glandular-pubescent above, to about 8 dm long. Lowest leaves with short, margined petioles, upper sessile and cordate-clasping, blades usually ovate, sometimes lance-ovate, apices acute to acuminate, variable in size, 1–8 cm long, 1–4 cm wide, margins entire. Cymes widely branched and open. Flower stalks 5–10 mm long, elongating to 2–4 cm as the fruits mature, glandular-pubescent. Sepals glandular-pubescent, ovate, acuminate apically, 5–6 mm long at anthesis, becoming 7–9 mm long in fruit. Petals white, somewhat longer than the sepals. Stamens 10. Styles 5. Capsule ovoid to oval, usually about equaling the calyx, glabrous. Seeds obliquely curved-reniform, dark reddish brown, papillose-tuberculate. (*Alsine aquatica* (L.) Britt., *Myosoton aquaticum* (L.) Moench.)

Marshes, meadows, wet woodlands, damp banks of streams, alluvial thickets. Local, Que. and Ont. to Minn., southward to n.w. N.C., W.Va., Ohio and Ill.; B.C.; native of Eur.

2. Stellaria graminea L.

Perennial, stems weak, reclining or weakly ascending, commonly diffusely branched, to about 6 dm long, glabrous or sometimes pubescent on the angles. Leaves sessile, linear to narrowly lanceolate, sometimes ciliate at the base, 1.5–5 cm long, margins entire. Cymes terminal, long-stalked, loosely divaricately branched, the flower stalks very slender, 2–4 cm long. Sepals elongate-triangular, 3–6 mm long, acute apically, green centrally on the outside and 3-nerved, with conspicuous, sometimes ciliate, hyaline margins. Petals white, exceeding the sepals, the lobes oblanceolate. Stamens 10. Styles 3. Capsule lance-ovate in outline, somewhat surpassing the sepals, glabrous. Seeds few, dark reddish brown, dull, surface strongly raised-reticulate and rough. (*Alsine longifolia* sensu Small, 1933)

Open wet grassy places, clearings, thickets, ditches, stream banks, meadows, shores. Nfld. and Que. to Minn., southward to n. S.C., W.Va., Ohio, Mo.; native of Eur.

4. Arenaria (SANDWORTS)

Low, often tufted, annual or perennial herbs. Leaves opposite, sessile or subpetiolate, without stipules. Flowers in terminal cymes or capitate clusters, rarely axillary and solitary. Sepals 5. Petals 5 (lacking in some), usually white, apically entire, notched, or bifid. Stamens 10. Ovary 1-locular, styles 2–5, 3 in most. Capsule 3- or 6-valved. Seeds numerous, smooth, muricate, or tuberculate.

1. Stems conspicuously leafy throughout, weak, prostrate below; flower stalks and sepals stipitate-glandular.   1. *A. godfreyi*
1. Stems filiform-wiry, erect, with few, widely spaced, inconspicuous leaves; flower stalks and sepals glabrous.
  2. Larger stem leaves mostly 10–20 mm long, linear; fruiting calyx broader than long; capsule scarcely if at all surpassing the calyx.   2. *A. glabra*
  2. Larger stem leaves 7 mm long or less, usually less than 5 mm, oblong; fruiting calyx longer than broad; capsule about ⅓ exserted from the calyx.
    3. Mature open corollas 5–12 mm broad, petals about twice as long as the sepals.   3. *A. uniflora*
    3. Mature open corollas 2–4 mm broad, petals scarcely exceeding the sepals.   4. *A. alabamensis*

1. Arenaria godfreyi Shinners.

Perennial. Stems glabrous or sparsely glandular-pubescent, prostrate below, weakly ascending above, usually diffusely branched, the ascending branches commonly intertangled with each other or with other vegetation, to about 4 dm long. Leaves narrowly linear or very narrowly linear-lanceolate, 1–3.5 cm long, sometimes glandular-pubescent along their proximal margins. Cymes diffuse and open, flower stalks filiform, glandular-pubescent, 1–3 cm long. Sepals 5, separate, lance-ovate to oblong-elliptic, acute apically, green, scarcely 1-nerved, glandular-pubescent, 3–5 mm long. Petals 5, white, mostly broadly and shallowly notched at the summit, nearly twice as long as the sepals. Styles 3, slender. Capsule ovate, glabrous, about equaling the calyx. Seeds plump, dark brown, dull, nearly orbicular, strongly papillose. (*Sabulina uniflora* sensu Small, 1933; *Stellaria paludicola* Fern. & Schub. not *A. paludicola* Robs.)

Seepage areas, wet woodlands near streams, clearings of flatwoods, adjacent ditches. Apparently very local, coastal plain, N.C. to n. Fla.

2. Arenaria glabra Michx.

Glabrous annual with an inconspicuous basal rosette of linear-oblanceolate leaves, 1 to several, erect, wiry stems 5–20 cm tall. Pairs of stem leaves few and widely spaced, linear, 10–20 mm long. Flowers in open cymes; flower stalks filiform, 5–20 mm long or a little more. Sepals oblongish, 3–4 mm long, blunt or rounded apically, with narrow hyaline margins. Petals white, obovate, 6–10 mm long, broadly and shallowly notched apically. Fruiting calyx broader than long, scarcely if at all exceeded by the capsule. (*Sabulina glabra* (Michx.) Small, *Minuartia glabra* (Michx.) Mattf., *A. groenlandica* (Retz.) Spreng. var. *glabra* (Michx.) Fern.)

Vernal pools and in seasonally wet moss-lichen mats, chiefly on granitic and siliceous rock outcrops. In our range, N.C. and Tenn. to Ga. and Ala.

3. Arenaria uniflora (Walt.) Muhl.

Glabrous annual with a basal rosette of inconspicuous oblanceolate leaves and 1 to several, erect, wiry stems 2–10 cm tall. Pairs of stem leaves few, widely spaced, oblongish, 7 mm long or less, usually less than 5 mm. Flowers few to several in open cymes, or more or less racemose; flower stalks filiform, 5–15 mm long. Sepals oblongish, 2–3 mm long, with narrow hyaline margins. Petals white, 3–7 mm long, obovate, broadly and shallow notched apically, about twice as long as the sepals. Fruiting calyx longer than broad, about ⅔ as long as the capsule. (*Sabulina brevifolia* (Nutt.) Small, *Minuartia uniflora* (Walt.) Mattf.; not *Sabulina uniflora* sensu Small, 1933)

4. Arenaria alabamensis (McCormack, Bozeman, & Spongberg) Wyatt.

Closely similar to *A. uniflora*. Flowers smaller, petals 1–3 mm long, broadly obovate, scarcely exceeding the sepals; mature open corollas 2–4 mm broad. (*Minuartia alabamensis* McCormack, Bozeman, & Spongberg)

Vernal pools and seasonally wet moss-lichen mats, granitic rock outcrops. Piedmont, Ala. (Randolph Co.) and N.C. (Rutherford Co.)

5. Sagina

Sagina decumbens (Ell.) T. & G. PEARLWORT.

Low, inconspicuous, winter annual, with few to many slender-wiry branches from the base, these from 2 to about 15 cm tall, glabrous. Leaves sessile, without stipules, dilated basally, scarious, and united around the node, linear-subulate, mucronate apically, 5–15 mm long. Flowers small, axillary, with filiform, usually stipitate-glandular stalks 3–25 mm long. Sepals 5, rarely 4, distinct, oblong-elliptic, obtuse apically, 1–2.5 mm long. Petals none, or

1–5 and rudimentary, or sometimes 5 and equaling the sepals, not notched apically. Stamens 3–10. Styles 4 or 5. Capsule ovoid to ellipsoid, 2–3 mm long, opening by 4–6 valves. Seeds numerous, obovoid, tiny, about 0.2 mm long, yellowish or reddish brown, dull, very minutely stipitate-glandular.

Wet to dryish fields, sometimes in vernal pools in fields, moist to wet roadsides and ditches, meadows, lawns, clearings, logging roads. Mass. and s. Vt., s. N.Y. to Ky., Ill., Mo., southward to Fla. and Tex.

Ceratophyllaceae (HORNWORT FAMILY)

Ceratophyllum (HORNWORTS, COON-TAILS)

Perennial, submersed aquatic plants, branching but with a single branch produced from a node. Roots lacking but leafy branches sometimes modified as "rhizoids"; stems breaking easily and pieces continuing growth separately. Leaves sessile, whorled, dichotomously divided into filiform or linear segments. Flowers unisexual, both staminate and pistillate on the same plant, very small, solitary in axil of one leaf of a given whorl, each subtended by an 8–12-parted involucre, perianth none. Staminate flowers with 4–10 stamens, filaments very short, anthers with a connective projecting distally and ending in 2 bristles. Pistillate flowers with 1 pistil, ovary superior, 1-locular. Fruit a 1-seeded, ovoid-oblong achene, warty, spiny, or spineless. (Plants of *Ceratophyllum* are difficult to identify with certainty to species without achenes.)

Reference: Lowden, Richard W. "Studies in the Submerged Genus *Ceratophyllum* L. in the Neotropics." *Aquatic Botany* 4: 127–142. 1978.

1. Base and lateral margins of achene with spines *or* lateral margins without spines and the base with spines.
  2. Base of achene with spines, lateral margins without.   1. *C. demersum*
  2. Base and lateral margins of achene with spines.   2. *C. muricatum*
1. Base and lateral margins of achene without spines, the surface warty.   3. *C. submersum*

1. Ceratophyllum demersum L.   Figs. 55 and 56

Leaves of principal stems 9–11 at a node, 1- or 2(3)-forked and with 2–4 ultimate leaf segments, the latter with 4 or 5 prominent teeth marginally. Involucral bracts of staminate flowers 1 mm long; stamens more than 15, anthers 1–1.1 mm long, dotted with red glands. Involucral bracts of pistillate flowers 1.2–2 mm long. Achene 4–6 mm long, with spineless lateral margins and 1 or 2 basal spines.

Quiet waters of streams, ponds, lakes, ditches, canals. More or less throughout temp. N.Am.; trop. Am.; Old World.

2. Ceratophyllum muricatum Cham.   Fig. 56

Similar to *C. demersum*. Leaves of principal stems 2- or 3-forked and with 3–8 ultimate segments, the latter without or with inconspicuous marginal teeth. Achene 5–7 mm long, the lateral margins and the base bearing spines. (*C. echinatum* Gray; incl. *C. floridanum* Fassett)

Habitats as for *C. demersum*. s.w. N.B., s. Maine to Minn., generally southward to s. Fla. and n.e. Tex.; Mex.

3. Ceratophyllum submersum L.   Fig. 57

Leaves of principal stems 7–10 at a node, 2- or 3(4)-forked and with 3–8 ultimate segments, the latter with 4–7 inconspicuous marginal teeth. Involucral bracts of staminate flowers 0.5–0.8 mm long; stamens 4–10, anthers pale green, 0.5–0.7 mm long. Pistillate flower with 7–11 involucral bracts 0.8–1.6 mm long. Achene about 4 mm long, without spines, surface warty.

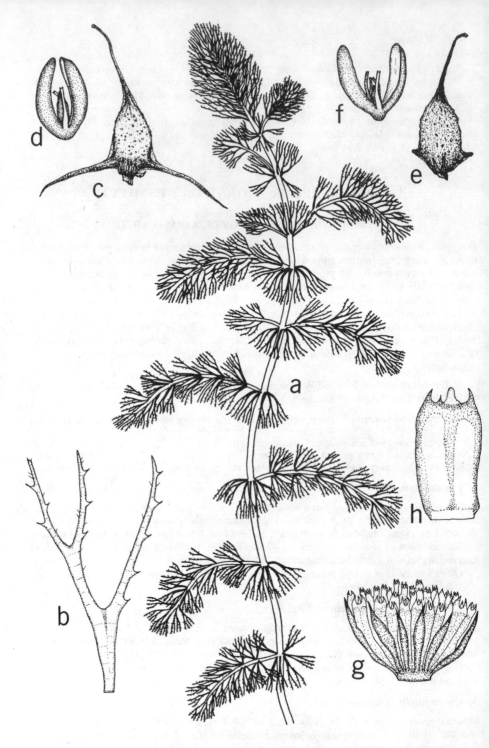

Fig. 55. **Ceratophyllum demersum:** a, habit; b, 1- or 2-forked dichotomous leaf; c, long reflexed basal spined achene and involucre; d, embryo showing cotyledons and a simple first node plumule leaf (Dom.Rep., Lowden 3171); e, short basal spined achene and involucre; f, embryo showing cotyledons and simple first node plumule leaves (Dom.Rep., Lowden 3378); g, staminate inflorescence and involucre; h, stamen. (From Lowden, "Studies on the Submerged Genus *Ceratophyllum* L. in the Neotropics." *Aquat. Bot.* 4: 127–142. 1978. Courtesy of and with the permission of the author)

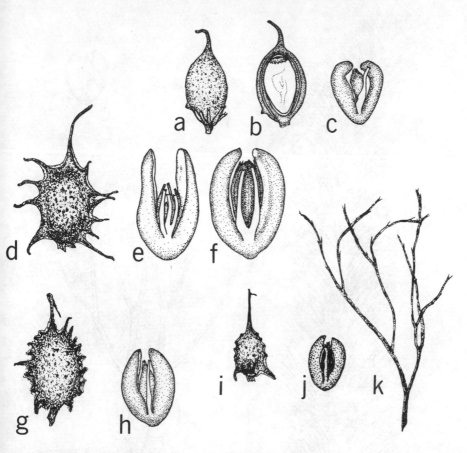

Fig. 56. a–c, **Ceratophyllum demersum** (Finland, Lindberg 1938): a, achene showing one basal spinal lobe; b, longitudinal section of achene; c, embryo showing cotyledons and simple first node plumule leaf; d–f, **Ceratophyllum muricatum** (U.S., Roberts 711): d, spiny-margined achene; e, embryo showing cotyledons and a 3-lobed first node plumule leaf; f, embryo showing cotyledons and a 2-lobed first node plumule leaf; g,h, **Ceratophyllum muricatum** (El Salvador, Fassett 28553, type of Fassett's *C. llernae*): g, spiny-margined achene; h, embryo showing cotyledons and a simple first node plumule leaf; i–k, **Ceratophyllum muricatum** (U.S., Killip 40732, type of Fassett's *C. floridanum*): i, spiny-margined achene; j, embryo showing cotyledons and a simple first node plumule leaf; k, 2- or 3-forked dichotomous leaf. (From Lowden, "Studies on the Submerged Genus *Ceratophyllum* L. in the Neotropics." *Aquat. Bot*. 4: 127–142. 1978. Courtesy of and with the permission of the author)

Native of Eur., known in the New World only from a single collection from s. Fla. (1883), and a recent one from the Dominican Republic.

Ranunculaceae (CROWFOOT FAMILY)

Herbaceous (or rarely woody), mostly perennial plants. Leaves usually alternate, sometimes opposite, with bases of petioles dilated, simple and often lobed, dissected, or compound. Flowers variable: in some, all parts present and distinct: in some. bisexual, in some, unisex-

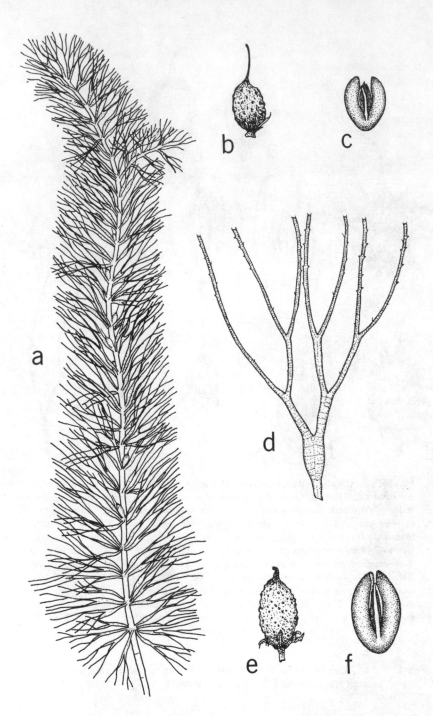

Fig. 57. **Ceratophyllum submersum:** a, habit; b, spineless achene and involucre; c, embryo showing cotyledons and a simple first node plumule leaf (Dom.Rep., Lowden 3414); d, 2- or 3-forked dichotomous leaf; e, spineless achene and involucre; f, embryo showing cotyledons and a simple first node plumule leaf (Tanzania, Richards 26962). (From Lowden, "Studies on the Submerged Genus *Ceratophyllum* L. in the Neotropics." *Aquat. Bot.* 4:127–142. 1978. Courtesy of and with the permission of the author)

ual; in some, petals none and the sepals petallike; sepals and/or petals spurred in some; stamens usually numerous; pistils few to many and distinct, somewhat united in some. Ovaries superior, 1-locular, with 1 to many ovules. Fruit an achene, a follicle, a berry, or a capsule.

1. Plant an upright shrub.   1. *Xanthorhiza*
1. Plant herbaceous, or if stems semiwoody, then a vine.
   2. Perianth in 1 series of similar parts (sometimes deciduous by the time flower is at full anthesis).
     3. Plant diminutive, with numerous basal, entire, linear to linear-spatulate leaves; pistils or fruits on an elongate, columnar receptacle.   2. *Myosurus*
     3. Plant not diminutive *and* with entire, linear leaves; pistils not on an elongate columnar receptacle.
       4. Leaves opposite.   3. *Clematis*
       4. Leaves alternate (although commonly closely set at the base of the stem).
         5. Leaf blades 2 or 3, ternately compound.   4. *Thalictrum*
         5. Leaf blades simple even if deeply divided.
           6. The leaf blade merely toothed, base cordate.   5. *Caltha*
           6. The leaf blade deeply palmately or pedately divided, base not cordate.   6. *Trautvetteria*
   2. Perianth parts in 2 dissimilar series.   7. *Ranunculus*

1. Xanthorhiza

Xanthorhiza simplicissima Marsh. YELLOW-ROOT.

A colonial shrub with the wood of stems and roots yellow. Stems slender, brittle, usually unbranched, to about 8 dm tall, the alternate leaves closely spiraled on short shoot of the current season. Leaves deciduous, with slender petioles as long as, usually longer than, the blades; blades odd-pinnately compound, leaflets 3–5, usually 5, bases cuneate, apices of leaflets or their lobes acute; lowermost pair of leaflets usually deeply cleft only on the abaxial side, the middle pair saliently cleft-toothed, very asymmetrical, the terminal one usually symmetrical and deeply 3-lobed, all margins strongly toothed except on the cuneate basal portion; blades of a given stem usually very unequal in size, the larger about 10–12 cm long overall and as wide across the basal pair of leaflets. Flowers in flexuous, drooping racemes or narrow panicles 5–15 cm long, 1 from the axil of each of several of the lowermost leaves, at full anthesis well before the leaves are fully grown; axes of inflorescence and the flowers commonly maroon or brownish maroon, sometimes greenish yellow, the axes and flower stalks pubescent, bracts linear. Sepals 5, somewhat unequal, more or less clawed, varying in shape, 3–5 mm long. Petals none, Staminodia 5, stalked, the summit with 2 diverging, rounded, nectariferous maroon lobes. Stamens usually 5, sometimes 10, usually shorter than the staminodes. Pistils usually 5–10. Fruit an obliquely oblong 1-seeded follicle, ripening, dehiscing, and falling early in the season.

     River and stream banks, moist thickets, springy places, usually where shaded. s.w. N.Y. to Va. and W.Va., southward to Ga., Ala., Miss., and Fla. Panhandle.

2. Myosurus

Myosurus minimus L. MOUSE-TAIL.   Fig. 58

Diminutive plant, with a short, relatively stout caudex usually bearing numerous closely set leaves. Leaves very narrow, filiform below, narrowly linear- or spatulate-dilated distally, variable in length, mostly 4–8 cm long. Flowers usually several, yellowish or whitish, solitary terminating a scape, the scape equaling the leaves or somewhat longer; bisexual; sepals 5, oblong, 2–3 mm long, spurred basally; petals 5, small, linear to narrowly spatulate, 2–3 mm long, clawed, a nectariferous pit at the summit of the claw; stamens 5–20; carpels free, numerous, borne closely set on an elongate receptacle, the fruiting "spike" 1–5 cm long at maturity, elongate-conical, extending beyond the leaves. Achenes closely packed on the axis, rectangular in outline, 1.5–2 mm long, 2–3 times as long as broad, a low keel on the rhombic summit.

125

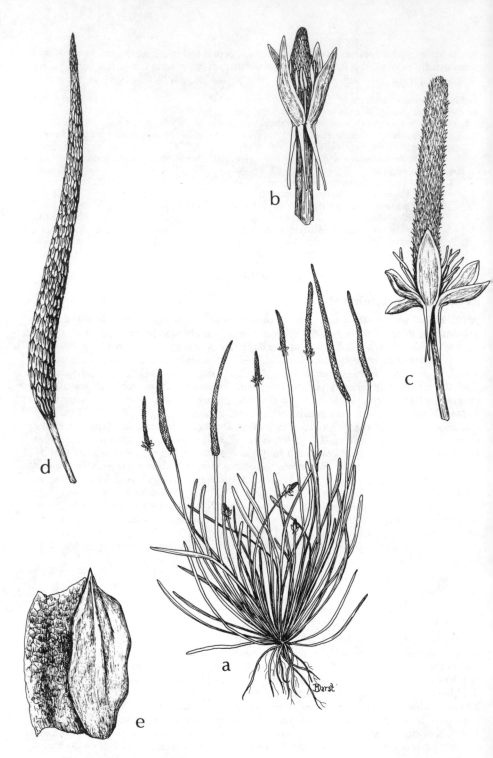

Fig. 58. **Myosurus minimus:** a, habit; b, flower, at anthesis; c, flower, postanthesis; d, fruiting "spike"; e, achene.

In moist to wet soils in early spring, commonly in fallow fields, in vernal pools, ditches, stream banks, edges of ponds, borrow-pits. s. Ont., Ill., Minn., Sask. to B.C., generally southward to Tenn., Ala., Miss., La., and Calif.; chiefly coastal plain and piedmont, s.e. Va. to S.C.; Euras.

3. Clematis

Herbaceous to slightly woody perennial vines, or upright perennial herbs; vines climbing by bending or twining of the stalks of the leaves or leaflets. Leaves opposite, 1- or 2-pinnately compound, ternately compound, or simple. Flowers in axillary panicled cymes, or axillary and solitary, bisexual or unisexual. Sepals usually 4, petallike, valvate in the bud, white, blue, violet, or purple. Petals absent but somewhat petaloid staminodes present in some. Stamens numerous, anthers fused. Pistils numerous, separate, ovaries superior, styles long. Fruit an aggregate of achenes, styles persistent, elongating as the fruits mature, often plumose.

Reference: Keener, Carl S. "Studies in the Ranunculaceae of the Southeastern United States. III. Clematis L." *Sida* 6: 33–47. 1975.

1. Flowers bisexual or unisexual, numerous, in axillary panicled cymes; sepals thin, spreading, whitish; filaments glabrous.
  2. Flowers usually bisexual; leaflets entire (rarely cleft), usually thickish-membranous; anthers 1.5 mm long or more.   1. *C. terniflora*
  2. Flowers usually unisexual; leaflets coarsely toothed or lobed; anthers less than 1.5 mm long.
    3. Leaves ternately compound; leaflets usually coarsely serrate but seldom conspicuously 3-lobed; achenes light to dark brown or greenish brown.   2. *C. virginiana*
    3. Leaves biternately compound or pinnate and with 5 leaflets; leaflets often conspicuously 3-lobed or -toothed, the lobes entire to coarsely serrate; achenes reddish brown to dark blackish purple.   3. *C. catesbyana*
1. Flowers bisexual, solitary in the leaf axils; sepals rose, blue, violet, or purple, rarely white, thickish-spongy, erect or ascending-erect; filaments pubescent.
  4. Plant an upright herb.   4. *C. baldwinii*
  4. Plant a vine.
    5. Leaflets green beneath, not glaucous; sepals broadened or dilated somewhat above the middle into a usually crisped margin of different texture, then narrowed to an acuminate tip; styles 1.5–3 cm long on the fruits, soft-pubescent, hairs ascending, not with a plumose effect.   5. *C. crispa*
    5. Leaflets glaucous beneath; sepals narrowed from the middle to the tip, of uniform texture; styles on the fruits 4.5–6 cm long, hairs spreading, with a markedly plumose effect.   6. *C. glaucophylla*

1. Clematis terniflora DC.

Climbing or sprawling vine, slightly woody near the base. Stem 3–5 m long, sparsely pubescent or glabrous. Leaves usually pinnately compound, the petioles to 5 cm long; leaflets usually 5, glabrous, thickish-membranous, ovate or triangular-ovate, apically obtuse or rounded, often apiculate or mucronate, margins entire, bases subcordate, truncate, sometimes cuneate; stalks of the leaflets elongate, that of the terminal leaflet longer than those of the laterals. Inflorescence an axillary panicled cyme. Flowers bisexual. Sepals white, oblong to oblanceolate, 6–18 mm long, margins pubescent. Filaments and anthers glabrous, anthers 1.8–2.8 mm long. Achenes elliptic, styles 2–5 cm long, plumose. (*C. dioscoreifolia* Levl. & Vaniot, *C. maximowicziana* Franch. & Savat., *C. paniculata* Thunb.)

Native of Japan; cultivated and sporadically naturalized in much of our range: vacant lots, dry to wet thickets and hedgerows, occasional in bottomland woodlands and thickets on shores of impoundments.

2. Clematis virginiana L. VIRGIN'S BOWER, DEVIL'S DARNING NEEDLE.

Sprawling or climbing herbaceous vine. Leaves ternately compound, rarely pinnately 5-foliolate, petioles to 10 cm long; leaflets thin, ovate, or occasionally lance-ovate, 2–10 cm long, 1–6 cm wide, bases truncate to subcordate, sometimes cuneate, apically acuminate, margins coarsely few-toothed, occasional leaflets entire, surfaces glabrous to sparingly pubescent.

127

Flowers in axillary cymes, plants functionally dioecious, either with all staminate flowers, all pistillate flowers, or flowers with functional pistils and abortive stamens. Sepals narrowly oblong, whitish, spreading, 6–12 mm long, pubescent on the back. Filaments glabrous, anthers 0.6–1.5 mm long. Achenes elliptic to elliptic-obovate, light to dark brown, or greenish brown, pilose or villous-hirsute, the styles 1–3 cm long, plumose.

Low woodlands, stream banks, thickets, hedgerows, borders of lowland forests, usually climbing over or into shrubs and trees. e. Can. to Man., southward to Ga. and e. Tex.; cen. pen. Fla. where probably introduced.

3. Clematis catesbyana Pursh.

Closely similar to *C. virginiana*. Leaves usually biternate to 5-foliolate, the leaflets 3-lobed, the lobes entire, or occasional leaflets unlobed, ovate to lance-ovate, 3–6 cm long, bases subcordate to rounded or truncate. Achenes reddish or purplish brown to dark blackish purple. (Incl. *C. micrantha* Small and *C. ligusticifolia* sensu Radford, 1968, not *C. ligusticifolia* Nutt., a related plant occurring west of our range)

Lowland and upland woodlands, river banks, moist to wet thickets and hedgerows. Va. to Ky. and Mo., southward to n. Fla. and Ark.

4. Clematis baldwinii T. & G. PINE-HYACINTH.

Upright perennial herb with slender, hard, horizontal, branched rootstocks from which arise 1 to several slender, ribbed, pubescent stems 2–6 dm tall, usually unbranched below and bearing 1 to several long-stalked flowers from the upper leaf axils. Lower portion of the stem with short internodes, leafless by flowering time, the old, scaly petiolar bases evident. Lower leaves (before shedding) smaller than the upper, very short petiolate, leaves becoming sessile upwardly; blades very variable in form and size, even on a single plant, unlobed, or shallowly to deeply 3–5-lobed or deeply 3–5-cleft, unlobed ones long-lanceolate to short-ovate, sometimes linear, if lobed or cleft, the divisions linear-oblong, their margins entire; upper surfaces glabrous, lower sparsely short-pubescent, principally along the major veins. Flowers bisexual, with stalks from about 15–25 cm long, erect, pubescent. Sepals thickish-spongy, variable in size, 2.5–4 (–5.5) cm long, narrowly oblong for half or more their length, pale to deep purple centrally on the exterior and with a white-tomentose band, crisped on its edge, on either side, this narrow proximally and widening somewhat upwardly, sometimes dilated above the middle and giving to an acuminate tip; sepals pale purplish to white within. Filaments and anthers pubescent, anthers sometimes sparsely so. Achenes ovate, silky-pubescent, about 5 mm long, plumose styles 6–10 cm long. (*Viorna baldwinii* (T. & G.) Small; incl. *C. baldwinii* var. *latiuscula* R. W. Long)

Moist to wet pinelands, adjacent roadsides and clearings, margins of swampy woodlands. Pen. Fla.

5. Clematis crispa L. LEATHER-FLOWER, BLUE-JASMINE.   Fig. 59

Climbing or weakly ascending glabrous vine, stem ribbed. Leaves petiolate, blades very variable from plant to plant, or to some extent even on a single plant, mostly compound, leaflets 3–5, occasionally 2, some leaves sometimes simple, lobed or unlobed; leaflets ovate, elliptic, lanceolate, or linear-lanceolate, 1.5–10 cm long, 0.3–5 cm wide, margins entire, bases broadly to narrowly cuneate, truncate, occasionally subcordate, apices obtuse, acute, or acuminate. Flowers bisexual, solitary, arising between a pair of leaflets terminating branchlets, their stalks bractless, mostly 5–10 cm long. Sepals thickish-spongy, lanceolate to lance-ovate, 3–5 cm long, variable in color, rose, bluish, violet, or occasionally nearly white, broadened or dilated somewhat above the middle into a usually crisped margin of different texture, then narrowed to an acuminate tip, glabrous without, whitish-tomentose on the inner margins. Achenes suborbicular to rhombic, 6–9 mm broad, corky-thickened marginally, faces depressed, flat, appressed-pubescent, styles 1.5–3 cm long, pubescent with ascending hairs 1–1.5 mm long, not appearing plumose. (*Viorna crispa* (L.) Small; incl. *V. obliqua* Small)

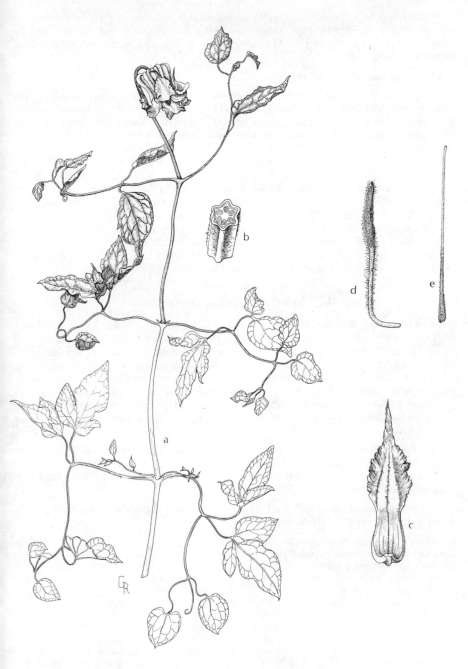

Fig. 59. **Clematis crispa:** a, small portion of plant; b, section of stem; c, sepal; d, stamen; e, style.

Swamps, wet woodlands, floodplain forests, marshes, marshy or thickety shores, wet pine savannas or flatwoods, river and stream banks, wet clearings and thickets. Chiefly coastal plain, s.e. Va. to cen. pen. Fla., westward to e. Tex., s.e. Okla., northward in the interior to s. Mo., s. Ill.

6. Clematis glaucophylla Small.

Climbing or sprawling vine, ribbed stems glabrous and glaucous. Leaves petiolate or sessile, blades on principal stems usually compound, with 3–5 leaflets, blades of branchlets often with 2 leaflets or simple; blades glaucous beneath, variable in shape and size, ovate with cordate bases, ovate with truncate, rounded, or asymmetric bases, broadly oval-elliptic, often with one large oblique lobe and somewhat mittenlike, margins entire; varying from suborbicular to about twice as long as broad, the latter mostly of the broadly oval-elliptic shape. Flowers bisexual, solitary in the leaf axils, or arising between the terminal pair of leaves of a branchlet, stalks bractless, 6–12 cm long. Sepals thickish-spongy, 1.5–2.5 cm long, ovate, acuminate apically, of uniform texture exteriorly, rich rose-purple, the inner margins whitish-tomentose. Stamens pubescent. Achene 8–10 mm long, rhombic, appressed-pubescent, corky-thickened marginally, faces depressed, flat, style 4.5–6 cm long, silky-villous, markedly plumose. (*Viorna glaucophylla* (Small) Small)

Floodplain forests and adjacent richly wooded bluffs, river banks, bottomland clearings and thickets. Local, Ga., Fla. Panhandle, Ala., Miss., s.e. Okla.

4. Thalictrum (MEADOW-RUES)

Perennial herbs with alternate, ternately compound or decompound leaves, lower, sometimes all, leaves petiolate, the principal divisions and the leaflets stalked. Inflorescence openly paniculate (in ours). Flowers (in ours) mostly unisexual (rarely all bisexual) and plants dioecious, occasional plants with a few bisexual flowers. Sepals 4 or 5, greenish, sometimes somewhat petallike, usually inconspicuous and falling quickly. Petals none. Stamens numerous, filaments relatively long and slender, sometimes dilated distally, anthers linear to globose. Pistils several to numerous, distinct, styles slender, often obliquely positioned at the summit of the ovary, stigmatic for most of their length. Fruit a longitudinally ribbed or nerved achene.

References: Boivin, Bernard. "American Thalictra and Their Old World Allies." *Rhodora* 46: 337–377, 391–445, 453–487. 1944.

Keener, Carl S. "Studies in the Ranunculaceae of the Southeastern United States. II. *Thalictrum* L." *Rhodora* 78: 457–472. 1976.

In obtaining specimens of thalictra for identification, it is important (if the plants are dioecious) to gather both staminate and pistillate plants.

1. Flowers all bisexual; fruit shaped like a curved sword.   1. *T. clavatum*
1. Flowers unisexual (rarely bisexual), plants mostly dioecious; fruit not shaped like a curved sword.
  2. Leaflets mostly narrow and unlobed, linear, narrowly lanceolate, narrowly elliptic, unlobed, 5–10 times as long as wide, only a few leaflets on a given plant short and lobed.   2. *T. cooleyi*
  2. Leaflets wider relative to their length, little if any over twice as long as wide.
    3. Plant not exceding 3 dm tall, stems slenderly wiry and lax; roots tuberous-thickened. 3. *T. debile*
    3. Plants much more than 3 dm tall, stems not slenderly wiry, stiffly upright; roots fibrous.
      4. Leaf subtending lowest inflorescence branch long-petioled.   4. *T. dioicum*
      4. Leaf subtending lowest inflorescence branch sessile or subsessile.
        5. Lower surfaces of leaflets with tiny amber or grayish beads of exudate and/or short hairs tipped with amber or grayish glands; lower surfaces of leaves occasionally glabrous, if so, then anthers linear, 1.8–3 mm long, the connective extending beyond the tips of the anther sacs as a distinct point.   5. *T. revolutum*
        5. Lower leaf surfaces not glandular and the anthers short-oblong or oblong-elliptic, 0.8–1.3 mm long, blunt apically.
          6. Filaments, fully developed, 3.5–6.5 mm long; leaflets predominantly 3-lobed; sepals longer than broad, oblong or obovate.   6. *T. pubescens*

6. Filaments, fully developed, 2–4 mm long; leaflets predominantly unlobed; sepals about as broad as long, suborbicular. 7. *T. macrostylum*

1. Thalictrum clavatum DC.

Glabrous plant with a cluster of fleshy, fusiform roots. Stems slender, 3–6 dm tall, with 1–3 basal leaves, the flowering stem arising from the caudex and usually naked to the flower-bearing portion where the leaves are biternate, flowers few. Leaflets variable in shape, size, and in lobing, ovate, obovate, suborbicular, or often broader than long, bases mostly broadly cuneate, occasionally truncate or subcordate, lobes rounded, 3–7 in number, mostly distally disposed, sometimes the broadly rounded apices merely crenate. Flowers bisexual, their slender stalks elongate. Sepals white and petallike, somewhat clawed basally, their blades elliptic to obovate. Filaments filiform below, distally broadly dilated and spatulate, white, broader than the anthers, anthers about 0.5 mm long, oblong-elliptic. Pistils 3–8. Mature achenes with stipes ½ to nearly as long as the achenes, widely diverging from the receptacle, falcate, upper margin convexly curved, lower concave, swordlike, strongly nerved, mostly 4–5 mm long.

Seepage slopes, wooded stream banks, mossy cliffs and ledges about waterfalls. Mts. of W.Va., Va., Ky., Tenn., Ga., Ala.

Thalictrum mirabile Small, chiefly inhabiting cliffs and ledges, often dripping wet, local in cen. Ky., cen. Tenn., and n. Ala., is apparently very closely similar and perhaps not distinct from *T. clavatum*. We have seen no specimens.

2. Thalictrum cooleyi Ahles.

Plant with a short, erect caudex. Stems sometimes slender and weak below, tending to be held erect by surrounding vegetation, sometimes more robust and stiffly erect, 4–8 dm tall. Lower and stem leaves petioled, the uppermost usually sessile. Leaflets varying, linear, lanceolate, elliptic, or oblong-oblanceolate, all except the latter usually entire and 4 or more times longer than wide, the oblong-oblanceolate ones 1–3-toothed-lobed distally and 2–3 times as long as wide; upper surfaces green and glabrous, the lower pale, glaucous, margins revolute. Plant dioecious. Sepals ovate, oblong, or obovate, 1.5–2 times as long as wide. Filaments slender below, tending to become curled, somewhat dilated distally, longer than the anthers; anthers linear-elliptic, or oblong when not fully developed, 1–2 mm long. Achenes sessile, ellipsoid, ridged and furrowed, the ridges sinuous, sometimes branching, the hooked, persistent style base about 2 mm long, body 4–6 mm long.

Savannas, rare. Coastal plain, s.e. N.C.; s.w. Ga., Fla. Panhandle.

3. Thalictrum debile Buckl.

Glabrous plant with a cluster of short-ovate to fusiform-attenuate tuberous roots. Stems delicate, flexuous and weak, often reclining, 1–3 (–4) dm tall. Lower leaves ternately decompound, upwardly 1-ternate. Leaflets mostly cup- or bell-shaped, with 3, short, rounded, erect, subequal lobes distally, 5–15 mm long, many a little longer than broad, some as broad as long. Dioecious. Sepals ovate to oblong or obovate, broadly rounded apically, pale grayish, sometimes suffused with purplish pigment. Filaments purplish, somewhat dilated distally, slightly longer than the anthers, anthers yellow or purplish, 2 mm long. Pistils few, sessile, the styles much longer than the ovaries at anthesis. Achene approximately oblong, 3–4 mm long, ribbed, longer than the persistent style.

In woodlands and prairies of the black belt, often in heavy, sometimes wet alluvial soils. Ala., Miss., Ark., Okla.

4. Thalictrum dioicum L.

Plant with a short, erect caudex bearing cordlike fibrous roots. Stem 3–7 dm tall, glabrous. Leaves all long-petiolate, 1–3 of them below the inflorescence. Leaflets thin, becoming fully grown after anthesis, glabrous, reniform to obovate in overall outline, many of them on an individual plant broader than long, with 3–7 rounded lobes distally, sometimes merely 3–7-crenate, larger ones about 4 cm broad and 3 cm long, varying to about 1 cm broad and a little longer. Plants dioecious. Sepals obovate, ovate, or suborbicular, those of the staminate flowers mostly at least ⅓ longer than broad, those of the pistillate little, if any, longer than broad. Filaments filiform proximally, somewhat dilated distally, anthers linear, about 3 mm long. Achenes short-stipitate or subsessile, ellipsoid, about 3–4 mm long, ridged and furrowed, the ridges often branched.

Rich woodlands, bluffs, ravines, and slopes, moist but not wet woodlands. Que. to Man., southward to Ga., Ala. (not in the coastal plain), and Mo.

5. Thalictrum revolutum DC.

Stem relatively stout and stiffish to slender and flexuous, to about 1.5 m tall, often much shorter, glabrous or sparsely pubescent. Basal leaves petioled, the middle and upper ones sessile; variable in size, blades of the lower 1–3 dm long and wide, ternately decompound. Leaflets variable, ovate, oblong, or obovate, some oblong-elliptic, many 3-lobed distally, some with 1–2 lobes, some unlobed, these often asymmetric, not infrequently some of each of the preceding on a single leaf; largest leaflets about 3 cm long and broad, smallest ones about 1 cm long, usually not as broad; upper leaf surfaces glabrous, the lower usually glaucous, with tiny amber or grayish beads of exudate and/or with short hairs tipped with amber or gray glands, sometimes glabrous, glaucous and nonglandular (the variation sometimes exhibited on plants in the same population, or even on leaflets of the same plant), margins weakly to strongly revolute. Dioecious. Sepals varying in shape, even on a single flower, linear-lanceolate to elliptic or ovate, most of them considerably longer than broad. Filaments 2–5 mm long, capillary proximally, dilated distally, anthers linear, 1.8–3 mm long. Pistils 6–12 or more, sessile or very short-stipitate. Achenes lanceolate or elliptic, 3–4 mm long, usually glandular-puberulent, style bases persistent as beaks 3–4 mm long, surfaces ridged and grooved.

Specimens (if collections include staminate plants) of this species without any of the glandularness on the lower surfaces of the leaflets may be distinguished from *T. pubescens* by characteristics of filaments and anthers. The filaments of *T. revolutum* are capillary proximally and little dilated distally, tend to become curled and entangled basally, the anthers are linear, 1.8–3 mm long, the connectives extending beyond the tips of the anther sacs as a distinct point; the filaments of *T. pubescens* are more stiffish and straight, more distinctly dilated from near their middles, the anthers short-oblong or elliptic-oblong, 0.8–1.3 mm long or less, blunt apically.

Well-drained woodlands and woodland borders, moist to wet thickets and clearings, prairies, barrens, meadows, (attributed to swamps by Small, 1933). Mass. to Ont., Ohio, Ill., Mo., southward to Ga., Fla. Panhandle, and La.

6. Thalictrum pubescens Pursh.

Plant with a short, erect caudex and fibrous roots. Stem glabrous (in our range), to 2.5 m tall. Lower leaves petiolate, upper sessile. Leaflets usually glabrous, sometimes minutely pubescent beneath but not glandular; variable in size, shape, and lobing, obovate, ovate, elliptic, with 3 obtuse, shallow or relatively deep lobes distally, or 3-toothed distally, or unlobed, frequently all three kinds on the same leaf; larger leaflets about 3 cm long and 2.5–3 cm wide, smaller ones about 10 mm long and 5 mm wide, some plants with leaflets predominantly in the larger range, some with them predominantly in the smaller. Plants polygamo-dioecious, usually some plants with wholly staminate flowers, others with pistillate and bisexual ones. Sepals obovate, oblong, or lanceolate. Filaments mostly dilated distally from near the base, or from below their middles, stiffish, white, 3.5–6.5 mm long, anthers short-oblong or elliptic, 0.8–1.3 mm long. Achenes short-stipitate or subsessile, with base of style 1–3 mm long persisting as a beak at the summit, the body elliptic, 3–4 mm long, longitudinally ridged and grooved, the ridges often branched, glabrous or pubescent. (*T. polygamum* Muhl. ex Spreng.) See note following description of *T. revolutum.*

Wet meadows, floodplain woodlands, marshes, bogs, wet clearings and thickets. Nfld. to Ont., generally southward to Ga., Ala., Miss.

7. Thalictrum macrostylum Shuttlw. ex Small & Heller.

Plant slender, about 1 m tall. Lower leaves petiolate, upper sessile, ternately decompound. Leaflets variable in shape and size on a given plant, 5–20 cm long, ovate, elliptic, or obovate, often asymmetrical, entire or shallowly 3-lobed distally, reticulate-veined beneath. Dioecious. Sepals subrotund to obovate, about 2 mm long. Filaments nearly straight, dilated distally, about 2–4 mm long, anthers short-oblong, about 1 mm long, blunt apically. Achenes sessile or subsessile, about 3 mm long, with low, inconspicuous ribs. (Possibly incl. *T. subrotundum* Boiv.)

Swampy woodlands, meadows, slopes, cliffs, limestone sinks. s. Va. to Ga.

5. Caltha

Caltha palustris L., var. **palustris. MARSH-MARIGOLD, COWSLIP.**

Glabrous perennial herb with a short rootstock bearing numerous cordlike roots, the stem at first short and with alternate, closely set, basal long-petioled leaves; flowering stem elongating to as much as 8 dm, its leaves having petioles progressively shorter upwardly, the uppermost nearly or quite sessile. Leaf blades mostly with broadly rounded, cordate basal lobes, the blade overall suborbicular to broadly ovate, or the uppermost often reniform, the margins shallowly and finely crenate or crenate-dentate. Flowers 1–3 in the axils of one or several uppermost bracteal leaves, their stalks varying in length from 1 to 6 cm. Sepals usually 5, but varying from 4 to 9, petallike, bright yellow, ovate to obovate, 10–25 mm long, 5–15 mm wide. Petals none. Stamens numerous, yellow. Pistils 4–12, ripening into follicles, these flattened, asymmetrically oblong, short-beaked, 8–20 mm long, 2.5–5 mm wide. Seeds several in each follicle, usually obovoid in outline, about 2 mm long, lustrous, reddish brown.

Marshes, swamps, wet meadows, wet woodlands, springy areas in woodlands, seepage bogs. In our range only in the mountains of N.C. and Tenn.; widespread in n. temp. N.Am. and Euras., with many "forms."

Reference: Smit, Petra G. "A Revision of *Caltha* (Ranunculaceae)." *Blumea* 21: 119–150. 1973.

6. Trautvetteria

Trautvetteria carolinensis (Walt.) Vail. **TASSEL-RUE, FALSE BUGBANE.**

Relatively coarse perennial herb, to 15 dm tall, glabrous except in the inflorescence. Leaves alternate, the basal close-set, long-petioled, blades when fully grown to 3–4 dm wide, considerably wider than long, deeply palmately or pedately 5–11 cleft-lobed, the primary lobes irregularly, saliently cleft-serrate; stem leaves few, much smaller, subsessile or sessile. Inflorescence irregularly corymbose-paniculate. Sepals 3–5 (deciduous by the time flowers are at full anthesis), white, obovate, concavely cupped, 3–5 mm long. Petals none. Stamens numerous, the filaments dilated distally. Pistils several, styles short-subulate. Fruit a 4-ribbed, 1-seeded, angled-inflated utricle 3–4 mm long, asymmetrically obovoid, with a hooked, persistent style.

Seepage slopes, bogs, springy areas in woodlands, wooded bluffs, prairies. s.w. Pa., W.Va., Ky. to Mo., southward mostly in the mts. to Ga. and Ala.

7. Ranunculus (BUTTERCUPS)

Perennial or annual herbs of varied habit. Leaves alternate (basal ones commonly closely set), petiolate or sessile, simple and entire, toothed, lobed, cleft, or divided; stipules laterally along the petioles of the basal leaves. Flowers radially symmetrical, bisexual. Sepals 3–5, rarely more, not petaloid, green or yellowish. Petals usually 5, fewer or more in some species, each with a nectariferous spot or pit on the upper side at or near the base, mostly larger than the sepals, smaller in some species, various shades of yellow, rarely white. Stamens numerous in most, as few as 5 in some species. Pistils several to numerous, ovaries superior; styles relatively long to very short, straight, curved, or hooked, persistent. Fruit an achene, usually somewhat flattened.

References: Benson, L. "A Treatise on the North American Ranunculi." *Am. Midl. Nat.* 40: 1–261. 1948.

Keener, C. S. "Studies in the Ranunculaceae of the Southeastern United States. V. *Ranunculus* L." *Sida* 6: 266–283. 1976.

In obtaining specimens for identification, it is important to have carefully extracted the main roots; fully developed flowers and fruiting heads with mature achenes are necessary.

1. Flowers sessile, borne on the stem opposite petiole bases.   1. *R. platensis*
1. Flowers stalked, borne from axils of leaves or bracteal leaves.
  2. Petals white.
    3. Leaves with blades shallowly 3-lobed, cordate to subtruncate basally, mostly floating.
                                              2. *R. hederaceus*
    3. Leaves dissected into filiform segments, mostly submersed.   3. *R. longirostris*
  2. Petals yellow.
    4. Faces of achenes papillose (sometimes the papillae with hooked spines distally), pebbled, tuberculate, spiny, longitudinally ridged, or rugose.
      5. Achenial faces papillose, the papillae with tiny hooked spines distally.   4. *R. parviflorus*
      5. Achenial faces variously pebbled, papillose, rugose, tuberculate, or spiny, tips of projections, if any, straight or curved (rarely hooked but if hooked then the projections stout).
        6. Pistils or achenes in a single whorl.   5. *R. arvensis*
        6. Pistils or achenes spirally disposed.
          7. Leaf blades entire, slightly wavy, or toothed, none of them deeply lobed.
            8. Plants spreading by stolons; perennials.
              9. Achenes longitudinally nerved.   6. *R. cymbalaria*
              9. Achenes pebbled on the faces.   7. *R. subcordatus*
            8. Plants erect or only the bases of the stems reclining, the latter sometimes rooting at the nodes; annuals, except *R. ambigens* (no. 10).
              10. Plants erect, with basal leaves, lower stems not reclining and rooting at the nodes.
                11. Petals minute, 1.5–2 mm long, slightly shorter than to slightly longer than the sepals; apex of achenes, viewed from above the tip, lenticular in outline, the style base decurrent across the top of the achene and forming an elliptical green strip.   8. *R. pusillus*
                11. Petals much more evident, twice as long as the sepals or more, 3–8 mm long; apex of achenes, viewed from above, broadly elliptic to nearly round, the style without a decurrent green base.   9. *R. laxicaulis*
              10. Plants with reclining lower stems, rooting at the nodes, without basal leaves; plants perennial.   10. *R. ambigens*
          7. Leaf blades markedly lobed, coarsely toothed, or if truly aquatic and with some submersed leaves then these divided into filiform segments.
            12. Plants aquatic, the submersed leaves divided into filiform segments.   11. *R. flabellaris*
            12. Plants terrestrial, or if the bases in shallow water, the leaves not divided into filiform segments.
              13. Achenes with stout spiny protuberances on the faces.   12. *R. muricatus*
              13. Achenes with low, rounded papillae on the faces, or the faces faintly horizontally rugose.
                14. The achenes with low rounded papillae on the faces.
                  15. Petals 5 mm long or a little less; stems, petioles, and leaf blades pilose.
                                             13. *R. trilobus*
                  15. Petals about 7 mm long; stems, petioles, and leaf blades glabrous or with few scattered hairs.   14. *R. sardous*
                14. The achenes horizontally rugose on the faces.   15. *R. sceleratus*
    4. Faces of achenes smooth or finely punctate-alveolate.
      16. Blades of basal leaves, some of them at least, unlobed, ovate to suborbicular or reniform, their

bases usually cordate, margins crenate, distinctly unlike the deeply pedately divided stem leaves.
    17. Petals 1.5–3 mm long, not longer than the sepals, barely evident.
        18. Style on the achene tiny, about 0.2 mm long, straight.   16. *R. abortivus*
        18. Style on the achene subulate, about 1 mm long (straightened out), hooked.
                            17. *R. alleghaniensis*
    17. Petals 5–8 mm long, about twice as long as the sepals, evident.   18. *R. harveyi*
16. Blades of basal and lower stem leaves markedly lobed or divided, bases rarely cordate.
    19. Achenial face convex, a relatively broadly corky-thickened zone around the face, green only across the rounded summit; style on the achene minute and barely evident.   15. *R. sceleratus*
    19. Achenial face flattened, sometimes slightly convex, the achenial margin distinctly differentiated into a green rim or keel; style on achene evident.
        20. Petals 2–4 (–5) mm long, equaling or shorter than the sepals.   19. *R. recurvatus*
        20. Petals 6 mm long or longer, longer than the sepals.
            21. Styles stigmatic laterally, the stigmatic portion nearly covering one side of the style.
                22. Sepals reflexed at full anthesis of flower.
                    23. Plant perennial, with a hard, bulblike or cormose base; sepals 5–8 mm long, petals 8–15 mm long.   20. *R. bulbosus*
                    23. Plants annual, with soft bases; sepals 2–4 mm long, petals 5–8 mm long.
                              14. *R. sardous*
                22. Sepals not reflexed at full anthesis of flower.
                    24. Principal leaves ternate, the terminal segment, or all the segments, stalked; receptacle pubescent.   21. *R. repens*
                    24. Principal leaves pedately (almost laciniately) divided into 5 principal, sessile segments; receptacle glabrous.   22. *R. acris*
            21. Styles stigmatic at or around their tips.
                25. Fascicle of roots below the stem including both tuberous and slenderly fibrous or cordlike ones.   23. *R. fascicularis*
                25. Fascicle of roots below stem uniform, all slenderly fibrous or cordlike.
                    26. Plants colonial, soon after flowering begins the branches become lax and reclining, often rooting at the nodes and there forming plantlets; sepals reflexed at full anthesis of the flower.   24. *R. carolinianus*
                    26. Plants erect to weak and reclining, not colonial; sepals spreading at full anthesis of the flower, sometimes tardily reflexed.
                        27. Blades of the earliest basal leaves all ternate and the divisions stalked; stipules usually rounded to truncate at the summit.   25. *R. septentrionalis*
                        27. Blades of the earliest basal leaves 3-lobed, or ternate and their lateral divisions sessile; stipules tapering at the summit.   26. *R. hispidus*

1. Ranunculus platensis Spreng.   Fig. 60

Low, very slender, soft annual, pilose throughout. Stems weakly ascending, 5–30 (–45) cm tall, sometimes a few from the base, sometimes numerous from the base. Principal leaves forming a basal rosette, laxly long-petioled, palmate, suborbicular to reniform, broader than long, the larger ones about 1.5 cm wide; some of them deeply 3-cleft, the divisions strongly dentate or crenate-dentate, some 3-lobed, the lobes dentate or crenate-dentate, some merely dentate or crenate-dentate, bases cordate to truncate; flowering stems with leaves reduced upwardly and shorter-petioled, few and widely spaced. Flowers solitary, sessile, positioned on the opposite side of the stem from the petiole, minute and barely visible to the unaided eye. Sepals 3, obovate, quickly deciduous. Petals 3, spatulate, pale yellow. Stamens 5. Carpels mostly 8–12, surfaces of the ovaries bearing short, pustular-based, hooked trichomes, styles curved. Achenes lenticular, somewhat oblique in outline, about 2 mm across, hooks of the trichomes mostly shed, the pustular bases rendering the surface papillose.

River floodplains, moist to wet fields, lawns, roadsides, ditches. Native from Brazil to Argentina. Naturalized in Fla. Panhandle, La., and s.e. Tex., perhaps elsewhere.

2. Ranunculus hederaceus L.

Glabrous aquatic or marsh plant, the creeping stems commonly floating or in mud, rooting at the nodes. Leaves with slender petioles 1–3 cm long; blades essentially reniform in outline, obtusely and shallowly 3–5-lobed, 5–20 mm wide, cordate to nearly truncate basally, the

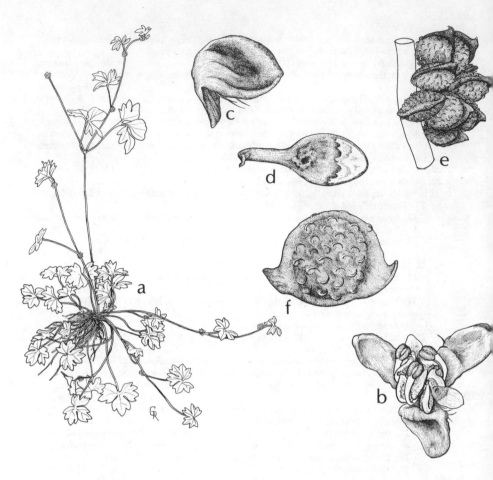

Fig. 60. **Ranunculus platensis:** a, habit; b, flower, face view; c, sepal; d, petal (drawn to a much larger scale than sepal); e, cluster of achenes from one flower; f, achene.

apical outline broadly rounded; stipules broad, tonguelike at the summit. Flower stalks 5–20 mm long. Sepals 5, spreading, ovate, 1.5–2 mm long, ⅔ as long as the petals, quickly deciduous. Petals 5, white, obovate, 1.5–2.5 mm long. Stamens 5–10. Fruiting head globose; achenes about 7–15, obovate, 1–1.5 mm long, horizontally rugose on the faces.

Pools, wet sandy depressions, marshes. Nfld.; coastal plain, s.e. Pa. to S.C.; Eur.

3. Ranunculus longirostris Godr. WHITE WATER-CROWFOOT.

Glabrous aquatic perennial with elongate stems. Leaves submersed, mostly relatively distant from each other, very short-petiolate or subsessile; blades mostly ternate, finely dissected into rather rigid filiform segments, in overall outline broader than long, mostly 1.5–3 cm broad. Flower stalks 1.5–2 cm long. Sepals spreading, narrowly elliptic, 3–4 mm long, half the length of the petals, quickly deciduous. Petals 5, white, obovate, 4–9 mm long. Stamens 10–20. Fruiting head subglobose; achenes 7–25, obovate, about 1.5 mm long, transversely ridged on the faces. (*Batrachium trichophyllum* sensu Small, 1933)

In streams, spring-fed ponds and lakes, springs and spring runs, sloughs, sometimes densely matted. s.w. Que. to Ore., southward to Del., Va., Ky., Tenn., Ala., Ark., Tex., N.Mex., Ariz.

136

4. Ranunculus parviflorus L. SMALL-FLOWERED BUTTERCUP.

Annual, pilose throughout. Lower stems with few to numerous closely set petiolate leaves, the petioles 3–6 cm long, blades mostly as broad as long or broader than long, in overall outline suborbicular to reniform, some with 3 principal dentate lobes, others irregularly lobed-toothed, about 1.5–2 cm long; flowering stems 1 to several, weakly ascending, 1–3 dm tall, their leaves petiolate and the blades much like the lower ones (except the uppermost, these deeply 2–5-cleft (or simple), the divisions lanceolate); stipules broadly rounded to truncate at the summit. Flowers axillary, solitary, with stalks 1–4 mm long at anthesis, elongating as the fruiting heads mature, becoming 5–20 mm long. Flowers small and inconspicuous, about 4–5 mm across. Sepals 5, 1 mm long, spreading, narrowly ovate, quickly deciduous. Petals 5, narrowly elliptic, pale yellow, 1–2 mm long. Stamens about 10. Fruiting head globose, 4–5 mm across; achenes 10–20, lenticular, mostly obovate and somewhat asymmetrical, the body 1.5 mm long or a little more, the styles triangular, shortly and softly attenuate, hooked apically, disposed somewhat to one side of the broadly rounded apex, margins smooth, faces with minute pustular-based, hooked trichomes, the hooks sometimes deciduous leaving the surfaces papillose.

Meadows, moist fields, lawns, and pastures, logging roads in bottomlands, swales, ditches, sometimes in upland woodlands. Native of Eur., widely naturalized in N.Am., throughout much of our range, in Fla. only in the Panhandle.

5. Ranunculus arvensis L.

Annual, usually with one erect main stem, to about 6 dm tall. Blades of both basal and stem leaves deeply 3-cleft (occasional ones 3-lobed), the divisions variously sublobed or toothed, linear, narrowly lanceolate, or oblanceolate; basal and lowermost stem leaves petiolate, sessile above; stems and leaves mostly sparsely to densely hirsute or pilose, sometimes glabrous. Flowers solitary in the axils of bracteal leaves, their stalks 1–2 cm long, elongating to about 4 cm as the fruiting heads mature. Sepals 5, spreading, 4–5 mm long, elliptic-oblong, hirsute over much of the outer surface but margins membranous and glabrous. Petals 5, pale yellow, obovate, 5–6 mm long. Stamens 5–10. Achenes mostly 5–8 in a single whorl, obliquely obovate in outline, the body 4–6 mm long, with a prominent green spiny or papillose-tuberculate rim, the faces flat but beset with coarse spiny protuberances, or papillose-tuberculate, the style triangular-subulate, essentially straight, 1.5–2 mm long.

Weedy, mostly in moist, low places in cultivated fields. Native of Eur., sporadically naturalized, in our range mostly in the piedmont, N.C. to Miss.

6. Ranunculus cymbalaria Pursh. SEASIDE CROWFOOT.

Glabrous or sparsely pubescent, tufted perennial producing slender, elongate leafy stolons. Leaves petiolate, petioles of the basal ones 2–5 cm long, those of the stolons about equaling the blades; blades ovate, suborbicular, or reniform, 1–3.5 cm long, mostly as broad as long, margins crenate-dentate. Flowers 1 or several on a branched axis, their stalks 1–3 cm long. Sepals 5, spreading, elliptic, 2–5 mm long, glabrous, quickly deciduous. Petals 5 (–12), bright yellow, narrowly obovate, 3–8 mm long. Stamens 10–30. Fruiting head oblong-cylindrical, 3–14 mm long; achenes many, wedge-shaped to oblong, about 1.5 mm long, surfaces longitudinally ridged, style short-triangular, at one side of the nearly truncate summit.

Muddy, mostly brackish or alkaline places, shores of streams and marshes, wet meadows, shallow water of pools and streamlets. A wide-ranging, variable species, s. Greenland, Labr., to Alaska, southward to N.J., Ill., Iowa, Kan., Ark., Okla., Tex.; widely distributed in the western states; Mex.; S.Am.; Euras.

7. Ranunculus subcordatus E. O. Beal.   Fig. 61

Subaquatic perennial producing elongate leafy stolons. Leaves petiolate, petioles of the lower ones to nearly 3 dm long, those of the upper about 1 cm long; blades entire to finely

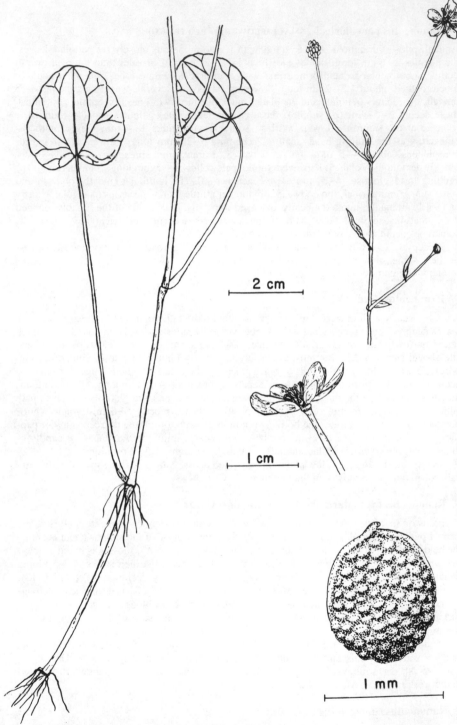

2 cm

1 cm

1 mm

Fig. 61. **Ranunculus subcordatus:** lower portion of a vegetative branch with adventitious roots (left); terminal flowering branch with flower and fruit (top right); flower (center right); fruit (bottom left). (From Beal, "A New Species of *Ranunculus* from North Carolina." *Brittonia* 23: 266–268. 1971)

dentate, to 3.5 cm long, upper gradually reduced and lanceolate. Flower stalks 1–4 cm long at anthesis, elongating somewhat as the fruiting heads mature. Sepals 5, ovate, about 2 mm long, sparsely pubescent dorsally, quickly deciduous. Petals 4–8, yellow, 3–4 mm long. Stamens 7–25. Fruiting heads globose; achenes numerous, about 1 mm long, faces finely pebbled, style minute.

Swamps and swamp margins. Coastal plain, Bladen and Halifax Cos., N.C.

8. Ranunculus pusillus Poir. SPEARWORT.

Annual, glabrous or rarely pubescent, the stem simple below or more commonly several-branched, weak, the lower stems sometimes reclining. Basal, lower, and usually the midstem leaves long-petiolate, the uppermost or bracteal ones sessile or subsessile. Blades of principal leaves ovate to narrowly lanceolate, the bases truncate to cuneate, apices obtuse, margins entire, irregularly shallowly wavy, or irregularly shallowly dentate; blades of bracteal leaves lanceolate to linear, margins entire. Inflorescence more or less paniculate. Flowers minute. Sepals 5, spreading, ovate, suborbicular, or obovate, 1–2 mm long. Petals 1–3 (–5), pale yellow, inconspicuous, 1–1.5 mm long. Stamens 5–10. Fruiting heads hemispherical, 2–4 mm long; achenes numerous, obovate, about 1 mm long, faces convex, papillate, styles in anthesis 0.1–0.2 mm long. (Incl. *R. lindheimeri* Engelm., *R. tener* Mohr)

In shallow water or mud, temporary pools, ditches, marshes, low places in cultivated fields, swamps, margins of small streams, wet meadows, floodplains, wet clearings. Chiefly coastal plain and piedmont, s.e. N.Y. to n. Fla., westward to Tex., northward in the interior to Mo., Ind., Ohio; Calif.

9. Ranunculus laxicaulis (T. & G.) Darby. MANY-FLOWERED SPEARWORT.   Fig. 62

In vegetative features closely similar to *R. pusillus*. Flowers larger and more conspicuous, sepals 2–3 mm long, petals 3–8 mm long, much exceeding the sepals and much more conspicuous than in *R. pusillus*, achenes more turgid, somewhat smaller, 0.6–0.7 mm long, more obliquely obovate, styles 0.5 mm long. (*R. texensis* Engelm.; *R. oblongifolius* sensu Small, 1933; incl. *R. mississippiensis* Small)

In habitats similar to those of *R. pusillus*. Coastal plain, Conn. to Ga., westward to s.e. Tex. and e. Okla., northward in the interior to Kan., Mo., Tenn., Ill., Ind.

10. Ranunculus ambigens S. Wats. WATER-PLANTAIN-SPEARWORT.

Relatively coarse perennial, the lower stems reclining and rooting at the nodes, 5–20 mm thick, upper stems weakly ascending, sometimes reaching 1 m long overall. Leaves with stoutish, dilated petioles, the blades lanceolate, to 15 cm long and 3 cm broad, margins entire to finely and irregularly dentate. Flowers solitary or in a forking inflorescence with few to about 20 flowers. Sepals 5, spreading, ovate, about 4 mm long, quickly deciduous. Petals 5 or 6, shiny yellow, oblong, 5–8 mm long. Stamens numerous. Fruiting heads globose-ovoid; achenes numerous, cuneate-obovate, 1.5–2 mm long, broadened distal portion with convex, longitudinally wrinkled or faintly ridged faces, styles thin, long-triangular. (*R. obtusiusculus* Raf. sensu Small, 1933)

Swamps, sloughs, marshes, marshy shores. Local, Maine to Minn., southward to S.C., Tenn., n. La.

11. Ranunculus flabellaris Raf. YELLOW WATER-CROWFOOT.

Glabrous to pubescent aquatic or marsh perennial with elongate, hollow stems. Leaves of submersed plants ternately decompound, finely dissected into linear-filiform segments, in overall outline suborbicular to reniform, to about 10 cm long, as broad or somewhat broader; leaves of plants in mud (or emersed leaves of plants in water) petiolate, the blades 3-cleft, the divisions 3-lobed, sometimes densely pubescent. Flowers usually several on a branched, essentially leafless sympodial stem, the flower stalks stout, 1–4 cm long at anthesis, elongating somewhat as the fruiting heads mature. Sepals 5, spreading, ovate to suborbicular, 5–8 mm long, quickly deciduous. Petals 5, golden yellow, broadly obovate, 10–15 mm long.

Fig. 62. **Ranunculus laxicaulis:** a, habit; b, flower; c,d, petal, two views; e, fruiting
"head." (From Correll and Correll. 1972)

Stamens numerous, 50–80. Fruiting head ovoid; achenes 50–75, obliquely obovate, 2.5–3.5 mm long including the flat, subulate style, prominently corky basally and along one margin, the faces rugose. (*R. delphinifolius* Torr.)

Quiet waters of ponds, sloughs, bayous, mud flats, marshes, swamps. Maine to B.C., southward to N.C., La., Okla., Utah, and Calif.

12. Ranunculus muricatus L.

Glabrous or pubescent annual with closely set basal leaves and usually several more or less paniculate flowering stems from near the base, these sometimes shorter than the basal leaves, the plants often only 5–15 cm tall, or sometimes plants much more robust with very long lower leaves and flowering stems to about 5 dm tall. Basal and lower stem leaves long-petiolate, the petioles varying in length, on some plants 2–3 cm long, on others to about 18 cm long; petioles gradually reduced upwardly, bracteal leaves sessile; blades of principal leaves commonly about as wide as long, sometimes wider than long, in overall outline reniform to suborbicular, bases cordate, truncate, or very broadly tapered, variably lobed or toothed, some 3–5-cleft, some 3–5-lobed, the divisions or lobes with obtuse to rounded coarsely dentate teeth. Sepals 5, reflexed, ovate, about 5 mm long. Petals 5, yellow, spatulate to obovate, about 8 mm long. Fruiting heads globose; achenes mostly numbering about 10–15, somewhat asymmetrical, obovate, the body 4–5 mm long, with a conspicuous, green, keeled, marginal rim, the faces flat but with prominent stiff-spiny or tuberculate protuberances, finely punctate-alveolate between the spines, styles prominent, continuous with the rim, triangular-subulate, usually curved and hooked at the extremity.

Low wet woodlands, wet pastures, wet alluvial outwash, moist to wet fallow fields. Native of s. Eur., sporadically naturalized, S.C. to n. Fla., westward to Ark., La., and Tex.

13. Ranunculus trilobus Desf.

Annual with a thickish short basal stem, numerous closely set basal leaves, the flowering stems usually several from near the base, 3–5 dm tall, glabrous or nearly so. Basal and lower to median stem leaves long-petiolate, the blades essentially similar, ternate, the lateral segments very short-stalked to sessile, the terminal one with a stalk 1–2 cm long; all segments ternately lobed or cleft, the divisions obtusely toothed-lobed. Overall the larger blades 5–6 cm long and about as wide across the lateral lobes. Flowers more or less panicled, their stalks 0.5–1 cm long, usually elongating to 3–4 cm as the fruiting heads mature. Sepals 5, ovate, reflexed, about 2 mm long. Petals 5, obovate, glossy yellow above, about 4 mm long. Stamens about 20. Fruiting heads ovoid or short-oblong; achenes numerous, somewhat asymmetrical, nearly orbicular, about 2 mm long, the faces papillose.

Wet clearings and ditches, low fallow fields. Native of s.w. Eur., apparently recently introduced, known thus far from Fla. Panhandle and coastal plain of Ala., La., and e. Tex.

14. Ranunculus sardous Crantz.

Somewhat like *R. bulbosus* (no. 20) but an annual, the base not bulbous, leaves closely similar in form. Tending to have more branches from near the base, these forking and more floriferous, the flower stalks long and the bracteal leaves much reduced, thus the flowers appearing panicled. Flowers similar to those of *R. bulbosus* but smaller, the sepals 2–4 mm long, petals 5–8 mm long, sulphur yellow. Achenes about 2 mm long, the margin narrowly green-keeled, faces smooth, or occasionally with a few low papillae, style short-triangular, about 0.3 mm long.

Weedy, fields, bottomland clearings, swales, waste places. Native of Eur., sporadically naturalized, sometimes locally abundant, various localities in our range, chiefly coastal plain and piedmont, N.C. to Ga., westward to e. Tex., northward in the interior to Mo. and Tenn.; in various places elsewhere in N.Am.

15. Ranunculus sceleratus L. CURSED BUTTERCUP.

Annual or short-lived perennial, usually glabrous below, pubescent distally on the branches. Stem erect, hollow, stout, often to 2 cm or more across near the base, usually much branched from somewhat above the base and with numerous axillary flowers distally on the branches. Blades of basal leaves (commonly absent by full anthesis) and lower stem leaves long-petiolate, blades mostly deeply 3-cleft, the divisions variously cleft or lobed as well, the apices of the lobes rounded; upwardly leaves becoming sessile, pedately divided, finally the bracteal leaves simple and lanceolate to oblanceolate, or linear. Flower stalks usually up to about 1 cm long at anthesis, elongating to as much as 3 cm as the fruiting heads mature. Sepals 5, spreading, ovate, 2–3 mm long, pilose exteriorly, quickly deciduous. Petals 5, yellow, obovate to oblong, mostly 3–4 mm long. Stamens 10–25. Fruiting heads oblong-cylindric, 5–15 mm long; achenes many, somewhat asymmetric, more or less wedge-shaped in outline, about 1 mm long, corky at the base and along the lateral margins, the faces within the corky zone faintly transversely rugose to smooth, style tiny.

Marshy shores and marshes, swamps, wet meadows, swales, sloughs, wet ditches. Nfld. to Alaska, generally southward to n. Fla., La., Tex. to Calif. Native of Eur.

16. Ranunculus abortivus L. KIDNEY-LEAF BUTTERCUP.

Annual or biennial with slender fibrous roots. Basal, petiolate leaves closely clustered, their blades 1–6 cm wide, ovate, suborbicular, or reniform, bases cordate, truncate, or broadly tapered, apices very broadly rounded, margins crenate, some blades sometimes deeply pedately divided; stipules gradually tapering to the summit. Flowering stems 1 to several from the base, to about 5 dm tall, their leaves sessile, variable, mostly deeply ternately divided, the divisions narrowly lanceolate to oblanceolate, often with 1 to several lobes, leaves sometimes not divided. Plants glabrous throughout, or with sparse pubescence of long, soft hairs on the petioles or blades of lower leaves. Flowers small, 6–7 mm across. Sepals 5, reflexed, usually short-clawed, their blades broadly elliptic, oval-oblong, or obovate, 3–5 mm long, longer than the petals. Petals 5, pale yellow, 2–3 mm long, elliptic. Head of achenes ovoid or short-oblong, mostly 4–6 mm long; receptacle glabrous or pubescent. Achenes lenticular, broadly obovate to suborbicular in outline, 1–1.5 mm long, glabrous or sparsely pubescent, the persistent style minute, soft, usually well to one side of the broadly rounded apex.

Floodplain woodlands, rich mesic woods, wooded stream banks and seepage areas, open moist to wet grassy banks, low areas in fields and pastures, ditches, muddy edges of streams, pools, ponds, or lakes. Labr. to Alaska, southward to n. Fla. in the East, to Wash., Colo., and Tex. westwardly.

R. micranthus Nutt. ex T. & G., in our range occurring in N.C., Tenn. and Ark., is closely similar to *R. abortivus*. Its roots, or some of them at least, are lance-attenuate and fleshy. The petioles and leaf surfaces are more densely pubescent with long, soft hairs. It occurs in dry to moist woodlands, on rocky slopes and calcareous banks, rarely, if at all, in wet lowland sites.

17. Ranunculus allegheniensis Britt.

In general features closely similar to *R. abortivus*. Basal petiolate leaf blades vary from simple and cordate-reniform, to 3-lobed, to divided and essentially ternately compound, the leafletlike divisions of the latter variously toothed-lobed. Achenes with a more pronounced, strongly curved or hooked beak 0.6–1 mm long (straightened out).

Mostly in well-drained woodlands, occasionally in grassy bottoms and on stream banks, probably not to be considered a wetland plant. e. Mass. and Vt. to s.e. Ohio, southward in the mts. to w. N.C. and e. Tenn.

18. Ranunculus harveyi (Gray) Britt.

Similar in general features to *R. abortivus*, the flowers larger and more showy. Flowering stems 2–4 dm tall. Plant usually glabrous, sometimes strongly pilose. Sepals spreading, not reflexed. Petals 5–7, commonly 7, oblong to oblong-spatulate, 6–8 mm long, about twice as long as the sepals, pale yellow, usually drying nearly white.

Ledges, bluffs, ravines, low woodlands, wet areas of thin soil in rocky glades. s. Mo., Ark., e. Okla., cen. Tex. and n.cen. Ala.

19. Ranunculus recurvatus Poir. HOOKED BUTTERCUP.

Perennial, the base somewhat bulbous or cormlike, bearing few to several closely set leaves, the pilose or villous flowering stems 1–3 from near the base. Leaves nearly all petioled, the uppermost bracteal leaves sometimes sessile, petioles usually pubescent, those of lower leaves of some plants 5–10 cm long, on others to 20 cm long, in either case reduced in length upwardly on a given plant; blades ovate to suborbicular, or reniform in overall outline, variable in size from plant to plant, those of the upper stem leaves often larger than those of the basal ones on a given plant, 3-cleft to or to somewhat below the middle, the 3 principal divisions irregularly lobed-toothed, the tips of lobes or teeth mostly obtuse. Flowers axillary, very short-stalked at anthesis, the stalks elongating to as much as 5 cm as the fruiting heads mature. Sepals 5, reflexed, pubescent, ovate, 3–7 mm long, longer than the petals. Petals 5, pale yellow, elliptic, 2–4 (–5) mm long. Stamens 10–25. Fruiting heads ovoid or globose-ovoid, 5–8 mm long; achenes from about 10–25, somewhat asymmetrical, obovate, about 2 mm long, thin and flattish, a narrow green keel on the edges, the face minutely alveolate, style subulate, hooked, about 1 mm long.

Rich, well-drained woodlands, wet woodlands, floodplain forests, borders of woodland streams. Nfld. to Man., generally southward to Fla. Panhandle and n.e. Tex.

20. Ranunculus bulbosus L. BULBOUS BUTTERCUP, MEADOWGOLD.

Perennial, the base bulblike or cormlike with several closely set leaves, flowering stems 1 to several from near the base, to about 6 dm tall, glabrous to villous. Basal and lower stem leaves long-petioled, ternate, the lateral divisions short-stalked to sessile, the terminal one usually stalked, the 3 primary divisions again 3-cleft or 3-lobed, their divisions variously toothed. Upper stem leaves few, mostly pedately divided. Flowers axillary, short- to long-stalked. Sepals 5, reflexed, ovate, 5–8 mm long, pubescent dorsally, quickly deciduous. Petals 5, golden yellow, much longer than the sepals, mostly 8–15 mm long, broadly obovate, apices broadly rounded. Stamens many. Fruiting heads globose-ovoid, 7–10 mm long; achenes few to numerous, 2.5–3 mm long, obovate to suborbicular, nearly flat, with a prominent green marginal rim, faces smooth, style triangular, about 0.5–0.8 mm long.

Pastures, meadows, lawns, fields, open stream banks, weedy. Nfld. to Ont., southward to Ga. and La.; w. N.Am. Native of Eur.

21. Ranunculus repens L. CREEPING BUTTERCUP.

Perennial, usually with several flowering stems from near the base, these to about 6 dm long, becoming repent and rooting at the nodes. Stems and leaves glabrous to pilose. Bracteal leaves sessile, others long-petioled; stipules broadly rounded at the tonguelike summit; blades ternate, the lateral segments sessile or short-stalked, the terminal with a stalk 1–5 cm long, the 3 principal segments of the blade again 3-cleft and irregularly lobed-toothed. Flowers with stalks 2–10 cm long, the stalks elongating somewhat as the fruiting heads mature. Sepals 5, spreading, nearly oblong, 4–7 mm long, quickly deciduous. Petals 5 (more in double-flowered garden forms), golden yellow, broadly obovate, very broadly rounded apically, 8–15 mm long. Stamens numerous. Pistils usually 20–25, the style stigmatic laterally its full length. Fruiting head globose or globose-ovoid; achenes flattish, obliquely obovate to suborbicular, 2–2.5 mm long, with a narrow keel on the edges, the faces finely punctate-alveolate, style nearly flat, triangular-subulate, 0.7–1.4 mm long, usually curved or hooked at the tip. Receptacles pubescent.

Wet open areas, meadows, lawns, stream banks, springy places. Nfld. to Minn., southward to S.C. and Ky.; Alaska to Colo. and Calif. Native of Eur.

22. Ranunculus acris L. TALL BUTTERCUP.   Fig. 63

Perennial, the base short and thickish. Stem slender, erect, to 10–12 dm tall, glabrous or pubescent, with several basal leaves with long, slender, pubescent petioles, blades subor-

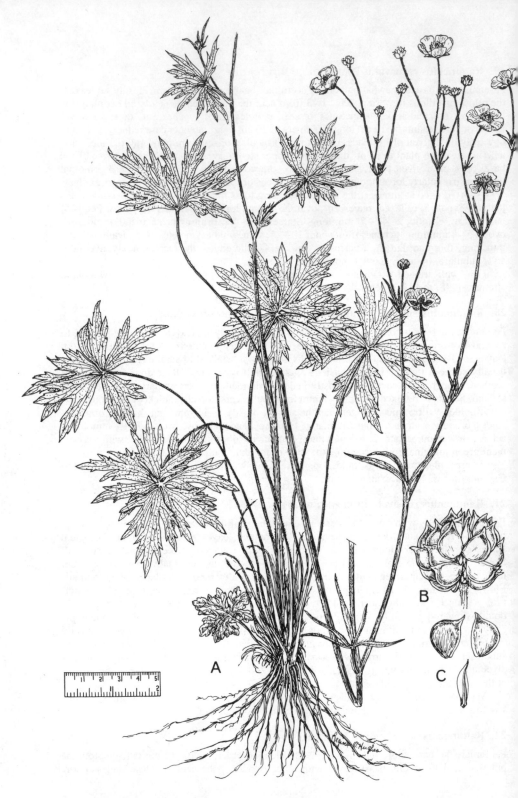

Fig. 63. **Ranunculus acris:** A, plant; B, fruiting head; C, achenes. (From Reed, *Selected Weeds of the United States* (1970) Fig. 91)

bicular in overall outline, deeply 5-cleft, the principal segments again deeply cleft and toothed-lobed, the divisions relatively narrow; stem leaves few, widely spaced, short-petioled to sessile, the bracteal leaves much reduced. Flowering branches pubescent, long, ascending-erect, usually bearing several flowers distally, their stalks elongating as the fruiting heads mature. Sepals 5, spreading, ovate to oblong-ovate, pubescent, 4–7 mm long. Petals usually 5, obovate to wedge-shaped, broadly rounded to truncate apically, deep yellow to cream-yellow, 6–15 mm long. Stamens numerous. Pistils few to numerous, the styles stigmatic laterally for about the full length. Fruiting heads subglobose; achenes flattish, obovate to suborbicular, 2–2.5 mm long, narrowly keel-margined, faces very finely punctate-alveolate, style triangular-subulate, hooked at the tip. Receptacles glabrous.

Wet meadows, wet clearings, open banks of streamlets or in the streams. Widespread across n. temp. N.Am., southward in our range to S.C., Ga., Tenn. Native of Eur.

23. Ranunculus fascicularis Muhl. ex Bigel. EARLY BUTTERCUP.

Perennial with a fascicle of roots including both tuberous-thickened and cordlike ones. Flowering stems usually several from near the base, weak, with long ascending to appressed hairs, 1–3 dm tall. Basal leaves several to numerous, closely set, long-petioled, blades mostly ternate (some sometimes undivided), the lateral segments short-stalked to sessile, the terminal one with a stalk 1–2 cm long; blades sometimes pinnately divided; lateral segments variously 3-lobed to 3-cleft or only toothed, the terminal one usually with 3 linear-oblong, ascending, blunt to rounded lobes; bracteal leaves much reduced, commonly undivided, linear to lanceolate. Stipules gradually tapered to the summit. Flower stalks variable in length, from about 1–6 cm long, usually elongating considerably as the fruiting heads mature. Sepals 5, spreading, ovate-attenuate, 6–8 mm long, pubescent with long, somewhat silky hairs. Petals 5, yellow, oblong to narrowly obovate, rounded apically, 6–14 mm long. Stamens numerous. Pistils 10–30, styles slender, stigmatic at the tip. Fruiting heads globose or globose-ovoid; achenes obliquely obovate-orbicular, body 2–3 mm long, narrowly keel-margined, faces somewhat convex, dull, style subulate, straight or curved, about 2 mm long.

Open, often rocky, woodlands, barrens, calcareous ledges, prairies, occasionally in semi-boggy places. Mass. to Minn., southward to Ga., Ala., La., and c. Tex.

24. Ranunculus carolinianus DC.   Fig. 64

Perennial with cordlike roots, pubescent to nearly glabrous. First stems of the season weakly erect, soon developing trailing or reclining ones that root at the nodes, plants usually colonial. Basal leaves few to numerous, closely set, long-petioled, blades varyingly 3-lobed, deeply 3-cleft, or ternate, the earliest ones usually 3-lobed, the lobes irregularly serrate, the later ternate ones varyingly 3-cleft to ternately compound, the divisions irregularly lobed or cleft and serrate. Stipules broadly rounded to truncate at the summit. Sepals 5, reflexed, ovate-attenuate, 3.5–5 mm long. Petals 5, bright to pale yellow, cuneate-oblong, rounded apically, variable in size, even on a single plant, 5–15 mm long, less than half as broad as long. Stamens numerous. Pistils 7–15, stigmatic at the tip. Fruiting heads subglobose; achenes obliquely ovate-orbicular or orbicular-obovate, prominently wing-keeled marginally, faces flattish, styles flat, triangular-subulate. (*R. palmatus* sensu Small, 1933)

Swamps, wet woodlands, in or on banks of woodland streams, floodplain forests, alluvial outwash, marshy shores, wet thickets and swales. Md. to Ky., s. Ind., s. Ill., Mo., Neb., generally southward to cen. pen. Fla. and e. fourth of Tex.

25. Ranunculus septentrionalis Poir. SWAMP BUTTERCUP.

Perennial, first stems erect, later ones repent, soft, hollow, hirsute to glabrous. Basal leaves long-petioled, blades ternately compound, divisions 2–3-cleft and sharply incised, cuneate to rounded-truncate at base, stalk of middle leaflet 1–5 cm long. Sepals 5, spreading, ovate, sometimes tardily reflexed, 6–10 mm long. Petals 5, golden yellow, cuneate-obovate, 7–15 mm long, broadly rounded at the apex where half as wide as long or more. Stamens numerous. Style stigmatic around the slender tip. Fruiting heads globose-ovoid; achenes obovate to

145

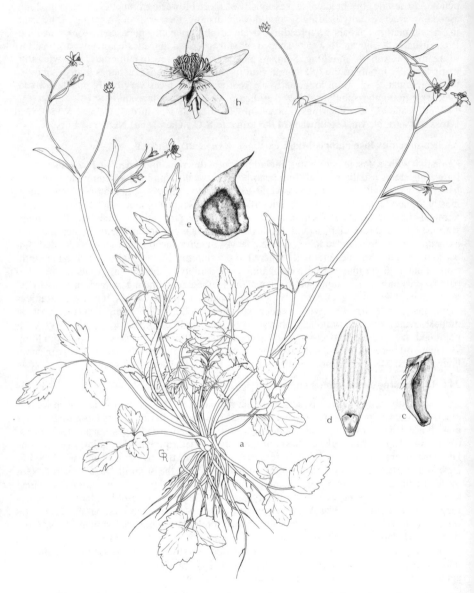

Fig. 64. **Ranunculus carolinianus:** a, habit; b, flower; c, sepal; d, petal; e, achene.

146

suborbicular, 3–4 mm long or a little more, narrowly keel-margined, faces finely pitted, style triangular-subulate.

Wet alluvial woodlands, wet meadows, thickets, stream banks. e. Que. to Man., southward to n. Ga. and Ark.

26. Ranunculus hispidus Michx. BRISTLY BUTTERCUP.

Perennial with cordlike roots, hirsute throughout. Flowering stems usually several from near the base, weakly erect. Early basal leaves variably 3-lobed, 3-cleft, or ternately divided, the latter with sessile lateral divisions; later basal leaves ternately to pinnately compound, the segments coarsely cleft to laciniate. Stipules short-tapered at the summit. Sepals 5, spreading, sometimes tardily reflexed, ovate-attenuate, 5.5–7.5 mm long. Petals yellow, obovate to oblong, 7–14 mm long, usually half as wide as long. Stamens numerous. Pistils 15–40, styles slender, stigmatic at their tips. Fruiting heads subglobose; achenes suborbicular, prominently wing-keeled, faces somewhat convex, minutely pitted, style subulate, straight, about 2 mm long.

Rich, moist woodlands, wooded, rocky creek banks, dryish woodlands, occasionally in alluvial woods or on seepage slopes. s. N.Y. to Ill. and Mo., southward to Ga., Ark., Okla.

Annonaceae (CUSTARD-APPLE FAMILY)

(Ours) shrubs or trees, the buds naked. Leaves deciduous (in ours), alternate, without stipules, simple, entire, pinnately veined, aromatic when bruised. Flowers axillary, usually more or less nodding, bisexual, the perianth of 3 or 4 sepals, and 6 or 8 petals in 2 whorls of 3 or 4 each, the inner smaller than the outer. Stamens numerous, spirally inserted on an elevated receptacle and forming a nearly globular mass; filaments and anthers poorly or scarcely differentiated from each other. Pistils (carpels) many, rarely few or even 1, free or sometimes the ovaries adhering or coalescing. Ovaries superior, each with a short style, a terminal stigma. Fruits developing from separate pistils becoming fleshy, pulpy berries; if fruit developing from numerous coherent or coalescing pistils and receptacle, then a fleshy syncarp. Ovules many, rarely few or even 1, in 1 or 2 rows or basal. Seeds large, (in ours) arillate.

• Pistils several and distinct, borne on the summit of a flat or globular receptacle, each maturing into a separate, fleshy or pulpy berry; petals thin-edged, not feltlike.   1. *Asimina*
• Pistils numerous, confluent with each other and coalescent with the elongate-conical receptacle, maturing into an aggregate fleshy fruit; petals thick-edged, feltlike.   2. *Annona*

1. Asimina (PAWPAWS)

(Species of wetlands) shrubs with stout-linear to fusiform taproots. Twigs glabrous or pubescent when young, with prominent lenticels. Flowers 1–4 in the leaf axils, enlarging with age, nodding to suberect, their stalks glabrate to tomentose. Sepals 3 or 4, equal, triangular to deltoid-ovate, nearly distinct, glabrate to persistently pubescent without. Corolla of 6 or 8 petals in two unequal series of 3 or 4 each, the inner smaller; petals veiny, spreading to erect, the inner saccate at the base and with or without an inner corrugated nectary zone. Pistils 3 to several, variously appressed-hairy, sessile on the summit of the receptacle. Fruit an oblong-cylindric (often irregular) pulpy berry, few to many seeded. Seeds bean-shaped, laterally compressed, with tough brown to chestnut brown coats.

Reference: Kral, Robert. "A Revision of *Asimina* and *Deeringothamnus* (Annonaceae)." *Brittonia* 12: 233–278. 1960.

• Flowers pleasantly fragrant, arising on wood of the previous season prior to or during the emergence of the current year's growth; outer petals 3–7 cm long, mostly more than 3, white.   1. *A. reticulata*
• Flowers with a fetid aroma, arising on current season's growth, axillary to leaves of the season; outer petals 1.5–3 cm long, pink with maroon streaks, or maroon.   2. *A. pygmaea*

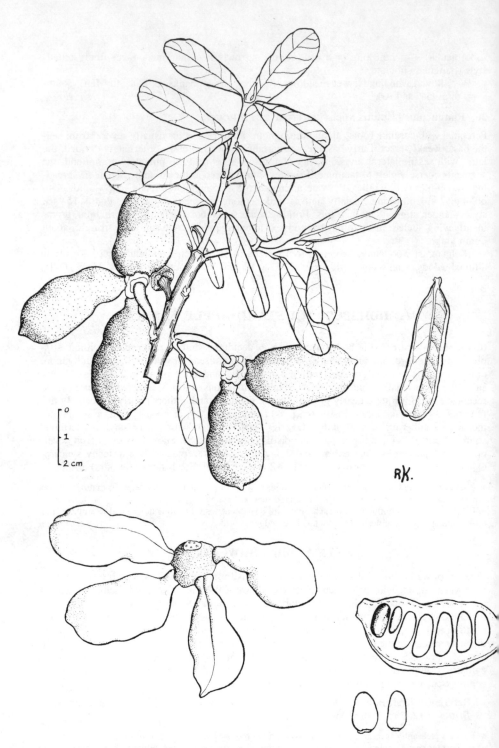

Fig. 65. **Asimina reticulata:** fruiting twig (upper center); lower leaf surface (center right); top view of receptacle with four attached fruits (lower left); longitudinal section of fruit and two seeds (lower right). (From Kral in *Brittonia* 12: 258. 1960)

1. Asimina reticulata Shuttlw. ex Chapm.   Figs. 65 and 66

Copiously and stiffly branched shrub to 1.5 m tall. Leaves leathery, oblong to elliptic or cuneate, 5–8 cm long, apically acute, obtuse, or rounded-emarginate, basally cuneate or abruptly rounded to the short petiole, margin strongly to moderately revolute; upper surface sparsely appressed-orange-pubescent, densely so on lower surface when young, aging to glabrous and pale green above, sparsely pubescent along the veins beneath and very pale green. Flowers 1–3 per node, fragrant, nodding on slender, moderately orange-hairy stalks 2–3.5 cm long, the flowers arising from the axils of prominent leaf scars on previous season's wood. Calyx 8–10 mm long, of 3 (or 4) triangular-deltoid sepals, reddish-pubescent

Fig. 66. **Asimina reticulata:** flowering twig (center); mature flowers (left and right).
(From Kral in *Brittonia* 12: 257. 1960)

149

externally, glabrous within. Outer petals 3–7 cm long, oblong to oval or obovate, wavy-margined, white, with tan or orange pubescence on the outer vein surfaces, glabrous and impressed-veiny within; inner petals ½–⅓ the length of the outer, lanceolate-hastate, revolute, fleshy, saccate-based, with the lower third of the inner face largely covered by deep purple corrugations. Pistils 3–8, fusiform, appressed-orange-hairy. Fruit 4–7 cm long, short-oblong, terete to irregularly bulging, yellow-green when ripe. Seeds 1–2 cm long, dark to pale brown, lustrous, in 2 irregular rows. (*Pityothamnus reticulatus* (Shuttlew. ex Chapm.) Small)

Seasonally wet pine savannas and flatwoods, Fla. scrub. Pen. Fla.

Fig. 67. **Asimina pygmaea:** flowering and fruiting twigs (center); flowers (upper right); upper leaf surface (lower right); longitudinal section of fruit and seeds (upper left). (From Kral in *Brittonia* 12: 263. 1960)

2. Asimina pygmaea (Bartr.) Dunal. Fig. 67

Dwarf shrub, 2–3 dm tall, from a stout, fusiform taproot, shoots 1 to several, sparingly branched or simple, reddish- to reddish brown–barked, with pale, raised lenticels, sparsely appressed-rusty-hairy distally when young, becoming glabrous with age. Leaves leathery, ascending-secund, of varying shapes, obovate to cuneate or oblanceolate, rarely linear-elliptic, 4–7 (–11) cm long, apically rounded to obtuse or emarginate, occasionally acute, bases tapering gradually to the short, twisted petiole, margins revolute; surfaces of young leaves sparsely stippled with small appressed reddish hairs, in age dark green and glabrous above, paler beneath and prominently reticulate. Flowers with a fetid aroma, nodding, secund on the side of the stem opposite the secund leaves, arising in the axils of leaves on new shoots of the season. Outer petals 1.5–3 cm long, oblong to ovate-lanceolate, pink with maroon streaks, or maroon, fleshy, margins revolute, tips reflexed; inner petals ⅓–⅔ the length of the outer, deep maroon, ovate-acute to lance-ovate, fleshy and with a saccate base which on the inside is densely corrugated, margins revolute, tips recurved. Pistils 2–5, narrowly fusiform. Fruit curved, oblong-cylindric, 3–4 (–5) cm long, yellowish green when ripe. Seeds about 1 cm long, brown, shiny, in 2 irregular rows. (*Pityothamnus pygmaeus* (Bartr.) Small)

Pine savannas and flatwoods, old fields and roadsides. s.e. Ga. to cen. pen. Fla.

2. Annona (CUSTARD-APPLES)

Annona glabra L. POND-APPLE. Fig. 68

Tree, attaining heights of up to 14 m, the base of the trunk buttressed; twigs glabrous. Leaves glabrous, short-petiolate, blades broadly oblong-elliptic or oval, 7–15 cm long; up to 6 cm broad, bases rounded to broadly short-cuneate, apices acute to shortly acuminate. Flowers solitary, usually borne on the internode, stalked, the stalks usually curved and dilated distally. Sepals 3, broader than long, reniform, 4–5 mm long, about 10 mm broad. Outer 3 petals creamy white, with a crimson spot at the base within, 2.5–3 cm long and 2–2.5 cm broad, ovate; inner 3 petals whitish without, dark crimson within, a little smaller than the outer. Fruit up to 12 cm long, 8 cm broad, exteriorly yellowish at maturity, pulp pinkish orange, rather dry, pungent-aromatic. Seeds light brown, about 1.5 cm long and 1 cm broad.

Swamps and ponds. s. Fla. (where no longer as common and abundant as it was formerly); trop. Am.; trop. w. Afr.

Illiciaceae (ANISE-TREE FAMILY)

Illicium floridanum Ellis. FLORIDA ANISE-TREE, STINKBUSH. Fig. 69

Shrub or small tree whose foliage and flowers emit a distinctive odor. Leaves evergreen, simple, glabrous, alternate (sometimes so closely approximate on distal portions of twigs as to give the impression of whorled), blades long-elliptic or lance-elliptic, mostly 8–15 cm long, narrowed below to short, narrowly winged petioles, margins entire, apices acute or short-acuminate, rarely blunt, glandular-dotted below. Flowers borne singly in the leaf axils on stalks 1–4 cm long, the stalks usually somewhat pendent so that the flowers are to a greater or lesser extent hidden below the foliage. Sepals 3–6, deciduous as the buds unfold, 8–10 mm long, roughly oblong, thin and their margins hyaline, obtuse or rounded apically, the upper ½–⅔ of the margins fringed with numerous minute pale cilia. Corolla showy and attractive, deep maroon, comprised of about 20–30 radially spreading, straplike petals in several series. Stamens 30–40, erect, in about 3 series, filaments colored as the petals, anthers darker, pollen white. Pistils 10–20, superior, in a single compact circle, the ovary of each laterally flattened, roughly ellipsoid, style stigmatic along most of one side; ovary 1-

Fig. 68. **Annona glabra:** flowering twig (above); fruiting twig (below). (From Little, Woodbury, and Wadsworth, *Trees of Puerto Rico and the Virgin Islands*. Vol. 2 (1974) Fig. 320.

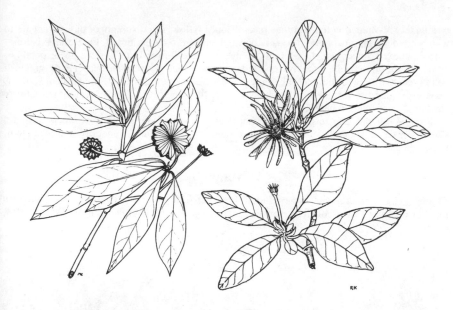

Fig. 69. **Illicium floridanum.** (From Kurz and Godfrey, *Trees of Northern Florida*. 1962)

locular, maturing into a hard follicle containing 1 seed. Seed roughly ellipsoid, somewhat flattened laterally, brown and glossy.

Lower slopes of richly wooded ravines, springy and swampy, densely wooded bottoms in ravines, swampy woodlands about streams. Fla. Panhandle from the Ochlockonee River westward, s.w. Ga., westward to La.

Magnoliaceae (MAGNOLIA FAMILY)

Shrubs or trees. Leaves simple, alternate, deciduous or evergreen, petiolate, blades pinnately veined, not toothed marginally, with stipules enclosing the bud, these when shed leaving conspicuous scars encircling the twig. Flowers solitary, terminal on the twigs, bisexual, all parts free (in ours). Perianth of 9–15 parts in whorls of threes, the outermost whorl sometimes partially differentiated as a calyx. Stamens numerous, spirally disposed on the elongated receptacle above the perianth and below the pistils. Pistils numerous, free, spirally arranged on the upper part of the receptacle, sometimes so compacted as to appear united; ovaries 1-locular, 1 or 2 ovules in each ovary. Fruits follicles or samaras, imbricated and collectively forming a sort of cone.

● Leaves about as broad as long, 4–6-lobed.   1. *Liriodendron*
● Leaves much longer than broad, not lobed.   2. *Magnolia*

1. Liriodendron

Liriodendron tulipifera L. YELLOW-POPLAR, TULIP-TREE.   Fig. 70

A tall deciduous tree, to 60 m tall and 3 m in diameter, mostly of rich, well-drained woodlands except in the southernmost parts of its range. Leaves glabrous, long-petioled, the blades about as broad as long, 4–6-lobed; upper pair of lobes, if present, forming the sides

153

of a widely **V**-shaped notch, another pair of lobes below, often sublobed; bases truncate to widely rounded or subcordate. Flowers borne in spring, 5–8 cm broad, the floral axis long, with 3 reflexed sepals which are shed by the time the flower is fully developed; petals 6, broad, erect, the corolla broadly inverted bell-like, that is, tulip-shaped. Petals greenish yellow, each with a large blotch of orange. Individual fruit a 2-seeded, indehiscent samara, these aggregated to form a spindle-shaped cone, the samaras falling separately at maturity leaving the persistent receptacle.

Rich woodlands and, especially in the southern part of the range, low woodlands, titi swamps or bays, wooded courses of small streams. Vt. to s. Mich., generally southward to n. Fla. and La.

2. Magnolia

Flowers relatively large and showy. Perianth parts 9–15, outer 3 partially differentiated (in ours) as a calyx. Pistils coherent, forming a conelike aggregate fruit comprised of follicles,

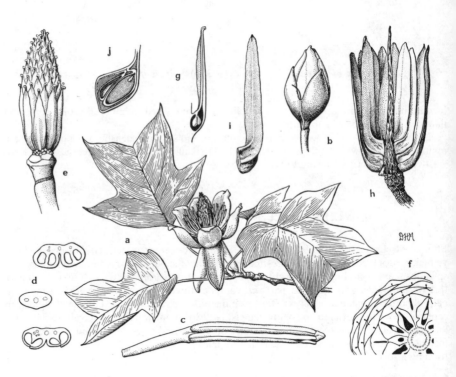

Fig. 70. **Liriodendron tulipifera:** a, flowering branchlet; b, flower bud with stipular bud scales; c, stamen, abaxial view; d, unopened anther, filament, and anther after anthesis, cross sections, pollen not shown; e, gynoecium, with sepals, petals and stamens removed from below; f, gynoecium, portion of cross section, with spirally arranged, imbricated carpels, ovaries adnate to axis to lower right, increasingly flattened styles toward outside, locules and stylar canals in black; g, carpel at anthesis, vertical section; h, mature gynoecium with many samaras already shed from axis; i, samara; j, lower part of samara, vertical section, with aborted ovule, left, and seed; d,f,g,j, semidiagrammatic. (From Wood in *Jour. Arn. Arb.* 39: 307. 1958)

each at maturity opening on the back, the seeds coming to hang from it by a slender thread for some time after dehiscence. Seeds with a fleshy scarlet to pink outer layer and a hard bony inner layer.

- Leaf blades dark lustrous green above, paler green·to rusty brown below.   1. *M. grandiflora*
- Leaf blades light green above, silvery below.   2. *M. virginiana*

1. Magnolia grandiflora L. SOUTHERN MAGNOLIA, BULLBAY.   Fig. 71

A large and handsome evergreen tree. Young twigs clothed with a brown, feltlike pubescence. Leaves short-petiolate, blades oblong, broadly elliptic, or oval, mostly 10–20 cm long or a little more, 6–10 cm wide, broadly cuneate basally, apices obtuse, rounded, or abruptly short-acuminate, upper surfaces dark, lustrous green and glabrous, lower usually rusty-tomentose, sometimes sparsely pubescent and dull green. Large, showy, very fragrant flowers appearing, a few at a time, in midspring. Petals creamy white, obovate, to spatulate,

Fig. 71. **Magnolia grandiflora.** (From Kurz and Godfrey, *Trees of Northern Florida.* 1962)

155

the corolla partially open urnlike in form, fully open the petals irregularly spreading, 15–20 cm across. Fruiting cones mostly oblong, 6–10 cm long, 5–6 cm in diameter, pubescent. Seeds bright lustrous red, 1–2 cm long.

Upland well-drained forests, ravine slopes and bottoms, floodplains where flooding is temporary, wet hammocks, often where surface water stands for considerable periods. Coastal plain, N.C. to cen. pen. Fla., westward to Ark., e. and s.e. Tex. Much cultivated for ornament both within and beyond the natural range, now becoming naturalized in woodlands in areas where not native.

2. Magnolia virginiana L. SWEETBAY.   Fig. 72

Shrubs or trees, the latter to about 25 m tall. Leaves evergreen southward, deciduous northward, short-petiolate, the blades mainly long-elliptic, oval, or oblong, 6–15 cm long and 2–6 cm wide, obtuse to acute at either extremity; upper surfaces light green and glabrous, lower surface at first pubescent, sometimes becoming glabrous and glaucous, sometimes permanently silvery-tomentose. Flowers borne in early summer, very pleasantly fragrant,

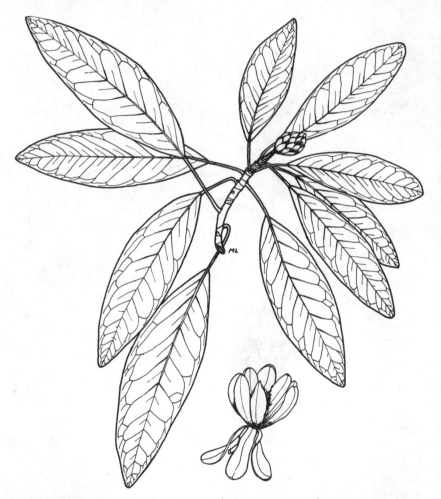

Fig. 72. **Magnolia virginiana.** (From Kurz and Godfrey, *Trees of Northern Florida*. 1962)

creamy white, 5–7 cm across when fully open. Sepals 3, reflexed as the flower opens but quickly deciduous. Petals 9–12, erect-ascending, elliptic to oblanceolate, 3–6 cm long. Individual fruits (follicles) closely compacted into an ellipsoid, oblong, or ovoid, glabrous cone 2.5–5 cm long and 2.5–3 cm in diameter. Seeds red, about 7 mm long.

Low wet woodlands, savannas, wet flatwoods, boggy wooded stream courses, common in titi bogs or swamps southward. Chiefly coastal plain, e. Mass., L.I., N.J. and Pa., southward to s. Fla., westward to e. Tex.

Cabombaceae (WATER-SHIELD FAMILY)

Rhizomatous, aquatic, perennial herbs. Flowers axillary, bisexual, radially symmetrical, with thickish stalks of variable length to 15 cm. Sepals 3 or 4, separate. Petals 3 or 4, separate. Stamens 3–20. Pistils 2–8, separate, ovaries superior, 1-locular, 1- to several-ovulate. Fruits indehiscent and nutlike, 1- to several-seeded.

- Plants with submersed leaves bisected into linear segments and with inconspicuous, simple, floating leaves (when flowering); parts not mucilaginous.   1. *Cabomba*
- Plants with all leaves floating and peltate; parts abundantly coated with mucilage.   2. *Brasenia*

1. Cabomba

Cabomba caroliniana Gray. FANWORT.   Fig. 73

Aquatic herb with numerous, elongate stems much branched from near the base, the roots numerous and fibrous. Plants submersed except for the floating leaves subtending the flowers, the flowers emergent only during full anthesis. Leaves of two kinds, those below the flowering shoot tips opposite, petiolate, the blade of each comprised of numerous, finely dichotomously dissected, linear segments palmately disposed; those at the flowering stem tips alternate, petiolate, simple, and floating, the blades elongate rhombic to linear elliptic, and peltate, mostly 1–3 cm long. Flowers mostly borne singly in the axils of the floating leaves, their stalks somewhat exceeding the subtending leaves. Perianth of 3 sepals and 3 petals, much alike in texture and color, white to pink or purplish, to about 12 mm long. Sepals obovate, narrowed abruptly to a short claw at base. Petals oval, clawed below, the blade with a yellow-spotted auricle on either side at the base. Stamens 3–6, usually 6, shorter than the petals. Carpels separate, 2–4, commonly 3, asymmetrically somewhat botlelike in form, ripening into a leathery, indehiscent fruit bearing 3 seeds.

Ponds, pools, lakes, quiet streams. Chiefly coastal plain, Va. to s. Fla., westward to e. Tex., northward in the interior, s.e. Okla. to Mo., s. Ill., w. Ky., w. Tenn.; naturalized from Va. to Mass.

Reference: Fassett, N. C. "A Monograph of *Cabomba*." *Castanea* 18: 116–128. 1953.

In the southeastern part of the range those plants with much purple pigment in both the vegetative and floral parts have been segregated as *C. pulcherrima* (Harper) Fassett = *C. caroliniana* var. *pulcherrima* Harper. We do not find characters other than color to distinguish *C. pulcherrima* from *C. caroliniana*. We have noticed locally that plants with much purple pigment are usually in quiet waters that become very warm during summer whereas those with little or no purple pigment are in relatively cool water.

2. Brasenia

Brasenia schreberi Gmel. PURPLE WEN-DOCK, WATER-SHIELD.

Aquatic herb with floating, peltate leaves. Rhizome small, creeping in substrate. Leaves alternate, very long-petioled, blades 3.5–11 cm long, broadly elliptical, the margins entire or slightly crenate; submersed leaves present only in seedlings. Submersed parts of plant and

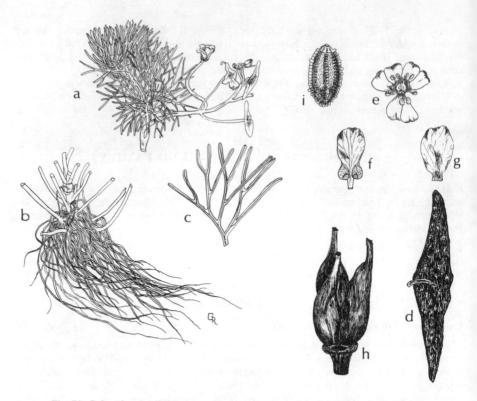

Fig. 73. **Cabomba caroliniana:** a, submersed leaves, to left, floating leaves and flowers, to right; b, base of plant; c, submersed leaf; d, floating leaf; e, flower; f, petal; g, sepal; h, carpels; i, seed.

lower leaf surfaces covered with a mucilaginous, jellylike substance. Flowers solitary, axillary, on stout stalks up to 15 cm long, emergent at anthesis. Sepals and petals 3 or 4 each, much alike in texture and color, dull purple, 12–20 mm long, linear-oblong, persistent. Stamens 18–36, filaments filiform. Carpels separate, 4 to numerous, club-shaped, tapering to linear stigmas, ripening into leathery, indehiscent, 1- or 2-seeded fruits.

Lakes, ponds, slow streams. Pr.Ed.I. to Ont., Minn., generally southward to s. Fla., Tex. and Okla.; B.C., Ore., Calif.; trop. Am.; Old World.

Nelumbonaceae

Nelumbo lutea (Willd.) Pers. YELLOW-LOTUS, WATER-CHINQUAPIN.   Fig. 74

Perennial aquatic herb with relatively spongy, long-cylindrical rhizomes about 1 cm in diameter, the leaves and flowers arising directly from the rhizome. Petioles long, stout, stiff; blades peltate, orbicular, some floating, some emergent, the larger 3–6 dm wide; blades of the floating leaves flat, those of the emergent ones somewhat elevated outward from the center, thus shallowly funnellike, the outermost portion undulating, the edge entire; upper surface dull bluish green, satiny, the lower pale and prominently veiny. Flowers solitary, showy, on long, stout, stiff stalks commonly overtopping the leaves, mostly 1–1.5 dm broad fully open. Perianth segments separate, numerous, imbricated, the outermost ovate, about 1 cm

Fig. 74. **Nelumbo lutea:** a, small portion of rhizome; b, leaf blade; c, part of undersurface of leaf; d, small section of petiole; e, flower bud; f, flower; g, stamen; h, petaloid stamen; i, section through a small portion of developing receptacle.

159

long and broad, more or less scarious-herbaceous, gradually increasing in size inwardly and becoming gradually more petaloid to about the middle of the series, these obovate, about 8 cm long and 4–5 cm broad; innermost becoming oblanceolate, some of them intergrading to petaloid stamens. Larger segments of the perianth (visible in face view of an open flower) pale yellow. Stamens numerous, their insertions in a tightly imbricate series; filaments slender, pale yellow, about 1 cm long, the anther tipped by a curved, linear, light yellow appendage 6–8 mm long. Pistils several to numerous, embedded in small pits near the top of a fleshy, spongy, obconical, flat-topped receptacle, the short styles protruding. Perianth segments and stamens deciduous leaving a series of scars below the obconical fruiting receptacle. Receptacle eventually becoming relatively hard and blackish or bluish, the flat summit with circular openings within each of which is a loosely seated, solitary, hard, indehiscent, nutlike or seedlike fruit 10–12 mm in diameter. (*N. pentapetala* (Walt.) Fern., name of uncertain application)

Ponds, lakes, sluggish streams, marshes. N.Y. and s. Ont. to Minn. and Iowa, generally southward to s. Fla., e. Okla., and e. third of Tex.

Nelumbo nucifera Gaertn., the SACRED-LOTUS, native of the Old World and cultivated for ornament, occurs locally and sporadically, spreading from cultivation, in our range. It is very much like *N. lutea* in general features, may be recognized by its pink flowers.

Nymphaeaceae (WATER-LILY FAMILY)

Aquatic, perennial herbs, rhizomatous, the long-petiolate leaves arising directly from the rhizome. Flowers borne singly from the leaf axils, with long stalks, floating or emersed, radially symmetrical, bisexual. Sepals 4 and petals numerous, the inner petals sometimes transitional to stamens, *or* sepals 6–9 (some of them petaloid) and "petals" scalelike or stamenlike. Stamens numerous. Gynoecium of several to many carpels, separate but compacted in a circle, or fully united, ovules numerous in each carpel or locule. Fruit a berry.

● Sepals and petals widely spreading at full anthesis; sepals 4, petals numerous, essentially flat and membranous; summits of the carpels extended as separate, incurved stylar projections.   1. *Nymphaea*
● Sepals 6–9, strongly concave and arched, stiff and tending to break easily; "petals" scalelike or stamenlike; summit of ovary surmounted by a disc, the stigmas sessile and radiating on the disc.

2. *Nuphar*

1. Nymphaea (WATER-LILIES)

Leaf blades ovate to orbicular, deeply cleft basally, some floating, rarely with some emersed, submersed blades often present, these thin and delicate. Flowers showy, varying from white to pink, blue, or yellow, at the water surface or somewhat emersed, perianth in most opening and closing over a period of several days, in some opening during daylight, in others during darkness; if opening several times, then the perianth enlarging somewhat from the time of first opening until the final opening. Sepals 4, greenish. Petals numerous, the inner usually transitional to stamens. Stamens numerous. Carpels compacted in a circle or fully united, sunken in a fleshy receptacle on the outer surface of which the petals and stamens are inserted, summits of the carpels prolonged into incurved stylar projections. Following final closing of the flower, the flower stalk spirals bringing the developing fruit to or near the substrate. Fruit surrounded by persistent perianth. Seeds enveloped by a membranous, saclike aril.

1. Petals bright yellow.   1. *N. mexicana*
1. Petals white, faintly pink, or blue to lavender.
  2. The petals blue to lavender (sometimes pale blue, almost white, but always with a bluish tinge which darkens upon drying).
    3. Leaf margins markedly sinuate-dentate.   2. *N. capensis*

3. Leaf margins entire or only vaguely sinuate.
    4. Leaves usually wine red beneath; upper surfaces with mound of fibrous tissue at point above junction of petiole and blade, papillose, the papillae densely and uniformly spaced over surface; petals medium to light blue.   3. *N. daubeniana*
    4. Leaves usually green beneath, upper surfaces without mound of fibrous tissue, papillose, the papillae more closely spaced over veins than elsewhere; petals usually very light blue.
      4. *N. elegans*
2. The petals white.
    5. Flowers opening nocturnally; outer surface of sepals marked by longitudinal reddish lines.
      6. Leaf blades green above and beneath, thin in texture and easily torn.   5. *N. blanda*
      6. Leaf blades green above, purple beneath (the purple pigment mainly restricted to many, short, sometimes forking lines), firm in texture.   6. *N. jamesoniana*
    5. Flowers opening during the day; outer surface of sepals without lines or spots, or marked with distinct, short, black lines.
      7. Leaf margins markedly sinuate-dentate; flowers emergent.   7. *N. ampla*
      7. Leaf margins entire or only vaguely sinuate; flowers not emergent.
        8. Petals rounded apically; flowers odorless or with scarcely perceptible fragrance.
        8. *N. tuberosa*
        8. Petals obtuse to subacute apically; flowers strongly fragrant.   9. *N. odorata*

1. Nymphaea mexicana Zucc. YELLOW WATER-LILY.   Fig. 75

Rhizome usually short and erect, bearing numerous fibrous roots, sometimes elongating to about a dm and becoming subhorizontal, perennating by very elongate, soft, spongy stolons about 1 cm in diameter, these producing new leaf- and root-bearing rhizomes from nodes at very long intervals; in autumn producing several descending, curved, fleshy, overwintering roots resembling tiny bananas. Petioles usually slender and flexuous; blades ovate-oval to orbicular in overall outline, margins entire to sinuate, mostly 8–15 cm long fully developed, some sometimes to 20 cm, the basal sinus about ⅓ of the length of the blade; green above, purple below. Sepals somewhat suffused with purplish pigment but this not in lines or dots. Corolla bright yellow, 6–10 cm across when fully developed and when fully open; petals obtuse to acute apically. Fruit about 3 cm long. Seeds grayish brown, oblong-oval, 5–6 mm long and 3 mm wide. (*N. flava* Leitn., *Castalia flava* (Leitn.) Greene)

Lakes, ponds, pools in marshes, sloughs, sluggish streams, ditches and canals. Fla. to e. and s. Tex.; Ariz.; introd. into e. N.C. and apparently persisting there; Mex.

2. Nymphaea capensis Thunb. CAPE BLUE WATER-LILY.

Leaf blades orbicular ovate in outline, 20–30 cm across, suffused with maroon beneath, margins markedly sinuate-dentate. Sepals obscurely many-nerved, without dark lines or spots externally, whitish interiorly. Petals blue to lavender, whitish basally.

Locally abundant as an escape from cultivation in sand-bottomed ditches. Native to S.Afr., naturalized in Indian River and Seminole Cos., Fla.

3. Nymphaea × daubeniana O. Thomas.

Leaf blades usually green on both surfaces; upper surfaces with a mound of fibrous tissue (rarely bearing a viviparous plantlet) at a point above the junction of petiole and blade, papillose, the papillae abundantly and uniformly spaced over surface. Sepals, at anthesis, 5–6 cm long. Petals medium to light blue.

Ditches and borrow pits. A horticultural hybrid occasionally naturalized from cultivation, mostly e. coast of Fla., notably Nassau to Brevard Cos.

4. Nymphaea elegans Hook. BLUE WATER-LILY.   Fig. 76

Rhizome short, erect, bearing fibrous roots, perennating by slender stolons late in the season, these bearing small tubers terminally. Petioles very slender, flexuous; blades oval-ovate to orbicular, relatively thin, green above, purple below, margins entire to sinuate-dentate, the sinus ½ the length of the blade or a little less, blades 8–20 cm wide. Flowers faintly fragrant. Sepals usually with dark purple streaks or dots on their outer surfaces, outer 1–4 pe-

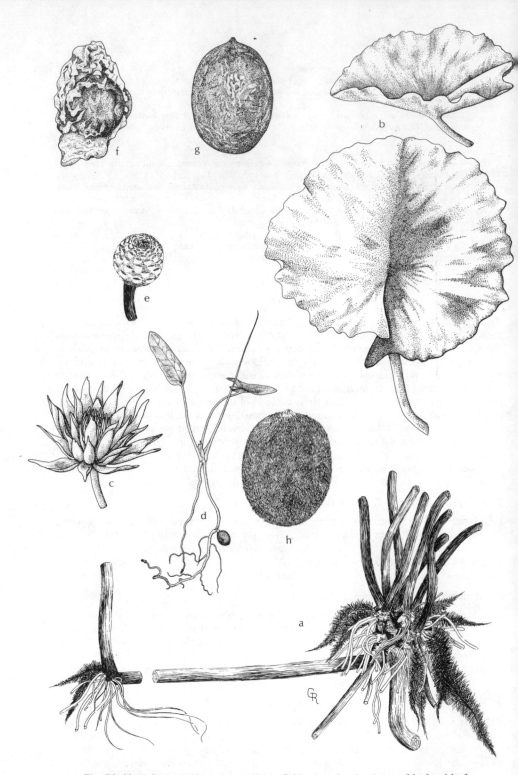

Fig. 75. **Nymphaea mexicana:** a, sections of rhizome, showing bases of leaf and leaf cluster; b, leaves; c, flower; d, seedling; e, fruit; f, developing seed with loosely enveloping membranous covering (aril); g, seed at later stage, membranous covering closely enveloping; h, seed, fully developed, membranous covering having taken the form of a dense pubescence.

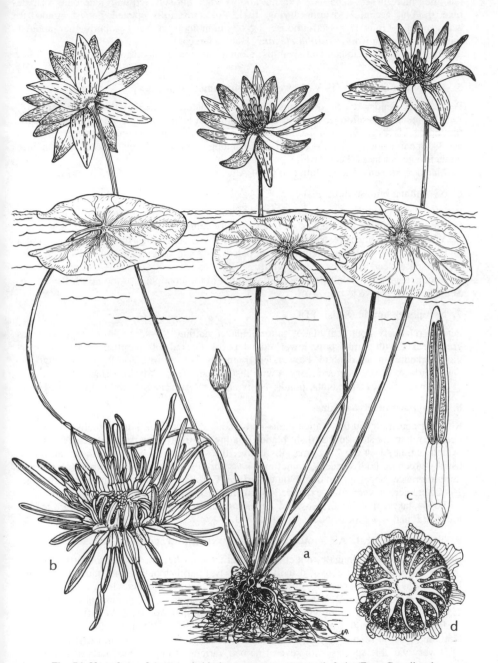

Fig. 76. **Nymphaea elegans:** a, habit; b, stamens; c, stamen; d, fruit. (From Correll and Correll. 1972)

tals often more or less sepallike. Corolla almost white but suffused with pale blue, or pale blue, the blue accentuated upon drying, 6–12 cm across fully opened; petals rounded to acute apically. Fruits depressed-globose, 1.5–3 cm in diameter. Seeds suborbicular, about 1 mm in diameter, brown. (*Castalia elegans* (Hook.) Greene)

Ponds, pools in marshes and swamps, sloughs, drainage ditches and canals. Pen. Fla.; La. and s. Tex.; Mex.

5. Nymphaea blanda G. F. W. Meyer var. fenzliana (Lehm.) Casp.

Petioles very slender, flaccid, glabrous; leaf blades orbicular-ovate in outline, 6–10 cm long, rounded apically, basal lobes blunt at their tips, thin in texture and easily torn, margins entire, both surfaces green, the upper minutely papillose and with short lines. Flowers 8–10 cm across, opening nocturnally. Sepals 3.5–4.5 cm long, with fine, closely spaced, reddish lines on their outer surfaces. Petals white.

Marshes, streams, lakes. Native to trop. Am., naturalized in Levy Co., pen. Fla.

6. Nymphaea jamesoniana Planch.

Petioles slender, subflaccid, glabrous; leaf blades 10–14 cm long, ovate-elliptic in outline, firm in texture, apices rounded, basal lobes broadly obtuse, upper surfaces green, lower purple (the purple pigment restricted to numerous, short, sometimes forking lines). Flowers 9–12 cm across, opening nocturnally. Sepals 1.4–2 cm long, outer surfaces with fine, closely spaced, reddish lines. Petals white.

In our range known only from a shallow pond in the floodplain of the Peach River, De-Soto Co., Fla. Native to trop. Am.

7. Nymphaea ampla (Salisb.) DC.

Petioles relatively stout; leaf blades suborbicular in outline, 15–45 cm long, upper surfaces green flecked with purple spots, lower reddish purple flecked with purplish black spots, margins markedly sinuate-dentate. Flowers emergent, 4–7 cm across, opening during daylight. Outer surfaces of sepals green marked with distinct, short, black lines. Petals white.

Drainage ditches and canals, ponds. Native to trop. Am., naturalized in Lee Co., Fla.

8. Nymphaea tuberosa Paine.

Rhizome constricted at the branch joints and readily disarticulating into tuberlike segments. Petioles relatively stout and stiffish; blades floating or slightly emersed, mostly orbicular in overall outline, 1–4 dm long, margins entire, the sinus about ⅓ the length of the blade, usually green on both surfaces, sometimes pale purplish below. Flowers odorless or with a faint fragrance. Sepals green on the outer surface. Corolla white or with a pale pinkish tinge, 10–20 cm across when fully expanded; petals rounded apically.

Ponds, lakes, sluggish streams. s.w. Que. and Ont. to Minn. and Neb., southward to Ky., n.e. Okla. and Ark.; local, N.Y., Vt., Pa., N.J.

9. Nymphaea odorata Ait. FRAGRANT WHITE WATER-LILY, POND-LILY.   Fig. 77

Rhizome horizontal, elongate, not constricted at the branches. Leaf blades mostly orbicular in overall outline, green above, usually (not always) purplish below, the sinus about ⅓ the length of the blade. Flowers very fragrant, usually floating. Corolla white, rarely pinkish. Fruit depressed-globose, 2.5–5 cm in diameter. Seed oblong-oval, grayish olive to orange, 1.5–2.0 mm long and 1 mm wide. (*Castalia odorata* (Ait.) Woodv. & Wood)

Ponds, lakes, sloughs, pools in marshes, sluggish streams, ditches and canals, swamps. Nfld. and Que. to Man., generally southward to s. Fla., Okla., e. Tex. and Ariz.

A very variable species with several named varieties from various parts of the overall range, two of which, var. *odorata* and var. *gigantea* Tricker (*N. lekophylla* (Small) Cory) have, historically, been attributed to our range. Plants of this complex, as we know them, exhibit diversity in size of rhizome, in diameter and texture of petioles and flower stalks, in size and texture of leaf blades, in size of flowers. This is further complicated because, as in

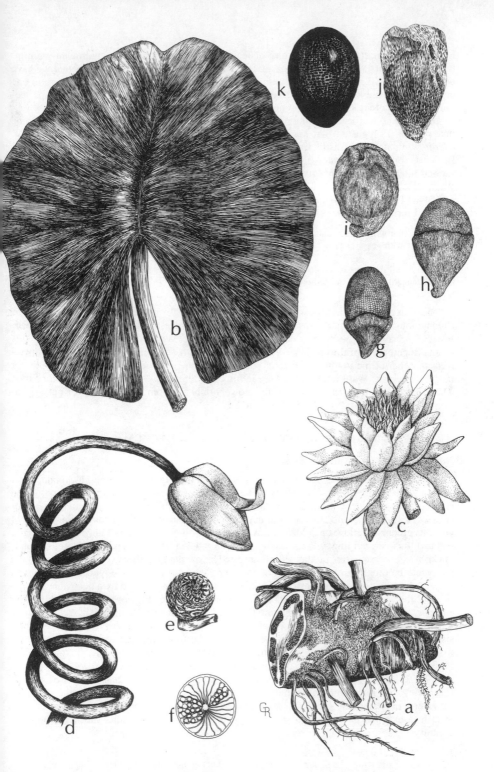

Fig. 77. **Nymphaea odorata:** a, small portion of rhizome; b, leaf blade; c, flower; d, portion of peduncle spiraled after fertilization of flower; e, fruit; f, diagrammatic cross-section of fruit; g–j, stages in development of aril around seed; k, seed.

water-lilies generally, size and texture of leaves and size of flowers is significantly diminished when plants are growing under adverse conditions. The result, for us, is that we have been unable satisfactorily to assign many specimens, either in the field or herbarium, to one or the other of the aforementioned varieties as described by authors. This is, perhaps, because of insufficient knowledge on our part.

Given the above, it is with not a little embarrassment that the senior author has for some time privately suggested that the Florida Panhandle harbors a white water-lily akin to *N. odorata* with attributes distinctive enough to warrant some recognition. Recently, Dr. Daniel B. Ward, focusing on water-lilies as a part of his ongoing studies of the flora of Florida, showed his concurrence by giving the plant the name *N. odorata* var. *godfreyi* Ward (see Ward, "Keys to the Flora of Florida — 4. *Nymphaea* (Nymphaeaceae)," *Phytologia* 37: 443–448 (1977)). He gives the following sets of contrasting characters for the white water-lilies of Florida:

● Flower often large, the sepals to 6–10 cm long; leaves medium to large (to 45 cm broad), maroon beneath but with veins usually greenish; petioles and peduncles thick, spongy; rhizome stout (2–4 cm thick). *N. odorata* var. *gigantea*

● Flower small, the sepals 3.5–6 cm long; leaves small to medium (8–20 cm broad), uniformly dark maroon beneath; petioles and peduncles thin, wiry; rhizome slender (1–2 cm thick).

N. odorata var. *godfreyi*

The above implies that the varieties *gigantea* and *godfreyi* are distinguishable from the tautonymous var. *odorata*; however Ward does not contrast either of them with the var. *odorata*, presumably because he does not consider it to be a part of the flora of Florida. This is not altogether helpful to us since we were unable with any definitiveness to assign specimens to either of the two varieties (var. *odorata* and var. *gigantea*) usually attributed to our range. Moreover, Ward does not include in his considerations *N. odorata* var. *minor* Sims. which, according to Gleason (1952), has flowers only 5–9 cm wide and fewer petals, stamens, and stigmas, and occurs from N.J. to La.

2. Nuphar (SPATTER-DOCKS, YELLOW COW-LILIES)

Nuphar luteum (L.) Sibth. & Sm.

Leaf blades broadly ovate, oval-ovate, orbicular, lanceolate or oblong-lanceolate, cleft basally, some floating, some sometimes emergent, some sometimes submersed, submersed blades, if present, very thin, flaccid, and usually crisped. Flowers at the water surface or emersed, not opening and closing from day to day. Sepals 6–9, some petaloid, varying from green to greenish yellow or golden yellow, rarely red-tinged, strongly concave and arched, stiff and thickish, tending to break easily, especially when dried. "Petals" numerous, small, thickish, yellow, scalelike or stamenlike, inserted below the numerous yellow stamens on the receptacle below the ovary, distally not extending beyond the ovary. Gynoecium of united carpels, thick, more or less constricted at the summit and surmounted by a persistent stigmatic disc, the separate, sessile stigmas radiating on its surface. Flower stalk not coiling after anthesis, the fruit thus maturing essentially in place. Fruit ovoid, greenish, yellowish- or reddish-tinged, usually with some or all of the perianth parts and shrivelled stamens persistent around it. Seeds numerous in each locule of the fruit, not arillate.

The most recent comprehensive treatment of *Nuphar* is Ernest O. Beal, "Taxonomic Revision of the Genus *Nuphar* Sm. of North America and Europe," *Jour. Elisha Mitchell Sci. Soc.* 72: 317–346 (1956). Beal recognizes but one species, *N. luteum*, comprised of nine subspecies, five occurring within our range. He professes that in areas where the ranges of subspecies overlap many specimens are intermediate between the overlapping subspecies and in such cases a decision to assign an intermediate plant to one or another subspecies must be arbitrary.

1. Leaf blades 2–3 times as long as wide or more.
  2. Sepals 6; stigmatic rays linear; leaf blades more than 3 times as long as wide.

1. *N.* subsp. *sagittifolium*

166

2. Sepals 6–9; stigmatic rays narrowly elliptic to oblanceolate; leaf blades not over 2½ times as long as wide. **2.** *N.* subsp. *ulvaceum*

1. Leaf blades less than twice as long as wide.

3. Upper and lower leaf surfaces both roughened-papillose, lower surface glabrous or sparsely pubescent. **3.** *N.* subsp. *macrophyllum*

3. Upper leaf surface roughened-papillose, lower densely pubescent (and, if papillose, the papillae hidden beneath the pubescence). **4.** *N.* subsp. *orbiculatum*

1. Nuphar luteum subsp. sagittifolium (Walt.) Beal.

Leaves with blades both submersed and floating; floating blades more than 3 times as long as wide, lanceolate to oblong-lanceolate or oblong, 1.5–3 dm long and 5–10 cm wide, firm, flat, the basal sinus usually not over 3.5 cm deep, surfaces glabrous and more or less papillose; submersed blades very thin, flaccid, crisped, about the same shape as the floating ones but usually larger. Flowers 2–3 cm across. Sepals 6, the outer oblong, the inner suborbicular. Stigmatic lines linear, 10–14 in number. (*Nuphar sagittifolium* (Walt.) Pursh, *Nymphaea sagittifolia* Walt.)

Rivers, estuaries, bayous, sloughs. Coastal plain, s.e. Va., e. N.C., n.e. S.C.

2. Nuphar luteum subsp. ulvaceum (Miller & Standl.) Beal. Fig. 78

Leaves with both submersed and floating blades, commonly fewer blades floating than submersed; floating blades ovate-lanceolate to nearly ovate, mostly 1½–2½ times as long as wide, 1.5–2 dm long and 8–10 cm wide, firm, flat, basal sinus ¼ the length of the blade or less, both surfaces glabrous and papillose; submersed leaves very thin, flaccid, crisped, similar in shape to the floating, usually larger, often twice as large or a little more. Flowers mostly 3 cm across. Sepals 6–9, irregular in shape, the inner usually short-clawed basally. Stigmatic rays narrowly elliptic to oblanceolate, usually 9–12 in number. (*Nymphaea ulvacea* Miller & Standl.)

Fresh waters of rivers and streams, mostly "black" waters. Fla. Panhandle, w. of the Apalachicola River, and s. Ala. (probably also s. Miss.)

3. Nuphar luteum subsp. macrophyllum (Small) Beal. Fig. 79

Leaf blades most commonly both floating and erect-emergent, sometimes all erect-emergent, sometimes all floating, usually closely similar in shape and texture whether floating or emergent; occasionally with what appear to be fully developed submersed leaves, the blades similar in shape to others, but very thin, flaccid, and crisped; blades mostly broadly ovate, the sinuses about ⅓ the length of the blades, variable in size, mostly 2.5–4 dm long and 1.5–3 dm wide; both surfaces roughened-papillose, the upper glabrous, lower usually glabrous, sometimes sparsely pubescent. Flowers about 2.5–3 cm across. Sepals 6, the outer broadly oblong, the inner obovate to oblate, commonly broader than long, short-clawed basally. Stigmatic rays linear or lanceolate, mostly 10–18 in number, sometimes fewer or more. (*Nymphaea macrophylla* Small, *Nymphaea advena* Ait., *Nuphar advena* (Ait.) Ait. f.; incl. *Nymphaea fluviatilis* Harper, *Nymphaea chartacea* Miller & Standl.)

Ponds, lakes, sluggish streams or backwaters, pools in marshes and swamps, ditches and canals. s. Maine to s. Wisc., generally southward to s. Fla. and Tex.; n.e. Mex., Cuba.

Nuphar luteum subsp. *ozarkanum* (Miller & Standl.) Beal is, according to Beal, questionably distinguishable from subsp. *macrophyllum*. It occurs in the Ozark region of Mo. and n.w. Ark., has sepals red-tinged within, and fruit variously red-tinged.

4. Nuphar luteum subsp. orbiculatum (Small) Beal.

Exposed leaf blades all floating, firm, flat; usually with some submersed leaves approximately of the same shape and size as the floating ones, but very thin, flaccid, and crisped; blades orbicular to broadly oblong-ovate, variable in size, mostly from about 1.5–4.5 dm long and wide, upper surface glabrous and roughened-papillose, lower densely pubescent. Flowers 4–6 cm across. Sepals 6, the outer broadly oblong or ovate-oblong, inner very broadly obovate, usually wider than long, short-clawed basally. Stigmatic rays linear or lan-

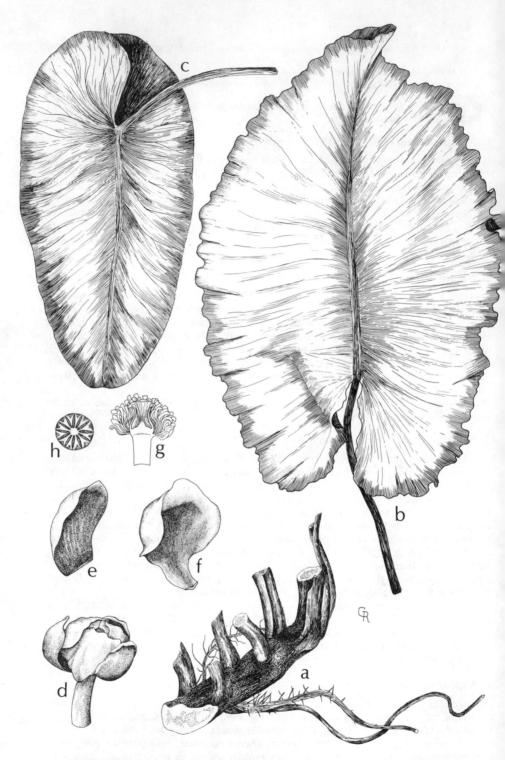

Fig. 78. **Nuphar luteum** subsp. **ulvaceum:** a, section of rhizome; b, blade of submersed leaf; c, blade of floating leaf; d, flower; e,f, sepals; g, longitudinal section through flower, calyx removed; h, summit of fruit, from above.

Fig. 79. **Nuphar luteum** subsp. **macrophyllum:** a, habit; b, flower; c, stamens; d, fruit, sectioned longitudinally. (From Correll and Correll. 1972)

ceolate, mostly 12–28 in number. (*Nymphaea orbiculata* Small; incl. *Nymphaea bombacina* Miller & Standl.)

Acid ponds and pools, pools in acid swamps, "black" waters of rivers and estuaries, ditches and canals. s.cen. and s.w. Ga., n.cen. Fla.

Cruciferae (MUSTARD FAMILY)

Annual, biennial, or perennial herbs with an acrid or pungent watery sap. Leaves alternate (in some approximate and appearing opposite or whorled), entire to lobed or pinnately divided, without stipules. Flowers bisexual, mostly radially symmetrical, without bracts, borne in racemes (infrequently solitary). Sepals 4, not persistent, usually oblong, erect, and more or less appressed to the corolla, sometimes spreading at anthesis. Petals 4 (rarely absent), yellow, white, or lavender. Stamens usually 6 (rarely fewer or more), in 2 series, the outer 2 shorter than the inner 4, the outer (and usually the inner) subtended by minute glands. Ovary superior, 2-locular, with 1 to many ovules in each locule, or less frequently 1-locular and with a single ovule. Fruit in most genera a silique, that is, a many-seeded capsule with 2 parietal placentae and 2 valves that separate from the septum on dehiscence, varying, however, to indehiscent, or 1-locular, or transversely septate.

There are numerous introduced, weedy members of this family occurring in our range which are seldom encountered in undisturbed natural vegetation. If they have no congeners native to the area that occur in other than disturbed places, we have omitted them from this account of wetland plants. Otherwise the account would tend to be nearly as inclusive (for the family) as manuals treating the total cruciferous flora for our range.

1. Fruit a silique, that is, a many-seeded capsule with 2 parietal placentae and 2 valves that separate from the septum on dehiscence.
  2. Petals yellow. 1. *Rorippa*
  2. Petals white, pinkish, or lavender.
    3. Leaves palmately parted into 2 or 3 divisions. 2. *Dentaria*
    3. Leaves simple or pinnately divided or compound.
      4. Plant aquatic; submersed leaves divided into filiform or capillary segments. 3. *Armoracia*
      4. Plant aquatic or terrestrial, if aquatic then the leaf divisions not filiform or capillary.
        5. Lower leaves pinnatifid below and with a large terminal lobe, thus essentially lyrate-pinnatifid; upper leaves ovate, lance-elliptic, to lanceolate, margins serrate. 4. *Iodanthus*
        5. Leaves not as described above.
          6. Plant aquatic, abundantly rooting at the nodes; silique terete or nearly so. 5. *Nasturtium*
          6. Plant terrestrial, or if sometimes aquatic then not rooting at the nodes; silique strongly flattened. 6. *Cardamine*
1. Fruit indehiscent, 2-jointed with transverse septae and breaking apart at the septae at maturity, 1 seed in each indehiscent joint. 7. *Cakile*

1. Rorippa (YELLOW CRESSES)

Annual (rarely biennial) herbs with a slender to thick vertical taproot, or perennial from spreading slender to thick subterranean stems and roots. Stems simple, or often several from the base, to 7 dm tall, infrequently taller, sometimes with a rosette of basal leaves. Stems and leaves glabrous or sparingly pubescent; if pubescent the hairs in some species vesicular, that is small, bladderlike and containing fluid, these when dry sometimes appearing scaly or as a whitish stubble. Leaves varying from simple with undulate or undulate-dentate margins to variously pinnately lobed, cleft, or divided. Racemes (or spikes) terminal, axillary, or terminal and axillary on the branches, blooming over a considerable period, the flowers small and closely set at the tips of the racemes, the raceme axes elongating as the fruits mature from the base upward. Sepals about 0.5 mm long, ascending or somewhat spreading at anthesis, green, often more or less suffused with purple, their margins narrowly hyaline, becoming yellowish or brownish in age. Petals (absent in some) yellow, sometimes whitish upon dry-

170

ing, from somewhat shorter than to a little longer than the sepals, usually clawed, their blades erect or spreading between the sepals. Stamens 6, the 2 shorter ones flanked by a pair of minute glands, the longer ones separated by a short, conic gland. Ovary cylindrical, style short and thick, stigmas capitate, persisting on the fruit. Seeds numerous.

Some species of the genus very variable, with several to numerous subspecies and varieties of the subspecies recognized.

Reference: Stuckey, Ronald L. "Taxonomy and Distribution of the Genus *Rorippa* (Cruciferae) in North America." *Sida* 4: 279–430. 1972.

1. Petals absent; fruits sessile or their stalks little if any exceeding 1 (–2) mm long.   1. *R. sessiliflora*
1. Petals present; fruits stalked, the stalks usually 4 mm long or more.
  2. Plant perennial; petals 4 mm long or more.
    3. Principal leaves of the stem pinnately divided, the sinuses reaching the midrib; upper sessile leaves without auricles.   2. *R. sylvestris*
    3. Principal leaves of the stem toothed to divided, if divided the sinuses not reaching the midrib; upper sessile leaves, at least, with auricled bases.   3. *R. sinuata*
  2. Plant annual; petals 3 mm long or less.
    4. Leaves variously toothed, lobed, or divided, if divided the sinuses not reaching the midribs and the leaves not at all giving the appearance of being bipinnatifid; racemes terminal and lateral; pubescence of stem or leaves, if any, of pointed hairs.   4. *R. palustris*
    4. Leaves pinnately divided, the sinuses reaching the midrib, the segments themselves markedly toothed or lobed, the leaves thus appearing almost bipinnatifid; racemes lateral; pubescence (usually sparsely and irregularly distributed on the lower stems and upper leaf surfaces) of small vesicular trichomes.   5. *R. teres*

1. Rorippa sessiliflora (Nutt. in T. & G.) Hitchc.

Glabrous annual or biennial, usually with a single principal stem. Lower and median leaves, at least, with decurrent-winged petioles, these not dilated-auriculate proximally, blades entire, or crenate, sometimes with oblong lobes with rounded apices on their lower halves, often somewhat crisped, in overall outline ovate, oblong, oblanceolate to obovate, or spatulate, variable in size, (2–) 3–10 (–12) cm long. Petals absent (rarely a single petal in 1 or 2 flowers on a single plant). Fruits sessile or with stalks not usually more than 1 mm long, the body cylindric or slightly compressed-cylindric, straight or only very slightly curved, 5–10 mm long and 2 mm wide. Seeds cordiform or somewhat kidney-shaped, about 0.6 mm wide, a bit wider than long, their surfaces finely alveolate-reticulate, buffish in color. (*Radicula sessiliflora* (Nutt. in T. & G.) Greene)

Gravel and sand bars of rivers, muddy banks of streams, floodplain forests, bottomland clearings and fields, wet exposed or marshy shores of ponds, lakes. Md. to Ohio, Ind., Ill., s. Wisc., Iowa, Neb., generally southward to n. Fla. and Tex.

2. Rorippa sylvestris (L.) Besser.

Rhizomatous perennial and colonial, stems to 5 dm tall, slender, often weak, glabrous or finely short-pubescent. Basal and lower leaves with slender petioles, the petioles little dilated proximally and not clasping, upper leaves sessile, not clasping, blades pinnatifid, the sinuses usually reaching the midrib, segments relatively narrow and irregularly toothed or lobed; in overall outline mostly oblong. Petals spatulate or spatulate-obovate, about 4 mm long, longer than the sepals. Fruiting stalks 5–10 mm long, very slender, widely divergent, ascending, or reflexed; fruits linear-falcate, sometimes nearly straight, 5–15 mm long, most of them about equaling the stalks, apparently rarely maturing seeds. (*Radicula sylvestris* (L.) Druce)

Gravels, sands, and muds along rivers and streams, bottomland fields, wet clearings, ditches. Native of Euras., naturalized principally in e. Can., n.e. and n.cen. U.S., sporadically elsewhere, in our range apparently chiefly along the Mississippi River.

3. Rorippa sinuata (Nutt. in T. & G.) Hitchc.

Perennial, usually several-branched from near the base, branches erect or decumbent, 1–4 dm long. Stem and leaves rather sparsely and irregularly clothed with nearly hemispherical, whitish, vesicular trichomes, these appearing scalelike when dry. Leaves strongly dentate-lobed, sinuate, or pinnatifid, if the latter the sinuses not reaching completely to the midrib, the segments entire to irregularly toothed; leaves in overall outline mostly oblanceolate, sometimes nearly oblong or lanceolate; median stem leaves, at least, with the narrow bases abruptly dilated-auriculate proximally and somewhat clasping the stem. Petals not clawed, oblong, oblanceolate, or spatulate, 3.5–5.5 mm long, longer than the sepals. Stalks of the fruits slender, mostly about 10 mm long, straight to variously curved; fruits cylindric, mostly 6–12 mm long, upwardly curved. Seeds irregularly angular, a little over 0.5 mm long.

River gravels and sands, wet river and stream banks, exposed and marshy shores of ponds and lakes, low fields, ditches. Ill., Minn., Mo., Ark., Okla., n.w. Tex., generally westward to Wash. and Calif.; s. Sask. and Alb.

4. Rorippa palustris (L.) Besser.

Annual or biennial, with a taproot. Stems single, or less commonly with several from the base, rarely decumbent, glabrous throughout or sparsely to densely hirsute (the trichomes pointed, not vesicular), to 14 dm tall. Basal and lower leaves usually with winged petioles, the petioles usually dilated proximally and auriculate, somewhat clasping, upper leaves sessile, their bases usually auriculate and somewhat clasping; blades glabrous or pubescent, varying from irregularly serrate to cleft, incised, or varyingly pinnately divided, the sinuses mostly not reaching the midrib, in overall outline nearly oblong to broadly oblanceolate. Racemes axillary and terminal, variable in length from 3–20 cm. Petals oblong to broadly spatulate, not clawed, shorter than, equaling, or slightly longer than the sepals, from about 1 to 3 mm long. Stamens 6. Fruiting stalks 3–10 mm long, divaricate, ascending, or reflexed; mature fruits ovate, oblong, or elliptic, 2–14 mm long, varying from as wide as long to about 8 times as long as wide. Seeds cordiform, reddish brown, mostly 0.5–0.9 mm long, surfaces finely pebbled or alveolate. (*Radicula palustris* (L.) Moench.; *Rorippa islandica* (L.) Borbas, name misapplied by authors to N.Am. plants)

Marshes, bogs, seepage and springy areas, sand, gravel, and mud banks of streams, floodplain forests, wet woodlands, wet clearings. Widely distributed in N.Am.; Eur.

A variable species for which Stuckey (1972) recognized four subspecies as well as varieties of the subspecies. Plants in our range are referable to *Rorippa palustris* subsp. *glabra* (O. E. Schulz) Stuckey var. *fernaldiana* (Butters & Abbe) Stuckey.

5. Rorippa teres (Michx.) Stuckey var. **teres.**   Fig. 80

Annual or biennial. Stems solitary and branched, or several from near the base, sometimes decumbent, 1–4 dm tall. In ours, lower stems and upper leaf surfaces with sparse and irregularly distributed, small vesicular trichomes. Lower leaves petiolate, upper sessile, pinnatifid, the sinuses reaching the midrib, the lobes themselves markedly but irregularly toothed or lobed, the leaves thus appearing almost bipinnatifid; leaf outline overall nearly oblong to very broadly lanceolate. Leafy branches, often even those low on the stem, bearing racemes, other racemes from axils progressively upward, no racemes truly terminal. Flower stalks short, mostly becoming only 2–4 mm long as fruiting stalks. Petals spatulate, minute, 1 mm long or a little more, shorter than, equaling, or slightly exceeding the sepals. Fruits linear-oblong, 1–2 cm long, slightly falcate, mostly ascending from the stalks. Seeds cordiform, tiny, reddish brown, surfaces very finely alveolate-reticulate, 0.4–0.5 mm in diameter. (*Rorippa obtusa* (Nutt. in T. & G.) Britt.; *Rorippa walteri* (Ell.) Small; *Radicula walteri* (Ell.) Greene)

Cypress-gum ponds, exposed or marshy shores of ponds and lakes, floodplain woodlands, wet clearings, wet banks of rivers and streams, moist to wet cultivated lands, ditches. Coastal plain, s.e. N.C. to s. Fla., westward to Okla., e. and s.w. Tex.

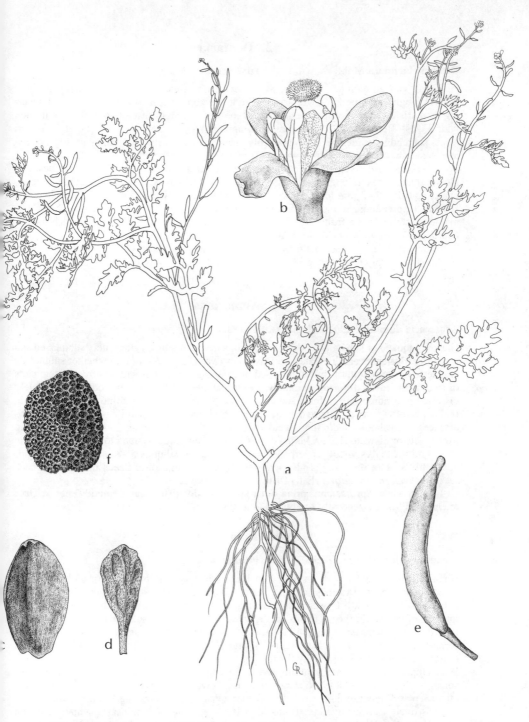

Fig. 80. **Rorippa teres:** a, habit; b, flower; c, sepal; d, petal; e, fruit (silique); f, seed.

2. Dentaria

Dentaria laciniata Muhl. ex Willd. TOOTHWORT.

Perennial with an elongate, jointed, horizontal, white rhizome with a pleasantly pungent taste. A single, long-petioled leaf arises from the rhizome but is absent at anthesis. Flowering stem arising from the rhizome, naked for most of its length, usually for 15–20 cm, then with 3, less frequently 2, short-petiolate, approximate, subopposite, or falsely whorled leaves subtending a raceme. Leaves deeply ternately divided, the segments mostly lanceolate, sometimes linear in outline, varying from nearly entire to laciniately toothed or divided again. Racemes relatively short, elongating somewhat as fruits mature, the axis pubescent. Flower stalks 1–2 cm long, fruiting stalks to 3 cm long. Sepals 4, erect, essentially oblong, obtuse apically, 5–8 mm long. Petals 4, 10–20 mm long, white to purplish, spatulate, ascending to spreading. Stamens 6. Silique narrowly cylindric, body 2–3 cm long, beak 5–8 mm long, stalks of the fruits 1.5–2.5 cm long, ascending. Seeds in 1 rank in each locule. (*Cardamine concatenata* (Michx.) Ahles)

Floodplain forests, creek bottoms and banks, mesic woodlands. Que. to Minn., generally southeastward from Que. to the Fla. Panhandle, and from Minn. to Neb., Kan., Okla., and La.

3. Armoracia

Armoracia aquatica (Eat.) Wieg. LAKE CRESS. Fig. 81

Glabrous aquatic or subaquatic, commonly submersed, sometimes alternately submersed and emersed as water levels fluctuate, the leaves very variable in relation to whether submersed or emersed or to length of time submersed or emersed, thus often with alternating segments of stem with leaves differing markedly. Leaves submersed for long periods 1–3-pinnately dissected into numerous filiform segments; other leaves oblong to elliptic or lanceolate in outline, varying from serrate to dentate to pinnatifid, mostly 3–7 cm long. Racemes of variable lengths to about 1.5 dm long, weak, often curved or bent. Sepals 3–4 mm long, elliptic to spatulate or obovate. Petals white, short-clawed, their blades oblong to obovate, 6–8 mm long. Fruiting stalks widely divergent, about 1 cm long; siliques 5–8 mm long, ellipsoid, apparently seldom maturing seeds; persistent style 2–4 mm long. Seeds in 2 rows in each locule. (*Neobeckia aquatica* (Eat.) Britt.)

In quiet water, springs and spring runs, lakes and sluggish streams, muddy shores. Que. to Ont. and Minn., generally southward to n. Fla., e. Okla. and e. Tex.

4. Iodanthus

Iodanthus pinnatifidus (Michx.) Steud. PURPLE ROCKET.

Perennial herb, usually with a single principal stem, this little branched or with several to numerous branches, racemes terminal on the branches. Stem leafy, glabrous or sparsely pubescent, 3–8 dm tall. Lowermost leaves with a short, winged petiole, blade pinnatifid the lower ⅔ of its length, the segments small, divergent or slightly ascending, increasing a little in size distally, with a very much larger, ovate or broadly lanceolate terminal lobe crenate-serrate on its margin, the shape essentially lyrate-pinnatifid, overall 10–15 cm long; median to upper leaves simple, gradually decreasing in size upwardly and becoming sessile, ovate, lance-elliptic, to lanceolate, margins serrate, mostly tapering to both extremities, broadest somewhat below the middle. Racemes elongate, flowers mostly relatively distant from each other. Sepals 4, erect, oblong to elliptic, obtuse apically, 3–5 mm long, usually purplish, the margins often scarious-hyaline. Petals 4, pale violet, lavender, or nearly white, spatulate with a long tapering base, the spatulate tips spreading at anthesis, 7–14 mm long. Stamens

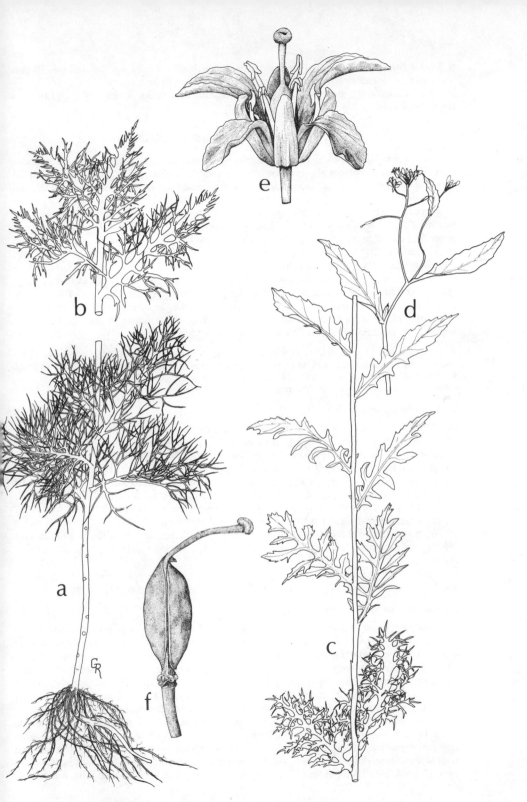

Fig. 81. **Armoracia aquatica:** a, basal portion of stem; b, midsectional portion of stem; c, upper midsectional portion of stem; d, flowering branch; e, flower; f, fruit.

6. Silique linear, straight, nearly terete, surface granular, 2–4 cm long, the stalks 4–10 mm long, widely spreading to ascending, seeds in 1 rank in each locule.

Floodplain forests, bottomland fields, river shores and banks. w. Pa., W.Va. to Ill. and Iowa, southward to Ala. and Tex.

5. Nasturtium (WATER CRESSES)

Aquatic or paludal, glabrous, perennial herbs with somewhat succulent stems. Leaves odd-pinnately divided to essentially compound, or undivided. Flowers in racemes terminating the branches, blooming over a considerable period of time, the flowers closely clustered and short-stalked at the tips of the racemes; raceme axes and stalks elongating considerably as the fruits mature. Sepals erect-ascending, the 2 inner ones saclike at the base. Petals white or whitish suffused with purple, clawed basally, blades obovate to oblanceolate. Shorter 2 stamens flanked by horseshoe-shaped, nectariferous glands. Ovary cylindric, style short and stout, scarcely differentiated, stigma capitate, slightly 2-lobed. Siliques essentially linear, often curved, seeds in 1 or 2 rows in each locule.

The identity of some populations of watercress occurring in large springs and flows of great magnitude from them, and in smaller springs and their rills, notably in Florida, has for a long time been very puzzling, chiefly because they flower and fruit rather infrequently (or collectors seldom happened to find them in flowering and fruiting stages). See Reed C. Rollins, "Watercress in Florida," *Rhodora* 80: 147–153 (1978).

● Mature siliques 2–3 mm broad and with seeds in 2 rows in each locule; each face of seed relatively coarsely alveolate-reticulate, with no more than about 50 relatively large alveolae.   1. *N. officinale*
● Mature siliques 1 mm broad and with seeds in 1 row in each locule; each face of seed very finely cancellate, with many more than 50 minute alveolae on each face.   2. *N. microphyllum*

1. Nasturtium officinale R. Br.   Fig. 82

Plant commonly much branched, rooted in the substrate, bases of the stems submersed, branch tips emersed, freely rooting from the nodes, often loosening from the substrate and forming dense and extensive floating mats. Stems variable in robustness, sometimes slender but in favorable situations up to 1 cm in diameter. Leaves seldom undivided, almost if not always with 3–11 segments, these variable in shape, oval, lanceolate, oblong, elliptic, or orbicular, the terminal segment usually, but not always, largest. Siliques 10–20 mm long, 2–3 mm broad. Seeds in 2 rows in each locule, amber-brown, lustrous, flattish to lenticular, mostly asymmetrical, nearly orbicular or short-oval in outline, about 1 mm across, surfaces of each face relatively coarsely alveolate-reticulate, with no more than about 50 relatively large alveolae. (*Sisymbrium nasturtium-aquaticum* L., *Rorippa nasturtium-aquaticum* (L.) Hayek)

Springs, spring runs, rills, clear water of sluggish streams, brooks. Native of Eur., cultivated for use as a green salad, naturalized locally more or less throughout temp. N.Am.

2. Nasturtium microphyllum (Boenn.) Reichnb.   Fig. 82

Plant relatively few-branched, roots from the nodes few and inconspicuous (relative to the condition on *N. officinale*), not known to us to form floating mats of any consequence. Stems usually slender, under favorable conditions attaining a diameter of 3–4 mm. Commonly growing only vegetatively and completely submersed in water 6–12 dm deep, leaf blades then always simple, orbicular or nearly so, or short-ovate. Also in relatively shallow water or wet seepage places about springs, stems erect or decumbent, lower submersed leaves often undivided, others with 3–5 (–7) segments, or if in very shallow water or entirely emersed, all leaves segmented. Terminal segment usually largest and often suborbicular or short-ovate, closely similar to the undivided leaves, other segments oblong to oblanceolate, sometimes all segments similar in size and shape. Plants wholly or at least partially emersed and with at least some divided leaves tending to be those on which flowers and fruits occur; however, plants with all simple leaves, having been at one time well covered with water, may

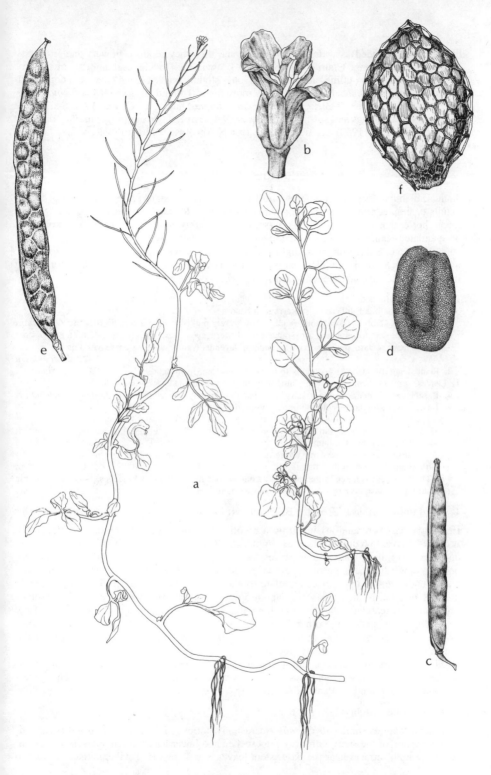

Fig. 82. a–d, **Nasturtium microphyllum:** a, habits; b, flower; c, silique; d, seed; e,f,
Nasturtium officinale: e, silique; f, seed.

177

have become stranded as water levels recede and then they produce flowers and fruit. Siliques 10–20 mm long, 1 mm broad. Seeds in 1 row in each locule, dull amber-brown, suborbicular to oblong in outline, about 0.5 mm long, slightly convex, each face often depressed centrally, very finely cancellate, with many more than 50 minute alveolae. (*N. officinale* var. *microphyllum* (Boenn.) Thell.; *Cardamine curvisiliqua* Shuttlw. ex Chapm.)

In and about springs and spring-fed streams. Native of Eur., in our range, in Fla., perhaps elsewhere; Fernald (1950) gives the range (for N.Am.) as: "Nfld. to Ont., s. to Fla. and Mich."

6. Cardamine (BITTER CRESSES)

Annual, biennial, or perennial herbs, most of relatively low stature. Leaves, some of them, petiolate, simple, pinnately lobed, or pinnately divided. Racemes simple or panicled. Sepals much shorter than the petals, erect, obtuse. Petals 4 (rarely none), white to lavender, obovate or spatulate. Stamens usually 6, 2 longer than the other 4. Ovary cylindric, style short, stigma truncate. Silique linear, straight, somewhat flattened parallel to the septum, tipped by a persistent style, dehiscing elastically from the base. Seeds numerous, flattened, wingless, in a single series or rank in each locule.

1. Leaves all undivided.
  2. Plant with bulblike, tuberous, subterranean base.
    3. The plant glabrous throughout; petals white (rarely pinkish); stem leaves 5–10.   1. *C. bulbosa*
    3. The plant with lower stems ashy-pubescent, lower leaves often sparsely pubescent on one or both surfaces near the base; petals pinkish lavender to lavender (rarely white); stem leaves 3 or 4.
                                                                                    2. *C. douglassii*
  2. Plant slenderly rhizomatous and with leafy stolons, mat-forming.   3. *C. rotundifolia*
1. Leaves, some of them on a given plant, pinnately lobed, divided, or compound.
  4. Basal leaves simple, cordate basally, stem leaves pinnately divided or compound. 4. *C. clematitis*
  4. Basal leaves all pinnately divided or compound.
    5. The basal leaves and lower to median stem leaves essentially similar; stamens 6.
      6. Leaf segments decurrent along or confluent with the midrib, the terminal segment usually considerably larger than the laterals.   5. *C. pensylvanica*
      6. Leaf segments not decurrent along nor confluent with the midrib, the terminal segment usually little if any larger than the laterals.   6. *C. parviflora*
    5. The basal leaves much larger and more conspicuous than the few stem leaves, thus the flowering or fruiting stems appearing essentially scapose; stamens 4.   7. *C. hirsuta*

1. Cardamine bulbosa (Schreb.) BSP. SPRING CRESS.   Fig. 83

Erect, glabrous perennial arising from a bulblike tuberous base. Stem simple or sparingly branched above, to 6 dm tall. Leaves simple, the basal long-petioled, their blades ovate-cordate, reniform, to orbicular; stem leaves 5–10, distant from each other, becoming shorter-petioled upwardly, the uppermost sessile, blades varying from ovate to lanceolate; margins entire, undulate, the upper ones undulate-toothed; lower surfaces sometimes purple. Sepals greenish or reddish, their tips white, about 5 mm long and 2 mm wide. Petals white, sometimes pinkish, spatulate, about 12 mm long and 5 mm wide. Stamens 6, the 4 longest nearly ½ as long as the petals, the other 2 about ¼ the length of the petals. Fruit slender, linear, 15–25 mm long, on ascending to divergent stalks up to 2 cm long, in some parts of the range, at least, infrequently maturing seeds.

Floodplain forests, seepage and spring areas, sometimes in shallow water, wet clearings, wet areas of pastures, wet meadows, moist to wet roadsides. Que. to Minn. and Kan., generally southward to the Fla. Panhandle and e. Tex.

2. Cardamine douglassii (Torr.) Britt.

In general features similar to *C. bulbosa*, usually not over 1–2 dm tall and not branched. Lower portion of the stem with ashy pubescence and lower leaves often sparsely pubescent on one or both surfaces near the base, stem leaves 3 or 4, sessile. Petals pinkish lavender to

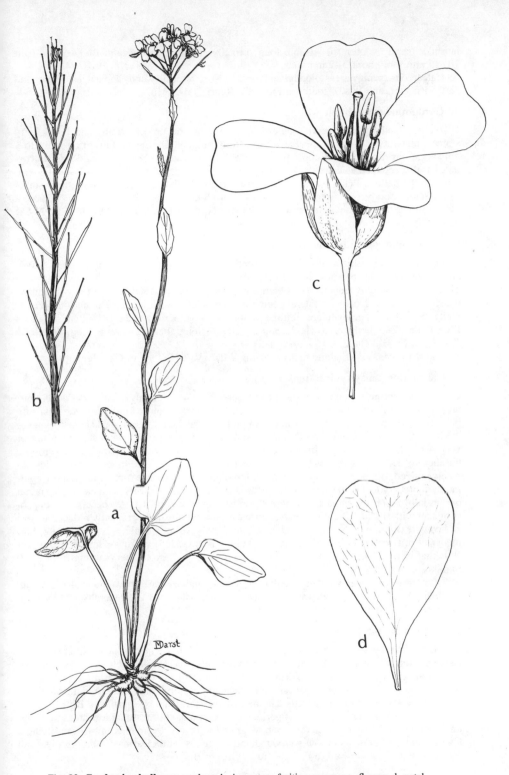

Fig. 83. **Cardamine bulbosa:** a, plant; b, immature fruiting raceme; c, flower; d, petal.

lavender (rarely white), 10–16 mm long, broadly spatulate to obovate. Body of the fruit 10–15 mm long, beak 3–7 mm long.

Calcareous springy areas, floodplain woodlands, rich mesic woods and ledges. Conn. and N.Y. to s. Ont. and Wisc., southward to S.C., Tenn., Ala., Mo.

3. Cardamine rotundifolia Michx.

Glabrous, slenderly rhizomatous perennial with weak decumbent stems and basal leafy stolons, mat-forming; stolons may also form from the flowering stems. Flowering stems to 3 dm long. Leaves petioled, or uppermost sessile, blades ovate to orbicular, rounded to subcordate basally, rounded apically, 2–4 cm long; lower leaves sometimes with 1 or 2 small lateral lobes on the petiole. Petals white, 5–10 mm long. Fruits divergent or ascending, their stalks 1–2 cm long, body 1–1.5 cm long, beak 2–3 mm long.

Springy places, brooksides, low woodlands. N.Y. to Ohio, southward to n.w. N.C., Tenn., Ky.

4. Cardamine clematitis Shuttlw.

Perennial, slenderly rhizomatous. Stems weak, 1–4 dm tall, pubescent below. Basal leaves simple, with slender petioles, blades orbicular or reniform, cordate basally, lower surfaces often purple, margins crenate, 1.5–2 cm long; stem leaves pinnately divided or compound, usually with 1 or 2 pairs of lateral segments or leaflets and a larger terminal one. Petals white, 6–8 mm long, spatulate. Fruiting stalks very slender, ascending, 1–2 cm long or a little more, fruit body narrowly linear, 2.5–4 cm long, beak 2–3 mm long. (Incl. *C. flagillifera* O. E. Schulz, *C. hugeri* Small)

Springy areas, in and along rocky streams. Mts., Va., N.C., n.w. Ga., Tenn.

5. Cardamine pensylvanica Muhl. ex Willd.   Fig. 84

Biennial, sometimes, perhaps, a short-lived perennial, of relatively small and slender stature, to 8 dm tall but commonly 1–4 dm, glabrous throughout or only the stem sparsely pubescent near the base. Varying from a single principal stem to numerous stems from near the base, the latter not infrequently decumbent or trailing, especially if in shallow water or very wet soil. Very variable with respect to leaves; first leaves usually forming a basal rosette but these often already disintegrated on flowering and fruiting specimens. Leaves odd-pinnately divided, often essentially compound, the segments or leaflets sometimes 3, more commonly more, up to 13, the size and shape as well as the number very variable, the terminal one usually much the largest, sometimes toothed or lobed, their bases decurrent along or confluent with the midrib; lower leaves, at least, petiolate, usually gradually reduced in size and complexity upwardly and becoming sessile. Petals white, spatulate, 2–4 mm long. Stamens 6. Fruit a silique, narrowly linear, 1–3 cm long, gradually narrowed into the persistent style that is 1–2 mm long; stalks of the fruits ascending, very slender, varying from 2–10 mm long.

Often in water, swamps, wet woodlands, springy and seepage areas, along streams, wet clearings and ditches, low cultivated fields and lawns. Nfld. to B.C., generally southward in the East to Fla. and Okla., in the West to Colo. and n. Calif.

6. Cardamine parviflora L.   Fig. 85

In general features much resembling *C. pensylvanica*, but a winter annual, usually not exceeding 3 dm tall, glabrous throughout, the leaves not nearly so variable. Leaf segments or leaflets not decurrent along the midrib, the lateral ones of the median stem leaves linear to oblanceolate, longer than broad, mostly not exceeding 2 mm broad, the terminal leaflet usually, but not always, as narrow as the laterals. Stamens 6. (Incl. *C. arenicola* Britt.)

Often in places wet only for short periods in winter and spring, small low areas in woodlands, pools, ledges, rock outcrops, sandy, gravelly, or rocky grounds in or near streams, lawns, fields, sometimes in low wet woodlands, ditches, and clearings. Nfld. and Que. to Minn., generally southward to Fla. and Tex.; Ore. to B.C.

Fig. 84. **Cardamine pensylvanica:** a, habit; b, flower; c, sepal; d, petal; e, stamens and pistil; f, fruit (silique); g, fruit, dehisced; h, seed.

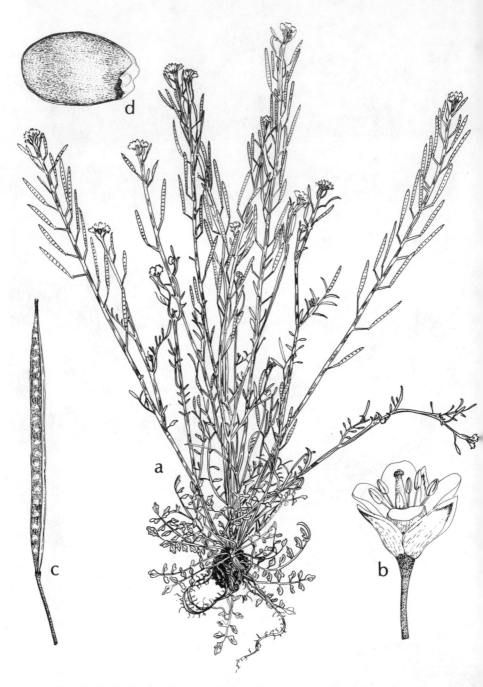

Fig. 85. **Cardamine parviflora:** a, habit; b, flower; c, fruit (silique); d, seed. (a, from Correll and Correll. 1972)

7. Cardamine hirsuta L.

In general features much resembling *C. pensylvanica* and *C. parviflora*, the larger leaves mostly in a basal rosette and relatively conspicuous, stem leaves few and much smaller, stems to 3 dm tall, glabrous. Petioles of the stem leaves ciliate, at least basally. Lateral segments or leaflets of basal leaves suborbicular or reniform to elliptic, often asymmetrical, irregularly shallowly crenate, terminal segment usually larger than the lateral and nearly always suborbicular or reniform; leaf segments or leaflets pubescent below, sometimes above. Stamens 4.

Usually in moist or wet soil, weedy fields, ditches, roadsides, lawns, clearings, stream banks. Native of Eur., naturalized from s.e. N.Y. to Ill., generally southward to Fla., Ark., and e. Tex.

7. Cakile (SEA ROCKETS)

Annual, fleshy, herbaceous, sometimes suffrutescent herb, (ours) glabrous, usually much branched basally, the branches spreading-ascending, sometimes prostrate, decumbent, or procumbent. Leaves alternate, entire to sinuate to irregularly pinnatifid, gradually narrowed below into the petiole, the lower leaves larger than the upper, ovate to spatulate or oblanceolate to obovate. Flowers borne in racemes terminating the branches, the racemes greatly elongating from first flowering to full development of the fruits, flower stalks 1–10 mm long, usually divergent. Sepals 4, green, erect at anthesis, deciduous, ovate, 3–5 mm long, 1.5–3 mm wide, margins hyaline. Petals 4, rarely absent, obovate to spatulate, sometimes distinctly clawed, blade reflexed, entire, sinuate, or emarginate, white, pinkish, lavender, or purple. Stamens 6, 4 long, 2 short. Pistil sessile, style none, stigma capitate, sometimes bilobed. Fruit dry and corky, of 2 indehiscent segments, each 1-chambered and each usually 1-seeded, upper beaked segment deciduous at full maturity by a transverse articulation; lower segment terete, vaguely angled, or ribbed, upper segment terete, angled, ribbed, or furrowed.

Reference: Rodman, James E. "Systematics and Evolution of the Genus *Cakile* (Cruciferae)." *Contr. Gray Herb.* 205: 3–146. 1974.

1. Axis of fruiting inflorescence zigzagged.   1. *C. geniculata*
1. Axis of the fruiting inflorescence straight or wavy, not zigzagged.
  2. Petals usually more than 3 mm wide; fruiting inflorescences more than 2 dm long.   2. *C. lanceolata*
  2. Petals usually less than 3 mm wide; fruiting inflorescences 1–2 dm long.
    3. Fruit 8-ribbed or 4-angled, the summit somewhat flattened and retuse or blunt.   3. *C. edentula*
    3. Fruit terete to 4-angled, the summit conical and acute.   4. *C. constricta*

1. Cakile geniculata (Robins.) Millsp.

Plant usually erect-spreading, 1–4 (–10) dm high. Leaves very fleshy, broadly ovate and petiolate to oblanceolate or spatulate, sinuately, dentately, or pinnately lobed. Fruiting racemes 1–2 dm long, their axes zigzagged at maturity, the stalks of the fruit of equal or nearly equal width to that of the inflorescence axis, fruits widely spaced. Sepals 3–4 mm long and 1–2 mm wide, sometimes with sparse, simple trichomes apically; petals obovate or spatulate, not distinctly clawed, 4–6 mm long and 1.2–2 mm wide, white to pale lavender. Mature fruits 20–27 mm long and 3–5 mm wide, upper segment usually distinctly 8-ribbed, terete, blunt to acute apically, conical, in outline lanceolate to oblong-lanceolate, often curved, to about twice as long as the lower segment; lower segment oblong-elliptic in outline, weakly 8-ribbed.

Coastal sands. La., w. of the Miss. Delta, to Tamaulipas and Veracruz.

2. Cakile lanceolata (Willd.) O. E. Schulz.

Plants herbaceous to suffrutescent, only moderately succulent, sometimes erect, usually straggling or prostrate, much branched, the branches often more than 5 dm long. First leaves up to 15 cm long and 4 cm wide, broadly ovate or ovate-deltoid and petiolate to ovate-lanceolate and attenuate basally, nearly entire to variously lobed, to deeply and finely pinnatifid; later leaves smaller, usually oblanceolate and less lobed. Fruiting racemes usually more than 2 dm long, linear or wavy. Sepals 3.5–5 mm long, dimorphic; petals clawed, 5–9.5 mm long and 1.7–4.3 mm wide, white to rarely pale lavender. Mature fruit from about 13 to 30 mm long, weakly 4-angled to terete, striate, or channeled; upper segment usually lanceolate in outline, the lower obconical, upper usually longer than the lower; without a constriction at the joint.

Coastal sands. Atlantic and Gulf coasts of pen. Fla.; Gulf and Caribbean coasts of s. Mex. and C.Am.; W.I. and n. S.Am.

3. Cakile edentula (Bigel.) Hook.

Succulent, rarely suffrutescent, usually erect, up to 8 dm tall, much-branched, the branches spreading-ascending, lowermost often decumbent. First leaves very fleshy, broadly ovate and petiolate to obovate or spatulate and attenuate at the base, crenately, sinuately, or dentately lobed, not pinnatifid, up to 15 cm long and 7 cm broad; later leaves smaller, usually oblanceolate, less lobed. Sepals ovate, apically often with sparse simple trichomes. Petals more or less obovate, not distinctly clawed, 4.6–9.7 mm long, 1.4–3.3 mm wide, white to pale lavender (petals frequently absent or present only as small white bristles). Mature fruit from about 12–30 mm long, 4-angled or 8-ribbed, the beak more or less flattened in one plane, usually blunt or retuse at the apex, sometimes acute, somewhat constricted at the joint of the two segments. Rodman (1974) recognizes two subspecies with distribution in our range, as follows.

C. edentula subsp. **edentula.** Upper fruit segment 4-angled, articulating surface of lower fruit segment more or less flat to concave and with 2 (–6) small teeth projecting upward.

Coastal sands. Lab. to N.C.; sporadic on shores of lakes Michigan, Erie, and Ontario; naturalized on the Pacific coast of N.Am.; shores of temp. Austl.; occasionally introduced as a ballast weed elsewhere.

C. edentula subsp. **harperi** (Small) Rodman. Upper fruit segment 8-ribbed, articulating surface of lower fruit segment flat and without teeth.

Coastal sands. N.C. to St. Lucie Co., Fla.

4. Cakile constricta Rodman.

Fleshy, 1–3 dm tall, much branched, branches commonly prostrate. First leaves ovate and petiolate to obovate or spatulate and attenuate; later leaves smaller, oblanceolate, more or less entire. Flowering racemes branched. Sepals about 4 mm long, 1.5–2.5 mm wide; petals somewhat clawed, 5.3–8 mm long, 1.3–2.6 mm wide, white to lavender. Fruiting racemes to 2 dm long and commonly straggly. Upper segment of fruit 9–15 mm long, lance-fusiform, terete to 4-angled, acute apically; lower segment 5–8.5 mm long, obtusely 4-angled to terete, oblanceolate in outline, its summit flat or with 2 apiculations within the margin; fruit distinctly constricted at the joint.

Coastal sands. Atlantic coast of Fla. and Gulf coast from Tampa Bay area to e. Tex.

Capparaceae (CAPER FAMILY)

Cleome spinosa Jacq. SPIDER-FLOWER.

Annual herb with a strong odor, 8–15 dm tall. Stem simple to freely branched, bearing long-stipitate glandular hairs and a stipular prickle at either side of the petiole bases. Leaves alternate, slenderly petiolate, petioles about as long as the blades, with both long and short gland-tipped hairs and usually bearing scattered, opaque, sharp prickles 1–2 mm long, as well; blades palmately compound, leaflets 5–7, the basal pair smallest, others increasing in size distally, terminal leaflets 4–10 cm long, all elliptic-oblanceolate or lance-elliptic, acuminate basally, acute to acuminate apically, margins reddish, ciliate-toothed, surfaces with a few scattered hairs and scattered prickles along the midrib beneath. Inflorescence a leafy-bracted raceme terminating the main stem or the branches. Flowers showy, bilaterally symmetrical, their slender, copiously glandular-pubescent stalks 2–4 cm long. Calyx slightly united basally, lobes 4, subulate, 3–5 mm long, glandular-pubescent exteriorly and marginally. Petals 4, pink, unequal, long-clawed, blades ovate, obovate, or oblanceolate, 1–3 cm long. Stamens 6, pink, the slender filaments longer than the petals, anthers long-filiform. Pistil 1, the ovary with a filiform stipe at first about 1 cm long, very quickly elongating to about 5–6 cm; ovary 1-locular, narrowly linear to fusiform, about 1 cm long, style barely perceptible. Fruit a narrow, 2-valved capsule 4–8 cm long, bearing numerous seeds on 2 parietal placentae. Seed looped, the extremities approximately meeting, about 2 mm long and broad, surface wrinkled-scaly. (*Neocleome spinosa*, as used by Small, 1933, *C. houtteana* Raf.)

Cultivated for ornament, naturalized in various disturbed places, often occurring on exposed shores and bars of rivers and in bottomland clearings or openings. s. N.Y. to Mo., southward to Fla. and Tex. Native of S.Am.

Droseraceae (SUNDEW FAMILY)

Annual or perennial, insectivorous herbs. Leaves alternate, infolded or circinate in vernation, the blades modified as active traps (as in *Dionaea*), or having mucilage-tipped tentacles (as in *Drosera*). Flowers bisexual, radially symmetrical. Sepals and petals each 5, distinct, sepals persistent. Stamens 5–15 (–20). Pistil 1, ovary superior, 1-locular, styles 3 and deeply bifid or divided, or single and with a fimbriate stigma. Fruit a 3–5-valved capsule, seeds small, numerous.

● Leaves with blades hinged lengthwise in the middle, the free margins with prominent bristles which interlock upon closing; stamens 15 (10–20); style 1.   1. *Dionaea*
● Leaves long-filiform or petiolate and with unhinged blades adorned on the upper surfaces and margins with gland-tipped, tentaclelike trichomes; stamens 5; styles 3, deeply divided and appearing as 6.
                                                                                            2. *Drosera*

1. Dionaea

Dionaea muscipula Ellis. VENUS FLYTRAP.   Fig. 86

Perennial carnivorous herb with leaves closely set on a short stem, the petiolar bases dilated and closely overlapping so that the short stem appears somewhat bulbous. Leaf blades hinged lengthwise in the middle, each of the two halves nearly kidney-shaped, the margin of each equipped with conspicuous bristles which interlock when the leaf is closed; petiole usually markedly winged and expanded distally. Inflorescence scapose, 1–3 dm tall, umbelliform-cymose. Sepals 5. Petals 5, white, spatulate, about 12 mm long. Stamens 15 (10–20). Ovary superior, 1-locular, style 1, stigma with numerous elongate papillae. Capsule ovoid. Seeds numerous, black and shiny, obovoid, tiny.

Bogs, between evergreen shrub bogs and longleaf pine savannas, flourishing best where burned periodically; wet, sandy-peaty soils of roadsides. Coastal plain, Beaufort Co., N.C. to s. of the Santee River, Charleston Co., S.C.

2. Drosera (SUNDEWS)

Annual or perennial insectivorous herbs. Leaves (circinate in bud), alternate. Upper surfaces and margins of the leaf blades covered with tentaclelike gland-tipped trichomes that secrete a sticky substance aiding in the trapping of small insects. Stipules scarious, fringed or di-

185

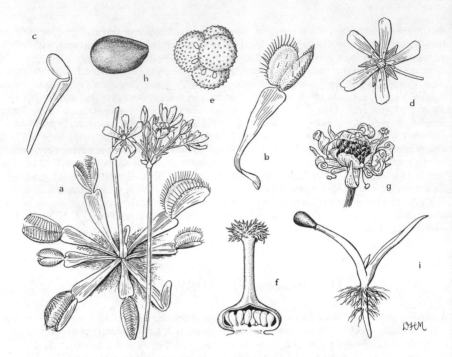

Fig. 86. **Dionaea muscipula:** a, habit; b, mature leaf; c, young leaf; d, flower; e, pollen tetrad; f, pistil, a portion of ovary wall removed to show basal placentation of ovules; g, open fruit, showing irregular dehiscence and shriveled stamens and petals; h, seed; i, seedling. (From Wood in *Jour. Arn. Arb.* 41: 158. 1960)

vided, adnate or free, or absent. Inflorescence scapose, a 1-sided raceme, nodding at the apex during development. Flowers bisexual, opening only during the day. Sepals 5. Petals 5, slightly united at the base. Stamens 5. Ovary superior, styles 3 or 5, each divided nearly to the base. Fruit a 3–5-valved capsule. Seeds minute, numerous, variously reticulated or ornamented.

1. Leaves filiform and bladeless, their expanded bases forming a cormlike structure.
  2. Petals 7–10 mm long; leaves to 25 cm long, slender-filiform, tentacles purple, drying to dark brown.   1. *D. filiformis*
  2. Petals 10–30 mm long; leaves to 40 cm long, coarsely filiform; tentacles green, drying to light brown.   2. *D. tracyi*
1. Leaves with blades expanded and petiolate, bases not expanded.
  3. Scapes with gland-tipped trichomes except toward the base.   3. *D. brevifolia*
  3. Scapes glabrous.
    4. Leaf blades suborbicular, at maturity as broad as long or broader; seeds about six times as long as broad.   4. *D. rotundifolia*
    4. Leaf blades spatulate to obovate; seeds twice as long as broad or less.
      5. Petioles glabrous; petals white; scapes arched outwardly at their bases; seeds uniformly papillose.   5. *D. intermedia*
      5. Petioles with at least a few long trichomes, not gland-tipped; petals pink; scapes straight at their bases; seeds short-papillose-ridged.   6. *D. capillaris*

1. Drosera filiformis Raf. DEW-THREAD, THREAD-LEAF SUNDEW.

Leaves slender-filiform, 8–25 cm long, less than 1 mm wide, covered with purple, glandular trichomes (drying to dark brown); stipules long-fimbriate and forming a woollike mat among the close-set leaf bases. Scape 6–22 cm long, glabrous, bearing 4–6 flowers, each about 1 cm across. Pedicels and calyces glandular-pubescent. Petals purple, 7–15 mm long, 5–8 mm wide. Seeds black, ellipsoid, coarsely crateriform.

Damp sands, sandy or gravelly pond margins. Coastal plain, Mass. to s.e. N.C.; Fla. Panhandle.

2. Drosera tracyi Macfarlane in L. H. Bailey. Fig. 87

Differing from *D. filiformis* in longer (to 40 cm) and wider (more than 1 mm) leaves bearing green tentacles. Scape 25–45 cm tall, with green, glandular trichomes (drying to light brown). Flowers 1.5–2 cm across, the petals 1.2–1.5 cm long and 15 mm wide, purplish rose in color. (*D. filiformis* var. *tracyi* (Macfarlane in Bailey) Diels in Engler)

Bogs, pine flatwoods and savannas, commonly on wet, sandy roadsides. Coastal plain, s. Ga., Fla. Panhandle to La.

3. Drosera brevifolia Pursh. Fig. 88

Basal rosettes seldom exceeding 3.5 cm broad. Leaf blades suborbicular to cuneate-spatulate, 4–10 mm long, usually about as long as the petiole. Stipules absent (or comprised of but a few setaceous segments). Scapes 1–8 cm tall, nearly uniformly clothed with gland-tipped trichomes, bearing 1–8 flowers; pedicels and calyces glandular-pubescent. Flowers to 1.5 cm across; petals white to pink or rose-purple, 4–5 mm long, 2–3 mm wide. Seeds black, obovate with crateriform markings. (*D. annua* Reed; *D. leucantha* Shinners)

Pine flatwoods and savannas, wet sand, commonly on roadsides. s. Va. to Fla., westward to Ark., Okla., Tex. and Tenn.

4. Drosera rotundifolia L. ROUND-LEAVED SUNDEW.

Basal rosettes variable, 10 (2–15) cm across. Leaf blades suborbicular, 4–10 mm long, often broader than long, much shorter than the petioles. Stipules adnate, fimbriate on the margins. Scape to 3 dm tall, glabrous, bearing 3–15 flowers. Pedicels and calyces glabrous or pubescent. Flowers 10–14 mm across. Petals white to pink, 5–7 mm long, 3–4 mm wide. Seeds light brown, oblique-fusiform, longitudinally striate, shiny.

Bogs and sandy swamps. n. N.Am., southward to n. Ga. in the East and southward to Calif. in the West.

5. Drosera intermedia Hayne in Schrad. Fig. 89

Leaves in a basal rosette or commonly along a stem up to 10 cm long, the stems sometimes proliferating. Leaf blades oblong-spatulate to obovate, 8–20 mm long, often very gradually tapering into the petiole; petioles usually 3–4 times as long as the blades. Stipules adnate at the base, then with many setaceous segments. Scapes 9–20 cm high, arching outward at the base, bearing up to 20 flowers. Pedicels and calyces glabrous or sparsely pubescent. Flowers 10–12 mm across. Petals white or pinkish, 5–8 mm long and 3–5 mm wide. Seeds reddish brown to black, ellipsoid-ovoid, densely and uniformly papillose.

Bogs, pond and stream or swamp borders, sometimes in wet ditches. Nfld. westward to Minn., southward to Fla. and Tex.; trop. Am.

6. Drosera capillaris Poir. PINK SUNDEW. Fig. 88

Basal rosettes very variable in size, to 10 (–12) cm across. Leaf blades obovate to spatulate, 5–10 mm long, usually somewhat shorter than the petioles. Stipules nearly free, fimbriate. Scape 4–20 cm tall, glabrous, bearing 2–20 flowers. Pedicels and calyces glabrous. Flowers about 1 cm across, petals pink, 6–7 mm long, 2–3 mm wide. Seeds brown, coarsely papillose-corrugated in 14–16 ridges.

Bogs, pine flatwoods and savannas, commonly in wet, sandy ditches. Coastal plain, Va. to Fla. and Tex.; trop. Am.

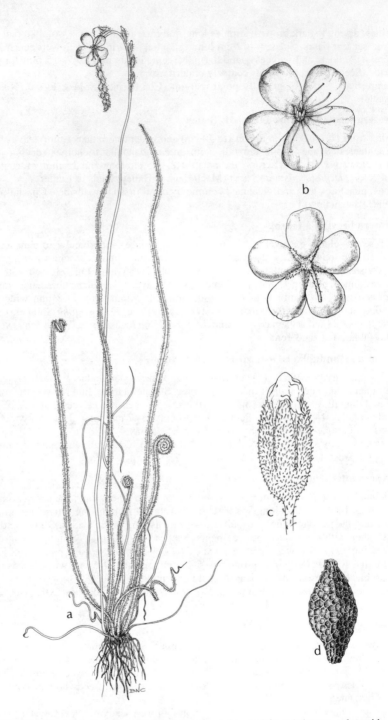

Fig. 87. **Drosera tracyi:** a, habit; b, flower, face view and from below; c, calyx with capsule protruding; d, seed.

Fig. 88. a–e, **Drosera brevifolia:** a, habit; b, leaf; c, flower, face view; d, calyx and capsule; e, seed; f,g, **Drosera capillaris:** f, calyx and capsule; g, seed.

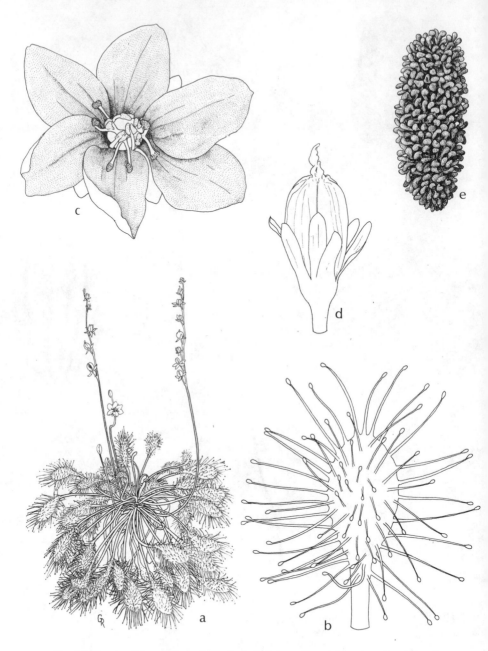

Fig. 89. **Drosera intermedia:** a, habit; b, leaf blade; c, flower, face view; d, calyx and capsule; e, seed.

190

Sarraceniaceae (PITCHER-PLANT FAMILY)

Sarracenia (PITCHER-PLANTS)

Long-lived, perennial, carnivorous plants, the stems wholly rhizomatous, rhizomes horizontal or vertical. Sessile, coarse, overlapping scalelike leaves with clasping bases and acuminate to acuminate-attentuate tips clothe the rhizomes; aerial leaves and flowers borne from amongst the scale leaves; in some species first aerial leaves of the season, and in some species aerial leaves produced late in the season and overwintering, differing from "summer" leaves generally taken to be characteristic for a species, and are referred to as phyllodia. Phyllodia, if present, shorter than the others, flattish, often twisted, falcate or somewhat swordlike. Summer leaves produced with or after the flowers, their bases dilated and clasping, narrowed above the base into a relatively slender, terete or basally winged, petiole, thence upwardly tubiform and urceolate, pitcherlike, or trumpetlike, laterally winged externally, the tube often conspicuously reticulate-veined distally, apically with an expanded, erect, to recurved hood to one side of an orifice terminating the tube, or in one species the margins of the hood fully folded-recurved and surrounding the orifice, the orifice thus appearing lateral. Flowers showy, borne singly, nodding at the terminus of a long scape arising from the rhizome, often becoming erect after the petals and stamens fall. Bracts 3, ovate, persistent, usually appressed to the calyx. Sepals 5, persistent. Petals 5, yellow, maroon, red, or pink, more or less fiddle-shaped, the apical portion ovate, obovate, or elliptic, pendulous over the edge of the style disc. Stamens numerous. Pistil 1, ovary superior, 5-locular, externally granular-tuberculate; style simple below, expanded above into a very large, persistent, umbrellalike, 5-lobed disc with a small stigma under the notch at the tip of each of its lobes. Fruit a loculicidally 5-valved capsule. Seeds numerous, irregularly clavate to obovate, reticulate-tuberculate, winged laterally.

Reference: McDaniel, Sidney. *The Genus* Sarracenia *(Sarraceniaceae)*. Bull. No. 9, Tall Timbers Research Station. 1971.

Where plants of two or more species grow together, especially where the substrate has been mechanically disturbed, hybrid individuals may occur. (See Bell (1952), Bell and Case (1956), and McDaniel (1971))

1. Leaves with areas of white or whitish tissue toward the summit of the tube and back of the orifice, or on both the distal portion of the tube and on the hood.
  2. Orifice of the leaf lateral beneath the hood, formed by partial union of the edges of the hood itself; leaves decumbent or sprawling, only occasional ones erect-ascending.   1. *S. psittacina*
  2. Orifice of the leaf terminal, to one side of the hood; leaves erect or strongly ascending.
    3. Areas of whitish tissue toward the summit of the leaf somewhat below and opposite the orifice, the areas of whitish tissue not enclosed within a conspicuous network of reddish venation; hood strongly and closely arching over the orifice; petals yellow.   2. *S. minor*
    3. Areas of whitish tissue all around the summit of the leaf and throughout the hood, the areas of whitish tissue enclosed by a conspicuous network of reddish venation; hood erect or somewhat arched but the blade held well aloft over the orifice; petals maroon.   3. *S. leucophylla*
1. Leaves not having areas of whitish tissue.
  4. Leaves erect.
    5. Petals yellow.
      6. Narrowed base of the hood purple or purple-spotted on the inside, its sides very strongly and loosely rolled back so that the edges are nearly contiguous in the back; blade of the hood broadly reniform to orbicular-reniform, broadly cordate basally.   4. *S. flava*
      6. Narrowed base of the hood not purple or purple-spotted on the inside, with revolute margins but not conspicuously rolled to the extent that the edges are anywhere nearly contiguous in the back; blade of the hood ovate, not at all or just barely cordate basally.
        7. Phyllodia usually present and conspicuous, recurved or falcate; plant occurring in cen. and n.e. Ala., n.w. Ga.   5. *S. oreophila*

7. Phyllodia, if present, swordlike, erect; plant occurring in s.w. Ala., s. Miss., s.e. La., s.w.
   La. and e. Tex.   6. *S. alata*
5. Petals maroon.   7. *S. rubra*
   4. Leaves decumbent or sprawling, occasional ones ascending, urnlike.   8. *S. purpurea*

1. Sarracenia psittacina Michx. PARROT PITCHER-PLANT.   Fig. 90

Leaves overwintering, 8–30 cm long, decumbent or sprawling, occasional ones ascending,
tubular portion very narrow, with a broad lateral wing; upper portion of the tube and the hood
with irregularly spaced small areas of pale greenish or whitish tissue and purple-reticulate;
hood strongly arched, keeled dorsally, the margins united and inrolled in such a way as to
surround a small orifice, the orifice thus lateral beneath the hood. Scape 15–35 cm long.
Bracts triangular-ovate, 6–8 mm long, blunt or rounded apically, reddish. Sepals ovate-triangular to rhombic, 1.5–2.5 cm long, blunt apically, maroon on the outside, green inside
with maroon edges and tips. Petals maroon, mostly 3–4 cm long, oblongish basally, apical
portion obovate. Style disc 2–3 cm broad, yellowish green.
   Wet pine savannas and flatwoods, open bogs, titi bogs and bays, adjacent boggy roadsides
and ditches, pineland seepage slopes. s. Ga., n. Fla., s. Ala., s.e. Miss., and s.e. La.

2. Sarracenia minor Walt. RAIN-HAT TRUMPET, HOODED PITCHER PLANT.   Fig. 91

Leaves erect, persistent through the winter, 12–35 cm long, occasionally longer, the tubular
portion gradually widening from the base to the orifice, sparsely and unevenly very short-
pubescent exteriorly, smooth and somewhat slick interiorly, green with prominent white
patches of tissue opposite and somewhat below the orifice, usually a faint flush of reddish
pigment in the tissue surrounding the white, wing on the tube 0.5–3 cm wide near the middle; hood narrowed very little basally, strongly arching-recurved over and essentially covering the orifice, the sides declined thus convex above, outer surface of hood bronze-red, or
somewhat streaked with bronze-red, inner surface strongly marked with reddish, reticulate
venation, with some glandular exudate and retrorsely short-strigose; phyllodia none. Scape
12–55 cm long, usually somewhat shorter than the longer leaves. Bracts narrowly oblong to
ovate-oblong, 5–10 mm long. Sepals broadly ovate-triangular, obtuse apically, mostly about
1.5–3.5 cm long, yellowish green. Petals yellow, usually oblongish basally and with an expanded obovate or ovate apex, sometimes gradually widening from the base and narrowly
obovate, mostly 3–4 cm long. Style disc about 3 cm across, pale yellow.
   Pine savannas, bogs, adjacent peaty-sandy ditches. Coastal plain, s.e. N.C. to cen. pen.
Fla., s.w. Ga. and to just w. of the Apalachicola River in the Fla. Panhandle.

3. Sarracenia leucophylla Raf.   Fig. 92

Leaves erect, not persistent through the winter, the larger to 9.5 dm long, tubular portion
gradually widening from the base to the orifice, usually narrowly winged or the wing sometimes to 2 cm wide; leaf green below, the upper portion of the tube all the way around and
both surfaces of the hood white and with conspicuous green to pink to red wine–colored
venation; narrowed basal portion of the hood short, nearly erect, the blade erect to angled
over and well above the orifice, ruffled on the edge, broadly rounded apically, white-hirsute
on the inner surface; phyllodia erect, 15–20 cm long, swordlike. Scape 3–8 dm long. Bracts
ovate, 5–10 mm long, rounded apically, purplish red. Sepals ovate or triangular-ovate,
3.5–5 cm long, upper surface maroon, lower greenish suffused with maroon. Petals 4–7 cm
long, obovate at the most basal portion and then constricted to an oblongish portion, finally
flared to a somewhat rhombic distal portion, deep maroon above the base, paler beneath.
Style disc 6–7 cm broad, reddish. (*S. drummondii* Croom)
   Bogs, wet pine savannas or flatwoods, boggy borders of branch bays and cypress depressions, boggy areas by small streams. s.w. Ga., Fla. Panhandle w. of the Ocklockonee River,
s. Ala., s.e. Miss.

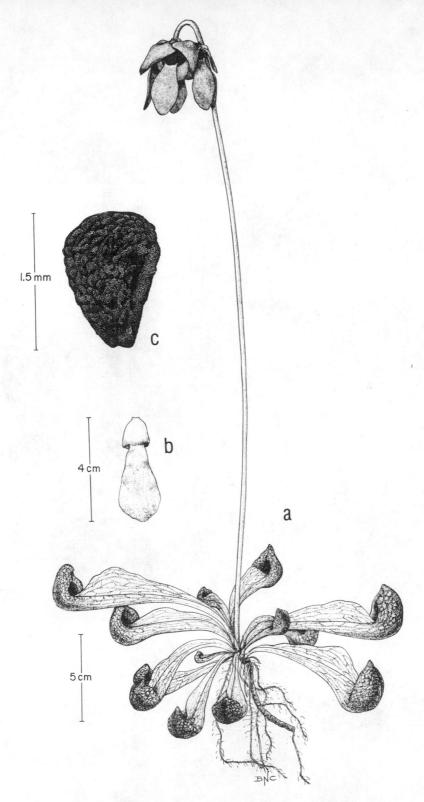

Fig. 90. **Sarracenia psittacina:** a, habit, plant in flower; b, petal; c, seed.

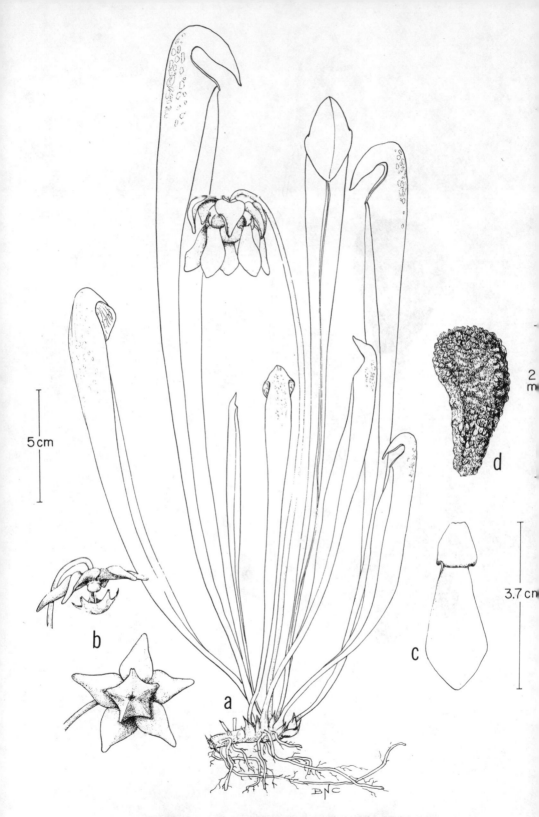

5 cm

2 m

3.7 cm

Fig. 91. **Sarracenia minor:** a, habit, flowering plant; b, fruit, lateral and face views; c, petal; d, seed.

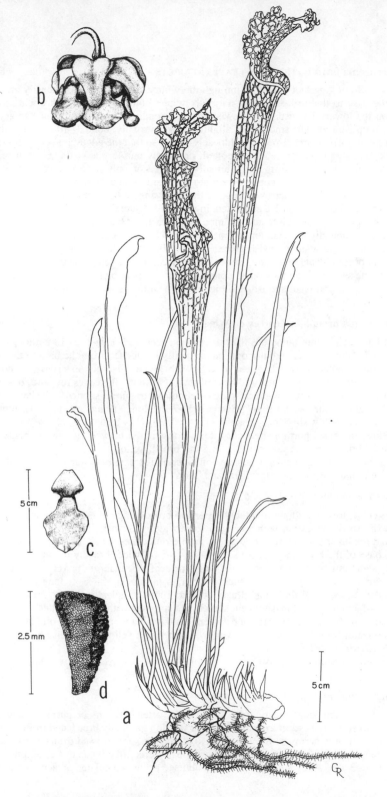

Fig. 92. **Sarracenia leucophylla:** a, habit, plant with early-season leaves (not shaded) and later "characteristic" leaves (shaded); b, flower, lateral view; c, petal; d, seed.

4. Sarracenia flava L. TRUMPET-LEAF, TRUMPETS.   Fig. 93

Leaves 3–9 dm long, not persistent through the winter, tubular portion gradually broadened from the base to the orifice, usually narrowly winged but sometimes with a wing to 3 cm wide on the lower half, usually yellowish green throughout but on some plants varying to wholly purplish red; tube smooth and slightly slick internally; hood 3–10 cm long, the narrowed basal portion very strongly and loosely rolled so that the edges are nearly contiguous in back, usually purple-streaked or -spotted, or wholly purple, the blade very conspicuous, reniform to orbicular-reniform, very broadly rounded or only occasionally lobed distally, very very finely and closely pubescent on the inner side, the pubescence becoming more or less sloughed eventually; narrowed base of the hood angled over the orifice and holding the blade well aloft over the orifice and extending beyond it laterally. Phyllodia 12–30 cm long, swordlike, produced in mid- to late summer. Scape to about 6 dm long. Bracts ovate-oblong, arched, blunt apically, 1–2 cm long. Sepals ovate to lance-ovate, rounded or obtuse apically, 3–5 cm long, yellow or yellowish green. Petals yellow to greenish yellow, 5–8.5 cm long, the apical portion ovate to elliptic. Style disc 6–8 cm broad, yellow or greenish yellow.

Wet pine savannas and flatwoods, pond-cypress swamps, bogs, pineland seepage slopes, titi thickets. Coastal plain and isolated piedmont localities, s.e. Va. to n. Fla., s. Ala., s.e. Miss.

5. Sarracenia oreophila Kearney ex Wherry.   Fig. 94

Phyllodia usually more numerous than the leaves, recurved-falcate, 5–18 cm long. Leaves erect, to 7.5 dm long, the tubular portion gradually broadened from the base to the orifice, rarely conspicuously winged, green to yellowish green, glabrous exteriorly, smooth and somewhat slick interiorly. Narrowed base of the hood short, the edges revolute, blade ovate to reniform, varyingly yellow-green, purple-reticulate, or purple-spotted at the base, glandular-pubescent on the inner surface. Scape 4.5–7 dm long. Bracts ovate, 6–12 mm long, blunt apically, yellowish. Sepals ovate to elliptic, yellow, 3–5 cm long, obtuse apically. Petals yellow, the basal portion short-obovate, a constriction between it and the blade, blade oblong-elliptic to obovate. Style disc 5–8.5 cm broad, yellow.

Wet thickets, boggy banks, wet sands on river and stream banks and shores, rich woodlands. cen. and n.e. Ala., adjacent Ga. and Tenn.

6. Sarracenia alata (Wood) Wood.   Fig. 95

Leaves erect, to 7.5 dm long, the tubular portion broadening gradually from the base to the orifice, narrowly winged or with a wing to 1 cm wide, yellowish green, distal portion of the tube and the hood commonly purple-lined, purple reticulate-veined, or strongly purple; narrowed base of the hood short, the blade ovate, 4–6 cm wide, broadly rounded apically and usually short-mucronate, strigose on the inner surface; tube glabrous or very short and finely pubescent exteriorly, smooth and slick interiorly; phyllodia ½–⅔ as long as the leaves, swordlike. Scapes to 6 dm long. Bracts oblong, rounded apically, 1–1.5 cm long. Sepals ovate, 3–6 cm long, yellowish green, sometimes suffused with purple. Petals 5–7 cm long, basal portion obovate, a constriction between it and the blade, blade ovate-orbicular to somewhat obovate, bright yellow, whitish yellow, or greenish yellow. Style disc 5–8 cm broad, yellow. (*S. sledgei* Macf.)

Wet pine savannas and flatwoods, bogs, pineland seepage slopes. s.w. Ala., s. Miss., s. La., e. Tex.

7. Sarracenia rubra Walt. SWEET PITCHER-PLANT.   Fig. 96

Leaves 0.8–7 dm long, erect, green to reddish, often reddish- or purple-veined above, glabrous to softly pubescent exteriorly, tubular portion gradually broadened from the base to the orifice, or little narrowed from the base to the orifice, or narrowed from somewhat above the middle; narrowed base of the hood short, nearly erect, the blade ovate, elliptic, or lanceolate, 1–3 cm wide, apically variable, obtuse, acute, apiculate, or acuminate, green,

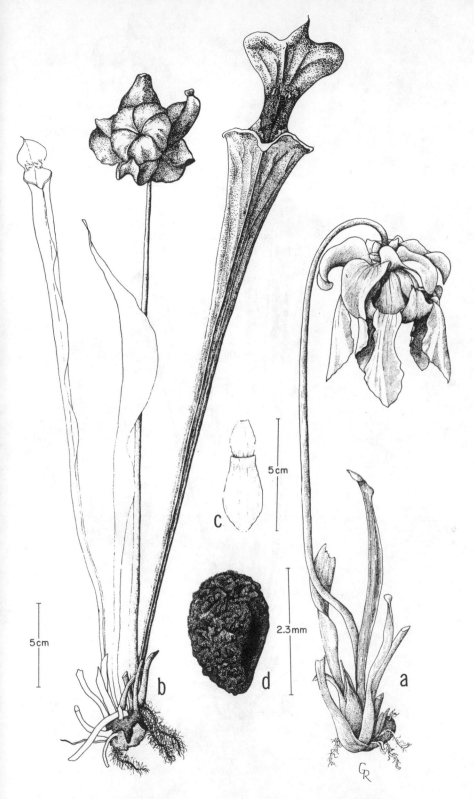

Fig. 93. **Sarracenia flava:** a, habit, with early-season leaves; b, habit, fruiting plant; c, petal; d, seed.

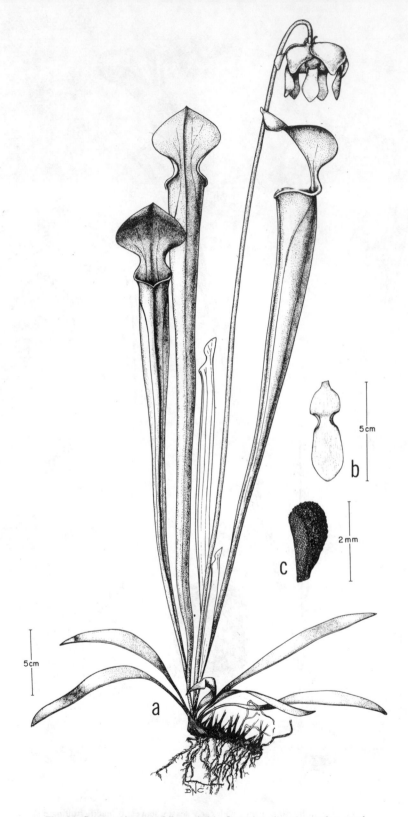

5 cm

5 cm

2 mm

Fig. 94. **Sarracenia oreophila:** a, habit, flowering plant; b, petal; c, seed.

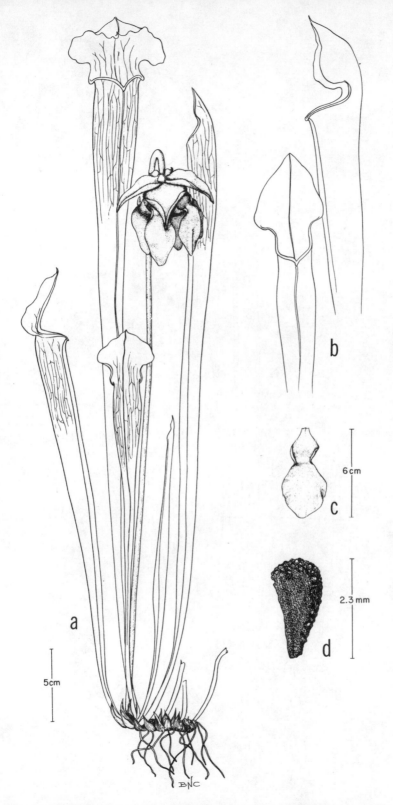

Fig. 95. **Sarracenia alata:** a, habit, flowering plant; b, tops of leaves, lateral and face views; c, petal; d, seed.

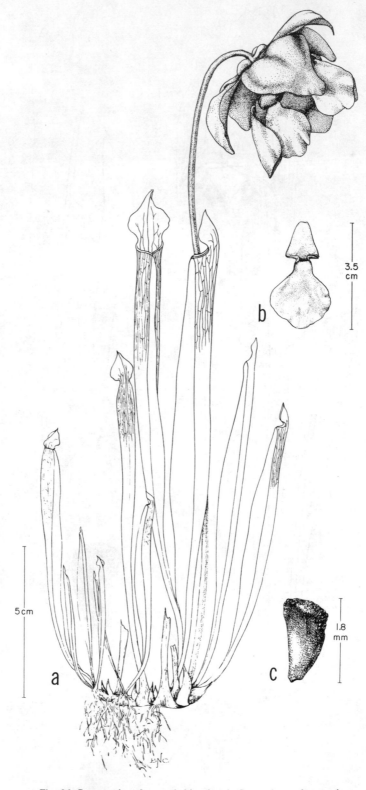

Fig. 96. **Sarracenia rubra:** a, habit, plant in flower; b, petal; c, seed.

faintly red- or purple-veined, or conspicuously and densely reticulate-veined, erect to some-what arching well above the orifice. Scape 1.2–7.5 dm long. Bracts 4–10 mm long, oblong or oblong-ovate. Sepals ovate-triangular, 1.8–2.7 cm long, purplish on the outer side. Petals maroon to red on the inner side, grayish to dull purple or red on the outer side, apical portion obovate. Style disc 2.8–4 cm broad, greenish.

Wet pine savannas, pineland seepage slopes, bogs, boggy stream margins. Mts. of s.w. N.C. and n.w. S.C.; coastal plain, s.e. N.C. to Ga. and s. Ala., w. Fla. Panhandle, s.e. Miss.; cen. Ala.

McDaniel (1971) stated that his broad concept of *S. rubra* included four regional variants with intergradation between them. Subsequently Case & Case (1974) described a segregate as *S. alabamensis* Case & Case. In a later paper (1976) the same authors treated the *S. rubra* complex as comprised of four taxa as follows: *S. rubra*, *S. jonesii* Wherry, *S. alabamensis* subsp. *alabamensis* and *S. a.* subsp. *wherryi* Case & Case. Schnell (1977, 1978) chose to distinguish the same taxa but all of them at the subspecific level. Schnell (1979) described an additional subspecies: *S. r.* subsp. *gulfensis* Schnell.

For purposes of this book, and until the dust settles, we leave it at that.

8. Sarracenia purpurea L. PITCHER-PLANT.   Fig. 97

Leaves overwintering, usually decumbent or sprawling, some leaves sometimes ascending, 0.5–4.5 dm long, urnlike, the tubular portion inflated distally, glabrous to hirsute externally, slick internally, with a very prominent lateral wing, variable in coloration from nearly green and with little suffused red or purplish red to variously reddish- or purple-veined, to nearly uniformly purplish red; hood 2–5 cm long, broader than long, reniform, sometimes notched apically, essentially erect, without a narrowed basal portion, the cordate basal lobes attached to the sides of the rim of the orifice, the orifice rim thus only from lobe to lobe of the hood, in extent half way or less of the circumference of the orifice, strongly revolute; inner surface of the hood markedly retrorsely pubescent. Scape 1.5–7 dm long. Bracts oblong to ovate-oblong, 5–8 mm long, rounded apically. Sepals ovate, 2–4 cm long, obtuse to rounded api-cally, usually more or less purplish red on the outside, more greenish within. Petals 3–6 cm long, deep maroon to rose or rose-pink, yellow in rare individuals, fiddle-shaped, the apical portion variably oblong, broadly elliptic, suborbicular, or obovate. Style disc 4–5 cm wide, yellowish green.

Bogs, swamps, wet pine savannas and flatwoods. Nfld. to Man., southward easterly, chiefly in the coastal plain, to the Fla. Panhandle and Miss., local in the piedmont and mts. of N.C., southward in the interior to Ohio, n. Ind. and Ill. to Iowa.

Crassulaceae (ORPINE FAMILY)

Annual or perennial, mostly succulent, herbs. Leaves simple, alternate, whorled, or op-posite, mostly persistent, without stipules. Flowers in cymes, less frequently in spikes or racemes or solitary, usually bisexual (rarely unisexual, the plants then dioecious), radially symmetrical. Sepals and petals 3–30 each (4 or 5 in ours), distinct in most, sometimes united basally. Stamens of the same number as the petals or twice as many, inserted on the base of the corolla in those having basally united petals. Carpels usually of the same number as the petals, superior, free or united basally, each subtended by a scalelike nectariferous gland, each maturing into a follicle. Seeds many (rarely few).

1. Leaves opposite, bases of the pairs united around the stem; flowers solitary in leaf axils. 1. *Tillaea*
1. Leaves alternate, sessile; flowers in small cymes terminating the branches.
  2. Carpels united basally (to somewhat below their middles); petals hooded apically; follicles dehisc-ing by a valvelike separation of nearly half of the lower portion of each carpel.   2. *Diamorpha*
  2. Carpels free; petals not hooded; follicles dehiscing along the upper suture.   3. *Sedum*

Fig. 97. **Sarracenia purpurea:** a, habit, plant in flower; b, petal; c, seed.

1. Tillaea

Tillaea aquatica L. PIGMYWEED.

Inconspicuous, glabrous annual, usually much branched and forming small mats, or, in water, with ascending stems to about 10 cm long or floating free. Leaves succulent, opposite, bases of the pairs united around the nodes, linear-oblong, 4–6 mm long. Flowers minute, about 1 mm wide, solitary in leaf axils, subsessile, stalks sometimes considerably elongating as the fruits mature. Sepals, petals, stamens, and carpels 3 or 4, usually 4. Petals twice as long as the sepals or a little more, greenish white, lance-elliptic. Follicles much exceeding the sepals, 8–10-seeded. Seeds minute, oblong, brown, striate and pitted between the striations (as seen with considerable magnification). (*Tillaeastrum aquaticum* (L.) Britt.)

Fresh to brackish tidal muds and shores, vernal pools, muddy banks, muds of depressions. Nfld. and Que. to Md.; inland across the n. states; Pacific coastal states; La. and e. and s.e. Tex.; Mex.; Euras.

2. Diamorpha

Diamorpha smallii Britt. ELF ORPINE.

Low, succulent, glabrous, reddish plants, mostly 3–7 dm high. Stem unbranched for 5–30 mm above the base, then with few to numerous short, leafy branches each bearing a cyme terminally, sometimes several-branched from the base, occasionally simple and with a single terminal cyme. Leaves alternate, sessile, oblongish to oblanceolate, 3–6 mm long, blunt apically. Cymes branched, each few-flowered. Sepals 4, minute, deltoid, 0.5 mm long or slightly more. Petals 4, pink or purple, elliptic-oblong, about 2–3 mm long, more or less persistent. Stamens twice as many as the petals, opposite to and alternate with them, filaments white, anthers black. Follicles usually 4, about 3 mm long, dehiscence of each by a valvelike separation of nearly half of the lower portion of the follicle. (*D. cymosa* (Nutt.) Britt. ex Small, name illegitimate; *Sedum smallii* (Britt.) Ahles)

In and about vernal pools on granite rock outcrops. Piedmont and lower mountains, N.C. to Ga., Ala., somewhat into Tenn. n. of n.w. Ga. and n.e. Ala.

3. Sedum

Sedum pusillum Michx. PUCK'S ORPINE.

Habit closely similar to that of *Diamorpha*, stem sometimes reddish below but the plant with an overall pale green aspect. Leaves alternate, sessile, nearly cylindric, 4–10 mm long. Sepals 4, deltoid, nearly 1 mm long. Petals 4, white, oblong or oblong-elliptic, 2.5–3 mm long, more or less persistent. Stamens 8, opposite to and alternate with the petals, filaments white, anthers purplish black. Follicles 4, 3 mm long or a little more, dehiscing along the upper suture.

In and about vernal pools on granite rock outcrops. Piedmont, N.C. to Ga.

Podostemaceae (RIVER-WEED FAMILY)

Podostemon ceratophyllum Michx. RIVER-WEED.   Fig. 98

Submersed aquatic herb, attached to rocks by fleshy discs, commonly in swiftly flowing water. Leaves rigid, alternate, 2-ranked, dissected into linear or filiform segments, sometimes simple, dilated into a pair of stipulelike appendages at base. Flowers axillary, solitary, bisexual, without perianth, subtended by a spathe from the leaf axil. Stamens 2, attached to the side of the ovary near the base, filaments fused below, free above; staminodia 2, short,

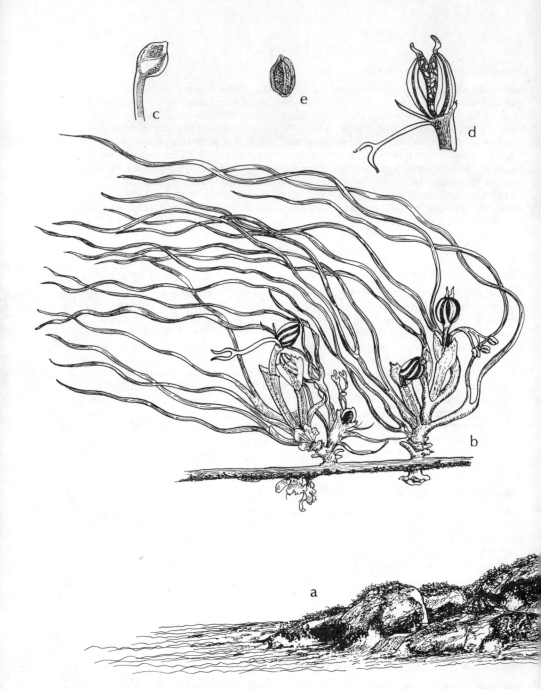

Fig. 98. **Podostemon ceratophyllum:** a, habitat, on rocks in river rapids; b, habit; c, bud; d, fruit; e, seed. (From Correll and Correll. 1972)

linear. Pistil 1, short-stipitate, 2-carpellate, ovary 2-locular, styles 2, 0.5–1 mm long, ovules numerous on a central placenta. Fruit a red capsule, stipitate (stipe elongating during maturation), obliquely oblong, 2-valved, longitudinally dehiscent, the larger valve persistent. Seeds sticky, red, oblong, about 5 mm long. (Incl. *P. abrotanoides* Nutt.)

Plants forming a low mat or crust on rocks in sluggish water, elongate, to several dm, in rapidly flowing water. Que. and Maine to Ont., generally southward to Ga., Ala., Miss., Ark., La., s.e. Okla. (Fla. Panhandle? e. Tex.?)

Saxifragaceae (SAXIFRAGE FAMILY)

Perennial, rarely annual, herbs, woody vines, or shrubs. Leaves opposite or alternate, without stipules. Inflorescence variable. Flowers usually bisexual, usually radially symmetrical, with a floral tube, the petals and stamens inserted on its rim. Sepals 4 or 5 (rarely 6 or 7). Petals mostly 4 or 5, sometimes more, rarely none. Stamens of the same number as the petals, twice as many, sometimes more numerous. Carpels fewer than the sepals, nearly distinct or the ovaries partially or wholly united, ovary or ovaries superior or partly inferior, styles usually as many as the carpels. Fruit a capsule, follicle, or berry.

1. Plant herbaceous.
  2. The plant annual, diminutive, forming a small rosette 1–3 cm broad, almost flat on the ground; leaves and sepals with red, glandular dots or short lines.   1. *Lepuropetalon*
  2. The plant perennial, habit not at all as described above; leaves and sepals not with red glandular dots or lines.
    3. Stems decumbent, rooting at the nodes at least below, mat-forming; leaves 5–15 mm long; sepals 4, petals none.   2. *Chrysosplenium*
    3. Stems not decumbent, not rooting at the nodes, not mat-forming; leaves (larger ones at least) much longer than 15 mm; sepals and petals 5 each.
      4. Leaves prominently and deeply palmately lobed.   6. *Boykinia*
      4. Leaves pinnately or palmately veined, marginally entire or toothed, not lobed.
        5. Stem leafy throughout.   3. *Penthorum*
        5. Stem with the principal leaves closely set near the base, essentially in rosettes.
          6. Leaves pinnately veined, margins toothed; flowers numerous, small, borne in scapose, panicled cymes.   4. *Saxifraga*
          6. Leaves palmately veined, margins entire; flowers showy, borne singly on a long scape usually having a single, sessile, clasping leaf at about its middle.   5. *Parnassia*
1. Plant woody.
  7. The plant a woody vine, the stem bearing numerous adventitious aerial roots that grow into the bark of the tree on which it climbs; flowers or fruits in compound cymes terminating branches of the current season.   7. *Decumaria*
  7. The plant an erect shrub; flowers or fruits in cylindrical racemes terminating branches of the current season.   8. *Itea*

1. Lepuropetalon

Lepuropetalon spathulatum (Muhl.) Ell.

Diminutive annual herb, forming a small rosette 1–3 cm broad and almost flat on the ground. Leaves alternate, glabrous, simple, sessile, spatulate, 3–10 mm long, margins entire, with red glandular dots or short lines. Flowers inconspicuous yet large considering the size of the plant, solitary at the end of the main stem and at the ends of branches. Floral tube campanulate. Sepals 5, ovate-triangular, 1–2 mm long, very unequal, glandular like the leaves, persistent. Petals 5, white, much smaller than the sepals, somewhat unequal. Stamens 5, filaments very short, Ovary inferior, 1-locular, styles 3. Seeds numerous, minute, oblong, reddish brown, reticulate.

Wet ditches, pond margins, depressions. s.e. N.C. to Ga., westward to e. half of Tex.; Mex.

2. Chrysosplenium

Chrysosplenium americanum Schwein.
WATER-MAT, WATER-CARPET, GOLDEN SAXIFRAGE.

Small, slender, glabrous, perennial herb, stems 5–20 cm long, decumbent or creeping, mat-forming, forking above. Leaves petiolate below, sessile above, mostly opposite, blades tender, somewhat succulent, ovate, suborbicular, or reniform, 5–15 mm long, margins entire or obscurely and irregularly crenate. Flowers small, terminal, solitary or in small cymes, bisexual, sessile or subsessile. Sepals 4, spreading, unequal, 2 ovate, 2 oblong, 1–1.5 mm long, greenish yellow to greenish red. Petals none. Stamens 4–8, inserted on a disc through which the 2 large-based, curved styles protrude. Carpels 2, united below the disc, ovary inferior, 1-locular. Seeds numerous, minute, globose, red, reticulate, short-pubescent.

Wooded or shaded seepage areas, margins and banks of small, shaded streams, about springs. Que. and Ont. to Sask., Minn., southward to Md., W.Va., Ohio, Ind., in the Appalachians to w. N.C., n. Ga., e. Tenn.

3. Penthorum

Penthorum sedoides L. **DITCH STONECROP.**   Fig. 99

Perennial rhizomatous herb. Stem leafy throughout, glabrous, decumbent at the base, then erect, simple or branched, to 7 or 8 dm tall. Leaves alternate, simple, lanceolate or elliptic, serrate, narrowed below to subpetiolar bases, acute to acuminate apically, 5–10 cm long, to about 4 cm wide. Flowers loosely spiked along the upper side of 2 to several branches of cymes terminating branches, branches of the cyme mostly 2–8 cm long; axes of the cyme stipitate-glandular. Flowers greenish, short-stalked. Sepals 5 (rarely 6 or 7), arising from a membranous floral tube, short-ovate-deltoid, obtuse to acute apically, persistent. Petals none. Stamens 10, inserted on the floral tube between the sepals, the shriveled filaments usually persistent. Ovary united to the floral tube only at the base thus nearly superior, the 5 carpels united to about the middle thence each abruptly narrowed to a short separate style, the styles incurved at anthesis; in fruit the basal united portion of the ovary forms a lobed-circular capsule, horizontally somewhat flattened, 5–6 mm across, the stylar summits separate and hornlike, erect or curved outwardly; dehiscence circumscissile laterally on each lobe. Seeds numerous, tiny, fusiform, rust-colored, surfaces reticulate and with many short, blunt protuberances.

Floodplain forests, swamps, marshes and marshy shores, ditches, wet clearings, and swales. Maine to Ont. and Minn., generally southward to n. Fla., e., s.e., and Panhandle of Tex.

4. Saxifraga (SAXIFRAGES)

Perennial herbs, the leaves alternate, principal larger ones closely set and essentially in a rosette (ours), the numerous small flowers in terminal cymes or panicled cymes on one or more branches bearing few bracteal leaves or none. Flowers bisexual, radially symmetrical or only the corolla slightly bilateral. Floral tube short, free from or fused to the base of the ovary or fruit. Sepals 5, persistent, in some ascending around the base of the fruit, in some reflexed. Petals 5, clawed basally in some. Stamens 10, filaments filiform or more or less clavate. Carpels 2 in most, united basally or nearly distinct, each maturing into a follicle dehiscent along its lower suture. Seeds numerous in each follicle.

For critical determination of identities, even for the few taxa treated in this work, it is important to have inflorescences, or portions of them, both with flowers at anthesis and with well-developed fruits. It is, perhaps, easier for one not familiar with the group to make identifications before specimens are dried.

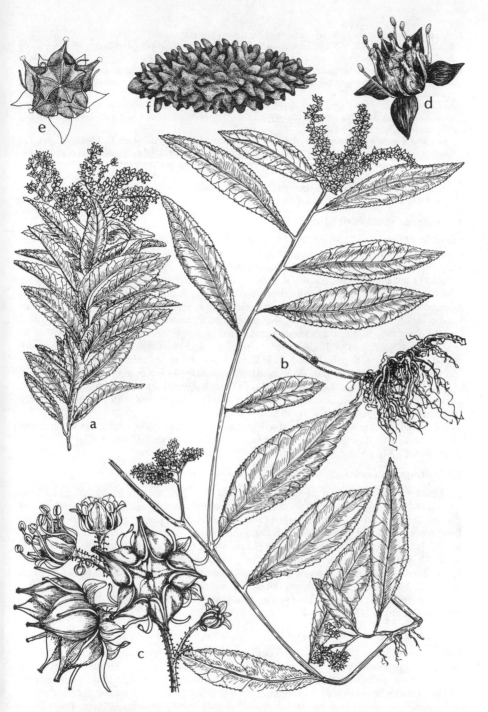

Fig. 99. **Penthorum sedoides:** a, tip of plant; b, part of procumbent stem of plant and roots; c, cluster of flowers and fruits; d, flower; e, fruit; f, seed. (a–c, from Correll and Correll. 1972)

1. Sepals ascending at anthesis and remaining so around the fruit.
　2. Carpels or follicles 2; inflorescence abundantly clothed with gland-tipped hairs.　1.　*S. virginiensis*
　2. Carpels or follicles 3 or 4; inflorescence essentially glabrous.　　　　　　　　　2.　*S. texana*
1. Sepals ascending or reflexed during anthesis but strongly reflexed following anthesis.
　3. Larger leaves with ovate blades twice as long as broad or less, abruptly contracted basally to the petioles.　　　　　　　　　　　　　　　　　　　　　　　　　　　　　　　　3.　*S. careyana*
　3. Larger leaves oblanceolate to long-oblong, the essentially bladed portions 3–8 times as long as broad, basally tapering very gradually into winged petioles (blades and petioles not distinctly differentiated).
　　4. Corolla with 3 distinctly clawed petals (somewhat larger than the other 2), blades of each minutely auriculate on either side at the base and each white with a pair of pale yellow spots proximally on the blade; other 2 petals wholly white, oblanceolate, tapered gradually from their widest places to the base.　　　　　　　　　　　　　　　　　　　　　　　　　　　　　　　4.　*S. michauxii*
　　4. Corolla with essentially similar petals, all short-clawed, all without basal auricles, each with a patch of pale yellow proximally on the blade.　　　　　　　　　　　　　5.　*S. micranthidifolia*

1. Saxifraga virginiensis Michx.

Larger leaf blades mostly ovate, but sometimes some of them oblong or elliptic-oblong, occasionally rotund, 1.5–6 cm long or a little more, most of them abruptly contracted to winged petioles about as long as the blades but sometimes tapering very gradually, apices obtuse to rounded, margins usually dentate or dentate-serrate, teeth few and blunt; petioles ciliate, more densely so proximally than distally, young leaf blades pubescent on both surfaces, most hairs sloughing later, the pubescence of long septate hairs. Scapes with their panicled cymes 1–2 dm high at anthesis, the inflorescences compact or relatively open, the axes usually elongating as the fruits mature and attaining length and breadth dimensions up to twice those obtaining at anthesis; inflorescence axes and short flower stalks bearing gland-tipped, septate hairs. Sepals deltoid, 1.5–2 mm long, obtuse apically, ascending at anthesis and remaining so. Petals white, oblanceolate or spatulate, 4–6 mm long. Carpels and follicles 2, fused to the floral tube basally, the follicles not much exceeding the sepals. Seeds falcate-elliptic or -oblong, about 0.5 mm long, reddish, striate and with papillae along the striations. (*Micranthes virginiensis* (Michx.) Small)

　Commonly inhabiting well-drained woodlands, rock ledges, mossy banks, in moist to wet shallow soil of rock outcrops. N.B. to Minn., generally southward (excluding the coastal plain) to Ga., Ala., Miss., Ark.

2. Saxifraga texana Buckl.

Leaf blades generally similar in shape to those of *S. virginiensis*, mostly smaller, to 4 cm long, mostly less, glabrous or nearly so, margins undulate. Scape with its inflorescence 5–15 cm tall, the inflorescence small, compact and remaining so; scape sparsely pubescent, inflorescence essentially glabrous. Sepals deltoid to oblong, ascending at anthesis and remaining so, 1.5–2 mm long, obtuse apically. Petals white, elliptic or obovate, 2.5–3 mm long, tapered basally. Carpels 3 or 4, united to the floral tube basally, follicles little exceeding the calyx. Seeds ovate-falcate, somewhat angled, about 0.4 mm long, reddish brown, surface roughish-ribbed. (*Micranthes texana* (Buckl.) Small)

　In seepage areas, rocky glades, prairies, and pastures. s.w. Mo. and s.e. Kan., southward through Ark. and e. Okla. to e. Tex.

3. Saxifraga careyana Gray in Hook.

Larger leaf blades mostly ovate, 5–10 cm long, 3–5 cm broad, abruptly contracted basally to margined petioles about as long as the blades, margins coarsely dentate, the teeth 6–10 on a side, obtuse; petioles and blade surfaces moderately pubescent with long, gland-tipped, soft hairs. Inflorescences with their open panicled cymes 1–4 dm tall, scape axes and flower stalks with pubescence like that of the leaves. Sepals oblong or ovate-oblong, about 1 mm long, obtuse apically, glabrous or nearly so, reflexed after anthesis. Petals equal in size, shortly clawed or not clawed, mostly oblong, white with 2 pale yellow spots proximally, about 4 mm long. Carpels and follicles 2, united basally, free from the floral tube. Seeds not seen (by us). (*Micranthes careyana* (Gray in Hook.) Small)

Mossy seepage areas, wet cliffs and ledges, rocky wooded slopes. Mts. of Va., Ky., N.C., Tenn.

4. Saxifraga michauxii Britt.

Leaves oblanceolate, proximally tapered very gradually into winged petioles, blades and petioles scarcely differentiated, 3–15 cm long overall, margins of bladed portion coarsely dentate or dentate-serrate, teeth 4–8 on a side, obtuse or acute; leaf surfaces, scape, inflorescence axes, and flower stalks copiously pubescent with long, soft, mostly gland-tipped hairs. Scape with diffusely open, lacy panicle 1.5–5 dm high, the panicle commonly as broad as high, more or less leafy-bracted. Sepals glabrous, mostly oblong, mucronate apically, 1–2 mm long, strongly reflexed after anthesis. Corolla with 3 distinctly clawed petals (somewhat larger than the other 2), the blades of each minutely auriculate on either side at the base and with a pair of yellow spots proximally on the blade; other 2 petals wholly white, oblanceolate, tapered gradually from their widest places to the base. Carpels and follicles 2, fused basally, free from the floral tube. Seeds ellipsoid, slightly falcate, about 0.5 mm long, dark brown, few-ribbed, with minute, hooked trichomes along the ribs. (*Hydatica petiolaris* (Raf.) Small)

Wet rock ledges and cliffs, rock crevices, rocky high mt. woodlands and grass balds. Mts., Va. and W.Va., southward to n. Ga. and Tenn.

5. Saxifraga micranthidifolia (Haw.) Steud.

Leaves similar in shape and form to those of *S. michauxii*, many of them attaining larger size, to 2 dm long or more, their marginal dentations more numerous, 12–40 on a side, the teeth on a given leaf commonly varying in size, some or all of them mucronate; the broadly winged petioles mostly moderately to copiously pubescent with long, soft hairs, the blade surfaces relatively sparsely pubescent with short hairs. Scapes with their open panicles 3–8 dm high, the panicles not as leafy-bracted and not so lacy as in *S. michauxii*. Scapes, inflorescence axes, and flower stalks pubescent. Pubescence of leaves and scapes usually not gland-tipped; that of inflorescence axes and flower stalks mostly gland-tipped but usually with some longer, nonglandular hairs intermixed. Sepals triangular to oblong or broadly oval, often varying thus in shape in a single inflorescence, obtuse to rounded apically, 1–2 mm long, glabrous, strongly reflexed after anthesis. Petals 2–3 5 mm long, clawed, blades oval to obovate, white with a patch of yellow proximally on the blades. Carpels and follicles 2, fused basally, free from the floral tube. Seeds falcate-fusiform, about 1 mm long, dark reddish brown, ribbed. (*Micranthes micranthidifolia* (Haw.) Small)

On rocks and logs of streams and stream banks, wet rock ledges and cliffs, seepage places. Mts., Pa., W.Va., Va., southward to n. Ga. and Tenn.

5. Parnassia (GRASSES-OF-PARNASSUS)

Glabrous, perennial herbs. Principal leaves closely set near the base, long-petioled, usually a single sessile, clasping leaf on the long scape bearing a single flower terminally, blades entire. Sepals 5, slightly united basally, persistent. Petals 5, spreading, white with conspicuous green or yellowish veins, often differing somewhat in size and shape in a single flower. Fertile stamens 5, alternate with the petals, 5 clusters of basally more or less united staminodia having gland-bearing tips alternating with the fertile stamens. Pistil 1, ovary superior, 1-locular, with 4 parietal placentae, stigmas 4, sessile or subsessile.

The species of *Parnassia* occurring in our range tend to look much alike so that in distinguishing them reliance is largely placed on a few "technical" characters.

1. Leaf blades reniform; petals short-clawed; staminodia shorter than or no longer than the fertile stamens. _____ 1. *P. asarifolia*

1. Leaf blades ovate to oblong; petals not clawed; staminodia considerably longer than the fertile stamens.
  2. Veins of the petals 5–9 (–11), the lower lateral pair with few, short, dead-ending branches; ovary
  white.   2. *P. grandifolia*
  2. Veins of the petals (9–) 11–18, the lower, lateral pair much branched and often with 1 to few re-
  ticulations; ovary green or white below and green distally.   3. *P. caroliniana*

1. Parnassia asarifolia Vent.   Fig. 100

Leaf blades reniform, 3–4 cm long, wider than long, the larger 5–10 cm broad. Petals abruptly contracted proximally to short claws, blades ovate to obovate, 10–18 mm long overall, main veins 11–15. Staminodia equaling or shorter than the fertile stamens.

Bogs, seepage areas, springy places. Mts., Va., W.Va., Ky., southward to n. Ga., Ala., Tenn.

2. Parnassia grandifolia DC.   Fig. 101

Leaf blades ovate to oblong, most of them longer than broad, larger ones 5–10 cm long, bases truncate, rounded, or cordate, apices broadly obtuse. Petals not clawed basally, elliptic, rhombic, or slightly obovate, 15–20 mm long (commonly each of these shapes and the range of lengths in the petals of a single flower), the lowest pair of lateral veins each with 4–7 ultimate branchlets terminating toward the margin, none of the veinlets anastomosing (see Fig. 101). Staminodia considerably longer than the fertile stamens. Ovary white.

Bogs, meadows, seepage areas, bases of dripping cliffs. W.Va., w. Va., westward to Mo., generally southward to n. Fla. and e. Tex.

3. Parnassia caroliniana Michx.   Fig. 101

Leaf blades essentially as in *P. grandifolia*. Petals not clawed basally, most of them ovate in a given flower, 15–20 mm long, the lowest lateral pair of veins much more dendritically branched, usually with 12–20 ultimate branchlets terminating toward the margin, often with 1 to several veinlets anastomosing (see Fig. 101). Staminodia as in *P. grandifolia*. Ovary green, or white basally and green distally.

Bogs, savannas, flatwoods. Coastal plain, s.e. N.C. to Fla. Panhandle.

6. Boykinia

Boykinia aconitifolia Nutt. BROOK-SAXIFRAGE.

Perennial, rhizomatous herb, 2–10 dm tall. Leaves alternate, the lower ones very long-petioled, few leaves on the stem, gradually shorter-petioled upwardly, uppermost sessile; blades deeply palmately lobed, the main lobes irregularly toothed-lobed, in outline more or less reniform, the larger ones 6–14 cm broad. Flowers small, in panicled cymes, inflorescence axis, short flower stalks, and floral tubes glandular-pubescent. Floral tube broadly urceolate, united to the lower half of the ovary but wholly enclosing the pistil. Sepals 5, triangular-subulate, 1.5–2 mm long, sparsely glandular-pubescent. Petals 5, white, short-clawed, obovate, spreading, a little over twice as long as the sepals. Stamens 5, opposite to and shorter than the sepals. Ovary half-inferior, 2-locular, styles 2, hornlike, stigmatic terminally, persistent; carpels maturing into follicles within the floral tube. Seeds numerous, about 0.4 mm long, more or less oval in outline, dark purplish brown, very finely papillose and somewhat lustrous. (*Therophon aconitifolium* (Nutt.) Millsp.)

Stream and river banks and their rocky shores, moist to wet woodlands and banks, seepage areas, wet rocks. Mts., Va., W.Va., Ky., southward to n. Ga. and n. Ala.

7. Decumaria

Decumaria barbara L. CLIMBING-HYDRANGEA, WOOD-VAMP.

A handsome, deciduous, woody vine, many adventitious roots growing from the stems and into the bark of trees on which it climbs to high levels, commonly with many short, usually

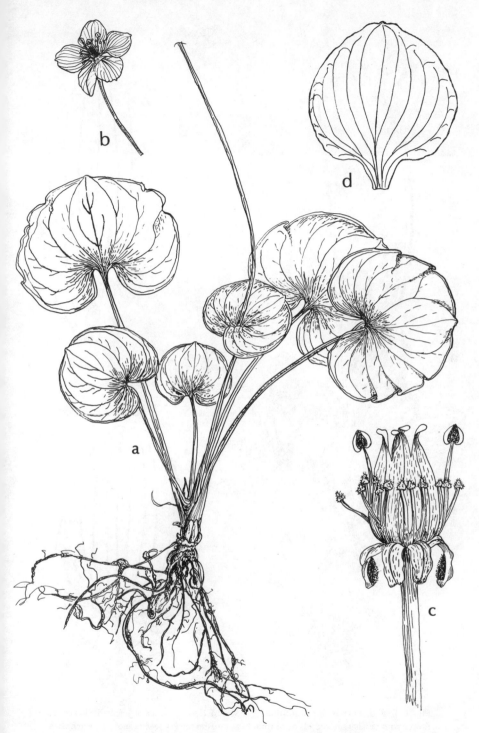

Fig. 100. **Parnassia asarifolia:** a, habit; b, flower; c, flower with petals removed; d, petal. (From Correll and Correll. 1972)

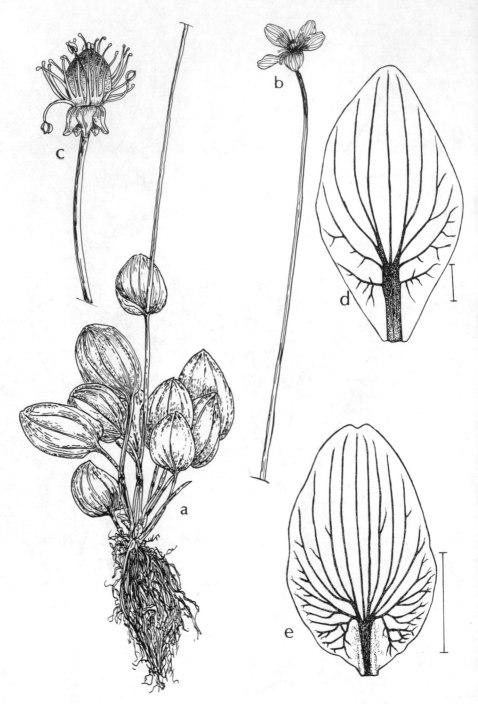

Fig. 101. a–d, **Parnassia grandifolia:** a, habit of lower portion of plant; b, flower; c, flower with petals removed; d, petal (bracket opposite lowest vein branch); e, **Parnassia caroliniana:** petal (bracket opposite lowest vein branch). (a–c, from Correll and Correll. 1972)

212

pendent branches, all along the tree trunk; sometimes with small branches bearing leaves much smaller than elsewhere on the vines creeping about the bases of trees and over rocks; young herbaceous portions of the stem glabrous, older stems to 1 cm or more in diameter, with roughish, brown bark. Leaves opposite, with slender petioles 1–4 cm long, blades ovate, oval, or nearly orbicular, usually varying in size and shape on a single plant, the larger ones 10–12 cm long, the wider ones to 8 cm broad; bases rounded to broadly short-tapered, apices abruptly short-acuminate, acute, or obtuse; upper surfaces dark green and glabrous, lower paler, usually sparsely short-pubescent when young, sometimes becoming glabrous or nearly so in age. Flowers small, numerous, borne in compound cymes, 4–10 cm broad, terminating branchlets of the current season. Floral tube conic, pale, thin, and narrowly low-ribbed at anthesis, becoming obpyramidal to campanulate, hard and strongly ribbed in fruit. Sepals 7–10, minute, triangular, persistent. Petals 7–10, white, oblanceolate, spatulate, or narrowly elliptic, 5–6 mm long. Stamens 20–30, filaments about as long as the petals. Ovary inferior, 7–10-locular, style 1, short and thick, surmounted by a 7–10-lobed stigma. Capsule 3–6 mm long, the strong ribs pale, dark brown between the ribs, the style and stigma persistent, tardily dehiscing along the ribs, the intervening tissue slowly sloughing away eventually leaving the bowed ribs, apical rim, and summit of the ovary with style and stigmas intact; then appearing lanternlike. Seeds numerous, angular, irregularly narrowly club-shaped in outline, about 2 mm long, tan and somewhat lustrous.

Floodplain forests, wet woodlands, river banks, sometimes in mesic woodlands on slopes above streams. Coastal plain, s.e. Va. and N.C., along rivers draining the eastern slopes of the Blue Ridge in s.w. N.C., throughout S.C., southward to cen. pen. Fla. and westward to Tenn. and La.

8. Itea

Itea virginica L. VIRGINIA-WILLOW, TASSEL-WHITE. Fig. 102

Deciduous shrub, to about 2.5 m tall, rather loosely branched, branchlets commonly arching, stems of young branchlets sparsely and minutely pubescent, quickly becoming glabrous. Leaves simple, alternate, without stipules, very short-petioled, blades very variable in shape and size even on a single branch, elliptic, oblong, oval, oblanceolate, or infrequently somewhat obovate, pinnately veined, cuneate basally, apices varyingly short-acuminate, acute, or obtuse, larger ones 5–10 cm long and 3–4 cm broad, (usually none of them fully grown at time of flowering); margins minutely serrate, the teeth pointed; petioles and lower leaf surfaces with few short hairs. Flowers small, borne on stalked, arching, cylindrical racemes 3–10 cm long and about 1.5 cm across, the short flower stalks diverging at right angles to the axis, the racemes terminating branchlets of the current season; axis of the raceme and flower stalks varying from sparsely to copiously short-pubescent, floral tubes sometimes similarly pubescent throughout or only at the base. Floral tube hemispherical, free from the ovary. Sepals 5, erect, subulate, 1 mm long or a little more, deciduous. Petals 5, white, linear-subulate, erect, pubescent at the base on the upper side, 6–7 mm long. Stamens 5, about half as long as the petals. Ovary superior, pubescent, 2-locular, style slender, persistent, stigma capitate, deciduous. Fruit conic-cylindric, hard, grooved along the two sutures, tardily splitting downward throughout the styles and along the sutures about half way, sometimes wholly, to the base; capsules more or less reflexed on the axis. Seeds several, irregularly oblongish in outline, a little over 1 mm long, a reticulum of isodiametric surface cells clearly evident, surface gold-colored, lustrous.

Swamps, wet woodlands, along wooded streams. s. N.J. and e. Pa., southward through e. and cen. Va., most of N.C., thence to pen. Fla. and westward to e. Tex., northward in the interior, s.e. Okla., Ark., w. Tenn., to s.w. Ky., s. Ill. and s.e. Mo.

213

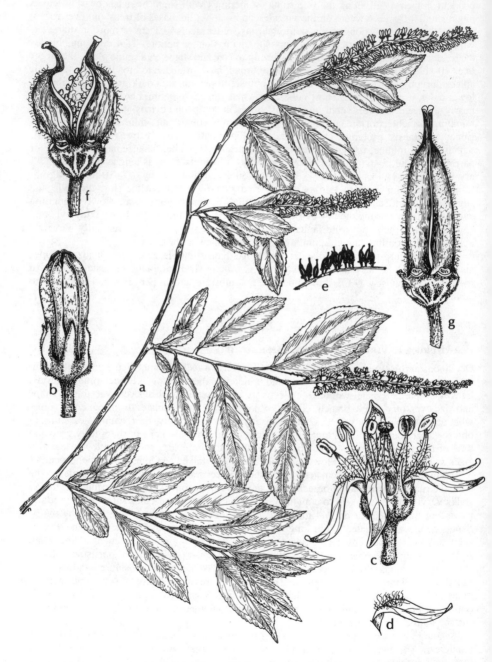

Fig. 102. **Itea virginica:** a, branch; b, bud; c, flower; d, petal; e, portion of fruiting inflorescence; f, opening fruit; g, empty fruit. (From Correll and Correll. 1972)

Hamamelidaceae (WITCH-HAZEL FAMILY)

Shrubs or trees with alternate, simple, deciduous leaves with quickly deciduous stipules. Flowers in heads or spikes, usually sessile, bisexual or unisexual, with a short floral tube adherent to the base of the ovary, calyx segments present or absent. Petals, if present, inserted on the floral tube. Stamens inserted on the floral tube, sometimes few and accompanied by scalelike staminodia, sometimes numerous. Ovary united below and 2-locular, styles 2, free, persisting on the woody capsular fruits.

1. Leaves palmately veined and lobed, glabrous; flowers and fruits in dense globose heads.
1. Liquidambar

1. Leaves pinnately veined, unlobed, stellate-pubescent at least below when young, usually sparsely so even in age; flowers in stalked, spicate, few-flowered clusters <i>or</i> in dense globose to oblong compact spikes.

   2. The leaves with undulate margins distally from about the middle of the blade or a little below; flowers few in short-stalked, loosely spicate clusters axillary to leaf scars on twigs of a previous season; petals 4; stamens 4. 2. <i>Hamamelis</i>

   2. The leaves crenate-dentate or dentate-serrate, usually distally from somewhat above the middle of the blade; flowers numerous in globose to oblong, compact spikes terminating branches or axillary to leaf scars; petals none; stamens numerous. 3. <i>Fothergilla</i>

1. Liquidambar

Liquidambar styraciflua L. SWEET GUM, RED GUM. Fig. 103

A handsome tree 40 m tall or more, its foliage aromatic when freshly crushed, with grayish brown bark and commonly (not always) with corky-winged or corky-ridged twigs. Leaves long-petiolate, glabrous, palmately veined and usually palmately 5-lobed giving much the effect of a star; margins sharply and evenly serrate, lobes acute and leaf base truncate to somewhat cordate. Stipules linear-oblanceolate, 8–10 mm long. Flowers unisexual, petals none, the plants monoecious. Staminate flowers greenish yellow, intermixed with scales and borne in small tightly congested globose heads on a stiff, stout axis; pistillate flowers pale green, in compact spherical heads one or more of which dangle on slender stalks to 5 cm long from below the staminate portion of the axis. Fruiting clusters mature into semiwoody, strongly echinate, pendent balls of fused capsules each of which opens between the free persistent styles and produces 1 or 2 winged seeds.

   Inhabits a variety of wooded mesic sites, in the southern part of the range commonly in wet or swampy woodlands as well, occasionally on sites where water stands almost continuously. s. Conn. to s. Ill., generally southward to s. Fla., e. and s.e. Tex., s.e. Okla.; Mex. and C.Am.

2. Hamamelis

Hamamelis virginiana L. WITCH-HAZEL. Fig. 104

Shrub or small tree, the young twigs and petioles densely stellate-pubescent, sometimes the surfaces of young leaves as well; older leaves usually with some stellate pubescence, at least on the veins below. Leaves short-petiolate, blades obovate, oval, or broadly elliptic, variable in size, to 15 cm long and 10 cm wide, bases broadly tapered to rounded, often oblique, apices rounded or obtuse, infrequently acute, margins distally from the middle or a little below with broad undulations. Stipules lanceolate, 6–8 mm long, densely stellate-pubescent. Flowers borne from the axils of leaf scars, single or usually in a short-stalked cluster of 3 and subtended by an involucre of 3 scales; bisexual or unisexual. Floral tube cuplike, with 2 or 3 bractlets at its base, 4 ovate calyx segments about 2.5 mm long on its rim. Petals 4, yellow, sometimes suffused with red, linear, 1–2 cm long. Fertile stamens 4, alternate with the petals, a scalelike staminodium between each fertile stamen. Ovary densely pubescent,

Fig. 103. **Liquidambar styraciflua:** a, twig with fruit; b, seeds. (From Correll and Correll. 1972)

partially inferior, styles 2. Capsule ovoid or thickly ellipsoid, its base fused with the floral tube, the persistent stylar beaks recurved, a single, large, hard, brown seed in each locule. Flowering occurs mostly in autumn, sometimes in winter or spring southward in the range, fruit maturation a year later.

In various upland woodland mixtures, wooded slopes and ravines, in the southeastern part of the range, at least, in floodplain forests, evergreen shrub bogs, stream banks. Que. and N.S. to n. Mich. and s.e. Minn., generally southward to cen. pen. Fla. and Tex.

3. Fothergilla

Fothergilla gardeni Murray. WITCH-ALDER.

Slender shrub to about 1 m tall, sprouting from subterranean runners and colonial. Young twigs, petioles, and both leaf surfaces densely stellate-pubescent; older leaves varying from pubescent to nearly smooth, usually a sparse pubescence on lower leaf surfaces even in age. Leaves short-petiolate, blades mostly obovate, less frequently broadly elliptic, rounded, truncate, or broadly cuneate basally, apices obtuse, rounded, or nearly truncate, margins usually crenate-dentate or serrate-dentate near their apices, sometimes entire. Stipules about 2 mm long, ovate-oblong, blunt apically, densely stellate-pubescent. Flowers in dense spikes terminal on twigs or axillary to leaf scars, appearing in spring before or as new shoots of the season emerge; mostly unisexual, petals none; staminate with numerous stamens borne on one side of an oblique, unlobed calyx; pistillate with a short cupular floral tube adherent to the base of the ovary, short, erect calyx segments at its summit; ovary pubescent, styles elongate at anthesis, in fruit becoming thick basally, hornlike, and curving outward at their tips. Base of capsule fused with the floral tube. (Incl. *F. parvifolia* Kearney)

Pine savannas and evergreen shrub bogs. Coastal plain, N.C. to Ga. and Ala.

Platanaceae (PLANE-TREE FAMILY)

Platanus occidentalis L. SYCAMORE, PLANE-TREE, BUTTONWOOD.   Fig. 104

A large tree, to 50 m tall, picturesque because of its bark, at first creamy white, becoming brown, later mottled or dappled as large platelike portions exfoliate. Old bark near the bases of trunks brown, furrowed and scaly. Leaves simple, alternate, deciduous, about as broad as long, sometimes broader, to 2 dm across (longer on vigorous sprouts); palmately veined and irregularly lobed, the margins irregularly coarsely toothed; bases truncate to cordate; thickly downy when young, becoming glabrous except along the veins beneath; petioles long, with enlarged bases, and with large leaflike, lobed or toothed stipules completely united around the twigs; stipules may drop early in the season or some may persist after leaf-fall giving the otherwise naked twigs a distinctive appearance. Flowers minute, borne in large numbers in long-stalked, unisexual, spherical balls. Calyx and corolla each of 3–6 very small sepals and petals or petals none. The fruiting head ("button balls") about 2.5 cm in diameter and comprised of numerous achenes each subtended by long, tawny hairs.

Mostly along streams, in floodplains and bottomlands. s. Maine to Ont., generally southward to Ga., w. Fla. and Tex.

Rosaceae (ROSE FAMILY)

Trees, shrubs, or (usually perennial) herbs. Leaves almost always alternate, simple or compound, usually with stipules (which are often adnate to the petioles). Inflorescences very varied. Flowers radially symmetrical, mostly bisexual, having a well-developed receptacular

217

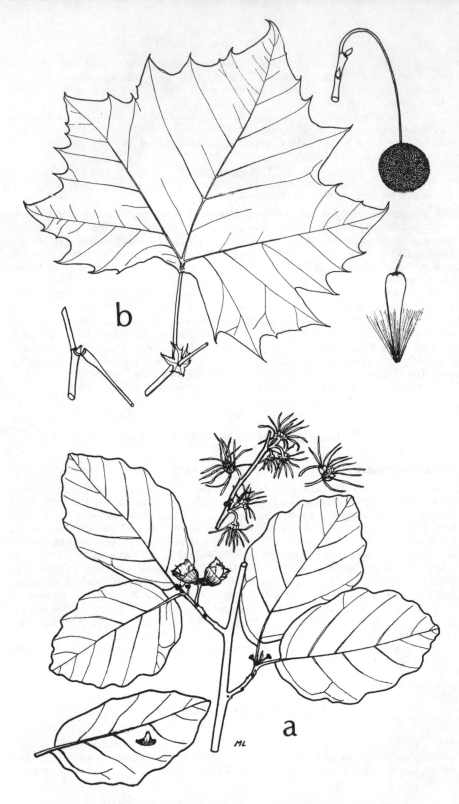

Fig. 104. a, **Hamamelis virginiana**; b, **Platanus occidentalis.** (From Kurz and Godfrey, *Trees of Northern Florida.* 1962)

disc, or a floral tube, the latter arising below the pistil or pistils and free from it or them, or fused to a greater or lesser degree, in cases where the carpels are fused then the ovary partially or wholly inferior. Sepals and petals most commonly 5 each, sometimes fewer or more, borne on the rim of the floral tube. Stamens distinct, commonly numerous (sometimes definite and then usually 5 or 10) and in 1 to several series, arising from the rim of the floral tube. Pistils 1 to many, distinct, or of 2–5 carpels united and these more or less fused with the floral tube. Type of fruit extremely various, a follicle, achene, pome, drupe, or aggregation of drupelets.

1. Plant woody, a shrub, tree, or bramble.
  2. Leaves simple.
    3. Pistils 3–5, superior.
      4. Leaf blades ovate, some on a given plant usually lobed; flowers in umbellike racemes.
                                                          1. *Physocarpus*
      4. Leaf blades elliptic or oblanceolate, unlobed; flowers (in those treated here) in many-flowered panicles.   2. *Spiraea*
    3. Pistil 1, wholly or almost wholly inferior.
      5. Thorns present on at least some of the branches.   11. *Crataegus*
      5. Thorns none.
        6. Flowers or fruits in corymbose clusters.   9. *Aronia*
        6. Flowers or fruits in racemes.   10. *Amelanchier*
  2. Leaves compound.
    7. Floral tube urceolate, wholly enclosing numerous separate pistils; leaves odd-pinnately compound.   8. *Rosa*
    7. Floral tube small, flat to hemispheric, the numerous separate pistils borne on a convex or conic receptacle elevating them above the floral tube; leaves pedately 5-foliolate or ternate.   5. *Rubus*
1. Plant herbaceous.
  8. Low perennial with slender rhizomes, to some extent mat-forming; leaves simple, blades orbicular to reniform; flowers axillary and solitary, some of them with petals and sterile, some without petals and fertile.   6. *Dalibarda*
  8. Plant without the above combination of characters.
    9. Flowers in dense spikes or heads; sepals 4, petaloid, white; petals none; pistil 1, ovary free from but enclosed by the floral tube.   7. *Sanguisorba*
    9. Flowers in many-flowered panicles, *or* solitary to few in leafy-bracted, irregularly branched, open inflorescences terminal on the branches; sepals 5, not petaloid; petals 5; pistils several to numerous, wholly or almost wholly elevated above the floral tube.
      10. Flowers numerous, in a terminal panicle; petals deep pink to rose; pistils 5–7, occasionally fewer or more, styles short, thickish, not jointed, not bent or looped; stigmas capitate.
                                                      3. *Filipendula*
      10. Flowers solitary to few in leafy-bracted, irregularly branched, open inflorescences terminal on the branches; petals white or yellow; pistils numerous, styles long and slender, 1-jointed, bent or with a loop near the middle or near the summit, stigmas linear.   4. *Geum*

1. Physocarpus

Physocarpus opulifolius (L.) Maxim. NINEBARK.

Shrub to about 3 m tall, the bark of older stems loose and peeling or shredding into long, thin layers or strips. Leaves simple, with petioles 1–2 cm long, glabrous or pubescent; blades palmately veined, mostly ovate in outline, many of them, sometimes all, 3-lobed, sometimes none lobed or both lobed and unlobed on a given branch, bases truncate, rounded, slightly cordate, or occasionally shortly tapered, apices broadly rounded, obtuse, or rarely acute, margins regularly or irregularly finely crenate or crenate-serrate; upper surfaces dark green, lower pale green, lower and sometimes the upper with sparse to dense stellate pubescence in the proximal vein axils, often with sparse pubescence along some or most of the principal veins; stipules narrowly lanceolate to linear-oblong, 6 mm long, quickly deciduous. Inflorescences umbellike racemes terminating branchlets; raceme axis, flower stalks, floral tubes, and sepals moderately to densely stellate-pubescent, sometimes nearly glabrous. Sepals 5, persistent, triangular, 3–4 mm long. Petals 5, white or white tinged with pink, suborbicular,

4–6 mm broad. Stamens numerous. Pistils 3–5, the ovaries maturing into 2–4-seeded, glabrous and lustrous to stellate-pubescent, 2-valved pods 5–10 mm long, beaked apically, extending beyond the floral tube and calyx for most of their length. Seeds essentially pear-shaped, tan or buff, lustrous, 1.5–2 mm long. (*Opulaster opulifolius* (L.) Kuntze; incl. *O. australis* Rydb., *O. alabamensis* Rydb.)

Creek banks, open woods and thickets astride creeks, rocky banks and slopes, thickets along shores and in seepage areas, bogs, moist cliffs. Que. to Minn., S.D., Colo., generally southward to w. half of N.C. and S.C., n. Ga. and Ala., Fla. Panhandle, Tenn., Ark.; s.w. Ga., cen. Panhandle of Fla.

2. Spiraea

Shrubs. Leaves simple, deciduous, short-petiolate, without stipules. Flowers bisexual, small, usually numerous in showy panicles or corymbs. Floral tube campanulate. Sepals and petals 5 each. Stamens numerous. Pistils usually 5, free within the floral tube, ovaries maturing into follicles dehiscent along the inner suture and finally also along the outer suture distally, several-seeded.

● Lower leaf surfaces entirely clothed with a reddish brown to whitish, dense, compact, cottony pubescence; petals pink or rose, very rarely white.   1. *S. tomentosa*
● Lower leaf surfaces green, glabrous or occasionally with a few hairs on the midrib; petals white, sometimes pinkish in bud.   2. *S. alba*

1. Spiraea tomentosa L. HARDHACK, STEEPLE-BUSH.

Shrub to 1 or 2 m tall. Stems simple or but sparsely branched below the inflorescence, grayish-tomentose, the tomentum usually more or less sloughed on older portions. Leaf blades oblanceolate or elliptic, rarely lance-ovate, 2–5 cm long, mostly cuneate basally, obtuse or less frequently acute apically; margins bluntly, less frequently acutely, serrate distally from a little below or from about the middle, blades glabrous or sparsely short-pubescent above, with a reddish brown to whitish, dense, compact, cottony pubescence beneath. Flowers usually many, crowded in an elongate, spirelike panicle, the panicle axes, flower stalks, floral tubes, and exterior surfaces of the sepals densely clothed with tomentum like that of the leaves. Sepals triangular, about 1 mm long, soon reflexed. Petals pink or rose, very rarely white, ovate to obovate, about 2 mm long, commonly bluntly few-toothed distally. Follicles tomentose, the tomentum sometimes sloughing, at maturity largely exserted from the floral tube, about 2 mm long, each usually beaked with the persistent style base.

Bogs, wet meadows, wet pastures, borders of swampy woodlands. N.S. and N.B. to Que. and Minn., generally southward through most of N.C., n.e. and n.w. S.C., Tenn., Ark.

2. Spiraea alba DuRoi. MEADOW-SWEET.

Shrub to 2 m tall. Stem, at least below the inflorescence, reddish brown, glabrous. Leaf blades oblanceolate or elliptic, 3–7 cm long, cuneate basally, obtuse or acute apically, margins mostly sharply and finely serrate from a little above the base, glabrous or nearly so on both surfaces. Flowers in panicles terminating the branches; panicles varying considerably in size, smaller ones sometimes oblong in outline, mostly pyramidal, the larger ones loosely branched below and often somewhat leafy. Sepals triangular, about 1 mm long, erect or becoming reflexed, apices obtuse or acute. Petals white (sometimes pinkish in bud), nearly rotund to broadly obovate, 2 mm long or a little more, often irregularly wavy on the margins. Follicles glabrous, half or more exserted from the floral tube, each usually beaked by the persistent style base.

Two intergrading varieties occur in our range. Those with the inflorescence axes pubescent may be designated var. *alba*; those with glabrous inflorescence axes may be designated var. *latifolia* (Ait.) Ahles (*S. latifolia* (Ait.) Borkh.).

Bogs, stream banks, wet meadows, wet thickets, shoreline thickets, pastures. e. Can. to Sask., generally southward to w. N.C. and Ark.

3. Filipendula

Filipendula rubra (Hill) Robins. QUEEN-OF-THE-MEADOWS.

Shortly rhizomatous, glabrous, perennial herb. Stem unbranched, 6–20 dm tall. Leaves pinnately divided, the major segments incised-lobed, often with very much smaller segments of blade tissue between them; larger leaves with terminal segment palmately 7–9-parted, the lobes again incised, 10–15 cm broad and long overall, lateral segments 7–12 cm long, often as wide, deeply or shallowly 3–5-lobed, the primary lobes again with small marginal lobing; all segments acute or acuminate and all edges serrate; stipules foliaceous, clasping. Flowers small, mostly bisexual, radially symmetrical, borne in long-stalked, panicled cymes. Floral tube saucerlike. Sepals 5, ovate, short-oblong, short-obovate, or nearly orbicular, broadly rounded apically, becoming reflexed, persistent, about 2 mm long. Petals 5, deep pink, very short-clawed, blades ovate, oblong, or obovate, their margins irregularly, shallowly erose and often crisped. Stamens numerous. Pistils 5–7, occasionally fewer or more, styles thickish and short, stigmas capitate, each ovary ripening into a folliclelike, indehiscent fruit, the style and stigma persisting, the fruits collectively almost wholly elevated above the floral tube.

Bogs, wet meadows, wet prairies. Pa. to n. Ill., Iowa, Minn., southward from Pa. to W.Va., Ky., w. N.C.

4. Geum (AVENS)

Perennial herbs. Leaves variable (on individual plants), the lower long-petioled, blades roundish and undivided, ternate, or pinnately divided, those of the midstem and above commonly but not always ternate, uppermost sometimes simple, becoming shorter-petioled upwardly; stipules variable, sometimes foliaceous, attached to the petiole basally. Flowers solitary or few in leafy-bracted, irregularly branched inflorescences terminal on the branches. Floral tube hemispheric to turbinate. Sepals 5, in most with smaller bractlets between them. Petals 5. Stamens numerous. Pistils numerous, borne on a conical or cylindrical receptacle, in fruit the "head" of achenes raised above the floral tube, ovaries maturing into achenes each bearing a persistent, conspicuous, 1-jointed style, it bent or looped near the middle or near the summit.

To facilitate identification of specimens of *Geum*, it is advisable that they have both flowers and at least some well-developed, preferably mature, fruiting "heads."

1. Bractlets none between the sepals; "head" of pistils raised on a stipe above the floral tube and calyx, this evident soon after, but usually not during anthesis; sepals 2 mm long or less.　　　　1. *G. vernum*
1. Bractlets between the sepals present; "head" of pistils sessile; sepals over 3 mm long.
　2. Fruiting "head" with no stiff pubescence from the receptacle exserted from between the achenes; stalks of the flowers or fruits with long, spreading hairs.　　　　2. *G. laciniatum*
　2. Fruiting "head" with stiff pubescence from the receptacle exserted from between the achenes; stalks of the flowers or fruits minutely velvety-pubescent.
　　3. Petals white, as long as or longer than the sepals.　　　　3. *G. canadense*
　　3. Petals cream-colored, shorter than the sepals.　　　　4. *G. virginianum*

1. Geum vernum (Raf.) T. & G.

Stems usually several from the basal crown, each arching below, 2.5–6 dm tall, sparsely pubescent to glabrous. Blades of basal leaves often roundish, mostly cordate basally, unlobed or few-lobed, sometimes some or all pinnately dissected, margins finely toothed; blades of median stem leaves pinnately dissected, the divisions irregularly laciniate-toothed-lobed, upper leaves mostly ternate; leaf surfaces usually sparsely pubescent, the margins of lobes or teeth mostly ciliate; stipules foliaceous, toothed or incised. Flowers small. Sepals triangular, soon reflexed, 1.5–2 mm long, pubescent distally; bractlets between the sepals none. Petals pale yellow or cream, oblanceolate, 1.5–2 mm long. "Head" of pistils, shortly after anthesis, elevated on a stipe above the floral tube and calyx. (*Stylipus vernus* Raf.)

Rich wooded slopes, banks and bottoms of woodland streams, fallow fields, roadsides. N.Y. to s. Ont. and Mich., southward locally to cen. N.C., Ark., s.e. Kan.

2. Geum laciniatum Murr.

Stems 4–10 dm tall, pubescent with longish, spreading or reflexed hairs. Basal and lower leaves with blades sometimes roundish and unlobed, sometimes with a large, roundish terminal segment and 1 or 2 pairs of very much smaller lateral segments, sometimes pinnately dissected, the segments incised; upwardly blades 3-lobed to pinnately dissected, the segments lobed or incised; upper leaves ternate to 3-lobed, sessile or nearly so; edges of all blades toothed, teeth mostly acute, lower leaf surfaces hirsute, especially along the principal veins, the upper glabrous or very sparsely pubescent; stipules foliaceous, toothed or incised. Sepals triangular-acuminate, soon reflexed, 5–8 mm long, hirsute exteriorly; small bractlets present between the sepals. Petals white, elliptic to obovate, shorter than the sepals. "Head" of pistils or fruits sessile, without stiff pubescence from the receptacle showing between them. Stalks of the flowers or fruits with long, spreading pubescence.

Wet meadows, thickets, bogs, ditches. N.S. to s. Ont. and Mich., southward to N.J., W.Va., n.w. N.C., Ill., Mo., and e. Kan.

3. Geum canadense Jacq.

Similar to *G. laciniatum* and *G. virginianum* in general features. Sepals triangular-acuminate varying to elliptic or ovate basally and abruptly acuminate apically. Petals cream-yellow, spatulate to narrowly obovate, distinctly longer than the sepals. "Head" of pistils or achenes sessile, with stiff pubescence from the receptacle exserted from them. Stalks of the flowers or fruits minutely velvety-pubescent.

Alluvial woodlands and moist woodlands generally, stream banks, marshy borders and clearings of lowland woods, bogs. N.S. to Minn. and S.D., generally southward to Ga., La., and Tex.

4. Geum virginianum L.

Similar to *G. laciniatum* and *G. canadense* in general features, the teeth of the leaf edges more rounded. Sepals triangular-acute. Petals cream-yellow, oblong-oblanceolate, as long as or longer than the sepals. "Head" of pistils or achenes sessile, with stiff pubescence from the receptacle exserted from between them. Stalks of the flowers or fruits minutely velvety-pubescent.

Alluvial woodlands, woodlands astride small streams, boggy meadows, moist upland woods and ravines, thickets, rocky banks. Mass. and N.Y. to Ind., generally southward to S.C. and Tenn.

G. aleppicum Jacq., widely distributed northward of our range, is known from our range only in boggy meadows in Avery Co., n.w. N.C., where it is in association with *G. virginianum*. It is distinguished from the latter (in part) by having bright yellow or golden yellow petals.

5. Rubus (BRAMBLES)

Perennial shrubs (those treated here), many bearing bristles, bristles and prickles, or prickles; sending up from the base biennial stems, these in the first year usually unbranched and not bearing flowers or fruits (termed primocanes), in the second year not elongating but forming relatively short, lateral branches most of which are flower-bearing (termed floricanes), the flowers solitary terminating the branchlets, or in racemes or panicles. Leaves mostly compound, those of the primocanes may differ in their complexity, in size, and in the form of the leaflets from those of the floricanes. Floral tube small, flat to hemispheric. Sepals usually 5, spreading to reflexed. Petals usually 5, erect or spreading. Stamens numerous. Pistils numerous, borne on a convex or conic receptacle which in some elongates as the ovaries mature into a cluster of drupelets, these falling separately or together.

Brambles (those comprising the blackberries), have been shown to hybridize, to exhibit various degrees of polyloidy, to have developed means of producing seeds without sexual reproduction; thus they exhibit an extraordinarily complex diversity of variable characteristics such that delimitation of species in the ordinary sense is apparently not possible. The number of "species" attributed to eastern North America by "specialists" is as high as 400 or more, the number varying greatly depending upon the particular "specialist," and "microspecies" perhaps as many as 10,000. Oversimplification appears to be the only way to achieve a practicable solution to the dilemma. Most occur in well-drained places.

1. Leaflets glabrous to relatively densely pubescent beneath but the pubescence not grayish and not felty.
   2. Stems trailing, prostrate, or sprawling, the primocanes rooting at their tips.
      3. Leaves of the primocanes with 5 leaflets, overwintering and some of them present along with the ternate leaflets of the new branchlets of the second year during flowering and fruiting; flowers all, or most of them, solitary terminating the floricane branchlets.   1. *R. trivialis*
      3. Leaves of the primocanes ternate, rarely some of them 5-foliolate, not overwintering, those of the floricane branchlets ternate; flowers several on each floricane branchlet.   2. *R. hispidus*
   2. Stems erect or high-arching, 1–3 m tall.   3. *R. argutus*
1. Leaflets with a dense, grayish, felty tomentum beneath.   4. *R. cuneifolius*

1. Rubus trivialis Michx. SOUTHERN DEWBERRY.

Stems trailing, reclining, prostrate, or sprawling, primocanes usually rooting at the tips, sometimes at some nodes, armed with slender bristles and large-based, stoutish, straight or hooked prickles, also on younger parts sometimes pilose and/or with sessile glands; similar armament and pubescence on petioles and stalks of the leaflets. Primocane leaves mostly 5-foliolate, those of the floricanes ternate, the former overwintering and some of them, at least, present until well after the new floricane branchlets with their flowers and fruits develop; leaflets elliptic or lanceolate, occasionally lance-ovate, rarely obovate, 2–6 cm long, upper surfaces glabrous, lower sparsely pubescent along the principal veins or glabrous, bases rounded or somewhat tapered, apices acute; margins sharply serrate, sometimes doubly serrate; some leaflets of ternate leaves sometimes oblique and with a lobe on one side. Flowers solitary, rarely 2, on the branchlets, their stalks mostly 2.5–4 cm long, sometimes bearing slender, longish, gland-tipped bristles or hairs, sometimes large-based prickles, usually softly pubescent distally as well. Sepals triangular, mucronate apically, becoming reflexed, 3–7 mm long, both surfaces tomentose, the upper more densely so than the lower. Petals commonly pink to rose in the bud, white or white tinted with pink at full anthesis, shortly subclawed basally, blades broadly elliptic to obovate, often with a lobe on one side, 1–1.5 cm long.

In both well-drained and poorly drained places, fields, waste places, along railroads, marshy swales, pineland depressions and clearings, stabilized dunes and interdune swales, banks and borders of marshes, bottomland clearings. Coastal plain and outer piedmont, s.e. Va. to s. Fla., westward to Tex., northward in the interior to Mo. and s. Ill.

2. Rubus hispidus L. SWAMP DEWBERRY.

Similar in habit to *R. trivialis*. Armature of straight to recurved prickles, the bases usually not enlarged, or sometimes some of them somewhat enlarged, sometimes very few or absent. Primocane leaves mostly ternate, some sometimes 5-foliolate, not overwintering; floricane leaflets ternate. Leaflets 1–4 cm long, the terminal one rhombic or obovate, wedge-shaped basally, obtuse apically, margins serrate or crenate-serrate distally from about the widest place or just below it, the lateral leaflets mostly inequilateral, more or less ovate, bases broadly tapered to rounded, apices obtuse to rounded, margins serrate or crenate-serrate from above the oblique taper on one side, nearly from the base on the other; upper surfaces glabrous, lower sparsely pubescent along the principal veins or glabrous. Flowers usually 2 to several or numerous, in racemes terminating the floricane branchlets, the stalk of the raceme and the stalks of the flowers very slender, more or less pubescent, sometimes with straight, very slender bristles as well. Flowers relatively small. Sepals elliptic-oblong, mucronate apically, 3–5 mm long, soft-pubescent. Petals white, 5–10 mm long, clawed basally, blades oblong-obovate.

Bogs, swamps, wet woodlands. N.S. and Que. to Wisc., generally southward to N.C. and closely adjacent S.C., Ky., Ill.

3. Rubus argutus Link. HIGHBUSH BLACKBERRY.

Stems erect or arching erect, to 3 m tall, perhaps somewhat taller in favorable habitats, coarse, ridged and grooved, sparsely pilose on young portions, older portions glabrous. Armature of stems, petioles, stalks of the leaflets, midveins of lower surfaces of leaflets, inflorescence axes, and flower stalks, consisting of hard, large-based, sharply pointed, straight or usually hooked prickles. Leaves of the primocanes 5-foliolate, petioles 6–10 cm long; stalk of the terminal leaflet longest, 2.5–4 cm long, those of the intermediate laterals mostly 1–2 cm long, those of the lower laterals varying from sessile to 5 or 6 mm long; blades elliptic, elliptic oblong, lance-ovate, or lance-obovate, bases rounded to cuneate, apices acuminate, margins sharply serrate; blade of terminal leaflet largest, 6–10 cm long or a little more, lowest pair smallest, 4–6 cm long. Leaves of the floricanes ternate, varying considerably in size but all much smaller in all respects than primocane leaves, leaflet shape similar. Leaflets glabrous above, varying from glabrous to densely pubescent beneath (but not felty-tomentose). Floricane branchlets bearing several-flowered cymes, these leafy-bracted below; axis of cyme and flower stalks bearing long, soft hairs, often with short, gland-tipped hairs intermixed. Sepals 4–6 mm long, ovate-triangular, becoming reflexed, densely tomentose interiorly, less so without, apex of each with a somewhat glandular mucro. Petals often pink to rose in the bud, white or white suffused with pink at full anthesis, obovate or rhombic-obovate, 2–2.5 cm long.

Wet or swampy woodlands and especially their borders, often where surface water stands much of the time, open stream banks, moist to wet clearings, shoreline thickets, margins of cypress-gum ponds and depressions; also in old fields and pastures, fence rows, and borders of rich woodlands. Md. to Mo., generally southward to s.cen. pen. Fla. and e. Tex.

4. Rubus cuneifolius Pursh. SAND BLACKBERRY.

Plants suckering from subterranean parts and commonly forming relatively dense stands. Stems mostly erect, sometimes the main stem sprawling, 3–15 dm tall, densely short-pubescent, the pubescence more or less sloughing on older portions, armed with stiff and sharp, large-based, straight, recurved, or hooked prickles. Primocanes not infrequently branching, leaves of lower portion of the main stem 5-foliolate, those of branches or tips of main stem commonly ternate; floricane leaves ternate; primocane leaves and their leaflets much larger than those of the floricanes; petioles and stalks of the leaflets armed like the stems. Leaflets with a grayish green cast, glabrous or pubescent above, if the latter, then not felty, lower surface wholly clothed with a grayish, compact, felty tomentum; terminal leaflet with a stalk 5–20 mm long, blade oblanceolate, spatulate, rhombic, or narrowly obovate, marginally serrate from near the base distally varying to serrate only somewhat above the middle; lateral leaflets sessile or subsessile, blades commonly inequilateral, oblanceolate, spatulate, obovate, or elliptic, varying in size but smaller than the terminal one. Branchlets of the floricanes with up to 10 flowers, rarely only 1. Sepals oblong, elliptic-oblong, or lance-ovate, 5–7 mm long, densely tomentose, with a thickened, glandular mucro at the tip. Sepals white, 1–1.5 cm long, clawed basally, blades obovate to oval.

In both well-drained and wet places, abundantly colonizing old fields, open mixed woodlands, clearings, frequently in burned-over places and places where the soil has been mechanically disturbed, including seasonally wet pine savannas and flatwoods, shrub-tree bogs and bays, edges of cypress-gum ponds or depressions. Conn., L.I., southward to s. Fla., westward to Miss.

6. Dalibarda

Dalibarda repens L. DEWDROP.

Low, herbaceous perennial with slender rhizomes from which leaves and short leafy shoots arise, to some extent mat-forming. Leaves simple, petioles slender and lax, 2–6 cm long, blades orbicular to reniform, 1.5–4 cm across, cordate basally, margins crenate, both surfaces and the petioles pubescent with long hairs; stipules subulate, about 5 mm long. Flowers axillary and solitary, some on straight stalks 5–10 cm long, petaliferous and usually sterile, some on curved or lax stalks 2–5 cm long, without petals and fertile. Flower stalks, floral tube, and sepals pubescent; sepals 5 or 6, unequal, ovate to oblongish, 3–5 mm long, the 3 larger ones usually toothed distally on the apetalous flowers, persistent; petals of petaliferous flowers 5, white, elliptic or oblong-elliptic, about twice the length of the sepals. Stamens numerous. Pistils 5–10, the ovary of each (in the apetalous flowers) maturing into a dry, 1-seeded drupe, its surface pale stramineus or whitish, somewhat wrinkled transversely, pubescent, 3–3.5 mm long.

Swamps, bogs, moist woodlands. In our range apparently known to occur in sphagnous bogs only in Transylvania Co., mts. of N.C.; W.Va., n. Ohio, Mich., northeasterly to s. Ont., Que., N.S.

7. Sanguisorba

Sanguisorba canadensis L. AMERICAN BURNET.

Glabrous, erect, perennial herb with stout rhizome, 9–15 (–20) dm tall, usually unbranched below, with few, stiffly erect branches above, these with long-stalked, densely many-flowered, cylindrical spikes. Leaves relatively few and widely spaced, odd-pinnately compound, the lower to 8 dm long including a long petiole, gradually becoming shorter, with shorter petioles and fewer leaflets upwardly. Lower leaves with 7–17 leaflets, lowermost pair the smallest, 1–2 cm long, then gradually increasing in size upwardly, the larger ones on leaves of some plants 3–4 cm long, varyingly on others up to 7–10 cm long, lowermost and uppermost pairs sessile or subsessile, all others with stalks 5–10 mm long; blades oblong, ovate-oblong, or lance-oblong, bases mostly cordate or rounded, apices rounded or obtuse, margins serrate; stipules foliaceous, their lower halves adnate to the petiole and somewhat clasping the stem. Flowers in dense spikes or heads. Floral bracts small, pubescent. Floral tube urnlike, contracted at the mouth. Sepals 4, petaloid, white, spreading, elliptic, 2–3 mm long. Petals none. Stamens 4. Pistil 1, ovary free from but enclosed by the floral tube, maturing into an achene.

Bogs, boggy meadows, seepage slopes, about waterfalls, moist prairies. Nfld. to Man., generally southward to N.J., Pa., Ohio, Ill., Ind., Mich., in the mts. to N.C. and n. Ga.

8. Rosa

Rosa palustris Marsh. SWAMP ROSE.

Shrub to about 2 m tall, forming clones from subterranean runners. Stems up to 1.2 cm thick near the base, bearing stout, hard, flattened-conical, sometimes hooked, prickles. Leaves odd-pinnately compound, leaflets 5–9, commonly 7, sometimes 3, lower pair smallest, increasing in size somewhat distally, the terminal one largest, mostly elliptic, short-stalked, 1–5 cm long, 0.5–2 cm wide; upper surfaces glabrous, lower varyingly soft-pubescent, sometimes only on the midrib, sometimes on all the principal veins, sometimes over all the surface; margins finely and sharply serrate nearly to the base; petioles mostly 1–2 cm long, stipules narrow, adnate to the petiole about ⅔ of its length, sometimes for its entire length, only the stipular tips free. Flowers solitary or in few-flowered clusters at or near the tips of branchlets. Floral tube urceolate to depressed-globose, outer surface stipitate-glandular, numerous pistils enclosed within. Sepals 1.5–2 cm long or a little more, lanceolate at base then narrowed to a linear portion above which the tips become somewhat dilated-foliaceous, lower surfaces densely short-pubescent and with stoutish stipitate-glandular trichomes proximally. Petals pink, 2–3 cm long, obovate. Stamens numerous. Pistils numerous, distinct, ripening into tan, bony achenes. Maturing or mature floral tube with persistent calyx and enclosed achenes referred to as a "rose hip."

Marshy or swampy shores of streams, sloughs, ponds, lakes, in swamps, wet thickets, commonly in shallow water. N.S. to Minn., generally southward to cen. pen. Fla., Miss., Ark.

9. Aronia (CHOKEBERRIES)

Shrubs, to 2–3 m tall, rarely taller, commonly spreading by subterranean offsets and colonial; when colonial, main stems commonly wandlike and with short lateral branches. Leaves simple, very short-petiolate, alternate, deciduous, with linear stipules adnate at base to the base of the petiole, margins finely toothed, the teeth tipped with red glands, upper surface glabrous save for scattered, narrow, red glands along the midrib, occasionally a few present on proximal parts of larger lateral veins as well. Flowering in early spring, the flowers in corymbose clusters of up to 25 terminating first branchlets of the season. Expanding flower buds usually pinkish, the outer surfaces of the petals then pink, usually white on both surfaces when fully expanded. Flowers about 1 cm across, the ovary inferior. Sepals 5, short-deltoid, at the summit of the floral tube. Petals 5, white, short-clawed, the claws inserted on the inner rim of the floral tube and spreading outward between the sepals, the blades rotund to obovate. Stamens 15–20, erect, filaments white, anthers lavender-pink, drying brown. Styles 5, united basally, stigmas capitate. Fruit a small subglobose berrylike pome.

Authors have differed in their interpretation of constituent taxa in *Aronia*. See James W. Hardin, "The Enigmatic Chokeberries (*Aronia*, Rosaceae)," *Bull. Torr. Bot. Club* 100: 178–184 (1973), wherein he states: "Two species (*A. arbutifolia* and *A. melanocarpa*) are fairly distinct. A third entity is intermediate and has variously been considered a species (*A. prunifolia* (Marsh.) Rehd.), a variety of each of the other two, or a hybrid." His analysis led him formally to recognize two "parental" species, the former two above, and intermediates of hybrid origin. The variability is apparently such that fruit color is the only reliable basis for distinguishing specimens as representing the "parental" taxa. In our range, the only area in which both species and intermediates occur is the southern Appalachians.

1. Aronia arbutifolia (L.) Ell. **RED CHOKEBERRY.**   Fig. 105

Young branchlets, inflorescence branches, flower stalks, floral tubes, and lower leaf surfaces usually (not invariably) densely pubescent. Mature fruit red, persisting into the winter. Leaves turning scarlet in autumn. (*Pyrus arbutifolia* (L.) L. f., *Sorbus arbutifolia* (L.) Heynhold)

Bogs, pine savannas and flatwoods, swamps, wet thickets, creek banks, moist rocky ledges. Nfld., N.S. to N.Y., e. Pa., W.Va., Ky. to s. and e. Ark., e. Tex., generally eastward to cen. pen. Fla.

2. Aronia melanocarpa (Michx.) Ell. **BLACK CHOKEBERRY.**

Young branchlets, inflorescence branches, flower stalks, base of floral tube, and lower leaf surfaces glabrous or sparsely pubescent. Fruits black, ripening and dropping early. Leaves not turning red in autumn. (*Pyrus melanocarpa* (Michx.) Willd., *Sorbus melanocarpa* (Michx.) Schneider)

In wet sites similar to those of *A. arbutifolia*, also in drier thickets, cliffs, bluffs, clearings. Nfld., N.S., s. Ont. to Wisc., n.e. Iowa, n. Ill., N.Y., middle and w. Pa., south in the southern Appalachians to n. Ga. and Ala.

Fig. 105. **Aronia arbutifolia:** a, flowering branch; b, flower cluster; c, pistil; d, fruiting branch; e, mature fruits after leaf-fall. (From Correll and Correll (1972), as *Pyrus arbutifolia*)

10. Amelanchier (SERVICEBERRIES, SHADBUSHES)

Trees or shrubs. Leaves simple, deciduous, stipules linear, quickly deciduous. Flowers bisexual, radially symmetrical, borne in racemes terminating short branchlets of the current season, appearing shortly before the foliage or as the leaves unfold. Floral tube campanulate or urceolate, ovary wholly or nearly wholly inferior. Sepals 5, persistent, spreading or reflexed. Petals 5, white, rarely pinkish, erect or spreading. Stamens 10–20, shorter than the petals, free, or the filaments united basally or to the middle. Ovary 5-locular, styles 5, free or united basally, ovules 2 in each locule but in development of the fruit separated by an ingrowing partition, thus the fruit, a berrylike pome, 10-locular, each locule 1-seeded.

Plants of eastern North American amelanchiers exhibit such a perplexing variability that authors, in both revisionary studies and in floristic treatments, have engendered a no less perplexing array of interpretations in delimiting its component taxa and in the application of names. Hybrids are allegedly common and one author has stated that about a third of the specimens in herbaria may be assumed to be from hybrid individuals. We ourselves have no appreciable first-hand knowledge of *Amelanchier* and have sought in vain to find much "common ground" in the most recent monograph (G. N. Jones, "American Species of *Amelanchier*," *Ill. Biol. Monogr.* 20, no. 2, 1946) and the treatments in subsequently published regional manuals: Fernald, 1950; Gleason, 1952; Radford et al., 1964. The two entities treated below are those we perceive to be inhabitants of wetlands within our range.

- Plant commonly a shrub with several upright stems to 8 m tall, and collectively with an overall vaselike form, sometimes with a single arborescent stem, not spreading by horizontal, suckering shoots and not colonial; racemes 2.5–6 cm long, flower stalks 10–20 mm long.   1. *A. canadensis*
- Plant spreading at or beneath the surface of the substrate by elongate, horizontal, suckering shoots and forming colonies of aerial stems to about 15 dm tall; racemes 1–3 cm long, compact, flower stalks 2–5 mm long.   2. *A. obovalis*

1. Amelanchier canadensis (L.) Medic.

Bushy shrub, usually with several stems from the base to 8 m tall, and collectively with an overall vaselike form, sometimes with a single arborescent stem. Leaves, at time of flowering, folded, half grown or less, heavily cottony-tomentose; at maturity glabrous, oblong or oblong-obovate, 3–6 cm long, rounded or slightly cordate basally, rounded and minutely mucronate apically, margins finely serrate. Racemes 2.5–6 cm long, flower stalks 10–20 mm long; raceme axis, flower stalks, floral tube, and sepals, densely tomentose. Sepals spreading, 1.5–3 mm long. Petals mostly oblong-obovate, varying to lance-linear, 7–10 mm long. Fruit dark purple or black. (*A. oblongifolia* T. & G.)

Swamps, bogs, pocosins, sandy upland woodlands. Chiefly coastal plain and outer piedmont, Maine and s.w. Que. to Ga.

2. Amelanchier obovalis (Michx.) Ashe.

Differing from *A. canadensis* in habit and stature, plant spreading at or beneath the surface of the substrate by elongate, horizontal, suckering shoots and forming colonies of aerial stems to about 1.5 m tall. Leaves scarcely evident at time of flowering, densely tomentose beneath; mature leaves glabrous, oblong, broadly elliptic, oblong-ovate or -obovate, or subrotund, 3–5 cm long, bases rounded, apices rounded and minutely mucronate, finely serrate distally from about their middles. Racemes dense, scarcely stalked, flower stalks 2–5 mm long; raceme axis, flower stalks, floral tubes, and sepals, loosely pubescent with long, soft hairs. Sepals spreading or reflexed, 2–2.5 mm long. Petals obovate, 4–8 mm long. Fruit purplish black.

Low, wet woodlands, pocosins, bogs, pine savannas and flatwoods. Chiefly coastal plain, N.J., s.e. Pa., southward to Ga.

11. Crataegus (HAWTHORNES, HAWS)

Small trees or shrubs, usually armed with thorns on at least some of the branches. Leaves simple, deciduous, serrate, dentate, or lobed, in some those on sterile shoots (produced after flowering) larger and tending to be more lobed or incised than on the flowering or fruiting branchlets; stipules small, very quickly deciduous. Flowers in simple or branched corymbs terminal on short branchlets produced early in the season, sometimes borne singly or 2 or 3 together. Flowers bisexual, floral tube cuplike, campanulate, or obconic, ovary wholly or almost wholly inferior. Sepals 5, persistent. Petals 5, often pink in the bud, white or pink at full anthesis. Stamens 5–20 (–25), the shriveled filaments often persistent at the summit of the fruit. Styles 1–5, free. Fruit a berrylike pome containing 1–5 1-seeded nutlets.

1. Leaf blades triangular in overall outline, essentially as broad at the base as long, all deeply cut-lobed, bases of the larger ones mostly broadly truncate or subcordate; inflorescence axis, flower stalks, and floral tubes copiously pubescent.   1. *C. marshallii*
1. Leaf blades oblanceolate, oval, elliptic, or ovate, most of them longer than broad, merely toothed, or some of those on the sterile shoots lobed, their bases mostly cuneate; inflorescence axis, flower stalks, and floral tubes glabrous or sparingly pubescent.
   2. Lower leaf surfaces with a tuft of pubescence in the axils formed by the midvein and the principal lateral veins, sometimes along the veins as well.
      3. Inflorescence branched.   2. *C. viridis*
      3. Inflorescence unbranched, essentially umbellate.   3. *C. aestivalis*
   2. Lower leaf surfaces wholly glabrous or rarely pubescent along the midvein, without tufts of hairs in the principal vein axils.   4. *C. crus-galli*

1. Crataegus marshallii Eggl. PARSLEY HAW.   Fig. 106

A small understory tree, to 6–8 m tall, with a thin, scaly bark, branches, if armed, with thorns 1–3 cm long; branchlets woolly-pubescent when young. Leaf blades alike on both flowering and later sterile shoots, membranous, not lustrous, triangular in overall outline, essentially as broad at the base as long, the smaller ones often broader than long, 1–3 (–5) cm long, most of them deeply cut-lobed, the main lobes irregularly sharply toothed or with smaller toothed lobes; petioles more or less woolly-pubescent, blades copiously pubescent beneath when young, somewhat less so above, much, but not all, of the pubescence slough- ing as the blades mature. Flowers few to numerous in usually branched clusters, the axes, flower stalks, and floral tubes copiously woolly-pubescent. Sepals triangular-subulate, 3–4 (–5) mm long, margins with gland-tipped teeth distally. Petals white or pale pink, sometimes white and pink-tinged, oblanceolate or obovate, 5–7 mm long. Anthers red. Fruit bright red, containing 3 nutlets.

Floodplain forests, adjacent wooded slopes, wet woodlands astride small streams, wooded ravine slopes, river banks. Coastal plain and outer piedmont, s.e. Va. to cen. pen. Fla., westward to e. Tex., northward in the interior to w. Tenn. and s.e. Mo.

2. Crataegus viridis L. GREEN HAW.   Fig. 107

Small tree, to 8 or 10 m tall, with gray, scaly bark sloughing in thin plates and exposing cinnamon-colored inner bark, branches, if armed, with thorns 1–4 cm long; branchlets mod- erately pubescent when young. Leaf blades variable, those of the flowering or fruiting branchlets obovate, spatulate, oblong-elliptic, ovate, or subrotund, 2.5–5 cm long, rela- tively finely and unevenly serrate above their middles, sometimes nearly to the base, mostly unlobed, those of the later, sterile shoots larger, more commonly ovate, serrations more sa- lient, those near the branch tips often lobed, glabrous save for tufts of pubescence in the principal vein axils beneath, petioles glabrous or sparingly pubescent. Flowers with a fetid odor, numerous, in branched corymbs, inflorescence axis glabrous or sparingly pubescent; flower stalks, distally, and floral tubes, usually with some long hairs. Sepals triangular-subul- ate, 3–5 mm long, glabrous or pubescent. Petals white, obovate, 5–8 mm long. Anthers yellow at first, turning brown or red. Fruit red or orange-red, containing 5 nutlets.

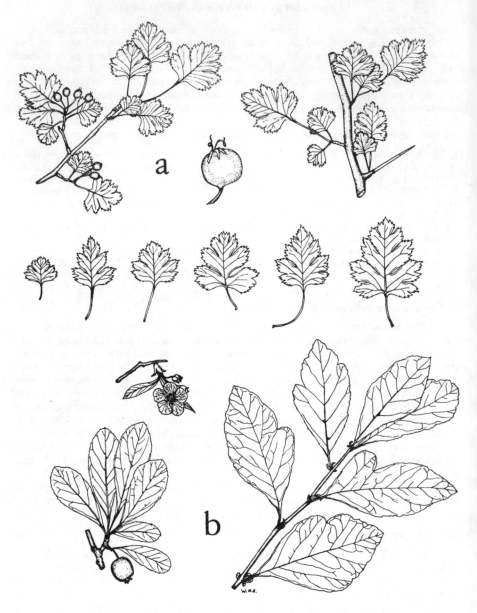

Fig. 106. a, **Crataegus marshallii**; b, **Crataegus aestivalis.** (From Kurz and Godfrey, *Trees of Northern Florida*. 1962)

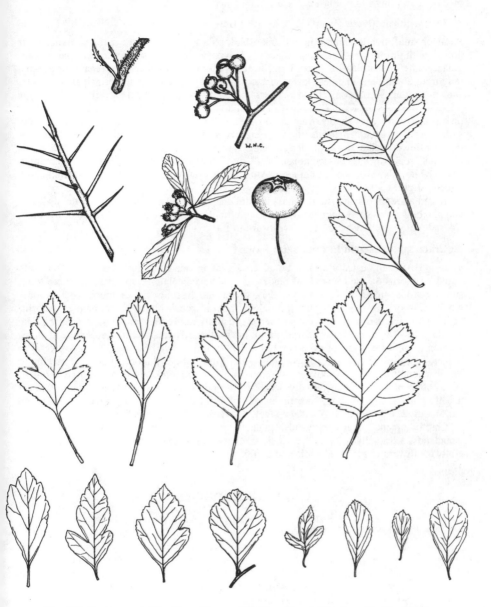

Fig. 107. **Crataegus viridis.** (From Kurz and Godfrey, *Trees of Northern Florida*. 1962)

231

Swamps, floodplain forests, wet woodlands, in and on the margins of pools and small ponds, sometimes in mesic woodlands. Va. to Mo., generally southward (excluding inner piedmont and mts.) to n. Fla., e. Okla., e. and s.cen. Tex.

3. Crataegus aestivalis (Walt.) T. & G. MAY HAW.   Fig. 106

Shrub or small tree to about 8 m tall, outer bark pale gray, inner bark reddish, branches, if armed, with thorns 1.5–4 cm long. Blades of leaves on flowering or fruiting branches mostly oblanceolate or spatulate, 2–5 cm long, lustrous above, subcoriaceous, crenate-serrate distally from about their middles, lower surfaces with tufts of hairs in the principal vein axils, sometimes pubescent along the veins, less frequently between them as well; blades of the proximal portions of sterile shoots usually similar but larger, those on their distal portions or on especially vigorous shoots ovate to obovate, 3–8 cm long, often with a lobe on one side, sometimes 3–5-lobed, margins unevenly crenate-serrate. Flowers single or 2–4 in an umbel, the flower stalks 5–12 mm long, shaggy-pubescent to glabrous; floral tube and sepals glabrous, sepals triangular-acuminate, 2–3 mm long, their margins sometimes glandular. Petals white, obovate, about 1 cm long. Anthers pink or purplish. Fruit red, containing 3–5 nutlets, slightly acid, prized for making jelly.

In and about pools and small ponds, floodplains, especially where regularly flooded, and about oxbows, wet or swampy woodlands, commonly where water stands much of the time. Coastal plain, s.e. N.C. to n. Fla., westward to La.

4. Crataegus crus-galli L. COCKSPUR HAW.   Fig. 108

Tree to 10 m tall, usually with a broad crown of intricate, thorny branches, often with branched thorns on the trunk. Leaf blades mostly oblanceolate or spatulate, 2–6 cm long, subcoriaceous, upper surface usually dark green and lustrous, lower much paler, usually wholly glabrous, infrequently pubescent along the midvein, without tufts of hairs in the principal vein axils, margins finely crenate-serrate, usually above their middles, sometimes from a little below them; blades of leaves on the later, sterile shoots similar but larger, in shape varying to obovate, oval, or even nearly rotund. Inflorescence branched, flowers several to numerous; inflorescence axis, flower stalks, and floral tube commonly glabrous, varying to hirsute. Sepals deltoid to subulate, 3–5 mm long, their margins entire or minutely toothed. Petals often pink in the bud, usually white at full anthesis, obovate, 5–7 mm long. Anthers usually pink or salmon-pink, sometimes yellow. Fruit dull red, rusty orange, or green and dark-mottled, containing 3–5 nutlets. (Incl. *C. pyracanthoides* Beadle)

Chiefly upland, open woodlands, thickets, pastures, fence rows; also in bottomland woodlands, seasonally wet pine flatwoods and flatwoods depressions. Que. to s. Minn., generally southward to Fla. Panhandle and e. Tex.

Chrysobalanaceae

Chrysobalanus icaco L. COCOA-PLUM.   Fig. 109

Evergreen shrub or small tree, sometimes with radially prostrate lower branches. Twigs glabrous, reddish brown, with conspicuous, raised, somewhat spongy, pale lenticels. Leaves simple, alternate, stipulate, very short-petiolate, blades obovate to orbicular, broadly obtuse basally, broadly obtuse, broadly rounded, or broadly rounded and shallowly emarginate apically, pinnately veined, coriaceous, glabrous, varying in size from about 2–6 (–8) cm long and 2–5 cm wide; stipules minute, very quickly deciduous and not leaving perceptible scars. Inflorescence of short racemes of small cymes or cymose throughout, sometimes racemose. Flowers small, bisexual, with a short, campanulate, densely silky-pubescent floral tube free from the ovary. Calyx segments 5, densely silky-pubescent, ovate, abruptly short pointed apically, 2–3 mm long. Petals 5, white, oblanceolate or spatulate, mostly dilated gradually from a narrow base, or short-clawed, about 6 mm long. Stamens numerous, sometimes unequal, exserted from the perianth, the filaments united up to half their length in several

Fig. 108. **Crataegus crus-galli.** (From Kurz and Godfrey, *Trees of Northern Florida.* 1962)

groups, densely pubescent except distally. Ovary pubescent, 1-locular, the slender pubescent style arising from the base of the ovary. Fruit a 2-seeded, ovate, orbicular, or obovoid, longitudinally ridged, yellow, red, or purple drupe 2–5 cm long. (Incl. *C. interior* Small, *C. icaco* var. *pellocarpus* (Meyer) DC.)

Cypress heads, hammocks, beaches, swamps. s. pen. Fla.; W.I., C.Am., e. S.Am., w. Afr.

Variation, chiefly in leaf shape and size and in fruit size, has led to the description of numerous segregate taxa in various parts of the range. Prance (1970) states that there is little correlation of the variable characters and discerns no basis for the recognition of more than one taxon.

Calycanthaceae

Calycanthus floridus L. CAROLINA ALLSPICE, SWEET-SHRUB, STRAWBERRY-SHRUB.

Fig. 110

Shrub, stems to about 3 m tall, extending by root sprouts and somewhat colonial, loosely few-branched, spicy-aromatic when crushed or bruised. Twigs reddish brown, at first quadrangular in cross-section, leaf scars conspicuously raised, more or less **V**-shaped, with 3(–5)

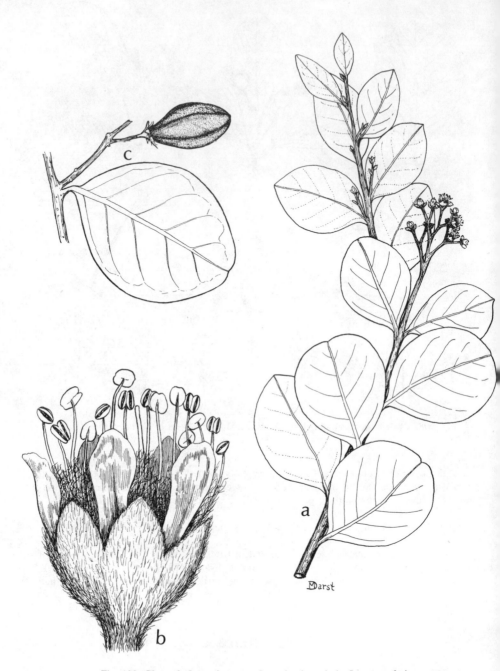

Fig. 109. **Chrysobalanus icaco:** a, flowering branch; b, flower; c, fruit.

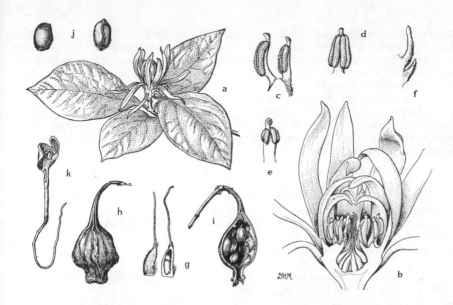

Fig. 110. a–h, **Calycanthus floridus** var. **floridus:** a, flowering branchlet; b, flower, vertical section to show carpels, stamens, staminodia, and cup-shaped "receptacle"; c, two stamens, lateral view; d, stamen, abaxial view; e, inner, reduced stamen, abaxial view; f, two staminodia from edge of cup, lateral view; g, carpels; h, mature dry pendulous pseudocarp; i–k, **Calycanthus floridus** var. **laevigatus** (*C. fertilis*): i, pseudocarp, vertical section, with mature carpels, some removed; j, mature carpels, lateral and abaxial views; k, seedling. (From Wood in *Jour. Arn. Arb.* 39: 324. 1958)

vascular bundle scars. Leaves deciduous, opposite, short-petiolate, the blades elliptic, oval, or ovate, rounded to broadly cuneate basally, acute to acuminate apically, margins entire, dark green and lustrous above, somewhat rugose, smooth and commonly glaucous, or short-pubescent beneath, 5–15 cm long. Flowers short stalked, borne singly and terminally on short branches of new growth; bisexual, about 3 cm long, fragrant. Receptacle cup-shaped, bearing on its outer surface and apex bracts numerous strap-shaped, somewhat fleshy, purplish green to purplish brown perianth parts (tepals). Stamens numerous on the edge of the receptacle, with stout filaments; inner stamens reduced to staminodia. Carpels numerous, distinct, inserted on the base or sides of the interior of the cup. Fruit an indehiscent pseudocarp resulting from the cup becoming fleshy and enclosing the achenes.

Rich woodlands of ravine slopes and bottoms, banks of rivers and small streams in woodlands. Local, s.cen. Pa., s. W.Va., e. Ky., more common from N.C. and e. Tenn. to the Fla. Panhandle, Ala. and s.e. Miss.

Nicely (1965) distinguishes two intergrading varieties: *C. floridus* var. *floridus*, with lower leaf surface, twigs, and petioles pubescent; *C. floridus* var. *laevigatus* (Willd.) T. & G. (*C. fertilis* Walt.), with lower leaf surface glabrous or with few scattered hairs, twigs and petioles glabrous to slightly pubescent.

Leguminosae (BEAN FAMILY)

Herbs, shrubs, vines, or trees. Leaves alternate, usually with stipules, mostly compound, relatively rarely simple; leaflets of compound leaves often with stipels. Flowers rarely solitary, usually in axillary or terminal racemes, panicles, spikes, or heads, usually bisexual, radially or bilaterally symmetrical. Calyx united below, 5-lobed, less frequently 4-lobed, or unlobed. Corolla basically of 5 petals (rarely of 1 petal or absent); petals distinct, or in some the 2 anterior ones wholly or partially coherent along 1 margin, in some all fused basally, papilionaceous in many. Stamens usually 10, distinct, or more or less united into a single tube (monadelphous), or the filaments of 9 more or less united and the 10th free (diadelphous), in some only 5, in some numerous. Pistil 1, ovary superior, 1-carpellate, 1-locular, ovules rarely 1 or 2, usually more and alternating in 2 rows from a parietal placenta along the ventral suture; style and stigma 1. Fruit a legume (pod) dehiscing by 2 valves or sometimes indehiscent, a samara in some (none of ours), or a loment (a pod constricted between the seeds and breaking up when mature into 1-seeded segments or joints).

References: Wilbur, Robert L. "The Leguminous Plants of North Carolina." *N.C. Agr. Exp. Sta. Tech. Bull.* no. 151. 1963.

Elias, Thomas. "The Genera of Mimosoideae (Leguminosae) in the Southeastern United States." *Jour. Arn. Arb.* 55: 67–118. 1974.

Robertson, Kenneth R., and Yin-Tse Lee. "The Genera of Caesalpinioideae (Leguminosae) in the Southeastern United States." *Jour. Arn. Arb.* 57: 1–53. 1976.

Isely, Duane. "Leguminosae of the United States: I. Subfamily Mimosoideae." *Mem. N.Y. Bot. Gard.* 25(1): 1–152. 1973.

Isley, Duane. "Leguminosae of the United States: II. Subfamily Caesalpinioideae." *Mem. N.Y. Bot. Gard.* 25(2): 1–228. 1975.

1. Plants woody, shrubs, trees, or vines.
  2. Leaves simple.   8. *Dalbergia*
  2. Leaves compound.
    3. The leaves odd-pinnately once-compound.
      4. Plant a twining woody vine; corolla papilionaceous, petals 5.   6. *Wisteria*
      4. Plant erect. Corolla with a single petal.   4. *Amorpha*
    3. The leaves even-pinnately once- or twice-compound.
      6. Plant a reclining or scrambling vinelike shrub; stem abundantly clothed with straight, sharp prickles; leaves bipinnately compound, their axes and the axes of the pinnae armed with large-based, hooked prickles.   2. *Caesalpina*
      6. Plant an erect shrub or tree; stem unarmed or with simple or branched, stout thorns arising from buds above the leaf axils or adventitiously, that is, branch thorns; leaves once-compound (*Sesbania*) or 1- or 2-compound, often both on an individual plant (*Gleditsia*).
        7. Shrubs, unarmed; flowering occurring over a considerable period in summer, flowers conspicuous, orange or yellow; corolla papilionaceous, petals 5; stamens 10, diadelphous, 9 and 1; ovary and pod strongly wing-angled.   7. *Sesbania*
        7. Trees, usually some of the branches and the trunks bearing stout, simple or branched thorns; flowering occurring during a very short period early in the season, flowers individually small and inconspicuous, green or yellowish green; petals 3–5, subequal, flower not papilionaceous; stamens 3–10, filaments free; ovary and pod not wing-angled, flat or nearly so.   1. *Gleditsia*
1. Plants herbaceous.
  8. Leaves with terminal tendrils.   12. *Vicia*
  8. Leaves without tendrils.
    9. The leaves simple (*P. virgata* in genus).   3. *Psoralea*
    9. The leaves pinnately compound.
      10. Leaves even-pinnately compound.
        11. Stipules and bracts subtending the stalks of the flowers attached peltately; fruit a legume with 2 longitudinally thickened sutures.   7. *Sesbania*
        11. Stipules and bracts subtending the stalks of the flowers attached basally; fruit a loment, transversely segmented, breaking up into 1-seeded joints.   11. *Aeschynomene*
      10. Leaves odd-pinnately compound.
        12. Plant a twining or trailing vine.
          13. Leaflets mostly 5–7 per leaf (an occasional leaf on a given plant with 3); corolla brownish purple.   9. *Apios*
          13. Leaflets 3 per leaf; corolla yellow.   10. *Vigna*
        12. Plant not a vine.
          14. Corolla not papilionaceous (petal 1; stamens 5).   5. *Petalostemum*
          14. Corolla papilionaceous; stamens 10.   3. *Psoralea*

1. Gleditsia

Trees or shrubs, twigs, branches, and trunks commonly armed with stout, simple or branched thorns developed from buds above the leaf axils or adventitiously. Leaves deciduous, alternate, often fascicled at the ends of short shoots, evenly 1- or 2-pinnately compound, both often on the same plant, sometimes a single leaf partly 1-pinnate, partly 2-pinnate; stipules small, quickly deciduous, stipels none; leaflets usually 9–18 pairs on 1-pinnate leaves or on individual pinnae of 2-pinnate leaves; pinnae of 2-pinnate leaves usually 6–8 pairs. Flowers small, both bisexual and unisexual, usually mostly staminate and a few bisexual on some individual plants, mostly pistillate and a few bisexual on others; flowers borne in axillary racemes, the largely staminate racemes densely many-flowered, the largely pistillate ones loosely relatively few-flowered. Calyx short, 3–5-lobed, lobes spreading. Petals 3–5, subequal, 4–5 mm long, yellowish or greenish yellow. Stamens 3–10, filaments free. Legume flat, 1- to many-seeded. Seeds flattened, ovate to suborbicular in outline.

● Mature leaves with their axes, stalks of the leaflets, and surfaces of leaflets glabrous or with but a few hairs; pod elliptic or oval, commonly oblique, tardily dehiscent, 5 cm long or less, 1–3-seeded, lacking pulp around the seed.   1. *G. aquatica*
● Mature leaves with their axes and stalks of the leaflets notably pubescent, lower surfaces of leaflets pubescent along their midribs at least proximally; pod elongate, 1–4 dm long, 2–3.5 cm broad, indehiscent, the margins usually contracting during maturation causing the pod to curve or coil, seeds numerous, a sugary pulp surrounding each.   2. *G. triacanthos*

1. Gleditsia aquatica Marsh. WATER-LOCUST.   Fig. 111

A tree to about 25 m tall, sometimes shrubby, with grayish brown to blackish, smoothish, narrowly furrowed, or warty bark. Thorns simple or few-branched, 7–14 cm long and up to 1.5 cm wide across the basal branches. Leaflets lanceolate, lance-oblong, or, infrequently, ovate-oblong, mostly 2.5–4 cm long, bases rounded or shortly tapered, often inequilateral, apices blunt or sometimes acute, margins obscurely crenate; axes of mature leaves usually glabrous, stalks of the leaflets glabrous or sparsely puberulent, surfaces of mature leaflets glabrous. Ovary markedly stipitate, glabrous or only sparsely ciliate along the sutures. Pod stipitate, elliptic or oval, commonly oblique, flat and thin, tardily dehiscent, 2–5 cm long, 2–3.5 cm broad, apiculate apically. Seeds 1–3 (not surrounded by pulp), nearly flat, suborbicular, 1–1.5 cm across.

River swamps and floodplains, river banks, wet hammocks astride small streams. Coastal plain, S.C. to s.cen. pen. Fla., westward to e. Tex., northward in the interior to s.e. Mo., s. Ill., s. Ind.

2. Gleditsia triacanthos L. HONEY-LOCUST.   Fig. 111

Similar to *G. aquatica*, attaining larger stature, presumably to 45 m tall, bark with deep fissures, long, narrow, scaly ridges between them. Thorns sometimes few or lacking, simple, or 3- to many-branched, the latter particularly on older stems or trunks, 6–15 (–40) cm long. Leaflets similar in size and shape to those of *G. aquatica*, sometimes somewhat more numerous; mature leaves with their axes and stalks of the leaflets notably pubescent, lower surfaces of the leaflets pubescent along their midribs at least proximally. Ovary sessile or nearly so, densely pubescent. Pod elongate, 1–4 dm long, mostly 2–3.5 cm broad, flat and thickish, indehiscent, the thickened margins usually contracting during maturation causing the pod to curve or coil. Seeds numerous (each surrounded by a sugary pulp), compressed but not flat, irregularly oblong in outline, dark brown, about 0.8 cm long.

In general inhabiting well-drained sites, upland woodlands and their borders, old fields, fence rows, less frequently on sites flooded for periods of short duration, river floodplains and hammocks, sometimes intermixed with *G. aquatica* where their ranges overlap. w. N.Y. and Pa. to e. S.D., generally southward to the Fla. Panhandle and e. Tex.

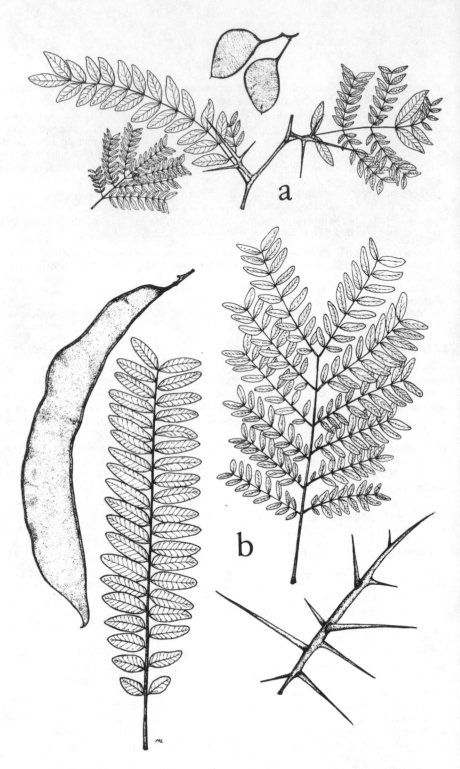

Fig. 111. a, **Gleditsia aquatica**; b, **Gleditsia triacanthos.** (From Kurz and Godfrey, *Trees of Northern Florida*. 1962)

2. Caesalpina

Caesalpina bonduc (L.) Roxb. GRAY NICKER, HOLD-BACK, WAIT-A-BIT VINE.

Robust, reclining or scrambling, vinelike shrub. Stems to 6 m long and 1 cm thick, forming impenetrable thickets, heavily armed with straight prickles (sometimes unarmed), the surface between the prickles and the prickles themselves clothed with short, curly, viscid pubescence. Leaves large, 4–6 dm long, evenly bipinnately compound, pinnae 4–9 pairs, each 4–12 cm long, ultimate leaflets 4–8 pairs on each pinna, very short-stalked, ovate to elliptic, 2–4 (–6) cm long and to 2.5 cm broad, bases truncate to rounded, apices obtuse to rounded and with a small apicule, sparsely short-pubescent along the veins, often glabrous in age; leaf axis and axes of the pinnae armed with stout, large-based, hooked prickles and pubescent with short, curly hairs. Flowers in axillary, simple or compound racemes. Calyx pubescent, tube 2–3 mm long, lobes 5, imbricate, 5–8 mm long, the outermost lobe slightly hoodlike. Petals 5, orange-yellow, free, clawed, nearly equal, oblong, 7–10 mm long. Stamens 10, filaments free. Legume short-stipitate, suborbicular to oblong or oval, compressed, 5–7 cm long, 3–5 cm broad, densely covered with straight, sharp prickles, tardily dehiscent. Seeds 1 or 2, gray, subglobose, 1.5–2 cm in diameter. (*Guilandina bonduc* L., not *G. bonduc* sensu Small, 1933, which is correctly designated *C. major* (Medic.) Dandy & Exell; *G. crista* (L.) Small)

Coastal strands, dunes, hammocks, borders of mangrove swamps, ditches and swales, often in disturbed places. Coastal pen. Fla.; pantropical.

3. Psoralea (SCURF-PEAS)

Perennial herbs with more or less tuberous-thickened rootstocks or rhizomes. Leaves alternate, mostly digitately or pinnately compound with 3 or 5 leaflets (leaflet solitary in one of ours); stipules usually persistent. Flowers in stalked spikes or racemes from the upper leaf axils. Calyx campanulate, in some swollen on one side, in some oblique, lobes 5, equal or unequal. Corolla lavender, blue, blue-purple, or purple (white on an occasional specimen), papilionaceous; standard usually clawed, blade orbicular to obovate, wings about as long as the standard, longer than the keel. Stamens 10, diadelphous, 9 and 1. Pod short, little if any longer than the calyx, 1-seeded, indehiscent, transversely wrinkled or corrugated.

1. Leaves unifoliolate.   1. *P. virgata*
1. Leaves trifoliolate.
    2. Floral bracts lanceolate, long-acuminate apically; corolla deep purple.   2. *P. simplex*
    2. Floral bracts ovate to suborbicular basally, abruptly narrowed distally to subulate tips; corolla lilac or lavender (white in a rare individual).   3. *P. psoralioides*

1. Psoralea virgata Nutt.

Rootstock slender, erect. Stem slender, solitary, sometimes with few, long, erect-ascending branches from near the base, sparsely pubescent below, more densely so above. Leaves unifoliolate, all petiolate, or the uppermost 1 or 2 sometimes subsessile, petioles of the larger lower ones to 5 cm long; blades of lowermost 1 or 2 (usually not present at flowering time) ovate to orbicular, others broadest at the rounded base (5–10 mm), gradually long-tapered to acute apices; upper surfaces sparsely pubescent to glabrous, darkly gland-dotted only on the uppermost, lower surfaces moderately pubescent. Bracts subtending flowers lanceolate, long-acuminate apically, pubescent, sparsely gland-dotted, quickly deciduous. Calyx purple or brownish, pubescent, sparsely gland-dotted, lobes acuminate, subequal. Corolla purple, about 8 mm long. Pod slightly convex, obliquely obovate or short-oblong, strongly transversely corrugated. (*Orbexilum virgatum* (Nutt.) Rydb.)

Pine savannas and flatwoods. Coastal plain, s.e. Ga., n.e. Fla.

2. Psoralea simplex Nutt. ex T. & G.

Rootstock short- or long-obconical. Stem solitary, sometimes with 2 or 3 erect branches from near the base, 3–9 dm tall, pubescent. Leaves pinnately 3-foliolate, lower petioled, petioles 0.8–7 cm long, upper sessile or subsessile; blades narrowly lanceolate, 2–4 (–7) cm long, tips mostly minutely mucronate, darkly gland-dotted above, both surfaces pubescent. Racemes oblong, 2–5 cm long, their stalks 4–10 cm long, copiously pubescent, relatively densely flowered, stalks of the flowers very short, copiously pubescent. Bracts subtending the flowers lanceolate and long-acuminate apically, quickly deciduous. Calyx purple, pubescent with spreading hairs, not gland-dotted, lobes acute, lower one a little longer than the others. Corolla deep purple, 7–10 mm long. Pod flat, obliquely orbicular, transversely wrinkled. (*Orbexilum simplex* (Nutt. ex T. & G.) Rydb.)

Seepage slopes of pinelands, pine savannas, adjacent ditches, clearings. Chiefly coastal plain, Ala. to e. and s.e. Tex., Ark., e. Okla.

3. Psoralea psoralioides (Walt.) Cory. SAMPSON'S SNAKEROOT.

Rootstock semiwoody, tuberous-thickened, obconic, usually with several stems from the crown. Stems 3–8 dm tall, usually sparsely clothed with stiffish, appressed-ascending hairs. Leaves pinnately 3-foliolate, lower petiolate; leaflets lanceolate or elliptic, 3–7 cm long, 0.5–2 cm broad, both surfaces gland-dotted varying to glandless, upper surface glabrous, lower and margins sparsely pubescent. Racemes slender, long-stalked, 2–10 cm long. Bracts basally ovate to suborbicular, abruptly contracted distally to subulate tips, quickly deciduous. Calyx 3–6 mm long, variously pubescent, with prominent glandular-dots, inconspicuous ones, or none, lobes triangular-acute, the lowest a little longer than the others. Corolla 4–7 mm long, lilac or lavender (white on an occasional individual plant). Pod obliquely suborbicular-obovate, strongly transversely corrugated, prominently gland-dotted varying to faintly so or without glands. (*Orbexilum pedunculatum* (Mill.) Rydb.) Two varieties may be distinguished, as follows.

P. psoralioides var. **psoralioides.** Leaflet surfaces, bracts, calyces, and pods bearing conspicuous glandular dots.

Chiefly in pine savannas and flatwoods, adjacent ditches, hillside bogs in pinelands, boggy meadows. Coastal plain and piedmont, Va. to s. Ga., Fla. Panhandle adjacent to s.w. Ga.

P. psoralioides var. **eglandulosa** (Ell.) F. L. Freeman. Leaflet surfaces sparsely gland-dotted or without glands; bracts, calyces, and pods with few inconspicuous, poorly developed glandular dots, or not glandular.

In habitats similar to those of var. *psoralioides*, or, more frequently, in open, well-drained woodlands. Upper piedmont and mts. (infrequently coastal plain of Ga., Ala.), Va. to Ohio, Ill., Mo., e. Kan., generally southward to Ga., e. and s.e. Tex.

4. Amorpha (LEAD-PLANTS, PLUME-LOCUSTS)

Low, suffrutescent to large bushy-branched shrubs. Leaves odd-pinnately compound, stipules setaceous to linear, quickly deciduous; leaflets stalked, each with a setaceous stipel to one side of its stalk, blades not punctate to conspicuously punctate, margins entire. Inflorescence a usually dense, spikelike, terminal raceme, the often-clustered flower stalks subtended by setaceous to linear, quickly deciduous bracts. Calyx persistent, the tube obconic, funnelform, or campanulate, strongly to barely 5-lobed. Corolla consisting of but one erect clawed petal, the standard, it purple, blue, violet, or white. Stamens 10, filaments united below, distinct above, extending beyond the calyx, usually beyond the petal. Fruit an indehiscent, 1-seeded pod, more or less oblique, straight to curved, without glands to strongly dotted with pustular, resinous glands.

Reference: Wilbur, Robert L. "A Revision of the North American Genus *Amorpha* (Leguminosae—Psoraleae)." *Rhodora* 77: 337–409. 1975.

1. Petiole much shorter than the width of the lowest leaflets.   1. *A. georgiana*
1. Petiole longer than the width of the lowest leaflet.
   2. Calyx lobes all acute or acuminate, half as long as the calyx tube or a little more; at anthesis the
   inflorescence axis, flower stalks, and calyces, with dense, whitish pubescence, the whole inflorescence
   markedly grayish to the unaided eye.   2. *A. paniculata*
   2. Calyx lobes, or some of them at least, with rounded lobes very much less than half the length of the
   calyx tube; at anthesis inflorescence axis, flower stalks, and calyces varying from glabrous to densely
   pubescent, the pubescence, if any, tawny and not giving the whole inflorescence a gray cast.
                                        3. *A. fruticosa*

1. Amorpha georgiana Wilbur.

Erect shrub with 1 to several slender stems from the base, mostly 3–6 dm tall, each simple to
few-branched above. Current season's branchlets glabrous to sparsely short-pubescent, with
scattered resinous glands. Leaves 6–15 cm long, petioles 1–4 mm long, shorter than the
width of the lowest leaflets, sparsely to moderately puberulent or glabrous; leaflets varying
from 15 to 43 per leaf, very short-stalked, blades oblong, elliptic-oblong, or ovate-oblong,
the terminal one sometimes obovate to suborbicular, mostly 6–35 mm long and 3–18 mm
broad, bases rounded to truncate, apices rounded or obtuse, the midvein exserted as a short
mucro, surfaces glandular-punctate, sparsely pubescent when young, the lower surfaces
sometimes remaining so. Racemes 1 to several, 5–20 cm long, subsessile or with a stalk to 8
mm long. Calyx tube turbinate to campanulate, about 2 mm high, orifice oblique, exteriorly
glabrous or sparsely short-pubescent, bearing few to numerous pustular glands except near
the base; lobes pubescent within, margins densely short-ciliate, upper lobes shortest, the
lower lobe longest and narrowest. Petal clawed, blade obovate to obcordate, reddish purple
or blue. Fruit 4.5–5 mm long, obliquely obovate, tapering proximally to a stipelike base,
glabrous, conspicuously punctate-glandular in the upper ½–⅔.

   Seasonally wet pine savannas, flats bordering streams, borders of swamps and wet woodlands, low pastures, sand ridges. Coastal plain, N.C. to Ga.

2. Amorpha paniculata T. & G.

Stout shrub, 2–3 m tall. Branchlets of the season densely gray-pubescent. Leaves 2–3.5 dm
long, petioles, axis of the leaf and stalks of the leaflets densely gray-pubescent; young, developing leaflets short-pilose above, becoming glabrous or nearly so, densely pubescent beneath, at maturity moderately pubescent beneath with grayish or tawny hairs; petioles 2–6
cm long, mostly longer than the width of the lowest leaflets; leaflets symmetrical, 11–19 per
leaf, oblong, oblong-elliptic, or ovate-oblong, most of them 3–8 cm long, 1.5–3 cm broad,
rounded at both extremities, rarely emarginate apically, short-mucronate. Racemes several
to numerous, mostly 1.5–3 dm long, their axes, stalks of the flowers, and calyces densely
gray-pubescent. Calyx with few, if any, resinous glands, tube funnelform, oblique at the orifice, lobes all acute to acuminate, half as long as the tube or a little more. Petal purple, 5–7
mm long, 3–4 mm broad. Fruit 6–8 mm long, curved, broadest somewhat above the middle, densely pubescent (pubescence sometimes sparse by full maturity), with few to numerous large, resinous, pustular glands.

   Bogs, swampy woodlands, thickets, ditches. w. La., s.w. Ark., e. Tex.

3. Amorpha fruticosa L. BASTARD-INDIGO.   Fig. 112

Shrub 1–4 m tall, often bushy-branched, very variable with respect to pubescence and size
and shape of leaflets. Branchlets of the current season glabrous to densely, tawny pubescent,
without glands or sparingly and inconspicuously punctate-glandular. Leaves 1–3 dm long,
petioles mostly longer than the width of the lowermost leaflets; petioles, leaf axes, stalks of
the leaflets, axis of the racemes, flower stalks, and calyces, (from plant to plant), varyingly
glabrous to densely pubescent, the hairs mostly tawny; leaflets 9–35, symmetrical or rarely
somewhat asymmetrical basally, oblong, oblong-elliptic, ovate-oblong, less frequently elliptic or lanceolate, 1–5 cm long, 0.5–3 cm broad, rounded to acute basally and apically, the
apices sometimes emarginate, sometimes with a mucro; surfaces varying from glabrous or
nearly so to densely pubescent, without glands or inconspicuously or rarely conspicuously

241

Fig. 112. **Amorpha fruticosa:** a, twig with inflorescence; b, portion of stem with infructescence; c, young leaflets; d, flower; e, pistil; f, cluster of fruits. (From Correll and Correll. 1972)

242

glandular-punctate. Calyx tube obconic, funnelform, or campanulate, 2–3 mm high, glabrous to moderately pubescent, distally punctate-glandular with small and inconspicuous to large, amber, resinous glands, occasionally without glands; lobes with short-ciliate margins, all much shorter than the tube, sometimes the upper and lateral ones hardly more than undulations, the lower one usually acute. Petal obovate, tapering gradually to an indistinct claw, dark reddish purple, purple, or blue. Fruit 5–8 mm long, moderately to strongly curved, glabrous, with few to numerous, conspicuous, resinous glands, rarely without them. (Incl. *A. bushii* Rydb., *A. croceolanata* P. W. Wats., *A. curtissii* Rydb., *A. dewinkeleri* Small, *A. tennesseensis* Shuttlw. ex Kuntze, *A. virgata* Small)

Shores and banks of rivers and streams, floodplain woodlands, wet woodlands and hammocks, marshes and thickets of shorelines, moist to wet clearings. Maine and s. Que. to N.D., s. Man., S.D., e. Wyo., generally southward to s. pen. Fla., e. Colo., N.Mex., Ariz., s. Calif.; n. Mex.

5. Petalostemum (PRAIRIE-CLOVERS)

Perennial herbs, usually with several stems from a basal, hard and somewhat woody crown, the several stems usually branched, branches terminated by globose to oblong, very dense spikes of many small flowers, flowering of a given spike occurring over a considerable period of time from the base upwardly. Leaves odd-pinnately compound, both their axes and lower surfaces of the small leaflets with punctate glands; stipules linear or subulate. Calyx tube obconic to campanulate, lobes 5. Petal 1 (the standard) clawed, inserted within the base of the calyx. Fertile stamens 5, filaments united below, free above; staminodes 4, their filament portions arising from the summit of the filament tube alternating with the free portions of the filaments of the fertile stamens, the blades petaloid, colored like the petal and little different in form from it, thus the flower "appearing to have" five petals. Fruit short, obliquely obovate to suborbicular, 1- to 2-seeded, usually indehiscent.

- Petal and petaloid staminodes white.   1. *P. gracile*
- Petal and petaloid staminodes lavender, lavender-pink, or rose.   2. *P. carneum*

1. Petalostemum gracile Nutt.

Branches usually radiating from the base and sprawling, to 1 m long or more, very slender, glabrous. Median stem leaves mostly 2.5–3 (–4) cm long, commonly but not always shorter than the internodes and not crowded, glabrous, petioles 4–6 mm long, leaflets 5–9, mostly 5, linear-oblong or narrowly elliptic, 8–15 mm long, lower surfaces minutely and often obscurely glandular-punctate; uppermost leaves mostly with 3 leaflets, sometimes 1. Spikes long-stalked, globose at first, becoming short-oblong, 1–2 cm long, about 1 cm thick in fruit; floral bracts strongly folded proximally, basal margins ciliate, tips subulate, about as long as the calyces. Calyx glabrous, about 3 mm high, tube campanulate, lobes erect, triangular, very much shorter than the tube, pubescent interiorly. Petal white, blade cuneate-obovate, gradually narrowed to a claw slightly longer than the calyx; filaments of petaloid staminodes extending to about the tip of the blade of the petal, their blades like the petal blade but extending beyond it for about their full length. Pod included within the calyx, compressed, obliquely obovate in outline, 1.5–2 mm long, the curved, hairy style persistent.

Seasonally wet places in pine savannas and flatwoods, sometimes on slopes, usually with seepage seasonally. Coastal plain, Ga., n. Fla., westward to Miss.

2. Petalostemum carneum Michx.

Stems much coarser and more rigid than in the preceding, usually erect, or decumbent only near the base, 5–10 dm tall, glabrous. Median stem leaves 2–3 cm long, numerous, longer than the internodes, usually with short, leafy branchlets in their axils, the leaves thus appearing crowded; petioles 5–8 mm long, leaflets 5–9 per leaf, narrowly oblanceolate to linear-oblong, 5–8 mm long, lower surfaces with conspicuous punctate glands. Spikes long-stalked, subglobose or short-oblong at first, becoming oblong and 2–4 cm long and 1 cm thick in fruit; floral bracts subulate, keeled on the back basally but not folded, about 4 mm long, glabrous. Calyx 2.5–3 mm long, tube glabrous, campanulate, lobes triangular-subulate, about ¼ as long as the tube, pubescent interiorly. Petal lavender, lavender-pink, or rose, blade obovate, abruptly narrowed to the claw, the claw barely exserted from the calyx; petaloid staminodes the color of the petal, their filaments short, blades elliptic, scarcely if at all exceeding the petal. Pod included within the calyx, obliquely obovate in outline, plump, pubescent distally, the curved, hairy style persistent.

Pine savannas and flatwoods, marl prairies. Coastal plain, s.e. Ga. to s. pen. Fla.

6. Wisteria

Wisteria frutescens (L.) Poir. AMERICAN WISTERIA.

High-climbing, twining, woody vine. Herbaceous branches of the current season at first densely clothed with silky, shaggy hairs, less frequently with short appressed pubescence, gradually becoming glabrous. Leaves odd-pinnately compound, 1–3 dm long overall, petioles 2–6 cm long; leaflets usually 9–15 per leaf, oblong to lance-ovate, their stalks 2–4 mm long, blades 2–6 cm long and 1–2.5 cm broad, bases broadly rounded or obtuse, apices blunt to acuminate, margins entire; in early stages of leaf expansion, all parts densely clothed with silky pubescence, eventually the upper surfaces sparsely pubescent or glabrous, the lower at least sparsely pubescent; stipules lanceolate, stipels setaceous, both about 2 mm long, very early deciduous. Racemes densely flowered, to about 12 cm long, mostly pyramidal at middevelopment. Flower stalks 4–6 mm long, subtended by elliptic to lance-ovate bracts 9–12 mm long, these deciduous by or before full anthesis of the flower. Flower stalks, bracts, and calyces densely pubescent, the hairs sometimes with few to numerous clavate glands intermixed. Calyx tube campanulate, 5–6 mm long, the limb somewhat 2-lipped, lobes unequal. Corolla blue-purple to lilac, papilionaceous, 1.5–2 cm long; standard short-clawed, blade suborbicular, reflexed; wing and keel petals obliquely obovate, each with a prominent auricle on one side at the base. Stamens 10, diadelphous, 9 and 1. Legume stipitate, glabrous, linear-oblong, 5–10 cm long, acuminate apically, often constricted between the seeds. Seeds few, reniform, 6–8 mm long, brown. (*Kraunhia frutescens* (L.) Greene; incl. *W. macrostachya* Nutt. = *Kraunhia macrostachya* (Nutt.) T. & G.)

Wet shores of streams, ponds, lakes, bayous, borders of wet woodlands, wet thickets, floodplain forests. Chiefly coastal plain, s.e. Va. to cen. pen. Fla., westward to e. and s.e. Tex., northward in the interior to s.e. Mo., s. Ill., s. Ind.

7. Sesbania

Annual or perennial herbs or shrubs. Leaves evenly pinnately compound, petioles short, stipules quickly deciduous; leaflets apiculate apically, stipels none. Flowers bisexual, bilaterally symmetrical, in axillary, stalked racemes. Calyx tube campanulate, broader than high, somewhat oblique, lobes 5, shorter than the tube. Corolla papilionaceous, petals all clawed; standard blade as broad as or broader than long, reflexed; blades of wing and keel petals obliquely oblong or oblanceolate, those of the keel united along their adjoining margins, the keel strongly arching. Stamens 10, diadelphous, 9 and 1. Pod with 2 thickened sutures, or 4-angled or -winged, dehiscent or indehiscent, stipitate at the base, beaked apically.

1. Annual herbs; wing and keel petals auricled at one side of the base of the blade; pod flattened, the 2 sutures thickened.
   2. Body of the pod oblong to elliptic, 2–5 (–8) cm long, 15–20 mm broad, 1–2-seeded (usually 2-), seeds not separated by cross-walls, valves of the pod separating at maturity into an outer, thicker layer and an inner, thin-papery layer; stipe of the pod distinct, slender, 1–2 cm long; flowers 6–9 mm long; standard clear yellow. 1. *S. vesicaria*
   2. Body of the pod elongate-linear, mostly 15–20 cm long, 3–4 mm broad, many-seeded, seeds separated by cross-walls; stipe of the pod thickish, about 3 mm long, not very distinctly differentiated from

the body; flowers 15–20 mm long; standard (especially noticeable in the bud just before full anthesis) strongly speckled with purplish brown.   2. *S. macrocarpa*
1. Shrub; body of the pod 4-angled or -winged; wing and keel petals not auricled at the base of the blade.
3. Corolla bright orange-red, becoming purplish in drying; calyx with much reddish purple pigment.   3. *S. punicea*
3. Corolla yellow, sometimes with reddish lines or speckles; calyx without reddish purple pigment, or with only a little at the base of the tube.   4. *S. drummondii*

1. Sesbania vesicaria (Jacq.) Ell. BAG-POD, BLADDER-POD.   Fig. 113

Annual herb, with a single main stem of very variable stature, from a few dm to 40 dm tall. Branchlets at first with long, white pubescence (this very dense as they unfold), becoming glabrous. Leaves 1–1.5 dm long, leaflets 20–40 (–52) per leaf, oblong to elliptic, the larger ones 1.5–3 cm long, 3–6 mm broad, bases rounded or tapered, apices rounded but apiculate, notably silky-pubescent as they expand, becoming glabrous; stipules linear-attenuate, 7–10 mm long, early deciduous. Racemes 8–14 cm long, their stalks 5–12 mm long, subtended by subulate bracts, a pair of bractlets below the calyx, both early deciduous; raceme axis, flower stalks, bracts, and calyces silky-pubescent as they expand, becoming glabrous. Flowers 6–9 mm long; standard usually clear yellow, its blade reniform, notched apically; wing and keel petals with an auricle on one side of the base of the blade. Body of the pod nearly flat, oblong to elliptic, 2–5 (–8) cm long, 15–20 mm broad, the 2 sutures thickened, 1- or 2-seeded (usually 2-), seeds not separated by cross-walls; valves of the pod at maturity separating into an outer, thicker and an inner thin-papery layer; stipe distinct, slender, 1–2 cm long, beak 3–8 mm long. Seed plump, nearly oblong in outline, about 1 cm long and 5 mm in diameter, reddish brown. (*Glottidium vesicarium* (Jacq.) Harper)

The leafless stems with some, at least, of the pods dangling from the branches usually persist over much of the winter.

Weedy, wet ditches, wet clearings, spoil banks and flats, low wet fields, disturbed places in marshes and on shores, drained ponds, exposed river bars, often in much abundance locally. Chiefly coastal plain, N.C. to s. Fla., westward to e. third of Tex., s.e. Okla., Ark.; W.I., whence probably introduced to our range.

2. Sesbania macrocarpa Muhl.   Fig. 114

Annual with stature and general appearance similar to that of *S. vesicaria*, stem glaucous. Developing shoots, leaves, and racemes with few, scattered, silky hairs, if any, these soon sloughed. Larger leaves 1–3 dm long or a little more, leaflets 20–70 per leaf, linear-oblong, 1–3 cm long, 2–6 mm broad. Flowers mostly 15–20 mm long; standard (especially noticeable in the bud just before full anthesis) strongly speckled with purplish brown, its blade suborbicular, notched apically; wing and keel petals with an auricle on one side of the base of the blade. Body of the pod elongate-linear, mostly 15–20 cm long, 3–4 mm broad, sutures thickened, many-seeded, seeds separated by cross-walls; stipe of pod 3 (–5) mm long, thickish, not very distinctly differentiated from the base of the pod body, beak 5–10 mm long. Seed compressed, oblong, 3–4 mm long, about 2 mm wide, more or less orange along the attachment side, the rest of the surface with a more or less olive-green background and strongly speckled or blotched with black. (*Sesbania exaltata* (Raf.) Cory, *Sesban exaltata* (Raf.) Rydb. in A. W. Hill)

In habitats similar to those of *S. vesicaria*, with which it is sometimes intermixed. Local, coastal plain and piedmont, s.e. N.Y. and s.e. Pa. southward to S.C. and becoming more frequent southward to s. Fla., westward to e. third of Tex., northward in the interior, locally, to Mo. and s. Ill.

Authors' Note: While this book was in proof, we encountered in Bay Co., Fla., large populations of *S. macrocarpa*, flowers of all plants of which had standards which were dark blood-red abaxially and golden yellow adaxially, sharply contrasting with the common condition of pale yellow, the abaxial surfaces speckled with purplish brown. James R. Burkhalter, Pensacola, Fla. (personal communication), informs us that all plants of the species known to him in the Pensacola area have flowers with standards dark blood-red abaxially.

Fig. 113. **Sesbania vesicaria:** branch.

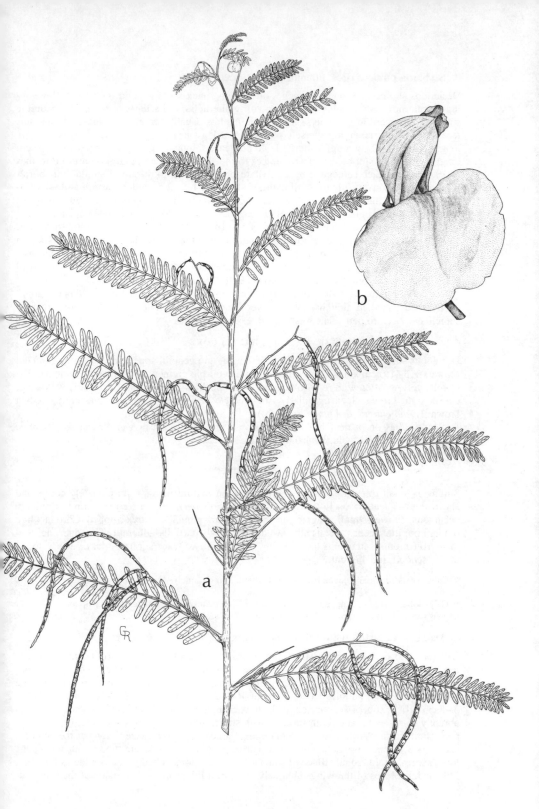

Fig. 114. **Sesbania macrocarpa:** a, branch, fruiting below, flowering at tip; b, flower.

3. Sesbania punicea (Cav.) Benth. in Mart. RATTLEBOX.

Deciduous shrub, 1−3 m tall, branches often dying back somewhat during winter. Developing shoots more or less silky-pubescent with white hairs; older parts of branches becoming sparsely pubescent to glabrous, stalks of the leaflets usually remaining pubescent, leaf surfaces commonly remaining sparsely pubescent at least near their margins. Leaves 1−2 dm long, petioles 1−2 cm long, leaflets 12−40 per leaf, oblong or oblong-oblanceolate, 1−3 cm long, 4−8 mm broad, bases tapered, apices rounded but apiculate, margins entire. Racemes 5−10 cm long, each subtended by a subulate bract about 2 mm long, 2 similar but shorter bractlets deciduous well before full anthesis. Flowers 1.5−2.5 cm long at full anthesis. Calyx usually with much reddish purple pigment. Corolla bright orange-red, becoming purplish in drying; blade of the standard suborbicular, 1.5−2.5 cm long, emarginate apically; wing and keel petals not auricled at the base of the blade. Pod dark brown or reddish brown, strongly 4-winged, edges of the wings usually undulating somewhat irregularly, oblong in outline, 5−8 cm long, 1−1.5 cm broad, with several to numerous seeds separated by crosswalls; stipe of the pod 5−15 mm long. Seed somewhat flattened, reniform in outline, 6−8 mm long, about 4 mm wide, reddish brown. (*Daubentonia punicea* (Cav.) DC.)

Native of S.Am., cultivated as an ornamental in the Atlantic and Gulf coastal plain, abundantly naturalized in sloughs, swales, wet clearings, vacant lots, sandy shores, marshy shorelines. N.C. to pen. Fla., westward to e. and s.e. Tex.

4. Sesbania drummondii (Rydb.) Cory. RATTLEBOX.

Deciduous shrub with stature, leaves, pubescence, and general appearance as for *S. punicea*. Flowers 1.3−1.6 cm long. Calyx without reddish purple pigment or with only a little basally. Corolla yellow, often with reddish lines and fine speckles. Pod similar but usually yellowish, tan, or light brown. Seed irregularly isodiametric in outline, 5−6 mm, puckered, reddish brown. (*Daubentonia drummondii* Rydb.)

Sandy areas on or near beaches, brackish flats, low, moist or wet places, waste areas, borders of woodlands. Fla. Panhandle to La., Ark., Tex.

8. Dalbergia

Shrubs or small trees. Leaves (in ours) simple on mature plants (odd-pinnately compound and with 2−7 leaflets on juvenile plants of *D. brownei*), margins entire, stipules quickly deciduous. Flowers small, fragrant, in short, axillary panicles. Calyx 5-lobed. Corolla white or pink, papilionaceous, petals all clawed, keel petals partially adherent along one side. Stamens 10 (in ours), filaments united below, free distally. Ovary stipitate, flat or nearly so, 1- to few-seeded, indehiscent.

- Calyx lobes essentially equal; fruit flat, suborbicular or obliquely suborbicular, 1-seeded.

　　　　　　　　　　　　　　　　　　　　　　　　　　　　　　　　1. *D. ecastophyllum*
- Calyx lobes unequal, 2-lipped, 2 lobes of upper lip broadest, central lobe of lower lip longest and the laterals shortest; fruit compressed, mostly oblong, 1−4-seeded.　　　　　　2. *D. brownei*

1. Dalbergia ecastophyllum (L.) Taub. in Engl. & Prantl. COIN-VINE.

Lax shrub to 6 m tall, with straggling or trailing branches, often forming impenetrable masses. Young twigs with dense, appressed, brownish to grayish pubescence, older ones brown, glabrous, with small, raised, circular, buff lenticels. Leaves simple on both juvenile and mature plants, petioles stoutish, about 10−15 mm long, blades 8−10 (−12) cm long, 3−8 cm broad, subcoriaceous and thickish when mature, rounded to truncate basally, apices abruptly short-acuminate, both surfaces with short, appressed pubescence when young, upper or both surfaces often becoming glabrous. Panicles to about twice as long as the petioles. Calyx 3−3.5 mm long, subtended by a pair of minute bractlets, the tube campanulate, the lobes short-deltoid, equal. Blade of standard 7−8 mm long, obcordate, as wide as long or a little wider, blades of the wings obliquely oblong, a little longer than blade of the standard, keel about 6 mm long. Fruit flat, 1-seeded, orbicular or nearly so, or short-oblong, often oblique, surfaces with short, appressed pubescence, sometimes becoming glabrous by full maturity. (*Ecastophyllum ecastophyllum* (L.) Britt.)

Borders of mangrove swamps, coastal dune scrub, thickets, hammocks. cen. pen. Fla., southward in Fla.; trop. Am.; w. Afr.

2. Dalbergia brownei (Jacq.) Urban.

Shrub up to 5 m high; juvenile plants with prostrate branches, 2–7-foliolate leaves; older plants with suberect, sprawling, or twining, often ropelike branches, simple leaves; often forming impenetrable entanglements. Young twigs brown-puberulent, older twigs dark brown, glabrous, striate, with raised, circular, brown lenticels. Petioles slender, 10–15 mm long, puberulent, blades ovate or lance-ovate, mostly 4–8 cm long, 2–4 cm broad, sub-coriaceous but thin, bases slightly cordate, truncate, or rounded, apices acuminate, upper surfaces glabrous, lower very sparsely puberulent. Panicles lax, with pubescent axes, often bearing one or more leaves. Calyx tube cylindric-campanulate, subtended by a pair of mi-nute, quickly deciduous bractlets, limb 2-lipped, the upper lip with 2 oblongish, obtuse lobes, lower with a linear-subulate central lobe somewhat longer than the lobes of the upper lip, and 2 short laterals, the calyx about 5 mm long overall. Blade of standard petal oblong, obcordate distally, 5–10 mm long, wings narrower than the standard, obliquely oblong, keel shorter than the wings. Fruit compressed, varying from short-oblong, little longer than broad to oblong and 2–3 times longer than broad, sometimes constricted medially, 1–3.5 cm long, sometimes short-apiculate apically, margins sparsely pubescent when young, glabrous at ma-turity, 1–4-, usually 2–4-, seeded. (*Amerimnon brownei* Jacq.)

On banks of streams, margins of mangrove swamps, coastal thickets. s. pen. Fla., Fla. Keys; trop. Am.

9. Apios

Apios americana Medic. POTATO-BEAN, GROUNDNUT.

Twining perennial vine, climbing to about 3 m high, rhizomatous, the rhizome bearing tu-bers 1–2 cm thick in a beadlike series; stem at first strigose, becoming glabrous or nearly so. Leaves odd-pinnately compound, leaflets mostly 5–7, 3 on occasional leaves of a given plant, 1–2 dm long at full maturity, petioles 2–7 cm long; leaflets with densely pubescent stalks 1.5–3 mm long, blades ovate or lance-ovate, 3–6 (–10) cm long and mostly 2–3 cm broad, bases rounded, apices acute to acuminate, margins entire; leaflets abundantly strigose as they unfold, especially along the veins, later sparsely strigose; stipules setaceous, pubes-cent, 3–6 mm long, deciduous much before the leaves are fully expanded, stipels similar but 1–2 mm long and they too deciduous very early in leaf expansion. Racemes oblong to ob-conical, axillary, often 2–4 in an axil, each often with short, few-flowered branches, thus "subpaniculate." Flowers with a grapelike fragrance. Calyx sparsely pubescent, subhemis-pheric, about 3 mm high, nearly truncate at the summit or with 5 broadly triangular, short, unequal lobes. Corolla brownish purple, papilionaceous, about 1 cm long; standard broader than long, rounded or retuse apically, with a pair of small auricles at the base; wings clawed and curved-ovate distally, becoming deflexed below the keel, an auricle on one side of the claw of each; keel petals approximately sickle-shaped, very short-clawed, an auricle on one side of each at the base of the blade. Stamens 10, diadelphous, 9 and 1. Style curved in conformity with the keel and further spiraled contrary to the arch of the keel, with approx-imately the shape of a question mark. Legume plump, linear-oblong in outline, stipitate, 5–12 cm long, obliquely long-apiculate apically, slightly constricted between the seeds, the valves dehiscing spirally. Seeds short-oblong, 5–7 mm long, wrinkled, purple. (*Glycine apios* L.)

Bottomland woodlands, stream banks, wet shoreline thickets, borders of swamps, wet clearings, meadows. N.S. and N.B. westward to Minn. and S.D., generally southward to s. pen. Fla. and Tex.

10. Vigna

Vigna luteola (Jacq.) Benth.

Perennial, herbaceous, trailing or twining vine, branches to 3 m long, often forming tangles. Stem somewhat striate, usually pubescent, the hairs long and shaggy-spreading, or stiffish and retrorse, rarely glabrous. Leaves trifoliolate, petioles 1.5–6 cm long, with long, retrorse, stiffish hairs; leaflets very variable in size and shape from plant to plant, ovate, lance-ovate, lanceolate, or linear-lanceolate, 2–8 cm long, up to 3.5 cm broad, surfaces sparsely strigose, truncate to cuneate basally, apices acute or obtuse, the short stalks of the leaflets and leaf axis usually antrorsely strigose, the axis sometimes glabrous; stipules lanceolate to lance-ovate, 3–5-nerved, 2–3.5 mm long, pubescent, stipels similar, 1–2 mm long, usually not early deciduous. Racemes axillary, of few flowers terminating stalks several times longer than the subtending leaves. Flower stalks short, shaggily retrorsely pubescent, subtended by a pubescent bract 1–1.5 mm long and with a pair of similar bractlets below the calyx, both quickly deciduous. Calyx tube broadly campanulate to subhemispheric, 2–2.5 mm high, glabrous or sparsely appressed-pubescent, limb 4-lobed, lobes ciliate, upper one notched, 2 mm long, laterals triangular, a little shorter, lower subulate, about 2.5 mm long. Corolla yellow, papilionaceous; standard very shortly and narrowly clawed, blade suborbicular, notched apically, a pair of minute descending auricles at the base; wings clawed, blades oblongish, each with an auricle on one side of the base; keel petals oblique, the keel itself strongly curved but not coiled. Stamens 10, diadelphous, 9 and 1. Mature legumes strongly reflexed from the axis; not stipitate, plump, linear-oblong, 4–6 cm long and about 5 mm broad, strigose. Seeds several, brown, about as broad as long, about 5 mm, outline irregular. (*V. repens* (L.) Kuntze)

Mostly near the coasts, in and about tidal marshes and flats, estuarine shores, borders of mangrove swamps, coastal flatwoods, ditches, swales, wet sands of roadsides and waste places. s.e. N.C. to s. Fla., westward to s.e. Tex.; trop. Am.

11. Aeschynomene (SENSITIVE JOINT-VETCHES, SHY-LEAVES)

Annual erect herbs with evenly pinnately compound leaves and stipules attached peltately (in those treated here). Leaves sensitive to light and usually to touch. Flowers in axillary, stalked, few-flowered racemes or panicles, the inflorescence stalk subtended by a bract similar to the stipules, each flower subtended by a nonpeltate bract, a pair of nonpeltate bractlets inserted below the calyx. Calyx (in ours) more or less persistent, strongly 2-lipped, the upper lip 2-lobed, lower 3-lobed. Corolla papilionaceous, yellowish to red or purple. Stamens 10, filaments united for about half their length, separating, sometimes tardily, into 2 groups of 5 each. Fruit stipitate, laterally compressed, forming a loment of 1-seeded segments.

● Leaflets (excepting the proximal pair) strongly 2- or 3-nerved beneath; loment strongly scalloped along one side.   1. *A. americana*
● Leaflets 1-nerved beneath; loment shallowly scalloped along one side.   2. *A. indica*

1. Aeschynomene americana L. var. americana.   Fig. 115

Annual erect herb, mostly 1–2 m tall, much branched only from well above the base, or often divaricately broadly branched at the base, upwardly the branches becoming erect-ascending. Main stem and branches with long, stiffish, pustular-based hairs, becoming glabrous below in age; pustular bases of the hairs dark purple; pubescence of the inflorescence axis and flower stalks of both short and long pustular-based hairs. Leaves short-petiolate, 2–7 cm long; stipules striate, above the point of attachment triangular to long-subulate, tips acute to setaceous-attenuate, prominently ciliate marginally, below the point of attachment usually oblong with oblique distal acuminations, margins entire; leaflets 20–60 per leaf, 4–15 mm long, 2- or 3-nerved beneath, linear or nearly so, inequilateral both basally and apically, minutely but sharply toothed on 1 margin distally from the middle or somewhat above, tip mucronate, petiole and axis pubescent. Inflorescence zigzagging, subtending bract like the stipules, axis and flower stalks with both short, stiff, purple-glandular, and long, straight, pale hairs. Floral bracts ovate-acuminate, conspicuously ciliate, flowers 5–10 mm long. Loment with a glabrous stipe 2–3 mm long, 3–9-segmented, lower margin strongly crenate; segments glabrous, each prominently moundlike centrally over the seed, the moundlike portion smooth or rough-tuberculate. Seed plump, approximately oval in outline but with a nipplelike protuberance at the side of one extremity, purplish black, about 2 mm long.

Weedy and sporadic in seasonally wet places, ditches, swales, clearings, roadsides, often in extensive almost pure stands during one year, absent there later. s. Ga. and Fla. (where probably introduced); trop. Am.

2. Aeschynomene indica L.
Fig. 115

Erect, often much branched annual of variable stature to 2.5 m tall. Stem often robust below, to 1 cm in diameter or a little more, forming numerous adventitious roots on submersed portions, if any, with a few scattered pustular-based hairs that are generally sloughed leaving reddish purple spots. Leaves short-petiolate, 5–12 cm long, petioles shortly glandular-pubescent, the axis similarly pubescent at least proximally; stipules striate, above the point of attachment linear-subulate, usually conspicuously ciliate marginally, below the point of attachment subulate, margins entire; leaflets mostly about 50–70, 5–10 mm long, glabrous, oblong, 1-nerved beneath, bases inequilateral, apices equilateral and rounded, with a minute apicule, margins entire. Inflorescence often bearing 1 or 2 leaves from the lower node or nodes, axis with sparse, short, glandular hairs, flower stalks glabrous or nearly so. Floral bracts ovate-acuminate, margins entire or toothed, flowers 8–10 mm long. Loment with glabrous stipe 3–8 (–10) mm long, 3–12-segmented, lower margin shallowly scalloped; segments somewhat convex or rounded over the seeds, the surface with scattered, purplish red dots where short, pustular-based hairs were present during early stages of development. Seed plump, reniform in outline, olive-green, about 3 mm long. (*A. virginica* sensu Small, 1933)

Ditches, sloughs, bayous, swales, estuarine shores, marshy shores of ponds and impoundments, floating mats of vegetation, wet clearings, borders of swamps, wet meadows. Coastal plain, N.C. to s.cen. pen. Fla., westward to Ark. and s.e. Tex.; P.R.; coastal Asia, Pacific Is., Austl., Afr.

Some North Carolina specimens may be referred to the dubiously distinct *A. virginica* (L.) BSP. See Radford et al., 1964.

12. Vicia (VETCHES)

Annual, biennial, or perennial, herbaceous vines, their stems erect, ascending, sprawling, or climbing. Leaves pinnately compound, those treated here with tendrils terminally; stipules persistent, usually with a spreading or descending lobe on one side at the base, thus "half-hastate" or "half-sagittate"; leaflets with very short stalks, entire marginally, stipels none. Flowers sessile or subsessile, single in the leaf axils, 2 to several in close axillary clusters, or in long-stalked axillary racemes or spikes (sometimes the stalk bearing a single flower). Calyx campanulate to somewhat turbinate, the limb symmetrical or asymmetrical, 5-lobed. Corolla papilionaceous, white, blue, violet, or yellow; standard folded downward, partially overlapping the wings, wings overlapping and somewhat adhering to the keel. Stamens 10, diadelphous, 9 and 1. Ovary sessile or short-stipitate, style with a tuft of hairs on one side below, or with a ring of hairs beneath the stigma. Legume flat to almost terete, 2- to several-seeded, the valves usually spiraling at dehiscence.

Reference: Hermann, F. J. *Vetches in the United States—Native, Naturalized, and Cultivated*. Handb. No. 168, U.S. Dept. Agr., 1960.

1. Ovary with 2 or 3 ovules or legume with 2 or 3 seeds; most leaflets elliptic or oval, few, if any, over twice as long as broad. 2. *V. floridana*

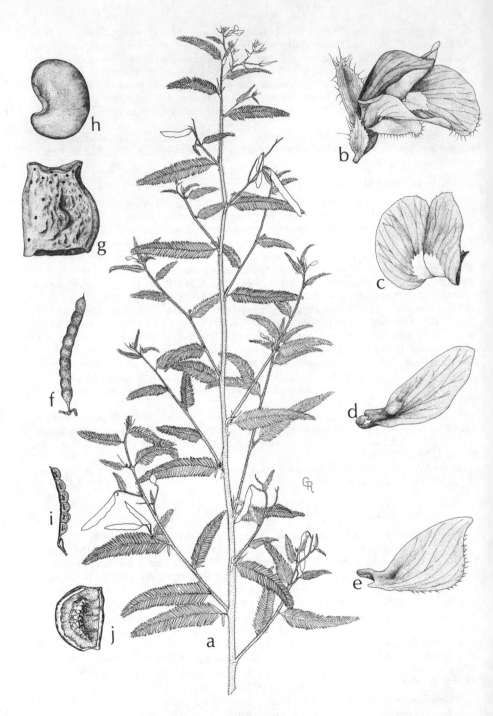

Fig. 115. a–h, **Aeschynomene indica:** a, branch; b, flower; c, standard; d, wing; e, keel; f, fruit (loment); g, segment of loment; h, seed; i,j, **Aeschynomene americana:** i, loment; j, segment of loment.

1. Ovary with 4–12 ovules or legume with 4–12 seeds; most leaflets linear to linear-oblong, considerably more than twice as long as broad.

   2. Stalk of the inflorescence shorter than the axis of the leaf (not including the tendril); flower solitary or rarely 2, 5–6 mm long.   1. *V. minutiflora*

   2. Stalk of the inflorescence as long as or a little longer than the axis of the leaf (not including the tendril); flowers mostly exceeding 8 per raceme, fewer than 8 on an occasional raceme, each 8 mm long or longer.

      3. Leaflets 2–4, mostly 4, narrowly linear, 1.5–3 cm long, 1–2 mm broad; flowers 8–9 mm long; legumes 2.5–3 cm long.   3. *V. acutifolia*

      3. Leaflets 4–6, mostly 6, linear-oblong or linear-elliptic, 3–5 cm long, 3–4 (–6) mm broad; flowers 10–12 mm long; legumes 4–4.5 cm long.   4. *V. ocalensis*

1. Vicia minutiflora Dietr.   Fig. 116

Winter annual. Stems angular, slender, usually several from the base, weakly erect, sometimes climbing, 3–8 dm long; young developing shoots and leaves pubescent, usually becoming glabrous by maturity. Lowermost leaves on the stem (early in the season at least) with 2 or 3 leaflets, these obovate or broadly elliptic, relatively short, 3–10 mm in length; leaves higher on the stem with 2–7 leaflets, mostly alternately disposed, linear to oblong or oblanceolate, 1–3.5 cm long, 1.5–6 mm broad, their apices acute to truncate, the latter often notched; tendril very slender, simple or few-branched. Flowers mostly solitary, rarely 2, the stalk shorter than the axis of the subtending leaf (not including the tendril). Corolla pale blue or purplish white. Legume 2–3 cm long, 5 mm broad, nearly flat, oblong, oblique at both extremities, 4–12-seeded. (Incl. *V. micrantha* Nutt. ex T. & G.)

    Mixed upland woodlands and fields, roadsides, also low areas in fields, bottomland clearings, wet woodlands, river and stream banks, thickets. Tenn. and Mo., generally southward to Fla. Panhandle, e. Okla. and e. and n.cen. Tex.

2. Vicia floridana S. Wats.   Fig. 117

Slender perennial. Stems angular to about 8 dm long, reclining or sprawling, often forming dense tangles; young developing branches and leaves pubescent, becoming glabrous by maturity. Leaflets opposite, 2–6, commonly 4, mostly elliptic or nearly so, 5–20 mm long, 2–8 mm broad, commonly twice as long as broad but some of them sometimes to 4 times as long as broad, shortly cuneate basally, mostly rounded apically, but the longer, narrower ones obtuse to acute, all minutely mucronate; tendrils very slender, unbranched. Stalk of the raceme usually shorter than the axis of the subtending leaf, sparsely short-pubescent; flowers 2–8 per raceme, about 6 mm long, corolla white or bluish white. Legume nearly flat, approximately oblong, 10–15 mm long, about 5 mm broad, tapered to a short stipe basally, obliquely tapered apically, 1–3(4)-seeded, mostly 2. Seed globose or subglobose, 4 mm in diameter, purplish brown and with a slight sheen.

    River and stream banks and shores, borders of wet woodlands, swales, ditches and moist to wet ditch banks and roadsides. Fla., westward to Dixie Co. in the north, southward to cen. pen. Fla.

3. Vicia acutifolia Ell.   Fig. 118

Slender perennial. Stems commonly several from the base, to about 12 dm long, reclining or climbing on neighboring plants, sometimes forming tangles, pubescent at first, glabrous or nearly so on older portions. Leaflets motly opposite, 2 or 4, linear or narrowly linear-oblong, 1.5–3 cm long, 1–2 mm broad, apices acute, mucronate; leaf stalks pubescent, leaf axis sparsely pubescent; tendril slender, unbranched. Stalk of the raceme usually somewhat longer than the axis of the subtending leaf (not including the tendril); raceme 2–10-, mostly 8–10-flowered, flowers 8–9 mm long. Corolla pale blue, the standard purple-lined and purple-tipped. Legume 2.5–3 cm long, 5 mm broad, oblique at both extremities, mostly 8–12-seeded. Seed orbicular in outline, flattened, about 2 mm in diameter, brown and with a slight sheen.

    Fresh and brackish marshes, borders of swamps and wet woodlands, marshy shores and clearings, ditches, moist to wet roadsides. Coastal plain, s.e. S.C. to s. pen. Fla., s. Ala.

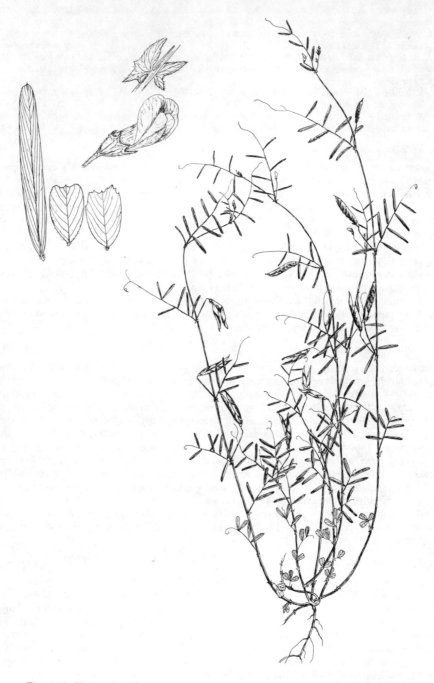

Fig. 116. **Vicia minutiflora.** (From Hermann, *Vetches of the United States* (1960)
Fig. 20)

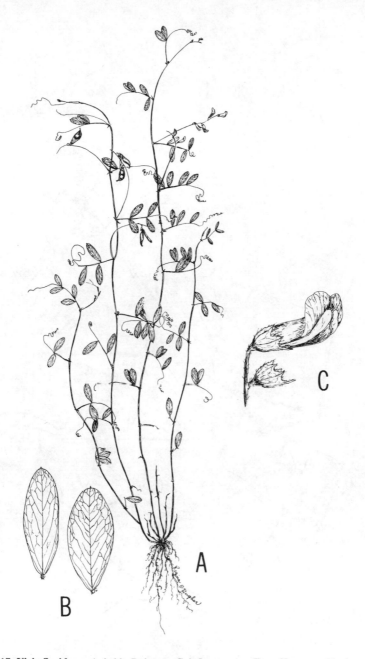

Fig. 117. **Vicia floridana:** A, habit; B, leaves; C, inflorescence. (From Hermann, *Vetches of the United States* (1960) Fig. 13)

B

A

C

Fig. 118. **Vicia acutifolia:** A, habit; B, flower; C, leaf. (From Hermann, *Vetches of the United States* (1960) Fig. 18)

4. Vicia ocalensis Godfrey & Kral.

Relatively robust perennial. Stems to 12 dm tall, reclining to suberect or climbing on surrounding plants, angled and striate; young developing shoots and inflorescences densely pilose; later, the stems, leaf axes, lower surfaces of leaflets, stalks of racemes, axes of racemes, flower stalks, and calyces moderately to sparsely pilose. Leaflets 4–6, mostly 6, linear-oblong, narrowly elliptic, or almost linear, 3–5 cm long, 2–4 (–6) mm broad, apices rounded or subtruncate, mucronate; tendril unbranched. Stalk of the raceme usually somewhat longer than the axis of the subtending leaf; racemes 12–18, mostly 15–18, flowered, flowers 10–12 mm long. Corolla bluish white, standard purple-lined and purple-tipped, wings white, keel white with purple tip. Legume nearly flat, 4–4.5 cm long, 7–8 mm broad, oblong, both extremities oblique, 8–12-seeded. Seed orbicular in outline, flattened, 3–3.5 mm in diameter, brown and with a slight sheen.

Marshy shores, banks and thickets along spring-fed streams, nearby ditches. Restricted, insofar as we know, to Ocala National Forest, n.cen. pen. Fla.

Linaceae (FLAX FAMILY)

Linum (FLAXES)

Slender annual or perennial herbs. Leaves simple, sessile, opposite on the lower part of the stems and alternate above, or opposite or alternate throughout. Inflorescence a terminal cyme or panicled cyme. Flowers radially symmetrical, bisexual. Sepals 5, separate. Petals 5, separate, yellow (in species treated here), falling very soon after anthesis. Stamens 5, filaments united basally, with intervening staminodia in some. Pistil 1, 5-carpellate, ovary superior, becoming more or less 10-locular by intrusion of false partitions; styles 5, wholly distinct, or united below and free above, stigmas capitate. Fruit capsular, dehiscing into 10 1-seeded or 5 2-seeded segments.

1. Style undivided to beyond the middle, usually to within 0.5–1 mm of the entire length; capsule dehiscing into five 2-seeded segments; sepals usually shed by full maturation of the capsule. 1. *L. carteri*
1. Styles wholly free; capsule dehiscing into ten 1-seeded segments; sepals persistent around the fully mature capsule.
  2. Leaves with a small, reddish purple, stipular gland at either side of the base, the glands usually persistent after leaf fall.   2. *L. arenicola*
  2. Leaves without stipular glands.
    3. The leaves opposite on about the lower half of the stem, sometimes to a higher level.
      4. Inflorescence of few, ascending, few-flowered branches terminating the stem; capsule subglobose, about 3 mm broad, apiculate apically; sepals thickish, blunt to rounded apically.
                                                  3. *L. westii*
      4. Inflorescence usually diffusely paniculate, the branches spreading, overall length commonly ½–⅓ the total height of the stem; capsule about 2 mm broad, usually a little broader than long, not apiculate; sepals thin, acute apically.   4. *L. striatum*
    3. Leaves alternate, or only a few near the base opposite.
      5. Capsule ovoid, longer than broad.   5. *L. floridanum*
      5. Capsule depressed-globose or oblate, broader than long.
        6. Medial stem leaves oblanceolate to spatulate, blunt apically; inflorescence branches laxly ascending or spreading; inner sepals entire marginally or less frequently with few, sessile glands.
                                      6. *L. virginianum*
        6. Medial stem leaves linear-lanceolate to linear-subulate, sharply acute apically; inflorescence branches stiffly ascending; inner sepals with stalked glands marginally.   7. *L. medium*

1. Linum carteri Small.

Annual, 1–6 dm tall. Stem wholly glabrous, minutely pubescent, or often finely scaly, narrowly wing-angled but the lower stem usually becoming terete in age. Leaves alternate, linear-subulate, 1–2 (–3) cm long, the lower and medial with entire margins, the upper or bracteal ones glandular-toothed; a small, purplish red, stipular gland present on either side of the leaf base in some individuals, absent in others. Cymes terminal, irregular, the branches often racemose, flower stalks thickish and rigid, angled, mostly 3–5 mm long. Sepals 5–7 mm long, lanceolate or lance-ovate, acuminate apically, margins glandular-toothed. Petals orange-yellow, 9–17 mm long, obovate. Style 6–9 mm long, undivided to within 0.5–1 mm below the stigmas. Capsule straw-colored, ovoid, 3–4 mm broad, dehiscing into 5 2-seeded segments, the sepals usually shed by full maturation of the capsule. Seed 3 mm long, thinly lenticular, lance-ovate in outline, brown, surface finely and faintly reticulate. (*L. rigidum* Pursh var. *carteri* (Small) C. M. Rogers, *Carthartolinum carteri* (Small) Small; incl. *L. carteri* var. *smallii* C. M. Rogers)

Shallow soils on lime-rock in open pinelands and glades, often in solution pits, clearings of pinelands, adjacent roadsides, outer margins of mangrove swamps. s. pen. Fla.

2. Linum arenicola (Small) Winkl. in Engl. & Prantl.

Slender, wiry perennial, 2–7 dm tall, usually with several stems from the base, each with a terminal cyme. Stem glabrous, irregularly striate-angled, the leaves, or most of them, usually shed by flowering time. Leaves opposite only at the base of the stem, alternate otherwise, linear-subulate, larger ones 7–10 (–15) mm long, margins entire or irregularly beset with minute glands, a small, purplish red, stipular gland at either side of the leaf base, these usually persistent after leaf-fall. Flower stalks 2 mm long or less. Sepals lanceolate to ovate, the inner ones sometimes obovate, with a prominent midrib, the outer 2.5–3 mm long or slightly more, the inner slightly shorter and with hyaline edges, all glandular-toothed, apices acuminate. Petals yellow, mostly 4.5–5.5 mm long, obovate. Styles separate. Capsule straw-colored, ovoid, 2–2.5 mm broad, short-apiculate apically, dehiscing into 10 1-seeded segments. Seed a little over 1 mm long, nearly flat, elliptic-ovate in outline, with a yellowish cartilaginous marginal band within which the surface is brown and very finely reticulate, lustrous throughout. (*Carthartolinum arenicola* Small)

Solution pits and shallow soil of ephemeral pools on lime-rock in open pinelands, pineland clearings, adjacent roadsides. s.e. tip of pen. Fla., w. Fla. Keys.

3. Linum westii C. M. Rogers.

Slender, glabrous perennial, 2–5 dm tall, often with 2 to several stems from near the base, each with a few-flowered terminal cyme. Stem narrowly wing-angled. Leaves opposite from the base to about midstem, the principal ones linear-oblong, narrowly elliptic, or narrowly oblanceolate, about 1.5 cm long and 2–4 mm broad, margins entire, bases very little narrowed, apices obtuse; leaves gradually diminishing in size upwardly, upper alternate ones with acute apices. Flower stalks 2–3 mm long. Outer sepals broadly elliptic to ovate, inner broadly elliptic, both thickish, with prominent midribs and blunt to rounded apices, about 3 mm long, margins of outer ones entire, inner usually minutely glandular-toothed. Petals pale yellow, about 6 mm long. Styles distinct. Capsule straw-colored, subglobose, apiculate apically, about 3 mm broad, dehiscing into 10 1-seeded segments. Seed nearly flat, narrowly obovate in outline, a little over 1 mm long, with a narrow, yellowish, cartilaginous band marginally and reddish brown within that, sublustrous.

Boggy depressions in pine flatwoods, margins of cypress ponds or depressions, *Hypericum* bogs, adjacent ditches. At present definitely known to occur in Jackson, Calhoun, and Franklin Cos., Fla. Panhandle, and in Baker Co., n.e. Fla.

4. Linum striatum Walt.

Glabrous perennial, 3–12 dm tall. Stem narrowly striate-angled, sometimes solitary, more commonly 2 to several stems from a basal crown, each usually with numerous, leafy and abundantly floriferous, spreading branches, the flowering portions commonly comprising ⅓–½ or more of the total height. Leaves opposite from the base to midstem or somewhat above, others alternate, the principal ones usually narrowly elliptic, sometimes oblanceolate, cuneate below, apices acute to rounded, 1–3.5 cm long and 2–11 mm broad, entire marginally. Flower stalks angular, broadened upwardly, 1–3 mm long or a little more. Sepals thin, elliptic, ovate, lance-ovate, or oblong, midribs faint, mostly 1.5–3 mm long, apices varyingly acute to mucronate, all entire marginally or the inner minutely glandular-toothed. Petals pale yellow, obovate, 2.5–4.5 mm long. Styles distinct. Capsule depressed-globose, about 2 mm broad, often with the style bases forming an apical point, dehiscing into 10 1-seeded segments. Seed about 1 mm long, nearly flat, falcate-elliptic in outline, brown, very faintly reticulate, sublustrous. (*Cathartolinum striatum* (Walt.) Small)

Seepage areas, marshy shores, banks of small streams, meadows, wet woodlands and their borders, wet clearings. Mass. to Pa., s. Ohio to s.e. Mo., generally southward to the Fla. Panhandle and e. Tex.; also interdunal swales near Lake Michigan in w. Mich. and n.w. Ind.

5. Linum floridanum (Planch.) Trel.

Glabrous perennial, 2–12 dm tall. Stem narrowly wing-angled but becoming terete below in age, commonly solitary, less frequently 2 to several stems from a basal crown, each with a terminal cyme with strongly ascending branches. Lower few leaves opposite, others alternate, usually numerous and closely set, linear-lanceolate to linear-subulate, relatively rigid, sharply acute apically, the larger ones 10–15 mm long and to 1 mm broad or a little more, entire marginally, gradually diminishing in size upwardly. Flower stalks angled, dilated distally, 2–3.5 mm long. Sepals lance-ovate, acute to acuminate apically, 2.5–3.5 mm long, outer entire marginally, inner with stalked glands at least distally, midribs usually faint. Petals lemon yellow, obovate, 5–10 mm long. Styles distinct, their bases commonly persistent. Capsule ovoid, longer than broad, 2–3 mm broad, yellow proximally and more or less suffused with purple distally, or wholly yellow, sometimes both on an individual plant, dehiscing into 10 1-seeded segments. Seed 2 mm long, nearly flat, elliptic-oblong in outline, reddish brown, dull, faintly and finely reticulate. (*Cathartolinum floridanum* (Planch.) Small, *L. virginianum* var. *floridanum* Planch.; incl. *L. floridanum* var. *chrysocarpum* C. M. Rogers)

Seasonally wet pine flatwoods and savannas, bogs, ditches, low fields, also on well-drained sands of longleaf pine-scrub oak ridges and hills. Coastal plain, Va. to s. Fla., westward to La.; Jam.

6. Linum virginianum L.

Glabrous perennial, 1–7 dm tall. Stem essentially terete, somewhat striate above, solitary, or sometimes with several stems from a basal crown, each with laxly ascending or spreading flowering branches, these often from leaf axils from midstem upwardly, sometimes only near the summit. Lower few leaves opposite, others alternate, relatively widely spaced, the larger ones mostly oblanceolate to spatulate, relatively thin and flexuous, apices blunt, 1–3 cm long and 3.5–6 mm broad, entire marginally. Flower stalks slender, obscurely angled to terete, dilated at the summit, mostly 2–6 mm long. Sepals lance-ovate, 2–4 mm long, the outer acute apically, margins entire, inner mucronate, margins sometimes with a few sessile glands distally. Petals yellow, obovate, 4–6 mm long. Styles distinct. Capsule oblate, broader than long, 2–3 mm broad, straw-colored, dehiscing into 10 1-seeded segments. Seed nearly flat, oblong-elliptic in outline, about 1.5 mm long, reddish brown, sublustrous. (*Cathartolinum virginianum* (L.) Reichb.)

Open upland woodlands, clearings, and thickets, rarely in low woods or bogs. Mass. to s. Ont., s. Mich. to Mo., southward to s.e. Va., thence chiefly inland, w. N.C., n.w. S.C., to cen. Ga., Ala. Tenn.; isolated localities in the coastal plain, S.C. and Ala.

7. Linum medium (Planch.) Britt. var. texanum (Planch.) Fern.

Glabrous perennial, 1.5–5 (–8) dm tall. Stem narrowly wing-angled, becoming terete below in age, solitary, or less commonly with 2 to several stems from a basal crown, each with stiffly ascending floriferous branches, these from midstem upwardly or only near the summit. Lower few leaves opposite, others alternate, usually numerous and relatively closely spaced, principal medial ones lance-linear to linear-subulate, sharply acute apically, 1–2.5 cm long, 2–3 mm broad, margins entire. Flower stalks 3–5 mm long, angled. Sepals ovate or lance-ovate, 2–3 mm long, acute or short-acuminate apically, outer ones with entire margins, inner with stalked glands marginally. Petals yellow, obovate, 4.5–8 mm long. Styles distinct. Capsule depressed globose, broader than long, 2–2.5 mm broad, more or less suffused with purple, dehiscing into 10 1-seeded segments. Seed thinly lenticular, narrowly elliptic in outline, 1–1.3 mm long, reddish brown, lustrous. (*Cathartolinum medium* (Planch.) Small var. *texanum* (Planch.) Moldenke, *L. virginianum* var. *medium* Planch.; incl. *Cathartolinum curtissii* Small)

Pine savannas and flatwoods, bog margins, moist to wet ditches, cypress depressions, marl prairies, interdune swales, infrequent in well-drained places. s.w. Maine to N.J., westward to Ohio, s. Mich., n. Ill., s.e. Iowa, Mo., s.e. Kan., generally southward to s. Fla. and e. Tex.; Bah.Is.

Balsaminaceae

Impatiens (TOUCH-ME-NOT, JEWEL-WEED, BALSAM)

Succulent annual herbs. Stems hollow. Leaves alternate, simple. Flowers bisexual, pendent, in stalked clusters in the leaf axils. Calyx petaloid, of 3 sepals, the two upper small, the lower one saccate, open in front, spurred below. Corolla of apparently 3 petals (actually 5), the upper often broader than long, each of the two lateral ones 2-lobed and considered as one. Stamens 5, with short flat filaments, the anthers more or less united around the stigma. Pistil 1, the ovary superior, 5-loculed, style very short or wanting. Fruit a fleshy, 5-valved capsule, explosively dehiscing, the valves coiling elastically, forcibly distributing the seeds.

- Flowers orange to reddish; spur 6 mm or more long bent parallel with the sac.   1. *I. capensis*
- Flowers pale yellow; spur 5 mm or less, bent at right angles to the sac.   2. *I. pallida*

1. Impatiens capensis Meerb. SPOTTED TOUCH-ME-NOT, JEWEL-WEED.   Fig. 119

Glabrous bright green herb to 1.5 m tall. Stems slightly glaucous. Leaves soft, pale or glaucous beneath, ovate to elliptic, 3–12 cm long, margins crenate, on petioles to 10 cm long. Flowers drooping on slender pedicels, orange and with crimson spots or variously colored, the saclike sepal conic, 6 mm or more long, its spur about 8 mm long and bent backwards parallel with the sac; minute cleistogamous flowers produced on small or poorly developed plants. Capsules about 2 cm long. (*I. biflora* Walt.)

Marshes, stream banks, alluvial woods. Nfld. to Alaska, s. to Fla. Panhandle and Tex.

2. Impatiens pallida Nutt. PALE TOUCH-ME-NOT, JEWEL-WEED.   Fig. 120

Glabrous herb to 2 m tall, in general aspect very similar to *I. capensis*. Leaves elliptic to elliptic-ovate, 5–13 cm long, margins crenate, on petioles to 6 cm long. Flowers in small axillary panicles, light yellow, the saclike sepal obtuse, 5 mm or less long, spur bent at right angles to the sac. Capsules green, 2–2.5 cm long.

Wet woods and meadows, Que. and N.S. to Sask., southward to Va., Ga., Tenn., W.Va., Mo. and Okla.

Surianaceae (BAY-CEDAR FAMILY)

Suriana maritima L. BAY-CEDAR.   Fig. 121

Commonly an openly branched, evergreen shrub 1–2 m tall, less frequently and in protected places attaining the stature of a small tree; in superficial vegetative appearance somewhat

Fig. 119. **Impatiens capensis:** a, flowering branch; b, base of plant; c, flower. (From Correll and Correll. 1972)

261

Fig. 120. **Impatiens pallida:** a, flowering branch; b, flower; c, stamens; d, ovary. (From Correll and Correll. 1972)

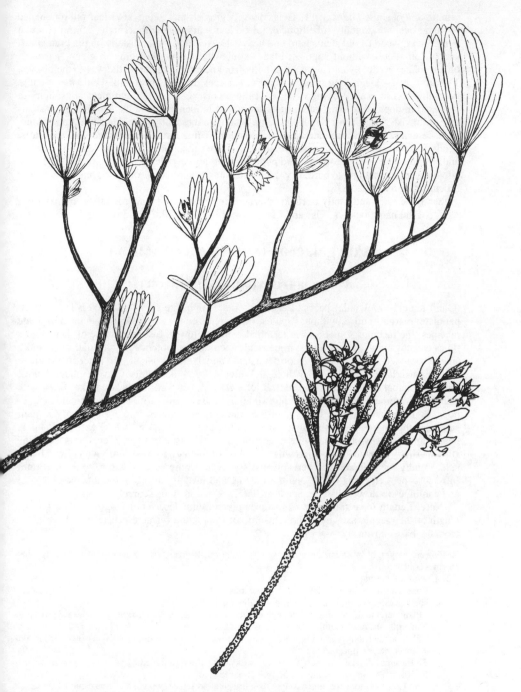

Fig. 121. **Suriana maritima.** (From Little, Woodbury, and Wadsworth, *Trees of Puerto Rico and the Virgin Islands*. Vol. 2 (1974) Fig. 412)

reminiscent of some *Podocarpus*. Twigs densely grayish-pubescent, after leaf-fall roughened by numerous, horizontally subelliptical, raised leaf scars each with a single, central vascular bundle scar; wood of old stems hard and heavy. Leaves closely set distally on the branchlets, simple, alternate, without stipules; blades somewhat fleshy, dull grayish green, narrowly oblanceolate, gradually narrowed proximally from near their apices all the way to their bases, 1–4, mostly 2–3, cm long, margins entire, surfaces densely short-pubescent when young, becoming less so in age, drying wrinkled. Flowers bisexual, radially symmetrical, solitary or clustered in upper leaf axils, oftentimes about half hidden amongst the leaves; flower stalks pale yellowish, pubescent. Sepals 5, densely short-pubescent, ovate, shortly elliptic-ovate, or lance-ovate, short-acuminate apically, mostly 6–7 mm long, persistent around the fruits. Petals 5, yellow, obovate, equaling or slightly shorter than the sepals. Stamens 10. Pistils 5 and separate, the ovaries rounded and pubescent, styles very slender, stigmas small and capitate. Ovaries maturing into hard, dry, 1-seeded, short and rounded-obovate fruits with rough-reticulate, hairy surfaces.

Sandy beaches and sandy or rocky shores, banks bordering mangrove flats, in mangrove flats, coastal hammocks. s. Fla. and Fla. Keys; trop. Am.; Old World tropics.

Polygalaceae (MILKWORT FAMILY)

Polygala (POLYGALAS, MILKWORTS)

(Ours) annual or biennial herbs. Leaves alternate, opposite or whorled, sessile or short-petiolate, margins entire, without stipules, rarely reduced to scales. Flowers in indeterminate racemes, the racemes sometimes in terminal cymes. Flowers bisexual, some of them sometimes cleistogamous, bilaterally symmetrical, each flower subtended by a small bract and 2 smaller bractlets, these deciduous or persistent. Sepals 5, persistent, the inner two (termed wings) much larger than the other three and often colored like petals. Corolla tubular below, with 3 principal lobes above, the central lobe clawed, often somewhat boatlike, often with sublobes, in some fringed apically, the two lateral lobes similar to each other; quickly deciduous. Stamens usually 8, less frequently 6, the filaments united into a tube nearly to their summits, the tube split on the upper side, united to the corolla tube in 2 groups. Pistil 1, ovary superior, 2-locular, one ovule in each locule; style usually slender, curved or bent, the stigma usually expanded and somewhat cuplike or tubelike, 2-lobed. Fruit a loculicidal capsule, usually rounded and notched apically, somewhat compressed contrary to the partition. Infructescences ripening basally on the raceme and usually falling as the raceme elongates and continues to flower distally. Seeds arillate, the aril in some scarcely evident.

Note: Length measurements for racemes given in the descriptions below are for the full length of the raceme axes even though some of the mature infructescences may have fallen from the bases upwardly.

1. Flowers varyingly lavender, purple, rose-purple, purplish green, or cream-white (if the latter then racemes solitary).
   2. Leaves all alternate.
      3. The leaves all reduced to small, inconspicuous scales.   1. *P. setacea*
      3. The leaves, some of them at least, not scalelike.
         4. Plant perennial, forming cleistogamous flowers on naked branches from the caudex; leaves predominantly obovate, spatulate-obovate, or spatulate.   2. *P. crenata*
         4. Plant annual, not producing cleistogamous flowers; leaves predominantly narrowly linear, linear-lanceolate, or linear-elliptic.
            5. Racemes delicate, loosely flowered, stalks of the fruits reflexed.   3. *P. leptocaulis*
            5. Racemes compact, stalks of the fruits not reflexed.
               6. Corollas elongate, much longer than the petaloid sepals (wings) and conspicuously exserted beyond them at anthesis; plant glaucous.   4. *P. incarnata*
               6. Corollas not elongate, not exserted beyond the petaloid sepals at anthesis; plant not glaucous.

7. Racemes not over 6–8 mm broad, narrowly oblong-obconical in outline. 5. *P. chapmanii*
7. Racemes 10–14 mm broad, broadly ovate to oblong in outline, truncate to rounded or merely abruptly short-pointed apically.
    8. Petaloid sepals (wings) about twice as long as the corolla; aril on the seed running along the seed body ⅔ of its length or more.   6. *P. sanguinea*
    8. Petaloid sepals (wings) only slightly longer than the corollas; aril reaching only ¼–⅓ the length of the seed body.   7. *P. mariana*
2. Leaves, some of them at least, opposite or whorled.
    9. Racemes compact, ovate to oblong, at most abruptly short-pointed apically.
        10. Bracts 2–3 mm long; petaloid sepals (wings) with cuspidate tips 1.5–3 mm long, the tips exserted at the periphery of the raceme giving it a somewhat brushy appearance.   8. *P. cruciata*
        10. Bracts 1 mm long; petaloid sepals (wings) merely short-mucronate at the tips, not exserted and the periphery of the raceme without a brushy appearance.   9. *P. brevifolia*
    9. Racemes loosely flowered, obconical apically.   10. *P. hookeri*
1. Flowers (when fresh) bright orange, yellow, or cream-white (if the latter then the racemes in panicled cymes).
    11. Racemes in panicled cymes.
        12. Fresh flowers yellow.
        13. Plants 4.5–12 dm tall; basal leaves numerous (present at time of flowering), linear, linear-lanceolate, or linear-oblanceolate; stem leaves linear-subulate.   11. *P. cymosa*
        13. Plants rarely exceeding 3 dm tall; basal leaves (often not present at time of flowering) spatulate to obovate; stem leaves narrowly spatulate to linear.   12. *P. ramosa*
        12. Fresh flowers cream-white to greenish white.   13. *P. balduinii*
    11. Racemes solitary.
        13. Fresh flowers bright orange.   14. *P. lutea*
        13. Fresh flowers yellow.
        14. Plants not exceeding 15 cm tall; bracts 4.5–6.5 mm long, their tips exserted at the periphery of the raceme; infructescences very tardily falling.   15. *P. nana*
        14. Plants 3–8 dm tall; bracts about 3 mm long, included within the periphery of the raceme; infructescences falling promptly from the base of the raceme upwardly as they mature.
        16. *P. rugelii*

1. Polygala setacea Michx.

Perennial, usually with a solitary, slenderly wiry stem 1–3.5 dm tall, with 1 to several short branches near the summit, these exceeding the central terminal raceme. Leaves alternate, reduced to short-subulate scales barely perceptible to the unaided eye. Racemes compact, obconical, or at length becoming oblong, 5–15 mm long fully developed, about 5 mm broad. Flowers cream-white or pinkish. Bracts subulate, about 0.5 mm long, quickly deciduous. Upper small sepal short-ovate, lower two sepals triangular-subulate, petaloid sepals (wings) short-clawed, blades elliptic-oval to obovate, rounded apically, about 2 mm long overall. Corolla united basally, crest with 4 narrow, erect, blunt lobes. Seed body nearly orbicular, plump, about 0.5 mm long, conspicuously white-pilose, a prominent 2-lobed, spongy aril barely overlapping the base and extending away from the base a distance about equal to the length of the seed proper.

Pine savannas and flatwoods. Coastal plain, Ga. to s. pen. Fla., westward to Miss.

2. Polygala crenata James.

Perennial. Stems usually several to numerous from the caudex, each usually unbranched and with a terminal raceme; short racemes of cleistogamous flowers produced from the caudex. Stems 1–3 dm tall, lax, sometimes reclining, commonly held more or less erect by surrounding vegetation. Leaves numerous on the stem, sessile, alternate, obovate, spatulate-obovate, spatulate, or linear-elliptic, the lower usually shorter and broader than those above; lowermost leaves mostly 5–10 mm long and obovate or spatulate-obovate, others variably spatulate to linear-elliptic, and although all relatively small, varying considerably in size, 3–15 mm long, sometimes the larger leaves more or less at midstem, sometimes uppermost. Racemes loose, at first pyramidal or obconical, becoming oblongish in outline, varying from 2–10 (–15) cm long fully developed. Bracts 1 mm long or a little more, boatlike, deciduous before the infructescences tardily fall. Flower stalks filiform, 3–4 mm long. Flowers usually bright purple, drying rose-purplish to pinkish. Small sepals ovate, obtuse apically, greenish on the back, the margins rose-purple, petaloid sepals (wings) short-clawed, the blades orbicular to oval-obovate, 4–5 mm long overall, greenish centrally on the back, margins rose-purple, evenly purple on the upper side. Corolla united to about the middle, the lateral lobes oblong, broadly rounded apically, crest fringed. Capsule oval-oblong to obovate, 3 mm long, longitudinally constricted medially, apically shallowly notched, the margins with narrow, crenate wings. Seed coat black, conspicuously white-pilose, body plump, oval, about 1.5 mm long, with a prominent spongy, basally pubescent aril around the basal rostrum and to about ½ the length of the seed body.

Wet pine savannas and flatwoods, hillside bogs, cypress depressions. Fla. Panhandle to s.e. La.

3. Polygala leptocaulis T. & G. Fig. 122

Plant annual, very slender. Stems solitary with a single terminal raceme or more frequently with 2 to several erect branches above the middle, to 5 dm tall. Leaves sessile, alternate, narrowly linear to filiform, lower ones 15–25 mm long, gradually reduced upwardly to 3–5 mm. Racemes delicate, loosely flowered, elongate-obconical, to about 8 cm long fully developed, 5–7 mm broad. Bracts subulate, about 1.5 mm long, quickly deciduous. Flower stalks very slender, 1–1.5 mm long, becoming reflexed after anthesis. Flowers lavender to rose-pink. Upper small sepal ovate, lower two sepals lanceolate to elliptic, petaloid sepals (wings) slenderly clawed, blades elliptic to elliptic-obovate, about 2 mm long overall. Corolla united at the base, lateral petals oblanceolate, 1.5 mm long, crest fringed. Capsule oblong, about 1.5 mm long, a row of glands on either side of the septum externally. Seed plump, widening somewhat from base to apex, club-shaped, about 1.2 mm long, dark brown to black, conspicuously white-pilose, the 2-lobed aril barely embracing the base.

Wet pine savannas and flatwoods, adjacent ditches, wet sands of exposed pond margins, bogs, prairies, fallow rice fields. Coastal plain, n. Fla. to e. Tex.; Mex., Cuba, S.Am.

4. Polygala incarnata L. Fig. 122

Plant annual, very slender, glaucous throughout. Stems solitary, commonly with a single terminal raceme, sometimes with 2 to several branches above the middle, mostly not exceeding 5 dm tall. Leaves sessile, alternate (opposite cotyledonary leaves sometimes persistent at time of flowering), the lower widely separate and scalelike, others subulate to linear-subulate, or linear and involute, often with their lower portions appressed to the stem, 5–15 mm long. Racemes compact, oblong in outline, 1–5 cm long fully developed, about 10–15 mm wide in portion at anthesis, 8–10 mm wide after corollas shed. Flowers lavender or pinkish. Bracts ovate-triangular, acute apically, about 1 mm long, somewhat tardily deciduous. Small sepals lanceolate to ovate-oblong, petaloid sepals (wings) narrowly elliptic-oblong, 3.5–3 mm long. Corolla slenderly tubular below, long-cylindric, 6–7 mm long, twice as long as the wings, lobes oblanceolate, central one conspicuously fringed. Capsule ovate-oblong in outline, 3–4 mm long, deeply depressed along the septum, plumply convex over the seeds, flattened distally. Seed oval, black, conspicuously white-pilose, nearly 2 mm long, with a prominent spongy, scarcely lobed aril below the base and ½ or more as long as the seed body. (*Gaypola incarnata* (L.) Small)

In a wide range of habitats, pine savannas and flatwoods, cypress prairies and depressions, bogs, fallow fields, open upland, mixed woodlands, pine–scrub oak woodlands on sand ridges, woodland borders and clearings, thickets. L.I. to Mich., Wisc., Iowa, and Neb., generally southward to s. Fla. and e. third of Tex.; Mex.

5. Polygala chapmanii T. & G. Fig. 123

Slender annual. Stems solitary, usually with several elongate, erect-ascending branches exceeding the main stem, mostly 3–6 dm tall. Leaves alternate, sessile, the lowermost scalelike, others subequal, linear to filiform, the larger 1.5–2 (–3) cm long. Racemes sub-

Fig. 122. a–e, **Polygala incarnata**: a, habit; b, inflorescence; c, flower; d, lobed crest of corolla, face view; e, seed; f, **Polygala leptocaulis**: habit; g–k, **Polygala ramosa**: g, habit; h, flower; i, capsule; j,k, seed.

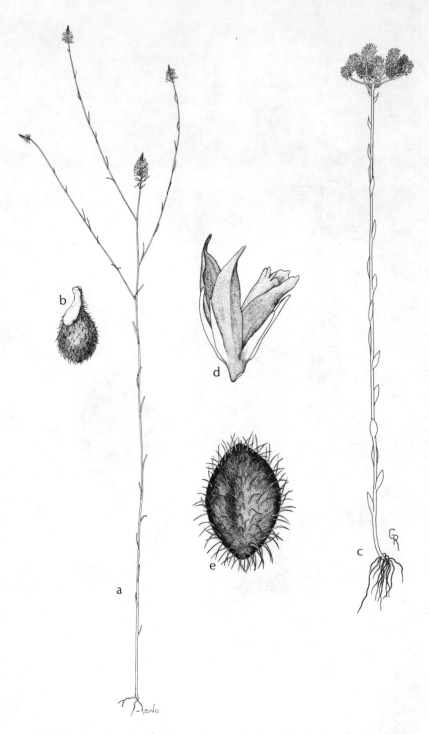

Fig. 123. a,b, **Polygala chapmanii:** a, habit; b, seed; c–e, **Polygala balduinii:** c, habit; d, flower; e, seed.

compact, slenderly oblong, obconical apically, 1.5–3 cm long fully developed, 6–8 mm broad. Bracts subulate, 1 mm long, somewhat tardily deciduous. Flower stalks 1–2 mm long. Flowers lavender. Lower small sepal larger than the other two, broadly ovate with conspicuous white-scarious or -hyaline margins, petaloid sepals (wings) clawed basally, then variably (even in one raceme) ovate, oval, or obovate, broadly rounded apically, about 3 mm long, slightly exceeding the corolla. Central lobe of the corolla crenate. Seed plump, obpyriform, 1 mm long, black, white-pilose, the two lobes of the aril clasping either side of the base.

Pine savannas, seepage slopes, bogs. Coastal plain, Fla. Panhandle to s. Miss.

6. Polygala sanguinea L.   Fig. 124

Low, erect annual. Stem solitary, sometimes with a single terminal raceme, sometimes with 2 to several branches distally, to 4 dm tall. Leaves numerous, sessile, alternate, linear to linear-elliptic, the lowermost smallest, gradually increasing in size upwardly, larger ones 2–3 cm long. Racemes dense, subglobose to oblong, rounded apically, 5–25 (–40) mm long, 10–14 mm broad, sessile or on stalks to about 3 cm long. Bracts subulate, 1 mm long or a little more, persistent to tardily deciduous after infructescences fall. Flower stalks 1–2 mm long. Flowers rose pink or rose purple, occasionally varying to whitish or greenish. Small sepals ovate to oval-elliptic, petaloid sepals (wings) scarcely clawed, ovate, ovate-oval, or broadly elliptic, rounded to obtuse apically, the midvein slightly exserted just behind or at the tip; wings 4–6 mm long, twice as long as the corolla and obscuring it. Capsule markedly humped over the seeds, deeply depressed along the septum between them, broader than long in the seed-bearing portion, flat distally beyond the seeds. Seed black, obpyriform, white-pilose, about 1 mm long, aril with 2 long, narrow lobes running flush against one side of the seed body about ⅔ of its length or a little more. (*P. viridescens* L.)

Wet meadows, bogs, boggy pastures, prairies, glades, moist old fields and roadsides. N.S. to s. Ont. and Minn., southward to N.C. (where rare in the coastal plain), n.w. S.C., cen. Ala., La., e. Tex. and Okla.

7. Polygala mariana Mill.   Fig. 124

In general appearance closely similar to *P. sanguinea*. Racemes averaging smaller, tending to be short-pointed apically early in development, always stalked above the bracteal leaves, stalks mostly 1–4 cm long. Bracts promptly deciduous as infructescences fall. Petaloid sepals (wings) 2.5–3.5 (–4) mm long, only slightly longer than the corollas. Aril on the seed with short lobes reaching only ¼–⅓ of the way along the seed body. (Incl. *P. harperi* Small)

Pine savannas and flatwoods, adjacent moist to wet roadsides and ditches, bogs. Coastal plain, s. N.J. to n. Fla., westward to s.e. Tex.

8. Polygala cruciata L.   Fig. 125

Low annual. Stems unbranched, or commonly with several to numerous branches from somewhat above the base, 1–5 dm tall. Leaves sessile, chiefly in whorls of 3 or 4, the lowermost smallest and often opposite, uppermost sometimes alternate, subequal upwardly from midstem or a little below, narrowly linear, linear-oblanceolate or linear-elliptic, punctate-dotted, 1–3 (–4) cm long. Racemes dense, mostly oblong in outline, abruptly short-pointed apically, 1–1.5 cm long fully developed, 1.5 cm broad, sessile or on stalks to about 2 cm long. Bracts subulate-attenuate above a shortly dilated base, 2–3 mm long, persistent after the infructescences fall. Flower stalks 1–3 mm long. Flowers usually purple or rose-purple, varying to purplish green. Small sepals ovate, apically obtuse to acutish, petaloid sepals (wings) broadly ovate, often with broadly rounded hastate lobes basally, narrowed above and acuminate with cuspidate tips 1.5–3 mm long, these tips giving the periphery of the raceme a somewhat brushy appearance. Corolla shorter than the wings and obscured by them, the central lobe crested with 3 or 4 usually bifurcated processes. Seed plump, oblongish, just slightly obovoid, black, sparsely short-pubescent, 1–1.5 mm long, aril with its linear lobes appressed to one side of the seed for about ⅔ of its length. (Incl. *P. ramosior* (Nash) Small)

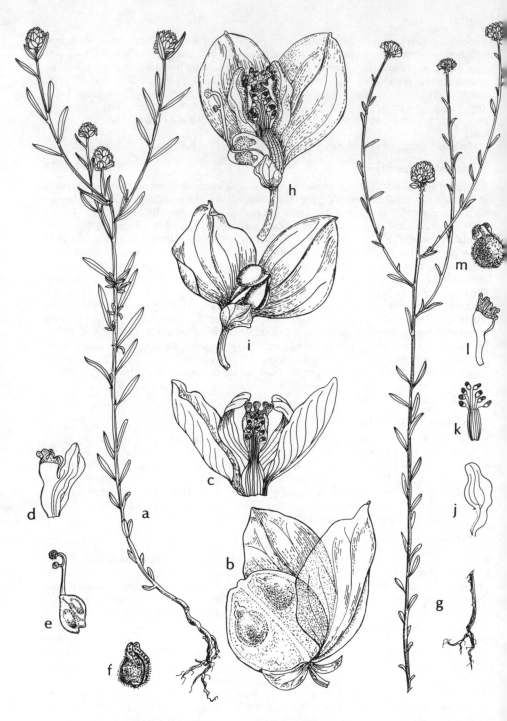

Fig. 124. a–f, **Polygala sanguinea:** a, habit; b, capsule with two inner sepals (wings); c, flower; d, keel; e, pistil; f, seed; g–m, **Polygala mariana:** g, habit; h, flower; i, capsule with two inner sepals (wings); j, petal; k, stamens; l, keel; m, seed. (From Correll and Correll. 1972)

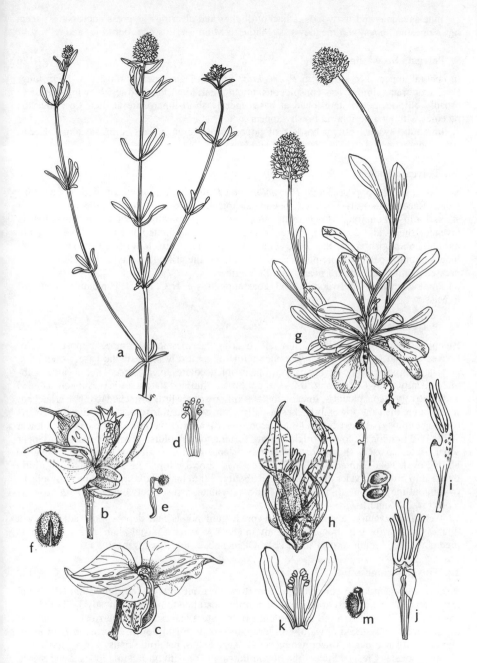

Fig. 125. a–f, **Polygala cruciata:** a, top of plant; b, flower; c, capsule with two sepals (wings); d, stamens; e, pistil; f, seed; g–m, **Polygala nana:** g, habit; h, flower; i,j, keel, two views; k, stamens connected to petals; l, pistil; m, seed. (From Correll and Correll. 1972)

271

Pine savannas and flatwoods, adjacent ditches and clearings, cypress depressions, seepage slopes and bogs, wet meadows. s. Maine to Minn., s. to s.cen. pen. Fla. and e. Tex.

9. Polygala brevifolia Nutt. Fig. 126

In general appearance similar to *P. cruciata*. Stalks of the racemes mostly 3–5 cm long. Persistent bracts similar, less conspicuous, about 1 mm long. Petaloid sepals (wings) ovate to broadly elliptic, not hastate-lobed at base, merely short-mucronate at their tips, thus the racemes without a peripheral brushy appearance.

Interdune swales, boggy borders of pocosins and branch bays, pine savannas. Local, coastal plain, s. N.J. to Fla. Panhandle and Miss.

10. Polygala hookeri T. & G.

Similar in general appearance to *P. cruciata* and *P. brevifolia*, somewhat more slender than either. Racemes loosely flowered, obconical apically, to 6 cm long fully developed, their stalks 1–3 (–6) cm long. Bracts ovate, short-acuminate apically, about 1 mm long. Flowers lavender, their stalks 2 mm long. Petaloid sepals somewhat dilated on each side basally (subhastate), oblongish above the base, acuminate apically, 3 mm long. Seed oblong-obovate, black, plump, sparsely short-pubescent, 1 mm long, the aril with its two elongate lobes appressed against one side for nearly its full length.

Pine savannas and flatwoods. Local, coastal plain, s.e. N.C. to Fla. Panhandle, westward to Miss.

11. Polygala cymosa Walt. Fig. 127

Biennial. Stem solitary, 4.5–12 dm tall, bearing racemes in a terminal cymose panicle. Leaves alternate, sessile, the larger ones numerous, closely set near the base, rosettelike, persistent through the flowering period, linear-oblanceolate, linear-lanceolate, or linear, subulate distally, 3.5–14 cm long, 0.2–0.6 cm broad, thin and flat; stem leaves much reduced and relatively inconspicuous, linear-subulate. Inflorescence little branched to much branched, if the latter then the lower branches usually elongate and floriferous only distally. Bracts ovate-acuminate, about 1.5 mm long, green, persistent after the infructescences fall. Flower stalks 1–2 mm long. Flowers bright yellow, turning to dark bluish green upon drying. Upper sepal ovate, sparsely short-ciliate, lower two lance-ovate, petaloid sepals (wings) oval to oblong-oval. Petals united only basally, crest with 2 or 3 bifurcating lobes. Capsules a little over 1 mm broad, broader than long. Seed about 0.8 mm long, dark brown to purplish black, oval in outline, plump, glabrous, very finely reticulate, aril scarcely evident. (*Psilotaxis cymosa* (Walt.) Small)

Wet pine savannas and flatwoods, cypress-gum ponds and depressions, adjacent wet ditches and borrow pits, bogs, commonly in shallow water. Coastal plain, rarely outermost piedmont, Del., southward to about Lake Okeechobee, Fla., westward to La.

12. Polygala ramosa Ell. Fig. 122

Annual. Very similar in general aspect to *P. cymosa* but rarely exceeding 3 dm tall. Basal leaves few (often not present by flowering time), spatulate to obovate, mostly 1–2 cm long, sometimes to 7 cm, narrowed to petiolelike bases; stem leaves sessile, narrowly spatulate to linear, gradually reduced upwardly. Inflorescence pattern and flower color as in *P. cymosa*. Upper sepal lance-ovate, lower two narrowly lance-ovate, petaloid sepals (wings) obovate or elliptic-obovate. Crest of lower lobe of corolla simple or with bifurcating lobes. Seed brown to nearly black, pubescent, oval, plump, about 0.5 mm long, aril usually minute and scarcely evident. (*Psilotaxis ramosa* (Ell.) Small)

Pine savannas and flatwoods, wet peaty sands of adjacent ditches and roadsides, open bogs, often on seepage slopes or hillside bogs. Coastal plain, s. N.J. to pen. Fla., westward to s.e. Tex.

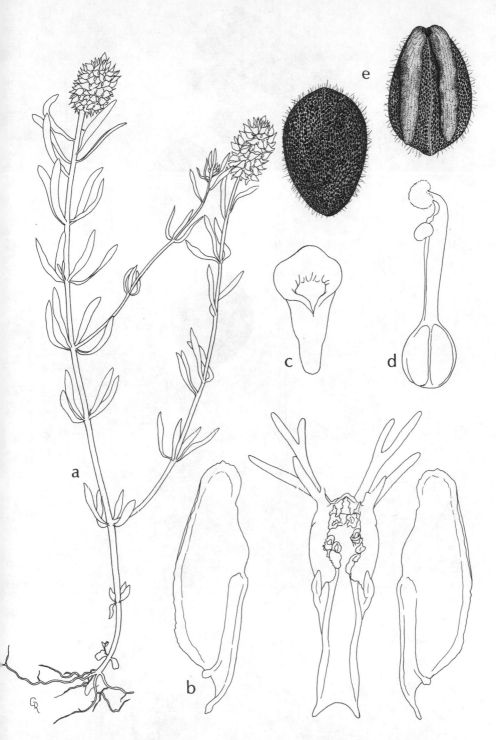

Fig. 126. **Polygala brevifolia**: a, habit; b, partial dissection of flower, petaloid wings of calyx either side, posterior part of corolla, center, with stamens adnate below; c, anther; d, pistil; e, seed, two views.

Fig. 127. a–d, **Polygala cymosa:** a, habit; b, flower; c, capsule; d, seed, two views; e,f,
Polygala lutea: e, habit; f, seed.

13. Polygala balduinii Nutt. Fig. 123

Annual or biennial, in general habit similar to *P. cymosa* and *P. ramosa*. Stems solitary or several from near the base, 1–7 dm tall. Leaves alternate, basal ones (commonly not present by flowering time) spatulate or oblanceolate to narrowly obovate, narrowed to short petiolelike bases, 5–25 mm long, 3–12 mm broad; stem leaves sessile, mostly narrowly elliptic, reduced upwardly. Inflorescence pattern as in *P. cymosa*. Flowers white, cream-white, or greenish white, usually drying somewhat buffish. Bracts ovate- or triangular-subulate, about 2 mm long, cream-white or green centrally with whitish margins. Upper sepal lance-ovate, two lower ones lanceolate or linear-lanceolate, petaloid sepals (wings) lanceolate or linear-lanceolate, narrowing distally to a cusplike extremity. Petals united only basally; crest with large lateral lobes and central bifurcating lobes. Seed dark brown, pilose, oval, slightly oblique, plump, aril minute, scarious, of 2 appressed oval lobes or a minute scale. (*Psilotaxis "baldwinii"* (Nutt.) Small)

Moist to wet pine savannas and flatwoods, adjacent ditches and roadsides, exposed wet sandy shores of ponds, prairies, marl prairies (in s. Fla.). s. Ga. and throughout Fla., westward to Miss.; e. Tex.; w. Cuba.

Respecting distinguishing *Polygala balduinii* var. *balduinii* and var. *carteri* (Small) Smith & Ward (*P. carteri* Small), and for treatment of related species, see Smith, Robert R., and Daniel B. Ward, "Taxonomy of the Genus *Polygala* series Decurrentes (Polygalaceae)," *Sida* 6: 284–310 (1976).

14. Polygala lutea L. RED-HOT-POKER. Fig. 127

Annual or biennial. Stems solitary with a single terminal raceme, or 2- to few-branched distally, each branch terminated by a raceme, or with several branches from near the base, stems 5–40 cm tall. Leaves alternate, the lower usually closely set in an irregular rosette, subsucculent, spatulate to obovate, 1.5–6 cm long, 0.5–2 cm broad, narrowed to petiolelike bases, apices rounded; stem leaves sessile, oblanceolate to elliptic, somewhat shorter than the basal ones. Racemes compact, at first small and ovate-acuminate, enlarging over a considerable period, becoming nearly oblong, to 2–4 cm long and about 1.5 cm broad. Flowers bright orange (very rarely yellow on an individual plant), drying dull pale yellow. Bracts subulate, about 3 mm long, mostly persistent after infructescences fall. Flower stalks mostly 1.5–2.5 mm long. Sepals all orange, three small ones short-ovate, tips acuminate, petaloid sepals (wings) oblong elliptic, cuspidate at the tips, 5–7 mm long. Petals united nearly their full length above the base, crest with 4 short bifurcated lobes, blunt apically. Seed dull black, pilose, oblong, plump, a little over 1 mm long, the aril nearly as long as the seed body and flush against one side of it. (*Psilotaxis lutea* (L.) Small)

Pine savannas and flatwoods, peaty sands of adjacent ditches and roadsides, bogs, boggy borders of cypress-gum ponds and depressions, seepage bogs. Coastal plain, L.I. and N.J., e. Pa., southward to s.cen. pen. Fla., westward to La.

15. Polygala nana (Michx.) DC. Fig. 125

Annual or biennial with general habital appearance of *P. lutea* but of lower average stature, 3–15 cm tall, commonly 5–10 cm. Basal leaves spatulate, narrowed to petiolelike bases, 1–5.5 cm long and 0.4–2 cm broad, subsucculent; stem leaves few, oblanceolate. Racemes compact, short-ovate at first, becoming oblong, 2–4 cm long, about 1.5 cm broad. Bracts subulate, 4.5–6.5 mm long, deciduous or persistent. Flower stalks about 1 mm long. Flowers lemon-yellow or greenish yellow. Upper small sepal lance-subulate, lower two lanceolate, petaloid sepals (wings) elliptic, apically long-acuminate to cuspidate, 5.5–7.5 mm long, the tips exserted at the periphery of the raceme. Corolla united about ⅔ of the length from the base, crest with narrow bifurcating lobes about 2 mm long and long-tapered apically. Seed nearly black, elliptic-oblong, plump, pilose, 1–1.6 mm long, a pronounced blunt rostrum at the base, the aril ⅓ to as much as the full length of the seed body. (*Psilotaxis nana* (Michx.) Raf.)

Pine savannas and flatwoods, usually on leached sands in small open spots, moist sands of roadsides and ditches, bog margins and seepage slopes, dried up to nearly dry cypress-gum ponds, openings in well-drained upland mixed woodlands. w. N.C. and middle S.C. to n.w. Ga., n. Ala., n. Miss., coastal plain, Ga. to s.cen. pen. Fla., westward to s.e. Tex.

16. Polygala rugelii Shuttlw. ex Chapm.   Fig. 128

Annual, biennial, sometimes perennial. Stems 1 to several from the base, the main stems usually with long branches, 3–8 dm tall. Basal leaves in an irregular rosette (often not present at time of flowering), spatulate, broadly obtuse to rounded apically, narrowed to petiolelike bases, 3–6 cm long, 0.5–1.5 cm broad; stem leaves numerous, oblanceolate below, grading to linear-lanceolate above. Racemes ovate to oblong, 1.5–3 cm long and 2.5 cm broad. Bracts subulate, about 3 mm long, included within the raceme, tardily deciduous. Flower stalks 2–3 mm long. Flowers bright yellow, drying bluish green or yellowish green. Upper small sepal broadly ovate, cuspidate apically, lower two subulate, petaloid sepals (wings) elliptic-oblong to narrowly obovate, somewhat oblique, 5–8 mm long, apically acuminate to cuspidate. Corolla united for most of its length, crest with 4 bifurcating lobes, their apices blunt. Seed dull brown, appressed-pilose, 1.2–1.6 mm long, oblong, plump, aril about as long as the seed body. (*Psilotaxis rugelii* (Shuttlw. ex Chapm.) Small)

Pine savannas and flatwoods, adjacent roadsides and ditches, margins of cypress depressions, bogs, occasionally on deep sands of scrub. Endemic to n.e. and pen. Fla.

Euphorbiaceae (SPURGE FAMILY)

A very large and diverse family. Annual or perennial herbs, shrubs, or trees, some succulent and more or less cactuslike, some kinds with milky or viscid sap. Leaves mostly alternate, in some species wholly or partly opposite or whorled, usually stipulate but stipules often small or falling early. Flowers unisexual, plants monoecious or dioecious. Inflorescences very variable (see description of *Chamaesyce* for very specialized one). Calyx and corolla present or absent, frequently different in staminate and pistillate flowers, parts free or rarely united. Stamens 1 to many, filaments in some branched. Pistil 1, ovary superior, usually 3-locular, rarely with fewer or more locules, ovules 1 or sometimes 2 per locule, styles as many as the locules, free or partially united, often branched. Fruit usually a 3-lobed capsule mostly splitting into 3 2-valved parts, rarely indehiscent.

1. Plant a shrub, subshrub, or tree.
    2. The plant restricted to maritime habitats; leaves opposite (*C. mesembryanthemifolia* in genus).
                                                                      7. *Chamaesyce*
    2. The plant not maritime; leaves alternate.
        3. Wood of the stem very light, approximately as light as cork; margins of leaf blades minutely appressed-crenate, the teeth tipped by tiny callosities.   5. *Stillingia*
        3. Wood hard, not extremely light in weight; margins of leaf blades entire.
            3a. Plant a shrub not usually exceeding 1.5 m tall; petioles 1 cm long or less; leaf blades lanceolate or elliptic-lanceolate, cuneate basally and tapering from their broadest places to acute tips; leaf nonglandular at junction of petioles and blade.   6. *Sebastiana*
            3a. Plant (potentially) a small to medium-sized tree; petioles 2–5 cm long on most leaves; leaf blades broadly rhombic-ovate or ovate, basally broadly rounded, truncate, or with a broad, short taper, mostly abruptly short-acuminate apically; leaf with a pair of glands at junction of petiole and blade.   9. *Sapium*
1. Plant herbaceous.
    4. Stem, leaves, calyces, and capsules notably stellate-pubescent.   2. *Croton*
    4. Stems and other parts glabrous, or if pubescent, then the pubescence not stellate.
        5. Ovary with 2 ovules or capsule with 2 seeds in each locule.   1. *Phyllanthus*
        5. Ovary with 1 ovule or capsule with 1 seed in each locule.
            6. Sap not milky; flowers clearly unisexual; perianth present in staminate or pistillate flowers or both.

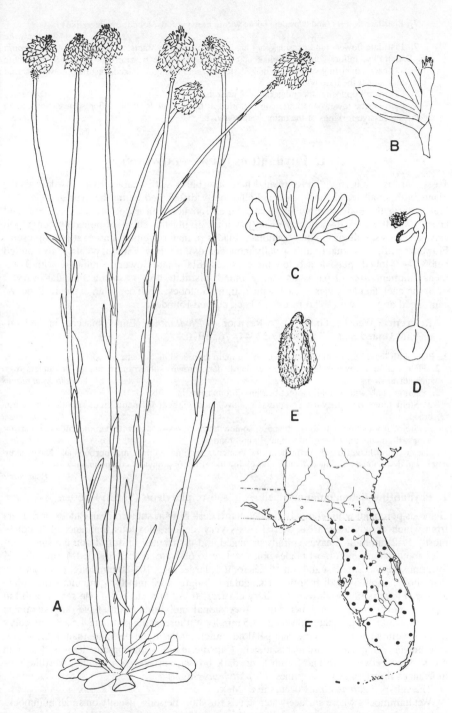

Fig. 128. **Polygala rugelii:** A, habit; B, flower; C, crest on keel (lower petal); D, pistil; E, seed; F, distribution. (From Smith and Ward in *Sida* 6: 303. 1976)

277

7. Pistillate flowers (and capsules) borne within conspicuous, lobed, involucrelike bracts.
3. *Acalypha*

7. Pistillate flowers (and capsules) not having involucrelike bracts. 4. *Caperonia*

6. Sap milky; inflorescence a cuplike or calyxlike involucre bearing few to several staminate flowers (each consisting of a single, naked stamen) and a single, naked, pistillate flower inserted centrally and at the base within.

8. Leaves all opposite, inequilateral or oblique basally. 7. *Chamaesyce*

8. Leaves, the lower ones, alternate, whorled at the first node of the inflorescence, opposite at the successive forkings of the inflorescence above. 8. *Euphorbia*

1. Phyllanthus (LEAF-FLOWERS)

Trees, shrubs, or herbs. (Those treated here low herbs with characteristics as described.) Plants with small leaves uniformly and alternately distributed along the main stem *and* its branches, *or* the main stem without leaves, the branches with naked proximal portions and bearing small leaves distichously distally, withal giving the effect of pinnately compound leaves; leaves sessile or minutely petiolate, blades pinnately veined, entire; stipules present. Flowers small, unisexual (plants usually monoecious), axillary, without petals. Calyx united below, 4–6-lobed, persistent below the capsules. Staminate flowers commonly with 3 stamens, filaments free or partly or wholly united; pistillate flowers with a 3-locular ovary, 2 ovules in each locule, styles 3, each with 2 stigmatic lobes. Fruit capsular, explosively dehiscent. Seed angled, with 2 flat faces, the back curved-rounded.

Reference: Webster, Grady L. "A Revision of *Phyllanthus* (Euphorbiaceae) in the Continental United States." *Brittonia* 22: 44–76. 1970.

1. Plant with leaves uniformly distributed on the main stem and its branches.
  2. Plant perennial, with relatively slender, hard, dark brown, subterranean caudex or runners from which aerial stems arise. 1. *P. liebmannianus*
  2. Plant annual, with a small taproot bearing fibrous roots laterally.
    3. Stem terete or somewhat compressed above, not winged; filaments free; capsule 1.5–2 mm broad. 2. *P. carolinensis*
    3. Stem, from somewhat above the base, somewhat compressed and with definite, although narrow, wings; filaments united; capsule about 3 mm broad. 3. *P. pudens*
1. Plant without leaves on the main stem, the branches with naked proximal portions and leaves alternately and distichously arranged distally thus giving the effect of pinnately compound leaves.
4. *P. urinaria*

1. Phyllanthus liebmannianus Muell.-Arg. subsp. platylepis (Small) Webster.

Glabrous perennial with relatively slender, hard, dark brown, subterranean caudex or runners from which aerial stems arise. Aerial stems very slender, flexuous, unbranched, reddish, mostly 1–2 dm high. Leaves spirally arranged and distributed uniformly on the stem, petioles about 1 mm long, most blades oblanceolate to obovate, apices rounded, rarely obtuse, lowermost leaves smallest and sometimes rotund, larger ones 10–20 mm long, paler beneath than above; stipules reddish brown, triangular-subulate, 1–2 mm long, minutely auriculate at base. Flowers in few-flowered axillary clusters, those in a cluster of one sex or both but pistillate never more than 1 per cluster (occasional individual plants have only staminate flowers), stalks of staminate flowers 2–3.5 mm long, those of the pistillate 4.5–8 mm; calyx lobes 6, yellowish green, somewhat petaloid, much larger in the pistillate than in the staminate flowers. Stamens 3, filaments united. Capsule about 4 mm broad, depressed at both poles. Seed between 1.5 and 2 mm long, dark brown, surface with minute, wartlike protuberances in irregularly wavy lines. (*P. platylepis* Small)

The subsp. *liebmannianus* occurs in e. Mex.

Wet hammocks where surface water stands for short periods, usually on small hummocks and about bases of trees, occasionally on thin soil of exposed lime-rock, banks of streams. Coastal counties of Fla. from Levy to Taylor.

2. Phyllanthus carolinensis Walt.

Annual, 1–3 dm high, the stem unbranched to diffusely branched, terete below, sometimes somewhat compressed above but not at all winged, usually glabrous, sometimes minutely scabrid, green to purplish. Leaves distichously arranged and uniformly distributed on the main stem and on the branches, if any, petioles 1 mm long or slightly more, blades obovate, oblong, or elliptic, apices rounded, lowermost leaves smallest, larger ones varying considerably in size from plant to plant, 5–20 (–30) mm long, 2–10 (–15) mm broad, glabrous; stipules triangular-acuminate, usually auriculate basally, minute, often quickly deciduous. Flowers in axillary clusters mostly with 1 or 2 staminate and 1 or 2 pistillate flowers, stalks 0.5–1 mm long, pistillate slightly the longer and reflexed. Staminate flower with (5)6 oblong to suborbicular calyx lobes 0.5–0.7 mm long, lobes somewhat longer than the tube, apices rounded to obtuse, stamens 3, filaments free; pistillate flowers with 6, rarely 5 or 7, linear or narrowly spatulate calyx lobes about 1 mm long, lobes considerably longer than the tube, calyces conspicuously persistent after the capsules fall. Capsule depressed at the poles, 1.5–2 mm broad. Seed about 1 mm long, cinnamon-brown, surface with minute, wartlike protuberances in lines. (Incl. *P. saxicola* Small)

Moist to wet, commonly alluvial soils, floodplain forests and their clearings, exposed sand and gravel bars of rivers, margins of flatwoods ponds, shores of impoundments, thin soil over lime-rock in pinelands (s. Fla.), frequently abundant on alluvium beneath highway bridges. Pa. to Mo. and s.e. Kan., generally southward to s. Fla. and e. Tex.; W.I.; Mex. to Urug. and Arg.

3. Phyllanthus pudens L. S. Wheeler.

Habit, leaf arrangement, petioles, and stipules like *P. carolinensis*, glabrous throughout. Stem, somewhat above the base, compressed and distinctly, although narrowly, winged. Leaf blades elliptic, oblong, some sometimes oblanceolate or spatulate, rounded or obtuse basally, mostly obtuse or sometimes a few acute apically, 10–15 (–20) mm long, 3–6 (–10) mm broad. Flowers in axillary clusters, each usually with 1–3 staminate and 1 or 2 pistillate flowers, stalks of the staminate about 0.5 mm long, those of the pistillate 1.4–1.8 mm long, reflexed. Calyx of staminate with 5 or 6 ovate lobes about 0.5 mm long, stamens 3, filaments united; calyx of pistillate with (5)6 ovate-oblong lobes 1 mm long or slightly more. Capsule depressed at both poles, about 3 mm broad. Seed 1.2–1.5 mm long, tan to brown, surface with minute, wartlike, dark purple protuberances in vague, irregularly wavy lines or uniformly distributed. (*P. avicularia* Small)

Wet depressions and marshy places in river bottomlands, coastal prairies, moist to wet fallow fields. s. La. and s. Tex.

4. Phyllanthus urinaria L.

Annual, 1.5–5 dm tall, usually with a single main stem, sometimes all branches ascending, commonly those near the base spreading at right angles, only the tips ascending, and about as long as the main stem; main stem glabrous, terete but with obscure wings, leafless, branches *resembling* stalked, pinnately compound leaves but actually branches with leafless proximal portions, distally bearing leaves alternately and distichously; naked portion of the branch more or less scabrid, with a conspicuous wing, leafy portion winged and angled, scabrid; leaves sensitive to touch, folding forward against the axis when stimulated. Leaves oblong or oblong-obovate, 6–20 mm long, 3–8 mm broad, bases rounded, commonly inequilateral, apices rounded and with a slight mucro; lower surfaces minutely stiff-pubescent on, and often near, the margins, same kind of pubescence visible only on the margin above; stipules conspicuously auriculate at the relatively broad base, then abruptly attenuate. Proximal leaves of a branchlet with solitary, sessile, pistillate flowers in their axils, distal leaves subtending minute clusters of short-stalked staminate flowers. Calyx of staminate flowers with 6 elliptic to oblong-obovate lobes about 0.5 mm long, stamens 3, filaments united; calyx of pistillate flowers with 6 linear-oblong or subulate lobes 0.5–1 mm long, their margins mi-

279

nutely toothed or entire. Capsule depressed at the poles, 2–2.5 mm broad, surface usually bearing flat, more or less appressed protuberances. Seed a little over 1 mm long, dull, light brown, surfaces prominently ridged and grooved. (*P. lathroides* misapplied by Small, (1933))

Weedy in both well-drained and poorly drained places, gardens, waste places, fallow fields, ditches, wet clearings, especially of floodplain forests, sand and gravel bars of rivers. An Old World species, relatively recently introduced to our range (ca. 1944), now very common and abundant in n. Fla., known to be in N.C., Ga., Ala., Miss., La., e. Tex., perhaps elsewhere; we have no knowledge of its frequency, abundance, or general distribution in our range as a whole.

A somewhat similar plant, *P. tenellus* Roxb., native of Mascarene Islands, widely naturalized in various tropical and subtropical areas, also relatively recently introduced to our range, is now very weedy in parts of the range. It does not, apparently, inhabit wet places but is not uncommonly seen growing with *P. urinaria* in well-drained, especially sandy, soils. It may be easily distinguished (in part) in that its pistillate flowers and capsules are on relatively long, filiform stalks.

2. Croton

Croton elliottii Chapm.

Annual herb 5–10 dm tall. Stem unbranched below, with few to numerous strongly ascending branches above; main stems terminated by a compact cyme (fruiting and falling early), several elongate, strongly ascending secondary branches arising from just below the first cyme, each of these terminate by a cyme, a tertiary set of branches arising just below it and each of these usually again 1 or 2 times cymosely branched; stems brown, clothed with paler appressed-stellate pubescence. Leaves alternate, simple, slenderly petiolate, lower ones with much longer petioles than the upper, petioles brown, stellate-pubescent; blades lanceolate, the larger ones 4–6 cm long and 1–1.5 cm wide, rounded basally, tapering somewhat from near their bases, apices bluntish, upper surfaces relatively sparsely stellate-pubescent, lower more densely so, lateral veins not or scarcely evident above, pinnate laterals evident below. Cymes compact, flowers unisexual, the few staminate on short stalked racemes arising from amidst the more conspicuous sessile pistillate ones. Staminate flowers with 5 short, ovate sepals clothed with tan stellate hairs, stamens 7–10, white-translucent. Pistillate flowers 3–5 per cyme, mostly with 5 or 6 oblongish sepals hooded at their apical extremities, stellate-pubescent on both surfaces, tan excepting the green, hooded, inwardly curving tips, enlarging after anthesis and persistently curving-erect around the fruit; style branches 3, spreading, each branch twice cleft, tan. Capsule ovate-orbicular in outline, with 3 rounded lobes, stellate-pubescent, about 5 mm long and broad, the withered styles persistent. Seeds 1 per locule, oval in outline, about 4 mm long, surface gray mottled with black, a bone-white caruncle over the micropylar end.

Exposed shores and bottoms of shallow sinkhole ponds, sometimes very abundant locally as water levels recede during drought periods. Coastal plain, s.w. Ga., s.e. Ala., cen. Fla. Panhandle.

3. Acalypha (THREE-SEEDED MERCURIES)

Annual or perennial herbs or shrubs (ours annual herbs). Leaves simple, alternate, petiolate, stipulate. Inflorescence spicate, terminal, axillary, or both. Flowers small, unisexual, plants monoecious (ours), the staminate below, pistillate above, or vice versa, sometimes the pistillate in separate spikes, petals none. Staminate flowers subsessile in clusters of several per minute bract, calyx united below, 4-lobed, stamens 4–8, filaments free or united basally; pistillate flowers sessile, subtended by an involucrelike, lobed bract which may enlarge after anthesis, in any case surrounding the ripe fruit, calyx lobes 3, ovary usually 3-locular, ovules 1 in each locule, styles 3, each usually divided several times. Fruit capsular. Seed ovoid, mottled, smooth, pitted, or tuberculate.

• Leaf blades lanceolate or lance-ovate, petioles ⅓ as long as the blades or less; hairs of the stem a mixture of short, incurved ones and longer, straight, spreading ones; bracts subtending the pistillate flowers 9–15-lobed, hairs on the bracts all pointed, none of them gland-tipped.   1. *A. virginica*
• Leaf blades ovate to somewhat rhombic-lanceolate or rhombic-ovate, petioles varying from about ½ as long to longer than the blades; hairs on the stem all short and incurved; bracts subtending the pistillate flowers 5–7-lobed, rarely to 9-lobed, hairs on the bracts all or some of them gland-tipped, rarely none of them glandular.   2. *A. rhomboidea*

1. Acalypha virginica L.

Stem erect, simple, or usually branched, commonly branched from the base upwardly, 1–7 dm tall, infrequently taller, sparsely to densely beset with a mixture of short, incurved and longer, straight hairs, occasionally the latter few in number. Petioles ⅓ as long as the blades or less, blades usually lanceolate, less frequently lance-ovate, those of the main stem generally larger than those on the branches, also varying considerably in size from plant to plant; midstem leaves 3–6 cm long; blades short-cuneate basally, acute apically, sparsely short-pubescent at least on the veins of both surfaces, margins crenate-serrate. Bracts subtending pistillate flowers irregularly 9–15-lobed, hairs all pointed, sparse to dense. Capsule pubescent. Seed with small pits in lines.

For the most part inhabiting well-drained sites, stream banks, open woodlands, disturbed places, sometimes where seasonally wet, swales, fields, open lowland woodlands. Maine to Ind., Ill., Mo., and Kan., generally southward to piedmont of Ga. and Tex.

2. Acalypha rhomboidea Raf.

Similar to *A. virginica* in habit and stature. Pubescence of the stem wholly of short, incurved hairs. Petioles ½ as long to longer than the leaf blades, blades ovate, rhombic-lanceolate, or rhombic-ovate, 2–9 cm long, broadly short-tapered basally, acute to acuminate apically, both surfaces sparsely pubescent, at least on the veins, with stiff, anteriorly appressed hairs, margins crenate-serrate. Bracts subtending the pistillate flowers with 5–7 (–9) lobes, hairs on the bracts all, or some of them usually, gland-tipped, rarely all pointed. Capsule pubescent distally. Seed with small pits in lines.

Commonly in well-drained places, moist, open woodlands, woodland borders, clearings, fields, also river banks, floodplain forests, swales. N.S., cen. Maine, s.w. Que., to Minn., generally southward to the Fla. Panhandle and e. Tex.

4. Caperonia

Annual or perennial herbs with hollow stems. Leaves alternate, stipulate, sessile or petioled. Inflorescences narrow, stalked, green spikes borne in axils of the upper leaves. Flowers unisexual, pistillate secund and interrupted on the lower part of the spike, staminate closer together and alternating above, each subtended by a short, broad bract. Staminate flower with an unequally 5-lobed calyx, 5 petals, and an androphore bearing 2 whorls of 5 short stamens. Pistillate flower with calyx deeply 5- or 6-lobed, petals 5 or 6, ovary 3-locular, each locule bearing 1 ovule, styles 3, each 4- or 5-parted. Capsule deeply 3-lobed, dehiscing loculicidally and septicidally.

• Stems bearing long, gland-tipped, spreading hairs.   1. *C. palustris*
• Stems glabrous.   2. *C. castaneifolia*

1. Caperonia palustris (L.) St.-Hil.   Fig. 129

Annual 3–10 dm tall, stem usually simple in the lower half, loosely and relatively few-branched above, bearing divergent, long, gland-tipped hairs, the latter sloughing on older, lower portions of the stem leaving it smooth. Lower, larger leaves with petioles 1–3 cm long, blades 6–15 cm long, lanceolate to lance-oblong, bases rounded, apices acute, margins sharply serrate, teeth small, prominently ascending-parallel lateral veins ending in the teeth; both surfaces sparsely pubescent with appressed-ascending, stiffish hairs; upwardly leaves

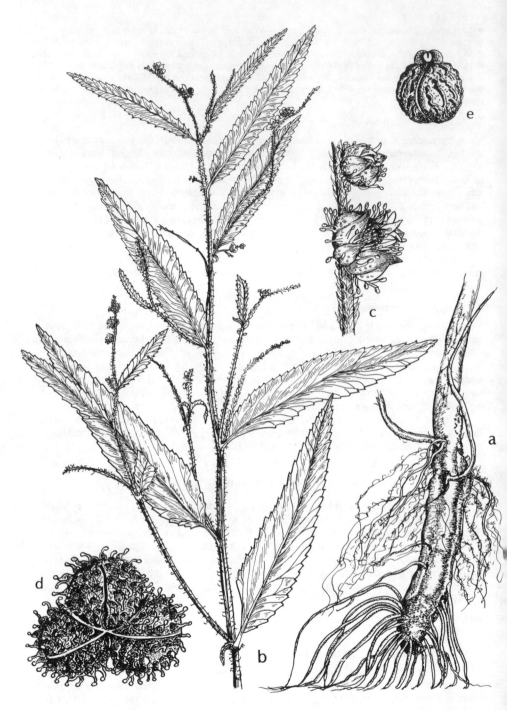

Fig. 129. **Caperonia palustris:** a, base of plant; b, portion of upper stem; c, pistillate flower; d, fruit; e, seed. (From Correll and Correll. 1972)

somewhat reduced in size, petioles progressively shorter, uppermost sessile; stipules subulate, 3–5 mm long. Petals oblanceolate or spatulate, those of the staminate flower exceeding the calyx, those of the pistillate shorter than the longest calyx lobes. Capsule about 3 mm high and 5 mm broad basally, persistent calyx subtending it, surface clothed with broad-based, flat, gland-tipped hairs. Seed globose, 2.5 mm in diameter, dark brown, surface minutely pitted and with a few flat pale trichomes appressed to it. (*C. castanaefolia* sensu Small (1933))

Marshy places, drainage and irrigation ditches, mostly in disturbed wet places in glades and pinelands, rice fields. s. Fla.; La. and s.e. Tex. Native to, and widespread in, warmer parts of Am. southward to Parag.

2. Caperonia castaneifolia (L.) St.-Hil.

Perennial(?), usually with several unbranched stems arising from the base, to 5 or 6 dm tall. Stems thickish-spongy below, glabrous throughout, by flowering time most of the leaves shed below the middle of the stem. Lowermost leaves smallest, short-petiolate, oblanceolate, apices rounded, 2–2.5 cm long, margins crenate-serrate; largest leaves at midstem, petioles about 2 cm long, blades lanceolate, short-tapered basally, long-acute apically, margins serrate, teeth triangular-ascending, obtuse, lateral veins not prominent, not conspicuously parallel to each other, ending in the teeth; surfaces glabrous above, glabrous or with a very few, irregularly scattered, hairs along the midrib beneath; stipules ovate, 1.5–2 mm long. Capsule like that of *C. palustris*. Seed slightly ovoid, almost globose, 3 mm in diameter, with a whitish, thin coating which sloughs readily. leaving a light brown, finely pitted surface.

Sporadic in shallow water of drainage ditches and canals, sloughs, pools, in s. pen. Fla. Native of trop. Am.

5. Stillingia

Stillingia aquatica Chapm. CORKWOOD.   Fig. 130

Glabrous shrub to 12 to 15 dm tall, with a short taproot bearing numerous small rootlets. Stem single, terete, tapering gradually from the base upwardly, leafy only on the branches; bark of older portions of the stem brown, with inconspicuous (to the unaided eye), raised, circular lenticels, the wood very light. Branches tend to be dichotomous or somewhat fascicled toward the top of the stem, branches reddish or purplish, more or less glaucous. Leaves alternate, closely set, some of them usually overwintering, petioles 1–4 mm long, blades lanceolate to linear-lanceolate, sometimes elliptic or rhombic, 3–8 cm long, cuneate at both extremities, margins minutely appressed-crenate, the teeth tipped by tiny callosities; stipules variable, sometimes minute glands, sometimes divided from very near the base into 2 or 3 subulate to filiform divisions 2–4 mm long. Flowers on a terminal, spikelike, green, yellow, or red inflorescence, unisexual, 1 per bract (pistillate) or 1 to several in a fascicle subtended by a bract (staminate), bracts with two marginal glands. Pistillate flowers borne on the lower portion of the inflorescence, sepals 3 or 2, usually separate, sometimes united, sometimes absent; petals none; ovary (2)3-locular, each locule with a single ovule, stigmas (2)3. Staminate flowers above the pistillate, calyx 2-lobed, always present; petals none; stamens 2, exserted, filaments joined at the base. Fruit a short-ovate capsule, about 1 cm broad, septicidally dehiscent. Seed about isodiametric, 4–4.5 mm across, silvery gray, with low, pinnaclelike tubercles or wrinkled-tuberculate.

Flatwoods ponds and depressions, drainage ditches and canals, in shallow water or where surface water stands much of the time. Coastal plain, s.e. S.C. to s. pen. Fla., westward to Miss.

6. Sebastiana

Sebastiana fruticosa (Bartr.) Fern. SEBASTIAN-BUSH.

Loosely branched, essentially glabrous shrub 1–2 m tall. Young stems slender, reddish brown, irregularly very narrowly winged; older stems becoming grayish, terete, bark irreg-

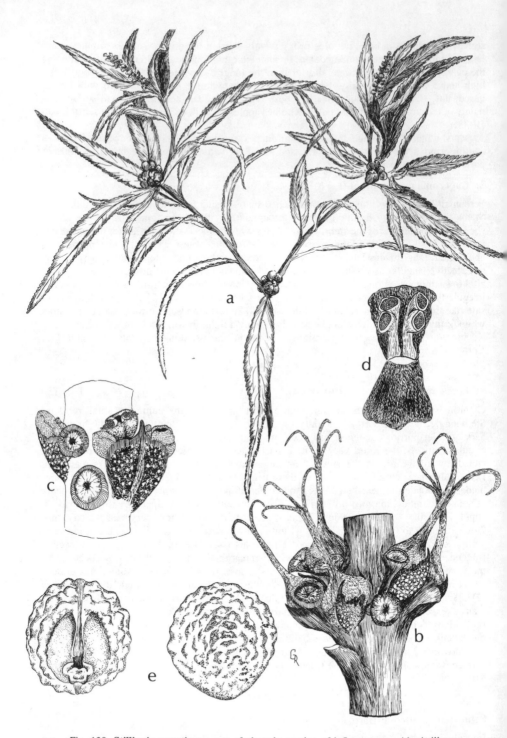

Fig. 130. **Stillingia aquatica:** a, top of plant; b, portion of inflorescence with pistillate flowers; c, portion of inflorescence with staminate flowers; d, staminate flower, sepals spread apart; e, seed, two views.

ularly slightly ridged, sloughing in small platelets. Leaves alternate, petioles 2–10 mm long, blades elliptic, oval, or lanceolate, cuneate basally, acute to acuminate apically, variable in size on most branches, 2–7 cm long, 1–2 (–3) cm broad, margins entire, pubescence, if any, minute on the petioles; stipules triangular, 1.5–2 mm long, glandular basally, apices acute to acuminate, margins minutely toothed. Inflorescences very narrow spikelike racemes 1–4 cm long, borne in the axils of uppermost leaves of branchlets, green or greenish yellow. Flowers small, unisexual, without petals, each flower subtended by a broad, short bract with a gland on either side. Pistillate flowers on the lower portion of the raceme, staminate above. Staminate flower with 3 small sepals united basally, 3 exserted stamens. Pistillate flower with 3 nearly separate sepals, sometimes 2 tiny additional ones in 2 of the sinuses between; ovary 3-locular, each locule with 1 ovule, stigmas 3, recurved. Capsule ovoid, about 8 mm long, dehiscing septicidally and loculicidally, the septae remaining attached to one side of each of the dehisced valves. Seed terete, ovoid-oblong, brown with a somewhat silvery cast, about 5 mm long. (*S. ligustrina* (Michx.) Muell.-Arg.)

River and stream banks, floodplain woodlands, richly wooded adjacent slopes. Coastal plain, s.e. N.C. to cen. pen. Fla., westward to e. and s.e. Tex.

7. Chamaesyce (SPURGES)

Herbs or subshrubs, rarely large shrubs or small trees (ours, excepting one, herbs), branching sympodial, that is, with an apparent main axis formed of successive secondary axes, each a fork of a dichotomy, the other fork being smaller or suppressed. Leaves opposite, entire or toothed, never lobed, usually inequilateral basally; stipules present, sometimes united. Plants usually monoecious. Basic inflorescence a cyathium comprised of a cuplike involucre with 4, rarely 5, nectariferous glands on its rim alternating with the lobes of the involucre, each gland usually with a somewhat petaloid appendage extending from beneath the gland; perianth none; involucre with a single, stipitate, pistillate flower centrally within it, its stipe exserting the pistil from the involucre; staminate flower consisting of a single stamen, its stalk jointed, borne on the sides of the involucre within, 5 staminate flowers singly or in clusters of 2 to several in a fascicle. Ovary 3-locular, each locule with one ovule, styles 3, free or basally united, each sometimes bifid distally. Each cyathium with its component parts simulating a flower, the cyathia borne in terminal and axillary cymes, or 1 or 2 in the leaf axils of short, congested, lateral branchlets. Fruit capsular, 3-lobed. Seeds terete or angular, smooth, wrinkled, pitted, or ridged and grooved.

To aid in critical identification, it is strongly recommended that care be taken to obtain specimens having mature capsules and mature seeds. Several of the species of *Chamaesyce* occurring within our range are notably weedy and inhabit sites having a wide range of moisture conditions. Others are weedy but restricted insofar as we know to well-drained places.

1. Capsule pubescent.
    2. Stems usually rooting at the nodes; faces of seed granular.   1. *C. humistrata*
    2. Stem not rooting at the nodes; face of seed with a few subregular, horizontal, low, rounded ridges, infrequently only with faint nearly circular to horizontally oblongish, shallow pits, not granular.
                                                                                                2. *C. maculata*
1. Capsule glabrous.
    3. Plant an annual herb.
        4. The plant with numerous, radiating, prostrate branches; cyathia 1 or 2 in the leaf axils of short, congested branchlets.   3. *C. serpens*
        4. The plant erect; cyathia in stalked, compound cymes.
            5. Cymes subtended by a pair of leafy bracts, otherwise not leafy-bracted; seed brown.
                                                                                            5. *C. hypercifolia*
            5. Cymes leafy-bracted; seed gray.
                6. Seed light gray, faces with 2 or 3(4) horizontal, low, blunt ridges, sometimes with 1 or more connecting cross ridges between them.   4. *C. hyssopifolia*
                6. Seed dark gray, faces without ridges, irregularly finely wrinkled.   6. *C. nutans*
    3. Plant an erect subshrub.   7. *C. mesembryanthemifolia*

1. Chamaesyce humistrata (Engelm. ex Gray) Small.

Annual, usually much branched, branches radiating, prostrate, commonly rooting at the nodes, sometimes ascending, 1–4 dm long. Stems sparsely to densely pubescent with incurled or crisped hairs, older portions sometimes becoming glabrous. Leaves very short petiolate, blades mostly oblong-obovate, varying to oblong, less frequently ovate or ovate-oblong, often larger on the main branches than on the branchlets, also smaller on plants growing in drier places than in wetter ones, 5–15 mm long, bases mostly inequilateral, apices broadly rounded, margins very minutely toothed or entire; petioles crisped-pilose, blade surfaces glabrous or with a few, scattered, longish hairs; stipules triangular-attenuate, or triangular in outline and with narrow, lacerate-fimbriate divisions, 1–1.5 mm long. Cyathia 1 or 2 in leaf axils of short, congested, lateral branchlets. Capsule strongly 3-lobed, ovate in outline, about 1.5 mm long and as broad at the base, appressed-pubescent. Seed somewhat unequally quadrangular, a little broader at one extremity than at the other, nearly 1 mm long, reddish-silvery, faces granular. (*Euphorbia humistrata* Engelm. ex Gray)

On sands, muds, and gravels of river banks, shores, and exposed bars and flats, openings and clearings of floodplains, adjacent ditches, alluvial outwash. Ohio to Kan., generally southward w. of the Appalachians to Fla. Panhandle and e. and s.e. Tex.

2. Chamaesyce maculata (L.) Small.

Annual, usually much branched, branches prostrate and radiating, less frequently erect-ascending, to 5 dm long. Stems shaggy-pubescent, hairs varying from curly to straight. Leaves very short-petiolate, blades variable in size and shape, even on a single plant, ovate, ovate-oblong, oblong, linear-oblong, less frequently oblong-obovate, those of the main branches commonly larger than those of branchlets, 4–15 mm long, inequilateral basally, rounded apically; most blades minutely serrate, some entire, surfaces mostly with few to numerous, scattered, longish hairs, some on a given plant sometimes glabrous; stipules deeply lacerate-fimbriate. Cyathia 1 or 2 in leaf axils of short, congested, lateral branchlets. Capsule strongly 3-lobed, ovate in outline, about 1.5 mm high and as broad at the base, pubescent with stiffish, ascending hairs. Seed somewhat unequally quadrangular, short-oblong in outline, silvery, the silvery coating sometimes partially or wholly sloughing leaving a rust-red surface, faces usually with few subregular, horizontal, low, rounded, ridges, sometimes only with faint, nearly circular to horizontally oblongish, shallow pits, not granular. (*Euphorbia maculata* L., *E. supina* Raf. ex Boiss., *C. supina* (Raf. ex Boiss.) Moldenke; incl. *C. matthewsii* Small, *C. tracyi* Small)

Weedy, commonly abundant locally, in a wide range of open or semiopen sites, well-drained and poorly drained, gardens, fields, waste places, roadsides, along railroads, sandy shores of bays and estuaries, exposed sand and gravel bars of rivers, dry, moist, or wet clearings, alluvial outwash, open woodlands of various mixtures, sandy flats about coastal marshes, stablized coastal dunes, exposed shores and bottoms of ponds. s. Que. to N.D., generally southward to s. pen. Fla. and more or less throughout Tex.; introd. in Calif., Ore., Eur., Japan.

3. Chamaesyce serpens (HBK.) Small.

Glabrous annual, with numerous, radiating, prostrate branches 0.5–4 dm long, sometimes rooting at the nodes. Leaves very short-petiolate, blades subcordate- or cordate-ovate, suborbicular or orbicular, or short-oblong, 2–7 mm long, bases both equilateral and inequilateral, apices rounded, margins entire; stipules connate around the node, with a lobed or lacerate scale at either side of the node. Cyathia 1 or 2 in leaf axils of branchlets. Capsule reflexed, glabrous, strongly 3-lobed, ovate in outline, about 1 mm high and 1.5 mm broad basally. Seed about 1 mm long, irregularly obtusely angular, ovate in outline, pinkish-silvery, faces smooth. (*Euphorbia serpens* HBK.)

Alluvial bottomlands, river bars, seasonally wet pools and flats, especially where calcareous, weedy in dry to moist places where an introduction. s. Ont. to Mont., s. to Ala.,

La., Tex., N.Mex.; sporadic introduction in the northeastern U.S.; Carib.Is.; Mex. to Arg.; introd. in s. France and Spain.

A weedy plant with the habit of *C. serpens*, also having cyathia 1 or 2 in the leaf axils of branchlets and glabrous capsules, occurs in dry to moist sands more or less throughout peninsular Florida and the Florida Keys. It is *C. blodgettii* (Engelm. ex Hitchc.) Small.

4. Chamaesyce hyssopifolia (L.) Small.

Erect annual, 1–8 dm tall, few- to multiple-branched, in the latter case branches commonly numerous from near the base, arching outward then ascending; stems wholly glabrous or less frequently with short pubescence along one side of the branchlets. Leaves with very short petioles, blades glabrous, oblong or oblong-elliptic, varying considerably in size from plant to plant or from one local population to another, 0.5–3 (–4) cm long, 3–15 mm broad, bases scarcely inequilateral and rounded to truncate, or strongly inequilateral with one side rounded or half-cordate, the other oblique, apices blunt; blades uniformly green or suffused with purplish red, paler beneath than above, margins minutely serrate, stipules more or less connate around the node, a short-deltoid tooth on either side of the node, stipular rim usually glandular, the teeth minutely lacerate or ciliate. Cyathia in stalked, compound cymes usually unequally branched, somewhat leafy-bracted. Capsule glabrous, ovate in outline, 1.5–2 mm long and as wide at the base. Seed angled, short-oblong in outline, a little under 1 mm long, light gray, faces mostly with 2 or 3(4) horizontal, low, blunt ridges, sometimes with 1 or more connecting ridges between them. (*Euphorbia hyssopifolia* L.; *E. maculata* (wholly or in part) of authors not L.)

Weedy, commonly abundant locally in both well-drained and poorly drained open places, mostly in sandy or peaty-sandy soil, roadsides, waste places, gardens, cultivated fields, ditches, spoil banks and flats, moist to wet clearings, along railroads, exposed sand and gravel bars of rivers. Coastal plain, S.C. to Fla. Keys, westward to La.; w. Tex. to s. N.Mex. and Ariz.; southward to Arg.; Carib.Is.

5. Chamaesyce hypericifolia (L.) Millsp.

Habit like that of *C. hyssopifolia*, glabrous throughout. Leaves much like those of *C. hyssopifolia*; stipules not or scarcely connate, triangular-acuminate, their tips frequently bifid, margins minutely toothed, tips pilose on the upper side. Cyathia in stalked, compound cymes, the branches usually subequal, each cyme subtended by a pair of leafy bracts, otherwise not leafy-bracted. Capsule glabrous, ovate and 1.5–2 mm long and as broad at the base, or broadest at about the middle and often a little broader than long. Seed about 0.5 mm long, angular, ovate to short-oblong in outline, brown, faces few-wrinkled. (*Euphorbia glomerifera* (Millsp.) L. C. Wheeler)

Weedy on roadsides, in gardens, waste places, clearings, also on borders of mangrove thickets, in saltbush flats, clearings of buttonbush-mangrove thickets, shallow soil on limerock and in solution pits. Cen. pen. Fla. southward to Fla. Keys; s. Tex.; Berm.; W.I., C.Am. and n. S.Am.

6. Chamaesyce nutans (Lag.) Small.

Habit like that of *C. hyssopifolia* and of *C. hypercifolia*. Stems pubescent, usually only on one side, with short, incurved hairs, becoming glabrous below in age. Leaves essentially like those of the two aforementioned in shape and marginal toothing, wholly glabrous, or pilose on the basal margins, or sparsely pilose on as much as the basal half of the upper surface, often red-blotched; stipules essentially connate around the nodes, a short, acute tooth on either side of the node, stipular rim more or less glandular, the teeth ciliate. Cyathia in stalked, compound, leafy-bracted cymes, the branches subequal or not greatly unequal. Capsule glabrous, ovate in outline, 2–2.5 mm long and as broad at the base. Seed angular, about 1.2 mm long, ovate or short-oblong in outline, dark gray, faces irregularly wrinkled. (*Euphorbia nutans* Lag.; *E. preslii* Guss.; sometimes included in *E. maculata* of authors not L.)

Dry, moist, or wet soil, weedy in open places. N.H. to Mich. and S.D., generally southward to Fla. and Tex.; warmer parts of the world.

7. Chamaesyce mesembryanthemifolia (Jacq.) Dugand.

Subshrub, commonly bushy-branched from the base, 2–12 dm tall, glabrous throughout, stems usually purplish, glaucous at first. Leaves usually grayish green, sessile, essentially alike on the principal stem and branches, sometimes those of the branchlets much smaller, especially on older, diffusely branched plants the leaves of main stems of which will have been shed; larger leaves ovate, broadly short-elliptic, oblong, or narrowly elliptic, mostly subcordate basally, obtuse apically, 8–15 mm long, margins entire; stipules triangular-acuminate, 1.5–2 mm long, lacerate-fimbriate marginally. Cyathia in axillary, leafy-bracted, short, simple or compound cymes, sometimes solitary in leaf axils. Capsule glabrous, deeply lobed, about 3 mm broad and 2 mm high. Seed nearly terete, about 1.5 mm long, slightly longer than broad, whitish, faintly wrinkled. (*Euphorbia mesembryanthemifolia* Jacq.; *C. buxifolia* (Lam.) Small)

Rocky shores, upper beach strand, dunes, coastal thickets, banks of lagoons and ditches. Fla., from about Flagler Co. southward on the Atlantic coast, from about cen. pen. Gulf coast southward; trop. Am.

8. Euphorbia

A very large and diverse genus (sensu lato) on a world-wide basis (and in the broadest sense including *Chamaesyce* here treated separately for convenience). For this book, it would not seem to serve any useful purpose to furnish a description that would encompass the variable and diverse elements of the genus since we are treating only two species. These have an inflorescence (cyathium) in a general way like that described for *Chamaesyce*; the species descriptions will serve to distinguish them otherwise.

1. Euphorbia curtisii Engelm. ex Chapm.

Perennial. Rootstock tan, elongate, 2–10 mm in diameter, more or less erect but crooked and gnarled, bearing 1 to numerous, little-branched to much-branched, erect, very slender, few-leaved stems; plant glabrous throughout, stem and leaves green, somewhat suffused with purple, or quite purple. Leaves low on the stem more or less bractlike, becoming larger upwardly, alternate; primary bracteal leaves of cymes of main stem or its branches paired or in threes, opposite and somewhat reduced above; larger leaves subsessile, linear or narrowly oblong, 1–4 (–6) cm long, 5–10 (–15) mm broad. Stalks of the cyathia very slender, mostly 5–10 mm long; cyathia about 1 mm high, campanulate, green on green plants varying to purplish or purple on the purplish or purple ones; glands appendaged distally and petaloid, spreading laterally, basal glandular portion green or greenish, appendage white or pink (pink on plants with much purple color throughout), the appendaged gland reniform, 1 mm high, 1.5 mm broad. Capsule exserted about 1 mm from the cyathium. (*Tithymalopsis curtisii* (Engelm. ex Chapm.) Small)

Pine savannas and flatwoods, bogs, often the rootstock of a given plant embedded in the tussock of wiregrass (*Aristida stricta*). Chiefly coastal plain, N.C. to Fla.

2. Euphorbia inundata Torr. ex Chapm.

Perennial. Rootstock brown, essentially erect but variably bent or gnarled, from about 1 to 4 cm thick, sometimes fusiform, usually bearing few to several stems from its crown. Plant glabrous throughout. Stems 1–3 dm tall or a little more, reddish purple below, green or slightly suffused with reddish purple above. Leaves sessile, those of lower 2–4 nodes bractlike, upwardly becoming abruptly longer, largest ones about midstem, alternate to near the summit and commonly bearing floriferous branches; bracteal leaves of cymes of main stem and its branches opposite; median stem leaves long-lanceolate or -oblanceolate, oblong, or linear, 2.5–10 cm long, 4–10 mm broad, reddish purple along the margins. Stalks of cyathia 5–10 mm long, stoutish; cyathial cup green or suffused with reddish purple, campanulate, 2 mm high, glands somewhat fleshy, spreading laterally, varying from green to maroon, in outline nearly cupform, 1.5 mm high and somewhat broader distally, distal edges shallowly undulate; lobes of cyathium extending a little above the bases of the glands, their distal margins lacerate. Stalk of mature capsule stout, exserted about 4 mm from the cyathium; capsule broader basally than long, 8–10 mm across the base, markedly 3-lobed, faces of the lobes nearly flat or moderately convex. Seed olive to grayish, ovoid, 2–3 mm in diameter at the base. (*Galarhoeus inundatus* (Torr. ex Chapm.) Small)

In sandy-peaty soils, seasonally wet pine savannas and flatwoods, pine-cypress savannas, bogs. Coastal plain, Fla., s. Ala.

E. floridana Chapm. (*Galarhoeus floridanus* (Chapm.) Small) is very similar to *E. inun-data* in general appearance of its above-ground parts. It inhabits deep, well-drained sands of pine–scrub oak ridges and hills. Its rootstock is buff-colored, slender, 2–4 (–6) mm in diameter, essentially straight to 3 dm or more into the soil, its crown usually bearing a single stem, infrequently 2 to several. Stem with few to numerous branches, 1–5 dm tall. Glands of the cyathium fleshy, with a flattish, semistalked, erect base and a flaring, more or less peltate summit sometimes depressed centrally, its outer (distal) margin shallowly crinkled with very short, fleshy lobes; lobes of the cyathium reaching about even with the bases of the glands, their distal margins minutely ciliate. Capsule markedly 3-lobed, faces of the lobes rotundly bulging. Seed brownish olive to grayish, spherical, 3–4 mm in diameter.

9. Sapium

Sapium sebiferum (L.) Roxb. CHINESE TALLOW-TREE.

Usually a fast-growing, small to medium-sized tree with milky sap, glabrous throughout, the young branches slender, commonly arching or somewhat drooping, glaucous. Leaves alternate, deciduous, stipulate. Stipules short-ovate; petioles slender, mostly 2–5 cm long; blades pinnately veined, broadly rhombic-ovate or ovate, as broad as or broader than long (excluding their acuminate tips), mostly 3–6 cm broad, bases broadly rounded, nearly truncate, or with but a short, broad taper, apices abruptly short-acuminate, margins entire, a pair of glands at the junction of petiole and blade. Inflorescences narrowly cylindrical, solitary terminating branches, 6–20 cm long. Flowers without petals. Pistillate flowers few, each solitary in the axil of a bract proximally on the inflorescence axis, calyx small, united only basally, with 3 triangular lobes, pistil with 3-locular ovary, 1 ovule in each locule, styles 3. Staminate flowers borne in numerous fascicles of up to 15 flowers per fascicle, flower stalks unequal, each fascicle in the axil of a bract having a pair of glands basally. Staminate flower with a small cuplike calyx about 1 mm across, stamens 2. Fruit a 3-lobed capsule about 1 cm long and as broad, upon dehiscing the capsular walls falling leaving 3 dull white seeds intact for a time. (*Triadica sebifera* (L.) Small)

Often in low, swampy or submarshy areas, shores of streams, ponds, lakes, and impoundments; also in upland, well-drained places, especially near human habitations. Native of Asia, cultivated as an ornamental, locally naturalized, in some places abundantly, chiefly coastal plain of our range, N.C. to n. Fla., westward to e. and s.e. Tex.

Callitrichaceae (WATER-STARWORT FAMILY)

Callitriche (WATER-STARWORTS)

Annual submersed or stranded diminutive herbs with slender and delicate stems. Leaves submersed, floating or emersed, opposite, entire, without stipules. Flowers minute and inconspicuous, scarcely perceptible to the unaided eye, borne in the leaf axils, unisexual (plants monoecious), perianth lacking. Staminate flower composed of a single stamen, 1–3 flowers in the axils of foliage leaves. Pistillate flower composed of a single pistil, ovary superior, styles 2, filiform, papillose, usually solitary in the leaf axil. Carpels splitting at maturity, usually forming a fruit 1 mm or less in diameter composed of 4 achenelike mericarps, the flattened mericarps winged, margined, or smooth, each with 1 seed.

A morphologically variable genus of plants identifiable with certainty in our area only by characteristics of the fruits. Habits vary greatly in relation to diverse habitats they occupy.

Reference: Fassett, N. C. "*Callitriche* in the New World." *Rhodora* 53: 185–194, 209–222. 1951.

1. Fruit as long as wide or a little longer, sessile.   1. *C. heterophylla*
1. Fruit wider than long, sessile or stalked.
  2. Mericarps bent at an angle with the face and appearing thickened on 1 side at base, i e , gibbous.
           2. *C. peploides*
  2. Mericarps not thickened on 1 side at base.
    3. Fruit stalked, margins with thin wings or wings curled over on themselves and thus appearing thick-margined when observed with magnification.   3. *C. nuttallii*
    3. Fruit essentially sessile, margins showing a narrow wing when observed with magnification.
           4. *C. terrestris*

1. Callitriche heterophylla Pursh emend. Darby.   Fig. 131

Leaves variable, often linear at lower nodes, with a rosette of floating, obovate leaves, or stranded on the substrate with the leaves all linear or all obovate or oblong. Linear leaves bidentate at the tips. Flowers with 2 inflated bracteoles at base. Fruit as long as wide or a little longer, sessile, slightly heart-shaped, convex on the face and thickest just above the base. Margins of the fruit wingless or with a very narrow wing. Grooves between the mericarps very narrow.

Quiet shallow waters and stranded on wet substrates. e. Can., throughout much of the U.S.; Mex. to Guat.

2. Callitriche peploides Nutt.   Fig. 133

Inconspicuous, often-overlooked plant. Stem prostrate, rooting below and with short erect branches. Leaves slightly crowded at the tips of the branches, cuneate to spatulate-obovate, rounded at tips. Flowers without bracteoles. Fruit wider than long, essentially sessile. Mericarps narrowed and elongated at base, pushing against each other so that each is bent at an angle with the face; fruit appearing greatly thickened at base. Grooves between mericarps narrow at top only.

On wet substrates in low places. S.C. to Fla. and Tex., Tenn., Ark.; e. Mex. to C.R.

3. Callitriche nuttallii Torr.   Fig. 132

Tiny plant with stems prostrate, rooting below, simple or much branched. Leaves oblanceolate-obovate, or spatulate, rounded at the tips. Flowers without bracteoles. Fruit wider than long, stalked. Mericarps with flat faces, margins with thin wings curled toward the face and thus appearing to be wirelike. Mericarps with a groove between them.

On wet substrate in low places, Ala. to Tex.; Ky., Ark.; Mex., C.Am.

4. Callitriche terrestris Raf. emend. Torr.

Tiny plant with simple to much-branched, slender stems. Leaves obovate-oblanceolate to spatulate. Flowers without bracteoles. Fruit wider than long, short-stalked. Mericarps with flat faces, usually equally rounded at both ends, occasionally vaguely heart-shaped, outer edges under high magnification showing a narrow wing. Mericarps with a groove between them except at the top.

Damp to wet open places, fallow fields, pathways, lawns. Local, N.Eng. southward to s.e. Va., Ala., La., e. Tex., northward in the interior, e. Okla., to Mo. and Ill.

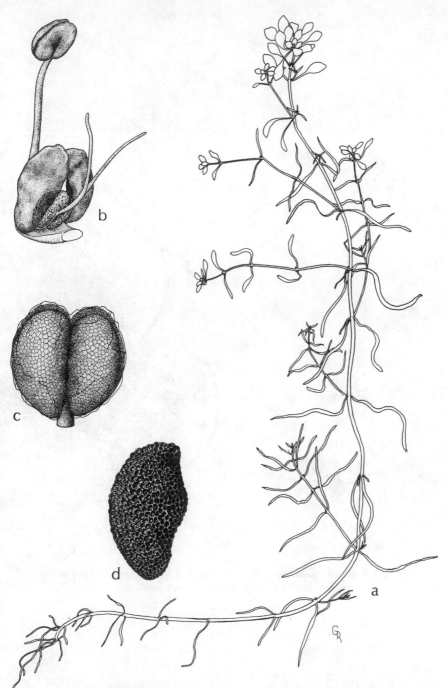

Fig. 131. **Callitriche heterophylla:** a, habit; b, flower; c, fruit; d, seed.

Fig. 133. **Callitriche peploides:** a, habit; b, fruit, two views; c, node with two fruits and aborted pistillate flower. (a,b, from Correll and Correll. 1972)

292

Fig. 132. **Callitriche nuttallii:** a, habit; b, node with staminate flowers (left) and node with fruits (right).

Anacardiaceae (SUMAC FAMILY)

Shrubs, small trees, or woody vines, with resin ducts in the bark, sometimes in the herbage, the sap milky or acrid. Leaves (in ours) alternate, simple, trifoliolate, or odd-pinnately compound. Inflorescences terminal or axillary, bracteate panicles. Flowers small, radially symmetrical, bisexual or functionally unisexual. Calyx usually 5-lobed, a glandular ring or cuplike disc within. Petals usually 5, free or united basally. Stamens as many as or twice as many as the petals, inserted at the base of the disc. Pistil 1, ovary superior, 1-locular, 1-ovulate, styles 3. Fruit a dryish, berrylike drupe.

1. Leaves evergreen, some present on twigs of the previous season (pinnately compound).
   2. Axis of leaf unwinged; leaflets manifestly stalked, 3–7, commonly 5, per leaf; stamens 5; drupes elliptic to oblong, about 1 cm long, brownish orange at maturity.   1. *Metopium*
   2. Axis of leaf narrowly winged distally; leaflets sessile or subsessile, 3–11, commonly 7–9, per leaf; stamens 10; drupe globose, about 6 mm across, bright red at maturity.   2. *Schinus*
1. Leaves deciduous, none on twigs of a previous season (pinnately compound or trifoliolate).
                                                     3. *Toxicodendron*

1. Metopium

Metopium toxiferum (L.) Krug & Urban. **POISON-TREE.**   Fig. 134

Evergreen shrub or small tree with widely spreading branches, all parts with sap irritating to the skin of many persons. Bark of older stems smoothish, light gray and mottled with yellow

Fig. 134. **Metopium toxiferum.** (From Little and Wadsworth, *Common Trees of Puerto Rico and the Virgin Islands* (1964) Fig. 132)

to brown spots; young twigs brown, with raised, orange-brownish lenticels, short-hairy at first, becoming glabrous. Leaves odd-pinnately compound, with petioles 2–8 cm long; leaflets with stalks 0.5–2.5 cm long, blades 3–7, commonly 5, pinnately veined, somewhat leathery, 3–9 cm long and 2.5–5.5 cm broad near their bases, bases broadly rounded to truncate, sometimes shortly inequilateral, apices blunt or bluntly short-acuminate, sometimes shallowly notched, surfaces glabrous, margins entire. Flowers small, in open axillary panicles 1–2 dm long, individual flowers short-stalked, mostly unisexual or functionally unisexual, the staminate and pistillate mostly on separate plants, sometimes some bisexual flowers intermingled. Calyx united basally, with 5 broad, short lobes broadly rounded apically. Petals 5, yellowish green, elliptic-oblong to oblong, considerably surpassing the calyx. Stamens 5. Pistil (rudimentary in functionally staminate flowers) seated on a disc, style very short, stigma shortly 3-lobed. Drupe elliptic to oblong, about 10 mm long, turning from green to brownish orange at maturity, the calyx persistent below.

Chiefly inhabiting pinelands on lime-rock and subtropical hammocks, to a lesser extent in seasonally wet marl prairies or submarshes of glades. s. pen. Fla. and Fla. Keys; W.I.

2. Schinus

Schinus terebinthifolius Raddi. BRAZILIAN PEPPER-TREE.

A bushy-branched shrub to about 3 m tall. Twigs nearly cinnamon brown, with numerous lenticels. Leaves evergreen, odd-pinnately compound (occasionally evenly pinnate by abortion of terminal leaflet in development), petioles mostly somewhat shorter than the lowest leaflets; leaflets glabrous, sessile or subsessile, 3–11, commonly 7–9, mostly oblong-elliptic, varying to lanceolate, dark green above, much paler below, mostly obtuse or rounded basally and apically, occasional ones acute, often slightly inequilateral, 2–5 cm long, rarely longer, 1–3 cm wide; axis of the leaf narrowly winged distally. Panicles axillary on shoots of the season, usually shorter than the leaves. Calyx minute, quickly deciduous, the lobes deltoid. Petals white, oblong, about 1.5 mm long. Stamens 10. Fruit globose, about 6 mm across, ripening through orange to bright red; plants in fruit very handsome.

Native of Brazil and Parag., introduced as an ornamental, now abundantly naturalized in a very wide range of habitats from about cen. pen. Fla. southward in Fla.; now generally considered a noxious weed there.

3. Toxicodendron

Shrubs, small trees, or woody vines, the sap poisonous to susceptible persons upon contact as are the fumes from burning parts. Leaves compound (in ours), trifoliolate or odd-pinnately compound, without stipules. Inflorescences axillary, bracteate panicles. Flowers bisexual and unisexual, plants dioecious or polygamous. Calyx united at base, 5-lobed, basal portion green, lobes cream-colored, often with purplish veins. Petals 5, separate, cream-colored, spreading at anthesis, about twice as long as the calyx. Stamens 5. Fruit whitish.

- Leaves trifoliolate; a low trailing shrub or more commonly a vine climbing by means of aerial, adventitious roots on the stem. 1. *T. radicans*
- Leaves odd-pinnately compound; plant an upright shrub or small, slender tree. 2. *T. vernix*

1. Toxicodendron radicans (L.) Kuntze. POISON-IVY. Fig. 135

A trailing shrub, or if support available, more commonly climbing; when climbing attached to support by numerous aerial roots from the stem, main stems on trees often attaining a diameter of 6–8 cm and upwardly bearing many vigorous lateral branches. Leaves petiolate, petioles to about as long as the blades, glabrous or variously pubescent; leaflets 3, the lateral subsessile or short-stalked, usually somewhat inequilateral, the terminal with a stalk varying in length from about 1 to 6 cm, usually equilateral; blades mostly ovate, rounded or short-tapered basally, acute or acuminate apically, unlobed and entire, or with 1 to several lobes or

Fig. 135. **Toxicodendron radicans:** A, habit; B, panicle; C, flowers; D, drupe; E, stones; F, aerial roots. (From Reed, *Selected Weeds of the United States* (1979) Fig. 126, as *Rhus radicans*)

coarse teeth, the lobing or toothing if present commonly very irregular, even on leaflets of a single leaf, lobes or teeth not infrequently restricted to one side of the leaflet; size of leaflets varying very greatly depending upon age or vigor of the plant, the larger to about 20 cm long and to 12 cm wide. Panicles ascending or spreading, not drooping, generally shorter than the petioles, mostly borne in axils of lower leaves of shoots of the season. Drupes grayish white, globose or subglobose, glabrous or pubescent, about 5 mm across. (*Rhus radicans* L.)

Essentially ubiquitous in terrestrial, nonacid habitats throughout our range. N.S. to B.C., generally southward to s. Fla. and Ariz.; Mex.; Asia.

Variant subspecific taxa within and beyond our range are recognized. For a very thorough account of this and closely allied species, see Gillis (1971).

2. Toxicodendron vernix (L.) Kuntze. **POISON SUMAC.** Fig. 136

Glabrous shrub or small slender tree to about 7 m tall. Leaves odd-pinnately compound, petioles 4–10 cm long, leaflet bearing axis 10–25 cm long, petioles and axes often reddish or maroon; leaflets 7–15, subsessile, 5–12 cm long, 2–5 cm wide, elliptic, oblong-elliptic, or ovate, bases cuneate to rounded, apices acuminate or acute, margins entire; leaf axis not winged. Panicles lax, loose, drooping, mostly borne in the axils of lower leaves on shoots of the season, from about half as long to as long as the leaves. Drupes yellowish white, glabrous, subglobose, 5–7 mm across, fruiting panicles persisting on leafless plants into the winter. (*Rhus vernix* L.)

Bogs, seepage slopes, evergreen shrub-tree bogs or bays, depressions in pine flatwoods, swamps, wet woodlands, thickets. Maine to Minn., generally southward to n.cen. pen. Fla. and e. Tex.

Cyrillaceae (CYRILLA FAMILY)

Shrubs or trees. Pith of the twigs continuous and homogeneous. Leaves alternate, simple, without stipules, thin but stiff-leathery, evergreen or semipersistent, entire, glabrous, sessile or short-petioled. Flowers small, in bracteate racemes, bisexual, radially symmetrical. Calyx united at the base or nearly free, 5-lobed (rarely 6- or 7-lobed), persistent. Petals of the same number as the sepals and alternating with them, separate, white or pinkish white. Stamens 5 or 10, when 5 alternate with the petals, when 10 those opposite the petals shorter. Ovary superior, 3–5-locular, each locule with 1–3 ovules; style short, stigma 2–5-lobed or entire. Fruit small, indehiscent, a dry drupe.

- Leaves with lateral veins scarcely if at all evident on either surface; flowers in terminal and axillary racemes prior to emergence of new shoot growth in early spring; calyx segments deltoid, short-oblong or crescent-shaped, obtuse or rounded at the summit, usually pinkish white; stamens 10; fruit markedly 2–5-winged. 1. *Cliftonia*
- Leaves with an evident reticulum of lateral veinlets either side of the midrib on both surfaces; flowers in clusters of racemes from the summit of twigs of the previous season and below leafy shoots of the season; calyx segments lance-ovate, sharply acute to acuminate apically; stamens 5; fruit not winged. 2. *Cyrilla*

1. Cliftonia

Cliftonia monophylla (Lam.) Britt. ex Sarg. **BLACK TITI, BUCKWHEAT-TREE.** Fig. 137

Shrub or small tree, generally forming dense stands. Leaves leathery, evergreen, blades mostly elliptic, varying to elliptic-oblanceolate or oblanceolate, sessile, or the cuneate bases of the blades narrowed to short petioles, 2.5–10 cm long and 1.2–1.8 cm wide, apically acute to obtuse, sometimes shallowly emarginate, upper surface dark green, the lower with a bluish-whitish bloom. Racemes 1 to several at the summit of twigs of the previous season, at full anthesis in early spring before new shoot growth commences. Axis of raceme ridged,

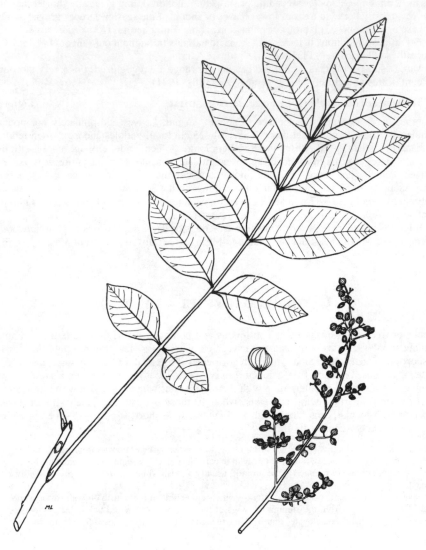

Fig. 136. **Toxicodendron vernix.** (From Kurz and Godfrey, *Trees of Northern Florida* (1962), as *Rhus vernix*)

298

Fig. 137. **Cliftonia monophylla.** (From Kurz and Godfrey, *Trees of Northern Florida.*
1962)

each ridge terminating in a small mound of tissue just below a flower stalk and subtending a bract which is deciduous before full anthesis. Calyx very much shorter than the corolla, usually united below, with 5 broadly deltoid, short-oblong, or crescent-shaped lobes obtuse to rounded at the summit. Petals 5, white or pinkish, spatulate to obovate, clawed at the base, rounded apically, 6–8 mm long. Stamens 10, the filaments dilated downwardly from the summit. Ovary 3–5-locular, one ovule in each locule. Fruit markedly 2–5-winged, 5–7 mm long, in outline little longer than broad

Acid shrub-tree bogs along stream courses and in flatwoods depressions. s.cen. Ga. and Fla. Panhandle to s. Miss.

2. Cyrilla

Cyrilla racemiflora L. TITI, HE-HUCKLEBERRY.   Fig. 138

Shrub or small tree, commonly reproducing vegetatively by sprouts from shallow horizontal roots and forming dense thickets. Leaves semipersistent or deciduous, short-petioled, blades very variable in size and shape, oblanceolate, narrowly obovate, elliptic-oblanceolate, or elliptic, with evident reticulate veinlets on both surfaces lateral to the midrib, cuneate basally, apices acute, obtuse, rounded, or emarginate, from about 1 to 10 cm long and 0.5–2.5 cm wide. Racemes clustered from the summit of twigs of the previous season and below shoots of the season. Flower stalks subtended by subulate bracts (usually persisting below the fruiting stalks) and with a pair of bractlets on the stalk. Calyx very much shorter than the corolla, segments lance-ovate, apices sharply acute or acuminate. Petals 5, white or creamy white, about 2.5–3.5 mm long, broadest at the base and gradually narrowed to an acute apex. Stamens 5, opposite the sepals, filaments broadest at the base and narrowed to the summit. Ovary 2 or 3 (rarely 4) -locular, ovules 1–3 in each locule. Fruit ovoid to subglobose, 2–2.5 mm long, often devoid of seeds, or not more than 1 seed maturing per locule. (Incl. *C. parvifolia* Raf.)

299

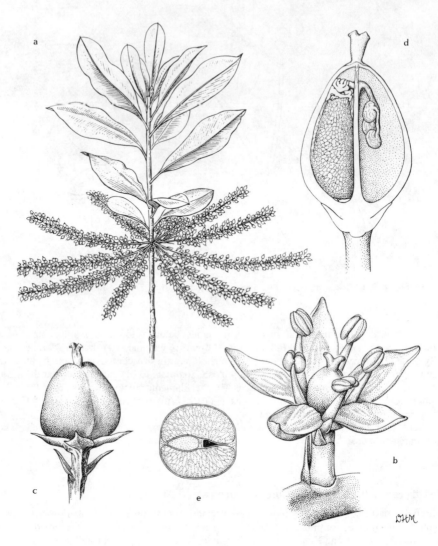

Fig. 138. **Cyrilla racemiflora:** a, fruiting branchlet; b, flower; c, fruit; d, nearly mature fruit in vertical section—note large seed (not sectioned) filling left locule, three ovules in right locule; e, fruit, semidiagrammatic cross-section. (From Thomas in *Jour. Arn. Arb.* 42: 99. 1961)

Acid shrub-tree bogs, pocosins, flatwoods depressions or stream courses, wet flatwoods, stream banks. Coastal plain, s.e. Va. to cen. pen. Fla., westward to s.e. Tex.; s. Mex., Brit.Hond., W.I., n. S.Am.

Thomas (1960), having studied *Cyrilla* extensively over a wide geographic range, concluded that numerous described taxa, showing local patterns of variation, intergrade to such an extent that but a single species can be recognized. In n. Fla., what has been known as *C. parvifolia*, in its extreme form, represents one of the patterns of variation; however, in the area in which it occurs, intermediates abound.

Aquifoliaceae (HOLLY FAMILY)

Ilex (HOLLIES)

Deciduous or evergreen shrubs or trees. Leaves simple, alternate, commonly varying considerably in size on a given plant, stipules minute and quickly deciduous. Flowers mostly functionally unisexual, staminate and pistillate on separate plants but often a few functionally bisexual flowers intermingled, all small, radially symmetrical, both staminate and pistillate with perianth of calyx and corolla; flowers in axillary or internodal cymes or fascicles or solitary, the pistillate ones more often solitary. Calyx usually persistent, united below, 4–9-lobed or -toothed. Corolla of 4–9 parts slightly united basally. Stamens as many as the corolla segments and alternate with them, free or inserted on the base of the corolla. Pistil 1, ovary superior, 4–8-locular, stigma lobed or capitate, sessile or nearly so. Staminodes and pistillodes reciprocally present in the functionally unisexual flowers. Fruit a berrylike drupe with as many 1-seeded nutlets as locules of the fruit, the nutlets usually with 2 flattish sides and 1 broadly rounded, the latter ribbed-grooved or smooth.

1. Leaves leathery, evergreen (as shown by the presence of some on twigs of the previous year).
  2. Leaf blades with marginal, distinctly spine-tipped dentations, or if margins entire, then always with a distinctly spine-tipped dentation apically.   1. *I. opaca*
  2. Leaf blades entire or toothed, if toothed, then the teeth not spine-tipped, the blades never distinctly spine-tipped apically.
    3. Margins of leaf blades appressed-crenate from base to apex.   2. *I. vomitoria*
    3. Margins of leaf blades entire, or if toothed, then the teeth absent from the basal margins.
      4. Lower surfaces of leaf blades punctate-dotted; drupe black.
        5. Margins of some, at least, of the leaf blades with 1–3 appressed-crenate teeth on a side near their apices.   3. *I. glabra*
        5. Margins of leaf blades entire or with a few, spreading, short, bristlelike teeth. 4. *I. coriacea*
      4. Lower surfaces of the leaf blades not punctate-dotted; drupe red, orange-red, or rarely yellow.
        6. Larger blades at least 1.5 cm wide, commonly much wider; branchlets mostly ascending at angles less than 45 degrees to the branch from which they arise.   5. *I. cassine*
        6. Larger leaf blades not exceeding 8 mm broad; branchlets mostly borne at angles of more than 45 degrees to the branches from which they arise, commonly at 90-degree angles.
                                  6. *I. myrtifolia*
1. Leaves membranous, deciduous (as shown by the lack of any on twigs of the previous year).
  7. Leaf blades predominantly oblanceolate, strongly cuneate basally, margins crenate distally.
                                  7. *I. decidua*
  7. Leaf blades predominantly elliptic, oval, oblong, or obovate, bases rounded or broadly tapered, margins serrate or entire.
    8. Lower surfaces of leaves prominently reticulate-veined, the veinlets forming the reticulations distinctly raised.   8. *I. amelanchier*
    8. Lower surfaces of leaves faintly reticulate-veined, the veinlets forming the reticulations impressed or only faintly raised.
      9. Calyx lobes ciliate marginally.   9. *I. verticillata*
      9. Calyx lobes entire.   10. *I. laevigata*

1. Ilex opaca Ait. var. **opaca.** AMERICAN HOLLY.   Fig. 139

Evergreen understory tree, to about 15 m tall. Young twigs puberulent, usually sparsely so, becoming glabrous, older twigs brown, roughish, with circular, raised lenticels. Leaves with puberulent petioles 5–12 (–18) mm long; blades variable in shape, oblong, elliptic, oval, slightly ovate or obovate, 3–10 (–12) cm long, 2–4 cm broad, stiff-leathery at maturity, usually with spine-tipped, marginal dentations, *always with a dentate-spinose tip*, marginal dentations varying from a single one on a side distally to all along each side to the base, usually not over 7 or 8 on a side at most, occasional trees with margins wholly entire; midrib on the upper surface of young leaves puberulent, usually glabrous in age, lower surface wholly glabrous or with a few short hairs along the midrib. Staminate and pistillate flowers borne similarly, in stalked, simple or compound cymes on leafless bases or in leaf axils of

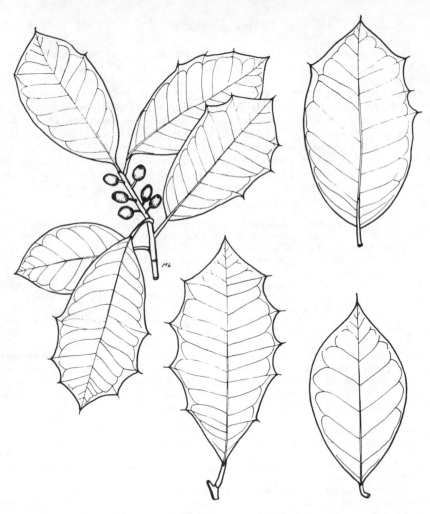

Fig. 139. **Ilex opaca** var. **opaca.** (From Kurz and Godfrey, *Trees of Northern Florida.* 1962)

developing shoots, or at leafless nodes or in leaf axils on wood of the previous season, inflorescence axes and flower stalks puberulent. Calyx lobes 4, triangular, acute or acuminate, their margins usually somewhat erose. Drupe red, without lustre, yellow on rare individual plants, globose to oval, 7–10 (–12) mm in diameter. Nutlets 4, 6–8 mm long, irregularly ribbed-grooved on the rounded side.

Mostly in mesic woodlands of various mixtures, also in floodplain forests, on river banks, in woodlands astride small streams and subject to flooding. s. Mass. to Md., s.e. N.Y., e. Pa., W.Va., s. Ohio to s.e. Mo., generally southward to n.cen. pen. Fla. and e. and s.cen. Tex.

2. Ilex vomitoria Ait. YAUPON.   Fig. 140

Evergreen shrub or small tree, to about 8 m tall, commonly sprouting from the roots and forming dense thickets, especially near the coasts. Young twigs sparsely to densely puberulent, the pubescence usually evident over the period of a year, older twigs gradually becoming glabrous and with a waxy gray coating that breaks up into a thin, more or less interlacing surface pattern. Leaves with puberulent petioles 2–3 mm long, rarely longer; blades stiff-leathery, oval, elliptic, or oblong-elliptic, rarely lanceolate, 0.5–3 cm long, 0.5–2.5 cm broad, upper surfaces dark green and lustrous, puberulent only along the midrib when young, lower surfaces paler, glabrous; margins appressed-crenate from base to apex, small glandular mucros at the sinuses and in a shallow notch of the blunt to rounded tip, bases rounded. Staminate and pistillate flowers borne similarly in 2- or 3-flowered, short-stalked or sessile fascicles in the axils of leaves or leaf scars of leafless nodes, the fascicles solitary or 2 to several glomerately at a given node; flower stalks mostly 1.5–3 mm long, more or less puberulent, those of the pistillate usually more evidently so than those of the staminate, calyces similarly pubescent. Calyx lobes 4, broadly short-triangular, obtuse to rounded. Drupe globose, 4–8 mm in diameter, clear, bright red, yellow on a rare individual plant. Nutlets 4, 3–4 mm long, irregularly shallowly ribbed on the rounded side.

Coastal dunes and interdune depressions, maritime forests, upland semiopen woodlands of various mixtures, fence rows, usually on sandy soils, pine flatwoods; for the most part the yaupon inhabits well-drained places but occurs as well on the edges of stream banks, in wet woodlands and floodplains. Coastal plain, s.e. Va. to cen. pen. Fla., westward to s.e. and s.cen. Tex., s.e. Okla., Ark.

3. Ilex glabra (L.) Gray. GALLBERRY, INKBERRY.

Evergreen shrub, to 2–3 m tall, commonly sprouting from subterranean runners and colonial. Twigs of the season green, very finely powdery-pubescent, older twigs glabrous, gray or grayish brown, the lenticels roundish and with vertical slits. Leaves with powdery-pubescent petioles 3–8 mm long; blades glabrous, leathery, elliptic to oblanceolate or subobovate, 2–5 cm long, upper surfaces lustrous green, lower paler, dull and with scattered punctate glands, usually reddish in color; margins somewhat thickish-banded, mostly with 1 or 2(3) small, appressed teeth on a side distally, sometimes entire, bases acute to obtuse, apices blunt, usually with a tiny mucro. Staminate flowers in axillary, stalked cymes, stalks of the cymes mostly 6–10 mm long, individual flower stalks of a given cyme variable in length to about 10 mm, minutely bracted at or near the base. Pistillate flowers mostly solitary in the leaf axils, sometimes 2 or 3, their stalks 4–10 mm long, with a pair of minute bracts varyingly from about midway their length to near their bases. Calyx glabrous, lobes 5–8, very short, broadly obtuse. Drupe black (very rarely white on an individual plant), dull or only sublustrous, globose, 5–7 mm in diameter, pulp dryish when fully mature, bitter; drupes persistent through the winter. Nutlets 5–8, 3–4 mm long, smooth on the rounded side.

Pine savannas and flatwoods, bogs, branch bays, seepage areas in woodlands, acid prairies. N.S. to s. Fla., westward to n.e. Tex., chiefly on the coastal plain.

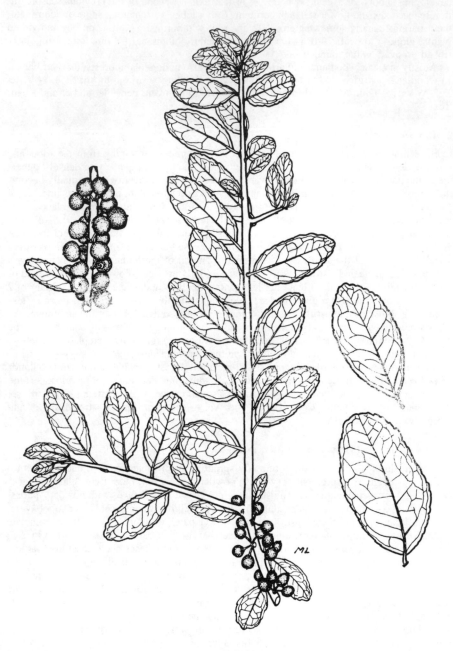

Fig. 140. **Ilex vomitoria.** (From Kurz and Godfrey, *Trees of Northern Florida*. 1962)

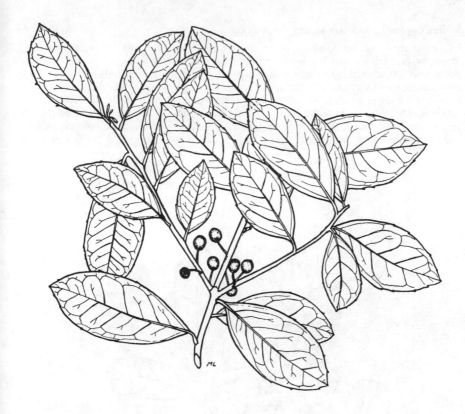

Fig. 141. **Ilex coriacea.** (From Kurz and Godfrey, *Trees of Northern Florida*. 1962)

4. Ilex coriacea (Pursh) Chapm. SWEET GALLBERRY.   Fig. 141

Evergreen shrub, rarely with the stature and dimensions of a small tree, to 5 m tall. Twigs of the season dark brown, finely powdery-pubescent, older twigs grayish to tan, lenticels nearly circular, with longitudinal slits. Leaves with powdery-pubescent petioles 5–10 mm long; blades leathery, elliptic, or oval, occasional ones lanceolate or subobovate, mostly 3.5–9 cm long, 1.5–4 cm broad, usually a few on a given branch smaller; upper surfaces dark green and lustrous, powdery-pubescent on the midrib; paler beneath and with scattered punctate glands, glabrous or pubescent, the pubescence often stellate; margins entire or with a few, almost bristlelike, usually spreading, teeth, bases acute, apices acute or short-acuminate, rounded or rounded and notched on occasional leaves. Staminate flowers sometimes in axillary fascicles, sometimes on leafless proximal portions of new shoots where fasciculate at the base and racemose above, stalks 4–6 (–10) mm long, minutely bracted at the base, glabrous or nearly so. Pistillate flowers sometimes solitary, more commonly in few-flowered fascicles at the bases of new shoots, their stalks 5–10 (–15) mm long, pubescent, without basal bracts. Calyx more or less tuberculate exteriorly, lobes 5–9, triangular, acute to acuminate, their margins entire, irregularly shallowly erose, or finely toothed. Drupe globose or subglobose, often somewhat broader than long, 6–8 (–10) mm in diameter, black and lustrous, pulp juicy and sweet at maturity, falling at or shortly after maturing. Nutlets 5–9, smooth on the rounded side.

Pine savannas and flatwoods, shrub-tree bogs and bays, open bogs, seepage areas in woodlands. Coastal plain, s.e. Va. to cen. pen. Fla., westward to s.e. Tex.

5. Ilex cassine L. DAHOON HOLLY, CASSENA. Fig. 142

Evergreen shrub, or tree to about 10 m tall, branches and branchlets strongly ascending. Young twigs, petioles, and leaf blades varying not a little as to vestiture; young twigs and petioles sometimes glabrous, more commonly shaggy hirsute, infrequently with short-curly pubescence; upper surfaces of leaf blades glabrous save along the midrib, the lower surfaces varyingly wholly glabrous, copiously shaggy-hirsute only along the midvein, copiously hirsute along the midvein and with longish, straight hairs relatively sparsely distributed over the remainder of the surface, or longish, straight hairs sparsely distributed on both the midvein and the remainder of the surface; inflorescence axes, flower stalks and calyces generally

Fig. 142. **Ilex cassine** (twig at lower right is probably from a plant intermediate between *I. cassine* and *I. myrtifolia*). (From Kurz and Godfrey, *Trees of Northern Florida*. 1962)

sparsely to copiously hirsute. Petioles 5–15 mm long, blades variable in shape and size from plant to plant or even on a single one, oblanceolate, nearly spatulate, oblong, oval, elliptic, or subobovate, 2–8 (–14) cm long, 0.8–4.5 cm broad, bases narrowly to broadly tapered, apices acute or obtuse, less frequently rounded, usually tipped with a small mucro, margins mostly entire, occasionally with a few, short, sharp dentations. Staminate flowers in axillary, usually diffusely panicled cymes to about 5 cm long. Pistillate flowers solitary, or in stalked 2–4-flowered, axillary cymes, stalks of the cymes about 5 mm long, sometimes with a pair of opposite or subopposite bractlets somewhat above the base, stalks of the flowers 2–3 mm long; occasionally pistillate flowers borne in diffusely panicled cymes similar to the staminate. Calyx lobes broadly triangular, margins minutely toothed or erose. Drupe globose, bright red, reddish orange, or yellowish red, rarely yellow, 5–8 mm in diameter. Nutlets 4, about 4 mm long, irregularly ribbed-grooved on the rounded side.

Edges of spring-fed rivers and streams and on their floodplains, cypress-gum ponds or depressions, depressions in flatwoods, wet hammocks, shrub-tree bogs, banks of brackish marshes. Coastal plain, s.e. N.C. to s. Fla., westward to s.e. Tex.

6. Ilex myrtifolia Walt. MYRTLE-LEAVED HOLLY, MYRTLE HOLLY.

Evergreen shrub or small, scrubby tree, usually not exceeding 5 or 6 m tall, the branches and branchlets stiff, many of them borne perpendicularly or nearly so to the branches from which they arise, almost all of them at an angle greater than 45 degrees. Young twigs and petioles puberulent, older twigs becoming glabrous and with a gray, waxy coating which breaks up into a thin, more or less interlacing, surface pattern. Leaves with petioles 1–3 mm long, blades 5–30 mm long, 3–8 mm broad, lanceolate, oblong, oblanceolate, or elliptic, stiff-leathery, dark green and lustrous above, glabrous or sometimes with sparse, minute puberulence on or near the midrib, paler beneath, usually wholly glabrous, occasionally with sparse, short pubescence on the midrib or both on the midrib and over the remainder of the surface, margins entire, tips mucronate. Staminate flowers rarely solitary, usually in stalked cymes, not over 8–10 per cyme, borne axillary to leaves on branchlets of the season, stalks pubescent, 2–8 mm long, without bractlets above their bases. Pistillate flowers axillary and solitary, their stalks pubescent, with 2 opposite or subopposite bractlets usually about half-way between their bases and midpoints. Calyx usually sparsely puberulent, lobes 4, short-triangular. Drupe globose, 5–8 mm in diameter, red, rarely orange or yellow, without luster. Nutlets 4, 4–5 mm long, ribbed-grooved on the rounded side. (*I. cassine* var. *myrtifolia* (Walt.) Sarg.)

Cypress-gum ponds and depressions, outer, sandy rims of ponds. Coastal plain, N.C. to n. Fla., westward to La. and near the mouth of the Brazos River, Tex.

Allegedly hybridizes with *I. cassine*, the hybrids intermediate in character.

7. Ilex decidua Walt. POSSUM-HAW.

Deciduous shrub or small understory tree, often attaining a height of about 10 m. Twigs of the season brown, glabrous, older ones gray, with raised, nearly circular lenticels. Leaves with copiously short-pubescent petioles 2–15 mm long; blades mostly but not always broadest somewhat above their middles, oblanceolate, spatulate, or elliptic, cuneate basally, apices rounded, obtuse, or obtusely subacuminate, variable in size from tree to tree, even on a single branch, 1–6 (–8) cm long, 8–30 (–45) mm broad, margins appressed-crenate, the teeth gland-tipped, upper surfaces sometimes sparsely short-pubescent on or near the midrib, at least when young, the lower sparsely to copiously shaggy-pubescent on the well elevated midrib, usually on the principal lateral veins as well, sometimes sparsely pubescent between the veins. Staminate flowers mostly in fascicles at about the junction of spur-shoots of the previous season and the twigs of the current season, also solitary in axils of developing leaves, stalks slender, 5–12 mm long. Pistillate flowers solitary or 2 or 3 in the leaf axils, or at nodes on twigs of the previous season, stalks 3–5 mm long. Calyx glabrous, lobes 5, deltoid, drupe 6–8 mm in diameter, globose or somewhat depressed-globose, scarlet, lustrous, yellow on a rare individual plant, persisting for a considerable period after leaf fall, often some on them throughout the winter. Nutlets 4, about 5 mm long, faintly to prominently ribbed on the rounded side.

Floodplain forests, alluvial swamps, low woodlands astride creeks, wet thickets, much less frequently on well-drained wooded slopes. Coastal plain, Md. to n. Fla., westward to e. and cen. Tex., e. Okla., northward in the interior to s. Ind., s. Ill., s. Mo.

Ilex longipes Chapm. ex Trel. (*I. decidua* var. *longipes* (Chapm. ex Trel.) Ahles) is variously considered by authors as specifically distinct from *I. decidua*, as an infraspecific variety of *I. decidua*, or not meriting any taxonomic distinction. Plants known to us in the field and identifiable as *I. longipes* (by us) appear to merit distinction at some level; where we have seen them, however, they were growing in well-drained places so we choose not to grapple with the problem in this work.

In the Suwannee River drainage, exclusively inhabiting well-drained places insofar as we know, occurs a holly which has been designated *I. decidua* var. *curtissii* Fern. (*I. curtissii* (Fern.) Small). This has not, to our knowledge, been recognized as meriting distinction from *I. decidua* except by Fernald and Small. We are of the opinion that further study of the matter is warranted.

8. Ilex amelanchier M. A. Curtis in Chapm. SARVIS HOLLY.

Deciduous shrub, usually with numerous main stems from near the base, to 3 m tall or a little more. Twigs of the season grayish brown, powdery pubescent or sparsely short-pubescent, with dotlike black lenticels; older stems smooth, gray-brown, the lenticels horizontally elliptical and pale buff. Leaves with puberulent petioles 3–15 mm long; blades oblong, oblong-obovate, ovate-oblong, or elliptic, the several shapes commonly on a single branch, larger ones 5–9 cm long and 3–4 cm broad, bases rounded or shortly tapered, apices rounded or obtuse, sometimes abruptly very short-acuminate; upper surfaces glabrous at maturity, lower surfaces and petioles persistently and rather uniformly clothed with soft, short pubescence, margins entire or with few small teeth. Staminate flowers in fascicles, the pistillate solitary, usually in the axils of both leaves and leaf scars on twigs of the previous season. Drupe cerise, without lustre, 5–10 mm in diameter, subglobose to slightly oblate, calyx lobes 4, calyx apparently deciduous well before maturation of the drupes. Nutlets 4, their rounded sides with 2 broad, longitudinal furrows, sometimes with a rib across one or both of the furrows.

Woodlands astride creeks, river floodplain forests, cypress-gum swamps. Coastal plain, (s.e. Va.?), N.C. to w. Fla. Panhandle, westward to s.e. La.

9. Ilex verticillata (L.) Gray. BLACK-ALDER, WINTERBERRY.

Deciduous shrub, 1–4 m tall, rarely treelike. Twigs of the current season sparsely pubescent at first, sometimes glabrous later, greenish older twigs grayish to brown, with nearly circular, tan lenticels. Leaves with pubescent petioles 5–20 mm long; blades variable in size from plant to plant, often even on a single branch, elliptic to obovate, 2–10 cm long, 1–5.5 cm broad, bases tapered, apices abruptly short-acuminate to prominently acuminate, margins appressed-serrate, upper surfaces sparsely short-pubescent, chiefly along the midrib, often becoming glabrous, lower varyingly glabrous, pubescent on the veins, or pubescent throughout. Staminate flowers usually in shortly stalked, rarely sessile, verticels in the leaf axils, stalks of the verticels, if present, 2–6 mm long, stalks of the individual flowers 2–5 mm long, both somewhat pubescent. Pistillate flowers solitary, or 2–4 in very short-stalked or sessile verticels, in the leaf axils. Calyx tube and lobes usually pubescent exteriorly, lobes 5–7, triangular, obtuse, ciliate marginally. Drupe 5–7 mm in diameter, globose, bright red, yellow on a rare individual plant. Nutlets 5 or 10, 3–4 mm long, smooth on the rounded side.

Swamps, wet woodlands, bogs, seepage areas in woodlands. N.S. to Minn., generally southward to Fla. Panhandle, Miss., s.w. Mo., Ark.

10. Ilex laevigata (Pursh) Gray. SMOOTH WINTERBERRY.

Similar to *I. verticillata*. Twigs of the current season glabrous, brown, with dotlike, dark red or brown lenticels, older twigs gray or grayish brown, with somewhat raised, nearly circular lenticels of the same color. Calyx glabrous, lobes acute. Drupe orange-red, yellow on a rare individual plant.

Shrub bogs, open bogs, swamps, wet woodlands. s. Maine to n. Ga., chiefly at low elevations.

Hippocrateaceae (HIPPOCRATEA FAMILY)

Hippocratea volubilis L.

Woody plant with milky sap, the main stem to 20 m long, with opposite, divaricate branchlets some of which loop or twist around other things and by which climbing is achieved. Leaves opposite, mostly somewhat leathery, persistent, petioles 5–8 mm long, blades oval-oblong, ovate, or obovate, 5–14 cm long, bases rounded to short-tapering, apices rounded to obtuse, margins entire or wavy; stipules minutely subulate, quickly deciduous. Stem, inflorescence axes, and flower stalks copiously short-pubescent. Inflorescences axillary on the divaricate branchlets, in stalked, dichotomously branched, open cymes to about 12 cm long and about as broad as long, flowers in irregular clusters, the very short flower stalks with 2 minute bractlets near the base. Flowers small, radially symmetrical, bisexual. Sepals 5, united only basally, lobes minutely deltoid-ovate, powdery-pubescent, persistent. Petals 5, distinct, thickish, white or yellowish near the base, much longer than the sepals, oblong-elliptic, spreading, their obtuse tips curved upwardly and inwardly. Stamens 3, seated on a conspicuous fleshy disc. Ovary 3-carpellate, 3-locular, the carpels fused basally and immersed in and fused with the disc, style short, subulate, stigma 3-lobed. Fruit deeply 3-parted, each capsular segment divergent, thinly lenticular, lanceolate, lance-ovate, or elliptic in outline, 2.5–3 cm long, loculicidally dehiscent along an inconspicuous median suture. Seeds 5 or 6 in each segment, each with a basal wing jointed with a fleshy caruncle.

Mangrove swamps, thickets, tropical hammocks. s. pen. Fla. and Fla. Keys; W.I.; cen. Mex. to Peru, Bolivia, s. Brazil, Parag., Arg.

Staphyleaceae (BLADDERNUT FAMILY)

Staphylea trifolia L. BLADDERNUT.   Fig. 143

Deciduous shrub, rarely becoming of small treelike stature, often colonial. Bark of twigs brown, dotted with pale, elongate to round lenticels, bark of older stems becoming gray to black, somewhat mottled. Leaves opposite, long-petiolate, with linear-subulate, quickly deciduous stipules, compound, usually with 3, rarely with 5, leaflets; lower pair of leaflets sessile or nearly so, the terminal one usually with a stalk 1.5–3 cm long; leaflets glabrous above, or short-pubescent when young, becoming glabrous, sparsely pubescent beneath, broadly elliptic to ovate-elliptic, 3–10 cm long and 2–5 cm wide, bases rounded or broadly tapering, sometimes oblique, apices short-acuminate, margins finely serrate, lateral veins adnate to the midrib for a short distance then abruptly divergent. Inflorescence a few-flowered panicle developing on the tips of branchlets as the new growth of the season is emerging; flower stalks slender, flexuous, 1–2 cm long, usually drooping, each jointed at or just above the middle, each subtended by a pair of long, linear-subulate, pubescent, membranous white bracts. Flowers bisexual, corolla campanulate. Calyx united only at base, lobes 5, erect, somewhat shorter than the petals, greenish white, linear-oblong, 6–8 mm

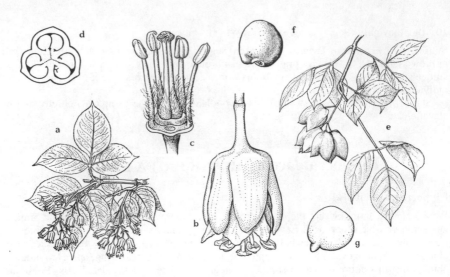

Fig. 143. **Staphylea trifolia:** a, branch with immature foliage and drooping panicles; b, individual flower with swollen area of pedicel and two-lobed anthers; c, same, perianth parts and one stamen removed—note pubescent filaments and ovary partly embedded in the disc; d, semidiagrammatic cross-section of three-loculate ovary with apotropous ovules in two rows; e, branch with mature foliage and immature, inflated capsules; f, seed; g, embryo with disc-shaped, flattened cotyledons. (From Spongberg in *Jour. Arn. Arb.* 52: 200. 1971)

long. Petals 5, erect, cream, striped with green. Stamens 5, inserted outside and below a fleshy disc, alternate with the petals and about equaling them in length. Ovary partially embedded in the disc, 3-locular, ovules 4–12 in each locule, styles 3, free below, united at the blunt stigmatic apex. Fruit inflated, bladderlike, thin-walled, 3–5 cm long and to 3 cm broad, the carpels separating at the summit as the fruit matures thus becoming 3-lobed and weakly 3-beaked by the persistent styles, fruits persisting after leaf fall. Seeds gray to brown, 5–7 mm long, subspherical, plump or somewhat flattened, seed coat hard. Often no ovules maturing into seeds; if seeds are produced, mostly 1 or 2 per locule.

Rich woods, floodplain forests, wooded stream banks, moist ravines, shores of lakes and ponds, pastures and fence rows. Que. to Minn., generally southward to Ga., n. Miss. and n. Ala., and Okla.; Apalachicola River, Fla. Panhandle.

Aceraceae (MAPLE FAMILY)

Acer

Trees or shrubs. Leaves simple and palmately veined, or ternately to pinnately compound, opposite, deciduous (in ours). Flowers small, radially symmetrical, borne in racemes, panicles, or fascicles, mostly functionally unisexual, the staminate and pistillate on separate plants but in some species occasional flowers structurally and functionally bisexual thus the plants polygamo-dioecious. Calyx usually of 5 sepals or united at the base and 5-lobed (rarely 4-parted or up to 12-). Petals none or of the same number as the calyx parts, inserted on the margin of a nectariferous disc. Stamens 3–12. Pistil 1, ovary superior, compressed, 2-

locular, each locule with 2 ovules, style 1, short, stigmas 2; each carpellate portion of the pistil bearing a wing, developing into a pair of conspicuously winged, 1-seeded samaras united basally and eventually separating.

1. Leaves compound. 1. *A. negundo*
1. Leaves simple.
    2. Terminal lobe of the leaf blade comprising more than half the length of the entire blade and the lobe narrower at the base than above; ovaries and young fruits pubescent, mature fruits pubescent at least basally. 2. *A. saccharinum*
    2. Terminal lobe of the leaf blade comprising half or less than half the length of the entire blade and widest at the base (or rarely leaves unlobed); ovaries and fruits glabrous. 3. *A. rubrum*

1. Acer negundo L. BOX-ELDER, ASH-LEAVED MAPLE. Fig. 144

Small to medium-sized, fast-growing tree, usually not exceeding 20–25 m tall, the wood brittle. Twigs green, sometimes glaucous, glabrous or pubescent, axillary buds prior to leaf-fall covered by the petiole bases, after leaf-fall surrounded by leaf scars. Leaves compound, long-petiolate, the blades ternate or with 5–9 leaflets pinnately arranged and pinnately veined, the lateral leaflets short-stalked, the terminal one usually (not always) with a stalk about twice as long. Blades of the leaflets mostly ovate, varying to elliptic, 5–10 cm long and to 5–7.5 cm wide, margins with few coarse, sometimes irregular, serrations, mostly from about the middle upwards, sometimes irregularly lobed-serrate; bases broadly tapered to rounded, apices acuminate; both surfaces copiously hairy as the leaves unfold, the upper surface becoming glabrous, the lower becoming glabrous or remaining pubescent. Flowers greenish, unisexual, the plants dioecious, flowering from buds on wood of a previous season just before or as new shoots emerge in spring. Staminate flowers in many-flowered fascicles, on long threadlike, pendent stalks; pistillate flowers in pendent racemes that elongate during fruit maturation. Calyx 5-lobed, corolla none. Samaras 2.5–3.5 cm long, yellowish, in large, loose pendent racemes. (*Negundo negundo* (L.) Karst.)

    Floodplain forests, river and stream banks, wet to moist hammocks and lowland woods, sometimes abundantly colonizing cut-over areas. Vt. to Man., generally southward to cen. pen. Fla. and e. half of Tex.; N.Mex., Ariz., Calif.

2. Acer saccharinum L. SILVER MAPLE, SOFT MAPLE. Fig. 145

Medium-sized to large tree, the bark gray and sloughing in large scales. Leaves simple, the petioles about as long as the blades; blades ovate in overall outline, about as broad as long, palmately 5-lobed, the primary lobes irregularly lobed-serrate, terminal lobe more than half the length of the entire blade and narrowed basally, sinuses between the lobes acute; unfolding leaves copiously pubescent beneath, sparsely so above, becoming green and glabrous above, markedly silvery below. Flowers greenish yellow to red, unisexual (plants dioecious), borne in dense clusters from buds on twigs of a previous season, appearing before new shoots emerge. On pistillate trees, as the fruits commence to develop the very short flower stalks start to elongate, eventually the mature samaras are in fascicles, the stalk of each samara 3–4 cm long. Samaras 4–6 cm long, yellowish, pubescent basally. (*Argentacer saccharinum* (L.) Small)

    Floodplain forests, river and stream banks, lowland woodlands. N.B. and Que. to Minn. and S.D., generally southward to the Fla. Panhandle, Okla. and La.

3. Acer rubrum L. RED MAPLE. Fig. 146

Medium-sized tree, young branchlets usually red, becoming reddish brown, then gray; bark of trunk brownish gray, with shallow fissures dividing the surface into irregular, vertical, flat ridges. Petioles mostly as long as or a little longer than the blades, usually reddish; blades simple, exhibiting much variation in size, lobing, and pubescence, the variation accounting for segregation by some authors into several species, varieties, or forms. In general, blades palmately 3–5-lobed (rarely unlobed and lanceolate or lance-ovate), the terminal lobe comprising half or less than half the length of the entire blade and widest at the base; blades

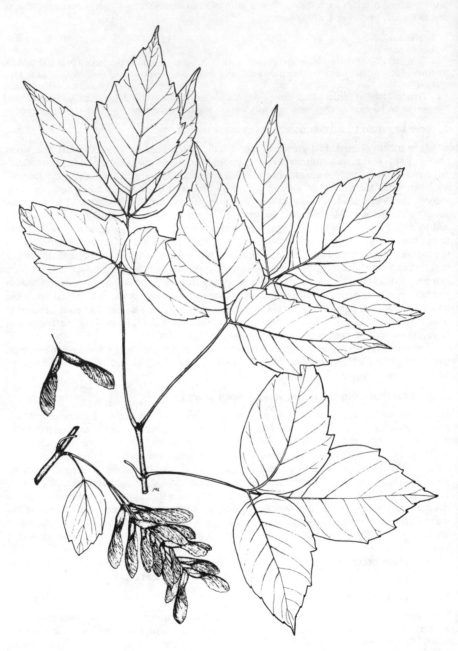

Fig. 144. **Acer negundo.** (From Kurz and Godfrey, *Trees of Northern Florida*, 1962)

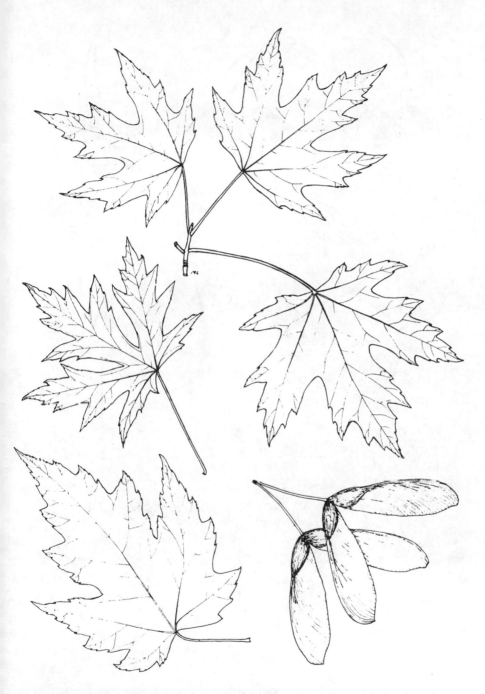

Fig. 145. **Acer saccharinum.** (From Kurz and Godfrey, *Trees of Northern Florida*. 1962)

313

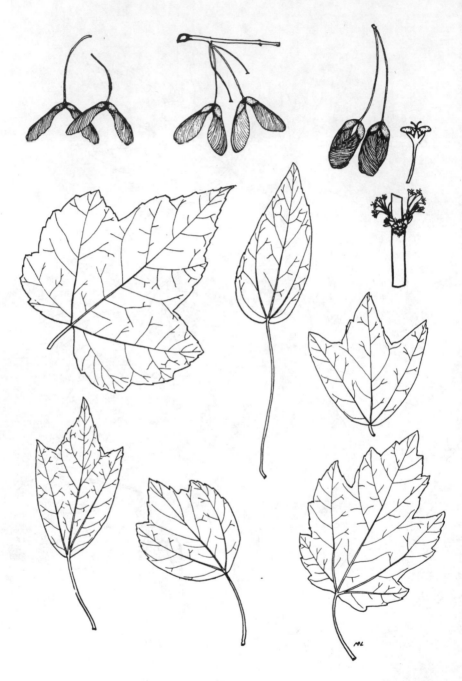

Fig. 146. **Acer rubrum.** (From Kurz and Godfrey, *Trees of Northern Florida*. 1962)

314

approximately ovate in overall outline, mostly from about 6–14 cm long, varying from as wide as long to considerably longer than wide, their bases subcordate, truncate, or U-shaped; primary lobes from irregularly and sharply serrate to lobed-serrate, crenate-serrate, or entire; upper surfaces glabrous and green, lower whitish to silvery, varying from glabrous to tomentose, usually pubescent at least on the veins. Flowers red, unisexual and plants dioecious, or sometimes some flowers bisexual, the plants polygamo-dioecious, appearing from clusters of buds on wood of a previous season, generally much before new shoot emergence, the fruits mostly maturing and falling before new shoot emergence. Staminate flowers in tight clusters, the pistillate (or bisexual) ones in nearly sessile clusters, their stalks elongating as the fruits develop, the fruits then in umbels. Sepals 5, distinct or nearly so, about 1 mm long, petals 5, linear-oblong, somewhat longer than the sepals. Samaras glabrous, 1.5–2.5 cm long, varying greatly in color from scarlet to bronze, more infrequently yellowish, generally making the fruiting trees very showy somewhat prior to new shoot development. (Many people think the fruiting trees are handsomely in flower when indeed they are in full fruit.) (*Rufacer rubrum* (L.) Small; *R. carolinianum* (Walt.) Small; *R. drummondii* (H. & A.) Small; *A. rubrum* var. *drummondii* (H. & A.) Sarg.; *A. rubrum* var. *trilobum* K. Koch; *A. rubrum* var. *trilobum* forma *tomentosum* (Tausch.) Siebert & Voss)

In both upland and lowland woods; much more frequent in lowlands in the coastal plain than elsewhere: swamps, floodplain forests, river and stream banks, wet woodlands. e. Can. to Man., generally southward to s. Fla., Okla. and e. Tex.

Rhamnaceae (BUCKTHORN FAMILY)

Shrubs, trees, or woody vines, unarmed or armed with thorns or stipular spines. Leaves simple, alternate (in ours), stipulate. Flowers small, radially symmetrical, bisexual, or functionally unisexual and the plants functionally polygamo-dioecious, usually borne in terminal or axillary cymes or panicled umbels. Flowers with a low floral tube or cup having a disc within, the tube bearing upon its rim 4 or 5 (rarely persistent) sepals, 4 or 5 petals, and 4 or 5 stamens, the latter opposite the petals and adnate to their bases. Pistil 1, sessile on the disc or somewhat immersed in it in early stages of its development, the floral tube and disc sometimes persisting and adherent to the base of the fruit; ovary 2–3-loculed, each locule with a single ovule (rarely with 2); style more or less 2- or 3-divided above or undivided. Fruit a drupe or capsule.

- Leaf blades with about 10 principal lateral veins to either side of the midrib, these essentially straight and angling-ascending parallel to each other.   1. *Berchemia*
- Leaf blades with 2 or 3, usually 3, principal lateral veins to either side of the midrib, these curvate-ascending.   2. *Colubrina*

1. Berchemia

Berchemia scandens (Hill) K. Koch. SUPPLE-JACK, RATTAN-VINE.

Deciduous, glabrous, flexible, tough, woody vine with short to widely divergent, reddish brown branches, to some extent twining but mostly scandent-scrambling on and over shrubs and low trees, in taller trees scrambling in the crowns. Leaves alternate, simple, short-petiolate; blades elliptic, oval, or ovate, their bases broadly short-tapered to rounded, apices obtuse or abruptly short-acuminate, usually tipped by a mucro, 3–8 cm long and to 4 cm wide, pinnately veined, the lateral veins about 10 to a side, conspicuous and angling-ascending parallel to each other, veinlets reticulate between them; margins slightly wavy to entire; upper surfaces bright green and shiny, the lower pale and dull; stipules lance-acute, eventually deciduous. Flowers small, about 2 mm across, greenish, functionally unisexual (plants functionally dioecious), borne on the branchlets in axillary and terminal panicles. Floral tube very short, a nectariferous disc within, this barely embracing the base of the ovary in the pistillate flowers, the floral tube surmounted by 5 deltoid sepals and 5 petals, petals in the staminate flowers obovate and each somewhat hooded about a stamen. Fruit an oblong-ellipsoid drupe 5–7 mm long, blue-black and glaucous when ripe, the thin disc below appearing more or less 2-flanged.

Floodplain forests, swamps, wettish pine flatwoods, pineland bogs, wet thickets; also in rich well-drained woodlands. Chiefly (not exclusively) coastal plain, s.e. Va. to s. Fla., westward to Tex., northward in the interior, Ark., s. Mo., Tenn.

2. Colubrina

Colubrina asiatica (L.) Brogn.

Glabrous, scandent-scrambling or spreading shrub with diffuse, elongate, smooth-barked branches (to 5 m long). Leaves persistent, alternate, slenderly petiolate, petioles 1–2 cm long; blades 4–9 cm long, 2.5–5 cm broad basally, ovate, bases truncate, rounded, or very slightly cordate, apices acuminate, margins finely crenate-serrate, upper surfaces dark glossy green, mostly with 3 curvate-ascending, principal lateral veins to either side of the midrib, the veinlets reticulate between them. Flowers small, functionally unisexual (plants functionally dioecious), borne in short-stalked, axillary clusters much shorter than the subtending leaves. Floral tube shallowly saucerlike, a prominent nectariferous disc within. Sepals 5, short-ovate, spreading. Petals 5, greenish white, each curvate and hooded about a stamen, each extending a little beyond the calyx. Fruit about 8 mm in diameter, globose, a dry drupe (the floral tube closely adhering to the base), 3-locular, each locule containing 1 seed, the endocarp yellow and chartaceous, dehiscing septicidally and partially loculicidally, this accompanied by a more or less regular breaking up of the exocarp. Seed about 5 mm long, slightly obovate or suborbicular in outline, rounded on the outer side and with 2 flat inner faces, surface smooth, dark purplish flecked with very tiny whitish papillae.

Mangrove swamps (where occasionally very thickly scrambling through and over the crowns of the mangroves), shoreline thickets, tropical hammocks. Native of E.Ind. and Pacific Is., naturalized in s. pen. Fla., Fla. Keys, and trop. Am.

Vitaceae (GRAPE FAMILY)

(Ours) unarmed, woody, or more or less fleshy vines, with tendrils opposite at least some of the leaves. Leaves alternate (rarely upper ones opposite), petiolate, blades simple or compound, stipules small and deciduous or absent. Inflorescences borne opposite the leaves, cymes or panicles, usually stalked, sometimes with tendrils, bracteate. Flowers very small, radially symmetrical, stalked, bisexual and/or unisexual (at least functionally), the plants sometimes polygamo-monoecious or -dioecious, with a nectariferous disc, cusp, or glands below the ovary. Calyx united, with 4 or 5 lobes, or the lobes absent. Petals 4 or 5, usually distinct, at anthesis spreading or cohering at their apices. Stamens of the same number as the petals and opposite them, distinct, sterile in the functionally pistillate flowers. Pistil 1 (rudimentary in the staminate flowers), 2-carpellate, ovary superior, 2-locular, style short or none, stigma sometimes 2-lobed. Fruit a 1–4 (–6) -seeded berry, the seed coat usually hard and bony.

1. Plant a semiwoody vine with more or less succulent branchlets and leaves; flowers 4-merous (except pistil).   1. *Cissus*
1. Plant a woody vine with nonsucculent branchlets and leaves; flowers 5-merous (except pistil).
   2. Petals free and spreading at anthesis; inflorescence a repeatedly bifurcate cyme; bark of the stem tight, with lenticels; pith white.   2. *Ampelopsis*
   2. Petals cohering at their apices and falling as a unit at anthesis; inflorescence a panicle; bark of older stems loosening and exfoliating in shreds (except in *Vitis rotundifolia*), without evident lenticels on the branchlets (except in *Vitis rotundifolia*); pith brown.   3. *Vitis*

1. Cissus

Cissus incisa (Nutt.) Des Moul. MARINE-IVY, MARINE-VINE.

A stoutish vine with warty stems up to 10 m long, scrambling or sprawling over marsh herbs, shrubs, small trees, or on banks or ledges. Young stems and leaves succulent or subsucculent, pith white. Leaves with petioles usually shorter than the blades; blades rarely undivided, usually deeply 3-parted or with 3 leaflets, to about 8 cm long overall and as wide, the divisions or leaflets ovate to obovate, cuneate below, the margins coarsely and irregularly toothed or lobed-toothed. Stalk of the cyme usually longer than the petiole of the leaf opposite, the divisions of the cyme more or less umbellate. Flowers 4-merous, bisexual or unisexual and the plants polygamo-monoecious. Calyx cuplike, the lobes short and rounded. Petals spreading at anthesis, free. Disc a deeply 4-lobed cup enveloping the ovary only basally. Berries ovoid to obovoid, black, 6–10 mm long, their stalks usually curved.

Salt marshes, usually on shell mounds, maritime woodlands, rocky ledges and rocky woodlands inland. Fla. to Tex., northward in the interior to Ark., Mo., Kan.; from Tex. to Ariz.; n. Mex.

2. Ampelopsis

Woody vines with nonsucculent branchlets and leaves. Bark tight and with evident lenticels, pith white. Tendrils bifurcate, without adhesive discs on their branch tips. Leaves petiolate, blades simple or compound. Inflorescence a bifurcate cyme, the ultimate divisions umbellike. Flowers 5-merous, bisexual. Calyx very small, saucerlike, with broad, shallow, rounded lobes. Petals green, free, oblong, rounded apically, exceeding the calyx, spreading at anthesis. Nectariferous disc cuplike, adherent to the base of the ovary, entire or lobed at the summit. Berries dryish, 1–4-seeded.

- Leaves simple, grapelike.   1. *A. cordata*
- Leaves twice ternately or partially tripinnately compound, not grapelike.   2. *A. arborea*

1. **Ampelopsis cordata** Michx. RACCOON-GRAPE, FALSE-GRAPE.

High-climbing vine, stems somewhat pubescent at the nodes at first, soon becoming glabrous. Leaves simple, petioles slender, about as long as the blades; blades ovate, often about as broad as long or a little longer than broad, their bases subcordate to truncate, apices short-acuminate, margins coarsely and often irregularly serrate, occasionally with 1 or 2 lateral lobes; petioles at first sparsely pubescent proximally, usually densely pubescent at juncture with the blade, blades at first sparsely pubescent, soon becoming glabrous. Berries somewhat flattened at both extremities, 5–8 mm across, not quite as long, ripening from green to orange, rose, purple, finally blue.

Floodplain forests, borders of wet woodlands, stream banks, sand and gravel bars, lowland thickets. Va. to Fla. Panhandle, westward to Tex., northward in the interior to s. Ohio, s. Ind., s. Ill., Mo. and Neb.

2. **Ampelopsis arborea** (L.) Koehne. PEPPER-VINE.

High-climbing vine, occasionally bushy and upright. Stem at first sparsely pubescent, soon becoming glabrous. Petioles short, much shorter than the blades; blades twice ternately or partially tripinnately compound; leaflets variable in size on a given leaf, the larger 3–4 (–5) cm long and about 2 cm wide, mostly ovate, their bases mostly cuneate, sometimes truncate, apices acuminate or acute, margins sharply and coarsely few-serrate; usually sparsely pubescent on the veins beneath and with tufts or webs of hairs in the vein axils. Berries obovoid or subglobose, 1–1.5 cm across, black when ripe.

Floodplain forests, stream banks, sand and gravel bars, lowland wet woodlands, marshes, woodland borders, wet thickets, sometimes in well-drained areas. Md. to s. Ill. and Mo., generally southward to s. Fla., Okla. and e. Tex.

3. Vitis (GRAPES)

(Ours) deciduous, woody vines or viny shrubs. Bark of older stems becoming loose and exfoliating in shreds (except in *V. rotundifolia*), without evident lenticels (except in *V. rotundifolia*), pith brown. Tendrils bifurcate (rarely simple), without adhesive discs at their branch tips. Leaves petiolate, blades simple, mostly ovate, bases mostly cordate. Inflorescence thyrsoid-paniculate, borne on branchlets of the season. Flowers bisexual or unisexual (at least functionally), plants mostly polygamo-dioecious. Calyx very small and saucerlike, or essentially absent. Corolla with 5 petals coherent apically, falling as a unit at anthesis. Stamens straight in the staminate and bisexual flowers, reflexed (or rarely absent) and sterile in the functionally pistillate flowers. Nectariferous disc of 5 more or less separate glands alternate with the 5 stamens. Fruit a pulpy 2–4-seeded berry.

1. Bark of branchlets with evident lenticels, that of older stems tight, not shredding and exfoliating.
　　　　　　　　　　　　　　　　　　　　　　　　　　　　　　　　　　　　　1. *V. rotundifolia*
1. Bark of branchlets without evident lenticels, that of older stems loosening and exfoliating in shreds.
　2. Lower leaf surfaces with a permanent, tight, felty tomentum.
　　3. Felty tomentum of lower surfaces of mature leaves rusty brown. 　　　　2. *V. labrusca*
　　3. Felty tomentum of lower surfaces of mature leaves markedly silvery. 　3. *V. shuttleworthii*
　2. Lower surfaces of mature leaves glabrous or variously pubescent, if densely pubescent then some of the pubescence cobwebby, loose, and raised.
　　4. Blades of mature leaves with lower surfaces more or less cobwebby-pubescent, or glabrous and glaucous.
　　　5. Branchlets terete; petioles more or less cobwebby-pubescent, without dense short hairs.
　　　　　　　　　　　　　　　　　　　　　　　　　　　　　　　　　　　　　4. *V. aestivalis*
　　　5. Branchlets angular; petioles densely short-pubescent, sometimes with some cobwebby pubescence as well. 　　　　　　　　　　　　　　　　　　　　　　　　　　　　5. *V. cinerea*
　　4. Blades of mature leaves green or grayish green, the hairs tending to be straight and mostly along the veins, cobwebby pubescence, if any, on young leaves disappearing except in the vein axils.
　　　6. Branchlets of the current season with a distinct purplish red cast. 　6. *V. palmata*
　　　6. Branchlets of the current season green to gray or brown.
　　　　7. Leaf blades unlobed or at most with but a suggestion of lobing. 　7. *V. vulpina*
　　　　7. Leaf blades (most of them at least) distinctly 3-lobed. 　　　　　8. *V. riparia*

1. Vitis rotundifolia Michx. MUSCADINE GRAPE, SCUPPERNONG.

Vigorous vines with tight, nonexfoliating bark, that of younger stems with evident lenticels. Petioles mostly as long as the blades, glabrous; blades orbicular to ovate, unlobed, lustrous, 5–12 cm long, as wide or nearly so, with a broad, shallow sinus basally, apices very short-acuminate, lower surfaces with cobwebby pubescence in the vein axils, margins prominently dentate-serrate, often irregularly so. Panicle 2–4 cm long. Fruits few per bunch, 1–2.5 cm across, purple-black to bronze, not glaucous. (*Muscadina rotundifolia* (Michx.) Small)

Inhabiting a wide variety of sites, both upland and well-drained and poorly drained or intermittently flooded bottomlands. Del. to Ky., s. Ind., Mo., generally southward to s. Fla., Okla. and e. Tex.

2. Vitis labrusca L. FOX GRAPE.

Vigorous vine, bark of older stems loose and exfoliating in shreds (as is the case in each of the following species). Young branchlets and leaves felty-pubescent, this persisting on lower leaf surfaces as a rusty, tight, felty tomentum. Petioles usually shorter than the blades, remaining cobwebby-pubescent; blades mostly ovate to suborbicular in overall outline, the larger 1–2 dm long, 3-lobed, bases cordate, the sinus varying from **V**- or **U**-shaped to very broad and shallow, margins shallowly dentate. Panicles 4–8 cm long. Fruits dark red to purplish black, 1.5–2.5 cm across.

Lowland woodlands, rich mesic woodlands, wet or dry thickets, woodland borders. s. Maine to s. Mich., southward to S.C., n. Ga. and Tenn.

3. Vitis shuttleworthii House. CALUSA GRAPE.

Vigorous vine, the young branchlets and leaves with a silvery, felty tomentum, this persisting on the petioles and lower leaf surfaces as a tight, felty tomentum, markedly silvery, and on the branchlets late in the season as a cottony covering. Petioles mostly shorter than the blades; blades ovate to reniform or suborbicular, the larger on most shoots 4–7 cm long, about as broad or a little broader, the margins shallowly, usually irregularly, crenate-dentate. Vigorous sprouts with much larger, shallowly 3-lobed blades to 12 cm long and about as wide and with margins coarsely and irregularly dentate. Panicles mostly 3–5 cm long. Fruits subglobose, 8–10 mm across. (*V. coriacea* Shuttlw.)

Mostly inhabiting well-drained places, hammocks, scrub, thickets, fence and hedge rows; also in woodlands bordering streams, or in clearings or partial clearings by streams, tropical hammocks. Pen. Fla.; W.I.

4. Vitis aestivalis Michx. SUMMER GRAPE.

High-climbing vine, the young branchlets terete, clothed with patches of cobwebby, whitish, or rusty pubescence, sometimes becoming glabrous. Petioles about as long as the blades, more or less cobwebby with whitish or rusty pubescence, without dense short hairs; blades rarely unlobed, usually shallowly 3-lobed, deeply 3–5-lobed on vigorous sprouts or vigorous lateral shoots; lower surfaces of mature leaves mostly with at least patches of cobwebby, whitish to rusty pubescence but sometimes whitish-glaucous and nearly glabrous; variable in shape and size, ovate to suborbicular in overall outline, more rarely reniform, about as broad as long, occasionally broader than long, the sinuses very variable, margins irregularly, usually shallowly, dentate-serrate. Panicles 5–15 cm long, usually relatively slenderly triangular in outline. Fruits 5–12 mm across, black or dark purple, thinly glaucous. (Incl. *V. argentifolia* Munson, *V. linsecomii* Buckl., *V. rufotomentosa* Small, *V. simpsonii* Munson)

Generally on well-drained sites, woodlands of various mixtures, woodland borders, thickets, fence and hedge rows, scrub, stabilized dunes; occasionally in floodplain or lowland woodlands, river or stream banks. Mass. to Ont. and s. Minn., generally southward to s. pen. Fla. and e. third of Tex.

5. Vitis cinerea Engelm. ex Millardet. PIGEON GRAPE.

Similar in general features to *V. aestivalis*. Branchlets angled, they and the petioles with dense short pubescence of grayish, straight hairs, these sometimes intermixed with tufts of grayish, cobwebby hairs. Panicles 1–2 dm long, often much branched and broadly triangular in outline. Fruits 4–9 mm across, purplish black, usually with a thin bloom.

Floodplain and lowland woodlands, stream banks, pond margins. s.e. Va. to s. Ohio, thence to s. Wisc., Iowa, and Kan., generally southward to n. Fla. and e. and n.cen. Tex.

6. Vitis palmata Vahl. RED GRAPE, CAT OR CATBIRD GRAPE.

Relatively slender, high-climbing vine, the branchlets of the season purplish red, somewhat angled or subterete, glabrous. Petioles slender, somewhat shorter than the blades; blades ovate in overall outline, mostly 4–8 cm long, nearly as wide, strongly 3-lobed (rarely with 2 additional small basal lobes), the 3 lobes narrowly tapered to longish, acuminate apices, margins sharply and irregularly dentate-serrate, basally cordate, the sinuses mostly broadly U-shaped, sometimes V-shaped, lower surfaces with tufts of hairs in the vein axils, sometimes hairy on the veins as well. Panicles 8–15 cm long. Fruits globose, mostly about 8 mm across, black, glossy, sometimes with a thin bloom.

Floodplain forests, stream banks, margins of ponds and sloughs, wet woodlands. s. Ind. to Iowa and Mo., southward to Fla. Panhandle and e. Tex.

7. Vitis vulpina L. WINTER GRAPE, FROST GRAPE, CHICKEN GRAPE.

Vigorous, high-climbing vine, the branchlets green to gray or brown, terete, glabrous. Petioles shorter than the blades, at first sparsely pubescent proximally, more densely so distally and densely pubescent at juncture with the blade, usually glabrous or nearly so at maturity, occasionally permanently pubescent; blades ovate, the larger 10–15 cm long, varying from a little longer than broad to about as broad as long, unlobed or occasionally with barely a suggestion of 1 or 2 lateral lobes; bases usually cordate and with a U-shaped sinus, sometimes with a broad shallow sinus or nearly truncate, apices short-acuminate, margins irregularly dentate-serrate and commonly, but not always, ciliate along the edges; lower surfaces with tufts or webs of pubescence in the vein axils, occasionally with patches of cobwebby hairs along the veins or even between them. Panicles 10–15 cm long. Fruits subglobose or flattened at both extremities, 5–10 mm across, black, shiny, slightly or not at all glaucous. (*V. cordifolia* Michx.; incl. *V. baileyana* Munson)

In upland well-drained woodlands of various mixtures, woodland borders, fence rows and hedges, also in floodplain and other lowland woodlands, stream banks, thickets. s.e. N.Y. to Mo. and e. Kan., generally southward to pen. Fla. and n.cen. Tex.

8. Vitis riparia Michx. FROST GRAPE.

Vigorous, high-climbing vine, the shoots of the season green to gray or brown, glabrous or nearly so at first and glabrous eventually. Leaf blades in overall outline much as in *V. vulpina* but definitely 3-lobed (deeply lobed on vigorous sprouts), usually short-acuminate, the margins more coarsely and sharply toothed than in *V. vulpina* and nearly always ciliate on the edges; lower surfaces with tufts of pubescence in the vein axils and usually short-pubescent along the veins as well. Fruits globose, 8–12 mm across, purplish black and strongly glaucous.

Stream banks, moist to wet thickets, moist woodlands. Que. to Man. and Mont., southward to Va., Tenn., Miss., Ark., (La.?) and Tex.

Malvaceae (MALLOW FAMILY)

Herbs, shrubs, or trees. Leaves alternate, simple, lobed or unlobed, stipulate. Flowers bisexual, radially symmetrical, solitary in leaf axils varying to racemose or paniculate, a calyxlike involucre subtending the calyx in some. Calyx usually united basally, 5-lobed, wholly united in some. Petals 5, free or barely united basally. Stamens numerous, filaments united into a tube below (monadelphous), free distally. Pistil 1, superior, 5- to many-carpellate, carpels loosely coherent and separating at maturity, or fully united; style 1, with as many branches as there are carpels, rarely twice as many; ovules 1 to several per carpel. Fruit a loculicidal capsule, rarely berrylike, or more or less ringlike and separating from the axis as separate dehiscent or indehiscent segments.

1. Plant a shrub or small tree.   1. *Pavonia*
1. Plant herbaceous, or at most subwoody at base.
  2. Blades of most leaves on a given plant having the central vein beneath, and often one or both veins to either side of it, dilated-glandular at or near the base; fruits bearing conspicuous bristles each with several peltately arranged, retrorse spicules at the tip.   2. *Urena*
  2. Blades of leaves without dilated glands on any veins beneath; fruits not as described.
    3. Low annual or biennial, stems mostly radiating, prostrate, and rooting at the nodes; involucral bracts 3; ovary of 15–25 carpels coherent in a ring.   3. *Modiola*
    3. Perennial, the stems strongly erect; involucral bracts 7 or more; ovary 5-carpellate, carpels fully united.
      4. Fruit a depressed or oblate, strongly angled-lobed capsule, each locule 1-seeded, dehiscing at maturity into 5 segments each comprised of halves of adjacent carpels.   4. *Kosteletzkya*
      4. Fruit a capsule, several to numerous seeds in each locule, neither depressed nor strongly angled-lobed, dehiscing loculicidally.   5. *Hibiscus*

1. Pavonia

Pavonia spicata Cav. MANGROVE MALLOW.

Shrub or small tree, 1–3 m tall, stellate-pubescent throughout. Leaves petiolate, petioles somewhat shorter than the blades; blades broadly ovate, cordate basally, strongly acuminate apically, 6–18 cm long, 4–10 cm broad basally, palmately veined, margins entire, irregularly wavy, or bluntly serrate; stipules early-deciduous leaving conspicuously raised bases. Flowers in loose racemes mostly bearing 3–12 flowers with stalks 1–2.5 cm long, those of lowest flowers longest. Involucral bracts lanceolate or lance-attenuate, 7–10 mm long. Calyx 5–7 mm high, tube bowl-shaped, lobes erect, deltoid, a little shorter than the tube. Petals greenish yellow, narrowly obovate, 2–2.5 cm long. Fruit approximately bowl-shaped in outline, 8–10 mm broad at the summit, a little broader than high, the carpellary portions with a median low crest on the truncated summit and two lateral cusps to either side, reticulate on the back and the summit. (*Malache scabra* B. Vogel)

Mangrove swamps, tropical hammocks, coastal thickets. s. pen. Fla. and Fla. Keys; W.I.; Mex. to n. S.Am.

2. Urena

Urena lobata L. BUR MALLOW.

Tough perennial, 3–15 dm tall, stellate-pubescent throughout; in southernmost part of our range, the lower part of the stem often overwintering and becoming subwoody, herbaceous branches arising from it in spring. Leaves petiolate, about as long as the blades; blades palmately veined, the larger ones usually 7-veined, the central vein beneath, often one or both veins to either side of it, dilated-glandular at or near the base, very variable in size, 2–8 (–12) cm broad, ours commonly somewhat broader than long and subreniform in outline, mostly subcordate basally, unlobed or more frequently with 3 short, broadly angulate lobes distally, margins irregularly and shallowly serrate, sometimes crenate or entire, upper surfaces dark green, relatively sparsely pubescent, sometimes becoming glabrous or nearly so, lower surfaces densely tomentose and grayish. Flowers very short-stalked, borne in leaf axils or in short racemes terminating branchlets. Involucre shallowly cuplike basally, with 5 subulate lobes 4–6 mm long, a little shorter than to a little longer than the triangular-acute calyx lobes. Petals 5, pink or rose, darker at the base, 1.5 cm long, obliquely obovate. Ovary 5-carpellate, 5-locular, style with 10 short branches near the summit. Fruit broadest at the summit, broader than long, 1 cm across, with 5 prominent, roundish lobes, each bearing prominent bristles each with several peltately arranged retrorse spicules at the tip; fruit eventually separating into 5 indehiscent segments each containing 1 dark brown, curved seed broadly rounded on the back.

Weedy and often abundant in a great variety of disturbed sites including thickets on edges of swamps and wet woodlands. n.e. to s. Fla.; native of E.Ind. and widely naturalized in tropical and subtropical areas.

3. Modiola

Modiola caroliniana (L.) G. Don.

Annual or biennial herb, usually much branched, branches commonly radiating, creeping, rooting at the nodes, to about 8 dm long, distal portions of branches sometimes ascending, less frequently weakly erect, hirsute throughout. Young stems shaggy-pubescent, becoming glabrous or nearly so. Petioles slender, about equaling the blades; blades 2–7 cm long, broadly ovate to subrotund in outline, palmately 3–7-lobed or -incised, lobes serrate or further incised. Flowers borne in leaf axils, their stalks 2–6 cm long, often as long as the petioles, sometimes equaling the blades. Involucral bracts 3, narrowly oblanceolate, about half as long as the calyx. Calyx 2.5–3.5 mm long at anthesis, enlarging somewhat as the fruits mature, united for about half its length, lobes ovate, apices acute or short-acuminate. Petals obovate, 4–8 mm long, salmon-colored to purplish red. Stamens numerous, staminal tube broad basally then narrowed to a necklike portion, free portions of filaments short and anthers aggregated. Ovary of 15–25 carpels coherent in a ring, each 2-locular, 1 ovule in each locule, style with as many divisions above the middle as there are carpels. Carpels separating at maturity of the fruit, each tardily bivalved at the top, the segments blackish, reniform in outline, 2-ridged and deeply channeled between, a prominent cusp dorsally on the curved ridges, a few long trichomes on the distal portion.

More or less weedy in fields, roadsides, lawns, gardens, sometimes in water on shorelines or in ephemeral pools. e. N.C. to n. pen. Fla., westward to s.e. Okla. and s. Tex.; trop. Am.

4. Kosteletzkya

Perennial herbs with 1 to several main stems from the base, each branched above, pubescent throughout (older, larger portions of stems becoming glabrous) with stellate or both simple and stellate hairs. Flowers solitary in axils of unreduced or reduced leaves, or, in addition, these grading into short racemes or panicles terminating branchlets. Involucre below the calyx of 7–10 bracts narrower than the calyx lobes. Petals pink or white. Ovary 5-carpellate, 5-locular, 1 ovule per locule. Fruit an oblate, strongly angled-lobed capsule dehiscing at maturity into 5 segments each comprised of halves of adjacent carpels. Seed smooth, plump, and strongly, often obliquely, curved.

- Corolla white, petals about 1 cm long; calyx lobes shorter than the radius of the mature capsule; pubescence of both simple and stellate hairs.   1. *K. pentasperma*
- Corolla pink (rarely white), petals 2–4 cm long; calyx lobes equaling or exceeding the radius of the capsule; pubescence stellate.   2. *K. virginica*

1. Kosteletzkya pentasperma (Bert. ex DC.) Griseb.

Plant to about 1 m tall. Stem bearing long, pustular-based, spiculelike hairs uniformly spaced over the surface and denser longitudinal bands of short, stellate hairs along one side of a given internode. Petioles very slender, to about 1 cm long; blades mostly 3–7 cm long, deltoid-ovate, some sometimes slightly lobed, truncate or rounded basally, acute apically, margins with blunt serrations; petioles with pubescence like that of the stems, both surfaces of the blades sparsely pubescent with both simple and few-branched stellate hairs. Flowers solitary in leaf axils of branchlets, their stalks slender, 5–10 mm long. Involucral bracts subulate, about ⅔ as long as the calyx lobes. Calyx lobes ovate-triangular, about 4 mm long, shorter than the radius of the fully developed capsule, pubescent. Petals white, about 1 cm long. Capsule 1 cm broad, strongly 5-angled-lobed, angles pubescent with simple spiculelike hairs.

Borders of mangrove swamps, coastal hammocks. s. Fla.; Mex. to Venez. and Ecuad.; Hisp., Cuba, Jamaica.

2. Kosteletzkya virginica (L.) Presl. SEASHORE MARSH-MALLOW.   Fig. 147

Plants of this species exhibit great variability in stature, density of pubescence, leaf size and form, and corolla size, the various characters intergrading. Stems 1 to several from the base, few-branched to much branched, from a few dm tall to 2.5 m tall. Stellate pubescence on all parts, including portions of the petals, varying from sparse to very dense and from harsh to soft-velvety. Leaves gradually reduced in size above midstem, blades of larger leaves, especially, varying greatly in size and form from plant to plant and variable to a lesser extent on some individual specimens; petioles of larger median stem leaves 4–10 cm long, stoutish, uppermost, reduced, or bracteal leaves becoming sessile; blades of median stem leaves cordate-ovate to long-triangular and unlobed, variously angulate-lobed, hastate, or sagittate, those of upper leaves mostly lanceolate or lance-ovate, some of them sometimes hastate; apices of most leaves acuminate, margins irregularly toothed. Flowers solitary in leaf axils, or in short racemes or panicles terminating branchlets. Bracts of the involucre subulate,

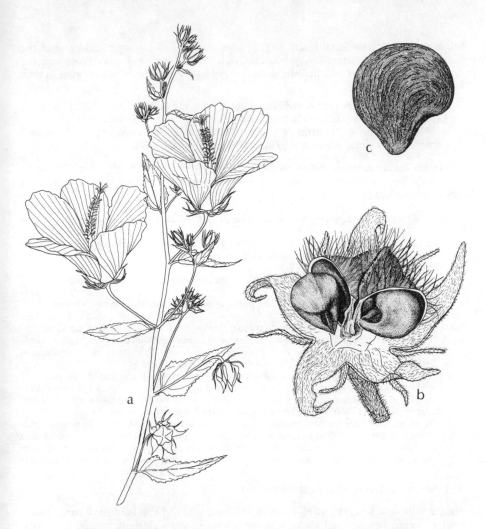

Fig. 147. **Kosteletzkya virginica:** a, flowering branch; b, capsule partially dehisced; c, seed.

much narrower and somewhat shorter than the calyx lobes. Calyx lobes triangular-acute, 10–12 mm long, equaling or usually considerably exceeding the radius of the mature capsule. Petals pink (rarely white), obovate, 2–4 cm long. Capsule strongly 5-angled-lobed, 10–12 mm broad, pubescent on and between the angles with both coarsely spiculelike and finely stellate hairs. (Incl. *K. altheaefolia* (Chapm.) Gray, *K. smilacifolia* Gray)

Salt, brackish, and fresh marshes, marshy shores, sloughs, ditches, borders of swamps and wet woodlands, wet clearings. Outer coastal plain, L.I. to s. Fla., westward to e. Tex.; W.I.

5. Hibiscus

Annual or perennial herbs or shrubs (ours perennial herbs). Leaves palmately veined, often palmately lobed or divided. Flowers borne in axils of unreduced or reduced leaves and/or in terminal panicled racemes, in some species their stalks with an articulation at, above, or

323

below the middle. Involucral bracts 7–15. Calyx united ⅔ of its length, enlarging as the fruits mature. Corolla large. Ovary 5-locular, style slender, with 5 short divisions near the summit, fruit a 5-valved, loculicidally dehiscent capsule, seeds several to numerous in each locule.

References: Wise, Dwayne A., and Margaret Y. Menzel. "Genetic Affinities of the North American Species of *Hibiscus* Sect. Trionum." *Brittonia* 23: 425–437. 1971.
Wise, Dwayne A. "Patterns of Genetic Differentiation in *Hibiscus* Sect. Trionum." Unpublished doctoral dissertation, Florida State University, 1972.

The key and descriptions below, as they involve taxa numbered 2–6b, are not drawn to account for the considerable number of hybrids between various of the pairs of taxa which may be encountered in nature.

1. Stems and leaves very harshly pubescent to touch; calyx lobes each with a nectary medially on the back.   1. *H. aculeatus*
1. Stems and leaves glabrous or variously pubescent, if the latter then the pubescence soft to slightly rough, not harsh, to touch; calyx lobes without nectaries on the back.
  2. Plant glabrous throughout.
    3. Corolla deep red; principal leaf blades deeply palmately divided, essentially compound.
                                              2. *H. coccineus*
    3. Corolla white or pink with a purple center; principal leaf blades unlobed or hastately lobed.
                                              3. *H. militaris*
  2. Plant with pubescent stem (younger parts at least), leaves pubescent on one or both surfaces.
    4. Involucral bracts copiously pubescent with short-stellate hairs and with longish, spreading simple hairs on or near the margins, thus more or less ciliate.   4. *H. lasiocarpos*
    4. Involucral bracts with uniform, tightly short-tomentose pubescence.
      5. Upper and lower surfaces of leaf blades densely pubescent; capsule copiously pubescent.
        6. Leaf blades, or most of the larger ones on a given plant, 3-lobed.   5. *H. grandiflorus*
        6. Leaf blades unlobed.   6c. *H. moscheutos* subsp. *incanus*
      5. Upper surfaces of leaf blades glabrous or very sparsely pubescent; capsule glabrous.
        7. Flower stalks nearly always bearing a leaf at, a little below, or a little above the middle; petals white or cream-colored and with a red to purple base.   6a. *H. moscheutos* subsp. *moscheutos*
        7. Flower stalks leafless or fused with the petioles only basally; petals usually pink, infrequently white, and with a red or crimson base.   6b. *H. moscheutos* subsp. *palustris*

1. Hibiscus aculeatus Walt. COMFORT-ROOT.

Plant usually with several angling-ascending to erect stems to 1 m tall from a hard, woody caudex. Stems, petioles, and blades very harshly pubescent with short, stellate trichomes, the trichome branches transparent and spiculelike. Principal stem leaves with petioles equaling or a little longer than the blades; blades mostly deeply palmately 5-lobed, lower 2 lobes oblongish, other 3 oblanceolate in outline but irregularly toothed or irregularly sublobed; upwardly blades deeply 3-lobed, uppermost reduced-bracteal and only toothed-lobed. Flowers solitary, mostly in the axils of upper 3-lobed or unlobed leaves, their stalks 5–12 mm long. Involucral bracts 8–10, linear-subulate and cleft apically, about half as long as the calyx. Calyx 10–15 mm long at anthesis, united about ⅓ of its length and campanulate, lobes triangular-acute or acuminate, exteriorly low-keeled, an elongate, purplish nectary somewhat back of the tip, keel and surface clothed with long, pustular-based, transparent, simple or 2- or 3-branched spiculelike trichomes, surface interiorly densely tomentose. Petals about 6 cm long, broadly obovate, cream-colored at first, becoming yellowish then pinkish, drying greenish yellow, dark purplish red basally. Capsule 12–14 mm long, exceeded by the calyx lobes, ovoid, tapered distally to an obtuse apex with a rigid beak about 2 mm long, surface abundantly clothed with long, spiculelike trichomes. Seed plump, obliquely obovoid to reniform, 3–4 mm long, dull brown, surface very minutely pitted, glabrous.
Hillside bogs in pinelands, pine savannas and flatwoods, adjacent ditches, swales. Coastal plain, N.C. to s.cen. pen. Fla., westward to La.

2. Hibiscus coccineus Walt.

Plant glabrous throughout. Stems 1 to several from a robust caudex, 1–2 (–3) m tall. Petioles variable in length to 15 cm; blades of lowermost leaves relatively small, unlobed, deltoid-acuminate to ovate-acuminate, bases more or less cordate, grading into palmately 3-lobed above with the terminal lobe much longer than the laterals (both of these leaf types mostly shed by flowering time); midstem blades largest, palmately 5-cleft to the junction with the petiole, essentially compound, segments or leaflets 4–25 cm long, mostly lanceolate, cuneate basally, long-acuminate apically, margins remotely, shallowly, and bluntly toothed to entire; uppermost leaves usually reduced, often deeply 3-cleft. Flowers solitary in axils of upper leaves, their stalks long, articulated 5–12 mm below the involucre. Involucral bracts usually 10, subulate, much shorter than the calyx. Calyx united for about ⅓ of its length, lobes lanceolate to ovate, acute or acuminate apically, 4–5 cm long. Petals deep red, cuneate basally, abruptly broadening to obovate, apices rounded or broadly, often obliquely, tapered, 7–10 cm long. Capsule ovoid or oblong-ovoid, 7–10 cm long, blunt apically, 2.5–3 cm long, striate and somewhat reticulate between the striations. Seed nearly globose, 3–4 mm in diameter, surface densely clothed with short, brown pubescence.

Swamps, marshes, sloughs, and ditches, commonly in water. s.e. Ga. to cen. pen. Fla., Fla. Panhandle, s. Ala.; cultivated as an ornamental.

3. Hibiscus militaris Cav. HALBERD-LEAVED MARSH-MALLOW. Fig. 148

Plant usually with several stems from the base, to 2 (–2.5) m tall, glabrous throughout. Petioles usually shorter than to equaling the blades; blades of lower leaves and often of lateral, short, sterile branchlets and bracteal leaves, unlobed, usually deltoid-acuminate; occasionally all blades unlobed on a given plant; midstem blades usually hastate, the central lobe narrowly to broadly long-deltoid-acuminate or lance-acuminate, much longer than the similarly shaped, divergent, basal lobes; very large leaves of particularly robust species sometimes with the basal lobes sublobed. Flowers solitary in axils of leaves or bracteal leaves distally on the branches, their stalks 1–6 cm long, articulated below or above the middle. Involucral bracts setaceous, about ⅔ as long as the calyx. Calyx 3–4 cm long, united ¾ of its length or a little more, lobes ovate or ovate-triangular, short-acuminate apically. Petals pink or white with a purple base, 6–8 cm long, cuneate-obovate, apices broadly rounded to truncate. Capsule nearly oblong in outline, 3.5–4 cm long, depressed-nerved, exceeded by the calyx. Seed globose, 3–4 mm in diameter, surface densely clothed with short, somewhat silky, brown pubescence.

Openings or clearings of river floodplains, river banks, marshy places along rivers and streams, ditches. s.e. Pa., s. Ohio to Minn., e. Neb., generally southward to the Fla. Panhandle and Tex.

4. Hibiscus lasiocarpos Cav.

Plant usually with several to numerous stems from the caudex, to about 2 m tall. Pubescence of stem sparse, somewhat roughly stellate-pubescent, older portions becoming glabrous. Petioles to 1 dm long; blades of larger leaves broadly to narrowly ovate, cordate to subcordate basally, sometimes to 12–15 cm broad basally, both surfaces pubescent, the lower gray or gray-green, the short-stellate pubescence relatively dense but not compactly felty-tomentose, pubescence of the upper surface sparse to abundant, the hairs longish and spreading, mostly simple or 2-forked from the base; upper surfaces usually green and drying brown, less frequently drying green; margins coarsely and irregularly dentate-serrate, rarely with a pair of low-angular lateral lobes. Involucral bracts 10–12, subulate, copiously pubescent with short-stellate hairs and with longish, spreading simple hairs on or near the margins, thus more or less ciliate. Calyx about 2.5 cm long at anthesis, united to somewhat above the middle, lobes lance-acuminate to ovate-triangular-acuminate, copiously short-stellate-pubescent exteriorly and interiorly on the lobes. Petals white to cream-colored or pale rose, deep purplish basally, cuneate-obovate, broadly rounded apically, 7–10 cm long. Capsule subglobose

325

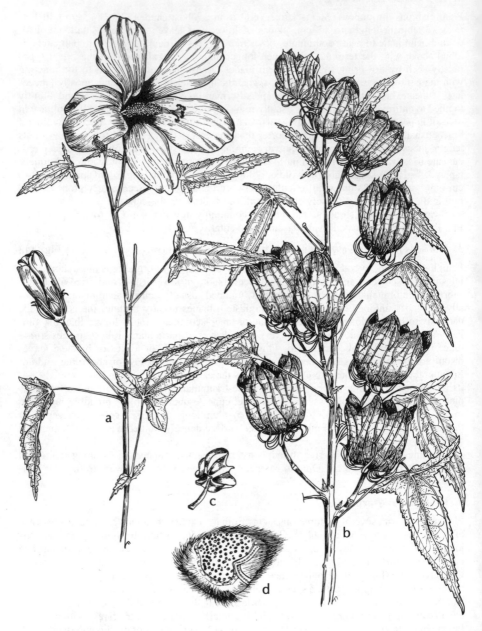

Fig. 148. **Hibiscus militaris:** a, flowering branch; b, fruiting branch; c, anther; d, seed.
(From Correll and Correll. 1972)

or short-cylindric, 2 cm long or a little more, surface with short-stellate pubescence more or less obscured by long, stiff hairs. Seed plump, obliquely broadly rounded-obovate, 4 mm broad, brown, surface very very finely cancellate and with scattered papillae.

Marshes, marshy shores, swamps, floodplains, swales, ditches. s. Ind., s. Ill., Mo., southward to Ala. and Tex.

5. Hibiscus grandiflorus Michx.

Plants to 3 m tall, with few to several stems from the basal crown. Stems, petioles, and flower stalks copiously pubescent with short, stellate hairs, velvety to touch, older portions of stems becoming glabrous. Petioles variable in length to about 10 cm on larger leaves; blades relatively large, larger ones 10–18 cm long and to 16 cm broad basally, ovate in overall outline, the larger ones mostly palmately 3-lobed, terminal lobe longer and broader basally than the laterals, margins irregularly and bluntly toothed, crenate or crenate-dentate, both surfaces densely and closely stellate-tomentose. Involucral bracts about 10, subulate, a little less than half as long as the calyx, closely and densely stellate-tomentose exteriorly and on the lobes interiorly. Calyx at anthesis 3–4 cm long, united to somewhat above the middle, lobes deltoid. Petals pink, infrequently white, red-purple basally, 12–14 cm long, cuneate-obovate, apices rounded. Capsule ovoid to subglobose, 2.5–3 cm long, abruptly contracted apically to a short beak, surface with dense short-stellate tomentum more or less obscured by long, slender, shaggy hairs. Seed obliquely obovate, 3 mm long, dark brown to almost black, smooth.

A few instances of hybridization between this and *H. coccineus* are known to us. *H. semilobatus* Chapm. probably represents a hybrid intermediate.

Marshes and marshy shores of ponds, lakes, rivers, swamps, glades, sloughs, ditches, canals, commonly in water. Coastal plain, s.e. Ga. to s. Fla., westward to s.e. La.

6a. Hibiscus moscheutos L. subsp. moscheutos.

Stems few to numerous from the basal crown, to 2.5 m tall, younger portions sparsely short-stellate-pubescent, older portions becoming glabrous or nearly so. Petioles 2–10 cm long, sparsely short-stellate-pubescent; blades very variable in size to about 20 cm long and 6 cm broad, mostly 2–2½ times as long as broad, ovate to lance-ovate or elliptic, usually unlobed, occasionally with 1 or 2 short angulate lobes medially or submedially, bases rounded or shortly and broadly tapered, apices acuminate, upper surfaces green and glabrous, infrequently with a few scattered trichomes, usually drying brown, lower surfaces gray with a dense, compact, felty tomentum, margins crenate, crenate-dentate, or irregularly and bluntly serrate. Flowers borne in the axils of leaves of the main stem or branches, often subracemose at branch tips, stalks of lower ones much longer than those of the uppermost, stalks usually bearing a leaf. Involucral bracts 10–15, subulate, felty-tomentose with stellate pubescence, shorter than the calyx. Calyx 2.5–3 cm long at anthesis, united for a little more than half its length, lobes triangular, apices acuminate; exterior of calyx varying from relatively sparsely pubescent with minute stellate hairs to felty-tomentose, lobes felty-tomentose interiorly. Petals white or cream-colored with a red or purple base, cuneate-obovate, 7–10 cm long. Capsule ovoid or ovoid-oblong, abruptly contracted to a short beak, 2.5–3 cm long, surface glabrous or sometimes sparsely pubescent on the sutures. Seed plump, obliquely obovoid, dark brown, surface very very finely cancellate and with scattered, small papillae.

Fresh and brackish marshes, swamps, swales, stream banks. Md. to s. Ind., southward to n. Fla. and s. Ala., thence to s.e. Tex.

Intergrading with subsp. *palustris* and with *H. lasiocarpus* where their respective ranges overlap.

6b. Hibiscus moscheutos subsp. palustris (L.) R. T. Clausen.

Midstem leaves usually broadly ovate, often 3-lobed, sometimes broader than long, or up to 1.5 times as long as broad. Flower stalks leafless or united with the petioles close to their respective bases. Corolla pink, rarely white, with or without a red center. Capsule blunt or barely beaked.

Salt, brackish, or fresh marshes. e. Mass. to e. N.C.; inland, w. N.Y., s. Ont., Ohio, n. Ind., s. Mich., n.e. Ill.

6c. Hibiscus moscheutos subsp. **incanus** (Wendl.) Ahles.   Fig. 149

Principal leaves similar in size and shape to those of subsp. *moscheutos*; blades densely and compactly felty-tomentose on both surfaces, the upper surface commonly gray and drying gray, sometimes green and drying brown. Calyx densely and compactly felty-tomentose. Flower stalks usually bearing a leaf. Capsule roughly stellate-pubescent. (*H. incanus* Wendl.)

Fresh and brackish marshes, marshy shores, pineland depressions, swales, swamps, wet thickets. Chiefly coastal plain, s.e. Md. to n.cen. pen. Fla., westward to e. Tex.

Sterculiaceae (CHOCOLATE FAMILY)

Melochia corchorifolia L. CHOCOLATE-WEED.

Annual herb, simple or commonly with several elongate, ascending branches from near the base, sometimes with several widely divaricate branches from near the base. Stems tough, usually more or less reddish, sparsely clothed with clusters of hairs or stellate hairs from pustular bases, the hairs usually deciduous leaving the pustular bases, the stems thus becoming roughish; if growing in shallow water the submersed portions of the stem with long, white, adventitious roots. Leaves simple, alternate, petiolate, petioles shorter than the blades, pubescence as on the stems; blades ovate to lance-ovate, venation pinnate, principal veins depressed above, raised beneath, finely reticulate between the principal veins, a few pustular-based simple hairs on the midvein above and on the main veins beneath, bases of larger leaves mostly truncate; stipules subulate, 4–6 mm long, margins long-ciliate. Flowers in compact headlike glomerules terminating axillary branchlets, the glomerule above the topmost leaf usually much larger than the others and appearing terminal. Flowers small, bisexual, each with 3 or 4 subulate, long-ciliate subtending bractlets, the bractlets collectively giving the heads a somewhat bristly appearance. Calyx united basally, with 5 ovate lobes. Petals 5, spatulate, pale purplish pink to nearly white, clawed below, 4–7 mm long. Stamens 5. Ovary lance-ovate, styles 5, distinct. Fruit a 5-locular capsule, subglobose to oblate in outline, shallowly 5-lobed, pubescent, 4–5 mm across, dehiscing loculicidally. Seeds 5, whitish-translucent, soft, 2 faces flat, 1 rounded.

In shallow water of ponds, or on wet exposed shores or bottoms of ponds, low wet areas in fields, rice fields. Native to Old World tropics, naturalized, coastal plain, N.C. to s. Fla. westward to s.e. Tex.; trop. Am.

Elatinaceae (WATERWORT FAMILY)

Small aquatic or marsh herbs. Leaves simple, opposite, with membranous stipules between them. Flowers small, solitary or clustered in the leaf axils, radially symmetrical, bisexual, perianth 2–5-merous. Sepals and petals free, persistent. Stamens of the same number as the petals and alternate with them or twice as many. Ovary superior, 2–5-locular. Seeds several to many, mostly oblong-cylindric, straight or curved, their surfaces more or less angular-pitted.

● Plant glabrous, mat-forming; flowers sessile; perianth 2–4-merous.   1. *Elatine*
● Plant glandular-pubescent, erect or with bases of branches decumbent, then erect-ascending; flowers short-stalked; perianth 5-merous.   2. *Bergia*

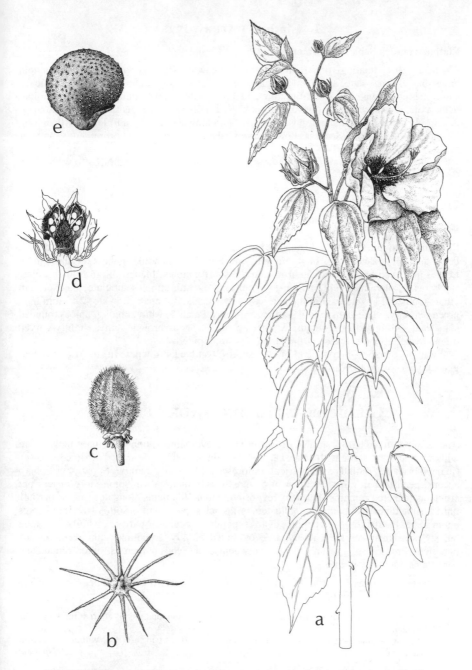

Fig. 149. **Hibiscus moscheutos** subsp. **incanus:** a, branch of plant; b, involucel of bractlets, from below; c, capsule, involucel and calyx removed; d, capsule, dehisced, involucel
and calyx intact; e, seed.

1. Elatine (WATERWORTS)

Elatine triandra Schk., var. **americana** (Pursh) Fassett.

Plants creeping, floating, or stranded in mud. Leaves flaccid-succulent, with subpetiolar bases, blades lanceolate, spatulate, or obovate, margins entire, rounded or obtuse apically, sometimes emarginate, to 7–10 mm long and 3 mm wide; stipules truncate. Flowers solitary in the axils, sessile. Perianth greenish or pinkish, usually with 3 sepals and 3 petals. Stamens 3. Capsule membranous, subglobose, 0.6–1.2 mm long and usually a little wider than long, loculicidally dehiscent. Seeds yellow, curved, about 0.5 mm long, prominently angular-pitted. (*E. americana* (Pursh) Arnott)

   Shallow water, shores of ponds and lakes, pools, ditches. Que. and N.B. to w. N.C., n. Ga., Miss., Tenn., Mo.

2. Bergia

Bergia texana (Hook.) Walp.

Plant simple to much branched from the base, the lower stems sometimes somewhat decumbent, then erect-ascending, to 4 dm tall, more or less glandular-pubescent throughout. Leaves narrowed basally and decurrent on the short petioles, blades elliptic, oblanceolate, rarely ovate, margins finely serrate, to 3 cm long overall; stipules subulate, their margins ciliate-toothed. Flowers 1–3 in the leaf axils, short-stalked. Sepals 5, ovate, 2–3 mm long, green centrally, white-margined, acuminate apically. Petals 5, white, oblong, apices rounded, somewhat shorter than the sepals. Stamens 5 or 10. Seeds brown, shiny, slightly curved, often broader at one end than at the other, obscurely pitted.

   Swamps, marshes, marshy and muddy shores, wet banks, ditches. Ill. to Ark. and Tex., generally westward to Wash. and Calif.

Guttiferae (ST.-JOHN'S-WORT FAMILY)

Annual or perennial herbs, shrubs, or subshrubs. Leaves simple, opposite, marginally entire, without stipules, in some with a basal articulation, in most punctate-dotted or -lined. Flowers (in ours) bisexual, radially symmetrical, inflorescence basically cymose. Sepals 4 or 5, herbaceous, persistent in most. Petals 4 or 5, mostly asymmetrical, in some more or less persistent as withered remains. Stamens few or numerous, filaments elongate, free or partially united, sometimes grouped in 3–5 clusters or fascicles, persistent in some, staminodia present in some. Pistil 1, 2–5-carpellate, ovary superior, 1-locular or partly or wholly 2–5-locular, styles as many as the carpels, free or united basally, sometimes connivent, usually persistent. Fruit capsular, usually septicidally dehiscent. Seeds numerous, pitted-reticulate or sometimes obscurely striate.

   Reference: Adams, Preston. "Clusiaceae of the Southeastern United States." *Jour. Elisha Mitchell Sci. Soc.* 89: 62–71. 1973.

● Petals flesh-colored to mauve-purple, or pinkish; 3 orange staminodial glands alternating with 3 clusters of 3 stamens each.   1. *Triadenum*
● Petals yellow or orange-yellow; stamens few or numerous, free or in fascicles, without intervening staminodial glands.   2. *Hypericum*

1. Triadenum (MARSH ST.-JOHN'S-WORTS)

Perennial, glabrous, rhizomatous herbs with erect stems. Leaves sessile or petiolate, without an articulation at the base, lower surfaces dotted with clear glands which may turn dark upon drying (punctate-dotted) except in *T. tubulosum*. Sepals 5, often somewhat unequal. Petals 5, flesh-colored, mauve-purple, or pinkish, quickly deciduous. Stamens 9, in 3 clusters of 3 each, the clusters alternating with 3 conspicuous, orange staminodial glands. Ovary 3-locular, placentation axile, styles 3, separate and divergent, stigmas capitate. Surfaces of seeds pitted-reticulate.

1. Lower leaf surfaces punctate-dotted.
    2. Leaves broadest at the base, cordate or subcordate, often clasping basally, all sessile; filaments united only close to their bases.   1. *T. virginicum*
    2. Leaves cuneate basally, the lower, at least, short-petiolate; filaments united to above the middle.
                                                             2. *T. walteri*
1. Lower leaf surfaces not punctate-dotted.   3. *T. tubulosum*

1. Triadenum virginicum (L.) Raf.   Fig. 150

Stems usually simple below, with several to numerous ascending branches from somewhat below or from about the middle, mostly 4–7 dm tall. Leaves sessile, ovate to oblong, bases cordate or subcordate, often clasping, apices rounded, occasionally shallowly notched, often short-apiculate, variable in size, mostly 2–7 cm long and 1–3 cm broad, lower surfaces pale green, punctate-dotted. Cymes sessile. Sepals oblong or oblong-elliptic, apically obtuse, acute, or acuminate, 4–7 mm long. Petals oblong or oblong-elliptic, apically obtuse, acute, or acuminate, 4–7 mm long. Petals 7–10 mm long, obovate, oblanceolate, or elliptic. Filaments of each stamen cluster united only basally. Capsule narrowly ovate-conical, 8–10 mm long. Seeds brown, short-oblong or sometimes elliptic-oblong, 0.5–0.8 mm long, very shallowly pitted-reticulate. (*Hypericum virginicum* L.)

    Swamps, marshy shores, bogs, sometimes in floating mats of vegetation, on submersed or floating logs. N.S. to N.Y., southward chiefly on the coastal plain to s.cen. pen. Fla., westward to e. Tex.; inland N.Y. to s. Ont. and Minn., southward to Ohio, Ind., Ill., W.Va., Tenn.

2. Triadenum walteri (Gmel.) Gl.   Fig. 151

In general aspect similar to *T. virginicum*, stems to 1 m tall. Leaves oblong-elliptic to oblong, the lower, sometimes all of them, short-petiolate, bases cuneate, apices rounded, to 15 cm long and 3.5 cm wide, lower surfaces pale, punctate-dotted. Cymes sessile or with stalks to 1.5 cm long. Sepals oblong to elliptic, rounded or obtuse apically, often unequal, 3–5 mm long. Petals narrowly obovate, 5–7 mm long. Filaments of stamen clusters united to the middle or above. Capsules narrowly elliptic, 7–10 mm long, persistent styles 1.5–3 mm long. Seeds dark brown, oblong, 1 mm long or slightly more, surfaces shallowly pitted-reticulate, somewhat lustrous, sometimes with a suggestion of iridescence. (*Hypericum petiolatum* Walt., *H. tubulosum* Walt. var. *walteri* (Gmel.) Lott., *T. petiolatum* (Walt.) Britt.)

    Swamps, floodplain forests, often on rotting logs or about bases of trees, wet seepage areas by woodland streams, marshy shores, amongst cypress and gum trees on pond or lake shores. Coastal plain, Md. to n. Fla., westward to e. Tex., northward in the interior, s.e. Okla., Ark., s.e. Mo., s. Ill. and Ind., Ky., W.Va.

3. Triadenum tubulosum (Walt.) Gl.

In general aspect similar to *T. virginicum* and *T. walteri*. Leaves usually all sessile, oblong, elliptic-oblong, elliptic, or oblong-oblanceolate, basally cuneate to rounded-truncate or sometimes subcordate, rounded apically and usually very short-apiculate, larger ones 5–15 cm long and 1–5 cm wide, lower surfaces pale, lacking punctate dots. Cymes sessile or on stalks to about 12 mm long. Sepals oblong, 5–7 mm long, apically obtuse or acute. Petals 5–8 mm long, elliptic-obovate. Filaments of stamen clusters united to the middle or above. Capsules elliptic or oblong-elliptic, 8–12 mm long, styles 1–1.5 mm long. Seeds essentially oblong, many of them curved or asymmetrical in outline, about 1 mm long, surfaces dark brown, somewhat lustrous, pitted-reticulate. (*Hypericum tubulosum* Walt., *T. longifolium* Small)

    Habitats as for *T. walteri*. Coastal plain, s.e. Va. to n. Fla., westward to s.e. Tex., northward in the interior to s.e. Mo., s. Ill., Ind., Ohio.

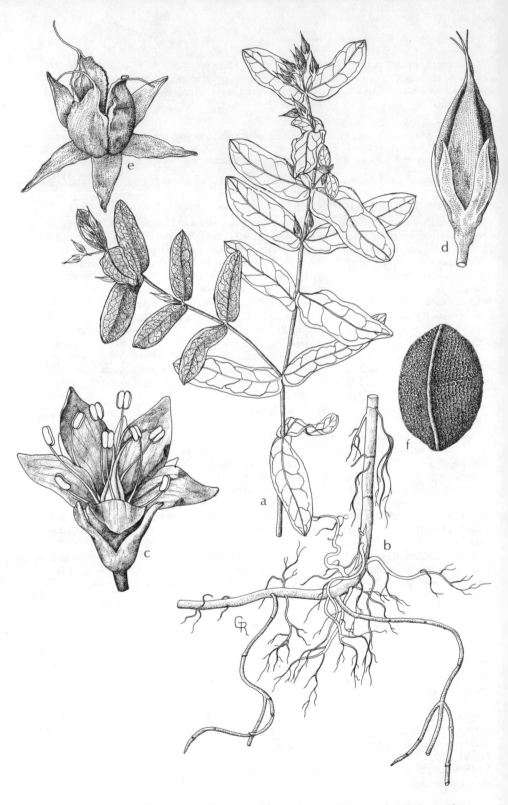

Fig. 150. **Triadenum virginicum:** a, top of plant; b, base of plant; c, flower; d, capsule; e, capsule after dehiscence; f, seed.

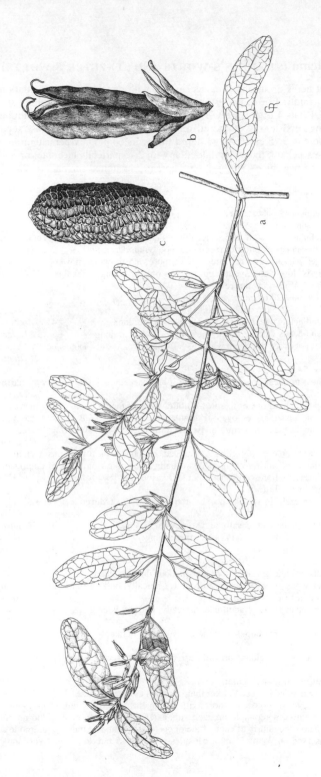

Fig. 151. **Triadenum walteri**: a, branch; b, capsule; c, seed.

2. Hypericum (ST.-JOHN'S-WORTS AND ST.-PETER'S-WORTS)

Annual or perennial herbs or shrubs. Leaves sessile, with or without an articulation at the base. Sepals 4 or 5 (rarely 3 or 6) persistent below the fruit, in some deciduous at time of capsule dehiscence. Petals 4 or 5 (rarely 3 or 6), yellow or orange-yellow, deciduous soon after anthesis or long persistent as withered remains. Stamens 5 to numerous, separate or filaments united below in 3–5 clusters, withered filaments commonly persistent around the fruits. Pistil 1, 2–5-carpellate, the ovary 1-locular with 2–5 parietal placentae, or 3–7-locular by intrusion of the placentae. Styles 2–5, separate, or united for part of their length, stigmas minute.

1. Plants herbaceous.
  2. Stem and leaves copiously pubescent.   1. *H. setosum*
  2. Stem and leaves glabrous.
    3. Leaves linear-subulate, 0.5–0.8 mm wide; flowers alternating along erect-ascending branchlets, thus the flowering branchlets essentially racemose, not cymose.   2. *H. drummondii*
    3. Leaves, the larger ones at least, 1.5 mm wide or more, mostly wider; flowers in cymes.
      4. Styles at anthesis closely connivent, appearing as one; leaf margins revolute. 3. *H. adpressum*
      4. Styles separate; leaf margins not revolute.
        5. Stamens numerous, 20 or more.
          6. Leaf blades 1.5–2.5 cm wide, rounded apically.
            7. Sepals and petals with black lines; styles 3–4 mm long.   4. *H. mitchellianum*
            7. Sepals and petals without black lines; styles 7–10 mm long.   5. *H. graveolens*
          6. Leaf blades not exceeding 1 cm wide, usually narrower, mostly acute apically.
               6. *H. denticulatum*
        5. Stamens few, 5–12.
          8. Leaves linear to linear-oblanceolate, little if any exceeding 3 mm wide, very narrow at the base, 1-nerved.   7. *H. canadense*
          8. Leaves ovate, deltoid-ovate, to short-elliptic, some of them clasping, 5–7-nerved.
            9. Middle and upper leaves ovate-triangular, tapered to a point apically; stem simple or branched from the base, the cymes above the foliage leaves thus appearing "naked."
               8. *H. gymnanthum*
            9. Leaves ovate-oblong to elliptic, rounded or blunt apically; stems usually diffusely branched, the lower floriferous branches usually emanating from axils of unreduced foliage leaves, the overall inflorescence having a leafy aspect, at least below.   9. *H. mutilum*
1. Plant woody, a shrub or subshrub.
  10. Sepals and petals 4 each (in species no. 14, the inner two so reduced as to be scarcely if at all evident).
    11. Leaves subcordate to cordate basally and somewhat clasping.   10. *H. tetrapetalum*
    11. Leaves not subcordate or cordate basally, not clasping.
      12. Styles 2.   14. *H. hypericoides*
      12. Styles 3 or 4.
        13. Outer sepals much larger and more conspicuous than the inner.
          14. The outer sepals broadly obtuse to rounded apically; bark of older stems reddish brown, exfoliating in thin strips or flakes.   11. *H. crux-andreae*
          14. The outer sepals acute or acuminate apically; bark of older stems grayish to nearly black, tight and smooth.   12. *H. edisonianum*
        13. Outer sepals somewhat larger than the inner but not markedly so.   13. *H. microsepalum*
  10. Sepals and petals 5 each.
    15. Midstem leaves ovate-triangular, subcordate to cordate basally and somewhat clasping.
               15. *H. myrtifolium*
    15. Midstem leaves not ovate-triangular.
      16. Leaves all linear-subulate, needlelike, their margins essentially parallel.
        17. Main stem leaves (not those of short axillary branches) not exceeding 11 mm long.
          18. Internode with a wing-angle on each side extending from the base of the internode to a position just below the midrib of each of the opposite leaves above, thus the internode 2-winged (this best observed on young stems); capsules mostly 4–5 mm long; seed very dark brown,

surface very finely alveolate, the alveolae nearly round and not obviously in longitudinal rows; plant with erect stems, usually bushy-branched above, to 10–15 dm tall. 16. *H. brachyphyllum*

18. Internode with a wing-angle on each side extending from the base of the internode to a position just below the midrib of each of the opposite leaves above *and in addition on each side of the internode* 2 wing-angles (one on either side of those described above) extending from the base of the internode to a position just below the two sides of the base of the opposite leaves above, sometimes forming a small auriculate appendage at that position, the internode then 6-angled; capsules mostly 6–7 mm long; seed usually black when fully ripe, surface relatively coarsely alveolate, the alveolae squarish and in longitudinal rows; plant usually bushy-branched from the base, commonly with many decumbent stems and mat-forming, to about 5 dm tall.

17. *H. reductum*

17. Main stem leaves (not those of short axillary branches) usually 13 mm long or more.

19. Young stems, leaves, and sepals strongly glaucous; bark of upper larger stems smooth, metallic-silvery, that of older lower portions of stems exfoliating in large, thin, curled plates.

18. *H. lissophloeus*

19. Young stems, leaves, and sepals not at all or scarcely glaucous; no part of the bark of the stem smooth and metallic-silvery.

20. Stem slender and wandlike, limber, usually less than 10 dm tall, unbranched or with few long-ascending branches from about the middle or above, these paniclelike in flower; stem not exceeding about 8 mm across near the base. 19. *H. exile*

20. Stem potentially 1 m tall or more, not limberly wandlike, eventually 1 to several cm thick at the base, bushy-branched above.

21. Bark of older stems thin and relatively tight, brown, reddish brown, or grayish, exfoliating in thin flakes or strips. 20. *H. nitidum*

21. Bark of older stems thin-corky to thick-corky and spongy, exfoliating in sheets, portions exposed buffish or cinnamon-colored.

22. Youngest internodes with a definite winged ridge on either side; cork layers of bark thin and smooth, with relatively small, inconspicuous, threadlike lactifers running between them; bark sloughing readily and not becoming thick on large old stems; flowers in terminal or terminal and lateral cymes mostly with 7–32 flowers. 21. *H. fasciculatum*

22. Youngest internodes terete; corky layers 3–4 mm thick, resin-filled lactifers running through them coarse, giving a striated appearance to the corky layers; bark cracking and disintegrating tardily on old stems and notably thick and soft, often as much as 3–4 cm thick; flowers solitary or in 3-flowered cymes in axils of leaves near the ends of branchlets. 22. *H. chapmanii*

16. Leaves not with parallel sides, narrowest basally, at least somewhat dilated distally.

23. The leaves with an articulation at the extreme base, this showing as a narrow horizontal line or groove.

24. Crown of plant bushy-branched, branchlets slender, flexuous, and usually spreading; larger leaves thinnish and flexuous, 2–3 cm long and 7 mm wide or less, their lower surfaces commonly, but not always, with evident lateral veins (observable with suitable magnification); placentae projecting little if any from the margin of the ovary toward the center; seeds 0.6–0.8 mm long. 23. *H. galioides*

24. Crown of plant with stiffish erect-ascending branches; larger leaves thickish and stiffish, more variable in size than the preceding (see description); placentae intruding conspicuously from the margin of the ovary toward the center; seeds 1 mm long or a little more.

24. *H. densiflorum*

23. The leaves without an articulation at the base.

25. Plants in low mats. 25. *H. buckleyi*

25. Plants erect.

26. Main stem leaves usually with short branches in their axils, successive pairs conspicuously at right angles to each other, lanceolate, linear-lanceolate, or linear-oblong, mostly broadest near the base and tapering to an acute or blunt point, thickish and stiffish, less than 10 mm wide; seeds pale brown, dull, faintly alveolate-reticulate, 0.4–0.5 mm long.

26. *H. cistifolium*

26. Main stem leaves without short branches in the axils, elliptic, lance-elliptic, or lanceolate, 1–3 cm wide, thinnish and flexuous, rounded to broadly obtuse apically; seeds dark brown and lustrous at maturity, conspicuously alveolate-reticulate, 1.5–2 mm long.

27. *H. nudiflorum*

1. Hypericum setosum L.

Annual or biennial. Stems solitary or 2 to several from the base, slender, unbranched below the terminal cyme, or less frequently with several to numerous strongly ascending branches each terminated by a cyme, 3–7.5 dm tall, copiously pubescent. Leaves copiously pubescent on both surfaces, glandular-dotted below, 1-nerved, those low on the stem spreading, usually shorter than the internodes, 3–10 mm long, broadly elliptic, oval, or oblanceolate; other leaves strongly ascending-erect, equaling the internodes in length, broadest at the base, the sides nearly parallel to above the middle then tapered to an acute tip, 5–15 mm long and 3–6 mm wide, varying little in size on a given stem. Branches of the cyme strongly ascending, flowers alternate on the lateral branchlets. Sepals 5, 3–4 mm long, somewhat variable in shape, usually broadest just above the middle, apically acute or short-acuminate, margins ciliate, glandular in lines on the outer surface. Petals 5, 5–8 mm long, obliquely oblong-elliptic to obovate, orange-yellow. Capsules ovate-conical or oval, 4–6 mm long, exceeding the calyx, dark reddish brown, styles 3, spreading. Seeds yellow, about 0.5 mm long, oblong, faintly reticulate.

Pine savannas and flatwoods, seepage slopes in pinelands, bogs, sometimes in open well-drained mixed woodlands. Coastal plain, s.e. Va. to cen. pen. Fla., westward to s.e. Tex.

2. Hypericum drummondii (Grev. & Hook.) T. & G.

Annual. Stem solitary, with strongly ascending wing-angled branches, punctate-dotted. Leaves ascending, linear-subulate, 1-nerved, 5–18 mm long and 0.5–0.8 mm wide, glabrous, punctate-dotted beneath. Flowers alternating on strongly ascending branches (not in cymes), their stalks about 2 mm long, the midribs of the sepals decurrent on them as ribs or wings. Sepals 5, lanceolate, 4–6 mm long, acute apically, the midrib elevated, their backs marked with both dots and lines. Petals 5, orange-yellow, 2.5–4.5 mm long. Capsules ovoid, equaling or little exceeding the calyx, styles 3. Seeds dull brown, short-oblong, about 0.8 mm long, shallowly pitted-reticulate. (*Sarothra drummondii* Grev. & Hook.)

In both well-drained and seasonally wet places, fields, open woodlands, clearings of bottomland woodlands and flatwoods, swales, along shores of ponds and lakes. Va. to Ohio, westward to Iowa and s.e. Kan., generally southward to n. Fla. and cen. Tex.

3. Hypericum adpressum Bart.

Rhizomatous, colonial, perennial herb, when growing in water the submersed stems spongy. Aerial stems usually simple below the inflorescence, 3–8 dm tall. Leaves numerous, narrowly oblong, lanceolate, or narrowly elliptic, acute or blunt apically, mostly 3–6 cm long, midrib depressed on the upper surface, strongly elevated beneath, margins revolute, lower surface pale green, conspicuously reticulate-veined (as observed with some magnification) upper surface glossy green. Lower portions of cymes leafy. Sepals 5, ovate to oblong, brown midribs strongly elevated on the outer surface, pale green either side of the midrib, margins strongly involute distally, sometimes their entire length. Petals 5, 6–8 mm long, long persisting and closely enveloping the maturing capsule. Styles closely connivent and appearing as one. Capsules ovoid, 3.5–4.5 mm long.

Marshy shores, wet meadows, bogs, swales, ditches. Local, coastal plain, e. Mass. to Ga.; also W. Va., Ind., Ill., cen. Tenn.

4. Hypericum mitchellianum Rydb.

Perennial. Stems erect, 1 to several from a short rhizomatous base, 4–6 dm tall, the lower and medial strongly ascending; inflorescence branches from the axils of unreduced leaves. Leaves oblong-elliptic, the larger ones 4–6 cm long and 1–2.5 cm wide, rounded, sometimes clasping basally, rounded apically, punctate-dotted on both surfaces, not revolute. Flowers 15–25 mm across. Sepals 5, unequal, long-triangular, lanceolate, or elliptic, 3–6 mm long, acute to obtuse apically, marked with black lines and dots. Petals 5, copper-yellow and marked with black lines and dots, oval, 6–12 mm long. Styles 3, 3–4 mm long. Capsules ovoid, 5–7 mm long. Seeds yellowish to brown, oblong, about 1 mm long, finely reticulate.

Seepage slopes, spray areas by falls, moist banks, moist to wet ditches. High elevations, mts. of Va., Tenn. and N.C.

5. Hypericum graveolens Buckl.

In general appearance closely similar to *H. mitchellianum*. Flowers larger, about 30 mm across. Sepals and petals without black lines or spots, petals 10–18 mm long. Styles 7–10 mm long.

Seepage slopes, wet ditches, balds. High elevations, mts. of N.C. and Tenn.

6. Hypericum denticulatum Walt.

Perennial, sometimes producing numerous stems from basal offshoots, exhibiting much variation in the leaves. Stems usually slender, unbranched below the inflorescence, 2–6 dm tall. Leaves numerous, ascending, commonly about equaling the internodes in length but on some plants much exceeding the internodes, 1-nerved; lanceolate, lance-subulate, narrowly elliptic, oblong-elliptic, or short-elliptic, bases tapering to truncate, mostly acute apically, punctate-dotted beneath or both beneath and above. Cymes open to dense. Sepals 5, oblong, lanceolate, or elliptic, acuminate apically, unequal, 2.5–6.5 mm long, punctate dotted distally. Petals 5, copper-yellow, 4–10 mm long, persistent and becoming inflexed around the capsule. Stamens numerous. Styles 3, 2–4 mm long. Capsules ovoid, 2.5–4 mm long. Seeds dark brown, semilustrous, faintly reticulate. (Incl. *H. acutifolium* Ell.)

Shallow soil on granite rock outcrops, pine savannas and flatwoods, cypress-gum ponds where often in shallow water, bogs, pineland seepage slopes, pine-oak barrens, clearings, pastures. N.J. to Ohio and s. Ill., southward to Fla. Panhandle and Miss.

7. Hypericum canadense L.

Annual, or sometimes perennating by short leafy stolons. Stem very slender, erect, 0.5–7 dm tall. Leaves linear or linear-oblanceolate, 1–3-nerved, 1–3 cm long, 1–3 (–4) mm wide, narrowed basally, blunt to acute apically. Cymes few- to numerous-flowered. Sepals 5, variable, ovate-triangular, somewhat triangular-subulate, lanceolate, or elliptic, acute or acuminate apically, 2.5–4.5 mm long, obscurely if at all punctate-dotted. Petals 5, yellow, 6–8 mm long, oblong to oblanceolate. Stamens few. Styles 3, 0.5–1 mm long. Capsules ovoid-conical, purplish, 5–6 mm long, somewhat exceeding the calyx. Seeds buffish, narrowly oblong, about 0.5 mm long, very numerous, very faintly reticulate, semilustrous.

Bogs, sandy or muddy shores, wet meadows, wet ditches. Nfld. and Que. to Minn., southward to n. Fla. and Ala.

8. Hypericum gymanthum Engelm. & Gray.   Fig. 152

Perennial. Stems solitary and simple or 2 to several simple ones from the base, erect, slender, 1–6 dm tall, the cymes usually well aloft of the foliage leaves, thus appearing "naked" (relative to the following species). Lower leaves much smaller than the medial and upper ones; medial and upper leaves firm, ovate-triangular, usually clasping, pointed apically, 5–7-nerved, about 1.5 cm long and 1 cm wide at the base, finely punctate-dotted, usually on both surfaces. Sepals 5, 3–5 mm long, lanceolate, long-acute to acuminate apically, oil tubes showing as lines. Petals 5, yellow, 3–6 mm long. Stamens few. Styles 3, 0.5–1 mm long. Capsules ellipsoid-conical, 3–4 mm long, equaling or slightly exceeding the calyx. Seeds buffish, minute, oblong, about 0.3 mm long, very faintly reticulate.

Bogs, sandy or muddy shores, marshy shores, savannas, cypress depressions in flatwoods, swales, glades. Coastal plain, s. N.J. to n. Fla., westward to s.e. and cen. Tex.; inland, Pa., W.Va., Tenn., Ohio to Ill., s. Mo., e. Kan.

9. Hypericum mutilum L.   Fig. 153

Perennial. Stems solitary or several to numerous from the base, each commonly diffusely branched, the lower floriferous branches, at least, borne from the axils of unreduced leaves,

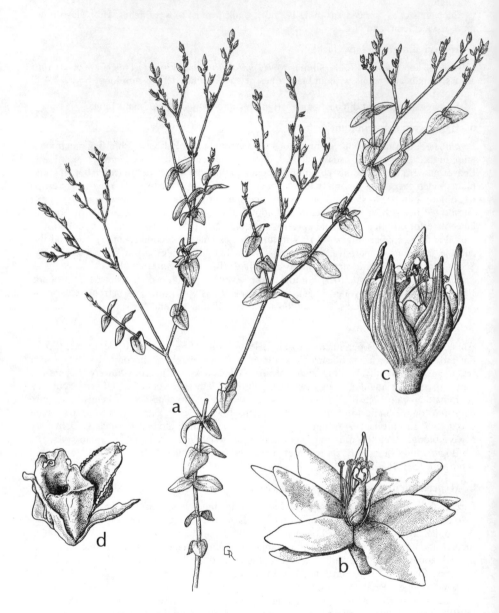

Fig. 152. **Hypericum gymnanthum:** a, top of plant; b, flower, at full anthesis; c, flower, after anthesis; d, capsule, after dehiscence.

338

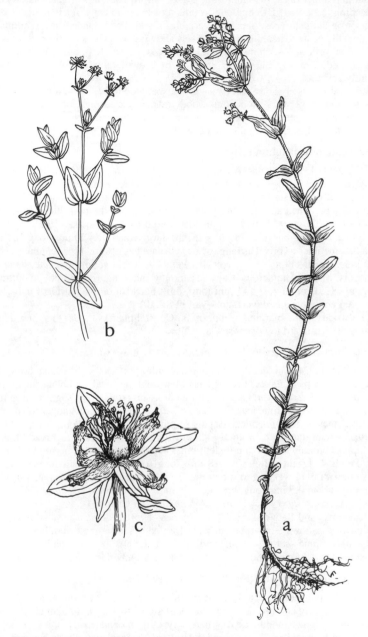

Fig. 153. **Hypericum mutilum**: a, habit; b, top of plant; c, flower. (From Correll and Correll. 1972)

the overall inflorescence thus appearing leafy at least below. Lower leaves much smaller than the medial and upper ones, the latter thin, ovate, elliptic, oblong, or lanceolate, sometimes partly clasping, 5-nerved, 1–5 cm long, rounded or blunt apically, finely punctate dotted on the lower or on both surfaces. Sepals 5, variable, linear, linear-lanceolate, or elliptic-oblanceolate, apically acute to acuminate, oil vessels showing as lines, sometimes as dots as well. Petals 5, light yellow, 2–3 mm long. Stamens few. Styles 3, 0.5–1 mm long. Capsules oblong-ellipsoid, equaling the calyx. Seeds light brown, minute, oblong, about 0.4 mm long, faintly reticulate.

Marshy and exposed shores, wet meadows, wet woodlands and swamps, open stream banks, sand and gravel bars, floodplain woodlands, wet clearings, on floating logs and in islands of floating vegetation, alluvial outwash, swales, ditches. Nfld. and Que. to Man., generally southward to s. Fla. and e. and cen. Tex.

10. Hypericum tetrapetalum Lam.

Shrub 1–10 dm tall. Stem slender, older portion with thin, reddish brown bark exfoliating in thin, grayish, irregular strips or flakes. Leaves ovate to oblong-ovate, the larger ones 1–2 cm long, subcordate or cordate basally, obtuse apically, punctate-dotted on both surfaces. Openly dichotomously branched above, solitary flowers terminating the branchlets. Flowers 2.5–3 cm across or a little more. Sepals 4, punctate-dotted on both surfaces, the outer 2 broader than the inner, varying considerably in size, the larger ones 15–20 mm long and about 12 mm broad, sometimes as broad as long, ovate, cordate basally and broadly obtuse to rounded apically; inner sepals elliptic, acute apically. Petals 4, bright yellow, obovate, much exceeding the sepals. Stamens numerous. Capsules about 6 mm long, oval-elliptic in outline, styles 3, diverging. Seeds oblong, 0.8–1 mm long, dark blackish brown, surfaces reticulate, sublustrous. (*Ascyrum tetrapetalum* (Lam.) Vail in Small)

Pine flatwoods, pond margins, wet prairies. Coastal plain, s.e. Ga. to s. pen. Fla., in the Fla. Panhandle westward to Okaloosa Co.; Cuba.

11. Hypericum crux-andreae (L.) Crantz. ST.-PETER'S-WORT.

Shrub 3–10 dm tall, not colonial. Stem slender, older portions with reddish brown bark exfoliating in thin strips or flakes. Leaves lanceolate, oblong, oval, or oblanceolate, the larger 2–3 cm long, bases mostly rounded or subtruncate, sometimes very slightly clasping, rarely cuneate, apices obtuse or rounded. Flowers 2–3 cm across, solitary and terminal or axillary, or in small cymes on the branchlets. Sepals 4, punctate-dotted on both surfaces, the outer 2 much broader than the inner, ovate, 1–1.5 cm long and commonly as broad basally, bases cordate, apices broadly obtuse to rounded; inner sepals lanceolate or narrowly elliptic, acute apically. Petals 4, bright yellow, 10–18 mm long, obovate. Stamens numerous. Capsules narrowly ovate or oval, varying to subglobose, 7–10 mm long, styles 3, rarely 4, diverging. Seeds oblong, about 0.8 mm long, brown, reticulate, sublustrous. (*H. stans* (Michx.) Adams & Robson, *Ascyrum stans* Michx.; incl. *A. cuneifolium* Chapm.)

Pine savannas and flatwoods, adjacent ditches, cypress depressions, swales, marshy or boggy shores, meadows, sometimes in well-drained upland mixed woodlands and on deep sands of longleaf pine–scrub oak hills and ridges. L.I., N.J., e. Pa., southward to cen. pen. Fla., westward to e. Tex., s. and cen. Ark., e. Okla.; Cumberland plateau, n. Ala. to s.e. Ky.

12. Hypericum edisonianum (Small) Adams & Robson.

Shrub to 15 dm tall, commonly with adventitious shoots arising from horizontal roots and becoming extensively colonial. Older portions of stems to 1.5 cm in diameter, bark grayish to nearly black, tight, smooth. Leaves narrowly elliptic or narrowly oblong-elliptic, acute apically, the larger ones 1.5–2.5 cm long, heavily glaucous above, less so beneath, punctate-dotted beneath, edges thickened and somewhat turned downward, a gland on each of the shortly auricled leaf bases. Flowers about 2 cm across. Sepals 4, glaucous, punctate-dotted, edges thickened, outer broader than the inner, ovate, acute or short-acuminate apically, mostly 10–12 mm long; inner sepals broadest at the base, tapering to subulate tips. Petals 4, yellow, 10–18 mm long, obovate. Stamens numerous. Capsules ovoid, 5–8 mm long, styles 3(4). Seeds oblong, about 0.8 mm long, buffish to brown, reticulate, dull. (*Ascyrum edisonianum* Small)

Pine flatwoods, flatwoods ponds and depressions, wet prairies. Endemic to Highlands, Glades, and Desoto Cos. of pen. Fla.

13. Hypericum microsepalum (T. & G.) Gray ex S. Wats.

A low, straggly to bushy-branched shrub, not exceeding 10 dm tall, commonly much lower, flowering in very early spring. Branches very leafy, flowers solitary at tips of branchlets or in leaf axils near the branch tips. Principal leaves mostly 5–10 (–15) mm long, 1.5–2 mm wide (commonly with very short branchlets bearing much smaller leaves in their axils), linear-oblong, oblanceolate, elliptic, or obovate, bright green above, pale green beneath, margins revolute, basally scarcely narrowed to cuneate, apices rounded to obtuse. Flowers showy, 1.5–2 cm across. Sepals usually 4, subequal, linear-elliptic to oblong-elliptic, obtuse to acute apically. Petals usually 4, yellow, subequal, spreading, usually asymmetric, obovate, broadly rounded apically (occasional flowers on some plants may have 5-merous calyces and corollas, or 3-merous calyces and 5-merous corollas, or 5-merous calyces and 3-merous corollas). Stamens numerous. Capsules conic, narrowly ovate, or elliptic-oblong, 3–4 mm long, styles 3. Seeds oblong, about 1 mm long, dark brown and lustrous, finely pitted-reticulate. (*Crookea microsepala* (T. & G.) Small)

Moist to wet pine flatwoods. Common and abundant in the Fla. Panhandle from Madison and Taylor Cos. westward to Walton Co. and in s.w. Ga.

14. Hypericum hypericoides (L.) Crantz. ST. ANDREW'S CROSS.

Plant erect, with a single main stem freely branched above, sometimes with several principal, erect stems from the base each freely branched above, commonly with short, leafy axillary branches, 3–15 dm tall; bark of older portions of stems reddish brown, sloughing or shredding in thin plates or strips. Leaves variable, narrowly oblanceolate, linear-oblong, or linear, the larger mostly 8–25 mm long, 1.5–6 mm wide (considerably larger if plants in deep shade), narrowed basally, obtuse to rounded apically, punctate-dotted on both surfaces. Flowers solitary above the pair of leaves terminating an ultimate branchlet, their stalks erect, 2–3 mm long, bearing a pair of narrowly subulate bractlets just below or very little below the calyx. Sepals 4, outer herbaceous, conspicuous, ovate to broadly elliptic, rounded or subcordate basally, obtuse apically, variable in size, 5–12 mm long, 4–10 mm wide, punctate-dotted on both surfaces; inner sepals minute. Petals 4, pale yellow, narrowly oblong-elliptic, mostly 8–10 mm long. Stamens numerous. Capsules ovate to elliptic in outline, 4–9 mm long, styles 2. Seeds oblong, 1 mm long, black, reticulate. (*Ascyrum hypericoides* L.; incl. *A. linifolium* Spach.)

In a wide variety of habitats, open, well-drained upland woodlands, pine barrens on sand hills and ridges, floodplain and wet woodlands, hammocks, moist to wet thickets, pine flatwoods, cypress-gum depressions, bogs. Va. to s.e. Mo. and e. Okla. generally southward to s. Fla. and e. third of Tex.; Berm., Bah.Is., Cuba, P.R., Haiti, Dom.Rep., Jam., e. Mex., Guat., Hond.

15. Hypericum myrtifolium Lam.   Fig. 154

Shrub to about 1 m tall. Stems slender, glaucous, mostly solitary, erect, loosely branched above, bark on older portions glaucous, grayish, sloughing in thin plates, becoming slightly corky if submersed. Leaves numerous, mostly ovate-triangular, 1–3 cm long, basally subcordate to cordate and somewhat clasping, obtuse apically, light green above, paler and slightly glaucous beneath, margins usually becoming narrowly revolute in drying, lower surface finely punctate-dotted. Flowers in cymes terminating the branchlets, 2–2.5 cm across. Sepals 5, glaucous beneath, subequal, ovate to elliptic, 8–10 mm long, acute apically, punctate-dotted on both surfaces. Petals 5, yellow, obovate, 10–12 mm long. Stamens numerous. Capsules ovate or ovate-conic, very dark brown to almost black, about 8 mm long, broadly

Fig. 154. **Hypericum myrtifolium:** a, base of plant; b, top of plant; c, flower; d, capsule, dehisced.

342

depressed along the 3 sutures and convex between the depressions, styles 3, erect. Seeds narrowly oblong, about 1 mm long, apiculate at both extremities, surfaces very dark brown, reticulate.

Pine flatwoods, cypress-gum depressions and ponds, borders of ponds (on exposed shores to high-water mark or in shallow water depending upon water level), bogs. Coastal plain, Ga. to s.cen. pen. Fla., westward to s.e. Miss.

16. Hypericum brachyphyllum (Spach) Steud.

Shrub, stem usually solitary, erect, bushy-branched above, usually 5–10, sometimes to 15, dm tall. Stem diameter near the base to about 1.5 cm, bark relatively thin and tight, exfoliating in small strips or plates. Principal leaves with very short axillary branchlets having smaller leaves giving a fasciculate appearance to the nodes. (See key for internodal characters.) Leaves needlelike, very strongly revolute, the margins usually folded tightly around and against the lower surface and together covering about ⅔ of the lower surface, the revolute margins (beneath) with relatively large punctate dots, these often appearing there in a longitudinal line, some punctate dots scattered on the rest of the upper surface; larger leaves 6–10 mm long. Flowers sometimes solitary in leaf axils, mostly in short, leafy cymes on branchlets, overall abundantly floriferous. Sepals 5, essentially like the leaves in form, 2–3.5 mm long. Petals 5, bright yellow, spatulate to obovate, mostly about 8 mm long. Stamens numerous. Capsules narrowly conical, narrowly depressed along the sutures, 4–5 mm long excluding the styles; styles 3 (rarely 4), usually erect, 3 mm long. Seeds short-oblong, about 0.4 mm long, very dark brown, surfaces very finely alveolate, the alveolae round or nearly so and not obviously in longitudinal rows.

Wet pine flatwoods and savannas, pond margins, cypress-gum ponds and depressions, borrow pits, wet ditches. s. Ga., to s. Fla., westward to s. Miss.

17. Hypericum reductum P. Adams. Fig. 155

Shrub, usually bushy branched from the base, often with numerous stems from decumbent bases and more or less mat-forming, a single plant as much as 1 m across; stems usually not exceeding 5 dm tall, often 1–2 dm. Stem diameter seldom reaching 1 cm, bark relatively thin and tight, exfoliating in thin plates and strips. Leaves as in *H. brachyphyllum*, the larger ones usually not exceeding 5 mm long, the fasciclelike appearance of the nodes generally more pronounced. (See key for characters of the internodes.) Branches abundantly floriferous, more flowers solitary in the leaf axils than in *H. brachyphyllum*, many of the branchlets thus appearing racemose, some flowers in short cymes as well. Sepals 5, 2–3 mm long. Petals 5, bright yellow, obovate, 7–8 mm long. Stamens numerous. Capsules narrowly oblong but shortly tapered at the tip, 4–8 mm long excluding the styles; styles 3, straight, 1–1.5 mm long. Seeds short-oblong, about 0.4 mm long, usually black when fully mature, alveolate, the alveolae relatively fewer and larger than in *H. brachyphyllum*, squarish, in longitudinal rows.

For the most part inhabiting well-drained sandy soils, longleaf pine–scrub oak hills and ridges, sand pine–scrub oak hills and ridges, old stabilized coastal dunes; also moist to wet interdune hollows or swales, outer sandy shores of ponds where sometimes in water at least for short periods. Local, coastal plain, s.e. N.C. to cen. pen. Fla., Fla. Panhandle, s.e. Ala.

18. Hypericum lissophloeus P. Adams.

Shrub to 4 m tall, the stems limber, to 4.5 cm thick near the base, the floriferous branches thin-flexuous, ascending, the flowers mostly solitary in the leaf axils, occasionally in 3-flowered cymes. Leaves articulate at the base, linear-subulate and needlelike, the larger ones mostly (10–) 12–17 mm long, strongly glaucous, the edges strongly curved downward, in drying flattened against the lower surface, punctate-dotted mostly along the revolute edges. Sepals 5, articulate at the base, subulate, 7–8 mm long, revolute, deciduous by the time the fruits are fully ripe. Petals 5, bright yellow, asymmetrical, obovate, with a tooth obliquely placed to one side of the summit, 10–12 mm long. Stamens many. Styles 3, rarely 4, con-

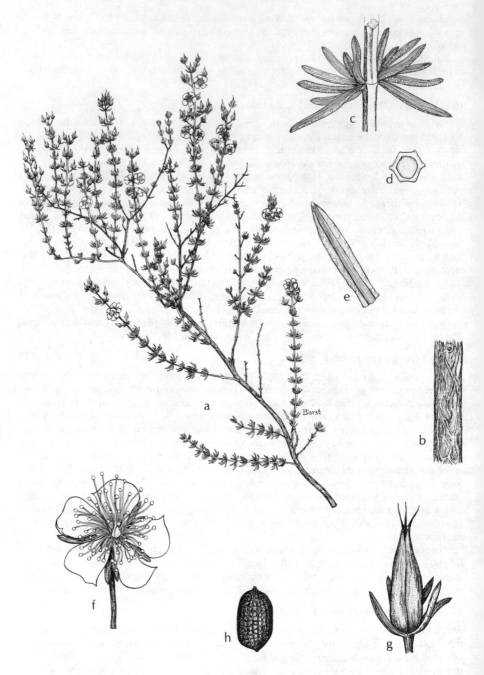

Fig. 155. **Hypericum reductum:** a, flowering/fruiting branch; b, portion of lower stem; c, node; d, diagrammatic cross-section of twig; e, lower surface of leaf; f, flower; g, capsule; h, seed.

344

nivent at anthesis. Capsules ovate-conic, narrowly depressed along the sutures, lustrous reddish brown, 6–8 mm long, styles slender, erect, about 4 mm long, often breaking off near the base by maturity of the capsules. Seeds oblong, tan to brown, 1.5 mm long or a little more, surfaces longitudinally markedly striate, with faint cross striae between.

Locally abundant, restricted to shores of sinkhole ponds and lakes, on exposed shores to the high-water mark and at high water levels in water to 1.5 m deep. Endemic to Bay and Washington Cos., Fla. Panhandle.

19. Hypericum exile P. Adams.

Slender shrub, stems wandlike, limber, usually less than 10 dm tall, with 2 to few ascending branches above. Bark thin, reddish brown, exfoliating in very thin, small, irregular flakes or strips. Leaves articulate at the base, spreading, linear-subulate and needlelike, (10–) 15–25 mm long, the edges strongly rolled downward, on drying tightly folded against the lower surface and punctate-dotted. Distally, the stems or main long branches with short floriferous branchlets, the flowers in 3–7-flowered cymes, the inflorescence overall having the aspect of a compact narrow panicle. Sepals 5, articulate at the base, subulate, not revolute, 6–7 mm long, usually deciduous at about the time the fruits are fully mature. Petals 5, bright yellow, asymmetrical, nearly elliptic, rounded at the tip, a tooth on one side somewhat below the tip, 6–7 mm long. Stamens many. Styles (2)3(4), connivent at anthesis. Capsules about 7 mm long excluding the beak, lustrous reddish brown, long-conical in outline, deeply depressed along the sutures, the styles persistent as a beak or spreading, 3 mm long. Seeds reddish brown, lustrous, oblong, 0.4–0.6 mm long, surfaces very finely reticulate.

Pine savannas and flatwoods, usually where soils remain water-saturated only for short periods, occasionally where wet for extended periods. Endemic to Liberty, Franklin, Gulf, Bay, and Washington Cos., Fla. Panhandle.

20. Hypericum nitidum Lam.

Plants commonly with several to numerous stems from the base, each bushy-branched above, overall relatively broad and dense, the main stems to 3–4 cm thick near the base, potentially 2 3 m tall. Bark thin and tight, brown, reddish, or gray, exfoliating in small thin flakes or strips. Leaves articulate at the base, linear-subulate, needlelike, spreading, the larger ones 15–25 mm long, the edges strongly curved around and in drying becoming flush against the lower surface, punctate-dotted. Sepals 5, articulate at the base, linear-subulate and needlelike, strongly revolute like the leaves, 3–4 mm long, usually deciduous by the time the fruits are fully mature. Petals 5, bright yellow, 5–6 mm long, obovate, asymmetrical, a tooth obliquely placed to one side of the rounded summit. Stamens many. Styles 3, sometimes connivent at anthesis. Capsules oblong-conic, 3–4 mm long, deeply depressed along the sutures, dull reddish brown, the slender styles about 4 mm long, usually breaking off about 1 mm from the base by maturity of the capsules. Seeds dull brown, 0.4–0.5 mm long, surfaces finely alveolate-reticulate, the alveolae roundish, not in obvious rows.

Wet pine flatwoods, open shores of blackwater streams draining flatwoods or shrub-tree bays or bogs, occasionally in broad, billowing masses in fresh water at shores of upper reaches of blackwater estuaries and their backwater basins, ditches, borrow pits, bogs. Local, Brunswick Co., N.C., Lexington Co., S.C., more general and sometimes abundant in s. Ga., Fla. Panhandle, s.w. Ala.

21. Hypericum fasciculatum Lam.

Erect shrub, to 1.5–2 m tall, much branched above (not conspicuously dwarf-tree-like as is usually the case with older specimens of the following species). Youngest internodes with a winged ridge on each side, not terete; bark on older stems corky, exfoliating in tissue-thin sheets, exposed portions buff or cinnamon-colored; bark sloughing freely and never attaining great thickness or notable soft-thickness on old basal portions. Principal leaves with short branchlets with smaller leaves in their axils giving a fasciculate appearance to the nodes. Leaves spreading to ascending, articulate at the base, linear-subulate and needlelike, the larger ones 13–26 mm long, revolute, punctate-dotted above and on the revolute portions beneath. Flowers mostly in 3–26-flowered cymes terminal or terminal and axillary on the branchlets, some sometimes solitary in the leaf axils. Sepals 5, similar to the leaves, (3–) 4.5–7 mm long, usually persistent below the fruits. Petals 5, bright yellow, 6–9 mm long, asymmetrical, obovate, a tooth obliquely to one side of the tip. Stamens many. Styles 3, sometimes connivent at anthesis. Capsules ovate-conic in outline, deeply depressed along the sutures, dark reddish brown, 3–5 mm long, the slender styles spreading, about 4 mm long, usually breaking off near the base by the time the capsules are fully ripe. Seeds oblong about 0.4 mm long, finely alveolate-reticulate, alveolae roundish, not in obvious lines.

Shores of ponds and lakes, cypress-gum ponds and depressions, wet flatwoods, ditches, borrow pits, bogs, open banks of streams, commonly in water. Coastal plain, e. N.C. southward from about the Neuse River, southward to s. Fla., westward to s. Miss.

22. Hypericum chapmanii P. Adams.

Erect shrub, to 2–3 (–4) m tall, usually with a single main stem, bushy-branched above, older plants with a markedly dwarf-tree-like appearance. Bark soft, spongy, with conspicuous vertical, resin-filled lactifers at discrete levels between cork layers 3–4 mm thick, large size of lactifers giving torn bark a striated appearance; cork buffish, reddish brown, or cinnamon-colored within, weathering grayish, bark often attaining 3–4 cm in thickness on old stems, soft-corky even on small stems. Leaves and sepals much as in *H. fasciculatum*. Flowers relatively few, solitary or in 3-flowered cymes in the axils of leaves near the tips of branchlets. Capsules ovate, about 6 mm long, depressed along the sutures, styles about 4 mm long, usually breaking off by the time capsules are fully ripe. Seeds oblong, 0.6–0.8 mm long, brown, surfaces very finely alveolate-reticulate, alveolae roundish, not in evident lines.

Wet pine flatwoods depressions, cypress-gum ponds or depressions, borrow pits. Endemic to coastal portion of Fla. Panhandle from a little e. of the Ochlockonee River to Santa Rosa Co.

23. Hypericum galioides Lam. Fig. 156

Shrub, mostly 1–1.5 m tall, bushy-branched, branchlets slender, flexuous, and usually spreading. Leaves articulated at the base, narrow, the larger ones mostly 2–3 cm long and 7 mm wide or less, narrowly oblanceolate (sometimes the sides rolling strongly downward in drying and then appearing linear), tapered below to a short-petiolate or subpetiolate base, apically rounded or blunt, upper surfaces punctate-dotted, lower sometimes with evident lateral veins. Main stem leaves with short branchlets in their axils. Flowers about 1.5 cm across, mostly in short axillary cymes, sometimes the distal portions of branches not very leafy, heavily floriferous, and paniclelike, often some flowers solitary in leaf axils. Sepals and petals 5. Sepals narrowly oblanceolate, 3–4 mm long, articulated at the base. Petals bright yellow, obovate, with a small tooth at one side of the rounded apex. Stamens numerous. Capsule obconic, 4–5 mm long, the 3 persistent styles 2 mm long. Seeds oblong, 0.4–0.8 mm long, dark brown, faintly reticulate. (Incl. *H. ambiguum* Ell.)

Cypress-gum swamps and depressions, floodplain forests, river and stream banks, wet woodlands, pine savannas and flatwoods, moist to wet clearings, shores of ponds and lakes, ditches. Coastal plain, N.C. to n. Fla., westward to s.e. Tex.

24. Hypericum densiflorum Pursh. Fig. 157

Shrub to 2.5 m tall, with numerous, erect-ascending, stiffish branches above. Leaves articulated at the base, very variable in shape and size, linear, linear-oblanceolate, linear-elliptic, linear-oblong, or narrowly elliptic, the larger ones 3–5 cm long and 3–12 mm wide, cuneate basally, sometimes subpetiolate, the broader ones rounded apically, the narrower ones acute, punctate-dotted above, very rarely with evident lateral veins beneath. Main stem leaves usually with short lateral branches in their axils. Flowers about 1.5 cm across, in compound cymes terminating the branchlets. Sepals and petals 5. Sepals articulated at the base, oblong,

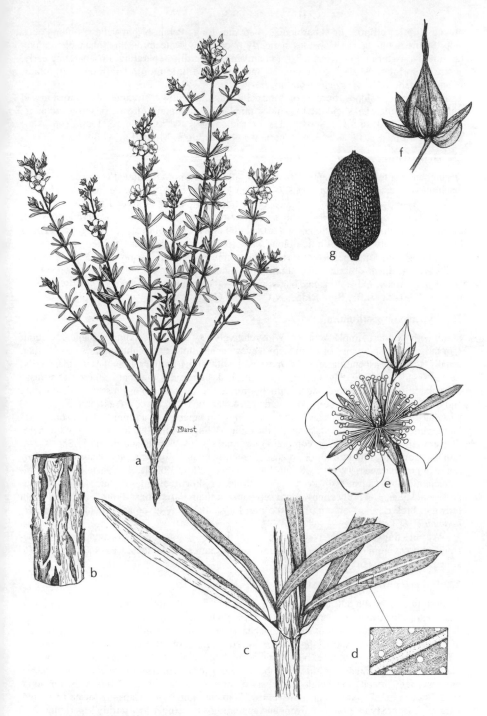

Fig. 156. **Hypericum galioides:** a, flowering/fruiting branch; b, portion of lower stem; c, node; d, small section of upper leaf surface; e, flower; f, capsule; g, seed.

347

oblong-elliptic, elliptic, or suborbicular, 3–5 mm long. Petals bright yellow, oblong-obovate. Stamens numerous. Capsules slenderly obconic to ovate in outline, channelled along the sutures, 4–5 mm long, styles 3–5, connivent and tardily separating, or separating early, about 3 mm long. Seeds linear-oblong, 1 mm long or a little more, dark brown, alveolate-reticulate. (Incl. *H. lobocarpum* Gattinger)

Bogs, seepage slopes, borders of ponds and lakes, meadows, stream banks, moist to wet thickets, roadside banks and ditches, moist pinelands, floodplain forests, mixed woodlands. Coastal plain, N.J. to S.C.; Appalachian mt. region, s.w. Pa. to n. Ga., e. Tenn., cen. Ala.; interior lowlands, s. Ill., s.e. Mo. generally southward to s. Miss., La., e. Tex.

25. Hypericum buckleyi M. A. Curtis.

Decumbent matted shrub, stems 0.5–4.5 dm tall. Leaves not articulated at the base, elliptic or elliptic-obovate, thin, larger ones 2–2.5 cm long and 5–12 mm wide, narrowed to subpetiolate bases, apices rounded, finely, often obscurely, punctate-dotted on both surfaces. Flowers 1.5–2 cm across, mostly solitary and terminal on the branchlets, sometimes in small cymes. Sepals and petals 5. Sepals not articulated at the base, obovate, 5–7 mm long. Petals bright yellow, oblong or elliptic-oblong, 10–15 mm long. Stamens numerous. Capsules ovate, 8–12 mm long, persistent styles 3 but connivent and appearing as 1, 3–4 mm long. Seeds brown, linear-oblong, about 2 mm long, finely alveolate-reticulate.

Seepage areas, moist crevices, balds, moist road embankments, ditches. Mostly at elevations of 3000–5000 ft., Blue Ridge, N.C., S.C., Ga.

26. Hypericum cistifolium Lam.

Shrub with slender, simple stems, or with relatively few, erect-ascending branches, to about 1 m tall. Successive pairs of main stem leaves conspicuously at right angles to each other, usually strongly ascending, for the most part with short axillary branches in their axils. Leaves sessile, not articulated at the base, 1-nerved, firm and somewhat leathery and stiffish, lanceolate, linear-lanceolate, lance-oblong, or oblong, the larger ones mostly 1.5–3 cm long and 6–7 mm wide, rarely somewhat wider, broadest near the base, the extreme base somewhat auriculate, the auricles declining, most of them tapering to a point distally, sometimes blunt, punctate-dotted on both surfaces, the midrib below continuous into a wing of the stem. Flowers 10–12 mm across, in compound terminal cymes. Sepals and petals 5. Sepals 2–3 mm long, oblong, oblong-elliptic, or less frequently ovate or obovate, the shape often differing in a single flower. Petals 5–8 mm long, bright yellow, obliquely cuneate-obovate. Stamens numerous. Capsules ovate to short-oblong, or shortly elliptic-oblong, 4–5 mm long, persistent styles 3, wholly connivent or separated at their extremities, 2 mm long. Seeds light brown to brown, short-oblong, 0.4–0.5 mm long, shallowly alveolate-reticulate. (Incl. *H. opacum* T. & G.)

Wet pine flatwoods and savannas, adjacent ditches, bogs, boggy seepage slopes of pinelands, cypress-gum depressions, interdune swales, banks of marshes. Coastal plain, N.C. to s. pen. Fla., westward to e. Tex.

27. Hypericum nudiflorum Michx.

Erect shrub, 5–20 dm tall, usually loosely branched. Leaves not articulated at the base, thin, lanceolate, lance-oblong, narrowly elliptic, or oblanceolate, the larger ones 3–6 cm long, 1–2 (–2.5) cm broad, without short branches in their axils; abruptly narrowed from the base of the green blade to a short, appressed, winged segment of different texture, apices mostly rounded, some broadly obtuse, both surfaces finely punctate-dotted, midrib beneath continuous into the wing of the stem. Flowers about 1.5 cm across, in compound cymes terminating the branches. Sepals and petals 5. Sepals oblong, 2–3 mm long. Petals 6–8 mm long, broadly cuneate-obovate, copperish yellow. Stamens numerous. Capsules ovate, 5–9 mm long, persistent styles 3 but connivent and appearing as 1, sometimes tardily separating, 2–3 mm long. Seeds very dark brown, almost black, linear-oblong, 1.5–2 mm long, finely but markedly alveolate-reticulate. (Incl. *H. apocynifolium* Small)

Fig. 157. **Hypericum densiflorum:** a, flowering branch; b, flower; c, stamen; d, fruiting branch; e, fruiting branch (enlarged); f, seed. (From Correll and Correll. 1972)

Along woodland streams, wooded river banks, wooded bottomlands, richly wooded, well-drained slopes. Coastal plain, s.e. Va., chiefly inner coastal plain and outer piedmont, N.C. and S.C., thence to Fla. Panhandle, westward to s. Ark., s.e. Okla., and e. Tex.; Cumberland plateau.

Theaceae (CAMELLIA FAMILY)

Gordonia lasianthus (L.) Ellis LOBLOLLY BAY.

A small to medium-sized tree with a relatively narrow, conical to columnar crown. Bark of trunk dark gray, roughened by interlacing, flat-topped ridges separated by rough, narrow furrows; twigs dark brown and somewhat glaucous when fresh, those of the season sparsely pubescent, the hairs mostly with 2–4 short, erect branches from the base, eventually becoming glabrous. Leaves alternate, simple, persistent, very short-petiolate, blades leathery, pinnately veined, mostly long-elliptic, 8–16 cm long and 3–5 cm broad, cuneate basally, apices obtuse to acute, sometimes notched, margins very shallowly appressed crenate-serrate, upper surfaces glabrous, dark green, the lower paler, more or less olive green, sparsely and usually persistently pubescent with hairs as on the twigs. Flowers handsome, solitary, axillary to close-set leaves on twigs of the season, their stiff stalks 5–8 cm long, usually several on a given twig but blooming one at a time, usually a relative few at one time on the tree as a whole, the flowering period extended over a number of weeks in summer. Flowers about 8 cm across; sepals short-clawed at base, their blades suborbicular, silky-pubescent exteriorly; petals 5, white, united basally, crinkly-fringed, their broadly rounded tips turned up, silky-pubescent exteriorly; stamens numerous, yellow, filaments basally forming a fleshy, deeply 5-lobed ring, each lobe of the ring adnate to or flush against the base of a petal; pistil 1, ovary superior, ovoid, 5-locular. Fruit a woody, ovate-oblong capsule about 1.5 cm long, its surface appressed silky-pubescent, dehiscing loculicidally, each locule with 4–8 flat, winged seeds about 1 cm long.

Evergreen shrub-tree bogs and bays, pond-cypress depressions, swamps. Coastal plain, N.C. to about n. of Lake Okeechobee, pen. Fla., westward to La.

Violaceae (VIOLET FAMILY)

Viola (VIOLETS)

Annual or perennial herbs (ours), most perennials rhizomatous or stoloniferous or both. Leaves alternate, in many very closely set on a basal, short caudex or at the tips of rhizomes, in others on aerial stems, petiolate, simple (in some lobed to deeply divided), stipulate. Early vernal leaves usually having shorter petioles and smaller blades than later vernal ones. Flowers in most species of 2 kinds, those of early season (vernal) with showy petals and mostly fertile, those of later season apetalous, self-pollinated in the bud and fertile (cleistogamous), their capsules usually bearing more numerous seeds than do those of vernal, petaliferous flowers. Flowers borne singly on 2-bracted stalks from leaf axils or directly from rhizomes or stolons, bisexual. Sepals 5, subequal to unequal, usually auricled at base, usually persistent. Petals (of vernal flowers) 5, strongly bilaterally symmetrical, the lowermost spurred at the base. Stamens 5, with short, distinct filaments, the 2 lowermost having nectaries extending into the spur of the lowest petal, anthers converging laterally around the ovary. Pistil 1, ovary superior, 1-locular, with 3 parietal placentae, each bearing several ovules; style 1, clavate, stigma obliquely positioned at the summit. Fruit a loculicidal, 3-valved capsule.

Violets in general have for many people a special aesthetic charm. For the taxonomist, specimens are frequently difficult to identify. Apparently interspecific hybridization is common and not a little introgression has occurred, blurring distinctions. For us, choosing what species to include for wetlands and how to circumscribe them was an agonizing chore. Our treatment, below, is perhaps overconservative.

1. Plant with leafy aerial stems.
  2. Petals pale violet.   1. *V. conspersa*
  2. Petals bright yellow with purple lines near the base, cream-white, or ivory.
    3. Petals bright yellow with purple lines near the base.   2. *V. pubescens*
    3. Petals cream-white or ivory.   3. *V. striata*
1. Plant with leaves closely clustered at the tips of the rhizome.
  4. Petals white (exclusive of the venation); plant stoloniferous at least during part of the season.
    5. Leaf blades lanceolate or linear-lanceolate.   4. *V. lanceolata*
    5. Leaf blades lance-ovate, ovate, suborbicular, or reniform.
      6. The leaf blades markedly cordate-ovate to reniform.   5. *V. pallens*
      6. The leaf blades mostly ovate or lance-ovate, their bases truncate, tapering, or barely subcordate.
                                                 6. *V. primulifolia*
  4. Petals deep to pale violet.
    7. Leaf blades (all of them) unlobed (apart from basal cordate-lobing); trichomes on petals knob- or clavate-tipped.
      8. Trichomes on petals knob-tipped; corolla with a dark eye centrally; sepals 8–12 mm long.
                                             7. *V. cucullata*
      8. Trichomes on petals clavate distally; corolla without a dark eye centrally; sepals 6–8 mm long.
                                               8. *V. affinis*
    7. Leaf blades both unlobed and lobed; trichomes on petals neither knob- nor clavate-tipped.
      9. Lateral petals bearded, spurred petal glabrous.   9. *V. esculenta*
      9. Lateral and spurred petals bearded, usually sparsely pubescent on the upper petals as well.
                                           10. *V. septemloba*

1. Viola conspersa Reichnb.

Perennial with a short, sometimes branched rhizome bearing several to numerous leafy, glabrous stems commencing to flower when very short, eventually reaching as much as 2 dm tall. Blades of lower leaves mostly reniform to reniform-orbicular, usually becoming cordate-ovate upwardly, 2–4 cm long, minutely crenate marginally, surfaces glabrous or the upper very sparsely short-pubescent; stipules lanceolate, with a few salient to fimbriate teeth marginally on the proximal ⅓–¾ of their length. Vernal flowers on slender, glabrous stalks surpassing the subtending leaves. Sepals lance-subulate, entire, glabrous. Petals pale violet (rarely white), with darker lines, the lateral pair bearded. Capsule ellipsoid, glabrous, 3–5 mm long. Seed buff, 1.5 mm long.

    Alluvial woodlands, meadows, seepage slopes. Que. and N.S. to Minn., relatively frequent southward to N.J., Pa., Ohio, n. Ind., n.e. Ill. and Wisc., in scattered localities southward; in our range, w. N.C., n.w. S.C., e. Tenn., n. Ala.

2. Viola pubescens Ait.

Perennial with an essentially erect rhizome bearing from its summit 1 to several leafy, pubescent or glabrous, erect or partially prostrate stems 1–3 dm long or a little more. Leaves 1 to several, long-petioled at the base, and 2–4, relatively short-petioled on the upper half of the stem; blades unlobed, orbicular-ovate to reniform or broadly ovate, 3–6 cm long, bases truncate to subcordate, apices short-acuminate, margins crenate-dentate or -serrate, upper surfaces glabrous or sparsely pilose, lower pilose on the veins or both on and between the veins; stipules lanceolate to lance-ovate, their margins entire to finely short-toothed, with or without short, marginal pubescence. Vernal flowers from the upper leaf axils, their sparsely to copiously pilose stalks (sometimes glabrous except on their distal extremities) shorter than to longer than the leaves. Sepals lanceolate, their margins pilose-ciliate throughout, only near their bases, or sometimes pubescent only across their bases. Petals bright yellow with purple lines near the base, the lateral ones bearded. Capsule glabrous to copiously woolly, ovoid,

10–12 mm long. Seed buff, ovoid to elliptic-ovoid, 2–2.5 mm long. (Incl. var. *pubescens* and var. *eriocarpa* (Schwein.) Russell, *V. pensylvanica* Michx.)

Alluvial woodlands and clearings, along streams, meadows, bogs, wooded slopes. N.S. to Man., generally southward to S.C., n. Ga. and Ala., Ark., n.e. Tex.

3. Viola striata Ait.

Perennial, rhizome slender, bearing several to numerous, reclining to erect, clump-forming, glabrous stems 0.6–3 dm long, leafy from the base upwardly. Leaf blades ovate, orbicular-ovate, or reniform, 1.5–5 (–7) cm long, bases cordate to truncate, apices broadly rounded, acute, or acuminate, margins shallowly crenate or crenate-serrate, glabrous or sparsely pubescent along the veins above; stipules more or less foliaceous, their margins saliently serrate or fimbriate. Vernal flowers from upper leaf axils, their glabrous stalks usually longer than the subtending leaves. Sepals with pronounced, descending auricles, lance-subulate above the auricles and with short-ciliate margins. Petals cream-white or ivory, purple-lined below on the spurred one, the laterals markedly bearded, the spurred one less so. Capsule ovoid, glabrous, 4–6 mm long. Seed pale brown, about 2 mm long.

Alluvial woodlands and clearings, stream banks and alluvium in streams, rocky slopes. N.Y. to Minn., southward in the piedmont and mts. to S.C., n. Ga. and westwardly to Ark.

4. Viola lanceolata L. LANCE-LEAVED VIOLET.   Fig. 158

Perennial, profusely producing slender stolons in summer (these sometimes bearing both apetalous and petaliferous flowers), often mat-forming. Vernal plants with closely set leaves on short caudices, first leaves usually with much smaller blades and much shorter petioles than later ones; blades lanceolate to linear-lanceolate, narrowed below to distally winged petioles, apices acute to blunt; smallest early leaves as little as 1–2 cm long (including petioles), later ones varying to as much as 15–20 cm long; margins of blades obscurely appressed-crenate, teeth callus-tipped; plants wholly glabrous or sparsely pilose on the petioles and proximal portion of the flower stalks. Stalks of vernal flowers shorter than to surpassing the leaves. Sepals 4–5 mm long, lance-subulate, apices acute or acuminate. Petals white, the spurred one with conspicuous purple veins on at least the lower half, the lateral ones without

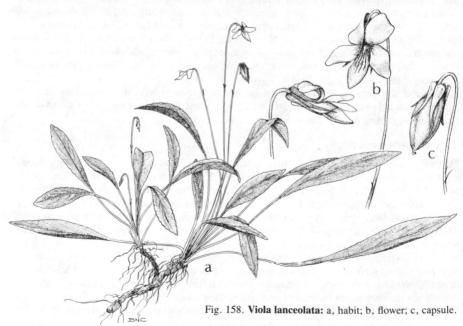

Fig. 158. **Viola lanceolata:** a, habit; b, flower; c, capsule.

352

purple venation or the venation much less conspicuous than that of the spurred one; corolla 10–14 mm across, lateral petals beardless or with a small tuft of trichomes near the base above. Capsules oblong, 6–8 mm long, glabrous.

Plants of this species in the southeastern coastal plain have, for the most part, longer, narrower leaves than elsewhere in the range; these have been designated subsp. *vittata* (Greene) Russell (*V. vittata* Greene).

Bogs, meadows, borders of ponds, lakes, open streams, seasonally wet pine savannas and flatwoods, wet ditches, cypress-gum depressions or borders of cypress-gum ponds, borders of evergreen shrub-tree bogs, often where shallow surface water stands, usually not in deep shade. s. N.S. to e. Minn., generally southward to s. pen. Fla. and n.e. Tex.

5. Viola pallens (Banks) Brainerd. SMOOTH WHITE VIOLET.

Plant with slender rhizomes, producing filiform, soft, flexuous stolons. Leaves tufted at the tips of rhizomes; earliest leaves usually with petioles 1–3 (–5) cm long, blades rotund to reniform in outline, some without basal lobes, most of them cordate basally, very broadly rounded distally, 1–2 cm broad; later leaves with petioles up to 10 (–12) cm long, blades cordate basally, ovate or suborbicular in overall outline, often a little longer than broad, distally rounded or tapered to a blunt tip, to about 5 cm long; petioles and blades glabrous or occasionally petioles sparsely short-pubescent and blades pubescent when first unfolding; margins of blades shallowly crenate. Stalks of vernal flowers usually surpassing the leaves. Sepals lanceolate to lance-ovate, about 3 mm long. Petals beardless, white, the 3 lower with brownish purple veins proximally, veins of the spurred petal most conspicuous. Capsule glabrous, ellipsoid, 4–5 mm long. Seed black. (*V. macloskeyi* Lloyd subsp. *pallens* (Banks) M. S. Baker)

Bogs, sphagnous shores, seepage areas in woodlands, banks of woodland streams, sometimes in slow-flowing water. N.B. and Maine to s. Ont., westward to Minn., southward to N.J. and Pa., cen. Ohio, n. Ind., n. Ill., e. Iowa, s.w. Mo., southward in the mts. from Pa. to n.w. S.C., n. Ga., n. Ala.

6. Viola primulifolia L. PRIMROSE-LEAVED VIOLET.

Perennial, producing slender stolons toward the end of or after the vernal flowering period. Leaves closely set on a short, erect caudex, very variable in size, in general the earliest leaves relatively small, petioles shorter than the blades, later ones with larger blades and petioles a little longer than the blades; blades ovate to lance-ovate (some of the earliest ones sometimes rotund or subrotund), bases subcordate to truncate, or tapering to wings on the distal part of the petiole, apices varyingly rounded, obtuse, or acute; smallest blades about 1 cm long, varying to 10 cm; margins mostly shallowly appressed-crenate, teeth callus-tipped, sometimes entire or nearly so; petioles, blades, and flower stalks vary from glabrous to shaggy-pubescent. Stalks of vernal flowers equaling to surpassing the leaves. Sepals subulate, mostly about 6 mm long, apices acuminate or acute. Petals white, the 3 lower with brown-purple veins near the base, the veins of the spurred petal much the more pronounced, the 2 laterals sometimes slightly bearded near the base above. Capsule oblong-ellipsoid, 7–10 mm long, glabrous.

Bogs, meadows, depressions in pine savannas and flatwoods, moist to wet ditches, banks of woodland streams, seepage areas in woodlands, sometimes in mesic woodlands. s. N.B. and s. Maine to Pa., Ohio, Ind., generally southward and southwestward to cen. pen. Fla., Ark., s.e. Okla., and e. Tex.

7. Viola cucullata Ait. BLUE MARSH VIOLET.

Perennial, glabrous, not stoloniferous. Leaves tufted at the rhizome tips; blades cordate-ovate to reniform, 4–10 cm long, apices of earliest ones rounded or bluntly tapered, those of later ones acute or short-acuminate, margins shallowly crenate to crenate-serrate, petioles to 2 dm long. Stalks of vernal flowers, long, often much surpassing the leaves. Sepals lanceolate, 8–12 mm long, sometimes ciliate, especially on the auricles. Petals violet-blue with darker veins usually forming a dark eye at the center of the corolla; lateral petals bearded proximally above with knob-tipped hairs. Capsule ovoid-cylindric, little longer than the calyx. Seed nearly black.

Wet meadows, springy areas in the open and in woodlands, bogs, along streams. Nfld. and Que. to Ont. and e. Minn., southward to S.C., n. Ga., Tenn., Ark., n. La.

8. Viola affinis LeConte.

Perennial, not stoloniferous. Leaves closely set in a tuft at the tip of a stout rhizome; blades of earliest leaves deeply cordate basally, ovate to reniform, broader than long, 2–3 cm broad, rounded from one basal lobe around the summit to the other basal lobe, petioles 2–4 cm long, blades of later leaves cordate basally varying to nearly truncate, gradually becoming nearly triangular in outline, sometimes longer than the width of the base, commonly broader across the base than long, apices blunt to acute, variable in size, many 3–4 cm broad, petioles 5–15 cm long; margins of blades shallowly crenate or bluntly serrate; surfaces glabrous, pubescent along the veins above or beneath or both, sometimes on parts of one or both surfaces between the veins as well, the pubescence very short, stiffish, and somewhat scalelike, both surfaces commonly (not always) finely rusty-red-dotted, dots, if present much more abundant beneath. Stalks of vernal flowers equaling or somewhat surpassing the leaves. Sepals lanceolate, acute to acuminate apically, 6–8 mm long, margins entire or occasionally minutely toothed. Bases of petals white or greenish white with violet veins much more pronounced on the spurred petal, violet distally, lateral and sometimes spurred petals bearded on the upper side basally with white, clavate hairs; in some populations, at least, corolla at first anthesis about 1.5 cm across, the exposed portions of the petals usually violet, becoming 3–3.5 cm across and usually pale violet by the time the petals commence to shrivel. (Incl. *V. chalcosperma* Brainerd, *V. floridana* Brainerd, *V. langloisii* Greene, *V. missouriensis* Greene)

Banks of woodland streams, low woodlands astride small streams, seepage areas in woodlands, rich, mesic woodlands. Vt. and Mass. to Wisc., generally southward to cen. pen. Fla. and e. Tex.

9. Viola esculenta Ell. Fig. 159

Perennial, glabrous or with few scattered hairs, not stoloniferous, with a thickish, somewhat erect or eventually horizontal rhizome. Leaf blades variable, both unlobed and lobed, but the lobing not so variable as in *V. septemloba*, from which it is considered indistinguishable by some authors; lobed leaves tending to come somewhat later during its vernal blooming season than is the case in *V. septemloba*. Unlobed leaf blades reniform, cordate-triangular, -ovate, or -orbicular, 2–6 cm long, margins crenate-dentate to shallowly crenate-serrate; lobed blades mostly with 3 major lobes, basal lobes variable, those of larger blades relatively broad, broader distally than proximally, one or both lobes bifurcate, the lower bifurcations divaricate and truncate-toothed apically, the upper bifurcations (if present) triangular to oblongish and slightly ascending; terminal lobe triangular, elliptic, or more frequently oblongish, obtuse apically. Sepals lanceolate, 6–7 mm long. Petals white or greenish white basally, the spurred one violet-lined on the distal portion of the white, otherwise violet; lateral petals bearded with pointed hairs, others glabrous, corolla 1.5–2 cm across, petals not enlarging appreciably after first anthesis, color not fading. Capsule elliptic-oblong, 8–10 mm long.

Alluvial, streamside woodlands and clearings, brooksides in ravines, adjacent wooded slopes. Chiefly coastal plain, s.e. Va. to pen. Fla., westward to s.e. Tex.

10. Viola septemloba LeConte. Fig. 160

Perennial, glabrous, not stoloniferous, with a short, soft, erect rhizome. Leaf blades very variable; some unlobed, triangular, ovate, or subreniform, 1–3 cm long, sometimes broader than long, bases cordate to subtruncate, margins obscurely crenate-serrate; other blades merely sharply toothed or toothed-lobed near their bases, with 2 smallish basal lobes, promi-

Fig. 159. **Viola esculenta:** plant and several leaf blades to show variation.

355

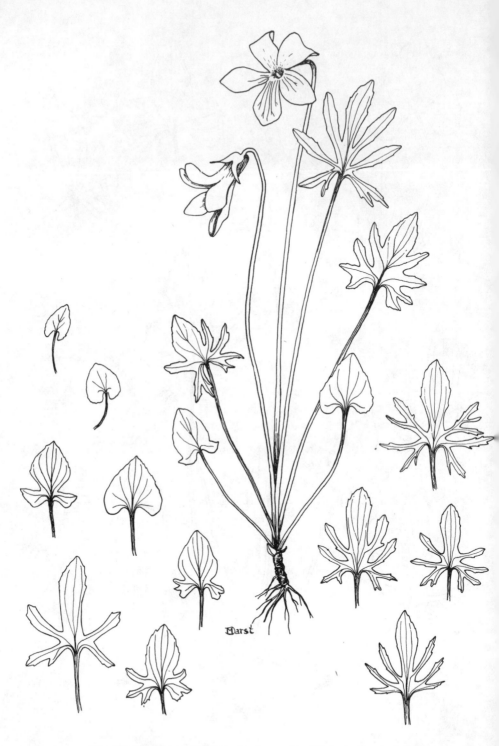

Fig. 160. **Viola septemloba:** plant and several leaf blades to show variation.

356

nently 3-lobed, or with 5 principal lobes of which the lowermost pair is often markedly bifurcated giving the overall effect of 7 lobes, or both lateral pairs bifurcated giving the overall appearance of 9 lobes; the lateral lobes and their sublobes of the most extremely lobed leaves much smaller than the terminal lobe and tending to have parallel sides; central, terminal lobe ovate, oval, oblong, or oblanceolate, 1–5 cm long and 1.5–2.5 cm broad at its broadest place. Stalks of vernal flowers very variable in length up to 2 dm. Sepals lanceolate to subulate, 8–12 mm long. At early anthesis lower part of petals greenish white, violet-lined outwardly, remainder of petals dark violet, the face of the open corolla about 18–20 mm across; over several days the petals enlarge, the portions at first deeply colored fade irregularly, often becoming essentially variegated, the corolla as a whole becoming up to 35 mm across; both lateral and spurred petals densely bearded, the upper one sparsely so. Capsule 10–15 mm long, elliptic-oblong. Seed ovoid, about 1 mm long, buff.

Seasonally wet to well-drained pinelands. Coastal plain, s.e. Va. to cen. pen. Fla., westward to e. and s.e. Tex.

Lauraceae (LAUREL FAMILY)

Trees or shrubs, or (in one not here treated) a parasitic vine. Fresh foliage notably aromatic when crushed. Leaves alternate, pinnately veined, usually with entire margins (scalelike in the parasitic vine), in some marked with tiny transparent dots. Flowers bisexual or unisexual, small, radially symmetrical, the perianth parts in two whorls of 3, similar or the outer smaller, free or somewhat united basally. Stamens 9–12, usually in 3 or 4 whorls, one or more whorls reduced to staminodia or glands, or absent. Pistil 1, ovary superior, 1-locular, style and stigma 1. Fruit a drupe.

1. Leaves persistent over winter; flowers all bisexual; persistent perianth evident below the fruit.
1. *Persea*
1. Leaves deciduous; flowers mostly unisexual and plants dioecious; perianth deciduous after anthesis but a small portion may remain as a disc below the fruit.
2. Branchlets zigzagging; mature leaves to 3 cm long and about 1 cm wide. 2. *Litsea*
2. Branchlets not zigzagging; larger mature leaves at least 6 cm long and 1.5 cm wide. 3. *Lindera*

1. Persea

Persea palustris (Raf.) Sarg. SWAMP-BAY, RED-BAY. Fig. 161

Shrub or small tree, the young twigs densely pubescent. Leaves with densely pubescent petioles 1–2 cm long, blades leathery, persistent over winter, long-elliptic, lanceolate, oblanceolate, or oblong, varying considerably in size on a given plant, the larger ones mostly 6–10 cm long and 2–3 cm wide or a little more, cuneate basally, acute or short-acuminate (infrequently rounded) apically, upper surfaces glabrous, the lower surfaces glaucous and from sparsely to densely pubescent, the principal veins tending to be rusty-colored and markedly raised. Inflorescences axillary, short cymose-paniculate, on stalks from 1 to 7 cm long, densely pubescent. Flowers bisexual, perianth densely pubescent, its segments free nearly to the base, the 3 outer shorter than the 3 inner, persistent below the fruit. Fertile stamens 9, in 3 outer series, and an inner series of 3 glandlike staminodia. Drupes dark blue to black with a thin whitish bloom, broadly ellipsoid to globose, about 1 cm long. (*Tamala pubescens* (Pursh) Small)

Swamps, wet woodlands, wet hammocks, wet pine flatwoods, banks in or at edges of marshes. Coastal plain, Va. to s. Fla., westward to s.e. Tex.

Persea borbonia (L.) Spreng. (*Tamala borbonia* (L.) Pax), in general features closely similar, considered by some authors to include (as a form) what is here described, is a plant of mesic to xeric woodlands and coastal dunes. Its young twigs, petioles, inflorescence stalks, and perianths are sparsely short-pubescent to glabrous, the leaf blades at most sparse-

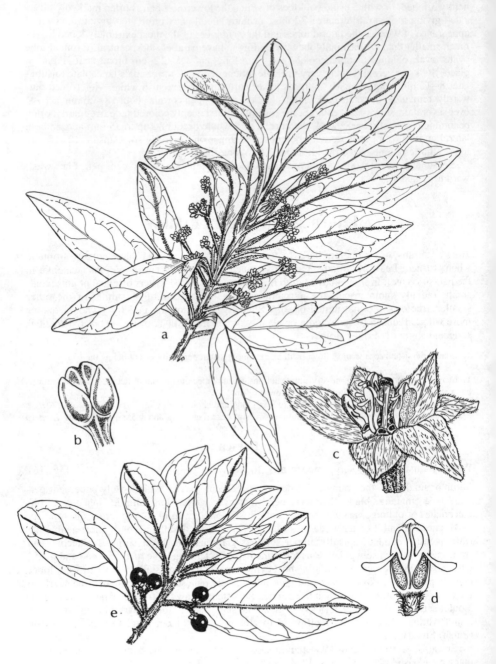

Fig. 161. **Persea palustris:** a, flowering branch; b, bud; c, flower; d, anther; e, fruiting branch. (From Correll and Correll (1972), as *P. borbonia*)

ly short-pubescent below, more commonly glabrous, thus the waxy bloom not so much obscured, the principal veins on their lower surfaces not markedly raised and not so rusty in color, usually buff.

Reference: Wofford, B. Eugene, and Ronald W. Pearman. "An SEM Study of Leaf Pubescence in the Southeastern Taxa of *Persea*." *Sida* 6: 19–23. 1975.

2. Litsea

Litsea aestivalis (L.) Fern. POND-SPICE.

Much-branched shrub to about 3 m tall. Branchlets zigzagging. Leaves deciduous, with very short, slender petioles, blades narrowly elliptic, lanceolate, or narrowly oblong, stiff-membranous, 1–3 cm long and 5–10 mm wide, glabrous. Flowers unisexual, plants dioecious; in few-flowered umbels axillary to leaf scars, the umbels subtended by 3–5 short scales, blooming prior to new shoot emergence; perianth yellow, 6-parted, segments nearly free to the base, not persistent. Staminate flowers with 9 or 12 fertile stamens in 3 or 4 series, those of series 3 (and 4 if present) with a pair of stipitate glands at the base, staminodia none. Pistillate flowers with 9 or 12 staminodia, those of series 3 (and 4 if present) with a pair of glands at the base. Drupe red, globose, 4–6 mm across. (*Glabraria geniculata* (Walt.) Britt. in Britt. & Brown)

Pond and swamp margins. Local, coastal plain, s.e. Va. (formerly), to n. Fla., westward to La.; Tenn.

3. Lindera

Shrubs, or in no. 1 sometimes attaining the stature of a small tree. Branchlets not zigzagging. Leaves short-petiolate, (deciduous in ours). Flowers in almost sessile, umbellike clusters from the axes of leaf scars, the clusters subtended by 2 pairs of short scales. Flowers mostly unisexual, the plants mostly dioecious; occasional plants with mostly pistillate flowers having an occasional staminate flower as well. Perianth yellow, with 6 similar parts in 2 whorls. Staminate flowers with 9 fertile stamens in 3 series, the innermost each with a pair of stipitate glands at the base; pistillate flowers with rudimentary stamens. Drupes bright red, about 1 cm long, borne on short, stubby stalks.

● Mature leaf blades (usually all of the larger ones) mostly obovate on any given plant but the smaller ones sometimes elliptic or oblanceolate, glabrous above, glabrous or pubescent below, spicy fragrant when crushed but not with the odor of sassafras.   1. *L. benzoin*
● Mature leaf blades elliptic-oblong, oval, or lance-ovate, pubescent on both surfaces, fragrant when crushed, the odor sassafraslike.   2. *L. melissaefolium*

1. Lindera benzoin (L.) Blume. SPICE-BUSH, BENJAMIN-BUSH.

Shrub to 5 m tall, sometimes with the stature and form of a small tree. Young twigs glabrous or pubescent. Larger leaf blades on a given plant mostly obovate, tapered to almost rounded basally, mostly short-acuminate apically; smaller leaves elliptic to oblanceolate, obtuse to rounded or even emarginate apically; leaves very variable in size, from 1 to about 15 cm long and from 1 cm or a little less to 6 cm wide, upper surfaces glabrous, lower glabrous or pubescent, usually pale below, margins usually pubescent, spicy-fragrant when crushed. Drupes glabrous or pubescent, elliptic-oblong. (*Benzoin aestivale* sensu Nees)

Plants with twigs and lower leaf surfaces permanently pubescent may be referred to *L. benzoin* var. *pubescens* (Palm. & Steyerm.) Rehd.

Rich, moist woodlands, beside small streams, floodplain forests. s.w. Maine, s. Ont., s. Mich. to e. Kan., generally southward to Fla. Panhandle and Tex.

2. Lindera melissaefolia (Walt.) Blume. JOVE'S-FRUIT.

Shrub, 3–20 dm tall or a little more. Twigs pubescent. Mature leaf blades elliptic-oblong, oval, or lance-ovate, obtusely tapered to rounded basally, acute or short-acuminate apically,

5–10 cm long, 1.5–3.5 cm wide, with a sassafraslike odor when crushed; both leaf surfaces pubescent. Drupe ellipsoid-obovoid. (*Benzoin melissaefolium* (Walt.) Nees)

Boggy margins of cypress-gum ponds, open bogs, swamps, sandy sinks. Coastal plain, N.C. to Fla. Panhandle, westward to La.; s.e. Mo.

Melastomataceae (MELASTOME FAMILY)

Rhexia (MEADOW BEAUTIES)

Perennial herbs, stems erect, from ascending subwoody caudices, slender rhizomes, tubers, or a combination of rhizomes and tubers. Stems hirsute and/or glandular-pubescent (save for *R. alifanus*), usually 4-angled, at midstem at least, in some species the angles winged, in some corky near the base. Leaves opposite, each pair at right angles to the preceding one (i.e., decussate), most (not all) with 3 principal veins, sessile or short-petiolate. Inflorescence cymose. Flowers few to many, radially symmetrical excepting the androecium. Bracts subtending the flowers like the leaves but much smaller, often deciduous. Flower with an elevated receptacle, forming with the perianth base a 2-layered, urceolate floral tube (hypanthium) the inner layer of which bears the stamens and petals at its tubular summit, the outer layer bearing 4 triangular calyx segments. Petals 4, separate, short-clawed, asymmetrical, blades disposed obliquely, ascending, or horizontally spreading, in color lavender, pink, purple, white, or (in one species) golden yellow, apices rounded or truncate, the midvein extending as a slender multicellular hair; surfaces of petals smooth above, in some species bearing glandular hairs on the part (abaxial surface) exposed in the bud. Stamens 8, subequal, in 2 whorls; filament long, slender, downwardly curved, with a small protuberance or hooklike appendage at the juncture with the anther, anthers shedding pollen through a terminal pore. Pistil 1, ovary inferior, 4-locular, style linear, stigma truncate. Fruit a loculicidal capsule enclosed by the hypanthium. Seeds curvate, except in *R. alifanus*, shaped like a snail shell, surface marked with concentrically oriented, regular or irregular ridges or tubercles, papillae, or laterally flattened domelike projections. (Surface sculpturing of seeds variable so drawings of seeds have limitations!)

Our treatment is, for the most part, adapted from R. Kral and P. E. Bostick, "The Genus *Rhexia* (Melastomataceae)," *Sida* 3: 387–440 (1969). These investigators give evidence of considerable hybridization between certain of the species of *Rhexia*. They did not find, however, evidence of hybridization involving the following: *R. lutea*, *R. nuttallii*, *R. petiolata*, *R. alifanus*, and *R. parviflora*. Since hybrids and hybrid swarms involving others of the species do occur, not a little difficulty is experienced in identification of specimens of varied intermediacy from certain localities.

1. Flowers yellow. 1. *R. lutea*
1. Flowers pink, lavender to purple, or white.
  2. Anthers about 2 mm long, almost straight; petals at anthesis ascending, or if slightly spreading, not flaring horizontally.
    3. Calyx segments apically acuminate-aristate; floral tube smooth save for a few hairs at and between the calyx segments; surface of seeds pebbled. 2. *R. petiolata*
    3. Calyx segments blunt or acute, not acuminate; floral tube glandular-pubescent; surface of seed marked with irregular, laterally compressed lines or ridges. 3. *R. nuttallii*
  2. Anthers 3 mm long or longer, curvate; petals at anthesis horizontally spreading.
    4. Stems glabrous throughout; leaf surfaces glabrous; seed wedge-shaped, definitely not shaped like a snail shell. 4. *R. alifanus*
    4. Stems pubescent, at least at the nodes; upper, or both upper and lower leaf surface pubescent; seeds shaped like a snail shell.
      5. Petals white.
        6. Calyx segments short-triangular, short-acuminate apically; corollas less than 2 cm across; bracts elliptic or broader, often nearly as broad as the floral tubes they subtend, giving the mature

inflorescence a "leafy" appearance; leaves at midstem with definite and distinct petioles and blades (a rare endemic in Franklin Co., Fla.).   5. *R. parviflora*

6. Calyx segments narrowly triangular, acute to acuminate; corollas over 2 cm across; bracts linear or narrowly triangular, distinctly narrower than the floral tubes they subtend, giving the mature inflorescence a relatively "naked" appearance; leaves at midstem gradually narrowed to the point of attachment (white-flowered form of widely ranging var.).

6a. *R. mariana* var. *mariana*

5. Petals pink, lavender, rose to purple.
 7. All four faces of stem at midstem approximately equal, the faces about flat, the angles sharp or winged.
  8. Petals dull lavender; hairs on the floral tube long, stiff, yellowish, chiefly crowded about the rim of the tube and on the bases of the calyx segments.   7. *R. aristosa*
  8. Petals bright lavender-rose; hairs of the floral tube scattered rather evenly over the surface.
   9. Leaf blades turned to a vertical position, i.e., with their plane surfaces at right angles to the substrate.   8. *R. salicifolia*
   9. Leaf blades not oriented vertically.
    10. Stem angles at midstem conspicuously winged; neck of floral tube about equal in length to the body.   9. *R. virginica*
    10. Stem angles at midstem wingless or narrowly winged; neck of floral tube usually longer than the body.
     11. Seeds papillate, the papillae in concentric lines.   6b. *R. mariana* var. *ventricosa*
     11. Seeds with irregular ridges in concentric lines.   6c. *R. mariana* var. *interior*
 7. All four stem faces not even approximately equal, one pair of opposite faces broader, darker green, convex to rounded, the narrower pair of opposite faces paler, flat to concave.
  12. Calyx segments broadly triangular; petals with glandular hairs near the margin on the abaxial (lower) side (best observed on buds before flower is at full anthesis).   10. *R. nashii*
  12. Calyx segments narrowly triangular to subulate; petals without glandular hairs.
   13. Floral tube 10–16 mm long, mostly 14–15 mm, the neck usually longer than the body.
11. *R. cubensis*
   13. Floral tube 6–9 (–10) mm long, the neck about equaling the body in length (lavender- or lavender-rose-flowered forms of).   6a. *R. mariana* var. *mariana*

1. Rhexia lutea Walt.   Fig. 162

Plant usually with several rigid, much-branched shoots 1–4 dm tall arising from a thickened, hard crown, the roots fibrous, nontuberous. Stem hirsute, at midstem and above somewhat 4-angled, the faces flat or convex. Leaves spatulate or oblanceolate to elliptic, all but the lowermost subsessile, the larger ones 2–3 cm long, 3-nerved, the two lateral nerves marginal on narrower leaves; leaf surfaces with stiff, yellowish pubescence, margins distantly low-serrate to subentire, teeth closely ascending, often hair-tipped. Inflorescence with "leafy" persistent bracts. Petals golden yellow, obovate to broadly cuneate, 1–1.5 cm long, ascending at anthesis, smooth except for hairlike projection at the tip. Mature floral tube 6–7 mm long, the body globose, abruptly contracted into a short-cylindrical lower neck then flaring-campanulate above, neck slightly shorter than the body; surface of neck and calyx segments with longish spreading pubescence, the calyx segments triangular, aristate apically. Seed about 0.7 mm long, with a few straight ridges of papillae along the crest, the sides with lower, more scattered papillae or smoothish.

Seasonally wet pine savannas and flatwoods, hillside seepage areas, and bogs. Coastal plain, N.C. to n. Fla., westward to s.e. Tex.

2. Rhexia petiolata Walt.   Fig. 163

Shoots solitary to several from hard caudices, roots nontuberous. Stem 1–5 dm tall, rigid, simple or few-branched, angular, the faces flat to rounded, internodes glabrous, sometimes a few hairs at the nodes. Leaves sessile or subsessile, 1–1.5 cm long, ovate to suborbicular, or short-elliptic, 3-nerved, lower surface glabrous, upper usually with a few long, nonglandular hairs, margins finely serrate, the ascending teeth mostly bristle-tipped thus the margins appearing ciliate. Inflorescence generally relatively compactly few-flowered, "leafy"-bracted, bracts persistent. Petals lavender-rose, somewhat asymmetrical, broadly oblong, elliptical,

Fig. 162. a–c, **Rhexia lutea:** a, habit; b, flower; c, seed; d–f, **Rhexia nuttallii:** d, habit;
e, mature floral tube (hypanthium); f, seed.

Fig. 163. **Rhexia petiolata:** a, base of plant; b, stems; c, flower; d, mature floral tube (hypanthium).

363

or ovate, 1–2 cm long, ascending to divergent, glabrous. Mature floral tube 5–7 (–9) mm long, a few long hairs around the rim and on the calyx segments, the body nearly globose, abruptly contracting into a short-cylindrical neck which flares into a spreading orifice. Calyx segments deltoid, acute apically, margins ciliate. Seed about 0.6 mm long, surface pebbled or with ridges of domelike processes. (*R. ciliosa* Michx.)

Seasonally wet pine savannas and flatwoods, bogs, wet peaty-sandy ditches, borders of cypress-gum ponds or depressions. Coastal plain, s.e. Va. southward into pen. Fla., westward to s.e. Tex.

3. Rhexia nuttallii James. Fig. 162

Similar in general features to *R. petiolata* but usually shorter and more slender, 1–3 (–4) dm tall, sometimes diminutive and flowering when 1 cm tall. Cyme to 8-flowered, commonly fewer than 8. Petals lavender-rose, 1–1.2 cm long, broadly cuneate to obovate, the abaxial surface with some gland-tipped hairs. Mature floral tube 5–7 mm long, glandular-pubescent, body nearly globose, abruptly narrowing to a short neck which flares distally to an orifice as broad as the body. Calyx segments broadly triangular, obtuse apically, margins entire, glandular-pubescent. Seed about 0.6 mm long, its surface sculptured with interrupted short, irregularly spaced ridges more or less longitudinally oriented. (*R. serrulata* Nutt.)

Pine savannas and flatwoods. Coastal plain, s.e. Ga. and in most of Fla.

4. Rhexia alifanus Walt. Fig. 164

Stems erect, wandlike, to about 1 m tall, arising from a short, spongy caudex. Stem glabrous, terete below, upwardly the internodes somewhat flattened in a plane parallel to that of the subtending leaf pair, longitudinally striate, the narrower bands paler and aligned with the leaf midribs. Leaves 3-nerved, predominantly lance-ovate, elliptic, or lanceolate, usually bluish green, subsessile, marginally entire or remotely low-toothed. Bracts ovate, glandular-pubescent, falling early. Petals lavender-rose, broadly oblong to suborbicular, 2–2.5 cm long, principal veins asymmetrical, purple, the lower surface with some hairs on the smaller half. Mature floral tube 7.5–10 mm long, glandular-pubescent, body subglobose to ovoid, longer than the abruptly constricting neck which flares above. Calyx segments with broadly triangular bases then narrowing to linear-oblong tips, the latter more or less disintegrating as the floral tube matures. Seed more or less wedge-shaped, 1–2 mm long, nearly smooth.

Pine savannas and flatwoods, bogs, adjacent ditches. Coastal plain, N.C. southward into pen. Fla., westward to s.e. Tex.

5. Rhexia parviflora Chapm. Fig. 165

Low plant, perennating by short, slender rhizomes. Stem 1–4 dm tall, 4-angled, faces subequal, internodes sparingly pubescent or glabrous, nodes pubescent. Larger leaves broadly ovate to elliptic, 1.5–3 cm long, short-petiolate, sparsely pubescent, margins finely toothed, the ascending teeth hair-tipped. Bracts "leafy." Petals white, asymmetrical, suborbicular to broadly obovate, mostly about 1 cm long, surfaces glabrous. Mature floral tube 5–7 mm long, body subglobose, neck short-cylindrical, shorter than the body, a few hairs near the orifice. Calyx segments short-triangular, apices short-acuminate. Seed 0.6 mm long, crested with irregular, roughly concentric interrupted lines of laterally flattened tubercles, these in turn vertically grooved.

Wet sands or peaty sands, borders of cypress ponds, borders of evergreen shrub bogs and *Hypericum* ponds. Franklin Co., Fla. Panhandle.

6a. Rhexia mariana L. var. mariana. Fig. 166

Plants forming extensive clones by shallowly set, slender, elongate rhizomes. Stems 2–10 dm tall, simple or freely branched from midstem upwards, spreading hirsute, stem faces unequal, one pair of opposite faces broader, rounded or convex, the narrower pair flat or concave. Leaves variable, linear, lanceolate, elliptic, or narrowly ovate, the longer ones 2–4 cm long, sparingly to copiously appressed-hirsute, apically acute to acuminate, basally acute to

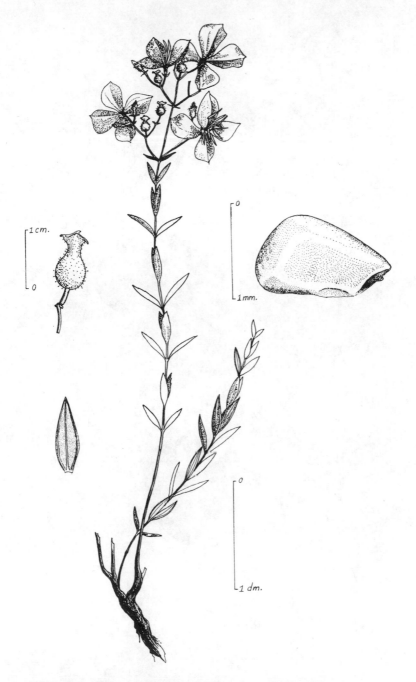

Fig. 164. **Rhexia alifanus:** habit (center); leaf (lower left); mature floral tube (upper left); seed (right). (From Kral and Bostick in *Sida* 3: 433. 1969)

Fig. 165. **Rhexia parviflora:** a, habit; b, flower; c, seed.

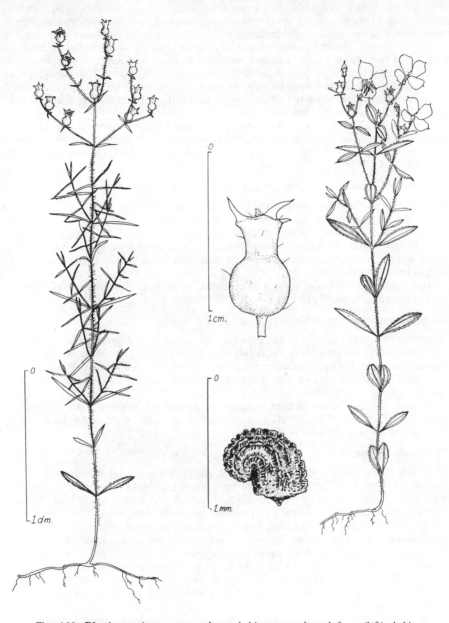

Fig. 166. **Rhexia mariana** var. **mariana:** habit, narrow-leaved form (left); habit, broader-leaved form (right); floral tube (top center); seed (bottom center). (From Kral and Bostick in *Sida* 3: 427. 1969)

367

attenuate, margins finely serrate, the teeth apiculate or hair-tipped. Cyme few- to many-flowered, the mature floral tubes often arranged secundly on the inside of the main branches. Petals broadly cuneate, oblong or obovate, 12–15 mm long, white to dull lavender, sometimes lavender-rose, surfaces glabrous. Mature floral tubes 6–10 mm long, body ovoid to subglobose, nearly evenly tapered into a cylindrical or upwardly slightly expanded neck as long as the body or very slightly longer, surface with frequent, sparse, or no pubescence. Calyx segments narrowly triangular, acute to acuminate. Seed about 0.7 mm long, surface longitudinally ridged, the ridges tuberculate, papillose, or with laterally flattened domes, most prominent ridges along the crest, rarely smooth. (Incl. *R. lanceolata* Walt., *R. delicatula* Small)

Pine savannas and flatwoods, bogs, margins of bays and cypress-gum ponds, or on exposed shores and bottoms of ponds, ditches, commonly weedy on moist to wet disturbed sites. Mass. to Va., Ky., s. Ind., s. Ill., generally southward to s. Fla. and southwestward to s.e. Okla. and e. Tex.

6b. Rhexia mariana var. ventricosa (Fern. & Grisc.) Kral & Bostick.

Similar to *R. mariana* var. *mariana* but the rhizomes stoutish, stem faces nearly equal, the angles sharp or narrowly winged. Leaves lance-ovate or lance-elliptic, the larger 4–6 cm long, 1–2 cm wide. Petals to 2.5 cm long, bright lavender-rose, lower petal surface with some glandular hairs. Mature floral tube 1–1.2 cm long, surface usually glandular-hirsute. Calyx segments deltoid, acuminate. Seed with concentric lines of close-set papillae. (*R. ventricosa* Fern. & Grisc.)

Pine barrens and savannas, bogs, marshy shores, openings in cypress-hardwood swamps, ditches and banks. Coastal plain, N.J. to S.C.

6c. Rhexia mariana var. interior (Pennell) Kral & Bostick.

According to Kral and Bostick (1969) the only striking character difference between this and var. *ventricosa* is that the seed of this is ridged, particularly along the crest, the ridges interrupted or laterally flattened domelike processes whereas the seed of var. *ventricosa* bears concentric lines of close-set papillae. The ranges of the two varieties are, however, markedly disjunct. (*R. interior* Pennell)

Moist to wet roadbanks, ditches, swamp forest clearings, prairies, glades, flatwoods. Lowlands of Ind., Ky., Tenn., interior prairies of Mo. and Kan., southward to La., e. Okla. westward to the prairie border in Tex.

7. Rhexia aristosa Britt.   Fig. 167

Plant with tuberous roots. Stem rigid, 4–7 dm tall, thickened-spongy below, subequally 4-sided at midstem, the angles sometimes narrowly winged, glabrous save for flaring yellowish hairs at the nodes. Leaves mostly lanceolate, 3-nerved, the larger 2–3 cm long, 3–5 mm broad, surfaces glabrous to sparingly appressed-pubescent, margins with low, regular, ascending, aristate teeth. Cymes broad. Petals asymmetrically oblong-obovate, 1–2 cm long, dull lavender, each aristate at the tip, abaxial surface with spreading glandular hairs. Mature floral tube 7–10 mm long, body ovoid, tapering into a cylindrical neck that flares at the summit, neck about the length of the body. Calyx segments narrowly triangular, aristate at the tip. Hairs of the floral tube stiff, slenderly tapering, yellowish, usually on the neck, rim, and calyx segments. Seed about 0.7 mm long, with irregular concentric ridges with a few isolated domes or papillae.

Wet pine barrens, savannas and flatwoods, bogs, ditches. Local, coastal plain, N.J. to Ga., and s.e. Ala.

8. Rhexia salicifolia Kral & Bostick.   Fig. 168

Plant forming tubers at the ends of some roots. Stem rigid, simple or more commonly bushy-branched, mostly about 2 dm tall, sometimes to 5.5 dm, at midstem 4-angled, angles nar-

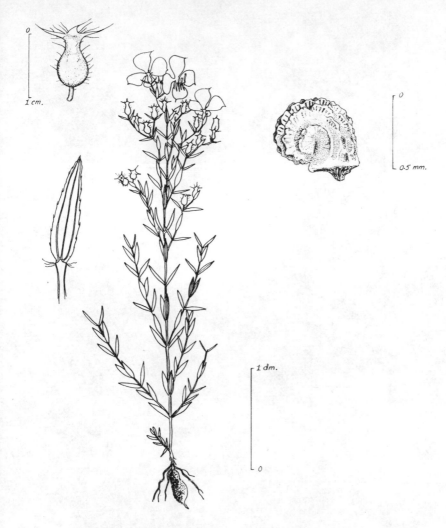

Fig. 167. **Rhexia aristosa:** habit (center); node and leaf (lower left); floral tube (upper left); seed (right). (From Kral and Bostick in *Sida* 3: 425. 1969)

rowly winged, the faces subequal, copiously glandular-pubescent. Leaves narrowly elliptic, oblanceolate, or linear, 1.5–4 cm long, 1–5 mm wide, 3-nerved, the lateral nerves often faint and extending little more than half the length of the blade, on narrower leaves the lateral nerves absent; leaf surfaces sparsely to copiously glandular-pubescent, margins with distant, low, ascending glandular hairs. Cymes few to densely flowered. Mature floral tube (4–) 5–7 (–'8) mm long, with scattered glandular pubescence, body globose, neck short-cylindrical, much shorter than the body. Calyx segments narrowly triangular. Petals broadly obovate to suborbicular, 11–12 mm long, deep lavender-rose, smooth above, glandular-pubescent below (in bud at least). Seed about 0.7 mm long, with 3–5 prominent, broad, symmetrical or tortuous longitudinal ridges or contiguous domelike tubercles in lines.

Sandy shores or exposed shores of limesink lakes, exposed bottoms of limesink cypress ponds, coastal interdune swales. Fla. Panhandle, s.e. Ala.

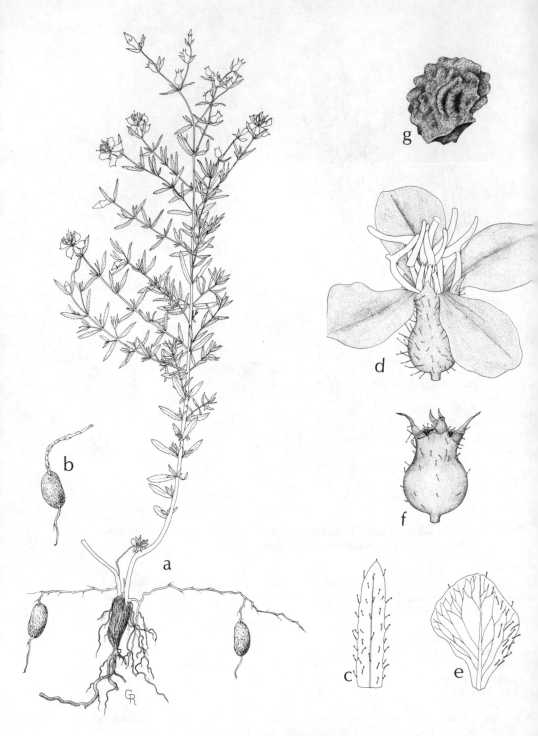

Fig. 168. **Rhexia salicifolia**: a, habit; b, germinating tuber; c, leaf; d, flower; e, petal, lower surface; f, mature floral tube (hypanthium); g, seed.

9. Rhexia virginica L.

Fig. 169

Plant with spongy-thickened or tuberiferous rootstocks. Stem rigid, simple or sparingly branched, to 1 m tall, 4-sided, wing-angled at midstem, sparsely glandular-pubescent (sometimes only at the nodes). Leaves ascending, ovate, elliptic, or more commonly lanceolate, 3–5 (–7) cm long, apically acute to acuminate, basally rounded to acute, margins finely serrate, the ascending tips of the teeth often hair-tipped. Cymes few- to many-flowered, open or contracted. Petals oblong to obovate, 1.5–2 cm long, lavender-rose to rose-purple, glandular hairs on the tips, some hairs usually present on the abaxial surface. Mature floral tube 7–10 mm long, glabrous to sparsely glandular-pubescent, body globose, with a narrow neck shorter than the body. Calyx segments narrowly triangular, acute to acuminate. Seed about 0.7 mm long, low muricate, papillose, or tuberculate in concentric lines, the sculpturing most prominent toward the crest. (Incl. *R. virginica* var. *purshii* (Spreng.) James = *R. stricta* Pursh not Bonpl.)

Shores of ponds, bogs, wet ditches, wet pine savannas and flatwoods. N.S. to s. Ont., and s. Wisc., e. Iowa, generally southward to n. Fla. and e.cen. Tex.

10. Rhexia nashii Small.

Fig. 170

Plants forming extensive clones from shallowly set, elongate, sometimes tuberiferous rhizomes. Stems of variable heights, 2–15 dm tall, at midstem hirsute, stem faces unequal, one pair of opposite faces broader, rounded or convex, the narrower flat or concave. Leaves ovate, lance-ovate, elliptic, or lanceolate, 3–7 cm long, both surfaces hirsute, acute apically, basally acute, often short-petiolate, margins finely to coarsely serrate, the teeth drawn out into hairlike tips and often with hairs on the tooth margins as well. Cyme symmetrical, open or contracted. Petals broadly obovate to suborbicular, 2–2.5 cm long or a little more, usually dull lavender, shortly hair-tipped, the lower surfaces usually glandular-pubescent. Mature floral tube 1–1.5 (–2) cm long, glabrous or glabrate, body ovoid to subglobose, abruptly or gradually passing into a cylindrical or narrowly funnelform neck at least equaling the length of the body. Calyx segments narrowly triangular, acute or acuminate. Seed about 0.7 mm long, surface concentrically lined with contiguous dome-shaped, sometimes laterally flattened, processes, the sculpturing most prominent toward the crest. (*R. mariana* var. *purpurea* Michx.)

Acid bogs and swamps, wet flatwoods and adjacent ditches, marshy shores, borders of cypress-gum ponds. Coastal plain, Va. to s. Fla., westward to La.

11. Rhexia cubensis Griseb.

Fig. 171

Plants forming clones from rhizomatous or tuberiferous bases. Stems 3–6 dm tall, glandular-hirsute, at midstem faces unequal (as in *R. nashii*). Leaves linear, linear-elliptic, oblong, or narrowly spatulate, only the midrib prominent, 2–4 cm long, both surfaces with scattered glandular-pubescence, margins regularly low-toothed, each ascending tooth terminated by a glandular hair. Cyme few- to several-flowered. Petals broadly cuneate to obovate, 1.5–2 cm long, horizontally spreading, bright lavender-rose to pale lavender-rose, almost white toward the clawed base, hair-tipped, glabrous or with very few glandular hairs abaxially. Mature floral tube 1–1.4 cm long, base ovoid or subglobose, tapering into a gradually widening neck at least as long as the body, glandular-pubescent, pubescence sometimes sparse or wearing away by full maturity. Calyx segments narrowly triangular or oblong, usually reflexed at full maturity. Seed about 0.7 mm long, concentrically evenly ridged, particularly along the crest.

Pine savannas and flatwoods, adjacent ditches, bogs, often weedy on roadsides. Coastal plain, N.C. to s. Fla., westward to s.w. Miss.; W.I.

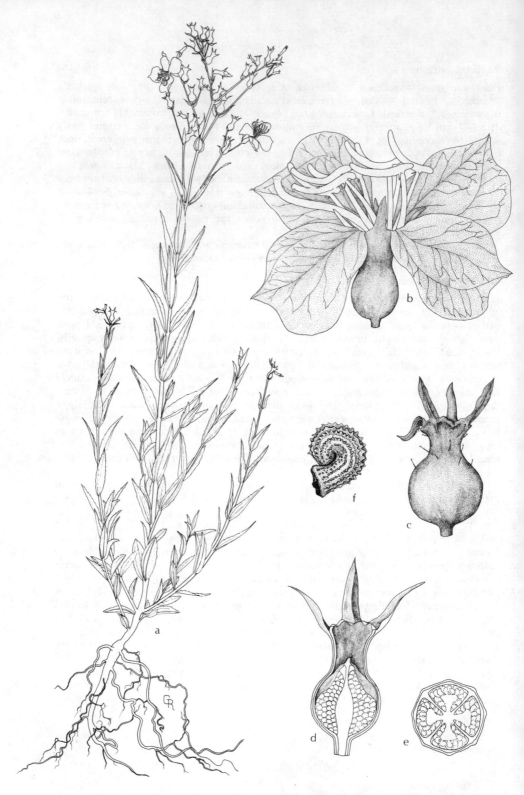

Fig. 169. **Rhexia virginica:** a, habit; b, flower; c, mature floral tube (hypanthium); d,e, longitudinal section and cross-section of floral tube; f, seed.

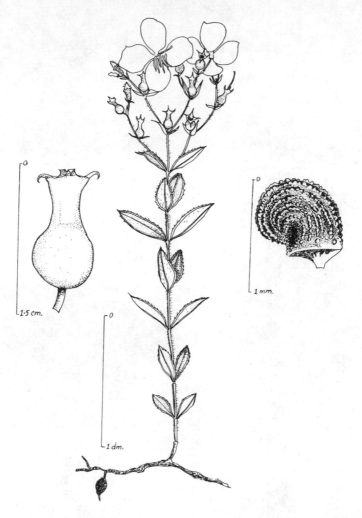

Fig. 170. **Rhexia nashii:** habit (center); mature floral tube (left); seed (right). (From Kral and Bostick in *Sida* 3: 432. 1969)

Lythraceae (LOOSESTRIFE FAMILY)

Herbs or shrubs (ours). Leaves simple, entire, opposite, whorled, or opposite below and alternate above, without stipules. Flowers bisexual, radially or bilaterally symmetrical, usually with 2 opposite bractlets either subtending the flower or on its stalk; ovary superior, borne free within a globose, campanulate, or cylindrical floral tube surmounted by 4–6 persistent calyx segments, in some with 3–5 alternating appendages. Petals sometimes absent, usually 4–6 inserted on the inner rim of the floral tube, alternating with the calyx segments, sometimes quickly deciduous after anthesis. Stamens as many as or twice as many as the petals (or more numerous), filaments often alternately unequal, inserted within the floral tube below the petals. Ovary often subtended by a disk, 2–4-locular; style 1, frequently dimorphic, stigma capitate, sometimes 2-lobed. Fruit a capsule seated within the floral tube. Seeds 3 to many.

373

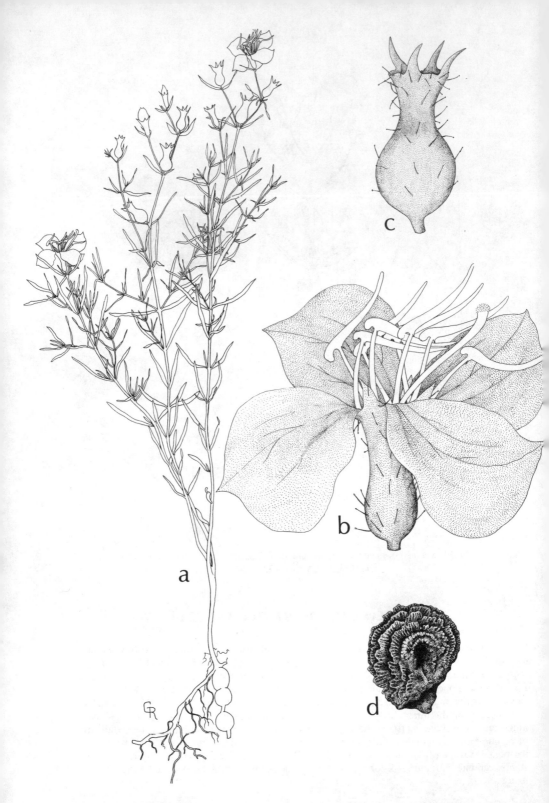

Fig. 171. **Rhexia cubensis:** a, habit; b, flower; c, mature floral tube (hypanthium); d, seed.

Reference: Graham, Shirley A. "Taxonomy of the Lythraceae in the Southeastern United States." *Sida* 6: 80–103. 1975.

1. Stems woody at the base and with conspicuously corky bark, distal portions of the branches herbaceous, some at least, arching and rooting at the tips.   1. *Decodon*
1. Stems (of ours) wholly herbaceous, not arching and rooting at tips.
  2. The stems pubescent; floral tube obliquely swollen on one side at the base.   2. *Cuphea*
  2. The stems glabrous; floral tube symmetrical.
    3. Floral tube cylindric to turbinate, considerably longer than wide.   3. *Lythrum*
    3. Floral tube globose to campanulate, about as wide as long to wider than long.
      4. Flowers or fruits solitary in the leaf axils.
        5. Appendages present between the calyx segments, *or*, if absent, then the length of the bractlets about equaling that of the floral tube; petals present; capsule septicidally dehiscent.   4. *Rotala*
        5. Appendages absent, the bractlets minute; petals none; capsule indehiscent.   5. *Didiplis*
      4. Flowers or fruits 2 to several in the leaf axils (on a given plant flowers or fruits may be solitary in a few leaf axils but not in most).   6. *Ammania*

1. Decodon

Decodon verticillatus (L.) Ell. WATER-WILLOW, SWAMP LOOSESTRIFE.   Fig. 172

Perennial, woody below, the bark very soft-corky under water, exfoliating in long cinnamon-colored strips above water, the ultimate branches herbaceous. Branches strongly arching, some of them at least, rooting at the tips. Young stems densely pubescent, older stems glabrate. Leaves opposite or whorled, short-petiolate, lanceolate or elliptic-lanceolate, the larger to about 20 cm long and 5 cm broad, acute at both extremities, glabrous above, closely pubescent with short, branched hairs beneath. Flowers axillary, about 2.5 cm broad, in short-stalked cymes. Floral tube usually surmounted by 5–7 bristle-tipped calyx segments and with spreading hornlike appendages in their sinuses. Petals 5 (–7), magenta, their bases stalked, the expanded portion lanceolate, crinkled, and with irregular margins. Stamens 8–10, with filaments of 3 possible lengths, 2 of the 3 occurring in any one flower. Fruit closely enveloped by the floral tube, 3–5-locular and dehiscent. Seeds irregularly obpyramidal, about 2 mm long, somewhat lustrous, the surface very finely reticulate, olive-green and with a large brown spot on one side.

Swamps, swamp clearings, swampy shores of streams, pools and bogs. N.S. to Ont. and Minn., generally southward to cen. pen. Fla. and e. Tex.

2. Cuphea (WAXWEEDS)

(Ours) annual or perennial herbs of relatively slender and low stature, the stems and floral tubes sticky, the pubescence in part, at least, of glandular hairs. Leaves opposite or whorled. Flowers 1–3 in the leaf axils or in racemes terminating the branches, their stalks with 2 bractlets; floral tube 12-nerved, obliquely swollen on one side at the base; calyx segments 6, short appendages alternating with them; petals 6 (–3), unequal, pale to deep purple, much exceeding the calyx segments. Stamens 10–12, usually 11, alternately unequal, the upper (adaxial) 2 inserted lower within the floral tube than the others. Ovary subtended by a glandular disk that extends free at one side; style 1, slender, capitate, rarely 2-lobed. Capsule septicidally dehiscent, the floral tube dehiscent as well on the adaxial side, the placenta extending out of both the capsule and floral tube.

1. Leaves mostly whorled (some on a given plant sometimes opposite).   1. *C. aspera*
1. Leaves opposite.
  2. Plant perennial; leaves sessile.   2. *C. glutinosa*
  2. Plant annual; leaves petiolate or at least the lower ones narrowed to subpetiolar bases.
    3. Floral tube 5 (–7) mm long, glabrous within; lower leaves narrowed to subpetiolar bases.
                                                                                        3. *C. carthagenensis*
    3. Floral tube (8–) 10–12 mm long, pubescent within; leaves with definite slender petioles.
                                                                                        4. *C. viscosissima*

Fig. 172. **Decodon verticillatus:** a, tip of a stem having arched, rooted at the tip, and from which new shoots are being produced; b, flowering branchlet; c, flower; d, seed.
(From Correll and Correll. 1972)

376

1. Cuphea aspera Chapm.

Fig. 173

Perennial with slender rhizomes some roots from which are fusiform-tuberoid distally. Stems simple to few-branched, 2–4 dm tall, glabrous near the base, medially with scattered, enlarged-based hairs; upper stems, flower stalks and floral tubes with elongate, spreading, enlarged-based, purple, glandular hairs and often with shorter, appressed, stiff, whitish hairs. Leaves sessile, whorled (or sometimes some opposite on a given plant), lanceolate to linear-elliptic, 1–2.5 cm long, acute apically, cuneate or occasionally truncate basally, prominently 1-nerved, more or less pubescent with short, stiff hairs. Flowers in bracteate racemes. Floral tube 7–9 mm long, purple, pubescent within. Calyx segments triangular, the intersepalar appendages bristle-tipped. Petals 6, lavender, the upper 2 largest. Seeds usually 3, slightly lenticular, broadly obovate in outline, dark brown, surface finely pebbled, 2 mm long and as broad, or a little broader basally. (*Parsonsia lythroides* Small)

Seasonally wet pine flatwoods, margins of shrub bogs. Endemic in Franklin and Gulf Cos., Fla.

2. Cuphea glutinosa Cham. & Schlect.

Perennial with slender, hard caudices, stems wiry, usually several from the base, erect-ascending to decumbent, 1–4 dm tall, pubescent with elongate, spreading, enlarged-based, glandular hairs and more numerous short, curly, whitish hairs. Leaves sessile, opposite, 1-nerved, mostly elliptic-lanceolate or lance-ovate, 1–1.5 cm long, acute at both extremities. Flowers solitary in the upper leaf axils or subracemose distally. Floral tube about 8 mm long, with both glandular and nonglandular hairs. Calyx segments very shortly and broadly deltoid, the intersepalar appendages minute. Petals 3–6, deep violet to purple. Seeds several, discoid, orbicular in outline, olivaceous, about 2 mm across.

Open moist to wet grassy areas, meadows, open woods by streams. Native of S.Am., naturalized in La. and e. Tex.

3. Cuphea carthagenensis (Jacq.) Macbr.

Fig. 174

Annual. Stems usually branched, stiffly to weakly erect, 1–6 dm tall, clammy-pubescent with slender, spreading, enlarged-based, glandular hairs and with shorter curly, nonglandular hairs intermixed on the upper portions. Leaves opposite, the lower narrowed to subpetiolar bases, elliptic, oval, ovate, or sometimes obovate, scabrous on the veins below, sometimes on or near the margins, acute apically, 2–6 cm long. Flowers solitary in the leaf axils or subracemose distally on the branches. Floral tube 5 (–7) mm long, often green, sometimes purplish, sparsely pubescent with long, spreading, weak, pale, glandular hairs with little enlarged bases, sometimes with sparse, short, stiffish hairs as well, glabrous within. Calyx segments minute, deltoid, the intersepalar appendages minute, shortly bristle-tipped. Petals 6, 2–3 mm long, linear-elliptic, greenish to purple, unequal. Seeds lenticular, obovate in outline, olive to brown, with pale banded edges, surfaces very finely reticulate, about 2 mm long. (*Parsonsia balsamona* (Cham. & Schlect.) Small)

Open, wet woodlands, swales, marshy shores, wet clearings, ditches. Coastal plain, N.C. to s. Fla., westward to s.e. Tex.; Mex. to S.Am.

4. Cuphea viscosissima Jacq.

Fig. 174

Annual. In general appearance very similar to *C. carthagenensis*. Leaves with definite slender petioles, the blades widest basally thence gradually tapering to acuminate tips. Floral tube (8–) 10–12 mm long, more purple than in *C. carthagenensis*, the glandular hairs dark purple, much more abundant, their bases much enlarged. Petals 6, deep purple, 4.5–5.5 mm long, the upper especially short-clawed and with ovate blades. Seeds flat, oval to suborbicular, brown, with more or less banded margins, surfaces very finely reticulate to nearly smooth, 2 mm long or slightly more. (*C. petiolata* (L.) Koehne; *Parsonsia petiolata* (L.) Rusby)

Usually on drier sites than those inhabited by *C. carthagenensis*, sometimes in moist to wet meadows, ditches, edges of ponds, lakes, and streams, wet gravel bars. N.H. to n. Ill.,

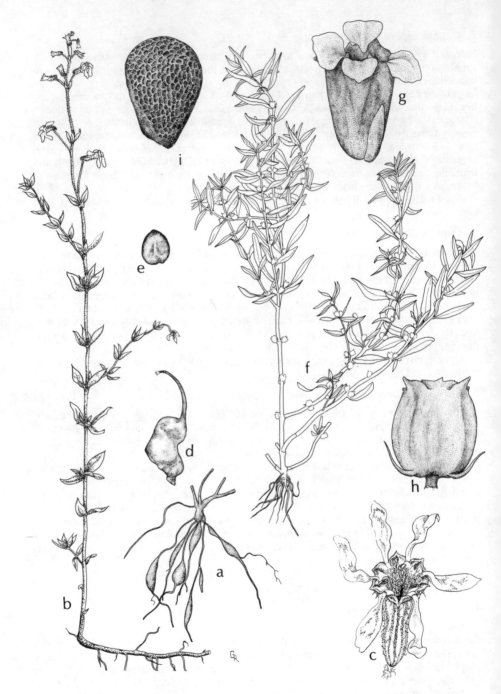

Fig. 173. a–e, **Cuphea aspera:** a, base of plant, showing tuberous roots; b, habit; c, flower; d, pistil; e, seed; f–i, **Rotala ramiosior:** f, habit; g, flower; h, fruit; i, seed.

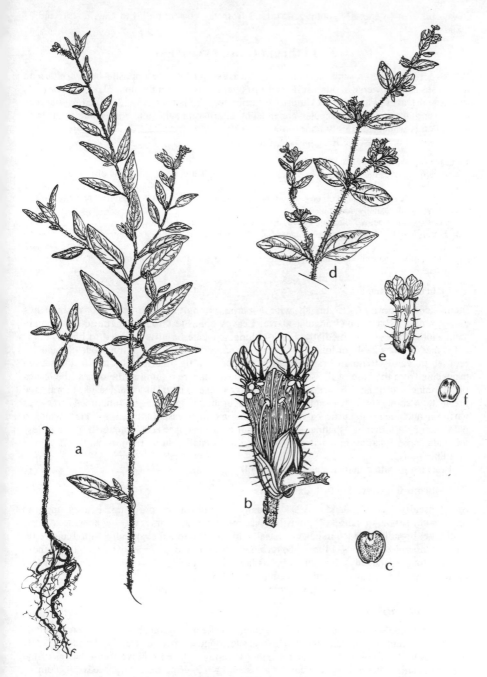

Fig. 174. a–c, **Cuphea viscosissima:** a, plant; b, flower, split longitudinally; c, seed; d–f, **Cuphea carthagenesis:** d, flowering branch; e, flower; f, seed. (From Correll and Correll. 1972)

Iowa, and Kan., generally southward (inland from the coastal plain) to Ga., n. La. and e. Okla.

3. Lythrum (LOOSESTRIFES)

(Ours) perennial herbs with slender stems. Leaves opposite, or opposite below, alternate above, sessile. Flowers (in ours) solitary (infrequently paired) in the upper leaf axils or axils of bracteal leaves. Floral tube cylindric to turbinate, symmetrical, 8–12-ribbed or -nerved. Calyx segments 5–7, intersepalar appendages alternating with and somewhat exterior to them. Petals usually 6 (rarely none). Stamens as many as the petals or twice as many. Capsule 2-valved or dehiscing irregularly. Seeds many.

1. Leaves opposite throughout.
   2. Principal leaves linear, much longer than broad, acute apically; flowering stems stiffly erect.
   1. *L. lineare*
   2. Principal leaves elliptic to oblong, 2–3 times as long as broad, rounded to bluntish apically; flowering stems weak, usually decumbent.
   2. *L. flagellare*
1. Leaves opposite below, alternate above, at least distally on the branches.
   3. Floral tube 3–4 mm long, appendages about the same length as the calyx segments; petals about 3 mm long.
   3. *L. curtissii*
   3. Floral tube 5–6 mm long, the appendages twice as long as the calyx segments; petals mostly 5–6 mm long.
   4. *L. alatum*

1. Lythrum lineare L.

Stems slender, mostly 6–15 dm tall, with numerous strongly ascending, floriferous branches above, terete at base, quadrangular above. Leaves opposite throughout, those of the main stem shorter below than medially, linear-oblong, cuneate basally, rounded to bluntish apically, medial ones 2–4 cm long, 3–4 mm wide, about as long as the internodes; gradually reduced upwardly to small bracteal leaves shorter than the internodes. Late in the season producing from the base of the stem long, prostrate runners rooting at the nodes, the leaves of the runners spatulate. Flowers usually solitary in the axils of bracteal leaves. Floral tube turbinate, 4 mm long; calyx segments deltoid-acuminate, appendages sometimes somewhat narrower and longer than the calyx segments, sometimes essentially alike. Petals white to pale lilac, 3–4 mm long, spatulate-obovate, not conspicuous. Seeds oblongish, more or less lenticular, somewhat asymmetrical, wing-margined distally, dull stramineous, 0.6 mm long or a little more.

Brackish to saline marshes and shores. Del. to s. Fla., westward to La.

2. Lythrum flagellare Shuttlw. ex Chapm.

Stems usually several, slender, weak, usually decumbent or sprawling. Leaves opposite throughout, sessile or subsessile, mostly 5–12 mm long, 2–3 times as long as wide, oblong or elliptic, broadly rounded to cuneate basally, bluntish to rounded apically, little reduced on the inflorescence branches. Floral tube turbinate, 4–5 mm long, appendages subulate, about twice as long as the calyx segments. Petals lavender to purple, 3–4 mm long, more or less spatulate, crinkled, considerably exceeding the calyx.

Low, open grounds, swamps, wet thickets. cen. to s. pen. Fla.

3. Lythrum curtissii Fern.

Stems simple or several from the base, slender, weak, much branched above, terete below, quadrangular above. Leaves sessile or subsessile, opposite below, usually alternate distally on the branches, those of the branches rather abruptly reduced relative to the leaves of the main stem although the latter are commonly shed by flowering time; larger leaves oblong to elliptic, 2–4 cm long, cuneate basally, bluntish to acute apically. Floral tube 3–4 mm long, narrowly turbinate to cylindrical. Sepals broadly deltoid, about 0.5 mm long, abruptly narrowed apically, appendages subulate to deltoid, about equaling the sepals in length. Petals about 3 mm long, lavender pink, elliptic, relatively inconspicuous, quickly deciduous. Seeds yellowish brown, somewhat flattened and asymmetrically twisted, spatulate in outline, slightly winged distally.

Floodplains and stream banks, boggy open areas in pinelands, wet thickets. Infrequent, coastal plain, s.w. Ga. and Apalachicola River region, Fla. Panhandle.

4. Lythrum alatum Pursh.

Stems slender to moderately robust, erect and wandlike, much branched above, terete to obscurely quadrangular. Leaves sessile, paler green below than above, opposite below, alternate on the inflorescence branches, gradually reduced upwardly; variable in shape, the principal ones lance-ovate with truncate to subcordate bases or linear-lanceolate to linear-oblong with strongly tapering bases. Floral tubes mostly 5–6 mm long, turbinate to cylindrical. Calyx segments deltoid or ovate-acuminate, the appendages subulate and usually much exceeding the calyx segments. Petals purple, obovate or spatulate-obovate, crinkled, mostly 4–6 mm long, much exceeding the sepals and relatively conspicuous, usually not quickly deciduous. Seeds yellowish brown, nearly fusiform, slightly asymmetrical and slightly winged distally, 0.5 mm long or a little more.

Moist to wet soil, marshes, marshy shores, swales, moist to wet clearings, flatwoods depressions, ditches.

Plants of this widely ranging species vary considerably, especially respecting leaf shape. Graham (1975), after extensive analysis of specimens from throughout the range, recognized two varieties, as follows.

L. alatum var. **alatum** (incl. *L. cordifolium* Niewl. and *L. dacotanum* Niewl.) Leaves ovate to oblong with rounded to subcordate bases.

Maine to s.e. N.D., e. S.D., southward to n. Va., Tenn., n.w. Ga., n. Ala., n. Ark., n.e. Okla., and Kan.

L. alatum var. **lanceolatum** (Ell.) T. & G. ex Rothrock in G. M. Wheeler (*L. lanceolatum* Ell.) Leaves lanceolate or linear-lanceolate, the bases tapering.

s.e. Va. and e. N.C. to Okla., generally southward to s. Fla. and e. half of Tex.

4. Rotala

Rotala ramosior (L.) Koehne in Mart. TOOTHCUP.   Fig. 173

Glabrous annual. Stem erect or branches decumbent, terete, or 4-angled above, simple to diffusely branched, if branched the lowermost branches sometimes prostrate below, the tips ascending, occasionally all branches prostrate. Plant usually not over 3 dm tall. Leaves opposite (rarely whorled), linear-lanceolate, lanceolate, or oblanceolate, the bases tapering, sessile or short-petioled, apices acute to obtuse, 1–3 (–5) cm long and 2–12 mm broad. Flowers tiny, about 2 mm long, solitary in the leaf axils, sessile, each subtended by a pair of subulate bracts less than half the length of the floral tube. Floral tube squarish in cross-section, closely enveloping the ovary or capsule. Calyx segments short, broad, inflexed and apiculate, the intersepalary appendages spreading, short, ovate-acuminate. Petals 4, white or pinkish, translucent, obovate, 3-lobed at their summits, about 0.6 mm long, scarcely exceeding the calyx segments, easily detached and quickly deciduous. Stamens 4 (–6), the tips of the anthers barely exserted. Style short. Capsule with microscopic dense transverse striations on the outer wall, septicidally dehiscent. Seeds numerous, tiny, about 0.3 mm long, igloo-shaped, the rounded surface very finely reticulate, the flat surface more coarsely reticulate and lustrous, yellow-brown.

In wet open sites, shores and exposed shores, pools, depressions, ditches. Mass. to Minn., generally southward to s. Fla. and e. half of Tex.; Pacific states; trop. Am.

See John H. Thieret, "*Rotala indica* (Lythraceae) in Louisiana," *Sida* 5: 45 (1972). Here Thieret reported *R. indica* (Willd.) Koehne as an inhabitant of rice fields in La. It is native to s. Asia and had hitherto been known (for the U.S.) as a naturalized introduction in rice fields in Calif. Distinguishing characteristics include leaves that gradually increase in size upwardly, the larger ones spatulate, 1–2 cm long, their margins thickened-translucent. Flowers on short, leafy, spikelike axillary branchlets. Bractlets subtending the flower about equaling the floral tube in length. Intersepalary appendages absent. Petals pale pink, ovate, acute apically, shorter than the calyx segments and persistent.

5. Didiplis

Didiplis diandra (DC.) Wood. WATER-PURSLANE.

Delicate annual, aquatic or amphibious herb. Stems weak, erect to procumbent, to 4 dm tall. Leaves opposite, numerous, sessile, 1–3 cm long, 0.5–4 cm wide, subelliptic and firm when emersed, very narrowly linear-lanceolate to linear-subulate and flaccid when submersed. Flowers solitary in the leaf axils, small and greenish. Floral tube enveloping the ovary but only the base of the capsule. Sepals 4, broadly triangular, intersepalary appendages absent. Petals none. Stamens 2–4, not exserted. Capsule globose, indehiscent, the wall transparent. Seeds numerous, about 1 mm long, oblanceolate or spatulate in outline, the summits enlarged and incurved, rounded, yellowish and minutely granular on one face, flat and greenish on the other. (*Peplis diandra* Nutt. ex DC.)

In shallow water, shores on ponds, streams, lakes, temporary pools. Va. to Ind. and Wisc., southward to N.C., Miss., La.

6. Ammania (TOOTHCUPS)

Subsucculent annual herbs. Leaves opposite, usually basally auricled. Flowers small, 4-merous, in short axillary cymes, 2–10 flowers per cyme, occasionally some solitary flowers on a given plant. Floral tube globose to campanulate, usually with 4 intersepalar appendages. Sepals short. Petals 4 or none, small, purple to pink, quickly deciduous. Stamens 4 or 8, enclosed or exserted. Capsule enclosed by the floral tube except at the summit, globose or nearly so, rupturing irregularly, incompletely 2–4-locular.

- Style 1.5–3.0 mm long, exserted in fruit; calyx lobes triangular with acute apices.   1. *A. coccinea*
- Style 0.5 mm long, included in fruit; calyx lobes obtuse, usually with minutely mucronate apices.

2. *A. latifolia*

1. Ammania coccinea Rottb.   Fig. 175

Stem simple to freely branching, to 10 dm tall, usually much less, quandrangular, sometimes slightly winged on the angles. Leaves linear to linear-lanceolate, sessile, the bases auriculate to cordate and somewhat clasping, to 10 or 11 cm long. Flowers in sessile to subsessile tight axillary cymes, (1–) 3–5 (–10) per cyme, often with the appearance of being whorled. Calyx segments very short and broadly triangular, apically acute, intersepalar appendages minute, protruding outwardly. Petals 4, orbicular to obovate, purple to rose-pink. Stamens 4 or 8, exserted. Style exserted, persistent on the fruit or disarticulating near the base. Floral tube enclosing all but the broad summit of the capsule, usually suffused with reddish pigment, the capsule dark red. Seeds numerous, minute, more or less triangular in outline, 1 face rounded and granular, the others flattish to depressed and irregularly ridged-reticulate, olive-brown.

Wet open sites, marshes and marshy shores, wet clearings, swales, ditches. N.J. to Ohio, Ind., Ill., Mo., generally southward to s. Fla. and Tex., sporadic in the western U.S.; trop. Am.

Ammania auriculata Willd. occurs outside our range in the cen. U.S. (and widely in the world elsewhere). Where it is sympatric with *A. coccinea* apparently hybrids or introgressants occur. Possible introgressants are infrequently found in our range in states bordering the Mississippi River. The cymes of *A. auriculata* are long stalked, a character shared by the introgressants. Occasional specimens of *A. coccinea* with elongated cymes occur elsewhere within our range but are not thought to be derived from hybridization with *A. auriculata*. (See Graham, 1975)

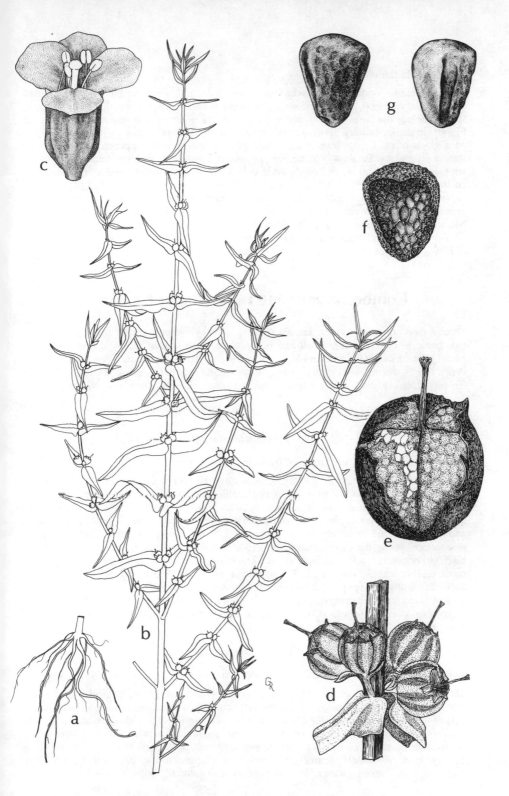

Fig. 175. a–f, **Ammania coccinea:** a,b, plant; c, flower; d, cluster of fruits; e, fruit, with wall partially disintegrated; f, seed; g, **Ammania latifolia:** seed, two views.

2. Ammania latifolia L.   Fig. 175

Stems simple to moderately branched, to about 9 dm tall, terete to barely quadrangular, rarely slightly winged. Leaves linear-elliptic to oblanceolate or spatulate, the lower sessile or short-petioled, the bases subauriculate to cuneate, the upper sessile and auriculate basally. Flowers in dense, axillary, short-stalked clusters, some sometimes solitary in a few leaf axils on a given plant. Sepals low, broad, rounded, the intersepalary appendages folded, outwardly protruding. Petals 4 or none, minute, about 1 mm long, quickly deciduous, pink to whitish pink. Stamens 4, enclosed. Style 0.5–1 mm long, usually persistent. Floral tube enclosing all but the broad summit of the capsule, green or with some reddish pigment, the capsule reddish brown. Seeds minute, numerous, more or less triangular, 1 face rounded and granular, the other irregular, somewhat convex, obscurely to definitely reticulate, nearly amber, somewhat lustrous. (Incl. *A. koehnei* Britt., *A. teres* Raf.)

Wet open places, tidal marshes and shores, borders of swamps, ditches. Coastal plain, N.J. to s. Fla., westward to s.e. Tex.; trop. Am.

Combretaceae (WHITE MANGROVE FAMILY)

Shrubs, trees, or woody vines. Leaves simple, leathery. Flowers small, bisexual or unisexual, borne in compact, often ball-like or conelike spikes, or in racemes. Ovary inferior. Floral tube surmounted by a deciduous or persistent, partially united, 4- or 5-lobed calyx. Petals 4 or 5, distinct or partially united, or none. Stamens twice, rarely thrice, as many as the calyx lobes. Pistil 1, ovary 1-locular, style 1. Fruit indehiscent, usually winged.

- Leaves alternate.   1. *Conocarpus*
- Leaves opposite.   2. *Laguncularia*

1. Conocarpus

Conocarpus erectus L. BUTTONWOOD, BUTTON MANGROVE.   Fig. 176

Shrub or small tree, to about 20 m tall. Leaves alternate, simple, evergreen, leathery, short-petiolate or sessile; blades with entire margins, elliptic to ovate (sometimes lanceolate), their apices acute to acuminate, a pair of glands on the petiole or at about the base of the blade; lower surface with small pits at juncture of midvein and lateral veins; stipules none. Twigs and leaves glabrous, or silvery-pubescent on some plants. Flowers in stalked, tight, ball-like or conelike spikes, the spikes axillary, in axillary clusters, or in terminal racemes. Flowers bisexual, or bisexual and staminate in the same inflorescence; floral tube greenish white surmounted by a cuplike calyx with 5 segments. Petals none. Stamens 5–8 (–10), exserted. Individual fruits (of the spike) 2-winged, flattened, scalelike.

Landward from red mangrove and usually above high tides, borders of streams, sandy or marly shores and hammocks. Coastal areas of cen. and s. pen. Fla. and the Fla. Keys; trop. Am.; w. Afr.

2. Laguncularia

Laguncularia racemosa (L.) Gaertn. f. WHITE MANGROVE.   Fig. 177

Shrub or small tree, sometimes 20 m tall. Leaves opposite, evergreen, glabrous, short-petiolate, oblong to obovate, apices usually obtuse, rounded, or notched, a pair of glands on the petiole just below the base of the blade. Inflorescence a terminal panicle or an axillary spike. Flowers greenish white, fragrant, bisexual. Floral tube more or less urnlike, surmounted by a short-campanulate calyx with 5 loosely triangular lobes, these persistent on the fruit. Petals 5, very small. Stamens 10, scarcely exserted. Fruit ellipsoid to obovoid, leathery, somewhat flattened, with 2 spongy wings, densely pubescent to glabrous, bearing 1 seed.

Fig. 176. **Conocarpus erectus.** (From Little and Wadsworth, *Common Trees of Puerto Rico and the Virgin Islands* (1964) Fig. 182)

Fig. 177. **Laguncularia racemosa.** (From Little and Wadsworth, *Common Trees of Puerto Rico and the Virgin Islands* (1964) Fig. 183)

Sandy or marly shores and borders of hammocks, somewhat intermixed with red and black mangroves and more or less between these and button mangrove on higher ground. Coastal areas of pen. Fla. southward from Brevard Co. on the east and Hernando Co. on the west, Fla. Keys; Berm.; trop. Am.; w. Afr.

Myrtaceae (MYRTLE FAMILY)

Melaleuca quinquenervia (Cav.) Blake.
CAJEPUT, PUNK-TREE, BOTTLE-BRUSH-TREE, PAPER-BARK TREE.

Evergreen tree with a slender, much-branched, somewhat columnar crown, the branches ascending on young trees, commonly somewhat drooping on older specimens. Bark thick, spongy, whitish at first, exfoliating in buffish to pale cinnamon-colored, many-papery layers giving it a very distinctive appearance. Leaves simple, very short-petiolate, arranged in 5 spiral rows; blades at first densely clothed with silvery, silky, appressed pubescence, becoming glabrous in age, dull green on both surfaces and dotted with reddish punctations, aromatic; narrowly elliptic, the principal veins more or less parallel, mostly 4–12 cm long. Flowers crowded in terminal spikes or panicles of spikes on woody axes, the stamens especially numerous and conspicuous, giving the inflorescence a bottle-brush aspect. Ovary inferior, the floral tube bearing 5 deciduous, ovate-deltoid calyx segments about 2 mm long, their apices obtuse. Petals 5, white, obovate to orbicular, 3–4 mm long. Stamens numerous, in 5 bundles opposite the petals, the filaments 1–1.5 cm long. Fruit short-cylindric to squarish, woody, 3-locular, dehiscing within and below the thick, circular rim of the floral tube. Seeds many, reddish brown and somewhat lustrous, asymmetric, long-angular, somewhat like tiny wedges, 0.5–1 mm long or a little more. (*M. leucadendra* of authors not L.)

The apices of the flowering twigs resume growth after flowering and during maturation of the fruits, as often as 3 times in a year on a given twig, thus segments of twig bearing sessile woody fruits may be sandwiched between leafy segments. Since the fruits are long persistent, older leafless branches are seen to bear interrupted fruiting segments.

Native of Austl., in cultivation and naturalized in s. Fla. in a great variety of habitats, some of them very wet, at least seasonally. In many places a pernicious weed tree and virtually replacing the native vegetation.

Rhizophoraceae (RED MANGROVE FAMILY)

Rhizophora mangle L. RED MANGROVE.   Fig. 178

Gregarious shrubs or trees, to 20 m tall. Bowed stilt roots numerous, arising from the trunk and branches. Leaves opposite, evergreen, dark green and shiny above, paler green beneath, leathery, stipulate; elliptic to elliptic-obovate or oblong, 4–12 (–15) cm long, 1.5–5 cm wide, apices obtuse to rounded, margins entire, short-petiolate. Flowers axillary in 2–8 (or more) -flowered cymes, bracts 2 and much reduced, bractlets 2, closely subtending each flower. Sepals 4(5), yellow, leathery, lanceolate, about 10 mm long. Petals 4, alternating with the sepals and often a little shorter than the sepals, leathery, pale yellow or yellowish white, densely pubescent with long, silky trichomes. Stamens 8, filaments very short, anthers elongate. Fruit cone-shaped, containing a single seed which germinates while the fruit is still attached, the first root becoming 2.5–3 dm long, fleshy and clublike.

Shallow water of coastal areas, shores of lagoons, creeks and rivers, mostly where water is salt or brackish, attaining its greatest size in wet flat places not inundated by the tides. Coasts of peninsular Fla., Volusia Co. on the east, Levy Co. on the west, southward through the Fla. Keys; trop. Am.

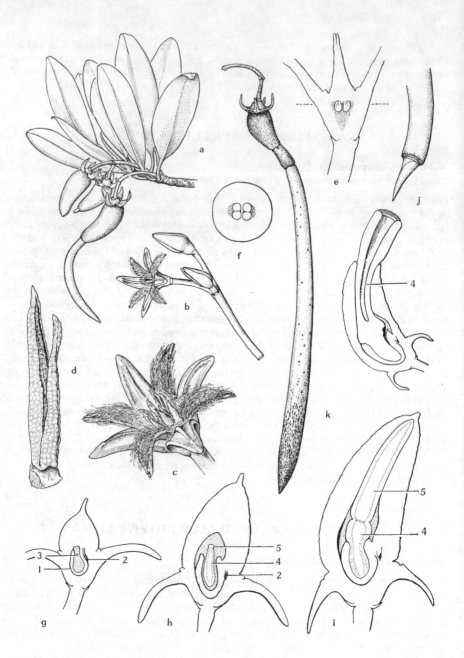

Fig. 178. **Rhizophora mangle:** a, fruiting branchlet, showing early stages of development of hypocotyl; b, inflorescence; c, flower; d, stamen; e, pistil and calyx in diagrammatic vertical section; f, two-locular ovary in diagrammatic cross-section at level of broken line in e; g–j, enlarging fruit in semidiagrammatic vertical section, showing progressive development from embryo to seedling—note endosperm (stippled), (1) seed coat, (2) aborted ovule, (3) embryo, (4) cotyledons, (5) hypocotyl (in j, hypocotyl disarticulated from cotyledonary tube revealing plumule); k, fruit with seedling showing fully elongated cotyledonary tube and hypocotyl. (From Graham in *Jour. Arn. Arb.* 45: 288. 1964)

Onagraceae (EVENING-PRIMROSE FAMILY)

Herbs or shrubs of diverse size and habit. Leaves simple, alternate or opposite; stipules minute and more or less glandular, or lacking. Flowers bisexual, mostly radially symmetrical, ovary inferior, the floral tube in some extending beyond the ovary. Calyx segments and petals 2–7 each (petals none in some species), distinct, borne at the summit of the floral tube. Stamens as many as the sepals and opposite them, or twice as many. Ovary 2–4-locular, style 1, stigma capitate, 2–4-lobed or -cleft. Fruit a many-seeded capsule, or an indehiscent nutlike structure.

1. Calyx segments, and petals if present, 3 each, 4 each, rarely 5 or 6; fruit a capsule, glabrous or pubescent but not beset with hooked bristles or hairs.
  2. The calyx segments persistent on the fruit; petals, if present, yellow; seed glabrous.   1. *Ludwigia*
  2. The calyx segments deciduous after anthesis; petals pinkish to rose, violet, purple, or white; seed bearing a terminal tuft of hairs.   2. *Epilobium*
1. Calyx segments and petals 2 each; fruit indehiscent, nutlike, beset with hooked bristles or hairs.
     3. *Circaea*

1. Ludwigia

Annual or perennial herbs or subshrubs. Leaves opposite or alternate, margins entire or minutely toothed, small stipules present. Floral tube not prolonged beyond the ovary. Calyx segments 4 or 5, rarely 6, persistent. Petals 4 or 5, rarely 6 (absent in some species), yellow, quickly deciduous and very easily detached. Fruit a many-seeded capsule, dehiscing longitudinally or terminally.

1. Leaves alternate.
  2. Stamens 8–10, in 2 series.
    3. Calyx segments 4 (rarely 5 in *L. peruviana*); seeds in several series in each locule, free of endocarp.
      4. Internodes conspicuously winged on the angles.   1. *L. decurrens*
      4. Internodes not at all or very faintly winged on the angles.
        5. Capsule broadly obconic or obpyramidal.   5. *L. peruviana*
        5. Capsule narrowly obconic or linear-oblong to long-clavate.
          6. Calyx segments 3–4 mm long; petals 4–5 mm long; capsule 10–16 mm long.   2. *L. erecta*
          6. Calyx segments 10–15 mm long; petals 15–30 mm long; capsule 2–5 cm long.
            7. Floral tube at anthesis markedly longer than the calyx segments; calyx segments at anthesis narrowly ovate, 3–5 mm wide near the base; petals 1–2 cm long, cuneate-obovate, corolla 2–4 cm across.   3. *L. octovalvis*
            7. Floral tube at anthesis the same length as the calyx segments or little longer; calyx segments at anthesis broadly ovate, 10 mm wide near the base; petals very broadly obovate, 3–5 cm long and as broad distally, corolla 6–7 cm across.   4. *L. bonariensis*
    3. Calyx segments 5, rarely 6 or 7; seeds in 1 series in each locule, seated within endocarp tissue.
      8. Stem erect; stalks of the flowers short, much shorter than the floral tube; seed embraced by a corky, horseshoelike segment of endocarp.   6. *L. leptocarpa*
      8. Stem's lower portions decumbent or creeping and rooting at the nodes; stalks of the flowers slender, much longer than the floral tube; seed embedded in a cube of woody endocarp.
        9. Bractlets at the base of the floral tube lanceolate; portions of the stem bearing flowers usually erect; stem and leaf surfaces glabrous or nearly so.   7. *L. peploides*
        9. Bractlets at the base of the floral tube or just below it on the flower stalk, deltoid or ovate; portions of the stem bearing flowers usually floating or creeping; stem, petioles, and leaf blades sparsely to densely pubescent with long, soft, shaggy hairs.   8. *L. uruguayensis*
  2. Stamens 4, in 1 series; calyx segments 4.
    10. Capsule little if any longer than broad.
      11. Flowers and capsules distinctly stalked.
        12. Bases of leaves cuneate-attenuate.   9. *L. alternifolia*
        12. Bases of leaves rounded or blunt.

13. Style 7–10 mm long; plant glabrous throughout, or, if pubescent, then the hairs very short.   10. *L. virgata*

13. Style not exceeding 3 mm long; plant pubescent throughout with curly-shaggy or long-spreading hairs.

   14. Calyx segments ovate; strongly reflexed after anthesis; pubescence of relatively short and spreading or curly-shaggy hairs.   11. *L. maritima*

   14. Calyx segments lanceolate to lance-ovate, spreading to erect after anthesis; pubescence of long, spreading hairs.   12. *L. hirtella*

  11. Flowers and capsules sessile or subsessile.

   15. Plant glabrous throughout.

    16. Leaves of the flowering stems oblanceolate, spatulate, or suborbicular, strongly cuneate tapered to subpetiolate bases.

     17. Calyx segments broadly ovate, about 1.5 mm long.   13. *L. microcarpa*

     17. Calyx segments triangular-acute, 2–2.5 mm long.   14. *L. curtissii*

    16. Leaves of the flowering stems lanceolate, oblanceolate, narrowly lanceolate, linear-oblong, or essentially linear, sessile and little if at all tapered basally.

     18. Capsule narrowly wing-angled.   15. *L. alata*

     18. Capsule obscurely quadrangular, the angles rounded.

      19. Flowers in globose headlike or oblong, bracteate spikes; leaves linear to linear-oblong, rounded basally.   16. *L. suffruticosa*

      19. Flowers in the axils of well separated leaves or bracteal leaves; leaves narrowly lanceolate, somewhat tapered basally.   17. *L. polycarpa*

   15. Plant pubescent throughout, or at least the calyx segments and floral tube pubescent.

    20. Plant copiously pilose throughout.   18. *L. pilosa*

    20. Plant with glabrous or sparsely short-pubescent stem and leaves; outer surfaces of calyx segments and capsule finely short-pubescent.   19. *L. sphaerocarpa*

 10. Capsule definitely longer than broad.

  21. Capsule shallowly grooved longitudinally below each calyx segment; petals none.
      20. *L. glandulosa*

  21. Capsule not longitudinally grooved; petals present.

   22. Calyx segments triangular-acute, 2–3 mm long, very much shorter than the capsule; seed buff.   21. *L. linearis*

   22. Calyx segments triangular-subulate, 5–6 mm long, ⅔ as long as the capsule or a little more; seed reddish brown.   22. *L. linifolia*

1. Leaves opposite.

 23. Flowers or capsules distinctly stalked, stalks 5 mm long or much more.

  24. Stalks of flowers or capsules markedly longer than the subtending leaves; petals somewhat longer than the calyx segments.   23. *L. arcuata*

  24. Stalks of the flowers or capsules shorter than the subtending leaves; petals equaling the calyx segments.   24. *L. brevipes*

 23. Flowers or capsules sessile, or (in ours) rarely with stalks 1–3 mm long.

  25. Plant with stem and both leaf surfaces, floral tube or capsule, and outer surfaces of calyx segments markedly hirsute.   25. *L. spathulata*

  25. Plant glabrous, or some parts very sparsely and minutely pubescent.

   26. Floral tube or capsule with a pair of bractlets borne somewhat above the base; floral tube or capsule without longitudinal bands of darker green.   26. *L. repens*

   26. Floral tube or capsule usually without bractlets, or, if a minute pair of bractlets present, then these borne below the base of the floral tube or capsule; floral tube with 4 longitudinal bands of a more definitely green color than that of the tissue between them.   27. *L. palustris*

1. Ludwigia decurrens Walt.   Fig. 179

Annual herb, of variable stature from 2 to 25 dm tall and from simple to widely and diffusely branched. Stem glabrous, with narrow wing-angles on the internodes, 2 running down from each leaf base. Leaves alternate, sessile or subsessile, lanceolate, 5–18 cm long, bases cuneate to rounded, apices acuminate, surfaces glabrous or sparsely short-pubescent, margins very finely scabrid. Flowers solitary in the axils of reduced leaves or bracteal leaves of the slender, flexuous branchlets, sessile or on stalks 1–5 (–10) mm long, 2 minute, scalelike bractlets at the base of the floral tube or on the flower stalks. Calyx segments 4, mostly triangular-subulate, 7–10 mm long. Petals 4, yellow, suborbicular to obovate, broadly rounded

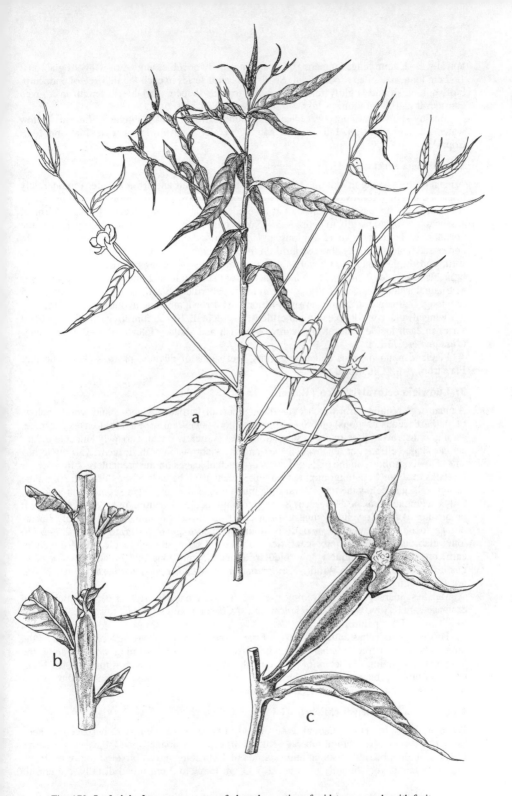

Fig. 179. **Ludwigia decurrens:** a, top of plant; b, portion of midstem; c, node with fruit.

apically, 8–12 mm long. Stamens 8. Capsule narrowly obconic, angled or winged, glabrous, 1–2 cm long, irregularly loculicidal. Seeds in several series in each locule, free of endocarp, buff, oblongish and slightly falcate, varying to pear-shaped, variable in length in a given capsule, 0.3–0.8 mm long. (*Jussiaea decurrens* (Walt.) DC.)

Marshy shores, swamps, wet clearings, muddy stream banks, ditches, often in shallow water. Va. and W.Va., s. Ind., s. Ill., Mo., generally southward to s. pen. Fla. and Tex.; trop. Am.

2. Ludwigia erecta (L.) Hara.

Annual herb, varying in stature from about 1 to 10 dm tall and from simple to moderately branched. Stem glabrous, internodes very narrowly, almost imperceptibly, angled by decurrent leaf bases, the angles usually not at all evident on larger, older stems. Leaves alternate, sessile or narrowed basally to petioles 2–15 mm long, lanceolate, cuneate basally, long-acute apically, the larger ones to 12 cm long and 2–3 cm broad medially, lateral veins prominent and raised beneath, finely short-scabrid on the midvein beneath, often on the lateral veins as well, and on the margins. Flowers solitary in the axils of leaves or bracteal leaves of the branchlets, sessile, 2 minute, scalelike bractlets at the base of the floral tube. Calyx segments 4, triangular, long-acute apically, 3–4 mm long, puberulent. Petals 4, yellow, obovate, 4–5 mm long. Stamens 8. Capsule narrowly oblong to narrowly obconic in outline, quadrangular, not winged, glabrous or sparsely puberulent, loculicidal, 10–16 mm long. Seeds in several series in each locule, free of endocarp, oblongish and slightly falcate, 0.3–0.4 mm long. (*Jussiaea erecta* L.)

Pineland pond margins and depressions, wet sands of ditches, prairies. Pen. and n.e. Fla.; trop. Am.; Old World tropics.

3. Ludwigia octovalvis (Jacq.) Raven.

Annual or perennial herb, herbaceous throughout, or herbaceous above and woody below. Stem (herbaceous portions) glabrous, sparsely short-pubescent, or hirsute. Leaves alternate, sessile, or narrowed proximally to subpetiolate bases, nearly linear, narrowly lanceolate, lanceolate, lance-elliptic, or oblanceolate, essentially glabrous on both surfaces, varying to hirsute. Flowers solitary in the axils of leaves or bracteal leaves on the branchlets, subsessile or on stalks to about 1 cm long, minute, scalelike bractlets at the base of the floral tube or on the flower stalk just below the tube; floral tube much longer than the calyx segments at anthesis. Calyx segments 4, ovate, 3-nerved, 8–14 mm long, acute to acuminate apically, puberulent or hirsute. Petals 4, yellow, cuneate-obovate, 1–2 cm long. Stamens 8. Capsule linear-oblong to obscurely clavate in outline, obtusely quadrangular, several-ribbed, 2.5–5 cm long, glabrous or sparsely pubescent. Seeds in several series in each locule, free from endocarp, brown, shiny, obliquely suborbicular, about 0.5 mm long and broad. (*Jussiaea angustifolia* Lam., *J. scabra* Willd., *J. suffruticosa* L. = *Ludwigia suffruticosa* Gomez Maza, not *L. suffruticosa* Walt.)

Marshes, marshy shores, swampy woodlands, wet clearings, stream banks, ditches and drainage canals, low, wet places in fallow fields. Coastal plain, s.e. N.C. to s. Fla., westward to e. half of Tex.; pantropical.

This is a herbaceous annual in parts of the range where frosts are severe; in parts of the range where frost-free or where and when frosts not severe becoming woody below, the lower stem attaining diameters of 1–3 cm, the wood very hard, with a thin, brown bark, upper portions herbaceous; subshrubby plants to 3 m tall. In our range, the subshrubby plants occur from about cen. pen. Fla. southward in Fla.

4. Ludwigia bonariensis (Micheli) Hara.

Perennial herb. In general habital features and in pubescence, similar to herbaceous specimens of *L. octovalvis*. Floral tube at anthesis little, if any, longer than the calyx segments. Calyx segments broadly ovate at anthesis, about 1 cm long and as broad at the base. Petals relatively large, very broadly obovate, 1.5–3 cm long, to 3 cm broad distally, the corolla about 6 cm across. Capsule obconic, 2–3.5 cm long, sparsely hirsute. Seeds as in *L. octovalvis*. (*Jussiaea neglecta* Lam.)

Marshes, swamps, wet clearings. Local, s.e. N.C.; Fla., perhaps elsewhere near the Gulf coast. Native of trop. Am.

5. Ludwigia peruviana (L.) Hara. PRIMROSE-WILLOW.

Plant very coarse, woody below, herbaceous above, commonly with numerous stems from near the base, to 3 m tall or more, lower stems often attaining a diameter of 3–4 cm in a single season, the wood becoming very hard; in areas where freezing occurs, depending upon the severity of the cold, tops of plant dying, dying back to near the base, or in extreme cold dying altogether. Herbaceous portions of stem copiously shaggy-pubescent, hairs tawny, older stems becoming glabrous, eventually the woody portions with thin, brown bark. Leaves alternate, lanceolate, lance-elliptic, or lance-ovate, cuneate basally to subpetiolate bases, acute to acuminate apically, 5–15 cm long and to 3 cm wide medially, both surfaces tawny, shaggy-pubescent, the lower more copiously so than the upper. Flowers solitary in axils of reduced leaves near the tips of branchlets. Flower stalks 1–2 (–3.5) cm long, shaggy-hairy, minute scalelike bractlets at the base of the floral tube or distally on the flower stalks. Calyx segments 4(5), triangular and with acuminate tips, 10–15 mm long and 5–7 mm wide basally, shaggy-pubescent. Petals 4(5), mostly bright yellow and showy, obovate with broadly rounded, truncate, or broadly and shallowly concave apices, mostly about 2.5 cm long and as broad distally. Stamens 8 or 10. Capsule broadly obconic in outline, quadrangular, 1–3 cm long, shaggy-pubescent, irregularly loculicidal. Seeds in several series in each locule, free of endocarp, buff to light brown, plump, elliptic to oblongish and slightly falcate, usually varying in size in a given capsule from about 0.3–0.8 mm long. (*Jussiaea peruviana* L.)

Weedy, commonly in shallow water, ditches and drainage canals, swales, sloughs, marshy shores, wet clearings. Common and abundant in cen. pen. and s. Fla., sporadic elsewhere in Fla. and s. Ga.; trop. Am.

6. Ludwigia leptocarpa (Nutt.) Hara.   Fig. 180

Annual herb, varying greatly in stature from about 1–20 dm tall and from simple to very widely and profusely branched; occasional large specimens with lower stems to 1 cm in diameter and very hard. Stem shaggy-pubescent with long hairs, the lower portions of large individuals often becoming glabrous. Leaves alternate, lanceolate to lance-elliptic, or oblanceolate, subsessile, or gradually narrowed basally into petioles to 2 cm long, acute apically, larger ones to 15 cm long and 3 cm broad medially, upper surfaces sparsely pubescent, lower usually pubescent mainly along the midrib, both surfaces finely papillose. Flowers solitary in the axils of leaves or bracteal leaves of the stiff branchlets, very numerous in larger specimens. Flower stalks 1–15 mm long, puberulent, at anthesis scarcely differentiated from the floral tube, bearing 2 minute, scalelike bractlets alternately. Calyx segments 5 (–7), lanceolate, acuminate apically, 5–8 mm long, their lower surfaces sparsely pubescent. Petals 5 (–7), yellow, broadly obovate, shallowly notched apically, 5–10 mm long. Stamens twice as many as the calyx segments. Capsule narrowly cylindric, 10–14-ribbed, sparsely pubescent with longish hairs, mostly 3–5 cm long, irregularly loculicidal. Seeds in 1 series in each locule, each seed tiny, buff-colored, seated within a brown corky, horseshoelike segment of endocarp, this obovate in its outline, flat on either side, somewhat over 1 mm long. (*Jussiaea leptocarpa* Nutt.)

Marshes and marshy shores, wet clearings, sand and gravel bars of streams, wet ditches, alluvial outwash. Coastal plain, N.C. to cen. pen. Fla., westward to e. Tex., northward in the interior to s.e. Mo. and s. Ill.; trop. Am.

7. Ludwigia peploides (HBK.) Raven subsp. peploides.   Fig. 181

Perennial herb, stems creeping and rooting at the nodes over wet substrate or in shallow water, sometimes forming floating mats, the flowering branches usually somewhat ascend-

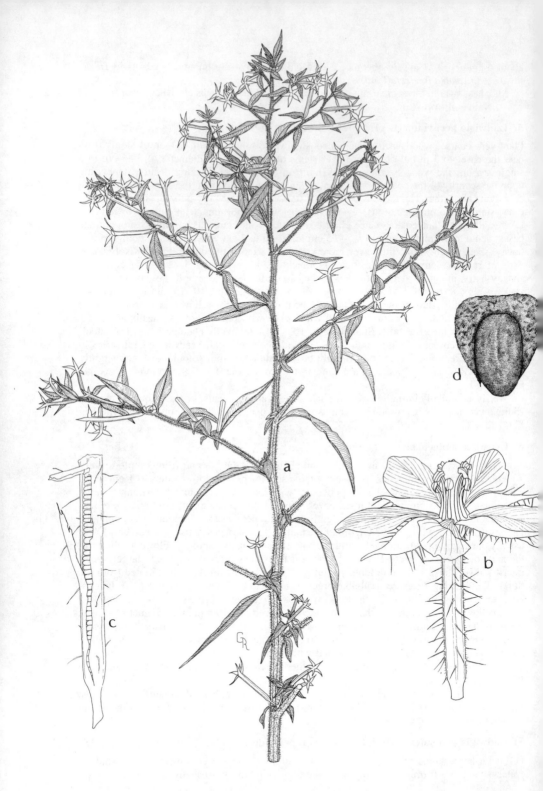

Fig. 180. **Ludwigia leptocarpa:** a, end of a branch; b, flower, from the side; c, part of capsule; d, seed with embracing segment of endocarp.

Fig. 181. **Ludwigia peploides** subsp. **peploides:** A, habit; B, flower; C, capsule; D, seeds with endocarp; E, seeds. (From Reed, *Selected Weeds of the United States* (1970) Fig. 133, as *Jussiaea repens* var. *glabrescens*)

ing, usually glabrous, sometimes sparsely pubescent. Leaves with petioles 1–5 cm long; blades of leaves on young shoots or those of lower portions of branches usually oblanceolate to spatulate, sometimes suborbicular, variable in size from 1–4 cm long; blades of leaves on distal portions of branches usually somewhat larger than those below, lanceolate to narrowly elliptic; surfaces usually glabrous, margins glabrous or sparsely short-pubescent. Flowers solitary in the axils of upper leaves, their stalks 1–6 cm long, the floral tube at anthesis scarcely thicker than the stalk below, two opposite or closely alternate, small, deltoid or ovate bractlets at or a little below the junction of the stalk and floral tube. Floral tube glabrous, or less frequently with a few longish hairs. Calyx segments 5, usually glabrous, sometimes with few longish hairs, mostly subulate, varying to lanceolate or ovate, 8–12 mm long. Petals 5, yellow, obovate, mostly 10–15 mm long. Stamens 10. Capsule hard, cylindric, glabrous, 2–4 cm long, 3–4 mm in diameter, tardily dehiscent. Seeds in 1 vertical series in each locule, each embedded in a cube of woody endocarp. (*Jussiaea diffusa* of authors not Forsk., *J. repens* L. var. *peploides* (HBK.) Griseb.; incl. *J. repens* var. *glabrescens* Kuntze)

Ponds, streams, sloughs, ditches, drainage and irrigation canals. Questionably native, sporadic in our range, chiefly on the coastal plain, and westwardly beyond the range to Ariz.; W.I., Mex., C.Am., S.Am. except Chile.

8. Ludwigia uruguayensis (Camb.) Hara. Fig. 182

Perennial herb, lower stems decumbent and rooting freely at the nodes, ascending portions simple to freely branched. Lower and median leaves petiolate, blades varyingly suborbicular, short-obovate, spatulate, or oblanceolate, apices usually rounded; upwardly leaves commonly becoming subsessile or sessile, lanceolate, oblanceolate, narrowly elliptic, acute at both extremities; stems, petioles, and blades varying from sparsely to densely pubescent with long, shaggy, soft hairs. Flowers solitary from upper leaf axils, their stalks 1–2 (–5) cm long, at anthesis scarcely differentiated from the floral tube, a pair of small lanceolate bractlets at the junction of the stalk and floral tube; stalks and floral tube copiously pubescent with long, shaggy, soft hairs. Calyx segments usually 5, sometimes 6, subulate to lanceolate, 10–12 mm long, long-acute, pubescent exteriorly. Petals usually 5, sometimes 6, bright yellow, obovate, broadly rounded to truncate apically, 1.2–2.2 cm long, 1–2 cm wide distally. Stamens 10. Capsules pubescent, obconic to short-cylindric, 1–2.5 cm long, 3–4 mm in diameter, tardily dehiscent. Seeds in 1 series in each locule, each embedded in a cube of woody endocarp. (*Jussiaea uruguayensis* Camb., *J. grandifolia* Michx. not R. & P.; incl. *J. michauxiana* Fern.)

Marshes, marshy shores, swamps, ponds and lakes, sloughs, ditches, irrigation and drainage canals, often forming dense, sometimes floating, mats. N.C. to Fla., westward to Okla. and Tex.; southward to n. Arg.

9. Ludwigia alternifolia L. SEED-BOX, RATTLE-BOX. Fig. 183

Plant to about 12 dm tall, commonly diffusely branched, usually having a fascicle of somewhat tuberous roots from the caudex. Stem usually subglabrous sometimes copiously pubescent with spreading hairs. Leaves alternate, the principal ones lanceolate, 5–10 cm long, narrowed below to subpetiolate bases, acute to obtuse apically, varyingly glabrous, sparsely short-pubescent on both surfaces and on the margins, sometimes copiously pubescent on both surfaces. Flowers solitary in the axils of somewhat reduced or bracteal leaves on the branchlets, their stalks 2–5 mm long and with 2 lanceolate bractlets distally. Calyx segments 4, long-triangular to ovate, 6–10 mm long, mostly finely short-pubescent marginally. Petals 4, yellow, about 6 mm long, obovate, apically broadly rounded to truncate and about as broad as long. Stamens 4. Capsule 4-angled, nearly cubical, the angles usually narrowly winged, 4–6 mm long and about as broad, glabrous or sparsely pubescent, rarely densely pubescent, dehiscent at first by a terminal pore, later loculicidal as well. Seed pale buff, smooth, about 0.5 mm long, plump, mostly asymmetrical-oblongish. (Incl. var. *pubescens* Palm. and Steyerm.)

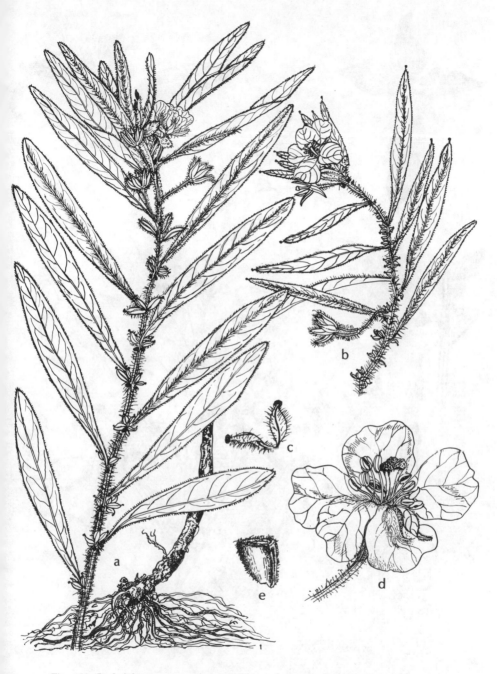

Fig. 182. **Ludwigia uruguayensis:** a, portions of plant; b, branchlet; c, stipules; d, flower; e, seed. (From Correll and Correll. 1972)

Fig. 183. **Ludwigia alternifolia:** a, base of plant; b, midsection of stem; c, tip of a branch; d, flower, from above; e, capsule, one-quarter cut away and turned back.

Banks of streams, marshes, marshy shores, bottomland woodlands and clearings, swales, sloughs, ditches. Mass. to s. Ont., s. Mich., Iowa, and Kan., generally southward to n. Fla. and e. Tex.

10. Ludwigia virgata Michx. Fig. 184

Perennial herb with horizontal, subterranean rootstocks from which arise tuberous roots singly or in a fascicle. Stems sometimes simple below, with few to numerous strongly ascending branches above, sometimes with numerous strongly ascending branches from near the base; some plants densely pubescent throughout with very short hairs, some glabrous or nearly so. Leaves alternate, sessile, the lowermost varying from oblong and about 2–3 cm long, 1.5 cm broad, to narrowly linear-oblong, bases and apices rounded to blunt, diminishing in size upwardly to bracteal leaves on the flowering branches. Flowering branches essentially racemose, wandlike, flower stalks 6–10 (–15) mm long and with 2 small, inconspicuous, opposite or closely alternate bractlets usually somewhat above their middles. Calyx segments 4, ovate to oblong-triangular, 5–10 mm long, strongly reflexed after anthesis, often completely hiding the flower stalks, usually deciduous about the time the capsules are fully mature, their lower surfaces very short-pubescent, the upper similarly pubescent or glabrous. Petals 4, yellow, quickly deciduous, obovate or oblong-obovate, 10–15 mm long, 6–10 mm broad. Style 7–9 mm long, with a pronounced disclike base, stigma umbrellalike. Stamens 4. Capsule hard, 5–7 mm long, 3.5–5 mm broad, 4-angled, short-oblong to globose in outline, surmounted by the style base (the style above the base quickly deciduous after anthesis) which during maturation of the fruit becomes 4-lobed and domelike; dehiscence occurring through a central pore of the style base. Seeds buff-colored to brown, smooth and somewhat shiny (somewhat reticulate seen with suitable magnification), 0.5–0.8 mm long, somewhat falcate-oblong, rounded at one extremity, apiculate at the other.

Seasonally wet pine savannas and flatwoods, bogs, not infrequently on well-drained sandy pinelands, especially after clearing. Coastal plain, s.e. Va. to s. pen. Fla., Fla. Panhandle and adjacent Ala.

11. Ludwigia maritima Harper.

Closely similar to *L. virgata* and *L. hirtella* in general features. Pubescent throughout, hairs of the stem, leaf surfaces, flower stalks, and capsules short, curly-shaggy. Calyx segments ovate, strongly reflexed after anthesis, usually deciduous by the time the capsules are fully mature. Style not exceeding 3 mm long. Capsule tending to fracture easily, 6–10 mm long, 4.5–8 mm broad. Seeds similar to those of *L. virgata* but usually not exceeding 0.5 mm long.

Pine savannas and flatwoods, bogs, ditches, banks of streams draining savannas or flatwoods, shores of open ponds, cypress-gum depressions. Coastal plain, N.C. to s. pen. Fla., westward to e. La.

12. Ludwigia hirtella Raf. Fig. 185

Closely similar to *L. virgata* and *L. maritima* in general features. Pubescent throughout, the hairs of the stem, leaf surfaces, flower stalks, and capsules long-spreading. Calyx segments lanceolate to ovate-lanceolate, spreading to erect after anthesis, sometimes deciduous by the time the capsules are fully mature, their lower surfaces pubescent with long, spreading hairs, upper usually glabrous. Style not exceeding 3 mm long. Capsule 5–8 mm long and broad, fracturing easily at maturity, dehiscent as in *L. virgata*. Seeds essentially as in *L. virgata*.

Pine savannas and flatwoods, burned-over shrub-tree bogs, ditches. Coastal plain, s. N.J. to Fla. Panhandle, westward to e. Tex., northward in the interior, s.e. Okla., Ark., Tenn., Ky.

13. Ludwigia microcarpa Michx. Fig. 186

Relatively diminutive perennial, glabrous throughout. Flowering stems simple to diffusely, often divaricately, branched, stiffly to weakly erect, or sprawling, 1–4.5 dm tall; producing

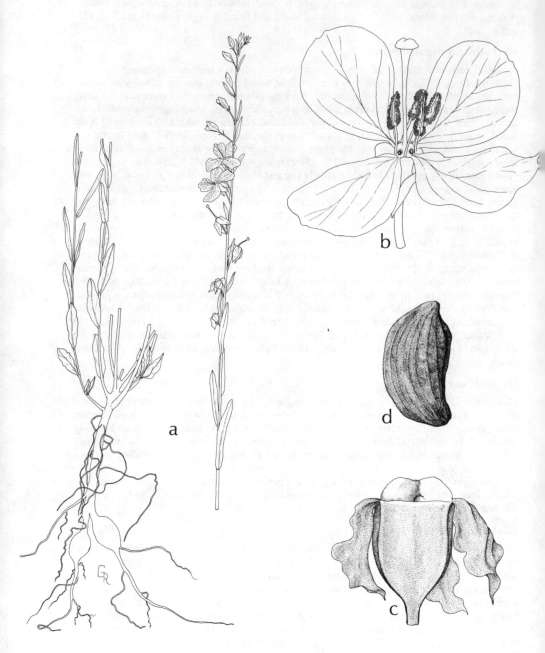

Fig. 184. **Ludwigia virgata:** a, plant; b, flower; c, capsule; d, seed.

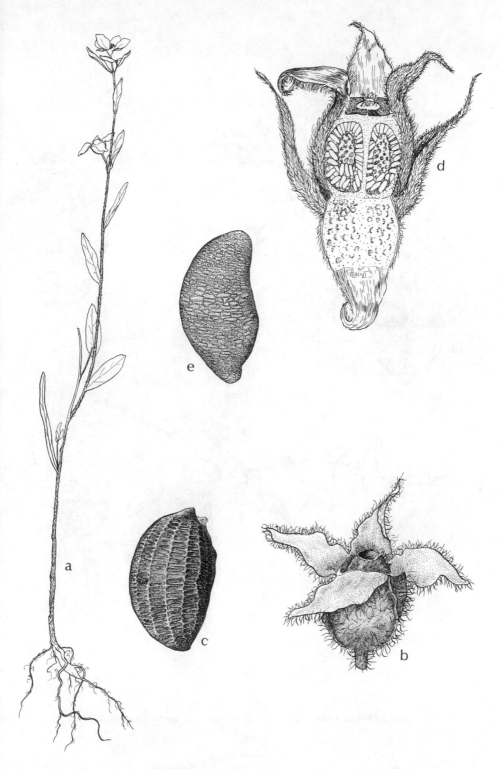

Fig. 185. a–c, **Ludwigia hirtella:** a, habit; b, fruit; c, seed; d,e, **Ludwigia pilosa:** d, fruit, slit down one side; e, seed.

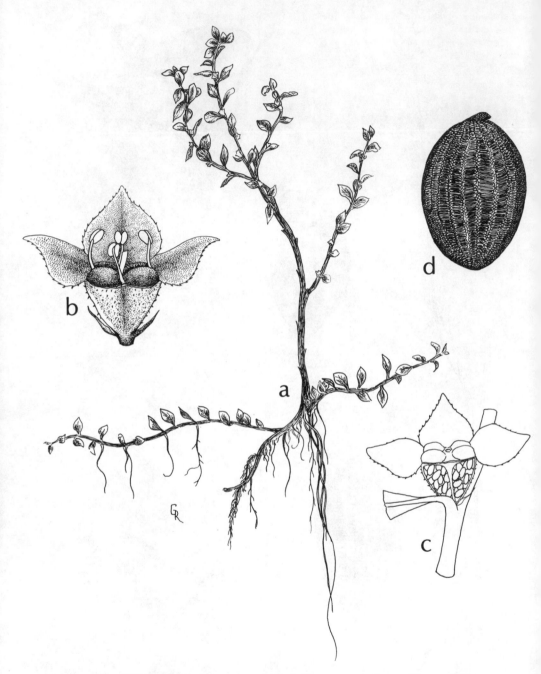

Fig. 186. **Ludwigia microcarpa:** a, habit; b, flower; c, capsule, one side removed; d, seed.

slender, leafy, autumnal stolons. Leaves alternate, oblanceolate, spatulate, or suborbicular, narrowed to barely petiolate bases, mostly blunt to rounded apically, 5–12 mm long or a little more, with a few sessile, marginal glands; leaves of the stolons similar. Flowers solitary in most of the leaf axils, sessile or subsessile; bractlets linear, inconspicuous, nearly as long as the floral tube. Calyx segments 4, broadly ovate, about 1.5 mm long. Petals none. Stamens 4. Capsule hemispheric to broadly obpyramidal in outline, somewhat quadrangular, the angles rounded, 1.5–2 mm long. Seed dark reddish brown, asymmetrically ovate-ellipsoid, about 0.4 mm long.

Marl prairies, bogs, wet sands or peaty sands of pond margins and marshy shores, pinelands, ditches, swales, marshes. Chiefly coastal plain, N.C. to s. Fla., westward to La.; cen. Tenn., s.e. Mo.

14. Ludwigia curtissii Chapm.   Fig. 187

Low perennial, glabrous throughout, flowering stem slender, erect or weakly erect, simple to few-branched, 1–5 dm tall, producing short, leafy, autumnal offshoots from the base. Leaves alternate, those of both the flowering stems and basal offshoots oblanceolate to spatulate, attenuate below to subpetiolate bases, rounded or obtuse apically, 8–20 mm long, 2–10 mm broad distally, with a few sessile glands marginally. Flowers solitary in the leaf axils more or less throughout; bractlets lanceolate, borne from the base of the floral tube, about half as long as the floral tube. Calyx segments 4, triangular-acute or -acuminate, 2–2.5 mm long. Petals none. Stamens 4. Capsule shortly obpyramidal in outline, obscurely quadrangular, angles rounded, mostly 2–3 mm long. Seed buff-colored, plump, mostly oblique, a little longer than broad, about 0.4 mm long. (Incl. *L. simpsonii* Chapm., *L. spathulifolia* Small)

Pine savannas and flatwoods, sandy-peaty depressions and ditches, pineland ponds, cypress ponds or depressions, marl prairies, solution pits in lime-rock of pinelands. Cen. pen. to s. Fla.

15. Ludwigia alata Ell.   Fig. 188

Perennial, glabrous throughout, flowering stem erect, relatively slender, simple to moderately branched, 4–8 dm tall, submersed portions, if any, spongy, producing leafy autumnal stolons from the base. Leaves alternate, the principal ones of the flowering stem sessile or subsessile, narrowly lanceolate, linear-lanceolate, or linear, 3–6 cm long and 2–8 mm broad, mostly acute at both extremities, margins with a few, small, sessile glands; leaves of the stolons short-petiolate, blades oblanceolate, obovate, or suborbicular, 1–2 cm long. Flowers solitary in the axils of leaves or reduced leaves of the branchlets, sessile, usually well separated from each other, rarely congested, the bracteal leaves being very closely approximate; bractlets lanceolate, inserted at the base of the floral tube, usually as long as the floral tube or slightly longer. Calyx segments 4, triangular, about equaling the length of the floral tube. Petals none. Stamens 4. Capsule 4-angled, the angles narrowly winged, shortly obpyramidal in outline, 3–4 mm long and as broad. Seed buff-colored, plump, oblong in outline, twice as long as thick or a little more, about 0.6 mm long. (Incl. *L. lanceolata* Ell., *L. simulata* Small)

Swamps, bogs, wet peaty soil in ditches and depressions, brackish or fresh marshes, interdune swales, wet stream banks, depressions in pine flatwoods, often in shallow water. Coastal plain, s.e. Va. to s. Fla., westward to Miss. (La.?)

16. Ludwigia suffruticosa Walt.   Fig. 189

Rhizomatous, colonial perennial and forming leafy, autumnal stolons. Flowering stems erect, simple or few-branched, 3–8 dm tall, glabrous. Leaves alternate, glabrous, the principal medial ones of the flowering stems sessile, linear to linear-oblong, 1–4 cm long, mostly 3–10 mm broad, rounded basally, long-acute apically; leaves near the base of the flowering stem sometimes like the medial ones, sometimes spatulate or oblanceolate, apices blunt to rounded, 5–15 mm long; leaves of stolons oblanceolate, spatulate, oblong, or suborbicular,

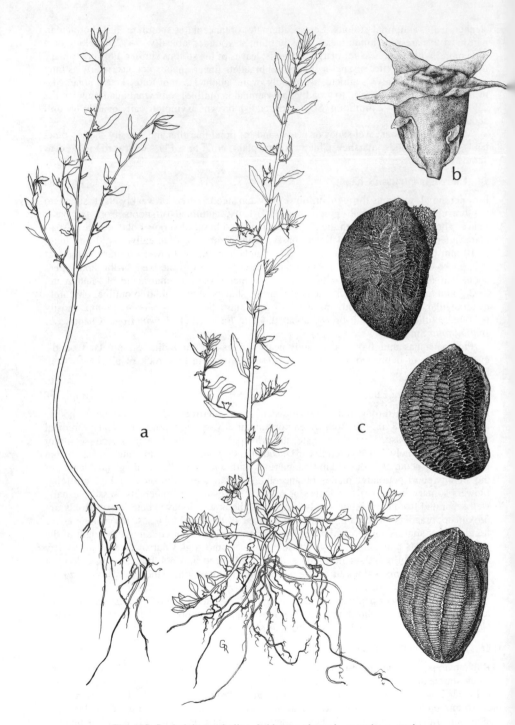

Fig. 187. **Ludwigia curtissii:** a, habit, two plants; b, capsule; c, seeds.

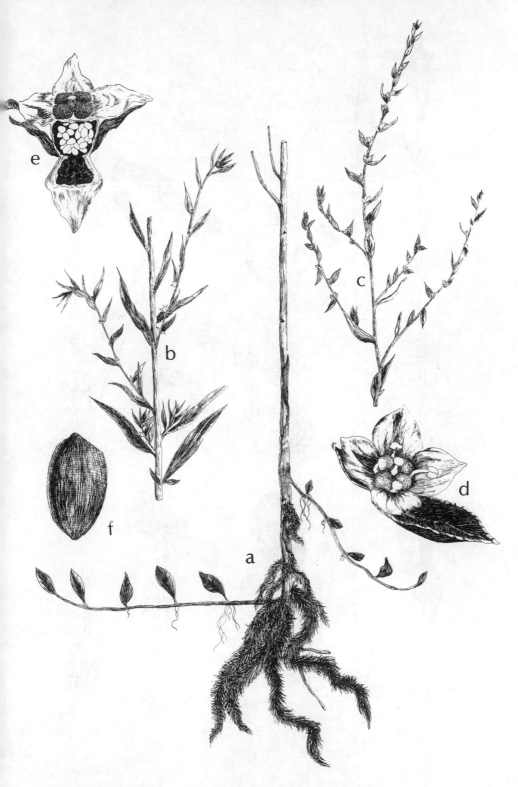

Fig. 188. **Ludwigia alata:** a, base of plant; b, midsection of stem; c, top of stem; d, flower, from above; e, capsule, with one-quarter cut away and pulled back; f, seed.

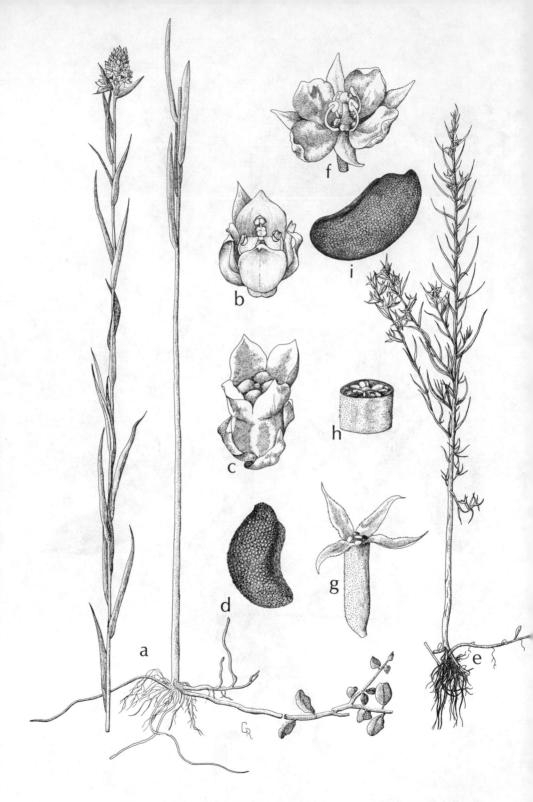

Fig. 189. a–d, **Ludwigia suffruticosa:** a, plant; b, flower; c, capsule; d, seed; e–i, **Ludwigia linifolia:** e, habit; f, flower; g, capsule; h, cross-section of capsule; i, seed.

sessile or very short-petiolate, 5–20 mm long. Flowers in globose to oblong, congested, headlike spikes terminating the main stem or its branches, the inflorescence axis pilose but the pubescence usually hidden by the congested flowers or fruits; bractlets relatively large, ovate to lance-ovate, inserted just below the floral tube, from about as long as the floral tube to as long as the calyx segments. Calyx segments 4, glabrous, broadly ovate-deltoid, 2.5–3 mm long, somewhat shorter than the floral tube. Petals none or minute. Stamens 4. Capsule obscurely quadrangular, the angles rounded, broadly obpyramidal in outline, 3.5–4 mm long, glabrous. Seed light brown, oblong to falcate-ellipsoid, very faintly reticulate, about 0.5 mm long.

Margins and exposed shores of sinkhole ponds, shallow pools, wet pineland depressions, wet savannas, brackish swales, ditches. Coastal plain, s.e. N.C. to s. Fla., westward to s. Miss.

17. Ludwigia polycarpa Short and Peter.

Perennial, glabrous throughout, flowering stem 1–10 dm tall, simple to diffusely branched, submersed portions, if any, spongy, producing leafy, autumnal stolons from the base. Leaves attenuate, the principal ones of the flowering stems sessile, narrowly lanceolate, 3–12 cm long, 5–8 mm broad, tapered to acute extremities, sometimes minutely scabrid marginally; leaves of the stolons crowded, lanceolate, oblanceolate, or nearly linear, 1–2.5 cm long, sessile or subsessile. Flowers solitary in the axils of foliage leaves, sessile or subsessile; bractlets lanceolate to subulate, inserted on the floral tube, sometimes near its base, varying to the middle, variable in length from plant to plant from about ¼ as long as the floral tube to about as long. Calyx segments 4, deltoid, about half as long as the floral tube. Petals none. Stamens 4. Capsule short-oblong to somewhat turbinate, not angled, 4–5 mm long, shallowly grooved longitudinally below each calyx segment. Seed buff-colored, surface somewhat mealy, falcate-oblongish, 0.4–0.6 mm long.

Marshy or wet shores of ponds and streams, swamps, wet prairies, ditches. s.w. Maine to Conn., westward to s. Ont., Minn., southward in the interior to Tenn., Mo., Kan.; s. Ala.

18. Ludwigia pilosa Walt.   Fig. 185

Perennial, flowering stems erect, 5–12 dm tall, usually bushy-branched, copiously pilose throughout; producing from the base elongate, leafy, autumnal stolons, their pubescence relatively sparse and short. Leaves alternate, the principal ones of the flowering stems lanceolate or elliptic, both surfaces pilose, 4–7.5 cm long and to about 1.2 cm broad, sessile or subsessile, tapered to both extremities; leaves of stolons short-petiolate, blades mostly obovate to suborbicular, 1–2 cm long. Flowers solitary and sessile in the axils of reduced leaves or bracteal leaves, usually a few crowded at the tips of each branchlet; bractlets subulate, inserted at or just above the base of the floral tube and extending to about the middle of the calyx segments. Calyx segments 4, ovate-triangular below, markedly acuminate apically, somewhat longer than the floral tube, pilose beneath. Petals none or minute. Stamens 4. Capsule pubescent, hemispheric to subglobose, terete or nearly so in cross-section, 3–4 mm long. Seed light brown, oblique-ellipsoid, obscurely reticulate, about 0.5 mm long.

Marshes, marshy shores, swamps, cypress-gum ponds and depressions, pools, borrow pits, drainage ditches and canals. Coastal plain, s.e. Va. to cen. pen. Fla., westward to s.e. Tex.

19. Ludwigia sphaerocarpa Ell.   Fig. 190

Perennial, flowering stem erect, glabrous or pubescent, 3–10 dm tall, little-branched to diffusely branched, producing from the base autumnal, leafy stolons; submersed portions of stems, if any, becoming very spongy. Leaves alternate, the principal ones of the flowering stems sessile, lanceolate to linear-lanceolate, 3–10 cm long, 3–8 (–10) mm wide, acute at both extremities, glabrous or sparsely short-pubescent, those of the flowering branchlets similar but reduced; leaves of the stolons oblanceolate to obovate, cuneate below to subpetiolate bases, obtuse to rounded apically. Flowers solitary in the axils of reduced leaves on

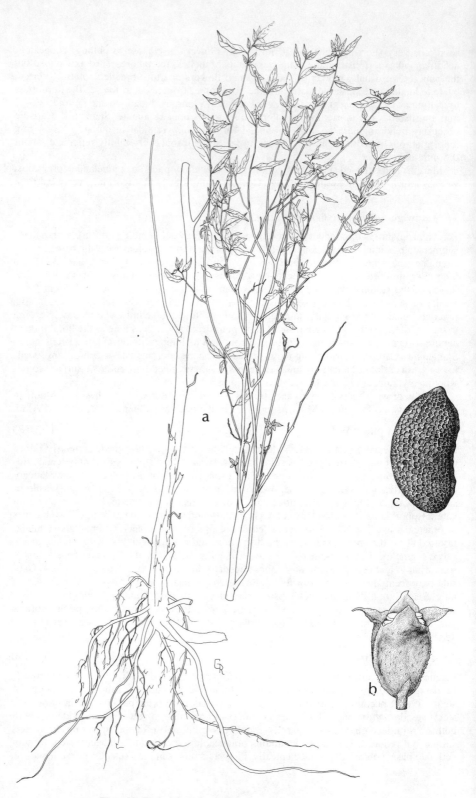

Fig. 190. **Ludwigia sphaerocarpa:** a, plant; b, flower; c, seed.

the branchlets, on a given branch varying from well separated from each other to congested, subsessile; bractlets from minute and scalelike and scarcely evident to lanceolate and up to ⅔ as long as the floral tube, inserted at the base of the latter. Calyx segments 4, triangular, their length about equal to that of the floral tube, short-pubescent beneath. Petals none. Stamens 4. Capsule usually hemispheric and wider than long, varying to subglobose and somewhat longer than broad, 2.5–4.5 mm long, terete in cross-section or with broadly rounded lobes, surface finely short-pubescent. Seed buff-colored, with a finely reticulate surface layer (seen with suitable magnification) which sloughs readily, about 0.5 mm long, falcate-oblongish.

Often in shallow water, marshes, marshy shores, cypress-gum depressions and swamps, depressions in pine savannas and flatwoods, bogs, drainage ditches and canals, borrow pits. Coastal plain, Mass. to s.cen. pen. Fla., westward to e. Tex.; n.w. Ind., and s. Mich.

20. Ludwigia glandulosa Walt.   Fig. 191

Perennial, flowering stem glabrous, erect, usually much branched, the branches often divaricate from near the base, submersed portions, if any, notably spongy, 2–10 dm tall, producing elongate, leafy, autumnal stolons. Leaves alternate, the principal ones of the flowering stems lanceolate, elliptic, or oblanceolate, 3–10 cm long and 6–12 (–20) mm broad, attenuate below to subpetiolate bases, acute apically, surfaces glabrous, often very finely toothed on the margins and with a few sessile glands; leaves of the stolons essentially similar. Flowers solitary in the axils of leaves more or less throughout, subsessile; bractlets minute at the base of the floral tube or none. Calyx segments ovate-acuminate, glabrous or sparsely short-pubescent, 1–2 mm long, very much shorter than the floral tube. Petals none. Stamens 4. Capsule oblong in outline, shallowly grooved longitudinally below each calyx segment, convex between the grooves, mostly 6–8 mm long, glabrous or sparsely short-pubescent. Seed pale buff-colored, falcate-oblongish, minutely reticulate, about 0.8 mm long.

Marshes, marshy shores, swamps, wet clearings, pools, cypress-gum ponds and depressions, sloughs, ditches, commonly in shallow water. Coastal plain and piedmont, Va. to n. Fla., westward to e. Tex., northward in the interior to s. Ind., Mo., Kan.

21. Ludwigia linearis Walt.   Fig. 192

Perennial, flowering stem erect, 3–10 dm tall, usually with numerous slender, erect-ascending branches, submersed portions, if any, notably spongy, producing leafy, autumnal, basal offshoots; angles of young stems often with minute, pale, somewhat scalelike trichomes, these sometimes scattered between the angles. Leaves alternate, the principal ones of the flowering stems oblanceolate, narrowly lanceolate, or narrowly linear-oblanceolate or -lanceolate, 3–7 cm long and 1.5–5 mm wide, sessile or narrowed to subpetiolate bases, acute apically, surfaces and margins usually with minute, pale, scalelike trichomes; leaves of the basal offshoots oblanceolate or spatulate, 1–3 cm long. Flowers sessile, solitary in the leaf axils more or less throughout the branches; bractlets minute, linear, arising at the base of the floral tube; floral tube densely covered with minute, pale, scalelike trichomes, these few on the outer surfaces of the calyx segments. Calyx segments 4, triangular-acute, 2–3 mm long, very much shorter than the floral tube. Petals 4, yellow, obovate, a little longer than the calyx segments. Stamens 4. Capsule turbinate or elongate-campanulate in outline, obtusely quadrangular, mostly 6–8 mm long. Seed buff-colored, falcate-oblongish, very finely and obscurely reticulate, about 0.5 mm long.

Marshy shores, swamps, wet clearings, cypress-gum ponds and depressions, pools, drainage ditches and canals, sloughs, in and on the banks of small, open streams, often in shallow water. Chiefly coastal plain, s. N.J. to n. Fla., westward to s.e. Tex., Ark., Tenn.

22. Ludwigia linifolia Poir. in Lam.   Fig. 189

Slender, low perennial, flowering stem erect, simple or few-branched, 1–4 dm tall, submersed portions, if any, scarcely spongy, producing leafy, autumnal stolons from the base. Leaves alternate, glabrous, those of the flowering stems narrowly linear-oblanceolate, 1–2

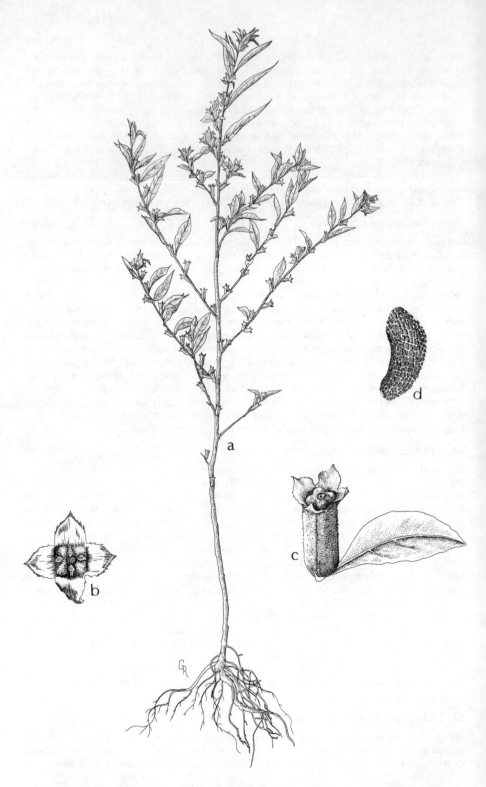

Fig. 191. **Ludwigia glandulosa:** a, habit; b, flower, from above; c, capsule; d, seed.

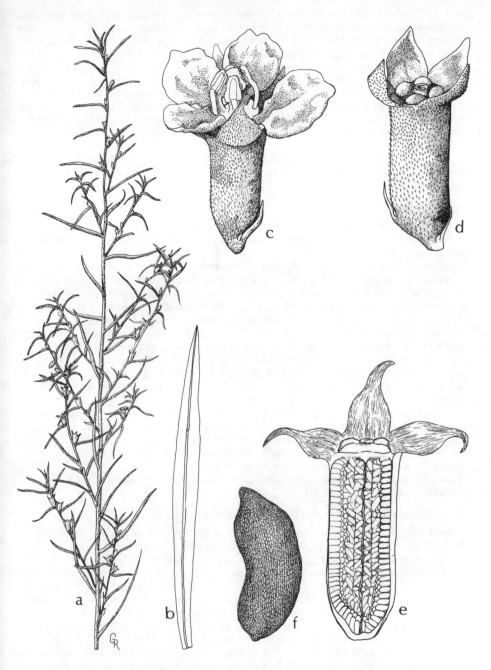

Fig. 192. **Ludwigia linearis:** a, top of plant; b, leaf; c, flower, side view; d, capsule; e, longitudinal section of capsule; f, seed.

411

(−3) cm long and 2–3 (−4) mm broad, attenuate-subpetiolate basally, acute apically; leaves of the stolons narrowly oblanceolate to spatulate, 2–6 mm broad distally. Flowers sessile, solitary in the leaf axils distally on the main stem or its branches, or branches sometimes floriferous almost throughout; bractlets linear-subulate, inserted at the base of the floral tube, very variable in length from plant to plant, sometimes 2–4 mm long and appressed to the floral tube, varying to nearly as long as the calyx segments and more or less spreading. Calyx segments 4, triangular-subulate, 5–6 mm long. Petals 4, pale yellow, narrowly obovate, 4–5 mm long. Stamens 4. Capsule essentially oblong to narrowly turbinate in outline, 8–9 mm long, obscurely quadrangular, surface with pale, minute, mostly scalelike, granular trichomes. Seed reddish brown, falcate-ellipsoid, very finely and obscurely reticulate, about 0.5 mm long.

Depressions in pine savannas and flatwoods, cypress-gum ponds and depressions, wet ditches, borrow pits, sometimes in shallow water. Coastal plain, s.e. N.C. to s. Fla., westward to Miss.

23. Ludwigia arcuata Walt.   Fig. 193

Perennial, branches prostrate and creeping, rooting at the lower nodes. Stem pubescent with short, hooked hairs, these sometimes shed on older portions. Leaves opposite, sessile, oblanceolate or some of them narrowly or broadly elliptic, the 3 shapes not uncommonly on a single plant, mostly 8–20 mm long, sometimes glabrous, sometimes minutely scabrid on and near the margins and with hooked hairs on the midrib beneath. Flowers solitary in leaf axils, usually only in the axil of one of the pair of leaves at a given node and relatively few nodes bearing flowers at all; flower stalks slender, markedly longer than the subtending leaves, sparsely clothed with short, hooked hairs as are the floral tubes and outer surfaces of the calyx segments; bractlets narrowly oblanceolate, 2–3 mm long, opposite or subopposite just below the floral tube or distally on the flower stalks. Calyx segments 4, elongate-triangular, 4–8 mm long. Petals 4, bright yellow, obovate, somewhat longer than the calyx segments. Capsule obconical in outline, obscurely quadrangular, usually curved, 6–8 mm long. Seed buff, ellipsoid or falcate-ellipsoid, about 0.5 mm long. (*Ludwigiantha arcuata* (Walt.) Small)

The description above applies to plants growing terrestrially or in very shallow water. Occasionally plants occur matted from the substrate in water to at least 1 m deep, many flaccid, nonflowering stems ascending to somewhat below the water's surface, the leaves flaccid, mostly linear-subulate or narrowly oblanceolate; very much resembling *Didiplis diandra*, for which it is sometimes mistaken.

In and on the shores of ponds and lakes, pools, marshes, wet ditches. Coastal plain, S.C. to s. Fla., Fla. Panhandle, s. Ala.

24. Ludwigia brevipes (Long in Britt. and Brown) Eames.

Closely similar to *L. arcuata*. Flower or fruiting stalks shorter than the subtending leaves. Sepals 4–6 mm long. Petals elliptic or spatulate, equaling the sepals.

Shores of ponds and lakes, pools, marshes, swamps. Coastal plain, s. N.J. to S.C.

The range of *L. brevipes* was given by Fernald (1950) as s. N.J. to n. Fla. There are some specimens at Fla. State University, collected in Fla., variously identified by their collectors, some of them as *L. brevipes*, which Peter Raven has annotated as *L. arcuata* × *L. repens*. The specimens look much alike superficially but exhibit various combinations of the characters of the putative parents save that no mature capsules are present. The flowers have slender stalks 6–10 mm long, much shorter than the subtending leaves (suggesting *L. brevipes*) but occur 2 at a node, unlike *L. brevipes*. It is possible that specimens of putative hybrid origin, like or similar to these, examined by Fernald, may account for the alleged occurrence of *L. brevipes* southward of S.C.

25. Ludwigia spathulata T. & G.   Fig. 194

Plant with prostrate and creeping branches, rooting at the nodes, often many plants with their branches intermingled and forming extensive mats, hirsute throughout. Leaves opposite,

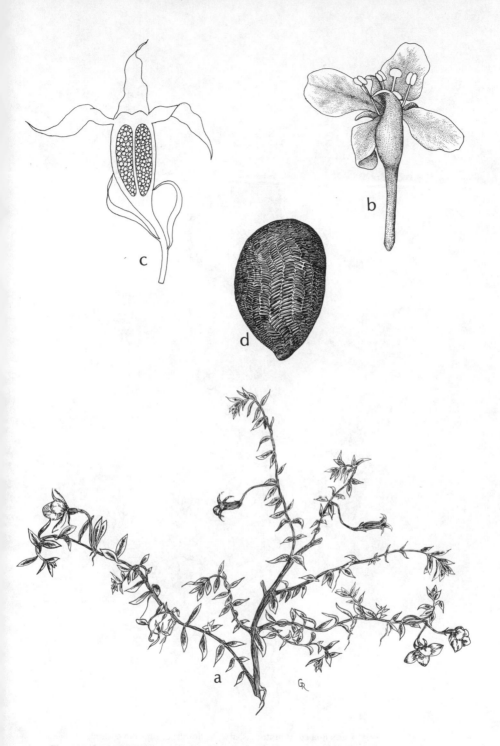

Fig. 193. **Ludwigia arcuata:** a, branch of plant; b, flower; c, longitudinal section of capsule; d, seed.

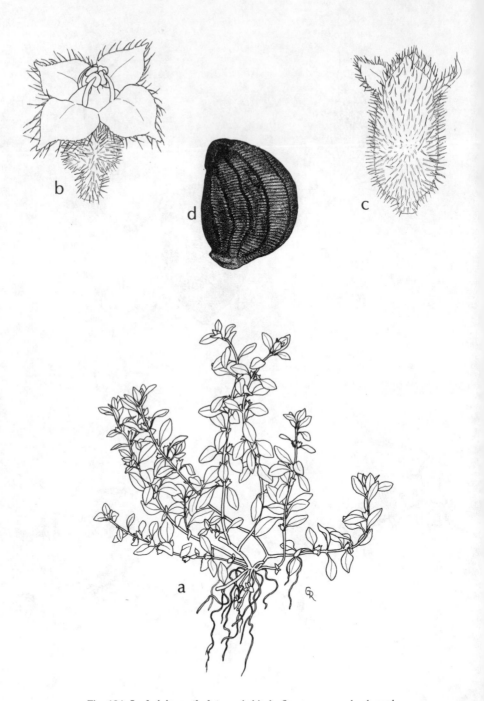

Fig. 194. **Ludwigia spathulata:** a, habit; b, flower; c, capsule; d, seed.

with broadly elliptic distal portions, these abruptly, then gradually, narrowed to petiolar or subpetiolar bases, 1–2 cm long overall, 3–5 mm wide at their widest places, apices obtuse. Flowers sessile, solitary in the axils of both leaves of many, sometimes most, nodes; bractlets none. Calyx segments 4, broadly ovate, abruptly short-acuminate apically, about 1 mm long. Petals none. Stamens 4. Capsule shortly obpyramidal in outline, vaguely quadrangular, 2.5–3 mm long. Seed obliquely ovate, plump, 0.3–0.4 mm long, dark reddish brown and with narrow, longitudinal stripes on the more rounded side. (*Isnardia spathulata* (T. & G.) Kuntze)

During times of low water, on exposed shores and bottoms of open sinkhole ponds, cypress-gum ponds, bogs; in a given place appearing at infrequent intervals, not necessarily at any particular time of low water levels. Local, coastal plain, S.C., s.w. Ga., Fla. Panhandle, s. Ala.

26. Ludwigia repens Forst. Fig. 195

Perennial. Plants growing terrestrially having prostrate or creeping branches, mat-forming; plants growing in water commonly completely submersed or only the tips floating, rooted in the substrate and with flaccid ascending stems; glabrous throughout, or more commonly with sparse, hooked hairs on parts of the stem, the petioles, on the floral tube and outer surfaces of the calyx segments; usually with a considerable amount of purple pigment throughout, especially in submersed plants. Leaves opposite, with elliptic to subrotund distal portions, these abruptly or gradually narrowed to petiolar or subpetiolar bases, apically blunt to rounded, sometimes cuspidate; leaves of plants growing terrestrially mostly 1–3 cm long overall and 5–10 mm broad at their broadest places; leaves of submersed plants (and sometimes those in dense shade) generally flaccid, varying in size up to 5 cm long and their distal portions 2.5–3 cm broad. Flowers mostly sessile, rarely with stalks not exceeding 3 mm long, solitary in leaf axils, sometimes in the axil of only one leaf at a given node, more commonly in the axils of both, sometimes flowers only at scattered nodes, sometimes at most nodes; bractlets 2, linear to linear-oblong, 2–4 mm long, borne somewhat above the base of the floral tube. Calyx segments 4, triangular, short-acuminate apically, 2–3.5 mm long. Petals 4, yellow, falling quickly (hence often not seen and said to be absent). Stamens 4. Capsule glabrous at maturity, 4–8 mm long, in outline obpyramidal, or more rarely oblong, quadrangular, relatively deeply furrowed longitudinally on two sides, very shallowly on the other two, without longitudinal bands of darker green. Seed stramineous, obliquely ovate, elliptic, or oblong, plump, 0.5–0.8 mm long. (*L. natans* Fll , *Isnardia repens* (Sw.) DC.; incl. *I. intermedia* Small and Alexander in Small)

In flowing or still, clear or "black," water of rivers and small streams, ponds and lakes, pools, swamps, drainage ditches and canals; also in wet, sandy, or sandy-peaty places. Coastal plain, N.C. to s. Fla., westward to Tex. and N.Mex., northward in the interior to Okla., Mo., Tenn.; Calif., southward to cen. Mex.; Berm.; W.I.

27. Ludwigia palustris (L.) Ell. Fig. 196

Perennial, branches prostrate and creeping, rooting at the lower nodes, mat-forming; in water, the branchlets laxly ascending or floating. Stem glabrous or the younger portions with a few, scattered, short and stiff, sometimes curved trichomes. Leaves opposite, their blades elliptic to ovate, abruptly narrowed below to petioles about as long as the blades, margins of the blades sometimes, not always, with a few sessile glands, sometimes with sparse, minute, stiff trichomes. Flowers sessile, solitary in leaf axils, usually one in the axil of each of the pair of leaves at a node and commonly at most of the nodes of the branchlets; bractlets none or minute and subulate, up to about 1 mm long, borne at the base of the floral tube. Calyx segments 4, ovate, 0.5–1 mm long. Petals none. Stamens 4. Capsule short-oblong in outline, quadrangular, the angles rounded and each with a longitudinal band of green darker than that of the tissue between, glabrous or with a few scattered, minute, scalelike trichomes. Seed stramineous, plump, usually asymmetrical, ovate-falcate to oblong, 0.4–0.8 mm long. (*Isnardia palustris* L.)

415

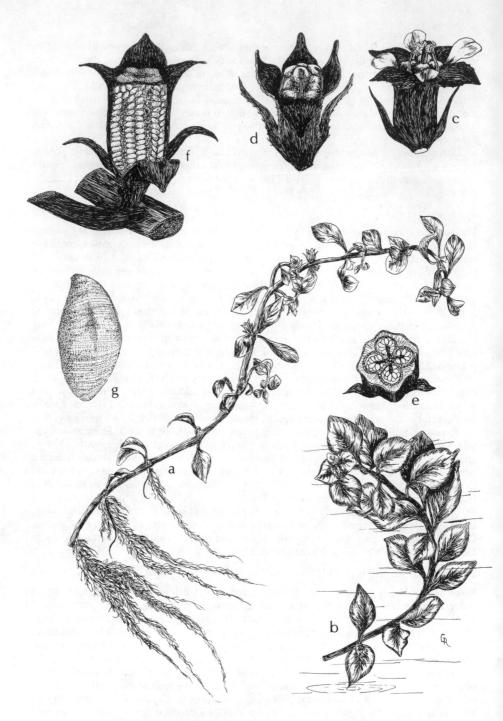

Fig. 195. **Ludwigia repens:** a, branch of plant, emersed form; b, tip of branch, submersed form; c, flower, with petals; d, flower, lacking petals; e, cross-section of capsule; f, longitudinal section of capsule; g, seed.

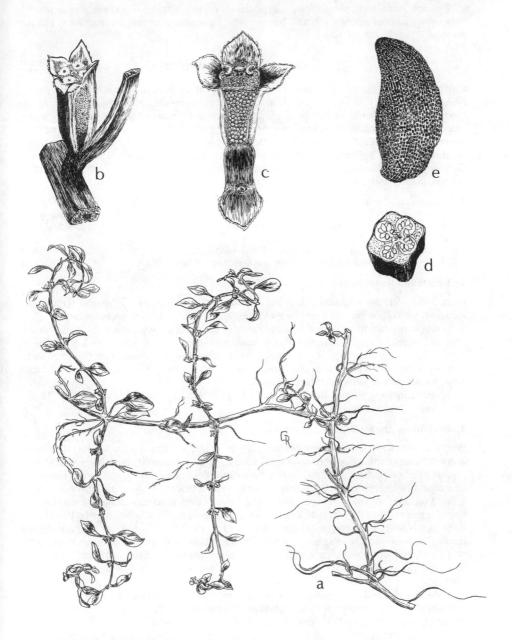

Fig. 196. **Ludwigia palustris:** a, habit, part of plant; b, capsule; c, capsule, one-quarter cut away and turned back; d, cross-section of fruit; e, seed.

417

Swampy woodlands, wet clearings, exposed shores and bottoms or wet margins of ponds, banks and shores of streams, ditches, borrow pits. Throughout our range excepting s. Fla.; in much of temp. N.Am.; Euras.; Afr.

2. Epilobium (WILLOW-HERBS)

Chiefly perennial herbs, often flowering the first year from seed. Stem erect, simple or branched. Leaves opposite or alternate, sessile or subsessile, margins entire or finely toothed. Flowers borne in the axils of little reduced or reduced leaves terminating the branches, or racemosely at the ends of branches. Floral tube little, if any, prolonged beyond the ovary. Calyx segments 4, deciduous after anthesis. Petals 4, pink to rose, violet, purple, or white. Stamens 8, filaments often unequal in length. Style 1, stigma capitate to 4-lobed. Capsule slender-elongate, 4-locular, loculicidally dehiscent. Seeds many, each with a terminal tuft of silky hairs.

1. Leaves linear or narrowly lanceolate, 2–3 (–5) mm wide, margins entire, lateral veins not evident.
1. *E. leptophyllum*
1. Leaves lanceolate to elliptic, exceeding 5 mm wide, margins entire to finely toothed, lateral veins evident.
    2. Petals 10–15 mm long, corollas showy; stigma 4-lobed.   2. *E. angustifolium*
    2. Petals 4–6 mm long, corollas relatively inconspicuous; stigma capitate.   3. *E. coloratum*

1. Epilobium leptophyllum Raf.

Plant 3–6 (–10) dm tall, stem, both leaf surfaces and flower stalks pubescent with short, incurved, whitish or grayish hairs. Leaves alternate, sessile, linear to linear-lanceolate, 2–3 (–5) mm wide, lateral veins not evident, margins entire; often with short branchlets in the axils of the primary stem leaves. Calyx segments oblong, blunt apically, 3–3.5 mm long. Petals white, 4–5 mm long. Stigma capitate. Capsule linear in outline, 3–6 cm long, more sparsely pubescent than the stem and leaves, stalks 1–5 cm long. Seed fusiform, about 2 mm long, brown, tuft of hairs pale, barely tawny.

Marshes, marshy shores, meadows, bogs. Que. to Alb., southward to w. N.C., e. Tenn., Mo., Kan., Colo.

2. Epilobium angustifolium L. FIREWEED.

Perennial with widely spreading roots that form buds freely. Stem erect, 1–2 dm tall, simple or much branched above, glabrous (ours). Leaves alternate, glabrous, essentially sessile, lanceolate to elliptic, varying greatly in size up to 2 dm long, margins entire, somewhat wavy, or irregularly finely dentate or serrate, lateral veins evident. Flowers in terminal, small-bracted racemes, or the lower ones from the axils of leaves in length exceeding the flowers. Calyx segments usually purplish, oval, oblong-oblanceolate, or elliptic, mostly 10–15 mm long. Petals rose, magenta, or purple, rarely white, 10–15 (–20) mm long, sharp-clawed basally, obovate distally. Stigma 4-lobed. Capsule linear in outline, 2.5–7 cm long, often purple, surface with matted pubescence, especially when young. Seed fusiform, 1 mm long or a little more, brown, tuft of hairs somewhat tawny. (*Chamaenerion angustifolium* (L.) Scop.)

Clearings, burned-over areas, moist ravines, in soils rich in humus, probably not really a wetland plant, in our range at least. Subarctic and boreal N.Am., southward in our range to higher elevations of mts., N.C. and Tenn., n. Ohio, cen. Ind., n. Ill., n. Iowa, S.D., N.Mex., Ariz., Calif.; Euras.

3. Epilobium coloratum Biehler.

Perennial, in autumn forming basal, rosettelike offshoots. Stem to 1 m tall, freely branched above except in small plants, short-pubescent above, lower stem glabrous in age. Leaves chiefly opposite, essentially sessile, glabrous, lanceolate or oblong-lanceolate, acuminate apically, the larger ones 4–15 cm long, 5–25 mm broad, margins finely toothed, lateral nerves evident. Flowers small, commonly collectively many, borne near the tips of the slender, leafy-bracted branchlets. Flower stalk and floral tube copiously short-pubescent. Calyx segments lance-oblong, 2.5–3 mm long, acute to acuminate apically, sparsely short-pubescent. Petals pink, 3–5 mm long, short-obovate, notched at the summit. Stigma capitate. Capsule very narrowly linear in outline, pubescent, 3–6 cm long, the stalks mostly 5–7 mm long. Seed fusiform, 1–1.5 mm long, brown, tuft of hairs brown.

Marshes, swamps, seepage slopes, meadows, bogs. Maine to Minn. and S.D., southward to w. N.C., n. Ga., n. Ala., Ark., Okla., Tex.

3. Circaea

Circaea alpina L. ENCHANTER'S NIGHTSHADE.

Low, perennial with tuberous rhizomes. Stem soft, weak, 1–3 dm tall, sparsely pubescent above with short, appressed hairs. Leaves opposite, thin, with slender petioles slightly shorter than the blades; blades deltoid to ovate, bases subcordate, truncate, or shortly tapered, short-acute to acuminate apically, 2–6 cm long, commonly as broad at the base as long, both surfaces sparsely pubescent with short, appressed hairs. Flowers small, in a few-flowered terminal raceme, the flower stalks 2–6 mm long, spreading to erect at anthesis, reflexed later. Calyx segments and petals 2 each, about 2 mm long, petals white. Stamens 2, alternate with the petals. Fruit turbinate, indehiscent, 1-locular, 1- or 2-seeded, its surface covered with soft, hooked hairs.

Seepage areas, mossy bogs, moist to wet woodlands. Nfld. to Alaska, southward in the Appalachian Mts. to N.C. and Tenn., and to Ind., Ill., Mich., Iowa, S.D., Colo., Utah, Wash., and to N.Mex., Ariz.

Haloragaceae (WATER-MILFOIL FAMILY)

Aquatic or amphibious, herbaceous perennials. Leaves without stipules, whorled or alternate, rarely opposite, simple, pinnately veined, or often pinnately dissected, frequently dimorphic. Flowers solitary or clustered in the axils of leaves or bracteal leaves, bisexual or unisexual, radially symmetrical; ovary inferior, the floral tube surmounted by 2–4 calyx segments, or these absent, and 2–4 small, often quickly deciduous petals, or petals absent. Stamens 3–8. Ovary 3- or 4-locular (in ours), each locule with 1 ovule; styles or sessile stigmas 1–4. Fruit indehiscent, a single nutlet, or the carpellate portions separating into separate nutlets.

- Leaves alternate; flowers bisexual, parts in threes.   1. *Proserpinaca*
- Leaves whorled, or occasionally some leaves on a given plant opposite or alternate; flowers mostly unisexual, some sometimes bisexual, parts in fours.   2. *Myriophyllum*

1. Proserpinaca (MERMAID-WEEDS)

Lower stems usually prostrate and rooting at the nodes, the upper weakly ascending-erect. Leaves alternate. Flowers usually solitary in the upper leaf axils, sometimes clustered, bisexual. Calyx segments 3, deltoid, persistent; corolla none; stamens 3; pistil 1, more or less 3-angled, stigmas 3. Fruit 3-locular, 3-angled.

- Leaves all pinnately divided.   1. *P. pectinata*
- Leaves merely toothed, or both toothed and pectinate ones on the same plant.   2. *P. palustris*

1. Proserpinaca pectinata Lam.   Fig. 197

Leaves all pinnately dissected, ovate or ovate-oblong in overall outline, 1–2.5 cm long, the linear-subulate divisions mostly 10–18, these sometimes bearing minute spiculelike teeth.

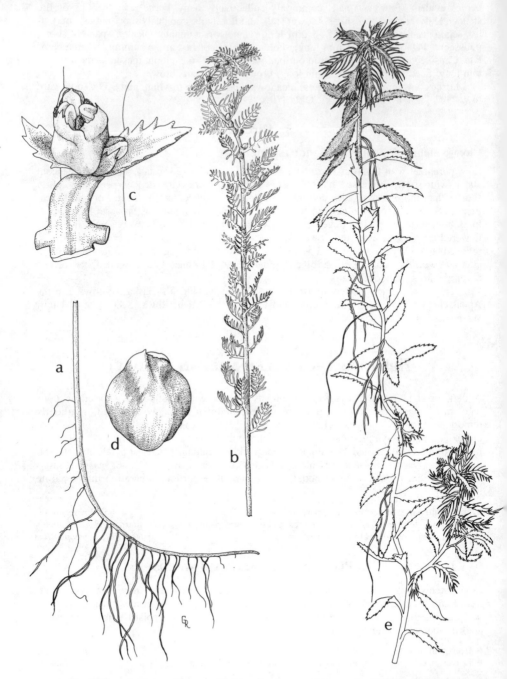

Fig. 197. a–d, **Proserpinaca pectinata:** a,b, plant; c, flower and subtending bracts; d, fruit; e, **Proserpinaca palustris:** vegetative portion of a plant, showing leaf dimorphism.

Fruit ovate, about as wide as long, 3–4 mm long, 3-angled, the faces flat or concave and usually with a few irregular ridges.

Cypress-gum ponds or depressions, swamps, wet pine savannas and flatwoods, ditches, wet clearings, often in shallow water, commonly matted. Chiefly coastal plain, N.S. and s.w. Maine southward to s. Fla., westward to s.e. Tex.

2. Proserpinaca palustris L.   Fig. 197

Leaves essentially sessile, very variable as to toothing or segmentation. In general, emersed leaves lanceolate and sharply serrate, submersed leaves from lanceolate to elliptic or ovate in overall outline and deeply pinnatifid or pectinate. Often they grade from lower to upper in respect to the degree of pinnasection. Apparently the polymorphism is related to the particular environmental conditions obtaining during the development of a given leaf or set of leaves. Not uncommonly a plant may have a set of deeply pinnatifid leaves below, then a set of lanceolate, serrate leaves, then another set of pinnatifid ones, presumably related to fluctuating water levels. Moreoever, in southern latitudes, completely emersed plants actively growing during the shorter days of winter (perhaps in deep shade in summer) may have all pectinate or all pinnatifid leaves, then as the days lengthen and growth continues there is a gradation to lanceolate, serrate leaves on the upper stems. Fruits mostly about as broad as long, varying from urceolate to trigonous, the trigonous ones very variable relative to the sharpness of the angles, whether or how much the angles are winged, whether the faces are smooth or irregularly pebbled.

Authors, using varying characters of the fruits, have recognized several varieties of this species; or some of the segregates have been accorded specific rank. (Incl. *P. palustris* var. *palustris*, *P. palustris* var. *crebra* Fern. & Grisc., *P. palustris* var. *amblygona* Fern., or *P. amblygona* (Fern.) Small; *P. intermedia* Mack.; *P. platycarpa* Small)

Shores of streams, lakes, ponds, in marshes and ditches, swamps. Que. and N.S. to Ont. and Minn., generally southward to s. Fla. and e. Tex.

2. Myriophyllum (WATER-MILFOILS)

In most species, plants submersed excepting the upper flower-bearing portions (unless stranded at times of low water levels). Leaves whorled, or sometimes some of them on a given plant opposite or alternate, the submersed ones pinnately dissected into capillary segments, the bracteal leaves or bracts subtending the flowers reduced. (*M. aquaticum*, having considerable portions of nonflowering stems emersed and flowers in the axils of unreduced leaves, an exception.) Lowermost flowers on an inflorescence usually pistillate, a few median ones bisexual, the upper staminate, each flower subtended by 2 bractlets in addition to the primary bract or bracteal leaf. Calyx segments 4. Petals 4 in the bisexual or staminate flowers, larger than the sepals, quickly deciduous, none in the pistillate flowers. Stamens 4 or 8. Stigmas 4, plumose, recurved. Fruit 4-lobed, eventually splitting into 4 nutlets.

1. Leafy vegetative branches in considerable part emersed; flowers (in ours) all pistillate (though rarely flowering).   1. *M. aquaticum*
1. Leafy vegetative parts mostly submersed (unless stranded at times of low water levels); lower flowers pistillate, median ones bisexual, upper ones staminate.
  2. Primary bracts of the inflorescence as long as, usually longer than, the internodes of the axis, easily perceived by the naked eye.
    3. Foliage leaves with a total of 8–10 capillary segments; primary bracts or bracteal leaves with longish serrations, usually subpectinately toothed.   2. *M. pinnatum*
    3. Foliage leaves with a total of 12–20 capillary segments; primary bracts or bracteal leaves with short serrations, or the lower ones sometimes entire.   3. *M. heterophyllum*
  2. Primary bracts of the inflorescence shorter than the internodes of the axis, scarcely perceptible to the naked eye.
    4. Foliage leaves with a grayish cast, the segments mostly disposed on the axis in pairs, or subopposite, those on each side extending outwardly nearly paralleling each other, thus the leaves with a markedly feathery appearance; upper primary bracts of the inflorescence obovate. 4. *M. spicatum*

421

4. Foliage leaves with a bright green to reddish cast, the segments mostly disposed on the axis alternately and irregularly, those on each side more or less intermingled, thus the leaves not markedly feathery in appearance; upper primary bracts of the inflorescence oblanceolate.   5. *M. laxum*

1. Myriophyllum aquaticum (Vell.) Verdc. PARROT-FEATHER.

Stem moderately elongate, relatively stout, partially submersed but with considerable portions of leafy branches emersed. Leaves all whorled, stiffish, usually with 20 or more linear-filiform divisions, appearing markedly featherlike, grayish green. Flowers (in ours) all pistillate (apparently flowering rarely), borne in the axils of essentially unreduced leaves. Bractlets filiform, 2- or 3-cleft. (*M. proserpinacoides* Gill.; *M. brasiliense* Camb.)

Sluggish waters, edges of streams, lakes, ponds, drainage and irrigation ditches and canals, backwaters, sloughs, lagoons, commonly in dense, sometimes floating, mats, often choking waterways. Native of S.Am., sporadically naturalized in much of our range and beyond.

2. Myriophyllum pinnatum (Walt.) BSP.   Fig. 198

Stems elongate, slender. Submersed leaves in whorls of 3 or 4, usually also some alternate and subopposite, mostly with a total of 8–10 capillary divisions. Inflorescence up to 15 cm long (occasionally longer). Bracts conspicuous, in whorls of 4, spreading-ascending distally, spreading horizontally in the medial region, usually reflexed proximally, all with longish serrations or subpectinately toothed; bractlets triangular-acute, their margins scarious-spinose. Petals of bisexual and staminate flowers translucent, whitish or slightly purplish. Stamens 4. Fruits about 1.5 mm long, green, surfaces tuberculate.

Ponds, lakes, streams, drainage and irrigation ditches and canals, sloughs, swamps, borrow pits. s. N.Eng. to Ohio, Iowa, and Kan., generally southward to s. Fla. and Tex.

3. Myriophyllum heterophyllum Michx.   Fig. 199

Stems elongate, relatively stout, reddish, occasionally much branched. Submersed leaves in whorls of 5 or 6, or 4–6 subapproximate, up to about 6 cm long, with a total of 12–28 capillary divisions. Bracts subfoliaceous, the lower ones ovate-lanceolate to lanceolate, entire or shallowly serrate, those above ovate to ovate-lanceolate, shallowly serrate (occasionally sharply serrate); bractlets triangular-acute, with spinose margins. Petals of bisexual and staminate flowers translucent, reflexed-spreading at anthesis. Stamens 4. Fruit about 1.5 mm long, green, segments rounded, the surfaces papillose.

Ponds, lakes, streams, ditches and canals, swamps, sloughs, borrow pits. s.w. Que. and Ont. to N.D., generally southward to cen. pen. Fla. and N.Mex.

4. Myriophyllum spicatum L. EURASIAN WATER-MILFOIL.   Fig. 200

Stem slender, commonly much branched. Leaves grayish green, all whorled, usually with a total of 24 or more filiform segments, these mostly disposed from the axis in pairs, the segments on each side of the axis extending outwardly and nearly paralleling each other, thus the leaves appearing markedly feathery. Bracts inconspicuous, scarcely evident to the naked eye, shorter than the internodes of the axis, the spikes appearing essentially naked and interrupted; lower bracts usually toothed, sometimes entire, mostly exceeding the flowers, the upper obovate, entire, equaling or shorter than the flowers. Petals of bisexual and staminate flowers reddish. Stamens 8. Fruit 4-lobed, subglobose, somewhat tuberculate.

Native of Eurasia. During recent years rapidly becoming established in both impoundments and natural waters, sometimes in brackish waters, even in clear, cool, spring-fed rivers, in various parts of our range; often forming very dense stands, displacing native, waterfowl food plants, and forming an impediment to navigation. Apparently being spread, in part, by transport of fragments.

5. Myriophyllum laxum Shuttlw. ex Chapm.   Fig. 201

Stems elongate, reddish, slender. Submersed leaves in whorls of 4 or 5, or 4 or 5 subapproximate, or 2 or 3 at a node, some sometimes alternate; with a total of 11–21 capillary divisions

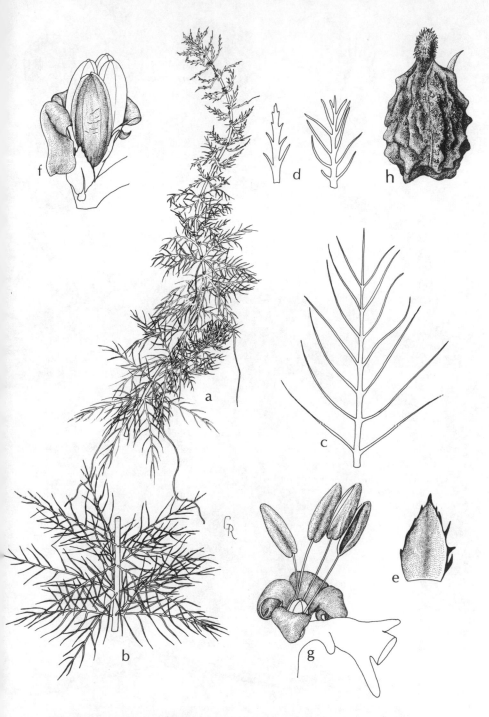

Fig. 198. **Myriophyllum pinnatum:** a, distal portion of branch; b, medial portion of branch; c, submersed leaf; d, leaves from distal flowering branch; e, bract; f, early stage of staminate flower; g, staminate flower; h, a single mericarp.

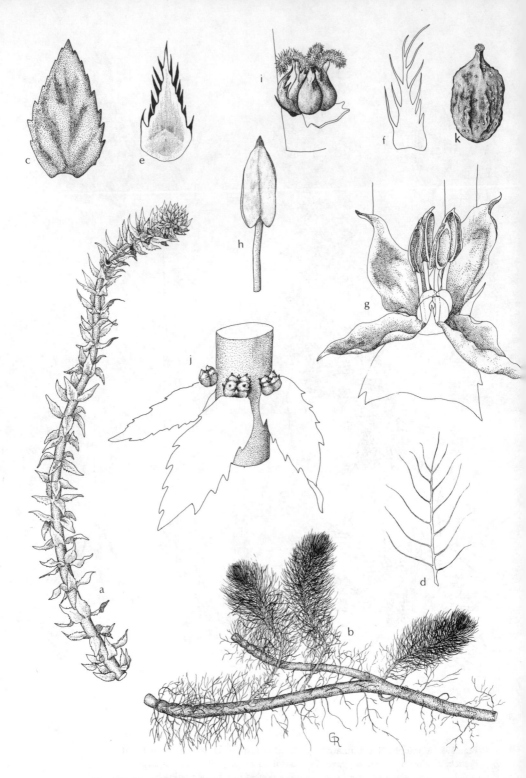

Fig. 199. **Myriophyllum heterophyllum:** a, distal portion of branch; b, medial section of branch; c–f, variation in leaves; g, staminate flower; h, stamen; i, pistillate flower; j, node, showing developing pistillate flowers in three leaf axils; k, a single mericarp.

Fig. 200. **Myriophyllum spicatum:** A, habit; B, whorl of leaves; C, part of flower spike, with pistillate flowers below, staminate above; D, immature fruits; E, mature fruit. (From Reed, *Selected Weeds of the United States* (1970) Fig. 136)

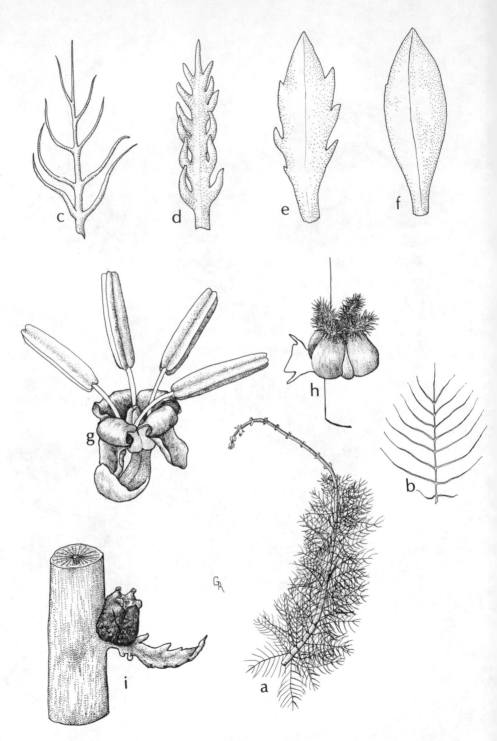

Fig. 201. **Myriophyllum laxum:** a, distal portion of a branch, with submersed vegetative portion and emersed inflorescence in developing stage; b, leaf; c, vegetative leaf just below inflorescence; d, a lower bracteal leaf; e, a median bracteal leaf; f, an upper bracteal leaf; g, staminate flower; h, pistillate flower; i, fruit at about maturity.

mostly alternately and irregularly disposed on the axis, more or less intermingling, the leaves thus not markedly feathery. Flowers minute, the bracts inconspicuous and scarcely evident to the naked eye, the spike thus slender, appearing naked, and interrupted. Lower bracts pectinate, upwardly gradually becoming few-toothed, then entire and spatulate. Petals of bisexual and staminate flowers translucent, pinkish or reddish, recurved at anthesis. Stamens 4. Fruits about 1 mm long, red, the segments broadly rounded, the surfaces smooth to warty.

Ponds, lakes, streams, backwaters, sloughs, ditches and canals. s.e. N.C. to n. Fla. and s. Ala.

Nyssaceae (SOUR GUM FAMILY)

Nyssa

Trees (sometimes flowering and fruiting when in a shrub stage). Pith with transverse diaphragms, homogeneous between the diaphragms. Leaves alternate, deciduous, without stipules, petiolate, blades simple, pinnately veined, entire, or with 1 to few coarse dentations. Flowers small and individually inconspicuous, radially symmetrical, green, unisexual or bisexual, plants mostly dioecious, sometimes polygamo-dioecious; staminate in axillary stalked heads, umbels, or racemes, with 5 minute sepals or none, petals 5–10, or none, stamens 5–12, commonly 10, inserted around the base of a nectariferous disc; bisexual or pistillate flowers axillary, solitary, or 2 to several in a bracted cluster at the end of a stalk, ovary inferior, the floral tube surmounted by 5 small sepals or none, by 5 slightly larger, often quickly deciduous, petals, or none, with 5–10 stamens, these fertile or infertile, or none. Flowers produced in spring on the newly emerging shoots of the season. Ovary 1-locular and 1-ovulate. Fruit a drupe.

A delicate and delicious honey is derived from plants of this genus. It is generally known in the trade as tupelo honey.

1. Mature leaf blades (some of them but not necessarily all on a given branch or tree) exceeding 10 cm long; staminate flowers numerous, sessile, in ball-like heads, pistillate or bisexual, solitary, subtended by bracts at the end of a stalk; drupes 20 mm long or more.
    2. Petioles 1–2 (rarely to 3) cm long; drupes red at maturity, usually somewhat longer than their stalks.   1. *N. ogeche*
    2. Petioles 3–6 cm long; drupes dark blue or dark purple at maturity, usually somewhat shorter than their stalks.   2. *N. aquatica*
1. Leaf blades 10 cm long or less; staminate flowers in short racemes, pistillate or bisexual, 2–4 in a sessile cluster subtended by bracts at the end of a stalk; drupes 10–15 mm long.   3. *N. sylvatica*

1. Nyssa ogeche Bartr. ex Marsh. **OGEECHEE-LIME, OGEECHEE TUPELO.**   Fig. 202

A small to medium-sized tree, frequently with numerous upright branches from a leaning trunk, or more commonly with several upright, often crooked, trunks from near the base; not infrequently many-stemmed shrubs, these often bearing flowers and fruits. Twigs of the season pubescent. Petioles mostly 1–2 cm long (rarely to 3 cm), pubescent; blades varying in shape and size, even on an individual branch, elliptic or subelliptic, narrowly obovate, or oblong-oval, to 1.5 dm long and 8 cm wide; bases mostly cuneate, infrequently rounded or subcordate, apices acute, barely acuminate, obtuse, or rounded; margins usually entire, very rarely with 1 to several coarsely dentate teeth; emerging leaves sparsely pubescent above, becoming glabrous in age, their lower surfaces very densely soft-pubescent, becoming relatively sparsely soft-pubescent in age, much paler beneath than above. Staminate flowers numerous, sessile, in compact ball-like clusters; pistillate or bisexual solitary, subtended by 2 to several unequal bracts at the ends of stalks 1.5–2 cm long. Drupes red at maturity, or at least with a red blush, oblong to obovate, 2–4 cm long, usually somewhat shorter than their stalks which are stiffish and not drooping. Stones of the drupes with papery wings extending outward to the skin of the fruit.

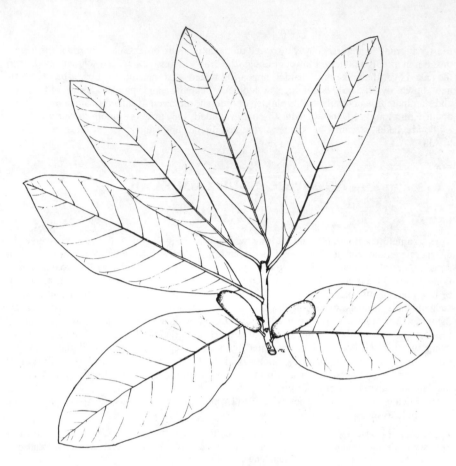

Fig. 202. **Nyssa ogeche.** (From Kurz and Godfrey, *Trees of Northern Florida*. 1962)

River banks, swamps, sloughs, bayous, pond and lake margins. Coastal plain, s.e. S.C., Ga., and n. Fla.

2. **Nyssa aquatica** L. WATER TUPELO. Fig. 203

A tree attaining about 30 m in height, with swollen or buttressed base. Twigs of the season pubescent or glabrous. Petioles 3–6 cm long, sparsely pubescent; blades variable in size, even on a single branch, ovate, or ovate-oblong, varying from about 6 to 30 cm long, some on a given branch 12 cm long or more; bases rounded to subcordate, or broadly cuneate, apices mostly short-acuminate, occasionally acute, rarely rounded; margins entire, or usu-ally some on a given tree with 1 to several coarsely dentate teeth or lobes; emerging leaves densely soft-pubescent beneath, becoming sparsely pubescent only along the veins in age, their upper surfaces at first pubescent on the veins, later glabrous, lower surfaces of mature leaves much paler beneath than above. Staminate flowers numerous, sessile, in compact, ball-like clusters; pistillate or bisexual solitary, subtended by 2 to several unequal bracts at the end of flexuous stalks 2–4 cm long. Drupes dark blue or dark purple at maturity, elliptic oblong to narrowly obovate, 1.5–3 cm long, usually somewhat shorter than their stalks. Stone of the drupe with 8–10 sharp, longitudinal ridges. (*N. uniflora* Wang.)

Usually growing where bases inundated much of the time, floodplain forests, swamps. Coastal plain, Va. to n. Fla., westward to e. Tex., northward in the interior to Mo., s. Ill., s. Ind.

428

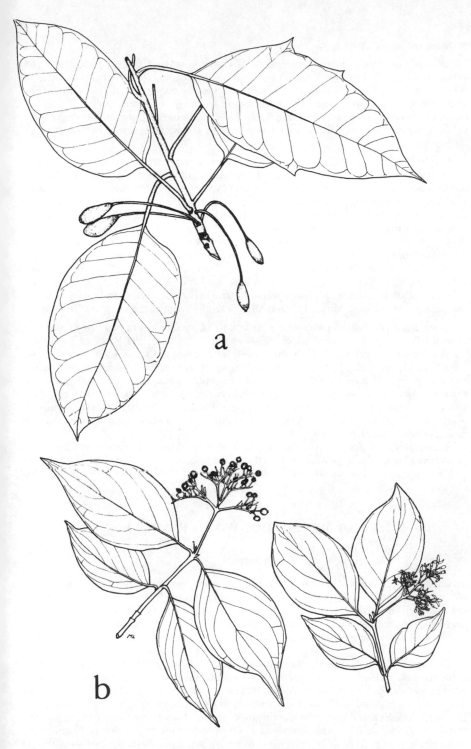

Fig. 203. a, **Nyssa aquatica;** b, **Cornus foemina.** (From Kurz and Godfrey, *Trees of Northern Florida*. 1962)

429

3. Nyssa sylvatica Marsh.

Variation in the "*Nyssa sylvatica* complex," as currently understood (see Eyde, 1963 and 1966), is such that, from the standpoint of a practical taxonomy, it is virtually a matter of personal preference as to whether to treat it as a species comprised of two varieties, or whether to recognize two species. For the most part, we think, specimens can be identified as one or the other of two morphological entities; however, since there are not infrequent individuals in parts of the range which are intermediate, we choose to recognize two varieties, as follows.

N. sylvatica var. **sylvatica. BLACK GUM, SOUR GUM.** In general a tree of forests on well-drained sites, occurring as single, scattered individuals in the forest. Leaf blades relatively membranous, obovate to elliptic-oblong, sometimes a few nearly orbicular, their bases usually cuneate, occasional ones rounded, apices mostly short-acuminate, margins entire, or with 1 to several dentate teeth, not uncommonly with at least some toothed leaves on a given tree; blades variable in size, even on an individual, from about 3 to 10 cm long; surfaces usually glabrous above at maturity, varying from glabrous to densely pubescent beneath. Staminate flowers in short racemes, pistillate or bisexual (2)3 or 4 in a bracted cluster at the end of a stalk; drupes about 1 cm long, elliptic to oblong-elliptic, blue-black. (Incl. *N. sylvatica* var. *dilatata* Fern., and var. *caroliniana* (Poir.) Fern.)

Well-drained woodlands of various mixtures, said to occur as well in lowland woodlands and swamps. Maine to s. Mich., Ill., and Mo., generally southward to Fla. and Tex.; Mex.

N. sylvatica var. **biflora** (Walt.) Sarg. **BLACK GUM, SWAMP TUPELO.** A gregarious tree of wetlands, commonly where water stands much of the time, the bases of the trunks swollen; where subject to periodic burning, with large subterranean bases and shrubby tops, these flowering and fruiting if the site not burned over every year. Leaf blades relatively thick and stiff, elliptic to lance-elliptic, occasional ones oblanceolate, cuneate basally, apically acute or obtuse, infrequently rounded; variable in size, 2–8 cm long, rarely to 10 cm, margins entire (some toothed on seedlings or sprouts). Pistillate or bisexual flowers usually 2, sometimes 1 or 3, in a bracted, sessile cluster at the end of a stalk. Drupes as in var. *sylvatica*, or rarely subglobose. (Incl. *N. ursina* Small)

Swamps, cypress-gum ponds, pond and lake margins, floodplain forests, wet flatwoods or savannas, shrub-tree bogs or bays. Chiefly coastal plain, N.J. to s.cen. Fla., westward to e. Tex.   Fig. 204

Cornaceae (DOGWOOD FAMILY)

Cornus

Trees, shrubs, or herbs (ours trees or shrubs). Leaves opposite (in those treated here), simple, pinnately veined, margins entire, deciduous. Flowers small, in cymes or heads, bisexual, radially symmetrical; ovary inferior, the floral tube bearing on its rim 4 minute calyx segments, 4 somewhat larger, spreading, white or creamy white petals, and 4 stamens; style slender with a terminal stigma. Ovary 2-locular, 1 ovule in each locule. Fruit a drupe, often only 1-seeded.

Cornus florida L., the familiar flowering dogwood, is the only species in our range whose inflorescences are capitate and subtended and surrounded by conspicuous white bracts. It inhabits well-drained upland sites, and is, therefore, excluded from this account. The dogwoods or cornels of our wetlands, those with which we are concerned, have given botanists, generally, considerable difficulty in identification; authors have, for the most part, treated the entities at the specific level with something less than uniformity in application of names. We have chosen to adapt from an analysis in the following publication: James S. Wilson, "Varia-

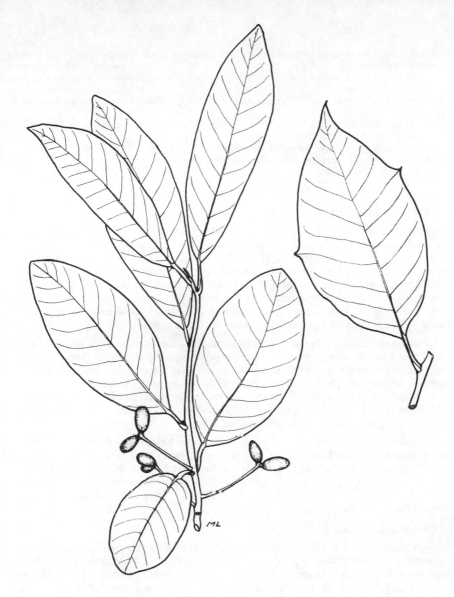

Fig. 204. **Nyssa sylvatica** var. **biflora.** (From Kurz and Godfrey, *Trees of Northern Flor-ida*. 1962)

tion of Three Taxonomic Complexes of the Genus *Cornus* in Eastern United States," *Trans. Kan. Acad. Sci.* 67: 747–817 ([1964] 1965). It is to be noted that Wilson considered hybrid-ization to be occurring between subspecies within a species and between species. It is to be presumed that identification of specimens as representing one or another of the named taxa, given the amount of intermediacy obtaining, will in many cases be arbitrary.

1. Twigs of the current season copiously pubescent.
  2. Upper leaf surfaces glabrous and smooth; styles abruptly dilated at about 1 mm below the stigma, thus with an enlargement below the stigma.

3. Leaf blades mostly ovate or broadly elliptic, not exceeding twice as long as wide, bases rounded to broadly cuneate.   1a. *C. amomum* subsp. *amomum*

3. Leaf blades mostly elliptic or oblong-elliptic, somewhat more than twice as long as wide, bases tapered.   1b. *C. amonum* subsp. *obliqua*

2. Upper leaf surfaces pubescent, often scabrous; styles barely dilated just below the stigma, or not at all dilated.

4. Leaf blades predominantly ovate, their bases broadly rounded, truncate, or broadly tapered; pith of twigs brown or tawny; mature fruits white.   2. *C. drummondii*

4. Leaf blades rarely ovate, varyingly lanceolate, elliptic, elliptic-oblong, or obovate, bases predominantly cuneate, some sometimes rounded (both often on the same branchlets); pith of twigs white; mature fruits bluish.   3b. *C. foemina* subsp. *microcarpa*

1. Twigs of the current season glabrous, or sparsely pubescent with appressed hairs.

3a. *C. foemina* subsp. *foemina*

1. Cornus amomum Mill.

Shrubs to about 5 m tall. Branchlets of the current season usually copiously pubescent, the hair usually rusty, infrequently silvery; pith of twigs brown or tawny. Petioles and lower leaf surfaces pubescent, upper surfaces glabrous and smooth. Cymes convex to flat-topped, to about 8 cm across, axes pubescent. Floral tube and calyx segments silvery-pubescent, a maroon nectariferous disc at the summit of the ovary. Style abruptly dilated at about 1 mm or somewhat less below the stigma, thus its summit about twice the diameter below. Drupe dark bluish with areas of cream color, usually subglobose, sparsely pubescent, about 8 mm in diameter. Two subspecies may be distinguished, as follows.

C. amomum subsp. **amomum.** Leaf blades mostly ovate, not exceeding twice as long as broad, their bases broadly rounded to broadly tapered, apically acuminate, variable in size, even on a single branch, the larger about 10 cm long and 5 cm broad, sometimes none exceeding 8 cm long and about 4 cm broad. (*Svida amomum* (Mill.) Small)

River and stream banks, borders of swamps, wet thickets and clearings, moist woodlands. s. Maine to Ind., Ill., southward mostly in the piedmont and mountains to s.w. Ga. and s. Ala.; extending into Gadsden and Jackson Cos., Fla.

C. amomum subsp. **obliqua** (Raf.) J. S. Wilson. Leaf blades mostly elliptic or oblong-elliptic, somewhat more than twice as long as broad, bases mostly tapered, apices acute or acuminate, variable in size, the larger 5–10 cm long and 2–4 cm broad, occasional specimens with all leaves smaller but with similar length/width proportions. (*C. purpusi* Koehne)

Similar habitats. s. Que. to Minn.; N.Y. to Ohio, Ky., cen. Tenn., and Ark.; Minn. to e. Neb., e. Kan., e. Okla.

2. Cornus drummondii C. A. Meyer.

Shrub or small tree to 6 m tall, sometimes spreading by root sprouts and colonial. Branchlets of the current season copiously pubescent; pith brown or tawny. Leaf blades predominantly ovate, sometimes broadly elliptic, about twice as long as broad, variable in size, the larger 6–10 cm long, bases mostly broadly rounded or truncate, varying to broadly tapered, apices acuminate; upper surfaces pubescent, commonly scabrous, the lower usually softly pubescent. Cyme convex, 4–7 cm across, pubescent on its axes. Floral tube densely silvery pubescent, calyx segments minute, a reddish nectariferous disc surmounting the ovary. Style barely, if at all, dilated below the stigma, the little enlargement there is being gradual. Mature drupes white, rarely faintly bluish, subglobose, about 5 mm in diameter. (*Svida asperifolia* sensu Small, 1933)

River and creek banks, floodplain woodlands, wooded bluffs, limestone outcrops and glades. Ohio, s. Mich., s. Minn., Iowa, e. Neb., generally southward to cen. Ga., Ala., and e. half of Tex.

3. Cornus foemina Mill.

Shrub or small tree to about 6 m tall. Pith of twigs white. Leaf blades variable, broadly to narrowly elliptic, oblong-elliptic, ovate, or lanceolate, bases broadly to narrowly cuneate, sometimes rounded, apices acuminate or acute. Cymes convex or flat-topped, 2 to about 10 cm across, their axes sparsely short-pubescent to essentially glabrous. Floral tubes covered with appressed silvery hairs, calyx segments sparsely pubescent, the nectariferous disc at the summit of the ovary reddish brown. Style not dilated at the summit. Mature fruit blue, globose, about 6–7 mm in diameter. Two subspecies may be distinguished, as follows.

C. foemina subsp. **foemina.** Branchlets of the current season glabrous or sparsely pubescent with appressed hairs. Leaf blades glabrous to sparsely pubescent above, not scabrous, glabrous to sparsely pubescent beneath, hairs if present loose or appressed. (*Svida stricta* sensu Small)

River and stream banks, pond and lake shores, wet thickets and clearings, floodplain forests, swamps, wet woodlands. s. Del., s.e. Va., s. Ind., s. Ill., s.e. Mo., generally southward except in the Appalachian highlands to s. Fla., Ark., s.e. Okla., e. Tex.

C. foemina subsp. **microcarpa** (Nash) J. S. Wilson. A shrub, mostly not over 2–3 m tall, rarely inhabiting lowland woodlands, differing from subsp. *foemina* in its lower stature, in its usually having copiously pubescent twigs of the current season, and in having scabrid upper leaf surfaces. (*C. microcarpa* sensu Small)

Mostly an understory shrub of well-drained mixed woodlands, occasionally in bottomland woodlands but not where water stands for any appreciable period. Coastal plain, s.e. N.C. to cen. pen. Fla., Fla. Panhandle, s.w. Ga., s. Ala.

Araliaceae (GINSENG FAMILY)

Aralia spinosa L. DEVIL'S-WALKING-STICK, HERCULES-CLUB.

A distinctive, deciduous shrub or small tree, the stem usually unbranched, or with a few ascending branches, the very large compound leaves on growth of the season giving an umbrellalike aspect. Producing elongate underground runners and clone-forming. Stem markedly prickly, the prickles coarse, stiff, and sharp, sloughing on older stems. Leafless nodes with raised ridgelike leaf scars nearly encircling the stem. Leaves alternate, 2- or 3-pinnately compound, up to 12 dm long, the axes sometimes irregularly prickle-bearing, the prickles commonly recurved. Petioles enlarged basally, clasping and nearly encircling the stem. Leaflets short-stalked, blades ovate, 3–10 cm long, their bases acute to rounded, sometimes oblique, apices acute or more commonly acuminate, margins sharply serrate. Frequently a pair of accessory leaflets is borne one on either side of the principal axis where two secondary axes arise. Inflorescence a very large, terminal compound panicle of small flowers in slender-stalked ball-like umbels, branches of the panicle and flower stalks pubescent. Ovary inferior. Calyx segments 5, very small, broader than long, obtuse or rounded apically. Petals 5, white, 2–3 mm long, spreading or reflexed. Stamens 5. Styles 4–6, usually 5, fused only basally. Fruit a globose drupe 4–6 mm in diameter, crowned by persistent style bases, purplish black. Fruiting panicles purplish throughout and showy.

Upland and lowland woodlands, shrub bogs. N.J. to s. Ind., Ill., and Iowa, generally southward to Fla. and e. Tex.

Umbelliferae (CARROT FAMILY)

Herbs (ours), many with stems hollow at maturity. Leaves alternate (opposite in only one of ours), simple in some, phyllodial in some, more commonly compound or divided. Flowers relatively small, mostly in simple or compound umbels; simple umbels or primary umbels of compound ones sometimes subtended by bracts (the involucre) and ultimate umbels of compound umbels sometimes subtended by bractlets (the involucel); flowers in some closely aggregated in involucrate heads. Ovary inferior, the floral tube in some surmounted by 5 sepals (these often minute, absent in some), by 5 petals, and by 5 stamens. Ovary 2-locular, styles 2, the stylar bases often swollen and spreading forming a stylopodium. Fruit dry, of 2 mericarps attached to each other by their faces (at the commissure) and at maturity dehiscing at the commissure, the two-mericarpal fruit called a schizocarp; schizocarp flattened to a greater or lesser degree dorsally (parallel to the commissural face) or laterally (at right angles to the commissural face), or terete or nearly so in cross-sectional outline; each mericarp may have 5 primary ribs, distinguished as lateral, 1 down each margin (next to the commissure on each side of the mericarp), as dorsal, the median rib (at the outer free edge of the mericarp), and as intermediate (on each side of the mericarp between the lateral and dorsal; rarely 4 secondary ribs are intercalated between the primary ribs; oil tubes present in some between or under the ribs, if present not always visible externally; mericarps 1-seeded.

In this family it is especially important for purposes of identification that specimens have at least some well-developed fruits.

1. Flowers closely sessile in compact bracteate heads, each flower subtended by a spine-tipped bractlet.
                                 1. *Eryngium*
1. Flowers in umbels or compound umbels.
 2. Stems elongate-rhizomatous, horizontal, bearing simple leaves and inflorescences (when present) from nodes of the rhizome.
  3. Leaves phyllodial, transversely septate.   2. *Lilaeopsis*
  3. Leaves petiolate and with definite blades.
   4. Leaf blades ovate to oblong; involucre of 2 ovate, oblong or obovate bracts mostly 3–4 mm long; ribbing on the fruit more or less reticulate.   3. *Centella*
   4. Leaf blades, in overall outline, round to reniform; involucre absent or its bracts minute; no reticulation on the fruit.   4. *Hydrocotyle*
 2. Stems erect, or if prostrate not rhizomatous.
  5. Leaves phyllodial, septate, hollow between the septae.
   6. Stems slender, weak; umbel with involucre of 1 or 2 short bracts or none. 5. *Ptilimnium*
   6. Stems relatively robust and stiff; umbel with an involucre of filiform bracts. 6. *Oxypolis*
  5. Leaves not phyllodial.
   7. The leaves opposite, blades palmately veined, palmately cleft, or palmately divided.
                              7. *Bowlesia*
   7. The leaves alternate, blades pinnately veined, pinnately divided, or ternately to pinnately compound.
    8. Petals yellow.   8. *Zizia*
    8. Petals white, greenish white, or pinkish.
     9. Stems more or less woolly; upper leaves with sheaths very broadly winging the petioles; leaf or leaflet blades soft-pubescent beneath.   9. *Heracleum*
     9. Stems not woolly; leaf or leaflet blades not soft-pubescent beneath.
      10. Principal leaves with 3 linear to linear-lanceolate, palmately disposed divisions or leaflets (sometimes reduced to 1).
       11. Fruit ovoid but narrowed to a neck distally, not winged dorsally. 10. *Cynosciadium*
       11. Fruit ellipsoid-oblong, not narrowed to a neck distally, winged dorsally. 6. *Oxypolis*
      10. Principal leaves with more than 3 leaflets or divisions (if ternately compound then the leaflets broader than linear-lanceolate, i.e., in no. 13 below, *Cryptotaenia*).
       12. Leaves once pinnately divided or compound.
        13. Leaf divisions or leaflets cross-septate.   11. *Limnosciadium*
        13. Leaf divisions or leaflets without cross-septae.
         14. Fruit strongly flattened dorsally, broadly winged dorsally.   6. *Oxypolis*
         14. Fruit scarcely flattened, all the ribs equal and prominently corky-winged. 12. *Sium*
       12. Leaves ternately compound, pinnately or ternately decompound, or pinnately divided.
        15. Principal leaves ternately once compound.   13. *Cryptotaenia*
        15. Principal leaves ternately or pinnately decompound or pinnately divided or dissected.
         16. Leaves with definite and discrete leaflets.   14. *Cicuta*
         16. Leaves decompound with the ultimate leaflets pinnately cleft or divided, or the entire leaf pinnately divided, the divisions linear to filiform.
          17. Leaves decompound, the ultimate leaflets pinnately cleft or divided. 15. *Conium*
          17. Leaves finely pinnately divided, the ultimate divisions linear to filiform.

18. Fruit lanceolate, or elliptic, much longer than broad.
    19. Bracts of involucel filiform.   16. *Trepocarpus*
    19. Bracts of involucel oblong to ovate.   17. *Chaerophyllum*
18. Fruit ovoid, oval, to suborbicular.
    20. Once-compound umbels sessile in the leaf axils from low on the plant to tips
of branches, stalks of the primary umbel shorter than to about equaling the leaves.
        21. Lower leaves biternate or ternate-pinnately dissected; ribs on fruit narrow,
edges acute.   18. *Ammoselinum*
        21. Lower leaves pinnately decompound-dissected; ribs on fruit corky, edges
rounded.   19. *Apium*
    20. Once-compound umbels long-stalked, terminating upper parts of the stem,
the primary stalks much exceeding the subtending leaves.
        22. Ovary and fruit warty-tuberculate.   20. *Spermolepis*
        22. Ovary and fruit not tuberculate.   5. *Ptilimnium*

1. Eryngium (ERYNGOS)

Biennial or perennial herbs. Stems erect or prostrate. Leaves membranous to coriaceous, often spinose on the margins or tips, petioles sheathing at the base, principal veins pinnate or parallel. Flowers borne in globose to cylindrical, involucrate heads, the involucral bracts often spinose on the margins or tips, heads borne singly from the leaf axils or in cymes or racemes. Sepals evident, rigid, persistent. Petals white to blue or purple. Styles slender, stylopodium none. Fruit globose to obovoid, obconic, or obpyramidal, not ribbed, the surface papillate, scaly, or tuberculate.

1. Plants low, the branches weak and commonly prostrate, sometimes prostrate below and weakly ascending distally; heads borne singly from the leaf axils.
    2. Heads, fully developed, oblong-cylindrical, subtending bracts conspicuously longer than the radius of the base of the head.   1. *E. prostratum*
    2. Heads, fully developed, subglobose, little if any longer than broad, subtending bracts not or barely extending beyond the base of the head.   2. *E. baldwinii*
1. Plants with erect stems; heads in cymes.
    3. Lower, larger leaves exceeding 1 dm long, several to many times longer than wide.
        4. Flowers blue; leaves without marginal bristles.   3. *E. aquaticum*
        4. Flowers white or greenish white; leaves with marginal bristles.   4. *E. yuccifolium*
    3. Lower, larger leaves (blades) less than 1 dm long, usually 7 cm long or less.   5. *E. integrifolium*

1. Eryngium prostratum Nutt. ex. DC.   Fig. 205

Plant low, commonly with several to numerous branches from the base, branches weakly erect, all prostrate, or prostrate below, the distal portions ascending; plants often growing in tangled mats. Basal leaves largest and longer petiolate than stem leaves, blades ovate, oblong, lanceolate, entire, dentate, cleft, or pinnatifid, variable in size, to about 7 cm long and to 2.5 cm wide; stem leaves reduced upwardly, sessile above. Heads borne singly from leaf axils, slenderly stalked, stalks usually 1–2 (–3.5) cm long; heads, fully developed, oblong-cylindrical, about 8 mm long, 3–4 mm wide. Involucral bracts 5–10, in length considerably exceeding the radius of the base of the head, linear, entire, usually somewhat reflexed; bractlets subtending the flowers scarcely extending outwardly from between the flowers, in fruit usually concealed. Flowers blue. Fruits subglobose to obconic, about 2 mm broad, usually broader than long, surface short-papillose.

Low, wet areas, often where temporarily flooded, floodplain forests, depressions, margin of ponds and lakes, wet woodlands, swamps, ditches. s.e. Va. to Mo., generally southward to cen. pen. Fla., e. Okla. and Tex.

2. Eryngium baldwinii Spreng.

Essentially similar in habit, foliar characters, color of flowers, to *E. prostratum*. Heads, fully developed, subglobose, little if any longer than broad. Bracts of the involucre shorter, not or barely extending beyond the base of the head. Bractlets subtending the fruits usually exserted

Fig. 205. a, **Eryngium prostratum** (From Correll and Correll. 1972); b–e, **Hydrocotyle verticillata** var. **verticillata:** b, habit; c, portion of inflorescence; d, flower; e, fruit; f, **Hydrocotyle verticillata** var. **triradiata:** portion of inflorescence.

somewhat from between the fruits. Surface of fruit with elongated, clavate papillae or tubercles.

In habitats similar to those of *E. prostratum*. s. Ga. and all of Fla. e. of the Apalachicola River drainage.

3. **Eryngium aquaticum** L.   Fig. 206

Plants slender to stout, to 18 dm tall, perennating by basal offshoots. Stems solitary, erect, striate. Basal leaves (usually absent at time of flowering) with long, broadly winged, channeled petioles sheathing the stem at base; margins of petiolar wings hyaline; blades linear, linear-lanceolate, or oblong, entire or finely toothed; leaves, including petioles, to 4 dm long, blades to 2.5 cm wide; leaves upwardly reduced, mostly without blades. Inflorescence openly cymose; heads, fully developed, ovoid, 15–30 mm long and about as broad at the base. Involucral bracts 6–10, reflexed, commonly blue, linear-attenuate, entire or with 2 to several sharp cusps or lobes, to 2.5 cm long; bractlets subtending the flowers oblong, with 1

Fig. 206. **Eryngium aquaticum:** a, plant; b, adaxial side of median node, with leaf; c, abaxial side of node; d, flower with subtending bractlet; e–g, variation in bractlets.

437

or 3 terminal spinelike cusps, if 3, then the central one sometimes longer than the laterals and extending outwardly well beyond the flowers or fruits, the heads thus bristly. Sepals lance-ovate, acuminate apically. Petals blue. Fruits short-oblong, 2–3 mm long, angled, with conspicuous flat scales on the angles. (*E. virginianum* Lam.; incl. *E. floridanum* Coult. & Rose, *E. ravenelii* Gray)

Wet soils or in shallow water, fresh to brackish marshes, wet pine flatwoods, swamps, bogs, depressions, ditches. Coastal plain, N.J. to n. and cen. pen. Fla.

4. Eryngium yuccifolium Michx. RATTLESNAKE-MASTER, BUTTON-SNAKE-ROOT.

Plant perennial, stems striate, stiffly erect, commonly clumped, to about 1 m tall. Lower, larger leaves scarcely petiolate, broadly linear-attenuate, to 10 dm long, 1–4 cm wide, principal veins parallel, marginally with prominent bristles; upper leaves similar but reduced in size. Inflorescence openly cymose; heads, fully developed, subglobose to ovoid, little longer than broad, 1–2.5 cm across. Involucral bracts 4–10, lance-ovate, spreading at first, later usually reflexed, usually not longer than the radius of the head, entire to spinulose-toothed or -lobed; bractlets subtending the flowers usually entire, with sharp, subulate tips, extending outwardly from between the flowers or fruits, the heads somewhat bristly. Sepals broadly ovate, obtuse and mucronate apically. Petals whitish. Fruit obpyramidal, about 3 mm long, with prominent flat scales on the angles. (*E. aquaticum* sensu Small (1933); incl. *E. synchaetum* (Gray ex Coult. & Rose) Coult. & Rose)

Bogs, pine savannas and flatwoods, prairies, ditches, also in well-drained openly wooded sand ridges and hills. Conn. to Minn. and Kan., generally southward to s. Fla., Okla. and Tex.

5. Eryngium integrifolium Walt.   Fig. 207

Plants slender, erect, 3–8 dm tall, stems striate, solitary to several, perennating from the base. Lower, larger, leaves (commonly not present at time of flowering) petiolate, blades elliptic, ovate, oblong, very variable, to about 8 cm long and 3–4 cm wide, entire or crenate marginally, bases cordate or oblique, obtuse apically; stem leaves also variable, reduced in size upwardly, mostly sessile, from short-ovate to linear-lanceolate, crenate to sharply spinose-toothed, or rarely laciniately cleft. Inflorescence openly cymose; heads, fully developed, subhemispherical to globose, to about 1.5 cm across. Involucral bracts 6–10, linear, rigid, 1–2 cm long, extending well outward from the base of the head, entire or usually with 3–5 spinose teeth, bracts subtending the flowers about 5 mm long, linear-oblong below, with 3 stiff, straight, equal spinose cusps terminally, the cusps usually protruding from between the flowers or fruits. Sepals lance-ovate, stiffly acuminate apically, 1.5–2 mm long. Petals greenish at first, then usually blue. Fruit about 2 mm long, more or less obpyramidal, the angles bearing short to clavate white papillae, these becoming flat and scaly on the mature fruit. (Incl. *E. ludovicianum* Morong)

Bogs, wet pine savannas and flatwoods, meadows. N.C. to n. Fla., westward to Okla. and Tex.

2. Lilaeopsis

Low, perennial herbs with rhizomatous, horizontal stems, the stems subterranean or matted in water, nodes of the rhizomes bearing 1 to several linear to spatulate, transversely septate, phyllodial leaves. Involucre of 3 to several minute bracts 1–1.5 mm long. Sepals minute or absent. Petals white. Fruit ovate to suborbicular, slightly flattened laterally or nearly terete, lateral ribs thick-corky, dorsal ribs narrow.

- Phyllodes mostly 1–5 cm long, occasionally to 7 cm; stalks of the umbels as long as the phyllodes or longer.   1. *L. chinensis*
- Phyllodes mostly 10–30 cm long, rarely shorter; stalks of the umbels much shorter than the phyllodes.   2. *L. carolinensis*

Fig. 207. **Eryngium integrifolium:** a, plant; b, leaves, to show variation; c, flower with bractlet.

1. Lilaeopsis chinensis (L.) Kuntze.

In tidal muds of brackish marshes and shores, usually inundated at high tides, commonly forming extensive sodlike mats. Phyllodes linear to narrowly clavate, usually not exceeding 5 cm long, with 4–6 transverse septae, mostly solitary-from the nodes. Stalks of the umbels equaling or somewhat exceeding the phyllodes, 4–10-flowered, flower stalks thickish, 3–4 mm long. (*L. lineata* (Michx.) Greene)

In mud of brackish marshes and tidal shores. N.S. to n. pen. Fla., westward to La.

2. Lilaeopsis carolinensis Coult. & Rose.   Fig. 208

Usually in shallow fresh water, or in mud, commonly in thick, dense, tangled mats. Phyllodes linear to conspicuously clavate, mostly 10–30 cm long, with 7–15 transverse septae. Stalks of the umbels much shorter than the phyllodes, 5–15-flowered, the flower stalks slender, 5–10 mm long. (*L. attenuata* (H. & A.) Fern.)

Fresh shallow water, pools, marshes, swamps, ditches, near and on muddy shores. Coastal plain, Va. to n. Fla., westward to La.; S.Am.

3. Centella

Centella asiatica (L.) Urban.

Perennial herb with slender horizontal stems from which arise a cluster of ascending petiolate leaves at each node. Stem and leaves glabrous to tomentose. Leaves simple, blades ovate to oblong, 1.5–5 cm long, palmately veined, apices rounded, bases cordate to truncate, margins denticulate, weakly sinuate, or dentate; petioles very variable in length, to 30 cm long, glabrous or tawny-pubescent. Inflorescence of 1–5 simple, loose to subcapitate, umbels per node. Peduncles very variable in length but usually much shorter than the petioles from the same node. Involucels of 2 oblanceolate bracts. Flowers 1–4 in each umbel, subsessile to short-pedicillate. Sepals absent. Petals white to greenish, often tinged with rose, spreading, triangular, often pubescent on their lower surfaces. Stamens and styles shorter than the petals. Anthers purple. Stylopodium depressed. Fruit strongly flattened laterally, broader than long, more or less pumpkin-shaped, prominently nerved with more or less raised reticulate venation between the nerves. (*C. repanda* (Pers.) Small; *C. erecta* (L.f.) Fern.)

Margins of lakes, ponds and streams, wet ditches, wet pine savannas and flatwoods, bogs. Coastal plain, Del. to s. Fla., westward to Tex.; pantropical.

4. Hydrocotyle (WATER PENNYWORTS)

Low perennial, (ours glabrous), the stems of most horizontal or arching, on or within the substrate, sometimes in floating mats, rooting at the nodes (flowering stems of *H. americana* commonly shortly erect or ascending); leaves solitary from the nodes, inflorescences from the nodes. In most species robustness of the stems and leaves varying considerably relative to conditions of moisture and fertility. Leaves long-petiolate, petioles nonsheathing, blades peltate or cordate, orbicular, reniform, or ovate, margins irregularly crenate or crenate-lobed. Flowers small, white, greenish or yellow, in simple to compound umbels or interrupted spikes, umbels of some proliferating. Calyx segments none or minute. Petals ovate. Fruit somewhat flattened laterally, broader than long, dorsal and lateral ribs present or absent, stylopodium conic to depressed.

1. Leaves not peltate, petioles attached at the sinus at the base of the blade.
  2. Inflorescences sessile or short-stalked, usually on shortly erect or ascending, very slender stems; petioles very slender, leaves thin; faces of the mericarps prominently ribbed.   1. *H. americana*
  2. Inflorescences distinctly stalked, borne from rather fleshy or stoutish horizontal stems; petioles stout, they and the leaves subsucculent; faces of the mericarps not ribbed.   2. *H. ranunculoides*
1. Leaves peltate.
  3. Flowers in a simple umbel.   3. *H. umbellata*

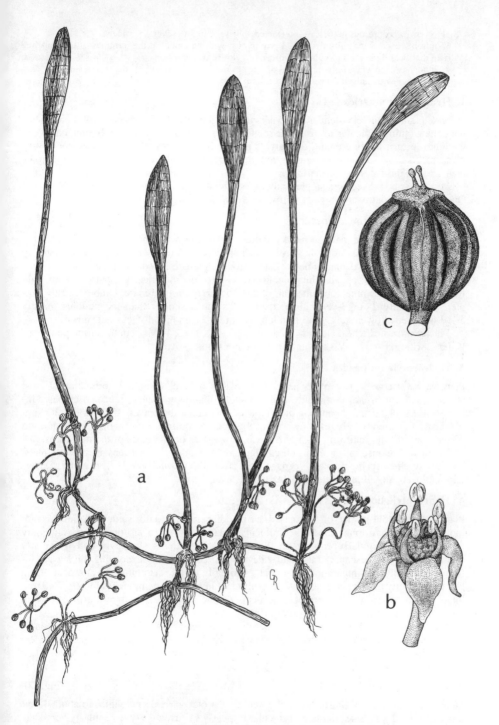

Fig. 208. **Lilaeopsis carolinensis:** a, habit; b, flower; c, fruit.

3. Flowers in proliferous umbels or interrupted spikes, or in 2 to several verticels.
4. Inflorescence when fully developed proliferous, with an umbel at the terminus of the stalk and from that bearing several to numerous ascending floriferous branches.   4. *H. bonariensis*
4. Inflorescences when fully developed of 1 to several interrupted spikes, the spikes simple or few-branched, or with stalked flowers in 2 to several verticels.   5. *H. verticillata*

1. Hydrocotyle americana L.

Delicate plants with very slender stems and petioles. Petioles 1–5 (–6) cm long. Leaf blades not peltate, thin, orbicular to orbicular-reniform, 1–4 cm across, shallowly crenate-lobed, the lobes in turn finely crenate, or margins merely irregularly crenate. Inflorescences very inconspicuous, sessile to short-stalked, few-flowered. Fruits about 1.5 mm wide, with prominent ribs on the faces of the mericarps.

Meadows, bogs, seepage slopes and ledges, about waterfalls, wet woodlands. Nfld. and Que. to Wisc., southward to w. N.C., w. S.C., W.Va., Ind.

2. Hydrocotyle ranunculoides L. f.

Plants usually in dense mats on mucky shores, in shallow water near shores, or in floating mats. Stems and petioles stoutish, leaves rather succulent. Petioles very variable in length, to about 35 cm long. Leaf blades not peltate, variably crenate-lobed, sometimes with a deep sinus terminally and opposite the pronounced basal sinus, thus nearly 2-parted, commonly with a prominent terminal lobe, margins of the lobes irregularly crenate. Umbels mostly simple, occasionally 2- or 3-verticillate, their stalks much shorter than the petioles, mostly 5–10-flowered, flower stalks 1–3 mm long. Fruit 2–3 mm broad, faces not ribbed.

Mucky shores, ditches, sloughs, floating mats. Pa., Del., W.Va., southward to pen. Fla., w. La., Ark. and Ariz.; Wash. southward to Pan.; Cuba; S.Am.

3. Hydrocotyle umbellata L.

Plant very variable in size of its parts. Leaf blades peltate, essentially orbicular, at most about 6 or 7 cm broad, commonly much smaller, shallowly crenate-lobed to crenate. Inflorescence stalks usually equaling or somewhat exceeding the leaves. Umbels usually simple, many-flowered (rarely few-flowered), subglobose in outline, flowers on slender stalks to 20 mm long. Fruits reniform, 2–3 mm broad, low ridges on the faces of the mericarps.

In small streams, on or near shores of ponds and lakes, sometimes in floating mats, swamps, ditches, spring runs and seepage areas, wet alluvial outwash. N.S. to Minn., generally southward to s. Fla., Okla. and Tex.; Ore., Calif.; trop. Am.

4. Hydrocotyle bonariensis Lam.

Plant very variable in size of its parts. Leaf blades peltate, usually somewhat fleshy, commonly broader than long, sometimes orbicular, maximally to 15 cm broad but commonly much smaller, irregularly crenate on the margins, sometimes shallowly and irregularly crenate-lobed. Inflorescence stalks, fully developed, considerably exceeding the leaves. Inflorescence at first a simple globose umbel, eventually with few to numerous branches from that umbel, fully developed essentially obpyramidal in outline, the branches mostly 7–10 cm long, flowers or fruits in interrupted verticels on the branches. Fruit reniform, 3–4 mm broad, faces of the mericarps with sharp ribs.

Wet sands, coastal dunes and upper beach strand, interdune swales, wet alluvial sands, sandy-peaty depressions and ditches, often matted in shallow water, blackland prairies. Outer coastal plain, N.C. to s. Fla., westward to Tex.; southward through Mex. to n. Arg., Urug., Chile; P.R.

5. Hydrocotyle verticillata Thunb.   Fig. 205

Stems, petioles, and inflorescence stalks slender. Petioles variable in length, to about 12 cm long. Leaf blades peltate, thinnish, orbicular or nearly so, irregularly to regularly very shallowly crenate-lobed or crenate, at most about 6 cm broad. Inflorescence stalks much shorter than to exceeding the leaves. Flowers sessile in few-flowered clusters forming interrupted spikes, *or* in 1 to few separated verticels, the flowers stalked. Fruits oblate, mostly about 2 mm broad, faces of the mericarps ribbed.

Plants with spicate inflorescences may be designated var. *verticillata*; those with stalked flowers in verticels may be designated var. *triradiata* (A. Rich.) Fern. (*H. canbyi* Coult. & Rose; *H. australis* Coult. & Rose)

Floodplain and wet woodlands, springy areas in woodland, swamps, in and along banks of woodland spring runs, sandy-peaty ditches. Coastal plain and piedmont, Mass. to s. Fla., westward to Calif.; Utah, Nev.; trop. Am.

5. Ptilimnium (BISHOP'S-WEEDS)

Glabrous annual herbs with fibrous, sometimes tuberous, roots, erect relatively few-branched stems. Leaves terete, hollow, septate phyllodia, *or* petiolate and decompound with filiform divisions, the petioles sheathing the stem, their borders narrow and hyaline. Inflorescences stalked, compound umbels, axillary and terminal. Involucre of entire or pinnatifid bracts; bracts of the involucel entire. Flowers white, or rarely pink. Sepals minute, lanceolate or deltoid. Carpophore 2-cleft about ½ its length. Fruit ovoid to suborbicular, compressed laterally, dorsal ribs rounded or acute, lateral ribs separate from the corky tissues on the commissural side of the rib. Corky tissue forming a longitudinal band around the fruit.

Reference: Easterly, N. W. "A Morphological Study of *Ptilimnium*." *Brittonia* 9: 136–145. 1957.

1. Leaves cylindrical, hollow, jointed phyllodia.
   2. Umbels 5–7 in an inflorescence; flower stalks less than 3 mm long; corolla 1–1.5 mm across.
                                                                                    1. *P. fluviatile*
   2. Umbels 6–15 in an inflorescence; flower stalks 3–6 mm long; corolla 3–3.5 mm across.
                                                                                    2. *P. nodosum*
1. Leaves petiolate, decompound with filiform divisions.
   3. Leaf segments crowded on the axis thus appearing nearly verticillate.   3. *P. costatum*
   3. Leaf segments not crowded on the axis and not appearing verticillate.
      4. The leaf segments usually 3 at a node on the axis; bracts of the involucre usually 3-cleft.
                                                                                    4. *P. capillaceum*
      4. The leaf segments usually 2 at a node on the axis; bracts of the involucre usually entire.
                                                                                    5. *P. nuttallii*

1. Ptilimnium fluviatile (Rose) Mathias.

Plants 2–5 dm high, erect or spreading. Leaves cylindrical, hollow, septate phyllodia 4–12 cm long. Involucre of small lanceolate bracts 1–4 mm long; involucel inconspicuous or lacking. Umbels 5–7 per inflorescence. Sepals lance-deltoid, 1–2 mm long. Petals white, rarely pink, the narrow tips inflexed. Anthers rose-colored. Fruit suborbicular. (*Harperella fluviatilis* Rose)

Rare, stream margins, rocky shoals, sandbars. n. Ala. through the Carolinas to the Potomac River valley.

2. Ptilimnium nodosum (Rose) Mathias.

Plants 4–10 dm tall, relatively robust if growing in wet places. Leaves cylindrical, hollow, septate phyllodia 8–30 cm long. Involucre of small lanceolate bracts 2–5 mm long; involucel inconspicuous or lacking. Umbels 6–15 per inflorescence. Sepals lance-deltoid, 1–2 mm long. Petals white, the narrow apex inflexed. Anthers rose-colored. Fruit suborbicular. (*Harperella nodosum* Rose)

Rare, savannas, wet ditches, shallow pond depressions in flatwoods. Presently known from Schley and Dooly Cos., Ga., and Aiken Co., S.C.

3. Ptilimnium costatum (Ell.) Raf.   Fig. 209

Plant 6–15 dm tall. Leaves pinnately dissected-compound, the ultimate divisions filiform, crowded on the axis and appearing verticillate. Involucral bracts usually entire; involucel of entire bracts shorter than the flower stalks. Umbels about 20–24 per inflorescence, flowers

Fig. 209. **Ptilimnium costatum:** a, flowering branch; b, midsection of stem; c, base of plant; d, leaf, enlarged; e, flower; f, fruit; g, one mericarp. (From Correll and Correll. 1972)

15–20 per umbel, their stalks 4–5 mm long. Sepals deltoid, acute to subacuminate, persistent. Petals white, the narrow apex inflexed. Anthers rose-colored. Styles 1.5–3 mm long. Fruit ovoid, 2–4 mm long, corky band around it flat.

Sporadic in occurrence, stream banks, swamps, pond margins, pine savannas. N.C., Tenn., Ill., Mo., generally southward to Ga., Ala. and e. half of Tex.

4. Ptilimnium capillaceum (Michx.) Raf.

Plant 1–8 dm tall. Leaves pinnately decompound-dissected, the divisions filiform and usually 3 at a node on the leaf axis. Involucre usually 3-cleft; involucel of filiform, entire bracts. Umbels 4–20 per inflorescence, their stalks subequal, 1–3.5 cm long. Flowers 5–20 per umbel, their stalks 3–12, usually 4–6, mm long. Sepals minute, deltoid, persistent. Petals white, rarely pinkish, their abruptly narrowed tips inflexed. Anthers purplish. Styles 0.2–0.5 mm long. Fruit broadly ovoid, 1.5–3 mm long, the lateral ribs conspicuous, corky band around the fruit prominent.

Brackish and fresh marshes, wet woodlands, swamps, ditches, shores of ponds and lakes, depressions in pine savannas and flatwoods. s.e. Mass., coastal plain and piedmont to s. Fla., westward to e. half of Tex., northward in the interior to Mo., Kan., s. Ill., Ky.

5. Ptilimnium nuttallii (DC.) Britt.

Plant 3–6 dm tall. Leaves pinnately decompound-dissected, the divisions relatively few, filiform, usually 2 at a node on the leaf axis. Bracts of the involucre filiform, mostly entire; involucel of filiform entire bractlets. Umbels 25–30 per inflorescence, their stalks subequal. Flowers 25–30 per umbel, their stalks 3–8 mm long. Sepals minute, deltoid-acuminate, persistent. Petals white. Anthers purple. Styles 0.5–1.5 mm long. Fruit ovoid, 1–1.5 mm long, lateral ribs conspicuous, corky band around the fruit conspicuous.

Moist sandy soil of open places, bogs, wet woodlands, wet prairies, ditches, abandoned rice fields, low sandy fields. Ky. to Kan., generally southward to La. and e. half of Tex.

6. Oxypolis (HOG-FENNELS)

Erect, glabrous perennials with fascicles of fibrous roots. Leaves pinnately or ternately compound or in the form of hollow, septate phyllodes. Pedunculate umbels compound, terminal and axillary. Bracts of the involucre and involucel several. Sepals minute. Petals white or garnet-maroon. Fruit dorsally flattened, with corky wings nearly as broad as the body or broader, dorsal and median ribs finely wirelike, stylopodium conic; oil tubes prominent, solitary in the intervals, 2–6 in the commissures.

1. Leaves phyllodial.
   2. Petals white; phyllodia not conspicuously jointed or brittle.   1. *O. filiformis*
   2. Petals garnet-maroon; phyllodia conspicuously jointed, the joints brittle and very easily segmented.
                                                                                       2. *O. greenmanii*
1. Leaves not phyllodial, blades pinnately or ternately compound or divided.
   3. Leaves odd-pinnately compound or divided.   3. *O. rigidior*
   3. Leaves ternately divided or simple.   4. *O. ternata*

1. Oxypolis filiformis (Walt.) Britt.   Fig. 210

Stems slender, many-striate, sparingly branched, to 1.8 m tall. Leaves phyllodial, the lower elongate-tapering, to 6 dm long, inconspicuously septate, not especially stout, not readily segmenting at the joints, markedly reduced upwardly. Inflorescence to 12 cm across in fruit, much smaller in flower; involucre of 4–10 lance-attenuate bracts with narrow scarious margins; involucels mostly with 6–8 linear-attenuate bracts. Flowers about 1.5 mm across, mostly 12–22 in the ultimate umbel, the pedicels subequal in fruit, about 1 cm long. Sepals minute, deltoid. Petals white, ovate-acuminate, strongly curled inward, tips declined. Fruit oval, elliptic, or narrowly obovate, 5–8 mm long and 3–4 mm wide, the lateral wings flat and thin.

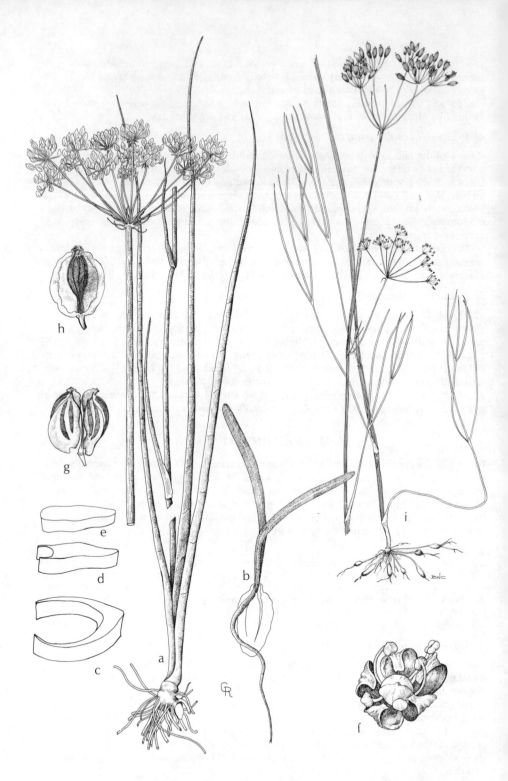

Fig. 210. a–h, **Oxypolis filiformis:** a, plant; b, seedling; c, section of sheath near base; d, section of sheath near apex; e, section at base of phyllode; f, flower; g, fruit, showing attachment of mericarps to carpophore; h, fruit from dorsal side, showing ribs; i, **Oxypolis ternata:** habit.

Open or semiopen wet sites, commonly in shallow water, bogs, pine flatwoods or cypress depressions, marshes, ditches, wet prairies. Coastal plain, N.C. to s. Fla., westward to Tex.; Cuba, Bah.Is.

2. Oxypolis greenmanii Math. & Const.   Figs. 211 and 212

Stems stout, many-striate, relatively few-branched, to 2 m tall or more, with a short, stout rhizome. Leaves phyllodial, the lower very stout, long-tapering, conspicuously septate-jointed, brittle and easily segmenting at the joints; markedly reduced upwards. Inflorescence to about 15 cm across in fruit, much smaller in flower; involucre of 6–10 lance-attenuate, usually unequal bracts; involucel of 6–10 bracts, one usually lance-attenuate, the others linear-attenuate. Flowers about 1.5 mm across, mostly 16–25 in the ultimate umbel, the pedicels subequal in fruit, about 1 cm long. Sepals minute, deltoid-attenuate. Petals garnet-maroon, ovate to orbicular, mostly broader than long, strongly curled inward, the abruptly acuminate tips declined. Filaments garnet-maroon, anthers white. Fruit oval, elliptic, narrowly obovate, sometimes ovate, 6–9 mm long and about 5 mm wide, the lateral wings flat and thin.

Stems and phyllodia are suffused with purplish pigment, flower stalks, sepals, petals, filaments, early stages of developing fruits are garnet-maroon. During maturation of the fruits, the pedicels become green and the ripe fruits are usually suffused with purplish pigment.

Locally abundant, usually in water, depressions of pine flatwoods, cypress ponds, *Hypericum* ponds or depressions, drainage canals and ditches, borrow-pit pools. Restricted to Calhoun, Gulf and Bay Cos. in Fla. Panhandle.

3. Oxypolis rigidior (L.) Raf. COWBANE, WATER-DROPWORT.   Fig. 213

Stems slender to moderately coarse, somewhat striate, to 1.5 m tall. Leaves petiolate, odd-pinnately compound, with 5–11 leaflets, the lower leaves to 30 cm long. Leaflets variable, broadly to narrowly lanceolate, oblong, oblanceolate, linear-lanceolate, sometimes 3-lobed terminally; rarely with entire margins, usually with irregularly distributed, salient teeth. Inflorescence to 15 cm across in fruit, much smaller in flower; involucral bracts several, long-attenuate; bracts of the involucel several and filiform. Flowers about 3 mm across, 15–45 in the ultimate umbel, the pedicels unequal in fruit, the longer about 1 cm. Sepals minute, deltoid. Petals white, orbicular, spreading, the abruptly acuminate tips strongly curved inwardly. Fruit oval or elliptic, about 5 mm long and 3 mm wide, the lateral wings thick and triangular in cross section. (*O. turgida* Small))

In and on the banks of small streams, bogs, wet woodlands, swamps, marshes and ditches. N.Y. and N.J. to Minn., generally southward to the Fla. Panhandle and Tex.

4. Oxypolis ternata (Nutt.) A. Heller.   Fig. 210

Stem relatively very slender, little-branched, with a few rounded ridges and grooves. Fibrous roots terminated by fusiform tubers. Leaves ternately compound, or the uppermost simple, slenderly long-petioled, the lower with 1–3, mostly 3, very narrow, entire leaflets; stem leaves few, the upper bracteal. Inflorescence to 10 cm across at maturity, much smaller in flower; involucral bracts several, linear-attenuate; bracts of the involucel several, unequal, filiform. Ultimate umbels mostly with 8–12 flowers, pedicels slender, subequal, to 15 mm long, flower about 3 mm across. Sepals minute, deltoid-acuminate. Petals white, broadly ovate to obovate, spreading, the short-acuminate tips strongly curled inwardly. Fruit elliptic to narrowly obovate, 3–5 mm long, the lateral wings prominent.

Bogs and wet pine flatwoods and savannas. Coastal plain, s.e. Va. to Fla. Panhandle.

7. Bowlesia

Bowlesia incana Ruiz and Pavon.   Fig. 214

Annual decumbent herb, usually with several to many subequal, lax branches from the base, the several stems sometimes reaching 6 dm long or more. Herbage stellate-tomentose to

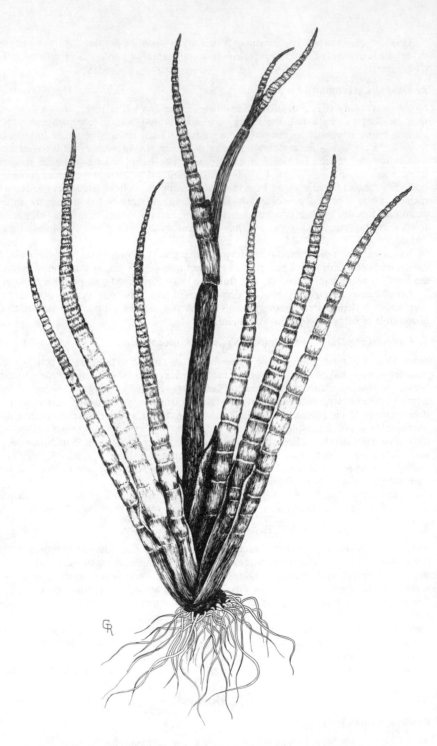

Fig. 211. **Oxypolis greenmanii:** plant with early stage of development of flowering stem.

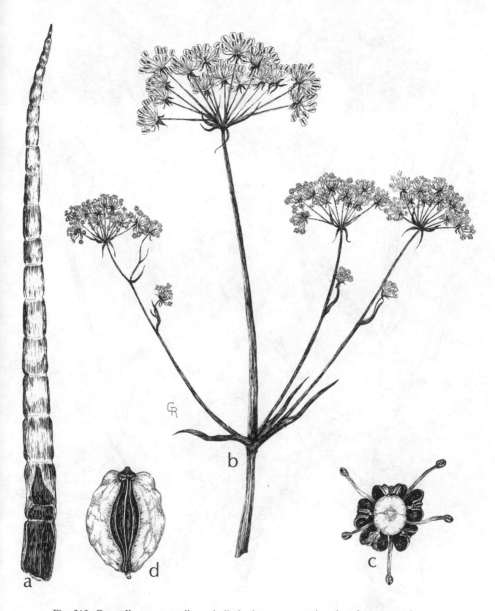

Fig. 212. **Oxypolis greenmanii:** a, phyllode, lower part sectioned to show septae; b, uppermost portion of flowering stem; c, flower, from above; d, fruit.

Fig. 213. **Oxypolis rigidior:** a, base of plant; b, inflorescence branch; c,d, leaves; e, flower; f, fruit. (From Correll and Correll. 1972)

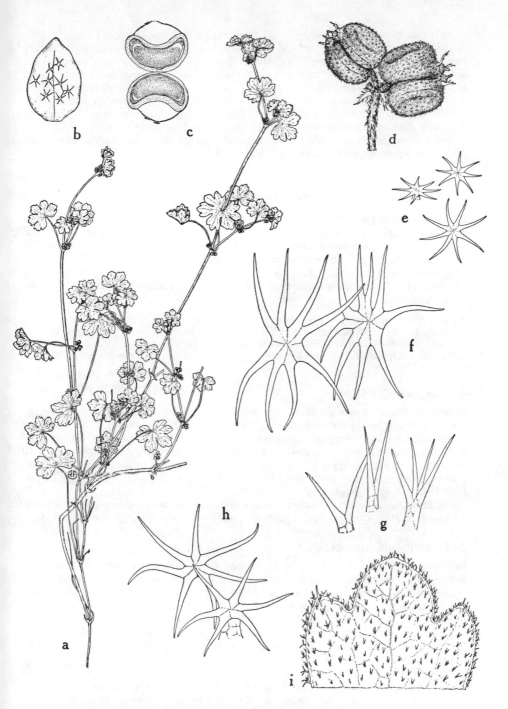

Fig. 214. **Bowlesia incana:** a, habit; b, dorsal surface of petal; c, fruit transsection; d, fruits; e, hairs from fruit; f, hairs from stem; g, hairs from upper leaf surface; h, hairs from lower leaf surface; i, upper leaf surface. (From Mathias and Constance in *Univ. Calif. Publ. in Bot.* 38: 32. 1965)

sparsely stellate-pubescent. Leaves simple, all except the first opposite, with slender lax petioles and hyaline stipules; blades orbicular-reniform in outline, shallowly to deeply palmately lobed, mostly broader than long, from 0.5–4.5 cm broad. Umbels usually paired at the nodes, their stalks short, rarely to 1 cm long, borne in the leaf axils; pedicels short; involucre of several lanceolate bracts. Flowers yellowish green to purplish; sepals triangular, hyaline, 0.3–0.7 mm long; petals usually lavender to maroon, oval, obovate, or suborbicular, 0.4–0.7 mm long. Fruit oval to globose, stellate-pubescent, much constricted between the mericarps, flattened dorsally, concave on the back, about 2 mm across.

In our range, mostly in floodplain woodland or clearings, sometimes weedy in low moist places. Fla. Panhandle (introduced?); La. and s.e. Tex.; s. Ariz., Calif., Mex.; s. S.Am.

8. Zizia (GOLDEN ALEXANDERS)

Glabrous perennials with a cluster of thickened roots. Stems and leaves not fully developed as flowering begins, becoming 6–10 dm tall. Leaf blades simple or ternately compound, some of both sometimes on the same plant. Umbels compound. Involucre none; involucel of several acicular bractlets. Central flower of each umbel sessile, others with stalks 1–3 mm long. Sepals minute, triangular. Petals golden yellow, obovate, abruptly narrowed apically to an inflexed tip. Stylopodium none. Fruit somewhat flattened laterally, with narrow, somewhat winged ribs.

● Blades of the leaflets finely serrate, mostly 5–10 teeth per cm of margin; stalks of the umbels stiffly ascending, 1–1.5 cm long in flower, 2.5–3.5 cm long in fruit; fruit oblong.   1. *Z. aurea*
● Blades of the leaflets crenate-serrate, mostly 2 to 3 teeth per cm of margin; stalks of the umbels slender, lax, and spreading, 2–4.5 cm long in flower, 4–5 (–8) cm long in fruit; fruit ovoid to orbicular.

2. *Z. trifoliata*

1. Zizia aurea (L.) Koch.

Leaves variable, the lower long-petiolate, gradually becoming essentially sessile upwardly, earliest leaves of the season rarely simple, the lower usually 1–3-, mostly 2- or 3-ternately compound, sometimes irregularly compound, leaflets sometimes cleft as well; leaflets ovate to lanceolate, usually acute to acuminate apically, rarely obtuse, finely serrate, mostly 5–10 teeth per cm of margin; upper leaves 1- or 2-ternately compound or irregularly compound. Umbels 6–20, their stalks stiffly ascending, 1–1.5 cm long in flower, 2.5–3.5 cm long in fruit. Fruit oblong, 3–4 mm long.

Meadows, moist to wet woodlands, swamp forests, thickets, moist roadsides and woodland borders. N.B., Que., to Sask., generally southward to Ga. (Fla.?) and e. third of Tex.

2. Zizia trifoliata (Michx.) Fern.

In general aspect similar to *Z. aurea*. Lowermost leaves more often with but 3 leaflets although sometimes biternate, the leaflets less tapered distally, mostly obtuse; margins with coarser and fewer teeth, usually 2 or 3 per cm of margin. Umbels 4–10, their stalks slender and lax, spreading, 2–4.5 cm long in flower, 4–5 (–8) cm long in fruit. Fruit ovate to suborbicular, 2–3.5 mm long. (*Z. bebbii* Coult. & Rose) Britt.; incl. *Z. arenicola* Rose)

Usually in mesic, well-drained woodlands; sometimes in moist to wet creek bottoms or floodplain woods, especially in southern parts of the range. Va. and W.Va., Tenn., Ark., to n. pen. Fla.

A third species, *Z. aptera* (Gray) Fern. (*Z. cordata* of authors), occurs in our range, inhabiting only well-drained sites insofar as we know. Its first, lowermost, leaves are simple, cordate-ovate.

9. Heracleum

Heracleum maximum Bartr. COW-PARSNIP. Fig. 215

Coarse perennial with a strong unpleasant odor, 1–2.5 m tall. Stem hollow, ribbed, woolly or pilose. Lower, larger leaves ternately compound, with long stout petioles 2–5 dm long, blades overall to nearly 3 dm long and broad; leaves reduced upwardly, the upper ternately lobed or divided (with form similar to leaflets of lower leaves) and with very broad (to 5 cm wide flattened out) sheaths decurrent the length of the petioles; leaflets of lower leaves and blades of upper leaves palmately cleft or divided, ovate to orbicular, sometimes broader than long, basally subcordate to cordate, margins irregularly lobed-toothed. Inflorescence of compound umbels 10–20 cm across, the peduncles 5–20 cm long, stout, pubescent, densely so below the umbel. Involucral bracts 1–6, usually dilated below, abruptly narrowed above to long-attenuate tips, sometimes linear-attenuate throughout, quickly deciduous; bractlets of the involucel similar but more persistent. Umbels 15–30 per inflorescence, their stalks unequal, 5–10 cm long, secondary umbels with numerous flowers, the pedicels 8–30 mm long. Sepals none. Petals white or purplish, those of the inner flowers of the umbel obovate with abruptly narrowed inflexed tips, those of the outer flowers of the umbel much more elongate, sometimes deeply 2-cleft, sometimes ½ of the petal truncate at the summit, the other half a tonguelike lobe. Fruit obovate, strongly flattened dorsally, 7–12 mm long, scarcely ribbed, oil tubes clearly visible. (*H. lanatum* Michx.)

Meadows, open stream banks, moist rich woodlands near streams, wet roadsides. Nfld. to Alaska, southward in the s. Appalachians to Ga.; Ohio, Ind., Ill., Mo., Kan., southward in the western mts. to N.Mex., Ariz., Calif.; e. Asia.

10. Cynosciadium

Cynosciadium digitatum DC. Fig. 216

Erect herbaceous annual, 3–5 dm tall, relatively few-branched, in the upper half more or less dichotomously branched. Basal leaf blades linear to lanceolate, septate, acute apically, tapering to a petiolelike base; stem leaves palmately parted, the 3–5 divisions narrowly elongate. Inflorescence of once-compound axillary and terminal umbels with unequal stalks 2–8 cm long. Involucre usually of a few linear-attenuate bracts of unequal length, to about 1 cm long; involucel usually of 2 very unequal bractlets. Umbels 2–10 per inflorescence, sometimes with 1 to few flowers as a primary umbel below; secondary umbels with 2 to several flowers, the pedicels usually unequal and spreading-ascending. Flowers 2.5–3 mm across. Sepals minute, deltoid-acuminate. Petals white, obovate, spreading. Fruit 2–3 mm long, ovoid below, narrowed into a short neck above, slightly flattened to subterete, dorsal ribs narrow, the lateral prominently corky-winged.

Lowland woods, coastal and blackland prairies, ditches, borrow pits. s. Mo. to e. and s.e. Tex., La. and Miss.

11. Limnosciadium

Limnosciadium pinnatum (DC.) Math. & Const. Fig. 217

Glabrous annual with single principal erect stem or several stems from the base, 1–8 dm tall. Basal leaves linear-lanceolate, 5–20 cm long, acute apically, tapering to petiolelike bases, septate, simple, or pinnate with the terminal segment elongate. Principal stem leaves pinnately divided, with 2–9 linear to linear-lanceolate divisions 3–10 cm long, acute basally and apically. Inflorescence mostly once compound, the peduncles 1–8 cm long. Involucre of several linear-filiform reflexed bracts 2–6 mm long; involucel of several filiform bractlets 1–5 mm long. Umbels 3–12 per inflorescence, their stalks 5–35 mm long, flowers 4–20 per

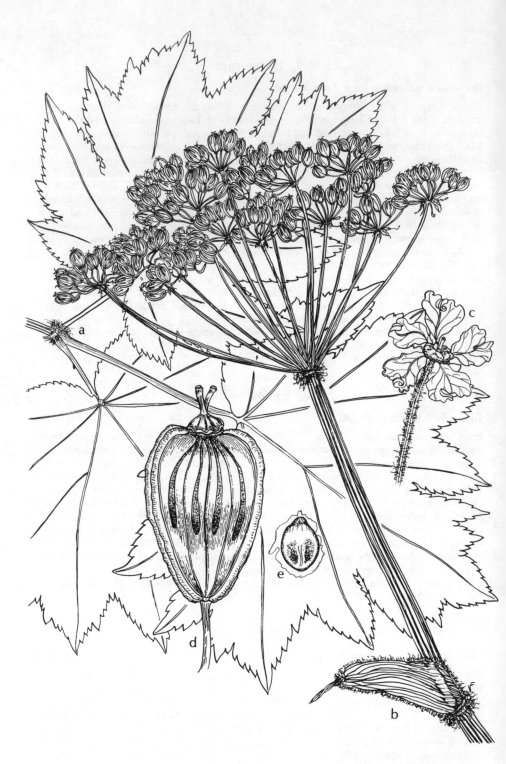

Fig. 215. **Heracleum maximum:** a, leaf; b, inflorescence branch; c, flower; d, fruit; e, seed. (From Correll and Correll (1972), as *H. lanatum*)

Fig. 216. **Cymosciadium digitatum:** 1, plant; 2, fruit, side view; 3, fruit, transverse section. (From Mathias and Constance in C. L. Lundell, *Flora of Texas*. Vol. 3, pl. 39)

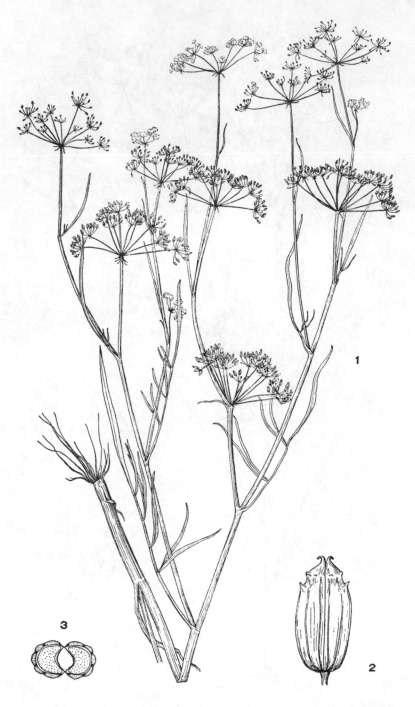

Fig. 217. **Limnosciadium pinnatum:** 1, plant; 2, fruit, side view; 3, fruit, transverse section. (From Mathias and Constance in C. L. Lundell, *Flora of Texas*. Vol. 3, pl. 40)

secondary umbel, pedicels 2–8 mm long. Flowers about 1.5–2 mm across. Sepals ovate-deltoid, about 0.4 mm long. Petals white, broadly ovate, abruptly short-acuminate apically, tips curved inward. Fruit broadly oval to suborbicular, 2–3 mm long, rounded basally and apically, slightly compressed dorsally, dorsal ribs low, lateral ribs corky-winged.

Low, wet open places, borders of ponds, ditches. s.w. Mo. and s.e. Kan. to Tex. and La.

12. Sium

Sium suave Walt. WATER-PARSNIP. Fig. 218

Slender to stout glabrous perennial with fascicles of fleshy fibrous roots. Stem to 12–15 dm tall, to 2 cm in diameter at the base, terete and hollow below, strongly corrugated or angled above, branching from about or above the middle; in water with the lower nodes bearing long feathery fibrous roots. Leaves long-petioled, compound, odd-pinnate; first leaves of the season, especially if submersed, 2- or 3-pinnately dissected into linear, linear-filiform, or lanceolate segments; lower stem leaves to at least 4 dm long, once compound and mostly with 7–15 leaflets, or in particularly stout specimens with about half of the primary divisions further divided; reduced upward, both in overall size and number of leaflets, the bracteal leaves of the inflorescence very much smaller and mostly with 3 (rarely 1) leaflets, sessile or subsessile. Leaflets ovate-lanceolate to lanceolate, often asymmetrical at base, the apices acute to acuminate; margins saliently toothed. Umbels twice compound, mostly becoming 5–8 cm across in fruit; bracts of involucres and involucels 6–10, linear- or lance-attenuate, with hyaline margins. Stalks of the ultimate umbels unequal, stiff and straight, sharply wing-angled, the edges of the angles white-opaque. Sepals minute, petals white. Fruit oval, slightly compressed laterally, 2–3 mm long, with prominently corky-winged ribs, constricted at commissure; oil tubes 1–3 in intervals, 2–6 on commissure.

In streams, swamps, bogs, floodplain forests. Nfld. to B.C., southward more or less throughout the U.S.; e. Asia.

The robustness of plants and their parts varies greatly, presumably as ecological conditions fluctuate. *S. floridanum* Small, of river bottomlands in the southeastern United States, may or may not be distinct. We suggest plants so named may be slender specimens of *S. suave* which have been inhibited in their growth by flood waters.

13. Cryptotaenia

Cryptotaenia canadensis (L.) DC. HONEWORT, WILD CHERVIL.

Glabrous perennial with slender fascicled roots. Stem relatively few-branched, 3–10 dm tall, striate. Leaves mostly long-petiolate, blades essentially ternately compound, the 2 lateral leaflets commonly (less frequently the central leaflet) irregularly cleft-lobed. Leaflets mostly ovate in outline, doubly serrate, the central one approximately symmetrical, acute basally, usually acute to acuminate apically; lateral leaflets asymmetrical, obtuse to rounded basally, acute or acuminate apically, rarely obtuse, not infrequently cleft-lobed, sometimes mittenlike; leaflets of principal leaves variable in size from about 8–15 cm long, 3–7 cm wide; upper leaves much reduced. Inflorescence rather loose and diffuse, the branching essentially paniculate. Ultimate umbels with few flowers, stalks of the umbels unequal, strongly ascending, the flower stalks very unequal. Flowers minute. Sepals lacking or extremely small. Petals white, very small, obovate, the narrowed tips inflexed. Stylopodium slender conic. Fruit glabrous, long-elliptic to fusiform, somewhat flattened laterally, ribs narrow, rounded, prominent, subequal.

Moist woodlands, river banks and floodplain woodlands, moist thickets. N.B., Que. to Man., generally southward, save on the coastal plain were infrequent, to Ga. and Tex.; along the Apalachicola River, Fla. Panhandle.

Fig. 218. **Sium suave:** a, small portion of flowering-fruiting branch; b, section of stem; c, lower stem leaf; d, submersed leaf; e, section of petiole; f, flower; g, fruit.

14. Cicuta (WATER-HEMLOCKS)

Stout, coarse perennials with a cluster of fleshy, tuberiform, fingerlike roots. Stems glabrous, hollow, to 2–2.5 m tall. Principal leaves petiolate, 1–3-pinnately compound, the upper ternately compound or simple; leaflets serrate, their principal lateral veins ascending to or toward the sinuses between the teeth. Inflorescence of axillary and terminal compound umbels. Involucre of 1 to several linear-attenuate unequal bracts, the longer to 15 or 20 mm, sometimes absent, if present usually quickly deciduous; bractlets of involucel several, similar. Petals white, short-clawed at base, blades ovate, the abruptly narrowed tips inflexed. Sepals deltoid to deltoid acuminate. Fruit ovate, oval, elliptic, or suborbicular, somewhat flattened laterally, with strong, flattish corky ribs; stylopodium depressed.

All parts of the plant are deadly poisonous to humans and livestock.

● Leaflets mostly lanceolate, or lance-oblong; fruit without a constriction at the commissure, the lateral ribs (next to the commissure) flush against each other.   1. *C. maculata*
● Leaflets ovate or ovate-lanceolate; fruit constricted at the commissure, thus with a furrow separating the lateral ribs.   2. *C. mexicana*

1. Cicuta maculata L.   Fig. 219

Stems erect, often mottled with purple below, variable in stature, to about 2 m tall. Lower, larger leaves long-petiolate, 2- or 3-pinnately compound, the blades overall as much as 30 cm long, 8–25 cm broad, upwardly becoming 1-pinnate, then ternate, bracteal leaves sometimes simple. Leaflets lanceolate to lance-oblong, thinnish, larger leaflets of lower leaves 0.6–3 cm broad, sometimes to 4 cm. Sepals deltoid. Fruit ovoid to ellipsoid, about 3 mm long, nearly 2 mm wide, not constricted at the commissure, the lateral ribs flush against each other; one mericarp often aborts, the fruit thus oblique.

Marshes, marshy shores of streams, swamps, wet meadows, ditches. Pr.Ed.I. and Que. to N.C. and Tenn., westward to the Dakotas and Tex.

2. Cicuta mexicana Coult. & Rose.   Figs. 220 and 221

Similar to *C. maculata*, usually stouter and coarser throughout. Stem to 2.5 m tall, up to 4 or 5 cm in diameter, surface commonly with alternating green and purple stripes, appearing essentially purple from a little distance. Plants in floating mats or in deep muck of marshy shores often with lower stems or branches growing horizontally, late in the season portions enlarge greatly, forming large fusiform tubers; eventually stems either side of the tuber disintegrate after which the tubers may float away, later becoming lodged, the next season germinating. Lower, larger leaves usually 3-pinnately compound, sometimes the overall blades to at least 40 cm long and 60–70 cm broad at the base, their leaflets usually ovate, sometimes lanceolate, commonly asymmetrical, usually unequal in size on a given leaf; prominently veiny, margins irregularly to regularly coarsely toothed. Sepals deltoid-attenuate. Fruits mostly ellipsoid, sometimes suborbicular, 2–3 mm long and 2–3 mm wide, constricted at the commissure thus a groove between the mericarps, the lateral ribs separated, not flush against one another. (*C. maculata* var. *curtissii* (Coult. & Rose) Fern.)

Commonly in water along marshy shores, in floating mats of vegetation, swamps, ditches. s.e. Va. to s. Fla., westward to Tex.; e. Mex.; (Tenn., Ky., Ark.?).

15. Conium

Conium maculatum L. POISON-HEMLOCK.   Fig. 222

Glabrous, stout biennial with stout taproot, coarse, hollow, spotted stem to 3 m tall. Leaves 3 or 4 times pinnately compound, in overall outline broadly triangular-ovate, the larger 2–4 dm long, ultimate divisions pinnately incised. Inflorescence a large compound cyme of compound umbels. Involucre of short-ovate bracts, their apices acuminate; bractlets of the involucel similar. Ultimate umbels numerous, their stalks 15–25 mm long, about 15-flowered, the

Fig. 219. **Cicuta maculata:** a, inflorescence branch; b, base of plant and a lower leaf; c, flower; d, petal; e, fruit (drawn to same scale as fruit of *C. mexicana* in Fig. 221).

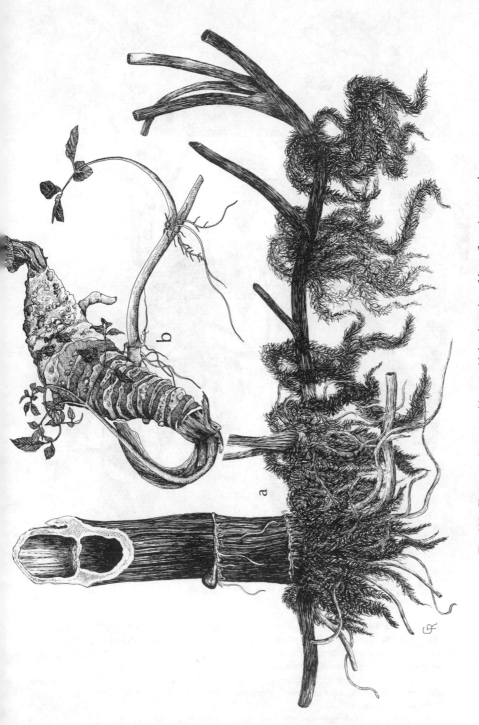

Fig. 220. **Cicuta mexicana**: a, rhizome with basal portion of large flowering stem; b, autumnal tuberoid rhizome which detaches by disintegration with connections to parent rhizome, usually floats away.

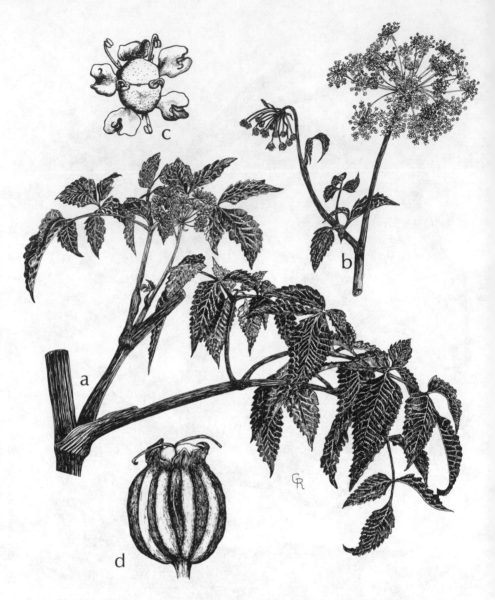

Fig. 221. **Cicuta mexicana**: a, median stem leaf; b, small branch of inflorescence; c, flower, from above; d, fruit.

pedicels 4–6 mm long. Flowers 2–3 mm across. Sepals none. Petals white, broadly obovate, the abruptly narrowed tips strongly inflexed. Stylopodium depressed-conic. Fruit glabrous, 2–2.5 mm long, broadly oval (mericarps separating at maturity), flattened laterally, ribs prominent, pale brown between the ribs.

Native of Eurasia and widely naturalized in N.Am. Waste places, open stream banks, open alluvial woodlands. In our range apparently rare and sporadic on the coastal plain, more common elsewhere.

All parts of the plant dangerously poisonous to humans and livestock.

Fig. 222. **Conium maculatum:** a, inflorescence branch; b, leaf; c, flower; d, fruit. (From Correll and Correll. 1972)

463

16. Trepocarpus

Trepocarpus aethusae Nutt.   Fig. 223

Erect, slender, glabrous annual with a slender taproot, 3–7 dm tall, with relatively few ascending branches. Leaves short-petiolate, pinnately decompound-dissected, the ultimate divisions flat, narrowly linear, 1 mm wide or less, leaves little reduced upward. Inflorescence of compound umbels, axillary and terminal, peduncles mostly 4–6 cm long. Each compound umbel bearing 2–4 ultimate umbels, umbel stalks 1–1.5 cm long, 2- to 8-flowered, the flower stalks subequal. Involucre of 1 to several linear-subulate, very unequal bracts, the longer usually about 10 mm long; involucel bracts similar. Flowers about 2 mm across in face view, the floral tube elongate. Sepals subulate, 1.5–2 mm long, somewhat unequal, persistent. Petals white, broadly obovate, their short abruptly narrowed apices inflexed. Anthers yellow. Styles very short, stylopodium short-conic. Fruit glabrous, linear-oblong, 8–10 mm long, slightly flattened laterally, strongly ribbed.

Floodplain forests, marshy meadows, moist to wet clearings, ditches. S.C., Tenn., Ala., Miss., La., Ark., s.e. Okla., e. half of Tex.

17. Chaerophyllum (CHERVILS)

Annual herbs. Leaves petiolate, 3-pinnately dissected, the ultimate divisions lobed. Umbels compound, the primary umbels sessile in axils of upper leaves, mostly 2–4 umbels per inflorescence, sometimes solitary, their stalks variable in length, at most to about 6 cm long. Involucres none; involucel of several ovate to oblong bractlets, erect at anthesis, reflexed or spreading when fruits fully mature. Flowers 3–10 per secondary umbel, sessile or very short stalked at anthesis, outer ones developing stalks to about 5–6 mm long as the fruits mature. Flowers very small. Sepals, if present, minute. Petals (in ours) white, obovate, about 1 mm long. Fruit much longer than broad, lanceolate to fusiform-elliptic, somewhat flattened laterally, ribbed.

● Fruit lanceolate, broadest near the base, narrowed distally to a neck; bracts of the involucel reflexed below the fully mature fruits.   1. *C. tainturieri*
● Fruit fusiform-elliptic, narrowed at the summit but without a neck; bracts of the involucel spreading below the fully mature fruits.   2. *C. procumbens*

1. Chaerophyllum tainturieri Hook.

Plants usually stiffly erect, very variable in stature, commonly slender, short, and unbranched when growing close together, more robust and with several branches when growing singly, maximally to 8–9 dm tall. Stems and leaves pubescent. Leaf blades 3-pinnately compound, the ultimate divisions flat, lobed. Fruits lanceolate, broadest near the base, narrowed to a neck above, 5–8 mm long, usually glabrous, occasionally pubescent, the ribs as broad as the intervals or broader.

Weedy in moist open places, in moist woodlands, sometimes abundant in floodplain woodlands. Va. to Mo. and Kan., generally southward to Fla. and Tex.

2. Chaerophyllum procumbens (L.) Crantz.

In general habital features very closely similar to *C. tainturieri*, the stems weaker, commonly with 2 to several branches from near the base, branches often somewhat spreading below, glabrous or sparingly pubescent. Bracts of the involucel spreading below the fully mature fruits. Fruits fusiform-elliptic, broadest at about the middle, often curved by full maturity, narrowed at the summit but without a neck, the ribs narrower than the intervals.

Moist to wet, usually alluvial woodlands, thickets and glades. N.Y. to s. Mich., Iowa, Kan., generally southward to Ga., Ark. and Okla.

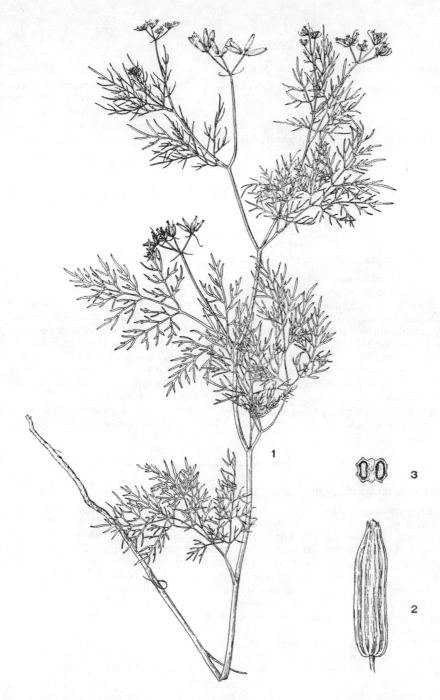

Fig. 223. **Trepocarpus aethusae:** 1, plant; 2, fruit, side view; 3, fruit, transverse section.
(From Mathias and Constance in C. L. Lundell, *Flora of Texas.* Vol. 3, pt. 5, pl. 43.
1951)

18. Ammoselinum

Ammoselinum butleri (Wats.) Coult. & Rose.   Fig. 224

Annual, branching from the base, the branches 4–5 (–12) cm high. Leaves biternate or ternate-pinnately dissected, the ultimate divisions linear, obtuse; overall the blades to 2.5 cm long and 1.5 cm wide. Inflorescences irregularly compound umbels, sessile in the leaf axils, not exceeding the leaves. Involucre lacking; involucel of several foliaceous bractlets. Umbels 2–6 per inflorescence, unequally short-stalked, or some sessile, 1- to 10-flowered, flower stalks unequal, 1–6 mm long. Sepals none. Petals white. Fruit ovoid, flattened laterally, 2.5–3 mm long, glabrous to roughened-callose, ribs narrow, edges acute, the lateral closely contiguous.

Principally in moist to wet floodplain woodlands. Tenn., Ark., s.e. Okla. Sporadically weedy in other parts of our range.

19. Apium

Apium leptophyllum (Pers.) F. Muell. MARSH-PARSLEY.

Slender, erect, glabrous annual 7–8 dm tall. Stems striate, with several alternate branches. Leaves petiolate, or the upper subsessile, petioles with abruptly expanded sheathing or subsheathing bases, margins of the sheath conspicuously white-hyaline; blades of the lower leaves pinnately decompound-dissected, the stem leaves ternate-pinnately dissected, the segments mostly filiform. Umbels sessile or stalked, the stalks mostly 1–2 cm long, borne from axils of stem and bracteal leaves from about mid-stem upward; umbel with the central flower sessile and about 10 unequally stalked flowers. Subtending bracts lacking. Flowers very small; sepals minute or lacking; petals very small, white or greenish. Fruit oval to ovate or orbicular, 2–4 mm long, compressed laterally, the ridges corky-thickened, oil tubes solitary in the intervals, 2 on the commissure. (*Cyclospermum ammi* (L.) Britt.)

Moist to wet open sites, ditches, wet clearings, borders of marshes, roadsides. N.C. to s. Fla., westward to Okla. and Tex.; scattered and local elsewhere as an adventive in coastal areas; trop. Am.

20. Spermolepis

Spermolepis divaricata (Walt.) Raf.   Fig. 225

Slender, glabrous annual, variable in stature, to about 7 dm tall, commonly with a single main stem, occasionally several-branched from near the base. Lower leaves short-petiolate, upper mostly sessile, blades ternately or ternate-pinnately dissected, the ultimate segments filiform. Inflorescence open, of irregularly compound umbels, the primary stalks filiform, 2–5 cm long, from the upper leaf axils. Involucre lacking; involucel of 1 or a few subulate bractlets. Ultimate umbels mostly 3–7 per inflorescence, their stalks 1–2 dm long, mostly 2- to 6-flowered, the central flower usually sessile, others unequally stalked, stalks to 15 mm long. Flowers minute. Sepals none. Petals white, ovate to oblong, nearly 1 mm long. Fruit ovate to oval, 1–1.5 mm long, somewhat flattened laterally, surface warty-tuberculate, weakly ribbed.

Weedy, commonly in well-drained sands, sometimes in moist to wet open places, ditches, borders or bottoms of ponds exposed at low water. Coastal plain and piedmont, Va. to s. Fla., westward to Tex., northward in the interior to Kan. and Mo.

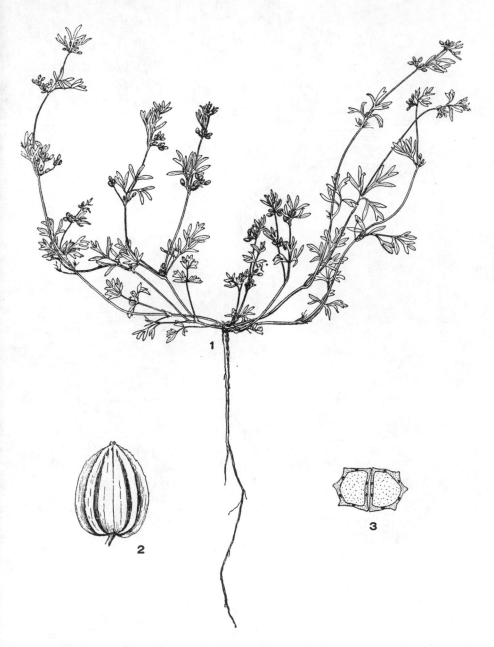

Fig. 224. **Ammoselinum butleri:** 1, plant; 2, fruit, side view; 3, fruit, transverse section. (From Mathias and Constance in C. L. Lundell, *Flora of Texas*. Vol. 3, pt. 5, pl. 44. 1951)

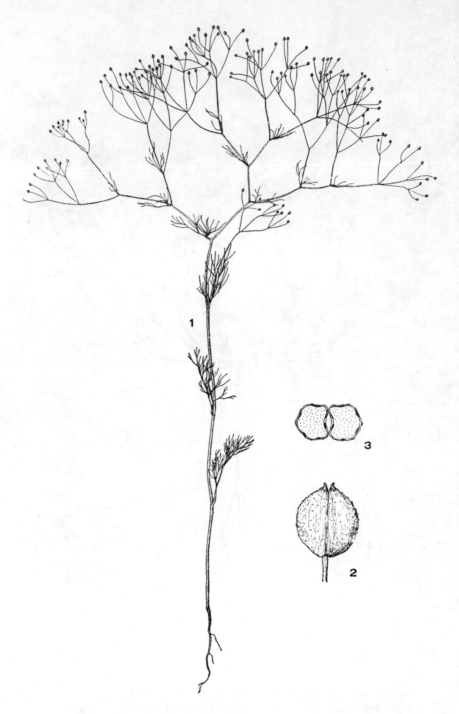

Fig. 225. **Spermolepis divaricata:** 1, plant; 2, fruit, side view; 3, fruit, transverse section. (From Mathias and Constance in C. L. Lundell, *Flora of Texas.* Vol. 3, pt. 5, pl. 46. 1951)

Clethraceae (WHITE-ALDER FAMILY)

Clethra alnifolia L. SWEET PEPPERBUSH.   Fig. 226

Shrub to about 3 m tall. Young branchlets tomentose, woody twigs glabrous. Leaves simple, alternate, deciduous, short-petioled; blades very variable in size, to about 8 cm long and 4 cm wide, averaging about ½ that; mostly narrowly obovate to oblanceolate, sometimes elliptic-lanceolate, cuneate basally, apically acute, acuminate, obtuse, or less frequently rounded, margins usually serrate on the distal halves, occasionally entire, glabrous above, lower surface varying from glabrous at full maturity to loosely tomentose or densely tomentose, pinnately veined. Flowers in terminal or terminal and axillary dense racemes on branchlets of the season; raceme axes, bracts, the flower stalks, calyces, and ovaries densely pubescent. Bracts subtending the flowers 2–10 mm long, linear-oblong to linear, acute apically, shorter than to about as long as the flowers, usually deciduous by the time the flowers open or shortly thereafter. Pedicels 2–5 mm long. Calyx 3–4 mm high, oblong or campanulate, united below, the 5 lobes triangular-ovate to oblong, apically rounded, obtuse, or acute. Petals 5, separate, white, oblong-obovate, glabrous, about 8 mm long. Stamens 10, 5 in each of two whorls, filaments glabrous, anthers sagittate, opening by terminal pores. Ovary superior, 3-locular, style slender and elongate, stigma capitate or slightly 3-lobed. Fruit a 3-valved, pubescent, subglobose capsule about 3 mm in diameter, calyx persistent and loosely surrounding it. Flowers emitting a pungent fragrance. (Incl. *C. tomentosa* Lam.)

Wet thickets, wet pine savannas and flatwoods, swamps, wet woodlands, bays and pocosins, stream banks, bogs. s. Maine, s. N.H., southward to n. Fla., westward to s.e. Tex.

Ericaceae (HEATH FAMILY)

Shrubs or trees, rarely vines. Leaves simple, alternate in most (of ours), pinnately veined, without stipules. Flowers bisexual, radially or bilaterally symmetrical, ovary superior in most, inferior in a few genera. Calyx united below, 4–7-lobed, or sepals distinct. Corolla of separate petals or tubular below, 4–7-lobed, tube often funnelform, campanulate, or urceolate. Stamens usually twice the number of petals or corolla lobes, distinct, usually inserted at the edge of a nectariferous disc, anthers inverted during development or at anthesis, dehiscing from the base (apparent apex) by slits, clefts, or pores, often with awns or spurs. Pistil 1, mostly 4–5-carpellate, locules usually as many as the corolla lobes, placentation axile, but sometimes loculate below, 1-locular above, or with twice as many locules as carpels by development of additional partitions; style 1, stigma truncate, capitate, or peltate. Fruit a loculicidal or septicidal capsule, a berry, or a drupe, the calyx usually persistent.

Reference: Wood, Carroll E., Jr. "The Genera of Ericaceae in the Southeastern United States." *Jour. Arn. Arb.* 42: 10–80. 1961.

1. Calyx lobes 7 and corolla of 7 separate petals.   1. *Befaria*
1. Calyx lobes 5 and corolla conspicuously tubular below, *or* calyx lobes 4 and corolla deeply 4-lobed, lobes reflexed.
  2. Calyx lobes 5 and corolla conspicuously tubular below, 5-lobed.
    3. Ovary superior; fruit a capsule.
      4. Corolla tube campanulate, or cylindric-tubular below, flaring above, in either case the opening broad.
        5. Corolla tube campanulate or cylindric-campanulate; capsule little, if any, longer than broad.
          6. Anthers, in the flower bud, seated in pouches of the corolla tube.   2. *Kalmia*
          6. Anthers, in the flower bud, not seated in pouches of the corolla tube.
            7. Anther-halves each with a pair of ascending terminal awns.   5. *Zenobia*
            7. Anther-halves without terminal awns (*L. mariana* in genus).   6. *Lyonia*
        5. Corolla tube cylindric-tubular below, flaring above; capsule much longer than broad.
                                3. *Rhododendron*

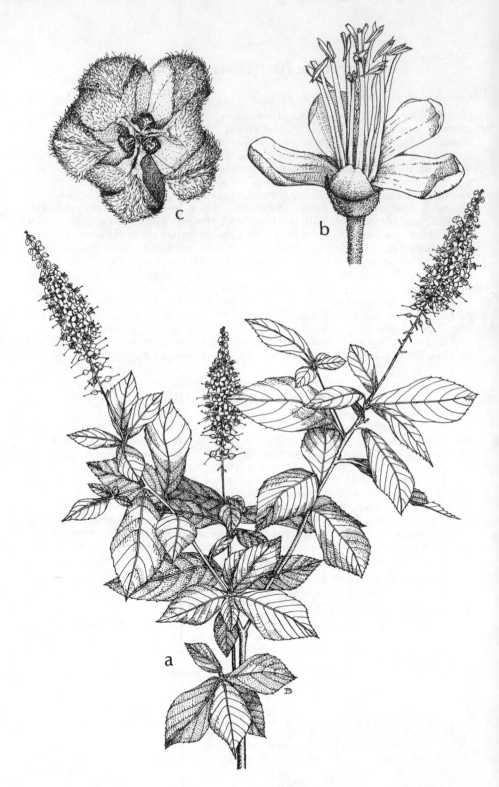

Fig. 226. **Clethra alnifolia:** a, branch with inflorescences; b, flower; c, fruit.

4. Corolla essentially urceolate, its opening narrow.
    8. Flowers or capsules solitary in the axils of reduced or bracteal leaves.   8. *Cassandra*
    8. Flowers or capsules in lateral or terminal racemes, in axillary umbelliform clusters, or in fasciclelike racemes.
        9. Stalks of flowers or fruits with a pair of subulate bractlets a little below the calyx. 7. *Pieris*
        9. Stalks of flowers or fruits with short-ovate bractlets near the base or short-ovate bractlets below the calyx.
            10. Flowers and capsules in definite racemes.   4. *Leucothoe*
            10. Flowers and capsules in umbelliform clusters or short, fasciclelike racemes. 6. *Lyonia*
3. Ovary inferior, fruit a berrylike drupe with 10 hard nutlets or a many-seeded berry.
    11. Leaves with sessile, amber, glandular dots beneath; fruit a berrylike drupe with 10 hard nutlets.
        9. *Gaylussacia*
    11. Leaves without sessile, amber, glandular dots beneath; if glandular-pubescent, then the hairs stipitate-glandular or clavate; fruit a many-seeded berry.   10. *Vaccinium*
2. Calyx lobes 4, corolla deeply 4-lobed, lobes reflexed (*V. macrocarpon* in genus).   10. *Vaccinium*

1. Befaria

Befaria racemosa Vent. TAR-FLOWER, FLY-CATCHER.

Evergreen, slender shrub with relatively few erect branches, 1–2.5 m tall. Twigs powdery-pubescent and with few to numerous, long, spreading hairs. Leaves sessile or subsessile, narrowly to broadly elliptic or ovate, occasional ones oblanceolate or obovate, mostly 2–4 cm long, 6–20 mm broad, cuneate basally, acute to rounded apically, margins entire, lateral veins usually faint, upper surfaces powdery-pubescent, lower pale, glabrous or with few to numerous longish hairs along the midrib. Flowers showy, fragrant, in racemes terminal on shoots of the season. Flower stalks 1–3 cm long, subtended by reduced or bracteal leaves, or the uppermost by subulate bracts, each stalk bearing 1, or usually 2, subulate bractlets closely below or at a little distance below the calyx. Calyx persistent, about 5 mm high, tube short-campanulate, surmounted by 7 erect, ovate-triangular lobes, inner surface glandular. Petals 7, separate, white and pink-tinged, sometimes wholly pink proximally, spreading, elongate, 2–3 cm long, tapering from a narrowly clawed base to a narrowly spatulate or oblanceolate blade, glutinous exteriorly. Stamens 12–14, filaments long, ½–⅔ as long as the petals, dilated below and cottony-pubescent, anthers short, dehiscing by 2 oblique terminal pores. Ovary superior, 7-locular, style slender, 1.5–2 cm long, stigma capitate. Fruit a subglobose, 7-valved, septicidal capsule 6–8 mm broad. Seeds narrowly elongate, somewhat falcate, 1–1.5 mm long, amber-colored.

In both poorly drained pine flatwoods and in well-drained sand-scrub. Coastal plain, s.e. Ga. to s. pen. Fla.

2. Kalmia (LAURELS)

Evergreen, rarely deciduous, shrubs, sometimes of small treelike stature. Twigs terete or 2- or 3-angled. Leaves alternate, opposite, or whorled, petioled or rarely sessile, blades usually leathery, margins entire, revolute in some. Inflorescence a lateral or terminal panicle, lateral raceme, or fascicle, or flower solitary, each flower subtended by a persistent bract, some with a pair of bractlets at the base of the flower stalk. Calyx persistent or deciduous in fruit, united below, deeply 5-lobed. Corolla rose-purple, pale rose, pink, or white, always with a red ring at the base of the limb, rotate to campanulate, tube short and narrow basally, limb shallowly 5-lobed with 10 pockets in which the anthers are held in the bud, the filaments strongly recurved and under tension, springing up suddenly when touched. Stamens 10, filaments slender, anthers short, opening by apical slits. Ovary superior, on a 5-lobed disc, 5-locular, style slender, straight, stigma flat, 5-grooved. Capsule septicidal, 5-valved, style persistent. Seeds numerous, minute, hard, light brown, longitudinally ridged, sometimes winged.

Reference: Southall, Russell M., and James W. Hardin. "A Taxonomic Revision of *Kalmia* (Ericaceae)." *Jour. Elisha Mitchell Sci. Soc.* 90: 1–23. 1974.

1. Leaves whorled or opposite; flowers in axillary racemes.   1. *K. carolina*
1. Leaves alternate; flowers in axillary fascicles, or solitary in the leaf axils.
  2. The leaves deciduous (as evidenced by absence of any on twigs of the previous season), mostly oblanceolate, cuneate-attenuate below; flowers in axillary fascicles near the tips of branches of the previous season.   2. *K. cuneata*
  2. The leaves persistent (as evidenced by the presence of some on twigs of the previous season), variable in shape but only an occasional one oblanceolate, bases mostly rounded or truncate; flowers solitary in axils of leaves on branches of the present season (rarely 2 or 3 in an axil).   3. *K. hirsuta*

1. Kalmia carolina Small. SHEEP LAUREL, LAMBKILL, WICKY.

Shrub to 1.5 m tall, sparsely branched, perennating by subterranean runners and sometimes forming low thickets. Young twigs terete, closely gray-pubescent, sometimes with glandular hairs intermixed, becoming glabrous in age. Leaves persistent, usually whorled, sometimes opposite, petioles 8–16 mm long, pubescent like the twigs, blades 2–8 cm long, 1–3 cm broad, oblong-elliptic, elliptic, or oblanceolate, light green becoming reddish in winter, usually cuneate basally, apices rounded or obtuse, often minutely mucronate, moderately powdery gray-pubescent above, sometimes becoming glabrous eventually, much more densely powdery gray-pubescent and usually with some short glandular hairs intermixed beneath. Flowers in axillary racemes on wood of the previous season, raceme axes to 10 cm long, usually much shorter; flowers alternate, opposite, or whorled, each subtended by a bract 5–6 mm long, a pair of shorter bractlets closely set above, axes and flower stalks densely short-pubescent and with some glandular hairs. Calyx persistent, lobes ovate, 2–2.8 mm long, obtuse apically, closely short-pubescent, not glandular. Corolla pink to rose-purple, nearly rotate, limb 4.6–5.6 mm long. Fruit globose, 2.5–3.5 mm across, closely gray-pubescent with intermixed glandular hairs elevated above. (*K. angustifolia* L., var. *caroliniana* (Small) Fern.)

  Rocky woodlands and bogs in the mts.; moist to wet thickets, bottomland woodlands, borders of swamps, pocosins, branch bays, bogs, in the coastal plain. Mts., s.w. Va. (Carroll Co.) to n.e. Tenn. (Johnson Co.), southwestward through N.C. to n. Ga.; coastal plain, s.e. Va. to S.C.

2. Kalmia cuneata Michx.

Shrub to about 9 dm tall, sparsely branched, perennating by subterranean runners. Young twigs terete, puberulent and stipitate-glandular, becoming glabrous in age. Leaves deciduous, alternate, 2–7 cm long, 5–20 mm broad, mostly oblanceolate, some sometimes elliptic-oblanceolate, cuneate-attenuate below, obtuse apically, dark green and glabrous above, hirsute and stipitate-glandular beneath and on the revolute margins. Flowers few in fascicles axillary to leaf scars near the tips of branches of the previous season. Flower stalks 2.5–3 cm long, oblongish and abruptly contracted at the tips, stipitate-glandular, persistent. Corolla greenish white, rotate, stipitate-glandular exteriorly, limb 7.5–8.5 mm long. Fruit oblate, 4–5 mm across, conspicuously stipitate-glandular.

  Sandy-peaty soils, borders of pocosins and branch bays. s.e. N.C. and n.e. S.C.

3. Kalmia hirsuta Walt. HAIRY WICKY.

Straggly shrub 1–6 dm tall, commonly with several to numerous stems from a hard, stout, subterranean base. Twigs light brown, bearing numerous long, spreading hairs. Leaves alternate, persistent, light green, sessile or subsessile, variable in shape, even on a single stem, ovate, elliptic, oblong, oblong-elliptic, oval, or oblanceolate, 5–15 mm long, 2–8 mm broad, bases rounded to truncate, apices acute or obtuse, infrequently rounded, surfaces glabrous to moderately pubescent with scattered long hairs, almost always with scattered long hairs on or near the revolute margins. Flowers solitary (rarely 2 or 3) in leaf axils along most of the present year's growth, their stalks 3–10 (–20) mm long, hirsute, each with a pair of elliptic-oblong, foliaceous bracts 4–6 mm long just above the subtending leaf. Calyx lobes lanceolate, acute apically, 5–8 mm long, margins markedly ciliate, often a few long hairs on the midrib above. Corolla campanulate, pale to deep pink or rose, less frequently white, marked with red near the stamen pockets and with a ring of red spots lower down, limb 5–6 mm long. Capsule oblate, 2–3 mm across, tan, the persistent sepals much exceeding it (before dehiscence). (*Kalmiella hirsuta* (Walt.) Small)

Pine savannas and flatwoods, borders of shrub-tree bogs, hillside bogs in pinelands. Coastal plain, s.e. S.C. to n.cen. pen. Fla., westward to s.e. La.

3. Rhododendron

Evergreen or deciduous shrubs, rarely of tree stature. Leaves alternate, usually petiolate. Buds with several to numerous imbricate scales; overwintering inflorescence buds solitary and terminal on branchlets, sometimes twinned and terminal, sometimes 2 or 3 subterminal as well; overwintering vegetative-shoot buds usually several just below the branchlet tips, these smaller than the inflorescence buds. Flowers in umbellate clusters. Calyx very small relative to the size of the corolla, united below, 5-lobed, persistent. Corolla showy, weakly bilaterally symmetrical, united below, campanulate to funnelform, in some the tube narrowly cylindrical, flaring above, the 5-lobed limb rotate or nearly so. Stamens 5–7 (–10), filaments slender and elongate, unusually unequal, anther-halves opening by terminal pores. Ovary superior, 5-locular, ovules numerous in each locule, style slender, curved, stigma capitate. Fruit a septicidal capsule.

1. Leaves evergreen (as evidenced by presence of some on wood of a previous season); stamens 10.

> 1. *R. minus* var. *chapmanii*

1. Leaves deciduous (none on wood of a previous season); stamens 5.
  2. Flower clusters appearing after leafy branches of the season are fully developed.   2. *R. viscosum*
  2. Flower clusters appearing before or with the leafy branches of the season.
    3. Flower stalks, calyces, and capsules not stipitate-glandular; plant to 5 m tall, not having elongate subterranean runners and widely colonial.   3. *R. canescens*
    3. Flower stalks, calyces, and capsules stipitate-glandular (nonglandular hairs may be present as well); plant rarely over 6 dm tall, forming elongate subterranean runners and widely colonial.

> 4. *R. atlanticum*

1. Rhododendron minus Michx. var. chapmanii (Gray) Duncan & Pullen.
FLORIDA RHODODENDRON.

Evergreen shrub to about 2.5 m tall, the branching relatively open and generally stiffly erect-ascending. Young twigs copiously scurfy-dotted. Petioles 3–8 mm long, scurfy-dotted, blades elliptic or oval, stiff-leathery, 1.5–5 cm long, 1–3 cm broad, slightly tapered basally, apices rounded to obtuse, rarely acute, dark green and sometimes with a few scattered scurfy dots above, abundantly dotted with brown scurfy scales beneath, margins entire, edges slightly to conspicuously rolled downward. Flower clusters appearing before the leafy shoots of the season. Flower stalks 5–8 mm long, scurfy-dotted. Calyx tube very shallowly saucer-like, lobes minute, mostly triangular, ciliate. Corolla light pink to rose-pink, irregularly gland-dotted exteriorly, tube funnelform, 1–2 cm long, limb 3–4 cm across, lobes ovate to oblong, rounded apically. Stamens 10, filaments glabrous, somewhat longer than the purplish red style. Capsule about 1 cm long, longitudinally lobed, slightly urceolate in outline, scurfy-dotted. (*R. chapmanii* Gray)

Seasonally wet pine flatwoods, borders of evergreen shrub-tree bogs or bays, transition zone between branch bays and pine ridges. Scattered localities from n.e. Fla. to cen. Fla. Panhandle.

2. Rhododendron viscosum (L.) Torr. SWAMP-HONEYSUCKLE, CLAMMY AZALEA.

Deciduous shrub to 5 m tall. Young twigs sparsely to densely shaggy-pubescent with long hairs. Exposed portions of scales of overwintering inflorescence buds densely short-ciliate marginally, otherwise glabrous, a stiff, stoutish mucro or short, spinelike awn arising somewhat back of the rounded or obtuse tip and extending a little beyond the tip. Flower clusters appearing considerably after leafy branchlets of the season. Leaves sessile or subsessile, mostly oblanceolate or narrowly obovate, varying to elliptic, 1–7 cm long, bases cuneate,

473

apices rounded, obtuse, or acute, minutely mucronate, both surfaces green, or green above, glaucous beneath, with appressed-ascending, stoutish hairs along the midrib beneath, margins finely bristly-ciliate, the trichomes usually curved beneath the leaf edge, scarcely if at all visible from above. Flowers fragrant, their stalks 5–20 mm long, usually with some sparse close-cottony pubescence and intermixed long-spreading, nonglandular and stipitate glandular hairs. Calyx stipitate-glandular. Corolla white or white with pink stripes or pink tint, rarely pink, stipitate-glandular exteriorly, tube cylindric except the abruptly expanded distal portion, 1.5–2.5 cm long, lobes spreading, shorter than the tube. Stamens 5, exserted, a little shorter than the style. Capsule lance-cylindric, 1–2 cm long, stipitate-glandular.

Swamps, wet woodlands and thickets, bogs, evergreen shrub tree bogs or bays, seasonally wet pine flatwoods.

A variable complex for which most authors distinguish two species: (1) *R. viscosum* (*Azalea viscosa* L.), ranging from s.w. Maine to n.e. Ohio, s. to S.C. and e. Tenn., (or to Fla. and Miss., depending upon the author); and (2) *R. serrulatum* (Small) Millais (*A. serrulata* Small), ranging on the coastal plain from s.e. Va. to cen. pen. Fla., westward to Miss. See Small (1933), Fernald (1950), Gleason (1952), and Radford et al. (1964) wherein are recognized *R. viscosum* var. *viscosum* and *R. viscosum* var. *serrulatum* (Small) Ahles.

3. Rhododendron canescens (Michx.) Sweet. HOARY AZALEA. Fig. 227

Deciduous shrub to 5 m tall. Young twigs with dense, shaggy, long hairs or, more frequently, a mixture of these and short, tangled, cottony hairs. Exposed portions of scales of overwintering inflorescence buds densely short-pubescent medially. Flower clusters appearing before or with the leafy shoots of the season. Petioles 2–10 mm long, pubescent like the twigs, blades mostly oblanceolate or narrowly obovate, varying to elliptic, 2–10 cm long, bases cuneate, apices obtuse to acute, minutely mucronate, sparsely to densely short-pubescent beneath, often with stouter, stiffer hairs along the midrib, margins bristly-ciliate with curved-ascending trichomes. Flowers fragrant, their stalks 4–12 mm long, densely soft-pubescent, rarely stipitate-glandular. Calyx densely soft-pubescent, the minute lobes pectinate-ciliate as well. Corolla deep pink to white, if the limb white, the tube white or pink, exteriorly densely pubescent with soft hairs and with sparse to dense stipitate-glandular hairs intermixed; lobes shorter than the tube, spreading, more or less undulate. Stamens 5, much exserted, about the same length as the style. Capsule about 1.5 cm long, lance-falcate, densely short-pubescent, not stipitate-glandular. (*Azalea canescens* Michx.)

Wet woodlands, springy places in woodlands, river and stream banks, pine flatwoods, borders of shrub-tree bogs or bays, well-drained wooded slopes, rocky open woodlands. Coastal plain and piedmont, Del., Md., southward to pen. Fla., westward to e. Tex., s.e. Okla., Ark., w. Tenn.

Rhododendron nudiflorum (L.) Torr. (*Azalea nudiflora* L.) is very similar to *R. canescens* in general habital features and aspect, apparently infrequently occurs in bogs, usually inhabits well-drained places. Exposed portions of the scales of overwintering inflorescence buds short-ciliate on the margins, otherwise glabrous or nearly so. Pubescence of young twigs relatively sparse, of stiffish, ascending hairs, that of flower stalks, calyces, and exterior of corolla tubes copious, of long, stiffish, usually ascending hairs, intermixed stipitate-glandular hairs, if any, only on the corolla tube. Capsule similarly pubescent, without stipitate glands. Range: Mass. and N.Y. to Ohio, southward to S.C., n. Ga., Tenn.

4. Rhododendron atlanticum (Ashe) Rehd. DWARF AZALEA.

Low, deciduous shrub, erect stems mostly 2–6 dm tall, occasionally to 10 dm or a little more, with extensive subterranean runners and forming colonies. Young branches more or less shaggy-pubescent with long hairs, less frequently stipitate-glandular or both. Flower clusters appearing before or with the branchlets of the season. Leaves cuneately narrowed below to subpetiolate bases or with definite petioles to 6 mm long; blades oblanceolate, obovate, elliptic, or oblong, 2–6 cm long, usually pubescent above when young, becoming glabrous or nearly so, with long, appressed-ascending hairs on the midrib beneath, bases

Fig. 227. **Rhododendron canescens.** (From Correll and Correll. 1972)

cuneate, apices rounded to acute, usually minutely mucronate, margins bristly-ciliate with curved-ascending hairs. Flowers fragrant, their stalks 5–15 mm long, hirsute, stipitate-glandular, or both. Calyx lobes 1–4 mm long, stipitate-glandular. Corolla white to deep pink or lavender, the largest lobe often marked with yellow, stipitate-glandular exteriorly, tube 1.5–2.5 cm long, cylindrical below, expanding to funnelform above, lobes spreading, shorter than the tube. Stamens 5, exserted, the style somewhat exceeding them. Capsule lance-cylindric, 1.5–2 cm long or slightly more, hirsute, stipitate-glandular, or both. (*Azalea atlantica* Ashe)

Pine savannas and flatwoods, borders of evergreen shrub-tree bogs or bays, bottomland clearings, sometimes in well-drained pinelands. Coastal plain, s. N.J. to Ga. and s.e. Ala.

4. Leucothoe (FETTER-BUSHES)

Evergreen or deciduous shrubs. Leaves alternate, short-petiolate. Flowers in racemes from axillary buds on wood of the previous season. Flower stalks each subtended by a bract and each having a pair of bractlets either basally or beneath the calyx. Calyx united basally, 5-lobed, persistent. Corolla white or white and pink-tinged, tube urceolate, with 5 small lobes at the summit. Stamens 10, not exserted from the corolla, anthers opening by terminal pores, awnless or 2–4-awned apically. Ovary superior, 5-locular, style straight, included or exserted from the corolla. Fruit a 5-valved, loculicidal capsule. Seed numerous.

1. Leaves evergreen (as evidenced by presence of some on wood of a previous season).
  2. Leaf blades finely reticulate-veined; bracts much shorter than the slender flower stalks; filaments long-pubescent.   1. *L. populifolia*
  2. Leaf blades not reticulate-veined; bracts half as long as the stoutish flower stalks; filaments short-pubescent.   2. *L. axillaris*
1. Leaves deciduous (as evidenced by absence of any on wood of a previous season).   3. *L. racemosa*

1. Leucothoe populifolia (Lam.) Dippel.*

Shrub with weakly ascending branches, to 4 m tall. Young twigs light brown, usually somewhat glaucous, sometimes sparsely short-pubescent. Leaves persistent, petioles short, 2–10 mm long, finely short-pubescent, blades ovate or lance-ovate, 4–10 cm long, mostly rounded basally, some broadly short-tapered, apices acute or short-acuminate, somewhat glaucous and grayish green, finely reticulate-veined, sparsely short-pubescent on the midrib beneath, margins narrowly cartilaginous-banded, entire or irregularly and obscurely wavy. Flowers in short, sessile or subsessile racemes borne in the leaf axils, raceme axes finely short-pubescent. Flower stalks slender, 7–10 mm long, finely short-pubescent, each subtended by a very small, ovate, persistent bract, with 1–3 minute bractlets somewhat above on the flower stalk. Calyx tubular for half its length or a little more, lobes very broadly low-triangular, 1 mm long or less, pubescent. Corolla 7–8 mm long, white, tube urceolate, broadest at base, usually with 2 constrictions, lobes short-oblongish. Filaments S-curved, long-pubescent, anthers awnless. Style tip exserted from the corolla. Capsule depressed-subglobose, 5–6 mm broad, dark brown. (*L. acuminata* (Ait.) D. Don)

Moist to wet hammocks, swampy woodlands about large springs and spring-runs. Coastal plain, s.e. S.C. to n. pen. Fla.

2. Leucothoe axillaris (Lam.) D. Don. DOG-HOBBLE.

Loosely branched shrub, leafy branches usually more or less arching, to 1.5 m tall. Young branchlets brown, usually finely short-pubescent. Leaves persistent, petioles 2–10 mm long, copiously short-pubescent, blades leathery, lustrous dark green above, duller and paler green beneath, ovate, elliptic, elliptic-oblong, or oval, 5–14 cm long, 1.5–5 cm broad, obtusely

---

*After this book had gone to press, it was indicated in a recent publication that this plant properly belongs to the genus *Agarista*, the binomial: *A. populifolia* (Lam.) Judd. See Walter S. Judd, "Generic Relationships in the Andromedeae (Ericaceae)," *Jour. Arn. Arb.* 60: 477–503 (1979).

short-tapered basally, apices short-acute, short-acuminate, or abruptly mucronate, upper surfaces glabrous, lower minutely pubescent when young, often glabrous in age, margins cartilaginous and variously cartilaginous-serrate, teeth few and mostly on the distal margins, or more or less throughout the margins and more numerous distally, teeth varying from (5–) 15–40 (–50) per side. Flowers in axillary, sessile racemes 2–7 cm long; axes of racemes pubescent, racemes not secund; flower stalks stoutish, 1–4.5 mm long, pubescent, each subtended by a broadly short-ovate or orbicular bract, 2 smaller bractlets closely set above the bract, both bracts and bractlets persistent. Calyx lobes ovate or lance-ovate, blunt-tipped, 1–2.2 mm long. Corolla white, 6–8 mm long, tube oblong-urceolate, lobes deltoid, very short, recurved. Filaments straight, short-pubescent, anther-halves awnless at the apex. Style pubescent, not exserted from the corolla. Capsule oblate, slightly if at all lobed, about 5 mm across, dark brown with buff sutures. Seeds 1.5–2 mm long, irregularly compressed-angular, lustrous amber, surface relatively coarsely reticulate.

Wet woodlands astride small streams, seasonally wet depressions in woodlands, floodplain forests, wooded stream banks, swamps, shrub-tree bogs. Coastal plain, s.e. Va. to Fla. Panhandle, westward to La.

A closely similar plant occurs in the mountains and piedmont, Va., N.C., S.C., n. Ga., e. Tenn. It inhabits cool, moist woodlands, chiefly understory heath thickets, and is not, insofar as we know, a wetland plant. It is *L. fontanesiana* (Steud.) Sleum. (*L. catesbaei* of authors, *L. editorum* Fern. & Schubert, *L. walteri* (Willd.) Melvin, misapplied).

3. Leucothoe racemosa (L.) Gray. FETTERBUSH.   Fig. 228

Deciduous shrub, to 4 m tall. Twigs of the season copiously short-pubescent. Leaf blades mostly elliptic, varying to lance-elliptic or oval, variable in size on a given branch, 1–5 cm long, 6–30 mm broad, cuneate to rounded basally, acute or short-acuminate apically, surfaces short-pubescent at least when young, usually remaining so, margins shallowly and obscurely crenate-serrate; petioles pubescent like the twigs. Racemes partially developing the summer before flowering, solitary, 2–10 cm long, axes densely short-pubescent, flowers secund but fruits not secund. Flower stalks about 2 mm long, subtending bracts mostly foliaceous, linear, 8–10 mm long, deciduous by full anthesis of the flowers, bractlets short, ovate-acuminate, borne just below the calyx, persistent. Calyx united only at the base, lobes lanceolate to lance-ovate, markedly acuminate apically, a little less than half as long as the corolla. Corolla white, sometimes pink-tinged, about 8 mm long. Each anther half 2-awned at the summit. Tip of style exserted from the corolla. Capsule oblate, not lobed, 4–5 mm broad, dark brown with buff sutures. Seeds angled, with 2 flat faces and the back rounded, about 0.3–1 mm long, light brown. (*Eubotrys racemosa* (L.) Nutt.; incl. *E. elongata* Small)

Swamps, shrub-tree bogs, open bogs, cypress-gum depressions, shoreline thickets. Chiefly but not exclusively coastal plain, e. Mass. to n.cen. pen. Fla., westward to s.e. La.; cen. Tenn.

5. Zenobia

Zenobia pulverulenta (Bartr. ex Willd.) Pollard.

Deciduous shrub, 1–2 m tall, forming colonies by subterranean runners, twigs glabrous, often glaucous. Leaves alternate, short-petiolate, blades leathery, elliptic, ovate-elliptic, occasional ones obovate, 2–8 cm long, 1–3 (–4.5) cm broad, cuneate to rounded basally, acute, obtuse, or rounded apically, usually short-mucronate, margins usually obscurely crenate, sometimes entire, upper surfaces glabrous, lower usually sparsely pubescent, often grayish-glaucous. (Individuals with leaves glaucous beneath also have glaucous flower stalks, bracts and bractlets, calyces, and ovaries.) Flowers in showy, umbellike fascicles from buds on wood of the previous season, racemosely arranged on leafless branches, the branches varying greatly in length; flower stalks glabrous, 1–2.5 cm long, each subtended by a short, broad, leathery bract and with a pair of bractlets at the base. Calyx persistent, leathery, united below, lobes 5, ovate-triangular, acute to acuminate apically, 2–3 mm long.

Fig. 228. a–d, **Leucothoe racemosa:** a, flowering branch; b, flower; c, fruiting branch; d, fruit; e–j, **Lyonia ligustrina:** e, flowering branch; f, flower; g, fruiting branch; h, fruit; i, persisting fruiting racemes of previous season; j, remains of old fruit. (From Correll and Correll. 1972)

478

Corolla white, tube broadly campanulate, 7–12 mm long, with 5, low, broad lobes. Stamens 10, filaments abruptly enlarged below, each anther-half surmounted with a pair of slender, ascending awns, each half opening by a pore at the base of the awns. Ovary superior, 5-locular. Fruit a depressed-globose or oblate, loculicidally 5-valved capsule. Seeds very irregularly angled, lustrous rust-colored, about 1 mm long. (Incl. *Z. cassinefolia* (Vent.) Pollard)

Shrub bogs and bays, pine savannas, borders of swamps. Coastal plain, s.e. Va. to S.C.

6. Lyonia

Evergreen or deciduous shrubs, rarely small trees. Leaves alternate, short-petioled. Flowers in umbelliform clusters in the axils of leaves or leaf scars, or the umbelliform clusters borne on leafy or leafless branches forming racemes or panicles; each flower stalk subtended by a bract and each with 2 bractlets near its base. Perianth 5-merous in ours. Calyx united basally, 5-lobed, persistent in most, deciduous in fruit in some. Corolla tube cylindric-campanulate, urceolate, or globose-urceolate, the 5 lobes short. Stamens 10; filaments flattened, often S-shaped, with or without a pair of spurlike appendages apically; anthers awnless, opening by terminal pores. Ovary superior, 5-locular, style straight, not exserted, stigma truncate or capitate. Fruit a 5-valved, loculicidal capsule, thickened along the paler sutures which appear as ridges before dehiscence. Seeds numerous, appearing like fine sawdust (scoboform).

1. Evergreen shrubs, some leaves present on wood of the previous season.
    2. Lower leaf surfaces (especially when leaves young) bearing stipitate-peltate or shieldlike, rust-colored scales.
        3. Flowers (and subsequent fruits) borne on twigs of the previous season essentially prior to production of any new shoot growth.   1. *L. ferruginea*
        3. Flowers (and subsequent fruits) borne on twigs of the season after the twigs are nearly fully developed.   2. *L. fruticosa*
    2. Lower leaf surfaces glabrous.   3. *L. lucida*
1. Deciduous shrubs, no leaves present on wood of the previous season.
    4. Young twigs terete; leaf margins minutely serrate; corolla ovoid-globose, 3–4 mm long.
                                                                                                            4. *L. ligustrina*
    4. Young twigs angled; leaf margins entire; corolla oblong-cylindric, 7–14 mm long. 5. *L. mariana*

1. Lyonia ferruginea (Walt.) Nutt. STAGGER-BUSH.

Evergreen shrub or small tree, often colonial, to 6 m tall; when of the stature of a relatively large shrub or small tree, then the trunks usually crooked, the crown spreading and the branchlets irregularly spreading. Young twigs with moderate to dense short, gray pubescence and stipitate-peltate or shieldlike scales, becoming less pubescent and less scaly with greater age and the scales becoming dirty gray. Leaf blades when only half or less than half developed wholly rusty-colored with stipitate-peltate scales on both surfaces, the scales quickly deciduous from the upper surfaces leaving them glossy green and with pale pubescence along the midribs; lower surfaces usually with persistent, short, and inconspicuous pubescence and with at least some rust-colored and gray scales, the scales usually of two size classes, the larger rust-colored and irregularly margined, the smaller entire-margined and gray, both more or less persisting or only the smaller ones persisting; leaves short-petioled, blades with entire margins, elliptic or oblanceolate to narrowly obovate, 1–7.5 cm long, 0.5–3 cm broad, at full maturity commonly with a downward curvature and revolute, principal lateral veins, of some leaves at least, usually impressed on the upper surface and in troughlike depressions. Flowers in axillary fascicles, on twigs of the previous season, their stalks loosely scaly and sometimes with nonglandular pubescence, each with a pair of bractlets at the base. Calyx persistent, lobes triangular, 1–2 mm long, outer surfaces scaly, inner glabrous or pubescent. Corolla subglobose-urceolate, 2–4 mm long, usually broader at the base than long, lobes very short, white. Capsule ovoid, scaly and pubescent, 5-angled, 3–6 mm long, with 5 thickened sutures which separate as a unit from the valves in dehiscence. (*Xolisma ferruginea* (Walt.) Heller)

In both poorly drained pine flatwoods (where commonly intermixed with *L. fruticosa*) and in well-drained sand pine–oak scrub (where *L. fruticosa* is absent). Coastal plain, s.e. S.C. to s.cen. pen. Fla., and in Fla. Panhandle to about Bay Co.

2. Lyonia fruticosa (Michx.) G. S. Torr. in Robins.

Evergreen shrub, usually colonial, to 1.5, rarely to 3, m tall, branches usually rigidly erect; young twigs often densely pubescent abundantly clothed with rust-colored scales, becoming glabrous or sparsely pubescent in age. Leaf blades when only half or less than half developed wholly rust-colored with stipitate-peltate scales of one size class; scales soon deciduous on upper blade surfaces leaving them dull green (relative to the color in *L. ferruginea*) and with grayish pubescence along the midrib; scales on lower leaf surfaces mostly eventually slough-ing or turning gray leaving the surfaces dull-grayish; mature leaf blades flat or their edges sometimes slightly curved upwardly, margins rarely revolute, obovate to oblanceolate, oval or elliptic, 0.5–5.5 cm long and 0.3–2.8 cm broad, principal lateral veins not depressed on the upper surfaces. Flowers in axillary fascicles, borne on newly developed twigs of the sea-son (rarely a few flowers on an occasional branch of an occasional plant on twigs of a previous season). Calyx persistent, lobes triangular, 1–1.5 mm long, the outer surfaces scaly-pubescent, the inner sparely to moderately non-glandular-pubescent. Corolla ovoid-urceolate, 2.5–5 mm long, usually about as broad at the base as long, scaly without, lobes very short, white. Capsule ovoid, 5-angled, 3–5 mm long, with thickened sutures which separate as a unit from the valves in dehiscence. (*Xolisma fruticosa* (Michx.) Nash)

Poorly drained pine flatwoods. Coastal plain, Ga. to s. pen. Fla., in the Fla. Panhandle west to the Apalachicola River drainage.

3. Lyonia lucida (Lam.) Koch. FETTERBUSH.

Evergreen shrub, commonly very showy and handsome when flowering; flowering speci-mens varying very greatly in stature from a few dm tall to at least 4 m, the large ones robustly branched from the base, their crowns about as broad as the height of the plant, usu-ally occurring where surface water is present much, nearly all, of the time. Young twigs strongly angled, green flecked with dark, loose, soon deciduous, narrow scales. Very young, partially grown, leaves minutely scaly on both surfaces (similar scales on flower stalks and calyces); mature leaves leathery, glabrous, dark glossy green above and with irregularly scat-tered, very minute, punctate dots, somewhat paler green beneath, the dots much more abun-dant and more evenly distributed (dots purple on fresh leaves, usually brownish on dried ones). Leaf blades narrowly to broadly elliptic, varying to oval or obovate, 2–8 cm long, 1–4 cm broad, bases cuneate, apices short-acuminate or acute, margins entire, a vein run-ning parallel to and near each margin, not revolute when fresh, revolute when dry. Flowers in axillary, nodding fascicles on wood of the preceding year. Calyx lobes lance-ovate, acute, 4–5 mm long, persistent. Corolla varying from white to deep pink, 6–10 mm long, tube dilated basally then cylindric-urceolate, constricted at the orifice, lobes minute, triangular, curved outwardly. Filament with a pair of descending, bristlelike appendages distally. Cap-sule ovoid, 4–5 mm long, about equaling the erect calyx lobes, truncate apically. (*Des-mothamnus lucidus* (Lam.) Small)

Shrub-tree bogs and bays, seasonally wet pine savannas and flatwoods, titi swamps, cypress-gum ponds, wet woodlands, also on deep, well-drained sands of scrub (in Fla. es-pecially). Chiefly coastal plain, s.e. Va. to cen. pen. Fla., westward to La.; w. Cuba.

4. Lyonia ligustrina (L.) DC. MALEBERRY, HE-HUCKLEBERRY.   Fig. 228

Deciduous shrub, 1–4 m tall. Young twigs terete, sparsely short-pubescent, soon becoming glabrous. Leaf blades lanceolate, elliptic, oval, or obovate, very variable in size, to 8 cm long or a little more and to 5 cm broad, short-cuneate basally, mucronate at the obtuse to rounded apices, margins minutely serrate, surfaces and petioles short-pubescent. Flowers in fascicles or short racemes on relatively short branches arising from buds in the axils of leaf scars on wood of the preceding season, the flowering branches often numerous on the old

480

wood giving the effect of a diffuse panicle; flowering branches more or less, or not at all, leafy-bracted. Calyx pubescent, persistent, lobes broadly triangular, 0.5–1 mm long. Corolla 3–4 mm long, whitish, tube ovoid-globose, the very small lobes curved outwardly. Filaments without appendages. Capsule sparsely short-pubescent, globose to oblate, the sutures prominently pale-cartilaginous, calyx only around the capsule base.

Thickets and woodlands along streams, shrub thickets in wet meadows, swamps, cypress-gum ponds and depressions, shrub-tree bogs and bays, pine flatwoods. Maine to N.Y., Ohio, Ky., Ark., s.e. Okla., generally southward to cen. pen. Fla. and e. Tex.

Plants with the inflorescence branches or branchlets without or with only a few foliaceous bracts may be designated *L. ligustrina* var. *ligustrina* (*Arsenococcus ligustrinus* (L.) Small). They occur chiefly in the mountains and piedmont south of Va., otherwise northward from Va. Plants with the inflorescence branches or branchlets conspicuously leafy-bracted may be designated *L. ligustrina* var. *foliosiflora* (Michx.) Fern. (*A. frondosus* (Pursh) Small). They occur chiefly on the coastal plain from s.e. Va. to cen. pen. Fla., westward to e. Tex., Okla., and Ark. (mts.)

5. Lyonia mariana (L.) D. Don. STAGGER-BUSH.

Shrub with subterranean runners and to some extent colonial, 2–10 (–20) dm tall, usually not branched below, branches above strongly ascending. Young twigs angled, glabrous, sparsely short-pubescent, or copiously short-pilose. Leaves deciduous (some sometimes overwintering, deciduous in spring), blades elliptic, oblong-elliptic, or narrowly obovate, bases broadly cuneate, apices obtuse to rounded, rarely acute, sometimes mucronate; varying considerably in size on a given plant, the larger ones 4–10 cm long, 1.5–4.5 cm broad, margins entire; upper surfaces usually glabrous, sometimes sparsely short-pubescent, lower varying from glabrous to short-pilose along the principal veins to uniformly and copiously short-pilose. Flowers nodding, in fascicles in axils of leaf scars on wood of the preceding season. Calyx lobes lanceolate, lance-subulate, or lance-oblong, 3-nerved, 6–8 mm long, as long as the developing capsule, deciduous by full maturity of the capsule. Corolla white or pinkish, 7–14 mm long, tube oblong-cylindric, lobes very short and broad, curved outwardly. Filaments with a pair of descending, bristlelike teeth distally. Capsule 5–7 mm long, ovoid-obpyramidal, truncate apically, sutures prominently pale-cartilaginous. (*Neopieris mariana* (L.) Britt.)

Pine savannas and flatwoods, hillside bogs in pinelands, edges of cypress-gum ponds or depressions, borders of shrub-tree bogs or bays, open, mixed, well-drained woodlands. Coastal plain and adjacent piedmont, s. R.I., s. Conn., to N.J. and e. Pa., thence to cen. pen. Fla. and to cen. Fla. Panhandle; La. to s.cen. Tex., northward in the interior, s.e. Okla., Ark., s.e. Mo., w. Tenn.

7. Pieris

Pieris phillyreifolia (Hook.) DC.   Fig. 229

Woody evergreen; infrequently an erect, leafy shrub with numerous stems from subterranean runners; more commonly ascending the trunks of *Taxodium ascendens* Brogn. (pond-cypress), its main, somewhat branched, stem (growing beneath the outerbark of the cypress) strongly flattened, bearing minute, brown, scale-leaves, the growing tips callused; lateral "normal" leafy branches exserted at intervals through the cypress bark, the branches more or less festooning the trunk of the cypress. Leaves alternate, leathery, very short-petiolate, blades glabrous, elliptic, oblong-elliptic, or lanceolate, 1.5–7 cm long, 5–20 mm broad, margins revolute, entire or with a few, small, very blunt teeth distally, less frequently toothed from somewhat below their middles. Flowers in 3–9-flowered racemes borne in the axils of leaves near the tips of branches well before new growth of the season commences; raceme axis and flower stalks powdery-pubescent, stalks 3–5 mm long, thickish, each subtended by a subulate bract about as long as the stalk and bearing a pair of subulate bractlets at a little distance below the calyx, both bracts and bractlets glandular-pubescent. Calyx persistent,

Fig. 229. **Pieris phillyreifolia:** a, habit of nonvining plant; b, portion of vining plant in bark of tree, bark partially torn away; c,d, stems from beneath bark of tree, showing scale leaves; e, enlargement of small section of d; f, flower; g, corolla opened out, with stamens and pistil; h, stamen.

482

united only basally, lobes 5, lance-triangular, about half as long as the corolla tube, margins stipitate-glandular above the middle. Corolla white, somewhat translucent, 7–9 mm long, tube ovoid-urceolate, narrowed at the throat and then with 5 very short, somewhat recurved lobes. Stamens 10, filaments somewhat **S**-curved, anthers with a pair of deflexed spurs on the back above the junction with the filament, each anther-half opening by an oval or **V**-shaped pore. Ovary superior, 5-locular, fruit a loculicidally 5-valved, subspheroidal capsule. Seeds brown, irregularly angled longitudinally, more or less obpyramidal in outline, about 1 mm long. (*Ampelothamnus phillyreifolius* (Hook.) Small)

Chiefly, if not exclusively, in cypress-gum ponds or depressions. Coastal plain, S.C. to cen. pen. Fla., Fla. Panhandle and s. Ala.

Reference: Harper, R. M. "A Unique Climbing Plant." *Torreya* 3: 21–22. 1903.

8. Cassandra

Cassandra calyculata (L.) D. Don. LEATHER-LEAF.

Shrub 5–10 dm tall, perennating by subterranean runners and often forming low, dense thickets. Young branchlets brown, pubescent with short, curly hairs, older stems dark reddish brown. Leaves leathery, persistent (except on fruiting branchlets), alternate, very short-petioled, blades elliptic, oblong-elliptic, or oblanceolate, 1–5 cm long, bases somewhat cuneate, apices mostly obtuse, varying to rounded or acute, often short-mucronate, both surfaces scurfy, the lower much more abundantly so than the upper, margins slightly revolute and obscurely toothed. Flowers small, nodding, short-stalked, solitary and descending-secund in the axils of bracteal leaves on horizontal or arching terminal branchlets, the flower buds having partially developed late in the season preceding flowering. Calyx persistent, united only basally, 5-lobed, lobes ovate-acute, 2–2.5 mm long, exteriorly scurfy except basally; calyx closely subtended by a pair of bractlets as broad as or broader than long, otherwise similar to the calyx lobes. Corolla 6–7 mm long, white, tube oblong-urceolate, with 5 short, recurved lobes. Stamens 10, not exserted from the corolla, filaments flat, tapering upwardly, anther-halves prolonged tubular, each opening by an oblique terminal pore. Ovary superior, oblate, somewhat 5-lobed, 5-locular, subtended by a 10-lobed, nectariferous, green disc; style straight, exserted from the corolla, stigma capitate. Capsule oblate, dark brown with buff, slightly thickened sutures, loculicidally 5-valved, but the wall splitting into 2 layers, inner layer 10-valved; style persistent on the capsule until about time of dehiscence. Seeds about 10 in each locule, lustrous brown, irregularly compressed-angular, about 1 mm long.

Bogs. Nearly circumpolar in distribution; in our range, in scattered localities of coastal plain of N.C. and in Henderson Co., s.w. N.C.

9. Gaylussacia (HUCKLEBERRIES)

Evergreen or deciduous shrubs, often with short to elongate subterranean runners and colonial. Leaves alternate, short-petioled. Flowers in axillary, usually few-flowered, bracteate racemes on wood of the preceding season, flower stalks with 1 or usually 2 bractlets. Floral tube adnate to the ovary. Sepals 5, persistent. Corolla white, greenish, greenish red or pinkish, tubular, 5-lobed, tube campanulate to urceolate. Stamens 10, inserted on the base of the corolla, filaments short, anthers without appendages on the back, each half narrowed upwardly into a tube, opening by a terminal pore or slit, awnless. Ovary inferior, 10-locular, each locule with 1 ovule. Fruit a berrylike drupe with 10 smooth, 1-seeded, bony nutlets.

1. Bracts of racemes not foliaceous, deciduous soon after flowering; floral tube and fruit glaucous, glands, if any, all sessile.   1. *G. frondosa*
1. Bracts of racemes foliaceous, persistent until the fruits mature; floral tube and young fruit stipitate-glandular.

2. Young twigs, raceme axes, flower stalks, and floral tubes with short curly hairs, with or without short stipitate-glandular ones intermixed.   2. *G. dumosa*
2. Young twigs, raceme axes, flower stalks, and floral tubes bearing long, spreading, silvery-silky hairs minutely glandular at their tips.   3. *G. mosieri*

1. Gaylussacia frondosa (L.) T. & G. DANGLEBERRY.

Deciduous shrub to 2 m tall, with sessile, amber, glandular dots on various of its parts. Young twigs often glaucous, glabrous, or sparsely short-pubescent varying to densely short-pubescent, glandular-dotted. Leaf blades oval, elliptic, oblanceolate, or obovate, cuneate basally, rounded to obtuse apically, sometimes retuse; margins entire, upper surfaces glabrous or very sparsely short-pubescent, lower usually moderately to strongly glaucous, less frequently without glaucousness, sparsely to densely short-pubescent. Racemes loosely 5–12-flowered, bracts not foliaceous, axes and flower stalks glabrous varying to copiously short-pubescent, sparsely gland-dotted. Floral tube glaucous, sometimes gland-dotted, calyx lobes triangular, about 1 mm long, obtuse to acute, gland-dotted. Corolla greenish white, usually suffused with purplish pink, dull and inconspicuous, tube cuplike, 2–4 mm long, a little longer than broad to broader than long, lobes very small, triangular, obtuse to acute, recurved. Stamens not exserted, stigmatic tip of the style about even with the mouth of the corolla. Fruit 5–8 mm in diameter, glaucous-blue. (*Decachaena frondosa* (L.) T. & G.)

Well-drained to moist woodlands and thickets, bottomland woodlands, poorly drained to well-drained pinelands, sphagnous bogs, shrub-tree bogs or bays.

Populations from N.H., Mass., s.e. N.Y., southward chiefly on the coastal plain to S.C. may be designated *G. frondosa* var. *frondosa*. Plant to 2 m tall, branches relatively widely spreading; twigs glabrous; larger leaves 3–6 cm long and 2–3.5 cm broad; corolla 3–4 mm long, usually a little longer than broad.

Plants closely similar but with densely short-pubescent twigs and lower leaf surfaces, apparently of infrequent occurrence from s.e. S.C. to cen. pen. Fla., Fla. Panhandle, may be designated *G. frondosa* var. *tomentosa* Gray (*Decachaena tomentosa* (Gray) Small).

Plants of most populations from s. Ga. to cen. pen. Fla., Fla. Panhandle to s. Miss. (La.?) are generally of smaller stature, 2–6 (–10) dm tall, the branches relatively short, the plant with a relatively columnar aspect; twigs copiously short-pubescent; larger leaves 2–4 cm long, 1–2 cm broad, sparsely short-pubescent and commonly strongly glaucous beneath; corolla 2–3 mm long, usually broader than long. These may be designated *G. frondosa* var. *nana* Gray (*Decachaena nana* (Gray) Small).

2. Gaylussacia dumosa (Andr.) Gray. DWARF HUCKLEBERRY.

Deciduous shrub with subterranean runners and colonial, 1–5 dm tall. Young twigs usually copiously pubescent with short curly hairs, with or without short stipitate-glandular hairs intermixed. Leaf blades oblanceolate or short-obovate, some sometimes elliptic, bases mostly strongly cuneate, apices rounded or infrequently acute, mucronate, variable in size on a given stem, the larger ones 2–3 cm long and 5–20 mm broad, margins sometimes with barely a suggestion of glandular crenulations; upper surfaces of young leaves usually short-pubescent at least on the veins and sparsely glandular-punctate, usually becoming glabrous and lustrous, edges often remaining pubescent, lower surfaces pubescent at least on the veins and gland-dotted. (Rarely in our range, leaf surfaces and margins more or less stipitate-glandular.) Racemes 5–40 mm long, mostly 4–10-flowered, bracts foliaceous and persistent, axes and flower stalks with somewhat matted curly hairs, short-stipitate glands intermixed. Floral tube usually abundantly clothed with both sessile and short-stipitate orange glands; sepals short-triangular, usually with only a few sessile glands and hairy on their margins. Corolla white or pinkish, campanulate when fully expanded, 6–9 mm long, constricted at the summit, lobes short-ovate. Stamens not exserted, stigmatic tip of style even with the mouth of the corolla. Fruit black, not glandular by full maturity, about 8 mm in diameter. (*Lasiococcus dumosus* (Andr.) Small)

Commonly in well-drained pinelands or pine-oak woodlands, also in areas transitional from high pinelands to shrub-tree bogs, and in sphagnous bogs. Nfld. to cen. pen. Fla., westward to Miss.; e.cen. Tenn.; also local in mts., W.Va., s.w. N.C., n.w. S.C.

3. Gaylussacia mosieri (Small) Small.

Deciduous shrub with colonial habit of *G. dumosa*, potential stature much greater, to 1.5 dm tall, in aspect very much resembling *G. dumosa*. Young twigs, raceme axes, flower stalks, and floral tubes bearing long, spreading, silvery-silky hairs minutely glandular at their tips, the glandular tips, at least some of them, red. Young twigs infrequently with short, nonglandular hairs. Leaf surfaces usually somewhat pubescent when young but the pubescence variable as to density and the relative number of stipitate-glandular and nonglandular hairs. Mature fruit black, bearing gland-tipped hairs. (*Lasiococcus mosieri* Small)

   Pitcher-plant bogs, borders of shrub-tree bogs and bays, seasonally wet pine savannas and flatwoods. Coastal plain, s. Ga., n. Fla., westward to La.

   G. mosieri and *G. dumosa*, where their ranges overlap, often grow in close proximity.

10. Vaccinium

Evergreen or deciduous, erect or trailing shrubs, rarely attaining tree stature. Leaves alternate, short-petioled, entire or serrate. Inflorescences from buds on wood of the previous season, flowers in the axils of bracts or leaves, inflorescences mostly racemose, rarely the flowers solitary or 2 together; flower stalks bearing 2 bractlets. Floral tube adnate to the ovary, the latter thus partly to wholly inferior. Calyx lobes 4 or 5, rarely none. Corolla cylindric, urceolate, or campanulate, 4- or 5-lobed, tube sometimes only basal. Stamens 8 or 10, anthers with or without spurs on the back, the halves narrowed upwardly into tubes opening by terminal pores. Ovary 4- or 5-locular, or falsely 10-locular above, 5-locular below. Fruit a (5- to) many-seeded berry crowned by the persistent calyx.

   A large genus exhibiting much polymorphism, some of which results from hybridization.

1. Stems trailing.
   2. Calyx lobes 4; corolla deeply 4-lobed, lobes strongly reflexed.   1. *V. macrocarpon*
   2. Calyx lobes 5; corolla tubular to the throat, the 5 lobes very small relative to the tube.
                                                                                      2. *V. crassifolium*
1. Stems erect.
   3. Evergreen shrub (leaves present on wood of previous year); leaves less than 1.5 cm long.
      4. Lower leaf surfaces with stalked or clavate glandular trichomes; floral tube and fruit not or only
         slightly glaucous.   3. *V. myrsinites*
      4. Lower leaf surfaces without pubescence; floral tube and fruit strongly glaucous.   4. *V. darrowi*
   3. Deciduous shrub (leaves shed in autumn, or if a few persistent over winter, these shed by the time
      new shoots are well developed in spring); leaves over 1.5 cm long.
      5. Leaves at maturity all less than 3 cm long, their margins (most of them) minutely serrate.
                                                                                      5. *V. elliottii*
      5. Leaves at maturity (most of them) over 3 cm long, their margins entire.
         6. Leaves and twigs pubescent; floral tube and fruit not glaucous, fruit black.   6. *V. fuscatum*
         6. Leaves and twigs glabrous; floral tube glaucous, fruit glaucous-blue.   7. *V. australe*

1. Vaccinium macrocarpon Ait. CRANBERRY.

Evergreen shrub with slender, long, forking, trailing stems with ascending flowering or fruiting branches. Young twigs with short, curly pubescence. Leaves with short, stoutish petioles, blades leathery, elliptic, elliptic-oblong, or oblong, 5–15 mm long, usually rounded basally and apically, margins entire and revolute, surfaces glabrous, the lower paler than the upper. Flowers 1–10 on short segment of stem with reduced or bracteal leaves, flowers solitary in their axils, a normally leafy branch above. Flower stalks filiform, 1–3 cm long, pubescent, a pair of subfoliose bractlets above the middle, flower nodding above the bractlets. Calyx lobes 4. Corolla pink, united only basally, lobes 4, 6–10 mm long, oblong, reflexed. Stamens exserted, anthers with long, terminal lobes. Berry red to pink, globose, ellipsoid, obovoid, or pyriform, 1–2 cm in diameter. (*Oxycoccus macrocarpus* (Ait.) Pursh)

Bogs, Nfld. to Man., southward to Va., W.Va., Ohio to Ill., isolated stations in coastal plain, N.C., mts. of N.C., e. Tenn., Ark.

2. Vaccinium crassifolium Andr. CREEPING BLUEBERRY.

Evergreen shrub with a stout, woody, blackish, subterranean caudex producing several to numerous slender trailing stems to 1 m long, forming mats or carpets to 2 m across. Stems densely short-pubescent. Leaves with slender, short petioles, blades leathery, elliptic or oval, 5–15 mm long, 3–10 mm broad, apices blunt to rounded, margins thickened, slightly revolute, with remote, low, gland-tipped crenations; upper surface very short-pubescent, lower glabrous. Flowers few in racemes shorter than the subtending leaves. Corolla pink, tube globular-urceolate, 2–4 mm long. Berry black or purple-black, about 5 mm in diameter. (*Herpothamnus crassifolius* (Andr.) Small)

Borders of shrub-tree bogs and bays, wet to dry pinelands. Coastal plain, s.e. Va. to s.e. Ga.

3. Vaccinium myrsinites Lam.

Low, evergreen shrub, to about 6 dm tall, with stout and elongate subterranean runners and colonial. Young twigs glabrous or sparsely to densely shaggy-pubescent. Leaves sessile or subsessile, commonly glossy green varying to glaucous and grayish green, oblanceolate, short-obovate, or elliptic, mostly 5–15 mm long, 2–10 mm broad, glabrous above, sparsely beset with stalked or clavate glandular hairs beneath, these usually shed on mature leaves, occasionally sparsely pubescent beneath with nonglandular hairs as well; margins not revolute, with minute, appressed-ascending, gland-tipped teeth, the glands eventually deciduous. Flowers 2–8 in axillary fascicles or short racemes, usually appearing before new growth of the season commences. Floral tube glabrous, sometimes glaucous, calyx lobes 5, broadly triangular-obtuse, deltoid, or very broadly rounded, about 1 mm long, usually glabrous, sometimes short-ciliate marginally. Corolla white to deep pink, urceolate, 6–8 mm long. Berry black, less frequently blue-black, 6–8 mm in diameter. (*Cyanococcus myrsinites* (Lam.) Small)

In seasonally wet to well-drained pinelands, scrub, prairies, borders of shrub-tree bogs or bays. Coastal plain, s.e. S.C., s. Ga., to s. pen. Fla., Fla. Panhandle, and s. Ala.

4. Vaccinium darrowi Camp.

Evergreen shrub with habit, leaf size and shape, and general appearance of *V. myrsinites* except that it is more markedly glaucous and bluish green; the two often sharing the same habitat where their ranges overlap. Young twigs moderately to densely pubescent with short, curly, gray hairs. Leaves commonly glaucous on both surfaces at first, usually remaining glaucous beneath, upper surface glabrous, short-pubescent along the midrib, or not infrequently the entire surface clothed with very fine, short, appressed, gray hairs; lower surface glaucous and glabrous, sparsely pubescent proximally along the midrib or along all of the midrib, without glandular hairs; thickened margins usually somewhat revolute, entire or with low, appressed-ascending, gland-tipped teeth. Fruit glaucous-blue, 4–6 mm in diameter.

Habitats essentially as for *V. myrsinites*. s. Ga. to s.cen. pen. Fla., westward to s.e. Tex.

5. Vaccinium elliottii Chapm. MAYBERRY.

Deciduous shrub to 4 m tall, often with a broad, bushy-branched crown. Young twigs varyingly glabrous, powdery-pubescent, or short shaggy-pubescent, the latter sometimes with a few stipitate-glandular hairs intermixed. Petioles short, pubescent, blades usually thin, glossy green above, narrowly to broadly elliptic, oblanceolate, or oval, 1–2 (–3) cm long, 5–10 (–15) mm broad, usually more or less pubescent along the midrib both above and beneath, sometimes sparsely pubescent over the entire surface beneath, margins minutely serrate, each serration of a developing leaf usually tipped by a quickly deciduous stipitate gland. Flowers in fascicles of 2–6, appearing before or as the shoots of the season develop; flower stalks 2–5 mm long, glabrous or pubescent. Floral tube and calyx glabrous, rarely glaucous, calyx lobes 5, broadly triangular, about 1 mm long. Corolla 5–7 mm long, narrowly urceolate, white or pinkish. Berry black, rarely glaucous, 5–10 mm in diameter. (*Cyanococcus elliottii* (Chapm.) Small)

In a wide variety of habitats, from wet to very well drained, bottomland forests, wet thickets and clearings, stream banks, various upland woodland mixtures. Chiefly coastal plain, s.e. Va. to n. Fla., westward to e. Tex. and Ark.

6. Vaccinium fuscatum Ait.

Straggly to bushy-branched shrub, to about 3 m tall. Young twigs shaggy pubescent. Leaves deciduous or some of them sometimes overwintering; petioles very short, usually copiously pubescent, blades lanceolate, elliptic, ovate, or obovate, cuneate basally, acute or short-acuminate apically, variable in size on a given plant or from plant to plant, 2–8 cm long, to about 3 cm broad; margins usually entire or glandular-toothed, upper surfaces glabrous to sparsely pubescent, the lower nonglandular, not glaucous, sparsely to densely shaggy-pubescent on the midrib, the midrib and principal veins, or over the entire surface. Flowers in few-flowered axillary fascicles or short racemes, appearing before or with the new shoots of the season. Floral tube glabrous, sometimes glaucous, calyx glabrous, lobes 5, about 1 mm long, broadly short-triangular, blunt-tipped. Corolla white to pink, 5–10 mm long, narrowly urceolate to cylindric-oblong. Berry black or blue-black, 5–10 mm in diameter. (*Cyanococcus fuscatus* (Ait.) Small; incl. *V. arkansanum* Ashe, *V. atrococcum* (Gray) Heller, *C. atrococcus* (Gray) Small, *C. holophyllus* Small)

Swamps, low woodlands, pine flatwoods, borders of shrub-tree bogs or bays, open bogs, cypress-gum depressions, also open upland woodlands. N.B. to s. Ont. and Ind., generally southward to s.cen. pen. Fla., Ark. and e. and s.e. Tex.

Note: *V. virgatum* Ait. (incl. *V. amoenum* Ait.), generally of upland, well-drained places, is similar. The lower surfaces of its leaves bear stipitate-glandular hairs intermixed with nonglandular ones.

7. Vaccinium australe Small.

Straggly shrub to about 4 m tall. Twigs glabrous. Petioles very short, blades mostly elliptic, 2.5–8 cm long, 1.5–3 cm broad, obtuse to acute apically and minutely mucronate, margins entire, surfaces glabrous, usually somewhat glaucous beneath. Flower clusters appearing before and with the new shoots of the season, in short axillary racemes and fascicles. Flower stalks mostly about 1 cm long, glabrous. Floral tube glaucous on its lower half, calyx lobes 5, deltoid, 1 mm long. Corolla white or pink-tinged, tube elliptic-oblong, 8–10 mm long, lobes minute, triangular-acuminate, recurved. Stamens and style not exserted. Berry blue, glaucous, 7–12 mm in diameter.

Shrub-tree bogs and bays, pine flatwoods, swamps, less frequently in upland woodlands. Coastal plain, N.J. to n. Fla. (chiefly Panhandle), westward to Miss.

Diapensiaceae (DIAPENSIA FAMILY)

Pyxidanthera barbulata Michx. PYXIE-MOSS.   Fig. 230

Prostrate, much branched, abundantly leafy, mat-forming, subshrub. Stem pubescent. Leaves sessile, alternate, approximate, oblanceolate, sharply acute apically, 3–8 mm long, upper surface pilose at or near the base, sometimes to the middle. Flowers usually numerous, each solitary and sessile in a leaf axil, bisexual, radially symmetrical. Sepals 5, pinkish, oblong, 3–5.5 mm long, concave, rounded apically. Corolla white to pink, about 10 mm across, united basally, lobes 5, spreading, spatulate-obovate, longer than the sepals. Stamens 5, filaments broad and flat, adnate to the corolla tube, shorter than and alternate with the corolla lobes, anthers short, bent inwardly. Pistil 1, ovary superior, 3-locular, with several

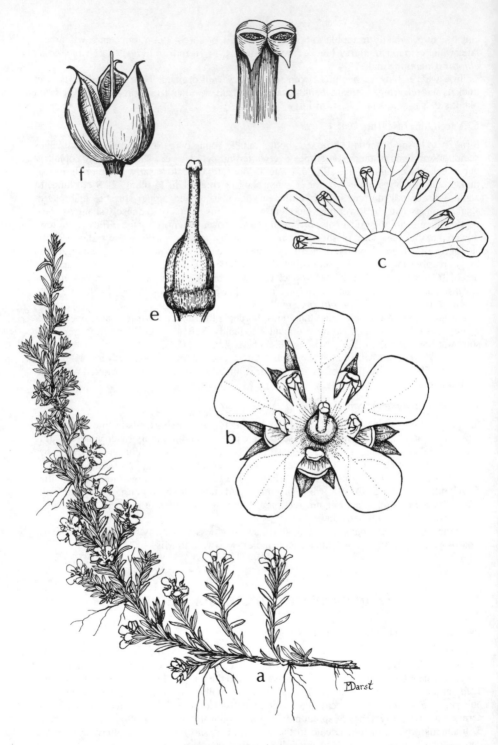

Fig. 230. **Pyxidanthera barbulata:** a, habit, portion of plant; b, flower, face view; c, corolla, cut and spread out; d, stamen; e, pistil; f, capsule.

488

ovules, style straight, exserted, stigma rounded-capitate. Fruit a globose, loculicidal capsule 2 mm in diameter. Seed nearly globose, pitted.

Sands or peaty sands of pinelands, often where seasonally wet. Coastal plain, s. N.J. to n.e. S.C.

Plumbaginaceae (LEADWORT FAMILY)

Limonium carolinianum (Walt.) Britt. SEA-LAVENDER, MARSH-ROSEMARY.

Perennial herb with a stout, often branched, woody subterranean rootstock, pink to red internally, the crown or each of the crowns bearing annually a rosette of alternately disposed leaves, the inflorescence a loose, open, small-bracteate, subdichotomously branched panicle of small, scarious flowers mostly borne along one side of the ultimate branchlets, the panicle fanlike in overall outline. The delicately textured and colored, lacy panicles much prized for use in dried floral arrangements. Leaves simple, alternate, with petioles equaling or more frequently exceeding the blades in length, commonly dilated and clasping the stem basally; blades thickish, somewhat leathery, punctate with salt glands, very variable in size and shape, linear-oblanceolate or -spatulate, narrowly to broadly elliptic, sometimes obovate, rarely suborbicular, 3–30 cm long and 4–75 mm wide; tapered below, apices acute or obtuse in narrower leaves, usually rounded apically, often retuse in the broader leaves. Panicle branches each subtended by a small, scarious, ovate to subulate, clasping bract. Flowers bisexual, radially symmetrical, solitary or 2 or 3 in short, scarious-bracted clusters, each flower subtended by 2 scarious bracts, the flowers or clusters usually somewhat separated from each other and mostly disposed along one side of the branchlets. Calyx cylindric-obconic to funnelform, scarious, glabrous or variously pubescent, often tinged reddish purple, with 5 erect, oblong to narrowly triangular, obtuse to acuminate, white lobes 0.4–2 mm long. Corolla pale to dark blue or lavender, scarious, tubular only at base, the 5 nearly distinct segments long-clawed, a stamen attached near the base of each. Pistil 1, ovary superior, 1-locular, 1-ovulate, styles 5, separate. Fruit membranous, indehiscent (a utricle), scarcely exserted from the persistent calyx. Seed smooth, shiny, red. (Incl. *L. nashii* Small, *L. obtusilobum* Blake, *L. angustatum* (Gray) Small)

Salt marshes, salt meadows, salt flats, interdune swales, mangrove swamps, saline ditches and shores. Coastal, Nfld. and Que. to s. Fla., westward to Tex.; n.e. Mex.; Berm.

Reference: Luteyn, James L. "Revision of *Limonium* (Plumbaginaceae) in Eastern North America." *Brittonia* 28: 303–317. 1976.

In considering *Limonium* in our range as comprising a single, polymorphic species, we follow the interpretation of Luteyn, which was based upon extensive field and herbarium studies; however, we are not at all confident that Luteyn's interpretation is a reasonable one.

Primulaceae (PRIMROSE FAMILY)

Annual or perennial herbs (ours). Leaves simple (dissected in one genus), without stipules. Flowers bisexual, radially symmetrical. Calyx and corolla united at their bases, usually 5-lobed, calyx persistent. Stamens 5, opposite the corolla lobes and inserted on the corolla tube or at the base of the lobes, sometimes staminodes present as well. Pistil 1, ovary superior or partly inferior, 1-locular, the placenta free-central, with few to many ovules. Style 1, stigmas capitate or truncate. Fruit a 2–6-valved capsule.

1. Plant aquatic, with finely dissected feathery leaves; inflorescence stalk (and to a lesser extent its axis) spongy-inflated.   1. *Hottonia*

1. Plant terrestrial, some in wet habitats but not aquatic, with simple leaves; without spongy-inflated inflorescence stalks.
  2. Annual, the leaves not over 20 mm long.   2. *Anagallis*
  2. Perennial, the leaves (the larger at least) well over 20 mm long.
    3. Ovary partly inferior; corolla white or pinkish.   3. *Samolus*
    3. Ovary wholly superior; corolla yellow to orange.   4. *Lysimachia*

1. Hottonia

Hottonia inflata Ell. FEATHERFOIL.   Fig. 231

Aquatic herb with finely divided, feathery, alternate leaves, those just below the inflorescence tending to be closely clustered, about oblong in overall outline, to 6–7 cm long. Inflorescences in a partly emersed floating cluster, each inflorescence stalk spongy-inflated, the axis inflated to a lesser degree. Flowers in separated whorls on the axis, each short-stalked, subtended by a lanceolate bract somewhat resembling the sepals. Sepals united only at the base, the 5 lobes somewhat foliaceous, linear-oblong, blunt apically, becoming larger as the fruit matures, then about 5 mm long. Corolla tube short, lobes white or whitish, somewhat shorter than the calyx tubes (at anthesis). Stamens not exserted. Capsule nearly globose to subpyriform. Seeds numerous, tiny, amber-brown, oblong-oval in outline, surface somewhat roughened.

Lakes, ponds, pools, swamps, ditches, and canals. Sporadic in occurrence. Chiefly coastal plain, Maine to n. Fla., westward to e. Tex., northward in the interior to Mo., Ill., Ind., Ohio, W.Va.

2. Anagallis (PIMPERNELS)

Low, erect, spreading, or procumbent annuals. Leaves small, not exceeding 20 mm long, alternate, opposite, or whorled. Flowers borne in the leaf axils. Calyx and corolla united only at their bases, (4)5-lobed. Stamens (4)5. Capsule circumscissile near the middle.

1. Flowers subsessile on very short, thickish, scarcely evident stalks.   1. *A. minima*
1. Flowers with evident, slender stalks.
  2. Leaves all opposite or whorled.   2. *A. arvensis*
  2. Leaves alternate above, sometimes some lower ones opposite or subopposite.   3. *A. pumila*

1. **Anagallis minima** (L.) Krause in Sturm. CHAFFWEED.   Fig. 232

Diminutive, somewhat succulent, glabrous herb. Stems simple or irregularly branched, erect, or often with several branches radiating from the base, prostrate below and rooting at the nodes, the tips ascending. Leaves very small, alternate, sessile, or blades narrowed proximally to subpetiolar bases; blades 3–5 mm long and to about 3 mm wide, elliptic, oblong, or obovate. Flowers tiny, in most of the leaf axils, with very short, thickish, scarcely evident stalks. Calyx lobes linear-oblong, apices attenuate. Corolla translucent, scarcely exceeding the calyx. Stamens not exserted. Seeds numerous, brown, obpyramidal, the summits flat and with circular outlines, surfaces finely pebbly. (*Centuculus minimus* L.)

Damp to wet sands and muds, often where water stands temporarily, pond and lake margins, wet clearings, ditches, alluvial outwash, low places in fallow fields. e. Can. to s. Fla., chiefly coastal plain southward, westward to Calif., northward in the interior to Ohio, Ill., Minn., Sask.; Mex.; Old World.

2. **Anagallis arvensis** L. SCARLET PIMPERNEL.

Low, glabrous plant, commonly with several branches from the base, these weakly erect or prostrate below, ascending distally, to 2–3 dm long. Leaves sessile, opposite (occasionally in whorls of 3), mostly ovate, occasionally lance-ovate or obovate, 5–20 mm long, apically acute or obtuse, punctate. Flower stalks slender, to 2.5 cm long, mostly exceeding the leaves, recurving in fruit. Calyx lobes lance-ovate, short-attenuate apically, somewhat

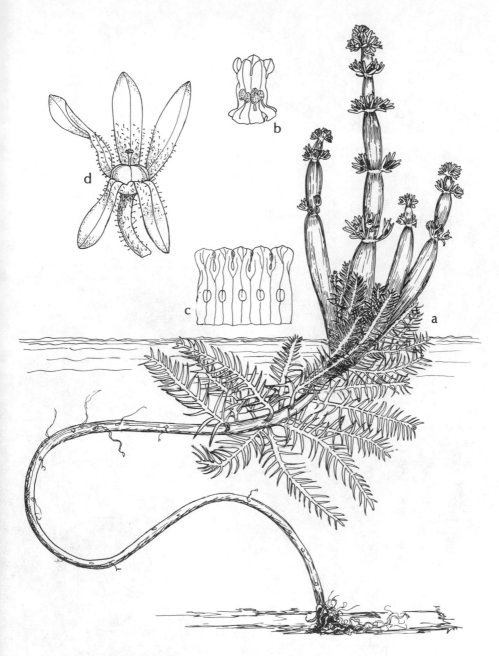

Fig. 231. **Hottonia inflata:** a, habit; b, corolla, diagrammatic to show stamens within; c, corolla, opened out; d, fruit with subtending calyx. (From Correll and Correll. 1972)

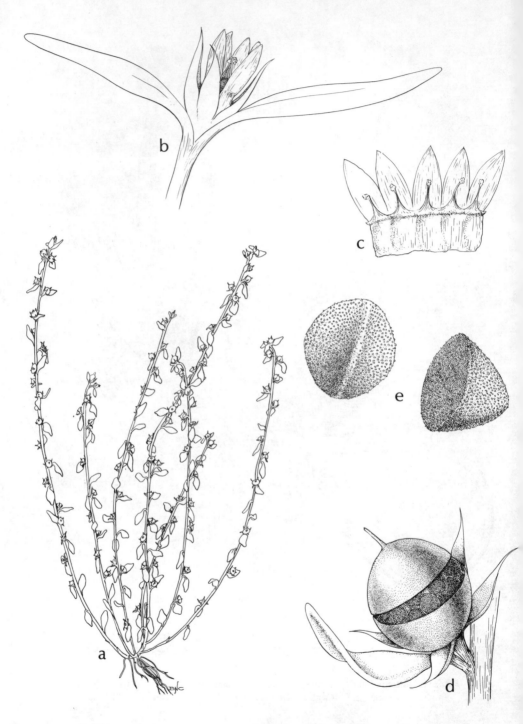

Fig. 232. **Anagallis minima:** a, habit; b, node, with flower; c, corolla, opened out to show insertion of stamens; d, capsule, to show circumscissile dehiscence; e, seeds.

492

keeled, margins hyaline, about 4 mm long. Corolla 5–15 mm across, lobes spreading in full light, closing in weak light, scarlet to salmon-colored or almost white, occasionally blue, obovate, their apices broadly rounded and usually fringed with minute stalked glands. Filaments pubescent. Seeds dark reddish brown, flat on one side, the other 2- or 3-angled, 1 mm across or a little more, with pale scurfy excrescences on the surface.

Usually in moist places, sand and gravel banks, fallow fields, lawns, ditches, depressions in prairies and flatlands. Native in Euras., widely but sporadically naturalized in much of temp. N.Am.

3. Anagallis pumila Sw.

Glabrous, usually much branched, branches commonly procumbent, to about 15 cm long. Some lower leaves usually opposite or subopposite, others alternate, sessile or short-stalked, elliptic, acute basally and apically, to 8 mm long, 4 mm wide. Flower stalks slender, to 6 mm long. Calyx lobes lanceolate, their tips attenuate, 2–3 mm long. Corolla greenish white. Filaments pubescent. Seeds brown, 3-angled, minutely rugose. (*Micropyxis pumila* (Sw.) Duby)

Low pinelands and low, open, moist to wet grounds. s. Fla.; pantropical.

3. Samolus

Caulescent or subscapose perennial herbs. Stems simple or with ascending branches. Leaves alternate, entire, the larger with bases tapering into straplike petioles; basal leaves commonly in a rosettelike cluster. Inflorescence a raceme or of branched racemes, the branches elongate-ascending, flower stalks wiry. Flowers 5-merous. Calyx segments triangular, persistent. Corolla white or pinkish, the lobes spreading and about equaling the tube. Stamens 5, not exserted, sometimes with alternating staminodia opposite the sinuses of the corolla lobes; filaments adnate to the corolla tube below, only short upper portions free. Ovary partly inferior. Capsule subglobose, 5-valved, the valves thick, the style persistent. Seeds numerous, angular.

- Flower stalks with small bracts about midway of their length; staminodia present; corolla white, the bases of the lobes glabrous within.   1. *S. parviflorus*
- Flower stalks without bracts; staminodia none; corolla pinkish, the bases of the lobes pubescent within.   2. *S. ebracteatus*

1. Samolus parviflorus Raf. WATER PIMPERNEL.   Fig. 233

Glabrous throughout, mostly 1–4 dm tall. Stems usually with rosettelike clusters of basal leaves. Leaves oblong-spatulate, the lower narrowed to winglike petioles, the upper sessile. Raceme simple or usually branched, the branches strongly ascending, bearing flowers from near the branch bases, flower stalks each with a small bract more or less midway of its length, ascending in flower or fruit, or spreading to recurved in fruit, the fruiting axis commonly (but not always) zigzagging. Calyx segments short-triangular, about as broad as long, shorter than the tube. Corolla white, the lobes oblong-obovate. Staminodia alternating with the stamens. Style short, stout. Seeds tiny, brown, flat at their summits, angled-tapering to their bases, the surfaces somewhat granular. (*S. floribundus* HBK.)

Shallow water and wet soil, stream banks, fresh-water shores, swamps, springy places, wet woodlands, ditches. Across s. Can., southward more or less throughout the U.S.; Mex., C.Am., W.I.; in S.Am. from Bol. and s. Brazil southward.

2. Samolus ebracteatus HBK.   Fig. 234

Glabrous below, more or less glandular-pubescent on the raceme axes and flower stalks. Rosettelike clusters of basal leaves present but these sometimes withered or shed by the time of flowering. Leaves broadly to narrowly oblong-spatulate to obovate, the lower narrowed to winglike petioles; foliage grayish green, both the stem and leaves usually somewhat suffused with reddish purple pigment. Stalks of racemes not leafy-bracted and not bearing flowers on

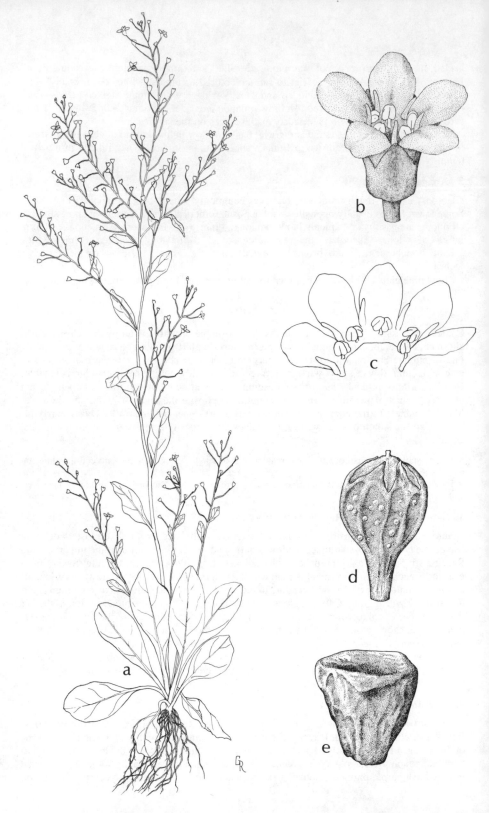

Fig. 233. **Samolus parviflorus:** a, habit; b, flower; c, corolla spread open to show insertion of stamens; d, capsule; e, seed.

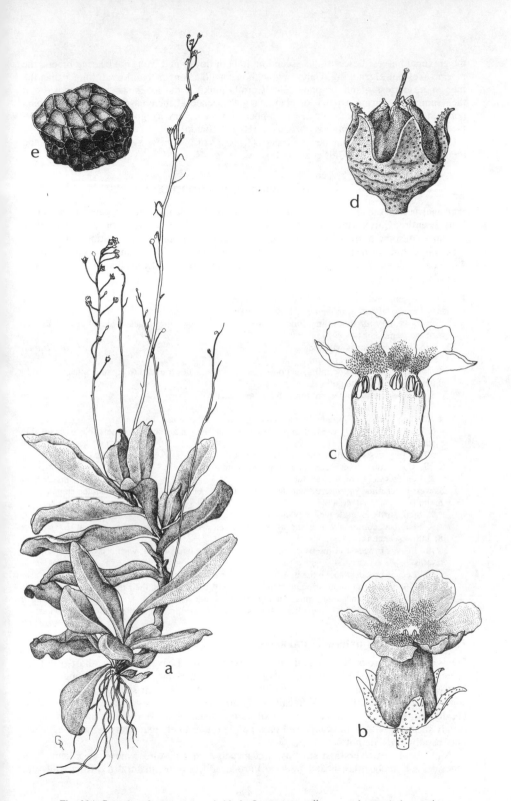

Fig. 234. **Samolus ebracteatus:** a, habit; b, flower; c, corolla, cut and spread; d, capsule; e, seed.

the proximal halves; flower stalks ascending in both flower and fruit, not bearing bracts, the raceme axes not zigzagging. Calyx segments triangular, longer than broad, longer than the tube, more or less glandular-pubescent. Corolla pinkish, the lobes short-oblong, wavy at their summits, bearing a patch of glandular pubescence at their bases within. Staminodia none. Style elongate. Seeds tiny, deep amber to brown, very irregularly angular, their surfaces reticulate, dull. (*Samodia ebracteata* (HBK.) Baudo)

Wet sands, brackish marshes, on lime-rock, edges of mangrove swamps. s. pen. Fla. and the Fla. Keys, along the Gulf coast to Tex., w. Okla., Nev.; Mex., W.I.

4. Lysimachia (LOOSESTRIFES)

Perennial herbs (ours). Leaves simple, opposite or whorled, glandular-punctate in some, margins entire. Calyx with a short tube and 5 or 6 lobes. Corolla 5- or 6-lobed, (ours) yellow to orange, in some purple-dotted. Stamens 5 (in some with as many alternating staminodia), filaments distinct or nearly so on a ring at the base of the corolla, or united basally. Capsule 5-valved, ovoid or subglobose, the style persistent on one valve. Seeds few to many, oblong, orbicular, or angled.

1. Stems creeping and trailing; leaf blades orbicular or nearly so.   1. *L. nummularia*
1. Stems erect, or if reclining then the leaves not round.
  2. Flowers axillary, *or* in terminal inflorescences with the subtending bracteal leaves essentially like those of the stem.
    3. Principal stem leaves whorled.   2. *L. quadrifolia*
    3. Principal stem leaves opposite.
      4. Median stem leaves definitely petiolate, bases of the blades rounded to broadly cuneate, sharply differentiated from the petioles.
        5. Petioles ciliate-pubescent throughout their lengths.   3. *L. ciliata*
        5. Petioles glabrous.   4. *L. radicans*
      4. Median stem leaves gradually tapered to sessile or subpetiolate bases.
        6. Leaves very finely toothed or at least rough on their flat margins (observable with suitable magnification).
          7. Calyx lobes faintly veined, acute apically.   5a. *L. lanceolata* var. *lanceolata*
          7. Calyx lobes with 3 conspicuous veins, acuminate apically.   5b. *L. lanceolata* var. *hybrida*
        6. Leaves smooth on the revolute margins.   6. *L. quadriflora*
  2. Flowers in terminal racemes or panicles, the subtending bracts much smaller than the stem leaves.
    8. Inflorescence paniculate.
      9. Stem stiffly erect; leaves in whorls of 3–5.   7. *L. fraseri*
      9. Stem lax, commonly reclining when fully developed; leaves opposite.   4. *L. radicans*
    8. Inflorescence racemose.
      10. Leaves broadest at the base, median ones ovate or lance-ovate, with 3 prominent principal veins.   8. *L. asperulaefolia*
      10. Leaves broadest above the base, lanceolate to linear, with one prominent nerve.
        11. Sepals with minute stipitate glands externally, especially on their margins; leaves linear to very narrowly lanceolate, mostly not exceeding 4 mm wide.   9. *L. loomisii*
        11. Sepals without stipitate glands; leaves lanceolate to elliptic, mostly 8 mm wide or wider.   10. *L. terrestris*

1. Lysimachia nummularia L. CREEPING-CHARLIE, MONEYWORT.

Stems repent, glabrous, rooting at the nodes, mat-forming. Leaves opposite, short-petiolate, 1.5–3.5 cm long, suborbicular, unevenly speckled with dark purple, round to elongate dots. Flowers solitary in 2 to several consecutive pairs of leaf axils, their stalks about equaling to somewhat longer than the leaves. Sepals barely united, 5–8 mm long, ovate, slightly cordate basally. Corolla 2–3 cm broad, lobes obovate or oblanceolate. Both calyx and corolla unevenly speckled with both orange and dark purple round to elongate dots. Filaments united only basally. Capsule not seen.

Moist banks of rivers and streams, wet meadows and pastures, moist woodlands, wet roadsides and ditches, lawns and gardens. Introduced from Eur., naturalized from Nfld. and

N.S. to Wisc., generally southward to n. Ga. in the East and Mo. and Kan. in the Midwest; Pacific states.

2. Lysimachia quadrifolia L.

Rhizomatous. Stems erect, 3–7 dm tall, glabrous or loosely pubescent with septate hairs. Leaves on about the lower ⅓ of the stem scalelike; other leaves sessile, commonly in whorls of 4 or 5, occasionally opposite, sometimes up to 7 in a whorl, the smallest just above the scale leaves, gradually increasing in size to somewhat above midstem, then diminishing gradually distally; lanceolate, lance-ovate, elliptic, to ovate, usually tapering basally, acute to acuminate apically, varying considerably in their dimensions from plant to plant, rather evenly dark purple–punctate, glabrous or pubescent. Flowers axillary on about the upper half of the plant, their stalks very slender, 1.5–6 cm long, usually about ⅔ as long as the leaves. Sepals united only at base, lobes linear-attenuate, 2–5 mm long, dotted and streaked with dark purple, their margins usually with stipitate-glandular hairs, sometimes a patch of such hairs near the base externally. Corolla united only at base, lobes oblong to ovate-oblong, 5–6 mm long, with a few scattered dark purple streaks of unequal size, a purple blotch at their bases. Filaments united below where densely yellow-glandular, free portions of unequal lengths, anthers purple. Capsule subglobose, 2.5–3.5 mm across, with few rather large dark purple punctations distally. Seeds few, dark brown, 1.5–2 mm long, their surfaces finely pitted.

Usually in well-drained sites, mesic to dry semiopen woodlands, thickets, open banks, sometimes on moist stream banks, occasionally in bottomland woods. Maine to Wisc., generally southward to S.C., n. Ga., n. Ala., Tenn., and Ill.

3. Lysimachia ciliata L.

Slenderly rhizomatous. Stems simple or branched above, glabrous below, sometimes glandular-puberulent above. Leaves opposite, petiolate, petioles markedly ciliate-pubescent, blades ovate to lance-ovate, bases broadly rounded, subcordate, or broadly short-tapered, apices acute to acuminate, rarely obtuse, 5–15 cm long, median leaves the largest. Flowers solitary in the leaf axils, their slender stalks 1.5–4.5 cm long, bracteal leaves often small but still essentially leafy. Calyx tubular at base, lobes 3.5–8 mm long, widest at base, tapering to their extremities, margins entire. Corolla yellow, not spotted, about 2 cm across, tube short, lobes obovate, mucronate apically, yellow glandular-dotted within the tube and proximally on the lobes. Filaments free, anthers yellow. Capsule globose to ovoid, 3.5–5.5 mm across. Seeds numerous, mahogany-colored, 3-angled, surfaces reticulate. (*Steironema ciliatum* (L.) Raf.)

Floodplain forests, marshy borders of streams and ponds, stream banks, wet depressions in prairies, moist to wet woodlands, swamps, thickets, open embankments. Across s. Can., Wash., generally southward to the Fla. Panhandle and Ariz.

4. Lysimachia radicans Hook.   Fig. 235

Plant with a short, erect, relatively slender caudex. Stem slender, lax, commonly arching-reclining, and rooting at distal nodes, laxly branched. Lowermost leaves not much smaller than (or equaling) the median leaves; leaves mostly petiolate, petioles usually few-ciliate at the base, bracteal leaves sometimes sessile; blades ovate to lanceolate, usually rounded at the base, sometimes tapering, apices acute, 5–9 cm long, margins entire or finely papillate. Flowers solitary, their very slender stalks mostly 1–2.5 cm long, the subtending bracteal leaves smaller than the principal ones but still essentially leafy. Calyx united only at the base, lobes lanceolate, acuminate apically, 3–4 mm long. Corolla yellow, not spotted, about 1 cm across, united basally, the tube yellow glandular-dotted within, lobes obovate, distally erose-dentate and mucronate. Filaments free, anthers yellow. Capsule subglobose, about 3 mm across. Seeds several, mahogany-colored, angled, surface minutely reticulate. (*Steironema radicans* (Hook.) Gray)

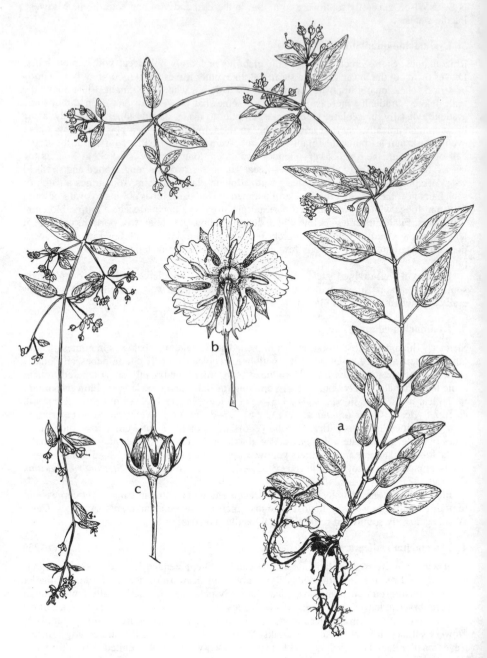

Fig. 235. **Lysimachia radicans:** a, habit; b, flower; c, fruit. (From Correll and Correll. 1972)

Floodplain forests, swamps, wet woodlands, shaded stream banks. Mo., Ky., generally southward to Miss., Ala., and s.e. Tex.; also e. Va.

5a. Lysimachia lanceolata Walt. var. lanceolata. Fig. 236

Slenderly rhizomatous. Stems weakly erect or reclining, to about 7 dm long, simple to much branched. Lowermost leaves usually definitely petiolate, petioles to as long as the blades, ciliate at the base or to about their middles, blades elliptic to nearly orbicular. Median and upper leaves commonly sessile, cuneate at the base, sometimes narrowed to a subpetiolar base; very variable in shape and dimensions, lanceolate, oblanceolate, narrowly elliptic, to linear. Flowers solitary or paired in the upper leaf axils, sometimes the stems diffusely short-branched and heavily floriferous; flower stalks slender, 2–4.5 cm long. Calyx united only at the base, lobes broadest at the base, tapering to their acute apices, mostly 3–5 mm long, usually glabrous, the midnerve indistinct, lateral nerves not usually evident. Corolla 1–1.5 cm across, yellow, not spotted, a purplish ring usually at the base of the tube, glandular-dotted on the tube within and proximally on the lobes, lobes obovate to suborbicular, erose-dentate apically. Filaments free, anthers yellow. Capsule subglobose, 3–4.5 mm across. Seeds several, angular, 1.2–2 mm long, dark brown. (*Steironema lanceolatum* (Walt.) Gray; incl. *S. heterophyllum* (Michx.) Raf.)

Wet banks, on rocks in streams, boggy meadows, wet ledges and cliffs, moist to wet woodlands, sometimes in mesic woodlands. Pa. to Wisc. and Iowa, southward to Fla. Pan-handle and Okla.

5b. Lysimachia lanceolata var. hybrida (Michx.) Gray.

Differing from var. *lanceolata* in its greater stature, to about 1 m tall, more rigid and erect stems, longer internodes, rather loose and wide branching. Sepals markedly acuminate api-cally, with 3 distinct, nearly parallel, principal nerves. (*L. hybrida* Michx., *Steironema hybridum* (Michx.) Raf.)

Swamps, meadows, sloughs, pond margins, thickets, wet prairies, ditches. Que. to Minn., N.D., Sask., Alb., Wash., generally southward to n. Fla. and Ariz.

6. Lysimachia quadriflora Sims.

Slenderly rhizomatous. Stem stiffly erect, 2–8 dm tall, shortly branched distally, glabrous. Lowermost leaves (usually shed by time of flowering) petiolate, blades mostly elliptical or oblong; median leaves sessile, linear, firm, margins smooth, distinctly and tightly revolute. Flowers solitary in the upper leaf axils, stalks slender, 2–3 cm long. Calyx united only at base, lobes lanceolate or lance-elliptic, acuminate apically, about 5 mm long. Corolla yellow, not spotted, 1.5–2 cm across, yellow glandular-dotted on the tube within and proximally on the lobes, lobes oblong to obovate, edges somewhat crinkled, apices toothed and often stipi-tate-glandular, conspicuously mucronate. Filaments free, anthers yellow. Capsule short-ovate or subglobose, 3.5–4 mm across. Seeds several, dark brown, with large deep pits separated by high narrow ridges. (*Steironema quadriflorum* (Sims) Hitchc.)

Calcareous bogs, swales, shores, stream banks, prairies, boggy meadows, springy places. Mass. to Man., southward to n.w. Ga., n. Ala. and Ark.

7. Lysimachia fraseri Duby.

Relatively stout, slenderly rhizomatous, rhizomes deeply seated in the substrate. Stems rigidly erect, 1–1.5 m tall, simple, or with few branches, glandular-pubescent above. Lowermost leaves scalelike; other leaves smallest below and above, largest medially, in whorls of 3–5, short-petiolate, blades mostly elliptic or oval, sometimes oblanceolate-obovate, cuneate basally, acute to short-acuminate apically, the larger ones 6–15 cm long, 2–6 cm wide, punctate with round to elongate dark purple or nearly black dots, much paler beneath than above, margins finely stipitate-glandular. Inflorescence paniculate, flower stalks abundantly stipitate-glandular. Calyx tube about 5 mm long, lobes linear-attenuate, 4–5 mm long, edged with a prominent maroon band which is stipitate-glandular on its edge. Corolla

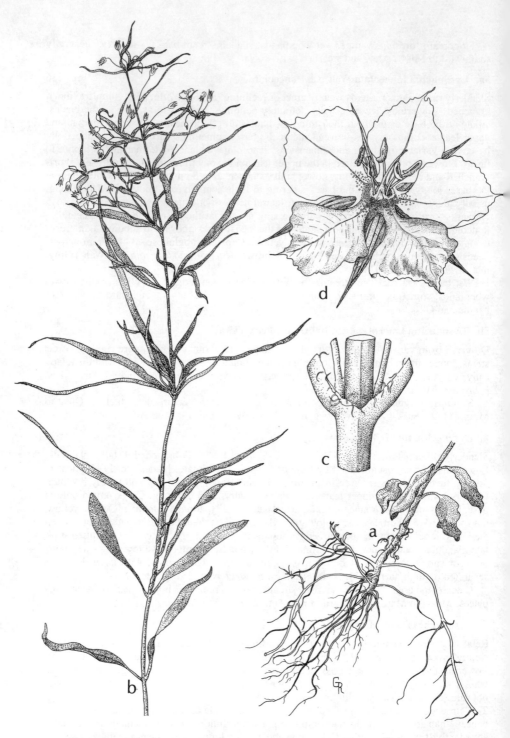

Fig. 236. **Lysimachia lanceolata** var. **lanceolata:** a, base of plant; b, flowering stem; c, node; d, flower.

yellow, not spotted, about 1.5 cm across, lobes elliptic to oblong, distally narrowed and with 1 or more largish teeth. Filaments united ⅓–½ their length, free portions of unequal length, anthers yellow. Capsule globose 3–4 mm across. Seeds several, angular, dark brown, surface finely pitted.

Meadows, stream banks and flats along streams, moist roadside banks, moist pastures. s. Appalachian mts., N.C., Tenn., n. Ga., and n. Ala.

8. Lysimachia asperulaefolia Poir. in Lam.

Stems slender, stiffly erect, 3–6 dm tall, simple or branched above, glandular-pubescent above. Lowermost leaves scalelike; other leaves sessile, in whorls of 3 or 4, with 3 distinct, palmate, principal veins, medial ones narrowly ovate, 2–4 cm long, stiffish, glaucous beneath, widest at the base, acute apically, margins revolute, upper surface with whitish punctate dots. Flowers in a terminal raceme 3–10 cm long, their stipitate-glandular stalks 5–10 mm long. Calyx glandular-pubescent, united only at the base, lobes linear-attenuate, 3–5 mm long. Corolla yellow, about 1.5 cm across, sometimes with reddish dots or streaks, lobes elliptic-lanceolate, acute apically, stipitate-glandular on both surfaces and the margins. Filaments united below. Capsule globose, 3–4 mm across. Seeds several, angular, somewhat winged, pale buff or gray.

Savannas and shrub bogs. Coastal plain, N.C., and S.C.

9. Lysimachia loomisii Torr.

Slenderly rhizomatous. Stems stiffly erect, 3–6 dm tall, with numerous leafy branches above, short branches in the leaf axils medially, mostly glandular-pubescent above. Leaves scalelike on about the lower ⅓ of the stem; medial and upper leaves opposite or subopposite, sometimes in whorls of 4, tapered to both extremities but essentially without petioles, linear to narrowly oblanceolate, punctate-dotted, 3–5 cm long, margins strongly revolute. Flowers in a terminal raceme 3–10 cm long, their usually smooth stalks about 1 cm long. Calyx tubular only at base, lobes 2–3 mm long, linear-oblong, acute to blunt apically, margins very finely stipitate-glandular, surface with elongate dark purple to black streaks in lines, sometimes the lines unbroken. Corolla yellow, about 1 cm across, lobes oblong-obovate, with several elongate maroon streaks and a ring of purplish color at the throat, yellow glandular-dotted proximally. Filaments united below, the tube and unequal free portions copiously glandular-dotted, anthers lavender. Capsule globose, about 3 mm across. Seeds few, angular, light brown, prominently pitted.

Savannas and shrub bogs. Coastal plain, N.C. and S.C. (one old record from Ga.)

L. loomisii is said to hybridize occasionally with *L. quadrifolia*.

10. Lysimachia terrestris (L.) BSP.

Relatively stoutly rhizomatous. Stems erect, 3–10 dm tall, simple or loosely branched above, without small fasciclelike branches in the medial leaf axils. Leaves of the lower stem scalelike; other leaves largest medially, sessile, opposite, the medial tapered to both extremities, mostly acute apically, lanceolate to narrowly elliptic, 3–10 cm long, 7–20 mm wide, margins flat to obscurely revolute, surfaces punctate with dark purplish dots and short streaks, somewhat glaucous beneath. Flowers in racemes 8–15 cm long, terminal on the main stem, sometimes on the branches, lowermost portions of the racemes sometimes leafy-bracted, the axis and slender flower stalks glabrous. Calyx united only at the base, lobes variable in shape, some widest basally and tapering to the summit, some lanceolate, some ovate, with or without elongate purple streaks, glabrous. Corolla about 1.5 cm across, yellow, within maroon at the throat overlaid with yellow glandular dots, lobes more or less oblong-elliptic, blunt apically, streaked with maroon. Filaments united at base, the tube and somewhat unequal free portions glandular-pubescent, anthers lavender-purple. Seeds few, angular to oval, with a tawny bloom overlaying a shiny black surface.

Swamps, marshy shores of ponds, lakes, streams, bogs, wet thickets. Nfld. and Que. to Man., generally southward to S.C., Tenn., Ill., Minn.

Lysimachia × *producta* (Gray) Fern. is a not infrequently occurring hybrid between *L. quadrifolia* and *L. terrestris* and has intermediate characteristics. It may be encountered where the two reputed parents occur together.

Myrsinaceae (MYRSINE FAMILY)

Myrsine guianensis (Aubl.) Kuntze. MYRSINE. Fig. 237

Evergreen shrub or small tree. Leaves simple, alternate, without stipules, dark glossy green above, dull and paler green beneath, glabrous, margins entire, mostly oblong-obovate to elliptic, 4–15 cm long, cuneate below to short-petiolate or subsessile bases, obtuse to rounded apically. Flowers small, bisexual, or unisexual and the plants polygamodioecious, borne in umbellate clusters on closely set, very short, thick-scaly, spur shoots from wood of the previous season, each flower with a thickish short stalk (elongating to about 4 mm in fruit). Calyx united below, with 5 short-deltoid or ovate-deltoid lobes. Corolla white, united below, the 5 oblong lobes spreading, blunt or rounded apically, considerably surpassing the calyx. Stamens 5, inserted on the corolla opposite and just below the corolla lobes, filaments very short, anthers about ⅔ as long as the corolla lobes or a little more. Pistil 1, style very short, stigma shortly lobed, persistent and forming a short apicule at the summit of the fruit. Fruit a subglobose, dryish, green to brown, 1-seeded drupe 3–4 mm in diameter, the calyx persistent about its base. (*Rapanea guianensis* Aubl.)

Well-drained to wet hammocks, scrub, cypress swamps, borders of mangrove swamps. cen. pen. to s. Fla. and Fla. Keys; trop. Am.

Ebenaceae (EBONY FAMILY)

Diospyros virginiana L. PERSIMMON. Fig. 238

Tree, reaching medium size, in open areas sometimes spreading by underground parts and thicket forming. Terminal buds frequently abort following which shoots develop from axillary buds, resulting in a bushy or spraylike branch pattern. Bark of older trunks brown to blackish, roughly broken into small blocks and cross fissures. Wood hard. Pith of the twigs homogeneous through most, if not all, of the first year, in year-old twigs becoming transversely diaphragmed, the diaphragms very thin, chambered between the diaphragms. Leaves deciduous, simple alternate, without stipules, short-petiolate; blades entire, broadly lanceolate, ovate-oblong, or widely elliptic, mostly 7–15 cm long and 3–7 cm wide, bases mostly rounded, sometimes broadly cuneate, apices usually short-acuminate, usually glabrous but sometimes pubescent beneath, upper surface dark green, often with blackish blemishes, the lower pale and somewhat whitish. Flowers inconspicuous, functionally unisexual, the plants functionally dioecious. Staminate flowers smaller than the pistillate, in few-flowered axillary clusters; pistillate solitary in the leaf axils. Calyx united below, 4- or 5-lobed above. Corolla 1–1.5 cm long, very much exceeding the calyx, greenish yellow, campanulate-urceolate, with 4 or 5 short, nearly triangular lobes at the summit. Staminate flowers usually with 16 fertile stamens, the pistillate with 8 sterile stamens. Ovary superior, 8-locular, styles 4, each 2-lobed at the summit. Fruit a several-seeded globose or depressed-globose berry to 4–5 cm in diameter, an enlarged persistent calyx at its base, in ripening its color turning from green to yellow, then orange, surface glaucous. Pulp of the fruit very astringent during maturation, becoming sweet and with a distinctively delicious flavor when fully ripe, the palatability varying, however, from tree to tree.

Inhabiting a wide range of sites of varying moisture conditions. Fields, upland woodlands of various mixtures, river bottomlands, pine flatwoods. s. N.Eng., e. N.Y., Pa., W.Va., Ohio, Ind., Ill., s.e. Iowa, e. Kan., generally southward to Fla. and e. Tex.

Fig. 237. **Myrsine guianensis:** a, fruiting branch; b, flower cluster; c, staminate flower; d, pistillate flower; e, fruiting cluster.

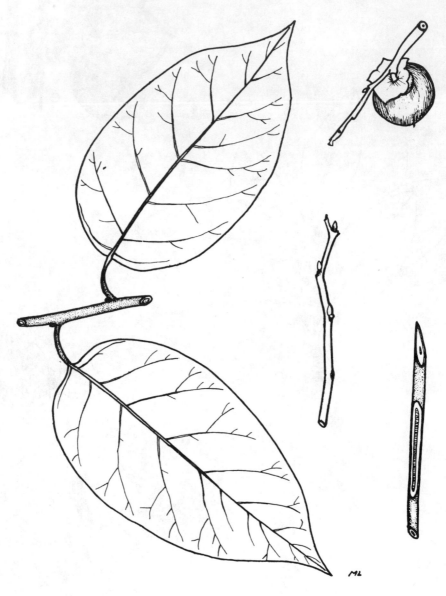

Fig. 238. **Diospyros virginiana.** (From Kurz and Godfrey, *Trees of Northern Florida.*
1962)

Sapotaceae (SAPODILLA FAMILY)

Bumelia (BUCKTHORNS)

Shrubs or small trees, sap milky, usually some of the branches with sharp thorns, the thorns sometimes naked, sometimes leafy branch-thorns. Leaves simple, sessile or short-petiolate, tardily deciduous, alternate—obviously alternate on elongated shoots resulting from rapid growth, approximate and appearing fascicled on stubby spur-shoots. Irregular and alternating periods of slow-growing spur-shoot and accelerated long-branch production yielding a distinctive crooked and rough branch system commonly inhabited by lichens, liverworts, and mosses. Leaves, under low magnification, exhibiting a marked reticulation of pale, raised veinlets. Flowers small, borne in umbels, mostly on the spurs; radially symmetrical, bisexual. Calyx united only basally, 5-lobed, persistent. Corolla united basally, 5-lobed, (in ours) each lobe with a lateral appendage on each side near the base. Stamens 5, opposite the corolla lobes, 5 petaloid staminodia alternate with the corolla lobes, both inserted on the corolla tube. Pistil 1, ovary superior, 5-locular, a single ovule in each locule. Fruit a drupelike berry.

- Larger leaves about 5 cm long, sometimes all somewhat shorter than 5 cm; shrub usually not exceeding 3 m tall; berry 5–7 mm long and about 5 mm wide.   1. *B. reclinata*
- Larger leaves 8–12 cm long; slender shrub or small tree, sometimes reaching 10 m tall; berry 10–12 mm long and 8 mm wide or a little more.   2. *B. lycioides*

1. Bumelia reclinata Vent.

Shrub, usually not over 3 m tall, often much less. Leaves oblanceolate, spatulate, or obovate, cuneately narrowed below to a short petiole, apices obtuse to rounded, sometimes emarginate, sparsely pubescent beneath when young, becoming glabrous, 1–5 cm long and 1–2 cm broad. Berry 5–7 mm long and about 5 mm broad. (Incl. *B. microcarpa* Small)

Inhabiting both well-drained and wet sites, upland hammocks, low, wet hammocks, depressions in pine flatwoods, swamps, floodplain forests, wiregrass bogs, glades. s. Ga. and Fla.

2. Bumelia lycioides (L.) Pers.

Shrub or small tree, sometimes attaining 20 m in height. Leaves elliptic, elliptic-oblanceolate, oblanceolate, rarely lanceolate, cuneately narrowed below to a short petiole, apically very variable, from acute to rounded, sometimes emarginate, the larger on a given plant 8–12 cm long, upper surfaces glabrous, the lower sometimes pubescent along the midrib. Berry 10–12 mm long and 8 mm wide or a little more.

Floodplain forests, often on natural levees, thickets, both low and well-drained woodlands. s.e. Va. to Fla. Panhandle, westward to s.e. Tex., northward in the interior to s. Mo., Tenn., s. Ind. and Ill.

Symplocaceae (SWEETLEAF FAMILY)

Symplocos tinctoria (L.) L'Her. COMMON SWEETLEAF, HORSE-SUGAR.   Fig. 239

A shrub or small tree, the leaves sometimes deciduous in autumn, commonly most of them persisting through the winter and shed in spring just before or during the flowering period. Younger twigs sparsely pubescent, grayish; older twigs brown but more or less covered with a waxy ash-colored bloom, the bloom eventually shed and the twigs brown, somewhat furrowed; pith chambered. Leaves simple, alternate, short-petioled, blades mostly elliptic, sometimes oblong or oblanceolate, cuneate at base, short-acuminate or acute apically, 5–15 cm long, margins entire or shallowly toothed, glabrous or pubescent above, moderately

Fig. 239. **Symplocos tinctoria:** a, flowering branch; b, flower; c, flower, another view; d, fruiting branch; e, fruit. (From Correll and Correll. 1972)

pubescent beneath, dull green, becoming yellowish green late in the season. Tissues of old leaves, especially near their midribs, sweet to the taste; leaves relished by browsing animals. Flowers yellow or creamy yellow, fragrant, close-set in sessile, almost ball-like clusters on twigs of the previous season before new growth commences. Ovary half-inferior (at anthesis, wholly inferior in the fruit), the floral tube surmounted by 5 very small, deltoid, persistent calyx segments. Petals 5, united at the base, the lobes oblanceolate or spatulate, 6–8 mm long. Stamens numerous, in fascicles inserted at the base of each of the corolla lobes, filaments free and variously united basally. Fruit an oblong, green drupe about 1 cm long, with small, erect, deltoid calyx segments at its summit.

Rich upland woodlands, moist ravines, floodplain forests, bottomland woodlands, palmetto marsh areas where seasonally wet, stream banks. Del. to Tenn., Ark. and s.e. Okla., generally southward to n. Fla. and e. Tex.

Styracaceae (STORAX FAMILY)

Deciduous shrubs or trees, various of their parts usually more or less pubescent with stellate hairs or peltate scales. Axillary buds superimposed, the lowest hidden by the petiole base prior to leaf-fall. Leaves alternate, simple, pinnately veined, without stipules. Flowers bisexual, radially symmetrical, floral tube with 4 or 5 small sepals (these sometimes lacking), the floral tube wholly or partially adnate to the ovary. Corolla united at least basally, usually 4- or 5-lobed, sometimes more. Stamens usually twice as many as the lobes of the corolla, filaments more or less united basally and inserted in a single series on the base of the corolla. Pistil 1, ovary 3–5-locular, ovules 1 to few in each locule, style slender, stigma terminal, usually minutely 3–5-lobed. Fruit (in ours) dry and indehiscent, or a capsule.

• Uppermost axillary bud thumblike in shape; flowers and fruits on short shoots of the current season; corolla lobes 5, rarely 6 or 7; fruit a 3-valved, irregularly dehiscent, subglobose capsule, the floral tube adherent to about its lower third.   1. *Styrax*
• Uppermost axillary bud ovoid; flowers and fruits in fascicles or short racemes axillary to leaf scars on wood of the previous season; corolla lobes 4; fruit dry and indehiscent, longitudinally winged, beaked apically.   2. *Halesia*

1. Styrax

Styrax americana Lam. STORAX, MOCK-ORANGE.   Fig. 240

Commonly a shrub, occasionally reaching small-tree stature, in spring bearing handsome white flowers. Young twigs sparsely to densely stellate-pubescent. Uppermost axillary bud (best developed after midsummer), thumblike in shape. Petioles 3–5 mm long, sparsely to densely stellate-pubescent, blades varying not a little in shape and size, elliptic, oval, ovate, obovate, or smaller ones lanceolate, up to 8 cm long and 4 cm broad, bases cuneate, apices obtuse, acute, or short acuminate; margins entire, wavy, or distantly low-toothed; surfaces varying from glabrous to sparsely to densely stellate-pubescent. Flowers borne on the developing shoots of the season, sometimes solitary in leaf axils, more frequently 2–5 semidrooping flowers on short lateral or terminal leafy shoots; flower stalks 2–8 mm long, lengthening as the fruits mature to 10–14 mm, sparsely to copiously stellate-pubescent as are the floral tubes and calyces. Sepals usually 5, small, surmounting the floral tube, considerably broader than long, broadly rounded but usually abruptly contracted to a mucro or short-subulate tip. Corolla white, united only basally, lobes usually 5, 8–12 mm long, oblong-acute or -acuminate, spreading or recurved, stellate-pubescent on both surfaces. Capsule subglobose, the floral tube usually adnate to its lower third, 6–8 mm in diameter, stellate-pubescent, usually filled with a single brown seed. (Incl. *S. pulverulenta* Michx., *S. americana* var. *pulverulenta* (Michx.) Rehd.)

507

Fig. 240. **Styrax americana:** a, flowering branch; b, flower; c, fruiting branch; d, fruit.
(From Correll and Correll. 1972)

Commonly in places where shallow surface water stands much or all of the time, floodplain forests, swamps, wet woodlands, wet shoreline thickets, cypress-gum depressions. s.e. Va. to s. pen. Fla., westward to e. Tex., northward in the interior, s.e. Okla. to s.e. Mo. and s. Ohio.

2. Halesia (SILVERBELLS)

Halesia diptera Ellis.   Fig. 241

A shrub or tree to about 10 m tall. Young twigs sparsely to densely stellate-pubescent. Uppermost axillary bud (best developed after midseason) ovoid. Petioles 5–15 mm long, blades variable in shape and size, broadly oval, obovate, suborbicular, or ovate, bases mostly rounded, varying to broadly short-tapered, apices abruptly short-acuminate to mucronate; margins irregularly and inconspicuously dentate-serrate; upper surface usually at least sparsely stellate-pubescent when young, sometimes but not always becoming glabrous in age, lower varying from soft or velvety stellate-pubescent to sparsely pubescent, often but not always glabrous in age. Flowers showy, pendent, in fascicles or short racemes of 2 or 3 to 6 or 7 from axils of leaf scars on wood of the previous season, prior to or as the new leafy shoots of the season develop. Flower stalks slender, they, the floral tubes, and sepals densely stellate-pubescent, the floral tube wholly adnate to the ovary, thus the ovary wholly inferior; floral tube gradually becoming nearly glabrous as the fruit matures, the pubescent sepals usually persisting. Corolla white, 1.5–3 cm long, united only basally, lobes erect, oval to obovate. Fruit dry, indehiscent, oblanceolate to elliptic or oblong, rarely oval or suborbicular, broadly 2-winged, beaked apically, 2.5–5 cm long.

Floodplain woodlands, adjacent slopes, river bluffs, ravine slopes, upland, well-drained, mixed woodlands. Coastal plain, s.e. S.C. to n. Fla., westward to e. Tex., Ark.

Note: *H. parviflora* Michx. infrequently occurs in floodplain woodlands of coastal plain streams; it generally inhabits well-drained woodlands. Its leaves are mostly oblong-elliptic, corollas fused over half their length, the fruits clavate in outline, narrowly 4-winged.

For a very detailed and very interesting account of nomenclatural vicissitudes relevant to taxa of *Halesia*, see James L. Reveal and Margaret J. Seldin, "On the Identity of *Halesia carolina* L. (Styracaceae)," *Taxon* 25: 123–140 (1976).

Oleaceae (OLIVE FAMILY)

Trees or shrubs with opposite leaves without stipules. Flowers bisexual or unisexual, radially symmetrical. Calyx small (sometimes absent), tubular, with 4 (rarely more) minute lobes or none. Corolla absent in some, in others united at least at the base and 4-lobed. Stamens 2 (rarely 3–5). Ovary superior, 1- or 2-locular, usually with 2 ovules in each locule, style 1 or lacking, stigma capitate or 2-lobed.

Reference: Hardin, James W. "Studies of the Southeastern United States Flora. IV. Oleaceae." *Sida* 5: 274–285. 1974.

1. Leaves pinnately compound; fruit a samara.   1. *Fraxinus*
1. Leaves simple; fruit a drupe.
  2. Flowers or fruits borne in panicles terminating leafy branchlets of the season.   2. *Ligustrum*
  2. Flowers or fruits borne in scaly glomerules, cymes, or panicles axillary to leaf scars on woody twigs, or axillary to leaves.
    3. The flowers in short, scaly-bracted glomerules or on short stiff panicles; corolla absent, or if present, then with short, broad, obtuse lobes.
      4. Flowers with minute calyx or none; corolla none; inflorescence a tight axillary glomerule developing just before or after new shoot emergence in spring.   3. *Forestiera*
      4. Flowers with evident calyx; corolla present, with short, broad, obtuse, creamy white lobes; in-

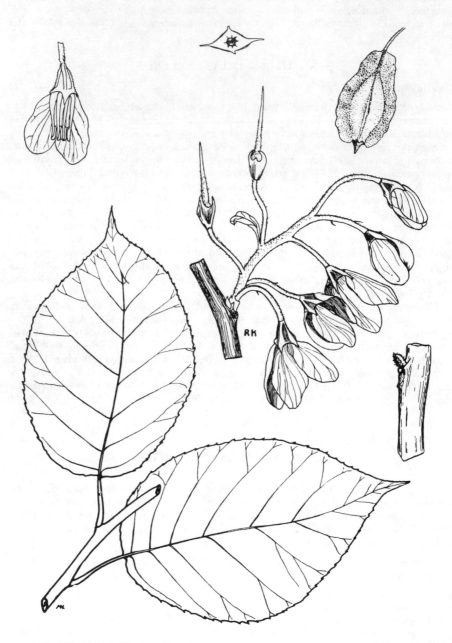

Fig. 241. **Halesia diptera.** (From Kurz and Godfrey, *Trees of Northern Florida*, 1962)

florescence a stiff, short panicle, developing partially in summer, overwintering thus, reaching full anthesis in spring.   4. *Osmanthus*

3. The flowers or fruits in large leafy-bracted, lax, drooping panicles; corollas with white or greenish white, linear lobes 1.5–3 cm long.   5. *Chionanthus*

1. Fraxinus (ASHES)

Deciduous trees with petiolate, 1-pinnately and odd-pinnately compound leaves. Flowers in dense fascicles, short compact racemes, or panicles, produced on old wood prior to the development of new shoots in the spring; unisexual (rarely bisexual), plants mostly dioecious. Calyx small, 4-parted, or none. Corolla of (2–) 4 (–6) petals, or absent. Stamens usually 2 in the staminate flowers. In the pistillate flowers ovary 2-locular, stigma 2-lobed. Fruit a 1- or 2-seeded samara.

1. Samaras with wings which extend to the base of the body or below; body flattish throughout.
     1. *F. caroliniana*
1. Samaras with wings which extend distally from the middle of the body or slightly above; body terete.
  2. Lateral leaflets mostly with a gradual narrowing at the base and a tendency to be narrowly decurrent along much of the stalk of the leaflet; stalks of the leaflet mostly 5–10 mm long. 2. *F. pennsylvanica*
  2. Lateral leaflets commonly oblique-rounded approximately at the base and decurrent only on the uppermost portion of the stalk of the leaflet; leaflet stalks very variable in length, 6–40 mm long.
     3. *F. profunda*

1. Fraxinus caroliniana Mill. POP ASH, CAROLINA ASH, WATER ASH.   Fig. 242

A small tree, commonly with several trunks. Unfolding branchlets commonly heavily pubescent, much or all of the pubescence later deciduous. Leaflets 5–7 (–9), rarely 3, very variable: lanceolate, ovate, oblong, oval, elliptical, or obovate, entire to variously toothed, usually narrowed and often somewhat oblique at the base, the apices broadly obtuse, acute, or acuminate; at maturity glabrous above, glabrous or sparsely short-pubescent beneath, the lower surfaces paler green than the upper, sometimes whitish. Samaras extremely variable in size and shape, commonly with very broad wings, these mostly decurrent to below the body; spatulate, linear-elliptical, broadly oblong or oval, ovate, or suborbicular, apically broadly rounded, obtuse, acute, sometimes shallowly notched. (Incl. *F. pauciflora* Nutt.)

Swamps, flatwoods depressions, wet shores, often with the bases inundated for long periods, even continuously. Chiefly coastal plain, Va. to Fla., westward to Ark. and Tex.

2. Fraxinus pennsylvanica Marsh. RED ASH, GREEN ASH.   Fig. 242

A large tree, the base buttressed if growing where inundated much of the time. Leaflets 5–9, commonly 7, their stalks mostly 5–10 mm long, the blades narrowly decurrent along much of their length; blades oblong-elliptical, ovate-lanceolate, lanceolate, rarely suborbicular, their bases gradually narrowed, the apices tapering, occasionally abruptly acuminate, entire or rarely toothed; lower leaf surfaces glabrous or pubescent. Samaras oblanceolate to spatulate, sometimes narrowly linear, sometimes notched apically, the wing usually less than 7 mm wide, the body less than 2 mm wide. (Incl. *F. darlingtonii* Britt., *F. smallii* Britt.)

River swamps, wet to moist woodlands, pineland depressions. Que. to Man., generally southward to n. Fla. and Tex.

3. Fraxinus profunda (Bush) Bush. PUMPKIN ASH.

A large tree, commonly with buttressed bases if inundated for long periods. Leaflets 5–9, commonly 7, their stalks very variable in length, 6–40 mm long, the blades decurrent only on the uppermost portion; blades ovate, ovate-oblong, or broadly elliptic, mostly broadly rounded at the base and abruptly narrowing to the short, broad wing on the upper portion of the stalk, commonly inequilateral, the apices broadly tapering or acuminate, margins entire or rarely toothed, pubescent beneath or becoming glabrous in age. Wings of the samaras narrowly decurrent to the middle of the body, elliptic, oblanceolate, oblong, to spatulate, the apices sometimes notched, the wing more than 7 mm wide, the body more than 2 mm wide. (*F. tomentosa* Michx. f.; incl. *F. michauxii* Britt.)

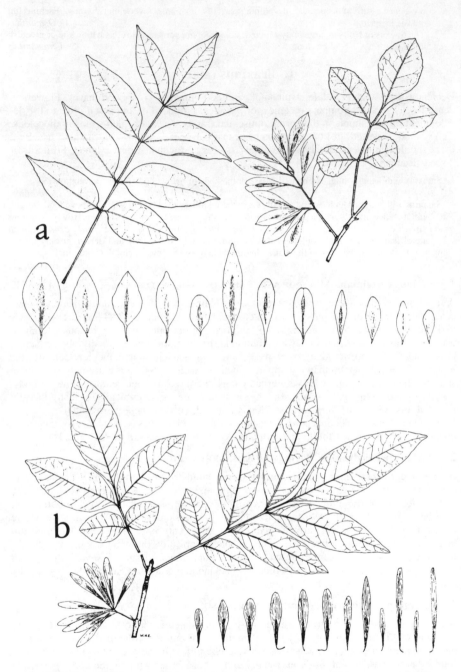

Fig. 242. a, **Fraxinus caroliniana;** b, **Fraxinus pennsylvanica.** (From Kurz and Godfrey, *Trees of Northern Florida*. 1962)

Swamps and wet woodlands. N.Y. to n. Fla., westward to La., northward in the interior to s. Ind. and Ill.

2. Ligustrum

Ligustrum sinense Lour. PRIVET.

Shrub or small tree. Twigs and branchlets densely short-pubescent. Leaves simple, short-petiolate, the petioles densely short-pubescent; blades mostly elliptic or oval, 2–3 cm long and 1–2 cm wide, apically obtuse or rounded, sometimes acute, glabrous above when mature, pubescent at least on the veins beneath. Inflorescences paniculate, terminating lateral pubescent branchlets of the season, their axes pubescent. Flowers white, with a somewhat disagreeable odor. Calyx persistent, cuplike, shallowly lobed at the summit. Corolla tube barely exceeding the calyx, the 4 lobes oblong to linear-oblong, about 3–4 mm long. Stamens exserted. Drupes dark blue or bluish black, ellipsoid to subglobose, mostly about 5 mm long.

Native of China, widely natrualized more or less throughout our range, fruits eaten by birds and their stones containing seeds dispersed by them. In both well-drained and poorly drained places, fence rows, wet thickets, bottomland woods. Plants of other species of privet are occasionally naturalized in our range; *L. sinense* is by far the most abundant, especially in wet places.

3. Forestiera

Forestiera acuminata (Michx.) Poir. in Lam. SWAMP-PRIVET.   Fig. 243

A shrub or small tree, its principal stems usually weak and leaning. Leaves deciduous, with petioles 1–2 cm long, blades commonly more or less rhombic, varying to oval or ovate, 4–8 cm long and 1–3 cm broad, cuneate at their bases, acuminate apically, margins usually shallowly, often indistinctly, serrate or crenate, at least on their upper halves (some leaves sometimes entire), surfaces usually glabrous, occasionally pubescent on the midrib above. Flowers mostly unisexual, the plants dioecious, appearing before the leaves, the staminate in dense, nearly sessile, lateral fascicles, the pistillate in short panicles, both from the axils of leaf scars. Calyx minute, 4-parted, quickly deciduous, or absent. Corolla none. In the staminate flowers, stamens 1–4, sometimes a rudimentary pistil, without a style, present. Ovary 2-locular, with 2 ovules in each locule. Fruit a drupe, commonly only 1 seed maturing, 7–15 mm long, more than twice as long as broad, oblong, oval, or elliptic in outline, somewhat wrinkled, dark purple.

River banks, sand or gravel bars, river swamps, wet woodlands. S.C. to n. Fla., westward to Tex., northward in the interior to s.e. Kan., Mo., s. Ind. and Ill., Ky., Tenn.

4. Osmanthus

Osmanthus americana (L.) Gray. WILD-OLIVE, DEVILWOOD.

Shrub or small tree. Leaves simple, opposite, evergreen, leathery, glabrous; variable in shape and size, the larger about 14 cm long, 5.5 cm wide; elliptic, oblong-elliptic, oblanceolate or obovate; margins entire and revolute; the bases cuneate, the apices generally acute, more rarely short-acuminate, obtuse, rounded, or notched. Flowers small, creamy white, in short, scaly-bracteate axillary panicles, mostly on twigs of the previous year—the developing flower clusters are evident from autumn and open in early spring. Calyx tubular with 4 minute lobes; corolla tubular, with 4 ovate spreading lobes. Stamens 2. Fruit a drupe, oval, 1–1.5 cm long, 1-seeded, dark bluish purple.

In a wide range of wooded habitats, well-drained uplands, bottomlands and flatwoods where there is standing water at some time. Coastal plain, s.e. Va. to Fla., westward to La.

5. Chionanthus

Chionanthus virginicus L. FRINGE-TREE, OLD MAN'S BEARD.   Fig. 243

A shrub or small tree to about 6 m tall, in spring before or as the leaves unfold with abundant dangling, showy panicles of white or creamy white flowers, thus a favorite subject as an ornamental. Leaves simple, deciduous, petiolate, somewhat leathery; variable in size, to 20 cm long and 10 cm wide; margins entire; broadly elliptic, oval, oblong, occasionally obovate; bases gradually tapering, the apices acute to acuminate, sometimes obtuse or rounded; upper surfaces glabrous, the lower copiously soft-pubescent when young, glabrous later, petioles commonly maroon or blackish purple as are the nodes which, together with the opposite leaves, is a useful distinguishing characteristic. Inflorescence a showy pendent panicle with numerous leaflike bracts, borne from twigs of the previous season. Flowers apparently bisexual but functionally unisexual and the plants functionally dioecious. Calyx minute, united at base and 4-lobed; petals 4, white, linear, separate nearly to the base; stamens 2, inserted in the short corolla tube. Style 1 with a 2-lobed stigma. Fruit a drupe, dark blue, oval-ellipsoid, 1–1.5 cm long.

In diverse habitats, well-drained upland hardwood or pine forests, rock outcrops, savannas, pine flatwoods, or shrub bogs where water may stand some of the time. N.J. to Ohio, generally southward to Fla., e. Tex., s.e. Okla.

Loganiaceae (LOGANIA FAMILY)

(Ours) herbs or vines. Leaves simple, opposite, with stipules (or if stipules quickly deciduous, a stipular line usually evident between the leaves). Flowers radially symmetrical, bisexual, sepals and petals partially united, 4- or 5-merous. Stamens 4 or 5. Ovary superior, 2-locular. Fruit a capsule.

1. Plant a woody twining vine.   1. *Gelsemium*
1. Plant herbaceous.
　2. Corolla 12–15 mm long.   2. *Spigelia*
　2. Corolla not exceeding 5 mm long.
　　3. Leaves linear-subulate, not over 2 mm wide; flowers solitary in the forks of the branches.
                                                                              3. *Polypremum*
　　3. Leaves much broader, lanceolate, elliptic, ovate, or orbicular; flowers in cymes terminating the branches.   4. *Mitreola*

1. Gelsemium (YELLOW-JESSAMINES)

Twining or trailing woody, evergreen vines. Leaves short-petioled, the petioles usually pubescent on the adaxial side, blades glabrous; with minute stipules which are quickly shed as the new shoot develops, leaving a stipular line between the leaves. Flowers axillary, solitary or in few-flowered cymes, showy. Sepals 5, very small relative to the corolla size, appressed to the base of the corolla tube, united only at the base. Corolla tube more or less funnelform, the lobes horizontally spreading, deeper yellow within the throat and the tube than on the lobes or without. Stamens 5, adnate to the corolla tube below, much exceeding the style and stigma on the flowers of some plants, the stigma being about halfway within the corolla tube, the anthers reaching the throat of the corolla; in the flowers of other plants, the style and stigma reach the throat and the stamens reach about halfway the length of the tube. Stigmas 2, each with 2 linear divisions. Fruit a 2-valved capsule, much flattened contrary to the septum. Seeds brown, several to many, flat.

● Sepals obtuse, blunt, or rounded apically, not persistent on the fruit; capsule oblong, very abruptly narrowed to a point 1.5–2 mm long; seed with a prominent membranous wing which is sharply distinguishable from the body of the seed; flowers delicately fragrant.   1. *G. sempervirens*

514

Fig. 243. a,b, **Forestiera acuminata:** a, fruiting branch; b, fruits; c–f, **Chionanthus virginicus:** c, flowering branch; d, leaves from flowering branch; e, fruiting branch; f, mature leaf. (From Correll and Correll. 1972)

• Sepals acuminate apically, persistent on the fruit; capsule elliptical, the tapering tip bearing a definite beak about 3 mm long; seed wingless; flowers without fragrance.   2. *G. rankinii*

1. Gelsemium sempervirens (L.) Jaume St. Hil.
CAROLINA-JESSAMINE, POOR-MAN'S-ROPE.   Fig. 244

Leaf blades mostly lanceolate, varying to ovate, occasional ones suborbicular, 6–9 cm long, to 1.5 cm broad. Flowers delicately fragrant. Sepals obtuse, blunt, or rounded apically, not persistent on the fruit. Corolla usually lemon yellow. Capsule broadly oblong, 1.5–2.5 cm long, 8–12 mm broad, abruptly narrowed to a short beak 1.5–2 mm long. Seeds to 1.5 cm long, asymmetrically winged.

Commonly in well-drained upland woodlands, thickets, fence rows; also in savannas, pine flatwoods, or lowland woods where water stands only for short periods. Coastal plain and piedmont, Va. to Fla., westward to e. Tex. and Ark.

Commonly festooning low-growing shrubs and trees, fences and hedge rows, dappling with yellow the crowns of trees very early in the spring.

2. Gelsemium rankinii Small.   Fig. 245

Leaf blades variable as in *G. sempervirens* but predominantly ovate-lanceolate; also ovate, suborbicular, oblong, and lanceolate. Sepals lanceolate, acuminate apically, persistent on the fruit. Corolla golden yellow. Capsule elliptical, 1–1.6 cm long, 6–8 mm broad, tapering to a tip about 3 mm long. Seeds to 4 mm long, wingless.

Swamps, wet woodlands, bogs, always in places with waterlogged soils much of the time. Coastal plain, s.e. N.C.; Ga., Fla. Panhandle, westward to La.

2. Spigelia

Spigelia loganioides (T. & G.) DC. PINK-ROOT, WORM-GRASS.

Glabrous, perennial herb 1–3 dm tall, stems 1 to several from a slender, short caudex, simple above the base or more frequently few-branched distally. Lowermost leaves smallest, 1–2 cm long, oblanceolate, others 3–4 cm long, lanceolate or elliptic, apices blunt, margins entire; stipules membranous, triangular-subulate, 2 mm long. Flowers few, in a terminal, leafy cyme. Calyx united at the base, with 5 linear-subulate, erect lobes about ⅓ as long as the corolla, persistent. Corolla white or white with pale lavender lines, 12–15 mm long, tube more or less funnelform, lobes 5, triangular, erect or slightly spreading, blunt apically. Stamens 5, filaments adnate to the corolla tube below, free above, anthers reaching about the middle of the tube. Capsule markedly 2-lobed, each lobe globose, 5–6 mm across the 2 lobes, the straight style 2 mm long, persistent. (*Coelostyles loganioides* T. & G.)

Seasonally wet hammocks. Levy and Marion Cos., Fla.

3. Polypremum

Polypremum procumbens L.

Low, glabrous herb, usually with numerous radially spreading-ascending or repent branches from a basal central crown, the branches not usually exceeding 3 dm long, commonly shorter. Leaves linear-subulate, 1–2 cm long, 0.5–2 mm wide, connected at the base by a whitish, hyaline, stipular membrane. Much of each branch essentially comprising a leafy cyme, the inconspicuous flowers sessile or subsessile in the forks of the branches. Calyx united only at the base, lobes 4, subulate, about 3 mm long. Corolla white, united about ⅔ its length, pubescent in the throat, lobes erect or somewhat spreading, short-oblong, apically rounded, about equaling the calyx. Stamens 4, inserted on the corolla tube and not reaching the throat of the corolla. Ovary superior, style very short, stigma capitate. Capsule ovate in outline, somewhat flattened and slightly lobed longitudinally, about 2 mm long, the shriveled style usually persisting in a shallow notch at the summit. Seeds dull amber, tiny, irregularly angled, many of them squarish.

Fig. 244. **Gelsemium sempervirens:** a, flowering branch; b, leaves, to show variation; c, flower; d,e, flowers, opened out, to show heterostyly; f, stamen; g, capsule; h, seed.

517

Fig. 245. **Gelsemium rankinii:** a, flowering branch; b, leaves to show variation; c, flower; d,e, flowers, opened out, to show heterostyly; f, flower bud; g, stamen; h, capsule, with calyx; i, seed.

Weedy in a variety of kinds of disturbed places, in cultivated and fallow fields, logging roads and logged-over woodlands, both upland and lowland, on exposed shores or bottoms of ponds during times of low water, exposed sand bars, river shores, stablized dunes and inter-dune swales, open places periodically flooded. L.I., N.J., e. Pa. to e. Mo., generally south-ward to s. Fla. and e. ⅔ of Tex.; trop. Am.

4. Mitreola (MITERWORTS, HORNPODS)

(Ours) annual, erect herbs. Leaves with firm, opaque, stipular membranes between the pairs. Flowers small, sessile along one side of the branches of a stalked cyme, the cymes terminal or terminating branches. Calyx united at the base, 5-lobed, about 1–1.5 mm long, per-sistent. Corolla white or pink, longer than the calyx, the tube globose-funnelform, the 5 lobes oblong-elliptic, about 1 mm long. Stamens 5, inserted near the base of the corolla tube and not reaching the throat. Ovary at anthesis subglobose, somewhat 2-lobed at the summit, the short style between the lobes, stigma capitate; during maturation of the fruit, the 2 car-pels separating except at the base, the basal part dilating, the distal parts strongly incurving, the whole fruit becoming much like a pair of horns, their inner faces grooved. Ripe capsule splitting longitudinally along the inner face of the separated portions. Seeds tiny, numerous.

1. Leaves mostly narrowed to petiolar or subpetiolar bases, sometimes some on a given plant sessile.
1. *M. petiolata*
1. Leaves sessile.
  2. The leaves narrow, lance-elliptic, oblanceolate, or almost linear, the larger ones four times as long as broad or a little more; mature capsule nearly smooth below, minutely tuberculate distally on both faces of the lobes; fully mature seed alveolate-reticulate, dark brown to black.   2. *M. angustifolia*
  2. The leaves ovate, suborbicular, or elliptic, the larger not exceeding 2 cm long, not over twice as long as broad; mature capsule markedly papillose or warty; surface of fully mature seed smooth, dark brown to black.   3. *M. sessilifolia*

1. Mitreola petiolata (J. F. Gmel.) T. & G.   Fig. 246

Stems 1–8 dm tall, simple or several-branched. Leaves mostly narrowed to petiolar or sub-petiolar bases, some sometimes sessile on a given plant, variable in size, 2–8 cm long, the larger usually over 3 cm long but shorter on occasional plants, 1–3.5 cm wide, acute apically (rarely obtuse), margins minutely toothed. Capsules smooth or sparsely papillate. Immature seeds yellowish and apparently papillate, fully ripe seeds black, iridescent, finely cancellate-reticulate, broadly oval to suborbicular in outline, nearly plano-convex. (*Cynoctonum mitreola* (L.) Britt.)

Swamps, floodplain forests, open boggy areas, wet pine savannas and flatwoods, wet marl prairies, wet hammocks, wet ditches and clearings, boggy shores, wet stream banks. s.e. Va., e. and cen. N.C. to s. Fla., westward to cen. Tex., northward to s.e. Okla., Ark., Tenn.; trop. Am.

2. Mitreola angustifolia (T. & G.) J. B. Nelson.   Fig. 247

Stems 1–5 dm tall, stiffly erect, simple to the inflorescence or with few erect branches. Leaves sessile, linear-oblong, lanceolate, or oblanceolate, the larger to 4 times as long as broad, with little taper to a truncate base, apically bluntish, margins minutely toothed. Cap-sule essentially smooth. Immature seed yellowish and apparently papillate, fully mature seeds dark brown or black, iridescent, finely cancellate-reticulate, broadly oval or ovate in outline, nearly plano-convex. (*Cynoctonum angustifolium* (T. & G.) Small; *C. sessilifolium* var. *angustifolium* (T. & G.) R. W. Long)

Cypress-gum depressions, wet pine flatwoods. s.e. Ga., n. Fla.

3. Mitreola sessilifolia (J. F. Gmel.) G. Don.   Fig. 248

Stems 1–5 dm tall, simple or with few to several erect branches. Leaves sessile, ovate to orbicular, or elliptic, basally rounded, occasionally obtuse, apically obtuse, rarely acute, not over 2 cm long, not over twice as long as broad, margins minutely toothed. Capsule usually

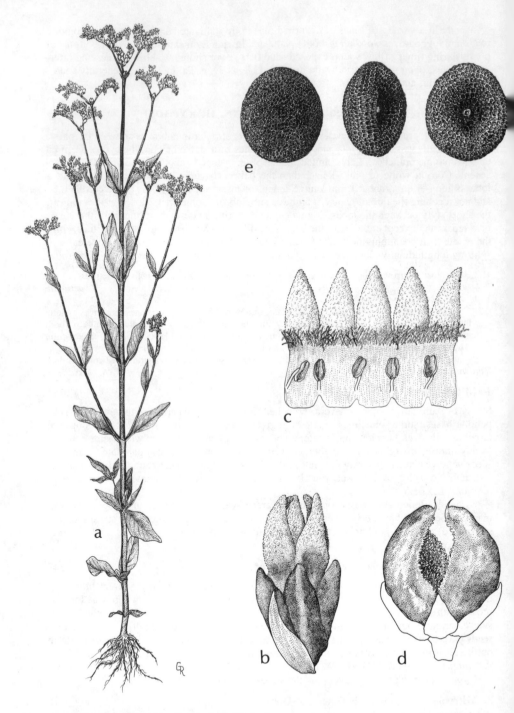

Fig. 246. **Mitreola petiolata:** a, habit; b, flower; c, corolla, opened out, showing insertion of stamens; d, fruit; e, seed, three views.

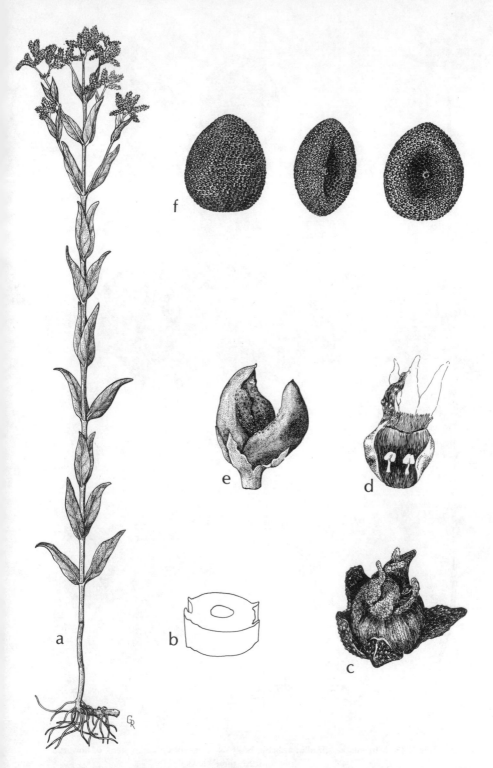

Fig. 247. **Mitreola angustifolia:** a, habit; b, outline of cross-section of stem; c, flower; d, corolla, partially cut away, showing insertion of two of the stamens; e, fruit; f, seed, three views.

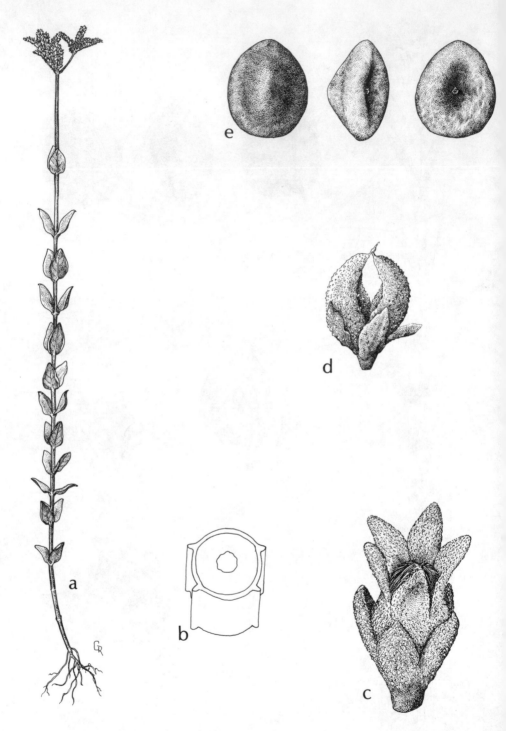

Fig. 248. **Mitreola sessilifolia:** a, habit; b, cross-sectional outline of stem; c, flower; d, fruit; e, seed, three views.

conspicuously papillose or warty. Immature seed yellowish, smooth, fully mature seeds dark brown to black, smooth, shiny, elliptic in outline, 1 side flattish and longitudinally shallowly furrowed, the other strongly convex or humped. (*Cynoctonum sessilifolium* J. F. Gmel.; incl. *C. sessilifolium* var. *microphyllum* R. W. Long)

Wet pine savannas and flatwoods, cypress-gum depressions, bogs, wet or seasonally wet marl prairies, pond margins, shrub bogs, borrow pits. Coastal plain, s.e. Va. to s. Fla., westward to e. Tex.; Bah.Is.

Gentianaceae (GENTIAN FAMILY)

Annual or perennial herbs, glabrous or nearly so, mostly with opposite or whorled (rarely alternate) sessile, entire leaves without stipules. Flowers bisexual, radially symmetrical, 4–12-merous. Calyx persistent, mostly tubular at least at base. Corolla tubular at least at base. Stamens as many as the corolla lobes and alternate with them, inserted on the corolla tube or at the throat. Ovary superior, 1-locular, with 2 parietal placentae; style 1, elongate to essentially absent, stigmas 1 or 2. Fruit a 2-valved capsule, septicidally dehiscent, many-seeded.

1. Leaves all scalelike, 5 mm long or less.   1. *Bartonia*
1. Leaves much larger, not scalelike.
  2. Corolla pink to rose or white; corolla lobes much longer than the corolla tube.   2. *Sabatia*
  2. Corolla blue to bluish white, greenish white, or white (if white, corolla lobes much shorter than corolla tube).
    3. Stamens inserted at the corolla throat, exserted; style slender.   3. *Eustoma*
    3. Stamens inserted on the corolla tube, not exserted; style short, stout, or absent.   4. *Gentiana*

1. Bartonia

Small and inconspicuous annual or biennial herbs. Stems filiform, in some twisted. Leaves reduced to subulate, acute, entire scales. Inflorescence a slender terminal panicle or raceme, or flowers terminal and solitary. Calyx of 4 sepals, free about to the base. Corolla of 4 petals, free nearly to the base. Stamens 4, inserted at the sinuses between corolla lobes. Style very short and stout, stigma slightly bifid. Fruit an ovoid to oblong capsule, abruptly beaked, slightly flattened. Seeds minute, very numerous.

1. Petals oblanceolate or spatulate to spatulate-obovate, tips rounded, 5 mm long or more; plants flowering in early spring.   1. *B. verna*
1. Petals oblong, or lance-oblong to ovate, acuminate apically or abruptly narrowed from a rounded tip to a mucro, mostly about 3 mm long; plants blooming in summer or autumn.
  2. The petals gradually narrowed to a long-acute or acuminate tip; scale leaves usually alternate below the inflorescence.   2. *B. paniculata*
  2. The petals apically abruptly narrowed from a rounded tip to a mucro; scale leaves opposite below the inflorescence.   3. *B. virginica*

1. **Bartonia verna** (Michx.) Muhl.

Stems usually purplish, 5–20 cm tall, rarely taller. Scale leaves usually opposite or subopposite but sometimes remote and alternate, 1–3 mm long. Flowers commonly terminal and solitary, varying to shortly racemose or shortly paniculate. Sepals mostly triangular-acute, much shorter than the corolla at full anthesis. Petals white, oblanceolate, spatulate, or spatulate-obovate, tips rounded, 5–10 mm long when fully developed. Stamens about ½ as long as the petals, filaments dilated below and gradually narrowing to the summit. Pistil ⅓ as long to as long as the corolla, ovary elliptic; style thickish, stigmas connivent.

Bogs, wet pine savannas and flatwoods, edges of shrub bogs, seepage slopes in pinelands, boggy roadsides and ditches. Coastal plain, N.C. to s. Fla., westward to s.e. Tex.

2. Bartonia paniculata (Michx.) Muhl. SCREW-STEM.

Stems commonly twisted, often sprawling, usually purplish, 1–4.5 dm tall. Scale leaves alternate, subopposite, opposite, or mixed, (mainly alternate), 1–3 mm long. Flowers in panicles or more rarely in racemes. Sepals lanceolate, acute or acuminate, in length nearly equaling the petals. Petals white or greenish, lance-oblong, long-acute or acuminate apically, 2–3 mm long. Stamens about as long as the petals, filaments dilated below, gradually narrowing to the summit. Pistil slightly longer than the corolla, ovary ovate or ovate-oblong, style short, stigmas slightly spreading. (*B. lanceolata* Small)

Swamps, bogs, wet meadows, marshes, edges of shrub bogs. N.S. to N.J., chiefly coastal plain southward to n. Fla., westward to e. Tex., northward in the interior to s.e. Okla., Ark., Ky.

3. Bartonia virginica (L.) BSP.

Stem erect, purplish below, greenish yellow above, 1–4.5 cm tall. Scale leaves chiefly opposite or subopposite, 1–4.5 mm long. Flowers borne oppositely in racemes or on strongly ascending subequal racemose branches. Sepals triangular-subulate, ⅔ as long as the petals or slightly more. Petals greenish yellow to creamy white, oblong, abruptly contracted at the rounded apex into a mucro, about 3 mm long. Stamens nearly as long as the corolla, filaments slender, little dilated below. Pistils about equaling the corolla, ovary lanceolate, style short, stigmas connivent.

Bogs, wet acid meadows. Que. and N.S. to Wisc., generally southward to n. Fla. and La.

2. Sabatia (ROSE-GENTIANS, MARSH-PINKS)

Erect, glabrous, annual or perennial herbs. Leaves below the branches opposite with successive pairs arranged at right angles to each other, sessile, margins entire, leaves or bracts of the inflorescence opposite or alternate. Inflorescence cymose or reduced-cymose. Calyx united below, 5–12 (–14) -lobed. Corolla tubular at the base, 5–12 (–14) -lobed, the lobes spreading. Stamens 5–12, filaments slender, inserted on the upper edge of the corolla tube. Ovary 2-carpellate, unilocular, margins of carpels intruding into locule and forming 4 parietal placentae. Style slender, stigmas 2, linear to slightly spatulate, spirally twisted at anthesis and bent to one side, later erect and exposing the densely papillate stigmatic surface. Capsule globose, ovoid, or cylindrical. Seeds numerous, pitted.

Our treatment, in considerable part, adapted from that of R. L. Wilbur, "A Revision of the North American Genus *Sabatia* (Gentianaceae)," *Rhodora* 57: 1–33; 43–71; 78–104 (1955).

1. Flowers with 7–12 (–14) corolla lobes (*S. calycina* may have occasional flowers with 7 corolla lobes but usually has 5 or 6 and is keyed under 2nd no. 1 of couplet).
  2. The flowers few to several, sessile or subsessile, subtending bracts markedly "leafy."
  1. *S. gentianoides*
  2. The flowers long-stalked, or if short-stalked then the subtending bracts not markedly "leafy."
    3. Upper stem leaves narrower than the diameter of the stem bearing them, or but little wider, narrowly linear; calyx lobes thickish, almost semicircular in cross-section. 2. *S. bartramii*
    3. Upper stem leaves noticeably wider than the portion of the stem bearing them; calyx lobes flat.
      4. Calyx lobes hyaline-margined; plants strongly stoloniferous; primary branches very often opposite; terminal flower short-stalked, usually its stalk exceeded by the first internode of the lateral branch arising at the same node. 3. *S. kennedyana*
      4. Calyx lobes not hyaline-margined; stolons usually lacking, if present then neither numerous nor strongly developed; primary branches usually alternate; terminal flower usually long-stalked, its stalk about equaling or exceeding the first internode of the lateral branch arising at the same node.
      4. *S. dodecandra*
1. Flowers with 5 or 6 corolla lobes, mostly 5 (7 lobes in occasional flowers of *S. calycina*).
  5. Inflorescence branches all or mainly opposite.
    6. Corolla pink (white on an occasional individual).

7. Stem leaves nearly oblong, sometimes slightly broader below the middle, three or more times longer than wide, not clasping. 5. *S. brachiata*
7. Stem leaves ovate, their bases clasping the stem, from about as wide as long at their widest places to little if any more than twice as long as wide. 6. *S. angularis*
6. Corollas white or creamy white.
   8. Stems quadrangular, the angles somewhat winged. 7. *S. quadrangula*
   8. Stems, lower portions at least, terete, not winged.
      9. Leaves and upper stem glaucous; stem terete throughout; corolla lobes mostly 5–7 mm long; bracts subtending upper inflorescence branches minute and scalelike, 1–4 mm long; corollas of dried specimens not changed in color appreciably. 8. *S. macrophylla*
      9. Leaves and stems not glaucous; upper stem angled or quadrate; corolla lobes mostly 10–15 (–20) mm long; bracts subtending upper inflorescence branches linear-subulate; corollas on dried specimens commonly changed in color to a saffron or orange hue. 9. *S. difformis*
5. Inflorescence branches mainly alternate.
   10. Calyx tube strongly 5-ridged-nerved.
      11. Leaves and calyx lobes membranous; leaves broadly rounded to truncate basally, clasping the stem or nearly so. 10. *S. campestris*
      11. Leaves and calyx lobes subsucculent; leaves broadest somewhat above the base, the bases at least somewhat tapered, not at all clasping. 11. *S. arenicola*
   10. Calyx tube not strongly 5-ridged-nerved.
      12. Calyx lobes foliaceous, oblong to oblanceolate, mostly the latter, equaling or mostly exceeding the corolla lobes in length. 12. *S. calycina*
      12. Calyx lobes linear-setaceous, if equaling the corolla lobes then very narrow and not foliaceous.
         13. Corollas pink (white on an occasional individual plant).
            14. Plant perennial, usually with several stems from the branched caudex; calyx lobes ¾ as long to as long as the corolla lobes, mostly exceeding ¾ as long. 13. *S. campanulata*
            14. Plant annual, stems solitary from a base with a tap root or several fibrous roots; calyx lobes not exceeding ¾ the length of the corolla lobes, usually less than ¾ as long.
               15. Corolla lobes 1.8–2.5 cm long, mostly 2–2.5 cm and corollas mainly 3.5 cm across or more; leaves succulent, on dried specimens becoming markedly rugose but remaining green. 14. *S. grandiflora*
               15. Corolla lobes 1–1.5 cm long, mostly less than 1.5 cm, and corollas mainly 1.5 cm across or less; leaves membranous, on dried specimens drying thin and smooth and usually becoming noticeably darkened if not dried quickly. 15. *S. stellaris*
         13. Corollas white. 16. *S. brevifolia*

1. Sabatia gentianoides Ell. Fig. 249

Erect annual 3–5 dm tall. Stem rigid, more or less terete but with fine and irregular internodal ridges. Very commonly unbranched below the terminal inflorescence, when branched, the branches usually restricted to upper quarter of stem. Branches opposite or alternate, mostly alternate, usually about 5–8 cm long. Leaves thick and somewhat succulent, drying rather thickly chartaceous and rugose, strikingly dimorphic; those of the basal rosette wide-spreading, oblong to orbicular-spatulate, typically long-persistent, 2–3 cm long, 8–12 mm wide, obtuse, rather strongly tapering to the base and appearing petiolate; the cauline leaves narrowly linear, 4–5 cm long, 1–3 mm wide, acute and slightly callose-tipped, sessile or clasping the stem. Flowers sessile or nearly so, borne singly at the principal apex or at that of short lateral branches, in robustly developed plants in few to several flowered compact clusters; each flower closely associated with 2 subtending, narrowly linear bracts 1.5–4 cm long. Calyx tube broadly campanulate, 4–6 mm long, smooth and unnerved; calyx lobes subulate, 5–10 mm long, spreading-ascending, arching outward especially at the tip, often unequal in size, 7–12 in number, 1 or 2 times as long as calyx tube, and about half or less as long as corolla. Corolla tube 6–10 mm long, pale greenish yellow, cylindrical. Corolla lobes 18–24 mm long, 6–8 mm wide, elliptical, spatulate to oblanceolate, acute to obtuse, pink to deep rose with an unlobed greenish yellow area at the base of the lobe 2–3 mm high. (*Lapithea gentianoides* (Ell.) Griseb.)

Seasonally wet pine savannas and flatwoods, adjacent ditches, boggy seepage slopes of pinelands. Coastal plain, N.C. to n. Fla., westward to s.e. Tex.

Fig. 249. **Sabatia gentianoides:** a, top of plant; b, base of plant; c, flower; d, pistil; e, fruit. (From Correll and Correll. 1972)

2. Sabatia bartramii Wilbur.   Fig. 250

A handsome perennial herb. Rhizome slender to thick, 3–5 mm in diameter, 4–6 cm long or more. Stems erect and rather rigid, terete. Branches typically alternate but occasionally opposite at one or two of the primary nodes; branches usually having but one flower, if further branched seldom with more than 2 flowers. Leaves thick, subsucculent, drying rather thickly chartaceous, the apices often callous-mucronate, strongly dimorphic with an abrupt transition from the basal to the cauline; basal leaves strongly spreading, conspicuously rosulate, oblanceolate, somewhat oblong, or more often broadly spatulate, strongly but gradually long-tapering to the almost petiolate base, apically obtuse or rarely acute, 4–8.5 cm long, 12–17 mm wide; stem leaves strongly ascendent and becoming closely appressed, the lower lanceolate but becoming gradually, or even abruptly, reduced to linear or very narrowly linear above. Inflorescence of reduced cymules of 1, rarely 2, flowers borne on slender, rigid, finely ridged stalks 4–8 cm long. Calyx tube ridgeless, crateriform to campanulate, usually broadly so, occasionally, especially in smaller flowers, somewhat turbinate, 3–4 mm long, usually straw-colored in strong contrast to the dark green lobes; calyx lobes usually subulate, 8–12 mm long, often subsucculent, usually drying somewhat rugose-thickened, ellipsoidal below to nearly rounded above in cross-section, often somewhat revolute, erect, strongly ascendent or but weakly spreading, dark green in color. Corolla tube cylindrical, 6–8 mm long, usually about 2 or 3 times as long as the calyx tube, pale yellow externally and darker within; corolla lobes 10–12 in number, 22–32 mm long, 7–10 mm wide, obovate-spatulate or rarely oblong to elliptic, usually broadly obtuse, deep rose-magenta to rose-pink (rarely white), with a slightly and irregularly toothed oblong yellow patch at the base of the lobe, this usually bordered by a dark red line. (*S. decandra* sensu Harper; *S. dodecandra* var. *coriacea* (Ell.) Ahles)

Cypress-gum ponds or their borders, wet pine flatwoods, adjacent ditches, often in shallow water. S.C., s. Ga., throughout Fla. except extreme southern tip, s. Ala., s.e. Miss.

3. Sabatia kennedyana Fern.

A handsome perennial herb, arising from slender rhizomes about 2–3 mm in diameter and up to 12 cm long. Stems usually solitary, occasionally 2 or more arising together, rigidly erect, terete. Branches mostly restricted to upper third or half of stem, the primary commonly opposite, additional branches when present alternate. Leaves thin, drying brittle, smooth, thinly chartaceous, only the midvein prominent, basal leaves, when present, and lower stem leaves similar in size and shape to those of rosettes; stem leaves strongly ascending, gradually reduced upwardly, lanceolate or linear, slightly callose-tipped, variable in size, averaging 3–5 cm long and 4–10 mm wide, mostly 2–4 times as long as the internodes. Inflorescence of complete or reduced cymules, the terminal flower usually short-stalked and greatly exceeded by the first node of the lateral branch arising from the same node. Calyx tube crateriform or shallowly campanulate, thin, smooth, or rarely with a few slightly elevated nerves; calyx lobes linear, mostly 6–10 mm long, thin, flat, acute apically, slightly hyaline-margined. Corolla tube 5–8 mm long, pale yellow without, darker within; corolla lobes 9–12 in number, mostly 18–24 mm long, 7–10 mm wide, obovate-spatulate, cuneate-obovate, or nearly oblong, rose-pink or pink (rarely white), corolla with a yellow but reddish-bordered "eye" at the base of the lobes. (*S. dodecandra* var. *kennedyana* (Fern.) Ahles)

Sandy and peaty margins of streams and ponds. s. N.S.; e. Mass., R.I.; s.e. N.C., n.e. S.C.

4. Sabatia dodecandra (L.) BSP.   Fig. 251

Perennial herb, 3–7 dm tall, with a slender to robust, somewhat branched rhizome, 4–10 cm long, 2–4 mm in diameter. Stems erect, terete and smooth or somewhat angular and ridged. Branches usually alternate. Basal rosettes absent from base of aerial stem, the lower cauline leaves neither densely clustered nor strongly contrasted in either size or shape with those

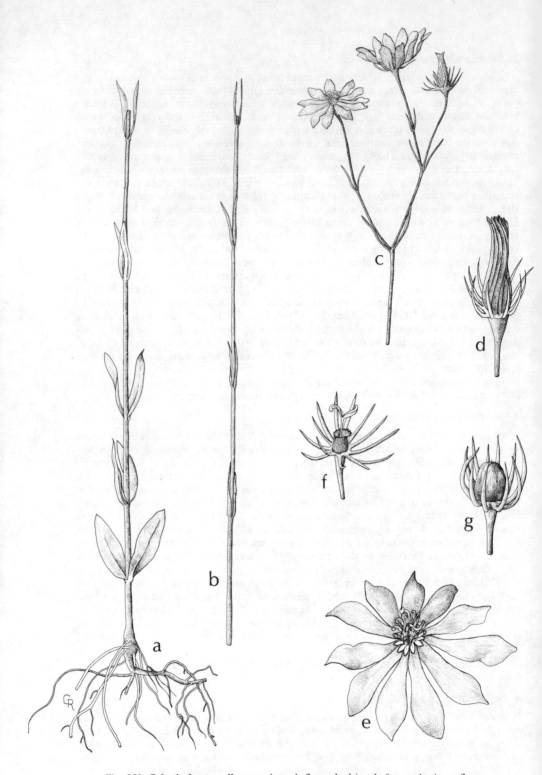

Fig. 250. **Sabatia bartramii:** a–c, plant; d, flower bud just before anthesis; e, flower, face view; f, flower, with corolla and stamens removed; g, capsule.

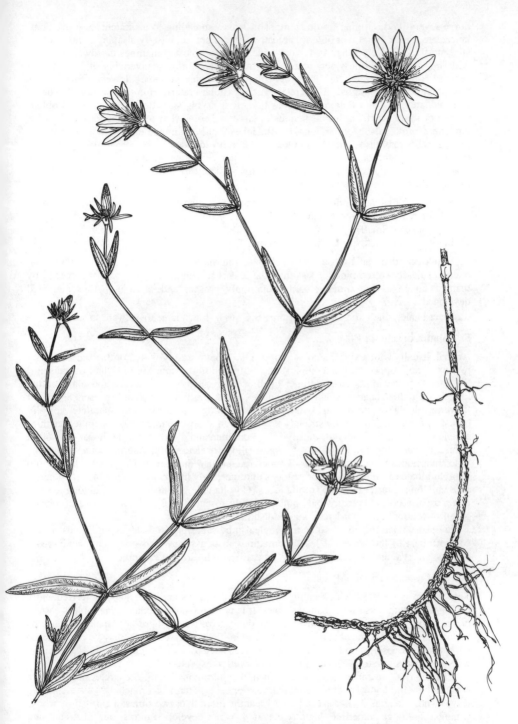

Fig. 251. **Sabatia dodecandra.** (From Correll and Correll. 1972)

borne several nodes higher up on stem; stem leaves spreading to ascendent, nonsucculent, lanceolate, linear, elliptic or oblong, apically acute to obtuse, basally clasping to merely sessile, mostly 2.5–4 cm long, 5–12 mm wide, the lowermost sometimes narrowly spatulate with long-tapering, almost petiolate bases. Inflorescence almost invariably of 1 to several reduced 1- or 2-flowered cymules, rarely complete cymules with three flowers. Flower stalks erect, rigid, slightly angled, 3–6 cm long. Calyx tube crateriform, somewhat turbinate or campanulate, 2–3 mm long; calyx lobes thin, flat, drying smooth, linear, narrowly oblanceolate or spatulate, from inconspicuous to large and somewhat foliaceous, acute, 0.8–1.8 cm long. Corolla tube 5–7 mm long; corolla lobes 9–12 in number, elliptic, oblanceolate, or oblong, acute to obtuse, 1.6–2.4 cm long, 5–8 mm wide, deep rose-purple, rose-pink, pink (rarely white), with oblong to somewhat triangular yellow patch at base of lobe, the patch sharply 3-lobed or irregularly toothed, usually bordered by a thin red line. Wilbur (1955) distinguishes two varieties, as follows.

S. dodecandra var. **dodecandra.** Internodes exceeding the leaves in length, commonly 1.5–3 times longer than the leaves; stolons rarely present, if present weakly developed. (Incl. *S. harperi* Small)

In saline and brackish marshes. Conn. to S.C.

S. dodecandra var. **foliosa** (Fern.) Wilbur. Internodes usually equaled in length by the leaves or nearly so, commonly less than the leaves in length and very rarely exceeded by them as much as 1.5 times; stolons commonly present and often numerous and well developed.

River banks, ditch and stream margins. S.C. to n. Fla., thence westward to s.e. Tex.

5. Sabatia brachiata Ell.

Annual, usually with a single stem arising from a rosette, rarely with 2 or 3. Branches, with rare exception, opposite, branching usually restricted to the upper third or half of the stem. Stem terete below although sometimes lined or finely ridged, subquadrate above and on the branches and there bearing slight wings on the angles. Basal rosette usually present at time of flowering, its leaves numerous, overlapping, spreading, usually broadly spatulate, apically obtuse or rarely acute, basally strongly tapered to an almost petiolate base; stem leaves thin, ascending, the larger 3-nerved, mostly 1.5–3 cm long and 4–10 mm wide, lower ones often obtuse, upwardly becoming acute, oblong to elliptic or lanceolate, tapering somewhat to the base. Internodes usually about 1.5–3 times longer than the leaves. Calyx tube 1.5–3 mm long, turbinate to campanulate; calyx lobes thin, narrowly linear, wide-spreading at anthesis, 3–8 mm long, shorter than the corolla lobes. Corolla tube greenish, lobes 5, oblong to spatulate, usually obtuse apically, pale pink to darkly roseate (rarely white), at base of lobes a greenish yellow "eye" bordered by a reddish line.

In open upland mixed woodlands, longleaf pine sand hills and ridges, old fields, in the southern part of the range, at least, sometimes in bogs, pine flatwoods, and sandy-peaty ditches. s.e. Va. and N.C. to s.e. Mo., generally southward to Ga. and La.

6. Sabatia angularis (L.) Pursh.

This species is the most wide-ranging member of the genus and is perhaps more frequently encountered within its range than any other. It is probably not a wetland plant unless certain of the heavy prairie soils are considered wetlands. We treat it here principally because its inclusion may ease somewhat the problem of identification given that it may be encountered so readily. Our description below is, however, restricted to a few diagnostic characters which will, we trust, serve to distinguish it from the wetland species.

Characteristic features include a conspicuously tetragonal stem, the angles of which are strikingly membranously wing-margined; an opposite pattern of branching; ovate-clasping leaves whose lengths do not exceed their breadths more than two times; a calyx tube whose venation, although sometimes slightly elevated, is not developed into strong ribs; and pink (rarely white) flowers. The combination of characters, not any one alone, is unique to the species.

Old fields, pastures, open upland woodlands, borders of upland woodlands, prairies, apparently occasionally in ditches and meadows. N.Y. to s. Mich. and Ill. to e. Kan., generally southward to the Fla. Panhandle and e. Tex.

7. Sabatia quadrangula Wilbur.

Annual herb 2.5–4.5 dm tall, usually with but one stem arising from each rosette. Stem strongly quadrate, conspicuously so below, finely membranously wing-angled. Branches typically opposite along stem and principal branches, ultimate ramifications sometimes alternate, the branches forming a flat-topped or convex crown. Basal rosette present or absent at anthesis. Stem leaves membranous, ascending, revolute, 1.6–3.2 cm long, 4–8 mm wide, obtuse to acute, typically apiculate, usually narrowly to broadly oblong or lanceolate, at least the lowermost, with strongly clasping bases. Inflorescence of cymosely to pyramidally arranged cymules, either one, both, or none of whose lateral branches may be suppressed, flowers usually closely associated in the cymule as the flower stalks are always short, not more than 1–2 mm long, always less than 4 mm long. Calyx tube thin, 5-ridged-angled, about half as long as corolla tube, or about 2–3 mm high, broadly turbinate; calyx lobes thin, narrowly linear, 4–8 mm long, usually about 0.5 mm or less in width, erect in bud, and apparently ascending at anthesis, usually exceeded by corolla lobes by 2–5 mm. Corolla tube 4–5 mm long, white but appearing greenish for the ovary is visible through the translucent wall; corolla lobes 5, usually oblong, spatulate, or somewhat elliptic, apically usually obtuse but not uncommonly acute, 6–12 mm long, 3–4 mm wide, spreading, white, occasionally with, more commonly without, a basal yellow patch, sometimes turning saffron yellow upon drying, especially along the reticulate veins. (*S. paniculata* of authors not (Michx.) Pursh)

Pine savannas and flatwoods, borders of shrub bogs, margins of limesinks, moist to wet ditches, shallow pockets on granite outcrops, seepage bogs on slopes of pinelands. e. Va., cen. and e. N.C. generally southwestward to n.w. Gulf coast of Fla. and Fla. Panhandle, s. Ala.

8. Sabatia macrophylla Hook.

Perennial herb 9–12 dm tall. Rhizome gnarled, stout, branched, up to 1 cm in diameter and often 10 cm or more long. Stems stiffly erect, terete throughout, strikingly glaucous above. Branches mainly opposite, producing a compact, flat-topped to somewhat convex inflorescence. Leaves thickish, somewhat succulent, drying thickly chartaceous, strongly ascending, very noticeably glaucous when fresh, ovate-lanceolate, lanceolate, oblong, to ovate-oblong or ovate, apically acute or rarely obtuse, scarious-mucronate-tipped, margins conspicuously scarious and often slightly revolute, strongly clasping basally, 3–6 cm long, 1–2.5 cm wide. Basal rosette none and the lowermost cauline leaves absent at time of flowering. Flower stalks slender, erect-ascending, 1–5 mm long. Calyx tube and lobes more or less colorless, tube campanulate, scarcely ridged or lined by elevated veins, 1–2 mm long; calyx lobes triangular-dentate, subulate, or linear, erect, slightly spreading, somewhat outwardly curved at the tip, or strongly recurved, 2–2.5 mm long, 1–3.5 times the length of the calyx tube. Corolla tube narrowly cylindrical, white, 3–3.5 mm long. Corolla lobes 5 (sometimes 6 in central flower of central cymules), oblong to oblong-spatulate, obtuse, widely spreading at anthesis, 5–7 mm long, 2–3 mm wide, entirely white or creamy white, color not changing or changing but little upon drying. (Incl. *S. recurvans* Small)

Seasonally wet pine savannas and flatwoods, cypress-gum ponds or depressions, borrow pits, ditches, seepage bogs on pineland slopes, often in shallow water. s. Ga., n. Fla., westward to e. La.

9. Sabatia difformis (L.) Druce.

Perennial herb 4.5–8 dm tall. Rhizome gnarled, stout-branched, 4–6 mm in diameter. Stems stiffly erect, more or less terete below, upwardly becoming angular and sometimes almost quadrate, branches mainly opposite, the inflorescence flat-topped to convex. Basal rosette none; lowermost stem leaves borne submersed or subterranean, at time of flowering usually represented only by scars, if present, then bractlike; upper stem leaves strongly ascending, linear to lanceolate, somewhat oblong, ovate-lanceolate to strongly ovate, usually acute, rarely obtuse, apically, somewhat revolute and both scarious-margined and -tipped, basally usually rather strongly clasping, or merely sessile in the narrower-leaved types, 1.8–4 cm long, 4–14 mm wide. Flower stalks slender, erect, finely ridged, 2–8 (–15) mm long. Calyx tube only slightly ridged, rather shallowly campanulate, crateriform or turbinate, 1–2 mm long. Calyx lobes linear to somewhat subulate, very slender, 4–10 mm long, 2–8 times the length of the tube, more or less outwardly arching at anthesis, erect in bud. Corolla tube narrowly cylindrical, 3–5 mm long; corolla lobes usually 5 (the centermost flower of the central cymules often 6-parted), oblong, weakly spatulate, elliptic, wide-spreading, 7–15 mm long, 2.5–6 mm wide, entirely white even at the base, usually turning orange or saffron upon drying, especially along the veins.

Pine barrens, pine savannas and flatwoods, borders of shrub bogs or burned-over shrub bogs, seepage bogs on slopes of pinelands, borrow pits, ditches. Coastal plain, s. N.J. southward throughout most of pen. Fla. and Fla. Panhandle, s. Ala.

10. Sabatia campestris Nutt.

Erect annual, usually 15–30 cm tall. Stems strongly quadrate throughout, the angles membranous-winged, branched mostly in the upper half. Leaves thin, the lowermost differing little from those above, narrowly to broadly lanceolate, ovate-lanceolate, sometimes oblong, generally clasping basally, apically the lower ones obtuse, becoming acute upwardly. Flower stalks wiry, rigid, pentagonal, mostly 2–5 cm long, with fine hyaline wings at the angles. Calyx tube broadly campanulate, strongly pentagonal with 5 elevated ribs; calyx lobes thin, strongly spreading at anthesis, linear, 1–2.2 cm long. Corolla tube usually 1–2 mm longer than the calyx tube, lobes 5, usually equaling or exceeding the calyx lobes, oblong, elliptical, or more commonly obovate or spatulate, rose to pale pink, rarely white; corolla with an "eye," yellow bordered by white and sometimes by a more densely roseate region.

Upland open woodlands, woodland borders, old fields, railway and highways embankments, prairies, sometimes in marshes and meadows, seasonally wet pinelands. Ill., Mo., s.e. Kan., Okla. and Ark., most of the e. half of Tex., eastward to s. Miss.

11. Sabatia arenicola Greenm.

Annual, stems erect or spreading, to 3 dm tall, simple or branching from the base to form a bushy-branched mass. Leaves thick-succulent, the venation obscured, widest above the base and tapered to the base, lanceolate, ovate-lanceolate, elliptic-obovate, apically obtuse, to 2.5 cm long, but mostly smaller. Calyx tube strongly pentagonal, the angles with strong ribs, lobes triangular-lanceolate, to 15 mm long. Corolla roseate, sometimes white, the lobes obovate to oblong, obtuse, equaling or a little shorter than the calyx lobes.

Beaches, interdune depressions, salt flats, wet savannas. Coastal, La., s. Tex.; n.e. Mex.

12. Sabatia calycina (Lam.) Heller.

Perennial herb. Rhizome slender, 1–3 mm in diameter, 2–4 (–10) cm long. Stems erect, rigid, smooth above or but slightly ridged. Branches mainly alternate, ascending to divaricate. Basal rosette lacking. Leaves thin, drying very thinly membranous; the lower stem leaves not conspicuously differentiated in shape from the median or upper leaves; upper stem leaves ascending or more often strongly spreading, 2.5–6 cm long, 10–18 mm wide, elliptical to broadly spatulate, mostly obtuse but still commonly acute, tapering into conspicuous, much-narrowed subpetiolate base. Inflorescence of usually reduced 1–2 flowered cymules; the flowers appearing loosely arranged and even solitary. Flower stalks slender, rigid, inconspicuously 5–7 angled, 2–5 cm long. Calyx tube thin, smooth or with veins but very slightly elevated, somewhat scarious or translucent, shallowly crateriform to broadly campanulate, 2–4 mm long; calyx lobes oblanceolate to spatulate, or rarely linear, often of unequal size, usually foliaceous, apparently enlarging after pollination, acute to obtuse, thick, membranous, 10–25 mm long, 1.5–6 mm wide. Corolla tube cylindrical, 4–5 mm long, colorless or white to pale pink; corolla lobes equaling the calyx lobes in number, usually 5- or 6-parted, but occasionally as many as 7, often exceeded in length by the calyx lobes which are sometimes as much as twice as long but not uncommonly equaling or even exceeding the calyx lobes by 1−2 mm, oblong to oblong-spatulate, oblanceolate or elliptic, obtuse or acute, 7−13 mm long, 3−5 mm wide, white or more commonly pale rose to pink, gradually giving away to white in the area above the triangular, pointed, yellow patch at the bases of the lobes.

Swamps, wet woodlands, springy areas in woodland, wet clearings, stream banks, wet ditches. Coastal plain, s.e. Va., southward throughout much of Fla., westward to s.e. Tex.; e. Cuba, Hisp.

13. Sabatia campanulata (L.) Torr.

Cespitose perennial 3−6 dm tall. Caudex erect, short, much branched underground, 1−4 cm long. Stems erect, terete but often strongly ridged-angled from fine elevated internodal lines. Branches generally alternate throughout. Leaves ascending to very commonly strongly spreading, margins somewhat thickened and slightly revolute, drying thick chartaceous, smooth or slightly rugose, obtuse to acute, broadly sessile or in the broader leaves somewhat tapering to the base; lower stem leaves 1.5−3 cm long, 2−7 mm wide, narrowly lanceolate, oblong, or linear, gradually reduced above to very narrowly linear or even filiform, those of the branches narrowly linear to filiform. Flowers apparently solitary, borne on alternate, rarely opposite, ascendent to widely divergent slender branches bearing 1 to several nodes, or in more obvious cymules which are often reduced. Flower stalks 4−7 cm long, slender, wiry, slightly angled. Calyx tube shallowly turbinate or more rarely somewhat campanulate, smooth or nearly so, not conspicuously nerved, scarious, usually about as broad as long, 1.5−2.5 mm long; calyx lobes linear to acicular or setaceous, erect to more commonly widely spreading, 7−17 mm long, usually less than 0.5 mm wide but occasionally as wide as 1 or even 2 mm, generally nearly equaling the corolla in length, rarely exceeding it. Corolla tube cylindrical, colorless to white or pale greenish yellow, 3−5 mm long; corolla lobes wide spreading, oblanceolate, oblong, or elliptic, acute to obtuse, 0.9−1.8 cm long, 4−7 mm wide, rose, pink, or rarely white, with an unusual unlobed yellow area 2−3 mm long at the bases of the lobes, often bordered by a dull red irregular line.

Pine savannas and flatwoods, bogs, seepage slopes of pinelands, semiopen boggy depressions in woodlands, ditches, borrow pits. Coastal plain, Mass. to n. Fla., thence westward to La. and Ark.; also s. Appalachians and locally in Ky., Ill.

14. Sabatia grandiflora (Gray) Small.

Erect annual 4−9 dm tall. Stems terete, although in dried specimens several fine lines evident running between the nodes. Branches invariably alternate, generally strongly divergent. Leaves erect, succulent, drying thick, rigid, rugose-roughened, the lower spatulate, elliptic, oblong-linear or linear, obtuse to acute, 2−4 cm long, 4−7 mm wide, gradually to abruptly reduced above to very narrowly linear or filiform, tips hyaline or callose-apiculate. Inflorescence with 2, 3, or rarely more flowers in reduced cymules; the flowers borne on slender, but rigid, elongate, terete stalks 4−8 cm long, the flowers thus appearing solitary. Calyx tube campanulate, usually broadly so, the sides more or less parallel, abruptly contracting to base or but gradually tapering in lower half, 2−4 mm long, generally very smooth or but very finely lined, the wall thin and somewhat hyaline-scarious especially with age. Calyx lobes erect to spreading, very narrowly linear or filiform, 1−2 cm long, usually about 3−6 times as long as the tube, the tips callose-apiculate. Corolla tube cylindrical, 5−7 mm long, exceeding calyx tube by 2−4 mm, apparently colorless; corolla lobes strongly spreading, oblanceolate, obovate, broadly spatulate, oblong, or elliptic, usually obtuse or more rarely acute, 1.8−2.5 cm long, 7−12 mm wide, deep rose to pale pink (rarely white) with oblong basal yellow area 2−4 mm long, rather irregularly lobed or toothed, and bordered by an intense red line or area (except in white forms).

Borders or exposed bottoms of cypress or cypress gum ponds and depressions, exposed shores of ponds and lakes, *Hypericum* ponds, pine flatwoods, wet prairies, borrow pits and ditches. pen. Fla., locally in the Fla. Panhandle, s.e. Ala.; Cuba.

15. Sabatia stellaris Pursh.

Erect annual 1.5–5 dm tall. Stem terete to strongly angled due to several somewhat irregularly disposed fine lines or ridges extending between nodes. Branches almost always alternate. Leaves ascending, rubbery in texture; after drying leaves commonly darken, usually thick membranous or occasionally the lowermost slightly rugose; lower leaves broadly to narrowly elliptic, or even linear, rarely spatulate or obovate, acute to rarely obtuse, often apiculate, usually tapering to both ends, 1.5–3 cm long, 3–8 mm wide; upper leaves more narrowly elliptic to linear, the uppermost sometimes very narrowly so, or filiform, 1–4 cm long, 1.5–3 mm wide, usually exceeding the diameter of the stem or at least equaling it. Flowers appearing solitary but usually arranged in very loose and reduced cymules and these sometimes aggregated in loose clusters. Flower stalks slender, straight, ascending to divergent, 4–10 cm long. Calyx tube turbinate, occasionally narrowly campanulate, usually rather gradually narrowing to the base, smooth or but very faintly nerved, thin, somewhat scarious, 2–4 mm long; calyx lobes narrowly linear to almost filiform, ascending or usually wide-spreading, 6–15 mm long, usually 3–6 times as long as calyx tube, slightly hyaline-scarious-margined. Corolla tube 2–3 times as long as wide, translucent, 4–6 mm long; corolla lobes strongly spreading, elliptic, oblong, spatulate, or obovate, obtuse to somewhat acute, 1–1.5 cm long, 4–8 mm wide, rose to pink (rarely white) with a usually irregularly 3-lobed yellow area at the base often bordered by a distinct bright or dark red line which, in turn, is sometimes bordered by a white area of variable width.

Salt and brackish marshes, wet marl prairies, marl pits and on marly spoil banks or flats, interdune depressions. s. Mass., southward along the coast, throughout much of coastal Fla., westward into La.; Bah.Is., w. Cuba, cen. plateau of Mex.

16. Sabatia brevifolia Raf.

Annual, stem erect, 3–6 dm tall, terete but finely ridged or lined. Branches mainly alternate, in large well-developed plants usually branching throughout the length of the stem, in smaller plants branching often restricted to the upper half of the stem. Lower leaves somewhat obovate, oblanceolate, oblong, elliptic, or linear, 1–2 cm long, 2–5 mm wide, usually obtuse apically, sometimes acute; upper stem leaves and those of the branches narrowly linear to filiform or subulate, gradually reduced to about 3–5 mm long and 0.5 mm wide. Internodes usually 1–2 times the length of the leaves, occasionally greater, sometimes less near the base. Flower stalks slender, wiry, 2–4 mm long. Calyx tube turbinate, smooth or finely nerved, lobes setaceous or subulate, 4–7 mm long, usually 2–4 times as long as the calyx tube, usually not exceeding half the length of the corolla. Corolla lobes spreading, elliptic, oblong, oblanceolate, or broadly spatulate, about 9–13 mm long, usually obtuse apically, sometimes acute, white with a greenish yellow to yellow patch at the bases of the lobes. (*S. elliottii* Steud.)

Pine savannas and flatwoods, seepage bogs on pineland slopes, borders of cypress-gum ponds, exposed shores of sinkhole ponds, sometimes open well-drained woodlands. Coastal plain, S.C. southward through most of Fla., s. Ala.

3. Eustoma

Eustoma exaltatum (L.) G. Don. CATCHFLY-GENTIAN.

Annual glabrous and glaucous herb of variable stature, commonly 7–10 dm tall, with few to several strongly ascending branches. Leaves sessile, most of them more or less clasping the stem; lowermost oblanceolate to obovate, usually smaller than midstem leaves, upwardly mostly oblong to oblong-elliptic, apically obtuse, uppermost often apically acute, variable in size depending upon robustness of specimens, up to 9 cm long and 3 cm wide, margins entire. Flowers on long stalks, solitary, or few to numerous, when numerous essentially in open panicles. Calyx fused only basally, the lobes 5 or 6, overlapping below, keeled, with a lanceolate basal portion and subulate tips, 1–2 dm long overall. Corolla campanulate-funnelform, blue, lavender, or almost white, the tube about 1 cm long, lobes 5 or 6, oblong-obovate, twice as long as the tube or a little more, mucronate apically. Stamens 5 or 6, inserted on the corolla at the throat. Ovary ellipsoid, style slender; stigmas 2, oblong to roundish. Capsule 2–3 cm long. Seeds very numerous, tiny, nearly round but irregular owing to being very markedly pitted, lustrous chestnut brown.

Variable open habitats including damp to wet calcarous and saline places, often where substrate is disturbed. pen. Fla. and Fla. Keys, local along the Gulf coast to s. and w. Tex.; Mex. to Venez.; W.I.

4. Gentiana (GENTIANS)

Annual or perennial relatively low herbs with fleshy roots or slender rhizomes; often several stems from the base. Leaves opposite, entire, sessile or short-petiolate, sometimes clasping. Flowers solitary or in terminal or axillary clusters, stalked or sessile, 4- or 5-merous (except pistil). Calyx tubular below, sometimes lined with an inner membrane that projects above the base of the lobes within. Corolla tubular below, the lobes ½ or less of the total length, salverform to funnelform, but usually closing quickly, persistent, often either plicate in the sinuses (the plaits notched, rounded, acute, lobed, or toothed) or with setacous scales basally on the inner surface of the lobes. Stamens united to the corolla tube for ⅓ of the length of the filaments or a little more, filaments free above, anthers free or coherent around the short style which ends in two stigmatic lobes (style sometimes absent). Seeds many, minute.

1. Flowers solitary and terminal.
  2. Calyx lobes widest at base, the sides parallel for some distance then gradually narrowed to an acute apex; free portions of the filaments widest at the base then gradually narrowing upwardly; corolla bright indigo-blue except in an occasional individual plant.   1. *G. autumnalis*
  2. Calyx lobes broadest somewhat above the base, narrowing somewhat from the widest point toward the base, and gradually narrowing upwardly from the widest portion to an acute apex; free portion of the filaments dilated at the base, then abruptly narrowing upwardly; corolla dull purplish green without.   2. *G. pennelliana*
1. Flowers terminal on axillary branches or in terminal and axillary clusters.
  3. Corolla without plaits, folds, or subsidiary lobing between the lobes.
    4. Corolla 4-lobed, the lobes broadly rounded and their margins fringed across the summit and part way down the sides.   3. *G. crinita*
    4. Corolla 5-lobed, the lobes acuminate, not fringed marginally.   4. *G. quinquefolia*
  3. Corolla lobes with alternating folds or plaits in the sinuses, these often of different color and texture than the lobes and toothed or fringed; in some species as long as the lobes.
    5. Stem, distally at least, scabrid with short, rough reddish pubescence or protuberances, these usually in lines.   5. *G. catesbaei*
    5. Stem smooth.
      6. Upper and bracteal leaves acute to obtuse; plaits deeply lobed or fimbriate at the summit; corolla lobes strongly ascending, summit of the corolla open at anthesis.   6. *G. saponaria*
      6. Upper and bracteal leaves acuminate; plaits shallowly toothed or lobed at the summit; corolla lobes incurved, summit of corolla scarcely open at anthesis.   7. *G. clausa*

1. Gentiana autumnalis L. PINE BARREN GENTIAN.

Perennial with one or few arching or erect, simple (rarely branched), glabrous stems, mostly 2–5 dm tall. Leaves sessile, linear to linear-oblanceolate, midstem leaves usually longest, these mostly 4–6 cm long. Flowers solitary and terminal. Calyx 5-lobed, lobes nearly linear, widest at the base then gradually narrowed to the summit. Corolla usually indigo blue externally but on individual plants may be white, greenish white, white and blue, or purple or lilac; corolla tube brown-spotted within; corolla lobes 5, spreading at anthesis, obovate, margins irregularly very shallowly toothed, plaited between the lobes, plaits somewhat fimbriate-laciniate. Free portions of the filaments widest at base, then gradually narrowed to an acute apex. Flowering in autumn. (*G. stoneana* Fern.; *G. porphyrio* J. F. Gmel.; *Dasystephana porphyrio* (J. F. Gmel.) Small)

Pine barrens, pine savannas. Coastal plain, cen. N.J. to S.C.

2. Gentiana pennelliana Fern. WIREGRASS GENTIAN.

Similar in general aspect to *G. autumnalis*. Calyx lobes widest somewhat above the base. Corolla externally dull purplish green. Free portions of the filaments dilated at base, abruptly narrowing above the dilated portion. Flowering in winter or very early spring. (*G. tenuifolia* (Raf.) Fern.; *Dasystephana tenuifolia* (Raf.) Pennell)

Seasonally wet pine savannas and flatwoods. (Individual plants usually remotely scattered and seldom observed since they mostly bloom in midwinter.) Fla. Panhandle.

3. Gentiana crinita Froel. FRINGED GENTIAN.

Annual, of varying stature, to about 8 dm tall. Stem glabrous, simple or with several elongate erect-ascending branches. Midstem leaves largest; lowest leaves spatulate and narrowed to petiolelike bases, those above sessile, oblong-linear, then the middle and upper subclasping, broadest at the base, lance-ovate, acute apically, to about 5 cm long. Flowers solitary on the branches, their stalks 4–12 cm long. Calyx ½ as long as the fully developed corolla or a little more, the tube and lobes keeled, the edges of the keels finely toothed, lobes 4, unequal, larger ones ovate-acuminate, smaller ones lance-attenuate. Corollas bright blue, broadly funnelform, 5–6 cm long, lobes 4, broadly obovate, apices broadly rounded, fringed at the summit and part way down the sides. (*Anthopogon crinitum* (Froel.) Raf.)

Bogs, wet meadows, brooksides, wet thickets, seepage slopes. Maine to Ont., s. Man., Mich., n. Minn., generally south to Pa. and n. Iowa, very local southward in the Appalachians to n. Ga.

4. Gentiana quinquefolia L. AGUE-WEED, STIFF GENTIAN.

Annual with glabrous stem freely branched, often from near the base upwardly, sometimes as much as 8 dm tall, more commonly 3 or 4 dm, the branches somewhat wing-angled, rigidly erect-ascending. Lower stem leaves smaller than the middle and upper, oblong and obtuse apically, middle and upper leaves ovate to ovate-lanceolate, basally somewhat clasping, acute or somewhat acuminate apically, the larger mostly 3–4 cm long. Flowers in clusters terminating the main stem and axillary branches, or in axillary clusters on the branches, virtually sessile or on stalks of varying lengths to 2.5 cm long. Lobes of calyx and corolla 5, those of the calyx nearly equal, longer than the calyx tube, linear-subulate to oblong or oblanceolate; calyx 3–6 (–9) mm long overall, very much shorter than the corolla. Corolla narrowly funnelform, pale lavender-blue to lilac, rarely greenish white, tube about 1.5 cm long, lobes short, ovate, acuminate apically and softly short bristle-tipped, without plaits in the sinuses. (*Gentianella quinquefolia* (L.) Small; incl. *Gentianella occidentalis* (Gray) Small = *Gentiana quinquefolia* var. *occidentalis* (Gray) Hitchc.)

Moist open somewhat boggy slopes, open moist stream banks, wet gravelly banks, damp fields, roadsides. Maine, s. Ont., to Mich. and Minn., southward in the Appalachians to N.C. and Tenn., southward in the western part of the range to La.

5. Gentiana catesbaei Walt.

Perennial with thickish fibrous roots. Stem simple, 3–8 dm tall, or with short branches near the summit, distally (or throughout) scabrid with short, rough, reddish pubescence or protuberances usually in vertical lines. Leaves smallest below, these variable in size, oblong to oblanceolate; midstem and upper leaves mostly elliptic, sometimes oblong-lanceolate, mostly 3–7 cm long, margins very finely toothed; tapered below, sometimes to a short-petiolate base, apically acute or obtuse. Calyx and corolla 5-lobed. Calyx about ½ as long as fully developed corolla, the lobes usually a little longer than the tube, linear to oblanceolate or narrowly elliptic, more or less foliaceous, very finely toothed on their margins. Corolla broadly funnelform to campanulate, deep blue to blue-violet, sometimes bluish white, 3–5 cm long, the tube much longer than the lobes; lobes more or less short-ovate, apically short-acuminate, plaits in the sinuses narrower and shorter than the lobes, lacerate at the summit. (*Dasystephana latifolia* (Chapm.) Small; incl. *D. parvifolia* (Chapm.) Small)

Pineland seepage slopes, borders of and clearings of wet woodlands, swales and ditches. Coastal plain, Del. to n. Fla., and s. Ala.

6. Gentiana saponaria L. SOAPWORT GENTIAN.

In general features similar to *G. catesbaei*. Stem smooth. Calyx a little less than ½ as long as the corolla, lobes about as long as the tube, lanceolate to oblong or oblanceolate, somewhat foliaceous, finely toothed marginally. Corolla blue to violet or bluish white, funnelform, 3–4 (–5) cm long, lobes short, ovate, acute to obtuse or rounded apically, plaits ¾ as long to nearly as long as the lobes and about as wide as the lobes, lacerate at the summit. (*Dasystephana saponaria* (L.) Small)

Bogs, marshes or marshy shores, wet woodlands, glades, swamps, moist thickets. N.Y. to Wisc. and Minn., generally southward to n. Fla. and e. Tex.

7. Gentiana clausa Raf. CLOSED GENTIAN.

In general features similar to *G. catesbaei* and *G. saponaria*. Stem smooth, middle and upper leaves ovate or lance-ovate. Calyx lobes shorter than the tube, ovate to obovate, finely toothed marginally, foliaceous, the calyx less than ½ as long as the fully developed corolla. Corolla nearly oblong in outline, whitish below, suffused with blue above and changing with age to blue-violet, 2.5–4 cm long; corolla very shortly lobed at the summit, lobes incurved, plaits whitish, about equaling the lobes in length, broad, truncate at the summit and short-toothed. Corolla essentially closed at anthesis.

Wooded slopes, banks of streams, meadows, thickets, in both mesic and moist to wet places. s. Maine, s. Que. to n.e. Ohio, N.J. and Pa., southward in the Appalachians to n.e. Tenn. and n.w. N.C.

Menyanthaceae (BOGBEAN FAMILY)

Nymphoides (FLOATING-HEARTS)

Perennial aquatic herbs with rhizomes in the submersed soil substrate, arising alternately from the rhizomes are several long, slender, petiolelike stems each of which bears 1 to several petiolate leaves, in some the petioles giving the appearance of being continuous with the stem, in some shortly branched from the node bearing the first leaf or leaves. Inflorescences umbellike in the leaf axils. In some, late in the season, the flowers or fruits admixed with or subtended by a somewhat bananalike cluster of fleshy, tuberlike roots. Flowers bisexual. Calyx united at base, 5-lobed. Corolla white, cream-colored, or yellow, united below the base of the 5 lobes, each lobe with a single scalelike gland internally at the base. Stamens 5, inserted on the short corolla tube. Ovary superior, many-ovuled, tapering to a short style. Stigma broad, 2-lobed. Capsule firm-walled, indehiscent or rupturing irregularly. Seeds smooth or papillate.

1. Node at end of petiolelike stem bearing 2 to several leaves, usually with 1 or more short branches from this node; without tuberlike roots admixed with or subtending the flower cluster; corolla golden yellow. 1. *N. peltata*
1. Node at end of petiolelike stem bearing a single leaf (rarely 2), without branches from this node; a cluster of tuberlike roots admixed with or subtending the flower cluster (late in the season at least); corolla white.
   2. Floating leaves cordate-ovate, lower surface relatively smooth; petiolelike stems less than 1 mm in diameter below the inflorescence, rarely red-punctate; capsule only slightly exceeding the calyx; seeds smooth (rarely papillate). 2. *N. cordata*

2. Floating leaves ovate to reniform, lower surface rough; petiolelike stems more than 1 mm in diameter below the inflorescence, conspicuously red-punctate; capsule considerably exceeding the length of the calyx; seeds conspicuously papillate.   3. *N. aquatica*

1. Nymphoides peltata (S. G. Gmel.) Kuntze. **YELLOW FLOATING-HEART.**

Node at the end of the petiolelike stem bearing 2 or more leaves, commonly with 1 or more short branches from this node. Leaves suborbicular in overall outline, the larger to 15 cm long and wide, the basal sinus about ⅓ the length of the blade, basal lobes cordate, margins entire or in oldest leaves mostly undulate-dentate, surfaces appearing smooth to the naked eye, the lower punctate as observed with suitable magnification. Flower stalks to about 6 cm long. Calyx lobes nearly oblong, about 1 cm long, tips obtuse. Corolla golden yellow, 2–3 cm across. Capsule oblongish, more or less asymmetrical, strongly beaked, 2.5–3.5 cm long including the beak. Seeds flat, narrowly winged, margin somewhat fringed.

Quiet waters of streams, ponds, impoundments, drainage ditches and canals. Native of s. Eur. and Asia Minor, introduced for cultivation, sporadically naturalized, s. N.Eng., N.Y. to Va., Ohio, Ind., Ill., Mo., Ark., La., Miss., Okla., n.e. Tex.; Ariz.

2. Nymphoides cordata (Ell.) Fern.   Fig. 252

Node at the end of petiolelike stem bearing 1 petiolate leaf (rarely 2), without branches from this node. Leaf blades cordate-ovate, 3–7 cm long, often variegated with purple above, smooth below. Petioles and petiolelike stems slender, the stems less than 1 mm in diameter below the inflorescence, rarely red-punctate. Inflorescence usually admixed with or subtended by a cluster of fleshy tuberlike roots about as thick as the stem, slenderly long-tapering distally. Flower stalks several, to 3 cm long. Calyx lobes oblong, 2–4 mm long, tips acute, sometimes faintly red-punctate. Corolla white or cream-colored, 0.5–1 cm across. Capsule ovoid to subglobose, about 4 mm long, only slightly exceeding the calyx. Seeds subglobose, smooth or rarely sparsely papillate. (*N. lacunosa* of authors not (Vent.) Kuntze)

Ponds, lakes, and quiet streams, Nfld. and Que., s. N.Eng., southward to Fla. Panhandle, westward to La.

3. Nymphoides aquatica (S. G. Gmel.) Kuntze.   Fig. 253

Leaf blades bluntly ovate to reniform, 5–15 cm long, green above, lower surface often deep purple and appearing rough to the naked eye. Petioles and petiolelike stems relatively stout, the stems over 1 mm in diameter below the inflorescence, conspicuously reddish purple – punctate. Inflorescence admixed with or subtended by a cluster of stout, fleshy, blunt-tipped, tuberlike roots late in the season. Flower stalks stout, conspicuously purple-red-punctate, to 8 cm long. Calyx purple-red-punctate, lobes 4–5 mm long, tips rounded to obtuse. Corolla white, 1–2 cm across. Capsule ovate, 10–14 mm long, considerably exceeding the calyx. Seeds globose, conspicuously papillate.

Ponds, lakes, quiet streams, drainage canals. Coastal plain, s. N.J. to Fla., thence to e. Tex.

Apocynaceae (DOGBANE FAMILY)

Herbs or twining woody vines (ours) with milky sap. Leaves simple, opposite or alternate, occasionally whorled, margins entire. Flowers bisexual, radially symmetrical. Calyx united basally, 5-lobed, in some with glandular appendages within. Corolla united below, 5-lobed, in some with appendages within the throat. Stamens 5, inserted on the corolla tube and basally adnate to it, alternate with the lobes of the corolla. Ovaries 2 (in ours) but the styles united, the stigma large, each ovary ripening into a follicle. Seeds naked or with a tuft of hairs on one end.

1. Leaves alternate or in some closely set and nearly whorled; stem erect, herbaceous.   1. *Amsonia*
1. Leaves opposite; stem twining, older portions woody.

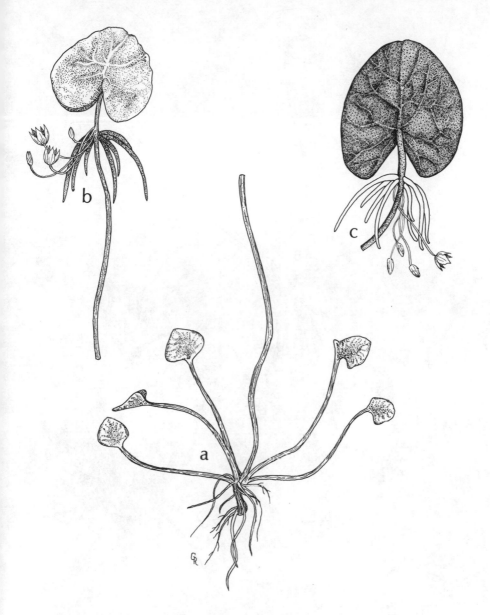

Fig. 252. **Nymphoides cordata:** a, base of seedling plant with part of stem from its distal node; b, cluster of fleshy roots, flower buds and flowers, and a leaf; c, lower leaf surface, roots, a flower and buds.

539

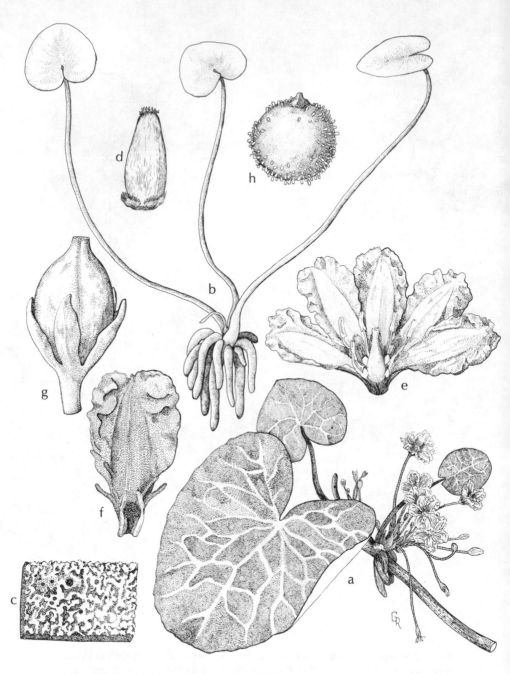

Fig. 253. **Nymphoides aquatica:** a, distal portion of petiolelike stem, bearing at the node a leaf, several tuberlike, spurlike roots, a flower cluster, and a pair of developing petiolelike stems each bearing a leaf; b, plantlet from a detached cluster of tuberlike roots; c, small portion of lower leaf surface; d, developing pistil; e, flower, corolla spread out; f, petal; g, fruit; h, seed.

2. Leaf blades abruptly short-acuminate apically; corolla tube 5–7 mm long, lobes 3–4 mm long.
2. *Trachelospermum*
2. Leaf blades rounded or merely mucronate apically; corolla tube and lobes much longer.
   3. Corolla yellow; calyx lobes 10–14 mm long.   3. *Urechites*
   3. Corolla white without, yellow within the tube; calyx lobes 4–5 mm long.   4. *Rhabdadenia*

1. Amsonia (BLUE-STARS)

Perennial herbs with a subwoody caudex and often several to numerous erect stems. Lowermost leaves scalelike, stem gradually becoming more "leafy" upwardly, the largest ones commencing about midstem, the uppermost little reduced if any. Inflorescence a compact, usually terminal, panicle. Corolla blue, funnelform-salverform, villous in the throat. Follicle linear-elongate, cylindric or with horizontal constrictions between the seeds.

1. Corolla glabrous exteriorly.   1. *A. rigida*
1. Corolla pubescent exteriorly.
   2. Leaf blades glabrous or sparsely pubescent beneath; follicle glabrous.   2. *A. tabernaemontana*
   2. Leaf blades permanently tomentose beneath; follicle pubescent at least distally. 3. *A. ludoviciana*

1. Amsonia rigida Shuttlw. ex Small.   Fig. 254

Stems usually several to numerous from the caudex, 5–15 dm tall, glabrous. Larger leaves short-petioled, blades varying considerably in size and shape from plant to plant, to 8 cm long and 1.5 cm broad, lanceolate, lance-elliptic, elliptic, ovate, or linear-lanceolate, cuneate basally, acuminate apically, both surfaces glabrous, the lower usually much paler than the upper, rarely ciliate on the margins. Calyx glabrous, lobes triangular-acute, 1.5 mm long. Corolla blue, glabrous exteriorly, tube 6–8 mm long, slightly dilated above the insertion of the stamens, not constricted at the orifice, lobes spreading, lanceolate, about 10 mm long, obtuse apically. Follicles slenderly cylindrical, 8–10 cm long, glabrous. (Incl. *A. glaberrima* Woods.)

Wet, semiopen or open depressions in pinelands, open swampy places, wet bottomland clearings, shores of ponds and impoundments. Coastal plain, s. Ga. to n. pen. Fla., westward to La.

2. Amsonia tabernaemontana Walt.

Stems solitary or several from the caudex, glabrous, 3–10 dm tall. Larger leaves with petioles 3–6 mm long, blades thinnish, ovate, broadly elliptic, varying to lanceolate, 8–15 cm long, 1–6 cm broad, broadly tapering to rounded basally, acuminate apically, upper surfaces glabrous or sparsely pubescent, lower glabrous, sometimes slightly glaucous, or usually sparsely pubescent. Calyx glabrous, lobes ovate or lance-ovate, 1–2 mm long. Corolla pale blue, tube 6–8 mm long, a little dilated above the insertion of the stamens, not constricted at the mouth, externally pilose at least on the upper part of the tube and on the lobes, lobes spreading, 4–7 mm long, lanceolate, blunt apically. Follicles erect, 8–10 cm long, narrowly cylindric, glabrous. (Incl. *A. salicifolia* Pursh, *A. tabernaemontana* var. *salicifolia* (Pursh) Woods; *A. amsonia* (L.) Britt.)

Mesic woodlands, river banks, floodplain woodlands and clearings, alluvial thickets. Va. to s. Ind., s. Ill., Mo. and e. Kan., generally southward to Fla. Panhandle and n.e. Tex.

3. Amsonia ludoviciana Vail.

Similar in general aspect to *A. tabernaemontana*. Stem pubescent at least when young. Leaf blades broadly elliptic, glabrous or nearly so above, densely and persistently matted-pubescent and whitish beneath. Calyx lobes and corolla tube softly short-pubescent exteriorly. Follicles pubescent distally.

Moist, open woodlands. Miss., La., Ark. (according to Small (1933)).

Fig. 254. **Amsonia rigida:** a, base of plant; b, top of plant; c, flower; d, fruit.

2. Trachelospermum

Trachelospermum difforme (Walt.) Gray. CLIMBING DOGBANE.   Fig. 255

Slender, deciduous, twining vine, stems glabrous, woody except at the herbaceous growing tips. Leaves short-petioled or subsessile, with minute stipules, blades, even on a single plant, varying in shape and size, lanceolate, elliptic, oval, obovate, or suborbicular, apices abruptly short-acuminate, to 14 cm long and 8 cm broad (mostly little more than half that size or less); upper surfaces glabrous or very sparsely short-pubescent, lower usually moderately short-pubescent, sometimes glabrous. Flowers small, in panicled, axillary cymes (only in the axil of 1 of a pair of leaves), cyme axes glabrous, flower stalks slender, glabrous, 4–10 mm long. Calyx glabrous, erect, lobes about a third as long as the corolla tube, lance-ovate basally, abruptly narrowed to long-acuminate or subulate tips. Corolla funnelform, pale yellow or greenish yellow, tube 5–7 mm long, lobes obliquely obovate, spreading, 3–4 mm long. Anthers adherent to the stigma, not exserted. Follicles very slenderly cylindric, 1–2 dm long, glabrous. Seeds angled, narrowly linear, truncate at both extremities, short-pubescent, bearing apically a conspicuous tuft of long, silky hairs.

River banks, floodplain and bottomland woodlands, clearings, and thickets, marshes, sloughs. Coastal plain and piedmont, Del. to n. Fla., westward to e. Tex., northward in the interior to Mo. and s. Ind.

3. Urechites

Urechites lutea (L.) Britt. RUBBER-VINE, WILD ALLAMANDA.

Twining, woody vine, sometimes pubescent, more often glabrous. Leaves opposite, variable in shape, oblong, obovate, ovate, oval, or elliptic, 3–7 cm long, mostly 2–3 cm broad, bases obtuse to rounded, apices mostly rounded, sometimes mucronate, bright, shiny green above, pale green beneath, venation raised-reticulate on both surfaces. Flowers in stalked, foliose-bracted, usually few-flowered cymes axillary to one of a pair of leaves, individual flowers occasionally sessile, usually on stalks 1–1.5 cm long. Calyx lobes lance-attenuate, 10–14 mm long. Corolla yellow, salverform, tube narrowly cylindric for about 1 cm at the base, flaring above into a campanulate throat 2.5–3 cm long and about 1.5 cm across distally; corolla 4–5 cm across the obliquely obovate, spreading lobes. Follicle narrowly cylindrical but tapering to pointed tips, usually incurved, 12–20 cm long.

Mangrove swamps, tropical hammocks. s. pen. Fla. and Fla. Keys; W.I.

4. Rhabdadenia

Rhabdadenia biflora (Jacq.) Muell.-Arg. in Mart. RUBBER-VINE.

Glabrous, woody, twining vine. Leaves opposite, short-petioled, blades oblong, elliptic-oblong, or obovate, 5–12 cm long, 1.5–5 cm broad, obtuse basally, rounded-mucronate apically. Flowers solitary or in 2–5-flowered, stalked cymes axillary to one of a pair of leaves; stalks of individual flowers usually 7–10 mm long. Calyx barely united basally, lobes elliptic-oblong, short-acuminate apically, 4–5 mm long. Corolla salverform, white without, yellow within the tube, tube 4.5–7.5 cm long, its basal half narrowly cylindrical, flaring above to a campanulate throat; corolla about 4 cm across the obliquely obovate, spreading lobes. Follicles narrowly cylindric-fusiform, straight or slightly curved, 10–14 cm long.

Mangrove swamps, canal banks, coastal thickets, tropical hammocks. s. pen. Fla. and Fla. Keys; W.I.; Mex. to n. S.Am.

Asclepiadaceae (MILKWEED FAMILY)

Herbs, shrubs, or vines, most with milky sap. Leaves simple, opposite or whorled in most, sometimes alternate. Flowers in umbels, bisexual, radially symmetrical. Calyx and corolla

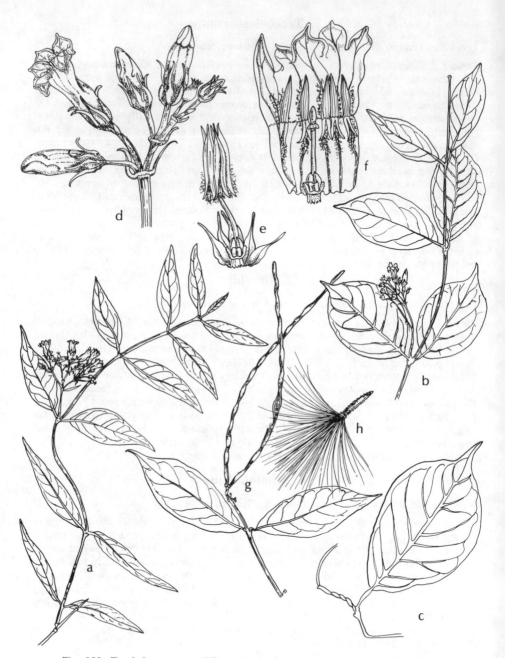

Fig. 255. **Trachelospermum difforme:** a,b, flowering branches with different leaf shapes; c, leaf; d, inflorescence; e, flower with corolla removed and calyx somewhat spread; f, corolla, opened out, and pistil; g, follicles; h, seed. (From Correll and Correll. 1972)

each united basally, each usually deeply 5-lobed. Stamens 5, alternate with the corolla lobes, filaments distinct or united into a tube called the column, anthers more or less connivent and adherent to the stigma, forming with the stigma a structure called the gynostegium; pollen grains united into waxy or granular pollinia. Five structures called hoods usually present between the corolla and stamens (collectively called the crown) and adnate to the column, to the corolla, or to both. Ovaries 2, superior, ovules numerous in each, style of each distinct, their stigmas united; if maturing, each ovary becoming a follicle, but both ovaries of a given flower seldom mature and in many flowers neither of them matures. Seed usually with a conspicuous tuft of long, silky hairs (a coma) at one end.

1. Stem erect or at most decumbent, not a vine.   1. *Asclepias*
1. Stem a twining vine.
  2. Corolla 12–15 mm across, lobes spreading at right angles to the floral axis; leaves oblong, oblong-elliptic, or lance-ovate.   2. *Sarcostemma*
  2. Corolla 5–7 mm across, lobes more or less erect and with tips arching outward; leaves linear or cordate-ovate.   3. *Cynanchum*

1. Asclepias (MILKWEEDS)

Perennial herbs, or a few annual. Leaves opposite or whorled in most, in some alternate or irregularly approximate. Flowers in stalked, terminal, or terminal and axillary umbels. Corolla lobes in most reflexed at anthesis concealing the smaller calyx, rarely erect. Hoods attached to the column just below the gynostegium, straight or curved, erect or ascending, diverse in form, usually saccate basally, in some bearing an internal horn.

1. Leaves alternate (often irregularly so; thus approximate leaves may be very close together).
  2. Umbels solitary at the apex of the main stem or its branches; tips of the hoods extending beyond the gynostegium each with an exserted horn curving over the top of the gynostegium.   1. *A. michauxii*
  2. Umbels 2–7 (rarely 1) distally on the main stem or its branches; tips of the hoods appressed against the base of the gynostegium, lacking horns.   2. *A. longifolia*
1. Leaves consistently opposite.
  3. At full anthesis, corolla lobes erect, their margins touching or nearly so, the corolla obscuring the hoods and gynostegium within.   3. *A. pedicellata*
  3. At full anthesis, corolla lobes reflexed, the hoods and gynostegium clearly showing above the corolla.
    4. Hoods each with a pair of triangular, erect lobes about 1 mm long on each side at the apex.
                                           4. *A. cinerea*
    4. Hoods without a pair of apical lobes.
      5. Corolla flame orange–red, bright crimson, or bright yellow; hoods bright orange.
        6. Perennial, rootstock bearing lateral, fusiform-tuberous roots; longer pairs of stem leaves 3–6, spreading at right angles to the stem, internodes elongate; umbels not leafy-bracted, borne on a stalk extending much above the uppermost pair of leaves.   5. *A. lanceolata*
        6. Annual, roots all fibrous; stem leaves numerous, ascending, internodes short; umbels (at least the lower) from axils of leaves.   6. *A. curassavica*
      5. Corollas and hoods variously colored but not as above.
        7. Leaves narrowly linear.   7. *A. viridula*
        7. Leaves not narrowly linear.
          8. The leaves sessile.
            9. Corolla greenish yellow; leaves mostly oblong, varying to oblanceolate, obovate, or oval, apices blunt; hoods greenish cream, arching and incurved, their tips meeting above the gynostegium, horns none.   8. *A. connivens*
            9. Corolla dull purplish red to lavender; leaves ovate to lanceolate, apices long-acute or acuminate; hoods orange-tinged, erect, surpassing the gynostegium and not incurved over it, a subulate horn exserted from each and loosely arching over the gynostegium.   9. *A. rubra*
          8. The leaves (some of them on a given stem at least) petiolate.
            10. Corolla white or white suffused with pale pink; stems slender, not exceeding about 6 dm tall; leaf blades long-acuminate at both extremities.   10. *A. perennis*
            10. Corolla bright rose purple; stem stout, 10–15 dm tall; bases of leaf blades rounded, truncate, subcordate, or short-tapered.   11. *A. incarnata*

1. Asclepias michauxii Dcne. in DC.

Perennial with a woody, more or less tuberous, medially fusiform rootstock bearing 1 to several stems 1–4 dm high. Stem purplish, clothed with short, upwardly curved hairs. Lowermost 1–3 leaves smallest, other abruptly larger and nearly equal, alternate or subopposite, tending to be secund, 5–10 cm long, linear-subulate or narrowly linear-oblong, sides folded upwardly, sparsely very short-pubescent. Umbels solitary and terminal on the main stem or on its branches, 3–4 cm broad, mostly 6–20-flowered, flower stalks 1–1.5 cm long. Calyx lobes lanceolate or lance-subulate, about 3 mm long. Corolla lobes 5 mm long, reflexed, their tips arching outwardly and upwardly, lower surfaces purple, upper greenish. Hoods purple, erect, with shape of a flour scoop, rounded sides outward, tips extending beyond the gynostegium, each with a purple horn exserted and curving over the top of the gynostegium. Follicle narrowly fusiform, 10–15 cm long, essentially glabrous.

Pine savannas and flatwoods, borders of cypress-gum ponds or depressions and shrub-tree bogs or bays, less frequently in mixed, well-drained woodlands. Coastal plain, s.e. S.C. to cen. pen. Fla., westward to La.

2. Asclepias longifolia Michx.

Perennial, with a woody, elongate, stout rootstock 0.5–3 cm in diameter, bearing 1 to several stems 2–7 dm tall. Stem purplish, nearly glabrous below, distally clothed with upwardly and inwardly arched, short hairs. (Hairs of other parts similar.) Leaves smallest at base of stem, linear-subulate, gradually becoming largest somewhat below midstem, larger ones 5–15 cm long, 2–12 mm broad, long-linear, linear-lanceolate, or lanceolate, relatively short-tapered basally, long-tapered distally; reduced lower leaves alternate, others below the inflorescence irregularly approximate, those of the inflorescence mostly opposite, sometimes whorled; upper surfaces with scattered hairs, mainly pubescent on the veins and along the margins beneath. Slenderly stalked umbels 2–7, rarely 1, each from one of a pair of leaves of the upper nodes; umbels 3–5 cm broad, with numerous small flowers on pubescent stalks about 2 cm long. Calyx lobes lanceolate to ovate, 2 mm long, pubescent. Corolla lobes 5–6 mm long, strongly reflexed, their obtuse tips arching outward, whitish on their lower halves, suffused with purple distally. Hoods with the form of a flour scoop, rounded sides outward, 2 mm long, unevenly suffused with purple, lacking horns within, their rounded tips appressed against the base of the gynostegium. Follicle lance attenuate in outline, about 10 cm long, pubescent. (*Acerates floridana* (Lam.) A. Hitchc.; incl. *Acerates delticola* Small)

Seasonally wet pine savannas and flatwoods, pine-cypress depressions, hillside bogs in pinelands. Coastal plain, Del. to s. pen. Fla., westward to e. Tex.

3. Asclepias pedicellata Walt.

Rootstock more or less elongate-cylindric but irregularly gnarled-knobby, 4–10 mm thick, distally (downwardly) abruptly expanded to a somewhat spongy, obconic, tuberous portion 3–4 cm in diameter. Stem usually solitary, 0.5–4 dm tall, purplish below, clothed with short, upwardly and inwardly curved hairs. Leaves opposite, sessile, lowermost 2 or 3 pairs bractlike, abruptly larger upwardly, largest medially, narrowly lanceolate to linear-oblong, 3–6 cm long, broadest near the base, long-tapered distally to acute apices, surfaces sparsely pubescent. Umbels short-stalked, few near the summit of the stem, lower ones in one of a pair of leaf axils, uppermost usually paired, each 2–8-flowered, flower stalks about 1 cm long. Calyx lobes lanceolate, 3 mm long. Corolla lobes 8–12 mm long, greenish yellow, erect, oblong, broadly channelled in the lower ⅔, their edges touching or nearly so, apices abruptly narrowed. Hoods inserted at the base of the column and base of the corolla, ¼ as long as the corolla lobes and scarcely ½ as long as the slender column, colored like the petals, narrowly saclike on the outer side, the tips ovate-acuminate and strongly inflexed-reflexed. Follicle not seen. (*Podostigma pedicellata* (Walt.) Vail ex Small)

Seasonally wet pine savannas and flatwoods, infrequently in well-drained sands of scrub. Coastal plain, s.e. N.C. to s. pen. Fla., in the Fla. Panhandle westward to at least the Apalachicola River region.

4. Asclepias cinerea Walt.

Rootstock woody, roughly cylindric, or with a portion fusiform-tuberous, sometimes with 2 or 3 fusiform-tuberous portions connected by narrow sections. Stem and leaves with aspect as in *A. viridula* (with which it sometimes grows) but the stem wholly glabrous. Umbels several, usually only one of a pair of upper leaves subtending an umbel, occasionally one in the axils of each of a pair, subtending leaves commonly shorter than the umbels, occasionally equaling or shortly surpassing them (all three conditions sometimes on a given plant). Flowers loosely 3–10 in an umbel, their stalks lax, 1.5–2.5 cm long. Calyx lobes lanceolate, blunt apically, 2 mm long. Corolla lobes 5–7 mm long, oblanceolate to obovate, reflexed, lavender with white margins, or whitish basally with lavender veins and lavender with white margins distally. Hoods variously suffused with lavender, with saclike basal portions 2–2.5 mm long, a pair of triangular, erect lobes about 1 mm long on the margins of each distally, the lobes not reaching the summit of the gynostegium, a horn protruding from within each hood, surpassed by the lobes of the hood. Follicle with a fusiform body, attenuate at the tip.

Pine savannas and flatwoods, often where wet or boggy, less frequently in well-drained sands of adjacent pine–scrub oak ridges. Coastal plain, s.e. S.C. to n. Fla., Fla. Panhandle.

5. Asclepias lanceolata Walt.   Fig. 256

Rootstock more or less cylindric, short to elongate, bearing laterally a cluster of fusiform, tuberous roots. Stem solitary, purplish at least below, 5–12 dm tall, glabrous. Leaves opposite, pairs few, internodes elongate, lowermost 2 or 3 pairs smallest, 1–3 cm long, linear, erect-ascending, abruptly larger above, largest ones medial, borne at right angles to the stem, linear-lanceolate to lanceolate, long-attenuate distally, 10–25 cm long, glabrous. Umbels 2–4, subapproximate, held well above the leaves, their subtending bracts very small, linear, quickly deciduous; flowers mostly 3–8 per umbel, their stalks about 1.5 cm long. Calyx lobes narrowly triangular-acute, 2–2.5 mm long. Corolla lobes flame red–orange (drying dull purplish), 10 mm long, lance-elliptic, reflexed and arching outwardly. Hoods orange, exteriorly saclike basally, distally flaring, tips rounded, spreading, much surpassing the gynostegium, an orange horn protruding from each at an angle over the gynostegium. Follicle narrowly fusiform, 8–10 cm long, glabrous and smooth.

Fresh to brackish marshes, bogs, wet savannas and pinelands, glades, cypress depressions, wet ditches. Coastal plain, s. N.J. to s. pen. Fla., westward to s.e. Tex.

6. Asclepias curassavica L.

Annual. Stem often tough and hard near the base, branched or unbranched, 3–12 dm tall, short-pubescent at first, soon becoming glabrous. Leaves opposite, petioles 5–10 mm long, blades lanceolate or long-elliptic-lanceolate, 5–12 cm long, 1–3 cm broad, at first minutely pubescent on the veins beneath and on the margins, usually becoming nearly or quite glabrous. Umbels several from one of each of a pair of leaves of upper nodes; flowers several to many per umbel, their stalks slender, 1.5–3 cm long. Calyx lobes lanceolate or lance-triangular, acute, 2–3 mm long. Corolla bright crimson or yellow (white on rare individuals), drying purplish, lobes reflexed, lanceolate or lance-elliptic, 7–10 mm long. Hoods orange, with the form of a flour scoop, rounded apically, surpassing the gynostegium, a similarly colored flat-subulate horn exserted from each and loosely curving above the gynostegium. Follicle narrowly fusiform, 6–10 cm long, glabrous and smooth.

Probably native to S.Am., widely distributed and weedy throughout the tropics and subtropics of both hemispheres. In our range, in low, open, moist to wet places, sometimes abundant in pastures in s. pen. Fla.; La.; s. Tex.

7. Asclepias viridula Chapm.

Perennial with a woody, tuberous, medially subglobose rootstock, usually bearing a solitary stem, infrequently more than one. Stem slender, purplish only at base, pubescent in longitudinal bands below the nodes with very short, spreading hairs. Lowermost 1–3 pairs of

Fig. 256. **Asclepias lanceolata:** a, plant; b, flower; c, hood with horn. (From Correll and Correll. 1972)

leaves minute and bractlike, others nearly uniform, opposite, narrowly linear, 4–10 cm long, glabrous. Umbels several, each in the axil of one of a pair of upper leaves, subtending leaves surpassing the umbels, mostly 6–10-flowered, flower stalks about 1 cm long. Calyx lobes lance-ovate, acute, 2 mm long. Corolla lobes 4–5 mm long, reflexed, their tips curving outwardly and upwardly, upper surfaces green, lower brownish purple. Hoods saclike basally, the erect free tips flaring, faintly brownish purple medially, otherwise cream-colored, each with a similarly colored horn slightly exserted and slightly arching over the gynostegium. Follicle narrowly fusiform, 8–10 cm long, glabrous.

Pine flatwoods, borders of shrub-tree bays or bogs. n.e. Fla. and lower Apalachicola River region.

8. Asclepias connivens Baldw. ex Ell.

Perennial, uppermost portion of rootstock tuberous-obconic, 2–4 cm thick, long-cylindric below and 5–8 mm thick. Stem stout, solitary, unbranched, 3–8 dm tall, more or less pubescent with short, curly hairs, becoming glabrous below. Leaves opposite, sessile, lowermost pairs smallest, often bractlike, others oblong, oblanceolate, narrowly obovate, or oval, 5–7 cm long, bases variable, rounded, truncate, or short-tapered, apices mostly blunt, infrequently acute, surfaces somewhat glaucous, sparsely short-pubescent beneath and minutely scabrid on the margins. Umbels usually several from axils of one of a pair of reduced leaves or small bracts, 3–6-flowered, flower stalks 1.5–2 cm long, copiously short-pubescent. Calyx lobes lanceolate to long-triangular, 6–7 mm long, acute apically. Corolla lobes greenish yellow, oblongish, 12–15 mm long, blunt apically, reflexed-rotate. Hoods prominent, strongly involute arching and incurved, the tips meeting above the gynostegium, horns none. Follicle 12–15 mm long, narrowly fusiform, attenuate distally, softly short-pubescent. (*Antherix connivens* (Baldw. ex Ell.) Feay ex Wood)

Bogs, wet pine savannas and flatwoods. Coastal plain, s. Ga., southward to s.cen. pen. Fla., Fla. Panhandle.

9. Asclepias rubra L.   Fig. 257

Perennial, rootstock fusiform-tuberous. Stem solitary, unbranched, 4–10 dm tall, glabrous or minutely and sparsely pubescent in longitudinal bands below the nodes. Leaves opposite, sessile, blades ovate to lanceolate, larger ones 6–14 cm long, 1.5–4 cm broad, rounded to truncate or subcordate basally, acuminate apically, glabrous, glaucous beneath. Umbels 2–4 from leafless upper nodes, flowers several to many per umbel, their stalks slender, mostly about 2 cm long. Calyx lobes narrowly triangular, 2–2.5 mm long. Corolla lobes dull purplish red to lavender, 8–9 mm long, oblong-acute, strongly reflexed. Hoods orange-tinged, lanceolate in outline, erect, much surpassing the gynostegium, their tips blunt, a subulate horn exserted about midway from each and loosely arching over the gynostegium. Follicle 8–10 cm long, narrowly fusiform. (Incl. *A. laurifolia* Michx.)

Bogs, pine savannas and flatwoods, borders of shrub-tree bogs or bays, swamps, wet meadows. Coastal plain, s. N.J., s.e. Pa., southward to Fla. Panhandle, westward to e. Tex.

10. Asclepias perennis Walt.   Fig. 258

Basal caudex relatively slender, bearing numerous fibrous roots. Stem solitary, or less frequently 2 to several from the caudex, each simple or few-branched, 3–6 dm tall, glabrous or sparsely short-pubescent in longitudinal bands below the nodes. Leaves opposite, glabrous, petioles 1–1.5 cm long, blades lanceolate, lance-ovate, or long-elliptic, mostly acuminate at both extremities, 6–12 cm long, 1–3 cm broad. Umbels 2–6 from axils of one of a pair of uppermost leaves or bracts, flowers several to numerous in each umbel, their stalks 8–12 mm long, pubescent with short upwardly and inwardly curved hairs. Calyx lobes oblongish, 2 mm long. Corolla lobes white or pinkish, elliptic, strongly reflexed then their distal halves arched outwardly, 4–4.5 mm long. Hoods colored like the corolla, with the form of a flour scoop, rounded side outward, about equaling the gynostegium, each with a prominently exserted horn loosely arching over the gynostegium. Follicle 6–7 cm long, body fusiform, attenuate distally, glabrous and smooth. Seed lacking a coma.

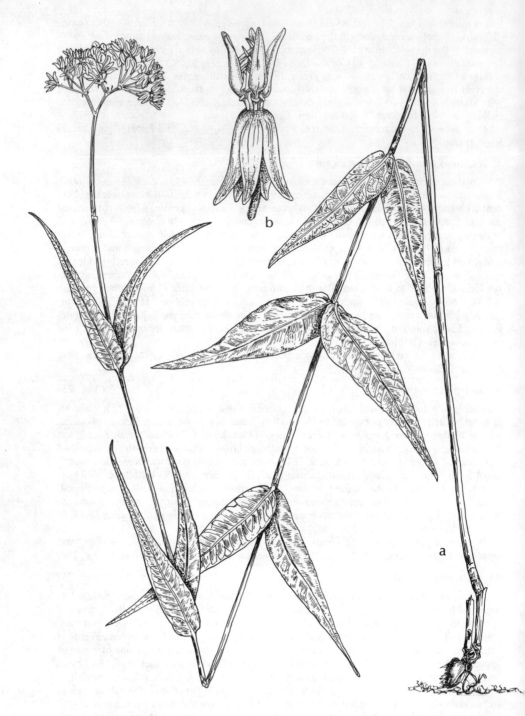

Fig. 257. **Asclepias rubra:** a, plant; b, flower. (From Correll and Correll. 1972)

Fig. 258. **Asclepias perennis:** a, basal portion of plant; b, upper part of plant with inflorescences and fruits; c, seed.

Fig. 259. **Asclepias incarnata:** a, base of plant; b, top of stem; c, flower. (From Correll and Correll. 1972)

Floodplain forests, swamps, wet woodlands, sloughs, ditches, muddy shores. Chiefly coastal plain, S.C. to cen. pen. Fla., westward to s.e. Tex., northward in the interior to s.e. Mo., s. Ill., s.w. Ind.

11. Asclepias incarnata L. SWAMP MILKWEED. Fig. 259

Perennial, rootstock short, bearing numerous fibrous roots. Stem usually stout, as much as 1–1.5 cm in diameter toward the base, to 15 dm tall, unbranched to much branched, glabrous, or with longitudinal bands of short, shaggy pubescence below the nodes, or copiously and uniformly short, shaggy pubescent. Leaves numerous, opposite, petioles 4–10 mm long, blades very variable from plant to plant, linear-lanceolate,.lanceolate, lance-elliptic, oblong, or long-ovate, 6–15 cm long, 1–4 cm broad, bases rounded, truncate, subcordate, or short-tapered, apices acute, gradually long-acuminate, or abruptly short-acuminate or acute; surfaces glabrous or nearly so, or moderately pilose-hispid, the lower more so than the upper. Umbels 2 to several distally on the branches, the lower often from axils of little-reduced leaves, other from axils of reduced leaves or small bracts, flowers several to many per umbel, their stalks, slender, mostly 1–1.5 cm long, pubescent. Calyx lobes oblongish, 2–2.5 mm long, sparsely pubescent. Corolla bright rose-purple, (white on a rare individual), lobes 3–4 mm long, elliptic-oblong, reflexed-rotate, tips blunt. Hoods colored like the corolla, saclike at their bases, oblongish in outline, rounded apically, barely surpassing the gynostegium, an acicular horn much exserted from each and curved over and well above the gynostegium. Follicle 6–10 cm long, body fusiform, attenuate distally.

Plants with relatively little pubescence on the stems (below the inflorescence) and leaves, leaf blades gradually long-acuminate apically, may be referred to *A. incarnata* var. *incarnata*; those with copiously and uniformly pubescent stems, leaves moderately pilose-hispid, acute or abruptly acuminate apically, may be referred to var. *pulchra* (Ehrh. ex Willd.) Fern.

Swamps, marshes, wet shores, meadows, ditches, wet prairies, low fields and pastures. N.S. to Man., generally southward to s. pen. Fla., and N.M.; Utah.

2. Sarcostemma

Sarcostemma clausum (Jacq.) Schult. in L. MILK WITHE.

Subsucculent, herbaceous, twining and clambering vine, commonly forming dense mats over shrubs and into and over small trees. Stems much branched, to several meters long, sparsely pubescent or glabrous. Leaves (sometimes absent at flowering time) opposite, with petioles about 5 mm long, blades mostly narrowly to broadly oblong, varying to elliptic-lanceolate or oval, 2–5 (–8) cm long, 0.5–3 cm broad, bases rounded to subcordate, apices acute, abruptly short-acuminate, or mucronate, lower surfaces sparsely pubescent. Umbels irregularly in the axils of one of a pair of leaves, their stalks longer than the subtending leaves; flowers fragrant, several to many per umbel, their copiously pubescent stalks slender, 7–20 mm long. Calyx lobes lanceolate to oblong, 2.5–3 mm long, short-acuminate apically, pubescent. Corolla greenish without, white within, lobes elliptic to ovate, rotate and with a 12–15 mm spread, pubescent without and on the margins. Hoods seated on a disc, erect and a little surpassing the gynostegium. Follicle 5–8 cm long, lanceolate in outline, pubescent. (*Funastrum clausum* (Jacq.) Schlecht.)

Weedy, swamps, marshes, canal and ditch banks, willow thickets, moist roadsides. s.cen. and s. pen. Fla.; native of trop. Am.

3. Cynanchum (SAND-VINES)

Twining herbaceous vines with opposite leaves. Flowers small, in umbels or short racemes axillary to one of a pair of leaves. Calyx segments triangular-acute. Corolla campanulate, white or greenish white.

- Leaves sessile, narrowly linear.   1. *C. angustifolium*
- Leaves petiolate, cordate-ovate.   2. *C. laeve*

1. Cynanchum angustifolium Pers.

Vine slender, subsucculent. Stem to 1 m long or more, sparsely short-pubescent when young, quickly becoming glabrous. Leaves sessile, commonly reflexed, narrowly linear, 2–8 cm long, glabrous. Stalks of the umbels a little shorter than to a little longer than the subtending leaves, flowers several per umbel, their stalks about 5 mm long, glabrous or nearly so. Calyx lobes narrowly triangular-acute, 1–2 mm long, short-pubescent on their margins. Corolla greenish white, sometimes tinged with rose or purple, lobes lance-ovate and acuminate, erect proximally, slightly spreading distally, 5–7 mm spread at the summit. Hoods seated on a disc, erect and somewhat surpassing the gynostegium. Follicle lance-attenuate in outline, 5–7 cm long, sparsely and minutely pubescent. (*C. palustris* (Pursh) Heller, *Lyonia palustris* (Pursh) Small)

Coastal salt and brackish marshes, moist, sandy or marly soil of adjacent hammocks, spoil banks and flats, shell middens. N.C. to s. pen. Fla., Fla. Keys, westward to Tex.; Bah.Is., W.I.

2. Cynanchum laeve (Michx.) Pers.

Vine slender, to 3 m long or more, pubescent in longitudinal strips. Leaves petiolate, petioles (1–) 3–5 cm long, usually pubescent on one side, blades cordate-ovate, acuminate or acute apically, 2.5–5 (–8) cm long and 2–5 cm across the base, margins entire, surfaces glabrous or, more frequently, short-pubescent along the veins. Flowers in umbels or short racemes, the umbel a little shorter than to a little longer than the petioles; stalks of the umbel and the flower stalks pubescent. Sepals triangular-acute, about 2 mm long, pale along their margins, pubescent at least proximally. Corolla white, about 6 mm long, deeply parted, lobes erect at anthesis; corona of 5 membranous, lanceolate, erect segments surpassing the gynostegium and about equaling the corolla lobes, each segment divided to the middle into 2 linear lobes. Follicle lance-attenuate in outline, 5–15 cm long and to 3 cm broad a little above the base. (*Ampelamus albidus* (Nutt.) Britt.)

In alluvial thickets, stream banks, floodplain and other low woodlands, lowland fields, fence rows. Pa. to Mo. and Kan., generally southward to the Fla. Panhandle and cen. Tex.

Convolvulaceae (MORNING-GLORY FAMILY)

Annual or perennial herbs (ours), shrubs or trees, many prostrate or climbing by twining, infrequently erect, stems of most with some milky sap; one genus (*Cuscuta*) without green pigment and parasitic. Leaves alternate, simple, without tendrils (reduced to scales or absent in *Cuscuta*). Flowers solitary or in cymes, bisexual, radially symmetrical, the calyx usually subtended by and often surrounded by 2 opposite or subopposite bracts. Perianth and androecium 4- or 5-merous. Sepals distinct or rarely united basally, somewhat imbricated. Corolla united, rotate, funnelform, or salverform, without lobes or variously lobed. Stamens inserted on the corolla near the base in most, in the throat or sinuses of the corolla in *Cuscuta*. Pistil 1, ovary superior, 1–4-locular, ovules 1 or 2 in each locule, styles 1 or 2, simple or forked. Fruit in most genera a capsule dehiscing longitudinally, circumscissile in some, irregularly dehiscent or indehiscent in some.

Dr. Daniel F. Austin of the Florida Atlantic University made available to us his unpublished manuscript treating genera of Convolvulaceae occurring in Florida. This was of much help to us and we acknowledge the courtesy most gratefully.

1. Plants without green pigment and parasitic; leaves scalelike.   1. *Cuscuta*
1. Plants green and autotrophic; leaves well developed.
  2. Corolla minute, shorter than the calyx, with 5 lobes longer than the tube; capsule deeply 2-lobed.
                                            2. *Dichondra*
  2. Corolla conspicuous, surpassing the calyx, unlobed, or if lobed the lobes shorter than the tube.
    3. Styles 2, free or fused only basally.

4. The styles free, each 2-cleft, stigmas 4, linear-filiform.   3. *Evolvulus*
4. The styles united basally, stigmas 2, capitate.   4. *Stylisma*
3. Style solitary.
    5. Flowers in dense, headlike clusters with interspersed, foliaceous, long-hirsute bracts.
                                                5. *Jacquemontia*
    5. Flowers solitary or in few- or many-flowered, loose cymes.
        6. Stigmas short-oblong.   6. *Calystegia*
        6. Stigmas capitate, entire or with 2 or 3 globose lobes.   7. *Ipomoea*

1. Cuscuta (DODDERS, LOVE-VINES)

Annual, twining, parasitic, yellow to orange vines without chlorophyll, roots absent on mature plants, attaching to host plants by haustoria. Leaves represented by minute scales. Flowers small, mostly white or yellowish, sessile or short-stalked, in few- to many-flowered cymose clusters, 5-merous (excepting pistil) in most, 3- or 4-merous in some. Both calyx and corolla united below and shallowly to deeply lobed above, or calyx of separate sepals in a few species. Corolla tube campanulate, cylindrical, or slightly urceolate, lobes erect, spreading, or reflexed, the tips inflexed in some; commonly with toothed, fringed, or fimbriated scale-like appendages below the stamens within and forming a corona. Stamens alternating with the corolla lobes and attached at or near the sinuses between the lobes. Ovary 2-locular, 2 ovules in each locule, styles 2, usually separate, somewhat united proximally in a few species, stigmas capitate to linear. Fruit a globose or ovoid capsule, bursting irregularly in some, or dehiscing with a regular or irregular line of circumscission near the base in some.

Fully developed flowers and fruits are necessary for purposes of accurate identification. If specimens are to be pressed and dried, it is best to have pieces of the host plant pressed with them, or to place appropriate-sized wads of paper about the *Cuscuta*, in either case so that some of the reproductive structures may dry with little or no pressing. If specimens are to be permanently retained, an ample amount of the material should be packeted, for the dried specimens are so fragile that parts essential for identification are very easily lost.

The treatment below is in large part adapted from that of Yuncker (1943).

1. Styles coherent nearly to the summit.   1. *C. exaltata*
1. Styles wholly separate and distinct.
    2. Capsule circumscissile, that is, easily separating near the base in a more or less regular line of
    cleavage.   2. *C. umbellata*
    2. Capsules not circumscissile, that is, not separating in a regular line of cleavage, when forcibly separated either coming away entirely or breaking very irregularly.
        3. Calyx manifestly united below; inflorescence not markedly bracteate and not markedly congested.
            4. Calyx and corolla each 3- or 4-merous.
                5. Lobes of calyx and corolla rounded apically.   3. *C. cephalanthi*
                5. Lobes of the calyx and corolla acute apically.   4. *C. coryli*
            4. Calyx and corolla each 5-merous.
                6. Calyx 5-angled by the projecting lobes where they overlap at the sinuses.   5. *C. pentagona*
                6. Calyx not 5-angled.
                    7. Lobes of the calyx rounded apically.
                        8. Calyx almost enclosing the corolla tube; capsule depressed globose, as broad as long or a little broader than long.   6. *C. obtusiflora*
                        8. Calyx shorter than the corolla tube; capsule ovoid, at least slightly longer than its breadth near the base.   7. *C. gronovii*
                    7. Lobes of the calyx pointed apically, obtuse to acute.
                        9. Corolla lobes spreading to deflexed, the tipmost portions upturned.   8. *C. campestris*
                        9. Corolla lobes erect, the tipmost portions inflexed.   9. *C. indecora*
        3. Calyx deeply divided into free or nearly free segments; inflorescence notably bracteate, flowers compactly clustered.
            10. Sepals and bracts nearly orbicular, with smooth margins; bracts closely appressed to the flowers.   10. *C. compacta*
            10. Sepals and bracts elongate and acute apically, margins somewhat toothed; tips of bracts spreading to slightly recurved.   11. *C. glomerata*

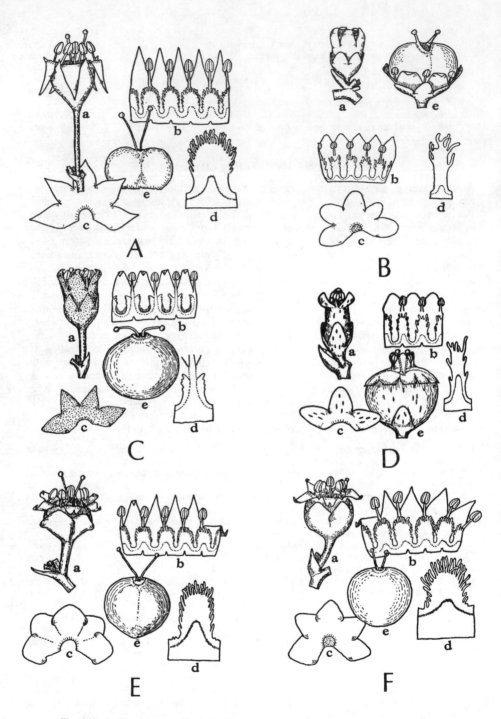

Fig. 260. A, **Cuscuta umbellata**; B, **Cuscuta obtusiflora** var. **glandulosa**; C, **Cuscuta coryli**; D, **Cuscuta cephalanthi**; E, **Cuscuta pentagona**; F, **Cuscuta campestris**. In each of the above: a, flower; b, opened corolla; c, opened calyx; d, infrastaminal scale; e, capsule. (From Yuncker in C. L. Lundell, *Flora of Texas*. Vol. 3, pt. 2. 1943)

1. Cuscuta exaltata Engelm.

Flowers about 4 mm long from the base of the flower to the corolla sinuses, sessile or sub-sessile, parts somewhat thick and fleshy, each flower subtended by an ovate bract. Calyx about enclosing the corolla, deeply divided, lobes 5, orbicular-ovate, overlapping, concave, rounded apically. Corolla tube cylindrical, lobes 5, ovate-orbicular, rounded or obtuse api-cally, erect or spreading; scales of 2 dentate or emarginate wings on either side of the fila-ment attachment. Stamens shorter than the corolla lobes, essentially sessile at the sinuses of the corolla. Styles partially or completely coherent (but easily separated). Capsule ovoid-globose, capped by the withered corolla, dehiscing circumscissily.

Common in Tex., known from a single collection in Fla. (Volusia Co.) Parasitic on sev-eral woody hosts including species of *Quercus, Vitis, Diospyros, Ulmus, Juglans,* and *Rhus.*

2. Cuscuta umbellata HBK.   Fig. 260

Flowers mostly 2–2.5 mm long from the base of the flower to the corolla sinuses, rarely a little larger, on stalks shorter than to much longer than the flowers, in dense compound cymes, the ultimate divisions with 3–7 flowers. Calyx turbinate, equaling or surpassing the corolla tube, the 5 lobes ovate-triangular, apically acute or acuminate, smooth or slightly papillate. Corolla lobes 5, as long as or longer than the tube, reflexed, lance-triangular, long-acute apically; scales oblong-subspatulate, reaching the filaments, moderately fringed mar-ginally. Stamens shorter than the lobes of the corolla, anthers shorter than to equaling the filaments. Styles slender, a little longer than the globose ovary. Capsule depressed-globose, with a ring of thickened knobs around the intrastylar aperture, tardily circumscissile, more or less surrounded by the withered corolla.

s. U.S.; W.I.; Mex.; n. S.Am. Parasitic on hosts in saline places, including species of *Sesuvium, Boerhavia, Portulaca, Atriplex, Suaeda, Alternantha, Euphorbia, Polygonum.*

3. Cuscuta cephalanthi Engelm.   Fig. 260

Flowers about 2 mm long from the base of the flower to the corolla sinuses, surfaces with numerous pellucid, glandlike cells, sessile or subsessile in spicate or paniculately cymose clusters, frequently arising endogenously in the region of the haustoria, the clusters at first open but soon becoming compact as the capsules mature. Calyx shorter than the corolla tube, deeply divided, the lobes 3 or 4(5) oblong-ovate, slightly overlapping at the base, the mar gins often minutely irregular. Corolla tube cylindric-campanulate, enlarging about the matur-ing capsule, lobes 3 or 4(5), much shorter than the tube, ovate, obtuse or rounded apically, erect to spreading. Stamens about as long as or somewhat shorter than the corolla lobes, attached in the sinuses or just below, filaments and anthers of about equal length; scales nar-rowly oblong, about reaching the filaments, fringed with short processes. Styles about equal to or slightly longer than the depressed-globose or globose ovary, stigmas capitate. Capsule depressed-globose or globose, commonly only 1 or 2 seeds maturing and the capsule becom-ing oblique, capped by the withered corolla.

Transcontinental from Maine to Ore. and Wash., southward to Va., Tenn., and Tex. Para-sitic on a variety of woody and herbaceous hosts including species of *Cephalanthus, Spi-raea, Dianthera, Teucrium, Physostegia, Vernonia, Solidago, Aster.*

4. Cuscuta coryli Engelm.   Fig. 260

Flowers mostly 1.5–2 mm long from the base of the flower to the corolla sinuses, somewhat fleshy, papillate, their stalks a little shorter or a little longer than the flowers, the clusters cymose-paniculate, sometimes originating endogenously in the region of the haustoria and forming dense, glomerulate clusters. Calyx lobes usually 4, about as long as the tube of the corolla, lobes ovate-lanceolate or triangular, scarcely overlapping at the base. Corolla tube cylindric-campanulate, lobes usually 4, about as long as the tube, lance-ovate or somewhat triangular-ovate, erect, tips acute, inflexed; scales represented by toothed wings. Stamens

557

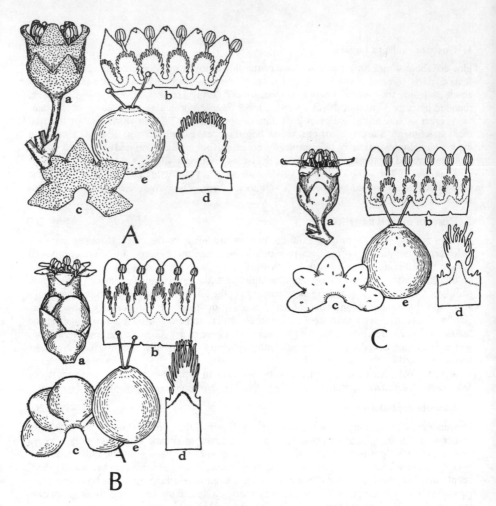

Fig. 261. A, **Cuscuta indecora;** B, **Cuscuta compacta;** C, **Cuscuta gronovii.** In each of
the above: a, flower; b, opened corolla; c, opened calyx; d, infrastaminal scale; e, capsule.
(From Yuncker in C. L. Lundell, *Flora of Texas.* Vol. 3, pt. 2. 1943)

nearly as long as the corolla lobes, filaments longer than the anthers. Styles mostly about as
long as the ovary. Capsule globose or slightly depressed-globose, the withered corolla cap-
ping the upper part and eventually splitting and falling off.

N.Eng. to Mont., southward to S.C., Miss., Tex., Ariz. Parasitizing a variety of woody
and herbaceous hosts including species of *Salix, Carya, Stachys, Solidago, Aster, Helianthus,* at least in part plants inhabiting low grounds.

5. Cuscuta pentagona Engelm.   Fig. 260

Flowers 1–1.5 mm long from the base of the flower to the corolla sinuses, the protruding
capsule soon causing the flowers to appear larger; commonly with pellucid, glandlike cells
on the flower stalks, the latter about as long as to shorter than the flowers; flowers borne in
loose, cymose clusters. Calyx usually about equaling the corolla tube and loose about it,
lobes 5, broadly ovate and broadly obtuse to rounded apically, overlapping at the sinuses to

form a conspicuously 5-angled calyx. Corolla tube campanulate, lobes 5, lance-triangular and acute apically, about as long as or slightly longer than the tube, spreading to reflexed, tipmost portions inflexed; scales short-oblong or oblong-ovate, prominently fringed. Stamens shorter than the corolla lobes, the filaments and anthers of nearly equal length. Styles slender, equal to or slightly shorter than the globose ovary, stigmas capitate. Capsule mostly depressed globose to ovoid, frequently a little longer than broad, the withered corolla remaining about the lower part.

s. N.Eng. to Mont., southward to Fla., Calif. Usually on herbaceous hosts of well-drained open places, perhaps not parasitizing wetland plants.

6. Cuscuta obtusiflora Kunth in HBK. Fig. 260

Flowers mostly 1.5−2 mm long from the base of the flower to the corolla sinuses, commonly with numerous, relatively large, pellucid, glandlike cells, subsessile in moderately open, glomerate-cymose clusters. Calyx nearly enclosing the corolla tube, lobes 5, commonly unequal, ovate-oblong, rounded apically, not overlapping at their bases. Corolla tube campanulate, becoming broadly so as it enlarges about the maturing capsule, lobes 5, shorter than the tube, triangular to broadly ovate, obtuse apically, spreading or reflexed; scales oblong-ovate, prominently fringed distally. Stamens a little shorter than the corolla lobes, filaments a little longer than the anthers. Styles stoutish, broadest basally and tapering a little to the apex, shorter than to about equaling the ovary, stigmas capitate. Capsule depressed-globose, the withered corolla often remaining for a time about the base. (Ours is the var. *glandulosa* Engelm., the var. *obtusiflora* being native to S.Am.; *C. glandulosa* (Engelm.) Small)

Ga. and Fla. to Tex.; W.I., N.M. Most frequently parasitic on plants of *Polygonum* species; on a number of other hosts as well.

7. Cuscuta gronovii Willd. in Roem. & Schult. Fig. 261

Flowers mostly 2−2.5 mm long from the base of the flower to the corolla sinuses, rarely smaller or larger, on stalks shorter than to about equaling the flowers, in loosely or densely panicled cymes, commonly with pellucid, glandlike cells on flowers or fruits. Calyx reaching to about the middle of the corolla tube, lobes 5, ovatish, rounded at the summit or very bluntly obtuse, overlapping a little at their bases. Corolla tube campanulate, lobes 5, shorter than the tube, spreading; scales about equaling the corolla tube, oblongish, conspicuously fringed distally. Stamens nearly as long as the corolla lobes, filaments a little longer than the anthers. Styles mostly shorter than the globose-conic ovary, stigmas capitate. Capsule globose-conic to obpyriform, surrounded at the base by the withered corolla.

Que. to Man., generally southward to Fla. and N.M.; W.I. Parasitic on a great many kinds of host plants, mostly those of low grounds.

8. Cuscuta campestris Yuncker.

Flowers 2−3 mm long from the base of the flower to the corolla sinuses, appearing larger as the fruit matures, on stalks mostly somewhat shorter than the flowers, borne in glomerate-cymose clusters. Calyx about as long as the corolla tube, lobes 5, overlapping at the base but not markedly angled at the sinuses, ovate to oval-ovate, bluntly obtuse apically, about as broad as long. Corolla tube campanulate, lobes 5, triangular to lanceolate, acute apically, spreading to reflexed, the tipmost portions turning upward; scales reaching the filaments, oblong or ovate-oblong, prominently fringed. Stamens shorter than the lobes of the corolla, filaments a little longer than the oval anthers. Styles as long as or a little longer than the globose ovary (but shorter than the mature capsule), stigmas capitate. Capsule globose or depressed-globose, the withered corolla about the lower portion. (*C. arvensis* Engelm. in Gray, not *C. arvensis* Beyr.; *C. pentagona* in part of Small (1933))

Transcontinental, across s. Can., southward to s. U.S. (possibly excluding Fla.); n.e. Mex. Parasitic on a great variety of hosts, mostly herbaceous, including species of *Justicia*, *Xanthium*, *Penthorum*, *Ludwigia*; also on many widely cultivated plants.

559

9. Cuscuta indecora Choisy.

Fig. 261

Flowers 2–2.5 (–3) mm long from the base of the flower to the corolla sinuses, somewhat fleshy, usually papillate, on stalks a little longer than the flowers, borne in loose to dense panicled cymes. Calyx mostly shorter than the corolla tube, lobes 5, ovate-triangular, obtuse to subacute apically. Corolla tube campanulate, lobes 5, lobes somewhat shorter than the tube, triangular, erect to spreading, tipmost portions portions inflexed; scales as long as or a little longer than the corolla tube, short-oblong, prominently fringed. Stamens shorter than or nearly as long as the corolla tube, filaments and oval anthers about equal in length. Styles as long as or slightly longer than the globose ovary, stigmas capitate. Capsule slightly depressed-globose, surrounded by the withered corolla which eventually splits by the enlarging capsule, dehiscing circumscissily.

Widely distributed in the New World. Parasitic on a wide variety of woody and herbaceous hosts including species of *Polygonum, Clematis, Sesbania, Cissus, Cephalanthus, Baccharis, Iva, Pluchea, Borrichia, Eryngium.*

10. Cuscuta compacta Juss. in Choisy.

Fig. 261

Flowers often greenish, up to 4.5 mm long from the base of the flower to the corolla sinuses, sessile or subsessile, in few- to compact-flowered clusters, these commonly closely clustered about the host's parts, frequently originating endogenously and forming a dense, ropelike compaction around the stem of the host, each flower subtended by 2 to several, orbicular-ovate appressed bracts. Calyx deeply divided or of 5 distinct sepals, these cupped, orbicular or oval, broadly overlapping. Corolla tube cylindrical, the basal part eventually enlarging about the developing capsule, lobes 5, spreading to reflexed, rounded or broadly obtuse apically, much shorter than the tube; scales reaching the filaments, oblongish, long-fringed. Stamens with short, approximately equal filaments and anthers, reaching to the sinuses of the corolla. Styles about as long as or somewhat shorter than the globose ovary, stigmas capitate. Capsule globose-conic, capped by a withered corolla.

s. N.H., Mass., to s. Ind., Ill., Mo., s.e. Okla., southward to Fla. and Tex. Parasitizing many kinds of host plants including species of *Alnus, Magnolia, Myrica, Cyrilla, Cephalanthus, Lycopus, Boehmeria, Clethra, Rubus, Rosa.*

11. Cuscuta glomerata Choisy.

Flowers 4–5 mm long from the base of the flower to the corolla sinuses, commonly originating endogenously and breaking forth in two parallel rows on opposite sides of the stem, forming ropelike floral masses winding tightly about the stem of the host; flowers sessile, subtended and surrounded by numerous, narrowly lanceolate, more or less boat-shaped bracts with recurved, acute tips. Sepals 5, distinct, similar to the bracts, tips spreading but not recurved. Corolla tube cylindrical, lobes 5, much shorter than the tube, long-triangular, acute apically; scales oblongish, reaching the filaments, prominently fringed around their apices, less so below. Stamens shorter than the corolla lobes, anthers and filaments of about equal length. Styles slender, considerably longer than the ovary, stigmas capitate. Capsule globose-flasklike, a collar about the base of the styles, with a withered corolla about the summit.

s.w. Mich. to S.D., southward to Miss. and Tex. Parasitizing many kinds of woody and herbaceous hosts, apparently with some predilection for members of Compositae including species of *Vernonia, Liatris, Solidago, Aster, Helianthus, Helenium.*

2. Dichondra

Dichondra carolinensis Michx. PONY-FOOT.

Perennial herb, stems spreading-prostrate, rooting at the nodes, sparsely pubescent, often becoming glabrous. Leaves petiolate, petioles 1–4 cm long, many of them longer than the blades, pubescent; blades cordate basally, orbicular to reniform in overall outline, mostly 1–2 cm broad, margins entire, both surfaces pubescent at least when young; stipules very small, subulate. Flowers minute, solitary or paired from leaf axils, their pubescent stalks varying from about half as long to as long as the petioles. Sepals 5, oblanceolate or spatulate, longer than the corolla, densely silky-pubescent on their backs, 1.5–3 mm long at anthesis, becoming 3.5–5 mm long in fruit. Corolla creamy white, deeply 5-lobed, lobes rotate, obtuse apically. Stamens 5, shorter than the corolla lobes. Ovary subglobose at first, 2-locular, each locule with 2 ovules, styles 2, stigmas capitate; capsule deeply 2-lobed, each lobe usually 1-seeded (one ovule aborting). Seed pyriform, 1.8–2.5 mm long, brown, pilose. (*D. repens* Forst. & Forst. var. *carolinense* (Michx.) Choisy in DC.)

In a wide variety of moist woodlands, often where flooded temporarily, moist, open banks, lawns. Coastal plain and piedmont, s.e. Va. to s. Fla., westward to Ark., s.e. Okla., e. and cen. Tex.; Berm., Bah.Is.

3. Evolvulus

Evolvulus sericeus Sw.

Inconspicuous perennial herb, often with branching leafless subterranean stems from which very slender, lax, leafy, prostrate or ascending, nontwining, glabrous or appressed-pubescent, aerial stems arise. Leaves sessile, narrowly elliptic, linear-elliptic, or linear-oblong, the lower usually shorter and broader than the upper, 1–3 cm long, 1–8 mm broad, acute apically, both surfaces glabrous, upper glabrous and lower appressed-pubescent, or both appressed-pubescent. Inflorescences mostly solitary in leaf axils, usually 1-flowered, a pair of subulate bracts at the juncture of inflorescence stalk and flower stalk, the latter the longer. Sepals 5, elliptic, 4–5 mm long, acuminate apically, silky-pubescent. Corolla white, wholly united, funnelform or funnelform-rotate, 1–1.5 cm broad distally and shallow notched-lobed. Stamens 5, inserted near the base of the corolla tube, somewhat exserted. Style united basally, each of 2 branches deeply cleft, stigmas 4, linear-filiform. Capsule subglobose, about 3.5 mm in diameter, somewhat surpassed by the calyx, 1–4-seeded. Seed dark brown or brownish black, dull, glabrous.

Several varieties have been recognized, chiefly distinguished by the presence or absence of pubescence on one or the other or both leaf surfaces.

Pineland bogs, on lime-rock in pinelands, prairies, grassy open places. Local, Fla. to Tex., northward to Tenn.; W.I. and S.Am.

4. Stylisma

Stylisma aquatica (Walt.) Raf.

Perennial herb. Stems usually several from the caudex, radiately spreading, prostrate or twining. Stem, petioles, blade surfaces, inflorescence stalks, flower stalks, bractlets, and calyces copiously soft-pubescent with short, silvery-silky hairs. Petioles short, 1–3 mm long, blades 2–3 cm long and 3–7 (–10) mm broad, mostly widest at their slightly cordate to truncate bases, most of them tapering but a little distally, thus essentially oblong, apices obtuse or obtuse with minute mucros. Inflorescences of 1–5 flowers, the inflorescence stalks long, 2 small, subulate, subopposite bractlets at the point of attachment of the flower stalks. Sepals 5, 5–8 mm long, 3–5 mm broad, their apices acute to acuminate. Corolla pink, lavender, or reddish purple, wholly united, funnelform, 1–1.5 cm long, notched-lobed distally, long-pubescent medially without. Stamens 5, inserted on the corolla tube near its base, partially exserted, filaments glabrous or with a few hairs near the base. Capsule oval or obliquely oval, 4–6 mm long, long-pubescent apically, 1(2)-seeded. Seed brown, short-oblong or -oval, 3 mm long. (*Breweria michauxii* Fern. & Schubert)

Pine savannas and flatwoods, pineland depressions and bogs, pond margins, wet prairies. Coastal plain, s.e. N.C. to Fla. Panhandle, westward to s.e. Ark. and s.e. Tex.

5. Jacquemontia

Jacquemontia tamnifolia (L.) Griseb.

Annual herb. Stem at first erect, later with long, prostrate or twining branches, with long spreading pubescence. Leaves with sparsely pubescent petioles about as long as the blades; blades ovate, 3–12 cm long, 2–6 cm broad basally, bases cordate, truncate, or rounded, apices acuminate, margins entire to irregularly and shallowly wavy, surfaces sparsely pubescent, usually becoming glabrous. Flowers in axillary, long-stalked, densely headlike cymose clusters, with lanceolate, foliaceous bracts, both bracts and calyces bearing conspicuous, long spreading hairs. Sepals 5, lance-subulate, 8–15 mm long. Corolla blue (rarely white on a given individual), wholly united, funnelform, 1–2 cm across, the edge shallowly wavy. Stamens 5, inserted near the base of the corolla, unequal, not exserted. Style filiform, stigmas 2, flattened. Capsule subglobose, 4–6 mm long, shorter than the sepals. Seed angled, with 2 flat and 1 rounded face, 2–3 mm long, dull brownish black, surface with minute tubercles. (*Thyella tamnifolia* (L.) Raf.)

More or less weedy in cultivated lands and waste places, in floodplain forests, exposed stream beds, river shores and bars, bottomland clearings, ditches. Chiefly coastal plain, s.e. Va. to s. Fla., westward to Ark. and e. Tex.; W.I., C. and S.Am. to Brazil; Afr.

6. Calystegia

Calystegia sepium (L.) R. Br. HEDGE BINDWEED.

Herbaceous, perennial, trailing or twining vine, the stems usually freely branching. Stems, petioles, leaf blades, and flower stalks glabrous to densely, softly, and finely pubescent. Leaves petiolate, petioles varying from shorter than to about as long as the blades; blades mostly hastate or sagittate basally, less frequently cordate, the basal lobes and the terminal portions varying greatly in breadth, margins entire. Flowers showy, mostly solitary, sometimes as many as 4, from leaf axils, their stalks shorter than to a little longer than the leaves. Calyx subtended by and essentially loosely surrounded by a pair of conspicuous, membranous bracts. Sepals 5, subequal, shorter than the subtending bracts. Corolla white, pink, rose, or rose-purple, funnelform, 4–7 cm long, 5–7 cm broad distally, the edges wavy or scarcely lobed. Stamens 5, inserted at the base of the corolla tube, about half as long as the tube. Ovary incompletely 2-locular, style 1, stigmas 2, short-oblong, blunt apically. Capsule ovoid to subglobose, enclosed by the bracts, 2–4-seeded. Seed subglobose to ovoid-trigonous, glabrous. (*Convolvulus sepium* L.)

Authors have segregated from this several specific or infraspecific taxa based on variation in leaf shape or form, pubescence, and corolla color. These include (at the specific level) *Convolvulus americanus* (Sims) Greene, *Convolvulus fraterniflorus* Mack. & Bush, *Convolvulus limnophilus* Greene, *Convolvulus nashii* House, and *Convolvulus repens* L.

Edges of marshes, marshy shores, swales, moist to wet thickets, moist to wet clearings, cultivated lands, waste places. More or less throughout temp. N.Am.; Euras.

7. Ipomoea (MORNING-GLORIES)

Herbaceous or shrubby, annual or perennial, many kinds twining vines, some with stems prostrate or trailing, infrequently erect. Leaves entire, lobed, or variously divided. Inflorescence axillary, flowers solitary or in stalked cymes. Sepals 5, commonly overlapping laterally, often unequal. Corolla usually radially symmetrical, funnelform to campanulate or salverform, with distinct midpetaline bands. Stamens 5, inserted low on the corolla tube, included in most, exserted in some. Style 1, stigma globose or with 2 or 3 globose lobes. Capsule globose or ovoid, 2- or 3-locular, 4–6-valved, 4–6-seeded (fewer by abortion).

1. Corolla salverform or nearly so, the tube cylindrical for most of its length, then abruptly spreading laterally near the summit.

2. Leaf blades pinnately divided to the midrib, the segments narrowly linear.   1. *I. quamoclit*
2. Leaf blades unlobed save for basal cordations, or toothed-lobed or palmately lobed.
 3. Corolla white, cylindrical portion of the tube 5–12 cm long.
  4. Sepals, at least the outer 2 or 3, abruptly narrowed distally to long, slender, taillike extremities; seed glabrous or with minute hairs only around the hilum.   4. *I. alba*
  4. Sepals broadly rounded and with minute mucros distally; seed pubescent over its entire surface.
   5. *I. violacea*
 3. Corolla red, or the tube yellowish and the limb red, cylindrical portion of the tube 1.5–3.5 cm long.
  5. Calyx 6–8 mm long; corolla 2 cm long; stalks of the fruits recurved.   2. *I. coccinea*
  5. Calyx 4–4.5 mm long; corolla 3–4 cm long; stalks of the fruits erect.   3. *I. hederifolia*
1. Corolla funnelform to campanulate, the tube gradually expanding from below the middle then gradually or abruptly flaring toward the summit.
 6. Stalks of the inflorescences and flowers retrorsely pubescent; ovary or capsule 3-locular.
  7. Sepals with acute to acuminate apices.
   8. Pubescence on the sepals, if any, slender and without swollen bases.   6. *I. indica*
   8. Pubescence on the sepals, especially proximally, with enlarged bases.   7. *I. purpurea*
  7. Sepals with long, taillike extremities much longer than the basal portions.   8. *I. hederacea*
 6. Stalks of the inflorescences and flowers glabrous, or, if pubescent, the pubescence not retrorse; ovary or capsule 2-locular.
  9. Stems not twining, elongate-prostrate, rooting at the nodes on older portions.
   10. Corolla white with yellow or purple center; some of the leaves lobed.   9. *I. stolonifera*
   10. Corolla lavender to rose-purple; all leaf blades unlobed.   10. *I. pes-caprae*
  9. Stems twining.
   11. Leaf blades deeply palmately divided, essentially palmately compound.
    12. Leaf segments entire, or one or both basal segments lobed.
     13. Corolla 1–2 cm long.   11. *I. wrightii*
     13. Corolla 5–8 cm long.   12. *I. cairica*
    12. Leaf segments sinuately lobed.   13. *I. sinuata*
   11. Leaf blades not deeply palmately divided.
    14. The leaf blades with conspicuous, narrowly sagittate basal lobes.   14. *I. sagittata*
    14. Leaf blades cordate to hastate basally.
     15. Corolla about 2 cm long.   15. *I. lacunosa*
     15. Corolla 4–8 cm long.
      16. Corolla lavender, lavender-pink, or rose-purple, 3.5–5 cm long; apices of sepals strongly acuminate.   16. *I. trichocarpa*
      16. Corolla white with purple-red center, 5–8 cm long; apices of sepals rounded sometimes slightly emarginate, sometimes with a small mucro, not acuminate.   17. *I. pandurata*

1. Ipomoea quamoclit L. CYPRESS-VINE.

Annual, glabrous, twining vine. Petioles 2–25 mm long, usually bearing a few, small, scale-like trichomes; blades 2–6 cm long, ovate to oblong in overall outline, pinnately divided to the midrib, the segments narrowly linear and nearly at right angles to the midrib. Inflorescences axillary to leaves, their stalks surpassing the leaf blades and bearing 1–4 flowers above a pair of minute bracts, the flower stalks dilated distally, variable in length, if more than 1, all much shorter than the inflorescence stalks. Sepals unequal, 4–6 mm long, oblong or ovate-oblong, minutely mucronate on their rounded apices. Corolla deep red, salverform, the tube 2–3 cm long, a little dilated from somewhat above the calyx, the limb 2 cm across, with short-triangular lobes, stamens and style well exserted. Capsule ovoid, 8–10 mm long, its stalk straight. Seed 5–6 mm long, angular, with 2 flat faces and 1 rounded, lance-ovate in outline, blunt at the tip, surface dark, dull brown and with sparse short, scaly trichomes. (*Quamoclit quamoclit* (L.) Britt. in Britt. & Brown)

 Cultivated as an ornamental, naturalized, on the coastal plain and piedmont in our range, occurring on fences, in moist to wet thickets and hedge rows, borders of wet woodlands and shrub-tree bogs, cultivated fields. Native of trop. Am.

2. Ipomoea coccinea L. SCARLET-CREEPER.

Glabrous, annual, twining vine. Petioles about as long as the leaf blades; blades broadly ovate, 2–7 cm long, cordate or angulate-lobed basally, markedly acuminate apically, mar-

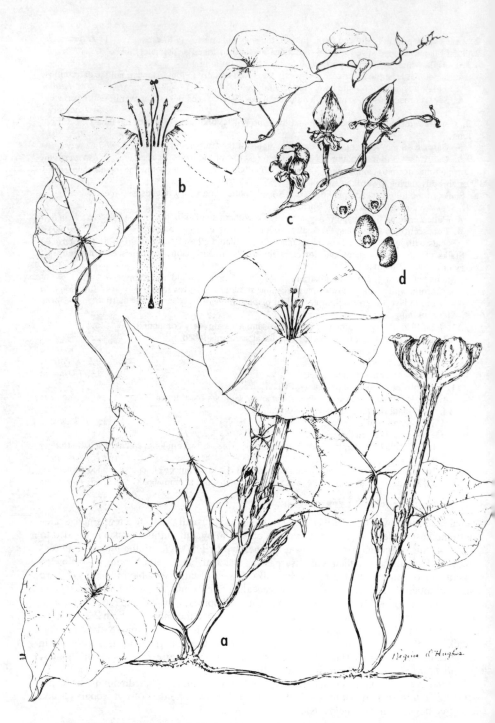

Fig. 262. **Ipomoea alba:** a, habit; b, cut-open corolla; c, capsules; d, seeds. (From Gunn, "Moonflowers, Ipomoea Section Calonyction in Temperate North America." *Brittonia* 24: 150–168. 1972)

gins entire or with few dentate teeth or small lobes. Inflorescence stalks variable in length, shorter than to surpassing the leaves, bearing a single flower or several-flowered cymes above a pair of minute bracts. Calyx 6–8 mm long, sepals unequal, each with an oblongish, round-tipped body 3–6 mm long and a taillike extremity 2–3 mm long, these mostly arising subapically. Corolla 2 cm long or little more, salverform or nearly so, scarlet, or the tube yellowish red, the limb scarlet, stamens and style exserted. Capsule subglobose, 4–5 mm high, at maturity its stalk recurved. (*Quamoclit coccinea* (L.) Moench.)

Alluvial thickets, muddy banks, sand and gravel bars of streams, fields, waste places. Said to be native to s.e. U.S., cultivated and naturalized elsewhere. (See note under *I. hederifolia*.)

3. Ipomoea hederifolia L.

Similar to *I. coccinea*. Leaf blades as in *I. coccinea*, but many of them, especially distally on the stems, prominently 3-lobed. Calyx smaller, 4–4.5 mm long, sepals 1.5–3 mm long excluding their taillike extremities. Corolla larger, 3–4 cm long. Stalks of mature fruits straight.

Habitats similar to those of *I. coccinea*. Native of trop. Am., cultivated for ornament and naturalized.

Daniel F. Austin (personal communication) states that *I. coccinea* is a temperate species of the southeastern United States which most authors have confused with the tropical *I. hederifolia* which has been widely introduced. It appears to us that *I. hederifolia* is now more common in parts of our range than is *I. coccinea*.

4. Ipomoea alba L. MOONFLOWER. Fig. 262

Perennial, twining vine, woody at base, sometimes producing long, prostrate runners. Stem with milky sap, to 30 m long, usually bearing, at least toward the base, numerous spinelike or warty protuberances 2–4 mm long. Petioles 5–20 cm long, smooth or bearing short, rough excrescences; mature blades 6–20 cm long, 7–15 cm broad, glabrous, thin, cordate-ovate, margins entire, apices acuminate, or variably angulate-lobed to prominently 3–7-lobed, basal lobes cordate to sagittate, other prominent lobes usually acuminate. Inflorescence stalks axillary, 1–24 cm long, each bearing 1 to several fragrant flowers, opening 1 at a time nocturnally, flower stalks 7–15 cm long, dilated distally, thickening and elongating somewhat as the fruits mature. Sepals unequal, 2 or 3 exterior ones shorter and thinner than the others, bearing long, slender, conspicuous, awnlike extremities, inner ones usually only minutely mucronate at their rounded apices; sepals at anthesis 5–15 mm long excluding the awnlike extremities, enlarging somewhat as the fruits mature, the awnlike extremities usually reflexed in fruit. Corolla salverform, the slenderly cylindrical tube 7–12 cm long, white, glabrous exteriorly, pubescent interiorly on the lower half; limb 11–14 cm across, white with greenish plaits, usually unlobed marginally, rarely 5-lobed. Stamens white, inserted on the corolla tube near its summit, they and the style exserted. Capsule 2.5–3 cm long, ovoid or obturbinate, 2-locular, 4-seeded, its stalk straight. Seed 10–12 mm long, ovoid-oblong, brown, glabrous or with minute hairs around the hilum. (*Calonyction aculeatum* (L.) House)

Widely cultivated as an ornamental and for screening, naturalized in s. Fla. in disturbed places, sometimes on borders of mangrove swamps. Native origin controversial, now pantropical.

5. Ipomoea violacea L. SEA-MOONFLOWER. Fig. 263

Trailing or twining perennial vine with milky sap. Stems to 10 m long, somewhat woody, not spiny or warty, with brown punctations. Petioles 3–16 cm long, a purple gland on each side at the junction with the blade; blades 5–16 cm long, 5–14 cm broad, broadly cordate-ovate or cordate-orbicular and with entire margins, or some angulate-lobed, thickish, acuminate apically. Inflorescences axillary, their stalks 0.7–7 (–12) cm long, 1–3(4)-flowered. Flower stalks 1.4–3 cm long, angular, somewhat thickened distally, especially in fruit; flowers not fragrant. Sepals subequal, 15–24 mm long, orbicular, broadly rounded apically, sometimes

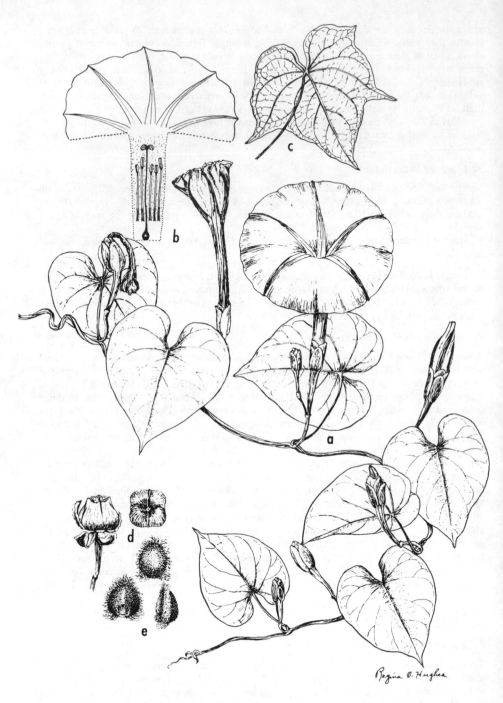

Fig. 263. **Ipomoea violacea:** a, habit; b, cut-open corolla; c, lobed leaf; d, capsules; e, seeds. (From Gunn, "Moonflowers, Ipomoea Section Calonyction in Temperate North America." *Brittonia* 24: 150–168. 1972, as *I. macrantha*)

notched, enlarging somewhat and becoming reflexed as the fruits mature. Corolla salverform, the cylindrical tube 5–12 cm long, white and glabrous exteriorly, white and minutely pubescent interiorly; limb 6–7 cm across, white with yellowish or greenish plaits. Stamens white, unequal, 2 long, 3 short, inserted near the base of the corolla tube, they and the style not exserted. Capsule globose with acuminate apex, 2–3 cm in diameter, 2-locular, 4-seeded, its stalk straight. Seed about 10 mm long, ovoid, dark brown, pubescent. (*I. macrantha* R. & S. in L., *Calonyction tuba* (Schlecht.) Colla)

In coastal vegetation. cen. pen. and s. Fla., Fla. Keys; pantropical, usually insular.

6. Ipomoea indica (Burm. f.) Merr.

Perennial, twining vine, low- to high-climbing, glabrous or pubescent. Petioles shorter than to about as long as the leaf blades; blades 4–16 cm long, 3–12 cm broad, cordate-ovate with entire margins and acuminate apices, or varyingly angulate-lobed to deeply 3–5-lobed, often some of each on a single plant, glabrous to softly and densely pubescent beneath. Inflorescence stalks 1–12 cm long, sometimes 1-flowered, usually bearing several to numerous flowers on short stalks; stalks of the inflorescences and flowers retrorsely pubescent. Sepals 1–2 cm long, lanceolate to lance-ovate, acuminate apically, glabrous or pubescent with slender hairs without enlarged bases. Corolla 6–9 cm long, 6–8 cm across the shallowly lobed limb, more or less funnelform, deep blue-purple with darker purple stripes along the plaits (commonly drying rose-pink or rose-purple). Capsule ovoid, 3-locular, 6-seeded. Seed angular with 2 flat faces and 1 rounded, 6 mm long, dark brown to black. (*I. acuminata* (Vahl) R. & S., *I. cathartica* Poir. in Lam., *I. congesta* R. Br., *Pharbitis cathartica* (Poir. in Lam.) Choisy in DC.)

Dry, moist, or wet thickets, borders of mangrove swamps, shoreline thickets, various disturbed sites and waste places; also cultivated for ornament. Fla. to s. Tex.; W.I.; Mex.; S.Am.; Old World.

7. Ipomoea purpurea (L.) Roth.   Fig. 264

Similar in habit to *I. indica*. Leaf blades all cordate-ovate with entire margins, or prominently 3-lobed, or on a given plant both entire and lobed leaves. Inflorescence and flower stalks retrorsely pubescent. Sepals lanceolate or lance-elliptic, acute apically, about 15 mm long, pubescent, at least proximally, with hairs having enlarged bases. Corolla 4.5–7 cm long, 4–5 cm across the limb, purple, red, deep pink, bluish, or white, sometimes the tube white and the limb colored, sometimes the color broken or streaked. Stamens and style not exserted. Capsule globose, about 1 cm in diameter, shorter than the sepals. Seed irregularly angular, dark brown, about 4 mm long. (*Pharbitis purpurea* (L.) Voight)

Weedy, cultivated fields, fence and hedge rows, dry to wet thickets and clearings. Native of trop. Am. and naturalized more or less throughout our range and beyond.

8. Ipomoea hederacea (L.) Jacq.   Fig. 265

Annual, twining vine. Stems pubescent with spreading, long and short hairs. Petioles shorter than to equaling the leaf blades; blades 5–12 cm long, about as wide across their bases, cordate-ovate with entire margins and short-acuminate apices, or prominently 3-lobed, lobes acuminate, occasionally both on a single plant, softly pubescent beneath. Inflorescence stalks shorter than the subtending leaves, bearing 1–3 flowers; inflorescence and flower stalks with some or all hairs retrorse. Sepals copiously long-pubescent, with basal portions broadest then narrowed to taillike extremities much longer than the basal portions, 1.5–2 cm long overall, enlarging somewhat as the fruits mature. Corolla tube white, limb at first blue with white center, changing to rose-purple, also drying rose-purple, 1.5–5 cm long, about 3 cm across the limb. Stamens and style not exserted. Capsule subglobose, 8–12 mm across, shorter than the sepals. Seed angular with 2 flat faces and 1 rounded, 4–5 mm long, brownish black, minutely pubescent. (*Pharbitis hederacea* (L.) Choisy; incl. *I. barbigera* Sims, *Pharbitis barbigera* (Sims) G. Don)

Weedy, fields, roadsides, thickets, stream banks, sand and gravel bars of streams, clearings. N.Eng. to N.D., southward to Fla. and Tex.; native of trop. Am.

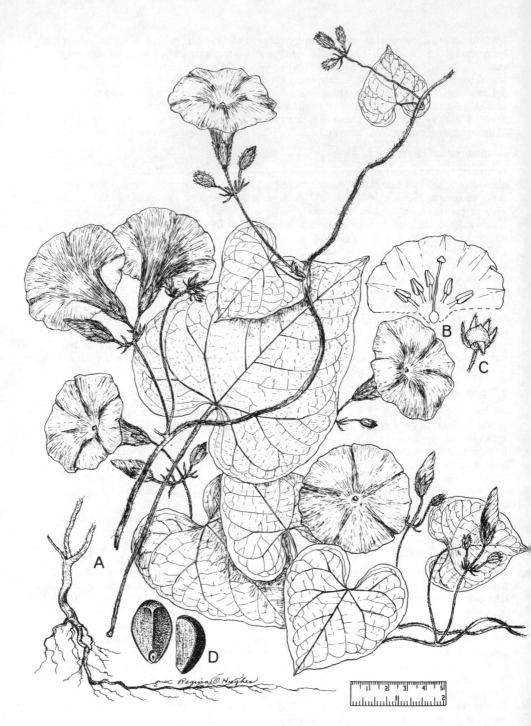

Fig. 264. **Ipomoea purpurea:** a, habit; b, diagram of flower opened up exclusive of calyx; c, capsule; d, seeds. (From Reed, *Selected Weeds of the United States* (1970) Fig. 149)

Fig. 265. **Ipomoea hederacea:** a, habit; b, calyx; c, capsule; d, seeds. (From Reed, *Selected Weeds of the United States* (1970) Fig. 148)

9. Ipomoea stolonifera (Cyrillo) J. F. Gmel. BEACH MORNING-GLORY.

Prostrate, glabrous perennial, stems commonly very long (up to 50 m), rooting at some of the nodes. Petioles 1–10 cm long; blades 2–12 cm long, 1–6 cm broad, somewhat succulent, proximal ones often oblong or ovate-oblong, unlobed, bases truncate to cordate, apices rounded or truncate, margins entire, later leaves variously 3–7-lobed. Inflorescence stalks 1–3 (–4) cm long, 1-flowered, the flower stalks gradually dilated distally, 1–2 cm long above a pair of minute bracts. Sepals subequal, 10–15 mm long, elliptic to elliptic-oblong, their apices rounded or blunt and apiculate. Corolla 6–7 cm long, the lower tube broad, the limb flaring-funnelform, white with yellow center, purplish at the base of the tube. Stamens inserted on the corolla tube near the base, they and the style not exserted. Capsule subglobose, about 15 mm across.

Upper strand of the beach and on dunes. s.e. N.C. to s. Fla., westward to Tex.; general on tropical shores.

10. Ipomoea pes-caprae (L.) R. Br. in Turkey. RAILROAD-VINE, GOAT'S-FOOT.

Trailing or prostrate, glabrous perennial, stems elongate (up to 80 m), rooting at some of the nodes. Petioles shorter than to much longer than the leaf blades; blades 4–10 cm long, somewhat succulent, oblong or ovate-oblong to orbicular or subreniform, margins entire, bases subcordate, truncate, or slightly tapered, apices very broad, mostly emarginate. Inflorescence stalks very variable in length, bearing 1 to several flowers above a pair of minute bracts, flower stalks of variable length to 5 cm, relatively coarse and somewhat dilated upwardly. Sepals unequal, ovatish, the larger ones 6–8 mm long at anthesis, all enlarging considerably both in length and breadth as the capsules mature. Corolla about 6 cm long, funnelform, whitish or pinkish near the base, lavender to purple upwardly, the center darker. Stamens and style not exserted. Capsule subglobose, about 15 mm across. Seed angular, with 2 flat faces and 1 rounded, slightly broader than long, about 1 cm broad, dark brown, copiously pubescent.

Upper beach strand and dunes, coastal flats, disturbed places along the coasts. Ga., Fla., westward to Tex.; general on tropical shores.

11. Ipomoea wrightii Gray.

Slender, glabrous, trailing or climbing vine. Petioles 1.5–5 cm long; leaf blades deeply palmately divided, essentially compound, segments or leaflets mostly 5, subequal to conspicuously unequal, lanceolate to linear-lanceolate, margins entire, 1.5–5 cm long, acute apically. Inflorescence stalks filiform, variable in length, the longer ones often twisted or coiled and used in climbing, usually bearing solitary flowers whose stalks, above a pair of minute bracts, are thicker than the inflorescence stalks below. Sepals subequal, 4–6 mm long, ovate or ovate-oblong, apices rounded or blunt. Corolla funnelform, 1–2 cm long, lavender-pink or rose, with darker center (drying purple). Stamens and style not exserted. Capsule subglobose, about 7 mm across. (*I. heptaphylla* (Rottb. & Willd.) Voight)

Alluvial thickets and clearings, borders of alluvial woodlands. Coastal plain, Miss. to cen. and s. Tex. Native of India, widely naturalized in warm regions.

12. Ipomoea cairica (L.) Sweet.

Twining vine. Petioles slender, 1–5 cm long; blades deeply palmately divided, essentially compound, segments or leaflets 5, mostly unequal, 1–5 cm long, lanceolate or narrowly elliptic, margins entire, one or both of the basal pair often lobed and mittenlike in form, short-pubescent on veins and margins. Inflorescence stalks 1–2 cm long, usually bearing a single flower on a somewhat dilated stalk above a pair of minute bracts. Sepals oblong, somewhat unequal, the longer ones about 1 cm long, apices rounded. Corolla funnelform, 5–8 cm long, 4–5 cm across the limb, mauve.

Cultivated for ornament and naturalized, fence and hedge rows, over shrubs and small trees, waste places, infrequently on wet sites. Native of Afr.

13. Ipomoea sinuata Ort.

Perennial, twining vine. Parts of stems, petioles, and inflorescence stalks bearing long, straight, spreading hairs with enlarged bases. Petioles 3–6 cm long; blades deeply palmately divided, essentially compound, segments or leaflets 5, 3–8 cm long, conspicuously and irregularly sinuate-lobed, mostly ovatish in overall outline, the lowermost pair often deeply 2-lobed (their sublobes sinuate), more or less mittenlike in form. Inflorescence stalks variable in length, usually as long as or longer than the petioles, bearing 1 or 2 flowers on short stalks above a pair of minute bracts, the flower stalks somewhat dilated distally. Sepals oblong or nearly so, somewhat unequal, their rounded apices mucronate, the larger at anthesis 2 cm long or a little more, becoming thickish and dryish as the capsules mature and at full maturity of the capsules spreading laterally below them. Corolla 3.5–5 cm long, funnelform below, spreading widely distally, the limb 4–5 cm across, white with a pink or striped pink center. Stamens inserted near the base of the corolla tube, unequal, 3 long and 2 short, not exserted. Capsule subglobose, 1.5 cm across. Seed about as broad as long, about 8 mm across, somewhat angled, the angles very blunt, glabrous, dull black. (*I. dissecta* (Jacq.) Pursh, *Merremia dissecta* (Jacq.) Hallier f. in Engl., *Operculina dissecta* (Jacq.) House)

Cultivated as an ornamental and sporadically naturalized in dry, moist, or wet thickets, stream banks, disturbed places. s. Ga., Fla., westward to s.cen. Tex. Native of trop. Am.

14. Ipomoea sagittata Poir. in Lam.   Fig. 266

Glabrous, perennial, low-twining vine. Petioles 1–3 cm long; leaf blades 3-lobed, 2 basal sagittate lobes and 1 terminal lobe, the 3 often nearly equal, or the terminal considerably larger than the basal ones; lobes from triangular-acuminate to nearly linear, 4–9 cm long overall, margins entire. Inflorescence and flower stalks subequal in length, together about equaling the petioles, flowers solitary. Sepals subequal, oblong, 1–1.5 cm long, rounded apically, purplish on the margins. Corolla 6–9 cm long, funnelform below, flaring distally, the limb 6–8 cm across, rose-pink, rose-lavender, or rose-purple, the center darker red-purple. Stamens inserted near the base of the corolla tube, stamens and style not exserted. Capsule subglobose in outline, somewhat lobed, about 1 cm broad. Seed 6–7 mm long, a little longer than broad, somewhat angled, angles rounded, dull dark brown, long-pilose chiefly on the angles.

Coastal marshes, interdune swales, sandy roadsides near marshes, wet prairies, glades, rarely inland. N.C. to s. Fla., westward to Tex.; W.I.

15. Ipomoea lacunosa L.

Annual, twining vine. Stems sparsely pubescent, the hairs spreading, with enlarged bases. Petioles mostly as long as the blades or little longer, sparsely pubescent; blades variable, cordate-ovate, margins entire, apices acuminate, varying to moderately angulate-toothed or -lobed to deeply 3–5-lobed, 2–10 cm long, 1.5–10 cm broad, surfaces sparsely pubescent. Inflorescence stalks 1 to several in a given leaf axil, usually not exceeding ⅓–½ as long as the subtending petiole, each usually bearing 1 flower above a pair of minute bracts, the flower stalks low-warty. Sepals 10–15 mm long, lanceolate, lance-ovate, or oblong, acuminate apically, long-ciliate on the margins. Corolla funnelform, about 2 cm long, white or white and pink-tinged. Capsule subglobose to hemispheric, pubescent toward the apex, about 6 mm across. Seed angular, angles rounded, 2 faces flat, the other strongly humped and broadest in the middle, tapering to the extremities, the outline 4–5 mm long and about as broad medially, dark purplish brown, glabrous.

Alluvial bottomland woodlands, openings, and clearings, borders of swamps, shoreline thickets, stream banks, low fields, meadows, exposed peaty shores and bottoms of ponds. N.J. to Ohio, Ill., Kan., generally southward to Fla. and e. Tex.

16. Ipomoea trichocarpa Ell.

Perennial twining vine (flowering the first year from seed). Stems sparsely pubescent with spreading hairs enlarged at their bases. Petioles from 2–12 cm long; blades variable in size

571

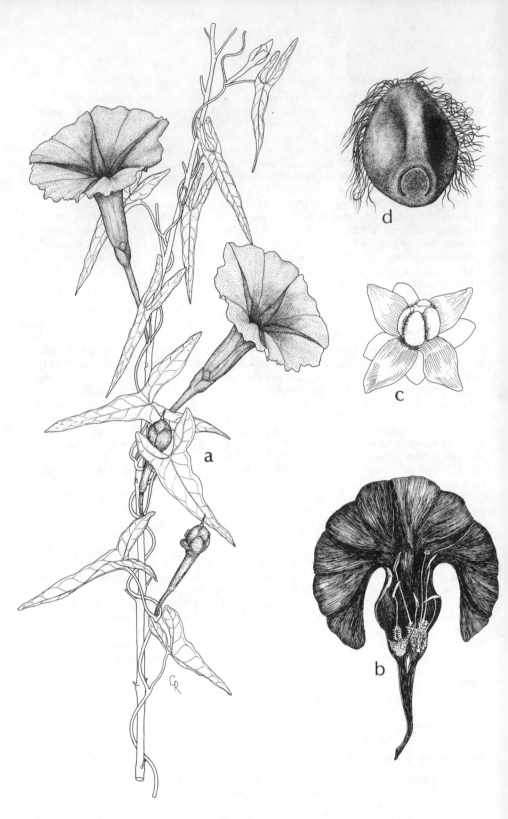

Fig. 266. **Ipomoea sagittata:** a, portion of vine twining on stem of another kind of plant;
b, flower, opened out; c, capsule, dehisced, seeds in center; d, seed.

and form, cordate-ovate with entire margins and acuminate apices, varying from angulate-lobed to deeply 3–5-lobed, to 10 cm long and about as broad basally, all often much smaller on a given plant, upper surfaces usually sparsely pubescent. Inflorescence stalks variable in length from about 5–20 cm long, usually 1, sometimes 2 in a given leaf axil, bearing 1–8 flowers distally in subumbellike clusters above a pair of minute bracts, the flower stalks 5–20 mm long. Sepals unequal, 10–15 mm long, more or less linear-oblong with acuminate apices, parts variably pubescent or glabrous exteriorly, sometimes ciliate marginally. Corolla funnelform, 3.5–5 cm long, lavender, lavender-pink, or rose-purple. Stamens and style not exserted. Capsule subglobose, 6–7 mm across, pubescent toward the apex. Seed angular, with 2 flat faces and 1 rounded (the latter not humped), glabrous, purplish black, 4 mm long.

Alluvial bottomland woodlands and clearings, stream banks, shoreline thickets, exposed peaty shores and bottoms of ponds, various disturbed and waste places. Coastal plain, N.C. to s. Fla., westward to Ark. and e. third of Tex.

I. lacunosa and *I. trichocarpa* hybridize and backcross where their ranges overlap so that given plants not infrequently exhibit various combinations of the characters of the two species and cannot be placed with either one or the other.

17. Ipomoea pandurata (L.) Mey. **WILD-POTATO, MAN-OF-THE-EARTH.**

Trailing or twining perennial with a much-enlarged root, stems glabrous or sparsely pubescent. Petioles shorter than to about equaling the leaf blades; blades 2–10 cm long, 2–9 cm broad, cordate-ovate and unlobed, varying to more or less fiddle-shaped, glabrous to copiously pubescent beneath. Inflorescence stalks usually equaling or exceeding the leaves, bearing 1 to several flowers above a pair of bracts, the flower stalks mostly about 1 cm long. Sepals oblong, unequal, the longer ones about 1.5 cm long, mostly rounded apically, glabrous or sparsely short-pubescent. Corolla funnelform-campanulate, 6–8 mm long and about as broad across the limb, white (and nearly always) with a purple-red center. Stamens and style not exserted. Capsule ovoid or oval, about 1 cm long. Seed 4 mm long, angled, with 2 flat faces and 1 rounded, dark brown and with shaggy, tawny hairs on the angles.

Borders of bottomland woodlands, clearings, river banks, dry to wet thickets, fence rows and hedge rows, fields. Conn. and N.Y., s. Ont., Ohio, s. Mich., Mo. and Kan., generally southward to s.cen. pen. Fla. and e. Tex.

Hydrophyllaceae (WATERLEAF FAMILY)

(Ours) herbaceous annuals, biennials, or perennials. Leaves alternate, or sometimes opposite below and alternate above, without stipules. Flowers bisexual, radially symmetrical, 5-merous (except pistil). Sepals distinct or united below, corolla united to about the middle. Stamens inserted on the corolla tube, alternate with the lobes of the corolla, usually with 2 scales at the base of each filament. Pistil 1, ovary superior, 1- or 2-locular, with few to many ovules, style 1, stigmas 2, or styles 2. Fruit a capsule.

1. Leaves entire.   1. *Hydrolea*
1. Leaves (some of them on any plant at least) lobed or divided.
  2. Leaves (except the lowermost) pinnately divided, the primary divisions 7–13.   2. *Ellisia*
  2. Leaves palmately lobed or divided, the primary divisions 2 or 3, or if more and subpinnate, the primary divisions then not usually exceeding 4.   3. *Nemophila*

1. Hydrolea

Perennial herbs of aquatic or wetland habitats. Leaves alternate, spines often present in leaf axils. Flowers in axillary or terminal cymes. Calyx 5-lobed. Petals 5, blue, rarely white, corolla bell-shaped to wheel-shaped. Stamens 5, filaments shorter or longer than the corolla,

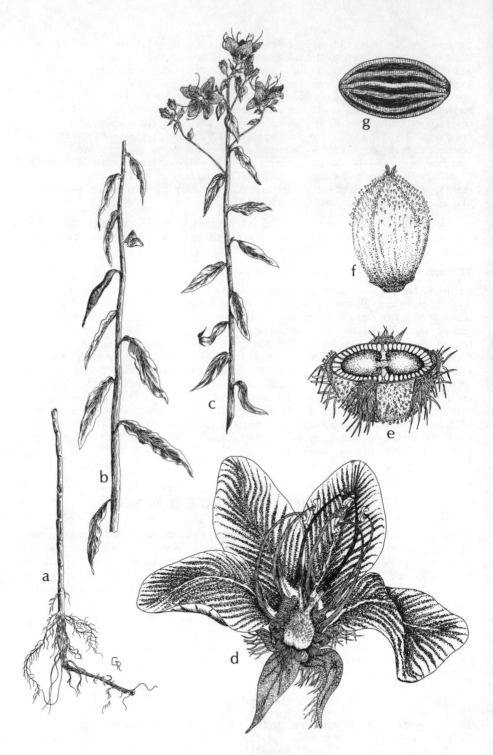

Fig. 267. **Hydrolea corymbosa:** a–c, basal, midsection, and top of plant; d, flower, petal removed; e, cross-section of fruit; f, pistil; g, seed.

dilated at base, inserted on the corolla tube. Styles 2, (rarely 3). Capsule globose, subtended by the persistent calyx. Seeds numerous, very small.

1. Flowers in a terminal cyme; style much longer than the ovary.
  2. Leaves elliptic to lanceolate; spines small or wanting.   1. *H. corymbosa*
  2. Leaves ovate to ovate-lanceolate; spines conspicuous.   2. *H. ovata*
1. Flowers in axillary cymes; styles about as long as the ovary.
  3. Calyx and stem sparsely pubescent with spreading hairs; calyx lobes linear or linear-lanceolate.
                                                                  3. *H. quadrivalvis*
  3. Calyx and stem glabrous or with minute short pubescence and/or sparse sessile glands, calyx lobes lance-ovate.   4. *H. uniflora*

1. Hydrolea corymbosa Macbr. ex Ell.   Fig. 267

Plants with slender underground rhizomes. Stems erect, up to 6–8 dm tall, glabrous below, puberulent above, the inflorescence branches puberulent and with a few long hairs intermixed. Leaves 2–3 cm long and 6–10 mm broad, of fairly uniform size to the inflorescence, sessile, glabrous, lanceolate to elliptic-lanceolate, bases obtuse, apices acute, margins very finely toothed. Leaf axils spineless except on an occasional plant. Inflorescence a terminal, panicled, leafy-bracted cyme, the flowers short-stalked. Calyx copiously white-hirsute and with shorter stipitate glands intermixed, lobes shorter than the corolla. Petals 10 mm long or slightly longer, free except at the very base, the short tube white-translucent, dilated just above the base to a glandular-puberulent inflated disc, then abruptly narrowed upwards; anthers at first pale salmon-colored, pollen yellow. Ovary stipitate-glandular from just above the base, the lowermost portions of the styles short-hirsute with intermixed stipitate glands. Capsule oval, glandular, completely enclosed by the persistent villous calyx. (*Nama corymbosum* (Macbr. ex Ell.) Kuntze)

Swampy woodlands, marshes, and ditches. Coastal plain, Fla. and Ga.

2. Hydrolea ovata Nutt.   Fig. 268

Plants with underground rhizomes. Stem erect, pubescent. Leaves 3–6 cm long, 15–25 mm wide, short-petioled, pubescent, ovate, apices acute, margins entire; spines to 1.5 cm long in axils of leaves. Inflorescence a terminal, panicled, leafy-bracted cyme, flowers stalked, to 2.5 cm broad. Calyx hirsute and with shorter stipitate glands intermixed, the lobes narrowly lanceolate, about equal, shorter than the corolla. Corolla about 6 mm wide, tube blue throughout; petals 10–14 mm long, 1 not fused. Ovary and lower part of style glandular, upper style glabrous. Pollen white. Capsule oval, glandular, enclosed by the hirsute calyx; short-beaked. (*Nama ovatum* (Nutt.) Britt.)

Edges of and in shallow water of swamps, streams, ponds, and ditches. Ga. to Tex., northward in the interior to Mo.

3. Hydrolea quadrivalvis Walt.   Fig. 269

Plants with lower portion of the stem decumbent and rooting at the nodes, the upper ascending. Stem thick, crooked, bright green, with spreading hairs, becoming smooth below with age. Leaves short-petioled, oblanceolate or lanceolate, up to 12 cm long, 2.5 cm broad, bases attenuate, apices acute; margins entire to shallowly and irregularly wavy; short stiff spines in the axils of some leaves. Flowers broad, borne in short-stalked, leafy-bracted, axillary cymes. Calyx lobes commonly unequal, the tube and the bases of the lobes long-hirsute. Corolla about 5 mm long and wide, tube green or greenish white within, with bright blue, spreading, ovate-obtuse lobes slightly shorter than to equalling the calyx lobes. Filaments attenuate at base, abruptly dilated upwards, then abruptly narrowed again, the dilated portion and that below white-translucent, the upper portion blue; anthers blue, the pollen white. Capsule broadly ovoid, calyx persistent. (*Nama quadrivalve* (Walt.) Kuntze)

Swampy woodlands, streams, marshy shores and ditches, commonly in shallow water. s.e. Va. to Fla., westward to La.

Fig. 268. **Hydrolea ovata:** a, top of plant; b, base of plant; c, corolla, opened out; d, stamen; e, flower, corolla removed. (From Correll and Correll. 1972)

Fig. 269. **Hydrolea quadrivalvis:** a, decumbent base of a branch; b, medial section of branch; c, top of branch; d, flower; e, flower, with part of calyx and corolla removed; f, cross-section of fruit; g, seed.

4. Hydrolea uniflora Raf.

Fig. 270

Very closely similar to *H. quadrivalvis* but glabrous or with minute short pubescence and/or sparse sessile glands throughout. The calyx lobes lance-ovate. Petals about 7 mm long, fused about ¼ their length. Pollen white. (*Nama affine* (Gray) Kuntze).

In similar habitats, Ind., Ill., Mo. southward to e. and s.e. Tex. and Miss.

2. Ellisia

Ellisia nyctelea (L.) L.

Low annual, the stem simple to diffusely branched, 1–4 dm tall, sparsely pubescent. Leaves petiolate, the petioles mostly shorter than the blades, petioles and both leaf surfaces pubescent; lowermost leaves opposite, the blades oblong-elliptic, entire; other leaves alternate and with blades pinnately divided, the primary divisions oblongish, 7–13 in number, oriented at right angles to the midrib, each segment usually 1–3-lobed or -toothed, overall leaf outline broadest at about the middle half, 3–8 cm long and 1–3 cm wide. Flowers stalked, solitary, arising opposite or above the petiole base, or sometimes in a small terminal cyme. Calyx lobes longer than the tube, enlarging as the fruit matures, pubescent. Corolla white or bluish, not exceeding the calyx, the lobes longer than the tube, a pair of minute scales at the base of each filament. Stamens not exserted. Style 1, stigmas 2. Ovary 1-locular, with 2 parietal placentae. Capsule globose, 5–6 mm across, exceeded by the calyx, usually 4-seeded. Seeds globose, reticulate. (*Nyctelea nyctelea* (L.) Britt.)

Floodplain woodlands, stream banks, prairies, cultivated fields, roadsides. Local, N.J. and Pa. to Va.; Man. to Alb., southward to Okla., Ark., Ind., Ill.

3. Nemophila

Nemophila aphylla (L.) Brummitt.

Low annual, with weak stems often much branched from the base, pubescent. Leaves petiolate, the petioles commonly as long as the blades, pubescent; lowermost leaf with a single orbicular or suborbicular blade, this commonly not present on plant at time of flowering; leaves alternate or sometimes the lower ones opposite; blades usually with 2–3 (4) palmate or subpalmate, irregular primary divisions, these in turn irregularly lobed, pubescent on both surfaces, in overall outline as broad as long or nearly so. Flowers small and inconspicuous, stalked, the stalks somewhat shorter than the petioles, solitary and arising opposite the leaves, sometimes in few-flowered racemes terminating branches. Calyx lobes longer than the tube, lanceolate, elliptic, or oblong, pubescent. Corolla white or pale bluish, somewhat exceeding the calyx, a pair of scales at the base of each filament. Style 1, stigmas 2. Ovary 1-locular, with 2 parietal placentae. Capsule ovoid or oval, 4–5 mm long, pubescent, 1–4 (–20) -seeded. Seeds roundish to oblong, pale brown, shallowly pitted, 1.5–2 mm long. (*N. microcalyx* (Nutt.) F. & M.)

Floodplain forests, moist lowland woodlands, mesic woodlands. e. Va., southward to Fla. Panhandle, westward to e. Tex., northward in the interior to Ark. Tenn., Mo.

Solanaceae (NIGHTSHADE FAMILY)

Lycium carolinianum Walt. CHRISTMAS-BERRY.

Sparingly bushy-branched, glabrous shrub to about 3 m tall, the main branches bearing shorter, more or less divaricate branchlets, many of which are rigid and sharply thorn-tipped. Leaves simple, alternate (commonly with very short, leafy branchlets in the axils giving a fascicled appearance), tardily deciduous (denuded very quickly in early winter when blocks

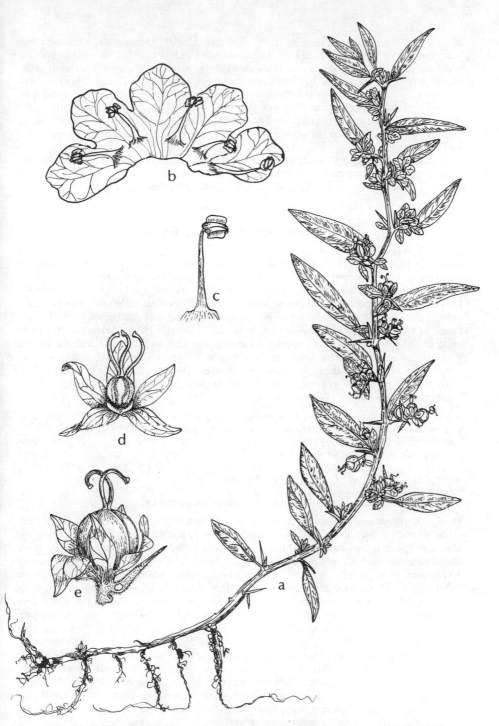

Fig. 270. **Hydrolea uniflora:** a, habit; b, corolla, opened out; c, stamen; d, flower with corolla removed (three styles unusual); e, fruit with subtending calyx (two styles usual). (From Correll and Correll. 1972)

of birds thrash about eating the ripe fruits), succulent, linear-oblanceolate or narrowly clavate, variable in length to 2.5 cm long. Flowers solitary in leaf axils, often in the axils of the small leaves of the short axillary branchlets and giving a fascicled aspect, the flower stalks slender, a little shorter then to equaling the subtending leaves. Calyx about 4 mm long, tubular below and with 4 short-deltoid lobes, persistent. Corolla blue or lavender, sometimes nearly white, tubular below, the limb with 4(5) rotate lobes about 5 mm long, slightly longer than the tube, pubescent in the throat and around the bases of the filaments which are inserted below the sinuses between the corolla lobes. Fruit a bright, lustrous red, elllipsoid berry 8–15 mm long.

Coastal sand spits, shell beaches and mounds, borders of, or in, salt and brackish marshes and mangrove swamps, coastal spoil areas, both where the bases may be inundated by high tides and somewhat above highest tide levels. Ga. and Fla., westward to Tex.; W.I.

Boraginaceae (BORAGE FAMILY)

Annual or perennial herbs (ours), or shrubs. Leaves alternate, simple. Flowers mostly in racemose or spicate cymes coiled at the tip before anthesis, in some elongating markedly during maturation, in some the axes short and flowers glomerate, rarely flowers borne singly. Flowers bisexual, radially symmetrical, or in some the calyx bilaterally symmetrical. Calyx united below, deeply or shallowly 5-lobed, persistent. Corolla tubular below, 5-lobed, commonly salverform, varying to funnelform or rotate, in some with appendages in the throat opposite the lobes. Stamens 5, inserted on the corolla tube and alternate with the corolla lobes. Ovary superior, 2-carpellate, usually 4-locular, a single ovule in each locule, in some deeply 4-lobed and with the style arising from between the lobes, in some unlobed or shallowly 4-lobed, the style terminal. Fruit in some breaking up into four 1-seeded nutlets, in some breaking up into two 2-seeded nutlets.

1. Flowers individually showy, the corollas trumpet-shaped and 25–30 mm long.   1. *Mertensia*
1. Flowers individually small, the corollas salverform, rotate, or funnelform, 5 mm long or less.
   2. Ovary unlobed or shallowly lobed, style terminal or essentially absent.   2. *Heliotropium*
   2. Ovary deeply 4-lobed, the style arising from between the lobes.   3. *Myosotis*

1. Mertensia

Mertensia virginica (L.) Pers. VIRGINIA-COWSLIP, BLUEBELLS.

Perennial, glabrous and glaucous herb with a thick, dark rootstock from which 1 to several stems arise, 2–7 dm tall. Leaves long-petioled below, gradually becoming sessile upwardly; blades of lower leaves tapering to margined petioles; blades of both petioled and sessile leaves punctate, variable in size and shape, larger ones sometimes 15–20 cm long and about 10 cm wide, ovate, elliptic, oblong-elliptic, or obovate, diminishing in size upwardly into the inflorescence, all rounded apically. Cymes axillary, relatively close together at first, often becoming distant as the main axis elongates; slender flower stalks to about 1 cm long. Calyx very much shorter than the corolla, about 5 mm long fully developed, tube campanulate, lobes somewhat longer than the tube, narrowly deltoid, blunt apically. Corolla showy, pink (rarely white) in bud, blue (white in some plants) when fully expanded, 1.5–3 cm long, long and slenderly funnelform below, abruptly expanded and cuplike distally, the cuplike portion with short, rounded lobes. Stamens inserted at about the base of the cuplike portion of the corolla and not exserted from it. Style slender, arising from between the lobes of the ovary, reaching the anthers or extending somewhat beyond them, remaining attached for a time after the corolla falls and then appearing long and threadlike beyond the calyx. Nutlets about 3 mm long, surfaces more or less pebbly or warty.

Alluvial soils of floodplain woodlands, stream banks, valley fields, sometimes on rocky slopes and bluffs. N.Y. to Minn., southward to N.C., Tenn., n. Ala., Ark., e. Kan.

2. Heliotropium (HELIOTROPES)

Annual or perennial herbs. Flowers small, in slender, crowded, 1-sided spikes or spikelike racemes circinate before anthesis, straightening after anthesis. Calyx shallowly to deeply lobed. Corolla small, salverform or funnelform. Anthers essentially sessile, included. Ovary unlobed or shallowly 4-lobed, style short, terminal on the ovary, stigma conical or capitate. Fruit separating into four 1-seeded nutlets or into two pairs of coherent nutlets.

1. Stem and leaves succulent, glabrous, usually glaucous.   1. *H. curassavicum*
1. Stem and leaves not succulent, not glaucous, pubescent.
   2. Flowers subtended by foliose bracts; leaves all sessile.   2. *H. polyphyllum*
   2. Flowers not subtended by bracts; lower leaves, at least, petiolate.
     3. Corolla limb pale to dark blue-purple with a white or yellow eye.   3. *H. indicum*
     3. Corolla limb white or white with a yellow eye (sometimes drying yellowish).
       4. Fruit much broader than long, separating into two 2-seeded nutlets.   4. *H. angiospermum*
       4. Fruit scarcely if at all broader than long, separating into four 1-seeded nutlets.
                                                      5. *H. procumbens*

1. Heliotropium curassavicum L.

Glabrous, usually glaucous, succulent annual or short-lived perennial, usually with several subequal, irregular branches from near the base, branches often decumbent, to about 6 dm long. Leaves mostly sessile, or a few sometimes narrowed to subpetiolate bases, oblanceolate to linear-oblanceolate, 1–6 cm long, blunt or rounded apically, almost veinless. Inflorescences commonly in stalked axillary pairs and in stalked pairs terminal on the branches, each of a pair 4–6 (–10) cm long fully developed, flowers sessile, alternating and approximate on the axes. Calyx united only basally, lobes deltoid, 1.5–2 mm long, blunt apically. Corolla salverform, tube about 2 mm long, limb white with a yellow eye. Fruit short-ovoid, shallowly 4-lobed, 2–3 mm across the base, splitting into 4 nutlets, persistent calyces ⅔ to about as long as the fruits. Nutlets tan, drying somewhat wrinkled-rugose.

Saline and brackish shores and flats, brackish or saline marshes, edges of mangrove swamps or flats, often weedy. Native of trop. Am., naturalized, somewhat sporadically and mostly near the coasts in our range, Mass. to s. Fla., and along the Gulf coast to Tex.; westwardly in inland Tex. and beyond.

2. Heliotropium polyphyllum Lehm.

Perennial herb, with a thickish, dark purplish brown, subterranean caudex from which 1 to several erect, weakly erect, or prostrate, simple or variously branched stems arise, stems very variable in length. Plant hispid throughout. Leaves sessile, little reduced upwardly to the inflorescences, narrowly lanceolate, linear, linear-elliptic, or narrowly oblanceolate, 1–2 cm long, only the midvein evident, this sometimes obscure, acute apically, margins entire. Racemes terminal on the branches, foliose-bracteate, flowers crowded on the axis at first, maturing fruits becoming somewhat separated as the axis elongates, fruiting racemes variable in length up to about 30 cm. Calyx tube short, lobes unequal, much longer than the tube, lanceolate, elliptic, or lance-ovate, 4–5 mm long. Corolla salverform, tube 3–4 mm long, limb white with a yellow eye or wholly yellow. Fruit ovoid, shallowly 4-lobed, brown, splitting into 4 nutlets, persistent calyx about 3 times as long as the fruits.

Dry to seasonally wet pinelands, pineland depressions, exposed margins or bottoms of pineland ponds, on lime-rock in pinelands, cypress and marl prairies, sand dunes, spoil banks or flats, low hammocks, brackish shores. Pen. Fla. and Fla. Keys; S. Am.

This "species" exhibits a complex variation of habit and flower color as yet not satisfactorily explained. Small (1933), for our range, treats three ill-defined species, *H. polyphyllum*, *H. leavenworthii* (Gray) Small [incorrectly attributed to Torr.], and *H. horizontale* Small. Long (1970), in a short note and without reference to the South American element, considers the complex to be comprised of *H. polyphyllum* var. *polyphyllum* and *H. polyphyllum* var. *horizontale* (Small) R.W. Long.

581

3. Heliotropium indicum L. TURNSOLE.

Erect annual herb, varying from simple, slender, and low, to coarse, much branched, and to 1 m tall. Stems and petioles sparsely clothed with stiff, pustular-based hairs. Leaves petiolate, blades abruptly narrowed proximally and decurrent on the distal portions of the petioles, mostly ovate or oval, obtuse to acute apically, pinnate venation more conspicuous beneath than above, upper surfaces glabrous or with a few scattered hairs on the midrib, lower surfaces usually with both long and short hairs on the principal veins and short-hispid between the veins, margins irregularly wavy. Stalked spikes single terminating branches becoming 8–30 cm long, flowers and fruits sessile, approximate, axes hirsute. Calyx united basally, lobes elongate-triangular, 1.5–2 mm long, sparsely hirsute. Corolla tube 2.5–3 mm long, narrowly urceolate, limb rotate, pale to dark blue-purple. Fruit a little broader than long, 2-lobed, lobes ribbed on the back, splitting into two 2-locular nutlets, one locule of each with 1 seed, 1 empty, rarely both seed-bearing. (*Tiaridium indicum* (L.) Lehm.)

Exposed sand and gravel bars of rivers and streams, alluvial muds, shores of impoundments, exposed shores or bottoms of ponds and lakes, floodplain forests, low wet places in fields. Va., W.Va., Ky., s. Ill. to Mo., generally southward to the Fla. Panhandle and e. half of Tex.; widely distributed in warmer parts of both hemispheres, supposedly native of Brazil.

4. Heliotropium angiospermum Murr.

Erect or sprawling annual, usually irregularly branched, up to 1 m tall. Stem sparsely hirsute, becoming glabrous below in age. Leaves short-petiolate, petioles hirsute, blades lanceolate, lance-ovate, or elliptic, mostly cuneate basally, acute to acuminate apically, both surfaces moderately hirsute, margins entire. Stalked spikes axillary and terminal on the branches, slender, becoming 4–6 cm long, axis hirsute, sessile flowers or fruits approximate. Calyx united only basally, lobes narrowly triangular, about 1.5–2 mm long, sparsely pubescent. Corolla tube short-cylindrical, 1–1.5 mm long, limb rotate, white with a yellowish eye, its lobes crinkled marginally. Fruit much broader than long, 2-lobed, about 2.5 mm across, surface scaly, separating into two 2-locular nutlets, each 2-seeded. (*Schobea angiospermum* (Murr.) Britt.)

Weedy on shell mounds, scrub, roadsides, citrus groves, saline shores, edges of mangrove swamps, only marginally a wetland plant. Near the coasts from cen. pen. Fla. southward and on the Fla. Keys; trop. Am.

5. Heliotropium procumbens Mill.

Annual, simple or usually bushy-branched, branches often sprawling, 1–5 dm long, strigose throughout, stems and leaves grayish green. Leaves petiolate, petioles to 1–1.5 cm long, blades oblanceolate to elliptic, cuneate basally, rounded to obtuse apically, margins entire, commonly only the midvein evident, pinnate laterals sometimes obscurely evident. Stalked racemes slender, solitary or paired, numerous, axillary and terminal on the branches; fruiting racemes varying greatly in length from about 3–15 cm, short-stalked flowers and fruits approximate. Calyx united only basally, lobes unequal, narrower ones subulate, broader ones lance-ovate, fully doubling in length after anthesis, becoming 2–2.5 mm long. Corolla tube funnelform, about 1.5 mm long, limb rotate, white (drying yellowish), lobes triangular. Fruit strigose (not tuberculate), somewhat broader than long, about 1.5 mm across, shallowly 4-lobed, calyx lobes somewhat exceeding the fruit, fruit splitting into four 1-seeded nutlets.

Commonly along rivers on muddy alluvium after flooding, river shores, exposed margins or bottoms of ponds, marshes, ditches. Fla. Panhandle to s.e. Tex., s.e. Okla., (Ark.?); trop. Am.

Specimens of this species from Fla., Miss., and La. deposited in the Fla. State University herbarium, were all identified by their collectors as *H. europaeum* L., a plant similar in general features, native of Europe and attributed to our range as a weed. Fruits of *H. europaeum* are minutely pubescent and tuberculate.

3. Myosotis (FORGET-ME-NOTS)

Annual, biennial, or perennial herbs with entire leaves. Flowers mostly in bractless racemes terminating the branches (sometimes the lower ones in the axils of little reduced leaves); racemes coiled at first, the flowers crowded, later the raceme axes elongating and straightening, and the maturing fruits well separated from each other. Calyx lobes narrow, unequal in some. Corolla salverform, tube short, lobes of the limb rounded apically, blue, white, or pink, sometimes yellow at first, then changing to blue, sometimes with a yellow eye, the throat partly closed by appendages opposite the lobes. Stamens with very short filaments inserted on the corolla tube. Ovary 4-lobed, style slender, terminal, stigma capitate. Fruit enclosed by the calyx, separating into four 1-seeded nutlets.

1. Hairs on the calyx (most of them) hooked at their tips.
    2. Calyx bilaterally symmetrical, lobes unequal; corolla white.   1. *M. macrosperma*
    2. Calyx radially symmetrical, lobes equal; corolla yellow at first, the limb (at least) turning blue to violet.   4. *M. discolor*
1. Hairs on the calyx appressed-ascending, straight.
    3. Calyx lobes shorter than the calyx tube: corolla limb 6–9 mm broad.   2. *M. scorpioides*
    3. Calyx lobes equaling or a little longer than the calyx tube; corolla limb 2–5 mm broad. 3. *M. laxa*

1. Myosotis macrosperma Engelm.

Annual or biennial. Stems laxly erect, few-branched, 2–6 dm tall, with copious spreading hairs. Both surfaces of the leaves pubescent, usually copiously, with stiff, pustular-based hairs. Basal leaves more or less in a rosette, tapering basally to margined petioles, blades rounded to obtuse apically, 2–8 cm long overall; stem leaves sessile, gradually reduced upwardly, lower ones with straplike bases and oblanceolate blades, becoming oblong or linear-oblong upwardly, the larger lower ones 5–20 mm wide. Racemes terminal on the branches, variable in length when fully developed, the flower number varying on a given branch from 2 or 3 to 20 or more, eventually relatively distant from each other, the lower ones 2–5 cm apart, the lowermost sometimes in the axils of bracteal leaves. Stalks of the fruits 3–10 mm long, spreading or recurved. Calyx with copious spreading hairs hooked at their tips, lobes triangular-acute, markedly unequal, the longest 2–3 mm long. Corolla white, limb 2–2.5 mm broad. Nutlets about 2 mm long, oval to ovate in outline, slightly lenticular, narrowly winged marginally, surfaces tan, smooth, lustrous.

Floodplain forests, adjacent bluffs and slopes, seepage areas in woodlands, moist to wet areas in fields, mostly where the substrate is somewhat calcareous. Md. to s. Ind. and Mo., generally southward to the Fla. Panhandle and e. Tex.

2. Myosotis scorpioides L.   Fig. 271

Perennial. Stems at first erect, often becoming decumbent, rooting below, sparsely pubescent with stiffly ascending hairs, 3–7 dm tall. Lower leaves with margined petiolar or straplike bases, oblanceolate, 2–8 cm long, median and upper ones mostly sessile, oblanceolate to oblong, somewhat reduced upwardly; all leaves with rounded to obtuse apices, surfaces sparsely pubescent with relatively short, stiff, appressed hairs. Racemes slender, lax, terminating the main stem or the branches, variable in length, eventually 2–3 dm long, the fruits 1–2 cm apart, stalks of the fruits spreading, 5–15 mm long. Calyx with short, stiff, appressed-ascending, straight hairs, in fruit 3–4 mm long, lobes nearly equal, short-deltoid. Corolla tube barely exceeding the calyx, limb blue with a yellow eye, 6–9 mm broad. Nutlets ovate in outline, 1.5–2 mm long, lenticular, margins keeled, surfaces dark brown to nearly black, shiny.

Open or wooded stream banks, marshy shores, wet meadows. Native of Eurasia, commonly cultivated, in our range sporadically naturalized chiefly in the mts., sometimes elsewhere.

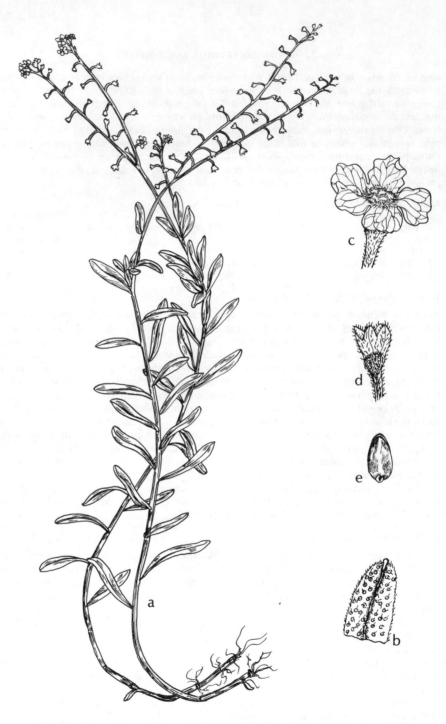

Fig. 271. **Myosotis scorpiodes:** a, habit; b, tip of leaf; c, flower; d, calyx; e, seed. (From Correll and Correll. 1972)

3. Myosotis laxa Lehm.

Perennial. Stems slender, single or several from the base, erect at first, sometimes becoming decumbent, 1–5 dm long, sparsely pubescent with short, stiff, appressed-ascending hairs. Lower leaves 2–5 cm long, narrowed to short winged petioles, oblanceolate to spatulate; most of the leaves sessile, narrowly lanceolate, or oblanceolate to oblong. Racemes terminal on the main stem or the branches, often branched on a given branchlet, eventually lax and the fruits relatively distant from each other, mostly 8–20 mm apart, stalks of the fruits spreading, slender, 5–10 mm long. Calyx with short, stiff, appressed-ascending, straight hairs, in fruit 2–3 mm long, lobes nearly equal, deltoid, as long as or a little longer than the tube. Corolla tube about as long as the calyx, limb blue with a yellow eye, 2–5 mm broad. Nutlets 1–1.5 mm long, ovate in outline, slightly lenticular, margins slightly keeled, surfaces brown, lustrous.

Marshy shores, stream banks or in shallow water of small streams, bogs, often in mud. Nfld. and Que. to Minn., s. to N.C., Ga., Tenn.

4. Myosotis discolor Pers.

Annual. Stem simple or often with several branches from near the base, 0.5–3 dm tall, with copious, spreading, straight, stiffish hairs, both leaf surfaces similarly pubescent. Basal leaves oblanceolate, blunt apically, narrowed to short petiolar bases, 0.5–3 cm long overall; other leaves sessile, oblanceolate to linear-lanceolate, some sometimes oblong-elliptic, gradually diminishing in size upwardly, none or but few on lowermost parts of the inflorescence branches, mostly blunt apically. Racemes terminal on the branches, loosely flowered, of varying lengths up to about 2 dm, flower stalks shorter than to about equaling the length of the calyx. Calyx symmetrical to asymmetrical, lobes equal or unequal; hairs on the calyx spreading, those on the tube with hooked tips, usually some of those on the lobes straight. Corolla pale yellow at first, the limb, at least, turning blue to violet (sometimes perhaps blue from the start), limb 1–2 mm across. Stalks of the fruits not exceeding 2 mm long. Nutlets a little over 1 mm long, ovate-lenticular, banded around the entire margin, surfaces brown, smooth and lustrous. (*M. versicolor* J. E. Sm.)

Wet meadows, fields, roadsides. Native of Eur., naturalized N.S., Que., southward to N.C.; s.e. Ala., perhaps elsewhere.

Verbenaceae (VERVAIN FAMILY)

Annual or perennial herbs, shrubs, or trees. Leaves opposite, simple or palmately compound, sometimes more or less pinnately divided, without stipules. Inflorescences axillary or terminating branches, spicate or in headlike spikes, racemose, cymose, or paniculate. Flowers bisexual or rarely unisexual. Calyx tubular at least below, mostly 4- (sometimes 2-, 5-, or 7-) lobed, or summit entire, persistent, in some enlarging as the fruit matures. Corolla tubular below, the limb essentially radially symmetrical or 2-lipped, 4- or 5- (rarely 7- or many-) parted. Stamens usually 4, in pairs, sometimes 2, or 4 or 5 and equal, inserted on the corolla tube, included or barely exserted. Ovary superior, 2–4-locular, ovules usually 1 in each locule; style terminal, stigma capitate or 2-lobed. Fruit dry and splitting into two or four 1-seeded nutlets or a drupe containing 2–4 nutlets.

- Flowers in spikes terminating the branches or branchlets; calyx unequally 5-lobed, the tube 5-angled or 5-ribbed; fruit splitting into 4 nutlets.   1. *Verbena*
- Flowers in congested headlike spikes terminating stalks borne from the leaf axils; calyx 2–4-toothed or -lobed, tube campanulate or somewhat flattened; fruit splitting into 2 nutlets.   2. *Phyla*

1. Verbena (VERVAINS)

Annual or perennial herbs with quadrangular stems. Leaves simple, pinnately veined, more or less incised or pinnatifid in some. Flowers in many-flowered, congested to loose and elongate bracteate spikes terminating the branches, the spikes sometimes more or less fastigiate or in umbellike clusters, or paniculate. Calyx 5-lobed, the tube 5-angled or 5-ribbed, the lower lobes usually longest. Corolla salverform, the tube straight or curved, the limb unequally 5-lobed, weakly 2-lipped. Stamens 4, in pairs, the short filaments of the pairs inserted at slightly different levels near the summit of the corolla tube, not exserted. Ovary 2-locular, mostly somewhat 4-lobed, style terminal, stigmas 2-lobed. Fruit enclosed by the persistent calyx, readily separating into 4 nutlets linear-oblong in outline.

1. Leaves sessile, or at most cuneate basally and subsessile, lanceolate, narrowly elliptic, or oblong-elliptic.
  2. Bases of the leaves little narrowed, subauriculate and subclasping.   1. *V. bonariensis*
  2. Bases of the leaves acutely narrowed.   2. *V. brasiliensis*
1. Leaves, lower and median stem ones at least, petiolate, the larger blades ovate or lance-ovate.
  3. Flowers and fruits ascending and overlapping each other on the spikes excepting sometimes the lowermost.   3. *V. hastata*
  3. Flowers and fruits scattered, diverging, and not or very little overlapping on the spikes.
    4. Lobes of the fruiting calyx long-acute, ascending and connivent, forming a point beyond the fruit; nutlets somewhat ribbed-reticulate on their outer faces.   4. *V. scabra*
    4. Lobes of the fruiting calyx short-deltoid and abruptly narrowed to a subulate tip, loosely arching over the summit of the fruit and not forming a point; nutlets smooth on their outer faces.   5. *V. urticifolia*

1. Verbena bonariensis L.

Coarse perennial. Stems erect, to 1.5 m tall or more, solitary or commonly several from a basal crown, with stiff, pustular-based hairs on and between the angles, some hairs retrorse, some antrorse, some spreading. Leaves sessile, lanceolate to oblong-elliptic, midstem leaves largest and mostly 8–15 cm long, 1–4 cm broad, bases subcordate and subclasping, apices acute, margins coarsely and sharply serrate along their margins from base to apex or from well below their middles, upper surfaces scabrid to hispid, lower hispid on and between the veins, the hairs on the veins longer than those between. Spikes individually dense, 1–4 cm long and 5–6 mm broad when fully developed, aggregated in fastigiate or umbellike clusters at the ends of opposite inflorescence branches, the lower branches longest. Bracts pubescent, lance-ovate, tips acuminate and stiff, as long as the calyces or slightly longer. Calyx 2.5–3 mm long, pubescent, lobes subulate, becoming connivent to a point beyond the fruit. Corolla purple, the tube nearly twice as long as the calyx. Nutlets oblong in outline, 1 mm long or slightly more, outer faces dull brown, smoothish and with few, usually faint, striate-reticulate elevations, inner angled faces covered with small buffish murications.

Weedy, mostly in moist to seasonally wet places, fields, clearings, swales, ditches, waste places generally. Coastal plain and piedmont, N.C. to s. Fla., westward to e. Tex., s.e. Okla., Ark., Tenn.; Calif. Native of S.Am.

2. Verbena brasiliensis Vell.

Closely similar to *V. bonariensis* in general aspect. Stems relatively sparsely and antrorsely hispid on and between the angles with pustular based hairs, the hairs above their bases commonly deciduous, the stems becoming relatively smooth in age. Leaves lanceolate or lance-elliptic, the larger midstem ones 4–6 (–10) cm long and 1–2 (–2.5) cm broad, acutely narrowed to sessile or subsessile bases, acute apically, sharply serrate distally above the tapered bases, upper surfaces evenly antrorsely hispid, lower more densely hispid on the veins and veinlets and between them. Spikes individually dense, 4–5 mm wide, disposed as in *V. bonariensis*. Corolla dull bluish purple. For the most part only 2 locules of the fruits mature, 2 abort, mature ones closely similar to those of *V. bonariensis* but fully 2 mm long.

Weedy, in similar places to those inhabited by *V. bonariensis* but more frequently on drier sites. Coastal plain and piedmont, s.e. Va. to s. Fla., westward to s.e. Tex., s.e. Okla., Ark.; Ore. and Calif., perhaps elsewhere. Native of S.Am.

3. **Verbena hastata** L. BLUE VERVAIN.

Coarse perennial. Stems to 1.5 m tall, rough-pubescent with short, usually antrorsely appressed or subappressed hairs. Leaves, excepting the uppermost, with short petioles, blades varying from ovate to lanceolate; lowermost leaves (usually shed by time of flowering), sometimes midstem ones as well, with a pair of narrow lobes at or somewhat above the base; bases of broader blades usually truncate or rounded, those of narrower ones cuneate, all long-acute apically, margins coarsely and sharply serrate, commonly doubly serrate, upper surfaces with stiff appressed pubescence, lower with more shaggy hairs, notably on the veins. Spikes individually relatively compact, flowers or fruits overlapping, variable in length to 10 cm, disposed in terminal panicles. Bracts triangular-subulate, keeled, about two-thirds as long as the fruiting calyx. Fruiting calyx appressed-pubescent, 2.5–3 mm long, lobes unequal, abruptly narrowed to points somewhat incurved over the tips of the fruits. Corolla blue or violet-blue, the limb 2.5–4.5 mm across. Nutlets linear-oblong in outline, about 2 mm long, outer rounded faces smooth, brown and sublustrous, inner angled faces finely muricate with pale grayish trichomes.

Shores and banks of rivers and streams, marshy shores, bogs, swales, meadows, prairies, wet woodlands, low fields, often forming large colonies on wetlands. N.S. to s. B.C., southward throughout much of the U.S. (according to most general references).

4. **Verbena scabra** Vahl. HARSH VERVAIN.

Relatively coarse perennial. Stem to 1.5 m tall, with spreading, pustular-based, stiff and sharp pubescence. Leaves petiolate, blades ovate to lance-ovate, largest leaves at midstem, 6–12 cm long and to 6 cm broad, bases rounded to broadly cuneate, margins crenate-serrate to coarsely, often doubly, serrate, sometimes some leaves on a given specimen with a pair of short lobes somewhat above the base, both surfaces pubescent with spreading, stiff, sharp, pustular-based hairs, the lower more densely so than the upper. Individual spikes slender, with spreading-ascending flowers or fruits, barely if at all overlapping; spikes of various lengths from 7–20 cm, usually disposed in diffuse panicles the lowermost branches of which arise from the axils of foliage leaves. Bracts triangular-subulate, hispid, mostly about half the length of the fruiting calyces and not appressed to them. Fruiting calyx hispid, mostly 2 mm long, the long-acute lobes ascending and connivent beyond the fruit forming a point. Corolla pinkish, bluish, or lavender, the tube barely exceeding the calyx, the limb 2–3 mm wide. Style and stigma remaining attached to the fruit until maturity and not exserted from the calyx. Nutlets oblong in outline, outer rounded faces rust-brown, somewhat ribbed-reticulate, inner angled faces with buffish murications.

Marshy shores, wet woodlands and their borders or clearings, floodplain forests, swales, stream banks, wet thickets, shell mounds. Coastal plain, Va. to s. Fla., westward to Calif.; n.Mex., W.I.

5. **Verbena urticifolia** L. WHITE VERVAIN.

In general aspect, leaf, spike, and inflorescence characteristics closely similar to *V. scabra*. Lobes of the fruiting calyx short-deltoid, abruptly narrowed to subulate tips that loosely arch over the summit of the fruit and not forming a point beyond the fruit. Nutlets smooth on their outer rounded faces.

Meadows, open or wooded stream banks, floodplain forests, marshy shores, swales, moist thickets, fence rows, fields, banks, open upland woodlands. Maine, s.w. Que., s. Ont. to S.D., southward to Fla. Panhandle and Tex.

2. **Phyla** (FROG-FRUITS)

Perennial herbs or subshrubs, main stems creeping or procumbent, flowering branches ascending. Flowers in dense, many-flowered spikes borne terminally on single stalks from the leaf axils, usually at alternating nodes, the spikes at first globular and headlike, at length becoming oblong-cylindric as flowering within a spike occurs over an extended period. Flowers small, borne singly in the axils of closely imbricated bracts. Calyx small, 2-lobed in ours, the tube campanulate or flattened. Corolla 2-lipped, upper lip notched, lower 3-lobed, the middle lobe often larger than the laterals. Stamens 4, in 2 pairs, filaments short, inserted on the corolla tube below the throat, 1 pair somewhat lower than the other, anthers not exserted. Ovary 2-locular, style 1, stigma obliquely capitate. Fruit separating into 2 nutlets, each broadly rounded on the outer side, flat on the inner.

1. Stems woody below; pubescence of straight appressed hairs attached basally; leaves with numerous pronounced lateral veins angling parallel to each other from the midrib to the sinuses between the marginal serrations, the veins deeply impressed and invisible on the upper surface, very strongly elevated below. 1. *P. stoechadifolia*
1. Stems herbaceous; pubescence of appressed hairs attached at their middles and tapering to both extremities; leaves with relatively few, relatively obscure lateral veins.
   2. Leaves mostly lanceolate to lance-elliptic (rarely oblanceolate), with a strong distal taper to an acute apex; most leaves marginally toothed distally from somewhat below the middle, teeth 7–11 on a side, rarely as few as 5. 2. *P. lanceolata*
   2. Leaves oblanceolate to spatulate or obovate, with a moderate distal taper or more commonly convex on either side of the narrowing distal portion, apices rounded to obtuse, rarely acute; most leaves marginally toothed distally from the middle or just above, teeth 3–5, mostly 5, on a side, rarely 7.
   3. *P. nodiflora*

1. Phyla stoechadifolia (L.) Small.

Plants woody below, the bark of older stems smooth, pale buff and with a whitish cast, that of younger stems reddish brown, clothed with antrorsely appressed, straight white hairs attached basally; both leaf surfaces, stalks of the spikes, and bracts with copious pubescence of the same kind. Stems decumbent below, branching in an open straggly fashion, herbaceous leafy flowering branches erect. Leaves lanceolate, stiff, 2–5 cm long and 5–10 mm wide, abruptly narrowed to subpetiolate bases, apices acute, margins evenly serrate from base to apex, with numerous pronounced lateral veins angling parallel to each other from the midvein to the sinuses between the serrations, the veins deeply impressed and invisible on the upper surface, very strongly elevated on the lower; the copious whitish pubescence on both surfaces gives the leaves a notably gray appearance. Stalks of the spikes varying from 2–10 cm long. Bracts obovate, abruptly short-acuminate apically, pubescent exteriorly on the upper half. Fruiting calyx 2.5 mm long, obovate in outline, flattened, the edges very narrowly wing-keeled, 2 short-triangular folded lobes apically. Corolla purple. Fruit 2 mm long, obovate in outline, summit nearly truncate, tardily splitting. (*Lippia stoechadifolia* (L.) HBK.)
   Marl prairies, glades. s. Fla.; W.I., Mex.

2. Phyla lanceolata (Michx.) Greene.

Main stems creeping, rooting at the nodes, often radially spreading and elongate, with weakly ascending flowering branches to 6 dm long; stems pubescent with appressed white hairs attached centrally and tapering to their extremities, the pubescence commonly soon sloughed leaving the stems glabrous. Pubescence of both leaf surfaces, inflorescence stalks, and bracts like that of the stems. Leaves lanceolate or lance-elliptic, rarely oblanceolate, 2–6 cm long and to 2.5 cm wide, tapering proximally to decurrent petioles to 10 mm long, or sessile; most leaves widest below their middles and tapering distally to acute apices, marginally toothed distally from somewhat below their middles, teeth 7–11 on a side, rarely as few as 5. Stalks of the spikes 4–10 cm long, from about as long as to 2.5 times as long as the subtending leaf. Bracts strongly folded-keeled proximally, abruptly flaring distally and broadly rounded but with a narrowed tip, pubescent exteriorly but the pubescence sometimes sloughed. Fruiting calyx campanulate below, with 2 broadly triangular-acuminate lobes whose tips arch toward each other over the summit of the fruit. Corolla pink, bluish, purplish, or white. Fruit plump, short-obovate, very nearly orbicular in outline, 1.5 mm long, splitting tardily. (*Lippia lanceolata* Michx.)

Fresh to brackish marshes, marshy shores, low woodlands along streams, ditches, sloughs. Ont. to Minn., southward on the Atlantic coastal plain to n.e. Fla., southward w. of the Appalachians to Ala., Miss., La., and w. of our range to Calif., n. Mex.

3. Phyla nodiflora (L.) Greene.

Similar in habital features to *P. lanceolata* but the ascending flowering branches shorter, rarely over 1 dm tall. Pubescence as in *P. lanceolata* but not sloughing from the older stems or bracts. Leaves sessile or tapered to short subpetiolate bases, oblanceolate, spatulate, or obovate, with a moderate distal taper or more commonly convex on either side of the narrowing distal portion, apically rounded or obtuse, rarely acute; most leaves marginally toothed distally from the middle or just above, teeth 3–5, mostly 5, on a side, rarely 7. Stalks of the spikes to 10 cm long, most of them 3 times as long as the subtending leaves or a little more. Bracts cuneate-obovate, the summit somewhat rounded laterally and coming to a point centrally. Calyx tube somewhat flattened, the edges with narrow pubescent keels each running into an erect traingular-subulate lobe about as long as the tube and extending beyond the fruit. Corolla rose-purple to white. Fruit plump, obovate in outline, nearly truncate at the summit, 1 mm long, readily splitting at maturity. (*Lippia nodiflora* (L.) Michx.)

Wet sands and peaty sands, mud flats, estuarine shores, beaches, interdune hollows, depressions in flatwoods, ditches, fields, thickets, clearings. Coastal plain, s.e. Va. to s. Fla., westward to Tex., Okla., Ark., s.e. Mo.; s. Calif.; trop. and subtrop. regions of New and Old Worlds.

Avicenniaceae (BLACK MANGROVE FAMILY)

Avicennia

Avicennia germinans (L.) L. **BLACK MANGROVE.**   Fig. 272

Shrub or small tree, sometimes attaining 20–25 m tall. Leaves opposite, evergreen, leathery, short-petiolate, the blades elliptic, oblong, lanceolate or somewhat oblanceolate, their apices obtuse to rounded; dark green and glabrous above, grayish and with a very tight felty pubescence beneath; stipules none. Inflorescence of stalked, bracteate cymes borne from the leaf axils or terminally on the twigs, pubescent. Flowers sessile, spicate on the cyme branches. Sepals 5, pubescent, nearly separate, 3–4 mm long, persistent on the fruit. Corolla white, tubular below then flaring and with one lobe above and a 3-lobed lip. Stamens 4, the filaments attached to the lower part of the corolla tube, anthers barely exserted. Fruit a flat, asymmetric, almost velvety 1-seeded pod much resembling a very short lima-bean pod. (*A. nitida* Jacq.)

Sandy or marly shores and hammocks, commonly intermixed with red mangrove, marginal to it, or with white mangrove. Coasts of Fla. southward from St. Johns Co. on the east and Levy Co. on the west, Fla. Keys; sporadic along the Gulf coast westward to La. but attaining little stature in this part of the range; Berm.; trop. Am.; W.Afr.

Labiatae (MINT FAMILY)

Annual or perennial aromatic herbs or low shrubs with square or 4-angled stems. Leaves simple, opposite. Flowers bisexual, bilaterally symmetrical, variously arranged, solitary or clustered in the leaf axils, racemose, paniculate, spicate, or in congested headlike cymes. Calyx tubular below, 2–5-toothed or -lobed, sometimes 2-lipped. Corolla tubular below, in most 2-lipped beyond the throat (upper lip obscure in *Teucrium*), lower lip 3-lobed. Stamens inserted on the corolla tube, 4 in 2 pairs or sometimes 2; anthers dehiscing longitudinally.

Fig. 272. **Avicennia germinans.** (From Little and Wadsworth, *Common Trees of Puerto Rico and the Virgin Islands* (1964) Fig. 249)

Pistil 1, 2-carpellate, ovary superior, deeply or shallowly 4-lobed, 4-locular, 1 ovule in each locule; style 1, arising from between the lobed summit of the ovary, stigmas 2. Each of the 4 lobed portions of the ovary maturing into a 1-seeded nutlet, the nutlets enclosed by the persistent calyx.

1. Calyx with a distinct protuberance, ridge, or helmetlike projection on the upper side of the tube.
   1. *Scutellaria*
1. Calyx tube not as above.
   2. Corolla limb with only a lower lip. 2. *Teucrium*
   2. Corolla limb 2-lipped, or nearly radial.
      3. Flowers in very tightly compacted, sessile, axillary clusters; stamens 2. 3. *Lycopus*
      3. Flowers arranged other than as described above; stamens 4.
         4. Plant with slender subterranean rhizomes or lower leafy stems creeping and rooting at the nodes, mat-forming; flowering stems weakly ascending, 5–40 cm high; leaf blades short-ovate to suborbicular, 3–15 (–20) mm long; flowers solitary in the leaf axils. 4. *Micromeria*
         4. Plant with other than the above combination of characters.
            5. Flowers in congested, involucrate cymes or heads.
               6. Congested cymes or heads themselves disposed cymosely at the summit of the plant.
                  5. *Pycnanthemum*
               6. Congested cymes or heads stalked from the axils of leaves or bracteal leaves.
                  7. Calyx slightly oblique at the orifice, 5-lobed, lobes nearly equal. 6. *Hyptis*
                  7. Calyx 2-lipped, the upper lip a single narrow lobe, lower lip with 2 broader lobes.
                     7. *Macbridea*
            5. Flowers not in involucrate, congested cymes or heads.
               8. The flowers in terminal racemes or panicled racemes; portion of the corolla extending beyond the calyx (10–) 20 mm long or a little more. 8. *Physostegia*
               8. The flowers in verticels axillary to bracts or in stalked to sessile congested cymes axillary to bracts; portion of corolla extending beyond the calyx 10 mm long or less.
                  9. Flowers in verticels axillary to bracts.
                     10. The flowers sessile. 9. *Stachys*
                     10. The flowers stalked. 10. *Calamintha*
                  9. Flowers in stalked to sessile, congested cymes axillary to bracts. 11. *Mentha*

1. Scutellaria (SKULLCAPS)

(Ours) perennial herbs, not aromatic, bitter to taste. Flowers axillary or in racemes, racemes commonly panicled. Calyx campanulate at anthesis, a conspicuous protuberance on the upper side of the tube, with 2 low, broadly rounded to almost truncate, or broadly triangular lobes distally, enlarging considerably after anthesis and ultimately splitting to the base along 2 sutures, in some the protuberances becoming enlarged or inflated following anthesis, upper portion usually falling away finally. Corolla tube in most long-exserted from the calyx, curved-ascending, dilated above, the limb 2-lipped, upper lip usually concave and arching (hoodlike), with a lobe on either side of its base, lower lip spreading; corollas pale to deep blue to violet (white in infrequent individuals). Stamens 4, ascending from the throat up and under the upper corolla lip.

1. Flowers 20 mm long or more.
   2. Blades of basal and lower stem leaves ovate, their bases truncate to subcordate. 1. *S. integrifolia*
   2. Blades of basal and lower stem leaves lance-ovate to narrowly ovate, their bases tapering.
      2. *S. arenicola*
1. Flowers 5–7 mm long.
   3. The flowers in stalked racemes axillary to leaves or bracteal leaves. 3. *S. lateriflora*
   3. The flowers solitary in the axils of leaves or bracteal leaves. 4. *S. racemosa*

1. Scutellaria integrifolia L.

Stems single or several from the base, 3–7 dm tall, variously pubescent, hairs short and curly to copiously pilose, on some specimens with some gland-tipped hairs. Lower leaves with slender, pubescent petioles, blades ovate, mostly 1.5–3 (–4) cm long, truncate to subcordate basally, rounded to obtuse apically, with few crenate-serrate teeth marginally; up-

Fig. 273. **Scutellaria lateriflora:** a, midsection of plant; b, branchlet; c, flower; d, corolla, opened out; e, fruiting calyx; f, calyx, opened out and showing nutlets on gynophore. (From Correll and Correll. 1972)

592

wardly leaves becoming narrower and longer, narrowly lanceolate to nearly oblong, tapering to margined subpetiolate bases, commonly with entire margins; leaf surfaces punctate-dotted, with sparse scattered pubescence on the upper surfaces, hirsute on the veins beneath, veins obscure on the upper surface, the midrib and laterals well elevated on the lower and without cross-veins. Flowers in terminal racemes or panicled racemes, flower stalks 3–4 mm long, pubescent. Calyx pubescent, sometimes some or all of the hairs stipitate-glandular, 2.5–3 mm long at anthesis, becoming 5–10 mm long in fruit and reticulate-veiny. Corolla pale to bright blue, sometimes pinkish, 2–2.5 cm long. Nutlets ovate-orbicular in outline, 1 mm or a little more in diameter, black or very dark brown, covered with imbricated, flattened, oblongish, nearly erect tubercles with blunt apices.

Bogs, meadows, pine savannas and flatwoods, borders of wet woodlands, wet thickets, sandy-peaty ditches, less frequently in well-drained open woodlands and their borders. s.e. Mass. to s.cen. pen. Fla. and within the area bordered westerly from s.e. Mass. to s. Ohio, Ky., Tenn., Ark., e. Tex.

2. Scutellaria arenicola Small.

Stems 3–6 (–7) dm tall, single or several from the base, shaggy-pubescent. Lower leaves with more or less winged petioles 1–1.5 cm long, blades narrowly ovate or lance-ovate, 1.5–3 cm long, tapered basally, usually obtuse apically, margins crenate-serrate; upwardly leaves gradually smaller, with shorter petioles, differing little in shape and toothing from the lower ones; leaf surfaces sparsely to densely pubescent, the lower surface especially rough-ish-punctate, veins obscure on the upper surface, the midrib, lateral and prominent cross-veins well elevated on the lower surface. Flowers in few-branched racemes, flower stalks 3–5 mm long, pubescent, some of the hairs gland-tipped. Calyx copiously glandular-pubescent, about 3 mm long at anthesis, becoming 6–7 mm long in fruit. Corolla 2–2.5 cm long, pale to bright blue. Nutlets suborbicular in outline, about 1 mm in diameter, brown, covered with flattish, imbricated, suborbicular, tubercles.

(Apparently) in both well-drained pinelands and in seasonally wet flatwoods. Endemic to cen. pen. Fla.

3. Scutellaria lateriflora L. Fig. 273

Plant forming very slender rhizomes and also surface runners with much reduced leaves. Stems to 1 m tall or more, usually branched, sometimes copiously branched, glabrous or pubescent on the angles. Leaves with slender, sparsely pubescent petioles 5–20 (–30) mm long, blades thin and flexuous, ovate, rounded to truncate, or shortly and broadly tapered basally, acuminate apically, margins serrate or crenate-serrate (rarely few-toothed or entire), 3–7 (–12) cm long and 1–5 cm broad, glabrous or short-pubescent on the veins beneath and on the margins. Flowers small, in stalked racemes 1–10 cm long in the axils of leaves or bracteal leaves, racemes often secund, flower stalks 1–2 mm long. Calyx 2 mm long or a little more at anthesis, becoming 3–4 mm long in fruit, powdery-pubescent, sometimes with some long stipitate-glandular hairs as well. Corolla blue, varying to pink or white, 5–8 mm long. Nutlets suborbicular in outline, flattish on 2 faces, 3rd face rounded, about 1 mm in diameter, tan-colored, surface pebbled.

Swamps, marshy shores of streams, ponds, or lakes, meadows, seepage slopes, wet thickets. Nfld. and Que. to B.C., generally southward to n. Fla. and s. Calif.

4. Scutellaria racemosa Pers.

Plant forming very slender rhizomes, these often moniliform-tuberous, colonial. Stems weakly ascending, to 3 dm tall, usually copiously branched, glabrous. Principal leaves short-petioled, blades triangular-ovate to lance-ovate, usually some with a blunt tooth or small lobe on one or both sides at the base, margins otherwise entire, to about 2 cm long and 1 cm wide at the base, apices blunt, glabrous or minutely pubescent on the margins and on the veins beneath; upwardly leaves becoming bracteal, the uppermost sessile or nearly so, lanceolate. Flowers small, borne singly in the axils of leaves or bracteal leaves, flower stalks about 3 mm long, softly short-pubescent. Calyx softly pubescent exteriorly, 1.5–2 mm long at anthesis,

Fig. 274. **Teucrium canadense:** a, topmost part of plant; b, base of plant; c, node; d, leaf venation; e, flower, opened out; f, node with fruits; g, nutlets. (From Correll and Correll. 1972)

becoming 4 mm long in fruit. Corolla softly pubescent exteriorly, the tube little if any exserted from the calyx, lavender, the back of the upper lip darker, lower lip purple-speckled. Nutlets broadly short-tapered basally and rounded domelike apically, tan, surface pebbly-alveolate.

Native of S.Am., apparently recently introduced, sporadically occurring in wet places, thus far known from the coastal plain of S.C., Fla. Panhandle, s.w. Ga., s.e. Ala.

2. Teucrium

Teucrium canadense L. GERMANDER.   Fig. 274

Rhizomatous perennial, colonial. Stems simple, terminated by a single narrow raceme, or branched only in the inflorescence, 5–15 dm tall, pubescent and the surface itself with tiny, beadlike, whitish, sessile glandular atoms amongst the hairs. Leaves petiolate, petioles 5–15 mm long; blades ovate, lance-ovate, lanceolate, or long-elliptic, margins saliently serrate, bases rounded to narrowly cuneate, apices acute to acuminate or rarely obtuse; larger leaves 6–10 cm long and 2–4 cm broad, longer than the internodes; leaves not much reduced below the inflorescence; upper leaf surfaces short-pubescent, to villous, to densely silvery-tomentose. Pubescence of inflorescence axes, bracts, and calyces much like that of the lower leaf surfaces on a given plant and surfaces sparsely to copiously clothed with tiny, beadlike, whitish, sessile atoms, these evident or to a greater or lesser degree obscured depending upon the density of the pubescence. Racemes commonly solitary, sometimes panicled, 5–20 cm long; flowers in pairs, whorled, or some of them alternate. Bracts lance-attenuate, equaling to considerably exceeding the calyces. Calyx 5–7 mm long, oblique at the orifice, lobes triangular, the upper 3 blunt, the lower 2 acute or acuminate. Corolla pink to lavender, 1–1.8 cm long, 1-lipped, the lip pinnately 5-lobed, 2 lowest lobes triangular-acuminate, recurved, and pointing toward the stamens arched above, next 2 lobes erect, oblong, tips obtuse, central lobe very much the largest, short-clawed basally then abruptly expanded to a declined suborbicular distal portion depressed centrally; corolla tube retrosely white-pubescent on the lower side interiorly; lip very minutely stipitate glandular exteriorly. Stamens 4, well exserted from the throat of the corolla and arched over the lip, the style arched with them and about equaling them. (Incl. *T. nashii* Kearney, *T. littorale* Bickn.)

Swamps, wet woodlands, floodplain forests, cypress depressions and prairies, wet thickets, wet clearings, meadows, marshes, marshy shores. Throughout our range and beyond.

A polymorphic complex for which various authors have, with little consensus, recognized segregate taxa on both the specific and infraspecific levels.

3. Lycopus (WATER-HOARHOUNDS, BUGLE-WEEDS)

Perennials, some stoloniferous and some of these bearing tubers at the tips of the stolons, some with rhizomes. Stems simple to much branched, angles rounded, ridged, or winged, faces between the angles flat to deeply grooved. Leaves glandular-punctate. Flowers in densely compacted axillary clusters, the opposite pairs of clusters often large enough to be contiguous around the nodes. Calyx tube campanulate, lobes 4 or 5, approximately equal or unequal, erect. Expanded portion of corolla white, nearly symmetrical or somewhat 2-lipped, 4- or 5-lobed, the throat more or less closed with pubescence. Stamens 2, exserted from the corolla in most, a pair of small staminodes present in some. Pistil borne on a well-developed disc, lobes of which alternate with the lobes of the ovary. Nutlets more or less triangular, obovate in outline, the summit smooth to variously corky-toothed.

References: Herman, Frederick J. "Diagnostic Characters in *Lycopus*." *Rhodora* 38: 373–375. 1936.

Henderson, Norlan C. "A Taxonomic Revision of the Genus *Lycopus* (Labiatae)." *Am. Midl. Nat.* 68: 95–138. 1962.

1. Calyx lobes obtuse to acute apically, shorter than or equaling the mature nutlets.
    2. Margin of leaf, proximally below the lowest tooth, with a dip or concavity in the taper; calyx and

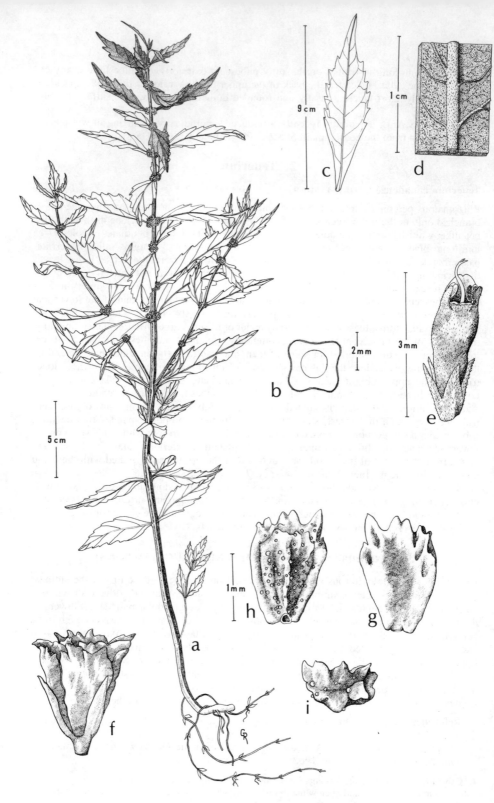

Fig. 275. **Lycopus virginicus:** a, habit; b, cross-sectional outline of stem; c, median stem leaf; d, portion of lower leaf surface; e, flower; f, nutlets with subtending persistent calyx; g, view of outer surface of nutlet; h, view of inner surface of nutlet; i, view of top of nutlet.

corolla 4-lobed, the corolla lobes essentially erect, stamens not exserted.   1. *L. virginicus*
2. Margin of the leaf (if tapered) proximally below the lowest tooth, with a straight taper; calyx and
corolla 5-lobed, corolla lobes spreading; stamens somewhat exserted.   2. *L. uniflorus*
1. Calyx lobes with acuminate to subulate apices, exceeding the mature nutlets.
   3. Stem sharply ridged on each angle.   3. *L. americanus*
   3. Stem angles rounded.
     4. Leaves sessile, their bases rounded or truncate, commonly subclasping.   4. *L. amplectens*
     4. Leaves markedly tapered basally.   5. *L. rubellus*

1. Lycopus virginicus L.   Fig. 275

Plant producing elongate, slender, cordlike runners, often branched, with small leaves, sometimes producing small tubers at their extremities, occasionally with runners from nodes on the upper part of the stem; stem angles rounded, the faces between them flat or shallowly and broadly grooved; stem appressed-pubescent, to 12 dm tall. Leaves dark green, often tinged with purple, sometimes purple, narrowly basally to a margined petiole, the margins of the basal taper with a dip or concave outline, lanceolate, lance-ovate, or elliptic, larger ones 6–9 cm long and 1.5–5 cm broad, apices acuminate, margins serrate distally from just below the middle, surfaces glabrous or with sparse pubescence on the principal veins. Calyx lobes usually 4, obtuse to acute, somewhat shorter than to about equaling the mature nutlets. Corolla 4-lobed, lobes erect. Stamens not exserted. Nutlets with a deeply toothed margin on the crest.

Floodplain forests, swamps, wet woodlands, stream banks, margins of wooded ponds, wet clearings and thickets, marshy shores. Mass. to N.Y., Pa., s. Ohio to s. Mo., generally southward to n. Fla. and e. Tex.

2. Lycopus uniflorus Michx.   Fig. 276

Habit similar to that of *L. virginicus*. Stems to about 10 dm tall, angles rounded, faces between the angles grooved, pubescent with short, usually appressed hairs, sometimes glabrous. Leaves rarely purple-tinged, sessile, the base sometimes rounded, if tapered to the base or to a short margined petiole, the sides of the taper usually straight, rarely concave; leaves lanceolate, lance-ovate, or oblong-elliptic, apices acuminate to blunt, margins serrate from somewhat below the middles upwardly. Calyx much shorter than to about equaling the mature nutlets, lobes 5, obtuse to acute. Corolla lobes usually 5, spreading. Stamens somewhat exserted from between the corolla lobes. Nutlets with an undulate to tuberculate crest. (Incl. *L. cokeri* Ahles)

Bogs, moist to wet woodlands, low open areas. Nfld. to B.C., southward to N.C., Ohio, Ind., Ill., Iowa; Okla.; southward in the West to n. Calif.

3. Lycopus americanus Muhl. ex Bart.   Fig. 277

Plant with subterranean runners, their tips slightly swollen but not tuberous. Stems simple or extensively branched, glabrous or sparsely short-pubescent, the angles sharply ridged. Leaves very variable in size, shape, and toothing; in outline ovate, lanceolate, narrowly oblong, to nearly linear; lower leaves usually, occasionally nearly all leaves, incised to pinnatifid, median varyingly serrate, uppermost shallowly serrate to nearly entire; lower ones petiolate or with margined subpetiolate bases, others cuneate basally. Calyx 5-lobed, lobes subulate-tipped, exceeding the mature nutlets. Corolla 4-lobed. Stamens exserted. Nutlets with a smooth or only slightly undulate crest.

Wet woodlands, marshes, marshy shores, stream and river banks, springy places, ditches. Widely distributed, Nfld. to B.C., more or less generally southward to n. Fla. and Calif.: much less frequent in our range than to the north of it.

4. Lycopus amplectens Raf.   Fig. 278

Plant producing long, slender, branching runners forming tubers at their extremities. Germinating tubers elongate and curved, attaining a crescent shape. Stems simple to freely branched, 2–10 dm tall, pubescent, the angles rounded, narrowly channelled between the

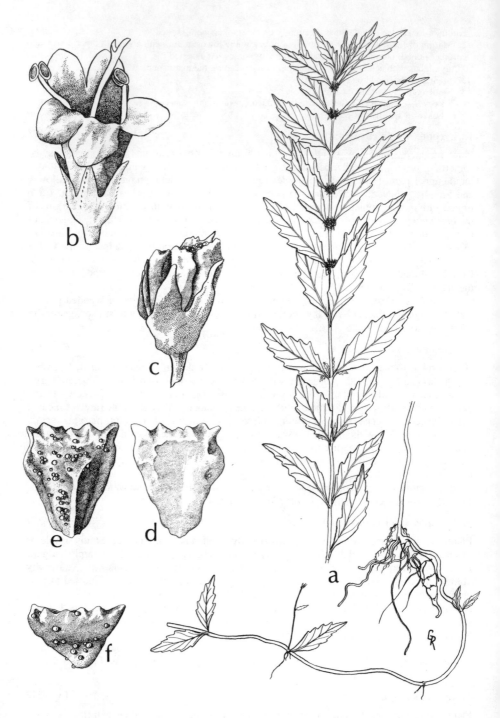

Fig. 276. **Lycopus uniflorus:** a, plant; b, flower; c, calyx with enclosed nutlets; d, view of outer face of nutlet; e, view of inner face of nutlet; f, view of top of nutlet.

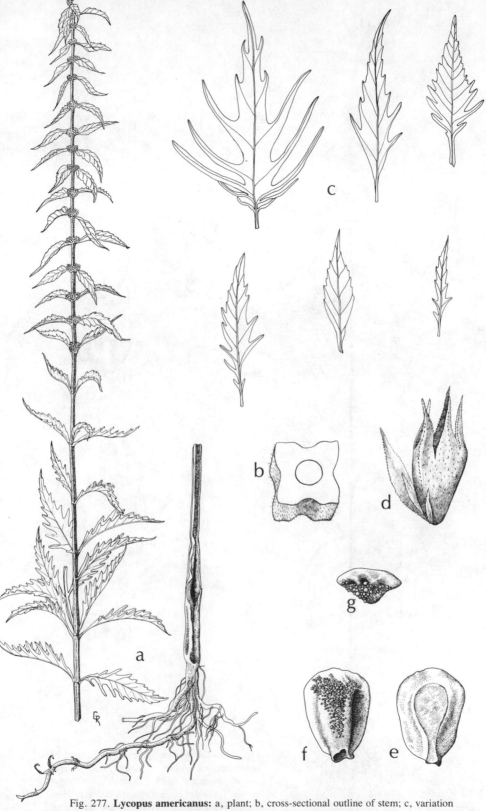

Fig. 277. **Lycopus americanus:** a, plant; b, cross-sectional outline of stem; c, variation in leaves on a single plant; d, calyx, with three subtending bracts; e, view of outer face of nutlet; f, view of inner face of nutlet; g, view of top of nutlet.

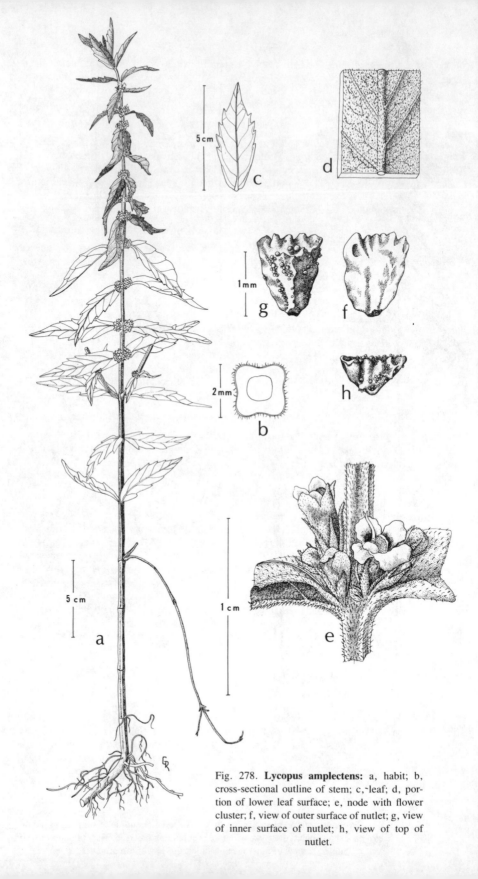

Fig. 278. **Lycopus amplectens:** a, habit; b, cross-sectional outline of stem; c, leaf; d, portion of lower leaf surface; e, node with flower cluster; f, view of outer surface of nutlet; g, view of inner surface of nutlet; h, view of top of nutlet.

angles. Leaves sessile, 3–6 cm long, bases rounded to subcordate and subclasping, sometimes with a somewhat tapered base, oblong-lanceolate, oblong, lanceolate, or occasionally oblanceolate, margins mostly with 4–6 low serrations on a side. Calyx lobes 5, attenuate-tipped, exceeding the mature nutlets. Corolla with 5 spreading lobes. Stamens exserted. Nutlets with a corky crest of 4 or 5 low, pointed teeth. (*L. sessilifolius* Gray; incl. *L. pubens* Britt. in Small)

Marshes, marshy shores, bogs, sandy-peaty ditches. Coastal plain, Mass. to n. Fla., s. Ala.

5. Lycopus rubellus Moench.

Plant producing long, slender, freely branching runners forming tubers at their extremities. Stems 4–12 dm tall, glabrous to densely pubescent, angles rounded, faces between the angles flat or broadly to narrowly channelled. Leaves variable, narrowed below to margined petioles, varying to sessile but with strongly tapering bases; lance-ovate to elliptic, to narrowly lanceolate, margins mostly serrate, surfaces glabrous or variably pubescent. Calyx lobes 5, attenuate-tipped, exceeding the mature nutlets. Corolla with 5 spreading lobes. Stamens exserted. Nutlet crests with low, rounded, occasionally acute, protuberances. (Incl. *L. velutinus* Rydb.)

A variable complex. We choose to recognize two intergrading varieties, as follows:

L. rubellus var. **rubellus.** Leaf blades lance-ovate to elliptic or lanceolate, the principal ones narrowed below to margined petioles, the margins of the taper with a dip or concavity, 4–12 cm long and 1–4 cm wide.

Stream banks, marshy shores of ponds and lakes, low wet woodlands, seepage areas in woodlands, floodplain forests. Maine to Ill. and Mo., generally southward to about Lake Okeechobee, Fla., and e. Tex., most abundant in the coastal plain and the interior woodlands. Fig. 279

L. rubellus var. **angustifolius** (Ell.) Ahles. Leaves essentially sessile, narrowly tapered basally, the margins of the taper usually straight or nearly so, mostly 4–8 cm long and 0.5–2 cm broad. (*L. angustifolius* Ell., *L. rubellus* var. *lanceolatus* Benner)

Cypress-gum ponds and depressions, hillside bogs in pinelands, marshy shores of ponds and lakes, in islands of floating vegetation, wet pine flatwoods. Coastal plain, Va. to about Lake Okeechobee, Fla., westward to La.; also n. Ga., Tenn., s.e. Mo. Fig. 280

4. Micromeria

Micromeria brownei (Sw.) Benth. var. **pilosiuscula** Gray. Fig. 281

Perennial, sometimes with slender subterranean rhizomes, mostly with leafy creeping stems rooting at the nodes, mat-forming. Flowering stems slender, weakly erect or ascending, 5–40 cm high, a small patch of short pubescence at the nodes between the petiole bases. Leaves short-petiolate, blades ovate to suborbicular, 5–15 (–20) mm long, margins obscurely crenate to entire, surfaces finely glandular-dotted, glabrous or with an occasional hair. Flowers solitary in the leaf axils from about midstem upwardly, their filiform, short-pubescent stalks exceeding the leaves. Calyx turbinate, 4–5 mm long including the lobes, tube slightly oblique at the orifice, 5-lobed, the lobes erect, triangular, the upper 3 obtuse or nearly so, about 1 mm long, lower 2 acute, 1.5 mm long or a little more. Corolla 8–10 mm long, the tube dilated above the calyx, then 2-lipped, pilose in the throat; lower lip whitish pink with darker splotches, 3-lobed, the middle lobe larger than the laterals, upper lip lavender-pink, unlobed, somewhat hoodlike, ciliate marginally. Stamens 4, 2 inserted basally on each lip of the corolla, those on the lower lip with anthers barely extending beyond the lobes. Nutlets tiny, dark purplish red. (*M. pilosiuscula* (Gray) Small)

Seepage areas in woodlands, spring runs, floodplain woodlands, muddy river and stream banks, swamps, meadows, ditches. From cen. pen. Fla. northward and westward in Fla., s.w. Ga., westward to s. Tex.; Mex., Guat.

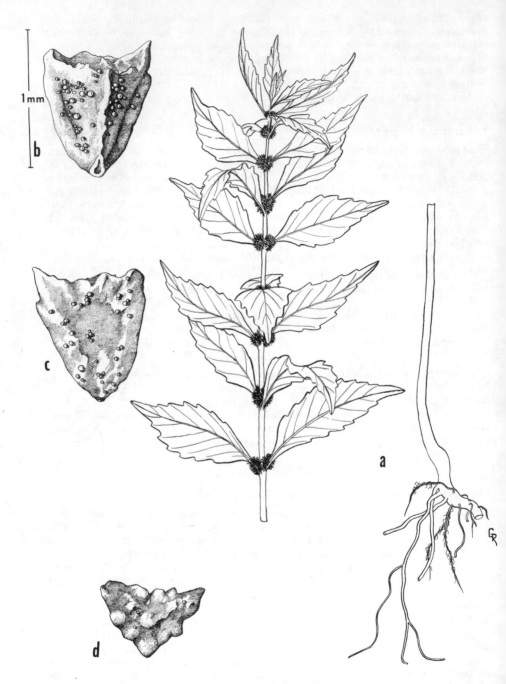

Fig. 279. **Lycopus rubellus** var. **rubellus:** a, plant; b, inner view of nutlet; c, outer view of nutlet; d, top view of nutlet.

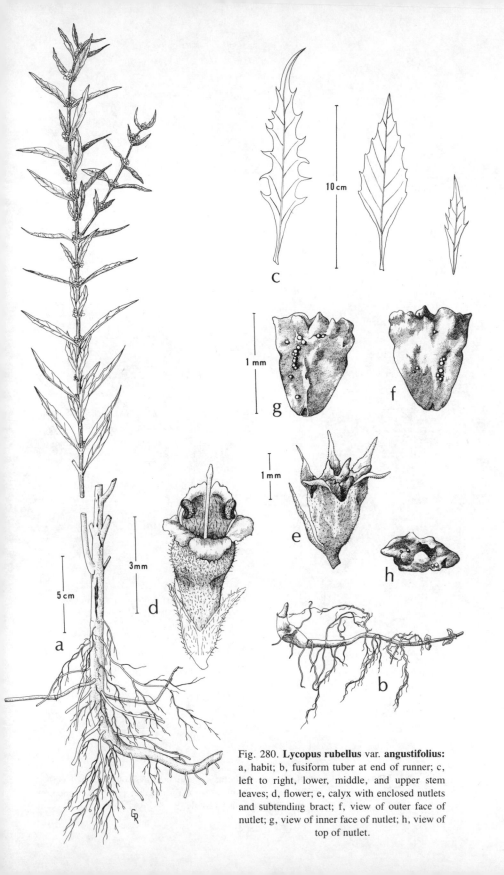

Fig. 280. **Lycopus rubellus** var. **angustifolius:**
a, habit; b, fusiform tuber at end of runner; c,
left to right, lower, middle, and upper stem
leaves; d, flower; e, calyx with enclosed nutlets
and subtending bract; f, view of outer face of
nutlet; g, view of inner face of nutlet; h, view of
top of nutlet.

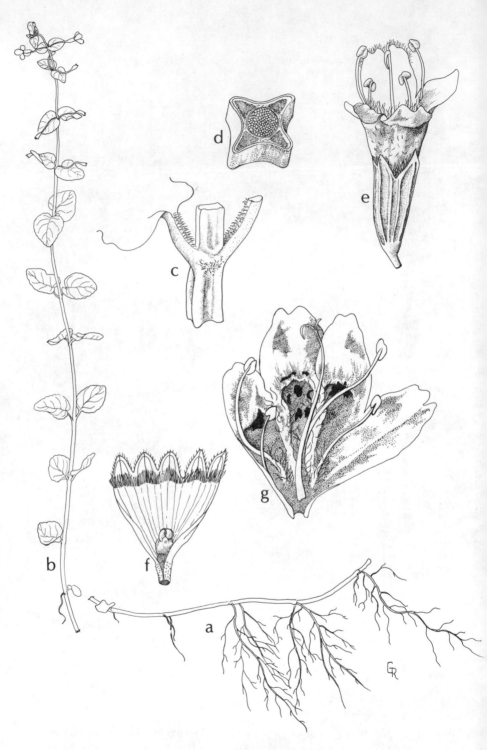

Fig. 281. **Micromeria brownei** var. **pilosiuscula:** a, decumbent base of plant; b, erect stem; c, node; d, cross-section of stem; e, flower; f, calyx, opened out; g, corolla opened out, with stamens and style.

5. Pycnanthemum (MOUNTAIN MINTS, BASILS)

Erect perennial herbs, stems simple below, branched in the inflorescences, infrequently below the inflorescences, in some the upper leaves and inflorescence bracts conspicuously whitened. Flowers small, in headlike clusters or in congested cymes; heads or cymes in turn disposed terminally on the branchlets or in open cymes. Calyx tube 10–13-nerved, lobes 5, erect, shorter than the tube, nearly equal or unequal and more or less 2-lipped. Corolla 2-lipped, white to purple, dotted or splotched with purple in most; upper lip 1-lobed with entire or emarginate apex, lower lip 3-lobed, lobes broadly rounded to obtuse. Stamens 4, inserted at the throat of the corolla, one pair longer than the other, exserted or included.

1. Principal leaves lanceolate to lance-elliptic or short-ovate, not exceeding 2 cm long.   1. *P. nudum*
1. Principal leaves variously shaped, much exceeding 2 cm long.
  2. Flowers so compacted in heads that no branching within the heads evident as viewed from the side.
    3. Stem glabrous.   2. *P. tenuifolium*
    3. Stem pubescent at least on the angles.
      4. Leaves linear-lanceolate to lanceolate, 3–10 mm wide, much longer than broad.
          3. *P. virginianum*
      4. Leaves ovate to lance-ovate, 15–40 mm broad, little if any more than twice as long as broad.
          4. *P. muticum*
  2. Flowers in close cymes but the branching within the cymes evident as viewed from the side.
    5. Calyx lobes short-triangular or deltoid, acute to obtuse apically, the tube and lobes felty-pubescent; principal leaves ovate or lance-ovate, acuminate or acute apically, whitish, felty-pubescent below.   5. *P. albescens*
    5. Calyx lobes with a very short, narrowly triangular base then stiffly long-awned, 7–8 mm long overall, tube and lobes densely short-pubescent but not felty; principal leaves lanceolate, narrowly elliptic, or oblong-elliptic, blunt apically, papillose or finely short-pubescent below, not felty.
          6. *P. hyssopifolium*

1. Pycnanthemum nudum Nutt.

Plant shortly rhizomatous. Stems glabrous, slender, mostly simple to the inflorescence, 3–6 dm tall. Leaves sessile, largest medially on the stem but not conspicuously smaller above and below, lanceolate, lance-elliptic, elliptic, or short-ovate, the larger ones usually not over 2 cm long and 10–12 mm wide, basally little tapered to rounded, apices blunt to acute, lateral veins strongly ascending, margins entire, surfaces glabrous, finely punctate-dotted. Inflorescences overall more or less turbinately cymose, varying in size from 3 to about 15 cm long and 3–10 cm broad at the top. Flowers not tightly compacted in the ultimate cymes, some of the branching within the cyme evident viewed from the side. Bracts lanceolate, their tips long-acute, glabrous, conspicuously punctate-dotted. Calyx turbinate, 4 mm long including the lobes, the tube nerved to just above the middle, glabrous and sparsely punctate-dotted, dots pale, distally and on the lobes conspicuously punctate-dotted, dots purplish, lobes about 1 mm long, triangular with sharply acute apices, copiously pubescent on their inner surfaces and margins. Corolla lips whitish, pubescent exteriorly and in the throat, purple punctate-dotted on both surfaces. Nutlets oblong in outline, pointed apically, dull tan, about 1.2 mm long. (*Koellia nuda* (Nutt.) Kuntze)

Wet pine flatwoods, hillside bogs in pinelands. s. Ga., n. Fla., s.e. Ala.

2. Pycnanthemum tenuifolium Schrad.

Plant slenderly rhizomatous. Stems glabrous, 5–8 dm tall, simple, terminated by a single cyme, or more commonly 3- to several-branched and the individual heads termminating branchlets reaching about the same level (rather than being storied). Leaves sessile, narrowly linear-subulate or narrowly linear-lanceolate, mostly 1.5–3 mm wide, rarely to 5 mm wide, longer ones 4–5 cm long, glabrous, both surfaces punctate-dotted, margins entire, often revolute, lateral veins paralleling the midrib. Heads compact, branching within them not evident viewed from the side, individually 5–10 mm across. Bracts lanceolate-attenuate, their midribs conspicuous and bonelike, outer surface varyingly powdery-pubescent to copiously white-pubescent, margins ciliate. Calyx 4–5 mm long including the lobes, tube turbinate, ribbed, glabrous and glandular below, more or less wooly-pubescent around the throat and on the lobes, lobes triangular-acute, about 1.3 mm long. Dilated portion of corolla tube and lips white with purple splotches, the tube and lips shaggy white-pubescent exteriorly and interiorly, hairs more sparse interiorly. Nutlets oblong in outline, blunt at both extremities, purplish black, about 1 mm long, somewhat granular on the surface. (*Koellia flexuosa* sensu Small (1933); *P. flexuosum* sensu Gl. (1952))

Bottomland fields, meadows, creek banks, bogs, pine savannas and flatwoods, thickets, low, wet pastures, prairies; also open, upland, well-drained woodlands and clearings. Maine to Minn., generally southward to Fla. Panhandle and e. fourth of Tex.

3. Pycnanthemum virginianum (L.) Durand & Jackson.

In general aspect similar to *P. tenuifolium*. Stem pubescent on the angles. Leaves sessile, linear-lanceolate, the larger ones 3–4 (–6) cm long and 3–6 (–10) mm broad, sometimes short-pubescent beneath, especially on the midveins, minutely scabrous on the margins, lateral veins 3 or 4 on a side, the uppermost arising at about midleaf, strongly ascending. Heads compact, the branching within them not evident as viewed from the side, 1–1.5 cm across. Bracts lance-ovate to lanceolate, their tips long-tapering to a sharp point, pubescent on the backs and copiously ciliate on the margins, hairs white. Calyx turbinate to cylindric, nerved, about 4 mm long, copiously short-pubescent and viscid-glandular distally and on the deltoid lobes, lobes less than 1 mm long. Corolla whitish to pink, splotched with purple, pubescent without. Nutlets about 1 mm long, triangular, oblong in outline, brown, smooth. (*Koellia virginiana* (L.) Britt.)

Banks of small streams, wet meadows, moist calcareous ledges, open upland woodlands, thickets. Maine to N.D., generally southward in mts. (rarely piedmont) to n. Ga. and to Okla. westwardly.

4. Pycnanthemum muticum (Michx.) Pers.

Relative coarse, stems to 12 dm tall, powdery-pubescent to pilose. Leaves sessile or very short-petiolate, ovate to lance-ovate, larger ones 4–6 cm long and 1.5–4 cm broad, bases rounded, apices short-acuminate, with distant low serrations marginally, punctate-dotted, surfaces glabrous above, glabrous or sparsely short-pubescent on the midvein beneath, petioles, if any, pubescent. Individual heads terminal and solitary or clustered on ultimate branchlets, each tightly compacted, branches within them not evident viewed from the side, 8–20 mm across; inflorescence copiously branched overall, the main branches conspicuously leafy, each head subtended by 2 or 3 ovate or lanceolate leafy bracts, these densely whitish-felty-pubescent above, sparsely short-pubescent below; bracts of the head itself linear-subulate to lance-subulate, copiously white-pubescent. Calyx about 3 mm long including the lobes, nerved, sparsely glandular, copiously pubescent distally and on the lobes, lobes deltoid, about 0.5 mm long. Corolla lips white splotched with purple, sparsely pubescent on both surfaces. Nutlets 1 mm long or a little more, one face rounded, 2 flat, oblong in outline, dull brown. (*Koellia mutica* (Michx.) Britt.)

Meadows, stream banks, low fields, bogs, dry to wet open woodlands. s.w. Maine to Mich., Ill., and Mo., generally southward to n. Fla., n.w. La., and n.cen. Tex.

5. Pycnanthemum albescens T. & G.

In general aspect similar to *P. muticum*. Stems copiously short-pubescent, becoming glabrous on lower portion in age, short-petioled, blades lanceolate to lance-ovate, larger ones 4–6 (–8) cm long and 1–2 cm broad, margins with few low serrations, some sometimes entire, surfaces finely punctate-dotted, upper glabrous and green, lower with tight, whitish tomentum. Individual cymes loose, the branching within them evident viewed from the side, 1–4 cm across. Inflorescence overall usually widely branched, leafy, the cymes subtended by a few leafy bracts of unequal size, these densely whitish-tomentose on both surfaces. Calyx 3–3.5 mm long including the lobes, densely and tightly whitish-pubescent throughout and glandular, lobes narrowly deltoid, 0.5–1 mm long, bluntish apically. Corolla lips whitish, pink, or lavender, purple-spotted. Nutlets about 1 mm long, oval in outline, one face rounded, others flat, brown, sparsely pubescent. (*Koellia albescens* (T. & G.) Kuntze)

Pine flatwoods and savannas, hillside bogs in pinelands, low, open woodlands, thickets bordering swamps, swales; also in well-drained, open, upland woodlands. s. Ga. and n. Fla., westward to e. Tex., northward in the interior, e. Okla., Ark., Mo.

6. Pycnanthemum hyssopifolium Benth.

Plant slenderly rhizomatous. Stems 3–10 dm tall, powdery-pubescent, older, lower portions becoming subglabrous. Leaves very short-petiolate, blades variable, narrowly oblong, lanceolate, lance-oblong, larger ones 2.5–4 cm long, 5–15 mm broad, usually recurved, apices blunt, bases tapered to rounded, minutely pubescent on both surfaces, margins with a few low serrations to entire, lateral veins few, strongly ascending. Cymes few to numerous, terminal on cymosely disposed branches, 1–3 (–4) cm across, the branching within the cyme evident viewed from the side, subtended by 2 to several reduced bracteal leaves, bracts of the head itself lanceolate to subulate. Calyx nearly cylindrical, 7–8 mm long including the teeth, tube markedly ribbed, glandular, minutely pubescent, surmounted by stiffly bristly, whitish, somewhat unequal awns 2–4 mm long or a little more, the cymes with a notably bristly aspect. Corolla lips white to lavender, purple-dotted or -splotched, short-pubescent exteriorly. Nutlets black, about 1 mm long, 3-angled, short-oblong in outline, white-pubescent apically. (*P. flexuosum* sensu Fern., 1950, and sensu Radford et al., 1964; *Koellia hyssopifolia* (Benth.) Britt.)

Pine savannas and flatwoods, bogs, evergreen shrub-tree bogs and their borders, swales, ditches. Coastal plain, s.e. Va. to n. Fla., Ala., Tenn.; also Henderson and Transylvania Cos. in w. N.C.

P. setosum Nutt. (*K. aristata* (Michx.) Kuntze) is closely similar to *P. hyssopifolium* but has ovate or lance-ovate leaves and outer bracts subtending the heads. Possibly of hybrid origin involving *P. hyssopifolium*.

6. Hyptis (BITTER MINTS)

(Ours) relatively coarse, erect herbs 1–2 m tall. Flowers in involucrate heads borne from the axils of foliar bracts or much-reduced bracts, 1–few flowers of a given head at anthesis at one time. Calyx tube slightly oblique at the orifice, 5-lobed, the lobes nearly equal, triangular-subulate or setaceous, stiffly erect. Corolla strongly 2-lipped, upper lip 2-lobed, the lobes short-ovate, rounded apically, lower lip with 2 lateral, rounded lobes and a central, longer saclike segment. Stamens 4, inserted at about the base of the lower lip.

• Principal leaves lanceolate, linear-lanceolate, to ovate- or rhombic-lanceolate, narrowed from their wide places to margined, subpetiolate bases; heads relatively large, 1.5–2.5 cm across fully developed (in fruit), borne on stalks 2–6 cm long; stems unbranched or few-branched only in the inflorescence.
                                                                              1. *H. alata*

• Principal leaves with definite slender petioles, blades broadly ovate, ovate-triangular, or rhombic-ovate, basally broadly cuneate to truncate; heads small, not exceeding 1 cm across fully developed, borne on stalks only 1–2 mm long; stems widely and diffusely branched, the distal portions of all the branchlets floriferous.
                                                                              2. *H. mutabilis*

1. Hyptis alata (Raf.) Shinners.   Fig. 282

Perennial with a hard caudex. Stems to about 2 m tall, simple or few-branched only in the inflorescence, often several stems from the base, sparsely to densely short-pubescent, lower portions becoming glabrous in age. Leaves lanceolate, ovate-lanceolate, rhombic-lanceolate, or linear-lanceolate, tapered from their broadest places to margined subpetiolate bases, acute to blunt apically, margins irregularly serrate, 4–10 (–15) cm long overall. Heads relatively few, mostly 3–12 pairs, borne singly from the axils of bracteal leaves, the latter gradually reduced upwardly, the fruiting heads 1.5–2.5 cm across, nearly globose, stalks of the heads

Fig. 282. **Hyptis alata:** a, top of stem; b, flower; c, fruiting calyx; d, nutlet. (From Correll and Correll. 1972)

2–6 cm long. Flowers many in each head, much compacted. Involucral bracts unequal, variable in shape, outer largest, lanceolate, oblanceolate, lance-oblong, or linear-oblong, densely pubescent with stiff hairs. Calyx densely white-pubescent, elongating as the fruits mature, about 8 mm long fully mature, stiffish-textured, with several strong longitudinal ribs basally, reticulate above. Corolla white spotted with lavender, the central lobe of the lower lip clawed basally, expanded distally and cuplike. Nutlets oblong in outline, dull black, smooth, 1–1.5 mm long. (*H. radiata* Willd.)

Pine savannas and flatwoods, pineland seepage slopes, bogs, wet thickets and clearings, swales, marl prairies, pond or lake margins, marshy shores of streams, ditches. Coastal plain, N.C. to s. pen. Fla., westward to s.e. Tex.; W.I.

2. Hyptis mutabilis (L. Rich.) Briq.

Perennial with a strong, unpleasant odor. Stems to 2 m tall, potentially widely and diffusely branched, rough-pubescent, the trichomes on the angles translucent, short, broad-based, and with hooked tips, those between the angles white, relatively thin and curly. Leaves with definite petioles, those of larger leaves 4–6 cm long; blades broadly ovate, triangular-ovate, or rhombic-ovate, broadly cuneate to truncate basally, acute apically, margins serrate or doubly serrate, the teeth obtuse to rounded; lower surfaces glandular-punctate, pubescent mostly on the veins, hairs on the larger veins of 2 kinds, long, scattered ones, many short ones, hairs on the smaller veins all short; larger blades about 8 cm long and 6–7 cm across their bases. Branchlets all floriferous distally, heads small, on stalks only 1–2 mm long, lowermost in axils of leafy bracts, most heads in axils of small bracts, the flowering branch tips appearing interrupted spikelike. Flowers in the lower larger heads about 10, fewer in the smaller upper heads and closely set but not compacted. Involucral bracts ovate acuminate, their margins short-ciliate. Calyx nearly urceolate, 6–7 mm long fully developed, half as long or less at anthesis, sparsely pubescent, longitudinally ribbed, with some cross-venation between the ribs on the neck. Corolla about 6 mm long, the tube pale lavender, pubescent, lips violet with whitish splotches, the upper lip recurved, the central lobe of the lower lip closed-saccate at first and enclosing the stamens, the style exserted above it, later the saccate portion snapping downward exposing the anthers as they dehisce. Nutlets oblong in outline, 1 mm long, purplish black, surface sublustrous.

Native of trop. Am. Naturalized in a wide range of habitats including lowland woodlands, floodplains, river and stream banks, marshy shores of ponds and lakes, brushy roadsides and thickets, moist to wet clearings, abundant chiefly where there is frequent disturbance. Throughout Fla., s.w. Ga., s. Ala., s. Miss., and s.e. La.

7. Macbridea

Rhizomatous perennials with erect stems 3–9 dm tall. Flowers in headlike clusters subtended by conspicuous closely appressed, empty bracts, the heads terminating solitary stems, or terminating branches. Calyx 2-lipped, the upper lip a single narrow lobe, the lower with 2 broader lobes obliquely notched distally. Corolla cylindric within the calyx or a little beyond then expanding to the throat, 2-lipped above the throat, upper lip hoodlike, arching, the lower spreading, 3-lobed. Stamens 4, extending from the corolla tube and arching under the upper lip, one pair longer than the other; anthers reniform, usually pubescent near the margins, the margins of the locules strongly pectinate-toothed. Style exserted, arching with the stamens, reaching the lower pair of anthers. Ovary seated upon a disc having a cylindrical lobe on one side that ascends beside the ovary and is higher than the ovary at anthesis, the mature nutlets exceeding the lobes.

• Corolla white with faint purplish markings in the throat; midstem leaves oblanceolate, their apices rounded or blunt.   1. *M. alba*
• Corolla pink to lavender, striped with purple and white, midstem leaves long-elliptic, their apices acute.   2. *M. caroliniana*

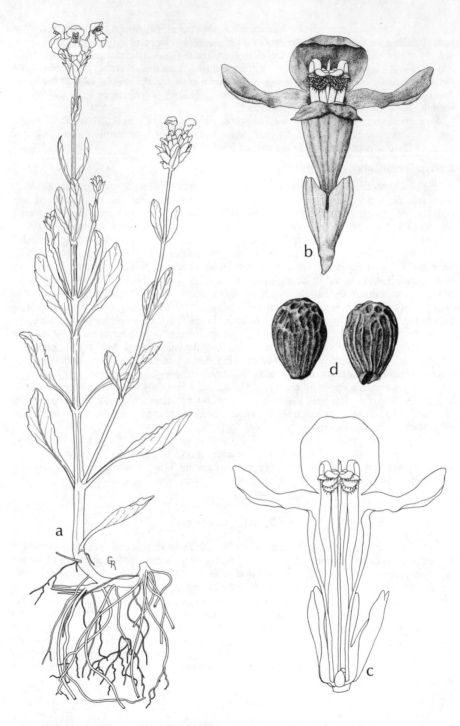

Fig. 283. **Macbridea alba:** a, habit; b, flower; c, longitudinal section of flower, semi-diagrammatic; d, seed, two views.

1. Macbridea alba Chapm. WHITE BIRDS-IN-A-NEST.   Fig. 283

Stem sparsely clothed with long, multicellular hairs and glandular-dotted, 3–4 dm tall, solitary and with a single headlike inflorescence or few-branched above, each branch terminated by a head. Leaves thickish, oblanceolate or spatulate, mostly in 6–8 pairs, gradually reduced in size from the base upwardly, usually all but the uppermost 2 pairs narrowed gradually from their widest places to margined subpetiolate bases, apices rounded; margins with a few very broad, very low, glandular-tipped protuberances thus appearing slightly wavy, both surfaces with scattered, long, straight, unicellular hairs and finely and abundantly glandular-punctate. Outer several bracts of the head empty, relatively large, varying in shape, the outermost oblong, usually becoming obovate then nearly orbicular; bracts subtending the flowers cuneate-obovate, clasping the calyces and reaching to somewhat beyond their throats. Calyx tube funnelform, striate-reticulate. Both bracts and calyces thick and becoming hard upon drying. Corolla 2.5–3 cm long, snow white excepting faint purple markings in the throat, with tiny atoms of yellow glandular exudate without. Filaments white, anthers purple basally. Nutlet 2–2.5 mm long, narrowly obovate in outline, light brown, longitudinally striate, in some part sometimes somewhat striate-reticulate, rounded summit wrinkled.

Pine savannas and flatwoods. Endemic to lower Apalachicola River region, Fla. Panhandle.

2. Macbridea caroliniana (Walt.) Blake.

Stems glabrous or becoming glabrous in age, with tiny dots of glandular exudate, 6–9 dm tall. Leaves petiolate to about midstem, upwardly narrowed to margined subpetiolate bases, uppermost sessile; blades thinnish, 6–12 cm long, 1.5–3 (–4) cm broad, narrowly elliptic-oval, apices acute, both surfaces sparsely pubescent to nearly glabrous, finely and abundantly glandular-punctate, margins as in *M. alba* but those of uppermost leaves, at least, ciliate. Heads essentially as in *M. alba*. Calyx tube faintly nerved, not striate. Corolla pink to lavender, striped with purple and white, 3–3.5 cm long. Nutlets obovate to oblanceolate in outline, light brown, irregularly ribbed. (*M. pulchra* Ell.)

Marshes, bogs, bottomland woodlands. Coastal plain, s.e. N.C. to n. Fla. and s. Ala.

8. Physostegia (OBEDIENT-PLANTS, FALSE DRAGON-HEADS)

Perennial herbs with straight, erect stems, simple below, commonly branched in the inflorescence, the branches usually nearly erect. Leaves sessile or the lower ones petiolate. Plant glabrous below, the inflorescence axes, bracts, flower stalks, and calyces pubescent, the hairs usually short and dense, in some the surfaces viscid-glandular as well. Calyx radially symmetrical or nearly so, tubular below, 5-lobed, lobes erect, the tube obscurely 10-nerved or the nerves not evident, enlarging somewhat as the fruits mature. Corolla much exceeding the calyx, in most narrowly cylindrical within the calyx tube then broadening upwardly, 2-lipped distally; upper lip concave, erect or arching, entire or slightly emarginate, the lower 3-lobed, spreading, the middle lobe largest, notched; corolla pink, rose, rose-lavender, magenta, or purple, rarely white, variously purple-lined, -dotted, or -splotched. Stamens 4, filaments adnate to the corolla tube for part of their length, upper pair shorter than the lower. Nutlets triangular.

We have had not a little difficulty in understanding the limits of the taxa comprising *Physostegia* as they occur in our range. After having wrestled with the problem an inordinate amount of time, it came to our attention that Mr. Philip Cantino of Harvard University Herbaria was engaged in the preparation of taxonomic revision of the genus. Via personal communication he has very graciously shared with us a generous amount of information accruing to him during the course of his studies. The key was furnished to us by him. For all of this we are very much in his debt.

1. Leaves, one or more of them, clasping the stem.
  2. Perennating buds borne at the ends of elongate. horizontal secondary rhizomes.

3. Calyx and rachis of inflorescence bearing minute, stalked glands, as well as nonglandular puberulence; leaves 2 or 3 times as long as wide.   1. *P. correllii*

3. Calyx and rachis of inflorescence puberulent but lacking stalked glands; leaves often more than 4 times as long as wide.

    4. Flowers 22–35 mm long; all or most of the larger stem leaves acute to attenuate at apex; axis of the raceme densely pubescent, always with some (and usually many) hairs 0.17–0.25 mm long; usually with half or more of the larger stem leaves sharply serrate.   7. *P. angustifolia*

    4. Flowers smaller, *or* half or more of the larger leaves obtuse at apex, *or* axis of the raceme minutely puberulent, few if any hairs exceeding 0.13 mm in length (mostly less than 0.1 mm); usually with most of the larger stem leaves bluntly toothed to subentire.

        5. Upper stem leaves, at least one pair, usually widest at base of blade and broadly clasping the stem; flowering calyx tube (1–) 2–4 mm long; flowers always less than 20 mm long.
                                        5. *P. intermedia*

        5. Upper stem leaves widest above to below the middle of blade but rarely at base; flowering calyx tube 3–7 (–8) mm long; flowers often longer than 20 mm.

            6. Upper leaves greatly reduced, those of uppermost pair below terminal spike often no larger than the floral bracts, those of second pair rarely more than a quarter as long as the internode directly above; primary rhizome unbranched or giving rise to 1 or 2(3) secondary rhizomes, usually no more than one of them greater than 5 cm long; usually no more than the 5 lowest nodes bearing petiolate leaves; principal stem leaves widest at to above the middle of the blade.   3. *P. purpurea*

            6. Upper leaves less reduced, those of uppermost pair below terminal spike usually considerably larger than the floral bracts, those of second pair ⅓ as long as to longer than the internode directly above; primary rhizome giving rise to 1–7 horizontal secondary rhizomes, most of them greater than 5 cm long; the lowest (3)4–9 nodes bearing petiolate leaves; principal stem leaves usually widest at to below middle of blade.   6. *P. leptophylla*

2. Perennating buds borne directly on primary rhizome or at the ends of short, vertical secondary rhizomes; horizontal secondary rhizomes lacking.

    7. Most or all of the larger leaves sharply serrate; largest leaves on dried specimens not exceeding 2.5 cm in width (wider on living plants) and rarely less than 5 times as long as wide.
                                        7. *P. angustifolia*

    7. Most or all of the larger leaves bluntly serrate to entire, or largest leaves more than 3 cm wide or less than 5 times as long as wide.

        8. Axis of raceme densely pubescent to tomentose, the hairs mostly 0.2–0.3 mm long; calyx lobes at anthesis generally 2–3 mm long, many or all of them attenuate or cuspidate; flowers tightly packed, adjacent calyces mostly overlapping half their length or more.   2. *P. digitalis*

        8. Axis of raceme puberulent to pubescent, hairs rarely reaching 0.2 mm long; calyx lobes at anthesis generally 1–2 mm long, mostly merely acute; flowers loosely spaced to tightly packed.
                                        3. *P. purpurea*

1. Leaves sessile or petiolate, but none clasping the stem.

    9. Larger leaves, all or most of them, sharply serrate.

        10. Axis of raceme densely pubescent, always with some (usually many) hairs 0.13–0.25 mm long; sterile bracts absent from inflorescence; nutlets 2–3 mm long; calyx lacking stalked glands; usually blooming April to early July.   7. *P. angustifolia*

        10. Axis of raceme minutely puberulent, few if any hairs more than 0.1 mm long; sterile bracts frequently present below flowers, but may be absent; nutlets usually 3–4 mm long; calyx with or without stalked glands; usually blooming July to October.   8. *P. virginiana*

    9. Larger leaves, half or more of them, bluntly serrate to entire.

        11. Calyx and rachis of inflorescence bearing minute, stalked glands, as well as nonglandular puberulence (glands barely visible with a good hand lens); nutlets 1.7–2 mm long, usually warty over all or part of surface.   4. *P. godfreyi*

        11. Calyx and rachis of inflorescence puberulent but lacking stalked glands; nutlets 2–3.6 mm long, smooth.

            12. Upper leaves greatly reduced, those of second pair below terminal spike rarely more than ¼ as long as the internode directly above; moist, open pinelands.   3. *P. purpurea*

            12. Upper leaves less reduced, those of second pair below terminal spike ⅓ as long as to longer than the internode directly above; wooded river swamps and fresh and brackish marshes, marshy shores.   6. *P. leptophylla*

1. Physostegia correllii (Lundell) Shinners.

Plant somewhat succulent, robust, rhizomatous, perennating buds borne at the ends of elongate, horizontal, secondary rhizomes. Stem simple or sparingly branched, to 2 m tall or a little more, the lower stem as much as 2.5 cm thick. Leaves firm and somewhat leathery, elliptic or oblong-elliptic, 5–13 cm long, 2.6–5 cm broad, only slightly reduced upwardly, slightly narrowed basally and somewhat clasping, margins conspicuously serrate-dentate. Inflorescence of spikelike, simple or compound racemes, short-leafy at the base, flower stalks very short, about 1 mm long in fruit. Calyx and axis of the inflorescence bearing minute stalked glands as well as nonglandular puberulence; calyx tube at anthesis subcylindric, 8–9.5 mm long, lobes slender, acuminate, about as long as the tube. Corolla lavender-pink, spotted or streaked with purple, about 3 cm long, sparsely pubescent. Nutlet a little over 2 mm long, sharply angled. (*Dracocephalum correllii* Lundell)

In water along streams, irrigation ditches. s. La.; s.w. Tex.; n. Mex.

2. Physostegia digitalis Small.

Coarse plant, rhizomatous, perennating buds borne directly on primary rhizome or at the ends of short, vertical secondary rhizomes, horizontal secondary rhizomes lacking. Principal leaves sessile, relatively large and conspicuous, thickish, 1–2 dm long and 2–7 cm broad, broadly long-elliptic, elliptic-oblong, or oblong, 1–2 cm broad at the clasping or subclasping base, half or more of the larger leaves acute to acuminate apically, margins entire, or the distal margins wavy or with few, blunt teeth. Flowers congested on the stout racemes, adjacent calyces mostly overlapping half their length or more. Bracts ovate, their tips acuminate to subaristate. Calyx at anthesis 8–10 mm long, the lobes generally 2–3 (–4) mm long, many or all of them attenuate to cuspidate. Corolla lavender to nearly white, with purple to lavender dots, portion beyond the calyx 2.5 cm long, 1–1.5 cm across at the throat. Nutlets brown, strongly 3-wing-angled, 2–3 mm long, 2 mm wide.

Sandy open pinelands, pineland depressions, priairies, borrow pits, highway and railway rights-of-way, w. La. and e. Tex.

3. Physostegia purpurea (Walt.) Blake.   Fig. 284

Plant usually with an erect, subterranean caudex with roots from its nodes, sometimes rhizomatous. Stems usually solitary, erect, simple or branched in the inflorescence, mostly 4–10 dm tall. Lowest 2 or 3 pairs of leaves usually longer than the internodes, petiolate or narrowed to wing-margined subpetiolate bases, blades elliptic, oblong-elliptic, obovate or oblanceolate, 4–13 cm long, rounded to obtuse apically, margins irregularly wavy, shallowly dentate, or dentate-serrate, teeth blunt, or margins sometimes entire; other leaves few, sessile, sometimes clasping the stem, thickish and stiffish, strongly erect-ascending, uppermost remote pairs bracteal, those of uppermost pair below the terminal spike often no larger than the floral bracts, those of second pair rarely more than a quarter as long as the internode directly above; medial leaves usually shorter than the internodes, oblanceolate, narrowly oblong, or linear-oblong, 4–6 cm long, 5–10 mm broad (occasionally broader and narrowly obovate), margins wavy, shallowly dentate, or dentate-serrate, teeth blunt; principal stem leaves widest at or a little above the middle. Lowermost pair of flowers 3–10 cm above topmost bracteal leaves. Racemes variable with respect to distance between pairs of flowers, varying from loosely spaced to tightly packed. Bracts broad-based and clasping, abruptly narrowed above the base and acute or acuminate. Inflorescence axis and calyx puberulent but lacking stalked glands. Calyx, including the lobes, 6–8 mm long at anthesis, lobes 1–2 mm long, mostly acute. Corolla pink to purple, portion above the calyx 2–2.5 cm long (rarely about 1 cm long), 10–12 mm wide at the throat. Nutlets brown, markedly 3-wing-angled, 2–3 mm long and broad, sometimes a little broader than long. (*P. obovata* (Ell.) Godfrey ex Weath.; *Dracocephalum purpureum* (Walt.) McClintock)

Wet pine savannas and flatwoods, often in depressions, marl prairies, ditches, margins of

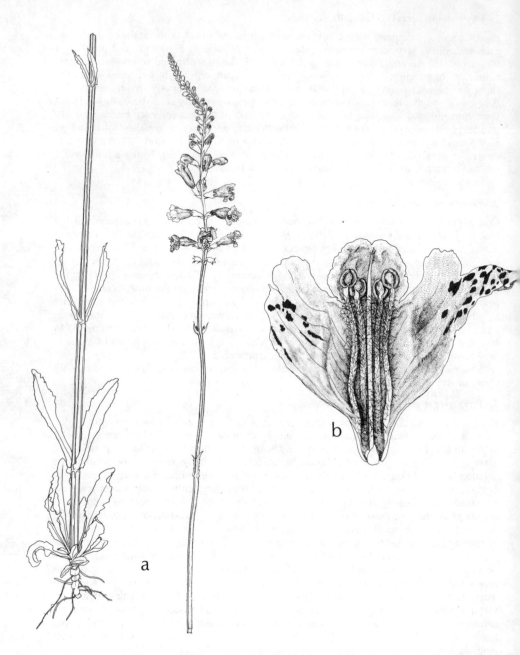

Fig. 284. **Physostegia purpurea:** a, habit; b, corolla, opened out.

cypress ponds and depressions. Coastal plain, N.C. to s. pen. Fla., s.w. Ga. and adj. Fla. Panhandle.

4. Physostegia godfreyi Cantino. Fig. 285

Perennial with rhizome 1–10 cm long. Stem erect, 6–10 dm tall or slightly more, unbranched or with few erect branches above. Leaves glabrous, those low on the stem (commonly shed by flowering time) petiolate, petioles slender, about as long as the blades; blades narrowly elliptic, narrowly elliptic-oblong, oblanceolate, or rarely lanceolate, sometimes slightly falcate, mostly 3–6.5 cm long, margins entire, slightly wavy, or occasionally with a few blunt teeth, bases tapered, apices blunt. Leaves becoming sessile upwardly, the uppermost bractlike and subulate. Stem glabrous below the raceme(s). Raceme axis or axes, the short flower stalks, and calyces bearing relatively abundant, short, pointed hairs with minute stipitate glands intermixed. Successive pairs of calyces not overlapping. Flower stalks 0.5–2 mm long; bracts inconspicuous, ovate-attenuate, about twice as long as the flower stalks. Calyx tube 3–5.5 mm long, somewhat obconic in outline, the short lobes triangular. Corolla 1–2 cm long or slightly longer, pale rose with the interior purple venation showing through, upper lip entire, the 3-lobed lower lip variously splotched with purple. Nutlets 1.7–2 mm long, trigonous, pebbly-warty over all or part of their surfaces.

Bogs, wet pine flatwoods and savannas, adjacent ditches, often in shallow water. Fla. Panhandle w. of the Ochlockonee River, in Liberty, Franklin, Calhoun, Gulf, and Bay Cos.

5. Physostegia intermedia (Nutt.) Engelm. & Gray.

Plant often with an erect subterranean caudex with fibrous roots at the nodes, perennating buds borne at the ends of elongate, horizontal, secondary rhizomes. Stem usually slender but upon occasion the basal portion submersed and spongy-swollen, mostly 3–10, sometimes to 15, dm tall, simple or the inflorescence few-branched. Lowermost or basal offshoot leaves petiolate (these usually shed by flowering time), others sessile; blades of petiolate leaves lanceolate to elliptic-oblong, mostly 2–3 cm long and to 1 cm broad, blunt apically, margins entire, wavy, or with few low blunt teeth; principal sessile leaves clasping the stem, linear, lanceolate, or lance-oblong, acute to attenuate apically, margins entire, wavy, or with few blunt teeth. Bracts broad basally then acuminate to attenuate. Calyx and inflorescence axis puberulent but lacking stalked glands. Calyx, including the lobes, 4–5 mm long at anthesis, tube (1–) 2–4 mm long, lobes triangular, 1–2 mm long. Corolla lavender-pink to almost white, portion extending beyond the calyx 1–1.5 cm long, 5–8 mm across the throat. Nutlets buffish to brown, strongly 3-angled or 3-wing-angled, 2–2.5 mm long and broad. (*Dracocephalum intermedium* Nutt.)

Swamps, marshes, wet meadows, open bottomland woodlands, prairies. s. Ill., Ky. to Kan., generally southward to La. and e. and s. half of Tex.

6. Physostegia leptophylla Small. Fig. 286

Plant slenderly rhizomatous, perennating buds borne at the ends of elongate, horizontal, secondary rhizomes, often forming dense colonies. Stem erect, to 1 or 1.5 m tall. Leaves thin and flexous, spreading below, angled-ascending above; basal and often those of leaves to midstem petiolate, petioles 2–4 cm long, blades lanceolate or narrowly elliptic, 4–8 cm long, 1.5–2.5 cm broad; blades of sessile leaves lanceolate, narrowly elliptic, or oblong-elliptic, to 14 cm long and 3.5 cm broad; leaf margins obscurely crenate, shallow-dentate, or dentate-serrate, tips of teeth blunt; principal stem leaves widest at to below middle of the blade; uppermost pair of leaves below the terminal spike usually considerably larger than the floral bracts, those of second pair ⅓ as long as to longer than the internode directly above. Bracts ovate or subhastate basally, abruptly narrowed above to attenuate tips. Calyx, including the lobes, 5–8 mm long at anthesis, lobes triangular, 1–2.5 mm long, their tips acute or acuminate. Corolla bright lavender-pink to purple, portion above the calyx 2–2.5 cm long, mostly 1.5 cm broad at the throat. Nutlets brown, markedly 3-angled, broadly elliptic in outline, 2–3 mm long. (Incl. *P. aboriginorum* Fern., *P. veroniciformis* Small, *P. den-*

615

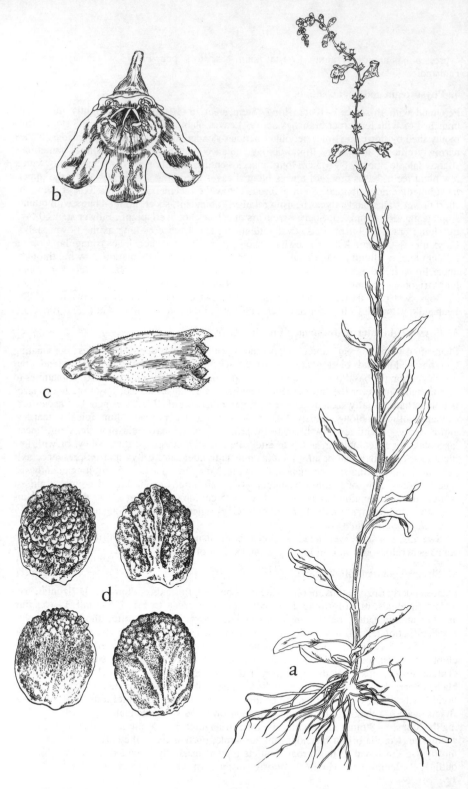

Fig. 285. **Physostegia godfreyi:** a, habit; b, flower; c, fruiting calyx; d, views of 4 nutlets (Drawn from photograph in Cantino, "*Physostegia godfreyi* (Lamiaceae), A New Species from Northern Florida" in *Rhodora* 81: 409–417. 1979)

Fig. 286. **Physostegia leptophylla:** a, habit; b, flower; c, fruiting calyx, opened out.

Fig. 287. **Physostegia angustifolia:** (From Correll and Correll, where attributed to Lundell, *Flora of Texas*, Vol. 2, pt. 2, pl. 18, 1969)

ticulatum of authors; *Dracocephalum leptophyllum* (Small) Small; *D. veroniciforme* (Small) Small)

In wet muck or peat, commonly in shallow water, edges of rivers and streams, river swamps, sloughs, marshes, and marshy shores. Coastal plain, s.e. Va. to s. pen. Fla., Fla. Panhandle.

7. Physostegia angustifolia Fern. Fig. 287

Plant with a short, erect caudex, rarely rhizomatous. Stems stiffly erect, 3–15 dm tall, simple or branched in the inflorescence. Leaves sessile, usually clasping the stem, stiff and rigid, strongly erect-ascending, lanceolate, linear-lanceolate, linear-oblong, or nearly linear, 4–18 cm long, mostly 3–25 mm wide, bases nearly truncate, apices of lower ones blunt, those of the median and upper ones usually acute to attenuate, margins of most or all of the larger leaves sharply serrate from somewhat below the middle. Racemes variable with respect to distance between pairs of flowers, often the lower pairs 1 cm or more apart, others tightly packed, adjacent calyces mostly overlapping half their length or more. Bracts variable, some broadest basally and rounded on each side, narrowing distally to long-acute or attenuate tips, some of them long-triangular, apices acute, sterile bracts absent from lower part of inflorescence. Calyx, including the lobes, 6–8 mm long at anthesis, lobes triangular-attenuate, 1.5–2.5 mm long, without stalked glands. Corolla pale lavender-pink to deep purple, sometimes nearly white, the portion beyond the calyx 2–2.5 cm long (rarely about 1.5 cm), 1–2 cm broad at the throat. Nutlets brown, 3-angled, 1 face often sunken, broadly oval in outline, 2–3 mm long. (Incl. *P. edwardsiana* Shinners)

Prairies, moist to wet swales and thickets, rocky glades, gravel bars, ditches, often in heavy, sticky, clayey soils. w. Mo. and e. Kan., generally southward to w. Ga., Ala., e. and cen. Tex.

8. Physostegia virginiana (L.) Benth.

Plant with basal offshoots, or slenderly rhizomatous, often colonial, perennating buds borne directly on primary rhizome or at the ends of short, vertical secondary rhizomes, horizontal secondary rhizomes lacking. Stem to 1.5 m tall, simple, or few-branched in the inflorescence. Leaves essentially sessile but the lowermost ones narrowed to broadly margined subpetioles; larger leaves lanceolate, oblanceolate, oblong-lanceolate, or lance-elliptic, spreading or spreading-ascending, thinnish, 5–18 cm long and 1–4 cm broad. Bracts broadest basally, tapering to long-acute or acuminate tips; sterile bracts frequently present on the axis below the flowers. Calyx, including the lobes, 4–8 mm long at anthesis, pubescent and with or without stalked glands. Corolla rose-lavender to purple, rarely whitish, portion beyond the calyx 2.5–3 cm long fully expanded, 12–15 mm wide at the throat. Nutlets oblong or oblong obovate in outline, strongly 3-wing-angled, 3–4 mm long. (Incl. *P. praemorsa* Shinners, *P. serotina* Shinners; *Dracocephalum virginianum* L.)

River banks and shores, lakes shores, wet thickets and swales, prairies, moist, open woodlands, ditches. Que. to Man., southward to n. Ga., n. Ala., s. La., Tex., local in s.e. N.M.; sometimes locally naturalized from cultivation elsewhere.

9. Stachys (HEDGE-NETTLES)

Annual or perennial herbs, the latter rhizomatous. Flowers in verticels on dense or interrupted, bracteate racemes, these terminal or on branches from the upper leaf axils. Calyx nerved, 5-lobed, the lobes nearly equal and erect or spreading. Corolla (in ours) whitish to pink, lavender, or purple, the tube usually exserted from the calyx, the limb 2-lipped, upper lip erect or arched and hoodlike, lower usually longer, spreading or deflexed, 3-lobed, the middle lobe larger than the laterals. Stamens 4, inserted just below the throat of the corolla, ascending under the upper lip of the corolla. Nutlets generally obovate in outline, rounded apically, 2 faces flat, 1 convex.

Plants of this genus are of local occurrence although they may be abundant in any one location. Characteristics within a species are highly variable and appear to intergrade between some species making it difficult to identify some specimens.

Our treatment is adapted, in part, from J. B. Nelson, "The Genus *Stachys* in the Southeastern United States," M.S. Thesis, Clemson Univ. (1975).

1. Leaves sessile, or if some of them petiolate then the petioles not over 5 mm long.
  2. Larger leaves linear-lanceolate to oblong, or elliptic-oblong, little if any over 10 mm wide.
    3. Calyx lobes evenly pubescent on the margins; leaf margins finely serrate, upper surface or both surfaces pubescent.   lb. *S. hyssopifolia* var. *ambigua*
    3. Calyx lobes glabrous or pubescent only on the midnerve; leaf margins entire or with small indentations, surfaces glabrous.
      4. Leaves linear lanceolate or lanceolate; stem glabrous or with very few, scattered, retrorse hairs on the angles.   la. *S. hyssopifolia* var. *hyssopifolia*
      4. Leaves mostly oblongish to elliptic-oblong; stems with long, spreading hairs on the angles, pubescent at the nodes between the leaf bases.   2. *S. lythroides*
  2. Larger leaves ovate-lanceolate to oblong-elliptic, 2–4.5 cm wide.   3. *S. eplingii*
1. Leaves definitely petiolate, some petioles 10 mm long or more.
  5. Longer petioles not over ¼ as long as the blades they bear; blades acuminate or acute apically, mostly sharply serrate marginally; stems to 1 m tall.   4. *S. tenuifolia*
  5. Longer petioles ⅓ as long to as long as the blades they bear; blades rounded to obtuse apically, crenate to crenate-serrate marginally; stems 1–4 (–6) dm tall.
    6. Perennial, rhizomatous, forming crisp, segmented tubers, colonial; calyx tube 4–5 mm long, about twice as long as the lobes; corolla 10–13 mm long.   5. *S. floridana*
    6. Annual or biennial, not forming rhizomes or tubers, not colonial; calyx tube 2–2.5 mm long; a little longer than the lobes; corolla 3–4 mm long.   6. *S. crenata*

1. Stachys hyssopifolia Michx.

Slenderly rhizomatous, the rhizomes becoming tuberous-thickened. Stems slender, ascending to sprawling, 3.5–8 dm tall or more, unbranched to much branched, glabrous to markedly pubescent. Leaves sessile or very short-petioled, usually not over 1 cm wide, linear-lanceolate to oblong, surfaces glabrous to copiously pubescent, margins entire or finely serrate. Raceme with 5–12 verticels of flowers, the verticels well set apart. Bracts lanceolate to lance-ovate, the lower exceeding the verticels they subtend. Calyx tube about equaling to somewhat longer than the triangular-attenuate, glabrous or variously pubescent lobes. Corolla pale purple to pink, mottled with purple and white, 10–12 mm long, upper lip hoodlike. Nutlets brown to nearly black, short-oblong in outline, 1.5–2 mm long. Two varieties may be distinguished, as follows.

S. hyssopifolia var. **hyssopifolia.** Stem glabrous or with few, scattered, retrorse hairs on the angles. Leaves sessile, linear-lanceolate or lanceolate, upper surface glabrous or slightly pubescent, lower glabrous, margins entire. Calyx usually glabrous, sometimes with a few hairs on the lobes.

Bogs, pine savannas, swampy or wet woodlands, marshy shores. Coastal plain and piedmont, Del., Md., Va., to s.e. Miss.; mts. of Va., N.C.; s. Mich., n.w. Ind.

S. hyssopifolia var. **ambigua** Gray. Stems copiously pubescent on the angles, sometimes on the faces as well, and at the nodes between the leaf bases. Leaves short-petioled, blades narrowly oblong, finely serrate on the margins, the upper surface, sometimes the lower, hirsute. Calyx tube glabrous to copiously pubescent, the lobes evenly pubescent on the margins. (*S. aspera* Michx.)

Swamps, marshes, prairies, pine savannas, open grassy places. Pa. to Iowa, southward in the eastern part of the range, chiefly on the coastal plain, to s.e. Ga., southward inland to Ky. and e.cen. Mo.

2. Stachys lythroides Small.

Stems usually vigorously and widely branched and somewhat sprawling, to about 8 dm tall, angles with long, spreading to somewhat retrorse hairs. Leaves sessile, oblongish or elliptic-oblong, the larger ones mostly about 1 cm wide, margins with very low, appressed teeth, surfaces glabrous. Calyx lobes with pubescence on the midveins, calyx otherwise glabrous. (Probably eventually to be considered as an infraspecific taxon of *S. hyssopifolia*.)

Wet places bordering Lake Iamonia and in nearby wet woodlands or their clearings, known only from n. Leon Co., Fla. (very close to the Fla.-Ga. state line).

3. Stachys eplingii J. B. Nelson.

Relatively coarse, long-rhizomatous. Stems erect, unbranched, 5–10 dm tall or a little more, the angles with spreading, pustular-based hairs, faces, especially on the upper portions, glandular-pubescent, somewhat pubescent at the nodes. Leaves sessile or subsessile, narrowly ovate to oblong-elliptic, 6–10 cm long, 2–4.5 cm broad, with a rank odor when fresh, rounded to subcordate basally, acuminate apically, margins finely crenate-serrate, upper surface canescent to scabrous, lower usually glandular-pubescent, the veins often bristly. Raceme with 4–25 verticels of flowers well set apart, verticels each with 8 or more flowers. Bracts foliose, pubescent. Calyx tube 3.5–4.5 mm long, glandular to hispid, lobes deltoid-attenuate, margins ciliate, about ½ as long as the tube. Corolla white with purple or lavender spots or lines, 1–1.3 cm long, upper lip hoodlike. Nutlets about 2 mm long, dark brown, obovate in outline, surface finely reticulate to irregularly bumpy.

Meadows, bogs, mountain forests. W.Va. to Ky. and N.C., n.w. S.C.; w.cen. Ark., e. Okla.

4. Stachys tenuifolia Willd.   Fig. 288

Plant with thickened-tuberous rhizomes. Stem 6–8 dm tall, sparingly to profusely branched, faces usually glabrous, angles variously roughened to bristly-pubescent. Leaves with petioles 1–2 cm long, blades variable in size, 5–10 cm long, 1–3 cm broad, ovate, elliptic-ovate, or lanceolate, bases truncate, rounded, or somewhat tapered, apices acuminate, margins serrate, surfaces glabrous or pubescent. Racemes with 9–12 interrupted verticels each with 6–8 flowers. Bracts foliose, reduced upwardly, glabrous or pubescent. Calyx tube 3–4 mm long, glabrous or bristly along the angles, lobes triangular-acuminate or -attenuate, 2–3 mm long, often divergent in fruit, glabrous or bristly on the margins. Corolla white to pale pink with dark purple spots, 1–1.2 cm long, upper lip hoodlike. Nutlets broadly obovate in outline, 1.5–2 mm long, dark brown, roughish. (Incl. *S. latidens* Small)

Swamps, wet woodlands, meadows, marshes and marshy shores, moist roadbanks, mountain woodlands. s. Que. to Minn., southward to S.C., n. Ga., Ala., Miss., La., e. fourth of Tex.

5. Stachys floridana Shuttlw. ex Benth. in DC.

Plant rhizomatous, rhizomes becoming segmented-tuberous, extensively colonial. Stems 1–5 (–6) dm tall, unbranched or variously branched from the lower nodes, angles usually with short, stiff, retrorse trichomes, sometimes with long spreading hairs below, copiously short-pubescent above. Basal and lower leaves with petioles ⅓ as long to as long as the blades, upper ones short-petioled or sessile, blades oblong to ovate-oblong, larger ones 2–4 cm long, mostly truncate or nearly so basally, obtuse apically, margins crenate-serrate, surfaces glabrous or sparingly short-pubescent. Racemes of 3–12 widely spaces verticels, each (3–) 6-flowered. Calyx copiously short-pubescent and with some sessile glands, tube 4–5 mm long, about twice as long as the triangular-attentuate lobes. Corolla whitish to pale pink, with purple spots and lines, especially on the lower lip, upper lip hoodlike. Nutlets 1.5 mm long, obovate in outline, flattened basally, plump distally, dark brown.

In a wide variety of habitats where open or semiopen, in areas with much moisture, even occasionally where water stands some of the time, or in well-drained places; often a noxious weed in moist lawns and gardens, difficult to eradicate, commonly distributed with nursery stock. s.e. Va. to s. Fla., chiefly coastal plain, westward to e. Tex.; occasional as a weed elsewhere in our range.

Fig. 288. **Stachys tenuifolia:** a, basal portion of plant; b, upper portion of plant; c, flower; d, corolla, opened out; e, node with fruits; f, nutlet. (From Correll and Correll. 1972)

6. Stachys crenata Raf.

Plant annual or biennial, not rhizomatous and not colonial, similar in stature and general appearance to *S. floridana*. Stems, petioles, and blades with long, spreading hairs. Petioles of basal and lowermost stem leaves as in *S. floridana*, becoming sessile from midstem upwardly; blades of larger leaves 2–4 cm long, mostly ovate or ovate-oblong, margins crenate, bases truncate to subcordate, apices rounded to obtuse. Calyx tube 2–2.5 mm long, a little longer than the lobes, lobes triangular below, tipped with stiff pointed awns, tube hirsute, many hairs gland-tipped, lobes similarly pubescent to the bases of the awns. Corolla 3–4 mm long (much less conspicuous than in *S. floridana*), upper lip hoodlike, pale purple, lower lip pale purple with darker spots fading to yellow. Nutlets 1 mm long, nearly suborbicular in outline, tan-colored to dark brown. (*S. agraria* Cham. & Schlecht.)

Sandy or muddy wet soil, shaded or open places, edges of ponds and lakes, stream banks, weedy in disturbed places. Coastal regions, Miss., La., e.cen. and s. Tex.; Mex.; reported as a weed in pasture, cen. Ala.

10. Calamintha

Calamintha arkansana (Nutt.) Shinners.   Fig. 289

Slender, low perennial, forming repent stolons or ascending sprouts from the base after flowering, herbage strongly aromatic. Flowering stems usually diffusely branched and abundantly floriferous, glabrous or sometimes somewhat pubescent at the nodes, 1–4 dm high. Leaves of stolons or basal offshoots elliptic to short-ovate, those of the flowering stems linear, narrowly oblanceolate, or narrowly elliptic, the larger ones to 25 mm long and 5 (–12) mm broad, margins entire or with 1–4 small, low teeth on a side, glabrous, punctate-dotted on both surfaces. Flowers 2 to several at each node, their filiform stalks 5–15 mm long, with a pair of linear bracts at the base. Calyx 3–4 (–5) mm long, glabrous, exteriorly strongly nerved, punctate-dotted, pubescent in the throat, lobes triangular-acuminate or -attenuate, much shorter than the tube. Corolla 7–10 mm long, tube dilated beyond the calyx, pubescent without, limb 2-lipped, upper lip erect, flattish, entire or notched, lower lip 3-lobed, middle lobe larger than the laterals. Stamens 4, unequal, ascending but barely, if at all, visible viewed from the side. Nutlets about 1 mm long, oblanceolate in outline, rounded apically, brown, surface finely alveolate and sometimes irregularly and loosely cobwebby. (*Satureja arkansana* (Nutt.) Briq.; incl. *Calamintha glabella* (Michx.) Benth. in DC., *Clinopodium glabellum* (Michx.) Kuntze)

In calcareous areas, pools and depressions on rocks in cedar glades, on rocks in stream courses, wet or damp banks, meadows, river and stream gravels, spring runs, seepage slopes. Ont. to Minn., southward to w. N.Y., Ohio, Tenn., Ark., Okla., e. and cen. Tex.

11. Mentha (MINTS)

Perennial herbs, rhizomatous or stoloniferous. Stems erect, sometimes diffusely branched. Leaves punctate-glandular at least beneath. Flowers small, borne in short, stalked cymes axillary to leaves or bracts, if the latter the inflorescences appearing interrupted or congested and spikelike. Calyx tube cylindric to campanulate, 10-nerved or -ribbed, slightly oblique at the orifice, 5-lobed, the lobes nearly equal, erect. Corolla tube little exserted from the calyx, with 1 entire or notched upper lobe or lip, a lower 3-lobed lip, not strongly bilateral. Stamens 4, equal, filaments straight, inserted at the throat of the corolla. Nutlets ovoid.

1. Stem and both leaf surfaces with abundant shaggy hairs; stamens much exserted. 1. *M. rotundifolia*
1. Stem glabrous or sparingly pubescent, especially near the nodes; lower leaf surfaces sparsely pubescent on the veins, upper surfaces glabrous, smooth; stamens not exserted.
   2. Leaves sessile, or rarely with petioles to about 2–3 mm long; bracts subtending the flower clusters with evenly ciliate margins. 2. *M. spicata*
   2. Leaves definitely petiolate, the petioles 5–8 (–15) mm long; bracts subtending the flower clusters glabrous or with few scattered hairs. 3. *M. piperita*

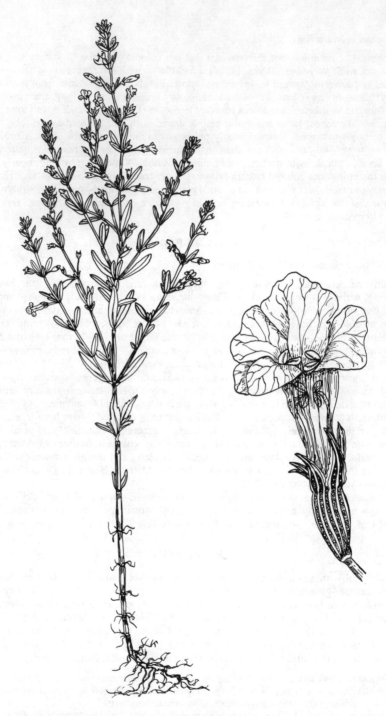

Fig. 289. **Calamintha arkansana.** (From Correll and Correll (1972), as *Satureja arkansana*)

1. Mentha rotundifolia (L.) Huds.

Plant with leafy stolons. Stems 5–8 (–15) cm high, abundantly shaggy-pubescent. Leaves sessile or subsessile, ovate to oblong or suborbicular, larger ones 3–6 cm long, bases rounded to subcordate, apices obtuse to rounded, both surfaces shaggy-pubescent, rugose-reticulate-veined, veins impressed on the upper surfaces, markedly elevated beneath, margins serrate. Inflorescence usually with several narrow, spirelike, spikelike branches, the flower clusters sometimes interrupted below, usually contiguous above. Bracts subtending the cymes lance-subulate, pubescent. Calyx, including the lobes, 1.5–2 mm long, rough pubescent throughout, the tube campanulate, lobes rigid-subulate, about as long as the tube, curved inwardly in fruit. Corolla usually white spotted with purple, 2.5–3 mm long. Stamens much exserted. Nutlets oblongish, about 0.5 mm long, brown, finely alveolate-reticulate.

Native of Eur., cultivated and sporadically naturalized in N.Am. in moist to wet open places, seepage areas, thickets, marshes.

2. Mentha spicata L. SPEARMINT.   Fig. 290

Plant with slender subterranean rhizomes. Stems erect, simple or variously branched, mostly 5–8 dm tall, glabrous or sparsely pubescent. Leaves sessile or less frequently with petioles 1–2 (–3) mm long, lance-oblong to oblong, not rugose, glabrous and smooth above, sparsely pubescent on the veins beneath, rounded to obtuse basally, acute apically, margins serrate. Inflorescence branches spikelike, spirelike, usually less than 1 cm broad across the pairs of flower clusters at anthesis, the lower clusters interrupted. Bracts subtending the cymes lance-subulate, glandular-dotted, their margins evenly ciliate. Calyx, including the lobes, 1.5–2 mm long, tube campanulate, ribbed and grooved, lobes subulate, pubescent, about as long as the tube varying to about ⅓ as long as the tube. Corolla lavender, pale violet, or pinkish. Stamens not exserted.

Native of Eur., cultivated and sporadically and widely naturalized in N.Am., in wet places, bogs, seepage areas, about ponds, creek banks, in ditches.

3. Mentha piperita L. PEPPERMINT.   Fig. 290

In general aspect similar to *M. spicata*. Leaves with definite petioles 5–8 (–15) mm long, blades ovate, lance-ovate, lanceolate, or elliptic, the larger ones 4–8 cm long, shortly and broadly cuneate or rounded basally, acute apically, margins sharply serrate, upper surfaces glabrous and smooth, lower sparsely pubescent on the veins, not rugose. Aspect of inflorescences much like those of *M. spicata*, usually over 1 cm across the pairs of cymes at anthesis, spikes dense except in the lowermost portions. Bracts subtending the cymes lance-attenuate, glandular, glabrous or sometimes with few irregularly spaced hairs on the margins. Calyx, including the lobes, 3–4 mm long, tube narrowly campanulate, nerved but not ribbed, glabrous and smooth, lobes subulate, much shorter than the tube, their margins ciliate. Corolla lavender. Stamens not exserted.

Native of Eur., cultivated and sporadically naturalized in much of N. Am., in wet places, brooksides, meadows, margins of ponds and lakes, thickets, ditches; probably the garden mint most frequently and abundantly naturalized in our range. Several other garden mints besides the three treated here may occur as escapes in the vicinity of habitations but apparently are less likely to "go wild."

Scrophulariaceae (FIGWORT FAMILY)

Annual, biennial, or perennial herbs, rarely shrubs or trees. Stems terete to quadrangular, smooth or ridged, occasionally winged. Leaves simple, alternate, opposite, or whorled, blades entire to deeply lobed, cleft, or divided. Flowers bisexual, variously arranged, axillary, racemose, spicate, paniculate, or thyrsoid, flower stalks often bearing 2 bractlets, these

Fig. 290. a–e, **Mentha spicata:** a, top of plant; b, tip of leaf; c, flower; d, corolla, opened out; e, pistil; f–i, **Mentha piperita:** f, top of plant; g, flower; h, calyx, opened out; i, corolla, opened out. (From Correll and Correll. 1972)

usually just below the calyx. Calyx of 4 or 5, equal or unequal, distinct, or partially united sepals. Corolla bilateral in most, in some nearly radial, of 4 or 5 partially or wholly united petals. Stamens 2 or 4, rarely 5, 1 or 2 staminodia present in some. Pistil 1, ovary superior, 1–3 (usually 2) -locular; style 1, terminated by 1 or 2 stigmas. Fruit a capsule. Seeds small, usually numerous.

1. Plant diminutive, with tufted linear leaves 2–5 cm long and 1–2 mm wide.   10. *Limosella*
1. Plant not as above.
  2. Leaves all alternate.
    3. Stem leaves pinnately lobed, cleft, or divided.   25. *Pedicularis*
    3. Stem leaves not pinnately lobed, cleft, or divided.
      4. Flowers subtended by conspicuous scarlet bracts.   24. *Castilleja*
      4. Flowers not subtended by scarlet bracts.
        5. Leaves serrate; corolla white, or white suffused with violet, about 10 mm long.   1. *Capraria*
        5. Leaves entire; corolla yellowish suffused with purple on the tube, 30–35 mm long.
           23. *Schwalbea*
  2. Leaves, the lower at least, opposite or whorled.
    6. Calyx lobes or sepals 4.
      7. Leaves pinnately lobed, cleft, or divided.   25. *Pedicularis*
      7. Leaves not pinnately lobed, cleft, or divided.
        8. Leaves punctate-dotted.   2. *Scoparia*
        8. Leaves not punctate-dotted.
          9. Flowers many, congested in long spirelike terminal racemes; leaves whorled, blades much longer than broad.   17. *Veronicastrum*
          9. Flowers solitary in the axils of leaves or bracteal leaves; leaves all opposite or opposite below.
            10. Flowers minute, less than 2 mm long, scarcely perceptible to the unaided eye.
               9. *Micranthemum*
            10. Flowers exceeding 2 mm long, easily perceived by the unaided eye.   18. *Veronica*
    6. Calyx lobes or sepals 5.
      11. Flowers bright orange.   19. *Macranthera*
      11. Flowers other than bright orange.
        12. Leaves whorled.   11. *Limnophila*
        12. Leaves not whorled.
          13. Leaves divided into long filiform segments, *or* pinnatifid, the segments not filiform.
            14. Leaves divided into filiform segments.   20. *Seymeria*
            14. Leaves pinnatifid, the segments not filiform.   13. *Leucospora*
          13. Leaves not divided.
            15. Plant aquatic (in vernal pools on granitic rock outcrops), with a basal rosette of lanceolate, acute, submersed leaves, lax stems arising from the base and distally bearing a pair of oval floating leaves with rounded apices; a single small flower produced between the pair of floating leaves, other flowers borne singly in the axils of the submersed basal leaves.
               6. *Amphianthus*
            15. Plant not having the above combination of characteristics.
              16. Calyx lobes shorter than to equaling the calyx tube.
                17. Flowers many, very small, congested in a terminal, spirelike raceme.
                   17. *Veronicastrum*
                17. Flowers not arranged as above.
                  18. Corolla tube narrowly cylindrical for its entire length, 5 lobes of the limb spreading at right angles to the tube, thus corolla salverform.   22. *Buchnera*
                  18. Corolla tube cylindrical only at the base, much dilated above the base.
                    19. Limb of the corolla very conspicuously 2-lipped; calyx tube angular or prismatic.
                       14. *Mimulus*
                    19. Limb of the corolla very weakly 2-lipped; calyx tube not angular. 21. *Agalinis*
              16. Calyx lobes distinctly longer than the tube, or the sepals distinct or essentially so.
                20. Flower with 1 staminode, this as long as or longer than the fertile stamens, distally broadening and flattening, bearded; inflorescence a terminal thyrsoid or panicled raceme; flowers 12 mm long or more.   16. *Penstemon*
                20. Flowers with or without staminodia, the latter if present not as described above; flowers not in terminal compound racemes or panicles.

21. Principal leaves all basal, lanceolate to linear-oblong, stem leaves scalelike.

12. *Dopatrium*

21. Principal leaves not all basal (or if essentially basal then as broad as long or nearly so), none of the leaves scalelike.

  22. Flowers in racemes whose axes zigzag. 7. *Mazus*

  22. Flowers not in racemes whose axes zigzag.

  23. Ovary with 4 sutures, capsule 4-valved.

  24. Flower stalks bearing a pair of small bractlets at their bases just above the much larger bracteal leaves. 4. *Mecardonia*

  24. Flower stalks without bractlets or these, if present, immediately below the calyx.

  25. Leaf margins entire, apices rounded or very blunt. 3. *Bacopa*

  25. Leaf margins toothed, even though weakly, and the apices mostly acute (leaves of *Gratiola hispida* entire, strongly acute). 5. *Gratiola*

  23. Ovary with 2 sutures, capsule 2-valved.

  26. Corolla 1 cm long or less; seed not winged. 8. *Lindernia*

  26. Corolla 2 cm long or more; seed broadly winged. 15. *Chelone*

1. Capraria

Capraria biflora L.

Erect perennial, usually with a single main stem and divergent branches. Stems 3–15 dm tall, the caudex thick and woody; lower part of stem with light grayish tan bark marked with a few low, irregular, vertical ridges; upper stems sometimes sparsely pubescent. Leaves alternate, sessile, oblanceolate, 2–7 cm long, to about 1.5 cm wide, progressively reduced upwardly, glabrous or with a few hairs along the margins and veins, apices acute to obtuse, bases cuneate, surface pebbly. Flowers 1 or 2 per node, each axillary to a leaf or foliar bract. Bractlets none. Flowers stalks slender, 5–20 mm long, sparsely pubescent. Sepals 5–7 mm long, distinct, linear, equal. Corolla white or violet-tinged, about 1 cm long, campanulate, the 5 lobes nearly equal, a little longer than the tube, apices acute, tube sparsely pubescent within, densely pubescent in the throat. Stamens 5, equal. Ovary ovoid to ellipsoid, obtuse, style short, somewhat dilated distally, terminated by 2 scarcely distinct stigmas. Capsules 4–6 mm long, ovoid or oblong-ovoid, dark brown, conspicuously glandular-dotted, septicidal and slightly loculicidal. Seed dark brown, a little longer than wide, irregularly tetrahedral, obtuse apically, oblique at the base, surface shallowly and coarsely alveolate-reticulate, the longitudinal reticular lines somewhat more pronounced.

Wet, sandy or marly soil, on or near beaches, swales, edges of mangrove swamps. s. Fla.; trop. Am.

2. Scoparia

Scoparia dulcis L. GOAT-WEED, SWEET-BROOM.

Perennial herb with a disagreeable odor. Stems erect, commonly solitary, sometimes 2 or more stems from the base, each with several ascending main branches, these subbranched; glandular-punctate, glabrous or short-pubescent, frequently more pubescent around the nodes than elsewhere. Leaves opposite or whorled, if the latter mostly 3 at a node, most of them cuneately narrowed to petiolar bases, larger ones 1–4 cm long and rhombic-lanceolate to rhombic-ovate, serrate distally from their broadest places, acute apically, both surfaces glandular-punctate; leaves gradually reduced upwardly. Flowers axillary, on slender stalks 3–5 (–9) mm long. Calyx 1.5–2 mm long, glandular exteriorly, united basally, 4-lobed, lobes ovate, oblong, or narrowly obovate, equal, 3-nerved. Corolla white, tubular basally, lobes 4, equal, ovate, horizontally spreading, a ring of long hairs at the base of the tube and extending to about the middle of the lobes. Stamens 4, alternate with the corolla lobes, exserted. Capsule ovoid to subglobose, 1.5–2.5 mm long, a little exceeding the calyx, dehiscing septicidally and slightly loculicidally. Seeds numerous, minute, irregularly angular, opaque brown, somewhat lustrous, coarsely and shallowly alveolate-reticulate.

Weedy; clearings of wet woodlands, floodplain forests, swamps, banks on edges of marshes, occasionally in brackish marshes, disturbed places in sand and gravel bars and exposed shores of rivers, spoil banks or flats, low pastures, ditches. Native of trop. Am., naturalized and often abundant in parts of our range, chiefly coastal plain, S.C. to s. Fla., westward to La.

3. Bacopa (WATER-HYSSOPS)

Low, mostly perennial, succulent or subsucculent herbs with prostrate or partially decumbent stems, some extensively mat-forming on wet substrates or in shallow water, some sometimes in floating mats. Leaves opposite, sessile or subsessile, glandular-punctate (sometimes obscurely so), not reduced distally on the stems. Flowers solitary (infrequently in pairs) in the leaf axils, commonly only one of a given pair of leaves subtending a flower; flower stalks in some with a pair of bractlets immediately below the calyx. Calyx of (4)5 nearly distinct, dimorphic sepals, equal or unequal in length, unequal in width, often appearing as if in 2 series, in some enlarging as the fruits mature; adaxial sepal largest, the 2 abaxial ones of intermediate size, and the 2 lateral ones smallest. Corolla tubular-campanulate below, (3–) 5 nearly equal, slightly spreading lobes above, lobes about as long as the tube. Stamens usually 4, in 2 pairs of unequal length, sometimes 3 or 2. Ovary globose, ovoid, or ellipsoid-cylindric, style with 2 terminal stigmas or stigma semicapitate. Capsule dehiscing septicidally, loculicidally, or both, enclosed by the persistent calyx. Seeds alveolate-reticulate.

1. Leaves tapering from their broadest places to narrow bases, not clasping.
   2. The leaves with only a midvein evident, margins not toothed.   1. *B. monnieri*
   2. The leaves with several evident veins ascending from their bases, margins crenate around the broadly rounded apices.   2. *B. egensis*
1. Leaves with broad clasping bases, or if obovate the basal taper broad proximally and subclasping.
   3. The leaves obovate to orbicular.
      4. Corollas 3.5–4.2 mm long, white or white tinged with pink, without a yellow throat; capsule 2–3 mm long.   3. *B. repens*
      4. Corollas 8–10 mm long, white with a yellow throat; capsule 5 mm long.   4. *B. rotundifolia*
   3. The leaves ovate.
      5. Plants not aromatic; corollas white, 3–4 mm long; without bractlets below the calyx; stamens 2.
                5. *B. innominata*
      5. Plants aromatic; corollas pale to bright blue or violet-blue, 9–13 mm long; a pair of subulate bractlets below the calyx; stamens 4.   6. *B. caroliniana*

1. Bacopa monnieri (L.) Pennell.   Fig. 291

Stems succulent, glabrous, (potentially) much branched, the principal branches extensively radiating and prostrate, rooting at the nodes, at length forming thick glossy green mats, the flowering branchlets decumbent or loosely ascending. Leaves oblanceolate, spatulate, or cuneate-obovate, succulent, glabrous, 5–15 mm long, margins entire or the upper margins obscurely finely wavy. Flowers solitary in the axils of one of a pair of leaves at some (not all) nodes; flower stalks at anthesis shorter than to somewhat exceeding the subtending leaves, fruiting stalks usually much longer, 1–2.5 cm long; bractlets 2, linear, 2–3 mm long. Corolla 8–10 mm long, tube yellowish green to yellow, limb pale lavender pink to white, a darker ring within the tube about midway its length. Stamens 4, the longer pair slightly exserted. Capsule 4–5 mm long, narrowly ovate. Seeds olive-colored, short-oblong or obliquely oblong, about 0.5 mm long, shallowly alveolate-reticulate, the alveolae longer longitudinally, the longitudinal striae the more conspicuous. (*Bramia monnieri* (L.) Pennell)

Fresh or brackish shores and marshes, sand flats, sandy alluvial outwash, ditches, pools, interdune swales, often in shallow water, most common near the coasts. Outer coastal plain, s.e. Va. to s. Fla. and Fla. Keys, westward to e.cen. and s. Tex.; subtropics and tropics generally.

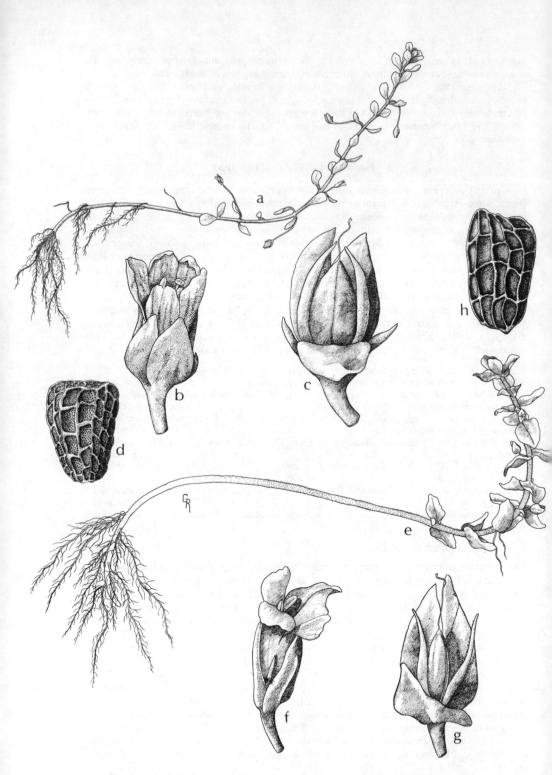

Fig. 291. a–d, **Bacopa monnieri:** a, habit; b, flower; c, capsule; d, seed; e–h, **Bacopa caroliniana:** e, habit; f, flower; g, capsule; h, seed.

2. Bacopa egensis (Poepp.) Pennell.

Stems lax, spongy, several radiating from the base, rooting at the lower nodes if these beneath the substrate. Leaves mainly in floating rosettelike clusters at the stem tips, broadly spatulate fully developed, narrowed below to a margined subpetiole, crenate on the margins of the broadly rounded apex. Flowers 1 or 2 in the axils of some leaves, their fully developed stalks 10–12 mm long, without bractlets. Calyx 2.5–3 mm long, lobes reflexed at full anthesis, erect in fruit. Corolla 3.5–4.5 mm long, lobes 3, white with bluish lines, spreading to reflexed at full anthesis. Stamens 3, inserted in the sinuses of the corolla lobes. Capsule globose, slightly exceeded by the calyx. Seeds shaped somewhat like a small banana, striate.

Shallow quiescent water, or on shores just above the water's edge, La. Native of trop. Am.; trop. Afr.

Reference: Depoe, Charles E. "*Bacopa egensis* (Poepp.) Pennell (Scrophulariaceae) in the United States." *Sida* 3: 313–318. 1969.

3. Bacopa repens (Sw.) Wettst.   Fig. 293

Plant aquatic, rooted in the soil beneath the water, sparingly branched, in deeper water stems growing toward the surface, their tips often lying just beneath the surface, to 5 dm long; in shallower water the branch tips emergent to about 10 cm; stems finely pubescent. Leaves ovate, obovate, less frequently broadly elliptic, 8–25 mm long, 5.5–15 mm broad, with several veins ascending from the base. Flowers solitary or in pairs from the leaf axils, their pubescent stalks 1.5–15 mm long, reflexed in fruit. Sepals 4 (or 5), some or all finely ciliate. Corolla 3.5–4.2 mm long, lobes 4, white or white tinged with pink, without a yellow throat, lobes spreading, slightly shorter to slightly longer than the tube. Stamens 4, anthers dark blue-purple. Capsule ovoid to ellipsoid, 2–3 mm long. Seeds cylindric to ellipsoid, brownish yellow, reticulate. (*Macuillamia repens* (Sw.) Pennell)

Description above adapted from that of John W. Thieret, "*Bacopa repens* (Scrophulariaceae) in the Conterminous United States," *Castanea* 35: 132–136 (1970).

Rice ponds, muddy pools. S.C. (where collected once in 1915); La.; trop. Am.

4. Bacopa rotundifolia (Michx.) Wettst.   Fig. 292

Plants with soft, spongy, prostrate stems, rooting at the nodes, usually on mud, in shallow water, or in floating mats, branch tips floating or laxly ascending, to 6 dm long; stems pubescent at least when young. Leaves thin, broadly short-obovate to orbicular, broadly short-tapered and subclasping or rounded and clasping basally, apically broadly rounded, with several conspicuous veins ascending from the base; larger leaves to 3.5 cm long and 2.5 cm wide. Flowers 1, or 2, in most of the axils of the distal leaves; flower stalks 1–1.5 cm long, usually not exceeding the leaves, pubescent, without bractlets. Corolla 8–10 mm long, white with a yellow throat, the lobes slightly shorter than the tube. Stamens 4, somewhat exserted. Capsule ovoid to subglobose, about 5 mm long. Seeds tan, cylindric-falcate, with a minute tail on either end, alveolate-reticulate. (*Macuillamia rotundifolia* (Michx.) Raf.)

Mostly in water or mud, in and about lakes, ponds, pools, pools in marshes, ditches. Ind. and Ill. to Mont., generally southward to Ala. and Ariz.

5. Bacopa innominata (Gómez Maza) Alain.   Fig. 294

Plant with decumbent, soft, pubescent stems, rooting at the nodes and mat-forming, weakly ascending branches 2–3 dm tall. Leaves glabrous or pubescent beneath, ovate, 5–10 (–15) mm long, mostly about 10 mm, 7–10 mm wide, clasping basally, obtuse apically. Flowers mostly in axils of one of a given pair of leaves at some distal nodes; flower stalks pubescent, 2–8 mm long, shorter than the subtending leaves, bractlets none. Corolla white, 3–4 mm long, lobes slightly shorter than the tube. Stamens 2, not exserted. Capsule ellipsoid-ovoid, 2–2.5 mm long. Seeds brownish black, iridescent, elliptic or oval, about twice as long as broad, irregularly angled, coarsely alveolate-reticulate, the longitudinal striae the more conspicuous. (*Herpestis rotundifolia* Gaertn. f., *B. cyclophylla* Fern.)

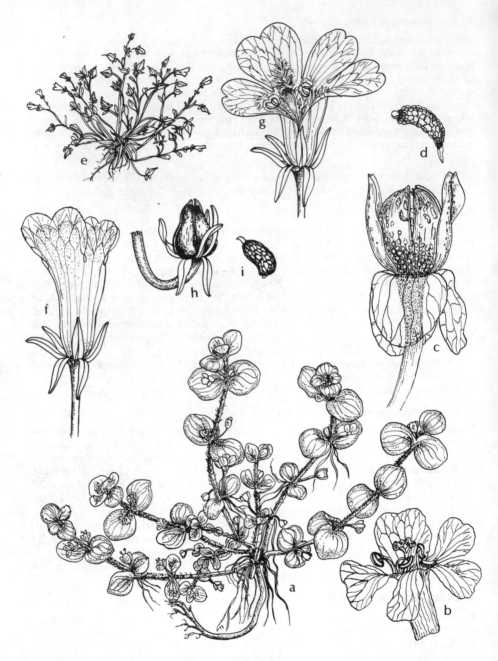

Fig. 292. a–d, **Bacopa rotundifolia:** a, habit; b, flower; c, capsule; d, seed; e–i, **Gratiola flava:** e, habit; f, flower; g, flower, another view; h, capsule; i, seed. (From Correll and Correll. 1972)

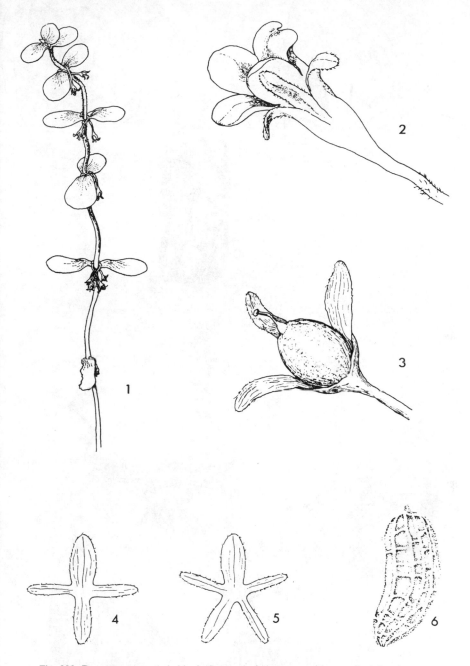

Fig. 293. **Bacopa repens:** 1, habit; 2, flower; 3, fruit (one sepal removed); 4,5, calyces; 6, seed. (From Thieret in *Castanea* 35: 132–136.' 1970)

Fig. 294. **Bacopa innominata:** a, habit; b, flower; c, corolla, split and opened, showing two stamens and pistil; d, capsule; e, seed.

634

Alluvial deposits along rivers and streams, wet ditches, muddy banks and shores. Coastal plain, s.e. N.C. to cen. pen. Fla.; W.I., Pan.

6. Bacopa caroliniana (Walt.) Robins. **BLUE-HYSSOP.**   Fig. 291

Plant aromatic, rhizomatous, forming very extensive mats, flowering stems commonly dense over extensive areas, lax, spongy, held erect by their density; often in shallow water with flowering stems emergent to 1–3 dm; sometimes extending from shore into water to 1 m deep or a little more, the branches very laxly ascending, tips emergent; often in low mats where surface water has receded. Stems short-shaggy-pubescent (becoming glabrous below), pairs of leaves close together. Leaves ovate, clasping at base, obtuse to rounded apically, 0.5–2 cm long and to 1 cm broad or a little more, dull green or olive green, abundantly punctate, veins pedately ascending from the base, sparsely pubescent on the veins of the lower surface. Flowers at anthesis scarcely if any longer than the subtending leaves, the pubescent stalks elongating as the fruits mature to as much as twice the length of the leaves; a pair of small subulate bracts just below the calyx. Corolla 9–13 mm long, pale or bright blue to violet blue. Stamens 4, the longer pair slightly exserted. Ovary with several minute fingerlike processes around its base, these disappearing as the fruits develop. Capsule 4–5 mm long, conic-ovoid. Seeds grayish brown, irregularly tetrahedral, slightly longer than broad, surface coarsely alveolate-reticulate, iridescent. (*Hydrotida caroliniana* (Walt.) Small)

Shallow water near shores of ponds, lakes, streams, cypress-gum ponds and depressions, bogs, drainage and irrigation ditches and canals. Coastal plain, s.e. Va. to s. Fla., westward to e. Tex.

4. Mecardonia

Erect to diffuse, glabrous, perennial herbs, green to dull yellowish green, turning almost black upon drying, especially if not dried quickly. Stems wing-angled. Leaves opposite, narrowed to a tapering sessile base or very shortly wing-petioled, more or less glandular-punctate, toothed at least on the upper margins. Flowers solitary in the axils of leaves or foliar bracts, their stalks bearing 2 small bractlets at their bases. Sepals 5, distinct or scarcely united, somewhat unequal. Corolla tube somewhat quadrangular, slightly enlarged upwardly, abruptly expanding into a slightly 2-lipped limb, the throat closed by a palate; upper lip 2-lobed, lower 3-lobed, lobes truncate and usually emarginate apically, shorter than the tube. Stamens 4, in 2 pairs of unequal length. Ovary cylindric to ovoid, surmounted by a short style terminated by a liplike stigma. Capsule glabrous, acute, septicidal, slightly loculicidal. Seeds numerous, cylindric, reticulate.

● Corolla yellow; outer 3 sepals broadly lanceolate to ovate and overlapping the slightly shorter, very narrow inner ones.   1. *M. vandellioides*
● Corolla white with purple veins; sepals lanceolate, nearly equal in length, the outer 3 only slightly wider than the inner 2.   2. *M. acuminata*

1. Mecardonia vandelliodes (HBK.) Pennell.

Very wide-ranging and variable, prostrate or decumbent to weakly erect herb, 1–4 dm tall. Leaves sessile, 1–3 cm long, ovate-elliptic to obovate-elliptic, apices obtuse to subacute, mostly serrate or crenate-serrate on the upper margins, entire on the lower, abruptly cuneate basally. Flower stalks slightly shorter than to 5 times as long as the subtending leaves or bracteal leaves. Sepals 4–9 mm long, the outer 3 ovate or broadly lanceolate, overlapping the 2 narrower inner ones. Corolla bright lemon yellow, 6–12 mm long. Capsule ellipsoid, 4–5 mm long. (*M. montevidensis* (Spreng.) Pennell, *M. procumbens* (Mill.) Small, *M. peduncularis* (Benth.) Small, *M. tenuis* Small, *M. viridis* Small, *Pagesia vandellioides* (HBK.) Pennell)

Mud and water of ditches, lagoons, ponds, streams, moist soils of cypress-marl prairies. s. Fla.; La.; cen., s., and w. Tex.; Ariz.; trop. Am.

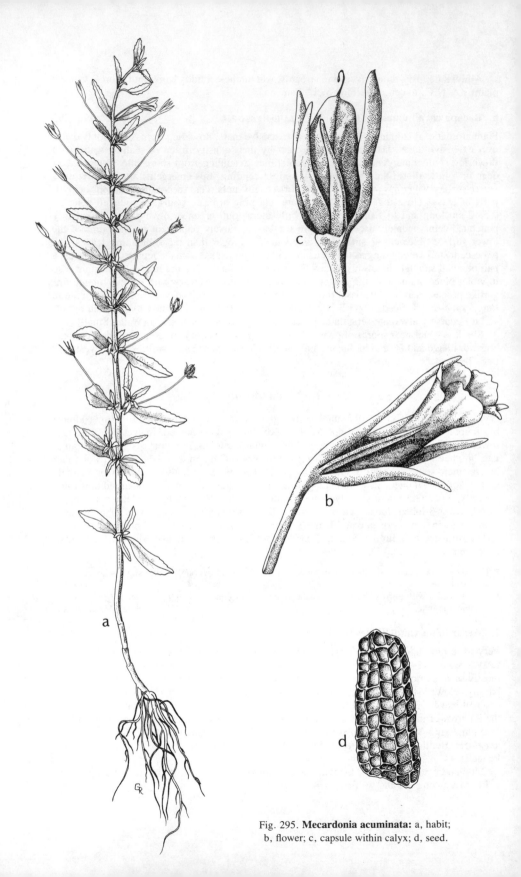

Fig. 295. **Mecardonia acuminata:** a, habit;
b, flower; c, capsule within calyx; d, seed.

2. Mecardonia acuminata (Walt.) Small.   Fig. 295

Usually erect, sometimes prostrate, herb 1–6 dm tall. Stems sometimes solitary and little if
any branched, sometimes solitary and diffusely branched, sometimes several stems from the
base, these simple or branched. Leaves 1–5 cm long, spatulate-elliptic to elliptic, serrate-
dentate on the upper margins, entire on the lower, tapering to a narrow base. Flower stalks
7–30 mm long, mostly longer than the subtending leaves or foliar bracts. Sepals lanceolate
to lance-ovate, the outer 3 equaling or slightly wider than the inner 2. Corolla 6–10 mm
long, white with purple veins, bearded within at the base of the upper lip. Capsule 4–6 mm
long.

   Pine savannas and flatwoods, bogs, marshy shores, wet clearings, floodplain forests,
cypress depressions and prairies, thin soil on or depressions in lime-rock, sand and gravel
bars of rivers, borrow pits, wet ditches. Coastal plain and piedmont, Del. and Md. to s. Fla.,
westward to e. Tex., northward in the interior to s. Mo., Tenn.

5. Gratiola (HEDGE-HYSSOPS)

Low, prostrate, or laxly erect, relatively inconspicuous, somewhat succulent, annual or perennial herbs, prostrate or decumbent stems commonly rooting at the nodes or producing
subterranean stems, some species, at least, producing extensive thick mats. Underwater portions of stems rather thick and spongy. Leaves opposite, sessile or with short pertioles,
blades toothed or entire, glabrous or glandular-pubescent, often glandular punctate. Flowers
sessile or borne on stalks axillary to leaves or foliar bracts, the stalks usually bearing 2 bractlets just below the calyx and very much like the sepals. Sepals 5, unequal, persistent. Corolla
tube somewhat quadrangular, cylindrical or slightly expanded upward then abruptly expanded into a limb, the limb in some nearly radial, mostly obscurely 2-lipped, the throat
closed by a palate; upper lip entire or 2-cleft, the lower 3-cleft; white to pale purple or
yellow-tinged, or bright yellow, often marked within and without by conspicuous darker venation, the deeper color of the veins gradually fading into the lighter background toward the
summit. Fertile stamens 2, sometimes with 2 sterile filaments as well. Ovary ovate or
globose, style slender, stigma dilated and flattened, liplike, recurved. Capsule 4-valved,
glabrous, dehiscing both septicidally and loculicidally nearly or quite to the base. Seeds numerous, elongate, reticulate.

1. Flowers and fruits sessile or subsessile.
   2. Leaves ovate; corolla equal to or only slightly exceeding the sepals.   1. *G. pilosa*
   2. Leaves linear-subulate; corolla at least twice as long as the sepals.   2. *G. hispida*
1. Flowers and fruits definitely stalked.
   3. Leaves narrowed toward the base, not clasping.
      4. Flower stalks stout, erect, usually less than 5 mm long; capsule globose, 4–7 mm long.
                                                                                  3. *G. virginiana*
      4. Flower stalks slender, divergent, 10 mm long or more; capsule ovoid to globose-ovoid, 3–5 mm
      long.
         5. Corolla 8–12 mm long; leaf blades elliptic, rhombic-lanceolate, or lanceolate. 4. *G. neglecta*
         5. Corolla 15–20 mm long; leaf blades oval or oblanceolate.   5. *G. floridiana*
   3. Leaves broadest at the base, clasping or subclasping.
      6. Corolla limb and tube golden yellow throughout, or the tube orange-yellow.
         7. Leaves and sepals linear-lanceolate or lanceolate; bractlets 2; capsule shorter than the sepals.
                                                                                  6. *G. aurea*
         7. Leaves linear-oblanceolate; sepals linear-spatulate; bractlet 1; capsule much longer than the se-
         pals.   7. *G. flava*
      6. Corolla limb pale lavender to white and usually marked with darker venation.
         8. Flower stalks bearing 2 sepallike bractlets just below the calyx.
            9. Leaves triangular to lanceolate, margins entire or with a few sharp teeth on the upper half;
            corolla with brown lines; sepals linear-lanceolate.   8. *G. brevifolia*
            9. Leaves oblong-ovate to ovate (linear-lanceolate if submersed), margins finely dentate; corolla
            with purple lines; sepals lanceolate to oblong-lanceolate.   9. *G. viscidula*
         8. Flower stalks bractless or rarely with a single small bractlet just below the calyx. 10. *G. ramosa*

1. Gratiola pilosa Michx.

Stiffly erect perennial producing 1 to several stems 2–6 dm tall from a short, hard caudex, villous-hirsute throughout. Stems often purplish green above. Leaves yellowish green, the lower ovate, mostly 1.5–2 cm long, apices obtuse or rounded, upwardly progressively smaller, ovate to lanceolate, acute apically; surfaces, besides the pubescence, pebbly above, glandular-punctate below, margins entire or few-toothed distally. Pubescence of multicellular, jointed, glandless, flattened, translucent hairs. Flowers sessile or subsessile, borne in the axils of the leaves. Bractlets 2, linear-lanceolate, about 9 mm long, borne just below and slightly exceeding the sepals. Sepals linear to linear-lanceolate, the longer ones 5–7 mm long. Corolla 5–9 mm long, the limb somewhat 2-lipped, tube yellowish, lobes white, pubescent within on the lower side, with conspicuous bluish purple veins within and without. Fertile stamens 2, the uppermost, 2 rudimentary filaments below. Ovary narrowly pyramidal, acuminate, sulcate when mature. Capsule conical, 4–5 mm long. Seeds brownish yellow, irregularly tetrahedral, obscurely lined. (*Tragiola pilosa* (Michx.) Small & Pennell)

Dry to wet sandy pinelands, bogs, cypress-gum depressions, pine savannas and flatwoods, sometimes in water. N.J. to s. Fla., westward to e. third of Tex., northward in the interior, e. Okla., Ark., Tenn., Ky.

2. Gratiola hispida (Benth. in Lindl.) Pollard.

Perennial herb, simple or frequently producing numerous stems from horizontal branches, the caudex rather thick and hard. Stems 0.5–2 dm tall, pubescent with stiffish, somewhat appressed, flattened, translucent hairs. Leaves sessile, 0.8–1.5 cm long, linear-subulate, firm, strongly revolute and entire marginally, pubescent along the midrib beneath, sometimes with a few scattered hairs besides, upper surface with a somewhat foamy aspect, lower pebbly. Axillary short-shoots with small leaves commonly present. Flowers sessile. Bractlets 2, somewhat longer than the sepals, revolute and hispid on the margins. Sepals 3–6 mm long, linear to linear-lanceolate, revolute, pubescent, the lower 3 longer than the upper 2. Corolla white, salverform, tube 10–13 mm long, pubescent within, the lobes much shorter than the tube. Stamens 2, staminodia 2, very small, or none. Ovary narrowly pyramidal, acuminate. Capsule 4–5 mm long, sulcate. Seeds dark brownish black or almost black, iridescent, oblique at one end, surface ribbed longitudinally, pebbly. (*Sophronanthe hispida* Benth. in Lindl.; *G. subulata* of authors)

Pine savannas and flatwoods, exposed shores of ponds, seepage slopes in pinelands, also well-drained deep sands of pine–scrub oak ridges and hills. Coastal plain, Ga. to s. Fla., westward to Miss.

3. Gratiola virginiana L.   Fig. 299

Annual, dark green, glabrous or nearly so below, somewhat glandular-pubescent above, simple or much branched, 1–4 dm tall. Lowermost branches commonly thickish, decumbent, rooting at the nodes and producing more slender vertical branches, mat-forming. Leaves 2–5 cm long, elliptic-lanceolate to ovate, narrowed at the base, sessile or the lower shortly wing-petioled, blades glandular punctate, apices rounded to broadly acute, margins with a few shallow teeth from the middle distally or entire. Flower stalks stout, 1–5 (–12) mm long, glabrous or slightly glandular-pubescent. Bractlets linear, slightly longer than the sepals. Sepals 4–6 mm long, linear. Corolla 9–14 mm long, the limb nearly white, often drying lavender, tube slightly greenish yellow to pale yellow, with purple or brownish purple veins, pubescent within at the throat. Capsule 4–7 mm long, globose. Seeds yellow or brownish yellow, nearly cylindric, oblique at the apex, conspicuously alveolate-reticulate, the longitudinal ribs the more pronounced.

In and on the edges of small woodland streams, seepage areas in woodlands, swamps and wet woodlands, pools in woodlands, alluvial outwash, ditches. N.J. to Ohio, thence to Iowa and Kan., generally southward to cen. pen. Fla. and e. third of Tex.

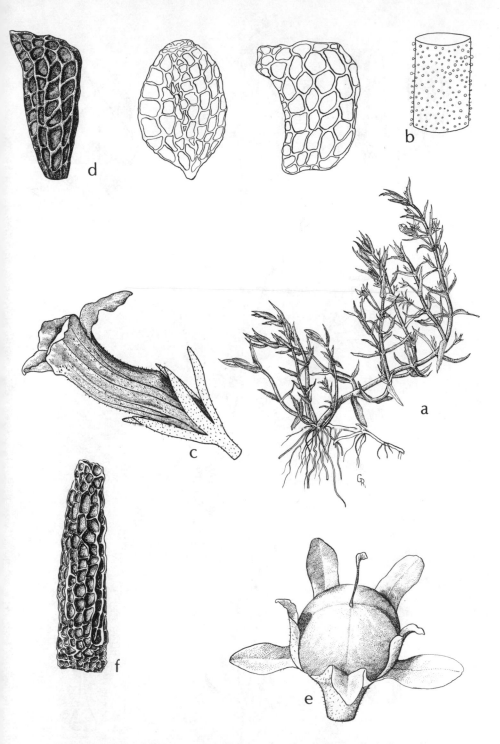

Fig. 299. a–d, **Gratiola ramosa:** a, habit; b, section of stem, diagrammatically, to show glandular pubescence; c, flower; d, seeds (figures to right outlined to show variation in shape and size); e,f, **Gratiola virginiana:** e, capsule, with subtending calyx and bracteoles; f, seed.

639

Fig. 296. **Gratiola floridana:** a, habit; b, flower, just prior to full anthesis; c, corolla; d, corolla, at late bud stage, opened out to show insertion of stamens, pistil, and pubescence within tube; e, capsule with subtending calyx and bracteoles; f, seeds.

4. Gratiola neglecta Torr.

Erect annual, 1–4 dm tall, branched or unbranched. Upper internodes, flower stalks, and expanding leaves more or less clammy-puberulent. Leaves elliptic, rhombic-lanceolate, or lanceolate, tapering to base and apex, 1–6 cm long, 2–12 mm wide, entire to irregularly finely serrate. Flower stalks filiform, 10–25 mm long, glandular-pubescent. Bractlets foliaceous, linear, equal to or exceeding the 3–5 mm long sepals. Corolla tube yellow, limb cream or white, 8–12 mm long, with clavate pubescence within. Staminodia minute or none. Capsule ovoid to globose-ovoid, 3–5 mm long. Seeds more or less cylindric, tan or yellowish, reticulate, the longitudinal reticular lines more prominent than the transverse ones.

Wet or muddy places, low woodlands, swamps, marshes, marshy shores, fields, ditches. Que. and Maine to B.C., southward in most of the U.S. except Fla.

5. Gratiola floridana Nutt.   Fig. 296

Annual, yellowish green, glabrous below, somewhat glandular-pubescent above, simple to much branched, 1–4 dm tall. Lowermost branches commonly thickish, decumbent, rooting at the nodes and with ascending branches, mat-forming. Leaves glabrous, to 4 cm long, mostly 2–3 cm, narrowed at the base, sessile or the lower shortly wing-petioled, blades ovate to oblanceolate, apices obtuse to acute, margins with a few shallow teeth from the middle distally or entire. Flower stalks slender, 20–45 mm long, finely glandular-pubescent to glabrate. Bractlets linear, as long as or a little longer than the sepals. Sepals linear, 3–6 mm long. Corolla 1.5–2 cm long, tube yellowish green with reddish purple to brownish yellow veins, limb nearly white. Capsule ovoid, acute, 3–5 mm long. Seeds yellow, 0.6–0.8 mm long, oblong to pyramidal, the alveoli longer than broad, slightly crescent-shaped.

Wet shaded places along streams and brooks, swampy woodlands, seepage slopes, often in shallow water. n. and cen. Ga. to Fla. Panhandle, Ala., Tenn., Miss.

6. Gratiola aurea Muhl. ex Pursh.   Fig. 297

Slender, laxly erect to sprawling, simple or branched, 1–4 dm tall; lowermost stems often decumbent, rooting at the nodes, mat-forming. Leaves sessile, 1–3 cm long, usually 3–7 (–10) mm wide, clasping at base, linear-lanceolate to ovate-lanceolate, apices rounded, margins entire or with a few shallow teeth distally. Flower stalks slender, 1–2.5 cm long, sparingly glandular-pubescent. Bractlets 2, slightly shorter than to exceeding the sepals and of about the same shape. Sepals 4–7 mm long, lanceolate to linear-lanceolate. Corolla golden yellow, without conspicuous veins, 10–15 mm long. Capsule about 3 mm long, globose or slightly longer than wide, shorter than the sepals. Seeds pyramidal, oblique at the apex, dark brownish yellow.

Nfld. and Que., southward in the coastal plain to Fla. Panhandle; inland, N.Y. and Ont. to Ill. and N.D.

7. Gratiola flava Leavenw.   Fig. 292

Glabrous, erect annual, less than 1 dm tall; stems clustered from the base. Leaves sessile, few, linear-oblanceolate, entire or with a few serrations, to 1.5 cm long and 5 mm wide, clasping at the base, obtuse apically. Flowers stalked, subtended by 1 linear-spatulate bractlet about as long as the sepals. Sepals 3–6 mm long, linear to linear-lanceolate, obtuse. Corolla about 12 mm long, limb golden yellow, tube orange-yellow. Capsule ovoid-pyramidal, almost twice as long as the sepals. Seeds brown, coarsely reticulate.

Wet sandy soil in prairies and fields. La., cen. and s. Tex.

8. Gratiola brevifolia Raf.   Fig. 298

Perennial, laxly ascending, somewhat spindly, simple or branched, 2–6 dm tall, often forming thick, tangled mats from decumbent lower branches which root at the nodes and produce vertical stems; stems covered with small, glandular hairs and the upper portions of the plants especially very sticky to the touch. Leaves triangular to lanceolate, 1–2 cm long, clasping at

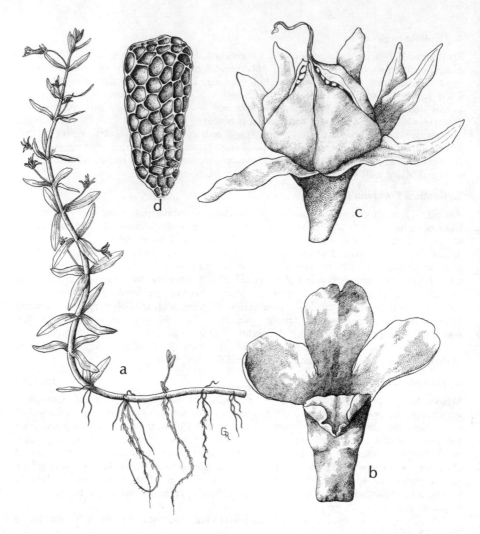

Fig. 297. **Gratiola aurea:** a, habit; b, corolla; c, capsule, with subtending calyx and bracteoles; d, seed.

the base, apices acute, margins entire or with a few teeth distally. Flower stalks slender,
12–20 mm long, finely and densely glandular-pubescent. Bractlets 2, just below and much
like the sepals. Sepals 5–7 mm long, glandular-pubescent, linear-lanceolate. Corolla 1–1.2
cm long, limb white, tube greenish yellow and with yellow or brownish yellow veins. Capsule globose, 1–2 mm long, much exceeded by the sepals. Seeds tiny, about 0.4 mm long,
obpyramidal, brown, reticulate.

Swamps and wet woodlands, seepage areas in woodlands, in and on the edge of water
about ponds and lakes. Ga., chiefly coastal plain, Fla. Panhandle, westward to e. Tex., s.e.
Okla.; s.e. Tenn.

9. Gratiola viscidula Pennell.

Perennial from a sparsely branched rhizome. Stem slender, erect or decumbent, usually
unbranched, 1–7 dm tall, pubescent. Leaves ovate to ovate-oblong (linear-lanceolate if

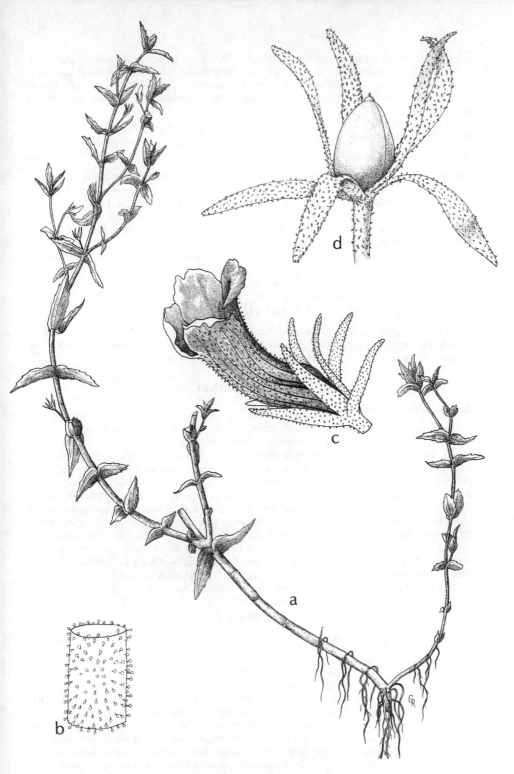

Fig. 298. **Gratiola brevifolia:** a, habit; b, section of stem, diagrammatically to show glandular pubescence; c, flower, with subtending bracteoles; d, capsule, with subtending calyx and bracteoles.

submersed), bases clasping, surfaces pubescent or glabrous, the upper usually glandular-punctate. Flower stalks filiform, 1 cm long or a little more, glandular-pubescent. Bractlets somewhat foliaceous, slightly shorter than the 3–7 mm long, ovate-lanceolate sepals. Corolla whitish with lavender lines, 1–1.3 cm long. Capsule globose, 2–2.5 mm long, shorter than the sepals.

Marshes, pond margins, swales, wet woodlands. Del., Md., e. Va., throughout much of N.C., upper S.C., n.e. Ga.; s. Ohio.

10. Gratiola ramosa Walt. Fig. 299

Perennial, more or less upright, branched or simple, 1–3 dm tall, often producing horizontal lower branches rooting at the nodes, forming small to large clumps. Herbage glandular-pubescent but not as markedly sticky to the touch as that of *G. brevifolia*. Leaves 0.7–2 cm long, blades sessile, clasping, subulate, lance-subulate, or lanceolate, apices acute, margins entire or with a few teeth on the upper half. Flower stalks 6–17 mm long, glandular-pubescent. Bractlets none, or 1 and minute. Sepals linear, 3–6 mm long, glandular-pubescent. Corolla 1–1.4 cm long, limb white, tube greenish yellow and with yellow or brownish yellow or purple veins. Capsule subglobose, 1–2 mm long, slightly longer than wide. Seeds tiny, about 0.3 mm long, obpyramidal, brown, finely reticulate.

Pine savannas and flatwoods, sandy-peaty pond margins or depressions, sandy-peaty ditches. Coastal plain, Md.; s.e. N.C. to s. Fla., westward to s.w. La.

This species and *G. brevifolia* are closely similar in aspect. In general the stems of *G. ramosa* are shorter, the branches more stiffly erect, and it inhabits more open, more acid habitats; *G. brevifolia* more commonly inhabits wooded places.

6. Amphianthus

Amphianthus pusillus Torr.

Diminutive, glabrous, aquatic annual, stems 5 (–10) cm long, branched only from the caudex, each lax stem held up by a pair of floating leaves. Basal, submersed, rosette leaves sessile, lanceolate, 5 mm long or a little more; filiform stems with a pair of opposite, ovate or oval leaves terminally, these 4–8 mm long, 3–5 mm wide, abruptly narrowed to short, sub-petiolar bases, rounded apically. Flowers axillary to both types of leaves. Flower stalks 0.5–2 mm long, bractlets none. Calyx united only at base, unequally 5-lobed, lobes about 1 mm long, rounded apically, the larger one obovate, others oblongish. Corolla of emersed (essentially floating) flowers white, 3–4 mm long, tube funnelform to narrowly campanulate, much longer than the 5, short, broad, emarginate, slightly spreading lobes; basal, submersed flowers not opening. Stamens 2, not exserted. Capsule broader than long, notched basally and apically, the somewhat flattened sides broadly rounded, loculicidal. Seeds oblong-subobpyramidal, about 1 mm long, dark brown, alveolate-reticulate.

Ephemeral pools on granitic rock outcrops. Piedmont, S.C., Ga.

7. Mazus

Mazus pumilus (Burm. f.) Steenis. Fig. 300

Low, lax annual, to about 1 dm tall, often much branched from near the base, branches sometimes decumbent. Larger leaves mostly closely set near the base of the stem, lowermost opposite or subopposite, wing-petioled, blades spatulate, oblanceolate, obovate, or oblong, up to 3 cm long, 1 cm wide, rounded apically, margins shallowly and often irregularly toothed, surfaces glabrous; the few leaves above the base alternate and inflorescence bracts alternate. Flowers in racemes terminating the branches, raceme axis usually zigzagging; flower stalks mostly about 1 cm long, sparsely pubescent. Calyx campanulate-tubular below, 5-lobed above, the lobes 4–5 mm long, slightly longer than the tube. Corolla about twice the length of the calyx, blue, 2-lipped, upper lip erect, slightly bifid, lower larger, 3-lobed, a pair

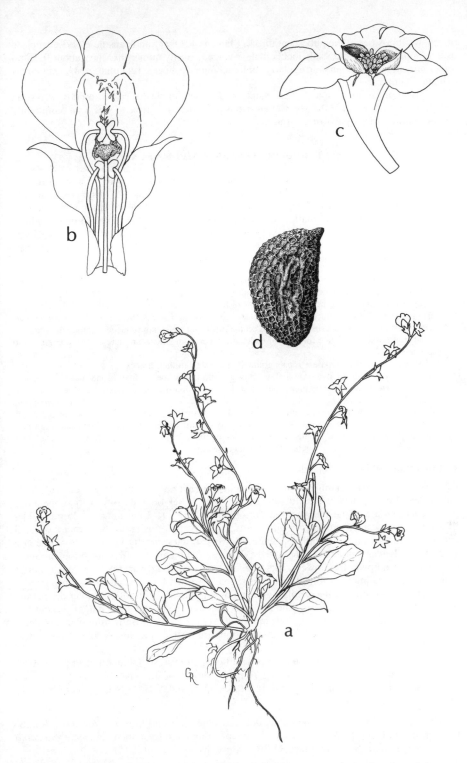

Fig. 300. **Mazus pumilus:** a, habit; b, flower, opened out; c, opened capsule surrounded by persistent calyx; d, seed.

of yellow gibbosities at the pubescent throat. Stamens 4. Ovary and capsule globose, capsule usually somewhat compressed, loculicidal, 2-valved. Seeds numerous, tiny, about 0.3 mm long, brown, mostly very irregularly oblique-angled, finely reticulate. (*M. japonicus* (Thunb.) Kuntze)

Weedy in moist lawns and gardens, sand, mud, or gravel bars of rivers, muddy banks, on wet alluvium in floodplain forests, wet clearings and ditches. Native to s.e. Asia, commonly naturalized in various parts of our range.

8. Lindernia (FALSE PIMPERNELS)

Annual, rarely perennial, low herbs. Leaves opposite. Flowers solitary in the axils of foliage or bracteal leaves, their slender stalks not bearing bractlets. Sepals 5, distinct or partially united. Corolla tube cylindric-funnelform, tube longer than the 2-lipped limb; upper lip narrower than the lower, erect, shallowly 2-lobed, lower lip somewhat deflexed or spreading, 3-lobed, 2 yellowish ridges within the throat. Fertile stamens 2, staminodia 2, or fertile stamens 4. Capsule glabrous, 2-valved, dehiscing septicidally. Seeds numerous, yellow, tan-colored, or brown, somewhat angled, obscurely to definitely alveolate-reticulate.

1. Flower stalks shorter than to about equaling the subtending leaves.   1. *L. dubia*
1. Flower stalks distinctly longer than the subtending leaves or bracteal leaves.
  2. Larger leaves essentially in basal rosettes or restricted to the lowermost portions of the flowering stems, the latter with very much reduced bracteal leaves and appearing "naked."   3. *L. monticola*
  2. Larger leaves not in basal rosettes, leaves on the flowering stems not reduced or gradually reduced.
    3. Calyx at anthesis connate for ½-⅔ of its length (lobes triangular-acuminate), later separating into separate sepals.   5. *L. crustacea*
    3. Calyx at anthesis of separate sepals gradually narrowed to their apices.
      4. Leaves orbicular or ovate-orbiculate, mostly little if any longer than broad, scarcely reduced upwardly; stems usually creeping, rooting at the nodes, mat-forming, ascending flowering stems sparsely floriferous.   4. *L. grandiflora*
      4. Leaves ovate to narrowly elliptic, many of them distinctly longer than broad; stems not creeping, occasionally some lowermost portions shortly decumbent and rooting at the nodes, not mat-forming; flowering stems usually abundantly floriferous from near the base upwardly and subtending leaves or bracteal leaves generally reduced upwardly.   2. *L. anagallidea*

1. Lindernia dubia (L.) Pennell.

Glabrous annual. Stems simple to much branched, weakly ascending, 1–2.5 dm tall, leafy throughout. Leaves mostly sessile, sometimes some lower ones narrowed to subpetiolate bases, oblanceolate, oblanceolate-ovate, or oblong-elliptic, 1–3 cm long, 5–25 mm wide, little reduced upwardly, cuneate basally, obtuse apically, margins entire or with a few small dentations, surfaces not evidently punctate. Flowers solitary, mostly in the axils of one of a given pair of leaves (sometimes one in the axil of each of the pair), often floriferous from the base of the stem upwardly; flower stalks slender, 1–2.5 cm long, varying from much shorter than to about equaling the subtending leaves. Sepals distinct, linear-attenuate, scabrous, 3–4 mm long, half as long as the corolla tube and as long as or a little longer than the mature capsule. Corolla 7–10 mm long, pale lavender, the margins of the lobes darker. Fertile stamens 2. Capsule oblong-elliptic, 4–6 mm long, style persistent until time of dehiscence. Seeds pale yellowish, oblong, truncate or oblique-truncate at the apices, obscurely reticulate. (*Ilysanthes dubia* (L.) Barnh.)

Commonly in alluvial muds of swamps and floodplain forests (sometimes in great abundance following subsidence of flooding), muddy shores of streams, bayous, ponds, wet meadows, tidal shores, alluvial outwash in ditches. Que. and N.H. to N.D., generally southward to n. Fla. and e. Tex.

This and *L. anagallidea*, where they occur together, are said to intergrade. (See Tom S. Cooperrider and George A. McCready, "On Separating Ohio Specimens of *Lindernia dubia* and *L. anagallidea* (Scrophulariaceae)," *Castanea* 40: 191–197 (1975).)

2. Lindernia anagallidea (Michx.) Pennell.

Glabrous, diffusely branched annual, stems slender, variable in height to about 2 dm. Leaves mostly sessile, sometimes a few lower ones narrowed to subpetiolate bases, ovate to narrowly elliptic, reduced at least on the distal portions of the flowering stems; larger lower leaves varying from about as broad as long to twice as long as broad, 5–20 mm long, bases varying from slightly clasping to shortly tapered, apices obtuse to rounded, margins entire to finely scabrid, surfaces not evidently punctate. Flowers mostly solitary in the axils of one of the pair of leaves at a given node, commonly stems floriferous from the bases upwardly; flower stalks filiform, much exceeding the subtending leaves or bracteal leaves. Sepals somewhat unequal, distinct, linear or linear-subulate, about half as long as the fully expanded corolla tube, varying from ½ as long to about equaling the mature capsule. Corolla 5–7 mm long, pale lavender to almost white. Fertile stamens 2. Capsule narrowly ovate to oblong-elliptic, 2.5–3.5 mm long, styles persistent until time of dehiscence. Seeds oblong or oblong-falcate, angled, buffish, the apices somewhat narrowed and with tiny darker apiculations, surface obscurely reticulate. (*Ilysanthes anagallidea* (Michx.) Raf.; incl. *I. inequalis* (Walt.) Pennell)

Sandy or sandy-peaty exposed shores or bottoms of ponds, depressions in pine savannas and flatwoods, adjacent ditches, cypress-gum depressions, marshy shores, low wet places in fields, wet meadows, ephemeral pools. N.H. to Minn., N.D., and Colo., generally southward to s. Fla. and Tex.; Pacific states.

3. Lindernia monticola Muhl. ex Nutt.

Plant perennial, usually purplish throughout, with a basal rosette of principal leaves or with 1 or 2 pairs of leaves similar to the rosette ones just above the base, other leaves very much reduced; rosette leaves elliptic to oval, oblanceolate, or spatulate, mostly narrowed basally and subpetiolate, obtuse apically, variable in size to about 2–3 cm long, margins entire or with a few low inconspicuous teeth, surfaces punctate; flowering stems 1 to several from the base, wiry, more or less dichotomously branched, the leaves essentially bracteal, oblanceolate, the stem appearing relatively "naked." Flower stalks filiform, 1–4 cm long, much exceeding the subtending bracteal leaves. Sepals distinct, lanceolate to subulate, 1.5–2 mm long, surrounding only the base of the fully expanded corolla tube, ⅔ as long as the mature capsule. Corolla violet-purple, sometimes with darker lines and blotches. Fertile stamens 2. Capsules 3–5 mm long, narrowly lance-elliptic. Seeds minute, brown, obscurely reticulate, longitudinally several-winged. (*Ilysanthes monticola* (Muhl. ex Nutt.) Raf.)

Shallow soil in dpressions or pools chiefly on granitic rock outcrops, N.C. to Ga. and Ala.; seasonally wet depressions in pine flatwoods, cypress-gum depressions, s.e. Ga., n.e. Fla.

4. Lindernia grandiflora Nutt.

Stems usually creeping, rooting at the nodes and forming glossy green mats with ascending-erect flowering branches 2–4 dm tall, the flowering branches usually sparsely floriferous. Leaves sessile, 5–15 mm long, orbicular or ovate-orbicular, as broad as long or little longer than broad, little if any reduced upwardly, bases mostly rounded to subcordate, apices rounded or obtuse, surfaces obscurely finely punctate. Flowers solitary in the axils of leaves, usually only one of a given pair of leaves subtending a flower and often several leaves not bearing flowers between those that do; flower stalks filiform, 1–3 (–4) cm long, much exceeding the subtending leaves. Sepals distinct, linear-attenuate, 3–4 mm long, surrounding only the base of the fully expanded corolla tube, about ⅔ as long as the mature capsule. Corolla 8–10 mm long, violet-blue, tube streaked with purple lines, some lobes mottled with purple. Fertile stamens 2. Capsule 4–5 mm long, oblong-elliptic. Seeds tan, a little longer than wide, alveolate-reticulate, reticules longer horizontally, the longitudinal striae 5–7, much more pronounced than the horizontal giving the surface a ridged appearance. (*Ilysanthes grandiflora* (Nutt.) Benth.)

Depressions in pine flatwoods, cypress depressions, swamps and floodplain forests, stream and river shores where sometimes matted in shallow water. s.e. Ga. to s. pen. Fla.

5. Lindernia crustacea (L.) F. von Muell.

Plant annual, commonly purplish, branched from near the base, stem sparsely pubescent, mostly on the angles, decumbent and rooting at the nodes, flowering branches weakly ascending, 5–20 cm long. Lower and often medial leaves with short-margined petioles, uppermost sessile; some blades on most plants ovate, varying to elliptic, 8–20 mm long, bases truncate to shortly and broadly tapered, apices obtuse to acute, margins variable, entire, undulate, or few-serrate, edges somewhat scabrid. Flowers solitary, usually in the axils of both of the pair of leaves at a given node, sometimes only in 1; flower stalks much exceeding the subtending leaves or bracteal leaves. Calyx 4–5 mm long, about ½ the length of the corolla tube and a little longer than the mature capsule, connate for ½–⅘ of its length at anthesis, free tips triangular-acuminate, later separating to the base into oblong sepals. Corolla light blue to violet blue, 6–7 mm long. Fertile stamens 4, in 2 pairs of unequal length. Capsule about 4 mm long, short-oblong, apically nearly truncate. Seeds buffish, irregular in outline, somewhat angled, with relatively few craterlike depressions on the surface.

Native of Indo-Malaya. Widely naturalized in tropical and subtropical areas; sporadic in the coastal plain of our range, S.C. to Fla., s.e. Ala., La., perhaps elsewhere; a weed in moist lawns, gardens, and nurseries, in sandy-peaty disturbed soils of flatwoods depressions and clearings, adjacent ditches.

9. Micranthemum

Low, matted, glabrous herbs. Leaves opposite, rarely in whorls of 3, sessile, margins entire. Flowers minute, scarcely perceptible to the unaided eye, solitary in the leaf axils (usually one of the opposite leaves with a flower in its axil, the other one without a flower, many pairs without any flowers), short-stalked, the stalks without bracts. Calyx tubular below, 4-lobed. Corolla bilaterally symmetrical. Stamens 2, curved or bent, appendaged. Ovary globose, style short and slender, cleft at the tip into 2 stigmas. Capsule 1-locular at maturity, several- or many-seeded, rupturing irregularly.

● Leaves orbicular or nearly so; calyx cleft nearly to the base, lobes oblanceolate; corolla with 3 subequal lobes and 1 larger lobe, lobes rotate.   1. *M. umbrosum*
● Leaves oblanceolate; calyx lobes more or less deltoid, 3 of the sinuses between them less than half the length of the calyx, the 4th sinus reaching nearly to the base of the calyx.   2. *M. glomeratum*

1. Micranthemum umbrosum (J. F. Gmel.) Blake.   Fig. 301

Low, extensively branched and usually matted herb, commonly forming dense, tight carpets a meter or more across, especially on alluvial soils along banks of streams and in shallow water along their shores; in deeper water, forming dense mats but the branches more elongate, flaccid, and with larger, more flaccid leaves. Leaves orbicular, 4–9 mm broad and long, or to 15 mm long when submersed. Hairlike roots produced at the nodes, especially when emersed. Flowers minute, usually in the axil of only 1 of the 2 leaves at a given node, their stalks 0.5–1 mm long. Calyx lobes about 1.5 mm long, equal, oblanceolate, sinuses reaching nearly to the base of the calyx. Corolla white, with 3 subequal and 1 larger lobe, the lobes rotate, slightly longer than the tube. Stamens exserted, filaments curved-appendaged near the base. Capsule about 1 mm long. Seeds cylindric, ribbed longitudinally and with fine transverse lines between the ribs, brownish yellow. (*Globifera umbrosa* J. F. Gmel.)

Wet places, edges of and in streams, pools, on mud or alluvial sands, ditches, swamps, sometimes on rotting logs or on cypress knees in swamps. Coastal plain, s.e. Va. to pen. Fla., westward to e. Tex.

2. Micranthemum glomeratum (Chapm.) Shinners.

Creeping, forming tangled mats in semiaquatic habitats, frequently producing hairlike, yellowish roots up to 5 cm long from the nodes. Leaves oblanceolate, variable in size up to 10

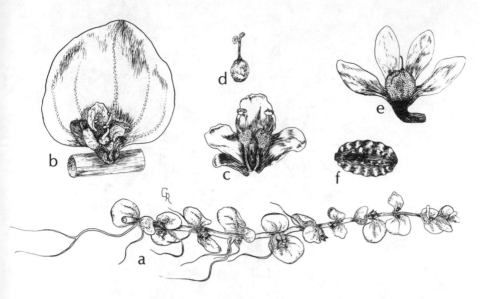

Fig. 301. **Micranthemum umbrosum:** a, habit, branch of matted plant; b, flower in a
leaf axil; c, flower, opened up; d, pistil; e, capsule; f, seed.

mm long, usually not exceeding 2–4 mm long. Flowers minute, their stalks 0.5–1 mm long
or a little more. Calyx 1–1.5 mm long, lobes nearly deltoid, 3 sinuses shallow, the 4th reaching nearly to the base of the calyx. Corolla white to pinkish, 1-lipped, lip 3-lobed, distal lobe
longest, about 2 mm long and curved upward, the 2 lateral lobes about ⅔ as long and turned
nearly at right angles to the long axis of the lip, a few hairs along the sinuses within. Filaments curved and distally almost paralleling the upcurved distal corolla lobe, a spurlike projection at the bend of each filament. Seeds cylindric, slightly ribbed longitudinally and with
fine transverse lines between them, brownish yellow. (*Hemianthus glomeratus* (Chapm.)
Pennell)

In and around swamps, shores of lakes, ponds, rivers. Endemic to pen. Fla.

10. Limosella

Limosella subulata Ives. **MUDWORT.**

Stoloniferous annual herb with tufted, linear-filiform, sessile leaves 2–5 cm long, 1–2 mm
wide. Flowers from the leaf axils, their stalks about half as long as the leaves, usually recurving after anthesis. Calyx tube campanulate, equally 5-lobed, lobes about ⅓ as long as the
tube, deltoid. Corolla white, 3.5–4 mm long, its tube barely if at all extending beyond the
calyx, equally 5-lobed, lobes shorter than the tube. Stamens 4. Capsule globose, dehiscing
septicidally.

Muddy or sandy intertidal flats and shores. Nfld. and Que. to n.e. N.C.

11. Limnophila

Soft perennial herbs, aquatic or on moist to wet shores. Leaves verticillate or subverticillate,
those submersed and those lying flat on wet substrate and the emersed ones differing markedly. If entirely terrestrial, stems branching somewhat beneath the substrate and mat-forming, flowering stems laxly or stiffly ascending-erect, 1–4 dm high; leaves oblanceolate or
lanceolate in outline, varying from clone to clone from merely serrate to pectinate-pinnatifid,

649

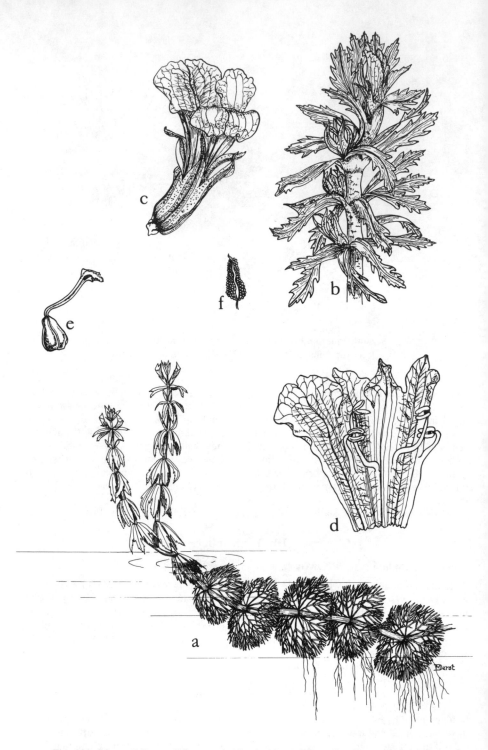

Fig. 302. **Limnophila sessiliflora:** a, habit, showing submersed and emersed parts; b, flowering shoot; c, flower; d, corolla, opened out; e, pistil; f, seeds, capsule removed from around them. (b–f, from Correll and Correll. 1972)

5–30 mm long, punctate; submersed stems lax, often in dense mats, to 1 m long or more, leaves 2 or 3 times dissected into linear-filiform segments, fanlike in outline (much like submersed leaves of *Cabomba*). Flowers solitary in the axils of emersed leaves or in the axils of submersed leaves as well. Calyx united below, nearly equally 5-lobed. Corolla nearly cylindric to beyond the calyx, somewhat dilated at the throat, limb weakly 2-lipped, upper lip a single entire or emarginate lobe, lower 3-lobed, lobes all spreading at full anthesis. Stamens 4, unequal, not exserted from the corolla tube. Style about equaling the longer stamens. Capsule 4-valved.

- Emersed stems powdery-pubescent below, hirsute distally; flowers sessile, bractlets none; calyx lobes distinctly longer than the tube.   1. *L. sessiliflora*
- Emersed stems glandular; flowers stalked, bractlets present; calyx lobes deltoid, equaling or somewhat shorter than the tube.   2. *L. indica*

1. Limnophila sessiliflora (Vahl) Blume.   Fig. 302

Emersed stems powdery-pubescent below, powdery-pubescent and with some longish hairs distally. Flowers sessile. Calyx 4–5 mm long, with a few longish hairs basally, the lobes long-triangular, their sides becoming rolled in drying and appearing subulate, distinctly longer than the tube. Corolla lavender or pale violet, with purple lines on the lower portion of the tube and extending into the lobes of the lower lip, tube about twice as long as the calyx.

Ponds, lakes, streams, impoundments. Native to the Old World tropics, sporadically naturalized, s.w. Ga. and Fla., perhaps elsewhere in our range.

2. Limnophila indica (L.) Druce.

Similar in general features to *L. sessiliflora*. Emersed stem glandular. Flowers stalked, bractlets present. Calyx glandular-pubescent, lobes short-deltoid.

Native to the Old World tropics.

Both *L. indica* and *L. sessiliflora* are grown, often side by side, by aquarists. It is possible that both may be naturalized in parts of our range. A fertile putative hybrid, *L.* × *ludoviciana* Thieret, naturalized in La., is presumed to have originated in cultivation. It is characterized as having the emersed stem glandular, stalked flowers, hirsute calyx, short calyx lobes. See D. Philcox, "A Taxonomic Revision of the Genus *Limnophila* R. Br. (Scrophulariaceae)," *Kew Bull.* 24; 101–170 (1970?).

12. Dopatrium

Dopatrium junceum (Roxb.) Hamilt. in Benth.

Small, slender, glabrous, annual emersed aquatic. Stems simple or branched, erect-ascending, 10–30 cm long, subsucculent. Leaves opposite, sessile, basal ones sometimes more or less in a rosette, lanceolate or linear-oblong, 1.5–2.5 cm long, 3–5 mm broad, obtuse apically; stem leaves gradually smaller, upper ones scalelike, obtuse. Flowers solitary in the leaf axils, stalks none or to 10 mm long, bractlets none. Calyx about 1.5 mm long, united basally, lobes 5, short, obtuse. Corolla purple, 5–6 mm long, much exceeding the calyx, cylindric to the slightly broadened throat, limb 2-lipped, upper lip a single erect somewhat obovate lobe broad at the apex, lower lip 3-lobed, lobes spreading or decurved, the central lobe largest. Fertile stamens 2, staminodia 2, minute. Capsule globose, 2.5–3 mm long, much exceeding the calyx. Seeds numerous, ovate to oblong in outline, reticulate.

Native of Asia. Naturalized in rice fields of La. and Calif.

13. Leucospora

Leucospora multifida (Michx.) Nutt.   Fig. 303

Low, diffusely branched annual herb, to about 2 dm tall and as broad, short-pubescent throughout. Leaves opposite, petioled, 2–3 cm long, deeply pinnately divided into 3–7

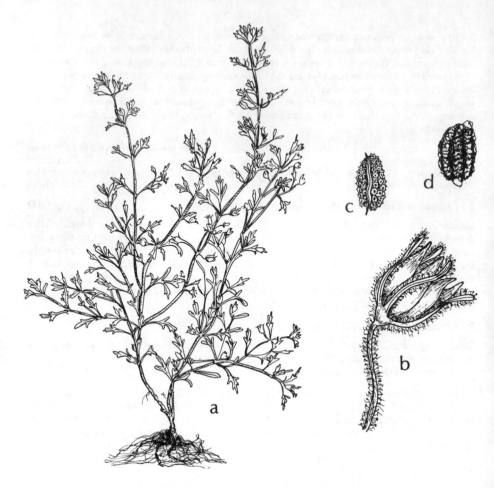

Fig. 303. **Leucospora multifida:** a, habit; b, flower; c, placenta (after shedding of seeds) removed from capsule; d, seed. (From Correll and Correll. 1972)

short-oblong to linear, ascending lobes. Flowers solitary in the leaf axils, their stalks 5–10 mm long, without bractlets. Calyx 4–5 mm long, united only basally, lobes subulate and equal. Corolla longer than the calyx, about 6 mm long, cylindrical to the throat, 2-lipped, lips much shorter than the tube, upper lip 2-lobed, lower 3-lobed, white with a pink tinge to pale lavender. Stamens 4, not exserted. Stigma 2-lobed. Capsule ovoid, thin-walled, septicidal. Seeds numerous, tiny, oval or oblong, pale-opaque, ridged and grooved (viewed with considerable magnification).

Exposed shores of ponds or lakes, gravel and sand bars of rivers and streams, river banks, depressions on rock outcrops in glades, mud flats, ditches. s. Ont. to Iowa and Kan., generally southward to n.w. Ga., Ala., La., and e. Tex.

14. Mimulus (MONKEY-FLOWERS)

Perennial, rhizomatous herbs, (ours) relatively coarse. Leaves opposite, serrate. Flowers solitary in the axils of leaves or bracts, their stalks without bractlets. Calyx with a relatively long, angled, or prismatic tube somewhat oblique at the orifice, with 5 lobes, uppermost lobe

652

largest. Corolla tube nearly cylindrical or somewhat dilated distally, limb strongly 2-lipped, the throat closed by a palate, upper lip 2-lobed, erect or reflexed-spreading, shorter and narrower than the lower, the lower 3-lobed, somewhat deflexed. Stamens 4, in 2 unequal pairs, inserted at about the middle of the corolla tube. Style elongate, stigma 2-lobed. Capsule dehiscing loculicidally, enclosed by the persistent calyx. Seeds many, minute.

• Angles of the stem rounded or obtuse; leaves sessile; stalks of the fruits mostly 2–3 (–4) cm long, in the upper part of the inflorescence, at least, longer than the subtending bracts.   1. *M. ringens*
• Angles of the stem with a narrow wing; leaves (except the uppermost) petiolate; stalks of the fruits 1 cm long or less, all of them shorter than the subtending leaves or bracteal leaves.   2. *M. alatus*

1. Mimulus ringens L.

Plant glabrous throughout. Stems 4–12 dm tall, unbranched or branched. Leaves sessile, lanceolate, lance-elliptic, oval, or oblanceolate, the larger ones 5–15 cm long, tapering to an acute base or relatively less tapered to a clasping or subclasping base, acuminate apically, margins with blunt serrations; leaves reduced upwardly and gradually becoming bracteate on upper parts of the inflorescence or its branches. Flowers in axils of foliage leaves below, on the main stem or its branches if any, in the axils of bracts above; flower stalks relatively long, elongating somewhat as the fruits mature, mostly 2–3 cm long, in the bracteate portions of the inflorescence much exceeding the subtending bracts. Calyx 10–15 mm long, cylindrical at anthesis, becoming somewhat bulged centrally in fruit, lobes short, triangular-subulate, sinuses between them narrowly **U**- or **V**-shaped. Corolla 2.5–3.5 cm long, violet blue, lavender, sometimes pinkish, or rarely nearly white, throat shaded with yellow and with reddish purple spots below. Capsule narrowly ovoid, 10–12 mm long. Seeds about 0.5 mm long, plump, terete, in outline somewhat falcate, a little larger at one end than at the other, pale amber with a few reddish very finely knobby striae, the red color sometimes interrupted, a tiny dark reddish knob on either extremity.

Bogs, seepage slopes, marshy shores, wet meadows, sloughs, borders of swamps or wet woodlands. Que. and N.S. to Sask., generally southward to Ga., La., Okla., Colo.

2. Mimulus alatus Ait.   Fig. 304

General stature as in *M. ringens*. Stem with narrow wings on the angles, glabrous. Leaves with distinct margined petioles (except the uppermost), petioles of the larger leaves 1–2 cm long, infrequently over 1 cm, blades lanceolate, lance-ovate, or oval-elliptic, cuneate basally, acuminate or acute apically, 5–15 cm long, margins usually more sharply serrate than in *M. ringens*, sometimes sparsely short, scabrid-pubescent on the edges and on the principal veins; leaves reduced upwardly but not becoming essentially bracteate. Flower stalks 1 cm long or less, much shorter than the subtending leaves. Calyx lobes deltoid below, abruptly narrowed to setaceous tips, intervals between the lobes broad, concave to nearly truncate, bristly short-pubescent on the edges as well as on the calyx lobes and on the upper portions of the angles of the tube. Corolla similar to that of *M. ringens*. Capsule ovate-oblong in outline, 1 cm long, seed about as in *M. ringens* but faintly reticulate, the knobby striae usually not reddish.

Floodplain forests, swamps, creek banks, marshy shores, ditches. Mass. and Conn. to s. Ont., s. Mich. to s. Iowa, e. Neb., south to Fla. Panhandle and e. Tex.

15. Chelone (TURTLEHEADS, SNAKEHEADS)

Perennial herbs. Stems erect, simple or few-branched above. Leaves opposite, relatively large, serrate, acuminate apically. Flowers in short spikelike racemes, their stalks very short and obscured by the subtending, closely imbricate, round-ovate, concave bracts and bractlets. Calyx united only at the base, 5-lobed, lobes broadly ovate, 6–10 mm long. Corolla purple, greenish white, or white tinged with pink or purple; tube abruptly expanded into an inflated throat, little open at the mouth, limb 2-lipped, upper lip broad and arched, keeled in the middle, notched apically, lower lip woolly-bearded at the throat, 3-lobed apically, middle

Fig. 304. **Mimulus alatus:** a, top of plant; b, base of plant; c, flower, opened; d, top of style.

654

lobe smallest. Fertile stamens 4, in 2 pairs of unequal length, not exserted, filaments villous; staminodium 1, usually glabrous. Capsule ovoid, septicidal. Seeds numerous, flat, nearly orbicular, broadly winged.

1. Leaves sessile, rounded basally; flowers 4-ranked in the racemes; staminodium purple.
1. C. cuthbertii

1. Leaves petiolate (some sometimes subsessile in C. glabra), blades tapered basally; flowers not 4-ranked in the racemes; staminodium white or green.
  2. Petioles less than 1.5 cm long; leaf blades widest at or near the middle, long-tapering basally.
    3. Corolla white or cream-white throughout, sometimes with a pink- or purple-tinged summit; staminodium green.
2. C. glabra
    3. Corolla purple throughout; staminodium white.
3. C. obliqua
  2. Petioles 1.5–4 cm long; leaf blades broadest below the middle, bases rounded or shortly and broadly tapered.
4. C. lyonii

1. Chelone cuthbertii Small.

Stem 4–10 dm tall. Leaves sessile, lanceolate, glabrous except along the veins, paler beneath than above, bases rounded or subcordate, 5–12 cm long and 1–5 cm wide. Flowers 4-ranked in the racemes. Bracts ovate, rounded apically, scarious-margined, finely ciliate. Corolla purple, 2–3 cm long. Staminodium purple, ½–¾ the length of the filaments of the fertile stamens.

Bogs, wet meadows, swampy woodlands. s.e. Va., mts. of N.C.

2. Chelone glabra L.

Stem 5–15 dm tall. Leaves short-petioled or subsessile, usually acutely tapered basally (in our range), elliptic-lanceolate, glabrous or pubescent along the margins and veins, to 18 cm long and 5 cm wide. Bracts ovate, rounded to acute apically. Corolla white throughout, greenish yellow, or tinged rose or pink distally, 2.3–3.5 cm long. Staminodium greenish, much shorter than the filaments of the fertile stamens, ⅓–½ the length of the latter. (Incl. C. montana (Raf.) Pennell & Wherry, C. clorantha Pennell & Wherry)

Stream banks, wet ditches, wet woodlands, low pastures, bogs. Nfld. to Minn., southward to Ga. and Ala. (Much more variable to the n. and n.e. of our range.)

3. Chelone obliqua L.

Stem 3–10 dm tall. Leaves tapering below to petioles 0.5–1.5 cm long, blades glabrous, paler beneath than above, lanceolate to elliptic-lanceolate, 6–14 cm long and 2–6 cm broad, long-tapered basally, broadest at about the middle. Bracts ovate, short acuminate or acute apically, ciliate. Corolla purple, 2.5–3.5 cm long, lower lip with pale yellow pubescence. Staminodium white.

Stream banks and in or on the margins of alluvial swamp forests. Chiefly coastal plain and mountains, Md. to Ala., Miss., Ark., in the interior, Mo., Tenn., Ky., Ind. to Minn.

4. Chelone lyonii Pursh.

Stems 4–10 dm tall. Leaves with petioles 1.5–4 cm long, blades glabrous, glaucous beneath, ovate to ovate-lanceolate, 8–14 cm long and 4–8 cm wide, bases rounded or shortly and broadly tapered. Bracts ovate, rounded apically, margins short-ciliate. Calyx lobes ciliate. Corolla purple, about 3 cm long. Staminodium white, sometimes tinged with rose at the tip, nearly as long as the filaments of the fertile stamens.

Open stream banks, rich woodlands, coniferous forests at higher elevations. Chiefly mountains, N.C., S.C., and Tenn. Cultivated in N.Eng. and naturalizing there, perhaps elsewhere.

16. Penstemon (BEARD-TONGUES)

(Ours) perennial herbs. Stems terete, simple or commonly several from a thickish hard caudex. Leaves opposite, basal ones closely set and forming rosettelike clusters, with winged or margined petioles; stem leaves (except sometimes the lowermost) sessile, often clasping, successive pairs at right angles to each other, gradually reduced in size upwardly from midstem. Flowers showy, borne in terminal thyrsoid or panicled racemes. Flower stalks without bractlets. Calyx of 5 sepals, free to the base or essentially so, somewhat overlapping at the base, not infrequently unequal, usually becoming longer after anthesis. Corolla tube much exceeding the calyx, in some cylindrical to somewhat beyond the calyx then dilated distally, in some dilated gradually above the calyx; limb more or less 2-lipped, upper lip 2-lobed, erect, lower 3-lobed, often somewhat longer than the upper, commonly thrust forward; orifice of the corolla open or partly closed by the up-arching of the lower lip. Fertile stamens in 2 unequal pairs, inserted at the base of the corolla tube, filaments more or less arching toward the lower lip; staminodium 1, as long as, often longer than, the fertile stamens, distally usually flattened and somewhat clavate, mostly pubescent on one side. Ovary and capsule ovoid-obconic, style slender, stigma capitate. Capsule dehiscing septicidally and partially loculicidally. Seeds numerous, angled.

1. Inflorescence axes glabrous or finely short-pubescent, not glandular-pubescent unless at the most distal portions.
   2. Stem glabrous; inflorescence axes glabrous except at their most distal portions.   1. *P. tenuis*
   2. Stem and inflorescence axes finely short-pubescent, not stipitate-glandular.   2. *P. laxiflorus*
1. Inflorescence axes stipitate-glandular, sometimes with some nonglandular hairs as well.
   3. Sepals spreading at anthesis, lance- or linear-attenuate, the attenuate tips usually considerably longer than the broader bases.   3. *P. calycosus*
   3. Sepals ovate, acute to acuminate apically, appressed at anthesis, the narrowed tips shorter than to about equaling the broader bases.
     4. Corolla white exteriorly; anther sacs usually with some stiffish trichomes dorsally.   4. *P. digitalis*
     4. Corolla purplish exteriorly; anther sacs glabrous.   5. *P. laevigatus*

1. **Penstemon tenuis** Small.

Stem 4–6 dm tall, essentially glabrous, hairs if present so short as to look powdery. Stem leaves oblong to lance-oblong or ovate, 6–10 cm long, somewhat clasping basally, acute to acuminate apically, margins mostly regularly and sharply but shallowly dentate-serrate. Inflorescence axes glabrous except on the most distal portions. Calyx 4–5 mm long, becoming 7 mm long in fruit, minutely glandular-pubescent, lanceolate-attenuate. Corolla pinkish or purplish, about 1.5 cm long, cylindrical to somewhat beyond the calyx then abruptly dilated and about 8 mm across at the orifice, lips short; base of tube rounded within and with violet lines. Staminodium bearded but not densely so above the middle, its tip reaching the orifice of the corolla; anthers purple, glabrous. Capsule 5 mm long, somewhat surpassed by the calyx.

     Low, poorly drained places, wet depressions, marshes and low prairies, alluvial thickets. w. Miss., La., Ark., s.e. Tex.

2. **Penstemon laxiflorus** Pennell.

Plants to about 7 dm tall. Stem and inflorescence axes finely short-pubescent, not stipitate-glandular. Midstem leaves sessile, somewhat clasping basally, glabrous or sparsely pubescent on the veins beneath, acute apically, triangular-oblong, 6–8 cm long, margins with irregular small dentations. Panicle laxly flowered, axes finely short-pubescent with nonglandular hairs; flower stalks and calyces usually with both stipitate-glandular and nonglandular hairs. Sepals 3–5 mm long, ovate to oval, mostly obtuse apically, margins with a narrow scarious band to the middle or just above. Corollas mostly about 20 mm long, cylindrical to somewhat beyond the calyx then dilated, about 6–8 mm across at the orifice; corolla white to pink or pale purplish exteriorly and stipitate-glandular, glabrous within, 2-ridged or pleated on the lower part of the tube within and finely violet-lined. Staminodium exserted from the orifice of the corolla, about equaling the lower lip, heavily bearded distally from above the middle with clavate, golden trichomes; anthers purple, finely papillate near the base. Capsule 8–9 mm long, somewhat exceeding the calyx. (*P. pauciflorus* Buckl., misapplied; possibly incl. *P. arkansanus* Pennell)

Sometimes in acid, seasonally wet pinelands, more frequently in well-drained, oak-pine or oak-hickory woodlands. cen. Ga. to e. Tex., e. Okla., s. Ark.; a single old record for Fla. Panhandle: 8 mi. w. of River Junction, Jackson Co., fide Pennell, 1935.

3. Penstemon calycosus Small.

Plant 6–12 dm tall. Stem usually glabrous, sometimes shaggy short-pubescent. Midstem leaves clasping, variable in size and shape, lanceolate to ovate, up to 15 cm long and 3 cm wide, apices acute to acuminate, surfaces glabrous or rarely minutely pubescent on the mid-veins finely dentate-serrate to subentire. Panicle loosely to closely.flowered, inflorescence axes, flower stalks, bracts, sepals, and outside of corollas relatively sparsely stipitate-glandular. Sepals about 8 mm long, spreading at anthesis, lance- or linear-attenuate, the attenuate portions usually longer than the wider basal parts. Corolla purplish violet externally, 2–3 cm long, usually cylindrical to somewhat beyond the calyx then abruptly dilated distally, sometimes dilated gradually above the calyx, 8–10 mm across at the orifice. Staminodium reaching the orifice of the corolla tube to as long as the lower lip, above the middle shaggy pubescent with yellowish, flat trichomes; anthers purple, appressed-pubescent on the surfaces (as observed with suitable magnification).

Most alluvial soils along streams, low thickets, meadows, wooded slopes, glades, moist roadsides and fields. Ohio to s. Mich., Ill., Mo., southward to Tenn., n. Ala.; from Ind. and Ohio said to have spread eastward, chiefly in fields, to Pa. and N.Y.

4. Penstemon digitalis Nutt.

In general features, stature, and size of corolla, much resembling *P. calycosus*. Stems commonly very finely short-pubescent. Leaves glabrous on both surfaces, margins entire to serrate or dentate-serrate with low inconspicuous teeth. Stipitate glandular-pubescent on the inflorescence axes, flower stalks, sepals, and outer corolla surfaces. Sepals ovate, 5–8 mm long, apices acute to acuminate, the narrowed distal portions shorter than to about as long as the broader basal portions. Corolla white without, with pink or purplish lines within the throat. Staminodium exserted from the corolla tube, about as long as the lower lip, distally relatively sparsely pubescent with yellowish, long, flat trichomes; anthers purple, usually (not always) with some straight trichomes from the dorsal surfaces of the anther sacs. Capsule 8–12 mm long.

In low places with poor drainage, open poorly-drained to well-drained woodlands and fields, meadows, marshy shores, stream bottoms. Maine to Mich., Minn., and S.D., southward to w. N.C., n. Ala., Okla., n.e. and e. Tex.

5. Penstemon laevigatus Solander ex Ait.

Similar to *P. digitalis*, generally somewhat smaller in stature, corollas smaller, 15–20 mm long, purplish exteriorly, purple-lined within. Anthers brown, glabrous. Capsule 4–8 mm long. (*P. penstemon* (L.) MacM.)

Low meadows, open bottomland woodlands, river and stream banks, rich woodlands, prairies, ditches. s. N.J., Pa., W.Va., southward to Fla. Panhandle and Miss.

17. Veronicastrum

Veronicastrum virginicum (L.) Farw. CULVER'S-PHYSIC, BOWMAN'S-ROOT.

Perennial herb. Stem 8–20 dm tall, usually unbranched to the inflorescence, glabrous, minutely pubescent, or rarely copiously clothed with long soft hairs. Leaves in numerous whorls of 3–7, the lower usually with short margined petioles, sometimes sessile, upper ones sessile; blades lanceolate to lance-ovate, 4–15 cm long, 1–3 cm broad, sharply serrate, acuminate apically, glabrous to pubescent with long soft hairs beneath. Flowers small, numerous, in terminal, panicled, narrowly spirelike racemes, the longer racemes to 15 cm long. Flower stalks 1 mm long or less, subtended by subulate-attenuate bracts about as long as the

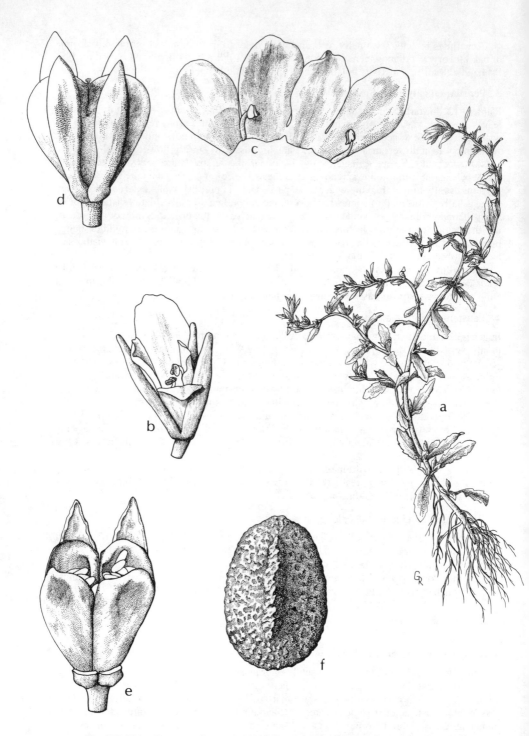

Fig. 305. **Veronica peregrina:** a, habit; b, flower; c, corolla, spread out; d, capsule, with persistent calyx; e, capsule, dehiscing, two sepals removed; f, seed.

calyx, bractlets none. Calyx about 2 mm long, united only basally, lobes 5, unequal, ovate or lance-ovate, acuminate apically. Corolla white, cream, pinkish white, or purplish, the narrowly funnelform tube about 5 mm long, much exserted from the calyx, pubescent within, with 5 short-deltoid, nearly equal, lobes. Stamens 2, long-exserted from the corolla as is the style. Capsule ovoid, narrowed apically, 3 mm long, opening by 4 terminal slits, style persisting until about time of dehiscence. Seeds numerous, minute, obscurely reticulate.

Bogs, wet meadows, marshy or boggy openings, stream banks, moist to wet clearings, prairies, moist open woodlands. Mass. and Vt. to Man., generally southward to w. S.C., westernmost Fla. Panhandle, and n.e. Tex.

18. Veronica (SPEEDWELLS)

Annual or perennial herbs. Foliage leaves opposite, bracteal leaves or bracts, except possibly the lowermost, alternate. Inflorescences racemose in the axils of opposite foliage leaves, or flowers solitary in the axils of alternate bracteal leaves or bracts terminating the main stem or its branches. Bractlets none. Calyx united only basally, 4-lobed, the lobes equal or the upper ones slightly longer. Corolla blue, violet or white, sometimes pinkish, the tube much shorter than the 4 spreading lobes; limb weakly 2-lipped, upper lip a single lobe, the largest, lower lip with 2 lateral lobes somewhat smaller, and a central much smaller one. Stamens 2, one arising at either side of the base of the upper lip. Ovary and capsule somewhat flattened in a plane contrary to the septum. Capsule loculicidal or both loculicidal and septicidal, style persistent at least until about the time of dehiscence. Seeds few to many, flat on one side, at least slightly convex on the other.

1. Flowers solitary in the axils of bracteal leaves or bracts, the inflorescences terminating the main stem and its branches.
    2. Style very short, thickish; plant annual, sometimes the lower stems short-decumbent and rooting at the nodes but not mat-forming.   1. *V. peregrina*
    2. Style slender, about as long as the ovary; plant perennial, lower stems (potentially) creeping, rooting at the nodes and mat-forming.   2. *V. serpyllifolia*
1. Flowers in racemes axillary to foliage leaves.
    3. Leaves all sessile and clasping or subclasping.   3. *V. anagallis-aquatica*
    3. Leaves distinctly short-petiolate.   4. *V. americana*

1. Veronica peregrina L. NECKWEED, PURSLANE-SPEEDWELL.   Fig. 305

Annual. Stem simple or much branched, if branched the lower stems sometimes shortly decumbent and rooting at the nodes but not mat-forming, branches laxly ascending, 1–3 (–4) dm tall, glabrous or glandular-pubescent on the inflorescence axes. Leaves below the inflorescence opposite, lowermost narrowed to petiolar bases, oblanceolate or spatulate, others sessile, lanceolate, oblanceolate, or linear-lanceolate, gradually reduced into the inflorescence, obtuse apically, few-toothed or entire marginally. Flowers solitary in the axils of mostly alternate leaves or bracts in spikelike racemes terminal on the main stem or branches. Flower stalks 1–2 mm long. Calyx lobes oblanceolate to linear-oblong, 3–4 mm long. Corolla white, 2–2.5 mm across the limb. Style very short and thickish. Capsule 2–4 mm long and broad, glabrous or occasionally glandular-pubescent, shallowly emarginate apically. Seeds numerous, about 0.6 mm long, ovate or short-oblong, only slightly convex on one side, golden brown, dull.

Weedy in moist to wet open places, often in shallow water of ephemeral pools, borrow pits, or ditches; low places in fields, meadows, about ponds, in mud along rivers and streams. Native of e. N.Am., now widely distributed throughout much of temp. N.Am. and in Eur.

2. Veronica serpyllifolia L. THYME-LEAVED SPEEDWELL.

Pubescent perennial with creeping stems rooting at the nodes and (potentially) mat-forming, flowering branches ascending, 5–20 cm high, short-pubescent. Leaves opposite below the inflorescences, some lower ones narrowed to short winged petiolar bases, mostly sessile, few and widely spaced, ovate to oblong, 1–1.5 (–2) cm long, broadly rounded apically, glabrous or nearly so, margins entire or with few fine teeth. Flowers in terminal, spikelike, lax, often arching racemes at first compact but elongating to about 10 cm, the bracts alternate. Flower stalks slender, 2–5 mm long, usually pubescent. Calyx lobes elliptic, rounded to obtuse apically, about 3 mm long. Corolla whitish or pale blue, 4–8 mm across the limb. Style slender, as long as or longer than the ovary or fruit. Capsule pubescent at least near the summit, in length about equaling the calyx, broader than long, notched apically. Seeds numerous, obovate to oblong, 0.5 mm long, only slightly convex on one side, yellowish brown, dull.

Wet meadows, seepage places about ponds or lakes, moist lawns, fields, roadsides, open woodlands. Native of Eur., naturalized from Nfld. to Minn., southward to n. Ga. (in piedmont and mts.) and Mo.; s. B.C. to n. Calif.

3. Veronica anagallis-aquatica L. WATER SPEEDWELL, BROOK-PIMPERNELL.

Erect, or decumbent below, sometimes sprawling, subsucculent, perennial, the stems often stout, rooting at the lower nodes, 2–10 dm tall, glabrous or glandular-pubescent above (in ours usually the latter). Flowers in axillary stalked racemes 5–15 cm long, flower stalks glabrous or (in our range) glandular-pubescent. Bracts linear-lanceolate, mostly a little shorter than the 4–8 mm long flower stalks. Calyx lobes elliptic-lanceolate to lance-ovate, acute apically, 4–5 mm long. Corolla bluish or lavender, the lobes violet-lined, 5–6 mm across the limb. Style slender, 3–4 mm long. Capsule nearly orbicular in outline, equaling or a little shorter than the calyx, obscurely notched apically, glabrous or (usually in ours) sparsely glandular-pubescent. Seeds numerous, minute, nearly orbicular, tan, dull. (Incl. *V. glandulifera* Pennell)

Commonly in shallow water, in or on the edges of streams and their banks, swamps, wet meadows. Native of Euras., naturalized from Maine to Wash., generally southward to w. N.C., n. Ala., cen. and n.cen. Tex., Calif.

Attributed to our range only in Tenn. and Ark. (much more widely distributed to the n., n.e., and n.w. of our range), a closely similar plant is recognized as specifically distinct by authors to which they variously apply one or the other of the following names: *V. catenata* Pennell, *V. comosa* Richt., *V. connata* Pennell, *V. salina* Schur. We fail to be able clearly to distinguish from *V. anagallis-aquatica* specimens at hand from various parts of the overall range and bearing one or the other of the above names.

4. Veronica americana (Raf.) Schwein. ex Benth.

Similar in general aspect to *V. anagallis-aquatica*. Glabrous throughout. Leaves short-petiolate, blades mostly lanceolate or lance-ovate, the lower sometimes nearly orbicular, truncate, rounded, or short-tapered basally, obtuse to acute apically, 1–9 cm long, 1–1.5 cm wide, margins serrate or entire. Corolla evenly light blue. Racemes lax, the flower stalks filiform, much longer than the linear subtending bracts.

Usually in shallow water, springs, spring runs, stream banks, marshy shores, swamps. Nfld. to Alaska, southward to w. N.C., e. Tenn., n.cen. Mo., cen. Okla., Edwards plateau of Tex., N.M., Ariz., Calif.; n.e. Asia.

19. Macranthera

Macranthera flammea (Bartr.) Pennell. Fig. 306

Large, coarse, stiff and brittle annual or biennial herb, mostly 15–30 dm tall, drying nearly black. Stem obtusely 4-angled, retrorsely pubescent above, glabrous below, essentially unbranched to the widely spreading inflorescence. Leaves opposite, sessile or with short winged petioles; blades glabrous or the margins sometimes minutely ciliate, 8–10 cm long, lanceolate to ovate, pinnately lobed, progressively reduced upwardly and the uppermost shallowly lobed, toothed, or entire; lower leaves usually shed by anthesis. Flowers borne on

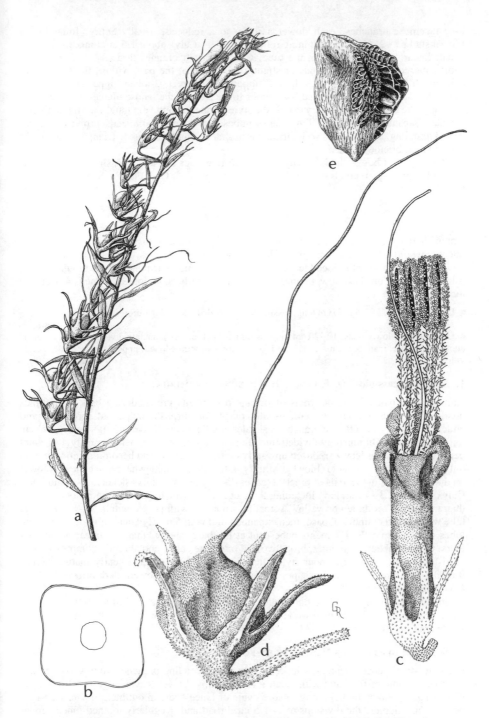

Fig. 306. **Macranthera flammea:** a, branch of inflorescence, fruiting below, flowering
above; b, cross-section of upper stem, diagrammatic; c, flower; d, capsule, e, seed.

long racemose branches, each flower axillary to a reduced, usually entire, foliar bract. Flower stalks 1–2 cm long, becoming reflexed in fruit. Calyx about 1.5 cm long, the 5 lobes equal, linear, much longer than the tube. Corolla bright orange, fleshy, 2–2.5 cm long, short-pubescent without, glabrous or slightly pubescent at the base within, the tube cylindric, much longer than the lobes; limb 2-lipped, lower erect, 2-lobed, upper spreading, 3-lobed. Stamens 4, equal, exserted contiguous to each other, the lanose filaments about twice as long as the corolla tube. Ovary ovoid, the style slender, about 3 cm long, the stigma linear-clavate. Capsule 1.5–1.8 cm long, short-pubescent, loculicidal. Seeds brownish black, 2.5–3 mm long, about half as wide, irregular in shape, angular-lunate to triangular, with 2 or 3 fluted membranous wings.

Bogs and wet boggy thickets, edges of shrub-tree bogs or bays, occasionally in shallow water of cypress-gum ponds or depressions. Coastal plain, s. Ga., Fla. Panhandle, westward to e. La.

20. Seymeria

Annual herbs. Leaves opposite, the lower, at least, pinnatifid or dissected, progressively reduced upwardly. Flowers, those of the lower portions of branches, borne in the leaf axils, upwardly in the axils of reduced leaves or foliar bracts. Bractlets none. Calyx tubular below, 5-lobed above. Corolla yellow, tubular below, with 5 nearly equal lobes. Stamens 4. Capsule loculicidal.

- Leaves all small, 1–1.5 (–2) cm long, sessile, finely divided into linear-filiform segments.

1. *S. cassioides*

- Leaves, larger lower ones, 10–20 cm long, wing-petioled, blades pinnately divided from the base to about the middle, main segments 1–1.5 cm broad, strongly toothed-lobed, upper leaves sessile, not divided.

2. *S. macrophylla*

1. Seymeria cassioides (J. F. Gmel.) Blake. SENNA SEYMERIA.

Annual herb, parasitic on the roots of pines, fresh plants green, drying dark, sometimes nearly black. Stems 5–10 dm tall, usually purple or purplish, pubescent, profusely and widely branched, the ultimate branchlets slender and flexuous. Leaves sessile 1–1.5 (–2) cm long, pinnately or bipinnately divided into short linear-filiform segments, mostly 4–7 pairs per leaf on the larger leaves; reduced upwardly on the branchlets and becoming entire or with only 1 or 2 pairs of segments. Flowers solitary in the axils of somewhat reduced leaves lower on the branchlets, in the axils of bracteal leaves distally. Flower stalks slender, 3–8 mm long. Calyx glabrous, 3–4 mm long including the 5 equal or unequal subulate lobes, lobes longer than the tube. Corolla lemon yellow, sometimes marked with purple within, 6–8 mm long, tube with a short cylindrical base, then expanded and with 5 nearly equal, oblong, spreading lobes rounded apically. Filaments pubescent at the base, shorter than the linear-oblong anthers. Capsule about 4 mm long, broadly ovoid or oblate below and abruptly compressed to a flattish summit, the long slender style often persistent until capsule nearly mature. Seeds 0.6–0.7 mm long, irregular in shape, somewhat ridged and furrowed, dark brown. (*Afzelia cassioides* J. F. Gmel.)

In both seasonally wet pine flatwoods, savannas, and swales, and in well-drained pine-oak ridges, hills, and open pine woodlands. Chiefly coastal plain, s.e. Va. to cen. pen. Fla., westward to e. Tex., inland, n. Ala., s.e. Tenn.; Bah.Is.

2. Seymeria macrophylla Nutt. MULLEN-FOXGLOVE.

Erect, relatively coarse annual, with relatively few ascending branches. Stems to 2 m tall, pubescent, some of the hairs with small glandular tips. Lower leaves large, 1–2 dm long, with winged petioles 1–3 cm long, blades ovate or lance-ovate in outline, pinnately divided below their middles, the divisions to 1–1.5 cm broad and irregularly toothed-lobed, upper portions of blades toothed; upwardly leaves becoming smaller, less divided, finally undivided, sessile, and lanceolate; petioles rough-pubescent, upper surfaces of blades sparsely short-pubescent, lower short-hispid on the veins. Flowers solitary in the axils of foliar bracts terminating the main stem and branches. Flower stalks stoutish, 2–3 mm long. Calyx 8–10 mm long including the lobes, pubescent, tube campanulate, lobes unequal, deltoid to oblong, the longer ones about as long as the tube. Corolla yellow, the tube campanulate, pubescent within, surmounted by 5 short, broad, rounded lobes. Filaments pubescent except at their apices, about as long as the oblong anthers. Capsule globose-ovoid, about 1 cm long, the hooked style often persistent. Seeds more or less falcate-fusiform, about 2 mm long, coarsely alveolate-reticulate, the alveolae nearly isodiametric, in lines. (*Dasistoma macrophylla* (Nutt.) Raf.)

River bottom woodlands, stream banks, moist woodlands. W.Va. to s. Wisc., Iowa, Neb., southward to n. S.C., n. Ga., cen. Ala., Miss., La. and n.cen. Tex.

21. Agalinis (GERARDIAS)

Mostly annual herbs (one of ours perennial), with erect, brittle, often much branched stems. Leaves and bracts all opposite, or opposite below and alternate on the flowering stems, infrequently all but the lowermost alternate or subopposite; sessile, entire, narrow, in some scalelike. Flowers with attractive pink to purple (infrequently white on an individual plant) corollas, borne singly in the axils of leaves or reduced bracteal leaves, in some forming racemes terminating the branchlets, sometimes solitary and appearing terminal on the branchlets. Calyx tube campanulate, hemispheric, or turbinate, lobes 5, nearly equal, in most shorter than the tube, in some much shorter or even reduced essentially to apiculations. Corolla pink to purple, the tube dilated above the calyx, often distended on the lower side, more or less 2-lipped, the 5 lobes nearly equal in most, rounded to truncate apically and commonly ciliate-fringed marginally; abaxial lip with 3 spreading, arched, or somewhat recurved lobes, adaxial lip with 2 reflexed or projecting lobes; corollas mostly with 2 yellow lines and purple spots in the throat, usually woolly-pubescent within at the base of the adaxial lobes. Stamens 4, in 2 pairs of unequal length, not exserted, filaments more or less woolly-pubescent at least toward the base, anthers woolly-pubescent. Capsule mostly globose or subglobose, in some obovate, dehiscing loculicidally. Seeds reticulate.

A sensible degree of understanding of what the taxa in the genus *Agalinis* are awaits, in our judgment, comprehensive study by someone possessed of a penchant for unsnarling a taxonomic-nomenclatural tangle and not afflicted with "splititis."

1. The flower stalks (or most of them on a given plant) twice as long as the calyx to much more than twice as long.
  2. Plant perennial, with slender rhizomes.   1. *A. linifolia*
  2. Plant annual.
    3. Leaves scalelike, 1 mm long or a little more.   2. *A. filicaulis*
    3. Leaves, the larger ones, 5 mm long or more.
      4. Main stems appearing conspicuously leafy, the primary leaves subtending "axillary fascicles" comprised of short axillary branches having short internodes bearing several leaves. 3. *A. filifolia*
      4. Main stems not conspicuously leafy, no "axillary fascicles" in the leaf axils.
        5. Flower stalks 15 mm long or less.
          6. Plant pale or yellowish green, not discoloring in drying; corolla tube glabrous exteriorly, corolla not discoloring in drying.   4. *A. obtusifolia*
          6. Plant dark green, drying blackish; corolla tube pubescent exteriorly, corolla blackening in drying.   5. *A. pseudaphylla*
        5. Flower stalks mostly 20–30 mm long.   7. *A. divaricata*
1. The flower stalks on a given plant mostly shorter than, equaling, or not exceeding 1.5 times as long as the calyx.
  7. Leaves scalelike, appressed to the stem, 2 mm long or less.   6. *A. aphylla*
  7. Leaves, the larger ones, 20 mm long or more, spreading.
    8. Plants restricted to saline habitats.   8. *A. maritima*
    8. Plants not occurring in saline habitats.
      9. Calyx lobes elongate triangular-subulate, 3–4 mm long, nearly as long as the calyx tube.
                                   9. *A. heterophylla*

9. Calyx lobes not elongate triangular-subulate, much shorter than the calyx tube.

    10. Stems with copious short-scabrid pubescence; primary leaves with well developed "axillary fascicles" comprised of short axillary branches with short internodes bearing several leaves, the stems thus appearing copiously leafy.   10. *A. fasciculata*

    10. Stems glabrous or with sparse and irregular short-scabrid pubescence; primary leaves without conspicuous "axillary fascicles" or these if present poorly developed and the stems not appearing conspicuously leafy.

        11. Sides of leaves flat or the margins somewhat revolute; calyx lobes with broad and low bases then abruptly narrowed into short points.   11. *A. purpurea*

        11. Sides of the leaves strongly involute, abruptly turned upward or folded over (upwardly) and the edges nearly meeting the midrib; calyx lobes triangular, mostly narrowed gradually from the bases to acute tips.   12. *A. pinetorum*

1. Agalinis linifolia (Nutt.) Britt.

Perennial with slender rhizomes, herbage and flowers tending to darken or blacken in drying. Stems glabrous, simple or with few erect or strongly ascending branches. Leaves and bracteal leaves opposite, linear, 3–5 cm long and to 3 mm wide, their surfaces irregularly pale-scaly, especially near the margins. Flowers relatively distantly disposed in about 8–20-flowered racemes, 1 or 2 flowers per node. Flower stalks glabrous, shorter than to longer than the subtending bracts. Calyx tube usually turbinate at anthesis, hemispheric on the fruit, the lobes little more than apicules. Corolla light violet-purple to pink, with paler dots internally, softly pubescent exteriorly. Capsule globose, 6–8 mm long, its apex extending beyond the calyx. (*Gerardia linifolia* Nutt.)

In water or waterlogged soils, cypress-gum ponds and depressions, depressions in pinelands, acid marshy shores, marshes. Coastal plain, Del.; s.e. N.C. to s. Fla., westward to s. La.

2. Agalinis filicaulis (Benth.) Pennell.

Annual with finely wiry, spindly, glaucous stems, irregularly, relatively few-branched, branches often one to a node, short and long on the same plant, divaricate to ascending; plant 1–5 dm tall; herbage and flowers not darkening in drying. Leaves opposite, whitish scale-like, triangular-subulate, 1 mm long or a little more, appressed, barely perceptible to the unaided eye. Flowers relatively few, mostly borne singly or in a pair from the highest node of a branchlet. Flower stalks filiform, 5–10 mm long, subtending bracts minute. Calyx 2 mm long, tube campanulate, lobes short-deltoid, broad shallow sinuses between them. Corolla lavender-pink, without lines or spots within, glabrous exteriorly, 10–13 mm long, abaxial lip projecting. Capsule globose, nearly half exserted from the calyx. (*Gerardia filicaulis* Benth.)

Seasonally wet pine-wiregrass savannas and flatwoods, bogs. Coastal plain, s. Ga., n. Fla., w. Ala. to s.w. La.

3. Agalinis filifolia (Nutt.) Raf.

Annual, commonly much branched, branches strongly ascending-erect, 3–8 dm tall; stem glabrous, appearing very leafy owing to presence of "axillary fascicles." Herbage at flowering time commonly strongly reddish purple, sometimes darkening in drying. Lowermost leaves opposite (mostly shed by flowering time), others and bracteal leaves alternate, linear-filiform, 1–2 cm long. Flowers 2–6, sometimes more, distally on the branchlets, rather irregularly racemosely disposed. Flower stalks slender but not filiform, 15–30 mm long, mostly about 20 mm, much longer than the subtending bracts. Calyx tube 4 mm long, campanulate at anthesis, subglobose on the fruit, lobes essentially apiculate, the sinuses between them broad and shallow. Corolla usually deep rose-purple, 25–35 mm long, pubescent exteriorly except near the base, lines and spots present within. Capsule subglobose, about equaling the calyx. (*Gerardia filifolia* Nutt.)

Generally on deep well-drained sands of pine–scrub oak ridges and stabilized dunes; also in seasonally wet interdune swales and flatwoods depressions. Coastal plain, s.e. Ga., throughout most of Fla., s.w. Ala.

4. Agalinis obtusifolia Raf.

Slender annual, simple or frequently with relatively few stiffly ascending branches, infrequently diffusely branched. Herbage pale green or yellowish green, it and the flowers not tending to darken in drying; stems squarish. Leaves opposite or occasionally the uppermost bracteal ones alternate, narrowly linear or somewhat broadened distally and linear-spatulate, 1–1.5 cm long, upper surfaces short scabrid-pubescent and sometimes purple-dotted, margins often drying either involute or revolute. Flowers 6–14 in racemes on the branchlets, or in few-flowered, more or less panicled racemes. Flower stalks mostly 5–15 mm long, some or all of them on a given plant considerably longer than the calyces. Calyx campanulate both at anthesis and on the fruits, about 2 mm long, tube somewhat reticulate-veined, lobes very low and broadly triangular, if any, but the principal veins of the tube thickened just back of the orifice and jutting out and forward as thickish apicules of the lobe, cartilagelike in texture. Corolla 1–1.5 cm long, lavender-pink, glabrous exteriorly, lobes spreading, spots and lines present or absent within. Capsule oblong to subglobose, about 3 mm long, very little exceeding the calyx. (*Gerardia obtusifolia* (Raf.) Pennell, *A. erecta* (Walt.) Pennell)

Seasonally wet pine savannas and flatwoods, hillside bogs in pinelands, in shallow soil on oolitic limestone in pinelands. Chiefly coastal plain, Del. to the Fla. Keys, westward to s.e. La.

5. Agalinis pseudaphylla (Pennell) Shinners.

Annual with slender, wiry, cartilagenous-ribbed stems to about 6 dm tall; stems glaucous or sometimes not glaucous but scabrid-pubescent, usually with several ascending-erect branches. Leaves opposite or subopposite, some of the bracteal leaves alternate, some opposite; larger leaves linear-subulate, to about 1 cm long, about 1 mm wide, spreading, notably scabrid-pubescent, more densely so on the upper than on the lower surfaces. Flowers relatively few and distant from each other on the raceme, commonly some of them aborting before developing very much. Flower stalks variable in length to about 15 cm long, some of them, at least, longer than the calyces. Calyx 2–3 mm long, campanulate at anthesis, hemispheric on the fruits, tube sometimes evidently netted-veined, sometimes not, lobes broad and low with apiculate tips, or little more than stout apicules. Corolla pink with darker pink dots and with or without yellow lines within, about 1.6 cm long, copiously pubescent exteriorly. Capsule globose, 5–6 mm long and wide, scarcely exceeding the calyx to about ⅓ longer than the calyx. (*Gerardia pseudophylla* (Pennell) Pennell, *A. oligophylla* Pennell var. *pseudophylla* Pennell)

Seasonally wet pine savannas and flatwoods, bogs, prairies. cen. to s. Ala., s. Miss., La.

6. Agalinis aphylla (Nutt.) Raf.

Similar in general aspect to *A. pseudaphylla*, stems similarly ribbed, apparently not glaucous, with some tendency to darken in drying. Usually only the lowest leaves opposite, others subopposite or mostly alternate, scalelike, larger ones to 2 mm long, mostly shorter, thickish, closely appressed to the stem, sparsely if at all pubescent. Flowers more numerous than in *A. pseudaphylla*, usually none of them aborting, relatively distant from each other, almost all alternately disposed on slender racemes 5–20 cm long. Flower stalks 1–2 mm long, none of them longer than the calyces. Calyx 1–2 mm long, campanulate at anthesis, hemispheric on the fruits, tube evidently veiny at anthesis but more conspicuously reticulate-veined on the fruits, lobes little more than stout apiculations. Corolla pink, usually with darker dots and yellow lines within, 1–1.5 cm long, glabrous exteriorly. Capsule globose, about 4 mm long, half exserted from the calyx. (*Gerardia aphylla* Nutt.)

Wet pine savannas and flatwoods, depressions in pinelands, bogs, edges of cypress-gum ponds and depressions. Coastal plain, s.e. N.C. to n. Fla., westward to s. La.

7. Agalinis divaricata (Chapm.) Pennell.

Annual with slender much-branched, glabrous or sparsely and minutely scabrous stems, 3–8 dm tall, sometimes darkening in drying, sometimes not at all. Leaves opposite, larger ones

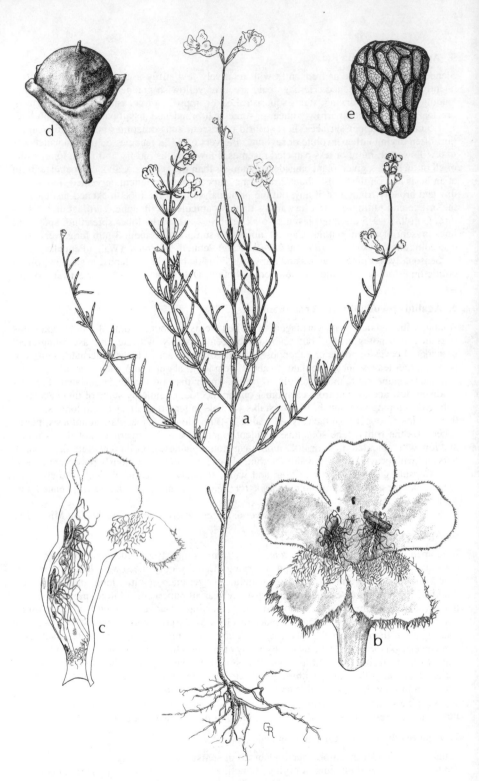

Fig. 307. **Agalinis maritima:** a, habit; b, flower; c, longitudinal section of flower; d, capsule; e, seed.

1.5–2.5 cm long, narrowly linear, mostly spreading at right angles from the stem. Flowers mostly in 6–12-flowered racemes, sometimes racemes somewhat panicled. Flower stalks filiform, 1–3 cm long, spreading at 45–80-degree angles from the axes, mostly much longer than the subtending bracts. Calyx 1.5–2 mm long, tube campanulate and faintly veined at anthesis, broadly turbinate to hemispheric and ribbed on the fruit, lobes sometimes short-deltoid, more frequently little more than apiculations. Corolla pink, with darker pink dots and sometimes yellow lines within, 1–1.5 cm long, lobes spreading, glabrous exteriorly. Capsule globose, 2.5–3 mm long, somewhat exceeding the calyx. (*Gerardia divaricata* Chapm.)

Generally an inhabitant of pine–scrub oak ridges with deep, well-drained sand and of open, upland, mixed woodlands on heavier soils; included here chiefly because it temporarily colonizes, in some abundance, mechanically much-disturbed, clear-cut sites formerly pine savannas or flatwoods; also in blackland prairies. Coastal plain, S.C. (fide Pennell, 1935 as *A. laxa*), to cen. pen. Fla., westward to Miss.

8. Agalinis maritima (Raf.) Raf.   Fig. 307

Subsucculent, glabrous annual, usually branched from the base upwardly, branches ascending, 5–40 cm tall. Leaves opposite (abruptly reduced on the inflorescence branches and some of them sometimes alternate there), those on the main stems linear or widening a little distally and very narrowly linear-oblanceolate, "axillary fascicles" weakly to moderately developed. Plants darkening in drying. Flowers in racemes terminating the branches, the raceme axes mostly without flowers in their lower halves, flowers or pairs of flowers relatively distant from each other, mostly 4–10 on each raceme. Flower stalks stoutish, 2–8 mm long, mostly not more than 1.5 times as long as the calyces, occasional ones on a given plant somewhat longer; bracts much shorter than to about equaling the flower stalks. Calyx 3–4 mm long, tube campanulate at anthesis, hemispheric on the fruits, lobes deltoid, obtuse apically. Corolla lavender-pink to purple, 1–1.5 cm long, finely punctate-dotted exteriorly. Capsule subglobose, 5–6 mm long, half exserted from the calyx. (*Gerardia maritima* Raf.; incl. *A. maritima* var. *grandiflora* (Benth.) Pennell)

Coastal salt and brackish marshes, salt flats, mangrove swamps. N.S. to the Fla. Keys, Gulf coast to Tex., Mex.; W.I.

9. Agalinis heterophylla (Nutt.) Small.

Annual with numerous strongly ascending branches, these essentially leafy throughout their flowering portions; stems glabrous or puberulent, 4–6 (–10) dm tall. Plants darkening in drying. Leaves opposite below, the lowermost usually 3-cleft but usually shed by flowering time; some leaves on the flowering branches often alternate; midstem leaves stiffish, rigid, ascending, nearly linear to linear-elliptic, 1.5–3 (–4.5) cm long; leaves on the upper portions of the branches commonly minutely scabrid on the margins and punctate or with pale surface scales on their upper surfaces. Flowers in the axils of only moderately reduced leaves of the branches. Flower stalks stoutish, mostly 1–2 mm long, much shorter than the calyces. Calyx tube 3.5–5 mm long, the principal veins extending on the lobes as midveins; lobes elongate-triangular-subulate, nearly as long as the tube, sometimes becoming a little longer than the tube, spreading by the time the capsules mature. Corolla deep pink to whitish and lavender-tinged, 2.5–3 cm long, pink-dotted and yellow-lined within, softly pubescent exteriorly. Capsule oval to subglobose, 6–8 mm long, its apex exserted from the tube of the calyx. (*Gerardia heterophylla* Nutt.)

Prairies, river banks, fallow fields, roadsides, ephemeral pools, wet gravelly soil on edges of ponds and lakes, open, usually moist woodlands. s. Mo., Ark., e. Okla., e. Tex., La., Miss.

10. Agalinis fasciculata (Ell.) Raf.

Relatively coarse annual, usually with a single main stem and numerous long ascending-erect branches, 6–12 dm tall, with well-developed "axillary fascicles" giving the stems a copiously leafy appearance. Plants tend to darken moderately in drying. Stems copiously short scabrid-pubescent. Leaves opposite, linear, 1.5–4 cm long, upper surfaces copiously scabrid, lower at least on the midribs. Flowers in racemes terminating the branches, mostly 12–30-flowered. Flower stalks stoutish, 1–4 mm long, in length about equaling the length of the calyx; subtending bracts as long as or longer than the flower stalks. Calyx 3–5 mm long, tube campanulate at anthesis, hemispheric on the fruit, lobes triangular-subulate, very finely scabrid on the margins and on the rim of the tube as well. Corolla purple, 2.5–3.5 cm long, short-pubescent exteriorly, darker spots and yellow lines within. Capsule 5–6 mm long, globose or subglobose, equaling the calyx to about half exserted from it. (*Gerardia fasciculata* Ell.)

Commonly somewhat weedy in semiopen to open moist to wet places: old fields, swales, interdune hollows, semimarshy places by ponds and lakes, pine savannas and flatwoods (especially where much disturbed), open woodlands. Chiefly coastal plain, s. Md. to s. pen. Fla., westward to e., s.e., and n.cen. Tex., northward in the interior to e. Neb., Ark., s.w. Mo.

11. Agalinis purpurea (L.) Pennell.

Annual, with much variation in branching pattern; sometimes diffusely and widely divaricately branched, varying to much branched, the branches arching-ascending or strongly ascending. Stems 3–12 dm tall, with short, broad-based trichomes more or less irregularly scattered on the narrow wing-angles, sometimes glabrous. "Axillary fascicles" weakly developed on some plants, completely absent on others. Plants usually only moderately darkening in drying. Leaves opposite, linear to linear-subulate, 1.5–4 cm long, about 0.5–1 mm broad, surfaces moderately short-scabrid, tips of the broad-based trichomes often shed leaving the surfaces pebbly; sides of leaves flat or edges somewhat revolute. Flowers disposed in longish racemes terminating ascending branches on some plants, on others a few flowers only on short, spreading branchlets, in general much variation in relation to overall branching pattern. Flower stalks stoutish, 2–6 mm long, few if any on a given plant more than 1.5 times as long as the calyces. Calyx 3–4 mm long, tube campanulate at anthesis, hemispheric on the fruits, lobes little more than apiculations shouldered basally by slightly elevated portions of the tube. Corolla pink to purple, (in our range) 2–3.8 cm long, pubescent exteriorly, darker spots and yellow lines within. Capsule globose, 5.6–6 mm long, surpassing the calyx tube by about ⅓. (*Gerardia purpurea* L.; incl. *A. virgata* Raf., *G. racemulosa* Pennell)

Moist to wet clearings, pine savannas and flatwoods, moist prairies, bogs, marshy shores, seepage areas of open woodlands, meadows, interdune swales, shallow soil on lime-rock in pinelands. N.S. to Minn., S.D., generally southward to s. Fla. and e. Tex.

12. Agalinis pinetorum Pennell.

Slender annual, with few strongly ascending branches, 4–7 dm tall. Stem only slightly angled, glabrous or with few and usually irregularly scattered, broad-based and very low trichomes. "Axillary fascicles" none or very weakly developed. Leaves opposite, narrowly linear, 1.5–2.5 cm long, spreading, upper surfaces and edges short scabrid-pubescent, lower surfaces smooth or nearly so; sides of leaves usually folded upwardly and over so that their edges nearly meet at the position of the midrib, the scabrosity of the edges thus showing down the middle on the upper side. Racemes few-flowered, flowers 2 at a node, sometimes 1, flowering nodes relatively distant from each other. Flower stalks stoutish, 1–2 mm long, shorter than the calyces; bracts, except the uppermost, equaling the calyces or longer. Calyx 3–4 mm long, tube campanulate at anthesis, lobes triangular, mostly tapered gradually from their bases to acute apices, sinuses between them V- or narrowly U-shaped. Corolla pink to purple, 1–2 cm long, not very broad at the throat, short-pubescent exteriorly. Capsule not seen by us. (Not *Gerardia pinetorum* Britt. & Wilson; *G. pulchella* Pennell *but not A. pulchella* Pennell)

Pine savannas and flatwoods, bogs. s. Ga., n. Fla., s.w. Ala., westward to La.

22. Buchnera (BLUEHEARTS)

Slender, erect, usually simple, perennial herbs with a hard caudex. Fresh plants dull bluish green, dried plants, including flowers, blue-black. Principal leaves opposite, sessile or very short-petioled, 3–6 pairs on the lower half of the stem, 1–3 pairs of reduced leaves above, one or more, usually alternate, distant bracts below the inflorescence. Flowers in a single terminal spike or on several spicate branches, each flower subtended by a small bract and 2 bractlets. Calyx tube cylindric, surmounted by 5 short lobes, the upper pair of lobes slightly longer than the lower 3. Corolla bluish purple (white on occasional plants), slenderly cylindric-tubular below, with 5 spreading lobes, the limb slightly bilateral. Stamens 4, in 2 pairs of slightly unequal length, not exserted from the corolla. Ovary ovoid, style short. Capsule included within or slightly exserted from the persistent calyx, dehiscing loculicidally. Seeds numerous, in general outline somewhat turbinate, with a short, right-angled bend at one end, obtuse or rounded at the other, caramel brown to brownish black.

The two species treated here overlap in parts of their respective ranges and appear to intergrade in areas of overlap. Measurements for calyx size, corolla tube and lobe size, appear to us to vary too much to be very useful diagnostically; moreover, size of these floral features varies with age of the individual flower.

- Leaves with 3 relatively conspicuous main veins, acute apically, the lower ones, at least, irregularly sharply toothed marginally; fully developed corolla tube usually 10–12 mm long.   1. *B. americana*
- Leaves obscurely or not at all 3-veined, some of them, at least, obtuse apically, entire or obscurely toothed marginally; fully developed corolla tube usually 6–10 mm long.   2. *B. floridana*

1. Buchnera americana L.

Stem scabrous, to 8 dm tall. Lower leaves lanceolate to lance-ovate, with 3 main veins, acute apically, margins irregularly sharply toothed; leaves gradually reduced upwardly. Calyx 6–7 mm long. Corolla purple, tube 10–12 mm long. Fruiting calyces ovoid-oblong, mostly about 7 mm long.

Moist soil of open woodlands, prairies, meadows, marshy places, stream margins, barrens. N.Y., s. Ont. to s. Mich., Ill., Mo., s.e. Kan., generally southward to Ga. and e. Tex.

2. Buchnera floridana Gand.

In general aspect similar to *B. americana*, the more robust specimens not as large as the more robust ones of *B. americana* but stature of both variable. Leaves obscurely if at all 3-veined, larger ones lanceolate to oblanceolate, sometimes linear-lanceolate, their margins entire or obscurely toothed, apices of some of them at least obtuse or rounded. Calyx 4–6 mm long. Corolla tube 6–10 mm long. Fruiting calyces mostly about 5 mm long. (Incl. *B. elongata* Sw., *B. brevifolia* Pennell)

Pine savannas and flatwoods, bogs, sandy-peaty ditches, grassy roadsides, marl prairies. Chiefly coastal plain, N.C. to s. Fla., westward to s. half of Tex., (Ark., Tenn., Mo.?); Bah.Is., w. Cuba.

23. Schwalbea

Schwalbea americana L. CHAFF-SEED.

Stiffly erect perennial. Stems densely short-pubescent, simple or 2- or 3-branched at or near the base. Leaves alternate, sessile, the lowest ones largest, 2–4 cm long and 6–10 mm wide, gradually reduced upwardly, lanceolate, or elliptic-lanceolate, mostly with 3 prominent veins, the midvein more pronounced than the laterals, apices obtuse to acute, margins entire, surfaces tomentose. Flowers in loose, terminal, bracteate racemes, the flower stalks 3–5 mm long. Bractlets 2, linear, arising just below the calyx. Calyx glandular-pubescent, tubular below, tube prominently 10–12-ribbed, 10 mm long or a little more, the limb 2-lipped, upper lip a short triangular lobe, lower lip very much larger, with 2 lateral lobes and a central, much larger lobe 3-notched at the summit. Corolla yellow or purplish yellow, about 3 cm long, 2-lipped, upper lip entire or merely emarginate, somewhat hooded, lower lip with 3 very short, broad, obtuse or rounded lobes. Stamens 4, extending from the corolla tube beneath the upper lip. Capsule hard, 10–12 mm long, about as long as the tube of the persistent calyx enclosing it, oblong-elliptic in outline, acuminate at the tip, 2-valved, shallowly channeled along the valves, brown. Seed body almost amber brown, linear-ellipsoid, with a prominent membranous wing of lighter color. (Incl. *S. australis* Pennell)

Moist to dry pinelands, oak woodlands, seasonally wet pine savannas. Local, coastal plain, e. Mass. to n. Fla., westward to e. Tex., also mts. of Ky., Tenn.

24. Castilleja

Castilleja coccinea (L.) Spreng. INDIAN PAINT-BRUSH, PAINTED-CUP.

Annual herb. Stem erect, usually simple, sometimes several from the base, 2–6 dm tall, pubescent, with a rosette of closely set, usually entire, oblanceolate, basal leaves, these often shed by the time of flowering. Leaves alternate, those of the stem sessile, rarely entire, usually 3–5-cleft, the segments to 4 cm long, linear to narrowly oblong, ascending. Flowers in conspicuously bracteate, compact, broad, terminal spikes 4–6 cm long, the spikes elongating to as much as 2 dm as the fruits mature, the fruits then well set apart from each other. Bracts more conspicuous than the flowers, wholly or mostly scarlet, rarely pale, commonly 3-lobed, sometimes 5-lobed. Calyx 2–3 cm long, thin-membranous, more or less scarlet, cleft to about the middle into 2 lateral halves, each half widened distally and rounded to truncate or barely emarginate apically. Corolla greenish yellow or pale yellow, about equaling the calyx or barely exserted from it, 2-lipped, the upper lip oblong, the lower larger, 3-lobed, deflexed. Stamens 4. Capsule included in the persistent, urceolate, fruiting calyx, ellipsoid, 8–10 mm long, loculicidal. Seeds straw-colored, more or less obpyramidal in outline, very prominently spongy-alveolate-reticulate.

Wet meadows, bogs, prairies. N.H. to s. Man., generally southward to S.C., n. Ga. and Ala., Miss., n. La. and Okla.

25. Pedicularis

Perennial herbs (ours). Leaves alternate or opposite, the lower usually closely set, pinnately or bipinnately lobed, cleft, or divided. Flowers in a spikelike, terminal, bracteate raceme. Bractlets none. Calyx tubular below, oblique at the orifice, longer on the upper side, in some 2–4-lobed. Corolla tube longer than the strongly 2-lipped limb, the upper lip hoodlike, arched and decurved, often beaked apically, lower lip shorter, 2-crested above, 3-lobed, lobes broadly rounded apically, the lateral lobes somewhat larger than the middle one. Stamens 4, ascending under the upper lip of the corolla. Capsule compressed, more or less falcate, glabrous, loculicidal. Seeds few.

- Leaves alternate, the lower petiolate, upper stem and inflorescence axis pubescent, the axis usually tomentose. 1. *P. canadensis*
- Leaves mostly opposite, sessile or subsessile; upper stem and inflorescence glabrous or nearly so. 2. *P. lanceolata*

1. Pedicularis canadensis L. COMMON LOUSEWORT.

Stems erect, 1.5–3 dm tall, relatively sparsely clothed below with soft white hairs, nearly tomentose above, especially on the inflorescence axis, commonly forming clones by means of subterranean branches. Leaves alternate, the larger ones closely set near the base, few or even none on the stem above the base and reduced upwardly; lower leaves with petioles 2–8 cm long (becoming sessile upwardly), blades elliptic or elliptic-oblong in outline, variable in length, 3–15 cm long, surfaces sparsely short-pubescent, pinnately divided, the segments short, with U-shaped sinuses between, their margins toothed. Raceme at first compact, elongating during flowering and as the fruits mature, at length flowers or fruits separated from each other, bracts subfoliose. Calyx 7–9 mm long, split on one side, sparsely pubescent. Corolla yellow or yellowish, sometimes shaded with lavender or reddish purple. Capsule lance-oblong, twice as long as the calyx. Seeds oblongish, plump, truncate at the base, rounded apically, about 2 mm long, finely reticulate.

In a wide variety of habitats, open, well-drained woodlands, prairies, creek banks, seepage slopes, moist open places. Maine and s. Que. to Man., s. to Fla. Panhandle, e. Tex., n. N.M.

2. Pedicularis lanceolata Michx. SWAMP LOUSEWORT.

Stems 4–8 dm tall, glabrous or sparsely pubescent throughout. Leaves mostly opposite, sessile or subsessile, 5–10 cm long, in overall outline lanceolate to lance-oblong, with short pinnate lobing, lobes with few short broad teeth marginally, sinuses between the lobes mostly V-shaped, sparsely pubescent, if at all, on the surfaces. Racemes much as in *P. canadensis*. Calyx about 1 cm long, 2-lobed, each lobe terminating in a foliaceous tip. Corolla pale yellow, upper lip incurved, with a short truncate beak apically, not much longer than the lower lip, the latter more or less erect and nearly closing the throat. Capsule obliquely ovate in outline, little exserted from the calyx.

Wet meadows, swampy woodlands, bogs, marshy shores. Mass., s. Ont. to Minn., southward to w. N.C., Tenn., Mo., Neb.

Lentibulariaceae (BLADDERWORT FAMILY)

Herbaceous plants, ours growing in water or wet places and comprised of two genera with very dissimilar vegetative bodies, both carnivorous but trapping mechanisms totally unlike. Flowers solitary on a scape or in scapose racemes, bisexual, markedly bilaterally symmetrical. Calyx united below, 2- or 5-lobed. Corolla united below, with a spur at the base, 2-lipped, the lower lip usually larger than the upper, with a palate in or extending from the throat or on the lower lip in front of the throat. Stamens 2, inserted at the base of the corolla tube, not exserted. Pistil 1, ovary superior, 1-locular, the placenta free-central, style short or none, stigmas 2. Fruit a capsule, 2- or 4-valved, usually rupturing irregularly.

- Plant with a short vertical stem bearing a compact rosette of soft, somewhat fleshy leaves, the roots fibrous; calyx 5 lobed.   1. *Pinguicula*
- Plant without leaves (scale-leaves in some), but sometimes some branches simulating leaves bearing bladderlike traps, roots absent; calyx 2-lobed.   2. *Utricularia*

1. Pinguicula

Plants with a short vertical stem bearing a compact rosette of alternately disposed, entire leaves, those of species whose rosettes are submersed mostly flat, those of terrestrial forms with their sides rolled upward, usually progressively more so toward the tips so that such leaves have a V-shaped outline. Flowers usually several from each rosette but the several developing and reaching anthesis over several weeks, each borne singly at the end of an elongate, nonbracteate scape. Glandular hairs, associated with a carnivorous function, occur on the upper leaf surfaces, the scapes, the outer surface of the calyx, to a lesser extent on the outer surface of the corolla tube and the ovary. Upper lip of the calyx with 3 lobes distinct nearly to the base, the lower lip 2-lobed and usually not incised more than halfway of the calyx tube. Corolla more or less cylindric-tubular at base, lowermost portion of the tube rather abruptly contracted into a spur; above the tube the corolla with 5 spreading lobes each notched or incised; from the lower side of the corolla tube just within the throat a hairy palate projects, usually obliquely outward from the throat (included within the throat in *P. pumila*). Hairs of the palate, of the ridge on the lower side of the corolla tube within and behind the palate, and on the inner side-walls of the tube, are more or less characteristic for each species. Filaments thick and stout, arching, they together with the anthers almost outlining a circle around the ovary. Ovary subglobose, with a bulbous free-central placenta bearing numerous ovules (in most species), a 2-lobed stigma at its summit. Capsule 2-valved. Seeds brown, their surfaces honeycombed.

References: Wood, C. E., Jr., and R. K. Godfrey. "*Pinguicula* (Lentibulariaceae) in the Southeastern United States." *Rhodora* 59: 217–230. 1957.

Godfrey, R. K., and H. Larry Stripling. "A Synopsis of *Pinguicula* (Lentibulariaceae) in the Southeastern United States." *Am. Midl. Nat.* 66: 395–409. 1961.

1. Expanded corolla not, or rarely, exceeding 1.5 cm across; palate not exserted from the throat of the corolla; rosettes rarely exceeding 3 cm broad.   1. *P. pumila*
1. Expanded corolla 1.8 cm across or more; palate markedly exserted from the throat of the corolla; rosettes 5 cm broad or more.
   2. Flowers sulphur yellow or golden yellow.   2. *P. lutea*
   2. Flowers violet to white.
      3. Pubescence on the lowermost portion of the scape consisting of elongated, pointed, multicellular, nonglandular hairs; that of the uppermost portion comprised of 1-celled glandular hairs (transition from one to the other very gradual); expanded portion of the corolla veiny, usually markedly so.
                                                                     3. *P. caerulea*
      3. Pubescence of the scape uniform throughout its length and all glandular; expanded portion of the corolla not at all veiny.
         4. Leaves dull red or reddish green; lobes of the corolla deeply incised.   4. *P. planifolia*
         4. Leaves bright green (although sometimes drying reddish); lobes of the corolla merely notched.
            5. Lobes of the corolla as broad as long or broader; outer ¾ of the corolla wisteria-violet and with a ring of white above the throat; corolla tube yellow, with reddish brown veins; trichomes on the inner side walls of the corolla tube yellow.   5. *P. primuliflora*
            5. Lobes of the corolla longer than broad; uniformly violet to white (no ring of white above the throat), deeper violet in the throat; corolla tube violet, with darker violet veins; trichomes on the inner side wall of the corolla tube white.   6. *P. ionantha*

1. Pinguicula pumila Michx.   Fig. 308

Rosettes mostly not exceeding 3–4 cm across. Leaves light green to olive green, ovate-oblong, the sides rolling markedly upward from just above the base, more so progressively toward the tip thus forming a **V**-shaped outline, the tips of the leaves mostly turning downward. Upper leaf surfaces glandular-pubescent and with some longer nonglandular hairs on either side of the sunken midrib near the leaf base, these hairs directed across the midrib, crisscrossing each other. Scapes to about 10 cm tall. Lobes of the upper lip of the calyx linear-oblong, blunt, about 3 mm long, those of the lower lip triangular, about 1.5 mm long, olive green to dull red. Corollas mostly 1.3–1.5 cm across, the lobes obovate, emarginate, violet to pale violet to white, rarely yellow; corolla tube violet, or yellow, violet mostly only in the deeper violet flowers, faintly to prominently reddish brown–veined; spur dull purplish brown in the more deeply violet flowers, varying to yellow, veined or not, mostly 3–5 mm long, nearly linear. Palate sulphur yellow, about 2 mm long, not exserted from the throat of the corolla tube, clothed with short, stubby, or slightly irregularly knobbed yellow hairs; hairs on the inner ridge of the tube below the palate chrome yellow to red, with short, stout stalks and knobby tips with very irregular surfaces; hairs on the walls of the tube yellow, more slender, but with very irregularly knobby tips. Filaments white, anthers and pollen very pale yellow or white. Stigma violet to white. Capsule 4–5 mm in diameter. Seed oblong to obpyramidal, mostly about 0.3–0.4 mm long, the alveolae mostly with 2 or 3 cross lines.

Moist to wet sandy soil, usually somewhat boggy, pine savannas and flatwoods, adjacent ditches and roadsides. Coastal plain, s.e. N.C. to s. Fla., westward to s.e. Tex.; Bah.Is.

2. Pinguicula lutea Walt.   Fig. 309

Rosettes to about 15 cm broad. Leaves yellowish green, ovate to oblong, the sides rolled upward beginning about ⅓ of the distance from the base, increasingly more so toward the

Fig. 308. **Pinguicula pumila:** a, habit; b, flower, longitudinal section; c,d, flower, face and lateral views; e, trichomes from palate; f, trichomes from ridge on corolla behind palate; g, trichomes from inner wall of corolla tube; h, capsule; i, seed.

Fig. 309. **Pinguicula lutea:** a, habit; b, two flowers, face view; c, flower, lateral view; d, trichomes from palate; e, trichomes from ridge of corolla tube behind palate; f, trichomes from inner walls of corolla tube; g, capsule; h, seed.

tips until the edges meet, thus the tips have an apparent acuminate tip owing to the curling. Leaves ciliate on the margins near the base, their upper surfaces glandular-pubescent, with a few longer nonglandular hairs near the midrib toward the base. Scapes mostly 15–25 cm tall. Lobes of the upper lip of the calyx oblong, broadly rounded at their tips, about 6 mm long, those of the lower lip ovate-triangular, broadly rounded at their tips, about 3 mm long. Corollas mostly 2.5–3.5 cm across, the lobes broader than long, notched or divided to about ⅓ their length, the edges of the lobes smooth to irregular, or occasionally each lobe irregularly 2- or 3-sublobed or toothed. Expanded portion of the corolla sulphur yellow, not veined, the tube greenish yellow, usually conspicuously purplish-veined on the lower side; spur slender, tapering, 5–10 mm long. Palate yellow, stout, blunt, well exserted from the throat, bearing abundant slender, yellow hairs with slightly enlarged clublike tips, hairs on the inner ridge of the tube, below the palate, with slender stalks and chrome yellow to orange-red clublike tips having very irregular surfaces; hairs on the walls of the tube yellow, somewhat more slender, with less irregular clublike tips. Filaments white, anthers pale yellow, pollen white. Stigma yellow on the center of the larger lobe, grading to white on the edges. Capsule 7–8 mm in diameter. Seeds mostly oblong, 0.5–0.8 mm long, the alveolae with 3 or 4 cross lines.

Moist to wet sandy-peaty soils of seepage bogs, pine savannas and flatwoods, adjacent ditches and roadsides. Coastal plain, s.e. N.C. to about Lake Okeechobee, Fla., westward to La.

3. Pinguicula caerulea Walt.   Fig. 310

Rosettes mostly 5–10 cm across. Leaves yellowish green, ovate-oblong, their sides curling upward from just above the base, progressively more so toward their tips thus forming a **V**-shaped outline, the tips of the leaves turned downward. Leaves glandular pubescent above and with some long nonglandular hairs near the base on either side of the sunken midrib these directed across the midrib, crisscrossing each other. Scapes to about 20 cm tall, densely pubescent, the lower ⅓ with long spreading, pointed, multicellular hairs; upwardly very gradually appear short, 1-celled hairs which produce a bulbous droplet of exudate on their tips; the long hairs very gradually diminish in number up the scape, the short ones become more numerous, the upper part of the scape having only the latter. Lobes of the upper lip of the calyx linear-oblong, about 8 mm long, those of the lower lip lance-triangular or lanceolate, about 5 mm long. Corollas mostly 2.5–3 cm across, the lobes cleft ⅓–½ their length, the subdivisions mostly broadly rounded, sometimes sub-notched or lobed. Expanded portion of the corolla varying from deep violet to pale violet, usually prominently veined, the veins deeper colored than the ground color, sometimes faintly veined. Corolla tube and spur violet to greenish yellow, distinctly violet-veined. Spur mostly 5–7 mm long, somewhat tapering. Palate broad and blunt, exserted from the throat, greenish yellow to cream-colored, abundantly clothed with long, slender, essentially pointed hairs. Hairs on the ridge of the tube within and on the walls of the tube essentially like those of the palate, with barely enlarged tips, or with somewhat branched tips. Filaments white, anthers very pale yellow, pollen pale yellow or white. Larger lobe of the stigma tinted yellow in the center, white nearer the margins. Capsule about 1 cm in diameter. Seeds oblong or obpyramidal, mostly about 0.5 mm long, the alveolae with 3 or 4 cross lines.

Moist to wet sandy and sandy-peaty soils of pine savannas and flatwoods, adjacent ditches and roadsides. Coastal plain, s.e. N.C. to about Lake Okeechobee, Fla., westward in the Fla. Panhandle to about the Apalachicola River.

4. Pinguicula planifolia Chapm.   Fig. 311

Rosettes to about 15 cm across. Leaves oblong, elliptic, or oblanceolate, thin and translucent, flat when fully developed, dull green and slightly to uniformly suffused with a dull purplish red pigment. Upper leaf surfaces with very short glandular hairs. Scapes relatively slender, to 25 cm high, sparsely glandular-pubescent below, more densely so near the summit. Calyx olive brown or reddish olive, the lobes of the upper lip oblong, about 4 mm long, those of the lower ovate-triangular, about 1 mm long, all rounded at their tips. Expanded

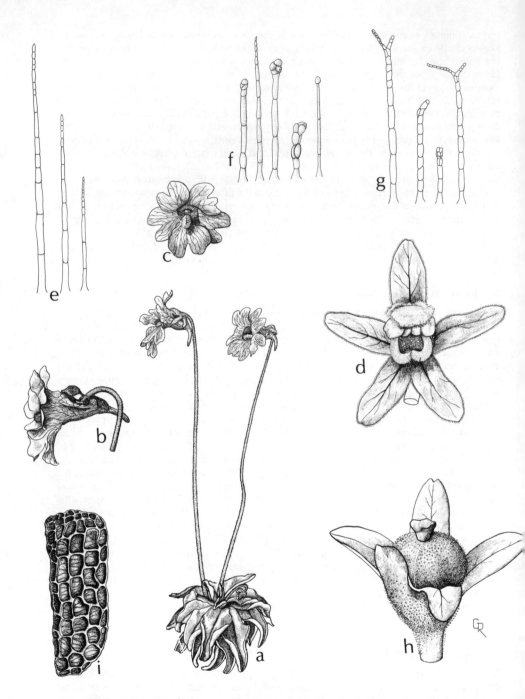

Fig. 310. **Pinguicula caerulea:** a, habit; b,c, flower, lateral and face views; d, face view of flower with corolla removed; e, trichomes from palate; f, trichomes from ridge of corolla behind palate; g, trichomes from inner wall of corolla tube; h, capsule; i, seed.

Fig. 311. **Pinguicula planifolia:** a, habit; b,c, flower, face view and lateral view; d, trichomes from inner walls of corolla tube; e, trichomes from ridge of corolla tube behind palate; f, trichomes from palate; g, capsule; h, seed.

677

portion of the corolla about 3 cm across, the lobes violet to almost magenta, occasionally nearly white, the throat deeper violet. Corolla lobes incised nearly ½ their length, lanceolate. Corolla tube somewhat veined. Spur oblong, 3−4 mm long, color of the calyx. Palate 4−6 mm long, exserted from the tube, slender, oblong, densely clothed with slender, yellow hairs with smooth to slightly irregularly surfaced clublike tips; hairs of the ridge of the corolla tube within and below the palate with sessile or very short-stalked, druselike hairs, yellow to orange or orange-red in color. Hairs on the wall of the tube within slender, stalked, white, and with smooth to very irregular clublike tips. Filaments pale to deep violet, anthers pale yellow, pollen white. Larger lobe of the stigma deep violet. Capsule about 5 mm in diameter. Seed obpyramidal, about 6−7 mm long, the alveolae with 2−4 cross lines.

In shallow water, margins of peaty ponds, bogs, boggy flatwoods, ditches and drainage canals. Fla. Panhandle from Leon and Liberty Cos. westward, thence to s.e. Miss.

5. Pinguicula primuliflora Wood & Godfrey. Fig. 312

Rosettes to about 15 cm broad. Leaves bright green, more or less oblong, fully developed ones essentially flat, the upper surfaces with short glandular hairs, propagating by forming plantlets along the leaf edges under certain conditions. Scapes sparsely glandular-pubescent, mostly 8−15 cm tall. Lobes of the upper lip of the calyx 5−6 mm long, broadly ovate-oblong, the lower about 3 mm long, broadly ovate-triangular, all rounded at their tips, dark olive green, sometimes suffused with a little reddish pigment. Expanded corollas 2.5−3 cm across, the lobes obovate to suborbicular, broader than long, shallowly notched. Open face of the corolla wisteria-violet and with a ring of white just at the base of the lobes and above the yellow throat, not veined. Corolla tube chrome yellow, especially within, and prominently reddish brown veined. Palate chrome yellow, exserted, cylindrical, 4−5 mm long, densely clothed with clublike yellow hairs. Ridge of the corolla tube behind the palate bearing stout, short-stalked hairs with obovoid or mitten-shaped orange to orange-red tips. Walls of the tube with chrome yellow, slender, clublike hairs. Filaments white, anthers very pale yellow, pollen very pale yellow to white. Larger lobe of the stigma white. Capsule depressed-globose, about 5 mm in diameter. Seeds 0.5−0.7 mm long, obpyramidal to subcylindric-truncate, the alveolae without cross lines.

In shallow, usually flowing, water of springy areas, boggy banks of small streams, swamps, rarely in shallow flowing water of ditches, usually in dense to partial shade. Western part of Fla. Panhandle, s.w. Ga., s. Ala. and Miss.

6. Pinguicula ionantha Godfrey. Fig. 313

Rosettes to about 15 cm across. Leaves bright green, oblong, rounded at their tips, essentially flat when fully developed, with just the edges rolled upward, the upper surface clothed with short glandular hairs. Scapes mostly 10−15 cm tall, rather sparsely short-glandular pubescent. Upper lip of the calyx with oblong, rounded lobes about 4 mm long, those of the lower ovate-triangular, obtuse, 3 mm long, all dark olive green and usually suffused with some purplish pigment. Expanded corollas mostly about 2 cm across, the lobes obovate, shallowly notched, longer than broad, pale violet to white. Throat of the corolla and the tube deeper violet than the limb and with darker violet veins. Spur 4−5 mm long, yellow to olive, linear-cylindric. Palate 4−6 mm long, exserted from the throat, conical or narrowly cylindric, the upper ⅔ yellow and clothed with yellow, slender, clublike hairs some of whose tips turn red as they age; lower ⅓ of the palate on the upper side violet like the throat and bearing few or no hairs, hairy to the base on the lower side. Ridge of the corolla tube within and below the palate with relatively distant hairs having short, stout stalks and spherical to broadly oblong, orange to orange-red tips. Walls of the tube with relatively few, white, clublike hairs. Filaments pale violet to white, anthers and pollen very pale yellow. Larger lobe of the stigma violet to pale violet. Capsule depressed-globose, about 5 mm in diameter. Seeds oblong-cylindric to obpyramidal, the alveolae with 1 or 2 cross lines.

In bogs, flatwoods depressions, adjacent ditches or drainage canals, the rosettes mostly submersed (at least when conditions are most favorable for the plant); frequently intermixed with *P. planifolia*. Fla. Panhandle: in Liberty, Franklin, Gulf and Bay Cos.

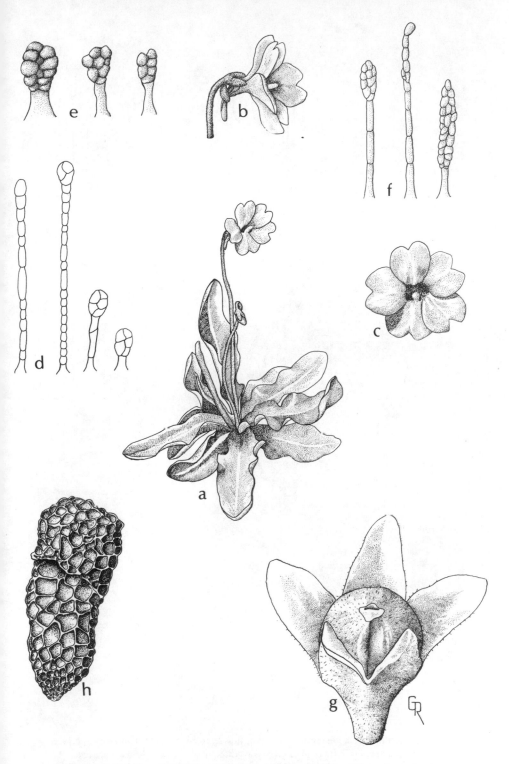

Fig. 312. **Pinguicula primuliflora:** a, habit; b,c, flower, lateral and face views; d, trichomes from palate; e, trichomes from ridge of corolla behind palate; f, trichomes from inner walls of corolla tube; g, capsule; h, seed.

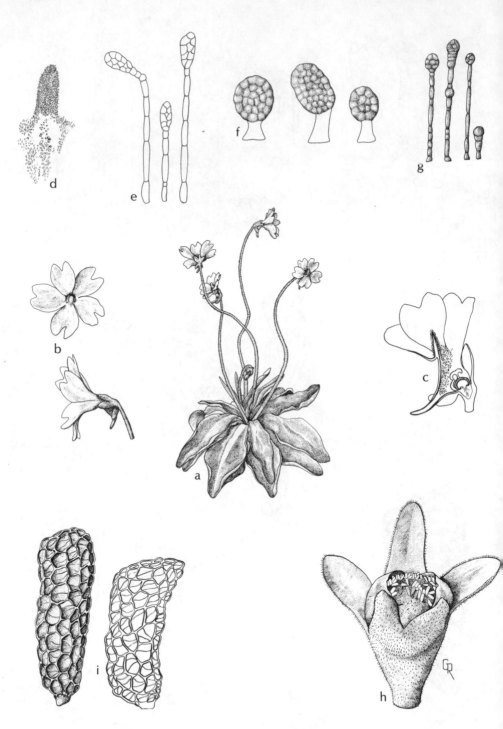

Fig. 313. **Pinguicula ionantha:** a, habit; b, flower, face view and lateral view; c, flower, longitudinal section; d, trichomes on palate, ridge of corolla tube behind palate, and on inner walls of corolla tube; e, trichomes, much enlarged, from palate, f, from ridge of tube behind palate, g, from inner wall of tube; h, capsule; i, seed.

2. Utricularia (BLADDERWORTS)

Aquatic or amphibious plants. Lacking roots. While most authors distinguish between stems and leaves in this group, there appears to be evidence that the vegetative portion is wholly comprised of a stem system, portions of which may be more or less leaflike. The system bears urnlike bladders upon at least some of its parts. They are highly specialized and unique structures whose homologies are not clear insofar as we know.

In some of our aquatic species the whole plant is free and unattached in the water, appears to grow indeterminately indefinitely, the older portions disintegrating as growth proceeds distally (e.g., in *U. foliosa*). In other aquatic species a characteristic portion is in the substrate, usually loose muck, and another characteristic portion extends into the water from the substrate and is more "foliar" (e.g., *U. floridana*). In our terrestrial species, the bulk of the branching system is subterranean although grasslike "leafy" portions emerge from the substrate when conditions are optimal. In only one of ours, *U. inflata*, do we know of the production of tubers: under special ecological conditions, long threadlike branches are produced in considerable numbers, each from an axil of 1 or more segments of leaflike branches and each bearing a tiny tuber at its tip. The tubers germinate under favorable conditions, giving rise to a new "generation." In north temperate regions some of our species form terminal winter buds (turions) of small, very crowded "leaves" by which the plants overwinter.

Small animals, especially small crustaceans, are trapped in the bladders and digested.

Flowers borne singly or more commonly in racemes on scapes from the nodes, a bract or a pair of bractlets sometimes on the flower stalks. Calyx 2-lobed, the upper lobe larger than the lower. Corolla 2-lipped, the lower lip usually 3-lobed and with a bearded palate that in some kinds closes the throat, the tube prolonged into a spur basally; upper lip usually erect and entire; corolla yellow, white, or from pink to purple. Stamens erect. Fruit a 2-valved capsule.

Corollas of some flowers in some species may be reduced, flowers cleistogamous, or under certain conditions a species which had chasmogamous flowers under favorable conditions may have some or all of them cleistogamous (presumably under unfavorable conditions).

1. Flowers deep pink, lavender, or purple (white on an occasional individual).
    2. Plants aquatic, unattached, branchlets verticillate.   7. *U. purpurea*
    2. Plants with principal branch system within a soil substrate, sometimes with either grass-blade-like branchlets emerging from the substrate *or* with a rosette of emergent obovate-spatulate to orbicular leaflike branchlets; branchlets not verticillate.
        3. Scape 1-flowered; bract on the scape tubular; emergent branchlets, if any, finely grass-blade-like.
                                                                               1. *U. resupinata*
        3. Scape usually with a several-flowered raceme at its summit (rarely 1-flowered); bracts on the scape several, not tubular; emergent branchlets, if any, leaflike, blades obovate-spatulate to orbicular.
                                                                               2. *U. amethystina*
1. Flowers white, yellow, or yellowish white.
    4. Bracts, bractlets, and calyx lobes fimbriate.   3. *U. simulans*
    4. Bracts, bractlets and calyx lobes not fimbriate.
        5. Flowers white, minute, about 1 mm long; plants diminutive, easily overlooked.   8. *U. olivacea*
        5. Flowers yellow, much longer than 1 mm.
            6. Flowering scape with a whorl of inflated branchlets about midway its length, these floating and thus holding the distal portion of the scape with its flowers elevated above the water surface.
                7. Racemes usually 9–14-flowered; stalks of fruits recurved; spur of the corolla notched at the tip; bracts of raceme longer than broad, unlobed; lateral leaflike branchlets with the primary basal divisions comprising 2 unequal forks; floats from the expanded portion tapering gradually toward the scape.   9. *U. inflata*
                7. Racemes mostly 3- or 4-flowered; stalks of fruits erect-ascending, rarely recurved; spur of corolla usually not notched at the tip; bracts of raceme as broad as long or broader than long, variably lobed; lateral leaflike branchlets with the primary basal divisions comprising 2 equal forks; floats from the expanded portion essentially parallel-sided until just at the scape axis.
                                                                               10. *U. radiata*

6. Flowering scape without a whorl of inflated, floating branches.
8. Plants aquatic (unless stranded at low water).
9. Principal axes of plant coarse, nearly flat and straplike.   11. *U. foliosa*
9. Principal axes of plant relatively slender and not flat and straplike.
10. Scapes to 10 cm long; plants usually in intricately tangled floating bunches or mats.
                                                                                 12. *U. biflora*
10. Scapes 15 cm long or longer; plants not usually in intricately tangled bunches or mats.
11. Scapes slender-wiry, about 1 mm across; "leafy" branches cylindrical in outline, 1–1.5 cm across.   13. *U. fibrosa*
11. Scapes subsucculent, commonly with 1 to several bows or bends, 2–3 mm across or a little more; "leafy" branches cylindrical in outline, 2–5 cm across.   14. *U. floridana*
8. Plants essentially terrestrial, the principal branch system within a sandy or sandy-peaty substrate (not uncommonly with surface water over the substrate during times of heavy rainfall).
12. Flower stalks each subtended by a single peltate bract.   4. *U. subulata*
12. Flower stalks each subtended by a nonpeltate bract and with 2 bractlets immediately above it.
13. Scapes green to yellowish green; all flowers chasmogamous; spur 10 mm long or a little more.   5. *U. cornuta*
13. Scapes greenish purple to purple; flowers both chasmogamous and cleistogamous; spur 7 mm long or less.   6. *U. juncea*

1. Utricularia resupinata B.D. Greene.   Fig. 314

Slender delicate branch system within a wet soil substrate, under optimal conditions with emergent, narrowly acicular, grass-blade-like branchlets; rarely in floating, densely tangled mats and then with abundant grass-blade-like branchlets. Descending branches with clawlike "rhizoids" from the base of the scape. Scapes 2–10 cm tall, slender-wiry, 1-flowered. Bract on the scape below the erect flower stalk (which appears like a continuation of the scape) tubular; flower stalks mostly 1–1.5 cm long, no bractlets subtending the flower. Corolla purple (rarely white), about 1 cm long, tipped backward and facing upward, the short-conic, blunt-tipped spur about 5 mm long, nearly horizontal. Seeds angled, somewhat obpyramidal, coarsely reticulate. (*Lecticula resupinata* (B. D. Greene) Barnh.)

Ponds, lakes, river shores, pools and ponds in pine flatwoods, adjacent ditches, borrow pits. N.S. and Que. to s. N.Eng., N.Y. to Del., n.w. Pa. to n.w. Wisc.; coastal plain, Ga., s. Ala., and throughout Fla.

2. Utricularia amethystina Salzm. ex St.-Hil. & Girard.

Relatively inconspicuous plant with branch system within a soil substrate, emergent leaflike branches in a rosette, these obovate, spatulate, or orbicular. Scapes filiform, 1.5–30 cm tall, raceme usually several-flowered (rarely 1-flowered), flower stalks filiform, mostly 10 mm long but varying from 5–20 mm; bracts several on the scape below the inflorescence, a bract and 2 bractlets subtending the flower. Corolla about 8 mm long, violet to mauve to nearly white, usually with a yellow spot on the plate; spur conic, acute at the tip. (Incl. *Calpidisca standleyae* Barnh.)

Wet pine flatwoods, pen. Fla.; Mex. to Arg.; Trin.

3. Utricularia simulans Pilger.   Fig. 315

Plant with a branch system within a soil subtrate, bearing bladders only and without leaflike branchlets; emergent linear-clavate leaflike branchlets evident when conditions are favorable. Scapes 5–15 cm tall, racemes usually several-flowered, sometimes 1-flowered; bracts fimbriate, several on the scape below the lowest flower; bract subtending the very short flower stalk and two bractlets below the calyx, the bracts, bractlets, and calyx lobes fimbriate. Corolla yellow, 6–8 mm across, spur subconic, about as long as the lower lip. Seed coarsely reticulate. (*U. fimbriata* of authors not Kunth; *Aranella fimbriata* sensu Barnh.)

Wet pine flatwoods. s. Fla.; C.Am., W.I.; much of S.Am. from Brazil northwards.

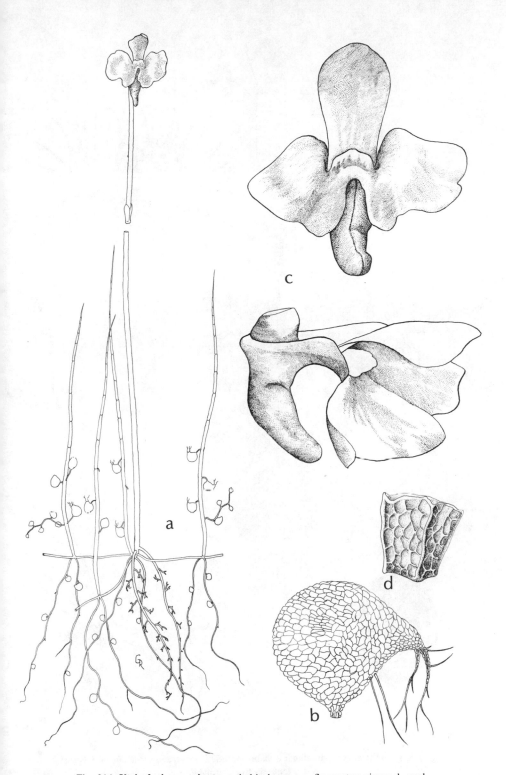

Fig. 314. **Utricularia resupinata:** a, habit; b, trap; c, flower, two views; d, seed.

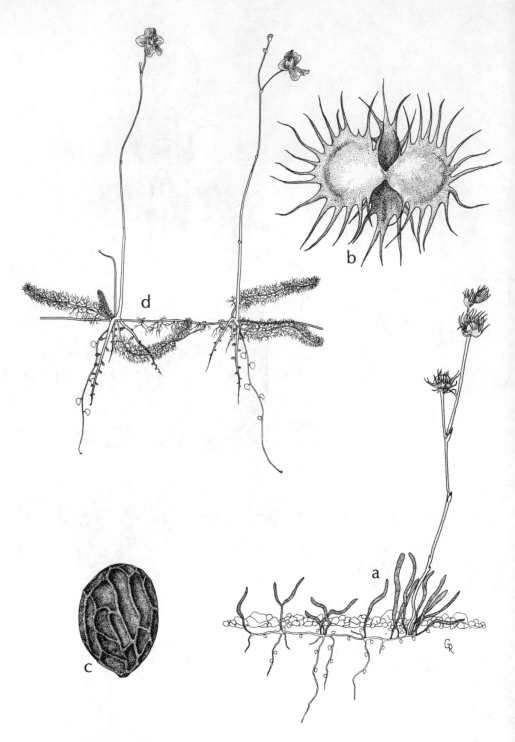

Fig. 315. a–c, **Utricularia simulans:** a, habit; b, bracts, outer, and sepals, inner; c, seed; d, **Utricularia fibrosa:** habit.

684

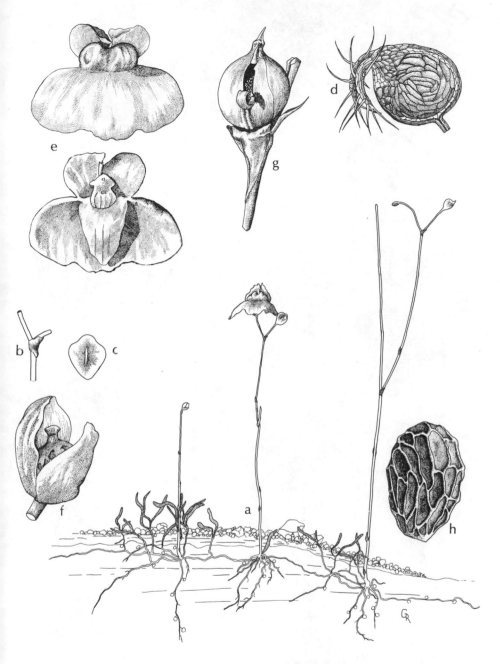

Fig. 316. **Utricularia subulata:** a, habit; b,c, bract, lateral and dorsal views; d, trap; e, corolla, two views; f, capsule, within calyx; g, capsule, split open; h, seed.

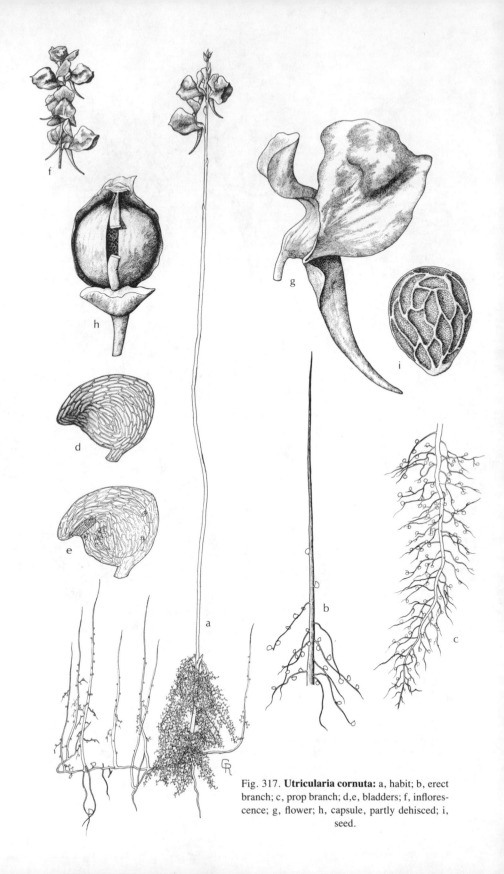

Fig. 317. **Utricularia cornuta:** a, habit; b, erect branch; c, prop branch; d,e, bladders; f, inflorescence; g, flower; h, capsule, partly dehisced; i, seed.

4. Utricularia subulata L.

Fig. 316

Vegetative body commonly entirely within a soil substrate, the short, linear-filiform, leaflike segments unbranched; under favorable conditions slenderly clavate, grass-blade-like branchlets emerge from the substrate. Scapes filiform-wiry, from about 4 to 18 cm tall, 1 to several widely separated peltate bracts on the scape, similar ones subtending the flower stalks. Flowers 1 to several, usually not over 8, commonly 2–4; flower stalks variable in length, to about 20 mm long. Corolla yellow, sometimes some or all flowers cleistogamous; corollas of chasmogamous flowers 6–12 mm long, 4–8 mm wide, lips very unequal, broader than long, the upper rounded, the lower more or less 3-lobed; palate prominent, 2-lobed; spur conic, about as long as the lower lip. Capsule 1.5–2 mm across. Seed about 0.2 mm long, coarsely reticulate. (*Setiscapella subulata* (L.) Barnh.; incl. *S. cleistogama* (Gray) Barnh.)

Bogs, wet pine savannas and flatwoods, clearings of swampy woodlands, boggy ditches and logging roads, commonly in wet sands and peaty sands in areas of soil disturbance. w. N.S., s.e. Mass., L.I., southward chiefly on the coastal plain to s. Fla., westward to Tex., northward in the interior to Ark. and Tenn.; W.I., C.Am., S.Am.; trop. Afr. and Madagascar; Thailand, Borneo.

5. Utricularia cornuta Michx.

Fig. 317

Delicate branch system within a soil substrate, bearing small linear-filiform leaflike lateral branches; under favorable conditions with soft, needle-shaped, leaflike branches emergent from the substrate. Scape erect, wiry, green to yellowish green, 1–4 dm tall, raceme terminal, 1–9-flowered, usually 3–5-, flowers crowded, their stalks short, each subtended by an oblong, usually yellowish, apically acute to acuminate bract with two narrow bractlets within it; similar widely separated bracts on the scape below the flowers. Corollas variable in size, 15–25 mm high, bright yellow to orange-yellow; spur long-conic, not appressed to the lip, mostly 8–12 mm long; palate a very broadly conical hump comprising much of the lower lip. Capsule globose, 1-valved, 2.5–3.5 mm across. Seeds tiny, reticulate, dull amber. (*Stomoisia cornuta* (Michx.) Raf.)

Bogs, alluvial outwash in pinelands, borrow pits, sandy-peaty margins of ponds, wet pine savannas and flatwoods, boggy ditches, commonly in shallow water. Nfld. and Que. to n. Ont. and Minn., generally southward to s. Fla. and e. half of Tex.

6. Utricularia juncea Vahl.

Fig. 318

In general habital features closely similar to *U. cornuta*. Emergent leaflike branches terete and gradually tapered to a long point if in water; if not covered by water, terete below, flattened distally and somewhat straplike. Scapes purplish green, bracts purple. Flowers both chasmogamous and cleistogamous in the same inflorescence, flowers not crowded. Corollas yellow, usually not exceeding 15 mm high; spur conic, 5–7 mm long; palate a very broadly conical hump comprising much of the lower lip. Seeds tiny, reticulate. (*Stomoisia juncea* (Vahl) Barnh.; incl. *S. virgatula* Barnh.)

Habitats as for *U. cornuta*. Chiefly coastal plain, L.I. and N.J. to s. Fla., westward to e. Tex. and s.e. Ark.; Bah.Is., Cuba, P.R.; Guat., Brit.Hond.; n. S.Am.

7. Utricularia purpurea Walt.

Fig. 319

Aquatic, free in the water, bladder bearing branchlets in whorls, stalked, the stalks mostly 1–1.5 cm long. Scapes emergent, borne at irregular intervals from the nodes, 5–10 cm long, slender, somewhat spongy, racemes 2–5-flowered; bracts none on the scape below the lowest flower; bracts subtending the flower stalks 2-lobed, the lower descending, much smaller than the ascending upper lobe. Corolla deep pink to purple, about 12 mm across, spur conic, blunt apically, 2–5 mm long. Capsule globose, about 3–4 mm across. Seed with conic, hornlike tubercles. (*Vesculina purpurea* (Walt.) Raf.)

Ponds, swamps, sluggish streams, sloughs, drainage ditches and canals. Que. and N.S. to Minn., southward to N.Y., n. Ind., s. Mich., Wisc., southward on the coastal plain to s. Fla., westward to s.e. Tex.; W.I., C.Am.

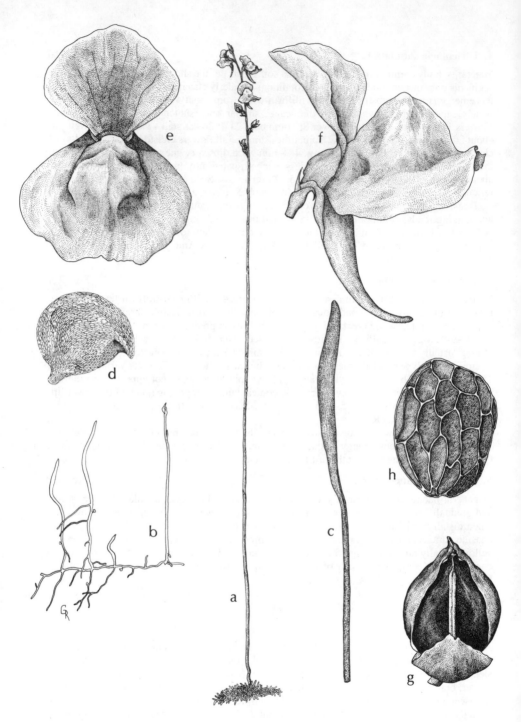

Fig. 318. **Utricularia juncea:** a, habit; b, small part of what is drawn at base of *a* cleared from the substrate and enlarged; c, phyllodelike branch, much enlarged, usually present only under very favorable conditions; d, trap; e,f, flower, face and lateral views; g, capsule, sepal pulled back; h, seed.

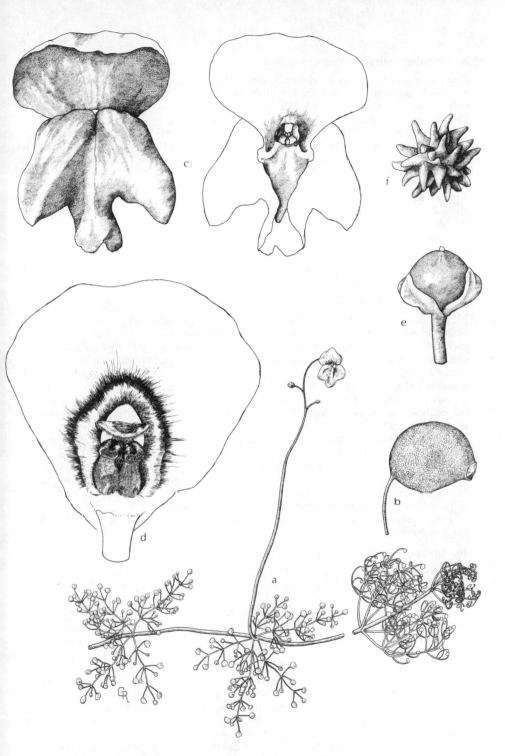

Fig. 319. **Utricularia purpurea:** a, habit, very small part of plant; b, bladder; c, flower, two views; d, view looking into flower, showing stamens and stigma draped over them; e, capsule; f, seed.

8. Utricularia olivacea Wright ex Griseb.

Diminutive aquatic (easily overlooked), free in the water, sometimes in matted bunches, commonly amongst vegetative parts of other aquatics, e.g., *Myriophyllum* spp., other utricularias, algae, sometimes stranded in muck as water levels recede. Slender stems bearing bladders only (although often with epiphytic filamentous algae), no leaflike branchlets. Flowers minute, borne mostly solitary from the nodes, the stalks soft, translucent, about 5 mm long, a bract at the point of origin. Corolla white, drying yellowish, the lower lip much longer than the upper, 1−2 mm long, saccate rather than spurred at the base, emarginate at the summit, upper lip short, broader than long, nearly heart-shaped. Ovary subglobose at anthesis, style short, ovules 2. Mature capsule with an elliptical body then narrowed gradually to a hollow beaklike summit, 1 mm long overall or a little more. Seed 1 per capsule, about 0.5 mm long, oblong-elliptic, olive green, smooth. (*Biovularia olivacea* (Wright ex Griseb.) Kam.)

Ponds and lakes. Coastal plain, N.J. to Fla.; W.I.; S.Am. In our range there are, insofar as we know, but a few scattered records of its occurrence. Its rarity is perhaps attributable more to the difficulty of perceiving the plant in the field than to its actual distribution. We find it abundant in numerous ponds and lakes in the general area of Tallahassee, Fla.

9. Utricularia inflata Walt.   Fig. 320

Plants free in the water. Main stems 2−3 mm wide. Lateral leaflike branches alternate, bushy-dendritic, variable in size, the larger 7 cm broad or more and 18 cm long or more, each with 2 unequal primary divisions, the two divisions spreading to opposite sides of the axis. Plants stranded in muck or mud during times of low water usually with long, slender, threadlike branches arising from axils of some of the dissections of the leaflike lateral branches and bearing tiny tubers at their summits. Scapes bearing a whorl of 5−10 spongy inflated branches which float and hold the inflorescence above the water surface; mature scapes, below the whorl of floats, 15−35 cm long, mature inflorescence stalks (above the floats) to 15 cm long; stalks of the floats taper gradually from the expanded portion to the axis. Racemes 4−18-flowered, usually 10- or 11-. Bracts longer than broad, oblong below, apically subacute to obtuse, unlobed. Corolla golden yellow, 1.5−2 cm broad, spur conic, notched at the tip, the lower lip about twice as long as the spur; palate saccate, streaked and dotted with red. Fruiting stalks mostly recurved. Capsule subglobose, 3−6 mm across. Seeds numerous, dark brown to black, their surfaces greatly roughened by laterally curved, irregularly hollowed out, or finely flattened, submembranous tubercles.

Ponds, lakes, swamps, drainage ditches and canals, sloughs. Coastal plain, N.J. to s. Fla., westward to e. Tex.

10. Utricularia radiata Small.   Fig. 320

In general habital features, similar to *U. inflata*. Primary divisions of leaflike lateral branches equal, the whole branch to 5 cm broad and 10 cm long. On plants marooned in muck or mud at times of low water, tiny vegetative buds sometimes produced in some of the axils of the dissections of the leaflike branch but no threadlike, tuber-bearing branches are produced. Mature scape, below the floats, to about 13 cm long; inflorescence stalks, (above the floats) maximally 6 cm long; stalks of the floats with parallel sides from the expanded portion toward the axis, tapering only very abruptly proximally. Racemes 1−7-, mostly 3−4-, flowered. Bracts variable, mostly as broad as long or broader than long, unlobed or variably lobed. Corolla dull yellow, 0.8−1.4 cm broad; spur conic, its tip rounded or with a small central papilla (rarely notched at the tip); palate streaked and spotted with red. Fruiting stalks usually stiffly erect-ascending, rarely slightly recurving. Capsule subglobose, 3−5 mm across. Seeds numerous, dark brown to black, surficially with tubercles much as in *U. inflata* but usually somewhat longer. (*U. inflata* var. *minor* Chapm.)

Ponds, lakes, swamps, drainage ditches and canals. N.S. southward chiefly on the coastal plain to s. Fla., westward to Tex.; w. Tenn., w. Va., n.w. Ind.; Cuba, S.Am.

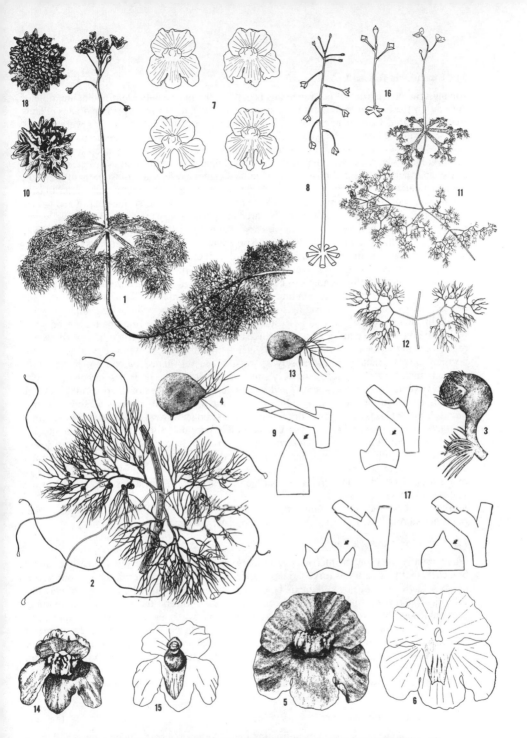

Fig. 320. 1–10, **Utricularia inflata:** 1, habit, small part of plant; 2, lateral foliar unit with tuber-bearing branches; 3, germinating tuber; 4, bladder; 5,6, corolla, two views; 7, variation in lower lip of corolla; 8, fruiting raceme; 9, node of raceme with subtending bract and outline of bract; 10, seed; 11–18, **Utricularia radiata:** 11, habit, small part of plant; 12, lateral foliar unit; 13, bladder; 14,15, corolla, two views; 16, fruiting raceme; 17, three nodes of raceme with subtending bracts and outline of bracts to show variation; 18, seed. (From Reinert and Godfrey, "Reappraisal of *Utricularia inflata* and *U. radiata* (Lentibulariaceae)." *Amer. Jour. Bot.* 49: 214. 1962)

11. Utricularia foliosa L.
Figs. 321 and 322

Amongst the utricularias of our range this is perhaps the most easily identified at any time of the year on the basis of a vegetative character, i.e., its flat, straplike principal axes. Plants free in the water, variable in extent but in favorable habitats not uncommonly several meters long and with numerous elongate branches, relatively very robust. In vigorous specimens principal flat stem may be 1 cm wide or a little more. Plants in relatively deep water tend to be much more vigorous and extensively developed than they are in shallow water, especially as the latter becomes heated in summer. Mucilage glands present and most parts of the plant are to a greater or lesser extent coated with mucilage, more so in autumn and winter than in summer. In deeper water, plants tend not to be nearly so close to the surface in fall and winter as they are in summer. Principal axes bear three morphologically relatively distinctive, vegetative, lateral branch systems: from a given node there is usually a pair, one the more leaflike, highly dissected, usually with its dissections spreading laterally in one plane, its outline commonly deltoid and (in the water) somewhat reminiscent of a fern frond; the other of the pair is usually 3–4 times as long, its dissections fewer and more spread, not oriented in one plane so that it is in overall outline tending to be cylindrical, the whole branch system usually more heavily bladder-bearing and descending in the water, sometimes its distal portion having silt or debris surrounding it (thus leading some to say the plant has a point of attachment but this is but an "inadvertence"); at some, usually not all, of the same nodes, or even between nodes, there occur ascending long, threadlike branches, (up to 60–120 cm long) whose tips reach the water surface and lie flat upon it; such a branch is naked below, but in its terminal portion bears alternately a few closely appressed, mussel-shaped bracts; not infrequently such a branch bears 1 or more relatively small dissected leaflike branches somewhere along its linear extent. Scapes borne singly and intermittently from nodes of the principal axes, the racemes emergent at anthesis, 10–30 cm long, the axis of the scape with shield-shaped bracts distally; bracts subtending the flower stalks similar; flowers to about 20 per raceme, the racemes developing distally over a considerable period of time; flower stalks about 1.5 cm long. Corolla yellow, 1.5–2 cm across, upper lip erect, nearly entire, margin wavy, lower lip spreading, slightly 3-lobed; spur slenderly conic, notched apically; palate saccate, streaked and spotted with red. Fruiting stalks usually recurved. Capsule subglobose, 5–7 mm across. Seeds discoid, circular in outline or nearly so, 3 mm across, appressed to the globose receptacle.

Ponds, lakes, sluggish streams, drainage ditches and canals, Everglades sloughs, sometimes in swamps. Fla. to La.; W.I., S.Am.; trop. Afr., Madagascar.

12. Utricularia biflora Lam.
Fig. 323

Plants usually very much intertwined, forming floating bunches or mats. Leaflike lateral branches small, 1 or 2 times forked. Scapes arise, singly or 2 or 3, intermittently from principal axes, 2–8 cm long; from the axis below the scape there is usually a cluster of clawlike "rhizoids." Growing tips circinate. Racemes with 1–3 flowers, the stalks erect-ascending, each subtended by a reniform bract 1 mm long, 2 mm wide, its margin wavy-transparent. Corolla yellow, lips about equal, 6–8 mm broad; palate prominent, copiously glandular pubescent, streaked with red; spur somewhat shorter than to a little longer than the lower lip, conic. Capsule subglobose, about 4 mm across. Seed dull brown, with a prominent thin, irregularly notched marginal wing. (*U. pumila* Walt.)

Ponds, pools, swamps, drainage ditches and canals. s.e. N.Eng., chiefly coastal plain southward to s. Fla., westward to s.e. Tex. and e. Okla.

13. Utricularia fibrosa Walt.
Fig. 315

Plant free in the water, sometimes in floating mats, commonly partially entangled and interwoven in loose debris from which narrow (1–1.5 cm wide) cylindrical foxtaillike branches extend, their leaflike lateral branchlets several times forked. Growing tips circinate. Descending "rhizoidal" branches below the scapes. Scapes slender-wiry, about 1 mm wide,

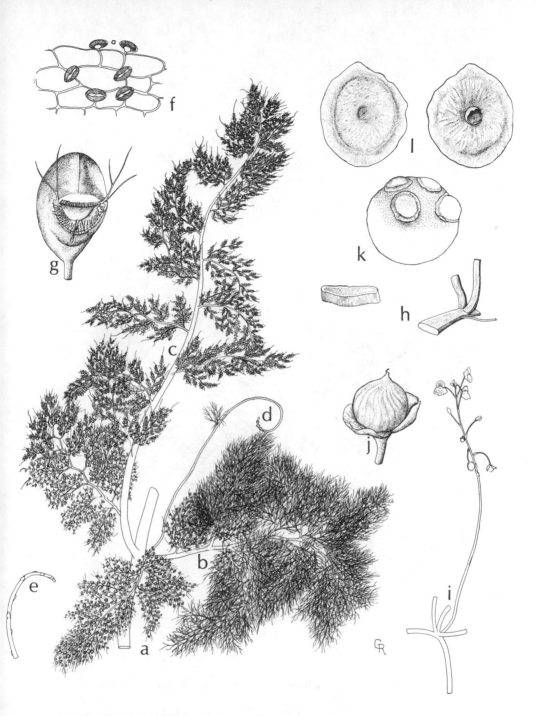

Fig. 321. **Utricularia foliosa:** a, single very short part of plant bearing three characteristic branches; b, usually nearly triangular in outline and extending horizontally in the water; c, an elongate bushy-branched part which usually descends in the water; d, a nearly naked branch the tip of which lies at the water surface; e, enlarged tip shown at d; f, very much enlarged surface of ultimate segment showing glands; g, trap; h, portions of principal axes, showing flattened cross-section; i, inflorescence; j, capsule; k, bulblike axis bearing flat seeds (not all shown); l, seed, two views.

Fig. 322. **Utricularia foliosa:** a, germinating seed; b, flower, two views.

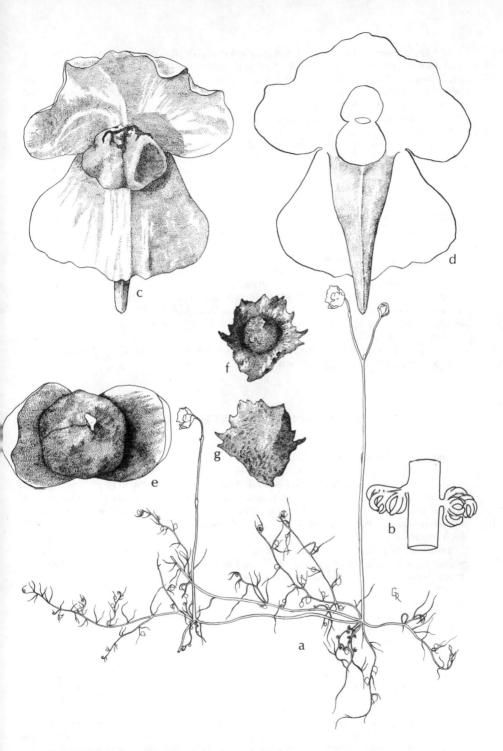

Fig. 323. **Utricularia biflora:** a, habit; b, rhizoid; c, flower, face view; d, diagrammatic view of opposite side to show spur; e, capsule; f,g, seed, two views.

1.5–3 dm in length below the 1–4-flowered inflorescence; flower stalks subtended by a usually 3-lobed reniform bract. Corolla pale yellow to greenish yellow, lips nearly vertical, nearly equal, about 12 mm broad and 9 mm long, their lower surfaces ribbed; palate saccate, puffed anteriorly, somewhat heart-shaped, pale yellow posteriorly, bright yellow and red-streaked anteriorly, centrally with copious colorless hairs, these extending into the corolla throat; spur conic, emarginate apically, ¾ as long as the lower lip or a little more. Capsule subglobose, 3–4 mm across.

Bogs, pools, swamps, ponds in flatwoods. s.e. Mass. to pen. Fla., westward to e. Tex. and e. Okla.

14. Utricularia floridana Nash.   Fig. 324

Portion of plant body usually nongreen, very loosely embedded in a soft silty or debris-laden substrate, from this chlorophyllous, cylindrical, foxtaillike branches ascend in the water, these mostly 2–5 cm across, the lateral leaflike branches several times forked. Growing tips circinate. Scapes arising intermittently from the branches within the substrate, elongate, to 1 m long, usually with some scattered, short, leaflike branches. Scapes sometimes flattened, subsucculent, 2–3 mm wide or a little more; racemes emersed, developing distally over a considerable period of time, their axes irregularly arched, bent, or twisted, commonly to about 20-flowered by full development. Corolla golden yellow, lips about equal, more or less undulate, each 1 cm long or a little more; palate saccate, red-streaked anteriorly; spur conic, nearly as long at the lip, blunt apically. Capsule globose, 7–8 mm across. Seed irregularly roughly tuberculate, dull brown.

Ponds, lakes (especially abundant in some karst lakes or ponds). Coastal plain, S.C. to cen. pen. Fla., Fla. Panhandle and s.w. Ga.

Acanthaceae (ACANTHUS FAMILY)

(Ours) perennial herbs, usually with mineral concretions (cystoliths) evident as short lines or dots on some or all of the green parts of the plant. Leaves simple, opposite. Flowers often with conspicuous subtending bracts, bisexual, bilaterally symmetrical to nearly radially symmetrical, solitary, in axillary glomerules, spikes, or capitate spikes. Calyx persistent, usually tubular near the base, 5-lobed. Corolla tubular below, commonly 5-lobed, sometimes 2-lipped (rarely 1-lipped). Stamens 4, in 2 pairs of unequal length, or 2, inserted on the corolla, sometimes 1–3 staminodes present in 2-stamened flowers. Pistil 1, ovary superior, 2-locular, 2–10 ovules in each locule, style slender, stigmas 1 or 2. Fruit a compressed 2-valved capsule, splitting and elastically spreading above the base at dehiscence, seeds flattened, borne on curved or hooked projections.

1. Leaves in a basal rosette.
  2. Scape with many hard, overlapping bracts from its base to the inflorescence.   1. *Elytraria*
  2. Scape without bracts below the inflorescence.   2. *Stenandrium*
1. Leaves not in a basal rosette.
  3. Corolla 2-lipped.
    4. Flowers subtended by obovate to oblong-elliptic foliaceous bracts.   3. *Dicliptera*
    4. Flowers subtended by narrow elongate-triangular or short, small bracts.
      5. The flowers in short axillary glomerules.   4. *Hygrophila*
      5. The flowers on long-stalked spikes.   5. *Justicia*
  3. Corolla not 2-lipped, lobes about equal.
    6. Corolla 10–14 mm across the open face, the tube not over 15 mm long from base to top of throat; leaves not exceeding 3 cm long.   6. *Dyschoriste*
    6. Corolla 30 mm or more across the open face, the tube 20 mm long or more from base to top of throat; leaves much longer than 3 cm.   7. *Ruellia*

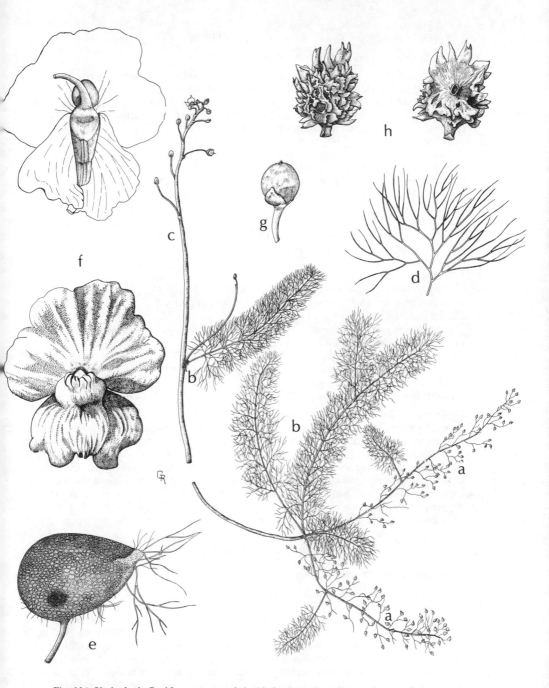

Fig. 324. **Utricularia floridana:** a, part of plant in loose mucky substrate; b, part of plant in water; c, emersed inflorescence; d, ultimate branch; e, trap; f, flower, two views; g, capsule; h, seed, two views.

1. Elytraria

Elytraria carolinensis (J. F. Gmel.) Pers.

Leaves essentially in a basal rosette, variable in size, maximally to about 20 cm long and 8 cm wide at their widest places, commonly smaller, oblanceolate to broadly oval or obovate, usually widest just above the middle, narrowing gradually proximally to a petiolar or sub-petiolar base, apically obtuse to rounded, margins irregularly shallowly wavy to entire, sparsely short-pubescent, sometimes becoming glabrous or nearly so. Inflorescence a con-gested, conical to oblong spike terminating a very much bracted scape 15–30 cm long fully developed; bracts of the scape overlapping somewhat, texturally hard, ovate, edges pale, margins more or less ciliate, sheathing the scape proximally, long-acuminate apically, the tips spreading. Spikes 2–6 cm long, 1–1.5 cm wide. Bracts of spike similar to those of the scape, marginally usually much more copiously ciliate. Flowers of a given spike bloom one at a time. Calyx scarcely tubular at base, sepals 5, unequal, 6–8 mm long, the narrower ones linear, the wider oblanceolate, acuminate apically, distal portions woolly. Corolla white to bluish, tube cylindrical-funnelform, exserted from the calyx, the 5 nearly equal lobes spread-ing, 10 mm across the face or a little more. Stamens 2, not exserted from the corolla. Cap-sule ovate, 4–5 mm long, not exserted from the persistent calyx. (*Tubiflora carolinensis* J. F. Gmel.)

Floodplain forests, wet woodlands, wet banks or seepage slopes, usually in calcareous areas or where limerock is close to the surface. Coastal plain, S.C. to s. Fla., s.w. Ga. and Fla. Panhandle.

In southernmost Florida, especially narrow-leaved *Elytraria* is designated *E. carolinensis* var. *angustifolia* (Fern.) Blake.

2. Stenandrium

Stenandrium floridanum (Gray) Small.

Low plant. Leaves in a basal rosette, petiolate, blades ovate, oval, elliptic, or obovate, abruptly narrowed to the petiole, apex rounded or obtuse, sometimes truncate, 2–4 cm long, 1–2 cm wide, margins entire. Flowers in a short spike terminating a scape 3–6 cm long. Bracts 2–2.5 cm long, foliaceous, elliptic, margins ciliate. Calyx tubular at base, the 5 lobes subulate, about 10 mm long. Corolla rose-purple, slenderly tubular below, the tube well ex-serted from the calyx, lobes obovate, unequal, spreading, about 2 cm across the open face. Capsule 9–12 mm long, somewhat exserted from the persistent calyx. Seeds hirsute. (*Gerar-dia floridana* (Gray) Small)

Seasonally wet pine-palmetto flatwoods. s. pen. Fla.

3. Dicliptera

Dicliptera brachiata (Pursh) Spreng. Fig. 325

Plant usually with widely spreading branches, often straggly, branches to about 8 dm long. Cystoliths usually evident as lines on the upper leaf surface. Stem angled, glabrous or sparse-ly short-pubescent, sometimes hirsute or villous, nodes conspicuously swollen. Larger leaves with petioles 3–6 cm long, shorter upwardly, blades ovate to lanceolate-ovate, varying in size, the larger to 14 cm long and 5–6 cm wide, broadly cuneate to rounded basally, acute to acuminate apically, glabrous or sparsely short-pubescent, margins entire. Flowers subsessile in axillary spikes or in terminal and/or axillary, stalked spikes; bracts mostly obovate, vary-ing to oblong-elliptic, ciliate. Calyx tubular only at base, the 5 segments subulate, 3–3.5 mm long, hirsute distally. Corolla pink, lavender, or bluish purple, tube cylindrical below, expanding to funnelform, then 2-lipped. Stamens 2, more or less enclosed by the upper lip of the corolla. Capsule obovate or oval, emarginate and apiculate in the notch, banded at the

Fig. 325. **Dicliptera brachiata:** a, top of stem; b, base of plant; c, flower; d, capsule; e, open capsule with seed; f, ovary and seed; g, cross-section of stem. (From Correll and Correll. 1972)

lines of dehiscence. Seeds 2–4, dark brown, flattened, nearly orbicular in outline, surface papillose-pubescent. (*Diapedium brachiatum* (Pursh) Kuntze)

Floodplain forests, stream and river banks, clearings of lowland woods. Coastal plain and outer piedmont, Va. to n. and cen. pen. Fla., westward to Tex., northward in the interior to e. Kan., Mo., s. Ind.

4. Hygrophila

Hygrophila lacustris (Schlecht. & Cham.) Nees.*   Fig. 326

Stems more or less quadrangular, surface with abundant cystoliths, glabrous or sparsely pubescent, decumbent below and rooting at the nodes, the branches ascending distally, to about 8 dm long. Plants often colonial. Leaves lanceolate or lance-elliptic, elongate-narrowed to a petiolelike base, acute apically, glabrous, margins entire; upper surface with abundant cystoliths both on and between the veins, cystoliths on lower surface mostly restricted to the veins. Flowers in sessile axillary glomerules, subtended by narrowly elongate-triangular bracts. Calyx lobes about equal, glabrous, narrowly elongate-triangular, 5–6 mm long, longer than the tube, with many cystoliths. Corolla (easily detached) bluish white (drying yellow), tube about 5 mm long, 2-lipped, upper lip 2-lobed, lower 3-lobed, lobes short, rounded apically. Stamens 4, inserted just above the middle of the corolla tube, one pair longer than the other, the longer barely reaching the throat of the corolla. Capsule linear-oblong, 6–12 mm long, acute apically.

Swamps, floodplain forests, in mud or shallow water of marshy shores, drainage ditches and canals. Fla., s.w. Ga., westward to s.e. Tex.; W.I.

5. Justicia

Plants rhizomatous and colonial. Stems more or less 4-angled, with evident cystoliths. Flowers solitary in leaf axils, or (in ours) in spikes or panicles, bracts small. Calyx 4- to 5-lobed, lobes narrow, equal or nearly so. Corolla tubular below, above 2-lipped, the upper lip bifid, more or less reflexed, lower spreading, 3-lobed; corolla white, red, lavender, or purple, with purple and/or white markings on the lower lip at the throat. Stamens 2, shortly exserted from the throat of the corolla. Capsule narrow below, abruptly widening to the seed-bearing portion, this irregularly oblong and abruptly narrowed again to a short-acuminate apex, 4-seeded, seeds disclike.

1. Stems, peduncles, bracts, calyces, and fruits with abundant glandular and nonglandular pubescence.
                                                                                               1. *J. cooleyi*
1. Stems, peduncles, bracts, calyces, and fruits glabrous, or if at all pubescent, then the hairs all nonglandular.
  2. Flowers in a congested spike terminating a stiffish peduncle.   2. *J. americana*
  2. Flowers not in congested spikes, solitary or in pairs on the spike, peduncle slender and flexuous.
    3. Calyx 12–15 mm long; leaves when fresh semisucculent, folded upward on each side of the midrib.   3. *J. crassifolia*
    3. Calyx 4–10 mm long; leaves membranous, flat.   4. *J. ovata*

1. Justicia cooleyi Monachino and Leonard.

Plant usually with several branches from near the base, each itself with ascending branches, 3–5 dm tall overall. Stems subquadrangular, abundantly clothed with white pubescence, some hairs gland-tipped, others not. Leaves mostly narrowed to petiolate bases, blades lan-

*After this book had gone to press, Donald H. Les (personal communication), reports that *H. polysperma* (Roxb.) T. Anderson is well established in riverine habitats in Lee Co., Fla., and is known from one collection from Pasco Co. It is native to the East Indies and was apparently introduced via the aquarium plant industry. While the flowers of *H. lacustris* are well distributed in distinct axillary clusters, those of *H. polysperma* are more or less hidden in crowded apical leaf axils.

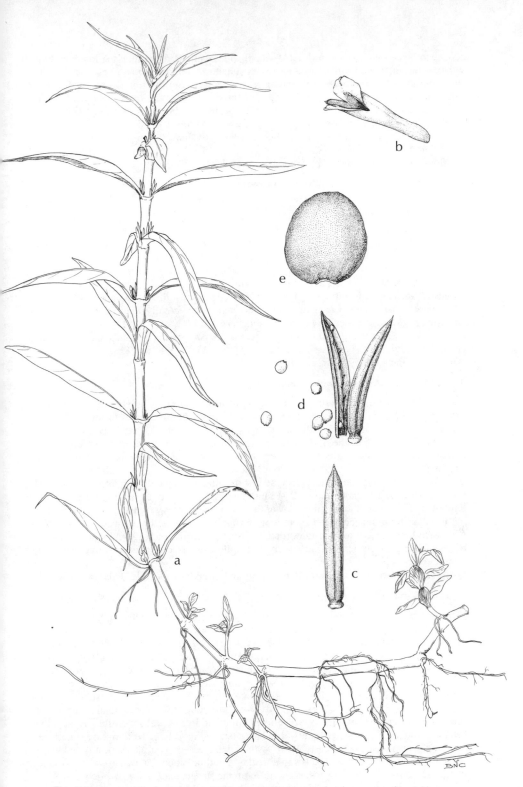

Fig. 326. **Hygrophila lacustris:** a, habit; b, corolla; c, capsule; d, capsule after dehiscence and seeds; e, seed.

701

ceolate to lance-ovate, 3–7 cm long, 1–3 cm wide, median larger than the lower, not much reduced upwardly, margins entire. Spikes terminal and axillary, sometimes branched, their stalks mostly shorter than the subtending leaves, flowers mostly, not always, borne on one side of the spike (secund). Peduncles, calyces, bracts, sepals, and fruits pubescent as on the stems. Bracts linear, 3–4 mm long. Calyx lobes 4, lance-subulate, 5 mm long. Corolla deep reddish purple with white herringbone markings on the lower lip, 8–12 mm long, sparsely pubescent externally. Fruit exceeding the persistent calyx. Seed obliquely orbicular, villous, about 2 mm across.

Moist to wet woodlands. Hernando Co., pen. Fla.

2. Justicia americana (L.) Vahl. WATER-WILLOW.   Fig. 327

Plant glabrous, strongly colonial. Stems sulcate-angled, relatively coarse, often decumbent below and abundantly rooting at the nodes, ascending above, to about 1 m long. Leaves lanceolate, linear, or linear elliptic, sessile, narrowly cuneate basally, acute apically, to 20 cm long, mostly not over 2 cm wide. Inflorescences terminal and axillary, each a dense, short-oblong spike terminating a stiffish stalk that is shorter than to a little longer than the subtending leaves. Bracts mostly short-ovate, apically acuminate. Calyx lobes 5, subulate, about 7 mm long. Corolla 1–1.4 cm long, violet to nearly white, with brownish purple markings on the lower lip, upper lip recurved. Fruit glabrous, exceeding the calyx, about 12 mm long. Seed densely warty, nearly orbicular in outline, subcordate basally, about 2 mm across. (*Dianthera americana* L.; incl. *J. mortuifluminis* Fern.)

In shallow water, margins and beds of streams, marshy shores of ponds and lakes, muddy or mucky alluvium of shores and ditches. w. Que. to Wisc., Mo. and Kan., generally southward to Fla. Panhandle, Okla. and Tex.

3. Justicia crassifolia (Chapm.) Small.   Fig. 328

Plants not in water usually with 1 to several stems from subterranean bases; in water forming extensive clones; glabrous. Stems sulcate-angular, simple below. Spikes terminal or terminal and axillary, most of their stiffly erect stalks exceeding the leaves. Lower leaves much smaller than the median and upper ones, principal leaves linear to linear-lanceolate or linear-elliptic, sessile, semisucculent and mostly with the sides folded upward, margins entire or irregularly shallowly wavy, 5–10 cm long. Flowers few, mostly 3–10, relatively distant on the spike, or approximate; bracts triangular, short-attenuate apically, obtusely keeled. Calyx lobes 5, linear or lance-attenuate, 12–15 mm long. Corolla rose-lavender to purple, 2–3 cm long, with a sculptured crest on the lower lip, crest rich purple with hairy white veins marking it, upper lip oblanceolate, 1.5 cm long, 8 mm broad, lower lip 3-lobed, lobes oval distally, each lobe 1.5 cm long, 1 cm broad. Capsule exceeding the calyx, about 23 mm long. Seeds suborbicular in outline, surface finely papillose, with a rounded, rimlike margin with minute setae on it.

Wet pine flatwoods, adjacent ditches. Gulf and Franklin Cos., Fla. Panhandle.

4. Justicia ovata (Walt.) Lindau in Urban.   Fig. 329

Plants often extensively colonial, sometimes solitary, glabrous. Stems sulcate-angled, 1–5 dm tall, simple or branched. Leaves membranous, flat, mostly sessile but sometimes narrowed to short subpetiolar bases, variable, linear, lanceolate-elliptic or lance-ovate, to about 10 cm long and 0.5–3 (–4) cm wide, margins entire to irregularly shallowly wavy. Spikes terminal or terminal and axillary, shorter than to considerably exceeding the subtending leaves, flowers in pairs on the spike or solitary, the latter mostly but not exclusively on one side of the spike (secund). Bracts short-ovate, short-triangular to short-subulate, short-pubescent on the margins. Calyx 5-lobed, 4–10 mm long, linear or subulate, short-pubescent on the margins of the lobes, sometimes on the midrib as well. Corolla size apparently heteromorphic, i.e., on some plants or in some populations about 10 mm long, in others 20 mm long. Corolla purple to nearly white, with variable white to purple markings on the lower lip. Capsule 1–1.5 cm long. Seeds suborbicular in outline, about 2 mm across, with rounded-banded margins, usually somewhat papillose on the surfaces. (*J. humilis* Michx.)

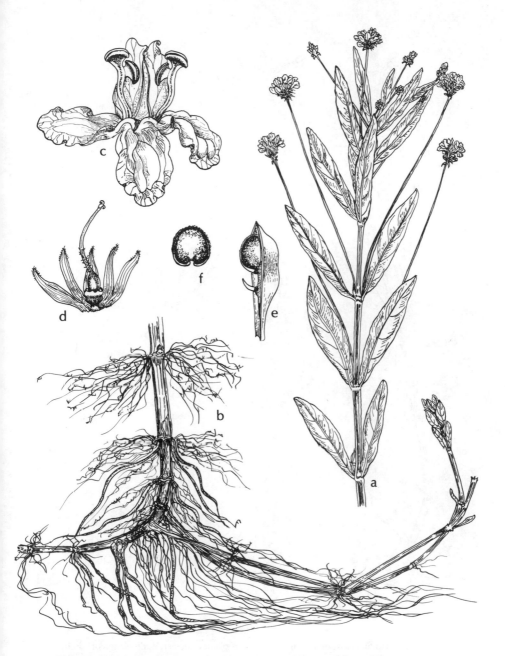

Fig. 327. **Justicia americana:** a, top of stem; b, base of plant; c, flower; d, flower, corolla removed and calyx somewhat spread; e, capsule, dehisced; f, seed. (From Correll and Correll. 1972)

703

Fig. 328. **Justicia crassifolia:** a, habit; b, top of plant; c, flower; d, dissection of corolla; e, capsule, dehisced.

704

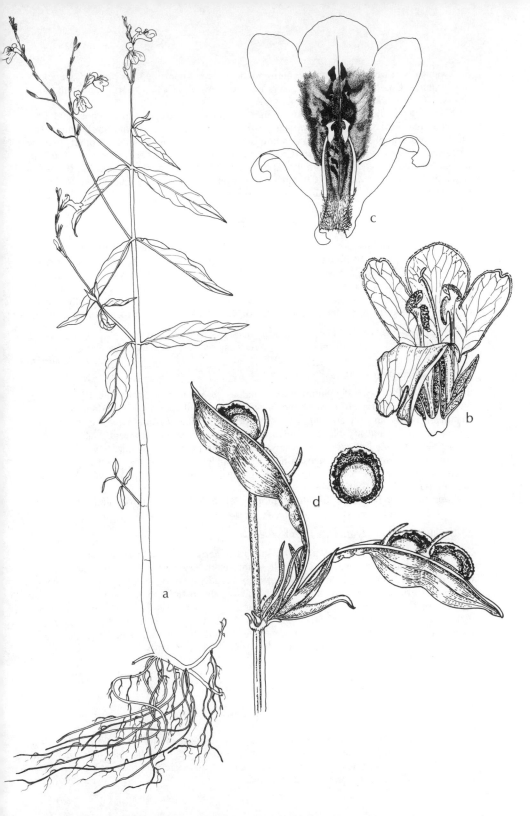

Fig. 329. **Justicia ovata** var. **lanceolata:** a, habit; b, flower; c, partial dissection of flower, showing insertion of stamens; d, capsule, dehisced, and seed. (b,d from Correll and Correll. 1972)

Swamps, wet woodlands, cypress-gum ponds or depressions, cypress prairies, wet clearings, ditches. Coastal plain, s.e. Va. to s. Fla., westward to e. Tex., northward in the interior, s.e. Okla., Ark., Tenn., s.e. Mo., w. Ky.

Plants having flowers borne mostly in pairs on the spike, the spike short and few-flowered, may be designated *J. ovata* var. *ovata*. Plants with flowers borne singly and mostly on the same side of the spike, more numerous and more distant from each other, may be designated *J. ovata* var. *lanceolata* (Chapm.) R. W. Long. A third entity (restricted to peninsular Florida) is segregated from the latter by some and is designated *J. angusta* (Chapm.) Small or *J. ovata* var. *angusta* (Chapm.) R. W. Long. We have not yet been able to discern the attributes by which authors distinguish *J. angusta*, either at the species or varietal level.

6. Dyschoriste

(Ours) herbs. Cystoliths not always evident. Flowers solitary or clustered in the leaf axils, sessile or nearly so, subtending bracts foliaceous, opposite. Calyx tube short, the segments 5, linear-aristate. Corolla blue, purple, or white, cylindric-tubular below, widened-funnelform in the upper ¾, the 5 lobes spreading, near equal or slightly 2-lipped. Stamens 4, the filaments united in pairs basally, one of each pair somewhat longer than the other, the anthers of the longer ones reaching the throat of the corolla. Ovules 1 or 2 per locule of the capsule.

- Leaves linear-spatulate, 2–3 mm wide at their widest places.   1. *D. angusta*
- Leaves lanceolate, elliptic, to suborbicular, mostly 8 mm wide or more at their widest places.
                                                                        2. *D. humistrata*

1. Dyschoriste angusta (Gray) Small.

Stem usually with several to numerous branches from the base, branches 1–2 dm tall, "leafy," sometimes branched within the substrate, short-pubescent. Leaves glabrous, linear-spatulate, mostly about 1 cm long, mostly tapering gradually from about the middle to a subpetiolar base, apically blunt though narrow, margins entire. Flowers usually solitary (sometimes 2) in the leaf axils. Calyx tube about 1 mm long, the 5 segments acicular, about 10 mm long and equaling the leaves or near so. Corolla blue or purplish, about 10 mm across the open face. Capsule 8 mm long, linear-oblong. (*D. oblongifolia* (Michx.) Kuntze var. *angusta* (Gray) R. W. Long)

In pinelands on lime-rock, usually in moist pockets, sometimes in wet sands of disturbed areas. s. Fla. and Fla. Keys.

2. Dyschoriste humistrata (Michx.) Kuntze.   Fig. 330

Stems usually short-branched within the substrate, aerial stems solitary and erect, or with few erect-ascending branches 1–3 dm tall, glabrous or very short-pubescent. Leaves essentially petiolate, blades variable in shape, mostly broadly elliptic, sometimes oval, some approaching orbicular; abruptly narrowed below to the petiolar base, obtuse to rounded apically, margins entire, mostly 2.5–3 cm long and about 8 mm wide at their widest places, glabrous. Flowers solitary (occasionally 2) in the leaf axils. Calyx tubular for 1–2 mm basally, the 5 segments acicular, 8–10 mm long. Corolla tube funnelform, 12–15 mm long, lobes short-oblong, broadly rounded apically, limb 10–14 mm across the open face. Capsule linear-oblong, about equaling the calyx.

Floodplain forests, wet pine flatwoods, wet mixed woodlands, sometimes in moist well-drained woodlands. Coastal plain, S.C. to cen. pen. Fla.

7. Ruellia

Perennial herbs. Leaves opposite. Flowers in axillary cymose clusters. Calyx tube short, the lobes long-acicular. Corolla (in ours) lavender-blue to purple, or white, long-tubular below, expanding to a throat shorter than to longer than the cylindric part of the tube, the lobes nearly equal. Stamens 4, not exserted. Stigmas 2.

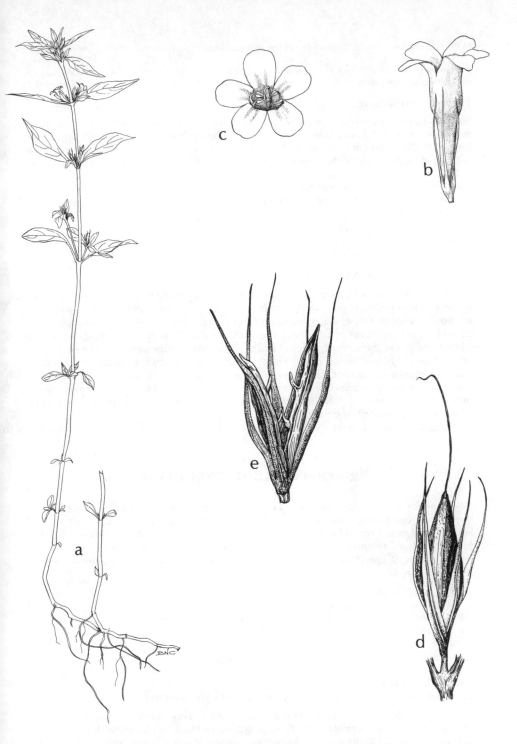

Fig. 330. **Dyschoriste humistrata:** a, habit; b, flower, lateral view; c, flower, face view;
d, capsule, with persistent calyx; e, capsule, dehisced.

- Flowers mauve to purple; midstem leaves linear-lanceolate or linear-elliptic, mostly 1 cm wide or less.

  1. *R. brittoniana*

- Flowers white, usually with inconspicuous very pale lavender-pink lines or dots above the sinuses; midstem leaves lance-ovate, elliptic, or oblong-elliptic, about 2 cm wide.

  2. *R. noctiflora*

1. Ruellia brittoniana Leonard ex Fern.

Stems to 1 m tall, purplish, decumbent at the base, hard, internodes glabrous. Leaves linear-lanceolate or linear-elliptic, narrowed to petiolelike bases, long-tapering to acute apices, the longer 10–20 (–30) cm long and to 1 cm wide or a little more, glabrous save for ciliate pubescence sometimes on the petiole bases and always short-ciliate at the nodes. Flower stalks and calyx usually stipitate-glandular. Calyx tube short, about 2 mm long, lobes subulate-attenuate, about 8 mm long. Corolla pale violet or mauve to purple, the cylindric base of the tube usually somewhat shorter than the expanded throat, 3–3.5 cm long overall, open face of the corolla 4–5 cm across. (*R. macrosperma* Greenm.)

Drainage ditches, shores of ponds or lakes, moist to wet wooded areas. Native to e. Mex., in parts of our range cultivated for ornament and sporadically naturalized: coastal plain, S.C. to Fla., w. to Tex.

2. Ruellia noctiflora (Nees) Gray.

Stems slender, erect, from a hard subterranean caudex with elongate cordlike roots, 3–4 dm tall or a little more, short-pubescent, purplish. Leaves lance-ovate, elliptic, or oblong-elliptic, broadly cuneate below, scarcely petiolate, apices bluntish, the larger 5–7 cm long, 2–3 cm wide, short-pubescent principally on the veins above and beneath, the margins with small whitish spiculelike pubescence; bracteal leaves more densely pubescent and with some stipitate glands. Calyx tube very short, the lobes acicular, about 4 cm long, with both glandular and nonglandular pubescence. Corolla white, usually with inconspicuous, pale lavender-pink lines or dots above the sinuses; corolla tube long-cylindrical, about 6 cm long, expanding to a throat about 1–2 cm long; corolla 7–8 cm across the face. Flowers open during the night, corollas falling about midmorning.

Wiregrass bogs, wet pine savannas. Coastal plain, s.e. Ga. and n. Fla., westward to La.

Bignoniaceae (BIGNONIA FAMILY)

Woody plants, shrubs, trees, or vines. Leaves opposite or whorled, rarely alternate, simple or pinnately compound, without stipules. Flowers large and showy, bisexual. Calyx united below, 2–5-lobed. Corolla largely tubular, very much exceeding the calyx, 5-lobed, 2-lipped to nearly radially symmetrical. Fertile stamens 2 or 4, inserted on the corolla tube. Pistil 1, style 1, stigma 2-lipped, ovary superior, 2-locular. Fruit a capsule. Seeds numerous, compressed, winged.

1. Trees; leaves simple.   1. *Catalpa*
1. Woody vines; leaves compound.
   2. Leaflets 2, margins entire, the leaf axis terminated by a branched tendril.   2. *Bignonia*
   2. Leaflets 7 or more, margins serrate, without a tendril.   3. *Campsis*

1. Catalpa

Catalpa bignonioides Walt.
INDIAN-BEAN, CIGAR-TREE, CATAWBA-TREE, CATERPILLAR-TREE.

A small to medium-sized tree, the bark thin and scaly. Leaves simple, deciduous, in whorls of 3 or opposite, with petioles about ½ the length of the blade, blades ovate, basally cordate to truncate, apically acuminate, 1–3 dm long, longest leaves twice as long as broad at the base, shortest leaves as broad basally as long, margins entire or somewhat wavy, pubescent on the veins above, pubescent beneath. Flowers in panicles 1–3 dm long and nearly as broad. Calyx 2-lobed, 6–10 mm long. Corolla white externally, marked within by yellow lines and brownish purple streaks and spots, tube campanulate, expanded at the throat, strongly 2-lipped, the lobes short, broader than long, rounded to truncate, crisped marginally. Fertile stamens 2 or 4, with 1 or 3 staminodia. Capsules long-cylindrical, 1–3 dm long or longer, 1–1.5 cm in diameter. Seeds winged on each side, the extremities of the wings cut into a fringe. (*C. catalpa* (L.) Karst.)

Floodplain forests, river and stream banks, low woodlands; cultivated widely for ornament and as a source of caterpillars used for fish bait and naturalized in uplands, chiefly in the vicinity of human dwellings, more or less throughout our range and beyond.

2. Bignonia

Bignonia capreolata L. CROSS-VINE, TRUMPET-FLOWER.

A high-climbing woody evergreen vine. Twigs show cross-shaped pith in cross-section. Leaves opposite, compound, each with 2 leaflets (sometimes 1) and a branched tendril which clings by means of adhesive disks at its branch tips. Leaflets glabrous, oblong-ovate, oblong-elliptic, lanceolate, or rarely obovate, their bases rounded to subcordate or auriculate, the apices acute to acuminate, rarely obtuse, margins entire. Flowers in axillary clusters of 2–5, with large showy orange or reddish orange corollas. Calyx small, cuplike, truncate at its rim or with 5 very short, blunt lobes. Corolla about 5 cm long, widely tubular above a narrow base, slightly bilabiate with 5 broad, short, and spreading lobes, orange or reddish orange, yellow to red within. Stamens 4, sometimes with a rudiment of a 5th, inserted on the corolla and not extending beyond it. Ovary superior, with a long style and bifid stigma. Fruit a flattened capsule, 1–2 dm long and about 2 cm wide. Seeds flat, oblong, broadly transversely winged. (*Anisostichus crucifera* (L.) Bureau; *A. capreolata* (L.) Bureau)

In a wide range of habitats relative to moisture condition of the soil, upland woodlands of various mixtures, floodplain and lowland woodlands, edges of evergreen shrub-tree bogs, thickets, fence rows. e. Md. to s. Ohio and s. Mo., generally southward to s. Fla. and La.

3. Campsis

Campsis radicans (L.) Seem. TRUMPET-CREEPER, COW-ITCH-VINE.

A woody vine, commonly trailing in fields, usually climbing by means of roots along the stem. Leaves opposite, deciduous, pinnately compound, leaflets mostly (5–) 7–11 (–15), mostly ovate but some lanceolate or nearly orbicular, 4–8 cm long, margins coarsely serrate, rounded at base, the apices acuminate. Flowers crowded in a terminal cluster, showy. Calyx tubular-campanulate, 1.5–2 cm long, with 5 erect, short-acuminate lobes about 5 mm long. Corolla funnelform, 6–8 cm long, slightly bilabiate and with 5 short, broad, spreading lobes. Stamens 4, borne on the corolla tube and about reaching its mouth. Ovary superior, with a long style and bifid stigma. Fruit a fusiform, falcate capsule, 1–2 dm long. Seeds broadly transversely winged. (*Bignonia radicans* L.)

In a wide variety of habitats, floodplain forests, lowland and upland woodlands of various mixtures, thickets, fence rows, old fields. N.J. to Iowa, generally southward to Fla. and Tex.

Martyniaceae (UNICORN-PLANT FAMILY)

Proboscidea louisianica (Mill.) Thell. UNICORN-PLANT, DEVIL'S CLAW.

Coarse annual, densely viscid-pubescent, strong-scented, the stems to 1 m long, prostrate-spreading, branches sometimes ascending. Leaves opposite or the upper subopposite, with stout petioles to 2 dm long, usually longer than the blades; blades reniform-orbicular to broadly ovate, usually cordate basally, to 25 cm long or more, often broader than long, margins entire or irregularly wavy. Flowers 8–20 in a loose, open raceme, their stalks 2–3 cm long at anthesis, to 4.5 cm long on the fruits, the corollas large and showy. Calyx (subtended by 2 bracts), partially united, somewhat membranous, split to the base along the lower side, 5-lobed, to 2 cm long. Corolla declined, tubular portion broadly campanulate, the limb spreading, 2-lipped, the 5 lobes broad, dull white to pinkish or purplish throughout, occasionally nearly clear reddish violet, usually mottled or blotched with reddish purple or yellow, internally with reddish purple spots the length of the tube. Stamens 4. Ovary superior, 1-locular with 2 parietal placentae, style 1, slender, stigma 2-lobed. Fruit a 2-valved capsule, usually crested on one side, with a very prominent curved beak, the beak splitting and elastically spreading, at maturity the split halves strongly incurving thus giving a markedly 2-horned effect. (*Martynia louisana* Mill.)

River banks and bars, wet ditches, fields, roadsides, gardens, waste places generally. According to Small (1933) native to the Mississippi valley; sometimes cultivated for its young pods, which are made into pickles, and spreading from cultivation; approximate present range, according to Fernald (1950), W.Va., s. Ohio, s. Ind., Ill., Minn., southward to Ga. and Tex.

Plantaginaceae (PLANTAIN FAMILY)

Plantago (PLANTAINS)

Annual or perennial herbs, those treated here all with few to numerous, alternate leaves closely set on a short stem. Inflorescences 1 to many, scapose, terminal, bracteate spikes borne from leaf axils. Flowers small, radially symmetrical, each sessile or subsessile in the axil of a bract, bisexual (in those treated here). Sepals 4, the two next to the subtending bract somewhat different from the two next to the axis, persistent. Corolla salverform, united half or more of its length, 4-lobed, chaffy and veinless, persisting after anthesis, the tube closely investing the maturing fruit, the lobes withering beyond the capsule and reflexed, spreading, or erect and more or less connivent, in some disintegrating by full maturation of the fruit. Stamens 4 or 2, inserted on the corolla tube. Pistil 1, ovary superior, 2–4-locular; style filiform, stigmatic for most of its length. Fruit a capsule dehiscing circumscissily. Seeds 1 to several in each locule.

1. Leaves narrowly linear.
  2. Capsule 3 mm long; seeds 10 or more per capsule; seed asymmetrically angular, 0.7–0.8 mm long.
                                                          1. *P. hybrida*
  2. Capsule 1–2 mm long, seeds 4 (or 2 by abortion) per capsule; seed oblong to lanceolate in outline, symmetrical, 1.2–1.3 mm long.   2. *P. elongata*
1. Leaf blades lanceolate to ovate.
  3. Spike distinctly interrupted throughout, portions of the axis freely exposed between flowers or groups of flowers.
    4. Leaf blades lanceolate, not over 4 cm broad at their broadest places, pubescent on both surfaces; bracts triangular-acute.   3. *P. sparsiflora*
    4. Leaf blades broadly ovate or oval, larger ones 6–20 cm broad at their broadest places, glabrous; bracts broadly ovate, obovate, or nearly orbicular.   4. *P. cordata*
  3. Spike, excepting the basal portion, densely flowered, the axis not or barely exposed.
    5. Capsule oblong-ellipsoid, broadest medially, very little tapered distally to a rounded summit, dehiscing at about the middle; sepals broadly rounded apically; surface of seed finely lined-reticulate.
                                                          5. *P. major*
    5. Capsule lance-ovoid, broadest below the middle, tapered to a blunt summit, dehiscing below the middle; sepals acute or narrowly blunt apically; surface of seed very minutely alveolate but not lined-reticulate.   6. *P. rugelii*

1. Plantago hybrida Bart.

Annual with 2 or 3 to many leaves and 1 to many inflorescences 6–20 cm tall. Leaves narrowly linear or linear-subulate, 5–10 (–15) cm long, bluntly callous-tipped, irregularly short-shaggy-pubescent. Scapes sparsely pubescent, spikes narrow, loosely flowered and portions of the axes exposed, to 10 cm long or a little more. Bracts glabrous, 2.5 mm long, distinctly longer than the calyces in the upper half of the spikes at least, somewhat saccate basally, a green central portion about as wide as the hyaline margin to either side of it. Sepals glabrous, the 2 nearest the bract notably inequilateral and falcate, with a narrow green midrib and hyaline otherwise, tips rounded; the 2 sepals away from the bract inequilateral only basally, green central portion keeled, margins hyaline, tips widely obtuse. Capsule ovoid, rounded apically, about 3 mm long and twice as long as the calyx, the corolla lobes erect beyond the capsule, dehiscence at the middle. Seeds 10 or more per capsule, dark brown, asymmetrically angular, mostly about 0.8 mm long, surfaces pitted. (*P. elongata* Nutt.)

Fallow fields, waste places, alluvial outwash, mud flats. N.J. and Pa. to Mo., generally southward to n. Fla. and Tex., westward to Calif.; southward to Arg.

2. Plantago elongata Pursh.

Closely similar to *P. hybrida* in general features. Bracts, sepals, and capsules similar but somewhat smaller. Bracts 1–2 mm long, varyingly shorter than to slightly longer than the calyces. Capsule 1–2 mm long, ⅓ longer than the calyx, dehiscence somewhat below the middle. Seeds 4 per capsule (or 2 by abortion), symmetrical, oblong to lanceolate in outline, 1.2–1.3 mm long, brown, surfaces pitted. (*P. pusilla* Nutt.)

Open places, in fields, on shallow soil and in pools of rock outcrops, generally in alkaline places. Mass. to s.e. N.Y., to Ore. in the northern tier of states, generally southward from N.Y. and Minn. to Ga. and Tex.

3. Plantago sparsiflora Michx. PINELAND PLANTAIN.

Perennial. Leaf blades long-lanceolate, long-tapered basally to winged petioles, acuminate apically, 8–20 cm long overall, blades 1–4 cm broad at their broadest places, pubescence scattered on both surfaces, margins entire. Inflorescences 1 to several, their axes more or less suffused with reddish purple, sparsely pubescent; spikes 20–40 cm long, loosely flowered and with portions of their axes visible between flowers or groups of flowers. Bracts glabrous, triangular-acute, 1–1.5 mm long, much shorter than the calyces. Sepals about 2.5 cm long, oblong or oblong-elliptic, inequilateral, with a somewhat keeled green midportion and either side scarious and more or less reddish purple. Capsule 3–3.5 mm long, ovoid or lance-ovoid, summit rounded to truncate, dehiscing a little above the base. Seeds 2 per capsule, 2.5 mm long, purplish black, one face rounded, the other deeply grooved, surface minutely pitted.

Seasonally wet pine savannas and flatwoods, adjacent ditches. Coastal plain, s.e. N.C. to n.e. Fla.

4. Plantago cordata Lam.

Glabrous perennial, bases usually in water, with a stout, corky, often partially hollow caudex. Leaves with petioles about as long as the blades; blades 1–3 dm long, 6–20 cm broad basally, ovate or oval, bases short-tapered to cordate, apices obtuse, margins entire. Inflorescences usually several, scapes hollow, 1–4 dm long, spikes 8–20 cm long, interrupted and the axes freely exposed between the flowers or groups of flowers. Bracts 2–2.5 mm long, little shorter than the calyces, broadly ovate, obovate, or nearly orbicular, broadly rounded at the summit. Sepals similar to the bracts, slightly longer. Capsule equaling or slightly surpassing the calyx, broadly ellipsoid to ovoid, dehiscing near the middle, the persistent corolla tube narrowed above and extending beyond the capsule, lobes reflexed. Seeds 2–4 per capsule, slenderly elliptic in outline, a pit on the inner face, about 3 mm long, dark brown.

Fresh-water marshes, small woodland streams and adjacent wet woodlands, ditches. s.e. N.Y. and s. Ont. to Mich. and Minn., generally southward to Ga. and La.

5. Plantago major L. COMMON PLANTAIN.

Fig. 331

Perennial. Leaves with petioles equaling to a little longer than the blades, varying greatly in size depending upon soil moisture and fertility levels; blades 5–30 cm long, ½–⅔ as broad near the base, broadly short-tapered basally, obtuse apically, surfaces (of plants in our range) glabrous to sparsely short-pubescent, margins entire, irregularly dentate or undulate. Inflorescences usually several, scapes not hollow, 1–2 dm long, sparsely pubescent, spikes 5–30 cm long, densely flowered except near the base. Bracts about equaling to a little surpassing the calyces, oblong to ovate, without keels, or keeled near the base, varying to keeled throughout their length, apices mostly rounded or obtuse. Sepals similar to the bracts. Capsule oblong-ellipsoid, mostly broadest at the middle, little tapered distally to a rounded summit, about twice as long as the calyx, dehiscing near the middle. Seeds 6–16 per capsule, 0.5–1.5 mm long, irregularly angular, brown, finely lined-reticulate.

Weedy, in lawns and gardens, roadsides, waste places, fields, clearings of bottomland woodlands, marshes, wet meadows, stream banks and exposed bars. Native of Eur., widely naturalized in much of temp. N.Am.

6. Plantago rugelii Decne.

Fig. 331

Closely resembling *P. major* in habit, leaves, and spikes (and often misidentified as *P. major*). Bracts mostly about 2 mm long and about half as long as the calyces, long-triangular, usually keeled, acute apically. Corolla tube little if at all surpassing the capsule, usually sloughed by the time the capsule is fully mature. Capsule lance-ovoid, broadest below the middle, tapered to a blunt summit, ½–⅔ longer than the calyx, dehiscing well below the middle. Seeds 4–10 per capsule, 1–2 mm long, irregularly angular, dark brown, finely pitted.

Rocky and gravelly shores of streams, creek beds, damp to wet shores, low woodlands and clearings, lawns, waste places. s.w. Que. to N.D., generally southward to Ga. and n.e. Okla.

Rubiaceae (MADDER FAMILY)

Trees, shrubs, or herbs, rarely climbing, ours mostly herbs. Leaves simple, entire, opposite or whorled and connected by interposed stipules, *or* stipules foliaceous, not distinguishable from the leaves, the stipules and leaves taken together giving the appearance of a whorled arrangement (in which case the leaves are said to be "appearing whorled"). Flowers bisexual, radially symmetrical in most, ovary inferior or partly so. Floral tube surmounted by 4 or 5 calyx segments, these sometimes lacking or nearly so. Corolla inserted on the rim of the floral tube within the calyx, (3)4- or 5-lobed, salverform, rotate, or funnelform. Stamens inserted on the corolla tube, as many as the lobes of the corolla and alternate with them. Pistil 1, 2–4-locular, each locule bearing 1 to many ovules. Fruit a capsule, berry, drupe, or separating at maturity into 1-seeded nutlets.

1. Plants woody, shrubs or small trees.
  2. Flowers in dense globose heads; calyx segments 4, very short, about 0.5 mm long, rounded apically; corolla white, lobes 4.   1. *Cephalanthus*
  2. Flowers in loose cymes; calyx segments 5, 1 of them, in outer flowers of the cyme at least, becoming very much larger than the others during anthesis and appearing like a large petaloid, pink to yellowish bract, collectively the petaloid sepals making the cyme very showy; other sepals lance-subulate, 10–15 mm long; corolla greenish yellow, mottled with brown or purple, lobes 5.   2. *Pinckneya*
1. Plants herbaceous.
  3. Leaves "appearing whorled."   3. *Galium*
  3. Leaves opposite.

Fig. 331. a–d, **Plantago major:** a, habit; b, flower; c, capsule with subtending calyx; d, seed; e–g, **Plantago rugelii:** e, flower; f, capsule with subtending calyx; g, seed.

713

4. Flowers in stalked pairs terminating the branchlets, the floral tubes of the pair fused near their bases; plant creeping and rooting at the nodes, evergreen; fruit a soft, bright red, berrylike drupe.
6. *Mitchella*

4. Plant without the above combination of characters.

   5. Ovules single in each locule of the ovary, or seeds solitary in each locule of the fruit.

      6. Flowers numerous in sessile, dense, axillary clusters.

         7. Fruit at maturity tardily splitting longitudinally, one part carrying with it the septum and therefore closed on the inner face, the other part open on the inner face; unopened capsule flat and circular at the summit.
7. *Spermacoce*

         7. Fruit at maturity readily splitting longitudinally in such a way that it gapes at the summit, both parts open on the inner face; unopened capsule without a flat and circular summit.
8. *Borreria*

      6. Flowers usually solitary in the leaf axils, infrequently 2 to several in each axil but not numerous and in dense clusters.
9. *Diodia*

   5. Ovules several to many in each locule of the ovary, or seeds several to many in each locule of the fruit.

      8. Calyx segments 4 and corolla 4-lobed.
4. *Hedyotis*

      8. Calyx segments 5 and corolla 5-lobed.
5. *Pentodon*

1. Cephalanthus

Cephalanthus occidentalis L. COMMON BUTTONBUSH.   Fig. 332

A scrubby shrub, rarely with the form of a small tree, to about 3 m tall. Twigs reddish brown, glabrous or short-pilose at first and becoming glabrous, with raised, corky lenticels; bark of older stems rough, ridged and furrowed, or bumpy. Leaves deciduous, opposite or in whorls of 3 or 4 (varying on a given plant), petiolate, petioles varying from 0.5–3 cm long, pubescent or glabrous; blades pinnately veined, oval, oblong-oval, elliptic, or ovate, very variable in size, 7–15 cm long, 3–10 cm broad, their bases broadly rounded to cuneate, apices acute or acuminate, upper surfaces glossy green, lower dull, sometimes glabrous, ours mostly short-pilose at least on the principal veins, sometimes uniformly soft-pubescent; stipules short-deltoid, these leaving stipular lines between the petioles after being sloughed. Flowers small, very numerous, in dense globose heads 3–4 cm in diameter, the heads long-stalked, solitary or in few-headed cymes borne both terminally and axillary on twigs of the season. Flowers sessile. Calyx segments 4, about 0.5 mm long, rounded apically. Corolla white, tube slenderly funnelform, 6–10 mm long, pubescent interiorly, with 4 short, rounded, spreading lobes, in bud a black gland present at the base of some or all of the sinuses, these sometimes persisting. Filaments short, inserted just below the sinuses of the corolla, anthers barely exserted. Style long, much exserted from the corolla, stigma capitate. Fruit 4–8 mm long, surmounted by the erect, persistent calyx segments, fruit eventually splitting into 2 or 4 obpyramidal nutlets.

The buttonbush has a dishevelled, scrubby appearance owing to the dying of leader shoots leaving dead and dying stumps.

Swamps, sloughs, shallow ponds, in small streams and on stream banks, marshes. e. Can. to Minn., generally southward to Fla. and Tex.; western states (as var. *californicus* Benth.); Mex., W.I. Ours is the var. *occidentalis* (incl. var. *pubescens* Raf.).

2. Pinckneya

Pinckneya bracteata (Bartr.) Raf. FEVER-TREE, PINCKNEYA, MAIDEN'S-BLUSHES.
Fig. 332

A shrub or small tree, very handsome and showy during its flowering season. Young twigs of the season, petioles, inflorescence axes, floral tubes, and perianth parts densely soft-pubescent. Twigs tawny in color at first, becoming reddish brown as some or all of the pubescence is sloughed and with pale, raised, corky or warty lenticels. Leaves deciduous, opposite, petiolate, petioles mostly 1–3 cm long; blades oval, elliptic, or ovate, varying in size on a given

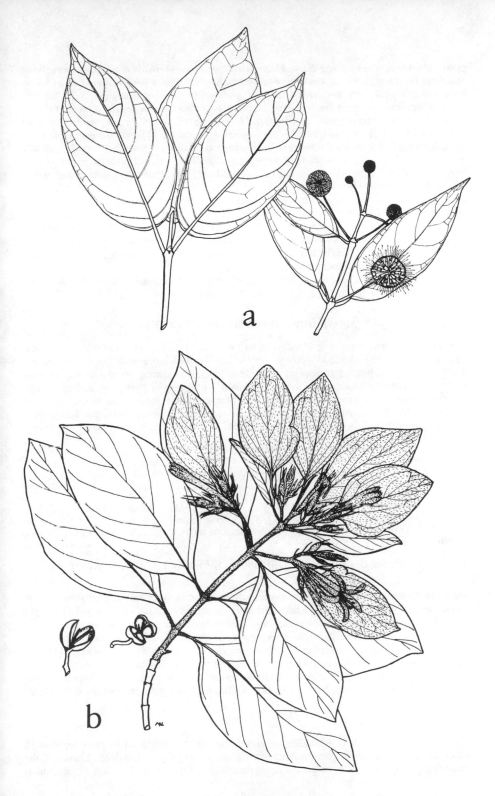

Fig. 332. a, **Cephalanthus occidentalis;** b, **Pinckneya bracteata.** (From Kurz and Godfrey, *Trees of Northern Florida.* 1962)

plant from 4 to 20 cm long and 2.5 to 12 cm wide, their bases broadly cuneate and somewhat decurrent on the petioles, apices obtuse to acute, upper surfaces with scattered short hairs, lower usually uniformly but moderately soft-pubescent; stipules triangular, very quickly sloughed leaving a stipular line between the petioles. Flowers in loose, few-flowered cymes terminally on the branches and often from the 1 or 2 nodes below. Calyx segments 5; at least 1 segment of some flowers of a cyme becoming greatly enlarged as the flower develops, these variable in size on individual flowers of a given cyme, leaflike in form but petaloid, mostly pink, some sometimes yellowish, largest ones ovate, 6–7 cm long and 4–5 cm broad, obtuse to rounded apically, lasting several weeks and rendering the cymes showy from a distance; other calyx segments lance-subulate, 1–1.5 cm long. Corolla greenish yellow, mottled with brown or purple, tube narrowly funnelform, 1.5–2.5 cm long, lobes long-triangular, shorter than the tube, usually somewhat curved-reflexed. Stamens inserted somewhat above the base of the corolla tube, exserted considerably beyond the throat. Style longer than the stamens, stigma capitate. Fruit a subglobose to ovoid, 2-valved, hard, brown capsule, its summit flattish and ringed by a scar left by the deciduous perianth. Seeds 2–3 mm long, numerous, flat, in a vertical stack in each locule. (*P. pubens* Michx.)

Bays, branch bays, seepage swamps, hillside bogs in pinelands, often associated with poison sumac. Coastal plain, s.e. S.C. to Fla. Panhandle.

3. Galium (BEDSTRAWS, CLEAVERS)

Annual or perennial herbs with slender, square or 4-angled stems. Leaves sessile or short-petiolate, "appearing whorled." Flowers very small, in single or branched, terminal and axillary cymes, sometimes 1 to several in the leaf axils. Floral tube ovoid globose, or subglobose at anthesis, becoming lobed or "twinned" and broader than long in fruit. Calyx segments none. Corolla rotate, 3- or 4(5)-lobed. Stamens short, inserted below the sinuses of the corolla. Styles 2, stigma capitate. Fruit dry or somewhat fleshy, each 1-carpellate portion globose or nearly so (sometimes one portion fails to develop), separating when ripe into 2 seedlike, indehiscent, 1-seeded parts.

1. Floral tube and fruit copiously pubescent with hairs hooked at their tips.   1. *G. aparine*
1. Floral tube and fruit without hooked hairs, smooth or at most scaly or bumpy.
  2. Leaves cuspidate at their tips.   2. *G. asprellum*
  2. Leaves blunt to rounded at their tips.
    3. Edges of the stem angles thin, in some places minutely retrorse-scabrid; leaves of the main stem mostly 5 or 6 per node; corolla lobes usually 3 (4 on some plants), obtuse apically. 3. *G. tinctorium*
    3. Edges of the stem angles markedly rounded, smooth; leaves of the main stem 4 per node; corolla lobes 4, acute apically.   4. *G. obtusum*

1. Galium aparine L. GOOSE-GRASS, SPRING CLEAVERS, CATCHWEED, BEDSTRAW.

Fig. 333

Weak, reclining or sprawling annual, commonly in dense more or less tangled stands. Stem angles armed with short, sharp, stiff, retrorsely hooked bristles. Leaves mostly 6–8 per node, linear-oblanceolate, oblanceolate, or linear, their tips cuspidate, margins and midvein beneath armed with bristles as on the stem, upper surface with sparse soft pubescence usually uniformly distributed. Flowers mostly 1–3 per cyme. Corolla white, 4-lobed, 2–3 mm across. Floral tube and fruit copiously clothed with long hairs hooked at their tips.

Floodplain forests and clearings, meadows, moist to wet thickets and waste places, usually in at least partial shade. Nfld. to Alaska, southward throughout most of the U.S.; Mex.; Euras.

2. Galium asprellum L.

Perennial, much branched, spreading or ascending, often leaning on or growing through other low plants. Stems retrorsely short-scabrid on the angles, to 2 m long. Leaves of the main stem 5 or 6 per node, oblanceolate or narrowly elliptic, 5–20 mm long, cuspidate at

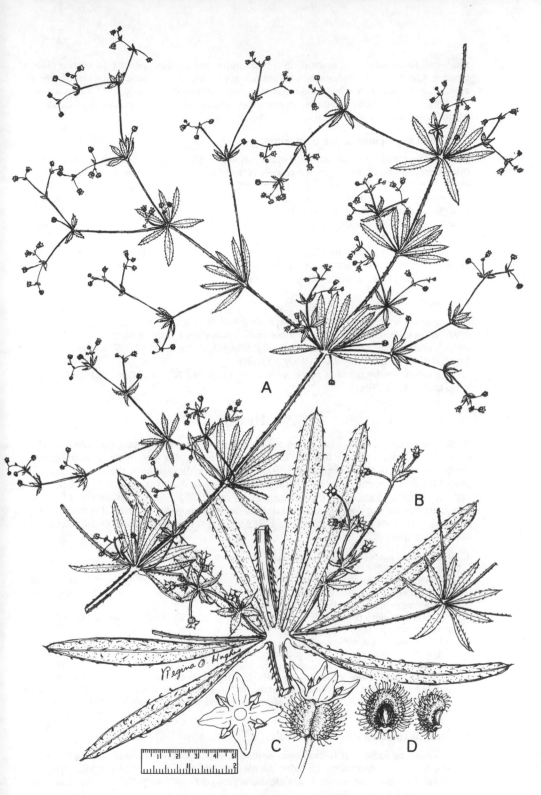

Fig. 333. **Galium aparine:** A, habit; B, enlarged leaf whorl; C, flowers, D, fruits. (From Reed, *Selected Weeds of the United States* (1970) Fig. 173)

717

their tips, finely retrorsely scabrid on their margins and usually sparsely so on the midrib beneath. Cymes compound, usually numerous, giving a diffusely paniculate-cymose overall effect. Corolla white, 4-lobed, about 3 mm across. Fruit smooth, black, 2–2.5 mm across.

Bogs, moist thickets, meadows, swamps, bottomland woodlands, springy places. Nfld. to Minn., southward to w. N.C., n. Ga., Tenn., Ill., Neb.

3. Galium tinctorium L. var. tinctorium.

Perennial with ascending or reclining stems, often in dense tangled stands. Stem angles with thin edges, in some places, at least, shortly retrorse-scabrid. Leaves on the main stems mostly 5 or 6 per node, 4 per node on some plants, oblanceolate to linear, 5–20 mm long, 1–3 mm wide, tips blunt to rounded. Cymes short, ours mostly 3-flowered, sometimes flowers single. Corolla white, usually 3-lobed, 4-lobed on some plants, less than 2 mm broad, lobes obtuse apically. Fruit dry, black, smooth or somewhat scaly, 2–3 mm broad. (*G. claytonii* Michx.)

Moist to wet thickets and clearings, boggy or marshy shores, floating islands, ditches, swamps and borders of swamps where often on rotting stumps and logs, or about the bases of trees. Nfld. and Que. to Ont., Mich., Neb., generally southward to Fla. and e. Tex.

4. Galium obtusum Bigel.

Similar to *G. tinctorium*, perhaps intergrading with it. Edges of the stem angles markedly rounded, smooth. Leaves similar to those of *G. tinctorium* but those of the main stem 4 per node. Corolla white, lobes 4, acute apically. Fruits 2.5–4 mm broad. (*G. tinctorium* sensu Small (1933); incl. *G. filifolium* (Wieg.) Small)

In habitats similar to those of *G. tinctorium*. Que. and N.S. to Minn. and S.D., southward to Ga. and e. Tex.

4. Hedyotis

Annual or perennial herbs. Leaves opposite, sessile or petiolate; stipules membranous between the leaf bases, distally truncate and entire, or with teeth or soft bristles emanating from the truncate apex of the membranous portion. Flowers small, solitary in the axils of each of a pair of leaves or sometimes solitary in the axil of 1 of a pair of leaves, in stalked umbels, or in compact axillary glomerules. Calyx segments 4; corolla 4-lobed. Ovary wholly inferior or extending beyond the floral tube (in fruit at least) and thus partly inferior, 2-locular, ovules several to numerous in each locule; style 1, stigmas 2. Fruit a capsule, loculicidally dehiscent across its summit.

1. Stipules membranous between the leaf bases, truncate and entire distally.
  2. Stems erect.
    3. Corolla pale bluish to lilac or white, with a yellow eye.   1. *H. caerulea*
    3. Corolla violet or purple, with a darker, sometimes reddish eye.   2. *H. crassifolia*
  2. Stems creeping, mat-forming.   3. *H. michauxii*
1. Stipules membranous between the leaf bases and with teeth or soft bristles emanating from the apex of the membranous portion.
  4. Flowers sessile or subsessile.
    5. Flowers solitary or 2 or 3 in the leaf axils; leaves narrowly linear-lanceolate, oblanceolate or linear-oblanceolate, several times longer than broad.   4. *H. boscii*
    5. Flowers occasionally solitary in the leaf axils, mostly numerous in compact axillary clusters; leaves elliptic to lance-ovate, little if any exceeding twice as long as broad.   5. *H. uniflora*
  4. Flowers slender-stalked, usually 2 to several in a slenderly stalked umbel, occasionally a single stalked flower in a leaf axil.   6. *H. corymbosa*

1. Hedyotis caerulea (L.) Hook. BLUET.

Low, solitary or somewhat tufted, erect annual or perennial; if perennial, then with short, filiform rhizomes connecting the tufts. Stems slender, glabrous, 5–20 cm tall. Principal leaves chiefly basal and on the lower stems, narrowed basally to petioles 3–10 mm long, blades ovate, oval, oblanceolate, or obovate, 3–8 mm long, blunt apically, margins irregularly short-ciliate; upper and bracteal leaves smaller, sessile or subsessile, mostly linear or linear-oblanceolate, acute apically. Flowers on slender stalks 2–7 cm long, solitary or in pairs above the uppermost pair of bracteal leaves, or from the upper axils as well. Calyx segments oblong, blunt apically, 1–2 mm long, surrounding only the base of the corolla tube. Corolla salverform, pale blue (rarely white) with a yellow eye, the tube narrowly cylindrical, 4–10 mm long, limb 10–14 mm across. Capsule extending much beyond the floral tube, somewhat flattened, 2-lobed, equaling or slightly shorter than the persistent calyx segments. (*Houstonia caerulea* L.)

Moist soil, grassy areas, often in lawns and fields, meadows, floodplain woodlands and their clearings, upland woodlands. N.S., N.B., Que., s. Ont. to Wisc., generally southward to Ga. and La.

2. Hedyotis crassifolia Raf. SMALL BLUET, STAR-VIOLET.

Annual, in general features resembling *H. caerulea*. Stems usually branched from near the base, sometimes simple, 2–10 cm tall. Flower stalks filiform, 1–3 cm long. Calyx segments triangular or lance-triangular, acute apically, varying from much shorter than to about as long as the corolla tube. Corolla salverform, violet to purple, with a darker, sometimes reddish eye, the tube 2–5 mm long, limb 4–8 mm across. (*Houstonia pusilla* Schoepf, *Houstonia patens* Ell.; incl. *Houstonia minima* Beck)

Moist soil, lawns, fields, roadsides, meadows, prairies, open upland woodlands. Va. to Ill., Iowa, Kan., southward to n. Fla. and e. third of Tex.

3. Hedyotis michauxii Fosb.

Perennial with creeping stems rooting at the nodes, mat-forming. Leaves of the creeping stems basally narrowed to margined petioles 3–7 mm long, blades shortly lance-ovate, ovate, or suborbicular, sometimes lanceolate, 3–6 mm long, apices obtuse to rounded, margins short-ciliate or glabrous. Flowering stems weakly erect, to 10 cm tall including the flower stalks, often numerous from the mats, collectively colorful and attractive; flowers usually solitary from the uppermost pair of reduced, sessile, bracteal leaves, their stalks filiform, 1–4 cm long. Calyx segments oblong, 1–2 mm long, blunt apically, about ⅓–½ as long as the corolla tube. Corolla a clear, often deep, blue with a yellow or white eye, the tube narrowly funnelform, limb rotate, mostly 10–14 mm across. Capsule broader than long, 2–3 mm broad, 2-lobed, somewhat shorter than the persistent calyx segments. (*Houstonia serpyllifolia* Michx.)

Moist to wet banks of streams, on rocks in streams, seepage areas on rock ledges or cliffs, adjacent ditches, about waterfalls. Mts. of Pa. and W.Va., southward to n. Ga. and e. Tenn.

4. Hedyotis boscii DC.   Fig. 334

Perennial. Stem usually diffusely branched from the base, the branches radially spreading, prostrate or nearly so, or weakly erect-ascending, 1–3 dm long, glabrous or short-scabrid on the angles. Leaves sessile or narrowed to short, subpetiolate bases, linear-lanceolate, narrowly oblanceolate, or narrowly linear-oblanceolate, 1–2.5 cm long and 1–3 mm wide, acute apically, glabrous. Flowers sessile or subsessile, solitary or 2 or 3 in the leaf axils. Calyx segments triangular-acute, 1.5–2 (–3) mm long, surfaces and edges granular (as is the floral tube). Corolla white, sometimes pink-tinged, tube very short-cylindrical, limb rotate, lobes 0.5–1 mm long, triangular to ovate-triangular, shorter than the calyx segments. Capsule not extended beyond the floral tube (wholly inferior), campanulate in outline, 2–2.5 mm long and broad. Seeds numerous, minute, strongly angular, dark purple. (*Oldenlandia boscii* (DC.) Chapm.)

Wet to dried-out sands or peats of shores or bottoms of ponds at times of low water levels, depressions in pine savannas and flatwoods, adjacent ditches, cypress-gum ponds and depressions, floodplain forests and clearings, sand and gravel bars of rivers. Coastal plain, s.e. Va. to Fla. Panhandle, westward to e. and s.cen. Tex., northward in the interior, s.e. Okla., Ark., s.e. Mo., Tenn.

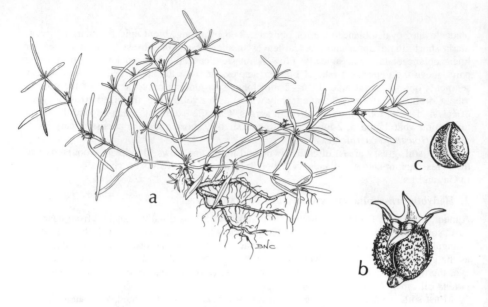

Fig. 334. **Hedyotis boscii:** a, habit; b, fruit; c, seed.

5. Hedyotis uniflora (L.) Lam.   Fig. 336

Annual. Stems simple to loosely branched, branches weakly ascending to decumbent, mostly 1–6 dm long (diminutive late-season individuals sometimes 1–2 cm high, cushionlike), varying from copiously to sparsely white-pilose on and between the angles to glabrous. Leaves sessile or subsessile, lanceolate, ovate, ovate-elliptic, 5–20 mm long and 4–10 mm broad, bases short-cuneate, occasionally nearly truncate and subclasping, apices blunt, surfaces and margins varyingly hispid to glabrous. Flowers sessile in the leaf axils, varying from solitary to several to numerous in compact clusters, mostly the latter. Calyx segments varying from narrowly triangular to deltoid or ovate-triangular, 1–1.5 mm long, the segments and the floral tube usually pilose, sometimes glabrous. Corolla white, tube very short-cylindrical, lobes rotate, a little shorter than the calyx segments. Capsule wholly inferior, about 2 mm broad and long. Seeds numerous, minute, strongly angular, from light to dark purplish. (*Oldenlandia uniflora* L.; incl. *H. uniflora* var. *fasciculata* (Bert.) W. H. Lewis)

Pools and depressions in pine savannas and flatwoods, moist to wet banks of streams, ditches, and canals, moist to wet clearings, marshy shores, interdune swales, alluvial outwash, sometimes in floating mats of vegetation. Coastal plain, L.I. to s. Fla., westward to e. Tex., northward in the interior to s.e. Mo.

6. Hedyotis corymbosa (L.) Lam.   Fig. 335

Annual. Stems glabrous, with several loosely spreading to weakly erect branches to 5 dm long. Leaves sessile or subsessile, lanceolate, linear-lanceolate, or narrowly elliptic, 1–2.5 cm long, 1–5 mm broad, bluntish to acute apically, margins glabrous or minutely scabrid. Flowers mostly in axillary, 2- to several-flowered, filiform-stalked umbels, each flower of the umbel filiform-stalked; occasionally flowers solitary and individually slender-stalked. Calyx segments triangular-subulate, about 1 mm long, minutely pubescent on the margins; floral tube glabrous. Corolla white, bearded in the throat, tube very short, lobes obovate, somewhat cupped-hooded distally, a little longer than the calyx segments. Capsule extending somewhat beyond the floral tube. Seeds numerous, minute, angular, dull brown, minutely alveolate-reticulate. (*Oldenlandia corymbosa* L.)

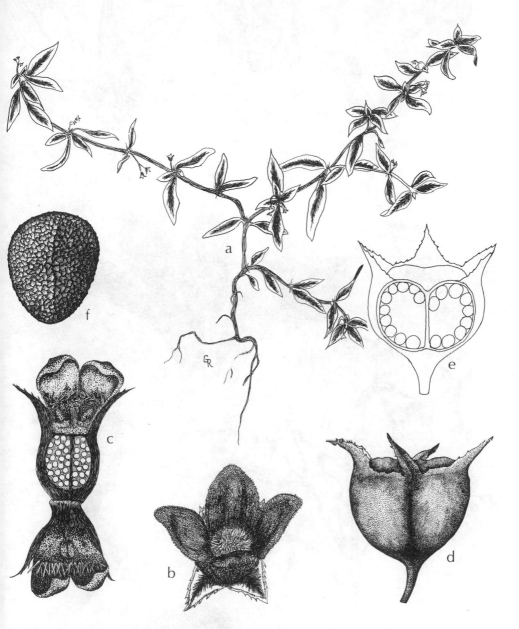

Fig. 335. **Hedyotis corymbosa:** a, habit; b, flower, from above; c, flower, cut longitudinally, cut portion folded back; d, capsule; e, diagram of longitudinal section through capsule; f, seed.

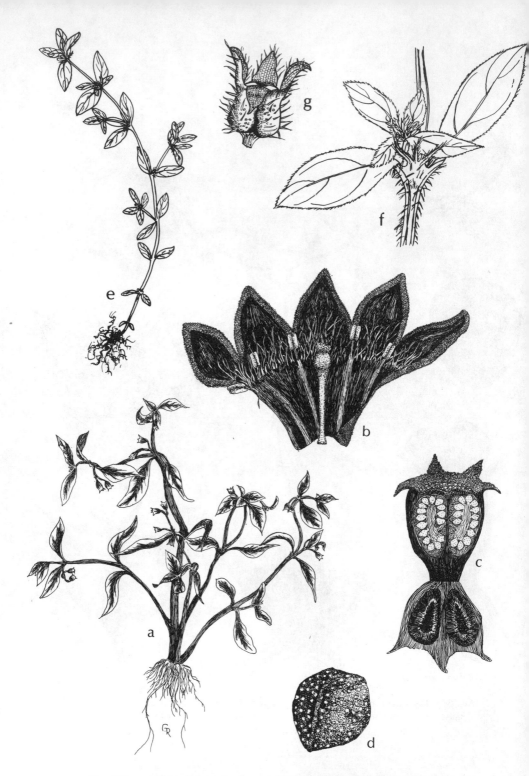

Fig. 336. a–d, **Pentodon pentandrus:** a, habit; b, corolla spread open; c, capsule, cut open longitudinally, cut portion spread back; d, seed; e–g, **Hedyotis uniflora:** e, habit (of small specimen with flower clusters not yet developed); f, node; g, fruit. (e–g, from Correll and Correll. 1972)

Pantropical. Introduced to our range, sporadic from Fla. to Tex. in moist lawns, gardens, moist to wet depressions in flatwoods, mucky shores, waste places.

5. Pentodon

Pentodon pentandrus (Schum. & Thonn.) Vatke.   Fig. 336

Low, soft annual, commonly diffusely branched from near the base, the branches sometimes prostrate and spreading radially, sometimes weakly erect-ascending. Stem 4-angled, glabrous. Leaves opposite, narrowed basally and decurrent on the distal portion of the short petioles, blades lanceolate, elliptic, or lance-ovate, 2–5 cm long, mostly obtuse apically, sometimes acute, surfaces glabrous, edges with very small, short, thickish, transparent trichomes which slough easily leaving the edges rough or with small cuplike depressions; stipules membranous between the petiole bases, irregularly few-toothed-appendaged distally. Flowers in short axillary cymes (including cymes in the axils of the uppermost pair of leaves on a branchlet) and thus these appearing terminal; flower stalks thickish, mostly 3–4 mm long. Floral tube campanulate to obovate in outline, surmounted by 5 deltoid to shortly triangular-subulate calyx segments 1–3 mm long, their margins and sometimes their outer surfaces bearing a few easily deciduous trichomes like those on the leaf edges. Corolla 5-lobed, white, quickly deciduous, tube funnelform, 2–3 mm long, lobes ovate-triangular, scarcely as long as the tube, margins thickish and involute; tube within and upper surface of corolla lobes with short, appressed pubescence, a villous ring at the throat. Free portions of the filaments very short, anthers reaching to about the bases of the sinuses of the corolla. Ovary 2-locular, ovules 10 or more in each locule. Fruit a 2-valved capsule. Seeds minute, about as broad as long, angled, surfaces alveolate-reticulate, reticules reddish brown, pits pale. (*P. halei* (T. & G.) Gray)

Moist to wet river and stream banks, swampy woodlands, marshy shores, coastal plain, Fla. to s.e. Tex.

6. Mitchella

Mitchella repens L. TWIN-FLOWER, PARTRIDGE-BERRY.

Perennial. Stems creeping, rooting at the nodes, often forming small mats, irregularly short-pubescent in lines. Leaves dark green, evergreen, opposite, short-petioled, blades 8–20 mm long, ovate, rounded to cordate basally, apices obtuse; stipules membranous between the petiole bases, and with triangular or ovate-triangular tips about 1 mm long. Flowers fragrant, in pairs on a short stalk terminating the branchlets, the bases of the two floral tubes joined (occasionally single by abortion of 1 flower). Floral tube 3–4 mm long, calyx segments 4, minute. Corolla white, often pink- or purple-tinged, tube 8–12 mm long, narrowly funnelform, only slightly dilated distally, pubescent within and on the upper surfaces of the 4 lance-oblong to ovate, spreading or recurved lobes (occasional flowers may have 3, 5, or 6 lobes). Stamens 4. All flowers of some individual plants with exserted stamens and included style and all flowers of other plants with included stamens and exserted style. Ovary wholly inferior, 4-locular, 1 ovule in each locule, style slender, stigmas 4, short-linear. Fruits twinned, each a bright red, rarely white, subglobose, berrylike drupe topped by the minute calyx segments, 4–6 mm in diameter, overwintering.

Rich deciduous or mixed well-drained woodlands, moist to wet mossy banks of woodland streams, seepage areas in woodlands where often mingled with sphagnum moss, sandy bogs. N.S. and Ont. to Minn., generally southward to pen. Fla., s.e. Okla. and e. Tex.

7. Spermacoce

Spermacoce glabra Michx. BUTTONWEED.   Fig. 337

Perennial herb. Stems glabrous, simple to much branched, weakly erect, spreading, or decumbent, 2–6 dm tall. Leaves opposite, sessile to subsessile, lanceolate, elliptic, or oblance-

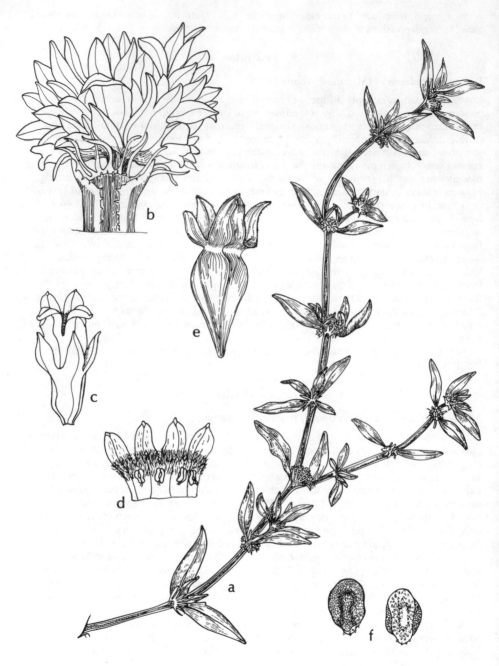

Fig. 337. **Spermacoce glabra:** a, branch of plant; b, cluster of flowers; c, flower; d, corolla, spread open; e, fruit; f, seed, two views. (From Correll and Correll. 1972)

olate, the larger ones 2–4 (–7) cm long and to 1.5 cm wide, sometimes rounded basally, mostly cuneate, if the latter some narrowed to short subpetioles, apices acute, surfaces glabrous or with sparse short pubescence on the midrib below; stipules membranous-sheathing and surmounted by a few long bristles. Flowers small, in dense axillary, sessile clusters. Floral tube glabrous. Calyx segments 4, triangular, 2–3 mm long, acute apically, finely toothed marginally, persistent. Corolla white, tube short-cylindric, equaling or slightly exceeding the erect calyx lobes, bearded in the throat, 4-lobed, lobes spreading, oblong, blunt apically. Filaments very short, inserted below the sinuses of the corolla, anthers not exserted. Ovary wholly inferior, 2-locular, a single ovule in each locule. Style short, stigmas 2. Fruit a somewhat turbinate capsule about 3 mm long, the summit, within the spreading calyx lobes, round and flat; capsule at maturity tardily splitting longitudinally into 2 separate parts, one part carrying with it the septum and therefore closed, the other part open on the inner face.

Floodplain forests and clearings, banks of rivers and sloughs. Coastal plain, s.e. S.C. to the Fla. Panhandle, westward to the e. third of Tex., northward in the interior to s.e. Kan., s. and cen. Mo., s. Ill., s. Ind., s. Ohio.

8. Borreria

Borreria laevis (Lam.) Griseb.

Annual. Stems simple or branched, erect, ascending, or spreading, to about 5 dm long, sparsely short-pubescent chiefly on the angles, pubescence commonly sloughing leaving the stem smooth. Leaves opposite, sessile, widely spaced, elliptic, tapering from about the middle to both extremities, 2–4 cm long and 8–15 mm wide, short-pubescent on or near their edges; stipules membranous-sheathing and with a few long bristles distally. Flowers small, in dense, sessile, axillary clusters (including a cluster above the terminal pair of leaves). Calyx segments 4, linear-subulate to linear, 0.5–2 mm long. Corolla white, pubescent in the throat, the tube short-funnelform or nearly cylindric, with 4 spreading, clavate lobes considerably exceeding the sepals. Stamens 4, filaments inserted at the bases of the sinuses of the corolla, anthers somewhat exserted. Ovary wholly inferior, 2-locular, 1 ovule in each locule; style slender, stigmas 2. Fruit a pubescent capsule, 3 mm long, turbinate in outline, somewhat flattened, the summit not circular and flat (as in *Spermacoce*), at maturity readily splitting longitudinally in such a way that the summit gapes and the inner faces are open.

Flatwoods depressions, wet sands on shores of flatwoods ponds, wet clearings, edges of marshes. s. Fla., local to La.; W.I., C.Am., S.Am.

9. Diodia (BUTTONWEEDS)

Annual or perennial low herbs. Leaves opposite, sessile, entire; stipules membranous-sheathing and with several elongate bristles distally. Flowers axillary, sessile, usually solitary, infrequently 2 to several in the axils. Calyx segments 2 or 4, rarely 3, persistent. Corolla tube cylindrical to funnelform, lobes 4, rarely 3, rotate, much shorter than the tube. Ovary wholly inferior, 2-locular, each locule with 1 ovule. Fruit splitting into 2 indehiscent parts.

● Calyx segments 2, rarely 3; corolla tube narrowly cylindrical; filaments inserted at the base of the sinuses between the corolla lobes, long and slender, well exserted from the corolla; stigmas 2, linear.

1. *D. virginiana*

● Calyx segments 4; corolla funnelform; filaments inserted on the corolla tube below the sinuses, very short, anthers barely if at all exserted from the corolla tube; stigma capitate or of 2 short, rounded lobes.

2. *D. teres*

1. Diodia virginiana L.
Fig. 338

Perennial herb. Stems usually branched from near the base, main branches 2–8 dm long, wholly prostrate, decumbent below, or weakly ascending, flexuous. Leaves variable, narrowly elliptic, elliptic, lanceolate, oblanceolate, or oblong, 2–6 (–10) cm long and to about

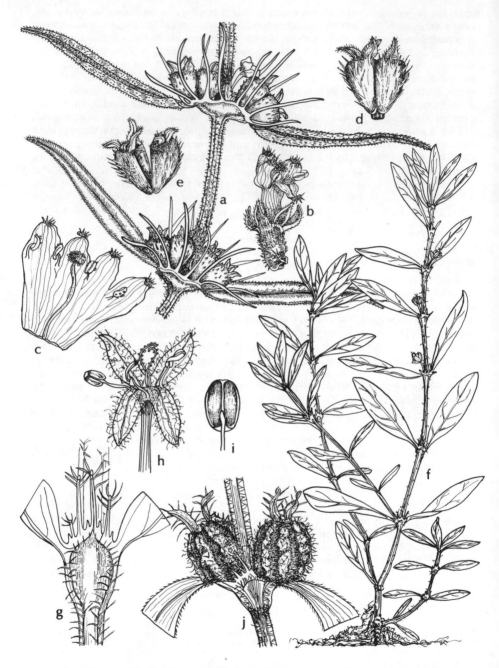

Fig. 338. a–e, **Diodia teres:** a, portion of stem; b, flower; c, corolla, opened up; d, fruit; e, fruit, splitting; f–j, **Diodia virginiana:** f, habit; g, leaf base showing stipules; h, flower; i, anther; j, node with fruits. (From Correll and Correll. 1972)

12 mm wide, mostly cuneate basally, sometimes nearly truncate, apically mostly acute, sometimes blunt, surfaces glabrous or pubescent, margins scabrid, stipular bristles flat, subulate. Calyx segments 2, rarely 3, 4–6 mm long, lance-subulate, triangular-subulate, lanceolate, or lance-ovate, pubescent. Corolla white, often faintly tinted with pink, tube narrowly cylindrical, 6–10 mm long, lobes 4, rarely 3, 3–4 mm long, ovate, obtuse apically, pubescent above. Stamens white, well exserted from the corolla, filaments slender and elongate, inserted at the bases of the sinuses between the corolla lobes. Style slender, stigmas 2, linear. Fruit 5–9 mm long, oblong-ellipsoid or oval, glabrous or varyingly pubescent, each half when fully ripe with 3 broadly rounded ribs on the back, very tardily splitting. (Incl. *D. tetragona* Walt., *D. hirsuta* Pursh, *D. harperi* Small)

Seasonally wet pine savannas and flatwoods, swamps, wet woodlands, borders of cypress-gum ponds and depressions, swales, shallow pools, marshy shores, wet prairies, ditches. s. N.J. to s. Ill. and Mo., generally southward to s. Fla. and e. third of Okla. and Tex.

2. Diodia teres Walt. **POOR-JOE, ROUGH BUTTONWEED.** Fig. 338

Annual. Stems stiff, erect, single to much branched from near the base, 1–8 dm tall, rarely more, variously pubescent (puberulent to hirsute), terete below, quadrangular distally on the branches. Leaves mostly lanceolate, sometimes linear, 2–4 cm long and to 8–10 mm wide, rounded to subclasping basally, sometimes very short cuneate, acute apically, upper surfaces scabrid, margins markedly scabrid, lower scabrid at least on the midvein; stipular bristles filiform, equaling to much exceeding the fruits. Calyx segments 4, triangular, 2–4 mm long, pubescent, transparent spicular trichomes on the margins, sometimes on the midvein, mucronate apically. Corolla pinkish purple to white, tube funnelform, 4–5 mm long, lobes 4, rotate, ovate, rounded to obtuse apically, glabrous. Filaments very short, inserted below the sinuses between the corolla lobes, anthers barely if at all exserted from the corolla tube. Stigma capitate, with 2 short, rounded lobes. Fruit about 4 mm long, obpyriform, pubescent, with a shallow longitudinal furrow on each side, readily splitting at maturity. (*Diodella teres* (Walt.) Small)

Commonly abundantly weedy on well-drained sandy soils of fields, open woodlands, pine–scrub oak ridges and hills, dunes, clearings, erosion areas, roadsides; also in seasonally wet interdune swales, wet sands of estuarine shores, shores of ponds exposed during times of low water, swales, ditches. s. N.Eng. to Wisc. and Iowa, generally southward to s. Fla. and e. two-thirds of Tex.; Ariz.

Caprifoliaceae (HONEYSUCKLE FAMILY)

(The genera treated here) shrubs, trees, or woody vines with opposite leaves with or without stipules. Flowers bisexual, ovary inferior, the floral tube constricted at the summit and bearing 3–5 calyx segments, these small and sometimes rudimentary or scarcely evident. Corolla tubular, at least below, 3–5-lobed, radially or bilaterally symmetrical. Stamens (in ours) as many as the lobes of the corolla, inserted on the corolla tube. Ovary 2–5-locular, style 1 or absent. Fruit a berry or drupe (in the genera treated here).

1. Leaves pinnately compound.   1. *Sambucus*
1. Leaves simple.
   2. Plant a twining or trailing vine.   2. *Lonicera*
   2. Plant a shrub or small tree.   3. *Viburnum*

1. Sambucus

Sambucus canadensis L. **ELDERBERRY.** Fig. 339

Soft-stemmed shrub, to about 4 m tall, the herbage with rather a rank odor when crushed or bruised, producing stout, elongate subterranean runners and colonial. Bark grayish brown, with prominent lenticels. Leaves opposite, pinnately once compound, sometimes the lower

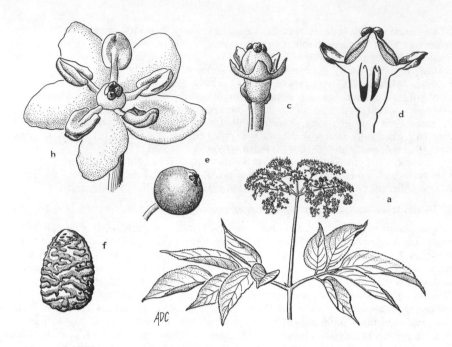

Fig. 339. **Sambucus canadensis:** a, flowering branchlet; b, flower; c, flower with corolla and stamens removed to show bracteoles and calyx lobes; d, semidiagrammatic vertical section of flower with corolla and stamens removed; e, fruit; f, stone. (From Ferguson in *Jour. Arn. Arb.* 47: 38. 1966)

leaflets further divided into three segments (and in the southeastern part of the range, in Florida and the West Indies especially, other leaflets often subdivided, to the extent that the leaf is sometimes bipinnate); petioles 3–10 cm long; leaflets mostly 5–11, lanceolate, elliptic, or ovate, variable in size, 5–18 cm long, 2–6 (–8) cm wide, serrate, glabrous or pubescent beneath, especially along the veins, sometimes downy beneath; bases tapering to rounded, apices acuminate; leaflets sessile or with stalks to 1 cm long, often (not always) with a small linear gland-tipped protuberance at the junction with the leaf axis. Flowers small, white, somewhat fragrant, 3–5 mm across, borne in large, flattish, many-flowered cymes to 2–4 dm broad, gland-tipped protuberances at the forks of the cymes. Floral tube with 2 bractlets at the base, surmounted by 5 (or 3) minute calyx lobes. Stamens 5. Ovary 3–5-locular, stigma sessile, 3–5-lobed. Fruit 4–6 mm long, a juicy, purplish black, berrylike drupe containing 3–5 stone-covered seeds. (Incl. *S. simpsonii* Rehd.)

Commonly in moist to wet open places, swamps, ditches, banks of canals and bayous, abundantly colonizing wet clearings and wet disturbed sites. N.S. and Que. to Man. and S.D., generally southward to Fla. and Tex.; W.I. and Mex.

Elderberry blossoms and fruits, one or the other, are used in the home in a variety of ways: for pies, preserves, jellies, wine, breadstuffs. The berries are relished by some songbirds. Pieces of the large, soft pith are often used in microtechnique for holding specimens in making freehand sections.

2. Lonicera (HONEYSUCKLE)

(Ours) woody twining vines. Leaves opposite, simple, without stipules. Flowers in axillary or terminal clusters. Floral tube surmounted by minute calyx segments. Corolla tubular below. Stamens 5. Stigma capitate. Fruit a several-seeded berry.

- Flowers in pairs on axillary stalks; corolla 2-lipped, whitish or cream-colored (often pinkish or tinged with purple), becoming yellow with age. 1. *L. japonica*
- Flowers in terminal interrupted spikes, usually a pair of flowers at each node of the spike; corolla long-tubular, with 5 short nearly equal lobes, coral red outside, often yellow within. 2. *L. sempervirens*

1. Lonicera japonica Thunb. JAPANESE HONEYSUCKLE.

Twining high-climbing or trailing woody vine. Younger stems pubescent. Leaves evergreen, short-petioled, the blades ovate, elliptic or oblong, mostly 4–8 cm long, entire (but leaves of vigorous new shoots in spring often lobed). Stalks bearing the pair of flowers mostly 2–10 mm long, the flowers subtended by a pair of bracteal leaves and a pair of minute bractlets. Flowers very fragrant. Calyx lobes 1–2 mm long. Corolla 3–5 cm long, whitish or cream, often pinkish or purple-tinged, becoming yellow, 2-lipped, the upper lip with 3 short lobes, the lower narrow and unlobed, the lips nearly as long as the pubescent tube. Stamens and style much exserted from the corolla. Berry black, globose or nearly so, 5–6 mm long. Seeds lustrous black, finely reticulate, 3–3.3 mm long. (*Nintooa japonica* (Thunb.) Sweet)

Native of eastern Asia and introduced. Perniciously weedy in woods, fields, thickets, roadsides, commonly overwhelming and eradicating native flora and difficult to control. s. N.Eng. to Mo. and Kan., generally southward to Fla. and Tex.

Fruits of honeysuckles are eaten by wild birds; herbage is said to be excellent deer browse.

2. Lonicera sempervirens L. CORAL HONEYSUCKLE, TRUMPET HONEYSUCKLE.

Twining or trailing woody vine, usually not diffusely branched and usually not exceeding about 5 m long. Branchlets glabrous or nearly so. Leaves tardily deciduous, mostly short-petiolate, the blades 3–7 cm long, variable in shape, oblong to elliptic, obovate or suborbicular; uppermost 1 or 2 pairs connate-perfoliate; apices obtuse to rounded, bases acute to rounded, margins entire, glaucous beneath. Flowers terminal on new growth, in interrupted spikes, 2–4 at a node, each cluster subtended by a small bract and each flower with a pair of bracteoles. Calyx none or of tiny lobes surmounting the floral tube. Corolla long-tubular, 4–5.5 cm long, with 5 short, nearly equal lobes, red without and often yellow within. Stamens and style not much exserted from the corolla. Berries red. (*Phenianthus sempervirens* (L.) Raf.)

For the most part in upland, well-drained places, open woodlands, borders of woodlands, thickets, fence rows; sometimes in floodplain woodlands and borders of wet woodlands. N.Eng. to Iowa, Neb., southward to Fla. and Tex.

3. Viburnum (ARROW-WOODS, BLACK-HAWS)

Some of the Viburnums in our area become small trees but all blossom and fruit when of the stature of shrubs. Leaves simple, opposite, deciduous. Flowers small, white or cream-colored, 5-merous, borne in terminal flattish cymes. Fruits 1-seeded drupes each of which is crowned with 5 minute persistent sepals and the stubby remains of the style.

Some of the species or species complexes of eastern North America exhibit much variation and there have been extremely differing interpretations in the delimitation of taxa.

1. Leaf blades mostly ovate, with conspicuous lateral veins, margins sharply dentate or dentate-serrate, bases subcordate, rounded, or truncate, mostly nearly as broad as long. 1. *V. dentatum*
1. Leaf blades elliptic, lance-elliptic, oval, oblanceolate, spatulate, or obovate, margins entire or weakly toothed, not sharply dentate or dentate-serrate, bases mostly tapered, rarely a leaf nearly as broad as long.
  2. Blades of larger leaves to 5 cm long, mostly oblanceolate to spatulate; cymes sessile.
2. *V. obovatum*
  2. Blades of larger leaves considerably exceeding 5 cm long, elliptic, lance-elliptic, oval, rarely broadly obovate; cymes stalked. 3. *V. nudum*

1. Viburnum dentatum L.

Shrub. Leaves with petioles 1–3 cm long, blades mostly ovate, an occasional leaf elliptic-ovate or lance-ovate, bases subcordate, rounded, or truncate, apices rounded to acute or short-acuminate, margins sharply dentate to dentate-serrate, the lateral veins conspicuous. Stalks of the cymes 2–12 cm long, cymes mostly with 6–8 principal branches, 4–10 cm across. Drupes elliptical to spherical in outline, sometimes obovate, 5–8 mm long, blue-black. Twigs, petioles, leaf blades, axes of the cymes, and fruits from glabrous to variously pubescent.

A polymorphic species complex variously interpreted by authors as comprised of several varieties or segregated into several species, some of them with varieties.

In floodplain forests, wet thickets, stream banks, wet woodlands, bogs; some of the variants in well-drained woodlands of various mixtures. Throughout our range and beyond.

2. Viburnum obovatum Walt.   Fig. 340

Shrub or small tree, stiffly branched. Leaves narrowed below to short, winged, petiolar or subpetiolar bases, blades chiefly oblanceolate, varying to spatulate, narrowly obovate, elliptic-oblanceolate, occasionally suborbicular; apically obtuse to rounded, margins usually somewhat revolute, entire, or irregularly serrate or crenate-dentate from about the middle upwards; varying in size, mostly 2–5 cm long and 1–3 cm wide, upper surfaces glabrous, the lower with many small glandular dots, similarly glandular-dotted on young twigs, petioles, and axes of the cymes. Cymes essentially sessile, with 2–5 primary branches, 4–6 cm across, flowering during emergence of new shoots in spring. Drupes elliptical to nearly spherical in outline, sometimes ovate, passing from red to black during maturation, 6–10 mm long. (Incl. *V. nashii* Small)

Stream banks, wet hammocks, floodplain woodlands, wet pine flatwoods. Coastal plain, S.C. to s. Fla., s.e. Ala.

3. Viburnum nudum L.   Fig. 341

Shrub or small tree. Leaves with winged petioles 5–20 mm long, blades varying in size and shape, to 15 cm long, narrowly elliptic, elliptic-oblong, lanceolate, lance-ovate, or obovate, bases acute to rounded, apices abruptly short-acuminate, acute, obtuse, sometimes rounded, margins mostly entire and somewhat revolute, varying to irregularly crenate-undulate, crenate-dentate, or finely serrate; upper leaf surfaces sparsely glandular-dotted, petioles and lower leaf surfaces relatively copiously glandular-dotted as are the young twigs and axes of the cymes. Cymes stalked, the stalks varying from about 5 to 25 mm long, mostly with 4 or 5 primary branches, to about 15 cm across, flowering after shoots of the season are fully expanded. Drupes mostly elliptical or ovate in outline, 6–10 mm long, during maturation varying from yellowish white to pink, then to deep blue, their surfaces with a waxy bloom. (Incl. *V. cassinoides* L., *V. nitidum* Ait.)

Swamps, flatwoods, shrub-tree bogs, wet woodlands, stream banks, creek bottoms, floodplain woodlands, heath and tamarack bogs or swamps, hemlock forests. Nfld. to Man. generally southward to n. Fla. and e. Tex.

Curcurbitaceae (GOURD FAMILY)

Annual or perennial herbaceous vines, trailing or climbing, usually with coiled, simple or branched tendrils opposite or to the side of the leaf bases. Leaves alternate, petiolate, blades simple, commonly lobed or divided. Flowers mostly unisexual (plants monoecious or dioecious), radially symmetrical or nearly so, ovary inferior. Calyx surmounting the floral tube, united below and (4)5(6)-lobed, or segments free. Corolla tubular below, the tube in some very short, 5-lobed. Staminate flowers with 3 or 5 stamens, if 5 then 2 pairs united by

Fig. 340. **Viburnum obovatum.** (From Kurz and Godfrey, *Trees of Northern Florida*. 1962)

the anthers, filaments distinct or partly or wholly united. Ovary of pistillate flowers 1- or 3-locular, placentae parietal, ovules 1 to many, style 1, stigmas 2 or 3, thick. Fruit indehiscent in most, fleshy and with an outer rind and spongy interior, or in some a paper-bladdery podlike structure.

1. Tendrils 3-forked distally; surfaces of the ovaries and fruits prickly.
   2. Stem and leaves glabrous or nearly so; pistillate flower or fruit solitary on a stalk from the leaf axil; fruit bladdery-inflated, 4-seeded, dehiscent by 2 pores at the summit.   1. *Echinocystis*
   2. Stem and leaves notably pubescent; pistillate flowers and fruits several in a headlike cluster terminating a stalk from the leaf axil; fruit firm, not inflated, indehiscent, filled with a single seed. 2. *Sicyos*
1. Tendrils not branched; surfaces of ovaries and fruits glabrous or pubescent, not prickly.
   3. Length of the larger leaf blades from point of attachment of the petiole to the tip of the central lobe 5 cm or more; pistillate flowers and fruits with very short (1–3 mm) stalks; fruit red or reddish.
                                                                                                3. *Cayaponia*

Fig. 341. **Viburnum nudum:** a, branch with inflorescence; b, flowers; c, twig with infructescence. (From Correll and Correll. 1972)

3. Length of the larger leaf blades from point of attachment of the petiole to the tip of the central lobe 5 cm or less, mostly 4 cm or less; pistillate flowers and fruit with slender stalks 1 cm long or more; fruit black or blackish green. 4. *Melothria*

1. Echinocystis

Echinocystis lobata (Michx.) T. & G. **BALSAM-APPLE.**

Monoecious annual. Stem trailing or high-climbing, essentially glabrous, tender. Tendrils 3-forked distally. Petioles about as long as the blades; blades 4–12 cm long and wide, glabrous but roughish, 5-lobed, lobes triangular-acute and short-aristate at their tips, finely serrate to entire marginally. Staminate flowers in flexuous axillary panicles much longer than the leaves. Pistillate flowers solitary, axillary, their stalks shorter than the petioles. Calyx segments 6, 1–2 mm long. Corolla rotate, 10–15 mm across, lobes 6, subulate, much longer than the tube. Stamens 3, filaments united into a column, anthers connivent, straight. Fruit bladdery-inflated, ovoid or ellipsoid, 3–5 cm long, 4-seeded, dehiscent by 2 apical pores, pulpy within but becoming dry and fibrous, surface covered by soft prickles. Seeds about 1.5 cm long, elliptic to obovate, dark gray, reticulate. (*Micrampelis lobata* (Michx.) Greene)

Floodplain forests, stream banks, alluvial thickets. N.B. to Sask., southward to Ga. and Tex.; also cultivated for arbors and freely escaping.

2. Sicyos

Sicyos angulatus L. **BUR-CUCUMBER.**

Monoecious annual. Stems trailing or climbing to several meters, viscid-pubescent. Tendrils 3-forked distally. Petioles varying from about ⅓ as long to as long as the blades; blades 6–15 (–20) cm wide, palmate, with basal more or less rounded-cordate lobes and 3 triangular lobes distally, the lateral ones acute, the terminal one acuminate, margins finely toothed, upper surfaces short-scabrid to nearly smooth, lower clammy-pubescent as are the petioles. Staminate flowers in short clusters terminating stalks as long as or somewhat longer than the leaves. Pistillate flowers sessile and clustered at the end of a stalk about equaling the petioles. Calyx segments 5, 1–2.5 mm long. Corolla white to cream, rotate, 8–12 mm across, 5-lobed, lobes deltoid. Filaments united into a column, anthers united, somewhat contorted. Fruit firm, not inflated, indehiscent, elliptic in outline, somewhat flattened, acuminate apically, 1–2 cm long, 1-seeded, surface covered with long, brittle prickles.

Stream banks, alluvial woodlands along streams and rivers, clearings, moist thickets and fields. s. Maine and w. Que. to Minn., generally southward to Fla. Panhandle and e. half of Tex.

3. Cayaponia

Monoecious (rarely dioecious) perennial vines. Tendrils (in ours) usually unbranched. Leaf blades palmately lobed or divided. Flowers solitary or few in the leaf axils, subsessile. Calyx segments 5. Corolla rotate or campanulate, 5-lobed. Stamens 3, distinct, filaments contorted. Fruit ellipsoid, pulpy, 3-locular, with few seeds in each locule.

- Terminal lobe of the leaf blade broadly triangular, usually not at all narrowed at the base.
  1. *C. grandifolia*
- Terminal lobe of the leaf blade narrowed at the base, broadest at about the middle. 2. *C. quinqueloba*

1. Cayaponia grandifolia (T. & G.) Small.

Leaf blades thin, palmately 5-lobed, the sinuses usually very broad and shallow; basal lobes rounded-cordate, lateral lobes broadly short-triangular, terminal lobe much larger than the laterals, broadly triangular, broadest at the base; margins with widely spaced minute dentations, both surfaces softly short-pubescent. Fruit 12–14 mm long.

Alluvial woodlands, clearings, and thickets along rivers and streams. Coastal plain, Miss., La., Ark.

2. Cayaponia quinqueloba (Raf.) Shinners.

Leaf blades somewhat firmer than in *C. grandifolia*, palmate, mostly with 3 principal lobes, varying from about half to almost wholly divided, the lateral lobes with a rounded, descending, basal portion and an oblongish ascending or spreading portion, the terminal lobe narrowed basally, broadest at about the middle, sometimes sublobed; margins with minute dentations, upper surface short-scabrid, lower pubescent but somewhat less scabrid. Fruit 16–20 mm long. (*C. boykinii* (T. & G.) Cogn.)

Wet alluvial woodlands and swamps along rivers and streams. Coastal plain, S.C. and Ga. to e. Tex.

4. Melothria

Melothria pendula L. MELONETTE, CREEPING-CUCUMBER.

Slender, trailing or low-climbing, perennial vine. Leaf blades mostly about as broad as long, 2–4 (–8) cm broad across the base, palmate, varying from scarcely to strongly 3–5-lobed, cordate basally; surfaces scabrid to hispid with short pustular-based hairs; petioles hispid. Tendrils unbranched. Flowers mostly unisexual, rarely bisexual. Staminate flowers in short racemes or clusters of 2–6 terminating slender stalks from the leaf axils, the pistillate or bisexual ones borne singly, one or more from the leaf axils, their stalks slender. Calyx segments 5, short-triangular. Corolla yellow, campanulate, 5-lobed. Stamens 3, filaments distinct, anthers distinct or barely connivent. Pistillate flowers with 3 stamens or 3 staminodes. Fruit 1 cm long or a little more, with numerous seeds, indehiscent, green to black, surface smooth.

Alluvial woodlands, clearings, and banks along rivers and streams, moist to wet thickets, edges of marshes and mangrove swamps, wet to well-drained hammocks. Va. to s. Ind., s. Mo., southward to s. Fla., Okla., e. and s. Tex.; Mex.

Segregates, treated either as separate species or as varieties of *M. pendula*, have been recognized from the s.e. portion of our range. We have been unable satisfactorily to distinguish them.

Campanulaceae (BELLFLOWER FAMILY)

(Ours) herbs, most with milky or watery sap. Leaves alternate, simple, unlobed, without stipules. Inflorescence basically a cyme but appearing spicate, racemose, or paniculate, sometimes flowers solitary in the leaf axils. Flowers radially or bilaterally symmetrical, bisexual, ovary wholly or partly inferior. Calyx segments 5 at the summit of the floral tube; corolla tubular at least basally, 5-lobed; stamens 5, alternate with the corolla lobes, inserted at the base of the corolla tube, or less frequently on the tube, free or variously united. Pistil 1, ovary 2–5-locular. Fruit a capsule bearing numerous seeds.

1. Flowers in a dense, cylindrical spike without bracts clearly evident exteriorly; stamens inserted about the middle of the corolla tube; capsule dehiscing circumscissily.   1. *Sphenoclea*
1. Flowers variously arranged, if in spikes, then the flowers with bracts clearly evident exteriorly; stamens inserted at the base of the corolla tube; capsule dehiscing loculicidally either on the sides or apically.
   2. The flowers radially symmetrical; stamens distinct; capsule dehiscing along the sides by valves or pores.   2. *Campanula*
   2. Flowers bilaterally symmetrical, 2-lipped; stamens with filaments and anthers united, or with only the anthers united; capsule dehiscing by apical valves.   3. *Lobelia*

1. Sphenoclea

Sphenoclea zeylandica Gaertn. CHICKEN-SPIKE.   Fig. 342

Coarse, glabrous annual. Stem hollow, branched above the base, to 1 m tall; when growing in water the base becoming spongy- or corky-thickened and with numerous fibrous roots from the nodes. Leaves short-petiolate, blades elliptic, margins entire, variable in size, to about 12 cm long and 5 cm wide. Inflorescences dense, nearly cylindric, stalked spikes terminating the branches. Flowers small, radially symmetrical, sessile, each subtended by a spatulate bract and 2 bractlets. Calyx segments short, nearly oblong, about 1.5 mm long, green excepting pale, opaque, minutely erose margins. Corolla white, slightly longer than the sepals. Stamens distinct, inserted on the corolla tube at about the middle of the tube; corolla together with the stamens quickly deciduous after anthesis. Capsule 2-locular, dehiscing circumscissily, the persistent calyx shed with the top of the capsule. Seeds numerous, minute, oblong or nearly so, about 0.4–0.5 mm long, buffish and somewhat lustrous, surface minutely roughened.

Native of the Old World tropics, widely naturalized in warm regions of western hemisphere. In our range, sporadically weedy in lowland wet areas of the coastal plain, S.C. to pen. Fla., westward to e. and s.e. Tex. and Ark.

2. Campanula (BELLFLOWERS, HAREBELLS)

Annual, biennial, or perennial. Flowers stalked, solitary to paniculate, radially symmetrical. Calyx segments persistent on the capsule. Corolla usually blue or bluish purple, varying to white on occasional plants of the same species, tubular basally or to somewhat beyond the middle, campanulate to rotate. Stamens distinct, inserted at the base of the corolla tube, filaments widened at the base, anthers linear. Ovary 3–5-locular. Capsule usually ribbed, dehiscing loculicidally along the sides by valves or pores.

1. Stem retrorsely scabrous on the angles or wing-angles and on the leaf margins and midribs below.
                                                                      1. *C. aparinoides*
1. Stem and leaves not scabrous.
  2. Plant a relatively coarse annual, 5–20 dm tall; leaves relatively large, 7–15 cm long, 2–6 cm wide; flowers about 2–2.5 cm across, in spikes terminating the branches.   2. *C. americana*
  2. Plant a low, slender, weak-stemmed perennial, with numerous stems from subterranean slender rhizomes; leaves small, 2–4 cm long and 0.5 cm wide or less; flowers 1–1.5 cm across, solitary in the leaf axils or terminating slender branchlets.   3. *C. floridana*

1. Campanula aparinoides Pursh. MARSH BELLFLOWER.

Perennial. Stem very slender, weak, 2–6 dm tall, usually sprawling or reclining on other plants, angled or wing-angled, irregularly retrorsely scabrous on the angles. Leaves sessile or subsessile, linear-attenuate, linear-lanceolate, or linear-elliptic, 2–6 cm long, entire or remotely and minutely toothed, the edges and midribs beneath usually retrorsely scabrid. Flowers terminating divergent, short, leafy branches. Calyx segments deltoid or deltoid-attenuate, 1–2 mm long. Corolla blue to whitish, tubular to somewhat beyond the calyx, funnelform, 5–8 mm long, lobes equaling to a little longer than the tube, oblong. Style about as long as the corolla. Capsule hemispheric, 1.5–2 mm long, opening by pores near the base.

Meadows, wet marshy shores, bogs, stream banks. N.B. and Maine to Sask., southward to w. N.C., n. Ga., Ky., Iowa, Neb., Colo.

2. Campanula americana L. TALL BELLFLOWER.

Annual. Stems erect, sometimes weakly so, simple to widely few-branched, glabrous, to 2 m tall. Lower and median leaves with short, margined petioles, upper sessile, blades elliptic, narrowly ovate, or ovate-oblong, margins serrate, apices acuminate, sparsely and irregularly scaly-pubescent on the principal veins beneath, 5–15 cm long and 2–6 cm broad. Flowers in

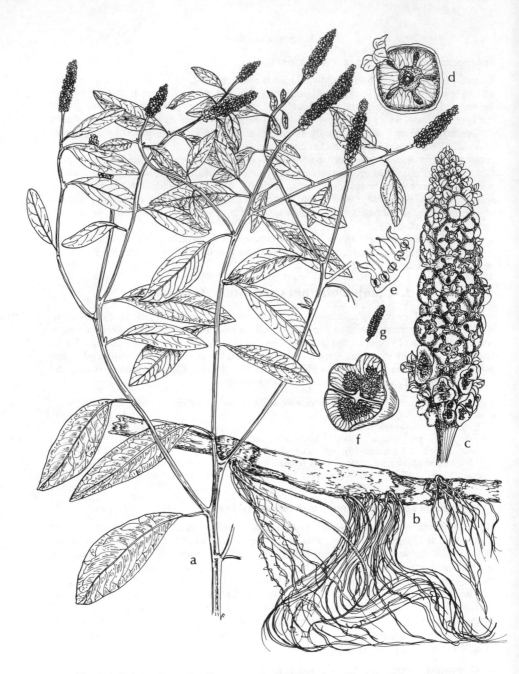

Fig. 342. **Sphenoclea zeylandica:** a, top of stem; b, portion of horizontal stem rooting at the nodes; c, spike; d, calyx and bracts; e, corolla, opened up; f, basal part of circum-scissile capsule; g, seed. (From Correll and Correll. 1972)

spicate racemes terminating the branches, solitary or in small clusters, lowermost flowers usually short-stalked, upper sessile, the lower subtended by reduced foliage leaves, these gradually reduced and more bractlike upwardly. Flowers 2–2.5 cm across. Calyx segments linear-subulate, 5–12 mm long. Corolla light blue, tubular basally, the elliptic-oblong lobes widely spreading at full anthesis (rotate). Style well exserted beyond the corolla, declined, then upwardly curved. Capsule obconic, 7–12 mm long, opening by pores just below the apex. Seeds broadly elliptical, flattish, lustrous brown, banded marginally, about 1.5 mm long. (*Campanulastrum americanum* (L.) Small)

Rich woodlands, river and stream banks, somewhat elevated places in floodplain forests, moist banks at roadsides bordering woodlands. N.Y. and Ont. to Minn., southward to Fla. Panhandle, Ala., Tenn., Ark., Okla.

3. Campanula floridana S. Wats. FLORIDA BELLFLOWER.

Low perennial, rhizomatous, commonly with numerous stems from the rhizomes, 2–4 dm tall, weakly ascending to reclining, usually glabrous but sometimes copiously short-pubescent. Lowermost leaves usually with short margined petioles, others sessile, blades variable, lowermost lanceolate, others elliptic-lanceolate, linear-lanceolate, or linear, 1–4 cm long, to 0.8 cm wide, mostly 0.5 cm wide or less, finely toothed. Flowers mostly solitary terminating very slender, short branchlets, about 1–1.5 cm across. Calyx segments subulate, 4–6 cm long, equaling or a little longer than the corolla. Corolla violet, tubular basally, narrowly oblong-elliptic lobes spreading at full anthesis (rotate). Capsule slenderly obconic, 3–4 mm long, opening by pores above or below the middle. Seeds about 0.5 mm long, nearly as broad, plump, buffish, reticulate-alveolate. (*Rotantha floridana* (S. Wats.) Small)

Pineland depressions, cypress depressions, adjacent ditches, moist to wet pastures, peaty-boggy margins of swamps or wet woodlands, calcareous stream banks. Endemic to pen. Fla. and easternmost Fla. Panhandle.

3. Lobelia (LOBELIAS)

(Ours) herbaceous, with acrid milky or yellow-milky sap. Stem usually erect, simple or branched. Flowers in bracteate racemes or panicled racemes, their stalks usually bearing 2 bractlets; flower stalks twist in anthesis thus inverting the flower with reference to the axis of the inflorescence, the apparent upper part actually being next to the bract. Calyx segments nearly equal, in some species with basal auriculate appendages. Corolla red, purplish, blue, or nearly white, bilaterally symmetrical, with a narrow tube, an upper lip of 2 nearly erect lobes, a lower spreading lip usually undivided basally and cleft into 3 lobes distally, lobes of the upper lip usually shorter and broader than those of the lower; tube split about to the base from between the 2 lobes of the upper lip, in some the tube on either side of the split separating incompletely forming windowlike openings and said to be fenestrate; inside base of lower lip pubescent or glabrous, sometimes with 2 somewhat elongate protuberances (tubercles). Filaments free at the base, united to a greater or lesser extent distally; anthers united into a tube around the style, 2 of them, the lower smaller ones (rarely all), with a tuft of white hairs apically. Ovary wholly inferior, or sometimes nearly free, 2-locular. Capsule dehiscing loculicidally at the apex. Seeds pitted-reticulate to tuberculate (in ours excepting *L. homophylla*).

Reference: McVaugh, Rogers. "Studies in the Taxonomy and Distribution of the Eastern North American Species of *Lobelia*." *Rhodora* 38: 241–263, 276–298, 305–329, 346–362. 1936.

1. Corollas crimson (very rarely white on an individual plant).   1. *L. cardinalis*
1. Corolla blue, violet, purple, or nearly white.
   2. Plants relatively very slender, lower stems usually trailing, rooting at the nodes and mat-forming, ascending flowering stems 1–3 dm tall; blades of lowermost or basal leaves orbicular; corollas bright blue or bright purple; restricted to n.e. and pen. Fla.   3. *L. feayana*
   2. Plants not with the above combination of characters; if spreading vegetatively, then rhizomatous; distribution not as above.

3. Larger leaves narrowly linear, not exceeding 0.5 mm wide.   13. *L. boykinii*
3. Larger leaves 2 mm wide or more.
  4. The larger leaves all basal and ascending-erect, elongate, linear-oblanceolate to oblanceolate, very long-tapering proximally into margined petioles.
    5. Calyx segments with small auricles at the base; flower stalks with a pair of very small and inconspicuous bractlets at the base; filament tube 6–11 mm long, anther tube 3 mm long or a little more.   17. *L. floridana*
    5. Calyx segments not auricled at the base; flower stalks without bractlets; filament tube 3–4.5 mm long, anther tube 2 mm long.   18. *L. paludosa*
  4. The larger leaves not all basal, or if basal, then not elongate and not very long-tapering proximally to margined petioles.
    6. Stem leaves linear to linear-oblanceolate, 4 mm wide or less, mostly not over 3 mm wide *and* flowers, including floral tube, 8–14 mm long.
      7. Lower lip of the corolla pubescent on the upper side near the base; flower stalks antrorsely scabrid; floral tube scabrid; corolla blue, without a white eye at the throat.   16. *L. canbyi*
      7. Lower lip of the corolla glabrous but with 2 tubercles on the upper side near the base; flower stalks and floral tube glabrous or with very few short-knobby hairs; corolla blue with a white eye at the throat.   19. *L. nuttallii*
    6. Stem leaves broader, the larger 5 mm wide or more; if less than 5 mm, then flowers, including floral tube, 15–24 mm long.
      8. Calyx segments with rather distantly disposed callus- or glandular-tipped teeth.
        9. Stem leaves not over 3 cm long, usually shorter, usually not over 5 (–7) mm broad at their broadest places, apically rounded or obtuse; calyx segments with prominent basal auricles, these declined and completely covering the floral tube.   4. *L. brevifolia*
        9. Stem leaves much longer than 3 cm and if as narrow as 5 mm, then their apices acute, *or* if leaves only a little over 3 cm long and with obtuse or rounded apices, then the calyx segments without basal auricles or with small auricles barely overlapping the summit of the floral tube.
          10. Lower lip of corolla pubescent on the upper side at the base; stem leaves linear, linear-lanceolate, or oblong-lanceolate.
            11. Larger leaves mostly linear, some sometimes linear-oblanceolate, mostly 5 mm wide or less, rarely to 7 mm wide, 5–15 cm long, apices acute; lower lip of corolla without a pair of tubercles at the base.   5. *L. glandulosa*
            11. Larger leaves mostly lanceolate or oblong-lanceolate, mostly 5–8 (–15) mm wide, 5–6 cm long, short-acute to bluntish apically; lower lip of corolla with a pair of tubercles on the upper side at or near the middle.   9. *L. flaccidifolia*
          10. Lower lip of corolla glabrous on the upper side at the base and with a pair of tubercles; stem leaves ovate, lance-ovate, or elliptic, rarely lanceolate.   6. *L. georgiana*
      8. Calyx segments with margins entire, or with short teeth not callus- or glandular-tipped, or ciliate.
        12. The calyx segments glabrous and with entire margins (sometimes with few marginal, remotely spaced, short stiff hairs in *L. homophylla* and *L. spicata*).
          13. Basal and lower stem leaves with petioles as long as the blades or longer, bases of blades truncate or subcordate.   20. *L. homophylla*
          13. Basal leaves, if any, and all stem leaves, sessile, or, if petiolate, then the petioles shorter than the blades and blades narrowed basally.
            14. Stem glabrous (sometimes with a very few chaffy hairs basally or on the angles below the leaves in *L. gattingeri*).
              15. Leaves above the base of the stem sessile, 1–4 cm long, bases rounded to very blunt, apices broadly tapered; flowers, including floral tube about 10 cm long; calyx segments about 3 mm long.   15. *L. gattingeri*
              15. Leaves, larger ones at least, long-tapered to margined petioles; flowers, including floral tube, 18–24 mm long; calyx segments 6–12 mm long or a little more.
                16. Leaves thin and flexuous, elliptic to ovate, rarely lanceolate, rounded, obtuse, or short-acute apically, 1.5–4.5 cm broad; plants of piedmont and mountains.
                                        7. *L. amoena*
                16. Leaves thickish, somewhat stiff, narrowly lanceolate, acute apically, to 1.5 (–2) cm broad; plants of outer coastal plain.   8. *L. elongata*
             14. Stem pubescent, at least below.
              17. Lower flowers of inflorescence ovate-leafy-bracted; capsule much inflated at maturity; calyx segments neither auricled at base nor flaring.   12. *L. inflata*
              17. Lower flowers of inflorescence with subtending bracts little, if any, larger than

those upwardly; capsule not inflated; calyx segments auricled at the base (very infrequently not auricled in *L. spicata*).

    18. Stems short-pubescent near the base, glabrous above; auricles at the base of calyx segments short-triangular to filiform, usually deflexed.   11. *L. spicata*

    18. Stems usually copiously pubescent throughout; base of calyx segments with auricles flaring along the summit of the floral tube.   10. *L. puberula*

   12. The calyx segments pubescent.

    19. Basal rosette of a few leaves usually present at anthesis, these petiolate or blades narrowed to winged petioles; leaf blades obtuse or rounded apically; raceme with a naked stalk below lowest flowers; bracts subtending the flowers linear-subulate, all similar and inconspicuous; corolla light blue or lilac; flowers 10–15 mm long.   14. *L. appendiculata*

    19. Basal leaves not in a rosette and not present at anthesis; leaf blades strongly acute to acuminate apically; raceme without a naked stalk separating it from uppermost leaves, the lower portion of the raceme leafy-bracted; corolla bright blue; flowers about 25 mm long.

                                              2. *L. siphilitica*

1. Lobelia cardinalis L. subsp. **cardinalis.** CARDINAL-FLOWER.   Fig. 343

Perennial, forming basal offshoots. Plants varying from slender to very coarse, 0.5–2 m tall, stem glabrous or sparingly short-pubescent. Lower leaves with margined petioles to 3 cm long, becoming sessile upwardly, blades lanceolate or elliptic, to 18 cm long and 5 cm wide, cuneate basally, acute to acuminate apically, margins with irregularly spaced, small knobby callose-teeth, or shallowly dentate-serrate, the teeth callus-tipped, surfaces glabrous or with short, stiffish pubescence principally on or near the veins. Raceme terminal, usually unbranched, few- to many-flowered, 1–5 dm long; flower stalks short-bristly-pubescent, with a pair of bractlets near the base, subtending bracts mostly linear-subulate, sometimes narrowly lanceolate, longer than the flower stalks, their margins callous-toothed. Floral tube smooth to bristly- or scaly-pubescent; calyx segments linear-subulate, flaring at the base, much exceeded by the corolla, glabrous to sparsely pubescent, auricles usually none. Corolla 3–4.5 cm long, deep red, crimson, or vermilion (white on a rare individual), tube fenestrate below. Filament tube 1.8–3.5 cm long, long-exserted from the corolla tube and of about the same color; anther tube 4–5.5 mm long, bluish gray. Capsule broadly campanulate to oblate, 8–10 mm long, often as broad. Seeds amber-brown, rough-tuberculate, elliptic to nearly linear in outline, 0.5–0.8 mm long.

    Swamps, low places in floodplain forests, river and stream banks, in streams, bogs, meadows, sometimes in mats of floating vegetation; in some parts of the range, locally very abundant and clogging waterways. N.B. and Ont. to Minn., generally southward to n. Fla. and e. Tex.

2. Lobelia siphilitica L. BIG BLUE LOBELIA, GREAT LOBELIA.

Perennial, forming basal offshoots. Stem usually unbranched, to 18 dm tall, erect, relatively coarse, smooth or chaffy-hirsute on the angles. Lower stem leaves narrowed into margined petioles; blades oblanceolate to elliptic, acute apically, variable in size to 18 cm long and 6 cm wide, margins mostly shallowly crenate-serrate, callosities at the extremities of the crenations thus seemingly in the sinuses, surfaces usually sparingly pubescent with short, stiffish hairs, sometimes glabrous. Racemes 1–5 dm long, leafy-bracted below, bracts reduced upwardly but often subfoliose-bracted even distally. Flower stalks loosely erect, 5–10 mm long fully developed, variously pubescent but commonly densely clothed with transparent, longish chaffy hairs, a pair of bracteoles near their summits. Flowers 2.3–3.3 cm long including the floral tube. Floral tube pubescent like the flower stalks, or sometimes glabrous; calyx segments conspicuous, with prominent lobes or auricles basally, thus the bases appearing ovate, subulate distally, margins markedly ciliate and usually also with callose teeth. Corolla blue, white-striped in the throat, lower lip inside at the base white with a pair of tubercles, tube fenestrate. Filament tube 12–15 mm long, anther tube 4–5.5 mm long. Capsule hemispheric, 8–10 mm across. Seeds linear-oblong in outline, about 0.8 mm long, amber-brown, lustrous, alveolate-reticulate and somewhat warty.

    Banks of small streams and in streams, moist woods and swamps, wet meadows, open

Fig. 343. **Lobelia cardinalis** subsp. **cardinalis:** a, basal decumbent part of plant; b, inflorescence at top of plant; c, medial section of stem, showing leaves; d, flower; e, anther tube and exserted upper portion of style; f, capsule, with persistent calyx segments; g, cross-section of capsule; h, seed.

wet places. Maine to Minn., Man. and Colo., southward to w. N.C., n.cen. Ala. and Miss., Ark., e. Okla. and n.e. Tex.

West of the Mississippi River, some plants have glabrous stems and leaves, smaller leaves, 15 mm wide or less, few-flowered racemes, usually fewer than 20 flowers, floral tube sparsely pubescent to glabrous, narrow, acute auricles at the bases of the calyx segments. Such plants may be designated var. *ludoviciana* A. DC.

3. Lobelia feayana Gray.

Leafy portion of stem very slender, decumbent, short, mostly 1–3 cm long, flowering stem weakly ascending to erect, 5–30 cm long, with 2 to several widely spaced, reduced leaves below a simple raceme. Lower leaves petiolate, petioles 5–20 mm long, blades orbicular to ovate, 8–13 mm wide, subcordate to truncate basally, very broadly rounded apically, margins with minute reddish callosities in shallow indentations; stem leaves lanceolate, lance-ovate, or oblanceolate, dentate, the teeth callus-tipped. Raceme lax, loosely flowered, with 2 tiny bractlets at the base of the smooth flower stalk. Floral tube glabrous; calyx segments triangular-subulate, glabrous, margins entire or with a few very minute callosities near the tip, basal auricles none. Corolla azure blue or purple, with a white eye, about 5–8 mm long spread out, with a pair of greenish tubercles inside at the base of the lower lip. Filament tube about 3 mm long, anther tube 1–1.5 mm long. Capsule campanulate, 3–4 mm long. Seeds tuberculate, about 0.4 mm long.

In wet, peaty, mucky, or sandy soils, seasonally wet pine flatwoods, cypress depressions, commonly very many plants matted together in ditches and on wet roadsides bordering flatwoods or marshy openings. n.e. and pen. Fla. southward to somewhat below Lake Okeechobee.

4. Lobelia brevifolia Nutt. ex A. DC.

Perennial with a small, short, erect caudex. Stems glabrous to sparsely very short-pubescent, unbranched, sometimes more than one from the base, often tinged purplish red near the base, 3–9 dm tall. Stem leaves numerous, mostly divaricately spreading or slightly ascending, essentially sessile although some lower ones often cuneately narrowed to subpetiolar bases, spatulate-oblanceolate, varying to oblong or linear-oblong, strongly short pectinately toothed, the teeth callus- or gland-tipped, mostly obtuse to rounded apically. Raceme few- to 30-flowered, flower stalks stoutish, short, 5–10 mm long fully developed, these and the floral tube more or less clothed with usually retrorse, transparent, spiculelike trichomes variable in length, stalks with a pair of bractlets near the base; subtending bracts linear-oblong, dilated at the base, about equaling the stalks, toothed like the leaves. Flowers 15–24 mm long including the floral tube. Calyx segments triangular, long-acute, strongly pectinately toothed, with broad, roundish basal auricles which collectively virtually cover the floral tube. Corolla azure to nearly purple, sometimes pale blue or grayish blue, pubescent exteriorly, usually but not always fenestrate, the lower lip smooth to pubescent inside at the base. Filament tube 6–11 mm long, anther tube 3.4–4 mm long. Capsule campanulate-hemispheric, 5–6 mm across. Seeds amber-brown, somewhat angular, ovate-elliptic in outline, about 0.5 mm long, alveolate-reticulate and somewhat warty.

Wet pine savannas and flatwoods, adjacent ditches and wet roadsides, bogs, banks of streams draining bays. Coastal plain, Fla. Panhandle to s.e. La.

5. Lobelia glandulosa Walt.

Fig. 344

Perennial with a slender, erect caudex. Stem usually unbranched, weakly erect, glabrous, mostly 7–10 dm tall, sometimes taller. Stem leaves relatively few, ascending, thickish, linear to narrowly oblanceolate, 8–15 cm long, mostly less than 5 mm wide, sometimes to 7 mm, essentially sessile, the wider ones narrowed to long, subpetiolar bases, apically long and narrowly tapered, margins callose-glandular, or sometimes without glands and entire to slightly wavy; often with very short, appressed, translucent trichomes on or near the leaf edges. Racemes with few to about 20 widely spaced flowers. Flower stalks stiff, erect, 5–12 mm

Fig. 344. **Lobelia glandulosa:** a, plant; b, median stem leaf; c, fruit; d, seed.

long fully developed, with a pair of bractlets at the base, the stalks and floral tubes varying from glabrous to densely pubescent with short, knobby, whitish trichomes, some of the trichomes sometimes spiculelike, subtending bracts linear-subulate, shorter than to exceeding the stalks in length, callose-toothed. Flowers 2–3.3 cm long including the floral tube. Calyx segments subulate, mostly with callose teeth marginally, basal auricles none or very small. Corolla blue, with a white eye at the throat, glabrous exteriorly, pubescent at the base of the lower lip within, fenestrate. Filament tube 7–10 mm long, anther tube 3.5–4 mm long. Capsule hemispheric, broader than long, about 8 mm across. Seeds amber-brown, lustrous, oblong-elliptic, about 0.8 mm long, alveolate-reticulate.

Wet pine savannas and flatwoods, adjacent ditches, bogs, bay margins, banks of streams draining bays, shores of sinkhole ponds. Coastal plain and outermost piedmont, s.e. Va. to s. Fla., Fla. Panhandle and s.w. Ala.

6. Lobelia georgiana McVaugh.

Perennial with a slender, erect caudex, perennating by basal offshoots. Stems erect to sprawling, 3–12 dm tall, sparsely pubescent to glabrous, usually unbranched. Stem leaves relatively few and widely spaced; leaves of basal offshoots and lower stem with margined petioles, blades ovate to elliptic; lower stem leaves smaller than those of midstem, above midstem becoming gradually reduced to or to somewhat below the raceme, sometimes the uppermost leaves little reduced; larger leaves lanceolate, elliptic, or oval, varying in size, to 12 cm long and 4 cm broad; margins irregularly finely dentate to wavy or entire; thickish in texture, dark green above, paler beneath, lower surface drying grayish. Raceme loose, flowers relatively distant, mostly few to 20, sometimes to 30. Flowers varying in size, 2–3 cm long including the floral tube. Flower stalks smooth, 2.5–4 mm long fully developed, with a pair of bractlets near the base, subtending bracts variable, the lower usually lanceolate, reduced upwardly, the upper linear to linear-subulate, marginally glandular-toothed. Floral tube glabrous; calyx segments dilated basally, subulate distally, margins glandular or glandular-toothed below the middle. Corolla blue, glabrous exteriorly, lower lip glabrous on the inside at the base and with a pair of tubercles, tube fenestrate. Filament tube 6–8.5 mm long, anther tube about 3 mm long. Capsule hemispheric, broader than long, about 6 mm across. (*Lobelia glandulifera* (Gray) Small, not *L. glandulifera* Kuntze)

Floodplain woodlands, along river and stream banks and in wooded areas astride small streams, hillside bogs in pinelands, meadows. e. Va. to e. Ky., southward to n. Fla. and s. Ala.

7. Lobelia amoena Michx.

Perennial, perennating by basal offshoots. Stem unbranched, 3–12 dm tall, glabrous or sometimes short-pubescent near the base. Leaves thin and flexuous, bright green above, paler beneath and drying grayish, larger ones elliptic to ovate, rarely lanceolate, rounded, obtuse or short-acute apically, 4–18 cm long, 1.5–4.5 cm wide, upper smaller leaves similar; lower leaves mostly tapered basally to short, margined petioles; margins of blades variable, sometimes crenate, the teeth callus-tipped and with sessile or embedded callosities between the teeth, sometimes finely callose-toothed, sometimes with smooth edges and with callosities embedded in the edges. Racemes 1–4 dm long, often secund, few- to about 40-flowered, lowest bracts usually reduced-foliaceous, lanceolate, longer than the flowers, reduced upwardly, the uppermost linear-subulate but relatively prominent, margins more or less callose-toothed. Flower stalks short, mostly 3–5 mm long, short-pubescent, with a pair of bractlets at the bases. Flowers 18–24 mm long including the floral tube. Floral tube glabrous or sparsely pubescent; calyx segments elongate-subulate, 5–12 mm long, glabrous, margins entire, basal auricles usually none or sometimes very small ones present. Corolla blue, with a whitish eye in the throat, the lower lip glabrous inside at the base, tube fenestrate. Filament tube 5–7 mm long, anther tube 2.5–3.5 mm long. Capsule campanulate-hemispheric, 5–6 mm across.

Seepage slopes, wet rocks, stream banks, wet woodlands, woodland borders. w. N.C. and S.C., n. Ga., e.cen. Ala.

Fig. 345. **Lobelia flaccidifolia:** a, plant; b, plant with differing leaves and somewhat larger flowers than in a; c, flower; d, stamens. (From Correll and Correll. 1972)

8. Lobelia elongata Small.

Similar to *L. amoena*, disjunct geographically. Differing in having thickish, stiff, narrowly lanceolate leaves, long-acute at both extremities, 0.5–1.5 (–2) cm broad, similarly colored on both surfaces. Filament tube 8–12 mm long, anther tube 4 mm long.

Swamps, low grounds, tidal marshes. Outer coastal plain, Del. to s.e. Ga.

Lobelia georgiana, *L. amoena*, and *L. elongata* apparently form a complex worthy of further intensive study. We note that Radford in Radford et al. (1964) does not distinguish between *L. elongata* and *L. georgiana* and suggests, in addition, that *L. amoena* may not be distinct from them either.

9. Lobelia flaccidifolia Small.   Fig. 345

Annual. Stems slender, erect, mostly 5–7 dm tall, sometimes a little taller, glabrous, usually unbranched. Lower stem leaves (sometimes shed by anthesis) narrowed to petiolar bases; blades lanceolate or oblong-lanceolate, blunt to acute apically, with minute callosities sunken in the edges, the margins sometimes slightly wavy; leaves becoming sessile somewhat below midstem, gradually reduced upwardly, the uppermost reduced leaf 6–12 cm below the lowermost flower; principal leaves mostly 5–8 cm long and 5–8 mm wide. Racemes loose, usually few-flowered (3–8). Flower stalks curved, 4–11 mm long fully developed, a pair of bractlets near the base, stalks and floral tube rough-pubescent, subtending bracts linear or long-triangular, pubescent to glabrous, the margins callus-toothed, shorter than to longer than the stalks. Flowers 14–20 mm long including the floral tube. Calyx segments triangular-subulate, pubescent, with a pair of prominent, declined basal auricles giving a sagittate effect. Corolla blue, lavender, or nearly white, with a white eye, glabrous to rough-pubescent exteriorly, fenestrate or not, the lower lip with a pair of tubercles and somewhat pubescent near the base inside. Filament tube 5–8 mm long, anther tube 2–3 mm long. Capsule hemispheric, 4–6 mm across. (Incl. *L. halei* Small)

Swampy woodlands along rivers and streams, pond margins, depressions in pinelands, prairies. Coastal plain, Ga., Fla. Panhandle to s.e. Tex.

10. Lobelia puberula Michx. DOWNY LOBELIA, PURPLE DEWDROP.   Fig. 346

Perennial with a slender, erect caudex. Stem usually unbranched, 3–16 dm tall, usually more or less purple-tinged, densely puberulent to short-hirsute throughout, occasionally becoming glabrous. Leaves sessile, the lower long-tapering to the base, variable in size to 12 cm long and 4 cm broad, oblanceolate to obovate, varying to oblong, elliptic, or lanceolate, midstem leaves largest, gradually reduced upwardly, more or less pubescent on both surfaces, rarely becoming glabrous, paler beneath than above, coarsely to finely dentate marginally, the teeth callus-tipped, or margins subentire with callosities on the edge. Racemes 4–50 cm long, dense or somewhat interrupted. Flower stalks short, stout, 3–5 mm long, pubescent, with a pair of bractlets basal or nearly so, subtending bracts usually subfoliose and lanceolate in the lower part of the raceme, gradually reduced upwardly, pubescent, margins toothed. Flowers varying in size from about 15 to 24 mm long. Floral tube more or less pubescent; calyx segments dilated at the base to flaring auricles, subulate distally, sometimes subulate or lanceolate, with small declined auricles, to 15 mm long. Corolla blue to purple, with a white eye in the throat, pubescent exteriorly at least on the veins, the lip inside at the base smooth to hirsute, tube fenestrate. Filament tube 6–15 mm long, anther tube 3–6 mm long. Capsule oblate to hemispheric, broader than long, to about 10 mm across. Seeds linear-oblong, about 1 mm long, amber-brown, lustrous, roughly alveolate-reticulate.

A widely distributed, morphologically variable complex for which some authors recognize segregate taxa at the varietal or specific level.

Bogs, meadows, thickets, roadsides, well-drained to moist woodlands. s. N.J. to Ill., southward to n. Fla. and e. Tex., e. Okla.

Fig. 346. **Lobelia puberula:** a, basal and middle portions of stem; b, inflorescence; c, flower; d, corolla, split open; e,f, two views of staminal column. (From Correll and Correll. 1972)

11. Lobelia spicata Lam.

Annual. Stem erect, usually unbranched, seldom with a few erect branches from near the base, roughly short-pubescent near the base. Very variable in leafiness; sometimes with a rosette of basal leaves, these narrowed to margined petioles usually shorter than the obovate to subrotund blades; plants sometimes without a rosette but lowermost leaves much as above; stem leaves generally few, sometimes only 2–4, mostly sessile, distant from each other, reduced upwardly, the uppermost 3–12 cm below the lowest flower, rarely some reaching the raceme, oblong, oblanceolate, or elliptic, mostly obtuse to rounded apically, uppermost rarely acute; largest leaves 4–11 cm long, 1–4 cm wide. Racemes loosely to relatively densely flowered, variable in length to about 4.5 dm. Flower stalks slender, 1–6 mm long, smooth to finely short-pubescent, sometimes with scaly trichomes intermixed, with a pair of small bractlets basally, subtending bracts subfoliaceous but not really conspicuous below, upwardly linear to subulate, surfaces sometimes bristly-pubescent, margins entire or callus-toothed, equaling or slightly longer than the stalks. Flowers 7–12 mm long including floral tube. Floral tube glabrous or pubescent; calyx segments 2–7 mm long, subulate, usually glabrous, sometimes with a few marginal hairs distally, basal auricles almost always present (in our range) but these very variable, triangular to oblong to filiform, usually declined. Corolla blue to lavender or whitish, glabrous exteriorly, lower lip pubescent at the base inside, tube not fenestrate. Filament tube 2–4 mm long, anther tube about 2 mm long. Capsule hemispheric, broader than long, 2–3 mm across. Seeds ellipsoid, 0.3–0.4 mm long, alveolate-reticulate, amber-brown.

A variable species, segregated by some authors into as many as five taxa, either at the specific or varietal level. Intergradation of characters used for distinguishing segregates and partial overlapping of ranges of the segregates is such that, for our purposes, we prefer to recognize but one variable entity. (Incl., for our range, *L. bracteata* Small, *L. leptostachya* A. DC., *L. spicata* var. *scaposa* McVaugh)

Open woodlands and bluffs, cedar glades and barrens, creek banks, bottomland clearings, meadows, prairies, only marginally in wet places. e. Can. to Minn., southward to Ga., Ark., e. Tex.

12. Lobelia inflata L. INDIAN-TOBACCO.

Annual. Stem erect, usually freely branched, usually hirsute throughout, rarely nearly glabrous, to about 1 m tall, dwarfed specimens in sterile places. Only the lowermost leaves narrowed to margined petioles, their blades short-oval to suborbicular; midstem leaves largest, varyingly ovate, lance-ovate, obovate, or ovate-oblong, 2.5–8 cm long and 1.5–3.5 cm broad, bases short-tapered to rounded, apices obtuse; upper leaves gradually reduced, acute apically; all serrate-dentate to crenate, sometimes with smaller teeth between the larger ones, teeth sometimes emarginate, small callosities on the extremities; lower surfaces usually pubescent; upper leaves becoming smaller toward and into the racemes, the latter thus leaf-bracted below, bracts upwardly becoming lanceolate, finally linear-acicular. Racemes loose, terminating branches (if plant branched), flowers variable in number. Flower stalks 5–10 mm long, minutely bristly-pubescent, basal bractlets minute. Flowers 8–10 mm long including floral tube. Floral tube glabrous; calyx segments subulate, about as long as the corolla tube, glabrous, or rarely sparsely pubescent, margins entire, basal auricles none. Corolla violet-blue to nearly white, glabrous exteriorly, lower lip pubescent within at the base, tube not fenestrate. Filament tube 2.5–3 mm long, anther tube 1.5–2 mm long. Capsule ovoid, oval, or subglobose, inflated. Seeds ellipsoid, about 0.5 mm long, amber-brown to dark brown, alveolate-reticulate, somewhat lustrous.

Dry to moist woodlands, woodland borders and clearings, bottomland woodlands and clearings, bottomland fields, pastures, open banks. Pr.Ed.I. to Sask., generally southward to cen. Ga., cen. Ala., Miss. and Ark. (rarely on the coastal plain).

13. Lobelia boykinii T. & G.

Rhizomatous perennial. Stem slender, sometimes spongy below, simple or few-branched, 5–8.5 dm tall. Leaves sessile, subulate or narrowly linear, to 2.5 cm long and 0.5 mm wide, entire or with a few minute callosities marginally; leaves on lower ⅓–½ of stem commonly shed before anthesis. Raceme loose, delicate, of 10–25 flowers, the flower stalks filiform, about 1 cm long, without bractlets, subtending bracts filiform, much shorter than the stalks. Floral tube glabrous; calyx segments filiform-subulate, glabrous, entire, without basal auricles. Corolla blue with a white eye at the throat, 10–12 mm long, lower lip usually pubescent inside and with 2 tubercles near the base. Filament tube 3–5 mm long, anther tube 1.5–2 mm long. Capsule campanulate-hemispheric, about 3 mm across. Seeds brown, irregularly turbinate, rough-tuberculate, about 0.4 mm long.

Cypress-gum depressions or ponds, wet pine savannas and flatwoods, adjacent ditches, often in shallow water. Coastal plain, s. Del.; S.C. to n. Fla.

14. Lobelia appendiculata A. DC.

Annual. Stem erect, unbranched or with few erect branches, 8–9 dm tall, glabrous or sparsely beset with short chaffy hairs. Usually with several basal leaves (sometimes shed by anthesis) with margined petioles, the blades lanceolate, oblanceolate, or elliptic, rounded to obtuse apically, about 2–6 cm long; stem leaves sessile, relatively few, variable in size to about 8 cm long and 1–2 cm wide, the lower oblanceolate, grading to oblong or elliptic and reduced upwardly, the larger ones cuneate to rounded basally, mostly rounded apically, margins obscurely wavy-dentate, with small callosities, the uppermost generally 4–10 cm below the lowest flowers. Racemes loosely flowered, 6–18 cm long. Flower stalks 4–6 mm long, with a pair of minute bractlets basally, these and the floral tubes roughly pubescent with short, white, chaffy hairs, the tube rarely glabrous, subtending bracts subulate, longer than the stalks, their margins with bristly teeth sometimes intermixed with callosities. Flowers 10–15 mm long including the floral tube. Calyx segments subulate, with bristly-toothed margins and prominent basal auricles, usually declined and obscuring the floral tube, the auricles drying lavender-blue. Corolla lilac or violet, glabrous exteriorly, lower lip pubescent on the inside at the base and with a pair of tubercles, tube not fenestrate. Filament tube 3–4 mm long, anther tube about 2 mm long. Capsule subhemispheric, nearly twice as long as broad, about 0.5 mm long, irregular in outline, amber-brown, lustrous, alveolate-reticulate.

Open pinelands, oak-pine woodlands, prairies, roadsides, marginally, if at all, a wetland plant. e. Kan., Okla., Ark., e. Tex., La., Miss., Ala.

15. Lobelia gattingeri Gray.

In general features closely similar to *L. appendiculata* but of smaller stature, to about 5.5 dm tall but not uncommonly diminutive. Flower stalks smooth or with short, knobby pubescence, subtending bracts glabrous, marginally with a few callosities. Floral tube glabrous; calyx segments glabrous, rarely minutely ciliate distally, with minute, spreading basal auricles not at all obscuring the floral tube. Capsule campanulate to hemispheric, mostly longer than broad, about 4 mm across. Seeds as in *L. appendiculata* but averaging smaller.

Calcareous rock outcrops, springy places or in seasonally wet places with shallow soil, bluffs, cedar glades and barrens. Narrowly endemic, cen. Tenn.

16. Lobelia canbyi Gray.

Annual, the stem slender, erect, simple throughout or sparingly branched above, 3–10 dm tall, glabrous below, sparingly scabrate on the inflorescence axes. Leaves sessile, linear or narrowly linear-lanceolate, 1–5 cm long, 1–4 mm wide, obscurely callus-toothed, sometimes very sparsely scabrid on the margins. Racemes lax, loosely 10–30-flowered. Flower stalks antrorsely scabrid, 3–8 mm long, usually with a pair of minute bractlets at the base, subtending bracts linear, usually equaling or a little longer than the stalks, with few minute callose teeth. Flowers about 10 mm long. Floral tube more or less antrorsely scabrid; calyx segments subulate, scabrid, or sometimes smooth, usually with a few low callose teeth, without basal auricles. Corolla blue to lavender, lower lip pubescent inside at the base. Filament tube 3–4 mm long, anther tube about 2 mm long. Capsule oblong-oval to campanulate, about 3 mm across. Seeds amber-brown, irregular in shape, minutely pebbled, about 0.5 mm long.

Depressions in pine savannas, bays, pocosins, coastal plain, N.J. to Ga.; swales, vernal pools, in oak barrens, cen. Tenn.

17. Lobelia floridana Chapm. Fig. 347

Glabrous perennial with a short, erect caudex. Stem simple or branched in the inflorescence, 5–15 dm, averaging 8–10 dm, tall. Larger leaves several, closely set near the base, 1–3 dm long, oblanceolate to broadly linear, the broader ones very long tapered into margined petioles, apices obtuse or blunt, acute in the linear leaves; margins entire or with callosities, sometimes crenate and with callosities in the indentations; leaves above the base few, remotely spaced, much reduced, 1–5 cm long, flower stalks stiff and thickish, usually rough, about 3–8 mm long, a pair of bractlets at the base, subtending bracts thickish-subulate, a little longer than the stalks, callosities on the margins. Flowers 13–20 mm long, including the floral tube, averaging 15–16 mm. Floral tube usually with a pebbled surface texture; calyx segments triangular to triangular-subulate, 2–6 mm long, glabrous, margins mostly finely callose-toothed, with small auricles basally. Corolla pale blue, sometimes with a pinkish tinge, to nearly white, variously pubescent exteriorly, the lower lip pubescent inside at the base, the tube rarely fenestrate. Filament tube 6–11 mm long, distally united from somewhat below the middle, anther tube 3 mm long or a little more. Capsule campanulate-hemispheric, 4–5 mm across. Seeds amber-brown, irregular in shape, about 0.8 mm long, tuberculate.

Wet pine flatwoods and savannas, adjacent ditches and borrow pits, cypress depressions, often in shallow water. Coastal plain, Fla. Panhandle to La.

18. Lobelia paludosa Nutt.

In habit and vegetative characters closely similar to *L. floridana* but smaller, averaging 5–6 dm tall, inflorescence less frequently branched. Basal leaves smaller, 5–12 cm long. Flower stalks more slender, less stiff, often somewhat nodding, bractlets none or very minute, subtending bracts, fully developed, half as long as the stalks or less. Calyx segments usually not exceeding 3 mm long. Corolla very light blue or nearly white, essentially glabrous exteriorly, the tube fenestrate. Filament tube 3–4.5 mm long, distally united from about the middle, anther tube 2–3 mm long.

Wet pine savannas, flatwoods, and prairies, adjacent ditches, cypress ponds and depressions, often in shallow water. s.e. Ga. to s. Fla., westward in the Fla. Panhandle to somewhat w. of the Apalachicola River.

19. Lobelia nuttallii R. & S.

Annual, stem very slender, simple or with long-ascending branches, 2–7.5 dm tall, pubescent below. Usually with a few small basal leaves (sometimes shed by anthesis) unlike those higher on the stem, ovate to elliptic, or oblanceolate, narrowed into margined petioles; lower stem leaves lanceolate, grading to linear upwardly, sessile, margins with few obscure callosities. Flowers relatively distantly disposed on the racemes or raceme branches, few to about 20 in a raceme. Flower stalks slenderly flexuous, usually glabrous, a pair of small bractlets near the base, subtending bracts narrowly linear, about equaling the stalks at anthesis, little if any more than half as long when the stalks are fully developed, usually with a few callose teeth marginally. Floral tube glabrous, mostly entire marginally, without basal auricles. Flowers 8–11 mm long, including the floral tube. Corolla blue with a white eye at the throat, glabrous without, glabrous at the base of the lower lip and with 2 tubercles. Filament tube 3 mm long, anther tube about 1.5 mm long. Capsule small, usually broader than long, hemispheric, 2–3 mm across.

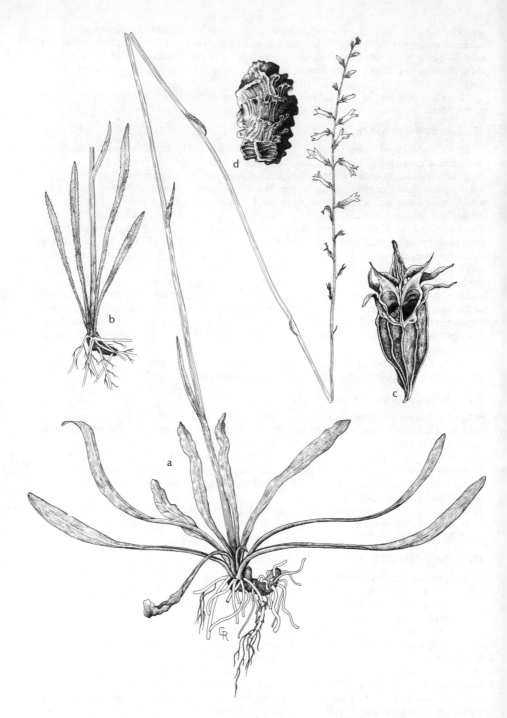

Fig. 347. **Lobelia floridana:** a, plant; b, base of another plant (probably extremes in size of a, an older plant, and b, a younger plant); c, fruit, after dehiscence; d, seed.

Savannas and adjacent savannalike roadsides and ditches, meadows, bogs, seepage slopes, low woods. Coastal plain and outer piedmont, L.I. to Fla. Panhandle; also inland, s.w. N.C., n.w. S.C., n. Ga., Ala., e.cen. Tenn. and s.e. Ky., s.e. La.

20. Lobelia homophylla F. E. Wimmer.

Annual. Stem simple, or several stems from near the base, often with few to several branches in the inflorescence. Leaves to about midstem petiolate, petioles of the lowermost longer than the blades, gradually reduced upwardly, the uppermost leaves sessile or with short, margined petioles; blades of petiolate leaves ovate to suborbicular, truncate to subcordate basally, obtuse to rounded apically, margins crenate, dentate, or dentate-serrate, the teeth mostly rounded to shallowly retuse apically and mucronate; margins of upper leaves sharply serrate or dentate, the teeth mucronate. Racemes loosely few- to many-flowered, delicate, flower stalks lax, filiform, glabrous, about 1 cm long fully developed, subtending bracts narrowly linear, margins ciliate, less than half as long as the flower stalks. Flowers 6–12 mm long. Ovary and capsule superior or very nearly so; calyx tube glabrous, subulate lobes glabrous or ciliate with a few short, stiff hairs, basal auricles none. Corolla blue, glabrous exteriorly, lower lip inside at the base pubescent, tube not fenestrate. Filament tube about 3 mm long, anther tube about 1.4 mm long. Capsule oval, longer than broad, 5–6 mm long. Seeds tiny, elliptic, brown, smooth and lustrous.

Weedy, low open woodlands, moist roadsides, ditches, canal banks, pastures. cen. pen. Fla.

Compositae (SUNFLOWER FAMILY)

A highly diversified and very large family. Our species annual, biennial, or perennial herbs or shrubs, sometimes herbaceous or shrubby vines. Leaves simple, entire to variously lobed or divided, the latter appearing essentially compound in some; stipules none. Flowers in heads, borne on a common receptacle, commonly mistaken for a single flower. Heads bearing few to very numerous individual flowers; in a few kinds few-flowered heads glomerated in such a way that the collective heads resemble a single one. Heads surrounded by an involucre of bracts (phyllaries), these few to numerous, varingly disposed in a single series to imbricated. Each flower of the head may be subtended by a bract (chaff or pale), the receptacle then said to be chaffy, or flowers without bracts, the receptacle said to be naked; in some the naked receptacle is pitted, a chaffy partition separating the pits, or the receptacle bristly or hairy. Flowers epigynous, the corolla surmounting the inferior ovary, calyx none but exterior to the corolla a pappus may or may not be present; if present commonly persistent on the fruit, sometimes deciduous. Pappus may be comprised of few to numerous, smooth, barbellate, or plumose capillary bristles in one or more series, or of scales, or a combination of the two, or of awns, teeth, or of a ring or crown. Corollas of three general types: (1) tubular, 4–5-lobed or truncate at the summit, radially symmetrical; (2) tubular only at base above which flat (ligulate or raylike), commonly bent to one side, sometimes toothed apically; (3) more or less intermediate between the first and second types, bilabiate, the outer side usually larger. Heads may be composed of all tubular, radially symmetrical flowers, usually bisexual, usually called disc flowers (heads discoid); or heads with all ligulate, bisexual flowers (heads ligulate); or heads may have disc flowers centrally, ray flowers peripherally (heads radiate). Stamens, if present, (4)5, filaments adnate to the corolla tube below but free above (lightly coalescent in a few genera), anthers coalescent and forming a tube around the style in most. Pistil solitary, comprised of 2 carpels, ovary 1-locular, with 1 ovule, style columnar, stigma bilobed. Fruit an achene.

1. Ligulate flowers present (in *Flaveria* a single, inconspicuous, ligulate flower per head in only some of the heads).
   2. Heads ligulate, flowers all ligulate and bisexual; sap milky.

2a. Heads in a raceme; involucral bracts markedly unequal; ligules pink or lavender. 43. *Prenanthes*
2a. Heads not racemose, borne singly terminating scapes or single stalks from leaf axils; involucral bracts equal or subequal; ligules yellow to orange-yellow (sometimes shriveling or drying with a lavender tint).
  44. *Krigia*
2. Heads radiate, having both ligulate (ray) and disc flowers; ligulate flowers pistillate or neuter; sap watery.
  3. Ray flowers yellow to orange-yellow, sometimes marked with purple or reddish brown basally.
    4. Stem woody at least below.
  4. *Borrichia*
    4. Stem herbaceous.
      5. Receptacle of the head with chaffy bracts united into cuplike tubules toothed or cleft at the summit, collectively honeycomb-like, each tubule surrounding the base of the flower.
  13. *Balduina*
      5. Receptacle of the head not as above.
        6. Pappus all, or at least in part, of capillary bristles.
          7. Involucral bracts equal and slightly imbricated *or* the larger ones of equal length in a single series and with a few very much-reduced bractlets around them at the base; ray flowers evenly disposed in a peripheral circle.
            8. Leaves opposite.
  16. *Arnica*
            8. Leaves alternate.
  17. *Senecio*
          7. Involucral bracts unequal, distinctly imbricated and graduated from the outside inwardly, outer smallest; ray flowers disposed irregularly around the periphery of the head.
            9. Heads disposed in panicles (in those treated here).
  21. *Solidago*
            9. Heads disposed in corymbs, the entire inflorescence more or less storied and flat-topped.
  22. *Euthamia*
        6. Pappus not even in part of capillary bristles, either crownlike, or of a few scales, or of awns or teeth, or none.
          10. Involucral bracts dimorphic, distinctly in 2 series.
            11. Pappus of 2–4 barbed or hispid teeth or awns; achene not winged.
  9. *Bidens*
            11. Pappus of 2 barbless teeth, a mere crown, or none; achene marginally winged (except sometimes in *C. tinctoria* in genus).
  8. *Coreopsis*
          10. Involucral bracts not in 2 distinct series, more or less imbricated.
            12. Receptacle of the head without chaff.
              13. Leaves alternate.
  14. *Helenium*
              13. Leaves opposite.
  15. *Flaveria*
            12. Receptacle of the head (of the disc at least) chaffy.
              14. The receptacle of the head strongly conical, domelike, or columnar.
                15. Leaves opposite.
  7. *Spilanthes*
                15. Leaves alternate.
                  16. Chaff subtending only the disc flowers.
  1. *Rudbeckia*
                  16. Chaff or pales subtending both ray and disc flowers.
  2. *Dracopsis*
              14. Receptacle of the head flat or convex.
                17. Disc achenes thin-edged or winged.
  6. *Verbesina*
                17. Disc achenes laterally compressed but neither thin-edged nor winged.
  5. *Helianthus*
  3. Ray flowers other than yellow or orange-yellow.
    18. Head with outer and medial involucral bracts foliaceous and conspicuously pectinately toothed.
  40. *Stokesia*
    18. Involucral bracts not as above.
      19. Leaves closely clustered basally; heads solitary on naked scapes; lower leaf surfaces with a dense, tight, grayish or whitish, felty tomentum.
  42. *Chaptalia*
      19. Leaves not all basal; heads not on naked scapes; leaves not felty-tomentose beneath.
        20. Stem winged by decurrent leaf bases.
  6. *Verbesina*
        20. Stem not winged.
          20a. Ray flowers deep maroon or orange-red (*R. graminifolia* in genus).
  1. *Rudbeckia*
          20a. Ray flowers other than deep maroon or orange-red.
            21. Pappus of scales, or of minute bristles above the achene body and short awns over the wings.
              22. Leaves opposite.
  3. *Eclipta*
              22. Leaves alternate.
  27. *Boltonia*
            21. Pappus of capillary bristles.

23. Involucral bracts nearly equal and scarcely imbricated, generally in 1 or 2 series; receptacle of the head flat, not alveolate.   25. *Erigeron*
   23. Involucral bracts imbricated in several series; receptacle of the head alveolate.
                                                                 24. *Aster*
1. Heads discoid, without ray flowers.
  24. Receptacle of the head bristly or chaffy.
    25. Pappus of capillary bristles.
      26. The pappus of soft, white, plumose, capillary bristles.   41. *Cirsium*
      26. The pappus of antrorsely short-barbellate, capillary bristles.   36. *Carphephorus*
    25. Pappus of awns or scales, or none.
      27. The pappus of awns or scales.
        28. Leaves opposite.   9. *Bidens*
        28. Leaves alternate, in some mostly closely set at or near the base of the plant.
          29. Corollas yellow.   6. *Verbesina*
          29. Corollas white, cream-white, or various shades of lavender to purple.   10. *Marshallia*
      27. The pappus none.
        30. Heads with staminate and pistillate flowers separately in heads on the same plant; involucre of the pistillate heads essentially united and forming a bur, but bracts free at their tips in the form of hooked spines.   12. *Xanthium*
        30. Heads containing both functionally staminate and fertile pistillate flowers; involucre without hooked spines.   11. *Iva*
  24. Receptacle of the head naked, that is without bristles or chaff.
    31. Plant a shrub.   26. *Baccharis*
    31. Plant herbaceous.
      32. Pappus of scales.
        33. Plant low, mat-forming, stems 1–4 dm tall, with whorled linear-subulate leaves 1–2 cm long.   34. *Sclerolepis*
        33. Plant not mat-forming, stem 6–12 dm tall, with elliptic-oblanceolate to elliptic leaves 5–15 cm long and 1–8 cm broad.   35. *Hartwrightia*
      32. Pappus, in part at least, of capillary bristles.
        34. Stem conspicuously winged by decurrent leaf bases; leaves persistently and tightly felty-gray-pubescent beneath.   29. *Pterocaulon*
        34. Stem not winged by decurrent leaf bases; leaves not felty-pubescent beneath.
          35. Heads with peripheral flowers pistillate, central ones bisexual.
            36. Involucre with principal bracts in 1 series, sometimes with a few small bractlets about the base; plants without a fetid odor.   20. *Erechtites*
            36. Involucre with bracts definitely imbricated; plants with a fetid odor.   28. *Pluchea*
          35. Heads with all bisexual flowers.
            37. Heads individually small and aggregated into a "compound" head subtended by a sub- or false-involucre of foliaceous bracts, the involucres of the individual heads of 4 pairs of non-foliaceous bracts, the pairs disposed at right angles to each other.   39. *Elephantopus*
            37. Heads not aggregated into a "compound" head.
              38. Pappus comprised of an outer (not easily perceived) series of minute bristles and an inner series of capillary bristles.   38. *Vernonia*
              38. Pappus bristles all similar but not necessarily of equal length.
                39. Plant a twining herbaceous vine.   33. *Mikania*
                39. Plant erect, not vining.
                  40. Corollas yellow.   23. *Bigelowia*
                  40. Corollas other than yellow.
                    41. Leaves alternate.
                      42. Flowers cream-white, sometimes with green, tan, or maroon tints, but not clearly lavender-pink to purple.
                        43. Flowers 5 per head.   19. *Arnoglossum*
                        43. Flowers numerous in each head.   18. *Cacalia*
                      42. Flowers lavender-pink to purple (white only on an occasional individual).
                      44. Heads corymbosely disposed, or (in *C. paniculatus*) the short, compact corymbs racemosely arranged.   36. *Carphephorus*
                      44. Heads spicately or racemosely disposed on the main stem or its branches, if any.   37. *Liatris*
                  41. Leaves, at least the lower ones, opposite or whorled.

45. Receptacle of the head conic; lobes of the corollas violet or violet-blue as are the conspicuously exserted stigmas.   30. *Conoclinium*
45. Receptacle of the head flat; lobes of the corollas white or whitish, or purple.
  46. Lobes of the corollas white or whitish, rarely with a bit of lavender tint; lower leaves opposite, or in whorls only on an occasional specimen.   32. *Eupatorium*
  46. Lobes of the corollas definitely lavender-pink to purple; lower leaves always in whorls of 3 or more.   31. *Eupatoriadelphus*

1. Rudbeckia (CONE-FLOWERS)

Annual or perennial herbs with alternate, entire to pinnatifid leaves. Heads radiate; ray flowers about 5–20, neutral, mostly yellow or orange, sometimes purple, red, or maroon; disc flowers numerous, bisexual and fertile, the corolla tube narrow at the base and with a longer, somewhat expanded distal portion, 5-lobed at the summit; disc corollas dark brown, purplish brown, purple, or almost black, yellow or greenish yellow. Involucre of 2 or 3 series of green, spreading or reflexed bracts. Receptacle of the head conic, columnar, globose, or hemispheric, a keeled and somewhat boatlike bract (pale) subtending and partly enfolding each disc flower. Achenes glabrous, quadrangular, obconic in outline; pappus a low, often irregular or toothed crown, or none.

1. Ray flowers deep maroon, rarely orange-red.   6. *R. graminifolia*
1. Ray flowers yellow or orange-yellow.
  2. Disc corollas yellow or greenish yellow.   1. *R. laciniata*
  2. Disc corollas dark purplish brown to nearly black.
    3. Largest leaves elongate and narrow, not exceeding 1 cm broad more or less medially, long-attenuate to both extremities.   7. *R. mohrii*
    3. Largest leaves not as above.
      4. Chaff pubescent marginally on the tip, or marginally on the tip and along the keels, the hairs not sticky.
        5. Larger stem leaves 5–10 cm broad, conspicuously auriculate-clasping basally; rays bright yellow; mature achenes purple or purple-brown; pappus of triangular-acute scales.
                                                                   4. *R. auriculata*
        5. Larger stem leaves less than 5 cm broad, not or barely clasping; rays orange-yellow; mature achenes gray; pappus a very low or barely perceptible crown.   5. *R. fulgida*
      4. Chaff sticky-pubescent at least distally.
        6. Stem glabrous.
          7. Stem and leaves glabrous, not glaucous; principal lateral veins of the leaf few, strongly ascending, wholly or nearly reaching the leaf tip.   2. *R. nitida*
          7. Stem and leaves glabrous and glaucous; principal lateral veins of the leaf numerous, arcuate from the midrib.   3. *R. maxima*
        6. Stem moderately to copiously shaggy-pubescent, at least above.   8. *R. subtomentosa*

1. Rudbeckia laciniata L. CUT-LEAVED CONEFLOWER.   Fig. 348

Perennial with a hard subterranean caudex from which lateral perennations arise. Variable in stature, sometimes only about 3–5 dm tall, commonly taller, to 3 m, freely branched above. Stem usually glabrous, often glaucous, infrequently sparsely pubescent, especially distally. Basal or rosette leaves with petioles 10–30 cm long, petioles of those above the base and to midstem or higher mostly not over 5–7 cm long, uppermost leaves sessile; in some populations blades of all leaves ovate, unlobed, and coarsely toothed; much more commonly the blades of basal and most stem leaves below the inflorescence variously pinnatifid or trilobed, the pattern of lobing extremely variable; uppermost blades generally ovate and unlobed. Heads few to numerous, long-stalked in a more or less paniculate pattern from axils of bracteal leaves. Involucral bracts relatively few, foliaceous, slightly imbricated, oblongish with acute apices, usually sparsely pubescent beneath. Ray flowers yellow, about 5–12 per head, 2–6 cm long, usually lanceolate, short-pubescent beneath, eventually drooping. Fully developed discs (in our range) mostly hemispheric to subglobose, infrequently elongating in age, 1–1.5 cm broad. Disc flowers numerous, their corollas yellow. Chaff strongly keeled, obtuse

Fig. 348. **Rudbeckia laciniata:** a, portion of inflorescence more or less surrounded by variable leaves; b, chaff; c, achene.

apically, exteriorly pubescent distally. Achene strongly angled, 3–4 mm long, dull brown; pappus a low crown. (Incl. *R. heterophylla* T. & G.)

Plants of this species vary not a little with respect to stature, leaf form, and head size.

Moist to wet woodlands, along streams, floodplain forests in places where flooding is intermittent, wet places in pastures, wet thickets and shores. Maine to Sask. and Idaho, generally southward to n.cen. pen. Fla. and Ariz.

2. Rudbeckia nitida Nutt.

Perennial with a short, stout, eventually few-branched caudex bearing 1 to few stems 5–12 dm tall. Stem glabrous, not glaucous, essentially terete but multistriate-ribbed, solitary and bearing a single head terminally on a few-bracted, longish stalk, or with relatively few erect-ascending similar branches. Basal and sometimes lower stem leaves largest, to 3 dm long overall, long-petioled, blades lanceolate or long-elliptic, somewhat decurrent on the petiole, margins entire, obscurely and irregularly toothed, or dentate-serrate, bases attenuate, apices rounded, obtuse, or acute; stem leaves gradually reduced upwardly, sessile; all leaves glabrous and sublustrous, principal lateral veins strongly ascending, intermediate reticulate venation conspicuous, especially beneath. Involucral bracts loose, more or less deflexed, subulate, glabrous. Rays mostly 7–10 per head, golden yellow above, paler beneath, deflexed, 3–4 (–5) cm long, mostly oblanceolate or linear-oblong, minutely 2- or 3-toothed apically, upper surfaces abundantly, very finely, and minutely papillate-granular, lower moderately very short-pubescent with yellow hairs. Fully developed discs mostly obconic, sometimes ovoid-hemispheric, 2–4.5 cm high. Disc corollas 3.5–4 mm long, purplish brown. Chaff 5–7 mm long, broadest near the summit, greenish to tan, with a purple line (dilated distally) paralleling each margin and usually one on the keel, backs short-pubescent on and along the keel proximally, usually more or less pubescent all over the distal third, tip obtuse, purplish but the purple obscured by paler pubescence. Achene 3–5 (–7) mm long, purplish and with a grayish metallic cast; pappus an irregularly toothed purplish crown.

Seasonally wet pine savannas and flatwoods, flatwoods depressions, clearings and pastures on former pineland sites. Local, s.e. Ga., n.e. Fla., w.cen. pen. Fla., cen. Fla. Panhandle; cen. La., e. and s.e. Tex.

3. Rudbeckia maxima Nutt.

In general aspect similar to *R. nitida*, taller, to 3 m, leaf blades larger, heads larger. Fresh stem and leaves glabrous and glaucous. Basal and lower stem leaves long-petioled, to 4 dm long overall; blades gradually narrowed and somewhat decurrent on the petioles or abruptly contracted to the petioles; blades broadly elliptic or ovate-elliptic, to 10 cm broad, apices obtuse, margins entire, undulate, or obscurely dentate; stem leaves sessile, somewhat clasping basally, ovate, obovate, or oblongish, sometimes irregularly and coarsely toothed; principal lateral veins of the blades numerous, pinnate, arcuate from the midrib. Stem sometimes unbranched and with a single long-stalked terminal head, or with few similar branches. Involucral bracts long-triangular-acute to linear-oblong-acute, sparsely pubescent on their backs and with short-ciliate margins, mostly eventually becoming deflexed. Rays as in *R. nitida* but not pubescent beneath. Fully developed discs oblong-obconic, 2.5–6 cm high. Disc corollas, chaff, and achenes like those of *R. nitida*.

Moist to wet open places, margins of swamps, swales, marshy places. n.w. La., s.w. Ark., e. Tex., s.e. Okla.

4. Rudbeckia auriculata (Perdue) Kral.

Robust perennial, to 3 m tall, perennating by short, stoloniferous offshoots eventually becoming 1 cm or more thick. Stem glabrous, terete but multiribbed, to 1 cm thick proximally. Rosette leaves to 6 dm long overall, long-petiolate, blades long-ovate, lance-ovate, or oblong, cuneate basally and shortly decurrent on the petiole, rounded, obtuse, or acute apically, margins entire, undulate, or crenate; principal stem leaves sessile and auriculate-clasping basally, some of them usually somewhat fiddle-shaped, others ovate, 15–20 cm long and to 10 cm broad, margins coarsely dentate-serrate, apices acute; surfaces of rosette leaves with scattered, stiff hairs somewhat enlarged basally, often breaking off near the base, stem leaves usually glabrous. Inflorescence more or less paniculate and with reduced bracteal leaves, or sometimes with branches from many of the larger stem leaves as well as from reduced bracteal leaves distally. Involucral bracts spreading, much shorter than the rays, the outer larger than the inner, lance-oblong to nearly linear, margins scabrid. Rays bright yellow, 1.5–4 cm long, lanceolate, oblanceolate, or oblong, with sparse short, appressed hairs on the backs, apices mostly minutely 2- or 3-toothed. Fully developed discs ovoid, about 1.5 cm high. Disc corollas 3–3.5 mm long, with greenish lower tube, the expanded portion and lobes purple. Chaff broadened from the base upwardly, the obtuse tip dilated, purplish, pubescent along the keel and exteriorly on the purplish tip and its margin. Achene somewhat curved, 4–6 mm long, purple or purple-brown; pappus a crown with several, prominent, unequal, triangular-acute scales. (*R. fulgida* var. *auriculata* Perdue)

Bogs, openings in and borders of swampy woodlands, swales, ditches. Ala., chiefly coastal plain.

5. **Rudbeckia fulgida** Ait.

Perennial, 3–10 dm tall, perennating by basal offshoots. Stem commonly solitary and bearing a single terminal head, varying to several main stems from the base or a single main stem few- to several-branched. Basal leaves long-petioled, blades lanceolate to ovate, 3–8 cm long, margins obscurely toothed to entire; lower stem leaves relatively short-petioled, upwardly a few leaves with winged subpetiolar bases, similar in shape to the basal; other leaves sessile, gradually reduced upwardly, lanceolate, oblanceolate, or narrowly oblongish, 3–6 cm long, apices mostly blunt; stem and both leaf surfaces usually with stiff, appressed hairs, less frequently with hairs spreading, sometimes glabrous. Stems below the heads naked for 3–10 cm. Involucral bracts green, 3–7 mm long, triangular-acute to nearly ovate and obtuse, pubescent, usually becoming reflexed. Rays golden yellow or orange-yellow, spreading, 1–2 cm long, broadest distally, the apices pointed, barely emarginate, or 3-lobed (often thus variable in a single head). Fully developed discs hemispheric to ovoid, mostly 1–1.5 cm broad. Disc flowers numerous, compacted, their corollas 3–3.5 mm long, pale basally, dark purplish brown to nearly black distally. Chaff obtusely keeled except at the triangular tip, a dark resinous line on either side from just back of the tip, attenuating proximally to below the middle, tip dark, pubescent marginally. Achene strongly angled, 2–2.5 mm long, grayish and smooth; pappus a low, sometimes scarcely perceptible crown. (Incl. *R. acuminata* Boynt. & Beadle, *R. chapmanii* Boynt. & Beadle, *R. foliosa* Boynt. & Beadle, *R. palustris* Eggert, *R. spathulata* Michx., *R. tenax* Boynt. & Beadle, *R. truncata* Small)

Dry to wet places, prairies, swales, open woodlands, fields, roadsides and ditches. N.J. to Ill., generally southward to s. Fla. and Tex.

6. **Rudbeckia graminifolia** (T. & G.) Boynton & Beadle.   Fig. 349

Perennial with a short, stoutish, basal caudex, the latter often shortly 2- or 3-branched, the caudex or each of its branches giving rise to a solitary, usually unbranched, stem 4–8 dm tall and bearing a single head on its elongate, naked terminal portion. Stem slender, lower portion glabrous, upper portion sparsely to copiously pubescent with mostly appressed-ascending, short, white hairs. Largest leaves basal and on the lower stem, narrow and elongate, some of them broadest somewhat above the middle and not exceeding 1 cm broad, attenuate to either extremity, blade and petiole scarcely differentiated, often some or all linear and 3–5 mm broad; only the midvein evident, this raised beneath, margins revolute, surfaces sparsely short-pubescent throughout, or only on and near the margin and on the midrib, sometimes becoming essentially glabrous. Involucral bracts subulate, their lower surfaces pubescent with white, stiff hairs, upper surfaces glabrous. Rays usually deep maroon, infrequently orange-red, darker above than beneath, soon becoming deflexed, 8–20 mm long, elliptic or oblong-elliptic, 2- or 3-toothed apically, upper surfaces evenly very finely and minutely granular-papillate giving a somewhat velvety sheen, lower surfaces dull, sparsely to moderately pubescent with short, white hairs. Fully developed discs ovoid, 1–2 cm high. Disc corollas 3–4 mm long, brownish red, at anthesis minutely atomiferous-glandular exteriorly, lobes

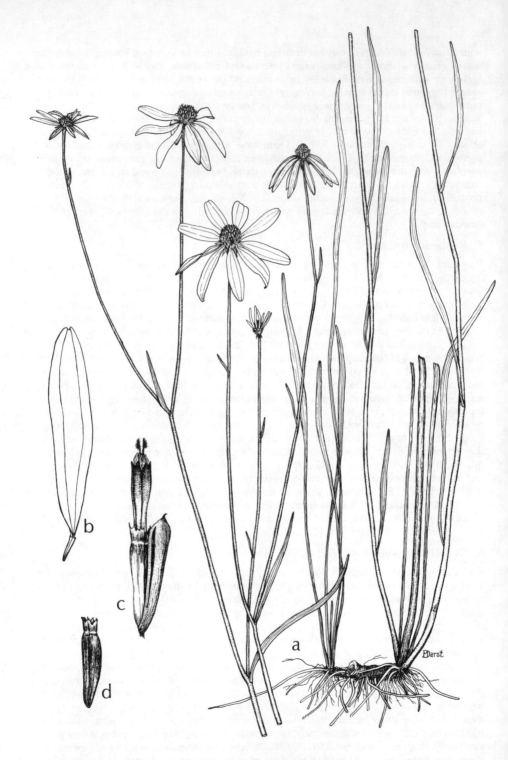

Fig. 349. **Rudbeckia graminifolia:** a, portions of plant; b, ray flower; c, disc flower with subtending pale; d, achene.

somewhat inflexed. Chaff brown or purplish, hard-leathery, broadest distally, a very dark, thin line, usually evident only interiorly, paralleling each margin, apex essentially hoodlike, tip obtuse, outer surface usually with a few hairs on the keel. Achene 2–2.5 mm long, surface with a grayish metallic sheen; pappus a low, brownish crown shortly toothed above the angles of the achene.

Wet pine savannas and flatwoods, borders of shrub-tree bogs or bays, open bogs, grassy cypress-gum depressions; often in shallow water. cen. Fla. Panhandle.

7. Rudbeckia mohrii Gray.

Perennial with a short, stout, basal caudex, 8–18 dm tall, glabrous throughout. Stem terete, multiribbed, commonly few-branched, infrequently multibranched, branches elongate, ascending-erect. Largest leaves on the lower stem, elongate and narrow, not exceeding 1 cm broad more or less medially, long-attenuate to either extremity, blade and petiole scarcely differentiated, 10–30 cm long, mostly with three well-defined, parallel ribs beneath, sometimes with faint ribs evident between them, edges tending to be thickened and closely rolled upwardly; other leaves gradually reduced upwardly, becoming narrowly linear, those on the elongate 1–3-headed, ultimate branches acicular-bracteate. Involucral bracts linear subulate, spreading fully across the base of the fully developed disc, the latter ovoid to hemispheric, 1–1.5 cm high and broad. Rays bright yellow, spreading at first, then drooping, 1–3 cm long, elliptic to linear-oblong, obtuse apically, sometimes with a small mucro, surfaces glabrous. Disc corollas 3 mm long, purple or purple-brown. Chaff 4 mm long, flared apically, a purplish mucro at the tip, otherwise tan with 1 purple line, broadest distally, paralleling each margin, sometimes with a second, less distinct or broken line. Achene 3–4 mm long, dark purple; pappus a distinct, similarly colored, shallowly 4-notched to 4-toothed crown.

Cypress ponds and depressions, flatwoods depressions, adjacent ditches, usually in shallow water. Coastal plain, s.w. Ga., cen. Fla. Panhandle.

8. Rudbeckia subtomentosa Pursh.

Perennial with a stout rhizome, 0.6–2 m tall. Stem terete but ribbed and channelled, moderately to copiously shaggy-pubescent at least above. Principal leaves petioled, blades variable, about 10 cm long, occasionally all on a given specimen ovate, lance-ovate, or lanceolate and unlobed, sometimes some unlobed, others irregularly and obliquely 2- or 3-lobed, commonly the lower and medial ones obliquely 3-cleft nearly to the base, margins of blades or lobes serrate, their apices acute or acuminate; leaf surfaces moderately to copiously short-pubescent, the hairs somewhat enlarged basally, more distinctly so above than beneath. Inflorescence branches few to relatively numerous, strongly erect-ascending, more or less leafy-bracted, the main branches usually from the axils of little, if at all, reduced leaves. Heads anise-scented. Involucral bracts usually reflexed, subulate, often involute, hispid. Rays bright to pale yellow, 2–4 cm long, linear-oblong to oblanceolate, mostly emarginate at the tip, copiously short-pubescent beneath. Fully developed discs ovoid or hemispheric, 1–1.5 cm high. Corollas of disc flowers about 3 mm long, purple or purple-brown except near the base. Chaff somewhat dilated near the tip, straw-colored with a purple line paralleling each margin, apically triangular-obtuse, purplish and glandular-pubescent. Achene 3 mm long, slightly curved, brown with a metallic sheen; pappus a barely perceptible crown.

Low meadows, prairies, open lowland woodlands, and thickets, stream banks, roadsides. Ind. to Wisc. and Iowa, southward to w. Tenn., w. Ala., La., s.e. Okla.

2. Dracopsis

Dracopsis amplexicaulis (Vahl) Cass.

Glabrous and glaucous annual 3–7 (–12) dm tall, usually with 3 to several strongly ascending branches from about midstem, each branch terminated by a single head, the stem below the head naked for 5–18 cm. Leaves alternate, pinnately veined, sessile and mostly cordate-clasping, the larger ones 6–12 cm long and to 5 cm broad, elliptic-oblong, uppermost smaller and ovate; margins of most leaves obscurely serrate, apices acuminate or acute. Involucral bracts relatively few in a single series, varyingly elliptic, lanceolate, ovate, or lance-subulate. Heads radiate. Ray flowers 5–10, neutral, yellow with a reddish brown to purple base, short-pubescent beneath, 1–2 (–3) cm long, broadest distally and shallowly and irregularly toothed, eventually reflexed. Receptacle of the head becoming columnar, elongating to 1.5–3 cm and broadening to 1.5–3 cm in fruit; chaff subtending both ray and disc flowers, nearly oblong in outline, 4–4.5 mm long, short-triangular-obtuse apically, the apex greenish exteriorly and pubescent on the margin, straw-colored or purplish proximally from the tip, a dark purple clavate marking on either side just below the tip. Disc flowers many and crowded, bisexual and fertile; corollas 3–3.5 mm long, with a short-cylindric tube basally, abruptly expanding to a longer cylindric throat, 5-lobed at the summit, upper half of the throat and the lobes brown to purple. Achene obscurely quadrangular, nearly oblong in outline, truncate apically, 2 mm long, surface with numerous very fine longitudinal lines and finely transversely rugose, brownish purple; pappus none. (*Rudbeckia amplexicaulis* Vahl)

In open, usually low places, bottomlands, prairies, along streams, ditches, lowland cultivated fields. Chiefly coastal plain, Ga. to Tex., northward in the interior, Okla., Ark., Mo., Kan.

3. Eclipta

Eclipta alba (L.) Hassk.

Annual (or perennial in warm regions), with a single few- to much-branched stem, or with several to numerous branches from a basal crown, branches weakly ascending to mostly prostrate, the latter sometimes rooting at the nodes, to 1 m long; stem and leaf surfaces relatively sparsely pubescent with stiff, antrorsely appressed hairs. Leaves opposite, simple, mostly lanceolate, varying to lance-elliptic or linear-lanceolate, often falcate, 2–10 cm long, 4–25 mm broad, tapered basally to sessile or short subpetiolate bases, apices acute or acuminate, margins remotely and obscurely serrate. Heads radiate, 1–3 borne in axils of one of a pair of leaves, their stalks unequal, of variable lengths, the longer ones to about 5 cm long. Involucre broad and low, bracts 10–12 in 2 series, subequal or the inner shorter and narrower than the outer, longer ones about 4 mm long, long-triangular-acute varying to ovate-acuminate. Receptacle flat to slightly convex, flowers subtended by bristlelike pales. Ray flowers numerous, pistillate and fertile, rays linear, white, inconspicuous, 1–2 mm long; disc flowers numerous, bisexual and fertile, corollas whitish, tube short and slender basally, expanded distally to a somewhat longer throat, with 4(5) short-deltoid lobes at the summit. Achene brown, 2 mm long, obconic in outline, somewhat 3- or 4-angled or laterally flattened, surface with rounded tubercles, summit minutely hairy. (*Verbesina alba* L.)

Weedy in diverse, usually open, wet places. Mass. to Wisc., Iowa, and Neb., generally southward to s. Fla. and Tex.; Calif.; general in tropics and subtropics.

4. Borrichia (SEA-OXEYES)

Rhizomatous, maritime herbs. Leaves opposite, gradually tapering basally to petioles or subpetioles. Petiole bases somewhat dilated in a **U**-shaped fashion against the stem, the upper edges of the pair meeting on either side of the node, an acutely triangular projection of stem tissue below and between where they meet; after the pair of leaves has fallen, the acutely triangular portions on either side of the node become free as sharp points directly below the next pair of leaves (or their scars) above. Heads radiate, stiffly stalked and solitary terminating the main stem and branches. Involucral bracts closely imbricated, outer ones of different form and texture from the inner, innermost sometimes transitional to the pales subtending the flowers. Receptacle flat. Ray flowers yellow, pistillate and fertile, corolla very short-tubular at base, abruptly expanded to a relatively small ray; disc flowers brownish yellow, bisexual and fertile, radially symmetrical, the tube broadening upwardly little if at all, lobes 5, essentially erect but slightly curved inwardly; in both ray and disc flowers, resin canals along the veins prominent. Achene 3- or 4-locular. Pappus crownlike.

- Involucral bracts, some of them at least, spine-tipped; pales hard and very rigid, with pronounced, very sharp, spinose tips.   1. *B. frutescens*
- Involucral bracts, none of them, spine-tipped; pales firm-papery, without spinose tips.
                                                                                              2. *B. arborescens*

1. Borrichia frutescens (L.) DC.

Plant freely rhizomatous, forming extensive colonies, 1.5–10 dm tall. Stem usually with relatively few strongly ascending branches, at first clothed with a dense, compact, gray pubescence, becoming glabrous and light brown, finally gray; older stems very irregularly and unevenly ridged and grooved. Leaves mostly narrowly to broadly oblanceolate, sometimes some of them elliptic, 2–6 cm long, margins usually entire, some sometimes minutely dentate, apices minutely mucronate; surfaces compactly and densely gray-pubescent like the young stems or sometimes glistening green. Outer involucral bracts mostly ovate, apices acute, short-mucronate, or shortly spine-tipped, in texture and pubescence like the leaves; inwardly bracts becoming firmer and harder, brown, much more sparsely pubescent, their tips often spinose. Pales oblong, hard and rigid, with hard, sharp, erect spine tips 1–3 mm long; in fruit the heads very "stickery" owing to the hardness, stiffness, and sharpness of the exserted, erect pales, the entire head hard and difficult to dissemble. Achene slightly obconic in outline, regularly 3- or 4-angled, metallic gray, about 3.5 mm long, the surmounting pappus a short crown, not as broad as the summit of the achene, with a sharp tooth above each angle.

Tidal marshes and mud flats, coastal and estuarine shores, edges of mangrove swamps, often weedy in vacant lots or on roadsides near the sea. Va. to Fla. Keys, westward to Tex.; c. Mex.; Berm.

2. Borrichia arborescens (L.) DC.

Plant usually bushy-branched, to about 12 dm tall, rhizomes short, scarcely colonial, the sharply triangular projections left between the leaf scars after leaf fall somewhat spreading and giving the stems at that stage a somewhat prickly aspect. Young stems and leaves of some plants with dense, compact, gray pubescence much as in *B. frutescens*; more commonly the pubescence of young stems much more sparse, the leaves mostly narrowly oblanceolate, glabrous and succulent. Outer involucral bracts lance-ovate or ovate, acute apically, of firm texture, grading to oblong, thin, inner bracts with rounded apices. Pales with a papery texture, oblong-oblanceolate, somewhat boatlike, rounded or obtuse apically, not at all spine-tipped. Achene unequally angled, somewhat compressed, slightly obconic in outline, 3 mm long, dark gray, the surmounting pappus a brownish, scarcely toothed, callus-thickened crown somewhat broader than the summit of the achene.

Saline shores, edges of or in disturbed coastal hammocks, mangrove swamps. s. pen. Fla. and Fla. Keys; W.I.

5. Helianthus (SUNFLOWERS)

Annual or perennial herbs, stems in most species erect and the inflorescence of heads more or less corymbosely disposed, individual heads mostly borne singly at the ends of naked, terminal stalks. Leaves simple, all opposite, or opposite below and alternate above; lower leaves, for the most part, with petiolar or subpetiolar bases. Heads radiate in most species. Involucre saucerlike to hemispheric, the bracts more or less herbaceous at least distally, subequal to strongly graduated, in 2 series or manifestly imbricated. Receptacle of the head flat to convex, chaffy, in the disc each chaffy bract embracing an achene. Ray flowers usually pistillate but always infertile, neutral in some, ligules yellow; disc flowers many, bisexual and fertile, their tubes yellowish in most species, 5-lobed at the summit, the lobes yellow to brownish, reddish, or purplish. Achenes somewhat compressed laterally, 4-sided and often subrhombic in cross-section, edges blunt, surfaces glabrous in a few species, in most having a few readily deciduous hairs near the apices of the achenes; pappus of 2 easily and quickly deciduous, scalelike awns, sometimes with additional, shorter scales on either side of the awns.

Reference: Heiser, C. B., Jr. "The North American Sunflowers (*Helianthus*)." *Mem. Torrey Bot. Club* 22 (3). 1969.

Many of the species of *Helianthus* exhibit considerable variation. Some of it can be attributed to plasticity, which renders some of the characters of individuals subject to environmental modification; not a little of the variation is a consequence of present or past hybridization and subsequent introgression.

There are approximately 30 species of sunflowers occurring in our area of coverage. For a few of them, ones with which we have slight acquaintance, we had difficulty in deciding, on the basis of references to habitat in the literature or reference to habitat on the labels of herbarium specimens, whether they should be included in this work. For the most part, when we were in doubt, we omitted them from the account.

1. Disc corollas with purple lobes.
  2. Plant annual.   1. *H. agrestis*
  2. Plant perennial.
    3. Principal leaves basal or very near the base.
      4. Ray flowers usually none, or only occasionally 1 to few present, these very short, inconspicuous, easily overlooked.   2. *H. radula*
      4. Ray flowers 12–20, conspicuous, 1.5–2.5 cm long.   4. *H. heterophyllus*
    3. Principal leaves extending high on the stem, flowering stem not at all scapiform.
                                                         5. *H. angustifolius*
1. Disc corollas with yellow lobes.
  4. Principal leaves basal or very near the base.   3. *H. carnosus*
  4. Principal leaves extending well up the stem.
    5. Stem leaves linear-lanceolate to linear, only the midvein evident, strongly revolute marginally.
                                                         5. *H. angustifolius*
    5. Stem leaves broader, or if linear-lanceolate to linear, then the lateral veins at least faintly evident, margins not at all or scarcely revolute.
      6. Leaves with a lateral vein either side of the midrib arising somewhat above the base, these laterals more evident than other lateral veins.
        7. Principal leaves oblongish, sessile and truncate to rounded basally, 2–3 times as long as broad.   6. *H. floridanus*
        7. Principal leaves lanceolate, tapered basally (usually to short subpetioles), much more than 3 times as long as broad.   8. *H. giganteus*
      6. Leaves with all lateral veins only faintly evident.   7. *H. simulans*

1. Helianthus agrestis Pollard.

Annual, 1–2 m tall, usually with several widely spreading branches above. Stem glaucous, glabrous below, usually sparsely and irregularly hispid on the inflorescence branches, and stalks of the heads copiously pubescent just below the heads. Leaves opposite below, alternate above, short-petioled, blades lanceolate, acute to acuminate apically, principal ones variable in size, 5–10 (–15) cm long and 0.6–3 (–4) cm broad, usually, but not always, ciliate marginally near the bases of the blades and on the petioles, short-hispid marginally from a little above the base and on the surfaces, most of the hairs on the surfaces with flat, scalelike bases. Involucral bracts imbricated in about 3 series, equal or subequal, lance-acuminate or -attenuate, glabrous on their backs and usually minutely hairy on their margins. Heads 5–6 cm across (ray tips to ray tips), rays usually about 12, mostly oblanceolate and often varying somewhat in length and width in a given head. Discs subhemispheric, 1.5 cm across or a little more; disc corollas pale below, their upper halves and lobes dark purple, thus the discs dark purple. Achene brown, about 4 mm long, surfaces glabrous and roughened by low tubercles or ridges.

Open marshy flats, marshy shores, wet flatwoods clearings, sometimes in extensive dense stands. Chiefly cen. pen. Fla. (and Thomas Co., Ga., according to Heiser (1969)).

2. Helianthus radula (Pursh) T. & G.

Perennial with a short, thickish crown, perennating by short, prostrate offshoots with rosettes of leaves. Stems 1 to several from a rosette, each shortly spreading laterally from the crown then arching upward, becoming erect, 5–12 dm tall, usually unbranched and bearing a single head, infrequently with 2 or 3 elongate branches; lower stem copiously hirsute, becoming sparsely so above. Largest leaves opposite and closely set in the rosette, with short, subpetiolar bases, blades obovate to subrhombic to orbicular, with 3 principal veins from the base, the lateral pair curvate, varying in size from about 4 to 12 cm long and 2 to 10 cm broad, margins entire or minutely dentate; lower stem leaves (above the rosette) opposite, sometimes 2–4 relatively approximate pairs rhombic to elliptic, only moderately reduced, then upwardly leaves sessile, few and much reduced, widely spaced, eventually bracteate, the stem naked for 2–5 dm below the head; oftentimes all leaves above the rosette very much reduced, lowermost pairs mostly oblanceolate, others narrow and bracteate; leaf surfaces rough to the touch, copiously clothed with long, pointed, transparent, pustular-based hairs. Involucral bracts imbricated, little graduated in size, dark purple, lanceolate, lance-ovate, or narrowly elliptic, appressed to and about equaling the disc, glabrous or hispid. Heads usually discoid, occasionally with 1 to few inconspicuous, short-oblong rays. Discs hemispheric, 1.5–2.5 cm broad. Chaff dark purple at least distally. Tubes of disc flowers usually straw-colored or sordid brown below, upper tube and lobes dark purple, the tips of the densely crowded flowers rendering the disc dark purple. Achene blackish purple, 4–4.5 mm long, short-pubescent near the apex.

Pine savannas and flatwoods, often where seasonally wet, less frequently in well-drained pinelands, often abundant on moist to wet roadsides and in moist to wet, sandy-peaty ditches, frequently intermixed with *H. heterophyllus*. Coastal plain, s.e. S.C. to cen. pen. Fla., westward to s.e. La.

3. Helianthus heterophyllus Nutt.

Perennial with a short, basal crown, the stem usually solitary and stiffly erect, sometimes with several long subequal branches from near the base, varying to few-branched more or less medially, or unbranched and with a single, terminal head, 4–12 dm tall; lower stem very rough to the touch, copiously pubescent with transparent, long, pointed, pustular-based hairs, the pubescence becoming shorter and much sparser upwardly, often becoming glabrous above. Basal and lower stem leaves opposite, the pairs approximate, the lowermost 1–3 pairs gradually narrowed basally to short or to relatively long subpetiolar portions, very variable in size and shape, bladed portions oblanceolate, lanceolate, or subelliptic, sometimes linear-lanceolate or linear-attenuate; other leaves few and widely spaced, alternate, much reduced, mostly bracteate and linear; leaf surfaces rough to the touch, the upper more so than the lower, pubescence like that of the lower stem, more copious above than beneath, margins entire, those of narrow leaves revolute. Stems below the heads generally long-scapiform. Involucral bracts imbricated, graduated in length, the outer shortest and acute, inner longest and with acuminate-attenuate tips, surfaces sparsely hispid to glabrous on their backs, scabrid to short-ciliate marginally. Heads very attractive at full anthesis, commonly 6–8 cm across (ray tips to ray tips), rays 12–20, ligules golden yellow and finely reddish brown–lined, mostly oblanceolate. Discs subhemispheric, 1.5–2.5 cm across; tubes of disc corollas light brown nearly or quite to their summits, the lobes purple, the compacted flower tips rendering the disc purple. Achene blackish purple, 4–4.5 mm long.

Seasonally wet pine savannas and flatwoods, hillside bogs, moist to wet, sandy-peaty ditches, often growing intermixed with *H. radula*. Coastal plain, s.e. N.C. to the Fla. Panhandle, westward to s.e. La.

4. Helianthus carnosus Small.

Glabrous perennial with a short, thick basal crown bearing buds from which a very short, leafy stem (essentially a rosette) arises on one side and from which a flowering stem or two, 6–8 dm tall, leafy only near the base, few-bracteate above, arches slightly outward, then upward, on the other side. Principal leaves opposite, gradually narrowed basally to short to long subpetioles, bladed portions mostly lanceolate to nearly linear, 10–25 cm long, 7–15 mm broad; bracteal leaves few, alternate, linear-subulate. Flowering stem usually unbranched and bearing a single, terminal head, infrequently with several elongate, subequal branches each bearing a terminal head. Involucral bracts imbricated, the outer not very much shorter than the inner, ovate and short-acuminate apically. Heads at full anthesis 6–10 cm across (ray tips to ray tips), rays 12–20; yellow and finely light brownish–lined, lanceolate to oblong. Discs subhemispheric, 2–2.5 cm across; tubes of disc corollas yellowish to stramineous, lobes yellow. Achene dark brown, about 3 mm long.

Seasonally wet pine flatwoods, adjacent roadsides. n.e. Fla.

5. Helianthus angustifolius L.   Fig. 350

Perennial with a short basal crown bearing crown buds, a rhizome sometimes extending from the crown, 5–15 (–20) cm tall. Stem simple and few-branched, varying to freely branched, the branches often widely spreading; pubescence of the stem varied, that of the lower, main stem, usually copious, sometimes comprised of long, spreading hairs, these sometimes intermixed with shorter appressed hairs, sometimes the hairs mostly short and either spreading or appressed, much of the pubescence often shed as the stem ages; pubescence of branches usually relatively sparse and short, spreading or appressed or both. Early leaves (occasionally present late in the season as shoots are newly developed from crown buds) small, narrowed to subpetiolar bases, the bladed portions oblanceolate to ovate; other leaves usually numerous, sessile, opposite below, alternate above, variable but mostly linear or narrowly lanceolate, to 20 cm long and 20 mm broad, generally 8–10 cm long and 3–8 mm broad, only the midvein evident, margins entire and strongly revolute, dark green above, paler beneath; upper surfaces usually hispid, the hairs usually enlarged and somewhat bulbous below, then abruptly flaring proximally into flat, closely appressed (as though glued), scalelike bases; lower surfaces sparsely to densely hispid, the hairs enlarged toward their bases or not and without the flaring, scalelike bases, also atomiferous-glandular. Involucral bracts imbricated, moderately graduated from the outside inwardly, outer shortest, subulate or lance-subulate, varying from glabrous to copiously hispid. Heads mostly 5–6 cm across (ray tips to ray tips), rays mostly 10–12, sometimes a few more, golden yellow and with rather faint brownish lines, mostly oblanceolate, lower surfaces sparsely and very finely atomiferous-glandular. Discs 1.5 cm across or a little more; tubes of disc corollas brownish or stramineous, lobes dark purple in most populations, brownish to yellow in some populations, chiefly in s. Ga. and Fla. where "genetically contaminated" from either *H. floridanus* or *H. simulans*. Achene brown, often mottled with black, 2.5–3 mm long, glabrous or with a few short hairs at the apex.

Bogs, low, open woodlands, pine savannas and flatwoods, swales and thickets, ditches, also open, upland woodlands and old fields. L.I., N.J., s.e. Pa., southward through e. Va., then more widespread across N.C. and Tenn. to s.e. Okla., generally southward to cen. pen. Fla., e. and s.e. Tex.; locally, s. Ohio, Ky., s. Ind., s. Ill., s.e. Mo.

6. Helianthus floridanus Gray ex Chapm.

Perennial, freely producing slender rhizomes late in the growing season. Stem short-hispid throughout, 8–12 dm tall, moderately branched above, the branches erect. Leaves opposite below, alternate above, sessile, the principal ones oblongish, truncate or rounded basally, 5–8 cm long and 2.5–3 cm broad, margins irregularly wavy, scarcely if at all revolute, a pair of arching-ascending lateral veins nearly as prominent as the midvein arising from the latter somewhat above the base; upper surfaces short-hispid, the hairs with bases as on the upper

Fig. 350. **Helianthus angustifolius:** a, plant; b, portion of lower leaf surface; c, head, viewed from below; d, disc floret; e, chaffy bract subtending floret; f, achene.

surfaces of leaves of *H. angustifolius*, lower surfaces atomiferous-glandular and subtomentose, the hairs short and without enlarged bases. Involucral bracts loosely imbricated, the outer deltoid to lanceolate, much shorter than the inner, their tips blunt to acute, medial ones oblong-acute, inner ones oblong-acuminate, surfaces glabrous or sparsely short-pubescent. Heads 5–6 cm across (ray tips to ray tips), rays golden yellow, faintly pale-lined, lower surfaces sparsely and very finely atomiferous-glandular. Discs hemispheric to subglobose, 1.5–2 cm across; lobes of disc corollas yellow. Achene black or nearly so, about 3 mm long, a few short hairs at its summit.

Pine savannas and flatwoods, borders of shrub-tree bogs, swales. s.e. S.C., s.e. and s.cen. Ga., n.e. Fla.

Populations of *H. floridanus* in the "pure" form are probably rare at present. Heiser (1969) suggests that the more aggressive *H. angustifolius* may be in the process of replacing *H. floridanus* through hybridization in the area of overlap. Chiefly in s. Ga. and e. Fla., plants in many populations have yellow discs but are otherwise very closely similar to *H. angustifolius* (indicating introgression in the direction of *H. angustifolius*).

7. Helianthus simulans E. E. Wats.

Perennial; when growing in wet, mucky soil freely producing slender rhizomes from its stout basal crown late in the season, rhizomes eventually becoming as much as 1 cm thick. Stem, except in depauperate individuals, coarse, freely branched, 2–3 m tall, rough-hispid, the hairs mostly curvate and with pustular bases; heads borne more or less corymbosely or sometimes racemosely disposed on branchlets. Leaves opposite below, alternate above, the larger ones narrowed to subpetiolar bases, blades linear-lanceolate to broadly lanceolate, 10–20 cm long overall, 1–3.5 cm broad, margins entire, not revolute or but scarcely so, midvein prominent, laterals faintly evident at most, upper surfaces with pubescence as for leaves of *H. angustifolius*, lower atomiferous-glandular and short-pubescent, the bases of the hairs little or not at all enlarged. Involucral bracts imbricated, the outer lance-subulate, varying inwardly to lance-attenuate, sparsely to copiously hispid on their backs. Heads (5–) 6–8 cm across (ray tips to ray tips), the rays 12–20 for the most part, oblanceolate, elliptic, or oblong-elliptic, larger ones 1–1.5 cm broad. Discs subhemispheric, about 2 cm broad; lobes of the disc corollas yellow to brownish but the discs given a purple aspect because of conspicuously protruding dark purple anthers. Achene dull brown, often mottled with black, glabrous, about 3 mm long.

Frequently cultivated for ornament in the Fla. Panhandle, perhaps elsewhere, and often abundantly naturalized in the vicinity of human habitations—on open banks, in swales, borders of pinelands, mostly in moist or seasonally wet places, occasionally on the banks of sloughs and in shallow water in the sloughs. Probably native only in s. La. where it is said to grow in wet, black muck. In Fla. it sometimes grows intermixed with *H. angustifolius* and in such places putative hybrid intermediates occur. It is often misidentified as *H. floridanus*.

8. Helianthus giganteus L.

Perennial with a short crown bearing crown buds and thickened, hard roots. Stems mostly 1–3 m tall, rough-pubescent (becoming glabrous below), usually reddish purple. Leaves opposite below, alternate above (rarely all opposite), the principal ones sessile or tapering basally to short subpetioles only 3–10 mm long, blades lanceolate, long-acute apically, 8–20 cm long, 1–3.5 (–6) cm broad, margins obscurely serrate to subentire, scabrous above, the hairs mostly with appressed, scalelike bases, pubescent beneath and atomiferous-glandular, the hairs spreading and without enlarged bases. Inflorescence of heads openly corymbiform, heads few to many. Involucral bracts loosely imbricated, subequal, narrowly lanceolate-acuminate to subulate, usually sparsely short-pubescent on their backs and bearing marginal cilia. Heads 4–5 cm across (ray tips to ray tips), rays 12–20, yellow and finely brownish-lined. Discs hemispheric, 1.5–2 cm across; disc corollas with yellow lobes. Achene dull brown, smooth, 3–4 mm long. (Incl. *H. alienus* E. E. Wats., *H. validus* E. E. Wats.)

Bogs, wet thickets, meadows, borders of swampy woodlands, ditches, low fields, occasionally in well-drained, open places. In our range, n.e. coastal plain, piedmont and mts. of N.C., upper S.C., n. Ga., e. Tenn.; widespread northeastward to N.S. and northwestward to Minn.

A somewhat similar sunflower, *H. grosseserratus* Martens, apparently inhabiting relatively dry or only marginally wet places, occurs locally in our range in Tenn., Ark., La., and s.w. Miss. Its stems are glabrous and glaucous, often strongly glaucous, at least below, its principal leaves more definitely petiolate, petioles to 5.5 cm long, the blades usually considerably larger and commonly more distinctly serrate.

6. Verbesina (CROWNBEARDS)

Perennial herbs (those treated here), stems winged from decurrent leaf bases in some. Leaves opposite or alternate, sometimes pinnately lobed or cleft, blades narrowed basally to winged petiolar bases, definitely petiolate, or sessile. Involucres campanulate, hemispheric, or cylindric-campanulate, bracts imbricated in 2 to several series. Receptacle conic, nearly globose to nearly flat, a chaffy bract subtending each flower. Ray flowers, if present, few, pistillate and fertile, pistillate and sterile, or neutral, yellow or white; disc flowers few to numerous, bisexual and fertile, corollas yellow or white, 5-lobed at the summit. Achene laterally flattened, with thin or winged edges; pappus usually of 2 persistent or easily deciduous awns or a crown.

1. Leaves all opposite, or the lower and median opposite, the upper bracteal leaves or bracts alternate.
   2. Heads radiate; stem winged from decurrent leaf bases.
      3. Principal leaves narrowed to definite petiolar bases, blades ovate; inflorescence with numerous to many heads.   1. *V. occidentalis*
      3. Principal leaf blades sessile, oblong, elliptic-oblong, or lanceolate; heads few. 3. *V. heterophylla*
   2. Heads discoid; stem not winged.   4. *V. chapmanii*
1. Leaves all alternate.
   4. Heads radiate.
      5. Ray and disc flowers white.   2. *V. virginica*
      5. Ray and disc flowers yellow.   5. *V. alternifolia*
   4. Heads discoid.   6. *V. walteri*

1. Verbesina occidentalis (L.) Walt.

Coarse perennial. Leafy stems 1–2 m tall, winged from decurrent leaf bases, short-pubescent, becoming glabrous below. Leaves opposite, the larger ones with a winged petiole less than half as long as the blade; blades broadly ovate, abruptly narrowed basally to the petiole, acuminate apically, to 18 cm long and 10 cm across the base, margins irregularly and relatively coarsely dentate-serrate, surfaces sparsely short-pubescent; upper leaves much smaller, blades lance-ovate to lanceolate. Heads radiate, usually many, cymose-panicled on the upper branches, inflorescence overall more or less flat-topped. Involucres cylindric-campanulate, 5–6 mm high; bracts erect, lanceolate to linear, very loosely imbricated, sparsely pubescent. Ray flowers 2–5 per head, yellow, lanceolate, tips blunt, 1–2 cm long, usually fertile; disc flowers several, corollas yellow. Achene obconical in outline, 4–5 mm long, somewhat angled, wingless, grayish, pubescent; pappus of 2 awns, 1 of them straight, or bent above the base then spreading laterally, the other arched-recurved. (*Phaethusa occidentalis* (L.) Small)

Floodplain forests subject to temporary flooding, bottomland woodlands, moist to wet clearings and thickets, coastal shell middens, wooded slopes. Pa. to Ill., generally southward to the Fla. Panhandle and Miss.

2. Verbesina virginica L. FROST-WEED.

Coarse perennial. Leafy stems 1–3 m tall, winged from decurrent leaf bases, relatively copiously short-pubescent, loose-tomentose on the inflorescence branches, becoming glabrous below. Leaves alternate, with winged petioles shorter than the blades; blades gradually or abruptly narrowed to the wings of the petiole; blades of the larger ones (most populations except southeasternmost part of the range), lance-ovate to ovate or elliptic, apices acuminate or acute, margins entire to irregularly coarsely toothed; blades of the larger ones (some populations or some individuals in some populations, chiefly southeasternmost part of the range) shallowly angulate-lobed to deeply, pinnately cleft-lobed; leaf surfaces sparsely to copiously short-pubescent, sometimes scabrid, sometimes soft to touch. Heads radiate, usually many in cymose panicles on upper branches, inflorescence overall more or less flat-topped. Involucres cylindric-campanulate, 6–8 mm high; bracts erect, imbricated (loosely so in fruit), mostly linear or linear-oblong, the outermost short, blunt-tipped, medial and inner about twice as long, acute apically, copiously pubescent. Ray flowers 1–5 per head, white, oval to oblanceolate, 3–10 mm long, fertile. Disc flowers few, white. Achene obovate (including wings), 4 mm long, body obovate in outline, unequally angled, gray, sparsely pubescent, wings straw-colored; pappus of 2 erect awns, one slightly longer than the other. (*Phaethusa virginica* (L.) Small)

Nonacid well-drained to wet woodlands of various mixtures, wet hammocks and their borders, stream banks, banks of marshes, coastal shell middens, meadows, rocky glades. Va. to Mo. and s.e. Kan., generally southward to s. Fla. and e., s.e. and n.cen. Tex.

Plants with few-lobed to deeply pinnately cleft leaves usually are designated *V. virginica* var. *laciniata* (Poir.) Gray (*Phaethusa laciniata* (Poir.) Small). It appears to us that the intergradation in leaf form, while not evident throughout the range of the species, is more nearly clinal than otherwise where it does occur.

3. Verbesina heterophylla (Chapm. in Coult.) Gray.

Perennial. Stems solitary or 2 to several from a basal crown or short rhizome, slender, 4–8 dm tall, winged by decurrent leaf bases, sparsely short- and rough-pubescent, leafy below the inflorescence branches, opposite- or alternate-bracteate on those branches. Principal leaves opposite, sessile, oblong, elliptic-oblong, or oblong-oblanceolate, 3–8 cm long and 1–2 (–3) cm broad, apices obtuse, margins bluntly low-toothed, surfaces scabrid. Heads few, radiate, borne terminally on bracteate branches. Involucres subhemispheric, 1–1.5 cm broad; bracts scabrid, imbricate in several series, lance-acute or -acuminate, medial ones 6–8 mm long. Ray flowers 5–10 per head, light yellow, ligules about 1.5 (–2) cm long, linear or lanceolate, mostly bifurcate apically; disc flowers few, corollas yellow. Achene obovate (including wings), 5 mm long, body dark brown, fusiform in outline, low-keeled, the wings purplish, narrow at base and broadening upwardly, at the summit even with and nearly obscuring a low, crownlike pappus. (*Pterophyton heterophyllum* (Chapm. in Coult.) Alexander; *V. warei* Gray)

Seasonally wet pine flatwoods. n.e. Fla.

4. Verbesina chapmanii J. R. Coleman.   Fig. 351

Perennial with a coarse, knotty, hard rhizome bearing coarse fibrous roots and few to many simple, unwinged stems 5–10 dm tall, the heads mostly solitary (rarely 2 or 3 close together) terminating the elongate, few-alternate-bracteate upper part of the stem; stem scabrid, becoming smooth below. Leaves opposite, sessile, the lowermost much reduced, gradually larger upwardly, those at about midstem largest; larger leaves lanceolate, elliptic, or oblong-elliptic, stiffish, blunt apically, 5–9 (–12) cm long and 1.5–2 (–3) cm broad, margins obscurely toothed; surfaces scabrid with pustular-glandular-based, short, spinelike hairs. Heads discoid. Involucres ovate to subhemispheric, 1–2 cm broad; bracts imbricated, the outer oblong and with blunt or rounded tips, inwardly becoming linear-acute, all thickish and purple, or the outer and medial thus, the inner thin and scarious, all with scabrid surfaces. Corollas yellow, conspicuously surpassing the involucre. Achene oblong-obovate (including wings), 1 cm long, body brown, elliptic in outline, irregularly low-ribbed, wings purplish to brown, slightly broader distally than at base, connecting with the knoblike or shallowly cuplike crown surmounting the achene. (*Pterophyton pauciflorum* (Nutt.) Alexander; *V. warei* Gray, misapplied by authors)

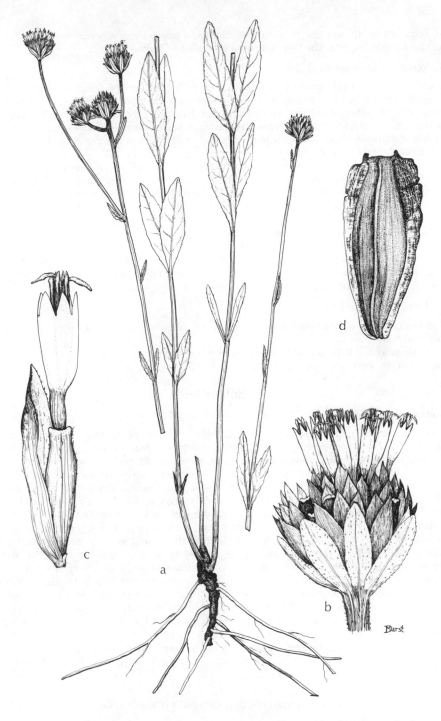

Fig. 351. **Verbesina chapmanii:** a, plant; b, head; c, floret with subtending scale; d, achene.

Bogs, seasonally wet pine savannas and flatwoods, grassy cypress depressions. Fla. Panhandle, Liberty and Franklin Cos. westward to at least s. Walton Co.

5. Verbesina alternifolia (L.) Britt.

Coarse perennial. Leafy stems to 3 m tall, more or less winged by decurrent leaf bases, rough-pubescent, becoming smooth below. Leaves alternate, gradually narrowed basally to short, winged, subpetiolar bases; larger ones lanceolate to lance-elliptic, 1–2 dm long, apices acute or acuminate, margins irregularly serrate to nearly entire, surfaces scabrid. Heads radiate, usually numerous, subhemispheric (excluding the rays), in open panicles. Involucres few-bracted; bracts mostly oblanceolate or some linear, loose, the outer soon deflexed. Ray flowers 2–10 per head, yellow, linear-oblanceolate or oblanceolate, 1–3 cm long, neutral; disc flowers numerous, loosely spreading, corollas yellow. Achene (including wings) broadly obovate, 6–7 mm long, body dark brown, oblanceolate to narrowly obovate in outline, lenticular or slightly angular, glabrous, wings straw-colored, broadening from the base upwardly, wing margins often irregularly lacerate, sometimes also ciliate; achene wingless on occasional individual plants; pappus of 2 erect or slightly arcuate awns. (*Actinomeris alternifolia* (L.) DC., *Ridan alternifolia* (L.) Britt. ex Kearney)

Floodplain forests where temporarily flooded, adjacent rich wooded slopes, wet woodlands and thickets, pastures. N.Y. and s. Ont. to Iowa, generally southward to Fla. Panhandle and La. (rarely on coastal plain in the Carolinas).

6. Verbesina walteri Shinners.

Apparently like *V. alternifolia* but the heads discoid, disc corollas white. (*Ridan paniculata* (Walt.) Small)

Habitats as for *V. alternifolia*. Piedmont of N.C., piedmont and coastal plain of S.C., thence to Ga., westward to La. and Ark.

7. Spilanthes

Spilanthes americana (Mutis) Hieron.   Fig. 352

Perennial, slenderly branching herb, branches weakly ascending or, more commonly, decumbent, to 1 m long, rooting at the nodes and colonial. Stem usually purple, sparsely pubescent to subglabrous, sometimes pubescent mainly at the nodes. Leaves simple, opposite, petiolate, petioles 1–2 cm long; blades of principal leaves deltoid-ovate, rhombic-ovate, or lanceolate, 2–4 cm long, bases truncate to cuneate, apices obtuse to acute, margins obscurely toothed, surfaces scabrid. Heads radiate, solitary on slender stalks 4–15 cm long from one of a pair of leaves on distal branches, the stalks naked or bearing 1 or 2 minute, solitary bracts. Involucre subtending only the base of the head; bracts loosely imbricated in 2 or 3 series, lance-oblong, subequal. Receptacle conic (high-conic in fruit), each flower with an oblanceolate pale about as long and partially enfolding it, pales persistent for a time after the achenes are shed. Ray flowers few, yellow, pistillate and sterile, 3–10 mm long, irregularly 2- or 3-toothed apically; disc flowers many, bisexual and fertile, corollas yellow, with 5 small, deltoid, spreading lobes at the summit. Achene 2 mm long, obconic in outline, lenticular or obscurely angled, with narrow, thin, sometimes minutely ciliate edges, each edge sometimes prolonged distally into a small bristle, surface dark brown with a somewhat grayish metallic sheen which sloughs readily; pappus none.

Swamps, wet woodlands, seepage areas and pools in woodlands, thickets, marshy places. Coastal plain, s.e. N.C. to s. Fla., westward to e., s.e., and n.cen. Tex., northward in the interior, s.e. Okla. to s.e. Mo. and s. Ill.

8. Coreopsis (COREOPSIS, TICKSEEDS)

Annual or perennial herbs. Leaves all opposite, all alternate, opposite on the lower part of the stem and alternate above, or alternate on the lower part of the stem and opposite above.

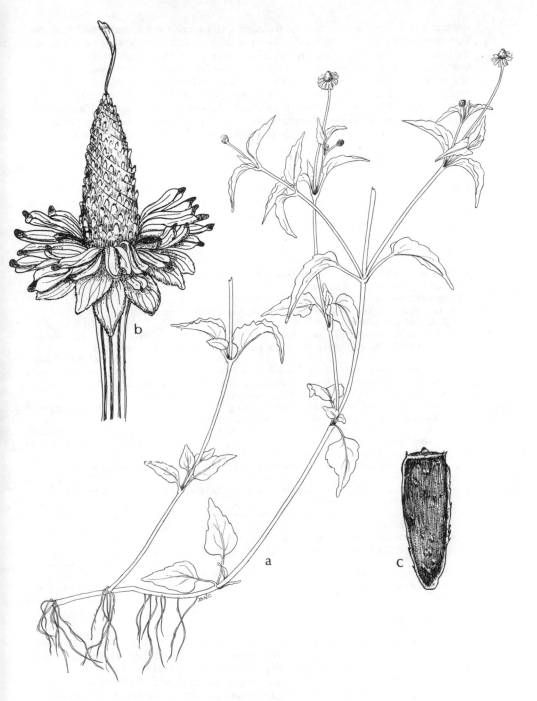

Fig. 352. **Spilanthes americana:** a, habit; b, head; c, achene. (b, from Correll and Correll. 1972)

Involucral bracts in 2 series, those of the outer series usually much smaller and more herbaceous than the differently textured inner series. Heads radiate, ray flowers spreading, neutral (ours) and infertile; rays usually 8, yellow in most, in some the base of the ray marked with reddish brown, rays in some lavender, pink, or white. Disc flowers bisexual and fertile, tube narrowly cylindrical below, abruptly expanded to cylindric-campanulate above, equally 4- or 5-lobed at the summit. Stamens 4 or 5. Receptacle of the head flat or nearly so, each flower subtended by a thin, scarious bract deciduous with the fruit. Achenes flattened in a plane parallel with the involucral bracts, marginally winged in most, faces warty-tuberculate in some. Pappus usually of 2 antrorsely setose awns or teeth, a low crown, or none.

Reference: Smith, Edwin B. "A Biosystematic Survey of *Coreopsis* in Eastern United States and Canada." *Sida* 6: 123–215. 1976.

1. Disc corollas 5-lobed at the summit; stamens 5.
  2. Plant rhizomatous and colonial; blades of principal leaves pedately trifoliolate (occasional specimens having the terminal leaflet again ternately divided); disc corollas yellowish at first, their throats and lobes becoming purplish to dark purple.   1. *C. tripteris*
  2. Plant not rhizomatous, not colonial; blades of principal leaves unlobed or, often, some of them with 1 or 2 pairs of similarly shaped but smaller pinnate basal divisions or leaflets; disc corollas yellow.
                 2. *C. pubescens*
1. Disc corollas 4-lobed at the summit; stamens 4.
  3. Rays lavender or pinkish lavender.   3. *C. nudata*
  3. Rays yellow or yellow with the base brownish red.
    4. Leaf blades unlobed, or some of them having 1 or 2 narrow auriculate lobes.
      5. Leaves all opposite.
        6. Leaf blades or their divisions linear to linear-oblanceolate, less than 1 cm broad (commonly much less); marginal wings of the achene entire, each as broad as the achene body.
                 4. *C. leavenworthii*
        6. Leaf blades broadly elliptic, lanceolate, or lance-ovate, 1.5–3 cm broad; marginal wings of the achene pectinate, each much less than half as broad as the achene body.   5. *C. integrifolia*
      5. Leaves all alternate, or 1 to few lower ones alternate, opposite above.
        7. Leaves alternate (sometimes excepting some of the bracteal leaves of the inflorescence); fresh blades not exhibiting tiny dark dots as observed with transmitted light.
          8. Leaves, at least some of the larger ones, with 1 or 2 narrow basal lobes; plants flowering May–June.   6. *C. falcata*
          8. Leaves all unlobed; plants flowering Aug.–Nov.
            9. Outer involucral bracts short-deltoid, less than 0.3 as long as those of the innermost series.
                 7. *C. floridana*
            9. Outer involucral bracts ovate-elliptic, mostly longer than broad, the larger ones half as long as those of the innermost series.   8. *C. gladiata*
        7. Leaves, 1 to few of those at or near the base of the stem alternate, those above all opposite; fresh blades exhibiting tiny dark dots as viewed in transmitted light.   9. *C. linifolia*
    4. Principal leaves, some of them at least, pinnately deeply divided or compound.
      10. Rays wholly yellow; pappus awns (0.3) 0.5–1 mm long; plants endemic to Fla.
                 4. *C. leavenworthii*
      10. Rays usually with a red-brown spot basally (rarely wholly yellow); pappus awns none or minute (not over 0.2 mm long); plants probably not occurring in Fla. (there being one doubtful record).   10. *C. tinctoria*

1. Coreopsis tripteris L.   Fig. 353

Perennial, perennating by elongate rhizomes and colonial. Stems solitary from the rhizomes, 1–2 (–3) m tall, relatively robust, usually glabrous and somewhat glaucous. Principal leaves with petioles 1–5 cm long, blades pedately trifoliolate (occasional specimens having the terminal leaflet again ternately divided), becoming reduced, simple, and subsessile to sessile above; leaflets (or simple leaves) mostly lanceolate, less frequently elliptic-oblong, oblong, nearly linear, or somewhat rhombic, 5–10 (–14) cm long, 6–25 (–30) mm broad, margins entire, surfaces glabrous or less frequently pubescent. Involucres hemispheric to campanulate, 7–10 mm high, often broader than high, united and more or less saucerlike basally (this

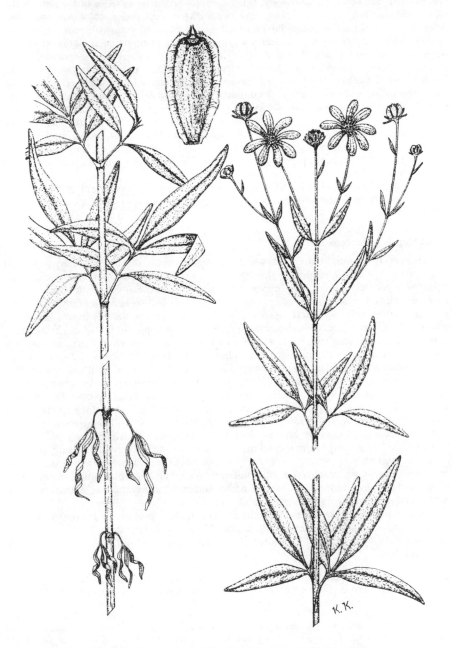

Fig. 353. **Coreopsis tripteris.** (From Smith in *Sida* 6(3). 1976)

more evident in mature fruiting heads); outer bracts herbaceous, in a single series, widely spaced from each other, linear-oblong, tips blunt, ½–¾ as long as those of the inner series when heads are fully developed; inner bracts somewhat overlapping, oblongish and somewhat tapered to blunt tips, much broader than the outer bracts, yellowish brown with thin-hyaline, yellowish, nearly translucent margins; all bracts glabrous or all short-pubescent. Rays 1–2.5 cm long, pale yellow with 10 or more slightly raised, brown, longitudinal lines; varying in shape (even in a single head), oblanceolate, oblong, or sometimes nearly oblong but broadest at the base, tips rounded, oblique, or irregularly lobed. Disc corollas 5 mm long, their cylindrical bases yellow, the throat at first yellowish brown and with 5 dark longitudinal lines, becoming purplish to dark purple distally, 5-lobed at the summit. Stamens 5. Chaff 3–4 mm long, narrowly linear, brown centrally and with yellowish margins. Achene body 4–7 mm long, narrowly obovate, oblanceolate, or oblong, dark brown, with narrow, scarious, straw-colored, entire, marginal wings, these often somewhat lacerate at either side of the summit of the achene; pappus of minute bristles, sometimes with 2 small, antrorsely setose awns as well.

In moist to wet lowland and well-drained upland woodlands, moist to wet thickets, blackland prairies. Pa. to s. Mich., Ill., s. Wisc., generally southward to s.w. Ga., adjacent Fla., and La.

2. Coreopsis pubescens Ell.

Perennial, 6–12 dm tall. Stems solitary or 2 to several from a basal crown, usually pubescent with spreading hairs, infrequently glabrous. Leaves opposite, the lower and medial commonly with petioles 1–4 cm long, upper or bracteal ones sessile or subsessile; blades sometimes all unlobed or undivided, often some of them with 1–2 pairs of similarly shaped but smaller basal divisions or leaflets, blades (or leaflets) varying considerably in size and shape, elliptic, ovate, lanceolate, or linear-oblanceolate, 3–12 cm long and 0.8–4 cm broad, margins entire, often pubescent, surfaces pubescent with short spreading hairs, infrequently glabrous. Involucres 7–10 mm high, hemispheric or campanulate, bracts in 2 series; bracts of outer series green and herbaceous, united basally, equal, triangular-subulate, from half as long to as long as those of the inner series; inner bracts yellowish brown to brown and submembranous, margins hyaline and yellow, mostly ovate with blunt apices. Rays yellow, 1–2.5 cm long, broadening upwardly from the base, irregularly 3–5-toothed apically. Disc corollas 4 mm long, yellow, 5-lobed at the summit. Stamens 5. Chaff linear-subulate, with 3 yellow lines centrally from the base to about the middle, margins whitish-hyaline, subulate upper half yellow grading to orange-yellow at the tip. Achene body usually broadly obovate, about as broad as long, strongly curved, purplish black, sometimes smooth, usually with prominent, purple, warty tubercles, margins with broad, buff to purple, entire, membranous wings, often with a horizontally oblong callus within the wing at either extremity of the achene body; pappus of 2 small, pale, chaffy scales.

Open or wooded banks of, and bottoms astride, streams, gravelly or sandy stream beds, alluvial thickets, meadows, open woodlands and bluffs, cliffs. Interior of Va. to s. Ill. and s. Mo., generally southward to Fla. Panhandle and La. Smith (1976) distinguishes three varieties, as follows.

1. Blades (or terminal leaflets) narrowly to broadly elliptical and acute or oblanceolate and acuminate; stout-stemmed glabrous or pubescent plants of mostly inland range, the stem little or not at all branched basally.
  2. Blades (or terminal leaflets) more or less broadly elliptical, ca. 1.5–4 cm wide, acute; stem, and often one or both surfaces of the leaves, rather densely hairy.   C. pubescens var. pubescens
  2. Blades (or terminal leaflets) narrowly elliptical to oblanceolate, ca. 0.6–2 cm wide, acuminate; stem and leaves glabrous.   C. pubescens var. robusta
1. Blades (or terminal leaflets) narrowly oblanceolate, acute; wiry-stemmed commonly glabrous plants of the Gulf Coast, the stem much-branched basally in mature specimens.   C. pubescens var. debilis

C. pubescens var. **pubescens.** Range as given above for the species.   Fig. 354

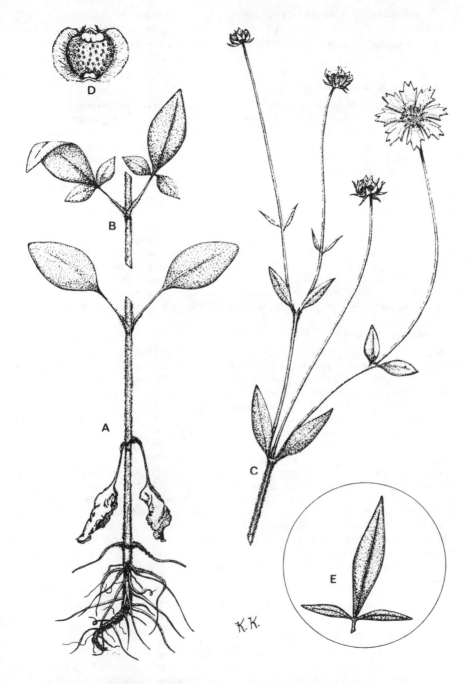

Fig. 354. A–D, **Coreopsis pubescens** var. **pubescens**; E, median leaf of var. **robusta.**
(From Smith in *Sida* 6(3). 1976)

C. pubescens var. **robusta** Gray ex Eames. w. Va., s. W.Va., w. N.C. and n.w. S.C., n.w. Ga.   Fig. 354

C. pubescens var. **debilis** (Sherff) E. B. Smith. cen. Tenn., Ala., n.e. Miss., s.e. La.
                                                                                        Fig. 355

3. Coreopsis nudata Nutt.   Fig. 356

Short-rhizomatous, glabrous perennial. Stem slender, simple to the sparsely few-branched inflorescence, mostly 6–12 dm tall. Leaves all alternate, terete, narrow and long-attenuate, those somewhat above but near the base largest, 1.5–2 (–4) dm long, few and much reduced in length upwardly. Heads few, 1–10 for the most part, on long, nearly erect stalks. Involucre campanulate, 8–10 mm broad; outer phyllaries usually much shorter than the inner, erect, narrowly elongate-triangular, inner ones usually about 10 mm long, nearly oblong but a little tapered distally, apices blunt, upper halves curved-reflexed, translucent-goldish and with numerous thin purplish red lines. Rays lavender or pinkish lavender, 1.5–2.5 cm long, gradually broadened from the base to a shallowly 3-lobed apex 8–15 mm broad. Tube of the disc corollas pale-translucent, the 4 short lobes yellow-translucent and conspicuously granular, recurved. Stamens 4. Chaff linear, nearly transparent, apically blunt or notched. Achene about 3 mm long, body oblong, purplish brown, the marginal wings pectinate with irregularly oblongish, membranous scalelike teeth; pappus a pair of upwardly setose, membranous awns 1–1.5 (–2) mm long.

   Usually in shallow water or wet boggy places, cypress ponds, depressions, or prairies, wet pine savannas, borrow pits, ditches. s. Ga., n. Fla., s.w. Ala., s.e. Miss., s.e. La.

4. Coreopsis leavenworthii T. & G.   Fig. 357

Annual or short-lived, glabrous perennial, mostly 5–8 dm tall. Stems solitary and few-branched in the inflorescence, solitary and copiously branched, or 2 to several from the basal crown and copiously branched. Leaves opposite, very variable, the lower usually petiolate or attenuate basally to subpetioles; some or all sometimes unlobed and entire, linear to oblanceolate, some specimens with lowermost leaves unlobed, several pairs above with 2, 3, or 4–6 pinnatifid lower segments and a larger, broader terminal segment, commonly specimens with several leaves with narrowly linear, pinnatifid segments, others simple and narrowly linear. Outer involucral bracts triangular-subulate, loosely spreading to reflexed, very much shorter than the erect, broadly ovate, inner ones which, collectively, have a subhemispheric form somewhat broader than long (5–15 mm broad in fruit). Rays yellow, 1–2 cm long, without a dark spot at base, broadest distally and with 3 short, blunt lobes. Disc corollas 3 mm long, dark distally, the 4 lobes blackish purple. Stamens 4. Chaff mostly about 5 mm long, acute apically, with a pair of narrow, dark lines from the base and running together somewhat back of the reddish brown tip, otherwise straw-colored. Achene body essentially oblong, warty-tuberculate, brown to purple, 2–3.5 mm long, with entire, thin-membranous, transparent, straw-colored or purple, marginal wings each as broad as the achene body; pappus a pair of soft, usually antrorsely setose awns (0.3–) 0.5–1 mm long. (Incl. *C. lewtonii* Small)

   Pine savannas and flatwoods, marl prairies, ditches, often abundant in roadside ditches and on roadside shoulders. Fla. (excepting Panhandle w. of Madison and Taylor Cos.)

5. Coreopsis integrifolia Poir. in Lam.   Fig. 358

Perennial, perennating by strong rhizomes and colonial, 4–7 dm tall. Stems very sparsely pubescent or glabrous, usually unbranched below the inflorescence, the latter usually with relatively few heads. Leaves opposite, the lower and medial with petioles 1–3 cm long; blades of principal leaves 3–7 cm long, 1.5–3 cm broad, elliptic, lanceolate, or lance-ovate, short-tapered basally, acutish apically, margins entire, varyingly short-pubescent to prominently ciliate, surfaces glabrous to moderately densely pubescent. Involucres hemispheric or

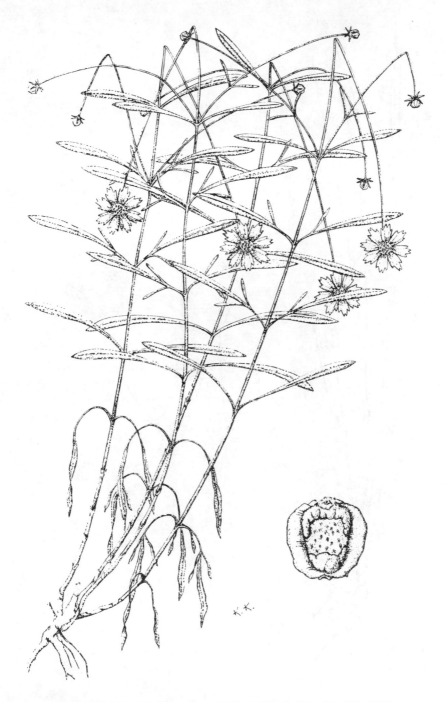

Fig. 355. **Coreopsis pubescens** var. **debilis.** (From Smith in *Sida* 6(3). 1976)

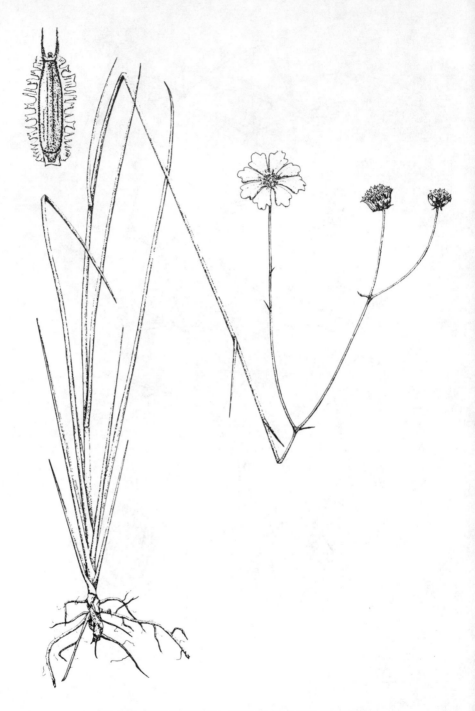

Fig. 356. **Coreopsis nudata.** (From Smith in *Sida* 6(3). 1976)

Fig. 357. **Coreopsis leavenworthii.** (From Smith in *Sida* 6(3). 1976)

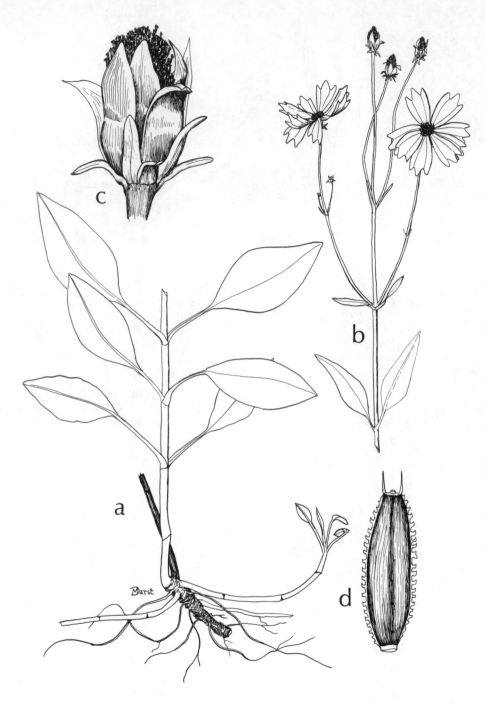

Fig. 358. **Coreopsis integrifolia:** a, lower portion of plant; b, upper portion of plant; c, fruiting head; d, achene.

campanulate, 1–1.5 cm broad and high; outer bracts loosely imbricated, their tips rounded, most of them oblong and much shorter than the innermost series but the uppermost (of the outer series) a little less than half as long as those of the innermost series and lance-ovate; bracts of the innermost series broadly ovate, abruptly narrowed to short, blunt tips apically. Rays bright yellow, 1–2 cm long, broadening upwardly from the base, with 3 blunt lobes apically. Disc corollas 3–4 mm long, purplish red, darkest distally, 4-lobed at the summit. Stamens 4. Chaff 4 mm long, linear-oblanceolate, with a pair of narrow dark lines centrally from the base to about ⅔ the length, sides straw-colored, the slightly dilated tip reddish purple. Achene body oblanceolate to oblong, 5 mm long, dark brown and with tan warty tubercles, margins pectinate with oblongish, often bifurcated, tan enations, much less than half the breadth of the body; pappus a pair of awns 0.5–1 mm long.

Banks and floodplains of rivers and creeks (subject to temporary flooding). Coastal plain, s.e. S.C., s. Ga., Fla. Panhandle (Chipola River drainage as now known).

6. Coreopsis falcata Boynt.   Fig. 359

Glabrous perennial 5–20 dm tall. Leaves all alternate, margins entire, the basal and lowermost stem leaves usually without lobes, lanceolate distally, gradually attenuate proximally to petioles 10–20 cm long, 10–25 cm long overall, apices blunt to acute; midstem leaves few, mostly with 1 narrow lobe or 2 narrow lobes more or less medially, lobes varying in length to 3 cm. Involucres about 1 cm high and 1.5 cm broad; outer bracts unequal, somewhat imbricated, lanceolate to lance-ovate, the longer ones about half as long as the innermost bracts; innermost ovate-elliptic to ovate, short-tapered to blunt extremities, 6–10 mm long. Rays bright yellow, 1.5–3 cm long, gradually broadened from the base upwardly, 7–12 mm broad at the 3-lobed apex. Disc corollas 4 mm long, dark purplish red or blackish red distally, 4-lobed at the summit. Stamens 4. Chaff 5–6 mm long, linear-acute, with narrow, dark, double stripes centrally and broader straw-colored margins on the lower ⅔, the distal ⅓ red. Achene body approximately oblong, 3 mm long, faces grayish brown and with scattered, small, tan tubercles, margins pectinate with pointed or bifurcated-pointed, tan teeth ¾ to fully as long as the breadth of the achene; pappus a pair of awns about 0.5 mm long.

Marshes, bogs, pine savannas, pocosins, wet ditches. Coastal plain and outer piedmont, N.C. to Ga.

C. falcata blooms in May and June (a rare individual in autumn) whereas *C. linifolia, C. gladiata,* and *C. floridana,* with which it is most closely allied, bloom (Aug.–) Sept.–Nov.

7. Coreopsis floridana E. B. Smith.   Fig. 360

Stoutish, glabrous perennial. Stem simple to moderately branched, 3–12 dm tall. Principal leaves alternate, unlobed, reduced upper bracteal ones linear, sometimes opposite; basal and lower stem leaves with narrowly elliptic, lanceolate, or oblong blades 4–15 cm long and to about 3 cm broad, gradually attenuate basally to petioles 5–20 cm long, rounded or obtuse apically. Involucres 1–1.5 cm high and 1–2.5 cm broad; outer bracts short-deltoid, rounded-obtuse apically, not exceeding 0.3 as long as the innermost, innermost oblong to elliptic, rounded or obtuse apically, 9–15 mm long. Rays yellow, 2–3 cm long, 1–2 cm broad distally, 3-lobed at the summit. Disc flowers 5–6 mm long, tube yellow below, purple-brown distally, 4-lobed at the summit. Stamens 4. Chaff linear, 9–10 mm long, a pair of dark stripes centrally for about ⅔ of their length, otherwise straw-colored, the tips somewhat dilated, dark purplish red. Achene grayish brown, body oblong, faces usually warty-tuberculate, 3.5–5 mm long, winged margin irregularly pectinate-lobed, about ¼–½ the breadth of the achene body; pappus a pair of antrorsely setose awns 0.5–2.5 mm long.

Bogs, marsh-prairies, cypress-prairies, glades, pine savannas and flatwoods, borders of cypress depressions, wet ditches. Fla. (excepting n.e. portion).

8. Coreopsis gladiata Walt.   Fig. 361

Perennial. Stem often with several subequal, erect-ascending branches from near the base or from about midstem, sometimes branched only in the inflorescence, main stem sometimes

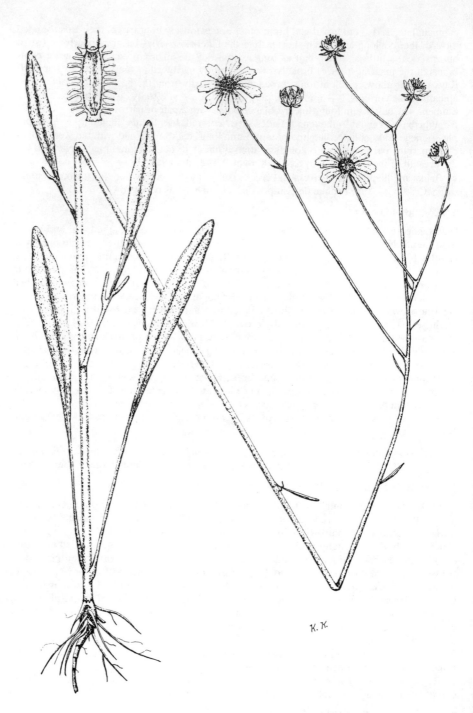

Fig. 359. **Coreopsis falcata.** (From Smith in *Sida* 6(3). 1976)

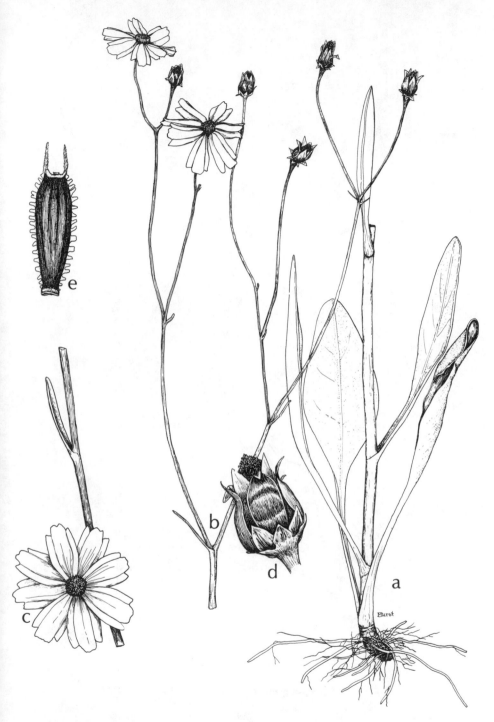

Fig. 360. **Coreopsis floridana:** a, lower portion of plant; b, inflorescence; c, head at full anthesis, face view, overlying midsection of stem; d, fruiting head; e, achene.

Fig. 361. **Coreopsis gladiata.** (From Smith in *Sida* 6(3). 1976)

zigzagging, 6–12 dm tall, glabrous or infrequently pubescent. Leaves all alternate, variable; principal ones of most specimens with well-differentiated petioles and blades; blades un-lobed, elliptic, varying to broadly or narrowly lanceolate, 6–15 cm long and 0.5–5 cm broad, one or both surfaces pubescent or glabrous. Involucres about 1 cm high and broad; outer bracts lanceolate to narrowly oblong, the longer ones about half as long as the inner-most, obtuse apically, the inner bracts about 10 mm long, elliptic-oblong, tapered to blunt tips. Rays yellow, sometimes with a minute reddish brown spot basally, broadening distally from the base, 3-lobed apically, 1–2 cm long or a little more. Disc corollas 4 mm long, dark brown-purple distally, 4-lobed at the summit. Stamens 4. Chaff linear-subulate, 5 mm long, with narrow, dark, double stripes centrally, yellowish otherwise. Achene dark brown, body oblanceolate, 3–4 mm long, margins setose, very narrowly winged, or pectinate with flat, oblongish, light brown, membranous enations mostly about ¼, sometimes to ½, as long as the breadth of the achene body; pappus a pair of antrorsely, minutely setose awns 0.5–2 mm long. (*C. angustifolia* Solander in Ait., *C. helianthoides* Beadle, *C. longifolia* Small)

Bogs, borders of shrub-tree bogs or bays, cypress-gum depressions, borders of swamps, wet ditches. Chiefly coastal plain, N.C. to s.cen. pen. Fla., westward to s.e. Miss.

9. Coreopsis linifolia Nutt.   Fig. 362

Glabrous perennial 5–10 dm tall. Stems usually solitary, occasionally 2 to several, from a basal crown. Leaves unlobed, margins entire; lower few leaves alternately disposed, mostly oblanceolate or linear-oblanceolate distally, gradually attenuate proximally to subpetioles of varied lengths to about 10 cm, 3–18 cm long overall, apices obtuse; other leaves opposite, gradually reduced upwardly, becoming narrowly linear and sessile; laminar portions of fresh leaves showing minute, dark dots in transmitted light (not always evident on dried leaves). Involucres about 8–10 mm high and broad; outer bracts ovate-elliptic, 2–3 mm long, obtuse apically, the inner oblongish with shortly tapered, blunt tips, 7–10 mm long. Rays yellow, 1–2 cm long, gradually broadened from the base to a 3-lobed summit 5–10 mm broad, the central lobe often notched. Disc corollas 3 mm long, yellowish below, the dilated distal portion dark purplish red, 4-lobed at the summit. Stamens 4. Chaff subulate, 4–5 mm long, purplish red at least distally. Achene body oblong or elliptic-oblong, 2 mm long or slightly over, faces grayish brown and often with scattered small tubercles, margin narrowly pecti-nate-winged; pappus a pair of antrorsely setose awns about 1 mm long. (*C. oniscicarpa* Fern., *C. gladiata* var. *linifolia* (Nutt.) Cronq.)

Bogs, pine savannas and flatwoods, shrub-tree bogs and bays or their borders, wet ditches. Coastal plain, s.e. Va. to n. Fla., westward to s.e. Tex.

10. Coreopsis tinctoria Nutt.   Fig. 363

Glabrous annual, mostly 3–8 dm tall. Stems commonly unbranched below the inflorescence, the latter few-branched to relatively diffusely branched, some plants with several stems from the basal crown. Leaves opposite, the lower petiolate, becoming subsessile above; blades of lower and medial ones 5–10 cm long, 1- or 2-pinnately divided, the segments or leaflets narrowly linear to linear-lanceolate or oblanceolate; blades of upper leaves undivided or with few divisions. Outer involucral bracts few, loosely imbricated, subulate to ovate-triangular, much shorter than the ovate-acuminate innermost which, collectively, have a subhemispheric form 8–10 mm broad, broader than high. Rays usually yellow with a conspicuous reddish brown base, varying (rarely) to all yellow or all reddish brown, 6–15 mm long, broadening from the base distally, mostly with 3 blunt lobes at the summit. Disc corollas yellow to red-dish purple proximally, reddish purple distally, 3–4 mm long. Stamens 4. Chaff linear, acute apically, about 4 mm long, with a pair of reddish lines about ⅔ of the length, margins straw-colored, upper ⅓ purplish red. Achene body elliptic-oblong, curved, 1.5–2.5 mm long, dark brown to black, one face or both faces often buffish-warty-tuberculate, wingless, nar-rowly winged, or broadly winged, wings straw-colored; pappus absent or a pair of minute awns. (Incl. *C. cardaminefolia* (DC.) Nutt., *C. stenophylla* Boynt.)

Meadows, open, wet, sandy or gravelly places, low fields, roadsides. Most abundant

Fig. 362. **Coreopsis linifolia:** a, plant; b, head; c, ray flower; d, disc flower. (From Correll and Correll. 1972)

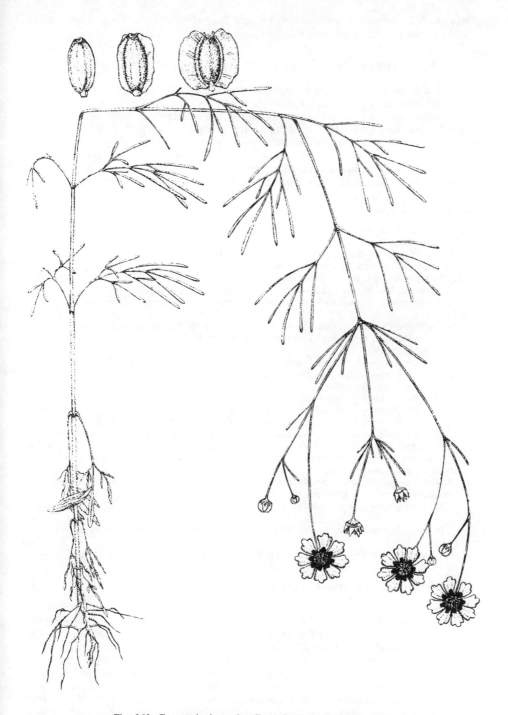

Fig. 363. **Coreopsis tinctoria.** (From Smith in *Sida* 6(3). 1976)

787

Wisc. to N.D., generally southward to La. and Tex.; also Wisc. to Wash., Calif., Ariz., N.M.; in much of our range cultivated for ornament and naturalized.

9. Bidens (BEGGAR-TICKS, STICK-TIGHTS, BUR-MARIGOLDS)

Annual or perennial herbs. Leaves opposite, simple or pinnately to ternately compound or dissected. Involucres of 2 series of bracts, the outer usually green and herbaceous, sometimes foliaceous, the inner yellowish to brownish and membranous, often striate, margins often hyaline. Receptacle of the head flat or slightly convex, chaffy, chaff flat or nearly so. Heads discoid or radiate, ray flowers when present neuter, less frequently pistillate and sterile, ligules yellow or less frequently white or pink; disc flowers bisexual and fertile, their corollas 5-lobed at the summit, usually yellow to orange. Achenes flattened parallel to the involucral bracts, not or scarcely winged, less frequently triangular or unequally 4-sided; pappus usually of 2–4 teeth or awns, these often retrorsely, sometimes antrorsely, barbed, sometimes not barbed, absent in a few species.

1. Leaves simple, unlobed or at most with 1 or 2 broad basal lobes.
  2. Heads conspicuously radiate and conspicuously showy.
    3. Leaves sessile or at most narrowed below to broadly winged bases.   1. *B. laevis*
    3. Leaves all, or at least the lower and medial, with slender and distinct petioles.   4. *B. mitis*
  2. Heads discoid, or if any ray flowers present then the rays inconspicuous and barely if at all showy.
    4. Leaves all sessile; stem usually short-pubescent at least above.   2. *B. cernua*
    4. Leaves mostly narrowed to narrowly winged subpetioles up to 3 cm long; stem glabrous.
                              3. *B. tripartita*
1. Leaves strongly pinnately lobed or divided, or ternately or pinnately compound.
  5. Heads radiate and showy, rays 1 cm long or more; outer, herbaceous involucral bracts about equal, not really foliaceous.
    6. Outer, herbaceous involucral bracts 6–11, little if any surpassing the inner bracts, their margins entire or finely short-ciliate, tips blunt to subacute.
      7. Achenes obovate or cuneate-obovate, 1.5–2 times as long as broad.
        8. The achenes with smooth edges.   4. *B. mitis*
        8. The achenes with scabrid, short-ciliate, or erose-notched edges.   5. *B. aristosa*
      7. Achenes relatively narrowly cuneate in outline, 2.5–5 times as long as broad. 6. *B. coronata*
    6. Outer, herbaceous involucral bracts (12–) 15–20 (–25), linear-attenuate and exceeding the discs, their margins often irregularly wavy or crisped, conspicuously and coarsely ciliate. 7. *B. polylepis*
  5. Heads discoid, or if rays present then these inconspicuous, less than 5 mm long; outer, herbaceous involucral bracts markedly unequal, some or all of them foliaceous.
    9. Outer, herbaceous bracts 5–20, their margins ciliate at least near the base; discs mostly 1 cm broad or more and 1–1.5 cm high.
      10. The outer, herbaceous bracts 5–10, mostly 8; discs mostly 10–15 mm broad and 10 mm high.
                              8. *B. frondosa*
      10. The outer, herbaceous bracts 10–20, mostly more than 10; discs to 2 cm broad and 15 mm high.   9. *B. vulgata*
    9. Outer, herbaceous bracts 3–5, not at all ciliate; discs mostly 3–8 mm broad and about 5 mm high.
                              10. *B. discoidea*

1. Bidens laevis (L.) BSP. BUR-MARIGOLD.   Fig. 364

Wholly glabrous annual, or perennial southward in the range. Flowering stems slender to relatively coarse, when in water decumbent below and rooting at the nodes, the flowering branches ascending, commonly forming extensive mats; flowering stems, when not growing in water, usually solitary and erect, 5–15 dm tall, branched above; in the southern part of our range (where perennial) plants perennating by basal offshoots, these sometimes very numerous from the base, essentially prostrate, radiating and forming mats as much as 2 m across, stems of such perennations slender and bearing numerous relatively small, thin leaves. Leaves of flowering stem thickish, often subsucculent, sessile, lanceolate to elliptic, acute apically, medial ones often tapered to very broadly winged, truncate bases, upper ones less tapered, larger ones 7–15 cm long, margins finely serrate. Inflorescence of heads openly to

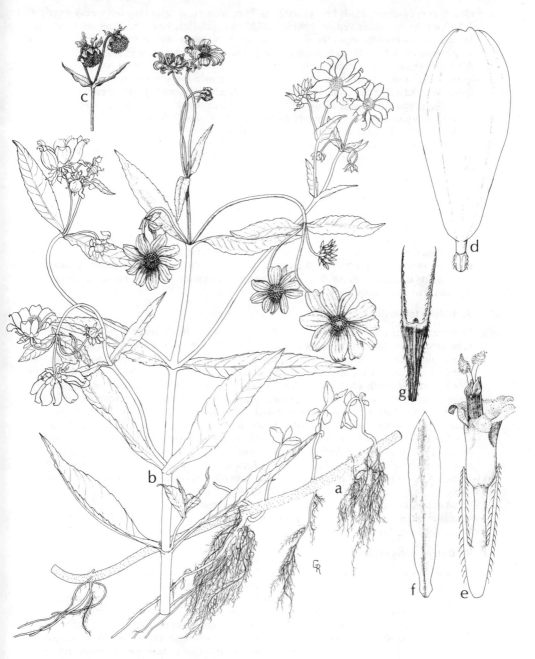

Fig. 364. **Bidens laevis:** a, decumbent lower part of stem; b, upper central branch of stem; c, two fruiting heads; d, ray flower; e, disc flower; f, chaffy scale subtending disc flower; g, achene.

closely corymbiform, heads few, radiate, 5–7 cm across (ray tips to ray tips), rays mostly 8, 1.5–3 cm long, golden yellow, broadly elliptic to spatulate; heads nodding after anthesis. Involucral bracts of the outer, herbaceous series 5–8, linear-oblong to oblanceolate, a little longer than the inner, obtuse to acute apically, finely serrate marginally; inner bracts yellowish to brownish. Discs hemispheric, about 1.5–2 cm across; chaff with orange to orange-brown or reddish tips, stramineous and sometimes brown-lined below. Achene strongly flattened, without a central rib on either face, edges thin, dull brown to purplish brown, 6–8 mm long, the pappus of 2 awns about half as long, infrequently with 2 shorter awns between, achenial margins and awns retrorsely barbed. (Incl. *B. nashii* Small)

In wet places, often in water, marshes and marshy shores, stream banks, in sluggish streams, sloughs, and ditches, wet meadows. Chiefly, but not exclusively, coastal plain, N.H. to s. Fla., westward to Calif., local in the interior; trop. Am.

2. Bidens cernua L.

Annual, closely similar to *B. laevis*, 1–10 dm tall. Leaves strongly tapered to a narrow base, subclasping, or subconnate, lanceolate to elliptic, usually somewhat more sharply serrate than in *B. laevis*. Heads radiate or discoid, rays, when present, 5–15 mm long, yellow. Achene compressed but 4-sided, 6–8 mm long, the margins tending to be thickened-cartilaginous, the pappus of 4 subequal awns about ⅓ as long as the achenial body, lateral margins and awns retrorsely barbed, the central ribs on the faces retrorsely barbed or smooth.

Habitats essentially as for *B. laevis*. Widespread in temp. N.Am., in our range in the s. Appalachian mts.; temp. parts of Euras.

3. Bidens tripartita L.

Glabrous annual of varied stature, 1–10 dm tall, stem simple or with strongly ascending branches. Leaves tapering to winged subpetioles up to 3 cm long, or sessile, blades (of plants in our range) unlobed and with serrate margins, lanceolate or lance-elliptic, sometimes with 1 or 2 broad basal lobes, apices acute to acuminate; principal leaves 6–20 cm long overall. Heads discoid or with a few inconspicuous rays not exceeding 10 mm long, usually not over 4 mm. Outer, herbaceous involucral bracts mostly 4–10, unequal, more or less foliaceous, sometimes the longer ones scarcely surpassing the discs, sometimes the longer ones really leaflike and much exceeding the discs. Discs hemispheric to campanulate, 1–2 cm across; chaff centrally streaked with brownish to reddish lines, the sides hyaline and pale, each side about as broad as the streaked portion. Achene flattened but 4-sided, narrowly cuneate to cuneate-obovate in outline, dull brown, edges smooth or antrorsely or retrorsely finely barbed, faces often tuberculate; pappus of 2, rarely 4, retrorsely, or rarely antrorsely, barbed awns 3–6 mm long. (Incl. *B. comosa* (Gray) Weig., *B. connata* Muhl.)

Marshes, swamps, ditches, wet meadows, cultivated and fallow bottomland fields. Widespread in the n.e. and n.cen. U.S., south in our range to piedmont and mts. of N.C., upper S.C., Tenn., n. Ga., and n. Ala.

4. Bidens mitis (Michx.) Sherff.   Fig. 365

Annual of variable stature, commonly 1–2 m tall, few-branched to diffusely and widely branched, stem erect, glabrous or unevenly and sparsely pubescent. Leaves highly variable, sometimes all petiolate, petioles 0.3–6 cm long, sometimes only the lower and medial ones petiolate, blades of principal ones unlobed and lanceolate to linear-lanceolate, their margins entire to sharply serrate, varying to lobulate-serrate, pinnately divided and with the segments linear to linear-oblong, or pinnately compound, sometimes ternate divided or compound, varying greatly in overall length/width dimensions. Outer, herbaceous involucral bracts 7–10, slightly shorter than to slightly longer than, usually narrower than, the inner ones, lanceolate, linear-oblong, or more commonly oblanceolate, mostly obtuse or rounded, infrequently acute, apically, their margins entire or finely and obscurely toothed; inner bracts yellowish, yellowish green, or reddish brown, with 3–7 pale to dark brownish or reddish lines and usually with purple speckles. Heads radiate, 3–6 cm across (ray tips to ray tips), rays

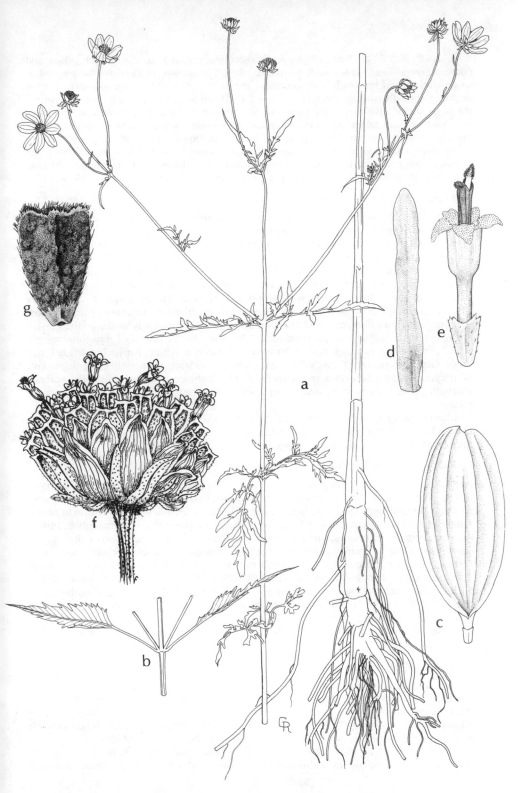

Fig. 365. **Bidens mitis:** a, plant; b, section of stem, showing undivided leaves characteristic of some specimens; c, ray flower; d, chaffy bract subtending disc flower; e, disc
flower; f, fruiting head; g, achene.

7–9, mostly 8, 1–3 cm long, elliptic to oblanceolate, rounded apically, golden yellow and purplish brown–lined. Discs subhemispheric, 1–1.5 cm across; chaff with a brown central stripe with several brownish lines centrally, the sides yellow, becoming scarious. Achene obovate to broadly cuneate in outline, 2.5–4.5 (–5) mm long, strongly flattened but 4-sided, the edges smooth, the ribs on the faces low and smooth, surfaces dark brown and smooth or with low, rounded tubercles, summit concave and without a pappus, or the edges merely elevated to short teeth, or to triangular scales to 1 mm long and finely antrorsely barbed.

Marshes, shallows in ponds or lakes, often in narrow to broad bands in shallow water outward from pond and lake shores, open, swampy woodlands and their borders, ditches, sloughs. Coastal plain, Md. to s. pen. Fla., westward to s.e. Tex.

5. **Bidens aristosa** (Michx.) Britt.

Annual of variable stature, 3–15 dm tall, stem erect, glabrous or sparsely pubescent. Leaves pinnately or bipinnately divided or compound, with petioles 1–3 cm long, segments of the blade lanceolate or linear-lanceolate, incised-serrate or pinnatifid, apices acuminate, 5–15 cm long overall. Outer, herbaceous involucral bracts 8–10, subequal, linear or nearly so, slightly shorter than to a little exceeding the inner, apices slightly tapered but the tips usually blunt, glabrous or sparsely short-pubescent on their backs, margins entire or finely short-ciliate; inner bracts elliptic, brown with yellow hyaline edges. Heads radiate, 3–5 cm across (ray tips to ray tips), rays 6–10, mostly oblanceolate or oblong-oblanceolate, golden yellow with purplish brown lines, apices obtuse or rounded, entire or minutely toothed. Discs 1–1.5 cm across; chaff yellow with purplish central lines and tip. Achene 4–7 mm long, mostly 1.5–2 times as long as broad, obovate or broadly cuneate in outline, flattish but 4-sided, ribs on the faces low, smooth, faces dull brown, glabrous and smooth, sparsely short-pubescent, or irregularly low-tuberculate and sometimes also pubescent, edges yellowish, scabrid, short-ciliate, or erose-notched; pappus of 2 short, antrorsely or retrorsely barbed awns or none.

Mostly low, moist to wet places, marshes, meadows, cultivated and fallow fields, ditches. Del. to Minn., generally southward to S.C., Ala., Miss., La., e. and s.e. Tex.; presumably introduced in e. part of the present range.

6. **Bidens coronata** (L.) Britt.

Annual, apparently similar in general features to *B. aristosa*, stems 3–15 dm tall, glabrous or sparsely pubescent. Leaves with petioles 3–15 mm long, up to 15 cm long overall, blades pinnately divided, segments 3–7, lance-linear to linear, incised-dentate or -serrate to entire, acute or acuminate apically. Outer, herbaceous involucral bracts 6–11, commonly 8, little if any surpassing the inner ones, linear to linear-spatulate, glabrous or nearly so on their backs, sometimes ciliate on their margins; inner bracts oblongish, subacute apically, brown with narrow, yellowish, opaque margins. Heads radiate, rays about 8, golden yellow, oblong, 1.5–3 cm long. Discs 8–15 mm broad; chaff with a brown to purple stripe centrally and narrow yellow margins. Achene dark purple, flattened but 4-sided, cuneate in outline, 5–9 mm long, 2.5–5 times as long as broad, edges smooth or sparsely antrorsely scabrid, medial ribs smooth; pappus of 2 short, antrorsely setose awns or scalelike awns.

Widespread in n.e. and n.cen. U.S.; in our range apparently restricted to coastal marshes of N.C. and S.C.

7. **Bidens polylepis** Blake.

Annual, in general aspect and most features closely similar to *B. coronata* and perhaps not specifically distinct from it. Outer, herbaceous involucral bracts (12–) 15–20 (–25), usually much surpassing the inner bracts, linear-attenuate, margins often irregularly wavy or crisped, conspicuously and coarsely ciliate, the ciliations often broad-based and collectively giving a toothed aspect to the margins. (*B. involucrata* (Nutt.) Britt. not Phil.)

Wet prairies and meadows, low, swampy woodlands, pond margins, borders of oxbow lakes, sandy flats and swales, cultivated and fallow fields, ditches. Presumably native in midwestern states; in our range occurring in N.C., upper S.C., Tenn., Ala., Ark., La., perhaps elsewhere.

8. Bidens frondosa L. Fig. 366

Annual of varied stature, up to 12 dm tall, stem simple to much branched, glabrous or sparsely pubescent on the internodes, nodes often pubescent. Leaves with petioles 1–6 cm long, blades pinnately compound, leaflets 3–5, commonly 3, firm, lanceolate, serrate, apices acuminate, the terminal one largest and tapered to a short stalk, laterals sessile, glabrous above, glabrous or sparingly pubescent beneath; leaflets variable in size up to about 10 cm long and 3 cm broad, occasionally 1 or more leaflets of a given leaf 1- or 2-lobed basally. Outer, herbaceous involucral bracts 5–8 (–10), commonly 8, unequal, some or all foliaceous, mostly oblanceolate, the longer ones surpassing the disc, margins irregularly and sparingly ciliate; inner bracts oblong with obtuse to acute apices, brown with narrow, pale, scarious margins. Heads discoid, or rarely with golden yellow ligules 2–3.5 mm long. Discs hemispheric to campanulate, (5–) 10–15 mm across, about 10 mm high; chaff with 3–5 brown lines centrally, the sides thin, pale, hyaline. Achene nearly flat and with an inconspicuous rib on either side, or the latter stronger and the achene 4-sided, cuneate in outline, 5–10 mm long, dark brown to purplish black, glabrous or pubescent, sometimes tuberculate; pappus of 2 strong, slender, retrosely, or, less frequently, antrorsely barbed awns 2–4 mm long.

In a wide variety of moist to wet, especially disturbed, places, borders of swampy woodlands, wet clearings, borders and banks of open streams, marshy shores, ditches, cultivated and fallow fields, waste places. Widespread in temp. parts of U.S. and adj. Can.; in most of our range except e. and pen. Fla.

9. Bidens vulgata Greene.

Similar to *B. frondosa* and perhaps not specifically distinct from it. Stem and leaves more often pubescent, sometimes rather copiously so. Outer, herbaceous involucral bracts 10–20, commonly more than 10, more conspicuously ciliate marginally, at least on their lower margins, sometimes pubescent on their backs. Discs averaging larger, to 2 cm across and 15 mm high. Achene nearly flat, oblong to cuneate-obovate in outline, 6–10 mm long, surfaces olivaceous to dark brown, usually smooth, edges usually minutely pubescent; pappus of 2 strong, retrorsely barbed awns 3–6 mm long.

Habitats as for *B. frondosa*. Widespread across s. Can. and n. U.S., reaching our range in mts. and piedmont of N.C. (and Tenn.?)

10. Bidens discoidea (T. & G.) Britt. Fig. 367

Similar to *B. vulgata* but more delicate. Leaves ternately compound (uppermost sometimes simple), leaflets thin, relatively more membranous. Outer, herbaceous involucral bracts 3–5, not ciliate or toothed marginally. Discs smaller, campanulate, 3–8 mm across and about 5 mm high. Achene narrowly cuneate, 3–6 mm long, pubescent, brown, or greenish yellow and mottled with brownish purple; pappus of 2 slender, antrorsely minutely setose awns 0.5–2 mm long.

Swamps, marshy shores and shallows of ponds and lakes, often on debris, around bases of trees, or on rotting logs or stumps, floodplain woodlands, shores of streams. N.S. to Minn., generally southward, excepting s. Appalachian mts., to n.cen. pen. Fla., s.e. Okla., e. and s.e. Tex.

10. Marshallia (BARBARA'S BUTTONS)

Perennial, erect herbs with nonwoody caudices or short rhizomes. Herbage obscurely gland-dotted or punctate, punctae often with resin droplets. Leaves alternate, all basal in some, simple, margins entire, usually with 1–3 (–5) paralleling, principal veins, glabrous or nearly so. Stem single, bearing a solitary, long-stalked head, or the heads long-stalked in cymes;

Fig. 366. **Bidens frondosa:** A, habit; B, head; C, flower subtended by chaffy bract; D, achenes. (From Reed, *Selected Weeds of the United States* (1970) Fig. 186)

Fig. 367. **Bidens discoidea:** a, branch of plant; b, head; c, chaff; d, achene. (From Correll and Correll. 1972)

stalks naked or leafy-bracted, channelled and grooved. Involucres hemispheric to campanulate, bracts subequal, in 1 or 2 series, imbricate or approximate. Receptacle of the head convex or conical, chaffy, in age pitted; chaff rigid, often hyaline-margined below, narrowly linear in outline, distally arched inward. Heads discoid, many-flowered, outer flowers blooming first, others inwardly over a considerable period of time, the heads expanding not a little during anthesis. Flowers all bisexual and fertile. Corollas white, cream-colored, or shades of lavender to purple, cylindric for most of their length, sometimes dilated a little below the summit, the 5 lobes long-linear, often twisting during anthesis, their tips bluntish. Achene somewhat 5-angled, usually more or less 10-ribbed, ribs pubescent, surfaces between the ribs usually atomiferous-glandular; pappus of 5(6) erect or spreading, scarious, translucent scales.

Reference: Channell, R. B. "A Revisional Study of the Genus *Marshallia* (Compositae)." *Contrib. Gray Herb.* 181: 41–132. 1957.

• Lower stem leaves erect, long-lanceolate or linear-lanceolate, attenuate basally to subpetioles and attenuate distally, firm in texture, evidently 3 (–5) -nerved; caudices usually fibrous with remnants of leaves of previous seasons; involucral bracts thick, ovate-attenuate.   1. *M. graminifolia*
• Lower stem leaves (and often some rosette leaves present at flowering time) spreading, oblanceolate or spatulate, tapered basally to subpetioles, rounded or obtuse apically, the 2 paralleling lateral nerves often faint, or occasional specimens with these leaves nearly linear, 1-nerved; caudices without fibrous remnants of leaves of previous seasons; involucral bracts thin, mostly linear-subulate.   2. *M. tenuifolia*

1. Marshallia graminifolia (Walt.) Small.

Plant with a short caudex, with coarse fibrous roots, and usually bearing fibrous remnants of leaves of previous seasons, perennating by basal offsets. Lower stem leaves erect, long-lanceolate, linear-lanceolate, or nearly linear, attenuate basally to subpetioles, attenuate distally, 1–2 dm long overall, to 1.5 cm broad, with 3 (–5) longitudinally paralleling veins; other leaves sessile, gradually diminishing in size upwardly, eventually to narrowly linear-acute bracts of the inflorescence or its branches, the latter sometimes bracteate very nearly to the summit, sometimes naked for a considerable distance below the heads. Stem, in age, glabrous below, upwardly becoming pubescent with incurving, moniliform, red to purple hairs, these sparse where they commence and becoming gradually more abundant upwardly. Stem sometimes simple, bearing a single terminal head, more commonly with 2 to several long, erect, cymose branches each bearing a head, the branching sometimes occurring not far above the base of the stem, more frequently more or less medially; upper bracteal leaves sometimes with a few hairs like those of the stem. Involucre hemispheric, enlarging as anthesis of the flowers proceeds, becoming 1.5–2.5 cm broad, occasionally broader, bracts often reddish, especially their margins and tips, thickish in texture, glandular-punctate, ovate-attenuate. Corollas pale lavender to purple. Achene unequally 5-angled, obconic in outline, 2–3 (–4) mm long, ribs sometimes obscure, surfaces between the angles atomiferous-glandular; pappus scales ovate-attenuate, finely toothed marginally, 1.5–2 mm long. (Incl. *M. laciniarioides* Small, *M. williamsonii* Small)

Pine savannas and flatwoods, adjacent ditches, bogs. Coastal plain, N.C. and S.C.

2. Marshallia tenuifolia Raf.

Closely similar to *M. graminifolia*. Differing in that the caudices are not clothed with fibers from old leaf bases; rosette leaves commonly still present at flowering time, these and the lower stem leaves mostly 5–6 cm long, of softer texture, usually spreading, oblanceolate or spatulate, narrowed basally to subpetioles, apices rounded or obtuse, with an evident vein paralleling the midrib on either side or more frequently these faint; some plants with linear, 1-nerved lower stem leaves; involucral bracts mostly linear-subulate, thin, with short, soft, attenuate tips.

Habitats like those for *M. graminifolia*. Coastal plain of Ga., southward to cen. pen. Fla., westward to e. Tex.

11. Iva (MARSH-ELDERS, SUMP-WEEDS)

Annual or perennial herbs or shrubs. Leaves opposite or opposite below, alternate above. Heads in a spicate, spicate-racemose, or paniculate arrangement, each bearing both staminate and pistillate flowers, the former 3–20 and central in the head, the latter 1–9 and peripheral; chaff subtending both the staminate and pistillate flowers or not one or the other. Involucres turbinate or hemispheric; bracts 3–9, free or united. Corolla of staminate flowers funnelform, 5-lobed; that of pistillate flowers tubular and truncate, or rudimentary or absent in some species. Achenes cuneate to obovate, somewhat compressed parallel to the involucral bracts; pappus none.

Reference: Jackson, R. C. "A Revision of the Genus *Iva* L." *Univ. of Kan. Sci. Bull.* 41: 793–876. 1960.

1. Plant a shrub.   1. *I. frutescens*
1. Plant an annual herb.
  2. Leaves sessile, narrowly linear, margins entire.   2. *I. microcephala*
  2. Leaves petiolate, blades lanceolate to ovate, margins irregularly serrate.   3. *I. annua*

1. Iva frutescens L. subsp. frutescens.

Bushy-branched shrub to 3.5 m tall, the distal portions of floriferous branches more or less herbaceous, atomiferous-glandular and pubescent with appressed-ascending hairs. Leaves opposite (excepting the bracteal ones of the flowering branchlets), short-petiolate, blades lanceolate, with 3 principal ascending veins, 4–10 cm long, 7–15 mm broad, acute at both extremities, margins serrate, surfaces appressed-pubescent, commonly olive green beneath. Involucre low, bracts 4 or 5, free, obovate, sparsely appressed-pubescent. Pistillate flowers 4,5(6), corollas about 1 mm long; staminate flowers 6–20, corollas about 2 mm long. Achene obovate in outline, 2–2.5 mm long, dark brown or dark purplish brown, resin-dotted, not hairy.

Coastal marshes, spits, bay and estuarine shores, mud flats, banks of dikes, causeways, and sloughs. Va. to s. Fla., westward to s.e. Tex.

Iva imbricata Walt., a succulent, low, bushy-branched subshrub, also occurs along the sea coasts in our range. It chiefly inhabits seaside dunes and sandy places near the shores of bays above the high-water marks.

2. Iva microcephala Nutt.

Annual, 4–10 dm tall, with few to many erect-ascending branches, often branching from near the base, the branches abundantly floriferous. Stems glabrous below, at least in age, usually sparsely short-pubescent on the inflorescence branches. Leaves opposite below, alternate above, linear, 2–6 cm long, 2–3 mm broad, punctate and sparsely short-pubescent. Heads numerous on all the branches, subtended by linear, foliaceous bracts, short stalked or sessile. Involucres low, bracts 4 or 5, free, obovate, margins whitish-scarious and ciliate, their backs green to purple, glandular-punctate, and sparsely short-pubescent. Pistillate flowers usually 3, corollas about 1 mm long; staminate flowers usually 5, corollas about 2 mm long. Achene obovate in outline, about 1.5 mm long, dark brown, dark purple, or nearly black, short-pubescent distally.

Exposed shores and bottoms of ponds, pine savannas and flatwoods, flatwoods depressions, swales, low places in pastures, often abundant in moist to wet disturbed places. Coastal plain, S.C. to s. pen. Fla., westward to cen. Fla. Panhandle and s.e. Ala.

3. Iva annua L.   Fig. 368

Annual of variable stature, 5–20 dm tall. Stem simple or with several to numerous spreading-ascending branches floriferous distally, glabrous below (at least in age), with stiffish, appressed-ascending hairs medially, and spreading-hirsute on the branches. Leaves mostly

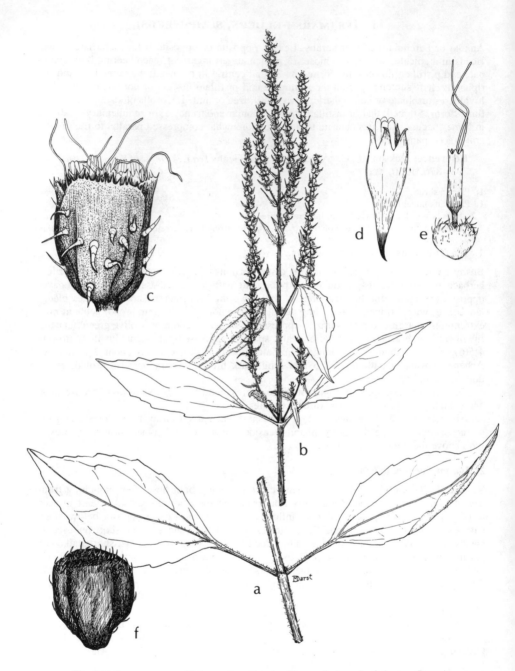

Fig. 368. **Iva annua:** a, midstem node with pair of leaves; b, branch of plant; c, flowering head; d, staminate flower; e, pistillate flower; f, achene.

opposite, petiolate, some of the upper bracteal leaves alternate, blades lanceolate to broadly ovate, with 3 principal, ascending veins from the base, 5–15 cm long and 2–7 cm broad, margins irregularly serrate, bases broadly short-tapered, apices acuminate, both surfaces pubescent, rough to the touch, hairs longer on the veins than between them. Heads sessile, crowded, solitary in the axils of relatively conspicuous, herbaceous, lanceolate to ovate bracts, the latter often with long, taillike tips, ciliate on the margins. Involucre low, its bracts 3–5, free, short and broad, apically truncate or rounded, a few long hairs on their backs and margins. Pistillate flowers 3–5, their corollas about 1.5 mm long; staminate flowers 8–16, their corollas about 2.5 mm long. Achene obovate in outline, about 2 mm long, dark brown, obscurely ridged longitudinally. (*I. ciliata* Willd.; incl. *I. caudata* Small)

More or less weedy and sporadic in much of our range (introduced from the prairie and plains region of cen. U.S.), sand and gravel bars of streams, stream banks, alluvial wood-lands, moist to wet clearings, waste places.

12. Xanthium

Xanthium strumarium L. COMMON COCKLEBUR.

Slender to coarsely branched, scabrid annual herb to 2 m tall. Leaves alternate (excepting those at the lowermost nodes), long-petiolate; blades to 15 cm long, broadly ovate to reni-form or suborbicular in outline, cordate, subcordate, or truncate basally, obtuse apically, margins irregularly toothed, often shallowly 3–5-lobed. Heads unisexual (plants monoe-cious), solitary or clustered in leaf axils. Staminate heads uppermost on the branches (smaller than the pistillate), many-flowered, involucre cuplike, of separate bracts in 1–3 se-ries, receptacle chaffy; staminate flowers with 5-lobed corolla, 5 separate stamens. Involucre of pistillate heads forming a conspicuous, tough and hard (at maturity) closed bur armed with hooked prickles and completely enclosing 2 flowers lacking a corolla, the 2 long styles of each protruding from the involucre between the terminal pair of prickles. Achene fusiform to obovoid; pappus none.

An extremely variable species (much of the variation involving the pistillate involucres or burs) from which numerous taxa have been segregated both on the specific and infraspecific levels. See Doris Love and Pierre Dansereau, "Biosystematic Studies of *Xanthium*: Tax-onomic Appraisal and Ecologic Status," *Canadian Jour. of Bot.* 37: 173–208 (1959).

Common, often extremely abundant weed inhabiting a great diversity of disturbed hab-itats and cultivated or pasture lands; sometimes in floodplain forests and their clearings, al-luvial banks and bars, shores. Cosmopolitan.

13. Balduina

Balduina uniflora Nutt.   Fig. 369

Perennial herb. Stem stiffly erect and bearing a single terminal head or with 2–20 stiffly erect branches each bearing a single terminal head, 4–8 dm tall; stem several-ribbed, sparsely short-pubescent and often atomiferous-glandular. Leaves sparsely pubescent and glandular-punctate, the basal rosette and lowermost stem leaves (often shed by flowering time) oblan-ceolate to linear-oblanceolate, attenuate basally to subpetioles, mostly 5–10 cm long overall; stem leaves upwardly ascending-erect and gradually reduced, those below the heads becom-ing remote and small-bracteate. Heads radiate. Involucre subtending the base of the head; bracts imbricate, firm, pubescent, the outer short-ovate and obtuse, gradually more elongate inwardly, the innermost acuminate apically. Receptacle somewhat convex, with chaffy bracts united into cuplike tubules toothed or cleft at the summit, collectively honeycomblike, each tubule surrounding the base of a flower. Ray flowers 5–20 per head, neutral, yellow, gradu-ally broadened from the base distally, 3–5-toothed at the summit, 1.5–2.5 (–4) cm long, their lower surfaces short-pubescent and atomiferous-glandular; disc flowers numerous, bisexual and fertile, their corollas 5–7 mm long, funnelform and with 5 erect, triangular

Fig. 369. **Balduina uniflora:** a, plant; b, flowering head, from below; c, portion of fruiting head showing connate bracts deeply seated within which are achenes; d, achene.

lobes at the summit, yellow. Achene deeply and firmly seated in the receptacular tubule, obconic, pubescent, about 3 mm long; pappus a ring of membranous scales 1 mm long or a little more, oblong basally and tapering to acute or acuminate apices. (*Endorima uniflora* (Nutt.) Barnh.)

Pine savannas and flatwoods, open bogs, borders of shrub-tree bogs and bays. Coastal plain, s.e. N.C. to n. Fla., westward to s.e. La.

B. atropurpurea (Harper) Small (*Endorima atropurpurea* (Harper) Small) is closely similar, its most conspicuous distinguishing feature being its purple disc flowers. It usually occurs with *B. uniflora* but has a much more restricted range and is of more local occurrence, chiefly s. Ga. and n.e. Fla., a disjunct locality in Darlington Co., n.e. S.C.

14. Helenium (SNEEZEWEEDS, BITTERWEEDS)

Annual or perennial herbs. Stems 1 to several, unbranched to paniculately or corymbosely branched above, sulcate or striate, unwinged or winged by the decurrent leaf bases. Leaves alternate, sessile to subpetiolate, margins entire to deeply toothed or lobed, impressed glandular-punctate. Heads 1 to numerous, most radiate. Involucral bracts flattish, herbaceous, in 2 series, the outer longer than the inner, usually spreading or reflexed at anthesis, usually impressed glandular-punctate, the outer sometimes united basally. Receptacle convex to ovoid or conic, alveolate, without chaff. Ray flowers (usually but not always present) pistillate and fertile or neutral, their ligules cuneate, 3 (–5) -lobed apically, yellow, variously tinged with red basally, varying to reddish brown or burnt orange, glabrous or pubescent beneath; disc flowers bisexual and fertile, their corolla tubes yellow or greenish yellow, exteriorly pubescent, 4- or 5-lobed at the summit, lobes yellow, brown, reddish brown, or purple. Achenes obconic, ribbed or angled-ribbed; pappus of 5 translucent scales each usually with an attenuate or awnlike tip, tip sometimes obtuse or rounded.

References: Berner, M. W. "Taxonomy of *Helenium* sect. Tetrodus and a Conspectus of North American *Helenium* (Compositae)." *Brittonia* 24: 331–355. 1972.

Rock, H. F. L. "A Revision of the Vernal Species of *Helenium* (Compositae)." *Rhodora* 59: 101–116, 128–158, 168–178, 203–216. 1957.

1. Disc flowers yellow throughout, the entire disc of the head thus appearing yellow.
  2. Stem leafy throughout, leaves not much reduced upwardly; heads few to numerous in a leafy inflorescence; ray flowers pistillate and most of them fertile. 1. *H. autumnale*
  2. Stem with all the larger leaves in a basal rosette and low on the stem, other leaves much reduced, stems thus subscapose; heads solitary and terminal on the single stems (rarely with 2 or 3 heads; ray flowers neutral).
    3. Stalk bearing the head decidedly pubescent, especially distally; midstem, reduced leaves barely decurrent on the stem, the wing extending only 1–5 mm below the insertion of the blade.
5. *H. pinnatifidum*
    3. Stalk bearing the head glabrous; midstem, reduced leaves decurrent on the stem rendering it winged 2 cm or more below the insertion of the blade, wings often extending from node to node.
      4. Achene pubescent; pappus scales deeply divided into numerous capillary segments ½ or more of the length of the scale. 6. *H. drummondii*
      4. Achene glabrous; pappus scales lacerate. 7. *H. vernale*
1. Disc flowers with their lobes reddish brown to purple thus the tips of the very much crowded flowers giving the disc a reddish brown to purple color (very infrequently only a sordid brown in *H. brevifolium*).
  5. Disc corollas predominantly 4-lobed. 2. *H. flexuosum*
  5. Disc corollas 5-lobed.
    6. Basal and lower stem leaves glabrous; stem glabrous below. 3. *H. brevifolium*
    6. Basal and lower stem leaves decidedly pubescent; stem pubescent below. 4. *H. campestre*

1. Helenium autumnale L.

Fig. 370

Perennial, stature various, little to much branched, up to 15 dm tall. Stem glabrous or finely short-pubescent, winged throughout, or lower portion unwinged at least in age. Leaves numerous, a little reduced upwardly, sessile or the larger one subpetiolate, lanceolate, narrowly

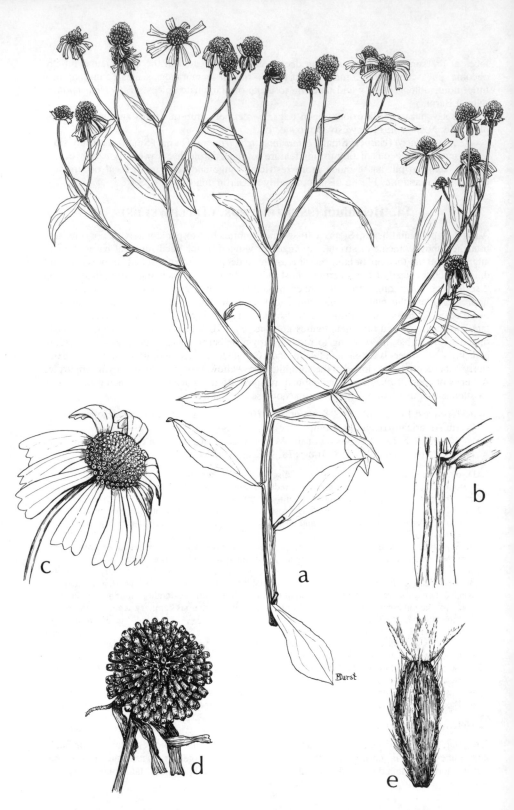

Fig. 370. **Helenium autumnale:** a, top of plant; b, node and small section of stem with decurrent petiolar wings; c, flowering head; d, fruiting head; e, achene.

elliptic, or narrowly oblong-elliptic, mostly cuneate to attenuate basally, acuminate to acute apically, 4–15 cm long, 1–3 (–4) cm broad, margins variable, entire, obscurely dentate or serrate, or less frequently sharply serrate. Heads few to numerous in a leafy inflorescence. Ray flowers pistillate and usually fertile, yellow, ligules varying in length from about 3 to 25 mm long, mostly 10–25 mm. Discs usually globose when fully developed, 0.6–2 cm broad; disc corollas yellow, 5-lobed at the summit. Achene brownish with pale ribs, pubescent on some or all of the ribs, 1.5–2 mm long; pappus scales short, ovate-acuminate or -attenuate, varying to subulate with attenuate tips, irregularly lacerate. (*H. latifolium* Mill; incl. *H. canaliculatum* Lam., *H. autumnale* var. *canaliculatum* (Lam.) T. & G., *H. parviflorum* Nutt., *H. autumnale* var. *parviflorum* (Nutt.) Fern., *H. virginicum* Blake)

Low, moist to wet, especially alluvial, places, pastures, meadows, woodlands, thickets, shores. Que. to B.C., generally southward to n. Fla. and Ariz.

2. Helenium flexuosum Raf.   Fig. 371

Perennial, stature various, commonly with a single stem from the short basal crown, sometimes with 2 to several stems, 4–10 dm tall, pubescent, winged, few- to much-branched above. Rosette and lowermost stem leaves (usually shed by flowering time), narrowed to winged subpetioles, blunt apically, mostly oblanceolate, varying to lanceolate or narrowly elliptic, entire or variously toothed marginally, up to 15 cm long, the lower stem leaves usually much longer than those of the rosette; other leaves mostly sessile, gradually diminishing in size upwardly, oblong, lanceolate, or linear-lanceolate, the lower ones with low dentations, upper entire. Heads few to numerous, disposed more or less paniculately. Ray flowers (rarely absent) neutral, ligules yellow or brownish, sometimes purplish basally, (5–) 10–20 mm long. Discs usually globose when fully developed, sometimes short-ovoid, 7–20 mm broad; disc flowers predominantly with 4 short-deltoid, glandular-pubescent, reddish brown to reddish purple lobes at the summit, tube yellowish, atomiferous-glandular. Achene about 1 mm long, prominently angle-ribbed, brownish and atomiferous-glandular between the angles, with pale-glistening, strongly ascending hairs on the angles; pappus scales mostly ovate or lanceolate and with attenuate or short, awnlike tips, margins irregularly few-toothed. (*H. nudiflorum* Nutt.; incl. *H. polyphyllum* Small)

More or less weedy in moist to wet places, pine savannas and flatwoods, adjacent roadsides and ditches, borders of shrub-tree bogs, borders of ponds, stream banks, meadows, prairies. N.Eng. to Mich., Ill., and Mo., generally southward to cen. pen. Fla., e. Tex., s.e. Okla.

3. Helenium brevifolium (Nutt.) Wood.

Perennial, 3–7 dm tall. Stem winged, commonly purple throughout, sometimes only near the base, usually solitary from a short, basal crown, sometimes bearing a single, terminal head, usually corymbosely 2–6-branched, each branch relatively long, erect, and bearing a single head; stem glabrous below, lower portions of the stalks of the heads thinly cottony-pubescent, the pubescence usually gradually more dense upwardly, each stalk naked or 1- or 2-bracteate. Rosette leaves commonly present during the flowering period, with oblanceolate bladed portions narrowed proximally to winged petioles, blunt to rounded apically, margins entire or undulate, surfaces glabrous, oftentimes suffused with purple, those of some rosettes 2–3 cm long overall and about 2 cm broad distally, those of other rosettes varying to 12 or 15 cm long and little if any broader; other leaves few, reduced, widely spaced, sessile, oblanceolate to narrowly oblong; uppermost stem and bracteal leaves usually thinly cottony-pubescent. Ray flowers neutral, ligules yellow, on most plants broadly cuneate, 7–15 mm broad distally, with 3–5 obtuse teeth; on occasional plants or an occasional head, ligules oblanceolate and not toothed and strongly bifid distally; ligules usually pubescent and atomiferous-glandular beneath, sometimes atomiferous-glandular above. Discs hemispheric, 1.5–2.5 cm broad; disc corollas with yellowish, atomiferous-glandular tubes and with 5 minute-deltoid, glandular-pubescent, red-purple to red-brown lobes. Achene 1–1.5 mm long, light brown, with low, rounded ribs, appressed-pubescent on the ribs and atomiferous-

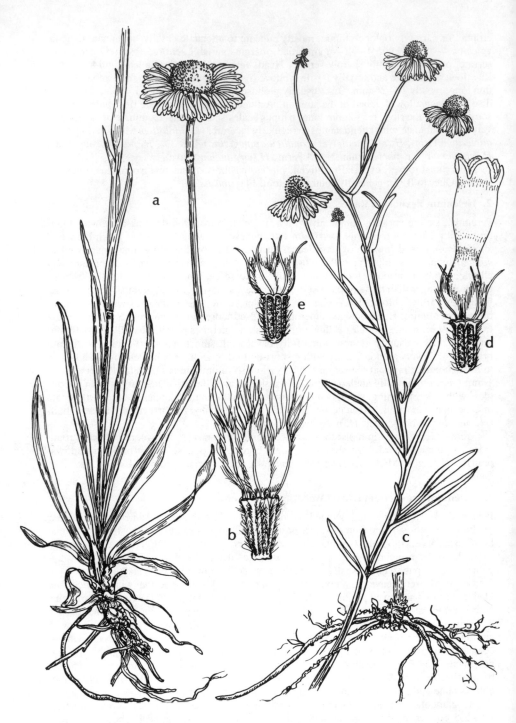

Fig. 371. a,b, **Helenium drummondii:** a, portions of plant; b, achene; c–e, **Helenium flexuosum:** c, base of plant and flowering branch; d, disc flower; e, achene. (From Correll and Correll. 1972)

glandular on and between the ribs; pappus scales 5–10, short-oblong to obovate, obtuse to rounded apically. (Incl. *H. curtisii* Gray)

Bogs, boggy clearings, boggy stream banks, seepage slopes, generally where the soil is saturated or even with standing surface water. Local, s.e. Va., coastal plain, piedmont, and mts., N.C. to w. Fla. Panhandle, Ala., Miss., La.

4. Helenium campestre Small.

Closely similar in habit and general inflorescence and floral features to *H. brevifolium*. Stem pubescent throughout, the hairs longish below, short above. Leaves pubescent on both surfaces. Discs more nearly subglobose than in *H. brevifolium*, 1–2 cm broad.

Low, open, pine woodlands, alluvial bottomland woodlands and thickets, moist places in prairies, rocky open woodlands. Arkansas, in a general north-south tier of counties on the piedmont of the Ozark and Ouachita Mts.

5. Helenium pinnatifidum (Nutt.) Rydb.

Perennial with a short basal crown bearing a rosette of closely set leaves, with a single stem from the crown or with 2 to several stems, in either case each stem simple and with a single terminal head, rarely branched above and with 2 or 3 heads; stem 3–8 dm tall, sparsely, if at all, pubescent below, becoming more definitely pubescent on the head-bearing portion, especially distally and on the base of the involucre, the summit of the stalk dilated and hollow. Rosette leaves usually present during the period of anthesis, variable, 3–8 (–20) cm long, 3–12 mm broad, with oblanceolate, elliptic, or linear bladed portions, narrowed proximally to subpetioles, the latter commonly expanded near their insertion and subclasping; leaf surfaces glabrous or with few, thickish, white trichomes, margins entire, undulate, few-dentate, or lobulate-pinnatifid; other leaves few and much reduced, gradually smaller upwardly rendering the stems subscapose, margins usually with a few low teeth; only the midstem leaves shortly and inconspicuously decurrent on the stem, the wings extending below the leaf insertion only 1–5 mm. Ray flowers neutral, their ligules yellow, 5–15 mm long, cuneate, 2–4-lobed apically, pubescent and atomiferous-glandular beneath. Discs hemispherical or merely convex, 1–2.5 cm broad; disc flowers yellow throughout, the tube atomiferous-glandular, the 5 minutely deltoid lobes glandular-pubescent. Achene about 1.5 mm long, light brown between pale ribs, almost always bristly-pubescent on the ribs (very rarely glabrous), atomiferous-glandular between the ribs; pappus scales 5–10, oblong, distal margins and sometimes the lateral edges unevenly, often deeply, lacerate. (*H. vernale* sensu Small, 1933, not Walt.)

Depressions in pine savannas and flatwoods, cypress ponds and prairies, borrow pits, borders of swamps, commonly in shallow water. Chiefly, almost exclusively, coastal plain, s.e. N.C. to s. pen. Fla., e. half of Fla. Panhandle.

6. Helenium drummondii Rock.   Fig. 371

Very closely similar to *H. pinnatifidum*, the rosette and lower stem leaves narrowly lanceolate, elliptic-lanceolate, infrequently oblanceolate, differing little from those of some specimens of *H. pinnatifidum*. Lower stem glabrous as in some specimens of *H. pinnatifidum*, the upper stem perhaps not so densely pubescent as on most specimens of *H. pinnatifidum*. The most significant features distinguishing *H. drummondii* from *H. pinnatifidum* would appear to be: the midstem leaves of the former are decurrent along the stem as manifest wings 2 cm long or more rather than only 1–5 mm in the latter; and the pappus scales of the former divided half or more of their length into a fringe of numerous capillary segments rather than being merely lacerate as in the latter. (*H. fimbriatum*, nomenclatural basis obscure)

Habitats as for *H. pinnatifidum*. e. Tex., s.w. La., and (perhaps) n.e. Fla.

7. Helenium vernale Walt.

Closely similar in general features to *H. pinnatifidum* and in the area of overlap in distribution sometimes growing intermixed with it. Stem glabrous throughout or very sparsely short-pubescent below the head and becoming glabrous. Rosette and lower stem leaves with their bladed portions narrowly oblanceolate to nearly linear, sometimes elliptic-oblanceolate, their margins entire, obscurely undulate, obscurely and remotely dentate, very rarely lobulate-pinnatifid, the latter much less frequently than in *H. pinnatifidum*, on the whole not dissimilar to such leaves on many specimens of *H. pinnatifidum*; midstem leaves decurrent along the stem rendering it winged for 2 cm or more below the point of insertion, commonly winged from node to node. Achene atomiferous-glandular between the ribs, not hairy. (*H. helenium* (Nutt.) Small)

Habitats as for *H. pinnatifidum*. Coastal plain, s.e. N.C. to n. Fla., westward to s.e. La.

15. Flaveria (YELLOWTOPS)

Annual or perennial herbs with sessile, opposite leaves, the bases of opposite pairs narrowly joining around the stem. Heads numerous, small, sessile or short-stalked, borne in dense glomerulate cymes sessile in leaf axils or in more open, terminal cymes, the heads more or less glomerate on the ultimate branches. Involucre (in those treated here) of 4–6, usually 5, yellowish bracts of subequal length, sometimes 1 or more bractlets below it on the stalks of the head. Receptacle of the head pitted. Some heads on a given plant (of those treated here) having a single pistillate and fertile ray flower and several bisexual and fertile disc flowers, other heads having only disc flowers. Achene narrowly obconic, 10-ribbed, grayish, 1–1.5 mm long; pappus none (in ours).

● Flowers mostly 5–10 per head; involucre linear-oblong in outline, bractlets below the involucre, if any, narrowly subulate, reaching to the base of the involucre or but little overlapping its base; larger leaves to 4 (–6) mm wide at their widest places.   1. *F. linearis*
● Flowers mostly 10–15 (–19) per head; involucre campanulate; bractlets below the involucre lance-acute, mostly extending half or more the length of the involucre; larger leaves mostly 8–15 mm wide at their widest places.   2. *F. floridana*

1. Flaveria linearis Lag.   Fig. 372

Stem simple or with 2 to several subequal, erect or decumbent branches from the base, to 8 dm tall, glabrous or rarely pubescent on the inflorescence branches. Leaves glabrous, linear or lance-linear, 5–10 (–12) cm long, 1–4 (–6) mm broad. Heads in a flat-topped or storied terminal cyme or in loosely glomerulate cymes terminating branches. Involucre narrowly linear-oblong (only slightly if at all broader basally than above), about 5 mm high, bractlets on the stalks of the head narrowly subulate, reaching about to the base of the involucre or scarcely overlapping its base. Flowers yellow, mostly 5–10 per head; ligules of ray flowers, if any, oval, about 2 mm long; disc flowers 3–4 mm long, tube cylindric below, flared to a funnelform throat, 5-lobed at the summit, lobes triangular, spreading.

Marshes and banks of marshes, marsh-prairies, cypress prairies, edges of mangrove swamps, marly flats, pinelands on limerock, floodplains, weedy in coastal areas. Pen. Fla. and Fla. Keys, extending around the Gulf coast of the Fla. Panhandle to Wakulla Co.; Bah.Is.; Yucatan.

2. Flaveria floridana J. R. Johnston.

Similar in general appearance to *F. linearis*, well-developed specimens more robust, to 12 dm tall. Larger leaves elongate-lanceolate, 5–15 cm long and 8–15 mm broad. Involucre campanulate, about 5 mm high, bractlets on the stalk of the head lanceolate-acute, commonly closely similar to the involucral bracts, some of them usually half or more as long as the involucre, sometimes surpassing it. Flowers bright yellow, 10–15 (–19) per head, individually like those of *F. linearis*. (Reputed to hybridize with *F. linearis*, the intermediates designated as *F. × latifolia* (J. R. Johnston) Rydb.)

Saline flats, interdune swales, borders of mangrove swamps or flats, waste places. Chiefly Gulf coast of pen. Fla. from Tampa Bay to Collier Co.

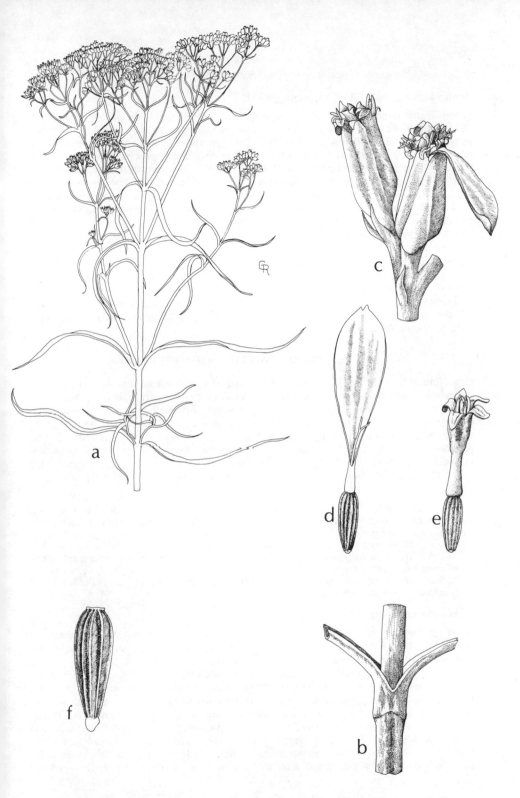

Fig. 372. **Flaveria linearis:** a, top of plant; b, node; c, two heads; d, ray flower; e, disc flower; f, achene.

16. Arnica

Arnica acaulis (Walt.) BSP. LEOPARD'S-BANE.

Perennial, the larger leaves basal on a simple, short, stout caudex, the inflorescence essentially scapose, the scape bearing relatively distant smaller opposite leaves, the bracteal leaves or bracts subtending inflorescence branches mostly alternate, sometimes subopposite or opposite. Scape 3–8 dm tall, abundantly clothed with gland-tipped hairs. Leaves sessile, broadly elliptic, ovate, or rhombic, 4–15 cm long and 1.5–8 cm broad, margins unevenly very shallowly wavy to entire, both surfaces with sessile, amber, tiny beadlike secretions in punctations and clothed with glandular hairs. Scape rarely unbranched and with a single terminal head, usually branched, the branches variable in number and very uneven in length, up to 2 dm long, each usually terminated by a single head. Heads campanulate, radiate, both ray and disc flowers yellow. Involucre 8–12 mm high and as broad or slightly broader; bracts somewhat imbricate, acute to acuminate, pubescent. Rays 10–15, pistillate, fertile, 1.5–2.5 cm long, apically with 3 or 4 short-obtuse teeth. Achenes black, striate, nearly fusiform but often with a longer taper basally, 4–5 mm long; pappus of numerous whitish, barbellate bristles 6–7 mm long.

Moist sandy-peaty soils, pine savannas and flatwoods, adjacent clearings, railroad savannas and moist grassy roadsides, sometimes in open upland pine woods. s. Pa., Del., southward to Fla. Panhandle.

17. Senecio (RAGWORTS, SQUAW-WEEDS)

Herbs (ours). Leaves alternate, unlobed to variously cleft or dissected. Inflorescence of heads corymbiform-cymose or paniculate. Heads radiate or discoid (usually radiate in those species treated here but those of occasional individuals discoid). Involucre of equal, elongate bracts in an essentially single series, this sometimes with additional subulate-setaceous, much smaller bracts about the base. Receptacle flat or slightly convex, without chaff. Ray flowers pistillate and fertile, the ligules of the corollas yellow (in ours) and often minutely 3-toothed terminally; disc flowers bisexual and fertile, their corollas yellow (in ours), 5-lobed at the summit. Achene columnar, 5–10-nerved; pappus of numerous, very soft, usually white, capillary bristles.

1. Plant annual, stem solitary; at flowering time all the principal leaves pinnately cleft. 1. *S. glabellus*
1. Plant perennial, usually with basal offshoots from a more or less horizontal caudex or rhizome.
  2. Basal and lowermost stem leaves mostly with cordate bases.   2. *S. aureus*
  2. Basal and lowermost stem leaves not having cordate bases.
    3. Stem persistently cottony-pubescent although the pubescence often becoming thin in age; achene minutely pubescent on the ribs.   3. *S. tomentosus*
    3. Stem cottony-pubescent only when young, in age glabrous or only with patches of pubescence about the lower nodes; achene glabrous.   4. *S. pauperculus*

1. Senecio glabellus Poir. BUTTERWEED.

Glabrous annual or winter annual, varying greatly in stature, sometimes slender and as little as 1.5 dm tall, the inflorescence with few heads, often coarse and as much as 15 dm tall, multibranched in the inflorescence; stem hollow. Usually a few lowermost small rosette leaves long-petioled and with unlobed ovate to orbicular blades about 2 cm long (such leaves rarely present at flowering time); other rosette leaves and lowermost stem leaves much larger, petioled, blades pinnately cleft, the lateral clefts opposite or subopposite, irregular in shape but tending to be blocky and with irregularly toothed margins, the terminal lobe usually largest and ovate to orbicular, sometimes sublobed; upwardly leaves becoming smaller and sessile, the blades similarly cleft; largest leaves of larger plants up to 25 cm long overall. Involucre 4–6 mm high, a few minute bracts on the dilated summit of the stalk of the head, the principal involucral bracts with acute to attenuate tips. Ligules of the ray flowers 5–12 mm long, golden yellow, with 3 minute teeth terminally, or the apices obscurely or not at all toothed. Disc corollas about 6 mm long, very narrowly tubular for about ⅔ their length then abruptly expanding to a funnelform throat. Achene about 3 mm long, strongly ribbed, constricted at both extremities, glabrous or minutely pubescent.

Swamps, wet woodlands, cypress ponds and lakes, wet places in fields and pastures, wet roadsides, often very abundant locally. Coastal plain of s.e. N.C., coastal plain and piedmont of S.C., southward to s. pen. Fla., westward to e. Tex., northward in the interior to Ohio, Mo., and S.D.

2. Senecio aureus L. GOLDEN RAGWORT.

Perennial, 2–8 dm tall, glabrous or cottony-pubescent in and near the axils of the basal leaves, on the inflorescence branches and involucres. Stems single or clustered from a branched, more or less horizontal rhizome, or clustered from stoloniferous basal offshoots. Basal and lowermost stem leaves long-petioled, blades mostly orbicular, round-ovate, or triangular-ovate, most of them basally cordate, sometimes blades of uppermost of the lower leaves cleft-lobed basally; all the lower leaves rounded apically, margins crenate, serrate, or serrate-dentate, variable in size, 1–6 (–18) cm long, commonly purple beneath; stem leaves, from a little above the base, reduced upwardly, varying from sublyrate to pinnatifid, becoming sessile above. Heads several to relatively numerous. Involucre 5–8 mm high, bracts often purplish-tipped or rarely black-tipped, linear, linear-oblong, or slightly oblanceolate, mostly acute apically. Ligules of the ray flowers orange-yellow, 5–15 mm long, mostly 3-toothed terminally. Corollas of the disc flowers 6–7 mm long, slenderly funnelform. Achene 2.5–3 mm long or a little more, strongly ribbed, not constricted at the extremities, glabrous. (Incl. *S. gracilis* Pursh)

Meadows, bogs, woodlands astride streams, stream banks, rocky shores of streams and on rocks in streams, floodplain woodlands, seepage slopes, spring runs, ledges, mesic woodlands. Nfld. to Minn., southward to n. Ga. and Ark., n.e. Okla.; a disjunct population along Little River, Gadsden Co., Fla. Panhandle.

3. Senecio tomentosus Michx.

Perennial, perennating by basal offshoots or stolons. Stem 2–7 dm tall, cottony-pubescent. Principal leaves in a basal rosette and on the lowermost portion of the stem, blades abruptly contracted to petioles or bases of the blades decurrent on the petioles where often bearing small lobes of blade tissue, blades varying in length, width, and shape on a given plant, 3–10 cm long and 1–4 cm broad, oblong, ovate-oblong, lance-oblong, elliptic, or lanceolate, surfaces at first cottony-pubescent, the upper usually, and sometimes the lower, becoming glabrous, margins usually crenate; leaves from a little above the base of the stem few, much reduced, mostly sessile, often minutely incised-lobed. Heads few to relatively numerous. Involucre cottony-pubescent, often becoming nearly glabrous, 5–8 mm high, bracts subulate. Ligules of the rays pale to deep yellow, 6–10 mm long, obscurely or not at all toothed terminally. Corolla of the disc flowers orange-yellow, about 6 mm long, slenderly funnelform. Achene 2–2.5 mm long, strongly ribbed, not constricted at the extremities, minutely pubescent on the ribs. (Incl. *S. alabemensis* Britt.)

Moist to wet sands and sandy clays of pinelands, sandy-peaty roadsides, moist to wet pastures, openings and clearings of mesic woodlands. Coastal plain, s. N.J. to w. Fla. Panhandle, westward to e. Tex., Ark., Okla.; also in shallow, seasonally wet soil of seepage on granite rock outcrops of piedmont, N.C. to e. Ala.

4. Senecio pauperculus Michx.

Slender perennial, 1–5 dm tall, unbranched below the few-headed inflorescence. Stem cottony-pubescent when young, in age sometimes glabrous or with patches of the pubescence persistent, especially at the lower nodes. Basal and lower stem leaves long-petioled, blades 1–10 cm long, lanceolate, narrowly to broadly elliptic, or oblanceolate, infrequently ovate, cuneate basally, blunt apically, margins finely serrate; other leaves few, widely spaced, becoming sessile above, more or less pinnatifid. Involucre 5–7 mm high, united basally and cuplike, often with a few minute bracts on the cup; principal bracts linear-oblong below and tapering to acute tips, glabrous (in ours). Ligules of the ray flowers yellow, 5–7 (–10) mm long. Tube of the disc corollas slenderly funnelform, 5–6 mm long. Achene 2–3 mm long, glabrous, not constricted at the extremities.

Bogs, wet meadows, peaty pinelands, wet rocks and gravels. Transcontinental, Can. and n. U.S.; occasional in our range southward to n. Ga. and Tenn.; a disjunct locality in Bay Co., Fla. Panhandle.

18. Cacalia

Cacalia suaveolens L.

Perennial herb, perennating by basal offshoots. Stem usually unbranched, glabrous, striate or grooved, 5–15 dm tall. Leaves alternate, most of them petiolate, blades pinnately veined, the lower and median with conspicuously sagittate-hastate bases, long-triangular above the junction of petiole and blade and to 2 dm long; upwardly the basal lobes of the blades becoming progressively less distinct, the uppermost often unlobed, their petioles becoming shorter and more winged, uppermost usually sessile; margins of blades sharply serrate, teeth numerous and unequal, upper surfaces dark green and glabrous, lower paler, sparsely short-pubescent chiefly along the veins. Heads discoid, few to numerous, in a terminal, more or less flat-topped, corymb. Involucre about 1 cm high, campanulate, 10–15 upper bracts closely imbricate, linear-attenuate, several lower bracts loosely imbricate, subulate, about half as long as the upper. Receptacle flat, without pales subtending the flowers. Flowers 20–40 per head, all bisexual and fertile, radially symmetrical; corolla cream-white, narrowly cylindric below, at about midway of the tube expanding to a nearly equally long, narrowly campanulate throat with 5 erect lobes shorter than the throat. Filaments with enlargements a very short distance below the anthers, otherwise filiform. Achene columnar, nerved. Pappus of numerous, white, minutely barbellate, capillary bristles. (*Synosma suaveolens* (L.) Raf. ex Britt. & Brown)

Wet woodlands, thickets, and clearings, bogs. R.I. and Conn. to s.e. Minn., generally southward to w. N.C., n. Ga., Tenn. and e. Mo.; disjunct in n.w. pen. Fla.

19. Arnoglossum (INDIAN-PLANTAINS)

Glabrous, perennial herbs, stem usually unbranched below the inflorescence. Leaves alternate, those of rosettes and lower stem petiolate, blades irregularly and coarsely toothed-lobed to entire. Heads discoid, borne in more or less flat-topped or storied corymbs. Involucre nearly cylindric, bracts of some with keeled or winged midribs, in a single series of 5, sometimes with a few very short, inconspicuous bractlets basally in addition. Receptacle with a short, conic, central projection, without pales subtending the flowers. Flowers 5 per head, bisexual and fertile, radially symmetrical, corolla cream-white, sometimes with green, tan, or maroon tints, or lavender, tube cylindric below, abruptly flaring above to a very short-campanulate throat, then deeply 5-lobed, lobes recurved. Filaments evenly filiform. Achene columnar or fusiform, ribbed. Pappus of numerous, very slender, soft, smooth or minutely barbellate, capillary bristles.

1. Involucral bracts with conspicuously winged midribs; lower stem angled or ridged and grooved.
   2. Blades of all or most stem leaves deltoid-ovate or cordate-ovate, margins angulate-lobed or angulate-dentate; corollas pale to bright lavender or cream-white with a lavender tinge. 1. *A. diversifolium*
   2. Blades of all or most stem leaves lanceolate to ovate, bases tapering, margins sinuate-crenate, crenate, or wavy-dentate; corollas cream-colored. 2. *A. sulcatum*
1. Involucral bracts without winged midribs, the latter not at all to only moderately evident; stem terete and striate throughout. 3. *A. ovatum*

1. Arnoglossum diversifolium (T. & G.) H. Robins. Fig. 373

Plant with a short, stout caudex, perennating by basal offshoots that form overwintering rosettes, the latter becoming detached from the parent plant by disintegrating of the connecting stems and not always forming a flowering stem the year immediately following. Stem purplish or pale, not glaucous, angled or ridged and grooved below, ribbed above, 7–15 dm tall. Leaves green on both sides, those of the rosettes and lower stem 15–30 cm long, the blade usually not more than ⅓ as long as the petiole, upwardly petioles progressively shorter, blades smaller, uppermost subsessile or sessile, all or most blades deltoid-ovate or cordate-ovate in outline, some sometimes broadly elliptic to rotund, margins of most of those on the stem irregularly and coarsely toothed-lobed, the basal ones more commonly shallowly dentate, wavy, or entire. Corymb varyingly narrow and short, with few heads, to broadly storied and with numerous heads. Involucre 6–10 mm long, bracts linear-oblong, apices obtuse, the midribs strongly winged, margins scarious. Corollas pale to bright lavender or cream-white with a lavender tinge. Achene fusiform, 4–5 mm long, inconspicuously ribbed. (*Cacalia diversifolia* T. & G., *Mesadenia diversifolia* (T. & G.) Greene)

Banks of woodland streams and seasonally wet places in richly wooded hammocks, swamps bordering rivers and streams. Levy Co., n.w. pen. Fla., then apparently disjunct to cen. Fla. Panhandle and adjacent s.w. Ga., s.e. Ala.

2. Arnoglossum sulcatum (Fern.) H. Robins. Fig. 374

Habit, stature, and surfaces of stems as in *A. diversifolium*. Lower leaves 20–25 cm long, petioled, blades comprising less than ½ the total length, mostly lanceolate to ovate and with tapered bases; blades of basal rosettes sometimes broadly ovate and with truncate or short-tapered bases; upwardly leaves with progressively shorter petioles, uppermost sessile; margins of blades entire, wavy or with few large dentations. Corymb diffuse and storied to congested, mostly the former. Involucre 8–10 mm high, bracts linear-oblong, acute or mucronate, rarely obtuse, apically, midrib conspicuously winged, margins scarious. Corolla cream-yellow. Achene 4 mm long, fusiform, ribbed. (*Cacalia sulcata* Fern., *Mesadenia sulcata* (Fern.) Small)

Peaty borders of swamps, sphagnous bogs, wet woodlands. s.w. Ga., Fla. Panhandle from Leon Co. westward, s. Ala.

3. Arnoglossum ovatum (Walt.) H. Robins. Figs. 375 and 376

Plant with a short and stout to somewhat elongate caudex, stem terete and striate throughout. Individuals of this species exhibiting much diversity in stature and in size and form of leaves, the extremes dissimilar but intergrading, even locally; in height varying from about 1 to 2.5 m tall or more; basal and rosette leaves varying from very long-petiolate, blades narrowly lanceolate and much shorter than the petioles, very gradually tapering into the petioles, to relatively short-petiolate, blades longer than the petioles, broadly ovate and abruptly contracting basally, to 15 cm wide; stems and leaves of individuals with narrowest leaves usually slightly glaucous, those of broadest-leaved ones conspicuously glaucous; margins of blades entire, wavy, or obscurely to sharply dentate; in all these respects grading from one extreme to the other; leaves usually sessile from about midstem upwardly. Corymbs varying greatly in size, from relatively few-headed and flat-topped, to multistoried, with many heads and to 2.5 or 3 dm broad. Involucre 8–11 mm high, bracts linear-oblong, acute, mucronate, or obtuse apically, margins hyaline, midrib not evident, evident, only proximally, or barely evident throughout, not at all winged. Corolla cream-yellow, sometimes tinged with pink or lavender. Achene 3–4 mm long, columnar or columnar-fusiform, ribbed. (*Cacalia lanceolata* (Nutt.), *Mesadenia lanceolata* (Nutt.) Raf.; incl. *C. elliottii* (Harper) Shinners, *C. lanceolata* var. *elliottii* (Harper) Kral & Godfrey, *Mesadenia maxima* Harper)

Pine savannas and flatwoods, wet sandy woodlands, ditches. Coastal plain, e. N.C., southward to s. Fla., westward to e. and s.e. Tex.

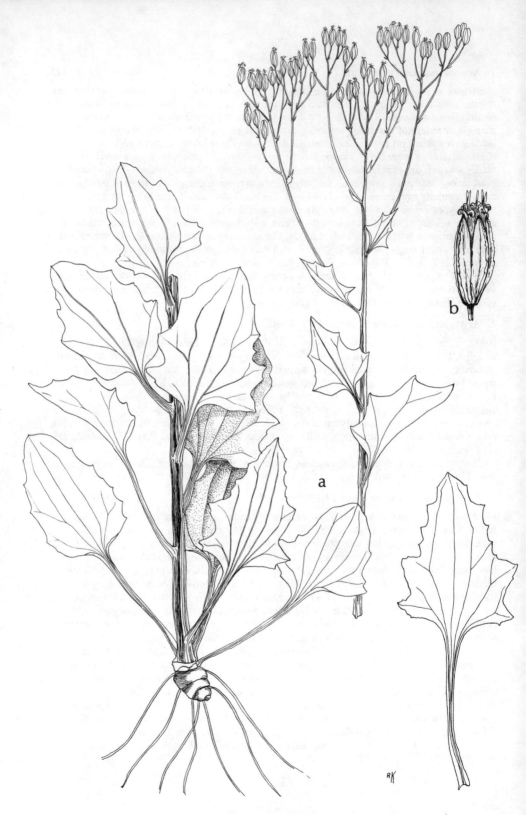

Fig. 373. **Arnoglossum diversifolium:** a, plant; b, head.

Fig. 374. **Arnoglossum sulcatum:** a, plant; b, head.

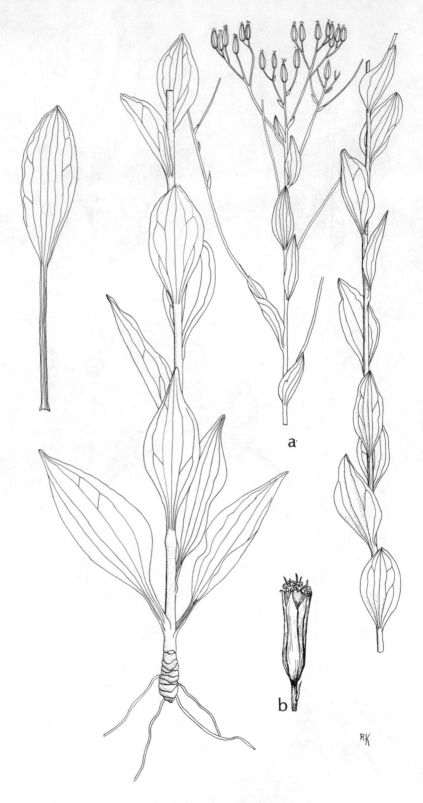

Fig. 375. **Arnoglossum ovatum:** a, plant (broader-leaved type); b, head.

Fig. 376. **Arnoglossum ovatum:** a, plant (narrow-leaved extreme); b, head.

20. Erechtites

Erechtites hieracifolia (L.) Raf. FIREWEED.

Annual, very variable in stature, in leaf form and size, and in pubescence, the variation commonly notable even in a small local population. Stems glabrous, setose, or pilose, from a few cm to 2–3 m tall, the bases of the former very slender and those of the latter 2–3 cm or more in diameter at the base, varying from simple and with a small cluster of heads at the summit to having numerous, elongate, strongly ascending principal branches from about the middle, these with heads disposed cymose-paniculately distally. Leaves alternate, variously pubescent, sometimes all on a given plant unlobed, irregularly and unevenly serrate or dentate-serrate, oblanceolate to elliptic, usually (not always) cuneately narrowed to winged petioles; commonly the lower unlobed and as described above, median leaves incised-serrate, sinuate-dentate, irregularly and coarsely pinnate-lobate, the lobes coarsely toothed or subpinnatifid, often the lower ½ or ⅔ of the leaf lobate to subpinnatifid, the distal portion merely toothed, acute apically, truncate to auriculate, sometimes tapered, basally; very variable in size, up to about 3 dm long and 6–8 cm wide across the medial lobes; leaves usually reduced toward the summit of the stem or its branches. Heads 1–2 cm long, campanulate basally, cylindric above, when fully developed (essentially cylindric in immature stages). Principal involucral bracts linear, glabrous or sparsely short-pubescent, a single series, connivent; additionally often with a few short, subulate bractlets around the campanulate base and on the stalks of the head; principal bracts reflexed after the achenes are shed. Receptacle of the head flat, not chaffy. Florets numerous, whitish to cream-colored, or yellowish, with an elongate-cylindric corolla somewhat dilated at the summit and 5-lobed; peripheral florets pistillate, others bisexual, all producing achenes, florets little exserted from the involucres. Achene brown, low-ribbed, nearly columnar, a little contracted at the summit, 2–2.5 mm long, surmounted by numerous, soft, white pappus bristles much longer than the achene.

In a wide variety of habitats relative to moisture conditions but not in excessively xeric places, commonly weedy, especially in disturbed and burned-over places; forest clearings and waste places, moist to wet ditches, swales, marshes and marshy shores. Nfld. to Sask., generally southward to s. Fla., e. and s.e. Tex.; W.I.

21. Solidago (GOLDENRODS)

Perennial, often rhizomatous herbs, mostly with stems erect and unbranched below the inflorescence; plants with short rhizomes forming close clumps of stems, those with elongate rhizomes often forming extensive clones, those with basal crowns sometimes perennating from crown buds and producing tufts of stems. Leaves alternate, the basal and lower stem leaves sometimes abruptly narrowed to petioles or gradually narrowed to subpetioles, upper stem leaves sessile, sometimes all sessile. Heads relatively small, usually numerous, disposed in axillary clusters, or in terminal, more or less elongate and cylindrical thyrses, heads not at all secund, sometimes in several such thyrsoid branches, or panicled, the branches spreading or often arching, heads secund, or rarely in corymbs, the heads not at all secund. Heads radiate (discoid in one species in our range, this not treated here), ray flowers pistillate and fertile, usually not disposed evenly around the periphery of the head; disc flowers bisexual and fertile, more numerous than the ray flowers; both ray and disc flowers yellow (white in one species of our range, this not treated here). Involucres cylindric or subcylindric to campanulate, bracts rarely subequal, usually imbricated and graduated from the outside inwardly, the outer smaller than the inner, yellowish-stramineous, often with a darker, more herbaceous median strip which is a little dilated and greener at the tip. Receptacle flat or slightly convex, without chaff. Achenes several-nerved; pappus a single series of numerous, equal or unequal, capillary bristles.

1. Stem glabrous below or throughout (sometimes lower portion of stem very sparsely and unevenly pubescent in *S. leavenworthii*).

2. Leaves at or near the base of the stem much larger than those at midstem, attenuated basally to winged subpetioles about as long as or longer than the bladed portions and much more than 4 cm long.
   3. Upper and lower surfaces of leaves glabrous save for the scabrid margins.
      4. Plant slenderly rhizomatous.   1. *S. stricta*
      4. Plant not rhizomatous but with a thick, branched caudex.   2. *S. sempervirens* var. *mexicana*
   3. Upper surfaces of leaves scabrid, lower glabrous.   3. *S. patula*
2. Leaves at or near the base of the stem essentially like those at midstem, not attenuated basally to subpetioles or, if subpetiolate, then the subpetioles not over 4 cm long.
   5. Most of the lower and midstem leaves 3–10 mm broad.   4. *S. leavenworthii*
   5. Most of the midstem leaves over 10 mm broad.
      6. Margins of leaf blades sharply serrate; a pair of lateral veins arising from the midrib well above the base of the blade, blade thus evidently triple-nerved.   5. *S. gigantea*
      6. Margins of leaf blades entire or obscurely and bluntly serrate; lateral veins not evident or, if present, then faint and not rendering the blade evidently triple-nerved.   6. *S. elliottii*
1. Stem evenly pubescent throughout (sometimes becoming glabrous in age on lower stem).
   7. Leaves triple-nerved well above the base of the blade.   7. *S. altissima*
   7. Leaves not triple-nerved.
      8. Midstem leaves mostly oblong or oblong-elliptic, subclasping basally, margins essentially entire.   8. *S. fistulosa*
      8. Midstem leaves variable in shape, none of them really oblong or oblong-elliptic, if subclasping basally then approximately ovate, margins sharply to bluntly toothed.   9. *S. rugosa*

1. **Solidago stricta** Ait.

Plant glabrous throughout, with a basal caudex to about 1 cm thick from which, late in the season, basal offshoots each bearing a terminal rosette of leaves are produced from crown buds, some of the offshoots with very short stems, some bracteate-rhizomatous and 2–10 cm long or more, usually both from the same caudex; flowering stems arising from the offshoots the following growing season thus resulting in 2 to several stems closely or loosely tufted. Stems very variable in stature, 7–20 dm tall, unbranched, strict and wandlike below the inflorescence, largest leaves basal or very low on the stem. Lowermost leaves with bladed portions attenuated to long subpetiolar portions dilated proximally and subclasping the stem, bladed portions lanceolate to long-elliptic, or oblanceolate, tapered distally and long-acute, varying to blunt or rounded tips, variable in overall length from about 6–30 cm long and in breadth from about 0.5–3 cm; other leaves usually erect, sessile, upwardly gradually or abruptly much reduced relative to the basal, becoming linear and smooth on the edges, the uppermost often very much reduced and bractlike; blades with only the midvein evident, glabrous save for scabrid margins. Inflorescence of heads paniculiform but the branches short, varying from few to very numerous, thus the panicle short to elongate, sometimes as much as 1 m long, heads sometimes secund on the branchlets; some specimens having the inflorescence with several elongate, erect branches, each branch thus elongate-paniculiform. Involucre 4–6 mm high, outer bracts short-subulate in outline, thick and rounded on the back, inwardly the bracts gradually flatter, longer, and broader, the medial usually blunt apically, the inner more nearly acute, the medial and inner often very finely and inconspicuously short-pubescent marginally. Achene subcylindric, about 2 mm long, sparsely short-pubescent; pappus about twice as long as the achene. (*S. petiolata* sensu Mack, in Small, not Mill.; incl. *S. pulchra* Small)

Bogs, seasonally wet pine savannas and flatwoods, coastal marshes (see note following description of *S. sempervirens*). Coastal plain, s. N.J. to s. pen. Fla., westward to e. and s.e. Tex.; W.I., s. Mex.

2. **Solidago sempervirens** L. var. **mexicana** (L.) Fern. SEASIDE GOLDENROD.

Plant glabrous throughout or with very sparse short pubescence on the inflorescence branches, with a stout, branched caudex, the branches to 2 cm thick and the caudex to 1–2 dm in overall diameter, producing, late in the season, very short-stemmed rosettes of leaves from crown buds on various parts of the caudex, flowering stems arising from each of the rosettes the following growing season thus resulting in clumps of stems. Stems 4–20 dm tall, commonly with numerous strongly ascending inflorescence branches, branching beginning low on a stem of short stature, relatively higher on stems of taller stature. Largest leaves basal or low on the stem, with bladed portions gradually attenuated to long, winged petioles dilated proximally and subclasping the stem, bladed portions long-elliptic or long-lanceolate, often, if not usually, blunt at their tips, mostly 10–20 cm long and 1–4 cm broad; other leaves gradually reduced upwardly; midvein of blade evident, other veins faint, glabrous save for minutely scabrid margins. Inflorescence of heads variable, commonly loosely and relatively broadly paniculiform, the main branches spreading and arching, heads mostly se-cund, varying to narrowly elongate and with few or none of the heads secund, often different inflorescence branches of a single stem thus variable. Involucre 3–5 mm high, outer bracts short, triangular to lanceolate, acute apically, thin and flat or a little rounded on their backs, bracts gradually longer inwardly and mostly linear-oblong or linear, tips blunt, sometimes finely and shortly pubescent marginally. Achene narrowly or slightly obconic, about 2 mm long, sparsely short-pubescent; pappus about 1.5 times as long as the achene. (*S. mexicana* L.)

Coastal marshes, estuarine and bay shores, coastal swales. s.e. Mass. to s. pen. Fla., westward to s.e. Tex.; W.I.

S. sempervirens var. *mexicana* and *S. stricta* appear to us to be reasonably distinct and to occupy different habitats for the most part; however in coastal marshes, perhaps especially of the Gulf coast, individual plants or entire local populations appear to us to be intermediate.

3. Solidago patula Muhl. ex Willd.

Plant of variable stature from 7–20 dm tall, with a relatively short, basal caudex, perennat-ing by basal offshoots. Stem irregularly ribbed-angled, or in part narrowly wing-angled, glabrous below, with short, rough pubescence on the inflorescence branches. Leaves of basal offshoots and lower part of the flowering stem narrowed to winged petioles often as long as the bladed portions, the latter elliptic, mostly short-acuminate apically, variable in size up to 15–20 cm long and 6–8 cm broad; leaves gradually reduced upwardly, variable in size and shape, those at about midstem sessile, elliptic to lanceolate, tapered to rounded basally, api-ces acute; leaf margins serrate or crenate-serrate, those of upper or bracteal leaves sometimes entire, upper surfaces antrorsely scabrid, the hairs short and stiff, lower surfaces glabrous. Inflorescence of heads paniculiform, sometimes with few short branches, more commonly with several strongly ascending branches leafy-bracted below, often curvate distally, heads not secund. Involucres 3–4.5 mm high, outer bracts short-triangular-acute, medial and inner ones oblong, apices usually rounded or obtuse, less frequently acute, margins of all bracts usually finely very short-ciliate. Achene 2–2.5 mm long, sparsely short-pubescent; pappus about as long as the achene. (*S. rigida* sensu Mack, in Small (1933) not L.; incl. *S. patula* var. *strictula* T. & G., *S. salicina* Ell.)

Bogs, swampy woodlands, shrub-tree bogs, stream banks, seepage slopes or ledges. Vt. to Wisc. or Minn., generally southward to w. Fla. Panhandle and e. Tex.

4. Solidago leavenworthii T. & G.

Rhizomatous, late in the season forming slender perennations from the rhizome and from the basal crown. Stem 5–12 dm tall, leafy, lower stem glabrous or sparsely pubescent, upwardly becoming pubescent with short, curvate hairs. Leaves little reduced to midstem or a little above then gradually reduced to the inflorescence, sessile, usually narrowly lanceolate or linear-lanceolate, less frequently lanceolate, larger ones 4–10 cm long, 3–10 (–12) mm broad, midvein evident, a pair of lateral veins usually much less pronounced, bases mostly narrowly rounded, apices acute, surfaces glabrous, often dotted beneath within the islets formed by the veinlets, margins antrorsely scabrid, often serrate as well. Inflorescence of heads paniculiform, the main branches usually arching and the heads for the most part se-cund, less frequently with the main axis having short branches and the heads more or less glomerate. Involucre 4–5 mm high, bracts often short-pubescent at or near their tips, the outermost short-triangular-subulate, gradually longer inwardly, the medial oblong, obtuse, the inner linear and blunt to acute. Achene 1 mm long or little more, short-pubescent; pappus about 3 times as long as the achene.

In both poorly and well drained places, pine savannas and flatwoods (especially in places of disturbance), borders of shrub-tree bogs, exposed shores of ponds and impoundments, clearings and borders of floodplain woodlands, interdune swales, thickets, sandy fields, open banks. Coastal plain, s.e. N.C. to s.e. and s.w. Ga., pen. Fla. and e. half of Fla. Panhandle, probably s.e. Ala.

5. Solidago gigantea Ait.

In general superficial aspect and stature similar to *S. altissima*. Late in the season perennating by basal offshoots from crown buds, thus sometimes forming clumps but not extensive clones. Stem leafy, glabrous and often glaucous below the inflorescence, inflorescence branches short-pubescent. Leaves of the basal offshoots attenuate below to subpetiolar bases, the bladed portions oblanceolate or spatulate; principal stem leaves mostly lanceolate, tapering basally and sessile or with a very short petiole, acuminate apically, 6–15 cm long and 1–2 cm broad; midvein most prominent, a pair of less prominent lateral veins arising from the midvein well above the base of the leaf; margins sharply serrate distally from a little below the middle (margins of uppermost, reduced leaves usually not serrate), the teeth with pointed callus tips, both surfaces glabrous and smooth, or upper surface minutely scabrid, the lower pubescent chiefly on the larger veins, occasionally on the surfaces between the veins as well; edges of all leaves antrorsely scabrid. (*S. serotina* Ait. not Retz; incl. *S. gigantea* var. *serotina* (Ait.) Cronq. = var. *leiophylla* Fern.)

Wet low woodlands and their borders, lowland clearings and fields, stream banks, shoreline thickets, meadows, commonly on alluvial soils. N.S. and Que. to Ore., generally southward to s.w. Ga. and Tex., westwardly to Utah and Colo. (doubtfully in Fla.; specimens from there so identified are probably *S. leavenworthii*).

6. Solidago elliottii T. & G.

Plant, late in the season, perennating by slender rhizomes. Stem (5–) 10–20 dm tall, glabrous throughout or very sparsely short-pubescent on the inflorescence branches. Lowermost leaves narrowed basally to subpetioles to 4 cm long, bladed portions oblanceolate; midstem leaves sessile, lanceolate, lance-elliptic, or elliptic, 6–15 cm long and 1.5–3.5 cm broad, short-cuneate to narrowly rounded basally, apices acute to acuminate, midvein evident but low and slender and not especially conspicuous beneath, lateral veins faint, or if evident then the blade rather obscurely 3-nerved from well above the base, margins entire or obscurely and bluntly serrate, the teeth with bluntly callused tips, edges scabrid, upper surfaces glabrous and smooth, lower glabrous or with sparse pubescence on the principal veins. Inflorescence of heads paniculiform, often leafy-bracted in about the lower half, the main branches slightly to strongly arching, relatively narrow to broad and open overall, heads mostly secund. Involucre 4–6 mm high, outer bracts short, more or less deltoid with blunt tips, medial and inner bracts graduated from oblong to linear, tips mostly rounded, the innermost often with minute pubescence on the edges of the tips. Achene 1.5–2 mm long or a little more, very sparsely pubescent; pappus about 1.5–2 times as long as the achene. (Incl. *S. mirabilis* Small, *S. edisoniana* Mack. in Small)

Wet woodlands and swamps and their borders, low hammocks, wet thickets, bogs. N.S.; coastal plain, e. Mass. to e.cen. pen. Fla.

7. Solidago altissima L.   Fig. 377

Plant with elongate rhizomes, often forming extensive clones. Stem 8–20 (–25) dm tall, leafy, copiously short-pubescent, in age usually glabrous or nearly so below and dull brown or reddish brown. Leaves little reduced to midstem or a little above, lowermost (usually shed by flowering time) narrowed to short, winged subpetioles, others sessile; principal midstem leaves lanceolate or oblanceolate, somewhat cuneate basally, acute to acuminate apically,

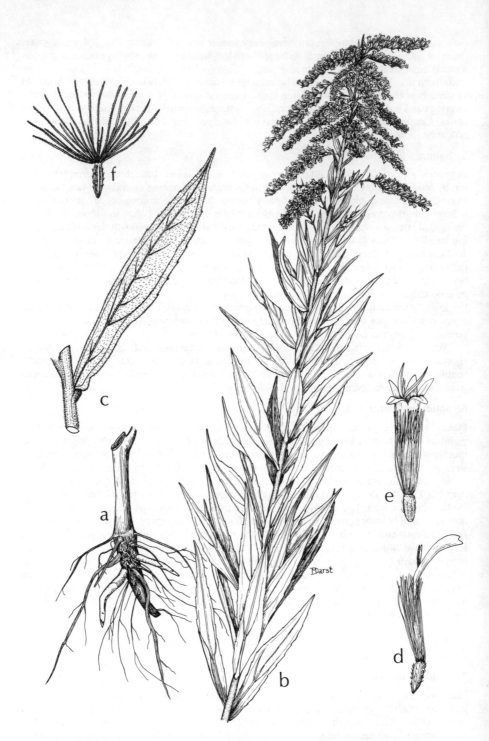

Fig. 377. **Solidago altissima:** a, base of plant; b, top of plant; c, short section of stem with leaf; d, ray flower; e, disc flower; f, achene.

6–15 cm long, midvein most prominent, especially beneath, a pair of less prominent lateral veins giving off from the midvein well above the base of the leaf, margins subentire or serrate, if serrate often obscurely so, scabrid above, sparsely to densely short-shaggy-pubescent beneath, oftentimes the pubescence beneath chiefly along the veins, edges scabrid. Inflorescence of heads pyramidal-paniculiform, the main branches usually spreading and arching, heads generally secund. Involucre 3–5 mm high, outer bracts short, triangular to subulate, acute, usually somewhat pubescent, inwardly bracts gradually longer, the medial more or less oblong, tips usually blunt or rounded, the inner linear, acute to rounded, medial and inner often with pubescent edges at the tips. Achene 1–1.5 mm long, sparsely, short-pubescent; pappus about 3 times as long as the achene. (*S. canadensis* L. var. *scabra* T. & G.; *S. hirsutissima* Mill.)

In diverse, open or openly wooded habitats, often in, but by no means restricted to, places at least seasonally wet, perhaps the most common and abundant goldenrod in our range. Que. to e. Mont., generally southward to Fla. (excluding s. pen.) and Ariz.

8. Solidago fistulosa Mill.   Fig. 378

Flowering stem from a short, erect crown terminating a horizontal caudex; late in the season several slender stolons produced from crown buds, the stolons radiating outward then arching upward, bearing several bracts along their length and a cluster of oblanceolate or spatulate leaves distally. Stem 7–20 dm tall, shaggy pubescent, usually copiously so in the inflorescence, often becoming glabrous below in age, abundantly leafy. Leaves usually relatively little reduced to somewhat above midstem then gradually reduced upwardly, sometimes gradually reduced upwardly from the base; lowermost leaves mostly 6–8 cm long, oblanceolate, with bladed portions narrowed to broadly winged subpetiolar bases; other leaves sessile, often subclasping, mostly oblong or elliptic-oblong, apices obtuse to acute, often minutely mucronate, margins essentially entire; only the midvein prominent, upper surfaces usually sparsely pubescent, sometimes copiously short-pubescent on the midvein, lower surfaces shaggy-pubescent on the midvein, sometimes also on other veins, less so on the rest of the surface, short-pubescent on and near the margins. Inflorescence of heads paniculiform, sometimes relatively narrow and compact, more frequently with long spreading and arching branches, the heads mostly secund. Involucre 4–5 mm high, bracts glabrous, outermost short and acute, inwardly becoming linear and blunt apically. Achene 1–1.5 mm long, short-pubescent; pappus 2–2.5 times as long as the achene.

In moist to wet places, borders of shrub-tree bogs, borders of cypress-gum ponds or depressions, pine savannas and flatwoods, borders of swamps and wet woodlands, ditches, swales. Coastal plain, s. N.J. to s. pen. Fla., westward to La.

9. Solidago rugosa Mill.

Plant with a short, basal crown, producing, late in the season, slender rhizomes from crown buds. Stem sparsely to densely pubescent, in either case the hairs long to short; varying greatly in stature, from 4–20 dm tall, and in respect to diffuseness or lack thereof of the inflorescence; inflorescence paniculiform, sometimes its branches short, spreading-arching, not at all leafy-bracted, or leafy-bracted, floriferous throughout, heads strongly secund; branching varying to elongate, spreading or ascending, straight or arching, to a greater or lesser extent leafy or leafy-bracted proximally, floriferous distally, or floriferous throughout, heads secund or not. Leaves sessile, the principal stem leaves various; elliptic to elliptic-oblanceolate, strongly to moderately tapered to both extremities, sharply to bluntly toothed marginally, surfaces scarcely if at all rugose, varying to lance-ovate or ovate, short-tapered to rounded basally, broadly short-tapered apically, strongly rugose; in general, the upper leaf surfaces scabrid with short, appressed hairs, lower densely to sparsely shaggy-pubescent, sometimes only on the veins; leaves 2–10 cm long, 1–4 cm broad. Involucres 3–5 mm high, outer bracts short, more or less triangular, apices blunt, bracts inwardly gradually longer, linear-oblong to linear, the inner ones varyingly acute to rounded, margins of some or all bracts glabrous to finely and shortly pubescent on their margins. Achene narrowly obconic,

Fig. 378. **Solidago fistulosa:** a, midsection of stem; b, top of plant; c, flowering head; d, disc flower; e, achene.

822

1–1.5 mm long, sparsely to moderately pubescent; pappus 2–2.5 times as long as the achene. (Incl. *S. aspera* Ait., *S. rugosa* var. *aspera* (Ait.) Fern., *S. celtidifolia* Small, *S. rugosa* var. *celtidifolia* (Small) Fern.)

Fernald (1950), Cronquist in Gleason (1952), and Ahles in Radford et al. (1964) distinguish two to four varieties from the material as here described. The senior author of this work fails to find sets of characters correlating clearly enough to warrant recognition of varieties.

In both poorly and well drained places, bogs, moist to wet thickets, swales, lowland woodlands, their borders, and clearings, shoreline thickets and banks, upland, open woodlands of various mixtures, their borders, and roadside banks. Nfld. to Mich. and Mo., generally southward to Fla. Panhandle, s.e. Okla., e. and s.e. Tex.

22. Euthamia (FLAT-TOPPED GOLDENRODS)

Perennial, rhizomatous herbs, some kinds viscid or glutinous. Stems erect, usually freely branching in the upper ½ or ⅓ and forming a more or less flat-topped, corymbiform, leafy inflorescence. Leaves sessile, linear to narrowly lanceolate, margins entire but often finely scabrid, surfaces punctate; stem below the inflorescence commonly devoid of leaves by flowering time, thus measurements of larger leaves (having importance for identification) difficult or impossible to assess in herbarium material. Heads small, radiate, sessile or short-stalked in fascicles or glomerules terminating branchlets; ray flowers usually more numerous than those of the disc and usually not disposed evenly at the periphery of the head. Involucral bracts imbricated, 1-nerved, more or less glutinous, yellowish with green tips (tips drying yellow to brown). Receptacle of the head alveolate, the alveolae fringed with hairs, the flowers not subtended by chaff; both ray and disc flowers yellow, the former pistillate and fertile, ligule short, disc flowers bisexual and fertile, corollas narrowly cylindrical below and abruptly dilated to a campanulate or funnelform throat of about equal length, 5 short lobes at the summit. Achenes several-nerved, short-hairy; pappus a single series of capillary bristles.

A North American genus exhibiting perplexing variability. Authors have named a plethora of taxa, most of which are not now considered to have sufficient distinctness to warrant their recognition at either the specific or varietal level.

Reference: Sieren, D. J. "A Taxonomic Revision of the Genus *Euthamia* (Compositae)." Ph.D. Diss., Univ. of Illinois at Urbana-Champaign. 1970.

1. Leaf surfaces multipunctate, the punctae giving the surfaces a rough appearance (observed with suitable magnification); inflorescence branchlets with at least a little scabrosity.
  2. Principal leaves narrowly linear, 1.5–2.5 mm broad, only the midvein evident; involucres 3.5–4.5 mm high.   1. *E. minor*
  2. Principal leaves narrowly lanceolate, (2–) 3–5 mm broad, some of them usually with a faint vein paralleling the midvein on either side; involucres 5–6 mm high.   2. *E. tenuifolia*
1. Leaf surfaces very faintly if at all punctate, often, but not always, minutely bubbly or pustulose, surfaces smooth (observed with suitable magnification); inflorescence branchlets without scabrosity.
                                           3. *E. leptocephala*

1. Euthamia minor (Michx.) Greene.

Plants mostly 5–7 dm tall, infrequently to 10 dm, the stem glabrous apart from a few very short hairs just above the leaf axils, or sometimes irregularly and sparsely very short pubescent on the internodes, especially those of the inflorescence branches, lower inflorescence branches usually commencing well above midstem. Principal stem leaves narrowly linear, 1.5–2.5 mm broad and 3–6 cm long, only the midvein evident, usually with very short pubescence on the upper surface near the base, minutely scabrid marginally, sometimes on the midvein beneath; leaves below the inflorescence usually with short, axillary branchlets bearing small leaves ("axillary fascicles"), both the leaves and the axillary branchlets usually shed by flowering time; leaves of the inflorescence branches shorter and narrower, essentially filiform, mostly not over 1 mm broad; both leaf surfaces abundantly punctate, one or both surfaces often, not always, glistening-glutinous. Involucre nearly cylindric (turbinate after drying), 3.5–4.5 mm high, usually glutinous. Ray flowers 7–11 and disc flowers 3 or 4 in most heads. (*Solidago microcephala* (Greene) Bush)

Old fields and pastures, both well and poorly drained, pine savannas and flatwoods, swales, marshy shores, ditches. Coastal plain and outer piedmont, e. Mass. to s. pen. Fla., westward to Miss.

2. Euthamia tenuifolia (Pursh) Greene.

Plants 3–12 dm tall, in our range mostly 6–12 dm, the stem often sparsely very short-pubescent, a little more so just above the leaf bases, usually sparsely short-hispid or scabrid on the inflorescence branchlets; lower inflorescence branches often commencing below the middle of the stem; principal stem leaves narrowly lanceolate, (2–) 3–5 mm broad and 4–8 cm long, usually some of them with a faint vein paralleling the midvein on either side, infrequently with short branchlets in the axils, usually shed by flowering time; leaves of the inflorescence branches sometimes filiform, more commonly narrowly lanceolate, up to 4 mm broad; all blades short-scabrid marginally, sometimes short-pubescent on the midrib beneath, surfaces finely punctate, not at all or inconspicuously glutinous. Involucre turbinate, (4–) 5–6.5 mm high. Ray flowers 10–16 and disc flowers 5–7 in most heads. (*Solidago tenuifolia* Pursh; incl. *Euthamia hirtipes* sensu Sieren but not *Solidago* × *hirtipes* Fern.)

Marshes and banks of marshes (fresh, brackish, and salt), swales, pine savannas and flatwoods. Chiefly but not exclusively near the coasts, (N.S. to Va.?) N.C. to cen. pen. Fla., westward to La.

3. Euthamia leptocephala (T. & G.) Greene.

Plants 5–10 dm tall, the stem wholly glabrous, even the inflorescence branchlets, not glutinous. Principal leaves without short branchlets in their axils, narrowly lanceolate or lance-oblong, to 8 cm long and 7 mm broad, mostly with a faint to clearly evident vein paralleling the midrib on either side, glabrous apart from the minutely scabrid margins, faintly if at all punctate, the surfaces smooth even if minutely bubbly or pustulose. Involucres turbinate to campanulate, 5–6 mm high, sparingly glutinous. Ray flowers 7–14 and disc flowers 3–5 in most heads. (*Solidago leptocephala* T. & G.)

Swales, wet openings, lowland fields, glades, floodplains, moist to wet roadsides, meadows, prairies. s. Ill. and s.e. Mo., to Miss., La., e. Okla., e. Tex.

23. Bigelowia (RAYLESS-GOLDENRODS)

Perennial herbs with rosettes of basal leaves from a small caudex. Stems glabrous, viscid when young, 2–8 dm tall. Leaves alternate, simple, margins entire, linear to oblanceolate or elliptic, progressively reduced in size above the rosettes, the upper linear, shorter than the internodes. Heads discoid, in open, flat-topped corymbs, 2–6-flowered. Involucres glutinous, cylindric, 4.5–9 mm high; bracts lanceolate, appressed, weakly keeled, spirally arranged and more or less in vertical ranks, yellowish with a darker tip. Receptacle pitted or with a central cusp, without bracts (pales) subtending the flowers. Flowers bisexual and fertile, corollas yellow, lower tube nearly cylindric, upper slightly dilated and with 5 recurved lobes. Achene turbinate or subcylindric, sparsely pubescent; pappus of a single series of capillary bristles.

Reference: Anderson, Loran C. "Studies in *Bigelowia* (Astereae, Compositae) 1. Morphology and Taxonomy." *Sida* 3: 451–465. 1970.

- Plants definitely rhizomatous; all leaves narrowly linear, 1–2 mm broad.   1. *B. nuttallii*
- Plants cespitose or weakly rhizomatous; lower leaves narrowly to broadly oblanceolate or elliptic, 2–14 mm broad.   2. *B. nudata*

1. Bigelowia nuttallii L. C. Anderson.

Plant with relatively short rhizomes and forming loose mats, stems 3–6 dm tall. Rosette leaves numerous, linear, 6–13 cm long and 1–2 mm broad; stem leaves 12–23 in number, linear, the lowest 4.5–10 cm long, 1 mm broad, progressively reduced upwardly. Involucres 6–9 mm high. Flowers 3–5 per head, corollas 4–5 mm long. Achene 3–3.4 mm long; pappus 4–4.4 mm long. (*B. virgata* (Nutt.) DC., *Chondrophora virgata* (Nutt.) Greene, names misapplied)

Mostly occurring in relatively dry places, sandstone and siltstone outcrops where sometimes in temporary seepage, stream banks subject to temporary flooding and on rocks in streams (n.e. Ala.), prairies. s.cen. Ga.; Ga. piedmont; Fla. (one locality in Tampa Bay area, one in the Panhandle); n.e. Ala.; s.w. La. and e. and s.e. Tex.

2. Bigelowia nudata (Michx.) DC. Fig. 379

Plants with stems solitary or tufted from a short crown or rhizome. Rosette leaves narrowly linear-oblanceolate to broadly oblanceolate, their bases tapering, sometimes long-tapering, to petiolar or subpetiolar bases, 4–14 cm long overall, 4–14 mm broad; stem leaves 6–15 in number, the lower oblanceolate, 3–9 cm long, progressively reduced upwardly. Involucres 4.5–7.5 mm high. Flowers 2–5 per head, 3–5 mm long. Achene 1.2–2 mm long; pappus 2.8–3.3 mm long. (*Chondrophora nudata* (Michx.) Britt.)

Bogs, pine savannas and flatwoods, in or on borders of cypress-gum depressions. Coastal plain, N.C. to s. pen. Fla., westward to s.e. La. Two subsp. recognized by Anderson, distinguished by him as follows.

B. nudata subsp. **nudata.** Basal leaves mostly 10 cm long, over 4 mm wide; flowers 3–4 mm long; head 4.5–6 mm tall.

N.C. to n.cen. pen. Fla., westward to s.e. La.

B. nudata subsp. **australis** L. C. Anderson. Basal leaves often over 10 cm long, less than 4 mm wide; flowers 4–5 mm long; heads 6–7.5 mm tall.

n.cen. pen. to s. Fla.

24. Aster (ASTERS)

Mostly perennial herbs, one (of ours) with a woody stem, one (of ours) annual; rhizomatous when perennial or with a basal caudex. Leaves alternate, simple, in some perennial species with basal or rosette leaves dissimilar to those of the stem, the head-bearing branchlets more or less leafy in most. Inflorescence of few to numerous heads (heads rarely solitary), the head-bearing branchlets disposed in corymbs, panicles, or racemes, or in not at all well-defined inflorescences. Heads radiate; ray flowers few to numerous, pistillate and fertile, white, bluish white, or various shades of blue, violet, or purple, sometimes roseate; disc flowers few to numerous, bisexual and fertile, their corollas red, purple, yellow, or nearly white, tube narrowly cylindric below and expanding above to a funnelform to campanulate throat, 5-lobed at the summit. Receptacle of the head flat or slightly convex, without chaff. Involucres hemispheric, turbinate, campanulate, or subcylindric, bracts imbricated in several series, the outer often green at the tip and sometimes also on part of the midrib. Achene more or less flattened, usually with one or more ribs on the faces; in most the pappus of numerous capillary bristles in a single series, in some, however, the pappus in 2 distinct series, the outer series very much shorter than the inner and difficult to perceive.

Probably 65 or more species of *Aster* occur in our range. There is a great deal of variation, some of which results from hybridization; identification of specimens, particularly among closely allied species, is difficult. It is an uncommonly onerous task to construct a minimally satisfactory diagnostic key.

We have chosen to exclude from our treatment species inhabiting wet cliffs and ledges and other such places. Other species were not included either because we did not know them well or because available information on moisture preferences was too meager.

1. Plant with older parts of stems definitely woody, much branched divaricately and scrambling over other vegetation or forming its own tangles. 1. *A. carolinianus*
1. Plants definitely herbaceous.
  2. Leaves closely set at the base of the stem or on basal offshoots very much the longest, linear or

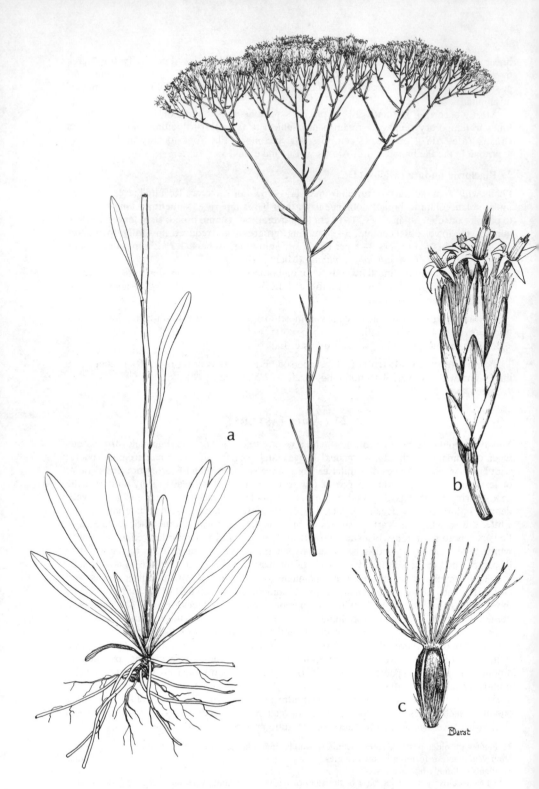

Fig. 379. **Bigelowia nudata:** a, plant; b, flowering head; c, achene.

nearly so and more or less grasslike, those above reduced and diminishing in size upwardly, all leaves persistent through the flowering period.

3. Involucre broadly hemispheric, 1.5–3 cm broad, tips of its bracts reflexed.   2. *A. eryngiifolius*
3. Involucre turbinate or turbinate-campanulate, 1–1.5 cm broad, its bracts erect.

4. Heads corymbosely disposed.

5. Involucral bracts conspicuously imbricated and graduated in length from the outside inwardly, the outermost shortest and subulate, all the bracts of about the same texture.   3. *A. chapmanii*
5. Involucral bracts imbricated but not markedly graduated in length, the outer and medial sub-equal, oblong and stramineous below then dilated to a lanceolate, lance-ovate, or ovate green tip, innermost bracts scarious throughout or tip barely greenish.   4. *A. paludosus*

4. Heads spicately or racemosely disposed.   5. *A. spinulosus*

2. Leaves not as described in first no. 2.

6. Pappus bristles in a single series; lowermost leaves not scalelike or bractlike.

7. Plant annual.   6. *A. subulatus*
7. Plant perennial.

8. Stem (falsely) subdichotomously branched, often zigzagging.   7. *A. tenuifolius*
8. Stem not at all giving the effect of dichotomous branching.

9. Leaves of midstem or upper stem (usually at least some of them) more or less clasping.

10. Bases of midstem or upper leaves (some of them at least) auriculate-clasping.

11. Margins of leaves entire, merely scabrid-ciliate.   8. *A. novae-angliae*
11. Margins of leaves at least obscurely toothed.   9. *A. puniceus*

10. Bases of midstem and upper leaves not auriculate at the clasping base.   10. *A. elliottii*

9. Leaves of midstem or upper stem not clasping.

12. Involucral bracts, at least the outer, commonly all, recurved or reflexed at their tips.   11. *A. novi-belgii*

12. Involucral bracts erect, their tips not at all recurved or reflexed.

13. Lobes of disc corollas ½–¾ as long as the expanded portion of the corolla tube and spreading or reflexed.   12. *A. lateriflorus*
13. Lobes of disc corollas scarcely more than ⅓ as long as the expanded portion of the tube, usually less, and erect.

14. Involucre 2.5–3.5 mm high; heads racemosely disposed and tending to be secund (this not usually evident in pressed and dried specimens); head-bearing branchlets 1–1.5 cm long.   13. *A. vimineus*
14. Involucre 4–8 mm high; heads racemosely disposed, not secund; head-bearing branchlets variable in length, mostly with some of them 2 cm long or more.

15. Stem usually with leaves persistent and little reduced nearly to the summit on the main axes through the flowering period; (viewed from beneath the leaf blade, the islets formed within the anastomosing veinlets irregular in shape but many or most of them considerably longer than broad).   14. *A. simplex*
15. Stem usually with few if any principal leaves persisting by the flowering period, leaves of upper axis or axes bractiform.

16. Islets formed within the anastomosing veinlets much longer than broad, or anasto-mosing veinlets not even evident.   15. *A. dumosus*
16. Islets formed by anastomosing veinlets evident (even on the upper bractiform leaves) mostly about as broad as long.   16. *A. praealtus*

6. Pappus bristles in two series, a relatively long inner series and a very short, not always easily perceived, outer series 1 mm long or less; lowermost stem "leaves" reduced to bractiform scales or bracts.

17. Stem and both leaf surfaces uniformly short-pubescent and atomiferous-glandular.   17. *A. reticulatus*
17. Stem glabrous below, often glaucous, sparsely short-pubescent on the inflorescence branches and sometimes a little below the branches; upper leaf surfaces glabrous or minutely scabrid, lower glabrous or pubescent only on veins.   18. *A. umbellatus*

1. Aster carolinianus Walt. CLIMBING ASTER.

Perennial, the principal stem and its branches woody, branchlets herbaceous at least distally, stem usually widely and diffusely branched, arching and scrambling over other vegetation to a height of 4 m or more, or forming a tangle of its own; larger woody stems to 1 cm thick, buff or tan, unevenly striate-ribbed; leafy flowering branches commonly divaricately spreading, copiously short-shaggy-pubescent. Larger leaf blades abruptly constricted to a suprabasal, shortly subpetiolar portion, the latter then flared on either side to a small but distinctly clasping auricle. Smaller leaves narrowed basally and similarly auriculate-clasping on either side; blades elliptic, acute apically, 2–6 cm long, to 1.5 (–2) cm broad, margins entire, surfaces sparsely to copiously short-pubescent. Involucre hemispheric or turbinate-campanulate, 7–12 mm high, bracts pubescent, strongly imbricated, the outer ones pale and linear-oblong below and with elliptic to subrhombic, dilated, often squarrose, green tips; inwardly bracts becoming gradually less dilated at their tips, the innermost linear and attenuated at their less conspicuously green tips, often lavender-pink along the scarious margins a little back of the tip. Rays numerous, ligules lavender-pink or lavender (often drying purple), 1–1.5 (–2) cm long. Achene columnar, brown, 2.5–3 mm long, glabrous.

Marshy shores, stream banks, marshes, edges of swamps and wet woodlands, mangrove thickets, often in water. Coastal plain, s.e. N.C. to s. pen. Fla., near-coastal portions of e. Fla. Panhandle.

2. Aster eryngiifolius T. & G.

Perennial with a short, basal crown which, in older plants, may attain a diameter of 4 cm or a little more, perennating by closely set basal offshoots, or the offsets sometimes shortly stemmed but scarcely rhizomatous. Stems solitary, or sometimes 2 to several from larger crowns, 3–8 dm tall, sometimes bearing a single head terminally, commonly with 2–4 heads in a terminal subcorymb, varying to 10–15 heads disposed more or less racemosely; stem shaggy-pubescent with relatively long hairs. Those leaves closely set basally much the largest, narrowly linear-oblanceolate, usually attenuate distally, a little tapered proximally from about the middle and not really petiolate, mostly 7–20 cm long and 3–10 mm broad; other leaves much shorter, usually fairly numerous, somewhat but not greatly reduced upwardly, erect or a little spreading, uppermost rigid and with spinose tips; leaf surfaces glabrous, venation parallel but veins other than the midvein and 1 or 2 to either side of it faint, margins entire or with widely spaced spinose to subulate teeth 1–6 mm long. Heads relatively large and hard. Involucre broadly hemispheric, 1.5–3 cm broad, bracts numerous, imbricated, thickish and stiff, green, subequal, subulate-attenuate, tips usually reflexed. Ray flowers numerous, their ligules 1–2 cm long, usually white, sometimes lavender-pink; disc flowers very numerous, their corollas yellow. Achene 1–2 mm long, narrowly fusiform in outline, brown, glabrous or sparsely pubescent, an inconspicuous rib on either edge and on either face.

Bogs, pine savannas and flatwoods, borders of cypress-gum depressions. cen. Fla. Panhandle, s.w. Ga., s.e. Ala.

3. Aster chapmanii T. & G.

Perennial with a short, basal crown, this sometimes bearing, besides the flowering stem, 2 or 3 closely offset rosettes of leaves. Stem glabrous, 6–10 dm tall, sometimes simple to a few-headed corymb, more commonly with a few elongate, erect, main branches each relatively little subbranched, long, slender, and flexuous branchlets bearing heads terminally. Those leaves closely set at the base of the stem much the largest, their nearly linear to narrowly linear-lanceolate or oblanceolate, thickish, bladed portions blunt to long-acute apically, gradually narrowed basally to petioles usually longer than the blades, 1–3 dm long overall, bladed portions with only the midvein evident; other leaves sessile, much reduced, gradually shorter upwardly, very narrowly linear to subulate, becoming bractiform on the branchlets, those just below the heads short-subulate and grading into the involucre; leaves glabrous. Involucre campanulate, 8–10 mm high, bracts strongly imbricated and appressed, sparsely pubescent on their backs and finely ciliate marginally, the outer short-subulate, grading longer inwardly, the inner linear-acute, all bracts with a green midstripe and scarious, usually purplish, margins, the tips of the innermost, even the green portion, suffused with purple. Ray flowers about 8–20, 1–1.5 cm long, violet; disc flowers yellow. Achene narrowly fusiform in outline, about 4.5 mm long, brown, glabrous or very sparsely pubescent, several-ribbed on each face.

Bogs, wet pine savannas and flatwoods, borders of shrub-tree bogs. cen. Fla. Panhandle, s.e. Ala.; e.cen. pen. Fla.

4. Aster paludosus Ait.

Perennial, producing slender, scaly rhizomes. Stems 2–8 dm tall, simple below the inflorescence, short-pubescent, pubescence denser and a little longer upwardly, lower stem sometimes becoming glabrous in age. Leaves of basal offshoots and lower stem with linear-lanceolate, somewhat grasslike, bladed portions gradually narrowed below to subpetiolate bases, to 20 cm long overall and 5–7 mm broad; upwardly leaves gradually reduced and becoming sessile; blades thickish and firm, glabrous, margins entire. Heads sometimes solitary and terminal, usually few in an open corymb. Involucre campanulate or turbinate-campanulate, 8–12 mm high; bracts loosely imbricated but little graduated in length, pubescent, the outer and medial with oblong, stramineous, thickish bases then dilated distally to lanceolate, lance-ovate, or ovate green tips, inner bracts essentially linear, tips obtuse or acute, scarious throughout or tips barely greenish. Ray flowers mostly 15–25, their ligules 1–2 cm long, lavender, violet, or purple; disc corollas yellow. Achene 3–4 mm long, slightly fusiform, glabrous or sparsely pubescent.

Pine savannas and flatwoods, lower slopes of sandy ridges, borders of and in branch bays. Chiefly coastal plain, N.C. to Ga.

5. Aster spinulosus Chapm.

Perennial with a short, basal crown or shortly rhizomatous, often bearing fibrous remnants of old leaf bases about the crown. Stem glabrous to spreading-pubescent, usually solitary, 3–7 dm tall, bearing a terminal spike or raceme of 2 or 3 to about 15 heads. Those leaves closely set at the base of the stem much the largest, stiff, with a linear, long-acute bladed portion 1.5–5 mm broad, only the midvein evident, gradually narrowed below and imperceptibly grading into a subpetiole, mostly 10–15 cm long overall, sometimes sparingly and irregularly ciliate marginally; other leaves much reduced, linear, gradually smaller upwardly. Heads all sessile or on stalks up to about 3 cm long, the uppermost at anthesis first. Involucre turbinate, or turbinate-campanulate, 6–8 mm high, bracts imbricated in several series, erect, subequal or graduated in size, the outer shortest, stiff, subulate, all or all but the innermost shortly spinose at the tip, all green except at their pale bases or the innermost sometimes scarious. Ray flowers 8–15, their ligules 1–1.5 cm long, violet to nearly white; disc corollas yellow. Achene fusiform in outline, about 2 mm long, pubescent.

Grassy openings in shrub-tree bogs, pine savannas, hillside seepage slopes in pinelands. Rare, perhaps only in Calhoun and Gulf Cos., Fla. Panhandle.

6. Aster subulatus Michx.

Glabrous, highly variable annual of diverse stature, 1–15 dm tall, few- to many-branched, branches widely spreading or strongly ascending, sometimes floriferous from the base of the plant upwardly, heads disposed more or less paniculately, subcorymbosely, or subracemosely. Principal leaves narrowly linear, lanceolate, or oblanceolate, up to 20 cm long and to 1 cm broad, rarely broader, margins entire; upwardly leaves reduced, most on the inflorescence branches subulate. Involucre turbinate, 5–8 mm high, bracts loosely imbricated, the outer subulate, grading inwardly to linear-attenuate, thin, greatly variable respecting how much green or purple coloration they have, the margins, save sometimes those of the outer bracts, scarious or hyaline. Ray flowers sometimes a little more numerous than the disc flowers, in general the proportionate number of each variable; ligules of the ray flowers variable in length, commonly circinate distally at least when dry, 0.2–6 mm long, although presumably not exhibiting that range of variation on a given plant, white, lavender, lavender-pink, or bluish; disc corollas yellow, their lobes often becoming purple after anthesis. Achene 1.5–2 mm long or a little more, fusiform in outline, light brown, sparsely pubescent, a pale rib on either edge and on either face. (Incl. *A. exilis* Ell., *A. inconspicuous* Less.)

In diverse moist to wet, usually open, places, saline to fresh marshes and marshy shores, spoil banks and flats, sandy shores, pools, mud flats, thickets, ditches, borders of wood-

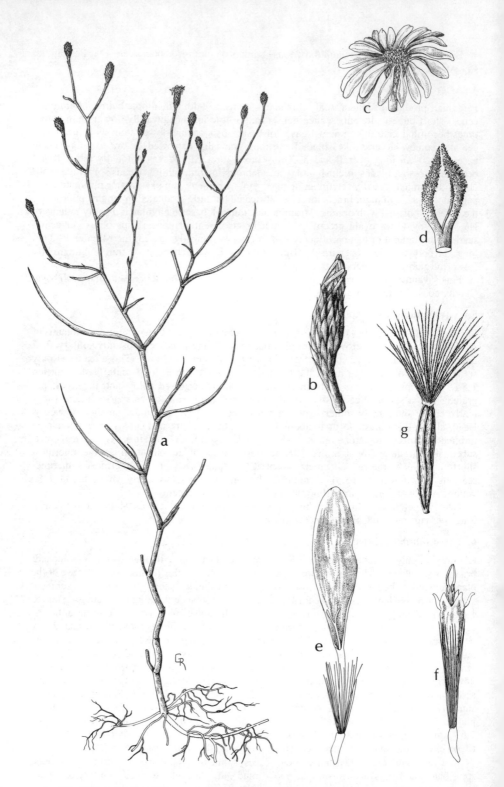

Fig. 380. **Aster tenuifolius:** a, habit; b, head in bud stage; c, head; d, stigmatic branches of style; e, ray flower; f, disc flower; g, achene.

830

lands, lowland fields. s. Maine to s. pen. Fla. and Fla. Keys, westward to Calif., in the interior to Mo. and Kan.; trop. Am.

7. Aster tenuifolius L.   Fig. 380

Glabrous, subsucculent perennial, sometimes slenderly rhizomatous, stems arising singly or 2 or 3 from the rhizomes, sometimes with a fibrous-rooted crown bearing a cluster of stems. Stem often somewhat zigzagging, (falsely) subdichotomously branched. Leaves few, linear or nearly so, 4–15 cm long and 1–5 mm broad, uppermost subbracteal to minutely bractlike. Heads terminating the few to numerous branches, distal portions of their stalks with short-subulate bracts grading into the involucres. Involucre turbinate, 4–9 mm high, outer bracts short-subulate, grading longer inwardly, the inner ones linear-acute or -acuminate, outer and medial usually green distally, the innermost scarious, sometimes with narrow, purplish margins. Rays 10–25, their ligules 5–8 (–10) mm long, often circinate distally at least when dry, white with a pale pink flush varying to lavender; disc corollas yellow, becoming reddish. Achene narrowly fusiform in outline, 2–4 mm long, tan, a rib on each edge and one on each face, sparsely pubescent. (Incl. *A. bracei* Britt. ex Small, *A. tenuifolius* var. *aphyllus* R. W. Long)

Salt and brackish marshes and sand-mud flats, lime-rock-pinelands, marl prairies, mangrove flats. Coastal, s. N.H. to s. pen. Fla. and Fla. Keys, Gulf coast to s.e. Tex.

8. Aster novae-angliae L.

Perennial with a short, stout, basal caudex or short, thickish rhizome. Stems usually somewhat clustered, 8–20 dm tall, copiously pubescent below with spreading, longish hairs, more shortly pubescent above, usually somewhat glandular-pubescent on the upper branches. Leaves numerous, only a little reduced upwardly, sessile and auriculate-clasping, mostly broadest basally and tapering to acute apices, varying to elliptic, larger ones 5–10 cm long and 5–15 mm broad, both surfaces uniformly short-pubescent, pubescence more scabrid above and somewhat softer beneath, margins entire but scabrid-ciliate. Heads in short, leafy-bracted panicles terminating branches. Involucre broadly hemispheric, 6–10 mm high and 15–20 mm broad, bracts somewhat unequal, subulate to lance-attenuate, commonly purplish, shortly glandular-pubescent. Ray flowers numerous, their ligules violet-purple, infrequently roseate or white, 1–2 cm long or a little more; disc corollas yellow to brown. Achene densely silky-pubescent.

Meadows, low thickets, shores, prairie swales, ditches, stream banks, low fields, open woodlands and their borders. Que. to s. Alb., generally southward to s. Appalachian highlands and prairies to Ala., Miss., Ark., e. Okla.

9. Aster puniceus L.

Perennial with a short caudex, sometimes perennating by slender rhizomes and colonial. Stems of variable stature, 4–25 dm tall, few- to much-branched above, branches angling-spreading, with coarse, stiffish pubescence on lower and midstem, hairs often tuberculate- or broad-based, shaggy-pilose on the branches (hairs sometimes shed on the lower stem by flowering time save for the bases which render the stems roughish). Lower stem leaves narrowed basally to winged subpetioles, others sessile, auriculate-clasping, lanceolate to oblong or elliptic-oblong, 7–20 cm long and 12–40 mm broad, short-scabrid above and along the edges, glabrous beneath or pubescent along the midvein, margins serrate to entire. Heads few to many, corymbiform-paniculate in a leafy inflorescence. Involucre hemispherical, 6–12 mm high, bracts loosely imbricated, all narrowly linear-attenuate or the outer ones foliose and narrowly oblanceolate with acuminate apices. Ray flowers numerous, their ligules blue, lilac, roseate, or rarely white, 1–1.5 cm long; disc corolla yellow at anthesis, becoming reddish.

Bogs, wet meadows, marshy shores, chiefly on the borders of swamps and wet woodlands, open stream banks, wet thickets, springy places. Nfld. to s. Man., southward (in our range) to N.C. and S.C. (chiefly piedmont and mts.), n. Ga., n. Ala., Tenn.

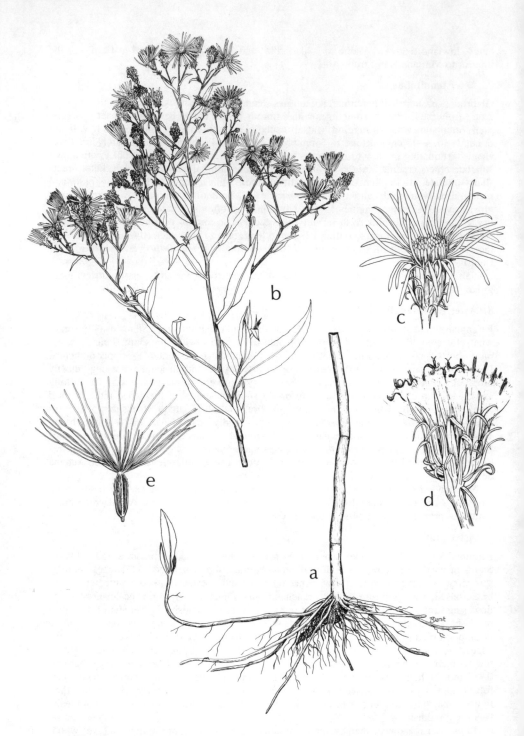

Fig. 381. **Aster elliottii:** a, base of plant; b, top of plant; c, flowering head; d, fruiting head; e, achene.

10. Aster elliottii T. & G. Fig. 381

Perennial with slender, often elongate, rhizomes and colonial. Stem of variable stature, commonly 1–2.5 m tall or a little more, up to 2 cm in diameter basally on the most vigorous specimens, glabrous below, short-pubescent above, the pubescence relatively evenly distributed or in lines or strips. Principal leaves gradually narrowed proximally to suprabasal, winged subpetioles then dilated at their extreme bases and sheathing the stem but not auriculate (lower and midstem leaves usually shed by flowering time, often, then, the sheathing bases persisting), 15–25 cm long overall; blades portions thickish, elliptic to elliptic-oblanceolate, acute apically, to about 5 cm broad, glabrous beneath, minutely scabrid above, margins serrate or dentate-serrate distally from their middles or a little below; leaves gradually reduced upwardly, becoming sessile and scarcely if at all sheathing. Inflorescence of heads corymbiform-paniculate, vigorous specimens with numerous strongly ascending branches and many heads, leafy at least below. Involucre turbinate-campanulate, 8–10 mm high, bracts loosely imbricate, narrowly linear-attenuate, green to purple distally, often irregularly short-pubescent marginally. Ray flowers numerous, their ligules lavender-pink, 1–1.5 cm long; disc corollas yellow at early anthesis, becoming red to purple. Achene purple, about 2.5 mm long, elliptic-oblanceolate in outline, nearly lenticular, a rib on either edge and one on either face, glabrous or sparsely pubescent.

Swamps and their borders, wet thickets, fresh to brackish shores and marshes, ditches. Outer coastal plain, s.e. Va. to s.e. Ga., throughout pen. Fla., coastal Fla. Panhandle to a little w. of Apalachicola.

11. Aster novi-belgii L.

Perennial with slender rhizomes. Stem usually slender, 2–10 dm tall, erect or sprawling, few- to much-branched, glabrous throughout varying to short-pubescent in lines or strips on the inflorescence branches, or pubescent just below the heads only. Leaves sessile, the principal ones elliptic, lanceolate, or lance-linear, infrequently oblong, some sometimes subclasping, 5–15 cm long and 4–25 mm broad, glabrous save for the minutely scabrid edges, margins obscurely toothed or entire; inflorescence, if few-branched and few-headed, with leaves much like those below but smaller, or, if multibranched and many-headed, the leaves numerous, much more bractiform, short and narrowly linear-oblong, arched-recurving. Involucre campanulate, 7–10 mm high, bracts loosely imbricated, oblanceolate, subequal, the outer usually, or sometimes all, with recurved tips, outer and medial ones green at least distally, the innermost usually purplish-scarious distally. Ray flowers numerous, their ligules 6–15 mm long, blue, violet, roseate, or nearly white; disc corolla yellow to reddish. (Incl. *A. elodes* T. & G.)

Meadows, shores, moist to wet thickets, borders of swamps and shrub-tree bogs, marshes, pine savannas. Nfld. to coastal plain of N.C. and S.C.

12. Aster lateriflorus (L.) Britt.

Perennial with a hard, basal, often shortly branched crown, perennating by basal offsets. Stems one or more from the crown, 3–12 dm tall, with few short branches or more commonly much branched, branches rather flexuous, arching to erect-ascending, very often with unreduced or little-reduced leaves on the main axis of the inflorescence and with numerous reduced leaves on the branchlets; stem commonly shaggy-pubescent, especially on the branches, sometimes glabrous throughout. Leaves of basal offshoots petiolate, petioles as long as the blades or a little longer, blades ovate, short-elliptic, or subrotund, 2–5 cm long, margins crenate-serrate distally from a little above the middle, apices broadly obtuse, bases shortly cuneate; principal leaves of the stem essentially sessile, lanceolate or lance-elliptic, tapered to both extremities (infrequently linear), 5–15 cm long and 5–30 mm broad, margins serrate, upper surfaces glabrous or short-scabrid especially toward the margins, wholly glabrous beneath or pubescent along the midvein; leaves or bracteal leaves along the axes of the flowering branches much smaller for the most part, tapered to both extremities, thin and flexuous. Heads on larger plants numerous, tending to be disposed subracemosely on the branchlets. Involucre turbinate-campanulate, 4–5 mm high, bracts loosely imbricated, erect, thin, essentially linear-oblong and little if at all broadened distally, apices mostly obtuse, a thin green midline on the proximal half, this dilating to an elliptic or rhombic green patch distally, sides rather broadly scarious except at the tip, often finely ciliate marginally around the tip. Rays relatively few, their ligules 4–6.5 mm long, often circinate distally at least when dry, usually white, varying to pale violet or lavender; throat and lobes of disc corollas yellow, commonly turning purplish, the lobes ½–¾ as long as the expanded portion of the corolla tube and usually spreading or reflexed. Achene about 2 mm long, slightly fusiform in outline, nearly flat and scarcely if at all ribbed, pubescent. (Incl. *A. agrostifolius* Burgess, *A. hirsuticaulis* Lindl., *A. spatelliformis* Burgess)

In diverse habitats, mostly in or associated with woodlands, commonly on the banks of and in floodplains of small woodland streams, river banks and floodplains, in wet woodlands and swamps and on their borders, thickets, bottomland fields, rich wooded slopes. N.S. to Minn., generally southward to cen. pen. Fla. and e. and s.e. Tex.

13. Aster vimineus Lam.

Rhizomatous perennial. Stems 4–15 dm tall, usually well branched, branches slender and relatively flexuous, divergent, generally floriferous throughout, short-pubescent. Basal leaves petioled, blades elliptic to lance-ovate, crenate-serrate marginally; lower and often median stem leaves with bladed portions narrowed to subpetioles, or sessile upwardly, bladed portions or blades lanceolate or linear-lanceolate, margins obscurely serrate, acute apically, principal ones 5–10 cm long overall and 3–10 mm broad, upper surfaces glabrous or short-scabrid, lower glabrous or short-pubescent; basal and principal stem leaves or most of them commonly shed by flowering time; leaves or bracteal leaves of the flowering branches and branchlets much reduced, narrowly oblanceolate, lanceolate, or nearly linear, acute apically. Heads racemosely disposed, commonly secund on the branchlets (this not very evident when specimens pressed and dried), the head-bearing branchlets mostly 1–1.5 cm long. Involucre 2.5–3.5 mm high, bracts imbricated, erect, all essentially linear, outermost shortest, at least the innermost acute or acuminate apically, most bracts with a green midstripe proximally, this gradually dilated to an oblanceolate-acute patch distally, sides scarious except at the tip, usually finely pubescent marginally on the distal halves. Ligules of the ray flowers usually white, sometimes purplish; expanded portions of the tubes and the lobes of the disc flowers yellowish, often turning purplish, the lobes about ⅓ the length of the expanded portion of the tube and erect. Achene narrowly fusiform in outline, 1.5–2 mm long, buffish, a faint rib on each edge and each face, pubescent. (Incl. *A. brachypholis* Small, *A. racemosus* Ell.)

River banks and floodplains, alluvial woodlands, fields and clearings, meadows, shores. Chiefly near the coast or on the coastal plain, Maine to n. Fla., westward to La., northward in the interior to s. Ohio, w. Mo., s.e. Okla.

14. Aster simplex Willd.

Rhizomatous perennial. Stems 6–15 dm tall, short-pubescent above in lines or strips, the main stem axis usually carrying about to the summit of the paniculate inflorescence and usually remaining leafy during the flowering period with leaves little reduced well toward the summit. Principal leaves lanceolate, infrequently nearly linear, 8–15 cm long, (3–) 8–20 (–35) mm broad, usually all much longer than broad, mostly acute at both extremities, margins varyingly sharply serrate, obscurely serrate, or entire, both surfaces glabrous, or the upper surfaces scabrous, edges scabrous; islets formed by the reticulum of veinlets (best viewed from beneath) irregular in shape but most or many of them considerably longer than broad. Branches and branchlets of the inflorescence only moderately floriferous, head-bearing branchlets 2–20 (–25) mm long. Involucre 4–6 mm high, bracts loosely imbricated, outermost shortest, outer and medial ones nearly linear, the inner slightly the broadest just back of the tip, outer and medial blunt apically, some or all of the inner ones usually acute, the extent of the green coloring varied and the green or nongreen portions, or both, sometimes suffused with purplish pigment. Ligules of the ray flowers usually white, occasionally lavender-pink or blue, especially near their tips; disc corollas yellowish or the lobes purple-tinged, lobes less than 1/3 to a little less than 1/2 as long as the expanded portion of the tube and usually erect. Achene narrowly fusiform, 2–3 mm long, light brown, faintly ribbed on each edge and on each face, pubescent. (*A. lamarckianus* Nees)

Wet meadows, river and creek banks, along sloughs, swales, ditches, alluvial thickets, shaded banks and woodland borders. N.S. to N.D., generally southward to Ga., Ark., Okla.

15. Aster dumosus L.

Perennial, with a short basal crown usually bearing a single stem, or freely and slenderly rhizomatous and somewhat colonial. Stem 3–15 dm tall, commonly freely and diffusely branched, sometimes relatively little branched, branches strongly ascending to subdivaricate, short-pubescent throughout or becoming glabrous below. Basal and lowermost stem leaves with oblanceolate bladed portions, gradually narrowed proximally to subpetioles, margins obscurely serrate, apices obtuse to acute; other principal stem leaves sessile, narrowly oblanceolate to linear, 3–12 cm long and 2–8 mm broad, margins entire or remotely toothed, both surfaces glabrous or the upper more or less scabrid; lower and principal stem leaves usually shed by flowering time; leaves of the branches and branchlets much reduced and bractiform, usually numerous, commonly thickish and stiff, oblongish, seldom over 1 cm long, all blunt at their tips and divaricate or reflexed on many plants, acute and ascending on others. Heads racemosely or subracemosely disposed, on the branches, the head-bearing branchlets usually variable in length, about 1–8 (–15) cm, generally at least some of them 2 cm long or more on any given plant. Involucre turbinate to campanulate, 4–8 mm high, bracts loosely imbricated, erect, graduated in length from the outside inwardly, outer shortest, all oblongish but some of the inner ones usually at least slightly broadened back of their tips, obtuse and sometimes minutely mucronate apically on most plants, infrequently the innermost acute; bracts with a thin green line on the proximal half, this broadening to an elliptic or oblanceolate patch of green distally, scarious on the sides except at the tip, sometimes more or less suffused with purplish pink pigment. Ligules of the ray flowers white, pale blue, or pale lavender, often circinate distally at least when dry; expanded portions of the tube and lobes of disc corollas yellow, usually turning purplish, lobes usually less than 1/3 as long as the expanded portion of the tube and erect. Achene 1.5–2 mm long, slightly fusiform, straw-colored, a rib on either edge and a less prominent, sometimes faint, one on either face, glabrous or pubescent. (Incl. *A. coridifolius* Michx., *A. gricilipes* (Wieg.) Alexander)

In both well-drained and poorly drained pinelands, adjacent ditches, borders of cypress-gum ponds and depressions and of shrub-tree bogs, thickets and clearings, shores, borders of upland woodlands, meadows, pastures, prairies, and glades. Mass. to Mich., generally southward to s. pen. Fla. and e. and s.e. Tex.

16. Aster praealtus Poir.

Perennial with elongate rhizomes and colonial. Stem 6–20 dm tall, pubescent in lines or strips or rather uniformly pubescent above, glabrous below at least in age. Principal leaves sessile, lanceolate, lance-elliptic, or linear-elliptic (usually shed by flowering time), thickish, 4–10 cm long and up to 10 mm broad, margins entire but scabrid, upper surfaces usually scabrid, glabrous or uniformly short-pubescent beneath, islets formed by the reticulum of veinlets (best viewed from beneath) irregular in shape but most of them about as broad as long. Inflorescence paniculate, often elongate and abundantly floriferous, branches strongly ascending or less frequently spreading, usually copiously bracteate-leafy, bracts mostly acute, on the head-bearing branchlets grading into the involucre; head-bearing branchlets 5–30 mm long. Involucre hemispheric (5–) 6–8 mm high, bracts loosely imbricated, outer shortest, subulate, grading inwardly to linear-acute, sometimes some of the inner ones broadest just back of the tip, the extent of their green coloring variable. Ligules of the ray flowers blue-violet to purple, rarely white; expanded portion of the tube and lobes of the disc corollas yellow, usually turning purplish, the lobes 1/4 as long as the expanded portion of the tube or less and erect. Achene curvate-fusiform, about 1.5 mm long, straw-colored, scarcely if at all ribbed, sparsely short-pubescent. (*S. salicifolius* sensu Small, 1933)

Marshes, wet meadows, lowland woodlands, fields, and clearings, prairies, swales, moist to wet banks and thickets along streams, ditches. Overall range not clear (to us); in our range, occurring in Tenn., n.w. Ga., Ala.(?), Miss. and La.; one known locality for Fla. Panhandle where probably introduced (in any event not persisting).

17. Aster reticulatus Pursh.

Perennial with a hard, branching rootstock from which several to numerous stems arise. Stems 4–8 dm tall, short-shaggy-pubescent and atomiferous-glandular, simple below the corymbose inflorescence. Lowermost stem bearing short, very broadly triangular bracts, upwardly bracts becoming longer and more leaflike, largest leaves at about midstem then gradually reduced upwardly, lower inflorescence branches leafy-bracted; principal leaves sessile, narrowly obovate, oblong-obovate, elliptic, or lance-ovate, mostly shortly and broadly tapered apically, sometimes acute, mostly 3–6 cm long and 1–3 cm broad, upper surfaces pubescent with very short hairs, the lower relatively coarsely reticulate-veined and copiously short-shaggy-pubescent, both surfaces atomiferous-glandular, margins entire or very rarely irregularly undulate or undulate-dentate. Involucre campanulate-hemispheric, 6–8 mm high, bracts erect, pubescent and often atomiferous-glandular, loosely imbricated, all linear-subulate, outer about half as long as the inner, gradually increasing in length inwardly, with a narrow, greenish midstripe, pale to either side, often partially and irregularly suffused with reddish pigment. Ray flowers about 8–20, rather unevenly disposed, their ligules 10–15 mm long, white or yellowish white, mostly minutely toothed at their tips. Lobes of disc corollas pale yellow, sometimes turning reddish purple. Achene narrowly fusiform to narrowly oblanceolate in outline, several-ribbed, about 4 mm long, brown and copiously pubescent, usually atomiferous-glandular; pappus of 2 series of bristles, the outer series very short, 1 mm long or less and not easily discerned, the inner 5–6 mm long. (*Doellingeria reticulata* (Pursh) Greene)

Boggy places in pinelands. Coastal plain, s.e. S.C., Ga., s.cen. pen. Fla., Fla. Panhandle westward to at least Walton Co.

18. Aster umbellatus Mill.

In general features similar to *A. reticulatus*. Stem 10–20 dm tall, glabrous and often glaucous below, sparsely short-pubescent on the inflorescence branches and sometimes somewhat below the inflorescence. Midstem leaves very short-petioled, blades lanceolate or narrowly elliptic, acute or acuminate apically, cuneate basally, 4–6 times as long as broad, varying to broadly elliptic, ovate-elliptic, or ovate, often broadly tapered at both extremities, or the base rounded or nearly so, 1.5–4 times as long as broad; upper surfaces glabrous or minutely scabrid, lower finely reticulate-veined, pubescent on the principal veins, margins entire and scabrid. Involucral bracts more or less pubescent marginally. Two varieties may be distinguished, as follows.

A. umbellatus var. **umbellatus.** Midstem leaves generally of the longer, narrower type as described above; involucral bracts thin and slender, usually most of them acute apically, little if any over 0.5 mm broad. (*Doellengeria umbellata* (Mill.) Nees.)

Meadows, thickets, wet to dry woodlands, bogs. Nfld. to Minn., generally southward to Va., Ky., in our range along the s. Appalachian mts. (rarely piedmont) of N.C., Tenn., n. Ga., n.e. Ala.

A. umbellatus var. **latifolius** Gray. Midstem leaves generally of the shorter, broader type as described above; involucral bracts relatively thick, obtuse to rounded apically, the broader ones 1 mm broad or a little more. (*A. umbellatus* var. *brevisquamus* Fern., *A. sericocarpoides* (Small) K. Schum., *Doellengeria humilis* (Willd.) Britt.)

Creek banks, low thickets in pinelands, bogs, open upland woodlands of various mixtures. Chiefly coastal plain, locally on the piedmont, N.C. to w. Fla. Panhandle, westward to Ark., s.e. Okla., e. Tex.

25. Erigeron (FLEABANES)

Herbs. Leaves alternate, all basal in some species, sessile or narrowed to subpetioles, entire, toothed, or lobed. Heads radiate, solitary terminating the stem or its branches, or few to numerous in corymbs. Involucres mostly hemispheric, bracts narrow, herbaceous, equal and in 1 series or little imbricated, without darker green tips (as in *Aster*). Receptacle flat or slightly convex, without chaff. Ray flowers pistillate and fertile, their ligules white, shades of blue, lavender, purple, or rose; disc flowers bisexual and fertile, their corollas yellow, with a short basal tube broadening to a slightly expanded throat 5-lobed at the summit. Achene slightly compressed, 2 (–4) -nerved; pappus of several capillary bristles, with or without an outer series of shorter bristles or chaffy scales (in some species the pappus of the ray and disc flowers unlike).

1. Stem and leaves, if pubescent, the hairs short; ray flowers mostly 20–30 per head; achene columnar or nearly so, scarcely flattened, 4-ribbed, without a pale, bony ring basally.   1. *E. vernus*
1. Stem and leaves shaggy-pubescent with longish hairs; ray flowers 50 or more per head; achene flattened, oblanceolate or narrowly obovate in outline and with each margin banded, with a pale, bony ring basally.
  2. Plant perennating by stolons; ray flowers 50–75 per head; achenes mostly 1.5 mm long or a little more.   2. *E. pulchellus*
  2. Plant without stolons; ray flowers 100 or more per head; achene about 1 mm long.
     3. *E. philadelphicus*

1. Erigeron vernus (L.) T. & G.

Perennial with a short, hard caudex or rhizome from which, usually, 1–3 simple, slender, flowering stems arise. Stem 1–6 dm tall, glabrous or sparsely short-pubescent below the inflorescence, branches of the latter sparsely short-pubescent. Leaves thickish, margins entire, undulate, or with a few dentations or serrations, glabrous or sparsely clothed with short pubescence, often minutely punctate and often with minute, grayish, scaly excrescences, especially above; larger leaves at or very near the base of the stem, oblanceolate to obovate, infrequently suborbicular, narrowed proximally to subpetioles, mostly obtuse apically, variable in size up to about 8 cm long, infrequently to 15 cm; stem leaves few, much reduced, and bracteate. Heads mostly 3 to about 20 in a loose, open corymb. Involucre about 5 mm high, 5–10 mm broad, bracts linear-subulate, somewhat glutinous, sometimes with a few small pustules exteriorly, glabrous or sparsely short pubescent, tips often pale pinkish or purplish. Ray flowers 20–30 per head, their ligules oblanceolate, 5–8 mm long, usually white, rarely pale lavender; disc corollas 2–3 mm long or a little more. Achene columnar, scarcely flattened, 4-ribbed, about 1.5 mm long, brownish, slightly lustrous, pubescent; pappus a single series of capillary bristles.

Wet, sandy-peaty soils of pine savannas and flatwoods, ditches, bogs. Coastal plain, s.e. Va. to s. Fla., westward to La.

2. Erigeron pulchellus Michx. ROBIN'S-PLANTAIN.

Perennial with a short caudex and perennating by slender stolons. Flowering stems shaggy-pubescent, usually solitary, 1–6 dm tall, bearing a single terminal head, or few heads in a loose terminal corymb. Leaves shaggy pubescent on both surfaces, the upper surfaces sometimes becoming glabrous or nearly so; larger leaves basal, sessile or narrowed to short, subpetiolar bases, oblanceolate, obovate, or suborbicular, apices rounded, 2–15 cm long, 1–3.5 (–5) cm broad, margins entire or crenate-serrate; other leaves few, usually widely spaced, diminishing in size upwardly, sessile and subclasping, oblanceolate to oblong, or the uppermost sometimes lanceolate to ovate. Involucre 5–7 mm high, 1–2 cm broad, bracts pubescent except at their acute, often purplish, tips. Ligules of the 50–75 ray flowers 6–10 mm long, about 1 mm wide, bluish to pink, or less frequently white; disc corollas 4.5–6 mm long. Achene flattened, banded on either margin, narrowly obovate in outline, 1.5 mm long or a little more, pale yellowish and sublustrous; pappus a single series of capillary bristles.

Rich wooded slopes, coves, stream banks, moist open places, rocky, often dripping, ledges and slopes. Maine to Minn., generally southward (rarely coastal plain), to Ga. and e. and s.e. Tex.

3. Erigeron philadelphicus L.

Biennial or short-lived perennial with a short caudex, commonly with a single stem, less frequently with 2 to several stems. Stems of variable stature, 2–12 dm tall; stems and leaves usually copiously to sparsely shaggy-pubescent. Basal and lower stem leaves oblanceolate to narrowly obovate, narrowed proximally to winged subpetioles, rounded to obtuse apically, margins variable, entire, crenate, crenate-serrate, or pinnately lobed, if lobed, lobes rounded to acute, varying in length from about 3 to 15 cm; other leaves sessile, commonly clasping, more or less reduced, oblongish, oblanceolate, or sometimes lanceolate, margins entire, crenate, or crenate-serrate, rarely, if ever, lobed. Heads few to numerous, in a loose, open corymb. Involucre 4–7 mm high, 8–15 mm broad, bracts linear-oblanceolate, acute apically, moderately pubescent with flat hairs, varying to subglabrous. Ray flowers mostly 100 or more per head, white to pink or rose-purple, their ligules 5–10 mm long, filiform below and a little broadened distally; disc corollas 2.5–3.2 mm long. Achene about 1 mm long, flattened, oblanceolate in outline, narrowly banded on each margin, tan and somewhat lustrous, narrowed at the base to a pale, shiny, bony ring; pappus a single series of capillary bristles.

Moist to very wet, grassy openings, wet meadows, floodplain and other lowland woodlands, stream banks, bottomland fields and pastures, moist to wet roadsides and thickets, seepage areas. Widespread in temp. N.Am., in our range absent from pen. Fla.

26. Baccharis (GROUNDSEL-TREES)

Shrubs (ours), often resinous. Leaves alternate, simple, petioled or sessile, 1–3-nerved, margins entire, toothed, or lobed. Heads discoid, relatively small, many-flowered, solitary or more commonly 2 to several in stalked clusters borne from axils of leaves or reduced bracteal leaves of the branchlets, often giving the appearance of being paniculiform. Involucres of imbricated, pale, papery bracts. Receptacle of the head pitted, minutely fringed, or smooth, flowers without subtending bracts (chaff). Flowers unisexual (pistillate) and functionally unisexual (staminate), plants functionally dioecious. Flowers of functionally staminate heads structurally bisexual, ovary aborting, corolla filiform below dilated a little below the summit and 5-lobed, a pappus of one series of crisped capillary bristles surmounting the abortive ovary and not evident beyond the involucre; flowers of pistillate heads with corollas only slightly dilated distally from about the middle, truncate at the summit or with minute teeth, the straight pappus bristles conspicuously extending beyond the involucre and the corollas, the collective pappuses of the numerous pistillate heads making the pistillate plants very much more conspicuous at flowering time than are the staminate plants. Achene cylindric or nearly so in outline, 5–10-ribbed.

1. Leaves nearly linear, almost needlelike, 1–3 mm broad.   1. *B. angustifolia*
1. Leaves broader, the larger ones of a given plant rarely less than 1 cm broad.
  2. Clusters of heads scattered along leafy branchlets and not giving the effect of a paniculiform inflorescence; all or almost all heads sessile; leaf blades without punctate glands or these inconspicuous.
         2. *B. glomeruliflora*
  2. Clusters of heads aggregated on distal portions of the branchlets giving the effect of a paniculiform inflorescence; many or all heads with short but definite stalks; leaf blades with conspicuous, pale amber, glandular punctations.   3. *B. halimifolia*

1. Baccharis angustifolia Michx. FALSE-WILLOW.

Plants much branched, to 4 m tall, glabrous. Leaves somewhat leathery (drying wrinkled), bright lustrous green, not or inconspicuously glandular-punctate, nearly linear and almost needlelike (tapering slightly to both extremities), 2–4 (–6) cm long, 1–3 mm broad, only the midvein evident, margins entire. (Leaves of an occasional plant somewhat broader and narrowly oblanceolate, having 1–3 salient teeth on a side, we suspect may be hybrids or hybrid derivatives of this and *B. halimifolia.*) Heads sessile and with stalks of variable lengths to about 15 mm, borne distally on branchlets. Involucres mostly about 4 mm high, campanulate. Achene slightly over 1 mm long, with 10 pale ribs, the intervals between the ribs brown; pappus about 10 mm long.

Borders and banks of salt and brackish marshes, mangrove swamps, and sloughs, coastal or near-coastal shores and spits, shrub-tree islands in coastal marshes, coastal hammocks. Seacoasts, N.C. to Atlantic and Gulf coasts of Fla.

2. Baccharis glomeruliflora Pers.

Plant loosely branched, the branches angling-ascending, to 2 m tall, glabrous. Leaves mostly cuneately narrowed to subpetiolar bases, the expanded portions usually bright green and sub-lustrous, thickish, the larger ones obovate to oblanceolate, mostly with 1–3 salient, dentate-serrate teeth on one or both sides from or from above the middle, the tip triangular to mucro-nate above the teeth; the larger leaves 3–4 cm long overall and 1.5–2 cm broad distally; leaves of flowering branches similar in shape, usually smaller, less distinctly toothed or with-out teeth. Clusters of heads scattered along the branchlets and not giving the effect of a panic-uliform inflorescence; heads all or nearly all of them sessile. Involucres 8–10 mm high, the bracts not resinous, their tips rounded or blunt. Achene 1.2–1.5 mm long, with 10 pale ribs, the intervals light brown; pappus about 8 mm long.

Moist to wet flatwoods hammocks, depressions in pine flatwoods, banks of marshes. s.e. N.C. to s. Fla., Fla. Panhandle.

3. Baccharis halimifolia L. GROUNDSEL-TREE, SILVERLING.

Plant freely branched, the branches strongly ascending, 1–4 m tall, glabrous. Leaves usually similar in shape and size to those of *B. glomeruliflora* but commonly a more dull, grayish green and with conspicuous, pale amber, glandular punctations. (Plants in what appear to us to be randomly scattered localities in the range may have leaves narrowly oblanceolate or spatulate and entire or barely toothed margins. These may represent what have been segre-gated as var. *angustior* DC.; we doubt that they are more than sporadic forms.) Clusters of heads aggregated on distal portions of branchlets giving the effect of paniculiform inflores-cences; many or all heads with short but definite stalks. Involucres 5–6 mm high, campanu-late, bracts mostly somewhat resinous, their tips blunt or the innermost and sometimes the medial acute. Achene about 1.2 mm long, with 10 pale ribs, the intervals honey-colored; pappus about 10 mm long.

Marshes and banks of marshes, shores, swales, old fields and various disturbed places. Coastal plain and piedmont, Mass. to s. Fla., westward to Tex., Ark., Okla.

27. Boltonia

Perennial herbs, perennating by basal offshoots, stolons, or rhizomes, glabrous or essentially so, few to diffusely branched. Leaves alternate, sessile, the lower and median (often not present by flowering time) narrowly elliptic to nearly linear, usually somewhat tapered to both extremities. Inflorescence of heads few-branched and with relatively few heads to dif-fusely branched and with many heads, the branching more or less paniculate. Heads radiate, with about 20–60 pistillate and fertile, white to pink or blue ray flowers and numerous, bisexual and fertile, yellow, disc flowers each with a narrow tube basally and an expanded throat 5-lobed at the summit. Involucre more or less hemispheric, bracts imbricated. Achenes of disc flowers laterally much compressed, narrowly to broadly winged on either side of the seed-bearing achene body, those of the ray flowers with the achene body thicker longitudinally on one side than the other, this bearing 2 wings, the thin edge similarly winged; pappus consisting of several minute bristles or scales above the achene body and usually short awns over the wings, the latter sometimes not developed.

Plants of *Boltonia* occur in wet places from N.Eng. to Man. and N.D., generally southward to s. pen. Fla. and Tex. Authors have differed not a little as to how many taxa may be distinguished, at what level, specific or varietal, the varieties variously shuffled amongst whatever species were recognized. The latest comprehensive treatment is Judy Tate Morgan, "A Taxonomic Study of the Genus *Boltonia* (Asteraceae)," Ph.D. diss., Univ. of North Carolina at Chapel Hill (1967).

The senior author of the present work has failed in his attempt to prepare a treatment for *Boltonia* for the geographic area of coverage. The number of specimens readily available to him is relatively meagre (approximately 100) and using them he has been unable to interpret what other authors have done in their treatments. Those using this book and seeking to identify specimens of *Boltonia* will have to refer to regional floras or manuals or the revisionary treatment of Morgan cited above.

28. Pluchea (MARSH-FLEABANES, STINKWEEDS)

Annual or perennial herbs (ours) or shrubs, emitting a strongly fetid odor. Stems glabrous, glandular-pubescent, or irregularly cottony-pubescent. Leaves alternate, simple, petioled or sessile, decurrent in some (none of ours) rendering the stem winged, blades unlobed. Heads discoid, borne in panicled cymes or corymbs, each head with few to numerous, heterogamous flowers; central flowers of the head relatively few in number, bisexual but sterile, their corolla tubes dilated in the upper third where cylindrical or slightly campanulate, with 5 deltoid lobes at the summit; outer flowers much more numerous, pistillate and fertile, their corollas filiform-tubular and only slightly dilated distally, with 3 slender lobes at the summit. Involucre of imbricated bracts. Receptacle naked. Achenes not exceeding 1 mm long, dark brown or reddish brown, 4–6-angled or prominently ridged and grooved, with a small whitish enlargement (caruncle) basally; pappus a single series of barbellate bristles.

1. Principal leaves definitely petioled or the blades strongly cuneately narrowed to winged subpetiolar bases.
    2. Central portion of the inflorescence of heads exceeding the lateral floriferous branches, paniculiform, not storied or flat-topped; involucral bracts conspicuously atomiferous-glandular on their backs, only the outer ones short-pubescent and -ciliate.   1. *P. camphorata*
    2. Central portion of the inflorescence of heads exceeded by some of the lateral floriferous branches, more nearly cymiform, storied and more or less flat-topped; outer and medial involucral bracts short-pubescent over their backs, atomiferous-glands usually sparse and inconspicuous.   2. *P. odorata*
1. Principal leaves sessile, mostly somewhat clasping but if tapered then still essentially sessile.
    3. Corollas rose-pink or rose-purple.   3. *P. rosea*
    3. Corollas creamy white.
        4. Fully developed heads 6–8 mm high, as broad as high; medial involucral bracts 2.5–3 times as long as broad, acute apically.   4. *P. foetida*
        4. Fully developed heads 10–12 mm high, twice as high as broad; medial involucral bracts 1–2 times as long as broad, broadly obtuse apically.   5. *P. longifolia*

1. Pluchea camphorata (L.) DC.

Annual or perennial, to about 15 dm tall. Stems and inflorescence branches short-pubescent, stems becoming glabrous below. Leaves (except sometimes the uppermost) narrowed to petioles 1–2 cm long, blades elliptic to oblong-elliptic, 6–15 cm long, 3–7 cm broad, acute or acuminate apically; margins dentate-serrate or crenate-serrate, both surfaces atomiferous-glandular, the lower more so than the upper, short-pubescent on the veins beneath. Inflorescence of heads in paniculiform clusters, the central portion exceeding the lateral floriferous branches, not storied or flat-topped. Heads campanulate, about 5 mm high, 3–4 mm broad. Median and inner involucral bracts conspicuously atomiferous-glandular, only the outermost short-pubescent and sometimes short-ciliate; outermost bracts ovate-obtuse the medial oblong-obtuse, varying to linear-oblong and obtuse or acute on the inside, all usually more or less purplish. Corollas lavender-pink to lavender-purple. (*P. petiolata* Cass.)

Moist to wet bottomland woodlands and clearings, river banks, swamps, marshy shores, wet meadows, swales and thickets, ditches. Chiefly coastal plain and piedmont, Del. and

Md., southward to n. Fla., westward to e., s.e. and n.cen. Tex., northward in the interior to s. Ohio and e. Kan.

2. Pluchea odorata (L.) Cass.

Annual, to 1.5 dm tall. Stem and inflorescence branches shortly glandular-pubescent, stem becoming glabrous below. Leaves (excepting the uppermost) usually cuneately narrowed to petioles from 5–20 mm long, blades short-ovate, ovate-lanceolate, lanceolate, or elliptic, mostly obtuse apically, 4–15 cm long and 1–7 cm broad; margins unevenly serrate, serrate-dentate, or crenate-serrate, surfaces atomiferous-glandular and sparsely to densely short-glandular-pubescent. Inflorescence of heads cymose, some of the lateral floriferous branches surpassing the central axis, more or less storied overall and flat-topped. Heads campanulate, 4.5–7 mm high and about as broad. Outer and medial involucral bracts short-pubescent on their backs, atomiferous glands sparse to numerous, the inner bracts usually variably pubescent to nearly glabrous; outer bracts ovate-obtuse, medial ones oblong-obtuse, inner linear-oblong and acute, all usually more or less rose-purplish. (*P. purpurascens* (Sw.) DC.)

In the range as a whole in diverse habitats: salt and brackish marshes, fresh marshes, wet marl prairies and glades, depressions in pine flatwoods, swales, wet hammocks, mud flats, saline and akaline flats. s.e. Mich. and s. Ill.; coastal plain, n.e. Mass. to s. Fla. and Fla. Keys, westward throughout the Gulf states, in Okla., Ark., Kan., from Tex. to Calif., s.w. Utah, Nev.; Mex., C.Am., n. S.Am.; Berm., W.I.

3. Pluchea rosea Godfrey var. rosea.

Perennial with 1 to several stems from basal crown. Stem and leaf surfaces with a mixture of sessile, atomiferous glands and sparse to dense viscid hairs, inflorescence branches viscid-tomentose. Leaves sessile, the lower with truncate to cuneate-truncate bases, median larger, oblongish, truncate to subclasping basally, obtuse apically, 2–7 cm long and 0.5–3 cm broad, upper leaves a little smaller but similar; margins shallowly apiculate-toothed. Heads few to many, in few compact corymbose clusters terminating branchlets, varying to many relatively loose clusters in a wide, more or less storied arrangement. Involucres campanulate to turbinate-campanulate, or turbinate, 4–6 mm high; outer bracts ovate-lanceolate and short-acuminate to lanceolate and long-acuminate, the medial ones oblong or lance-oblong and obtuse or short-acuminate varying to lance-acuminate, inner ones linear and acute to attenuate, all more or less rose-purplish. Corollas rose-pink or rose-purple.

Wet pine savannas and flatwoods, shores and bottoms of ponds exposed at times of low water level, dune swales, bogs, marl prairies, borrow-pits, ditches. Coastal plain, N.C. to s. Fla. and Fla. Keys, westward to s.e. Tex.; Bah.Is., Cuba.

4. Pluchea foetida (L.) DC.

Perennial, 5–10 dm tall, with 1 to few stems from a hard basal crown, similar in vegetative features to *P. rosea*. Stems often somewhat spongy-thickened basally, commonly dark purplish, viscid-pubescent, becoming glabrous below. Leaves 3–10 (–13) cm long, 1–3 cm broad, commonly sessile and somewhat clasping, sometimes cuneate basally; median and upper leaves oblong, oblong-elliptic, or the upper ones short-deltoid-ovate; margins shallowly and unevenly apiculate-toothed, surfaces atomiferous-glandular, the upper usually short-pubescent on the veins, lower sparsely to densely pubescent. Heads in loose to congested corymbose clusters terminating branchlets, tending to be storied and flat-topped. Involucres broadly campanulate, 6–8 mm high and about as broad; bracts pubescent and atomiferous-glandular, the outer lance-ovate with obtuse or obtuse-apiculate tips, medial ones oblong-acute, inner linear-acute or -attenuate. Corollas creamy white. (Incl. *P. imbricata* (Kearney) Nash, *P. tenuifolia* Small, *P. eggersii* Urban)

Wooded or open wet areas, swamps, pineland depressions, shrub-tree bogs or bays, marshy shores, ditches. Chiefly coastal plain, s. N.J. to s.cen. pen. Fla., westward to e. and s.e. Tex.; Dom.Rep.

Fig. 382. **Conoclinium coelestinum:** a, branch of plant; b, flowering head; c, flower.

5. Pluchea longifolia Nash.

Similar vegetatively and in inflorescence structure to *P. foetida* but usually coarser and attaining greater height, to 2.5 m tall. Leaves similar but to 20 cm long and 6 cm broad. Heads evidently much larger, 10–12 mm high, mostly twice as long as broad. Involucres turbinate or campanulate-turbinate; bracts powdery-pubescent, not at all or very sparsely atomiferous-glandular, the outermost smallest ones beginning somewhat down on the stalks, those just above short-ovate-obtuse, inwardly becoming longer and more oblong, the inner ones linear-oblong to linear-attenuate. Corollas creamy white.

Brackish to fresh marshes, lake shores, swamps, drainage ditches and canals, wet hammocks, commonly in shallow water. n. Fla. from Franklin Co. eastward, then to s.cen. pen. Fla.

29. Pterocaulon

Pterocaulon pycnostachyum (Michx.) Ell. BLACK-ROOT.

Perennial herb with 1 to several stems from a cluster of hard, tuberous thickened, dark brown or blackish roots. Stem 2–8 cm tall, densely, persistently, and tightly gray-felty-pubescent, conspicuously winged by long, decurrent leaf bases. Leaves alternate, simple, sessile, 3–10 cm long, 1–4 cm broad, cuneate basally, apices rounded, obtuse, or acute, margins entire, irregularly and shallowly undulating, or dentate, upper surfaces dark green, glabrous or thinly cottony-pubescent, especially along the midrib, lower surfaces pubescent like the stem. Heads discoid, few to numerous in dense, thickly oblongish (when fully developed) spikes 3–10 cm long terminating the branches or branchlets. Involucre campanulate, 4–5 mm high, densely gray-pubescent, bracts imbricate, the inner longer and narrower than the outer and acuminate. Receptacle without chaff. Peripheral flowers of the head pistillate and fertile, their corollas filiform-funnelform, very minutely toothed at the summit; central flowers bisexual but functionally staminate, their corollas narrowly tubular below, with a campanulate throat bearing 5 lobes each about as long as the throat; corollas creamy white; pappus of both kinds of flowers comprised of a single series of capillary bristles. Achene dark brown, minute, ribbed, silky-pubescent, with a smooth, shiny, bony ring basally. (*P. undulatum* (Walt.) Mohr)

Both poorly and well-drained pinelands, adjacent ditches, pineland depressions, interdune swales. Coastal plain, N.C. to s. pen. Fla., Fla. Panhandle, adjacent Ala.

P. virgatum (L.) DC., occurring in the U.S. chiefly in Tex., extends into s.w. La. Its leaves are linear-lanceolate, long-acute distally, the inflorescence of heads in elongate spikes. We do not know whether it inhabits poorly drained places. Correll and Correll (1972) do not include it as a wetland plant.

30. Conoclinium

Conoclinium coelestinum (L.) DC. MISTFLOWER.   Fig. 382

Slenderly rhizomatous perennial. Stem simple below the inflorescence varying to relatively few-branched, branches spreading-ascending and often commencing low on the stem, the main stem or the branches terminated by compact corymbs of heads; stem green to purple, clothed throughout with short-shaggy pubescence, this more copious on the inflorescence branches where usually also glandular. Leaves opposite, with slender, sulcate, pubescent petioles usually considerably shorter than the blades; blades mostly deltoid or ovate-deltoid, the larger ones 5–10 cm long and 3–6 cm broad basally, bases commonly truncate, sometimes subcordate or shortly tapered, apices acute or obtuse, margins crenate or crenate-serrate, upper surfaces sparsely appressed-pubescent, hairs short, lower glandular-punctate and short-pubescent on the principal veins and larger veinlets. Heads discoid, relatively small, bearing numerous flowers. Involucre campanulate, about 5 mm high and 5–7 mm broad distally, bracts erect, loosely imbricate at the base of the involucre, spreading away from each other distally, all subulate, outer shortest but not greatly lengthening inwardly, each usually purple-tipped, glabrous or sparsely pubescent, without glands or sparsely glandular-punctate. Receptacle conical, without chaff. Corolla tube narrowly funnelform, 5-lobed at the summit, tube whitish or sometimes violet-striped, lobes violet or blue-violet as are the conspicuously exserted stigmas which give the "mist effect"; lobes and stigmas often purple when dry; lobes inconspicuously papillose on both surfaces, tube sparsely glandular. Achene 1.5 mm long or slightly longer, 5-angled, narrowly obconic in outline, dark brown and sublustrous, glabrous or scantily hairy or glandular; pappus of relatively few stiffish capillary bristles about 3 mm long, these fused basally. (*Eupatorium coelestinum* L.)

Occupying a diversity of habitats, in general more common and abundant in moist to wet places, borders and banks of small streams, floodplain woodlands and clearings, thickets, woodland borders, meadows, borders of ponds and impoundments, sloughs, ditches, and fields. N.J. to Kan., generally southward to s. pen. Fla., e., s.e. and n.cen. Tex.; w. Cuba.

31. Eupatoriadelphus (JOE-PYE-WEEDS)

Perennial herbs, sometimes rhizomatous and colonial. Stems unbranched below the inflorescence. Leaves in whorls of 3–8, sometimes opposite on the upper stem, blades gradually or abruptly tapered proximally to short petioles. Heads discoid, relatively small and relatively numerous, in terminal panicled corymbs, ultimate inflorescence branchlets alternate, the whole inflorescence more or less flat-topped or more commonly domed. Involucral bracts imbricated, very unequal in length, faintly to strongly suffused with purple, inner ones usually glabrous or only pubescent marginally. Receptacle convex, without chaff. Flowers bisexual and fertile; corolla tubes slenderly funnelform, 5-lobed at the summit, tube and lobes glandular exteriorly, occasionally short-pubescent near the base, lavender-pink to purple; exterior of the corolla lobes with stomates. Achene narrow, tapered from base to apex, usually with a small calluslike thickening at the extreme base, 5-ribbed-angled, surface atomiferous-glandular; pappus of numerous, scabrid, purplish, capillary bristles.

• Leaves 3 or 4 at a node, commonly 3, their blades ovate or lance-ovate, the lower pair of arching-ascending lateral veins longer and more prominent than the other laterals; blades shortly and often abruptly narrowed to the petioles; lower stem not glaucous.   1. *E. dubius*
• Leaves 4–7 at a node, commonly 5 or 6, their blades narrowly to broadly lanceolate, the lateral veins about equally long and equally prominent; blades gradually tapered to the petioles; lower stem glaucous.
                                                                                        2. *E. fistulosus*

1. Eupatoriadelphus dubius (Willd. ex Poir.) King & H. Robins.

Stem 7–15 dm tall or a little more, sometimes hollow, usually purple-dotted or -flecked, glabrous below at least in age, sparsely short-pubescent upwardly, gradually becoming more copiously short-shaggy-pubescent and atomiferous-glandular on the inflorescence branches. Leaves usually 3 or 4 at a node, sometimes in pairs on the upper stem; blades ovate or lance-ovate, 5–18 cm long and 2–6 (–9) cm broad, the lower pair of arched-ascending lateral veins usually more prominent than other laterals, shortly and often abruptly narrowed proximally to the petioles, apices acute or short-acuminate, margins serrate or crenate-serrate, upper surfaces glabrous or sparsely short-pubescent, lower atomiferous-glandular, pubescent at least on the principal veins and larger veinlets. Involucre subcylindric, about 8 mm high; bracts unevenly graduated in length inwardly, the outer short, obtuse, triangular to elliptic, somewhat greenish and suffused with purple, pubescent and glandular on their backs, the medial ones a little longer, purple, scarcely if at all pubescent on their backs and not glandular, obtuse, the inner ones unequal but considerably longer than the medial, purple, wholly glabrous or finely short-pubescent marginally, obtuse or subacute. Achene 3–3.5 mm long, brown with amber glandular droplets; pappus bristles usually purplish, 5–7 mm long. (*Eupatorium dubium* Willd. ex Poir.; *Eupatorium purpureum* sensu Small, 1933)

Moist to wet thickets, swales, open stream banks, ditches, marshes, meadows. N.S. and s. Maine to e. N.Y., southward on the coastal plain and piedmont through S.C., perhaps into Ga.

2. Eupatoriadelphus fistulosus (Barratt) King & H. Robins.

In general features similar to *E. dubius*, usually stouter and taller, mostly 1.5–3 m tall. Stem hollow, commonly purple or purplish, glaucous when fresh, glabrous below, pubescent and glandular above much as in *E. dubius*. Leaves mostly in whorls of 4–7, usually 5 or 6, blades narrowly to broadly lanceolate, 1–3 dm long and 2–6 cm broad, gradually tapered proximally to the petioles, acute or acuminate apically, lateral veins all about equally prominent, margins crenate-serrate, upper surfaces glabrous or nearly so, lower atomiferous-glandular and sparsely short-pubescent on the principal veins and larger veinlets. Involucres like those of *E. dubius*. Achene 5–6 mm long, dark brown and with minute glandular atoms; pappus bristles buff to purplish, about 6 mm long. (*Eupatorium fistulosum* Barratt; *Eupatorium maculatum* sensu Small, 1933)

Moist to wet thickets and swales, open alluvial woodlands, meadows, often very abundant along brooks through fields and along ditches. s. Maine to Iowa, generally southward to Fla. Panhandle and e. Tex.

32. Eupatorium (BONESETS, THOROUGHWORTS)

Perennial herbs, rarely rhizomatous, usually perennating from crown buds and commonly with clumps of stems from the basal crown. Stems unbranched below the inflorescence, pubescent throughout. Leaves opposite throughout or sometimes alternate on the inflorescence branches, in some species an occasional specimen having them in whorls of 3 or 4; leaves sessile or petioled, blades entire, toothed, lobed, or dissected. Heads discoid, relatively small, usually numerous, relatively few-flowered, disposed corymbosely or in more or less flat-topped or storied cymes, heads in corymbs terminating the ultimate, usually alternate, branchlets, or diffusely paniculate in a few species. Involucral bracts very unequal, imbricated in 2 or 3 series, usually green save for scarious margins, rarely slightly purplish, pubescent and atomiferous-glandular exteriorly. Receptacle flat, without chaff. Flowers bisexual and fertile; corolla tube slender below, with an expanded throat above, 5-lobed at the summit, tube and lobes whitish, glandular exteriorly; stomates none on the corolla lobes exteriorly. Achene 5-ribbed or ribbed-angled, dark brown or blackish, glandular, not pubescent, without a stalk or enlargement basally; pappus of numerous, antrorsely minutely barbed, whitish bristles.

References: Sullivan, V. I. "Investigations on the Breeding Systems, Formation of Auto- and Alloploids and Reticulate Pattern of Hybridization in North American *Eupatorium* (Compositae)." Ph.D. Diss., Florida State Univ. 1972.

———. "Diploidy, Polyploidy, and Agamospermy among Species of *Eupatorium* (Compositae)." *Canad. Jour. Bot.* 54: 2907–2917. 1976.

———. "Putative Hybridization in the Genus *Eupatorium* (Compositae)." *Rhodora* 80: 513–527. 1978.

The base chromosome number for *Eupatorium* is $x = 10$. All plants of certain species apparently are always diploid and sexual; plants of certain other species may be diploid and sexual or polyploid and apomictic (agamospermic), the diploids often with a very much more restricted distribution than the polyploids. Hybridization is not uncommon, and if hybrid derivatives of diverse morphology happen to be or become polyploid, apomictic and ecologically successful, they may produce significantly large and wide-ranging populations.

1. Blades of principal leaves 1–3 times pinnately divided into linear-filiform or filiform segments; inflorescence of heads paniculate.

2. Stem (except lower portion in age) shaggy-pubescent; heads diffusely paniculately disposed, not secund on branchlets. 1. *E. capillifolium*
2. Stem glabrous throughout save for short hairs sometimes present between the lower leaves; heads in secund, short fascicles on spreading, arching, or ascending branches of an essentially pyramidal terminal panicle. 2. *E. leptophyllum*
1. Blades of principal leaves not pinnately divided; heads disposed corymbosely on ultimate branchlets, the main branches of the inflorescence corymbose or of more or less flat-topped, storied cymes.
  3. Leaves with sessile, strongly clasping, or perfoliate bases. 3. *E. perfoliatum*
  3. Leaves sessile or petiolate, not strongly clasping or perfoliate.
    4. Leaves distinctly petiolate, petioles 1–5 cm long.
      5. Stem glabrous or very scantily short-pubescent below, somewhat glaucous; leaves subsucculent, their petioles usually twisted so the blades are held more or less vertically; margins of blades crenate. 4. *E. mikanioides*
      5. Stem copiously short-shaggy-pubescent (except lower portion in age); leaves not subsucculent, petioles not twisted, blades held horizontally, margins of blades serrate. 5. *E. serotinum*
    4. Leaves sessile or minutely petiolate.
      6. Blades of midstem leaves ovate or triangular-ovate.
        7. The blades of principal leaves short-ovate, little if any longer than the breadth of the broadest portion near the base. 6. *E. rotundifolium*
        7. The blades of principal leaves more or less long-triangular-ovate, mostly nearly twice as long as the breadth of the widest portion near the base, often a little longer proportionately.
          7. *E. pilosum*
      6. Blades of midstem leaves of a more narrow and elongate type, lanceolate, oblanceolate, linear-lanceolate, or narrowly elliptic.
        8. Median and inner involucral bracts with acuminate-attenuate white tips; leaf surfaces harsh-scabrid to the touch. 8. *E. leucolepis*
        8. Medial and inner involucral bracts obtuse, to rounded apically; leaf surfaces not harsh-scabrid.
          9. Leaf surfaces finely and densely tomentose (as well as atomiferous-glandular), much paler beneath than above. 9. *E. resinosum*
          9. Leaf surfaces pubescent but not tightly tomentose, lower surfaces not significantly, if at all, paler than the upper.
            10. Herbage olive green in color; midstem leaves mostly definitely oblanceolate; perennation from small, definitely nontuberiform crown buds. 10. *E. semiserratum*
            10. Herbage not olive green; midstem leaves lanceolate, linear-lanceolate, or narrowly elliptic; perennation by conspicuous, ovate to conical, tuberiform crown buds which eventually become shortly rhizomatous. 11. *E. recurvans*

1. Eupatorium capillifolium (Lam.) Small. DOG-FENNEL. Fig. 383

Plant perennial in much, if not all, of its range (some authors describe it as annual), blooming the first year from seed, perennating by basal offsets, eventually forming several stems, each again perennating basally, the central portion of the clump commonly disintegrating so that finally stems often more or less in a ring. Stem shaggy-pubescent, becoming glabrous below in age, varying greatly in stature, those of young plants often slender and only a few dm tall, those of older plants to 3 m tall, widely and very diffusely branched, flowering branchlets slender and lax, abundantly floriferous, heads paniculately disposed, not at all secund. Lower leaves opposite, 5–8 cm long overall, with petioles 1–3 cm long, blades 2 or 3(4) times pinnately divided, the segments 1–1.5 mm broad; upwardly leaves becoming alternate and sessile, progressively with fewer divisions, the divisions filiform and 0.5 mm broad or less, those distally on the flowering branchlets mostly filiform and undivided; all leaves bright green and glabrous, sometimes sparsely punctate but not viscid. Involucre about 2 mm high, with 2–4 short, lance- or triangular-acute outer bracts, an inner series of about 5, much longer, oblong-oblanceolate inner ones with abruptly acuminate or apiculate apices; bracts glabrous, not glandular, varying from green to reddish purple, with narrow hyaline margins. Corolla creamy white, tube narrowly cylindric below, abruptly dilated to a more broadly cylindric or funnelform throat of about equal length, lobes minute. Achene 1–1.5 mm long, narrowly obconic in outline, grayish black, glabrous, not glandular; pappus bristles 2 mm long.

Fig. 383. **Eupatorium capillifolium:** A, basal portion and top of plant; B, small portion of panicle of heads; C, small section of stem with leaves; D, flower; E, achenes, one with pappus removed. (From Reed, *Selected Weeds of the United States* (1970) Fig. 199)

Lowland fields or low places in fields, bottomland clearings, moist to wet thickets, borders of ponds and lakes, exposed shores and bottoms of ponds and lakes, in upland areas where soil mechanically much disturbed but there not usually persisting, aggressively weedy and often in dense and extensive stands. Chiefly coastal plain and piedmont, N.J. to s. pen. Fla., westward to s. Ark., La., and e. Tex.; Berm., Bah.Is., Cuba.

E. compositifolium Walt. is similar in general features. It is restricted to the coastal plain and outermost piedmont from e.cen. N.C. to s.cen. pen. Fla., westward to e. and s.e. Tex., inhabiting, for the most part, well-drained sites. It grows intermixed with *E. capillifolium* only where the latter has colonized exposed and then dry shores or bottoms of ponds or where soils mechanically much disturbed. Stems of older plants usually not over 1.5 m tall. Leaves a dull grayish green, pubescent, abundantly glandular-punctate, and viscid, leaf segments very variable but generally broader, often conspicuously broader, than those of *E. capillifolium*. Where the plants of the two species grow intermixed they hybridize to some extent but the hybrids generally do not persist.

Plants of both *E. capillifolium* and *E. compositifolium* hybridize with those of *E. perfoliatum* in those places where, or at those times when, their blooming seasons overlap. The hybrid intermediates of both crosses are closely similar; generally they are few in any one place but on occasion they are abundant. The applicable name for the hybrid is *E.* × *pinnatifidum* Ell. (Incl. *E. eugenei* Small, *E. pectinatum* Small) (See c–f of Fig. 386.)

2. Eupatorium leptophyllum DC. Fig. 384

Plant perennating by basal offsets or by rhizomes to about 1.5 dm long. Stems 8–20 dm tall, commonly relatively slender, glabrous save for some short hairs sometimes between the bases of the lower leaves. Lower leaves opposite, with slender petioles 1–2 cm long, blades bipinnately divided into narrowly filiform segments; leaves becoming alternate upwardly and 1-pinnately divided, the upper ones simple and filiform; leaves glabrous and bright green. Inflorescence of heads terminal, paniculate, not conspicuously leafy-bracteate, more or less pyramidal in overall outline, the heads mostly in secund, short fascicles on spreading, arching, or ascending lateral branches, the terminus of the panicle narrow and compact. Involucre about 3 mm high, bracts in essentially 2 series, glabrous, outer few subulate, the inner few, lanceolate to linear-oblong with acuminate to short-attenuate apices, narrow hyaline margins. Corolla creamy white to white, not glandular, tube narrowly cylindrical below, with an abruptly more broadly cylindrical to funnelform throat, lobes minute. Achene 1–1.5 mm long, nearly columnar or slightly broadened upwardly from base to apex, purplish black and slightly iridescent, glabrous; pappus bristles 2–3 mm long. (*E. capillifolium* var. *leptophyllum* (DC.) Ahles)

Borders of, and exposed shores or bottoms of, open ponds, cypress-gum ponds, or cypress-titi ponds, or in s. Fla. more uniformly distributed over wet marl prairies. s.e. N.C. to s. Fla., Fla. Panhandle, adjacent Ga. and s. Ala.; Bah.Is., Cuba.

3. Eupatorium perfoliatum L. THOROUGHWORT, BONESET. Fig. 385

Plant perennating by rather long basal offsets eventually yielding clumped stems. Stem relatively little to widely branched, 6–15 (–20) dm tall, copiously shaggy pubescent below with long hairs, hairs shorter above and on the inflorescence branchlets where interspersed with glandular atoms. Leaves opposite, sessile, bases mostly perfoliate, lanceolate, tapering distally to long-acuminate tips, larger ones 5–20 cm long and 2–4.5 cm broad, margins crenate or crenate-serrate, lateral veins forming a more or less rugose reticulum, upper surfaces relatively sparsely short-pubescent, lower softly subtomentose and atomiferous-glandular. Involucre subcylindric, about 4 mm high, bracts imbricated, pubescent and atomiferous-glandular, outer acute, considerably shorter than the medial and inner, these acuminate, tips and margins of medial and inner ones often white. Corolla whitish, rarely purple-tinged, a few atoms of glandular exudate usually on the lobes, often both surfaces. Achene narrowly obconic to fusiform in outline, 2–2.5 mm long, black, usually bearing a few atoms of glandular exudate; pappus bristles 2.5–3 mm long.

Fig. 384. **Eupatorium leptophyllum:** a, midsection of stem; b, top of plant; c, flowering head; d, flower; e, achene.

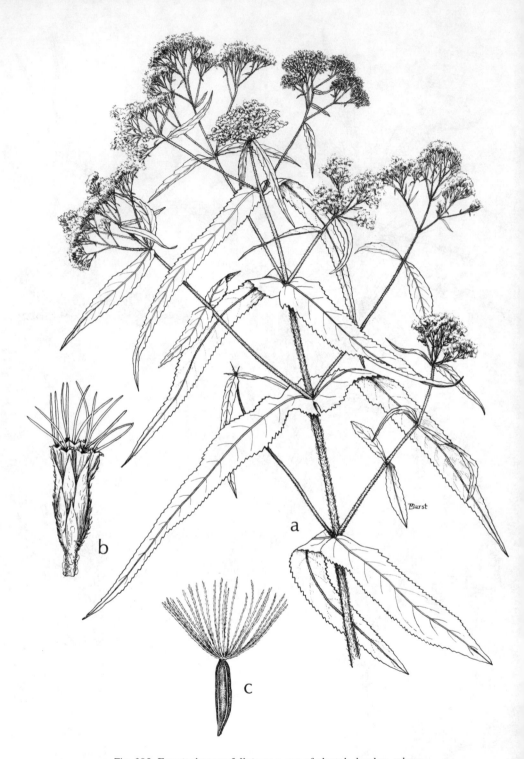

Fig. 385. **Eupatorium perfoliatum:** a, top of plant; b, head; c, achene.

Low, moist to wet places, woodlands, bogs, meadows, swales, clearings, thickets, pastures, banks and borders of streams, fields. N.S. and s. Que. to s.e. Man., generally southward to n. Fla., Kan., and e. Tex.

Plants of *E. perfoliatum*, sporadically and apparently relatively rarely, hybridize with those of *E. serotinum*, perhaps only northward of our range where their blooming seasons overlap. Some names which have been applied to the hybrids are: *E. truncatum* Muhl. ex Willd., *E. perfoliatum* var. *truncatum* (Muhl. ex Willd.) Gray; *E. cuneatum* Engelm. ex T. & G., *E. perfoliatum* var. *cuneatum* (Engelm. ex T. & G.) Gray; *E. serotinum* var. *polyneron* F. J. Herm. See also discussions under *E. capillifolium*, *E. recurvans*, and *E. rotundifolium*.

4. Eupatorium mikanioides Chapm.

Plant producing long, slender rhizomes or stolons and colonial. Stems erect to more or less reclining, 5–10 dm tall or long, glabrous or very scantily short-pubescent below, somewhat glaucous, progressively more pubescent upwardly, tomentose on the inflorescence branches. Leaves subsucculent, opposite, distinctly petiolate, petioles pubescent, 1–2.5 cm long, usually twisted so that the leaf blades are held more or less vertically, lowermost leaves (usually shed by flowering time) small, with lanceolate blades; blades of midstem leaves mostly somewhat deltoid-ovate, broadly short-tapered basally, obtuse to rounded apically, 2–7 cm long and 1.5–5 cm broad near their bases, margins crenate distally above the tapered bases, both surfaces glandular-punctate; emerging blades white or silvery with dense silky-cottony pubescence on both surfaces, quickly becoming glabrous above and sparsely pubescent beneath. Involucre 4–5 mm high, bracts few, not very much imbricated, outer much shorter than the inner, all acute apically, pubescent and atomiferous-glandular on their backs. Corollas whitish, usually with a few resinous atoms on the tube and lobes exteriorly. Achene narrowly obconic in outline, about 2 mm long, brown and with a few glandular atoms on the surface; pappus bristles about 5 mm long.

In and on borders of salt, brackish, or fresh marshes, pine flatwoods, adjacent ditches and roadsides, spoil banks and flats, interdune swales, marl prairies, glades, solution holes in lime-rock. Fla., coastal areas, from Volusia Co. southward on the Atlantic side, northward and westward along the Gulf of Mexico to Gulf Co.

5. Eupatorium serotinum Michx.

Clump-forming, commonly relatively coarse plant. Stem 0.6–3 m tall, few- to much-branched, lower stems of particularly robust specimens to 2 or 2.5 cm in diameter, often hollow in the center, copiously short-shaggy-pubescent, lower stem becoming glabrous in age. Leaves opposite, distinctly petiolate, petioles to 5 cm long; blades of principal leaves chiefly long-triangular-ovate, bases truncate to shortly and broadly tapered (lowermost usually subcordate), tapering from the broadest portion near the base to acuminate apices, larger ones 6–20 cm long and 3–8 cm broad basally, two lateral, angled-ascending veins arising basally or just above the base usually more prominent than other laterals, margins coarsely serrate, both surfaces glandular-punctate, upper essentially glabrous, lower pubescent along the larger veins, more sparsely so on the veinlets. Involucre about 3 mm high, bracts few, sparingly white-pubescent below, above more copiously so and ciliate marginally, also atomiferous-glandular, outer ones considerably shorter than the medial and inner, apices truncate or broadly rounded. Corollas dirty whitish, pale roseate, or pale lilac, sparsely atomiferous-glandular exteriorly on the tube and the lobes. Achene slightly obconic in outline, 1.5–2.5 mm long, black, slightly glutinous; pappus bristles 3–3.5 mm long.

Chiefly in moist to wet places, fields, meadows, pastures, thickets, clearings, in and on the borders of fresh to brackish marshes and glades, marshy shores, floodplain woodlands and their clearings, sloughs, open stream banks. s.e. N.Y. to s. Wisc. and e. Kan., generally southward to s. Fla., e., s.e., and n.cen. Tex.; n.e. Mex.

See discussion under *E. perfoliatum* concerning hybridization.

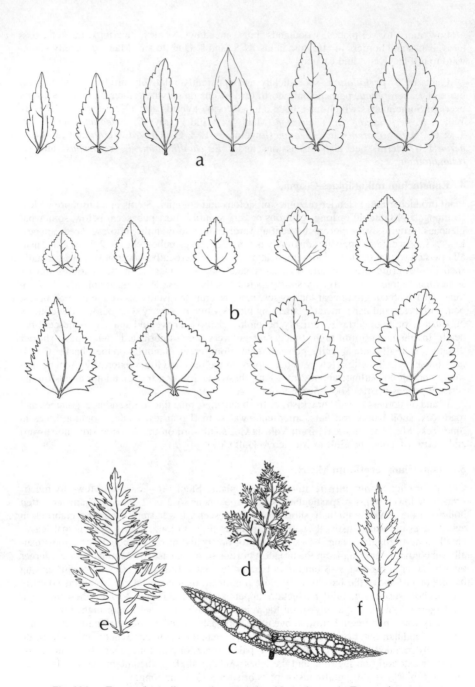

Fig. 386. a, **Eupatorium pilosum:** characteristic midstem leaves; b, **Eupatorium rotun-difolium:** characteristic midstem leaves; c, **Eupatorium perfoliatum:** pair of midstem leaves; d, **Eupatorium capillifolium:** lower midstem leaf; e,f, **Eupatorium × pin-natifidum:** midstem leaves.

6. Eupatorium rotundifolium L. Fig. 386

Plant with a knotty basal crown often bearing several stems. Stem 4–12 dm tall, uniformly and densely short-pubescent throughout. Leaves minutely petiolate or sessile, opposite save for alternate bracteal ones on the upper inflorescence branches, or the principal ones whorled on an occasional individual; lowest leaves smallest, progressively increasing in size to somewhat below midstem, diminishing in size from somewhat above midstem; blades of principal leaves ovate, variable in size, 2–8 cm long and 1.5–6.5 cm broad near the base, bases commonly truncate, sometimes cordate or shortly and broadly tapered, margins serrate, crenate-serrate, or shortly crenate-lobulate, lower ones sometimes incised-lobed; blades with a pair of evident, ascending lateral veins arising at or very near the base, both surfaces short-pubescent and minutely atomiferous-glandular. Involucre 5–6 mm high, bracts imbricated, outer subulate, becoming longer inwardly, the inner linear-oblong and with rounded to acute or short-attenuate, whitish tips, copiously pubescent and atomiferous-glandular on their backs. Corolla dull whitish, irregularly atomiferous-glandular exteriorly on the tube, densely so on the lobes. Achene 2–2.5 mm long, narrowly obconic in outline, brown to nearly black and atomiferous-glandular; pappus bristles 3 mm long or slightly more.

Bogs, pine savannas and flatwoods, borders of shrub-tree bogs or bays, open moist banks, low places in pastures and old fields. s. Maine to N.C., Tenn., s. Ark., generally southward to s.cen. pen. Fla. and e. Tex.

Plants of *E. rotundifolium* hybridize, or have perhaps long ago hybridized, with those of several other species, hybrid derivative populations of varied ancestry and showing a diversity of intermediacy sometimes occurring in abundance locally; if in local abundance and persisting, the plants usually polyploid and apomictic. Hybridization is known or suspected with the following: *E. lancifolium* (T. & G.) Small, *E. mikanioides*, *E. perfoliatum*, *E. pilosum*, *E. recurvans*, *E. semiserratum*, *E. serotinum*, and *E. sessilifolium* L.

7. Eupatorium pilosum Walt. Fig. 386

Plant with a knotty basal crown, perennating from crown buds and eventually stems often clumped. Stem 0.5–1.5´(–2) m tall, commonly purplish, shaggy-pubescent below, densely pubescent and atomiferous-glandular on the inflorescence branches. Leaves sessile or very very short-petiolate, opposite (whorled on an occasional specimen) to the inflorescence, all or nearly all bracteal leaves of the inflorescence alternate (and branches alternate); lowermost pairs of stem leaves very small, progressively increasing in size to about midstem then gradually diminishing upwardly; midstem blades of most plants somewhat ovate-triangular, less frequently oblong or lance-oblong, 3–10 cm long and 1–5 cm broad, bases truncate, subcordate, or broadly short-tapered, apices usually blunt, sometimes acute, lower pair of lateral veins arising from the midvein at the base or only slightly above, arching-ascending, more prominent than other laterals but not so evident as the midvein, margins usually unevenly and irregularly toothed, teeth rounded or blunt, rarely sharp save on upper, reduced leaves, 3–12 on a side; surfaces sparsely pubescent, hairs on the upper surface sometimes pustular-based, one or both surfaces glandular-punctate. Involucre 4–5 mm high, bracts somewhat imbricated, outer short, triangular-acute, becoming linear-oblong inwardly, inner ones with acute to attenuated white tips and pale hyaline margins. Corollas white, atomiferous-glandular exteriorly on tube and lobes. Achene narrowly obconic in outline, about 2 mm long, dark brown, sparsely atomiferous-glandular; pappus bristles about 4.5 mm long. (*E. verbenaefolium* Michx., *E. rotundifolium* var. *saundersii* (Porter) Cronq.)

Pine savannas and flatwoods, bogs, in and on borders of shrub-tree bogs, seepage slopes in pinelands, moist to wet banks bordering mt. woodlands, swales. s. N.Eng. to s. N.J. and e. Pa. to cen. pen. Fla., the boundary from s.e. Pa., roughly through W.Va., e. Ky., e. Tenn. to s.w. Miss.

8. Eupatorium leucolepis (DC.) T. & G.

Plant with a knotty basal crown commonly bearing a single stem. Stem 4–8 (–10) dm tall, short-pubescent throughout and atomiferous-glandular at least above. Principal leaves opposite, sessile, commonly curvate, divaricately spreading or deflexed, those low on the stem small, progressively larger to midstem then diminishing in size upwardly; midstem leaves variably lanceolate, lance-oblong, oblanceolate, linear-lanceolate, or linear-oblong, only the midvein evident or with two ascending, relatively prominent lateral veins from the base or a little above, margins obscurely to sharply serrate, both surfaces scabrous and glandular-punctate. Involucre obconic to subcylindric, 6–10 mm high, bracts imbricated, gradually increasing in length inwardly, the outer short-subulate, grading to linear with acuminate-attenuate white tips, all copiously white-pubescent and atomiferous-glandular, with narrow white-hyaline margins. Corollas dull white, tube and lobes atomiferous-glandular exteriorly. Achene slightly obconic in outline, about 2.5 mm long, dark gray to black, sparsely atomiferous-glandular; pappus about 4 mm long.

Bogs, pine savannas and flatwoods. Coastal plain, e. Mass. to n.cen. pen. Fla., westward to e. and s.e. Tex.

9. Eupatorium resinosum Torr. ex DC.

Plant with a knotty basal crown, perennating by long-scaly offshoots from crown buds. Stem 6–10 dm tall, softly short-pubescent, densely so above. Leaves opposite, sessile, those lowest on the stem smallest, gradually increasing in size to midstem and scarcely diminishing in size to or into the inflorescence; midstem leaves lanceolate, 6–10 (–12) dm long and about 1 cm broad, little tapered basally, the extreme base essentially truncate, apices acute-attenuate, margins entire to finely serrate more or less medially, both surfaces atomiferous-glandular and finely and closely tomentose, the lower surface much paler than the upper. Involucre 3–4 mm high, bracts varying from essentially in two series, the outer much shorter than the inner, to imbricated and graduated in length from outside inwardly; all bracts oblongish and with broadly short-obtuse apices, sparsely to densely pubescent on their backs and atomiferous-glandular. Corollas dull white, scantily and irregularly atomiferous-glandular on the tube and lobes exteriorly. Achene narrowly obconic in outline, about 2 mm long, dark brown and atomiferous-glandular; pappus bristles 3 mm long.

Sphagnous bogs in pinelands and shrub bogs. Local, coastal plain, s. N.J. and Del.; inner coastal plain, N.C.

10. Eupatorium semiserratum DC. Fig. 387

Plant commonly with several clumped stems arising from crown buds, herbage olive greenish. Stem 7–10 dm tall, rarely taller, uniformly short-pubescent, more copiously so above than below. Stem, leaf surfaces, involucres, exterior of corolla lobes, and achenes atomiferous-glandular. Leaves opposite (whorled in a fair number of individuals) save for some alternate bracteal ones in the inflorescence, usually some, at least, of the principal ones short-petiolate, blades mostly oblanceolate, less frequently some of them lanceolate, larger ones mostly 5–7 cm long and 1.5–2 cm broad, mostly triple-nerved from well above the base, margins finely serrate distally ½–¾ of the length of the margins, bases cuneate, apices acute, surfaces copiously short-pubescent but not tightly tomentose, both surfaces the same shade of olive green. Involucre subcylindric, 2.5–3.5 mm high, bracts few, essentially in two series, the outer short-oblong, obtuse, subimbricated, the inner oblong-obtuse, 2–3 times as long as the outer, all tomentose. Corolla dull or dirty white. Achene columnar in outline, 1.5–2 mm long, dark brown; pappus bristles about 3 mm long. (E. glaucescens sensu Correll & Johnston, 1970, not Ell.; not incl. E. cuneifolium Willd. and E. lancifolium (T. & G.) Small = E. semiserratum var. lancifolium T. & G.)

Depressions in pine savannas and flatwoods, borders of cypress-gum ponds, potholes in upland woodlands, swales, wet prairies, alluvial woodlands, their borders and clearings, ditches. Coastal plain, Va. to cen. Fla. Panhandle, westward to e. and s.e. Tex., northward in the interior, Ark., Tenn., s.e. Mo.

Perhaps the most significant hybridization involving plants of E. semiserratum is that with plants of E. rotundifolium. In places where plants of each are diploid, hybrids are occasionally produced at the present time but they do not seem to persist. Chiefly in s.e. S.C. and

Fig. 387. **Eupatorium semiserratum:** a, top of plant; b, leaf; c, flowering head; d, flower.

in s. Ark. and La., a not inconsequential proportion of *E. rotundifolium*-like specimens (apparently polyploid and apomictic) have a leaf morphology closely approximating that of specimens known to be F₁ hybrids of the cross, *E. rotundifolium* × *E. semiserratum*. In general, the leaves of such plants are more nearly like those of *E. rotundifolium* than they are those of *E. semiserratum*, but they have strongly cuneate bases and the lateral pair of nerves, if evident, are usually much less so than is the case for leaves of *E. rotundifolium*. A name which has been applied to such intermediates is *E. scabridum* Ell. The variability exhibited by plants surely having "*E. rotundifolium* in their ancestry" is perhaps much greater in s. Ark. than anywhere else.

11. Eupatorium recurvans Small.

Plant perennating during or just after anthesis by forming 1 to several, ovate to conical, tuberiform crown buds 2–3 cm long and about 1 cm thick; over winter each tuberiform bud produces fibrous roots and becomes essentially rhizomatous, in spring an erect stem of the season emerges from its tip. Stem simple below the inflorescence or often bearing 2 to several erect branches from near the base; sometimes each of up to 4 or 5 of the "rhizomatous buds" of a single crown produces an unbranched stem in which case the stems appear clumped; sometimes each produces a stem with several branches from near the base thus giving a conspicuously bushy-branched effect. Stems 3–6 (–7) dm tall, short-pubescent throughout, rather more densely so above than below. Leaves opposite, sessile or very short petiolate; blades of midstem leaves lanceolate, linear-lanceolate, or narrowly elliptic (tending to have short, smaller-leaved branchlets in their axils), 2–4 cm long and 3–10 mm broad, usually curved downwardly or deflexed; margins sometimes entire, usually irregularly serrate, teeth low and bluntish, both surfaces pubescent with short, appressed hairs and conspicuously glandular-punctate. Involucre 3–4 mm high, bracts glandular-punctate and short-pubescent, few in essentially two series, outer much shorter than the inner, ovate to oblong, tips rounded, inner linear-oblong, with white hyaline margins and tips, tips rounded and somewhat ciliate-fringed, margins below often sparsely ciliate. Corollas off white, irregularly atomiferous-glandular on the tube and lobes exteriorly. Achene narrowly obconic in outline, 1.5–2 mm long, dark brown and atomiferous-glandular; pappus bristles about 4 mm long.

Pine savannas and flatwoods, adjacent ditches, swales. Coastal plain, s.e. and s.cen. Ga., southward to s. pen. Fla. and s.e. Fla. Panhandle.

Plants described above as *E. recurvans* are, insofar as we are aware, always diploid and sexual. A "race" of closely similar plants that are usually triploid (rarely tetraploid) and apomictic (agamospermic) occurs commonly and abundantly within the range of, often intermixed with, or near *E. recurvans* (as described); they occur as well beyond the range of *E. recurvans*, to e. Va. to the northeast, and to e. Tex. to the west. This race (and another alluded to in the discussion below) is presumed to be derivative from the cross, *E. recurvans* × *E. rotundifolium* (see Sullivan, 1972). In any event, plants comprising it closely resemble those of *E. recurvans*, generally having a greater stature, mostly 10–15 dm tall, seldom, if ever, having the stems branched from near the base. Where these and plants of *E. recurvans* grow together or near each other, the latter are at full anthesis well in advance of the apomictic ones, usually having most or all heads with fully developed achenes by the time the apomictic ones reach anthesis. Involucres of *E. recurvans* are smaller, 3–4 mm high, and the bracts all rounded apically; involucres of the apomictic plants are 5–7 mm high and some, at least, of the inner bracts have acute apices.

The apomictic race is certainly not easily and not always distinguishable from the diploid "race." If one wants to grace it with a name, the applicable one is *E. mohrii* Greene, and we shall use that name in the discussion below.

Another race, usually tetraploid, rarely triploid, and apomictic, is more nearly like *E. mohrii* than *E. recurvans*, especially respecting stature, lack of low-stem branching, and characters of the involucre. It is apparently much more sporadic in its occurrence, although we have it in the Florida State University collection from e. N.C. to n. Fla., westward to s. Ala. It differs from both *E. recurvans* and *E. mohrii* in its uniformly broader and more definitely and sharply serrate or incised-serrate leaves, the principal ones of these on a given plant being lanceolate to narrowly lance-ovate, 1.5–3 cm broad at their broadest places. To one accustomed to seeing plants of this assemblage in the field, the character of the leaves makes the plants conspicuously contrasting. Again, if one wants a name for it, the applicable one is *E. anomalum* Nash.

It is perhaps useful to emphasize that all plants of this assemblage produce, at about the end of the flowering period, ovoid to conical, tuberiform, crown buds that later become essentially short-rhizomatous; this is an attribute exclusively its own among eupatoria.

33. Mikania

Mikania scandens (L.) Willd. CLIMBING HEMPWEED.

Twining, climbing, or sprawling, herbaceous vine, commonly densely matted over herbs, shrubs, or small trees. Stem glabrous or infrequently sparsely to densely pubescent. Leaves opposite, petiolate, petioles 1–10 cm long; blades palmately veined, ovate to deltoid-ovate in overall outline, bases mostly cordate or hastate-cordate, apices strongly acuminate, very variable in size, 2.5–10 (–14) cm long and commonly as broad basally as long, margins entire, irregularly wavy, or with few, broad dentations, surfaces usually finely atomiferous-glandular, sometimes sparsely, rarely densely, pubescent. Inflorescences axillary, their stalks mostly much exceeding the leaves, heads compactly to loosely aggregated in corymbs. Involucre cylindric, 3.5–6 mm long, of 4 principal bracts in one series, mostly with one smaller outer one; bracts short-pubescent to glabrous, usually atomiferous-glandular. Heads discoid, flowers 4 per head, bisexual. Corolla white, pink, or bluish, 5-lobed at the summit. Receptacle naked. Achene columnar or slightly obconic in outline, sharply 5-angled, atomiferous-glandular, 1.5–2.5 mm long, surmounted by a pappus of numerous capillary bristles 3–4 mm long in a single series. (Incl. *M. batatifolia* DC.)

Marshy or shrubby shoreline thickets, marshes, borders of swamps and wet woodlands, marl prairies, swales, bogs, commonly in water. Locally inland from Maine to s. Ont., Mich., Ill., Mo., generally southward, excepting mts., to s. Fla., s.e. Okla., s. and s.e. Tex.; trop. Am.

34. Sclerolepis

Sclerolepis uniflora (Walt.) BSP.   Fig. 388

Low glabrous herb perennating by slender whitish rhizomes, mat-forming. Stems usually unbranched, occasionally few-branched above, weak, decumbent and rooting for a short distance near the base, 1–4 dm tall. Leaves in whorls of 3–6, sessile, linear-subulate, margins entire, tipped by a very small callus, 1–2 cm long, 1–1.5 mm wide at the base. Heads solitary and terminal, to about 1.5 cm across at full anthesis. Flowers numerous, corollas all tubular and bisexual, pink or lavender, rarely white. Phyllaries subequal in 2 series, sometimes with a few much smaller bracts below, their margins pubescent, the upper portions especially purplish. Receptacle of the head naked. Pappus of 5 short, broad, straw-colored scales, their apices broadly rounded to truncate. Achene 1–1.3 mm long, narrow, tapered from below, sharply whitish-ribbed, the interrib surfaces dark brown to black, with an irregular calluslike stipe.

Usually in mats on a soft sandy to peaty substrate, often in still, shallow water at least part of the year; cypress-gum ponds or depressions, wet flatwoods, swamps, bogs, ditches. N.H., coastal plain southward to Fla. and Ala.

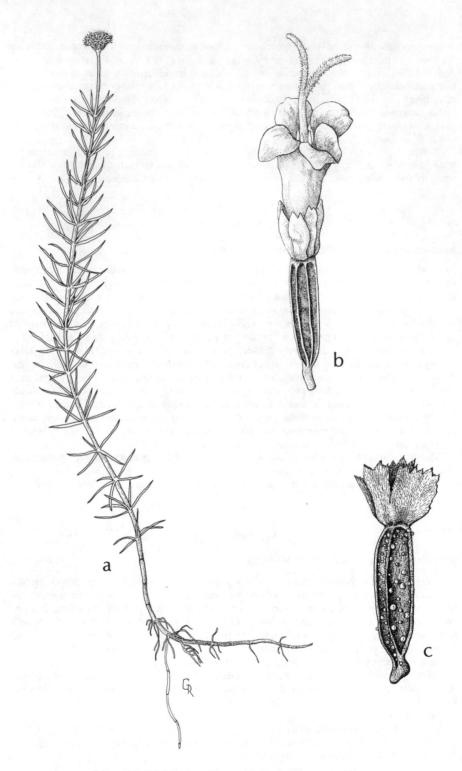

Fig. 388. **Sclerolepis uniflora:** a, habit; b, flower; c, achene.

35. Hartwrightia

Hartwrightia floridana Gray.

Perennial with a short rhizome; atomiferous- or impressed-glandular and viscid throughout, the involucral bracts, corollas, and achenes especially sticky. Stem 6–12 dm tall, openly corymbose-branched in the inflorescence. Leaves alternate, pinnately veined; basal and lowermost stem leaves much the largest, blades elliptic-oblanceolate to elliptic, variable in size, 5–15 cm long, 1–8 cm broad, margins entire or irregularly wavy, apices blunt to rounded, tapering to an elongate petiole; stem leaves few, reduced, narrowly lanceolate to linear, uppermost sessile. Heads discoid, each bearing relatively few flowers. Involucre of few loosely imbricated, linear-oblong bracts. Flowers all radially symmetrical, bisexual and fertile, without subtending chaff. Corollas about 3 mm long, pinkish purple, with a very short, narrow tubular base then flared to a funnelform throat surmounted by 5 short-triangular, erect lobes. Achene 3–3.5 mm long, obconic in outline, with 5 prominent wing-angles; pappus usually none or, if present, then of 4 or 5 fragile, quickly deciduous awns.

Boggy places in pinelands, edges of cypress depressions and shrub-tree bays, adjacent wet pastures. Local and infrequent, n.e. to s.cen. pen. Fla.

36. Carphephorus

Perennial herbs with 1 to several stems from a stoutish basal crown or short rhizome; stems glabrous in some, varying to densely pubescent in others, some with copious amber-colored, resinous, atomiferous glands. Leaves alternate, simple, the basal and sometimes those of the lower stem largest (the basal sometimes in rosettes), with an expanded, more or less bladed distal portion tapering below to petiolar or subpetiolar bases (linear and without an expanded blade in one species), upwardly leaves smaller and sessile. Inflorescence of heads disposed loosely to compactly in corymbs or thyrsoid panicles. Heads discoid, campanulate or hemispheric, flowers few to many per head. Involucral bracts in 2–6 series, from few, subequal, and scarcely imbricated to numerous and much imbricated. Flowers bisexual and fertile, receptacular chaff subtending all the flowers, sometimes only those near the periphery of the head, or absent. Corollas funnelform, 5-lobed at the summit, lavender-pink, lavender, or purple, rarely white. Achene obconic, 10-angled or -ribbed, short-pubescent; pappus of numerous, somewhat unequal, antrorsely barbellate, capillary bristles.

Reference: Correa, Mireya D., and Robert L. Wilbur. "A Revision of the Genus *Carphephorus* (Compositae, Eupatorieae)." *Jour. Elisha Mitchell Sci. Soc.* 85: 79–91. 1969.

1. Rosette or basal leaves sessile, narrowly linear, needlelike, and involute, many times longer than broad.   3. *C. pseudo-liatris*
1. Rosette or basal leaves with bladed distal portions narrowed below to petiolate or subpetiolate bases, not linear, not involute.
  2. Heads relatively large, the involucres 7–10 mm high, composed of 15–30 strongly imbricated bracts and containing 15–30 flowers.
    3. Involucral bracts with scarious margins, their tips mostly erose; corollas not atomiferous-glandular exteriorly.   1. *C. corymbosus*
    3. Involucral bracts without scarious margins, their tips not erose; corollas atomiferous-glandular exteriorly.   2. *C. tomentosus*
  2. Heads relatively small, involucres not exceeding 6 mm high, composed of 5–12 not strongly imbricated bracts and mostly containing 4–10 flowers.
    4. Stem glabrous, often slightly glaucous; leaves slightly glaucous.   5. *C. odoratissimus*
    4. Stem manifestly pubescent, not glaucous.
      5. Heads in a more or less storied and flat-topped corymbiform panicle.   4. *C. carnosus*
      5. Heads in short corymbs racemosely arranged or corymbiform-paniculate, the inflorescence overall short- to long-cylindric and neither storied nor flat-topped.   6. *C. paniculatus*

1. Carphephorus corymbosus (Nutt.) T. & G.

Stem solitary from a stoutish basal crown, 3–15 dm tall, moderately short-pubescent throughout. Lowermost leaves in a basal rosette, mostly or all of them often shed by flowering time; rosette and lowermost stem leaves with spatulate bladed portions gradually narrowed to subpetiolar bases, mostly 5–15 cm long and 0.8–2.5 cm broad distally; leaves from a little above the base of the stem numerous, ascending to erect, gradually smaller and much reduced upwardly, sessile, oblong-spatulate or oblong; all leaves with entire, thickened margins, surfaces inconspicuously punctate and atomiferous-glandular. Heads several to numerous, relatively compactly to loosely corymbose-paniculate, their short-hirsute stalks with relatively numerous, alternately disposed bracts grading into those of the involucres. Involucres campanulate, 7–10 mm high and as broad at the summit; bracts 15–20, imbricated, margins scarious, mostly oblong with rounded tips, at least the inner with distal margins finely erose, medial and inner ones 6–10 mm long. Flowers mostly 15–20 per head; receptacle of the head chaffy at least toward the margin, chaff 6–9 mm long, linear-oblong below and slightly dilated apically, tips rounded and minutely erose marginally, usually bearing irregular resinous lines on their distal halves. Corollas mostly lavender-pink, not atomiferous-glandular exteriorly. Achene 3–4 mm long, dark brown with pale pubescence, finely dark punctate-dotted when immature, the dots usually not discernible at maturity; pappus 2.5–3 times as long as the achenes, bristles straw-colored below, tips usually lavender-pink.

Often abundant in well-drained, open pinelands, occurring with less frequency and abundance in seasonally wet, boggy places in pinelands. s.e. Ga., s. in Fla. to about s. end of Lake Okeechobee, in the Panhandle westward to Taylor and Madison Cos.

2. Carphephorus tomentosus (Michx.) T. & G.

Stem solitary from a basal crown, unbranched below the inflorescence, 4–7 (–10) dm tall, sparsely to densely pubescent with spreading hairs, these longest on the lower part of the stem, also atomiferous-glandular. Leaves near the base of the stem largest, with oblanceolate bladed portions gradually narrowed to petioles, 5–14 (–20) cm long overall; leaves markedly reduced above the lowermost and upwardly gradually becoming bracteate, grading from short-oval to short-oblong; all leaves with entire margins, usually densely to sparingly pubescent, punctate, and copiously atomiferous-glandular, sometimes glabrous or nearly so. Heads in open storied and flat-topped, thyrsoid corymbs, their stalks with few, short, broad bracts. Involucre campanulate, 6–10 mm high; bracts relatively numerous, imbricated, margins not scarious, the outer deltoid or deltoid-ovate, obtuse apically, grading inwardly to oblong and acute, purple, densely pubescent and gland-dotted, tips not erose. Flowers mostly 20–30 per head; chaff usually fewer than the flowers, mostly 8–12 mm long, linear and straw-colored but with slightly dilated tips, the latter purple, pubescent, and gland-dotted exteriorly. Corolla purple or lavender, atomiferous-glandular exteriorly. Achene 3–4 mm long, dark brown and sparingly to moderately short-pubescent, somewhat viscid; pappus about twice as long as the achene, bristles markedly unequal, usually lavender-pink throughout.

Pine savannas and flatwoods, s.e. Va. to S.C. and just into Ga.

3. Carphephorus pseudo-liatris Cass.   Fig. 389

Plant usually with a solitary stem or infrequently with 2 to several stems from a basal crown, the latter bearing numerous, slender fibrous strands remaining from disintegrated bases of leaves of previous seasons. Stems stiffly erect, unbranched below the inflorescence, mostly 3.5–7 dm tall, shaggy-pubescent throughout. Leaves sessile, margins entire and ciliate; larger leaves basal and on the lower stem, narrowly linear and involute, 12–34 cm long and 1–2 mm broad; midstem and upper leaves very much shorter than the lower, numerous, 1.5–2.5 cm long, closely appressed to the stem, subulate. Inflorescence of heads compactly corymbose, heads numbering 5–15, mostly 7–12, obconic to campanulate, the distal halves of their stalks with several alternately disposed, small subulate bracts grading into those of

Fig. 389. **Carphephorus pseudo-liatris:** a, base of plant; b, midsection of stem and upper section of stem and inflorescence; c, head, at anthesis; d, flower; e, scale or chaff that subtends flower; f, achene.

the strongly imbricated involucres, the latter 6–9 mm high. Involucral bracts about 15–25, moderately to densely grayish pubescent and inconspicuously resin-dotted, mostly lance-ovate or lanceolate or the innermost linear, apices acute, usually purplish. Flowers 2–35 or more per head; chaff about as numerous as the flowers, linear-oblong, 5–9 mm long, straw-colored excepting the slightly dilated, triangular, purplish, pubescent tips. Corollas lavender. Achene about 3 mm long, grayish, pubescent; pappus about twice the length of the achene, bristles straw-colored.

Seasonally wet pine savannas and flatwoods, open bogs, seepage slopes of pinelands. s.w. Ga., cen. and w. Fla. Panhandle, westward to s.e. La.

4. Carphephorus carnosus (Small) James.

Stem solitary, unbranched below the inflorescence, 3–5 (–9) dm tall, densely short-pubes-cent throughout. Largest leaves in a basal rosette, with oblanceolate bladed portions acute apically and gradually narrowed below to short subpetiolate bases, 3–7 (–9) cm long, linear-oblong, gradually shorter upwardly, uppermost reduced to bracts. Heads in a relatively open, more or less elongate, storied and flat-topped, thyrsoid panicle, axes 1- to few-bracted. In-volucre campanulate, 4–6 mm high; bracts 6–9 (–11), loosely imbricated, moderately to densely short-pubescent, ciliate marginally, and atomiferous-glandular, green or suffused with a little purplish pigment, middle and inner ones oblong to elliptic, rounded or obtuse apically. Flowers 4–10 per head; pales none or few, if present, 3–3.5 mm long, linear, resin-dotted and with pale pubescence. Corollas lavender or lavender-purple, atomiferous-glandu-lar exteriorly. Achene about 2.5 mm long, brownish gray, pubescent and resin-dotted; pappus about twice as long as the achene, the bristles usually straw-colored below, lavender-pink distally. (*Litrisa carnosa* Small)

Seasonally wet pinelands and prairies. cen. pen. Fla.

5. Carphephorus odoratissimus (J. F. Gmel.) Hebert.

Plant with a short, stout, subterranean caudex, sometimes bearing basal offshoots but usually only the central caudex bearing a solitary, glabrous, often slightly glaucous stem to 1.5 (–2) m tall, unbranched below the inflorescence. Rosette leaves largest but variable in size, with narrowly to broadly oblanceolate to spatulate bladed portions narrowed below to petiolar or subpetiolar bases, 5–30 (–40) cm long overall, 1–6 (–12) cm broad at their broadest places; leaves low on the stem (above the rosette) strongly erect-ascending, often, but not always, large-bladed, others gradually reduced upwardly, all but the lowermost sessile, subclasping to clasping, from midstem to the inflorescence much reduced, the uppermost bractlike; mar-gins of rosette and lower stem leaves usually entire, reduced leaves often with a few undulate or crenate-serrate teeth; leaf surfaces glabrous and slightly glaucous, minutely punctate. Heads relatively numerous to very numerous, in an open, more or less storied and flat-topped corymbiform cyme, the latter variable in size up to 5–7 dm high, axes and stalks of the heads glabrous or somewhat glandular. Involucre narrowly campanulate, 3–4 (–5) mm high; bracts usually about 10–12, purplish, imbricate, the outer short and deltoid to oblong, grad-ually longer inwardly, the innermost spatulate to lance-spatulate, markedly punctate and sparingly to densely beset with atomiferous glands, the glands sometimes short-stalked. Re-ceptacle of the head naked or with few pales. Corollas lavender to purple, atomiferous-glan-dular exteriorly. Achene 2–3.5 mm long, buffish, pubescent and atomiferous-glandular; pappus about ⅓ longer than the achene, bristles somewhat unequal, straw-colored. (*Trilisa odoratissima* (J. F. Gmel.) Cass.)

Savannas and pine flatwoods, less frequently in well-drained open pinelands. Coastal plain, e.cen. N.C. to s. pen. Fla., westward to s.e. La.

6. Carphephorus paniculatus (J. F. Gmel.) Hebert.   Fig. 390

Plants with a stoutish subterranean caudex, the latter eventually with several branches each bearing a rosette, a single, erect, spreading-hirsute stem 3–12 dm tall borne from one or more of each rosette in a given season. Rosette and few lowermost leaves with an oblanceo-

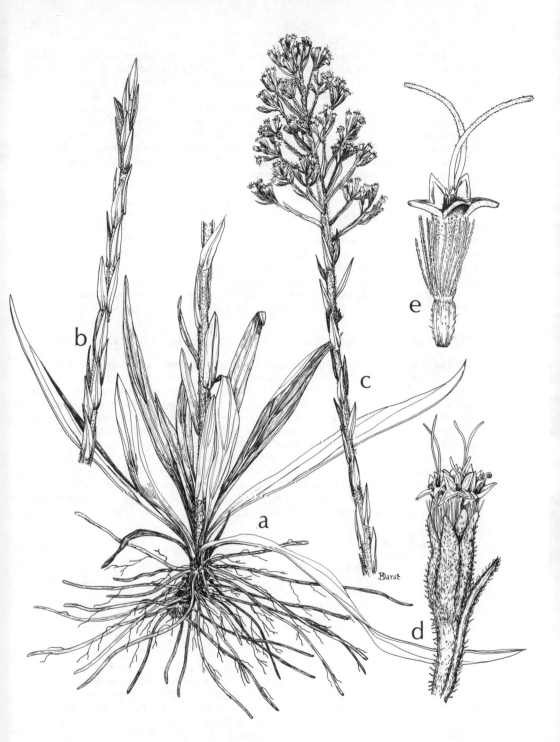

Fig. 390. **Carphephorus paniculatus:** a, base of plant; b, midsection of stem; c, top of plant; d, flowering head; e, flower.

late to spatulate bladed portion gradually narrowed below to a winged petiolar or subpetiolar base, varying greatly in size, 5–20 (–30) cm long and 1–2 (–3) cm broad distally, surfaces glabrous and minutely and inconspicuously punctate; leaves from just above the base numerous, markedly reduced and gradually smaller upwardly, mostly strongly ascendent or even appressed-erect, varying from oblanceolate below, to oblong, to linear-subulate above; surfaces glabrous or pubescent near the base and on the midrib beneath, sometimes on the margins; margins of all leaves usually entire, infrequently obscurely toothed. Heads commonly in short, compact corymbs racemiformly disposed on a short to elongate main axis thus the inflorescence of heads short- to long-cylindric overall, varying to corymbiform-paniculate and more or less cylindric overall, never storied and flat-topped unless the terminus of the axis damaged during development; Inflorescence axes and stalks of the heads usually purplish, hirsute and often glandular. Involucre purple, campanulate, 4–6 mm high and as broad distally; bracts 6–11, subequal, loosely imbricate, mostly elliptic to oblong, apices rounded to acute, sparsely to moderately short-pubescent, the hairs often gland-tipped, or mainly atomiferous-glandular. Flowers 4–10 per head; pales none or (reportedly) few. Corolla bright purple to lavender, rarely white, atomiferous-glandular exteriorly. Achene 2–3 mm long, sparsely pubescent, atomiferous-glandular; pappus about twice as long as the achene, bristles unequal, straw-colored. (*Trilisa paniculata* (J. F. Gmel.) Cass.)

Pine savannas and flatwoods, open bogs, borders of or burned-over shrub-tree bogs or bays. Coastal plain, e.cen. N.C. to s.cen. pen. Fla., Fla. Panhandle, s.e. Ala.

37. Liatris (BLAZING-STARS, GAY-FEATHERS, BUTTON-SNAKEROOTS)

Perennial herbs with a tuberous thickened, usually cormlike, rootstock. Stems solitary or 2 to several from the tuberous base, each unbranched, the inflorescence a spike or raceme, sometimes diffusely branched, notably, but not necessarily, if the terminal bud aborts or is damaged during development; uppermost heads at anthesis first. Leaves alternate, simple, margins entire, surfaces more or less punctate, narrow and sessile, or the basal and lower stem leaves petiolate, lower largest, diminishing in size upwardly. Heads discoid, each bearing few to numerous flowers, chaff none. Involucre of imbricated, often punctate, bracts; bracts wholly herbaceous and green, suffused with pink to purplish pigment throughout, or with thin margins and tips, the latter commonly pink to purple. Flowers radially symmetrical, bisexual and fertile. Corollas pink to purple (white in a rare individual), tube usually narrowly funnelform, lobes 5. Achene about 10-ribbed, pubescent; pappus of 1 or 2 series of barbellate or plumose bristles about 12–40 in number.

1. Subterranean rootstock cormlike.
    2. Heads in racemes. 1. *L. gracilis*
    2. Heads in spikes.
        3. Involucral bracts appressed, not recurved at their tips.
            4. Medial and inner involucral bracts broadly rounded apically. 2. *L. spicata*
            4. Medial and inner involucral bracts acuminate apically. 3. *L. acidota*
        3. Involucral bracts, at least the medial and inner ones, reflexed or spreading at their tips.
            4. *L. pycnostachya*
1. Subterranean rootstocks a cluster of tuberous thickened roots. 5. *L. garberi*

1. Liatris gracilis Pursh. Fig. 391

Subterranean base usually cormlike, sometimes gnarled or irregular. Stem 4–10 dm tall, often purplish, unbranched or less frequently diffusely short-branched (usually in the lower half and even without abortion of, or damage to, the terminal bud), copiously to sparsely shaggy-pubescent. Basal and lower stem leaves with lanceolate to linear-lanceolate bladed portions gradually attenuate to slender petioles of varied lengths, 6–15 cm long overall, apices acute to blunt; reduced stem leaves lanceolate to linear-subulate, mostly sessile; surfaces moderately punctate, glabrous or with a few long hairs on the midrib and near the margins beneath. Heads usually numerous, racemose, their stalks 3–12 (–18) mm long, divaricate,

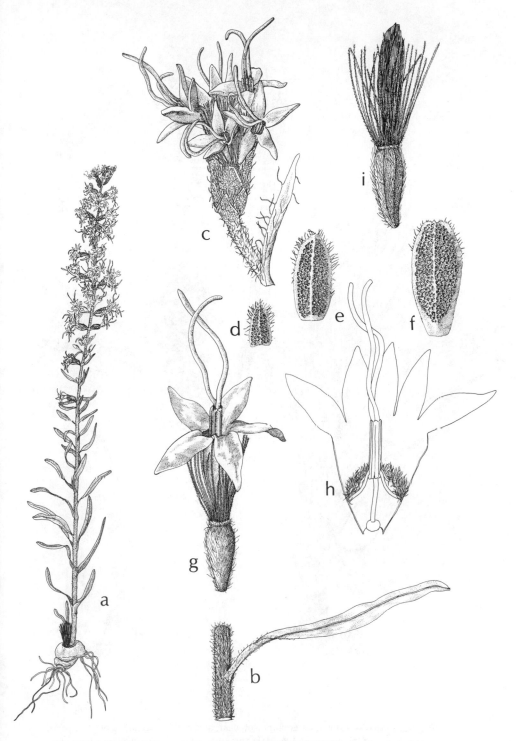

Fig. 391. **Liatris gracilis:** a, habit; b, small portion of stem with leaf; c, head; d–f, involucral bracts; g, flower; h, corolla opened up; i, achene.

Fig. 392. **Liatris spicata:** a, lower portion of plant; b, part of midsection of plant; c, part of inflorescence; d, head, at anthesis; e, flower; f, achene.

ascending, or deflexed. Involucres turbinate, 5–7 mm high; bracts glandular-punctate, variously suffused with purple, edges sometimes scarious, margins ciliate, sometimes pubescent on the backs as well, tips abruptly short-pointed to rounded. Flowers mostly 3–6 per head. Corollas lavender-pink to purple, tube about 5 mm long, atomiferous-glandular without, lobes 3 mm long. Achene obconic, 3–5 mm long, pappus about 5 mm high, usually purplish, bristles barbellate. (*Laciniaria gracilis* (Pursh) Kuntze, but not sensu Small, 1933; incl. *Liatris laxa* (Small) K. Schum., *Laciniaria laxa* Small)

Seasonally wet pine savannas and flatwoods, grass-sedge bogs; also, and more frequently and abundantly, in well-drained pinelands. Coastal plain, s.e. S.C. to s. pen. Fla., westward to s.e. Miss.

2. Liatris spicata (L.) Willd.   Fig. 392

Subterranean base cormlike, commonly with few to numerous fibers from the summit. Stem usually wandlike and unbranched (infrequently branched, the terminal bud having been destroyed at a relatively early stage of growth), very variable in stature, 6–25 dm tall. Stem and leaves usually glabrous, but, in specimens of some populations, having both more or less pilose. Leaves alternate, punctate, lower longest and with winged, subpetiolar bases, their broader portions linear or linear-lanceolate, 1–4 dm long overall and 5–20 mm broad (mostly 10 mm or less); leaves gradually reduced upwardly, the uppermost subulate. Inflorescence of heads spicate, varying from about 0.6 to 8 dm long, heads sometimes well separated from each other, varying to densely overlapping. Heads varying in width and length, subcylindric to campanulate or turbinate-campanulate, 8–15 mm long, 4–18-flowered. Involucral bracts appressed, imbricated in several series, all except the outermost oblong, apices rounded or nearly truncate, herbaceous centrally and atomiferous-glandular, scarious marginally and around the apices, sometimes wholly purple, varying to purplish or straw-colored on the scarious margins. Corollas deep rose-purple, 6.5–9 mm long, lobes triangular-subulate. Achene narrowly obconical, 4–6 mm long, sparsely short-pubescent, dull grayish black; pappus 5–7 mm high. (Incl. var. *resinosa* (Nutt.) Gaiser; *Laciniaria spicata* (L.) Kuntze)

Bogs, seepage slopes, wet meadows, pine savannas and flatwoods, borders of shrub-tree bogs or bays, ditches. s. N.Y. to Mich., generally southward to s. Fla. and La.

3. Liatris acidota Engelm. & Gray.

Subterranean rootstock cormlike, usually bearing from its summit fibrous remnants of previous years' leaves. Stem slender, 5–8 dm tall, glabrous. Leaves glabrous, the basal and lower stem leaves elongate, narrowly linear, others reduced, erect, linear-subulate. Heads few to numerous, spicate, ascending. Involucres 8–10 mm high, mostly turbinate; bracts few, a little suffused with purple to wholly purple, the outer mostly short-triangular-acute, becoming oblong inwardly, medial and inner ones with acuminate apices, the very tips stiffish. Corollas purple, tube 6–7 mm long, lobes mostly about 2 mm long, atomiferous-glandular exteriorly, sometimes glandular near the summit of the tube exteriorly. Achene 4–5 mm long, obconic, minutely glandular between the ribs; pappus 6 mm high, bristles purplish, strongly barbellate.

In moist to wet soils, pine savannas, prairies. Gulf coastal regions of La. and Tex.

4. Liatris pycnostachya Michx.   Fig. 393

Subterranean rootstock cormlike or more or less cylindrical and rhizomelike, usually bearing at its summit fibrous remnants of previous years' leaves. Stems 1 to several from the rootstock, each 6–15 dm tall, glabrous, sparsely to copiously shaggy-pubescent throughout, or thus pubescent mainly in the inflorescence. Basal and lowermost stem leaves with narrow, elongate, bladed portions 4–5 mm broad and long-attenuate basally to petiolar portions, attenuate distally; other leaves markedly reduced, sessile, gradually diminishing in length upwardly; leaf surfaces punctate, wholly glabrous or sparsely pubescent beneath. Heads usually numerous, crowded in a spike, each bearing 5–7 (–12) flowers. Involucres cylindric to turbi-

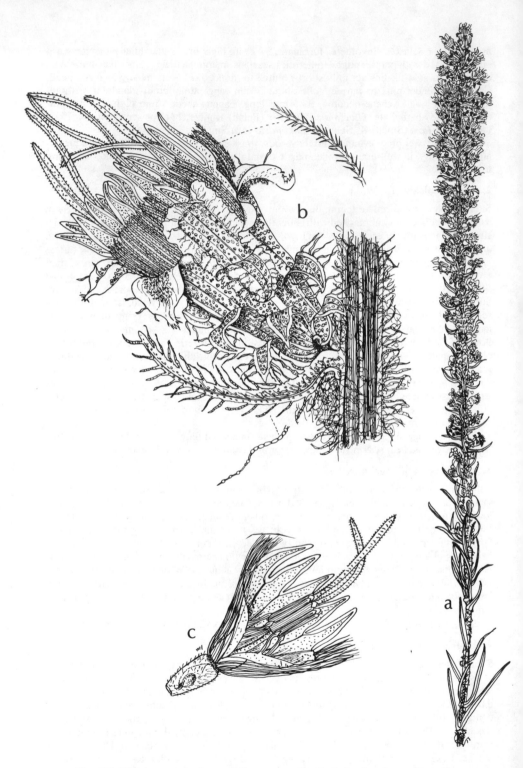

Fig. 393. **Liatris pycnostachya:** a, inflorescence of heads; b, head; c, flower. (From Correll and Correll. 1972)

nate, 8–10 mm high; bracts wholly green, wholly bright purple, or green centrally on the backs and with scarious margins and tips bright purple to pink-purple, bracts all acuminate apically, the medial and inner with reflexed or spreading tips; exterior surfaces conspicuously atomiferous-glandular, margins mostly but not always ciliate, often sparsely pubescent on the backs. Corolla bright purple (white in rare individuals), tube about 6 mm long, lobes 2–3 mm long, distal portions of the tube and the lobes atomiferous-glandular exteriorly. Achene obconic, 6 mm long or a little more, brown; pappus 7–8 mm high, straw-colored or pale purplish, bristles barbellate. (*Laciniaria pycnostachya* (Michx.) Kuntze)

Moist to dry prairies, prairielike rights-of-way, in and around bogs, boggy ditches and depressions, open well-drained woodlands and clearings. Ind. to S.D., generally southward to Miss., La., Tex.

5. Liatris garberi Gray.

Subterranean base usually a cluster of slenderly tuberous-thickened roots. Stem 2–8 dm tall, green, sparsely shaggy-pubescent below, more densely so in the inflorescence. Basal leaves narrowly linear to linear-oblanceolate, attenuate basally to subpetioles of varied lengths, 7–30 cm long overall and 2–8 mm broad, the attenuate bases often more or less ciliate on their margins; leaves much reduced above the base and markedly diminishing in size upwardly, mostly linear-setaceous; surfaces conspicuously punctate. Heads few to numerous, racemose, their stalks pubescent and punctate, 3–8 mm long. Involucres 7–14 mm high, turbinate-campanulate; bracts conspicuously glandular-punctate, green or purplish throughout, or purplish only along the margins, with or without narrow scarious margins; outer bracts short-triangular-acute, medial elliptic, inner oblong, tips of both of the latter usually mucronate, on some specimens broadly rounded and without mucros, margins conspicuously ciliate or without cilia, the backs not pubescent or pubescent centrally. Flowers 5–10 per head. Corolla tube 6–7 mm long, lobes 2–3 mm. Achene 3–4 mm long, narrowly obconic, dark brown; pappus about 6 mm high, faintly purplish, bristles barbellate. (*Laciniaria garberi* (Gray) Kuntze; incl. *Laciniaria chlorolepis* Small)

Low pinelands, grass-sedge bogs. cen. and s. pen. Fla.

38. Vernonia (IRONWEEDS)

Perennial herbs (those treated here). Leaves alternate, pinnately veined, mostly much longer than broad. Heads discoid, few to very numerous in a terminal, often more or less leafy corymb. Involucre campanulate to hemispheric, bracts imbricated, the outer shortest, progressively longer inwardly. Receptacle of the head flat to subconvex. Chaff none. Flowers all radially symmetrical, bisexual and fertile. Corollas (in ours) pale to deep purple or rose-purple, tube narrowly cylindrical or narrowly funnelform, surmounted by 5 lanceolate to linear or subulate lobes. Achene ribbed (in ours), often resin-dotted between the ribs; pappus an outer (not easily perceived) series of minute bristles and an inner series of numerous, purple, brown to buff-colored, or nearly white, capillary bristles.

Reference: Jones, Samuel B., and W. Zack Faust. "Compositae Tribe Vernonieae." *North American Flora* Ser. II. Part 10: 180–196. 1978.

There is considerable hybridization between pairs of species in places where their ranges overlap; the identification of many specimens is thus difficult if not impossible except for those persons having a very intimate knowledge of the group.

1. Medial and inner involucral bracts broad-based and with filiform tips.
  2. Leaf surfaces both punctate.   1. *V. arkansana*
  2. Leaf surfaces not punctate.   2. *V. novaboracensis*
1. Medial and inner involucral bracts acute, short-acuminate, or short-mucronate.
  3. Midstem leaves linear, less than 1 cm broad.
    4. Stem glabrous; plants of Ouachita River drainage, Ark. and Okla.   3. *V. lettermannii*
    4. Stem sparsely short-pubescent; plants of s. Fla., Fla. Keys, and Bah.Is.   4. *V. blodgettii*
  3. Midstem leaves broader, 2–4 cm broad or more.

Fig. 394. **Vernonia arkansana:** a, base and top of plant; b, flower. (From Correll and Correll (1972), as *V. crinita*)

5. Leaves tomentose beneath; heads bearing 30–60 flowers.   5. *V. missurica*
5. Leaves short-pubescent beneath, not tomentose; heads bearing 12–30 flowers.   6. *V. gigantea*

1. **Vernonia arkansana** DC.   Fig. 394

Stem 7–12 (–20) dm tall, leafy, sparsely short-pubescent below and becoming glabrous or nearly so in age, persistently short-pubescent with crinkly hairs on the inflorescence branches. Middle stem leaves sessile or often attenuate to subpetiolar bases, mostly narrowly elongate-lanceolate, long-attentuate to acute apices, 7–20 mm broad at their broadest places; both surfaces punctate, upper sparsely pubescent with pointed hairs, lower usually sparsely to copiously clothed with a mixture of short-pointed and minute clavate hairs; margins usually with small callus-teeth. Corymb usually leafy, irregular, the number of heads varying from 3 to about 30, each bearing numerous flowers. Involucres broadly hemispheric, mostly 10–15 mm broad and high, bases of bracts tightly imbricate excepting the outermost which grade into bracts of the stalks; outermost bracts long-subulate, others mostly with oval bases then rather abruptly contracted to long-filiform, spreading tips; outer bracts pubescent and gland-dotted throughout, others with the oval bases ciliate distally and pubescent and conspicuously gland-dotted centrally and distally, the filiform tips pubescent. Corollas about 11–12 mm long, bright purple, tube narrowly funnelform, lobes linear-acute. Achene about 6 mm long, cylindric, sparsely short-pubescent on the ribs, gland-dotted between the ribs; pappus dull purplish, the capillary bristles 5–6 mm long. (*V. crinita* Raf.)

Sand and gravel bars of streams, rocky stream bottoms, open bottomland woodlands, low pastures and wet meadows, prairies, rocky slopes and glades. Ozark region of Mo., Kan., Ark., Okla.; naturalized in Ill. and Wisc.

2. **Vernonia novaboracensis** (L.) Michx.

Relatively robust plant, 0.9–2 m tall or a little more. Stem leafy, green to dark purple (more commonly the latter), short-pubescent throughout but usually becoming glabrous below in age. Leaves short-petiolate, blades at midstem long-lanceolate to long-elliptic, mostly attenuate to both extremities, 12–28 cm long, 2–4 (–6) cm broad, glabrous to short-pubescent and slightly scabrous above, sparsely to copiously short-pubescent beneath, margins with short, callus-tipped teeth. Corymb leafy proximally, irregular, loose and spreading, heads relatively few to numerous, each bearing 30–50 (–65) flowers. Involucre campanulate or hemispheric, 6–10 mm high and about as broad; bracts usually purple, the outer loose, filiform, those inwardly with oval closely imbricated bases and acuminate to filiform tips; filiform outer bracts short-pubescent, basal portions of medial bracts conspicuously ciliate, those of innermost usually glabrous, tips of both mostly short-pubescent. Corollas 8–11 mm long, bright purple, tubes slightly flared distally, lobes linear-acute. Achene cylindric, about 4 mm long, with few glandular dots between the ribs; pappus dull brownish purple, the capillary bristles 6–7 mm long. (Incl. *V. harperi* Gl.)

Wet meadows, stream banks, moist to wet pastures, low open woodlands, thickets, and swales. Mass., N.Y., Pa., thence coastal plain, piedmont, and mountains southward to cen. Fla. Panhandle and Ala.

In the central Fla. Panhandle where this species is infrequent and *V. gigantea* is frequent and abundant, there are, perhaps, more specimens intermediate between them than there are those clearly identifiable as *V. novaboracensis*.

3. **Vernonia lettermannii** Engelm. ex Gray.

Stem 2–6 (–10) dm tall, leafy, glabrous. Leaves sessile, narrowly linear, to about 10 cm long and 1–2 mm broad, glabrous, both surfaces punctate, margins entire or with sessile callosities. Corymb leafy, with few to numerous heads, each bearing 10–15 flowers. Involucres campanulate-cylindric to cylindric, 7–10 mm high and 4–5 mm broad, bracts closely imbricate, more or less suffused with purple, the outer short-acicular, outer and medial short-pubescent on the margins, grading to oblong and acute or short-acuminate inwardly, the in-

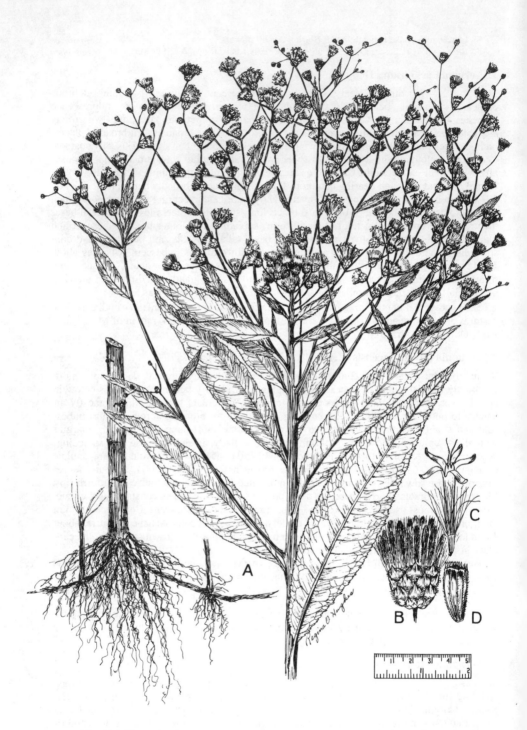

Fig. 395. **Vernonia gigantea** subsp. **gigantea:** A, plant; B, flower head; C, flower; D, achene. (From Reed, *Selected Weeds of the United States* (1970) Fig. 217, as *V. altissima*)

nermost glabrous. Corolla 10–11 mm long, bright purple, tube only slightly flaring distally, lobes linear-acute. Achene slightly curvate, obconic-cylindric, 3–4 mm long, gland-dotted between the glabrous ribs; pappus dull brownish to dull purplish, the capillary bristles about 6 mm long.

Gravel bars, rocky stream beds, rock ledges, edges of lakes or impoundments. Ouachita River drainage, s.w. Ark., s.e. Okla.

4. Vernonia blodgettii Small.

Slender plant, 2–5 dm tall. Stem usually purple, sparsely short-pubescent, leafy only near the base or leafy to the inflorescence. Principal leaves sessile, linear, narrowly linear-lanceolate, or oblanceolate, 6–7 cm long and 2–5 (–10) mm broad; lowermost leaves sometimes smallest, with a short blade and very short petiole; commonly only the midvein evident, this raised beneath, other veins faint if evident at all, margins entire, often revolute; upper surfaces sometimes punctate, usually sparsely pubescent with short, pointed hairs, commonly glabrous or nearly so in age, lower surfaces with longish pointed hairs on the midrib, at least when young, and minute, white, clavate trichomes either side of the midrib. Corymb with 3 to about 50 heads, each head bearing about 20 flowers. Involucre campanulate, 6–7 mm high and as broad distally; bracts loosely imbricate, purple, more or less ciliate, the outer and medial ones with acute tips, the inner very short-acuminate. Corollas about 10 mm long, purple, tubes narrowly funnelform, lobes subulate. Achene about 3 mm long, obconic-cylindric, sparsely pubescent on the ribs, gland-dotted between the ribs; pappus straw-colored, the capillary bristles 5–8 mm long. (Incl. *V. insularis* Gl.)

Seasonally wet pinelands and cypress prairies, usually where lime-rock outcrops or is near the surface. s. pen. Fla. and Fla. Keys; Bah.Is.

5. Vernonia missurica Raf.

Stem leafy, 1–2 m tall, sparsely, or more frequently, copiously short-pubescent with crinkly hairs. Leaves at midstem very short-petiolate to sessile, blades lanceolate to long-elliptic, 8–20 cm long, 2–4 (–6) cm broad, shortly tapered basally, acuminate apically, margins short-serrate, the teeth mostly callus tipped, upper surfaces scabrid, the lower tomentose. Corymb irregular, loose, leafy-bracted at least proximally, heads few to numerous, each bearing 30–60 flowers. Involucres ovoid-campanulate, 5–8 (–10) mm high, bracts closely imbricated and appressed, brownish or tips of the inner ones purple, outermost deltoid-acute, medial ovate and mucronate, inner oblong and mucronate, ciliate marginally, sometimes pubescent on their backs. Corollas 9–12 mm long, purple, tube narrowly cylindric for more than half its length from the base, then dilated, lobes elongate, linear-acute. Achene 4 mm long, somewhat flattened, oblong, with few glandular dots between the ribs; pappus dull brown or infrequently dull purple, the capillary bristles 6–7 mm long.

Low woodlands and their clearings, along streams, bottomland pastures, bogs, wet meadows, wet ditches, prairies. s.w. Ont. and Mich. to Iowa and Neb., generally southward to Ala. and e. and s.e. Tex.

6. Vernonia gigantea (Walt.) Trel. ex Branner & Coville subsp. gigantea.   Fig. 395

Stem leafy, 1–3.5 m tall (the largest specimens with an inflorescence to 1 m high and as broad, with a multitudinous quantity of heads), green to purple, puberulent but becoming glabrous below. Midstem leaves attenuate to subpetiolar bases 2–4 cm long, blades lanceolate, acute to acuminate apically, 15–25 cm long and 3–4 cm broad, margins sharply serrate to entire, upper surface slightly scabrous near the margins, lower short pubescent, sometimes only along the veins. Heads 12–30-flowered. Involucres campanulate to hemispheric, mostly about 5 mm high, bracts closely imbricated and appressed, usually more or less suffused with purple, margins mostly short-ciliate, the outermost subulate and acute, inwardly becoming oblong and the apices acute to mucronate. Corollas 5–6 mm long, bright purple, tube cylindric, somewhat broader in the distal ⅓ than below, lobes very short relative to the tube, varying from 1–1.5 mm long and nearly deltoid to linear-oblong and 2–3 mm long.

Achene about 3 mm long, oblong, sparsely hispid on the ribs, often lacking glandular dots between the ribs; pappus dull purple to light brown, the capillary bristles about 5 mm long. (Incl. *V. altissima* Nutt.)

Gravel and sand bars of streams, floodplain woodlands and their clearings, low thickets, ditches, stream banks, wet pineland clearings, wooded bluffs. w. Pa. and W.Va., Ohio, Ind., s.w. Mich., Ill., Mo., s.e. Neb., e. Kan., southward to Fla. Panhandle and e. Tex., apparently very infrequent in the Carolinas.

V. gigantea subsp. *ovalifolia* (T. & G.) Urbatsch (*V. ovalifolia* T. & G.) occurs in s.w. Ga., adjacent Fla. Panhandle, eastward in n. Fla. and southward to s. cen. Fla. Insofar as we know, it inhabits well-drained places.

39. Elephantopus (ELEPHANT'S-FOOT)

Perennial herb with 1 to several, subdichotomously branching stems from a basal crown; stems, leaves, bracts, and involucres pubescent and atomiferous-glandular. Leaves alternate, pinnately veined, unlobed, sessile or tapering to winged, subpetiolate bases. Heads discoid, 1–5-flowered, aggregated together into a compound head subtended by foliaceous bracts; involucre of individual heads of 4 pairs of bracts, the pairs arranged at right angles to each other, the 2 outer pairs about half as long and narrower than the inner; outer pairs subulate to linear-acute, thin, inner stiff, linear-oblong, acute or mucronate. Flowers bisexual. Corolla lavender, lavender-pink to deep purple, with 5 narrow, elongate lobes extending unilaterally at the summit. Receptacle naked. Achene narrowly obconic, 10-ribbed, antrorsely pubescent and atomiferous-glandular; pappus of 5–10 (sometimes more) antrorsely barbellate bristles dilated at the base.

- Principal, larger leaves in a basal rosette or near the base of the stem, markedly reduced above.
  1. *E. nudatus*
- Principal, larger leaves medial on the stem.
  2. *E. carolinianus*

1. Elephantopus nudatus Gray.   Fig. 396

Stem 2–6 dm tall, usually, but not always purple or purple-brown, sparsely and irregularly pubescent with stiffly ascending hairs, bearing the largest leaves in a basal rosette or near the base, markedly reduced above. Basal leaves narrowly to broadly oblanceolate, 8–20 cm long, 2–6 cm broad distally, cuneately narrowed below to subpetiolar bases, apices mostly obtuse, margins mostly shallowly crenate-serrate, both surfaces sparsely pubescent with stiff, ascending or spreading hairs. Compound heads terminating ultimate branches 1–8 cm long. Inner involucral bracts of individual head 7–8 mm long. Corolla deep purple. Achene brown with pale ribs narrower than the intervening spaces, 3 mm long or a little more; pappus bristles 5, 3–5 mm long.

Seepage slopes and depressions in pinelands, pine flatwoods, borders of cypress-gum depressions and shrub-tree bogs or bays, creek bottoms in woodlands, moist hammocks. Coastal plain, Del. to n.cen. pen. Fla., westward to s.e. Tex., Ark.

2. Elephantopus carolinianus Raeusch.

Stems 2–10 dm tall, usually green, infrequently purplish, with few to numerous branches, shaggy-pubescent, leafy about ⅔ of their length, the largest leaves medially on the stem, more or less reduced on distal portions. Leaves gradually to abruptly narrowed to subpetiolar bases, the larger ones 10–25 cm long, the dilated portions ovate to elliptic, mostly obtuse apically, 4–10 cm broad, margins crenate-serrate, surfaces sparsely pubescent. Inner involucral bracts of individual heads 8–10 mm long. Corollas lilac to purple. Achene 3.5–5 mm long, narrower than to equaling the intervening spaces; pappus bristles 4–5 mm long.

Bottomland forests and their clearings, river and stream banks, forested river bluffs, meadows, moist to wet thickets, rich woodlands. N.J. to Mo. and Kan., generally southward to s. Fla., e., n.e., and n.cen. Tex.

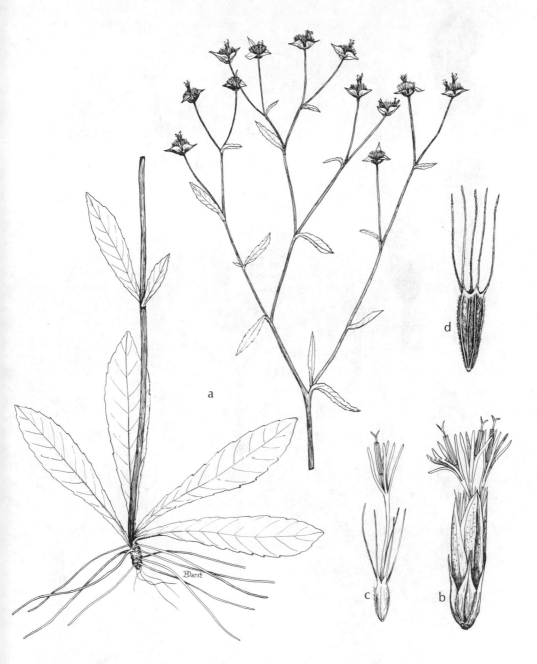

Fig. 396. **Elephantopus nudatus:** a, plant; b, individual head taken from conglomerate of heads; c, floret; d, achene.

Fig. 397. **Stokesia laevis:** a, plant; b, head, from above; c, involucral bracts, outer (left) to inner (right); d, flowers, outer (left) to inner (right); e, style and stigmas; f, achene.

40. Stokesia

Stokesia laevis (Hill) Greene. STOKESIA.   Fig. 397

Perennial herb with a stout caudex bearing thick roots and from which 1 to several stems arise. Stems up to 6 dm tall, bearing one head terminally (rarely), usually with several branches of uneven length each bearing a single head. Lower stems sparsely grayish-lanate, the upper stems and flower stalks densely lanate. Leaves alternate, glabrous, bright green, the lowermost approximate, the upper more distant; lowermost leaves with long-elliptic, entire blades narrowed to broadly winged petioles, the apices obtuse. Upward, the leaves soon becoming sessile, oblanceolate, some entire, others with a few aristate teeth near the base; upper bracteal leaves of the peduncles ovate-lanceolate, aristate-toothed often several to numerous near the summit of the peduncle and gradually merging with the outer involucral bracts. Outer and median involucral bracts foliaceous, sparsely lanate, oblanceolate, conspicuously, softly, pectinately, spinose-toothed, some of the teeth branched; innermost series of bracts linear, submembranaceous, their surfaces smooth, the margins ciliate near the acuminate tips. Heads showy, 7–10 cm across, of many tightly compacted flowers, the corollas bright violet-colored. Large 5-cleft ray flowers are just within much smaller cleft-lobed radially symmetrical tubular flowers like those which comprise the bulk of the head within the largest flowers, but the outer several series of flowers with varying degrees of development of the ligules and very variable in size. All flowers bisexual and fertile. Achenes greenish white, smooth, tightly compacted on a naked receptacle, 4 mm long, 4-sided, the angles ridged, somewhat broadened upward but the summit contracted and with a depression around the short persistent style base. Pappus not present on the achenes (falling with the other flower parts when they are shed), consisting of linear-attenuate, soft, white scales, 3 each on the outer flowers of the head, 4 or 5 on the inner.

Seasonally wet, sandy-peaty soil, pine-savannas and flatwoods, boggy borders of evergreen shrub bogs, hillside seepage bogs, adjacent sandy-peaty roadsides, railroad savannas, clearings. Coastal plain, S.C. and Ga., Fla. Panhandle, westward to La. Cultivated for ornament and sometimes escaped beyond the natural range.

41. Cirsium (PLUME THISTLES)

Mostly biennial or perennial, often stout and coarse, herbs. Leaves alternate, their margins entire, more commonly with marginal spines or spine-tipped teeth, or spinose-lobed, or pinnatifid, lobes or segments of the latter irregularly toothed, teeth ending in spine-tips. Heads terminal and solitary on the main stem or its branches or branchlets, or variably clustered, discoid, many-flowered. Flowers bisexual and fertile (rarely functionally unisexual). Involucres ovoid, obconic, obpyramidal, campanulate, urceolate, subcylindric, or nearly globose, becoming very much broader distally as the whole involucre spreads during maturation of the fruits; bracts numerous, markedly imbricated, generally some or most of them spine-tipped, in some species some of them with a glutinous dorsal ridge distally. Receptacle flat or somewhat conic, abundantly clothed with soft bristles or hairs. Corollas very slenderly tubular below, expanded above to a slightly broader, cylindric throat with 5 narrowly linear lobes at the summit, pigmented part of the corollas usually restricted to the throat and lobes, variably lavender to purple or garnet, less frequently pale yellow to nearly white. Achene glabrous, somewhat compressed or obscurely angular, often curvate, not ribbed; pappus of numerous plumose, soft, white, capillary bristles united basally into a ring, the entire pappus deciduous as a unit.

1. Heads subtended by a series of conspicuously spiny-toothed, erect bracts collectively more or less investing the head and forming a false or subinvolucre.   5. *C. horridulum*
1. Heads without such a series of bracts subtending and investing the head.
  2. Leaf bases, those of at least some leaves, decurrent on the stem, sometimes extending from node to node.

3. Lower stem conspicuously winged from node to node by decurrent leaf bases, the wings of uniform width the length of the nodes; involucre about 2 cm high.   2. *C. nuttallii*
3. Lower stem nodes partially winged by decurrent leaf bases, the wings inconspicuous and strongly tapering from the node to their endings below the node; involucre 3–3.5 cm high.   1. *C. lecontei*
  2. Leaf bases, none of them, decurrent on the stem.
    4. Lower leaf surfaces densely gray to white felty-pubescent, the pubescence easily perceived; outer and medial involucral bracts acute apically and spine-tipped.   3. *C. virginianum*
    4. Lower leaf surfaces pale green, thinly and so tightly tomentose as to appear smooth, or sometimes without pubescence in age; none of the involucral bracts spine-tipped, apices of the outer and medial ones rounded to obtuse.   4. *C. muticum*

1. Cirsium lecontei T. & G.   Fig. 398

Biennial with stout, subtuberiform roots. Stem 4–10 dm tall, usually unbranched and with a single terminal head, less frequently with 2 to several elongate branches each terminated by a head. Stem and lower surfaces of leaves clothed with a grayish, loose cottony web of pubescence, this sometimes partially or almost wholly disappearing in age, the upper leaf surfaces glabrous or with a few long hairs. Leaves of the lower stem long, the bladed portion gradually narrowed to a long, undifferentiated subpetiole, 15–20 cm long overall, bladed portion narrowly elliptic-oblong, 1–1.5 cm broad, dentately spine-lobed or dentately spine-toothed; leaves becoming sessile not far above the base of the stem, gradually reduced upwardly, eventually becoming bractiform, stem sometimes naked 5–10 cm below the head, leaf bases to some extent decurrent on the stem, sometimes from node to node. Involucre rounded basally then tapered to broadly obconic in outline, 3–3.5 cm high, 2.5–3 cm broad at the summit at anthesis; outer bracts narrowly short-subulate, gradually becoming a little longer and a little wider to about midinvolucre, shortly spine-tipped, with a glutinous dorsal ridge distally, the inner bracts then much longer, essentially linear below, gradually long-tapering distally to relatively soft points. Corolla tube about 3 cm long, lobes erect, 5 mm long, the lower tube pale, upper tube and lobes pinkish lavender or pinkish purple. Achene brown, 2–3 mm long; pappus bristles 3 cm long. (*Carduus lecontei* (T. & G.) Pollard)

    Wet pine savannas and flatwoods, sphagnous bogs, frequently associated with anthills. Coastal plain, s.e. N.C. to Fla. Panhandle, westward to La.

2. Cirsium nuttallii DC.

Biennial of variable stature, varying from slender and about 7 dm tall to coarse and to as much as 3.5 m tall, the low, slender plants with few heads in a loose, terminal corymb, larger plants with a much more branched, loose, open corymb, largest and most vigorous ones with long inflorescence branches cumulatively comprising 1.5 m or more of the height; ultimate head-bearing branchlets few-bracteate. Stem strongly and irregularly ribbed, thinly cobwebby-pubescent, lower internodes wholly spiny-winged by decurrent leaf bases, progressively less so upwardly. Leaves sessile, deeply pinnatifid (medial segments largest), larger ones 10–25 cm long and 3–10 cm broad (across the longest segments), primary segments irregularly and coarsely sublobed or toothed, lobes or teeth spine-tipped; upper surfaces essentially glabrous, at least at maturity, the lower cobwebby-pubescent, pubescence more or less sloughing in age. Involucre campanulate at anthesis, about 2 cm high and as broad distally; outer bracts subulate, grading to broader and a little longer to about midinvolucre, acute apically, with glutinous dorsal ridges distally, tipped with stiff, abruptly spreading, short spines; inner bracts longer, linear-acute to -attenuate, extreme tips purplish and usually crisped. Corolla tube about 2 cm long, lobes 6–7 mm long, erect, upper tube and lobes lavender or lavender pink, often pale, even nearly white. Achene 4–5 mm long, oblongish-curvate, dull pale purplish and finely speckled or streaked with dark purple save for a dull buffish or scarcely lustrous distal band. (*Carduus nuttallii* (DC.) Pollard)

    Marshes, banks of marshes, clear-cut or much-disturbed pinelands, moist to wet thickets and roadsides, also in uplands particularly where the soil mechanically disturbed or where pastured. Chiefly coastal plain, s.e. Va.; S.C. to s. pen. Fla., westward, chiefly along the Gulf coast, to Miss. (and La.?)

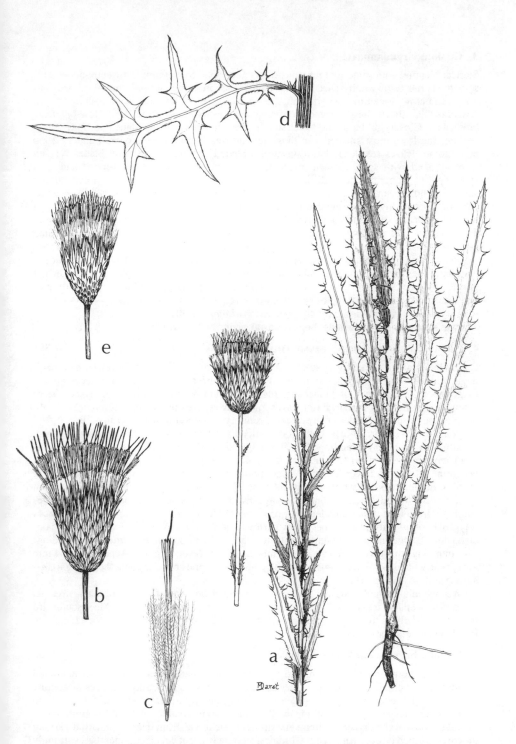

Fig. 398. a–c, **Cirsium lecontei:** a, plant, cut into three sections; b, head, at anthesis; c, flower; d,e, **Cirsium muticum:** d, midstem leaf; e, head, at anthesis.

3. Cirsium virginianum (L.) Michx.

Slender biennial with somewhat tuberous-thickened roots. Stem thinly cobwebby-pubescent, sometimes partly or wholly shedding the pubescence in aging, irregularly low-ribbed, unbranched below, sometimes wholly unbranched with a single terminal head, usually at least few-branched above, heads vaguely corymbosely disposed on rather long, few-bracteate branchlets. Closely set basal and lower stem leaves subpetiolate, the bladed portions oblanceolate, the lowermost smaller than those just above, 5–15 cm long overall and 1–2 cm broad; other leaves sessile, reduced upwardly; bladed portions of leaves or blades varying from unlobed with spiny-toothed margins, to irregularly dentately short-lobed and with spine-tips, to pinnatifid and with simple or forked or toothed spine-tipped segments; upper surfaces bright green and glabrous or with few cobwebby hairs, lower surfaces densely gray-felty-pubescent. Involucre rounded at the base and spreading upwardly, 1.5–2 (–2.5) cm high; outer bracts short-subulate, grading to linear-subulate at about midinvolucre, with relatively narrow and moderately glutinous dorsal ridges distally and with erect to spreading spine-tips 1–3 mm long; inner bracts gradually longer, linear-attenuate, without glutinous ridges, the tips long-pointed but not spiny, the extreme tips purplish. Corolla tube about 1.5 cm long, lobes about 4 mm long, tube pale below, upper tube and lobes rose-purple to bright purple. Achene 3–4 mm long, somewhat oblanceolate in outline and usually slightly curvate, dull buff-colored below a distal band that is lustrous yellowish; pappus 1.5–1.8 cm long. (*Carduus virginianus* L.; incl. *Cirsium revolutum* Small)

Pine savannas and flatwoods, bogs. Coastal plain, N.J. to s. Ga.

4. Cirsium muticum Michx. SWAMP THISTLE.   Fig. 398

Coarse biennial with a relatively stout basal caudex. Stem to 2 (–3) m tall, in vigorous specimens with numerous branches from about midstem, heads borne terminally on relatively long, bracteate branches or branchlets, the main stem several-ribbed, cottony-pubescent when young, glabrous in age. Basal leaves gradually tapered below to undifferentiated subpetioles, other sessile; blades of larger leaves roughly oblong-elliptic in overall outline, 1.5–3 dm long, 2–20 cm broad (across the lobes), deeply pinnately lobed, the spine-tipped lobes usually with 1–2 (–3) spine-tipped, coarse dentations, often with some short, slender, marginal spines as well; upper leaf surfaces dark green and sparsely short-pubescent, sometimes becoming glabrous, lower surfaces paler, thinly and so tightly tomentose as to appear smooth. Involucre broadly turbinate at anthesis, 2 (–3) cm high, thinly cobwebby; bracts not spine-tipped, the outer shortly linear-oblong, inwardly to about midinvolucre gradually becoming a little longer and broader, with prominent, glutinous, dorsal ridges distally, apices rounded to obtuse, inner bracts becoming narrower and acute, the innermost linear-attenuate, distal dorsal ridges narrow or faint. Corolla tube about 2 cm long, lobes somewhat spreading, 5–6 mm long, the lower tube pale, upper tube and lobes lavender-pink. Achene about 4 mm long, nearly black, obscurely angled, slightly curvate; pappus bristles about 2 cm long. (*Carduus muticus* (Michx.) Pers.)

Wet woodlands and marginal thickets, in wooded borders of small tributary streams, meadows, seepage areas. Nfld. to Sask., generally southward to Del. and Mo., southward from Md. chiefly in the Appalachian mts. but with scattered localities in the coastal plain to Fla. Panhandle, westward to e. Tex.

5. Cirsium horridulum Michx. (complex).

Biennial, to about 1 m tall, sometimes to 1.5 m, varying greatly as to robustness of stems and amount of branching in the inflorescence of heads. Stems with varying densities of cottony pubescence. Rosette and larger stem leaves varying greatly in size and form from plant to plant, sometimes strongly pinnatifid, the divisions variably sublobulate and spiny-toothed, the main body of the divisions often very strongly puckered; from deeply pinnatifid varying to scarcely lobed or only with marginal spines; stem leaves not decurrent, their bases strongly clasping to truncate; leaf pubescence highly variable. Heads subtended and more or less invested by erect, spiny-toothed, foliaceous bracts that may be construed as a subinvolucre or false involucre. Outer and medial involucral bracts spine-tipped and variously spinose-margined and pubescent. Corolla color, from plant to plant, deep garnet-purple, bright rose-lavender to pale lavender or essentially white, or cream-yellow; plants of local populations having corollas predominantly of one or another of the colors, or wholly of one or another of the colors; analysis of color variance may show it to be significantly different geographically and seasonally. Respecting seasonality, in parts of the overall range there appears to be an early spring surge of flowering and fruiting, a late spring surge, and a late summer and autumn surge.

Meadows, pine savannas and flatwoods, upland, open pine woodlands, pastures, marshes and marsh-prairies, cypress-prairies, shores, ditches, roadsides. Coastal plain and piedmont, s. Maine to s. pen. Fla. and Fla. Keys, westward to e. and s.e. Tex., e. Okla.

This is an enigmatic complex of thistles which in our judgment is very poorly understood. Plants belonging to it share one very easily discernible character which sets them apart from other kinds of plume thistles occurring in our range: the heads are subtended by a series of conspicuously spiny-toothed, erect bracts more or less investing the head and forming what may be construed as a false or subinvolucre.

Small (1933) recognized the complex as comprised of three species (*Cirsium horridulum*, *Cirsium smallii* Britt., and *Cirsium vittatum* Small) but his characterization of them does not serve to distinguish one from another adequately. Ahles in Radford et al. (1964), for the Carolinas, recognized two species (plume thistles being placed by him in the genus *Carduus*), *Carduus spinosissimus* Walt. = *Cirsium horridulum* Michx., and *Carduus smallii* (Britt.) Ahles = *Cirsium smallii* Britt. Ahles's characterization of the two does little, if anything, to distinguish one from the other. Long and Lakela (1971), for south Florida, recognized two varieties of *Cirsium horridulum*, var. *horridulum* and var. *vittatum* (Small) R. W. Long. It is doubtful whether the two diagnostic characters they used to distinguish one from the other have any utility.

The senior author suggests that an extensive and intensive study of this complex of thistles might be rewarding. His own observations of them in the field, chiefly but by no means exclusively in northern Florida, lead him to believe that no one will achieve any basic understanding of the group without first gaining an extensive field knowledge throughout the entire range of distribution and across the seasons.

42. Chaptalia

Chaptalia tomentosa Vent. SUN-BONNET, NIGHT-NODDING BOG-DANDELION.   Fig. 399

Perennial, subrhizomatous herb with abundant long fibrous roots, a short stem with alternate leaves in a basal rosette. Leaves oblanceolate to elliptic, narrowed below to a short, flat subpetiolar base, rounded or obtuse apically, variable in size in a given rosette, mostly 4–10 cm long, 2–3 cm wide at the widest places; upper surface of very young leaves with a thin, white, cottony pubescence which is quickly sloughed leaving the surface dark green and glabrous, lower surface permanently clothed with a dense, whitish, felty pubescence; margins with remote, very small, blunt, retrorse teeth. Heads solitary, in the early bud stage almost sessile in the leaf axils, the scape greatly elongating during further development, ultimately reaching 15–30 cm long, usually of uneven lengths on a given plant; flowering in early spring, inflorescences on a single plant in varying stages of development over a period of several weeks; scapes purplish red beneath a covering of rather loose cottony pubescence; heads closely nodding until first full anthesis when they become erect, the rays spreading; after first opening of the head during days of bright light, heads close again and nod at night, open again and become erect the next day (but remaining closed and nodding on days of low light intensity), this repeated for several days, finally remaining closed and nodding. Mature heads in the closed phase 2–3 cm long, the rays about twice the length of the involucre, lower sides of the rays pink to purple; rays mostly white on the upper surface, thus the heads

Fig. 399. **Chaptalia tomentosa:** a, habit; b, ray flower; c, disc flower; d, apex of larger lobe of corolla of disc flower; e, achene.

appearing white, seen from above, when fully open. Involucres campanulate, bracts numerous, closely appressed-imbricate, more or less cottony-pubescent. Flowers discoid and radiate; rays about 20, pistillate and fertile, their apices truncate, rounded, or with 1–3 small rounded lobes. Corollas of disc flowers ivory-colored, bilabiate, one lip 1-lobed, the lobe broadly oblong, 3-notched at the summit, curled. Anther tube yellow below, ivory-colored distally, anther bases long-tailed. Pappus of numerous, soft, white capillary bristles 4–6 mm long. Ovaries of disc flowers sterile. Achene glabrous, body elliptic, narrowed to a neck above, body 5–7-ribbed, the pappus tawny.

Bogs, pine savannas and flatwoods, borders of evergreen shrub bogs, sometimes abundant in wettish sand-peaty mowed areas alongside roads. Coastal plain, N.C. to s. Fla., westward to e. Tex.

43. Prenanthes (RATTLESNAKE-ROOTS)

Prenanthes autumnalis Walt.

Slightly glaucous, perennial herb with milky sap, glabrous throughout, 5–12 dm tall, unbranched. Leaves alternate, principal ones on the lower part of the stem, abruptly reduced above and mostly linear-bracteate below the inflorescence; larger leaves mostly incised-lobed, lobes spreading or directed downward, 5–30 cm long, those on plants with largest ones of the range of size narrowed basally to long, winged subpetioles. Heads borne in a narrow, terminal raceme, or 2 to several heads on very short branches, usually unilaterally disposed from the main axis, spreading or deflexed at anthesis, deflexed in fruit. Involucre cylindric, mostly 10–12 mm long; bracts commonly pinkish, glabrous; stalks of the heads with spirally arranged bracts, smallest ones proximally, gradually becoming a little larger distally and grading into the several small basal bracts of the involucre; inner bracts (abruptly much longer than the outer) about 8, equal, narrowly elongate-lanceolate to linear. Flowers all ligulate and fertile, 8–12 per head, pink or lavender (often drying rose), oblanceolate, usually, with 5 small teeth apically. Achene cylindric or slightly obconic, light brown, about 5 mm long; pappus of numerous, minutely antrorsely barbellate, straw-colored, readily deciduous, capillary bristles about 7 mm long. (*Nabalus virgatus* (Michx.) DC.)

Seasonally wet pine savannas and flatwoods, open bogs, shrub-tree bogs. Coastal plain, s. N.J. to s. Ga.

44. Krigia (DWARF-DANDELIONS)

Small annual or perennial herbs. Leaves alternate, in some closely set on a short stem thus essentially in rosettes; if leaves present more or less throughout the length of the stem, the uppermost sometimes in approximate pairs, one of the pair usually longer than the other; commonly on a given plant, or from plant to plant in some species, varying from entire to variously toothed or pinnatifid. Heads solitary on naked scapes from axils of congested basal leaves or leaves of congested basal branches, or stems more or less elongate and leafy, unbranched or branched, one or more single-headed stalks from a given leaf axil. Involucre a single or double series of nearly equal bracts. Receptacle naked. Flowers bisexual, all with yellow to orange, ligulate corollas, thus heads ligulate. Achenes narrowly turbinate to oblongish, 10–20-nerved or -ribbed; pappus absent, or of 5 prominent, short scales and 5 much longer bristles, or of 5–10 scarcely perceptible, short scales and 5–40 much longer bristles.

Note: Species of *Krigia*, other than those treated here, occur in our range, presumably not in wetlands. For a full treatment of the genus for the range, see Cronquist (1980).

1. Plant perennial, having a slender, vertical rhizome from a tuber, a short, rosulate stem above the rhizome; pappus bristles numerous.   1. *K. dandelion*
1. Plant annual; pappus of 5 broad, short scales and 5 longer bristles, *or* pappus none or a vestigial scaly crown.

2. Stem short, unbranched or with short branches, scapes solitary from axils of leaves; pappus of 5 broad, short scales and 5 longer bristles; involucral bracts reflexed after achenes are shed.

2. *K. virginica*

2. Stems elongate and leafy-branched, scapes 1–3 (–5) from common leaf axils at various levels on the leafy-branch system; pappus none or a vestigial scaly crown; involucral bracts remaining erect after achenes are shed.

3. *K. cespitosa*

1. Krigia dandelion (L.) Nutt. POTATO-DANDELION.

Perennial with a very short aerial stem bearing a rosette of closely set, alternate leaves; having a slender, vertical rhizome 0.5–6 cm long from a tuber; eventually producing from the crown threadlike, simple or branching stolons or rhizomes developing tubers at their tips, these giving rise to the flowering plants of the succeeding year. Lowermost early leaves much shorter than later ones and commonly soon withering, with oblanceolate to elliptic, entire blades narrowed to subpetiolar bases; later leaves narrowly linear to oblanceolate, margins entire or with few, remote, dentate teeth or lobes, bases less distinctly narrowed or sessile; leaves glaucous, glabrous, or with few, appressed, whitish hairs on both surfaces. Scapes 1 to several, each axillary to a leaf of the rosette, fully developed usually much exceeding the leaves, 1–5 dm long, each terminated by a single head; glaucous, glabrous below, commonly bearing just below the heads translucent, jointed, spreading hairs with dark capitate glands at their tips. Heads at anthesis (open during the morning) 3–4.5 cm across. Involucres mostly 10–14 mm high, united basally, bracts linear-subulate, in 2 equal series; united basal portion of the involucre and proximal portions of bracts often purple or purple-black, the dark pigmentation usually truncated distally, bracts otherwise green with narrow, hyaline margins and often purplish tips. Corollas much exceeding the involucres, golden yellow to orange-yellow, tubes pubescent, ligules oblongish, truncately 5-toothed apically, sometimes purplish on the back as they wither or dry. Achene about 2.5 mm long, dark reddish brown, columnar-turbinate, very short-pubescent; pappus of about 10 very inconspicuous scales and 25–45 prominent, scabrous, unequal, pale bristles, the longer ones about 10 mm long. Involucral bracts reflexed after achenes are shed. (*Cynthia dandelion* (L.) DC.)

Low, moist to wet woodland borders, fields, roadsides, meadows, shallow, seasonally wet soils of rock outcrops, ephemeral pools; also well-drained open or openly wooded places. Chiefly coastal plain and piedmont, N.J. to Fla. Panhandle, westward to e., s.e., and n.cen. Tex., northward in the interior to Ind. and Kan.

2. Krigia virginica (L.) Willd.

Annual, at first with a very short stem bearing a few very closely set leaves, 1 to few scapes, later producing several to numerous short branches with distinct but short internodes, commonly numerous scapes, the leafy portion of the plant eventually cushionlike. Few early leaves having small, oblanceolate to elliptic blades with few dentate teeth, narrowed to petioles shorter than to as long as the blades; later more numerous leaves larger, to 12 cm long, oblanceolate or oblongish in outline, blades mostly pinnately divided, narrowed below to winged subpetioles dilated basally, glabrous or irregularly pubescent. Scapes glabrous throughout, pubescent only below the heads, or often pubescent, usually irregularly so, throughout, the pubescence of spreading, jointed, translucent hairs with dark glandular tips; scapes at first short, mostly rapidly elongating during anthesis and as fruits mature, fully developed ones much exceeding the leafy bases, from a few cm tall on young plants to 35–40 cm tall on older, more luxuriant ones. Heads solitary terminating the scapes, at anthesis (open during the mornings) 1–1.5 cm across. Involucres 4–9 mm high, usually united basally, bracts about 9–15 (rarely 5), in 2(3?) series, outer linear-subulate, inner lanceolate to lance-ovate, green or variously lined or blotched with purple, the inner with broadly hyaline margins. Corollas yellow (ligules often becoming lavender as they shrivel or dry), ⅓–½ longer than the involucres, tube pubescent, ligules truncately 5-toothed apically. Achene turbinate, nearly 2 mm long, dark reddish brown, ribbed, minutely scabrous; pappus of 5 short-obovate, membranous scales 1 mm long or less, slightly erose at their broadly rounded apices, and 5 antrorsely scabrid bristles about 5 mm long. Involucral bracts spreading as achenes mature, reflexed after achenes are shed. (*Cynthia virginica* (L.) D. Don sensu Small (1933))

In general a weedy plant of both poorly and well-drained places, pastures, lawns, meadows, open woodlands; often in ephemeral pools, thin soils of seasonally wet places on rock outcrops, alluvial sands and muds. s. Maine, s. N.H., s. Ont., Mich., Wisc., and Iowa, generally southward to cen. pen. Fla. and Tex.

3. Krigia cespitosa (Raf.) Chambers in Vuilleumier.

Annual with a few, alternate leaves closely set on a very short stem, the latter eventually producing few to numerous leafy branches whose upper leaves are in remote subopposite pairs or in approximate 3's, the approximate leaves commonly unequal in length; heads borne singly on 1, 2, or 3 (−5) stalks from common leaf axils, thus heads appearing at various levels of the branch system their stalks 2–6 cm long, often glandular below the heads. Lower leaves with linear, linear-oblanceolate, oblanceolate, or elliptic blades (first ones rarely suborbicular), narrowed below to petiolar bases of unequal length, broader leaves blunt apically, narrower ones acute; leaf margins entire to remotely dentate-toothed or -lobed, to 15 cm long; upper leaves sessile, lance-linear to oblong, mostly 3–5 cm long, subclasping basally, acute apically; leaves glabrous, often glaucous. Involucres 3–5 mm high; bracts 6–10 in 2 series of nearly equal length, outer lanceolate to lance-ovate, acute apically, inner somewhat broader, with hyaline margins, tips blunt. Heads at anthesis 8–15 mm across. Corollas yellow, golden yellow, or orange, about twice as long as the involucral bracts, ligules truncately 5-toothed apically. Achene oblong to elliptic-oblong, 1–1.5 mm long, longitudinally ribbed and minutely cross-ribbed, dark reddish brown; pappus none or a minute scaly crown. Involucral bracts remaining erect after achenes are shed. (*K. oppositifolia* Raf., name invalid; *Serinea oppositifolia* (Raf.) Kuntze)

Usually moist to wet, low places, ephemeral pools, floodplain clearings, shallow, seasonally wet soil of rock outcrops, fallow fields, pastures, ditches, alluvial outwash. Coastal plain and piedmont, s.e. Va. to Fla. Panhandle, westward to Tex., northward in the interior to s. Ill. and cen. Kan.

Glossary

Abaxial. Pertaining to the side of an organ away from the axis, such as the lower surface of a leaf.

Acaulescent. Seemingly without a stem, actually applied to a plant which has a very short or subterranean stem.

Achene (Akene). A hard, dry, indehiscent, 1-seeded fruit, 1-locular.

Acicular. Needlelike.

Actinomorphic. Descriptive of a flower or set of flower parts which can be cut through the center into equal and similar parts along 2 or more planes; radially symmetrical.

Acuminate. Tapering to a pointed apex in such a way that the sides of the taper are more or less concave.

Acute. Sharply angled, the sides of the angle essentially straight.

Adaxial. Pertaining to the side of an organ toward the axis, such as the upper surface of a leaf.

Adnate. Fusion of unlike structures or parts.

Adventitious. Said of buds, roots, etc., which grow in irregular or unusual places.

Aggregate. Crowded into a cluster; a number of separate fruits from a single flower aggregated together.

Alternate. Said of leaves occurring one at a node; said also of members of adjacent whorls in the flower when any member of one whorl is in front of or behind the junction of two adjacent members of the succeeding whorl.

Alveolate. Pitted, honeycombed, as are the surfaces of some seeds or achenes.

Anastomosing. Connecting by cross-veins and forming a network.

Androecium. A collective term applied to all structures of the stamen whorl or whorls.

Androgynous. Staminate flowers above the pistillate in the same inflorescence.

Androphore. A support or column, formed by fusion of filaments, on which the stamens are borne.

Annual. A plant which completes its life history within a year.

Anther. The pollen-bearing part of the stamen.

Anthesis. The time at which a flower is open; the time during which a plant is in bloom.

Antrorse. Directed forward or upward.

Apetalous. Without petals.

Apical. At the tip or summit.

Apiculate. Terminated abruptly by a small, distinct point, an apiculus or apicule.

Apomictic. In general, reproducing without sexual reproduction; often used to denote seed production without a sexual process having been involved.

Arachnoid. Covered with long hairs so entangled as to give a cobwebby appearance.

Arcuate. Curved, bowlike.

Aril. An appendage growing at or about the hilum of a seed.

Aristate. Awned; provided with a bristle at the apex.

Articulate. Jointed; provided with places where separation may take place.

Ascending. Directed or rising upward obliquely.

Asepalous. Without sepals.

Attenuate. Gradually narrowed to a long point at apex or base.

Auricle. An earlike appendage.

Awn. A stiff, bristlelike appendage, usually at the end of a structure.

Axil. The angle found between any two organs or structures.

Axillary. In an axil.

Barbellate. Provided, usually laterally, with fine, short, reflexed points or barbs.

Bayou. A creek, usually slow-moving, usually near the coast.

Biennial. A plant requiring two years in which to complete its life cycle, the first year growing only vegetatively, the second flowering, fruiting, then dying.

Bifid. With 2 lobes or segments, as at the apices of some petals or leaves.

Bilaterally symmetrical. Said of corolla or calyx (or flower) when divisible into equal halves in 1 plane only; zygomorphic.

Bisexual. Pertaining to an individual flower having both male and female parts.

Blade. The expanded portion of a leaf, petal, or other structure.

Bloom. The white powder, dust, or waxy covering sometimes on stems, leaves, flowers, or fruits.

Bract. A reduced leaf or small leaflike structure, particularly one subtending a flower or an inflorescence branch.

Bristle. Stiff, strong but slender hair or trichome.

Bulb. A short underground stem surrounded by fleshy leaves or scales.

Caducous. Said of a part, such as a sepal or petal, that falls off quickly or early.

Callus. A hard protuberance or callosity; new tissue covering a wound.

Calyx. The outer series of perianth parts of a flower, commonly green in color, frequently enclosing the rest of the flower in the bud stage, occasionally colored or petallike (petaloid), in some groups greatly reduced or lacking.

Calyx tube. Tube formed by wholly or partially fused sepals. Not the floral tube of a perigynous or epigynous flower.

Campanulate. Bell-shaped, usually applied to calyx or corolla.

Cancellate. Latticed, or resembling a latticed construction, usually said of a surface such as that of an achene or seed.

Capillary. Very slender, threadlike.

Capitate. Aggregated into a dense, compact cluster or head.

Capsule. A dry, dehiscent fruit originating from 2 or more fused carpels.

Carinate. Provided with a ridge or keel extending lengthwise along the middle.

Carpel. The unit of structure of the female portion of a flower.

Carpophore. A prolongation of the receptacle (axis) that projects above the attachment of the perianth parts and bears the carpel or pistil; in Umbelliferae, a wiry stalk that supports each half (carpel) of the dehiscing fruit.

Castaneous. Chestnut brown.

Catkin. An erect, lax, or pendent, scaly-bracted spike bearing unisexual, apetalous flowers.

Caudex. The thickened persistent base, usually of an herbaceous perennial.

Caulescent. Having a well-developed stem above ground level.

Centrum. Central portion.

Cespitose (caespitose). Growing in tufts.

Chaff. A small, dry, membranous scale or bract; in the Compositae, the receptacular bracts subtending the individual flowers of the head.

Channeled. Having a deep longitudinal groove.

Chartaceous. Having the texture of thin but stiff paper.

Chasmogamous. A "normal" open flower (see **cleistogamous**).

Choripetalous. Corolla with petals distinct from one another.

Ciliate. With marginal hairs that form a fringe.

Circinate. Coiled from the top downward with the apex as the center of the coil.

Circumscissile. Opening or dehiscing by a line around the middle, the top coming off as a lid.

Clavate. Club-shaped, gradually broadened upward.

Claw. The narrowed, stalklike base of some sepals or petals.

Cleft. Cut about halfway to the midvein.

Cleistogamous. Small flowers, self-pollinating without opening; often in addition to the "normal" flower, sometimes in a different place on the plant as well (as on an underground part).

Coherent. Having like parts united.

Colonial. Usually used to describe cloning by vegetative reproduction, the seemingly separate plants having arisen from rhizomes, stolons, or roots of a single or of neighboring "parent" plants.

Colony. A stand, group, or population of neighboring plants of one species, the origin having been clonal, from seeds, or both.

Coma. A tuft of soft hairs, as at the apices or bases of seeds.

Commissure. A place of joining or meeting, as where one carpel joins another in the Umbelliferae.

Comose. With a tuft of soft hairs, a coma.

Connivent. Approximate but not organically united.

Convoluted. Said of parts rolled or twisted together when in an undeveloped stage, as in some corollas in the bud stage.

Cordate. With a sinus and rounded lobes at the base, the overall outline usually ovate; often restricted to the base rather than to the outline of the entire organ; heart-shaped.

Coriaceous. Leathery.

Corm. A bulblike structure in which the fleshy portion is predominantly stem tissue, the covering of membranous scales.

Corolla. The inner, usually colored or otherwise differentiated, whorl or whorls of the perianth.

Corymb. A racemose type of inflorescence in which the lower pedicels are successively elongate forming a more or less flat-topped inflorescence, the outer flowers opening before the inner.

Crenate. Shallowly rounded-toothed; scalloped.

Crown. That part of a stem at or just below the surface of the ground; an inner appendage of a petal or the throat of a corolla; an appendage or extrusion standing between the corolla and stamens, or on the corolla; an outgrowth of the staminal part or circle as in milkweeds.

Cuneate. Wedge-shaped, tapering toward the point of attachment.

Cusp. A strong or sharp point.

Cuspidate. Tipped with a short, rigid point.

Cyathium. A type of inflorescence characteristic of some members of Euphorbiaceae: consisting of a cuplike involucre bearing unisexual flowers, staminate on its inner face, pistillate from the base.

Cyme. A type of inflorescence in which each flower is terminal either to the main axis or to a branch.

Cystoliths. Intercellular concretions, usually of calcium carbonate.

Deciduous. Falling after completion of the normal function.

Decumbent. Reclining on the ground, with ascending distal portions.

Decurrent. Extending downward, applied usually to leaves in which the blade is apparently prolonged downward as two wings along the petiole or along the stem.

Dehiscent. Opening and shedding contents; said of stamens and fruits.

Deltoid. Equilaterally triangular.

Dentate. Having marginal teeth pointing outward, not forward.

Diadelphous. In two sets as applied to stamens when in two, usually unequal, sets.

Dichotomous. Forking in pairs.

Digitate. Fingered, several members arising from one point.

Dimorphic. Having two forms.

Dioecious. Said of a kind of plant having unisexual flowers, the male and female flowers on different individual plants.

Disc flowers. The radially symmetrical flowers of the head in Compositae, as distinguished from the ligulate ray flowers.

Distal. Farthest away from the center, the point of attachment, or origin.

Distichous. Two-ranked, on opposite sides of a stem and in the same plane.

Divaricate. Widely divergent.

Divided. Referring to the blade of an appendage when it is cut into distinct divisions to, or almost to, the midvein.

Dorsal. Pertaining to the back; the surface turned away from the axis.

Downy. Covered with short, weak hairs.

Drupe. A fleshy or pulpy fruit with the inner portion of the pericarp hard or stony and enclosing the seed; usually 1-locular and 1-seeded, sometimes more than 1-locular and more than 1-seeded.

Echinate. Bearing prickles.

Elliptic. An outline that is oval, narrowed to rounded at the ends and widest at about the middle (as the outline of a football); ellipsoid, a solid with an elliptical outline.

Emarginate. Said of leaves, sepals, or petals, and other structures that are notched at the apex.

Emersed, emergent. Standing above the water surface.

Enation. An outgrowth of the surface of an organ.

Endocarp. The inner layer of the wall of a matured ovary.

Entire. Having a margin devoid of any indentations, teeth, or lobes.

Ephemeral. Referring to an organ living a very short time, usually a day or less; in general, lasting but a short time, as an ephemeral pool.

Epigynous. Pertaining to a flower having sepals, petals, and stamens borne at the summit of a floral tube, the ovary being within and fused with the floral tube, the ovary thus said to be inferior.

Epipetalous. Said of stamens when they are inserted on the corolla.

Erose. Uneven; said of margins that give the appearance of having been torn, or of margins with very small teeth of irregular shape and size.

Even-pinnate. Said of compound leaves having an even number of leaflets, this usually easily determined because there is a pair terminally.

Excrescence. An outgrowth; a disfiguring addition.

Excurrent. Projecting beyond the edge or margin, as the midrib of a leaf or bract.

Exfoliating. Peeling off in thin layers, shreds, or plates, as the bark of some trees.

Exocarp. The outer layer of the wall of a matured ovary.

Exserted. Extending beyond (some enclosing part).

Falcate. Curved like a sickle.

Farinaceous. Containing starch or starchlike substance; also applied to a surface with a mealy coating.

Farinose. Term applied to a surface with a mealy or scurfy coating.

Fascicle. A bundle or close cluster.

Fastigiate. Clustered, parallel erect branches.

Fenestrate. With windowlike, perforated or transparent areas.

Filament. The stalk bearing the anther.

Filiform. Threadlike, long and very slender.

Fimbriate. Cut into regular segments and appearing fringed.

Flabellate. Fan-shaped.

Flaccid. Weak, limp, soft, or flabby.

Floccose. Said of pubescence which gives the impression of irregular tufts of cotton or wool.

Floricane. The stem at flowering and fruiting stage (of a bramble, *Rubus*).

Foliaceous. Leaflike.

Follicle. A fruit developing from a single carpel and dehiscing along one suture.

Fusiform. Tapering from approximately the middle to both extremities.

Gamopetalous. Corolla with petals to some extent united. Same as sympetalous.

Gamosepalous. Calyx with sepals to some extent united. Same as synsepalous.

Geniculate. Bent abruptly at the nodes.

Gibbous. A distended, rounded swelling on one side, as on a calyx or corolla tube or segment.

Glabrate. Becoming glabrous with age.

Glabrous. Without pubescence.

Gland. A secreting part or appendage.

Glandular. Having or bearing secreting organs, glands, or trichomes.

Glandular-pubescent. Hairs or trichomes capitate and secretory.

Glaucous. Having a frosted or whitish waxy bloom or powdery coating.

Glomerate. In compact groups.

Glutinous. With a gluey or sticky exudation.

Gynecandrous. Having staminate and pistillate flowers in the same spike or spikelet, the latter above the former.

Gynoecium. The pistil or pistils of a flower, taken collectively.

Gynophore. Stalk or stipe of a pistil.

Gynostegium. Anthers more or less connivent and adherent to the stigma, as in *Asclepias*.

Habit. Term used for the growth form of a plant.

Hastate. Arrow-shaped, with basal lobes that spread nearly or quite at right angles.

Herbage. Above-ground, nonwoody parts of plant.

Hilum. The scar at the point of attachment of an ovule or a seed.

Hirsute. Clothed with long, shaggy hairs, rough to the touch.

Hispid. Clothed with stiffish hairs.

Hood. In some Asclepidaceae, structures in a circle present between the corolla and stamens (collectively called the crown).

Hyaline. Of thin membranous texture, usually transparent or translucent.

Hypanthium. An expansion of the receptacle of a flower that forms a saucer-shaped, cup-shaped, or tubular structure (often simulating a calyx tube) bearing the perianth and stamens at or near its rim; it may be free from or united to the ovary; in this book usually referred to as the floral tube.

Hypogynous. Pertaining to a flower having sepals, petals, and stamens with their point of origin below the ovary, the ovary thus said to be superior.

Imbricated. Overlapping like shingles on a roof.

Immersed. Growing under water.

Incised. Having deeply cleft margins.

Included. Not projecting beyond (an enclosing part).

Indehiscent. Said of fruits that remain closed and do not shed their seeds.

Inflorescence. An aggregation of flowers occurring clustered together in a particular manner usually characteristic of a particular kind of plant.

Infructescence. The inflorescence in a fruiting stage; collective fruits.

Internode. The portion of a stem between nodes.

Involucel. A secondary involucre, as the bracts subtending the secondary umbels in the Umbelliferae.

Involucre. A group of closely placed bracts that subtend or enclose an inflorescence.

Involute. Said of margins that are rolled inward (toward the adaxial side).

Keel. The folded edge or ridge of any structure.

Labiate. Lipped, as in a calyx or corolla.

Lacerate. Said of a margin torn irregularly.

Laciniate. Cut into narrow lobes or segments.

Lacunate. With air spaces or chambers in the midst of tissue.

Lamellate. Made up of thin plates.

Lamina. The blade or expanded part of a leaf, petal, etc.

Lanate. Woolly, with long, intertwined, curled hairs.

Lanceolate. In outline, broadest toward the base and narrowed to the apex, several times longer than wide.

Legume. A 1-locular fruit, usually dehiscent along two sutures, bearing seeds along the ventral suture; a leguminous plant.

Lenticel. Corky spots on young bark, arising in relation to epidermal stomates.

Lenticular. Lens-shaped, biconvex.

Ligulate heads. In the Compositae, heads having all ligulate flowers.

Ligule. In Compositae, the flattened, straplike portion of the ray flower.

Limb. The spreading part of a synsepalous calyx or sympetalous corolla; usually referring only to the calyx or corolla lobes.

Linear. Long and slender with parallel or nearly parallel sides.

Lip. The upper or lower part of a bilabiate calyx or corolla.

Lobulate. Divided into small lobes.

Locule. A compartment of an anther or an ovary.

Loculicidal. Said of capsules that dehisce along the back of the locule.

Loment. A fruit of some legumes, contracted between the seeds, the 1-seeded segments separating at fruit maturity.

Lunate. Crescent-shaped.

Lyrate. Lyre-shaped; pinnatifid, with the terminal lobe larger than the other lobes.

Marcescent. Withering but remaining persistent.

Marsh. A tract of wet land principally inhabited by emergent herbaceous vegetation.

Membranous. Having a thin, soft, pliable texture.

Mericarp. The portion of the fruit which splits away as a seemingly separate fruit, as the two carpels in the Umbelliferae.

Mesic. Pertaining to conditions of medium moisture supply.

Monadelphous. Stamens united in one group by union of their filaments.

Moniliform. Constricted laterally and appearing beadlike.

Monoecious. Said of a kind of plant having unisexual flowers, the male and female flowers on the same individual plant.

Mucro. A sharp, abrupt point or tip.

Muricate. Having a rough surface texture owing to small, sharp projections.

Naturalized. Of foreign origin, but established and reproducing as though native.

Neutral flower. Said of a sterile flower composed of a perianth without any sexual organs.

Node. The place upon a stem which bears 1 or more leaves or bracts.

Nodose. Nodular, knotty.

Nutlet. Diminutive of nut; loosely applied to any small, dry, nutlike fruit.

Oblanceolate. In outline, broadest above the middle, shortly tapered to the apex, long-tapered to the base, much longer than broad.

Oblate. Flattened at the poles (as is a tangerine).

Oblique. Slanting; unequal-sided.

Oblong. In outline, longer than broad, the sides nearly parallel for much of their length.

Obovate. Inverted ovate.

Obsolete. Rudimentary or not evident; applied to a structure that is almost suppressed; vestigial.

Obtuse. Bluntly angled, the angle more than 90 degrees, less than 180.

Ocrea. A sheath around the stem just above the base of a leaf (derived from the stipules); used chiefly in Polygonaceae.

Ocreola, ocreole. Diminutive of ocrea; usually applied to the small sheaths in the inflorescences of Polygonaceae.

Odd-pinnate. Said of compound leaves having an odd number of leaflets, this usually easily determined because there is a single terminal leaflet.

Opposite. Said of leaves or bracts occurring two at a node on opposite sides of the stem. Said of flower parts when one part occurs in front of another.

Orbicular. Approximately circular in outline.

Ovate. In outline, broadest at or near the base, not greatly longer than broad (roughly the outline of an egg).

Ovoid. A three-dimensional object with an ovate outline.

Palate. In a 2-lipped corolla, the projecting part of the lower lip which closes the throat, as in a snapdragon.

Pale. In Compositae, the chaffy receptacular scale or bract subtending the individual flower of a head; in general, a small, thin scale or bract.

Palmate. With 3 or more lobes or veins or leaflets arising from one point.

Pandurate. Fiddle-shaped.

Panicle. A loose irregularly compound inflorescence with stalked flowers, i.e., the branches racemose.

Papilionaceous. Butterflylike; said of the flowers of those members of the pea family having the corolla composed of an upright banner or standard and two lateral wings, each representing a petal, and a keel comprised of two petals variously united.

Papillose. Descriptive of a surface beset with short, blunt, rounded, or cylindric projections.

Pappus. In Compositae, designation for the collective scales, bristles, crown, etc., at the summit of the ovary (or achene) and outside the corolla.

Parietal placentation. Said of the attachment of ovules or seeds when borne on the ovary wall or on structures raised from the ovary wall.

Pectinate. Said of a structure which is cleft into divisions in such a way as to resemble a comb.

Pedicel. The stalk of a flower in an inflorescence.

Peduncle. The stalk of a flower borne singly or the stalk of an inflorescence.

Peltate. Said of a plane structure that is attached at a point on its surface, not attached marginally.

Perennial. Living 3 or more seasons.

Perfoliate. Said of opposite or whorled leaves or bracts that are united into a collarlike structure around the stem that bears them.

Perianth. The calyx and corolla taken together, or either one if one is absent.

Perigynous. Pertaining to a flower having the sepals, petals, and stamens borne at the summit of a floral tube, the ovary (ovaries) being within but free from the tube.

Persistent. Remaining attached after the normal function has been completed; long-continuous.

Petal. The unit of structure of the corolla.

Petaloid. Said of a floral part (not a petal) having the form of a petal, usually showy.

Petiole. The stalk of a leaf.

Phyllode. An expanded, bladeless petiole.

Pilose. Pubescence comprised of scattered long, slender, soft hairs.

Pinna. A leaflet or a primary division of a compound leaf.

Pinnate. Having a common elongate axis, with branches, lobes, veins, or leaflets arranged divergently on either side.

Pistil. The unit of female function of a flower, may be comprised of a single carpel or two or more carpels united.

Pistillate. Said of a flower bearing a pistil or pistils but not stamens; may refer also to a plant having only pistillate flowers.

Placenta. The structure or tissue within the ovary bearing the ovules.

Plano-convex. Flat on one side, convex on the other.

Plicate. Folded into plaits, usually lengthwise.

Plumose. With hairlike branches, feathery.

Pocosin. A bog that has formed in a shallow, undrained depression, the surrounding land being somewhat elevated, the vegetation predominantly evergreen shrubs or small trees. Pocosins vary greatly in size.

Pollinium. A mass of coherent pollen characteristic of orchids and milkweeds.

Polygamo-dioecious. Polygamous but chiefly dioecious.

Polygamo-monoecious. Polygamous but chiefly monoecious.

Polygamous. Having bisexual, pistillate, and staminate flowers on the same individual plant.

Polypetalous. With distinct or separate petals.

Pome. A fruit in which the ovarial portion is relatively small and the fleshy portion is derived largely from the floral tube, as in an apple or pear.

Primocane. The first-year, nonflowering or nonfruiting shoot (of a bramble, *Rubus*).

Procumbent. Trailing or lying flat, but not rooting.

Proliferous. Bearing supplementary structures such as buds or flowers, either in an abnormal manner, or in a manner that is normal but from adventitious tissues.

Puberulent. Minutely pubescent, the hairs soft, straight, erect, scarcely visible to the unaided eye.

Pubescent. A general term for hairiness.

Punctate. With depressed dots scattered over the surface.

Pustulate hair. Hair with an enlarged base.

Pyriform. Pear-shaped.

Raceme. An inflorescence with a single axis, the flowers stalked.

Rachis. The central prolongation of the stalk (peduncle) through an inflorescence, or of a leaf stalk (petiole) through a compound leaf.

Radially symmetrical. Said of a flower or set of flower parts which can be cut through the center into equal and similar parts along 2 or more planes; actinomorphic.

Radiate head. In Compositae, a head having marginal, ligulate flowers.

Ray flower. In Compositae, the flowers of the head having a ligulate corolla.

Receptacle. The more or less expanded apex of a floral axis which bears the floral parts.

Remote. Considerably separated from one another.

Reniform. Kidney-shaped.

Repent. Said of a stem that is prostrate and rooting at the nodes.

Reticulate. Netted.

Retrorse. Having hairs or other processes turned toward the base.

Retuse. With a shallow, rounded notch at the apex.

Revolute. Said of margins that are rolled backward (toward the abaxial side).

Rhizome. A horizontal underground stem.

Rhombic. An outline like a rhomboid, a parallelogram with equal sides, having two oblique angles and two acute angles.

Roseate. Rose-colored.

Rosette. A group of organs, such as leaves, clustered and crowded around a common point of attachment.

Rotate. Radially spreading in one plane.

Rugose. With a wrinkled surface.

Saccate. Having a saclike swelling.

Sagittate. Shaped like an arrowhead with the basal lobes pointing downward.

Salverform. Said of a corolla in which the tube is essentially cylindrical, the lobes abruptly spreading.

Samara. An indehiscent, winged fruit, as in *Ulmus, Fraxinus, Acer*.

Scabrous, scabrid. Rough and harsh to the touch.

Scape. A leafless (or only bracteate) flowering stem arising from below the substrate or from a very short leafy stem.

Scarious. Thin and dry, usually more or less opaque.

Schizocarp. A fruit that splits into 1-seeded portions, often referred to as a nutlet.

Scorpoid. Said of a circinately coiled determinate inflorescence in which the flowers are 2-ranked and borne alternately at the right and left.

Secund. Disposed on one side of a stem or axis.

Sepal. The unit of structure of the calyx.

Septate. Partitioned by walls.

Septicidal. Said of capsules that dehisce along the junction of their carpels opposite the septae.

Serrate. Having marginal teeth pointing forward.

Sessile. Not stalked.

Setaceous. Bristlelike.

Sheath. The basal part of a lateral organ that closely surrounds or invests the stem.

Silique. In Cruciferae, a narrow, 2-locular fruit with two parietal placentae, in dehiscence the two halves usually separating from the partition.

Slough. A wet place of deep mud or mire; a sluggish channel.

Spadix. A spike with a fleshy axis, the flowers more or less embedded in the axis.

Spathe. A sheathing lateral organ, usually open on one side, enclosing a flower bud, a developing inflorescence, or an inflorescence.

Spatulate. Shaped like a spatula, i.e., gradually widening distally and with a rounded tip.

894

Spicule. A small, slender, sharp-pointed piece, usually on a surface.

Spike. An inflorescence with a single axis, the flowers sessile.

Spur. A tubular or saclike projection from a petal or sepal; also refers to a very short branch with closely spaced leaves.

Stamen. The male organ of a flower; usually consisting of a stalk or filament and an anther producing the pollen.

Staminate. Said of a flower bearing a stamen or stamens but not pistils; may refer also to a plant having only staminate flowers.

Staminodium. A sterile organ in the stamen whorl, presumed to be evolutionarily derived from a stamen.

Stellate. Starlike; said of hairs that branch in such a manner as to radiate from a central point.

Stigma. That part of the pistil that receives the pollen.

Stipe. The stalklike basal part of an ovary, or of a fruit such as an achene.

Stipel. Stipule of a leaflet of a compound leaf.

Stipule. An appendage at the base of a petiole or a leaf or on each side of its insertion.

Stolon. A stem that trails along the surface of the ground, usually rooting at the nodes.

Stramineus. Straw-colored.

Striate. Marked with fine parallel lines.

Strigose. Surface clothed with stiff, often appressed hairs, these usually pointing in one direction.

Style. The part of the pistil (if present) connecting the ovary and the stigma.

Stylopodium. An enlarged style base more or less covering the top of the ovary, as in Umbelliferae.

Subulate. Relatively narrow, with a long taper from base to apex.

Suffrutescent. Obscurely shrubby, usually woody only basally.

Sulcate. Grooved.

Suture. A seam or line or groove; usually applied to the line along which a fruit dehisces; any lengthwise groove that forms a junction between two parts.

Swale. An open hollow or depression, usually wet at least seasonally.

Swamp. A wooded area having surface water much of the time.

Syncarp. A multiple or aggregate fruit derived from numerous separate ovaries of a single flower.

Tendril. A slender twining or clasping process, modified stem, leaf, or part of a leaf, by which some plants climb.

Tepal. Denoting a unit of the perianth when the sepals and petals are essentially alike and not readily differentiated.

Terete. Circular in cross-section.

Ternate. In threes.

Testa. The outer covering of a seed.

Throat. Term applied to an expanded part of a corolla tube below the lobes.

Thyrse. Denoting a compact, usually narrow, panicle.

Tomentose. Densely covered with short, matted hairs.

Trichome. Any hairlike outgrowth of the epidermis.

Trigonous. Three-angled in cross-section.

Truncate. Cut squarely across, either at the base or apex of an organ.

Tuber. A thickened, solid portion of an underground stem, with many buds; as an adjective, tuberous, also referring to a fleshy, enlarged root.

Tubercle. A surficial nodule; a thickened, solid, spongy crown or cap, as on an achene.

Turbinate. Inversely conical.

Umbel. A flat or convex flower cluster in which the flower stalks arise from a common point (like the rays of an umbrella).

Unisexual. Pertaining to an individual flower having only male or only female parts, not both.

Urceolate. Urn-shaped.

Utricle. A small, thin-walled, often somewhat bladdery, 1-seeded fruit.

Vernal. Of the spring season.

Verrucose. Warty.

Versatile. Turning freely on its support, as an anther attached near the middle and capable of swinging freely on the filament.

Verticil. A ring of organs, e.g., leaves, flowers, flower parts, at a given position on an axis; a whorl.

Vesicular. Bladderlike or saclike cavity or cyst.

Vestiture. That which covers a surface, as hairs, scales, etc.

Villous. Covered with fine long hairs, the hairs not matted.

Viscid. Sticky.

Wet woodland. A wooded area having markedly fluctuating surface water.

Zygomorphic. Said of the corolla or calyx when divisible into equal halves in one plane only; bilaterally symmetrical.

References

Arber, A. *Water Plants: A Study of Aquatic Angiosperms*. Cambridge: University Press, 1920. Reprint with introd. by W. T. Stearn as *Historiae Classica, 1963*.

Beal, Ernest O. *A Manual of Aquatic Vascular Plants of North Carolina with Habitat Data*. North Carolina Agr. Exp. Sta., Tech. Bull. 247. 1977.

Bell, C. Ritchie. "Natural Hybrids in the Genus *Sarracenia*. I. History, Distribution and Taxonomy." *Jour. Elisha Mitchell Sci. Soc*. 68: 55–80. 1952.

_____, and Frederick W. Case. "Natural Hybrids in the Genus *Sarracenia*. II. Current Notes on Distribution." *Jour. Elisha Mitchell Sci. Soc*. 72: 142–152. 1956.

Bogle, A. Linn. "The Genera of Portulacaceae and Basellaceae in the Southeastern United States." *Jour. Arn. Arb*. 50: 566–598. 1969.

_____. "The Genera of Molluginaceae and Aizoaceae in the Southeastern United States." *Jour. Arn. Arb*. 51: 431–462. 1970.

Brizicky, George K. "The Genera of Violaceae in the Southeastern United States." *Jour. Arn. Arb*. 42: 321–333. 1961.

_____. "The Genera of Anacardiaceae in the Southeastern United States." *Jour. Arn. Arb*. 43: 359–375. 1962.

_____. "The Genera of Sapindales in the Southeastern United States." *Jour. Arn. Arb*. 44: 462–501. 1963.

_____. "The Genera of Celastrales in the Southeastern United States." *Jour. Arn. Arb*. 45: 206–234. 1964.

———. "The Genera of Rhamnaceae in the Southeastern United States." *Jour. Arn. Arb*. 45: 439–463. 1964.

———. "The Genera of Vitaceae in the Southeastern United States." *Jour. Arn. Arb*. 46: 48–67. 1965.

———. "The Genera of Sterculiaceae in the Southeastern United States." *Jour. Arn. Arb*. 47: 60–74. 1966.

Case, Frederick W., Jr., and Roberta B. Case. "*Sarracenia alabamensis*, A Newly Recognized Species from Central Alabama." *Rhodora* 76: 650–665. 1974.

———. "The *Sarracenia rubra* Complex." *Rhodora* 78: 270–325. 1976.

Channell, R. B., and Carroll E. Wood, Jr. "The Genera of the Primulales of the Southeastern United States." *Jour. Arn. Arb*. 40: 268–288. 1959.

———. "The Genera of Plumbaginaceae of the Southeastern United States." *Jour. Arn. Arb*. 40: 391–397. 1959.

———. "The Leitneriaceae in the Southeastern United States." *Jour. Arn. Arb*. 43: 435–438. 1962.

Cook, C. D. K. *Water Plants of the World*. The Hague: Dr. W. Junk b.v., 1974.

Correll, Donovan S., and Helen B. Correll. *Aquatic and Wetland Plants of Southwestern United States*. Environmental Protection Agency, Washington, 1972. Reprint (in 2 vols.). Stanford: Stanford University Press, 1975.

Cronquist, Arthur J. *Vascular Flora of the Southeastern United States*. Vol. 1: *Asteraceae*. Chapel Hill: University of North Carolina Press, 1980.

Davis, John H., Jr. "The Natural Features of Southern Florida." Florida Geological Survey, Geological Bull. 25. 1943.

Duncan, Wilbur H. *Woody Vines of the Southeastern United States*. Athens: University of Georgia Press, 1974.

Elias, Thomas S. "The Genera of Ulmaceae in the Southeastern United States." *Jour. Arn. Arb*. 51: 18–40. 1970.

———. "The Genera of Fagaceae in the Southeastern United States." *Jour. Arn. Arb*. 52: 159–195. 1971.

897

————. "The Genera of Myricaceae in the Southeastern United States." *Jour. Arn. Arb.* 52: 305–318. 1971.

————. "The Genera of Juglandaceae in the Southeastern United States." *Jour. Arn. Arb.* 53: 26–51. 1972.

Ernst, Wallace R. "The Genera of Hamamelidaceae and Platanaceae in the Southeastern United States." *Jour. Arn. Arb.* 44: 193–210. 1963.

Eyde, Richard H. "Morphological and Paleobotanical Studies of the Nyssaceae. I. A Survey of the Modern Species and Their Fruits." *Jour. Arn. Arb.* 44: 1–54. 1963.

————. "The Nyssaceae in the Southeastern United States." *Jour. Arn. Arb.* 47: 117–125. 1966.

Fassett, N. C. *A Manual of Aquatic Plants.* 2nd ed. (revised appendix by E. C. Ogden). Madison: University of Wisconsin Press, 1957.

Ferguson, I. K. "The Genera of Caprifoliaceae in the Southeastern United States." *Jour. Arn. Arb.* 47: 33–59. 1966.

————. "The Cornaceae in the Southeastern United States." *Jour. Arn. Arb.* 47: 106–116. 1966.

Fernald, Merritt L. *Gray's Manual of Botany.* 8th ed. New York: American Book Co., 1950.

Gillis, William T. "The Systematics and Ecology of Poison-Ivy and the Poison-Oaks (*Toxicodendron*, Anacardiaceae)." *Rhodora* 73: 72–159, 161–237, 370–443, 465–540. 1971.

Gleason, Henry A. *The New Britton and Brown Illustrated Flora of the Northeastern United States and Adjacent Canada.* 3 vols. New York: New York Botanical Garden, 1952.

Godfrey, Robert K., and Jean W. Wooten. *Aquatic and Wetland Plants of Southeastern United States: Monocotyledons.* Athens: University of Georgia Press, 1979.

Graham, Shirley A. "The Genera of Lythraceae in the Southeastern United States." *Jour. Arn. Arb.* 45: 235–250. 1964.

————. "The Genera of Rhizophoraceae and Combretaceae in the Southeastern United States." *Jour. Arn. Arb.* 45: 285–301. 1964.

————. "The Genera of Araliaceae in the Southeastern United States." *Jour. Arn. Arb.* 47: 126–136. 1966.

————, and C. E. Wood, Jr. "The Genera of Polygonaceae in the Southeastern United States." *Jour. Arn. Arb.* 46: 91–121. 1965.

————, and C. E. Wood, Jr. "The Podostemaceae in the Southeastern United States." *Jour. Arn. Arb.* 56: 456–465. 1975.

Jones, Samuel B., Jr. "Mississippi Flora. IV. Dicotyledon Families with Aquatic or Wetland Species." *Gulf Research Reports* 5: 7–22. 1975.

Kral, Robert. "Some Notes on the Flora of the Southern States, Particularly Alabama and Middle Tennessee." *Rhodora* 75: 366–410. 1973.

Kurz, Herman, and Robert K. Godfrey. *Trees of Northern Florida.* Gainesville: University of Florida Press, 1962.

Little, Elbert L., Jr., and Frank H. Wadsworth. *Common Trees of Puerto Rico and the Virgin Islands.* U.S. Dept. Agr., Agr. Handb. 249. 1964.

————, Roy O. Woodbury, and Frank H. Wadsworth. *Trees of Puerto Rico and the Virgin Islands.* Vol. 2. U.S. Dept. Agr., Agr. Handb. 449. 1974.

Long, Robert W. "The Genera of Acanthaceae in the Southeastern United States." *Jour. Arn. Arb.* 51: 257–309. 1970.

————. "Additions and Nomenclatural Changes in the Flora of Southern Florida." *Rhodora* 72: 16–46. 1970.

————, and Olga Lakela. *A Flora of Tropical Florida.* Coral Gables: University of Miami Press, 1971.

Mason, Herbert L. *A Flora of the Marshes of California.* Berkeley: University of California Press, 1957.

Mathias, Mildred, and Lincoln C. Constance. "Umbelliferae." In C. L. Lundell, *Flora of Texas*, vol. 3, pt. 5. Renner: Texas Research Foundation, 1951.

Miller, Norton G. "The Genera of the Urticaceae in the Southeastern United States." *Jour. Arn. Arb*. 52: 40–68. 1971.

———. "The Polygalaceae in the Southeastern United States." *Jour. Arn. Arb*. 52: 267–284. 1971.

Mitchell, R. S. "Variation in the *Polygonum amphibium* Complex and Its Taxonomic Significance." *Univ. of Calif. Publ. in Bot*. 45: 1–65. 1968.

———. *A Guide to Aquatic Smartweeds* (Polygonum) *of the United States*. Virginia Polytechnic Institute and State University, Water Resources Research Center, Bull. 41. 1971.

Muenscher, W. C. *Aquatic Plants of the United States*. Ithaca, N.Y.: Comstock Publishing Co., 1944.

Nicely, Kenneth A. "A Monographic Study of the Calycanthaceae." *Castanea* 30: 38–81. 1965.

Pennell, Francis W. *The Scrophulariaceae of Eastern Temperate North America*. Acad. Nat. Sci. Phila., Monograph 1. 1935.

Prance, Ghillean T. "The Genera of Chrysobalanaceae in the Southeastern United States." *Jour. Arn. Arb*. 51: 521–528. 1970.

Reed, Clyde. *Selected Weeds of the United States*. U.S. Dept. Agr., Agr. Res. Ser., Agr. Handb. 336. 1970.

Robertson, Kenneth R. "The Linaceae in the Southeastern United States." *Jour. Arn. Arb*. 52: 649–665. 1971.

———. "The Genera of Rosaceae in the Southeastern United States." *Jour. Arn. Arb*. 55: 303–332, 344–401, 611–662. 1974.

Schnell, Donald E. "Infraspecific Variation in *Sarracenia rubra* Walt.: Some Observations." *Castanea* 42: 149–170. 1977.

———. "*Sarracenia rubra* Walt.: Infraspecific Nomenclatural Corrections." *Castanea* 43: 260–261. 1978.

———. "*Sarracenia rubra* Walter ssp. *gulfensis*: A New Subspecies." *Castanea* 44: 217–223. 1979.

Sculthorpe, C. Duncan. *The Biology of Aquatic Vascular Plants*. London: Edward Arnold, 1967.

Small, John Kunkel. *Manual of the Southeastern Flora*. 1933. Reprint. New York: Hafner, 1972.

Sponberg, Stephen. "The Staphyleaceae in the Southeastern United States." *Jour. Arn. Arb*. 52: 196–203. 1971.

Steyermark, Julian A. *Flora of Missouri*. Ames: Iowa State University Press, 1962.

Thieret, John W. "The Martyniaceae in the Southeastern United States." *Jour. Arn. Arb*. 58: 25–39. 1977.

———, Robert R. Haynes, and David H. Dike. "Aquatic and Marsh Plants of Louisiana: A Checklist." *Louisiana Soc. for Hort. Res*. 13: 1–45. 1972.

Thomas, J. L. "A Monographic Study of the Cyrillaceae." *Contr. Gray Herb*. 186: 1–114. 1960.

———. "The Genera of the Cyrillaceae and Clethraceae of the Southeastern United States." *Jour. Arn. Arb*. 42: 96–106. 1961.

Vuilleumier, Beryl Simpson. "The Genera of Senecioneae in the Southeastern United States." *Jour. Arn. Arb*. 50: 104–123. 1969.

———. "The Tribe Mutisieae (Compositae) in the Southeastern United States." *Jour. Arn. Arb*. 50: 620–625. 1969.

———. "The Genera of Lactuceae (Compositae) in the Southeastern United States." *Jour. Arn. Arb*. 54: 42–93. 1973.

Webster, Grady L. "The Genera of Euphorbiaceae in the Southeastern United States." *Jour. Arn. Arb*. 48: 303–430. 1967.

Wells, B. W. *The Natural Gardens of North Carolina*. 1932. Reprint. Chapel Hill: University of North Carolina Press, 1967.

Wilson, Kenneth A. "The Genera of Hydrophyllaceae and Polemoniaceae in the Southeastern United States." *Jour. Arn. Arb.* 41: 197–212. 1960.

———. "The Genera of Myrtaceae in the Southeastern United States." *Jour. Arn. Arb.* 41: 270–278. 1960.

———. "The Genera of Convolvulaceae in the Southeastern United States." *Jour Arn. Arb.* 41: 298–317. 1960.

———, and Carroll E. Wood, Jr. "The Genera of Oleaceae in the Southeastern United States." *Jour. Arn. Arb.* 40: 369–384. 1959.

Wood, Carroll E., Jr. "The Genera of Woody Ranales in the Southeastern United States." *Jour. Arn. Arb.* 39: 296–346. 1958.

———. "The Genera of Nymphaeaceae and Ceratophyllaceae in the Southeastern United States." *Jour. Arn. Arb.* 40: 94–112. 1959.

———. "The Genera of Theaceae of the Southeastern United States." *Jour. Arn. Arb.* 40: 413–419. 1959.

———. "The Genera of Sarraceniaceae and Droseraceae in the Southeastern United States." *Jour. Arn. Arb.* 41: 152–163. 1960.

———. "The Genera of Ericaceae in the Southeastern United States." *Jour. Arn. Arb.* 42: 10–80. 1961.

———. "The Saururaceae in the Southeastern United States." *Jour. Arn. Arb.* 52: 479–485. 1971.

———. "The Balsaminaceae in the Southeastern United States." *Jour. Arn. Arb.* 56: 413–426. 1975.

———, and R. B. Channell. "The Empetraceae and Diapensiaceae of the Southeastern United States." *Jour. Arn. Arb.* 40: 161–171. 1959.

———, and R. B. Channell. "The Genera of the Ebenales in the Southeastern United States." *Jour. Arn. Arb.* 41: 1–35. 1960.

———, and Preston Adams. "The Genera of Guttiferae (Clusiaceae) in the Southeastern United States." *Jour. Arn. Arb.* 57: 74–90. 1976.

Wright, A. H., and A. A. Wright. "The Habitats and Composition of the Vegetation of Okefenokee Swamp, Georgia." *Ecol. Monogr.* 11: 110–232. 1932.

Yuncker, T. G. "Convolvulaceae. 1. Cuscuta." In C. L. Lundell, *Flora of Texas*, vol. 3, pt. 2. Renner: Texas Research Foundation, 1943.

Index to Common Names

907

Index to Scientific Names

Avicenniaceae, 589
Azalea
 atlantica Ashe, 476
 canescens Michx., 474
 nudiflora L., 474
 serrulata Small, 474
 viscosa L., 474

B

Baccharis, 838
 angustifolia Michx., 838
 glomeruliflora Pers., 839
 halimifolia L., 839
  var. *angustior* DC., 839
Bacopa, 629
 caroliniana (Walt.) Robins., 635
 cyclophylla Fern., 631
 egensis (Poepp.) Pennell, 631
 innominata (Gómez Maza) Alain, 631
 monnieri (L.) Pennell, 629
 repens (Sw.) Wettst., 631
 rotundifolia (Michx.) Wettst., 631
Balduina
 atropurpurea (Harper) Small, 801
 uniflora Nutt., 799
Balsaminaceae, 260
Bartonia, 523
 lanceolata Small, 524
 paniculata (Michx.) Muhl., 524
 verna (Michx.) Muhl., 523
 virginica (L.) BSP., 524
Bataceae, 107
Batis maritima L., 107
Batrachium tricophyllum sensu Small, 136
Befaria racemosa Vent., 471
Benzoin
 aestivale sensu Nees, 359
 melissaefolium (Walt.) Nees, 359
Berchemia scandens (Hill) K. Koch, 315
Bergia texana (Hook.) Walp., 330
Betulaceae, 36
Betula nigra L., 38
Bidens, 788
 aristosa (Michx.) Britt., 792
 cernua L., 790
 comosa (Gray) Weig., 790
 connata Muhl., 790
 coronata (L.) Britt., 792
 discoidea (T. & G.) Britt., 793
 frondosa L., 793
 involucrata (Nutt.) Britt. not Phil., 792
 laevis (L.) BSP., 788
 mitis (Michx.) Sherff, 790
 nashii Small, 790
 polylepis Blake, 792
 tripartita L., 790
 vulgata Greene, 793
Bigelowia, 824
 nuttalii L. C. Anderson, 824

 nudata (Michx.) DC., 825
  subsp. australis L. C. Anderson, 825
  subsp. nudata, 825
 virgata (Nutt.) DC., 825
Bignonia
 capreolata L., 709
 radicans L., 709
Bignoniaceae, 708
Bilderdykia
 convolvulus (L.) Green, 79
 cristata (Engelm. & Gray) Greene, 79
 dumetorum (L.) Dum., 79
 scandens (L.) Greene, 78
Biovularia olivacea (Wright ex Griseb.) Kam., 690
Boehmeria
 decurrens Small, 56
 drummondii Wedd., 56
 cylindrica (L.) Sw., 56
Boltonia, 839
Boraginaceae, 580
Borreria laevis (Lam.) Griseb., 725
Borrichia, 760
 arborescens (L.) DC., 761
 frutescens (L.) DC., 761
Bowlesia incana Ruiz and Pavon, 447
Boykinia aconitifolia Nutt., 210
Bramia monnieri (L.) Pennell, 629
Brasenia schreberi Gmel., 157
Breweria michauxii Fern. & Schub., 561
Brunnichia
 cirrhosa Banks ex Gaertn., 65
 ovata (Walt.) Shinners, 65
Buchnera, 669
 americana L., 669
 brevifolia Pennell, 669
 elongata Sw., 669
 floridana Gand , 669
Bumelia, 505
 lycioides (L.) Pers., 505
 microcarpa Small, 505
 reclinata Vent., 505

C

Cabomba
 caroliniana Gray, 157
  var. pulcherrima Harper, 157
 pulcherrima (Harper) Fassett, 157
Cabombaceae, 157
Cacalia
 diversifolia T. & G., 811
 elliottii (Harper) Shinners, 811
 lanceolata Nutt., 811
  var. *elliottii* (Harper) Kral & Godfrey, 811
 suaveolens L., 810
 sulcata Fern., 811
Caesalpina
 bonduc (L.) Roxb., 239
 bonduc sensu Small, 239

915

virginiana L., 725
Dioanea muscipula Ellis, 185
Diospyros virginiana L., 502
Doellingeria
    humilis (Willd.) Britt., 836
    reticulata (Pursh) Greene, 836
    umbellata (Mill.) Nees, 836
Dondia linearis (Ell.) Millsp., 99
Dopatrium junceum (Roxb.) Hamilt. in Benth., 651
Dracocephalum
    correllii Lundell, 613
    intermedium Nutt., 615
    leptophyllum (Small) Small, 619
    purpureum (Walt.) McClintock, 613
    veroniciforme (Small) Small, 619
    virginianum L., 619
Dracopsis amplexicaulis (Vahl) Cass., 759
Drosera, 185
    annua Reed, 187
    brevifolia Pursh, 187
    capillaris Poir., 187
    filiformis Raf., 186
        var. *tracyi* (Macfarlane in L. H. Bailey) Diels in Engler, 187
    intermedia Hayne in Schrad., 187
    leucantha Shinners, 187
    rotundifolia L., 187
    tracyi Macfarlane in L. H. Bailey, 187
Droseraceae, 185
Drymaria cordata (L.) Willd. ex R. & S., 118
Dyschoriste, 706
    angusta (Gray) Small, 706
    humistrata (Michx.) Kuntze, 706
    oblongifolia (Michx.) Kuntze
        var. *angusta* (Gray) R. W. Long, 706

E

Ebenaceae, 502
Ecastophyllum ecastophyllum (L.) Britt., 249
Echinocystis lobata (Michx.) T. & G., 733
Eclipta alba (L.) Hassk., 760
Elatinaceae, 328
Elatine
    americana (Pursh) Arnott, 330
    triandra Schk. var. americana (Pursh) Fassett, 330
Elephantopus, 874
    carolinianus Raeusch., 874
    nudatus Gray, 874
Ellisia nyctelea (L.) L., 578
Elytraria
    carolinensis (J. F. Gmel.) Pers., 698
        var. angustifolia (Fern.) Blake, 698
Endorima
    atropurpurea (Harper) Small, 801
    uniflora (Nutt.) Barnh., 801
Epilobium, 418
    angustifolium L., 418

coloratum Biehler, 418
leptophyllum Raf., 418
Erechtites hieracifolia (L.) Raf., 816
Ericaceae, 469
Erigeron, 837
    philadelphicus L., 838
    pulchellus Michx., 837
    vernus (L.) T. & G., 837
Eryngium, 435
    aquaticum L., 436
    aquaticum sensu Small, 438
    baldwinii Spreng., 435
    floridanum Coult. & Rose, 438
    integrifolium Walt., 438
    ludovicianum Morong, 438
    prostratum Nutt. ex DC., 435
    ravenelii Gray, 438
    synchaetum (Gray ex Coult. & Rose) Coult. & Rose, 438
    virginianum Lam., 438
    yuccifolium Michx., 438
Eubotrys
    elongata Small, 477
    racemosa (L.) Nutt., 477
Eupatoriadelphus, 844
    dubius (Willd. ex Poir.) King & H. Robins., 844
    fistulosus (Barratt) King & H. Robins., 845
Eupatorium, 845
    anomalum Nash, 856
    capillifolium (Lam.) Small, 846
        var. *leptophyllum* (DC.) Ahles, 848
    coelestinum L., 843
    compositifolium Walt., 848
    cuneatum Engelm. ex T. & G., 851
    dubium Willd. ex Poir., 844
    eugenei Small, 848
    fistulosum Barratt, 854
    glaucescens of authors not Ell., 854
    lancifolium (T. & G.) Small, 854
    leptophyllum DC., 848
    leucolepis (DC.) T. & G., 853
    maculatum sensu Small, 845
    mikanioides Chapm., 851
    mohrii Greene, 856
    pectinatum Small, 848
    perfoliatum L., 848
        var. *cuneatum* (Engelm. ex T. & G.) Gray, 851
        var. *truncatum* (Muhl. ex Willd.) Gray, 851
    pilosum Walt., 853
    pinnatifidum Ell., 848
    purpureum sensu Small, 844
    recurvans Small, 856
    resinosum Torr. ex DC., 854
    rotundifolium L., 853
        var. *saundersii* (Porter) Cronq., 853
    scabridum Ell., 856
    semiserratum DC., 854
        var. *lancifolium* T. & G., 854

917

sagittata Poir. in Lam., 571
sinuata Ort., 571
stolonifera (Cyrillo) J. F. Gmel., 570
tricocarpa Ell., 571
violacea L., 565
wrightii Gray, 570
Iresine
paniculata misapplied, 107
rhizomatosa Standl., 105
Isnardia
intermedia Small and Alex. in Small, 415
palustris L., 415
repens (Sw.) DC., 415
spathulata (T. & G.) Kuntze, 415
Itea virginica L., 213
Iva, 797
annua L., 797
caudata Small, 799
ciliata Willd., 799
frutescens L., subsp. frutescens, 797
imbricata Walt., 797
microcephala Nutt., 797

J

Jacquemontia tamnifolia (L.) Griseb., 562
Juglandaceae, 21
Jussiaea
angustifolia Lam., 392
decurrens (Walt.) DC., 392
diffusa of authors not Forsk., 396
erecta L., 392
grandifolia Michx. not R. & P., 396
leptocarpa Nutt., 393
michauxiana Fern., 396
neglecta Lam., 393
peruviana L., 393
repens L.
    var. glabrescens Kuntze, 396
    var. peploides (HBK.) Griseb., 396
scabra Willd., 392
suffruticosa L., 392
uruguayensis Camb., 396
Justicia, 700
americana (L.) Vahl, 702
angusta (Chapm.) Small, 706
cooleyi Monachino & Leonard, 700
crassifolia (Chapm.) Small, 702
humilis Michx., 702
mortuiflumensis Fern., 702
ovata (Walt.) Lindau in Urban, 702
    var. angusta (Chapm.) R. W. Long, 706
    var. lanceolata (Chapm.) R. W. Long, 706

K

Kalmia, 471
angustifolia L., var. caroliniana (Small) Fern.,
    472
carolina Small, 472
cuneata Michx., 472

hirsuta Walt., 472
Kalmiella hirsuta (Walt.) Small, 472
Koellia
albescens (T. & G.) Kuntze, 607
aristata (Michx.) Kuntze, 607
flexuosa sensu Small, 606
hyssopifolia (Benth.) Britt., 607
mutica (Michx.) Britt., 606
nuda (Nutt.) Kuntze, 605
virginiana (L.) Britt., 606
Kosteletzkya, 322
altheaefolia (Chapm.) Gray, 323
pentasperma (Bert. ex DC.) Griseb., 322
smilacifolia Gray, 323
virginica (L.) Presl., 322
Kraunhia
frutescens (L.) Greene, 244
macrostachya (Nutt.) T. & G., 244
Krigia, 883
cespitosa (Raf.) Chambers in Vuilleumier, 885
dandelion (L.) Nutt., 884
oppositifolia Raf., 885
virginica (L.) Willd., 884

L

Labiatae, 589
Laciniaria
chlorolepis Small, 869
garberi (Gray) Kuntze, 869
gracilis (Pursh) Kuntze, 867
laxa Small, 867
pycnostachya (Michx.) Kuntze, 869
spicata (L.) Kuntze, 867
Laguncularia racemosa (L.) Gaertn. f., 384
Lapithea gentianoides (Ell.) Griseb., 525
Laportea canadensis (L.) Wedd., 53
Lasiococcus
dumosus (Andr.) Small, 484
mosieri (Small), 485
Lauraceae, 357
Lecticula resupinata (B. D. Greene) Barnh., 682
Leguminosae, 235
Leitneria floridana Chapm., 27
Leitneriaceae, 27
Lentibulariaceae, 671
Lepuropetalon spathulatum (Muhl.) Ell., 205
Leucospora multifida (Michx.) Nutt., 651
Leucothoe, 476
acuminata (Ait.) D. Don, 476
axillaris (Lam.) D. Don, 476
catesbaei of authors, 476
editorum Fern. & Schub., 477
fontanesiana (Steud.) Sleum., 477
populifolia (Lam.) Dippel, 476
racemosa (L.) Gray, 477
walteri (Willd.) Melvin, misapplied, 477
Liatris, 864
acidota Engelm. & Gray, 867
garberi Gray, 869
gracilis Pursh, 864

922

Nymphoides, 537
  aquatica (S. G. Gmel.) Kuntze, 538
  cordata (Ell.) Fern., 538
  lacunosa of authors, 538
  peltata (S. G. Gmel.) Kuntze, 538
Nyssa, 427
  aquatica L., 428
  ogeche Bartr. ex Marsh., 427
  sylvatica Marsh., 430
    var. biflora (Walt.) Sarg., 430
    var. *caroliniana* Poir., 430
    var. *dilatata* Fern., 430
    var. sylvatica, 430
  uniflora Wang., 428
  ursina Small, 430
Nyssaceae, 427

O

Oldenlandia
  boscii (DC.) Chapm., 719
  corymbosa L., 720
  uniflora L., 720
Oleaceae, 509
Onagraceae, 389
Operculina dissecta (Jacq.) House, 571
Opulaster
  alabamensis Rydb., 220
  australis Rydb., 220
  opulifolius (L.) Kuntze, 220
Orbexilum
  pedunculatum (Mill.) Rydb., 240
  simplex (Nutt. ex T. & G.) Rydb., 240
  virgatum (Nutt.) Rydb., 239
Osmanthus americana (L.) Gray, 513
Oxycoccus macrocarpus (Ait.) Pursh, 485
Oxypolis, 445
  filiformis (Walt.) Britt., 445
  greenmanii Math. & Constance, 447
  rigidior (L.) Raf., 447
  ternata (Nutt.) A. Heller, 447
  turgida Small, 447

P

Pagesia vandellioides (HBK.) Pennell, 635
Parietaria, 53
  floridana Nutt., 56
  floridana of authors not Nutt., 55
  nummularia Small, 56
  pensylvanica Muhl. ex Willd., 56
  praetermissa Hinton, 55
Parnassia, 209
  asarifolia Vent., 210
  caroliniana Michx., 210
  grandifolia DC., 210
Parsonsia
  balsamona (Cham. & Schlecht.) Small, 377
  lythroides Small, 377
  petiolata (L.) Rusby, 377
Pavonia spicata Cav., 321

Pedicularis, 670
  canadensis L., 670
  lanceolata Michx., 671
Penstemon, 655
  arkansanus Pennell, 656
  calycosus Small, 657
  digitalis Nutt., 657
  laevigatus Solander ex Ait., 657
  laxiflorus Pennell, 656
  pauciflorus Buckl., 656
  penstemon (L.) MacM., 657
  tenuis Small, 656
Pentodon
  halei (T. & G.) Gray, 723
  pentandrus (Schum. & Thonn.) Vatke, 723
Penthorum sedoides L., 206
Peplis diandra Nutt. ex DC., 382
Persea
  borbonia (L.) Spreng., 357
  palustris (Raf.) Sarg., 357
Persicaria
  hirsuta (Walt.) Small, 85
  hydropiper (L.) Opiz, 85
  hydropiperoides (Michx.) Small, 89
  lapathifolia (L.) Small, 82
  longistyla (Small) Small, 85
  muhlenbergii (Meisn.) Small, 79
  opelousana (Ridd. ex Small) Small, 89
  pensylvanica (L.) Small, 85
  persicaria (L.) Small, 93
  portoricensis (Bertero) Small, 82
  punctata (Ell.) Small, 89
  setacea (Baldw. ex Ell.) Small, 89
Petalostemum, 243
  carneum Michx., 243
  gracile Nutt., 243
Phaethusa
  laciniata (Poir.) Small, 768
  occidentalis (L.) Small, 767
  virginica (L.) Small, 768
Pharbitis
  barbigera (Sims) G. Don, 567
  cathartica (Poir. in Lam.) Choisy in DC., 567
  hederacea (L.) Choisy, 567
  purpurea (L.) Voight, 567
Phenianthus sempervirens (L.) Raf., 729
Philoxerus vermicularis (L.) R. Br., 104
Phyla, 587
  lanceolata (Michx.) Greene, 588
  nodiflora (L.) Greene, 589
  stoechadifolia (L.) Small, 588
Phyllanthus, 278
  avicularia Small, 279
  caroliniensis Walt., 279
  lathroides, misapplied by Small, 280
  liebmannianus Muell.-Arg.
    subsp. liebmannianus, 278
    subsp. platylepis (Small) Webster, 278
  platylepis Small, 278
  pudens L. S. Wheeler, 279
  saxicola Small, 279

925

pensylvanicum L., 82
persicaria L., 93
punctatum Ell., 85
sagittatum L., 76
scandens L., 76
    var. cristatum (Engelm. & Gray) Gl., 78
    var. scandens, 78
setaceum Baldw. ex Ell., 89
virginianum L., 79
Polypremum procumbens L., 516
Populus, 33
deltoides (Bartr.) ex Marsh. var. deltoides, 34
heterophylla L., 34
Portulaca, 116
oleracea, 116
pilosa, 116
Portulacaceae, 116
Prenanthes autumnalis Walt., 883
Primulaceae, 489
Proboscidea louisianica (Mill.) Thell., 709
Proserpinaca, 419
amblygona (Fern.) Small, 421
intermedia Mack., 421
palustris L., 421
    var. *amblygona* Fern., 421
    var. *crebra* Fern. & Grisc., 421
    var. palustris, 421
pectinata Lam., 419
platycarpa Small, 421
Psilotaxis
    "baldwinii" (Nutt.) Small, 275
    cymosa (Walt.) Small, 272
    lutea (L.) Small, 275
    nana (Michx.) Raf., 275
    ramosa (Ell.) Small, 272
    rugelii (Shuttlw. ex Chapm.) Small, 276
Psoralea, 239
    psoralioides (Walt.) Cory, 240
    var. eglandulosa (Ell.) F. L. Freeman, 240
    var. psoralioides, 240
simplex Nutt. ex T. & G., 240
virgata Michx., 239
Pterocaulon
pycnostachyum (Michx.) Ell., 843
undulatum (Walt.) Mohr., 843
virgatum (L.) DC., 843
Pterophyton
heterophyllum (Chapm. in Coult.) Alexander, 768
pauciflorum (Nutt.) Alexander, 768
Ptilimnium, 443
capillaceum (Michx.) Raf., 445
costatum (Ell.) Raf., 443
fluviatile (Rose) Mathias, 443
nodosum (Rose) Mathias, 443
nuttallii (DC.) Britt., 445
Pycnanthemum, 605
albescens T. & G., 606
flexuosum sensu Gl., 606

flexuosum sensu Fern., and sensu Radford et al., 607
hyssopifolium Benth., 607
muticum (Michx.) Pers., 606
nudum Nutt., 605
setosum Nutt., 607
tenuifolium Schrad., 605
virginianum (L.) Durand & Jackson, 606
Pyrus
arbutifolia (L.) L. f., 226
melanocarpa (Michx.) Willd., 226
Pyxidanthera barbulata Michx., 487

Q

Quamoclit
coccinea (L.) Moench., 565
quamoclit (L.) Britt. in Britt. & Brown, 563
Quercus, 40
bicolor Willd., 45
falcata Michx.
    var. falcata, 46
    var. pagodaefolia Ell., 46
geminata Small, 45
laurifolia Michx., 49
lyrata Walt., 42
macrocarpa Michx., 41
michauxii Nutt., 42
nigra L., 50
nuttallii E. J. Palmer, 46
obtusa (Willd.) Ashe, 50
oglethorpensis Duncan, 45
pagoda Raf., 46
palustris Muenchh., 46
phellos L., 48
prinus sensu Small, 44
similis Ashe, 42
stellata Wang. var. *paludosa* Sarg., 42
virginiana Mill., 45
virginiana var. *maritima* (Chapm.) Sarg., 45

R

Radicula
palustris (L.) Moench., 172
sessiliflora (Nutt. in T. & G.) Greene, 171
sylvestris (L.) Druce, 171
walteri (Ell.) Greene, 172
Ranunculaceae, 123
Ranunculus, 133
abortivus L., 142
acris L., 143
allegheniensis Britt., 142
ambigens S. Wats., 139
arvensis L., 137
bulbosus L., 143
carolinianus DC., 145
cymbularia Pursh, 137
delphinifolius Torr., 141
fascicularis Muhl. ex Bigel., 145